Lecture Notes in Civil Engineering

Volume 309

Lecture Notes in Civil Engineering (LNCE) publishes the latest developments in Civil Engineering—quickly, informally and in top quality. Though original research reported in proceedings and post-proceedings represents the core of LNCE, edited volumes of exceptionally high quality and interest may also be considered for publication. Volumes published in LNCE embrace all aspects and subfields of, as well as new challenges in, Civil Engineering. Topics in the series include:

- Construction and Structural Mechanics
- Building Materials
- Concrete, Steel and Timber Structures
- Geotechnical Engineering
- Earthquake Engineering
- Coastal Engineering
- Ocean and Offshore Engineering; Ships and Floating Structures
- Hydraulics, Hydrology and Water Resources Engineering
- Environmental Engineering and Sustainability
- Structural Health and Monitoring
- Surveying and Geographical Information Systems
- Indoor Environments
- Transportation and Traffic
- Risk Analysis
- Safety and Security

To submit a proposal or request further information, please contact the appropriate Springer Editor:

– Pierpaolo Riva at pierpaolo.riva@springer.com (Europe and Americas);
– Swati Meherishi at swati.meherishi@springer.com (Asia—except China, Australia, and New Zealand);
– Wayne Hu at wayne.hu@springer.com (China).

All books in the series now indexed by Scopus and EI Compendex database!

Gian Paolo Cimellaro
Editor

Seismic Isolation, Energy Dissipation and Active Vibration Control of Structures

17th World Conference on Seismic Isolation (17WCSI)

Springer

Editor
Gian Paolo Cimellaro
DISEG
Politecnico di Torino
Turin, Italy

ISSN 2366-2557 ISSN 2366-2565 (electronic)
Lecture Notes in Civil Engineering
ISBN 978-3-031-21186-7 ISBN 978-3-031-21187-4 (eBook)
https://doi.org/10.1007/978-3-031-21187-4

This Springer imprint is published by the registered company Springer Nature Switzerland AG
The registered company address is: Gewerbestrasse 11, 6330 Cham, Switzerland

Preface

Many new seismic protection systems have been developed in the last decades, either to reduce the earthquake forces acting on the structure or to absorb a part of the seismic energy. Among them, base isolation in one of the most widely accepted and implemented seismic protection system. It consists of a method in which the structure (superstructure) is separated from the base (foundation or substructure) by introducing a suspension system between the base and the main structure.

The World Conference on Base Isolation (WCSI) is an international event that started right after the Anti-Seismic Systems International Society (ASSISi) was founded on October 5th, 2001 in Assisi (Italy). The WCSI Conference represents the major event of the ASSISI's activity and is focused on fostering development and acceptance of seismic protection solutions and techniques. After 21 years in 2022 the conference has been organized in Italy by the Politecnico di Torino and it has been hosted in the Department of Structural, Geotechnical & Building Engineering (DISEG).

The 17th edition (WCSI2022) has brought in Turin prestigious speakers from all over the World for days of exciting lectures on various aspects of the current seismic protection technologies, addressing the most important advances of our discipline in major countries around the World. The conference has allowed a constructive sharing of ideas, opportunities, and knowledge to contribute to build disaster resilient societies and to create new directions of research and technology implementation.

This book features chapters based on accepted papers from the World Conference on Base Isolation, WCSI 2022, held in Turin, Italy, from 12 to 16 September 2022. The variety and quality of the papers are a testament of the very stimulating event that the WCSI 2022 represents. Overall, the Proceedings include the latest advances in the widely popular research areas of seismic protection techniques, such as base isolation and are organized in one volume and several chapters. Each chapter represents a special session organized by international experts in their respective fields. The first chapter includes the invited and keynote lectures.

The book is of interest to a wide range of professionals in earthquake engineering, including designers, researchers, and graduate students working in the field of seismic protection systems.

September 2022

Gian Paolo Cimellaro
WCSI 2022 Chair

Acknowledgments

The organization of the World Conference and the success of this volume would not have been possible without the contributions of many people who have assisted or provided guidance over the last year. As such, we would like to acknowledge:
Dr. Alessandro Cardoni, Ph.D. Politecnico di Torino, Department of Structural, Building and Geotechnical Engineering (DISEG), Italy
who helped me during the organization of the entire event and the development of this book. I also would like to thank the local organizing Committee:

Marco Domaneschi, Ph.D., Politecnico di Torino, DISEG, Italy
Melissa De Iuliis, Ph.D., University of Rome Sapienza, DISG, Italy
Raffaele Tarantini, Ph.D., Politecnico di Torino, DISEG, Italy
Alessio Vallero, Ph.D., Politecnico di Torino, DISEG, Italy
and the International Scientific Board, which include the following members:
Aiken Ian—SIE Inc., Berkeley, California, USA
Benzoni Gianmario—Università di Salerno, Treasurer ASSISI, Italy
Boroschek Rubén—University of Chile, Santiago, Chile
Bubis Alexander—EERC, TsNIISK, 16WCSI organizer, Moscow, Russia
Buffarini Giacomo—ENEA, Italy
Cameron Black—University of California, Berkeley, USA
Castellano Maria Gabriella—R&D FIP MEC srl, Italy
Castino Chiara—Somma Int., Italy
Cimellaro Gian Paolo—Politecnico di Torino, Italy
Constantinou Micheal—University at Buffalo SUNY, USA
Clemente Paolo—ENEA (Presidente ASSISi)
Dall'Asta Andrea—Università degli Studi di Camerino, Italy
de la Llera Martin Juan Carlos—Pontifica Universitad Catolica de Chile, Chile
Dal Pedri Roberto—Hirun, Taiwan
Doglioni Carlo—Universitá di Roma Sapienza, Italy
Domaneschi Marco—Politecnico di Torino, Italy
Douglas John—University of Strathclyde, UK
Du Yongfeng—Lanzhou University of Technology, China

Dutu Andreea—University of Bucharest, Romania
Erdik Mustafa—Bogazici University, Turkey
Filiatrault André—University of New York at Buffalo (SUNY), USA
Fossetti Marinella—Università degli Studi di Enna "Kore", Italy
Fragiacomo Massimo—Università degli Studi dell'Aquila, Italy
Fragiadakis Michalis—National Technical University of Athens (NTUA), Greece
Giovinazzi Sonia—Enea, Italy
Kelly James—University of California, Berkeley, USA
Leoni Graziano—Università degli Studi di Camerino, Italy
Lomiento Giuseppe—California State Polytechnic University, Pomona, USA
Lopez-Garcia Diego—Pontifica Universitad Catolica de Chile, Chile
Martelli Alessandro—First President of ASSISI, Italy
Medeot Renzo—Seismic Engineering Consultant, Italy
Mele Elena—Università degli Studi di Napoli Federico II, Italy
Modena Claudio—Università degli Studi di Padova, Italy
Monti Giorgio—Sapienza Università di Roma, Italy
Nakashima Masayoshi—Kyoto University, Japan
Nuti Camillo—Università degli Studi Roma III, Italy
Occhiuzzi Antonio—Direttore ITC, CNR, Italy
Ponzo Felice Carlo—University of Basilicata, Italy
Pampanin Stefano—Sapienza University of Rome, Italy
Pavese Alberto—University of Pavia, Italy
Prota Andrea—Università degli Studi di Napoli Federico II, Italy
Ricciardi Giuseppe—University Messina, Italy
Ryan L. Kery—University of Nevada, Reno, USA
Saito Taiki–Toyohashi Università of Technology, Japan
Sadan Bahadir—MEF University, Istanbul, Turkey
Sextos Anastasios—University of Bristol, UK
Sorrentino Luigi—Sapienza University of Rome, Italy
Spacone Enrico—Università degli Studi G. D'Annunzio Chieti -Pescara, Italy
Takayama Mineo—Fukuoka Università, Japan
Tan Ping—Guangzhou University, China
Wada Akira—Tokyo Institute of Technology, President of JSSI, Japan
Warn Gordon Patrick—Penn State University, USA
Whittaker David—Beca Ltd, Christchurch, New Zealand
Zhou Ying—Tongji University. Shanghai, China
Zhou Fu Lin—University of Guangzhou, China

September 2022

Gian Paolo Cimellaro
WCSI 2022 Chair

Contents

Seismicity and Seismic Isolation Design

New Isolation and Energy Dissipation Devices

Latest Applications on Seismic Isolation in Different Countries

Applications and Recent Studies on Seismic Isolation in Italy

Paolo Clemente(✉)

ENEA, Casaccia Research Centre, 00123 Rome, Italy
paolo.clemente@enea.it

Abstract. The development and application of seismic isolation in Italy are presented in this paper. The absence of a suitable technical code limited the use of seismic isolation until the Molise earthquake of October 31st 2002. A great impulse was given by the 2009 L'Aquila earthquake, when the Italian Civil Protection Department realised buildings with seismic isolation for the homeless, the so-called C.A.S.E. project, using sliding pendulum isolators. After that, seismic isolation has always been considered, especially for public buildings, in the reconstruction after seismic events but not only. The evolution of seismic isolation application in Italy is analysed and represented in different diagrams. The structural type, the new or retrofit application and the type of isolation device are distinguished. A summary of the Italian technical code is also provided. Despite the mentioned problems, research has always been very active and advanced in Italy, thanks to the work of several universities and research institutions as well as producer companies. Among the most interesting subjects, here are reported the application for the retrofit of existing buildings, keeping them operational during the works, and the observed behaviour under far fault earthquakes, when the onset of motion could not be guaranteed, especially for friction based devices.

Keywords: Seismic isolation · Application of seismic isolation · Technical code · Evolution of seismic isolation · Observed seismic behaviour

1 Introduction

After the pioneering applications in the fields of bridges (the Somplago Viaduct, 1975) and buildings (the Fire Brigade building in Naples, 1981, and the Telecom Building in Ancona, 1991), the use of passive anti-seismic systems in Italy remained limited for a long time [1–4]. This was mainly due first to the lack of a specific design code (as a matter of fact, the first guidelines appeared in 1998 [5]), and then to the very complicated and time-consuming approval process required by the mentioned guidelines.

Therefore, a significant number of applications of seismic isolation started only after the 2002 Molise earthquake, which shook public opinion because of the collapse of the Francesco Jovine School at San Giuliano di Puglia, where twenty-seven children died together with a schoolteacher. A new technical code was rapidly issued [6], which allowed the use of base isolation. The new school in San Giuliano di Puglia, open in

G. P. Cimellaro (Ed.): WCSI 2022, LNCE 309, pp. 3–16, 2023.
https://doi.org/10.1007/978-3-031-21187-4_1

2008, was the first school in Italy to be protected by an isolation system composed of 61 high damping rubber bearings of two different types and 13 slider devices [7]. In the successive year, two residential buildings in San Giuliano were also designed with seismic isolation and, especially, several schools in different Italian regions. Seismic isolation was also considered the best solution for strategic buildings.

A powerful impulse was given by the 2009 L'Aquila earthquake, mainly thanks to the use of base isolation in buildings for temporarily hosting the home-less residents. The so-called C.A.S.E. project, performed by the Italian Civil Protection Department, consisted of pre-fabricated buildings, each placed on an isolated reinforced concrete slab supported by curved surface sliders (CSS), installed at the top of circular columns [8]. This was the first time that CSSs were used in Italy. In fact, until then, they had been used widely, but almost exclusively, in the USA. With a significant diffusion of seismic isolation, the recommendations for the correct use became important [9].

Seismic isolation was also considered in the reconstruction after the 2012 Emilia earthquake, while the organization of the reconstruction after the 2016–2017 Central Italy seismic sequence is still in progress [10].

Nowadays, Italy is the fifth country in the world (after Japan, China, Russia and the USA) and the first in Western Europe for the overall number of passive anti-seismic system applications. Seismic isolation has been used both for new and existing structures, both for bridges and buildings. Among these, it has been used mainly in reinforced concrete and masonry buildings, for which the economic convenience was also demonstrated [11, 12].

In this paper, the available data on the seismic isolation applications in Italy, received by the producers, are represented in suitable diagrams. The basic concepts of the present technical code are also described. Despite the mentioned problems, research has always been very active and advanced in Italy, thanks to the work of several universities and research institutions as well as producer companies. Among the most interesting subjects the following ones are reported: the application in the retrofit of existing buildings, keeping them operational during the works, and the observed behaviour under far fault earthquakes, when the onset of motion could not be guaranteed, especially for friction devices.

2 Applications of Seismic Isolation in Italy

It is very tough to have a complete list of seismically isolated structures in Italy nowadays. Actually, the data used in the following analysis are to be considered only indicative of the structures protected with seismic isolation in Italy. They are relative to the time interval from 1976 to 2022, but the list relative to 2022 is obviously not complete.

In Fig. 1, the total number of applications versus the year is plotted for the different structural types, classified as bridges, buildings and others. As one can see, the first applications referred mainly to bridges. Actually, the excellent behaviour of the Somplago Viaduct during the events that struck the Friuli region on September 1976 (maximum magnitude $M = 6.0$), with epicentres only a few kilometres from the viaduct, caused an immediate and rapid extension of the application of anti-seismic systems to new bridges and viaducts [13]. This ensured worldwide leadership to Italy for the number and importance of anti-seismic systems applied to bridges and viaducts at that time. A very significant increase occurred after 2004 for bridges and after 2009 also for buildings.

In Fig. 2, the total number of applications in buildings is reported again and compared with the number of applications of seismic isolation in residential buildings. Actually, the first seismically isolated buildings were mainly strategic or relevant structures, then seismic isolation was also used for residential buildings, which now represent a large percentage of the total number of applications.

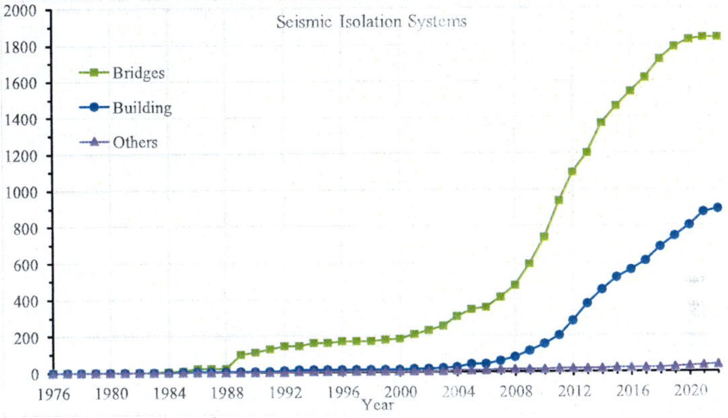

Fig. 1. Seismic isolation application for each structural type.

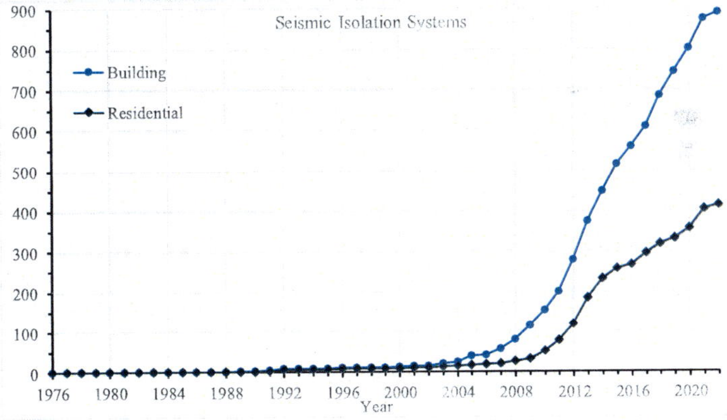

Fig. 2. Seismic isolation application in buildings.

The use of seismic isolation in new structures is almost always possible and not difficult, except for a limited number of situations. The present challenge is the use of seismic isolation in the retrofit of existing structures. In Fig. 3, the number of applications of seismic isolation in new buildings in Italy is compared with the number of the existing ones retrofitted by means of this technique. The evolution of the application in the retrofit of existing buildings has been quite similar to that of the application in new buildings and now the buildings retrofitted with seismic isolation represent an important percentage of the total number of seismically isolated buildings. The increase of applications in the retrofit of existing structures occurred during the reconstruction after the most severe earthquakes, as for new buildings.

The same analysis for bridges is shown in Fig. 4. The total number of applications in existing bridges is similar to that of the existing buildings, but the percentage with reference to the new structures is quite low.

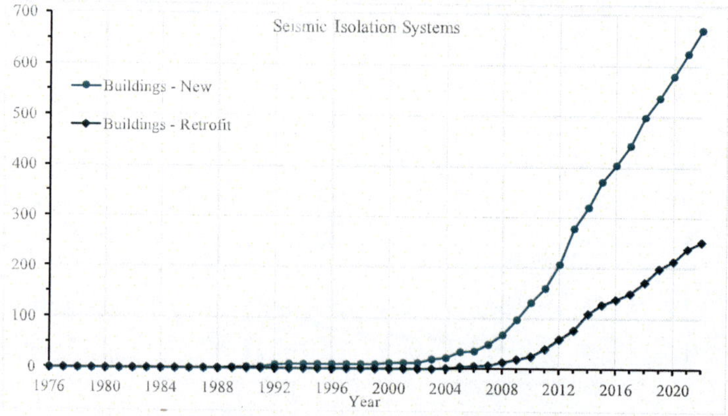

Fig. 3. New and retrofitted buildings with seismic isolation.

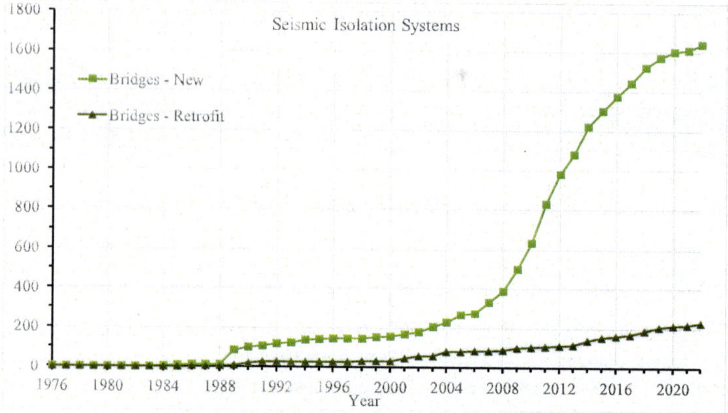

Fig. 4. New and retrofitted bridges with seismic isolation.

Finally, in Figs. 5 and 6, the number of the different device types used during the years for buildings and bridges are shown, respectively. Four types of devices are considered: high damping rubber bearings (HDRB), curved surface sliders (CSS), lead rubber bearings (LBR) and sliding devices (SD) associated with dampers of different types. Other systems, such as energy dissipation systems, sacrificial devices and buffers are not considered.

With reference to buildings (Fig. 5), initially, HRDBs were predominantly used, often with SDs to obtain the design stiffness and its optimum distribution in plan. CSSs have been used in Italy, for the first time, in the already mentioned C.A.S.E. project, after the 2009 L'Aquila earthquakes. The use of LRBs has been almost limited in buildings. For bridges (Fig. 6), in the first applications, dampers associated with sliding devices or similar were mostly used. CSSs have been also considered since L'Aquila earthquake but their use remained quite limited compared with that of HDRBs. The use of LRBs has been limited but not negligible for bridges.

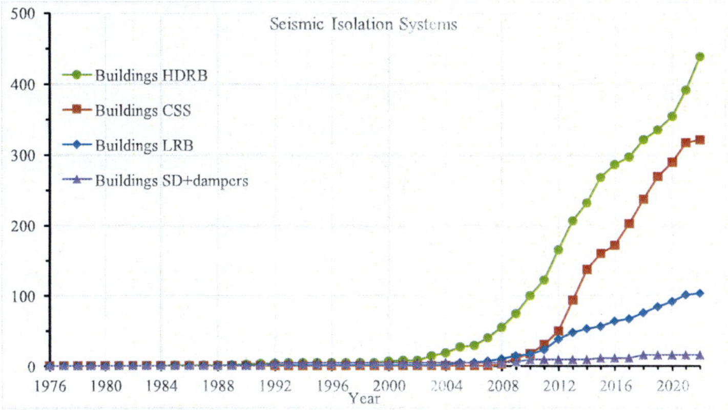

Fig. 5. Seismically isolated buildings for different types of devices.

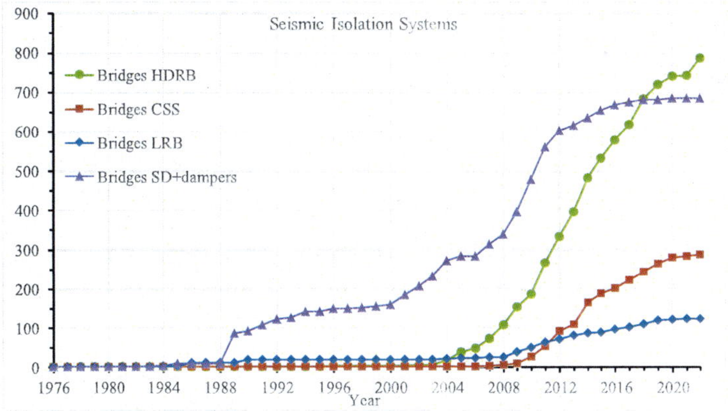

Fig. 6. Seismically isolated bridges for different types of devices.

3 The Seismic Design According to NTC-2018

3.1 The Design Earthquake

In the Italian technical code [14], the acceleration elastic response spectrum is defined by (Fig. 7):

- The horizontal peak ground acceleration a_g, the maximum amplification F_0, and the value $T_C{}^*$ of the period, given for any site and for different exceedance probability values P_{50} in 50 years (related to the return period T_R). The minimum value $P_{50} = 2\%$ corresponds to $T_R = 2475$ years. The values are relative to a rigid soil (type A) with a horizontal surface and a damping ratio $\xi = 5\%$. A different value of ξ is accounted for by means of the parameter η (=1 for $\xi = 5\%$).
- The soil type (four other types of soil are considered, namely B, C, D and E) that influences the soil parameter S, which modifies the spectrum amplitudes, and the value of T_C (Table 1).
- The topographical conditions that influence the soil parameter S.

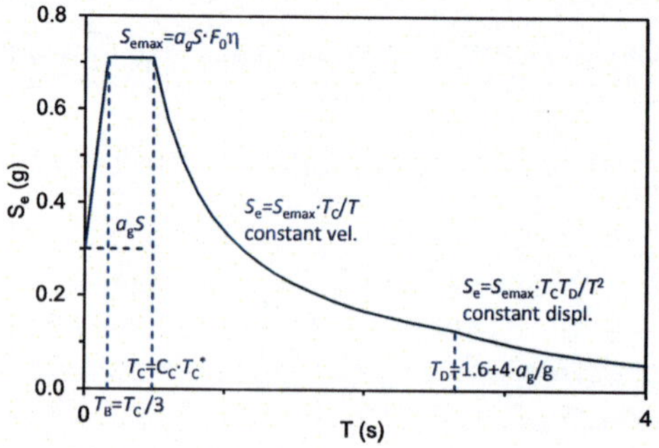

Fig. 7. Acceleration response spectrum according to the NTC-2018.

Table 1. Hazard parameters for NTC-2018.

Parameters	Notes
a_g, F_0, $T_C{}^*$	Given for any site and different P_{50} (or T_R)
S	Function of ground type, a_g, F_0, topographical conditions
T_C	$= c\cdot(T_C{}^*)^d$, (c, d = f(ground type))
T_B	$= T_C/3$
T_D	$= 1.6 + 4\, a_g/g$

The elastic response spectra to consider for the different limit states depend on the nominal service lifetime of the structure V_N (= 50 years, usually, but higher values should be considered for important infrastructures), and the use factor C_U (=1.0, 1.2 and 1.4 for ordinary, relevant and strategic structures, respectively). The product $V_R = V_N \cdot C_U$ is the reference time, from which the return period can be deduced, for a fixed exceedance probability P_{VR} in V_R ($T_R = -V_R/(\ln(1 - P_{VR}))$).

Four limit states are considered, two of them corresponding to service limit states (the operational limit state SLO, and the damage limit state SLD), and two corresponding to ultimate limit states (the life safeguard limit state SLV, and the collapse limit state SLC). A value of the exceedance probability P_{VR} is associated with each limit state (Table 2). The safety checks of the substructure and the superstructure must be carried out with reference to the SLV, while the check of the isolation devices must be done with reference to the SLC.

Table 2. Limit states and exceedance probability P_{VR}.

Limit State	SLO	SLD	SLV	SLC
P_{VR} (%)	81	63	10	5

In order to make a comparison with technical standards of other countries, the usual case of $V_N = 50$ years is considered for the reference spectrum. The adoption of a use factor C_U translates into assuming for the SLV a lower value of the exceedance probability P_{50}. Analogously, for the isolation devices, for which the SLC check is required, the exceeding probabilities to be considered are further lowered (Table 3).

Table 3. Values of P_{50} to be assumed in the design for substructure and superstructure, and isolation devices, for structures of different importance.

	Structure	P_{50} (%)
Reference Spectrum	Ordinary	10
Substructure and superstructure	Ordinary	10
	Relevant	6.8
	Strategic	5.1
Isolation devices	Ordinary	5
	Relevant	3.4
	Strategic	2.5

3.2 Seismic Action and Load Combination

The vertical component must be considered, in general, if $a_g \geq 0.15$ g, and if $K_v/K_{esi} < 800$ (also if $a_g < 0.15$ g) for seismically isolated bridges and buildings (K_v = total vertical stiffness, K_{esi} = total horizontal stiffness of the isolation system).

The effects of the seismic actions in the three orthogonal directions are summed according to the relation:

$$E = E_i + 0.3 \cdot E_j + 0.3 \cdot E_k \tag{1}$$

where x, y and z are i, j and k, respectively, and must be permutated. The seismic effects are combined with those of the gravitational loads at the SLV with the law:

$$G_1 + G_2 + E + \sum_j (\psi_{2j} Q_{Kj}) \tag{2}$$

where G_I, G_2 and Q_{kj} represent the structural loads, the permanent non-structural loads and the variable loads, respectively (ψ_{2j} = variable load factor, < 1).

3.3 Other Main Rules

- Isolation devices must be positioned in order to allow their inspection and replacement. They should be protected from fire, chemical and biological attacks.
- To minimise torsional effects, the effective stiffness centre and the centre of damping of the isolation system should be as close as possible to the projection of the centre of mass on the isolation interface.
- The compressive stress induced in isolating devices by the permanent actions should be as uniform as possible, to minimise different behaviour of them.
- Tension in the devices should be avoided ($V \geq 0$). Only for HDRBs a maximum tension stress not higher than 1.0 MPa (or $2G$ if lower, G = shear rubber modulus) can be accepted.
- The structural elements located above and below the isolation interface should be sufficiently rigid in both horizontal and vertical directions so that the effects of differential seismic ground displacements are minimised. In buildings, this condition is considered satisfied if: (i) rigid diaphragms (reinforced concrete slab or a grid of tie-beams) are provided above and under the isolation interface and (ii) the relative horizontal displacement of any vertical element, between each device and the rigid diaphragm should be lower than 1/20 the relative displacement of the isolation system.
- A suitable gap must be considered between the superstructure and the ground or the surrounding buildings, in order to allow the seismic displacements.

3.4 Modelling, Analysis and Checks

Superstructure and substructure should be modelled as linear elastic systems. The isolation system can have a linear viscous elastic behaviour if certain conditions are satisfied. In this case, it is characterised by an equivalent stiffness K_{esi} (sum of the stiffness of single devices) and an equivalent viscous damping ξ_{esi}; if K_{esi} and ξ_{esi} depend strongly on the design displacement, an iteration procedure must be used.

Linear and time-history analyses can be considered. Linear static and dynamic linear analyses can be applied if the isolation system can be modelled as linear. Usually, a response spectrum analysis is used but also a step-by-step procedure can be performed. The values of the elastic spectrum are relative to the damping ratio of the isolation system (usually $\geq 10\%$) for $T > 0.8 \cdot T_{is}$. A non-linear model should account for the influence of the deformation and velocity on the behaviour of the devices during the cycles.

The capacity of the superstructure and the substructure are evaluated by using the same partial load factor γ_M of non-isolated buildings. In buildings, the resistance condition of the structural elements of the superstructure may be satisfied considering the seismic action effects divided by a behaviour factor $q \leq 1.5$. The non-dissipative parts of the devices should keep in the elastic range. Suitable gaps and flexible connections for the pipes should be realized to absorb the seismic relative displacements.

The isolation devices should be able to support the displacement at the SLC. The connection should be able to support the same displacement with the same safety degree. The level of protection against the SLD of the substructure can be considered achieved if the checks against the SLV are satisfied. For the superstructure, the inter-storey drifts obtained from the analysis must be less than 2/3 of the limits indicated for the SLD of non-isolated structures.

For HDRBs, the checks refer to the angular strains, instability and stress in the steel. A suitable graphic procedure for the optimum design of isolator devices, shown in Fig. 8, was proposed in [15]. For CSSs, the checks are very influenced by the friction, which depends also on the vertical load [16]. This is different for each device and varies during the quake.

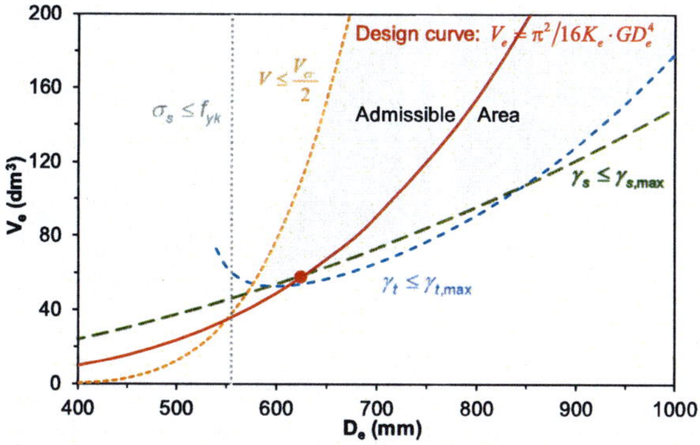

Fig. 8. Graphic procedure for the optimum design of HDRBs. The four inequalities define an admissible area on the plane V_e-D_e (V_e = rubber volume, D_e = rubber diameter). The optimum solution is given by the lowest intersection point with the design curve. The admissible area can be modified by changing the geometrical parameters of the devices [15].

4 Seismic Isolation in the Retrofit of Existing Buildings

The application of seismic isolation in new buildings is quite easy and almost always possible. It is also economically convenient, at least in medium and high seismicity areas both for reinforced concrete and masonry buildings. The challenge of seismic isolation is the protection of existing buildings, especially in countries where maintenance of old structures is an important issue.

The design issues for the application of base isolation in existing buildings were discussed in previous papers, with particular reference to the positioning of the isolation devices in plan and elevation and the transferring of the vertical actions during the building phases [17, 18].

Base isolation is particularly suitable for existing buildings. One can choose the period of vibration so as to have only low energy inputs in the superstructure for which it remains in the elastic range. As a result, the conventional retrofit intervention in the superstructure is limited or even avoided.

The intervention can be concentrated at the lowest level, while no intervention is necessary in the superstructure, so it is not necessary to demolish and rebuilt non-structural elements. As a result, an "operational seismic retrofit" would be possible, that is the superstructure could remain operative during the retrofit works. Obviously, a safety degree not lower than the existing one should be guaranteed during the works. This can be obtained by using temporary fixing systems for the isolation devices or temporary structures for the columns [19].

Particular attention is due to cultural heritage structures, for which seismic isolation allows the retrofit without modifying the architectural characteristics of the superstructures. A suitable system is the "Seismic Isolation Structure for Existing Buildings" [20], which consists of an isolated platform to be inserted under the foundations of the building, without touching it (Fig. 9). A discontinuity between the foundations and the soil is created by inserting horizontal pipes by means of pipe jacking and positioning the isolation devices at their horizontal diametric plane. In order to facilitate successive placement operations, the pieces of pipe have a particular shape and are composed of two portions, the lower and the upper cylindrical sectors, respectively, which are connected by means of removable elements. Then, the building is separated from the surrounding soil, in order to allow horizontal displacements during an earthquake. In this way, the structure is seismically isolated without modifying its architectural characteristics. Even the underground levels are not modified, but can be part of the seismically protected structure [21].

5 Seismic Behaviour Observed During Far Fault Earthquakes

The observed behaviour of seismically isolated buildings in different countries testified the suitability of this technique to protect structures against strong earthquakes. In Italy, up to now, base isolated buildings have been subjected only to low energy earthquakes at the site.

Seismically isolated buildings are characterised by low resonance frequencies when subject to their design earthquake. Instead they exhibit vibrations of low amplitudes and higher frequencies under low intensity dynamic actions. In order to study these effects, the actual behaviours of seismically isolated buildings subject to low intensity earthquake were analysed. The results obtained can be summarized as follows [22]:

- A building seismically isolated by means of HDRBs only, under a low energy earthquake at the site, showed a fundamental frequency much higher than that assumed in

the design. This could become close to the resonance frequencies of the superstructure. In this case, a suitable decoupling of motion between the superstructure and the soil-foundation system could not be guaranteed [23–28].
- In a building seismically isolated by means of CSSs, the seismic behaviour is governed by the static friction. If this has been designed for the vertical loads relative to the static ultimate limit state, the actual friction coefficient could be quite greater under the vertical seismic loads and the isolation system could not be put in action [29, 30].
- In a building seismically isolated by means of HDRBs + SDs the onset of motion is governed by the static friction of SDs. After that, the stiffness of the HDRBs determines the effective resonance frequencies, which could become close to that of the superstructure [31].

The observed behaviour should be accounted for when designing seismically isolated buildings.

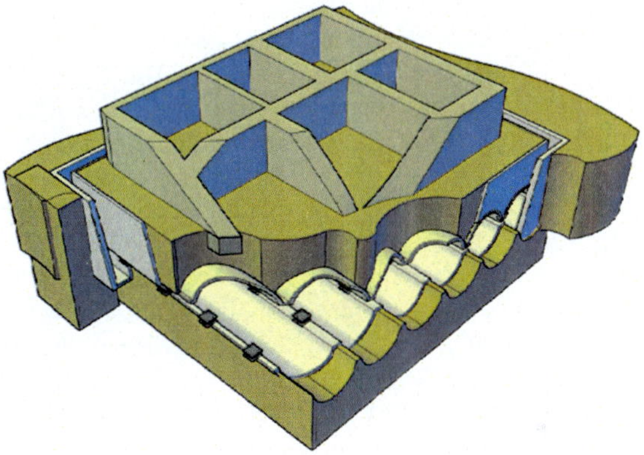

Fig. 9. Seismic isolation structure for existing buildings.

6 Conclusions

The development of seismic isolation in Italy has been influenced by the absence of a suitable technical code up to the Molise earthquake of October 31st 2002. Before that, specific guidelines had been issued in 1998, according to which the design with seismic isolation should undergo an approval process that was complicated and time-consuming. The distrust of technicians and citizens towards modern technologies had a terrific shock after the earthquake of 2009, when the Italian Civil Protection Department realized buildings with seismic isolation for the homeless. While until then the most widely used isolation devices in buildings were HDRBs, with the so-called C.A.S.E. project CSSs were introduced in Italy for the first time. After that, seismic isolation has been always considered, especially for public buildings, in the reconstruction after seismic events but not only.

In this paper, the evolution of application of seismic isolation in Italy has been represented in suitable diagrams based on the data of the producers and the main characteristics of the Italian technical code have been presented.

Among the most interesting research subjects, the application of seismic isolation in the retrofit of existing buildings and the observed behaviour under far fault earthquakes have been presented in detail. For the first one, the possibility of keeping the buildings operational during the works has been analysed, for the second one, it has been pointed out that the onset of motion could not be guaranteed, especially for friction based devices.

Acknowledgements. The statistical analyses shown in this paper were possible thanks to the contribution of Fip Mec, Freyssinet, Hirun, Maurer and Somma.

References

1. Clemente, P., Martelli, A.: Anti-seismic systems: worldwide application and conditions for their correct use. In: Proceedings of the 16th World Conference on Earthquake Engineering (16WCEE, Santiago, Chile, Jan 9–13), Keynote lecture, IAEE & ACHISINA (2017)
2. Clemente, P.: Seismic isolation: past, present and the importance of SHM for the future. J. Civ. Struct. Heal. Monit. **7**(2), 217–231 (2017). https://doi.org/10.1007/s13349-017-0219-6
3. Clemente, P., Martelli, A.: Seismically isolated buildings in Italy: state-of-the-art review and applications. Soil Dyn. Earthq. Eng. **119**, 471–487, Elsevier Science Ltd (2019). https://doi.org/10.1016/j.soildyn.2017.12.029
4. Clemente, P., Martelli, A.: Recent applications of seismic isolation in Italy. In: Proceedings of the 16th World Conference on Seismic Isolation, Energy Dissipation and Active Vibration Control of Structures (16WCSI, St. Petersburg, July 1–6), Keynote lecture, pp. 21–36 (2019). https://doi.org/10.37153/2686-7974-2019-16-21-36
5. LG-1998: Linee guida per progettazione, esecuzione e collaudo di strutture isolate dal sisma. Presidenza del Consiglio Superiore dei Lavori Pubblici – Servizio Tecnico Centrale, Ministero dei Lavori Pubblici (1998)
6. OPCM-3274: Primi elementi in materia di criteri generali per la classificazione sismica del territorio nazionale e di normative tecniche per le costruzioni in zona sismica. Ordinanza Presidenza del Consiglio dei Ministri, Supplemento Ordinario n. 72 della Gazzetta Ufficiale n. 105 dell'8.05.2003 (2003)
7. Bongiovanni, G., Buffarini, G., Clemente, P., Saitta, F., De Sortis, A., Nicoletti, M., Rossi, G.: Behaviour of the base isolated Jovine School in San Giuliano di Puglia, Italy, under the December 20th 2013 earthquake. In: Proceedings of the 14th World Conf. on Seismic Isolation, Energy Dissipation and Active Vibration Control of Structures (San Diego, 9–11 Sep) (2015)
8. Turino, R. (ed.): L'Aquila, il progetto C.A.S.E.. IUSS Press, Pavia (2010)
9. Martelli, A., Clemente, P., De Stefano, A., Forni, M., Salvatori, A.: Recent development and application of seismic isolation and energy dissipation and conditions for their correct use. In: Ansal, A. (ed.) Perspectives on European Earthquake Engineering and Seismology. GGEE, vol. 34, pp. 449–488. Springer, Cham (2014). https://doi.org/10.1007/978-3-319-07118-3_14
10. Bongiovanni, G., Buffarini, G., Clemente, P., Saitta, F., Tripepi, C.: Retrofit of existing buildings with seismic isolation: design issues and applications. In: Proceedings of the 16th World Conference on Seismic Isolation, Energy Dissipation and Active Vibration Control of Structures (16WCSI, St. Petersburg, 1–6July), pp. 111–122 (2019). https://doi.org/10.37153/2686-7974-2019-16-111-122

11. Clemente, P., Buffarini, G.: Base isolation: design and optimization criteria. Seismic Isolation Protection Syst. **1**(1), 17–40, Mathematical Science Publisher (2010). https://doi.org/10.2140/siaps.2010.1.17
12. Clemente, P., Bontempi, F., Boccamazzo, A.: Seismic isolation in masonry buildings: technological and economic issues. In: Modena, C., da Porto, F., Valluzzi, M.R. (eds.) Brick and Block Masonry: Trends, Innovation and Challenges (Proceedings of 6th International Conference IB2MAC, 26–30 Jun 2016, Padua), pp. 2207–2215. Taylor & Francis Group, London, UK (2016). https://doi.org/10.1201/b21889
13. Infanti, S, Castellano, M.G.: Seismic protection of bridges through isolation and energy dissipation devices. In: Proceedings of the Indian National Conference, IIBE, May 13–14 2004, Chennai (2004)
14. NTC-2018: Aggiornamento delle Norme Tecniche per le Costruzioni, Ministero delle Infrastrutture e dei Trasporti, Supplemento Ordinario n. 40 della Gazzetta Ufficiale n. 105 dell'20.02.2018 (2018)
15. Tripepi, C., Clemente, P.: Graphic procedure for the optimum design of elastomeric isolators. In: Practice Periodical on Structural Design and Construction, vol. 26, no. 1, 04020058-1_9, ASCE American Society of Civil Engineers (2021). https://doi.org/10.1061/(ASCE)SC.1943-5576.0000547
16. Saitta, F., Clemente, P., Bongiovanni, G., Buffarini, G., Salvatori, A., Grossi, C.: Base isolation of buildings with curved surface sliders: basic design criteria and critical issues. In: Advances in Civil Engineering, 2018, Article ID 1569683, 14 pages, Hindawi (2018). https://doi.org/10.1155/2018/1569683
17. Buffarini, G., Clemente, P., De Flaviis, A., Ormando, C., Salvatori, A.: Seismic retrofit of r.c buildings with base isolation: techniques and evaluation. In: Cimellaro, G.P. (ed.) WCSI 2022, LNCE, vol. 309, pp. 1–10. Springer (2023). https://doi.org/10.1007/978-3-031-21187-4_85
18. Buffarini, G., Clemente, P., Mancinelli, G.: Experimental analysis and seismic rehabilitation of an earthquake damaged building. In: Proceedings of the First Middle East Conference on Smart Monitoring, Assessment and Rehabilitation of Civil Structures (Smar2011, Dubai, 8–10 Feb 2011), Paper 171 (2011)
19. Saitta, F., Clemente, P., Buffarini, G., Bongiovanni, G.: Vulnerability analysis and seismic retrofit of a strategic building. J. Performance Constr. Facilities **31**(2), 04016085, ASCE American Society of Civil Engineers (2017). https://doi.org/10.1061/(ASCE)CF.1943-5509.0000948
20. Clemente, P., De Stefano, A.: Application of seismic isolation in the retrofit of historical buildings. In: Brebbia, C.A. & Maugeri M. (eds), 8th World Conference on Earthquake Resistant Engineering Structures, ERES 2011 (Sept. 7–9, Chianciano, Italy), vol. 120, pp. 1746–4498. WIT Press, Southampton, UK (2011). https://doi.org/10.2495/ERES110041
21. De Stefano, A., Clemente, P., Invernizzi, S., Matta, E., Quattrone, A.: Innovative technique for the base isolation of existing buildings. In: Papadrakakis, M., Papadopoulos, V., Plevris, V. (eds.) Computational Methods in Structural Dynamics and Earthquake Engineering (5th Int. Conf. COMPDYN 2015, Crete Island, Greece, 25–27 May 2015), vol. 1, pp. 377–387, NTUA, Athens, Greece. ISBN: 978-960-99994-7-2 (2015)
22. Buffarini, G., Clemente, P., Ianniruberto, U., Ormando, C., Scafati, F.: Observed seismic behaviour of base isolation systems in Italy. In: Cimellaro, G.P. (ed.) WCSI 2022, LNCE, vol. 309, pp. 1–12. Springer (2023). https://doi.org/10.1007/978-3-031-21187-4_50
23. Clemente, P., Bongiovanni, G., Benzoni, G.: Monitoring of seismic isolated buildings: state of the art and results under high and low energy inputs. In: Proceedings of the New Zealand Society for Earthquake Annual Conference and 15th World Conference on Seismic Isolation, Energy Dissipation and Active Vibration Control of Structures (NZSEE2017 and 15WCSI,

Wellington, 27–29 Apr), Invited lecture. New Zealand Society for Earthquake Engineering NZSEE (2017)
24. Clemente, P., Bongiovanni, G., Buffarini, G., Saitta, F., Castellano, M.G., Scafati, F.: Effectiveness of HDRB isolation systems under low energy earthquakes. Soil Dyn. Earthq. Eng. **118**, 207–220, online 15.01.2019, Elsevier Science Ltd (2019). https://doi.org/10.1016/j.soildyn.2018.12.018
25. Clemente, P., Bongiovanni, G., Buffarini, G., Saitta, F., Scafati, F.: Monitored seismic behavior of base isolated buildings in Italy. In: Limongelli, M., Çelebi, M. (eds.) Seismic Structural Health Monitoring, pp. 115–137, Springer Tracts in Civil Engineering. Springer, Cham (2019). https://doi.org/10.1007/978-3-030-13976-6_5
26. Clemente, P., Di Cicco, A., Saitta, F., Salvatori, A.: Seismic behaviour of base isolated civil protection operative centre in Foligno, Italy. J. Performance Constr. Facilities **35**(4), 04021027, ASCE American Society of Civil Engineers (2021). https://doi.org/10.1061/(ASCE)CF.1943-5509.0001589
27. Salvatori, A., Di Cicco, A., Clemente, P.: Monitoring of base isolated building subjected to far fault earthquake. In: Rainieri, C., Fabbrocino, G., Caterino, N., Ceroni, F., Notarangelo, M.A. (eds.) CSHM 2021. LNCE, vol. 156, pp. 311–338. Springer, Cham (2021). https://doi.org/10.1007/978-3-030-74258-4_21
28. Salvatori, A., Di Cicco, A., Clemente, P.: Seismic monitoring of buildings with base isolation. In: Papadrakakis, M., Fragiadakis, M. (eds.) Computational Methods in Structural Dynamics and Earthquake Engineering (7th ECCOMAS Thematic Conf., COMPDYN 2019, Crete Island, Greece, 24–26 June 2019), ID 19221, Institute for Structural Analysis and Antiseismic Research, National Technical University of Athens (2019). https://2019.compdyn.org/proceedings/
29. Scafati, F., Ormando, C., Clemente, P., Bongiovanni, G.: Observed behavior of buildings seismically isolated with CSSs under a low energy earthquake. J. Civ. Struct. Heal. Monit. **12**(2), 225–243 (2021). https://doi.org/10.1007/s13349-021-00539-z
30. Ormando, C., Clemente, P., Iannirubertο, U., Scafati, F.: Onset of motion of curved surface sliders used in seismic isolation systems. Pract. Periodical Struct. Des. Constr. **26**(3), 04021015, ASCE American Society of Civil Engineers (2021). https://doi.org/10.1061/(ASCE)SC.1943-5576.0000580
31. Salvatori, A., Bongiovanni, G., Clemente, P., Ormando, C., Saitta, F., Scafati, F.: Observed seismic behaviour of a HDRB and SD isolation system under far fault earthquakes. Infrastructures 2022 **7**(2), 13, MDPI Multidisciplinary Digital Publishing Institute (2022). https://doi.org/10.3390/infrastructures7020013

State of the Art in Application of Seismic Isolation and Energy Dissipation in Turkey

Bahadir Sadan(✉)

MEF University, Istanbul, Turkey
sadanb@mef.edu.tr

Abstract. This paper summarizes the passive structural control system applications and other related developments in Turkey, emphasizing the period between 2019 and 2022. The country hosts state-of-the-art seismic isolation applications, relatively greater in size, and use more isolators in each project (415 isolators per building on average) compared to the other seismic isolation projects worldwide. Construction of the world's largest seismic isolated building, Istanbul Başakşehir Pine and Sakura City Hospital, covering more than 1 million square meters of area and employing more than 2000 seismic isolators, was completed and has been in service since May 2020. 1915 Çanakkale Bridge, constructed in memory of the great war of Çanakkale during World War I, is now the world's longest suspension bridge with a 2023m main span length. Eight massive hydraulic dampers were used at the main deck and 48 at the approach viaducts in combination with 72 seismic isolators. Construction of a residential building complex in Istanbul consisting of 16 isolated blocks covering more than 170,000 m^2 area and having 454 seismic isolators is coming to an end very soon. Historical Nusretiye Clock Tower in Istanbul was relocated a few meters over seismic isolators. Historical Goztepe Railway Station in Istanbul was retrofitted to accommodate an additional railway line using seismic isolation technology. Seismic codes for buildings and bridges now include rules for the seismically isolated design of structures. Additionally, all seismic isolation designs must be peer-reviewed by law. A new seismic isolator testing facility was established in Eskişehir to respond to the demand for the large number of isolators that need to be tested.

Keywords: Turkey · Earthquake · Seismic · Isolation · Buildings · Bridges

1 Introduction

Earthquakes worldwide threaten human life and pose a great risk to the economy of countries and property owners. Life safety performance is not adequate nowadays, especially for the structures that need to continue their functions immediately, even after a major earthquake, such as hospitals, schools, museums, critical infrastructure, etc.

The codes limit the damage to the structural system by assigning an importance factor based on its occupancy. In force-based design, increasing the importance factor increases the force the structures are designed for, assigning higher strength to the structural

G. P. Cimellaro (Ed.): WCSI 2022, LNCE 309, pp. 17–25, 2023.
https://doi.org/10.1007/978-3-031-21187-4_2

members. Meanwhile, the damage to the structural members is controlled by applying special ductile solutions at the locations where damage can not be avoided.

However, such an approach does not reduce the seismic forces acting on the non-structural members but the contrary, increases. Higher strength assigned structures may avoid seismic damage to the structural members; still, they can not avoid damage to the non-structural members such as false ceilings, pipings, partition walls, claddings etc. After each major earthquake, several healthcare facilities were reported undamaged structurally but can not continue service due to non-structural damages.

Seismic isolation is the best solution to avoid damage to both structural members but also the contents, and non-structural members. The slight increase in the investment cost in applying seismic isolation to structures returns safe and resilient structures reducing the indirect post-earthquake repair/replacement costs.

Passive structural control systems have been widely applied to bridges, buildings, and industrial structures during construction or later as a retrofitting method in Turkey after the 1999 earthquakes. The Ministry of Health regulation which mandates the use of seismic isolation at all state hospitals to be constructed at seismic zones with more than 100-bed capacity increased the pace of applications exponentially. Seismic codes for buildings released in 2018 and for bridges in 2020, include special rules for the design of seismically isolated structures and testing of the seismic isolators.

2 Brief History and Recent Developments Regarding Seismic Isolation and Energy Dissipation

The history of passive structural control applications in Turkey until 2019 was presented during the previous World Conferences on Seismic Isolation, Energy Dissipation and Active Control of Structures (WCSI) organized by the Anti-Seismic Systems International Society (ASSISi). The last four of these were published as conference proceedings by Yenidogan et al. in 2013 in Sendai, Japan [1], Erdik et al. in 2015 in San Diego, USA [2], Erdik in 2017 in Wellington, New Zealand [3] and Erdik in 2019 in St. Petersburg, Russia [4].

Passive structural control system applications according to the structural type and other developments after 2018 until 2022 are summarized in this paper.

2.1 Bridges and Viaducts

The history of structures with passive structural control in Turkey dates back to the construction of Bolu Viaducts in 1991 with steel hysteretic energy dissipation units (later substituted with the curved surface friction isolators after the Düzce Earthquake in 1999) [5]. Since then, there have been various applications of seismic isolators and energy dissipating devices for bridge type of structures. The most recent ones are the 400 km long North Marmara Highway viaducts and the 1915 Canakkale Suspension Bridge.

1915 Çanakkale Bridge was constructed in western Turkey with a record-breaking 2023 m main span length and 318 m high towers. Seventy-two curved surface friction isolators were used in the approach viaducts of this suspension bridge, each having a displacement capacity varying between ±610 and ±1,090 mm. Additionally, 48 viscous dampers with 4,225 kN maximum force were employed at the approach viaducts. Eight massive hydraulic dampers with 5,000kN axial load capacity were used to sustain the movements of the main deck in the longitudinal direction, having a displacement capacity of ±1,250 mm. The bridge was opened to traffic in March 2022.

2.2 Retrofitting Applications

After the 1999 earthquakes, Ataturk Airport International Terminal building which was in construction at the time of the earthquakes, was retrofitted by placing curved surface friction isolators below the steel roof, which can be considered the first retrofitting application with seismic isolators in Turkey [6]. Antalya Airport terminal building [7] and Grand Tarabya Hotel [8] in Istanbul were retrofitted with seismic isolators in 2005, while the latter was demolished and reconstructed conventionally afterward. Marmara University Basibuyuk Research and Training Hospital (Fig. 1a), retrofitted by seismic isolators, was finally in service in April 2020 [9]. A residential building (Fig. 1b & 1c) in the popular neighborhood of Moda in the Kadikoy district of Istanbul was also retrofitted as one of the urban renewal projects going on in Istanbul.

(a) (b) (c)

Fig. 1. Photos (a) from inside of Marmara University Başıbüyük Research and Training Hospital, (b) installed seismic isolator at the retrofitted residential building in Kadıköy, Istanbul (c) outside view of the retrofitted residential building

2.3 Industrial Facilities

In 1998, LNG tanks of Ege Gaz in Izmir were constructed using rubber isolators, while an ammonia tank in Samsun (Fig. 2) was retrofitted with seismic isolators in 2015. These applications can be listed as one of the few applications of seismic isolation among industrial facilities in Turkey.

Fig. 2. Photos during the construction stage of the retrofitting of the ammonia tank in Samsun

2.4 Data Centers

The interest of the investors in seismic isolation technology at the data center projects is high in Turkey. Turkcell, as a leading telecommunication operator, requires new data center investments to take advantage of seismic isolation technology. Turkcell data centers in Tekirdag Corlu, Kocaeli Gebze, and Izmir were constructed with seismic isolators. The data centers of Akbank in Kocaeli Gebze, Is Bankasi and Star of Bosphorus in Istanbul Tuzla were also constructed seismically isolated.

2.5 Hospitals

Kocaeli University Hospital, constructed in 2001, was Turkey's first seismically isolated hospital. The pace of seismically isolated hospitals in Turkey increased dramatically after the regulation of the Ministry of Health of Turkey, which mandated the use of seismic isolators for the state hospitals that will be constructed in seismic zones with more than 100-bed capacity. There are 65 seismically isolated hospitals as of 2021, and a lot more of them are being constructed or in the design phase at the moment.

Basaksehir Pine and Sakura City Hospital in Istanbul (Fig. 3) now holds the world record for being the world's largest seismically isolated building. The hospital has a total area of more than one million square meters and features more than 2,068 triple friction pendulum isolators with a minimum displacement capacity of 772 mm [10].

2.6 Residential Buildings

The first residential building with seismic isolators was constructed in the Silivri district of Istanbul in 2013. In the following years, the contractors could not be convinced to construct another residential building utilizing seismic isolators until 2020.

In March 2020, a residential building complex project, Mavera Comfort Apartments, was announced with 16 isolated apartment blocks (Fig. 4a). The building complex covers more than 170,000 m^2 and employs 454 lead rubber isolators with a calculated maximum displacement of 33 cm. A maintenance floor was created over the ground floor of each building as the main isolation interface (Fig. 4b). The elevators and stairs in the buildings' core were isolated at the foundation level. The required gap to accommodate the isolator displacements can be seen in Fig. 4b.

(a) (b)

Fig. 3. Photos (a) during the construction stage and (b) inner car park floor of Başakşehir Pine and Sakura City Hospital

(a) (b)

Fig. 4. (a) Aerial view photo of Mavera Comfort Apartments and (b) photo from the maintenance floor as the main isolation interface

2.7 Historical Structures

The seismic isolation technology was first used in the rehabilitation of historic structures during the translocation of the Zeynel Bey Tomb in 2017 [11]. The tomb was moved 1935 m away from the original location due to the flooding caused by the construction of the Ilısu Dam in the Hasankeyf district of the city of Batman in the southeastern part of Turkey.

Similarly, Nusretiye Clock Tower in Istanbul was relocated some meters away from the original location and placed over seismic isolators during the construction of the Galataport in 2019 (Fig. 5). Four curved surface sliding isolators were used, each having a displacement capacity of ±660 mm.

Göztepe Railway Station, constructed in 1915 on the railway line between Haydarpaşa (Istanbul) and Izmit, was retrofitted in 2019 with seismic isolators during the rehabilitation of the railway line of Istanbul as part of the Marmaray Project. The primary railway system had two rail lines passing under the station structure between the masonry piers (Fig. 6a). The new railway line required three lines which were not possible due to the position of the existing masonry piers. A solution was created by

(a) (b)

Fig. 5. Photos (a) during the installation process of seismic isolators and (b) current stage of the Nusretiye Clock Tower in Istanbul

demolishing the existing piers, temporarily suspending the structure, creating a new single reinforced concrete pier in the midspan, and inserting seismic isolators underneath the station structure (Fig. 6b). Nine curved surface friction isolators were used, each with a displacement capacity of $\pm$ 300mm. The retrofitted station building is no longer used as a railway station but for social purposes.

(a) (b)

Fig. 6. (a) Old railway station resting on two mid piers accommodating two rail lines (b) retrofitted station resting on single mid pier accommodating three rail lines

2.8 Seismic Isolation in Numbers

According to an ongoing study conducted by the members of the Turkish Association for Seismic Isolation, there are 102 isolated structures in Turkey, of which 65 are hospitals, 8 are bridges, and 3 are industrial (Fig. 7a). Approximately 44,826 isolators were used in those structures, of which 5,856 are for bridges, and 33,932 are for hospitals. 58% of isolators are curved surface friction isolators, while 41% are elastomer-based isolators (Fig. 7b). 76% of all isolators were used in hospital projects, while 13% were used in bridge projects (Fig. 7c). The numbers and percentages are due to change since the study is not official yet finalized.

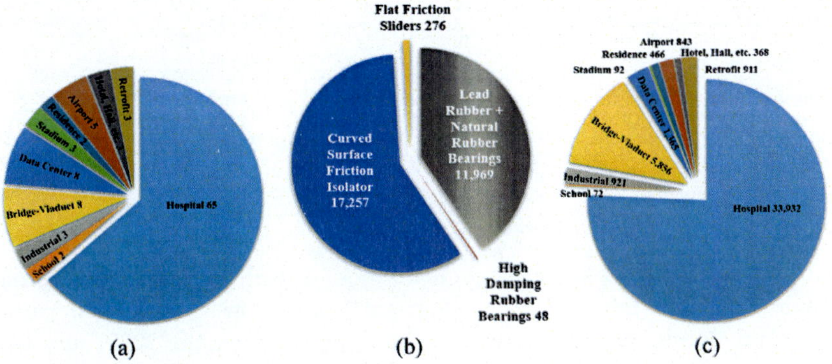

Fig. 7. Charts for (a) number of isolated structures (b) number of isolators (c) number of isolators per structure, as of 2021

2.9 Code Developments

The first seismic code of Turkey, which contains specific rules for the design of seismically isolated buildings, was released in 2018 [12]. Seismic isolation-related rules of the Turkish Seismic Code were described and compared with the other international seismic codes in detail by Erdik et al. [13]. Turkish Seismic Code for Bridges with seismic isolators and energy dissipating devices was also released in 2020 [14].

The Ministry of Environment, Urbanization and Climate Change announced a new declaration regarding a mandatory peer reviewing system for special structures, including seismically isolated buildings. Since 2019, licensed peer reviewers have been reviewing seismic hazard studies, soil-structure interaction analysis, seismic isolation and seismically isolated building designs.

2.10 Testing

Turkey's first dynamic seismic isolator testing facility, ESQUAKE was established on the premises of Eskisehir Technical University in 2018. Table 1 demonstrates the capacity of the facility.

3 Conclusions

The number of passive structural control system applications, especially seismically isolated structures, is increasing yearly in Turkey. Still, the majority of applications are at hospitals or bridges. After the launch of the Mavera Comfort Apartments project in March 2020, the interest in seismic isolation for other residential building projects going

Table 1. Testing capacity of ESQUAKE, seismic isolator testing facility

Maximum static vertical load	20,000 kN
Maximum dynamic vertical load	15,000 kN
Horizontal load	2,000 kN
Maximum displacement	±600 mm
Maximum velocity	1000 mm/s
Maximum size of the test specimen	1.5 × 1.5 m

on increased but could not find a way of realization. As well, the superior performance of seismically isolated Elazığ Hospital compared to other conventionally built buildings in the vicinity during Mw 6.5 Elazığ Earthquake on January 24th, 2020, attracted the attention of the public and the media but did not make the expected effect. The contractors still consider the system as a cost-increasing application.

Furthermore, seismic isolated building design is more controlled and of higher quality than previous years because of the mandatory peer reviewing system. Meanwhile, there are some problems in putting the peer reviewing system on legal grounds; the peer reviewing system was temporarily suspended at the end of July 2022. The author believes that the peer reviewing system is essential and somehow needs to be transformed into professional engineering in the near future.

Acknowledgments. The author would like to express his sincere gratitude to Prof.Mustafa Erdik and the members of the Turkish Association for Seismic isolation for providing the necessary data for this paper. Moreover, the author would like to thank Kerim Efe Özcanlı, Mehmet Emre Özcanlı, Dr.Fatih Sütçü, and Dr. Ian Aiken for their contributions to the preparation of the inventory of the seismically isolated structures in Turkey.

References

1. Yenidogan, C., Erdik, M.: The state of art in seismic isolation and energy dissipation applications in Turkey (「各国の最近の免震構造」 紹介シリーズ アルメニアとトルコ). Menshin **83**, 49–59 (2014)
2. Erdik, M., Tuzun, C., Ulker, O.: Evaluation of seismic isolation applications of health care facilities in Turkey. In: Proceedings of the 14th World Conference on Seismic Isolation, Energy Dissipation and Active Vibration Control of Structures, San Diego (CA), USA (2015)
3. Erdik, M. O.: State of the art on application, R&D and design rules for seismic isolation and energy dissipation for buildings, bridges and viaducts, cultural heritage and chemical plants in Turkey. In: Proceedings of the NZSEE Conference (2017)
4. Erdik, M.: State of the art on application, R&D and design rules for seismic isolation and energy dissipation in Turkey. In: Proceedings of the 16th World Conference on Seismic Isolation, Energy Dissipation and Active Vibration Control of Structures, St.Petersburg, Russian Federation (2019)
5. Roussis, P.C., Constantinou, M.C., Erdik, M., Durukal, E., Dicleli, M.: Assessment of performance of seismic isolation system of Bolu Viaduct. J. Bridg. Eng. **8**(4), 182–190 (2003)

6. Constantinou, M. C., Whittaker, A. S., Velivasakis, E.: Seismic evaluation and retrofit of the Ataturk international airport terminal building. In: Proceedings of the 2001 Structural Congress and Exposition, Washington, DC, USA , pp. 21–23 (2001)
7. Yilmaz, C., Booth, E., Sketchley, C.: Retrofit of Antalya airport international terminal building, Turkey using seismic isolation. In: Proceedings of the 1st European Conference on Earthquake Engineering and Seismology (2006)
8. Coşkun, N.: Rehabilitation of a reinforced concrete building in Turkey under the risk of earthquake. Eng. Sci. **5**(2), 300–309
9. Kubin, J., Kubin, D., Özmen, A., Şadan, O. B., Eroğlu, E., Sucuoğlu, H.: Seismic retrofit of an existing multi-block hospital by seismic isolators. In: Proceedings of the 15th World Conference on Earthquake Engineering, Lisbon, Portugal (2012)
10. Zekioglu, A., Gogus, A., Binzet, S., Chok, K.: Performance-based seismic design succeeds in Turkey. Civil Eng. Magaz. Arch. **91**(2), 56–63 (2021)
11. Yılmaz, M.: Hasankeyf Zeynel Bey Türbesinin Koruma ve Kurtarma (Taşıma) Projesi (The Conservation and Salvage (Relocation) Project of Zeynel Bey Tomb in Hasankeyf) (2017)
12. TBDY-2018, Türkiye Bina Deprem Yönetmeliği (2018). https://www.resmigazete.gov.tr/eskiler/2018/03/20180318M1.pdf
13. Erdik, M., Ülker, Ö., Şadan, B., Tüzün, C.: Seismic isolation code developments and significant applications in Turkey. Soil Dyn. Earthq. Eng. **115**, 413–437 (2018)
14. Deprem Etkisi Altında Yalıtımlı ve Sönümleyicili Köprüler ve Viyadüklerin Tasarımı için Esaslar (2020). https://www.resmigazete.gov.tr/eskiler/2020/10/20201006M1-5-1.pdf

Seismic Isolation Applications in Romania

Andreea Dutu[1(✉)], Adrian Iordachescu[2], Dragos Mocanu[3], and Lucian Soveja[4]

[1] Technical University of Civil Engineering Bucharest, Bucharest, Romania
andreea.dutu@utcb.ro
[2] Ion Mincu University of Architecture and Urbanism, Bucharest, Romania
[3] European Business Consult SRL, Bucharest, Romania
[4] Exprosov Iasi SRL, Iaşi, Romania

Abstract. Romania is one of the most seismic prone countries in Europe, as it has the Vrancea point source producing intermediate depth earthquakes and affecting a vast territory. The last big earthquake was in 1977 with a magnitude of 7.5 M_W on Richter scale. While the building stock is gradually aging, since most of it was built during the communist period (more than 40 years ago), many buildings, heritage or not, are now considered vulnerable to strong earthquakes and need urgent interventions. The conventional strengthening solutions, such as adding shear walls or jacketing, are usually being used due to economic reasons, but there are situations in which the disruption of the building's functionality is the most important factor when deciding how to strengthen/retrofit a building. For such situations, recent technology such as seismic isolation is suitable. In Romania, there are few applications of seismic isolation, only applied at existing heritage buildings. The case studies will be briefly presented, with a special focus on the latest one, which is now under construction, the Fire Tower (Foisorul de Foc) Building, where 18 rubber bearings and 5 viscous dampers were used to seismically isolate the masonry tower from the foundation.

Keywords: Rubber bearings · Friction pendulum · Dampers · Seismic · Masonry · Heritage

1 Introduction

Seismic isolation is a technology applicable everywhere in the world with the aim to protect the buildings from lateral forces, which can be seismic or from wind. This technology nowadays is not new, but in developing countries the engineers are still getting familiar with the specific practices corresponding to its application.

Japan is the leader in applying the seismic isolation system in the world and based on JSSI data in 2021, it had already 12,000 applications in buildings (according to Dr. Matsutaro SEKI, Fig. 1). However, there are also other countries following the good example: China (Fig. 3) has more than 10,000 applications on buildings (according to Prof. Zhe QU), Russia had 700 applications, Italy had 400 applications in 2013 [1] (Fig. 5), while USA had approx. 250 applications in 2013 [2]. Other countries in which the number of applications of new anti-seismic technologies is significant are South

G. P. Cimellaro (Ed.): WCSI 2022, LNCE 309, pp. 26–46, 2023.
https://doi.org/10.1007/978-3-031-21187-4_3

Korea, Taiwan, Armenia, New Zealand, France [3], Turkey [4], Mexico, Canada and Chile. In many cases Italian devices have been used [2].

Armenia, had 55 completed isolated buildings in 2020 [5]. In New Zealand, one of the motherlands of anti-seismic devices (in particular of those based on the use of lead, like LRBs and LDs) the isolated structures had an excellent behaviour event (M = 6.3). Similarly, the isolated structures in Santiago, Chile, showed an excellent behaviour during the Maule earthquake of February 27th, 2010 (M = 8.8) [2]. Examples of seismically isolated buildings or devices from China, Chile and Russia are shown in Figs. 2, 3 and 4.

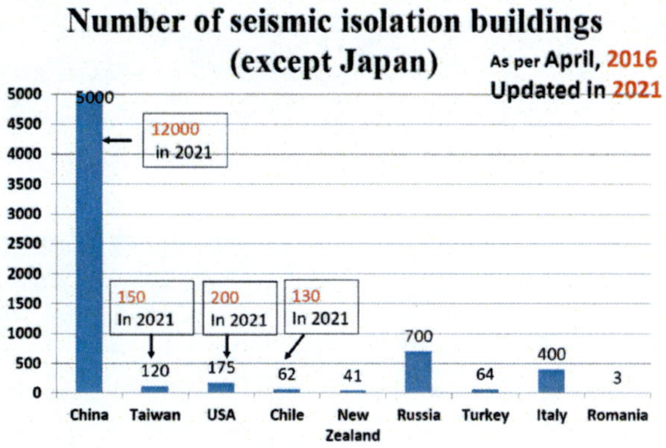

Fig. 1. Seismically isolated buildings in the world according to Dr. Matsutaro SEKI

Fig. 2. Seismically isolated building in China (Pingchuan Hospital), Gansu province

Fig. 3. Experimental test on an innovative isolator in Sankt Petersburg

Fig. 4. The mid-story isolated building Innovacion Tecnologica UC, Anacleto Angelini (CIT) building, Macul Santiago

2 Seismic Isolation in Romania

Romania is one of the countries in Europe with high seismic hazard (Fig. 5), which is mainly coming from the Vrancea seismic source. Vrancea can be characterized as a point source producing intermediate depth earthquakes, high peak ground accelerations, long predominant periods of the records (about 1.4–1.6 s in Bucharest), and undesirable side effects due to the soft soil conditions. All these have led to heavy losses and great damages in several catastrophic earthquakes (e.g. the seismic event of March 4 1977 of 7.5 M_W, as recorded in Bucharest);

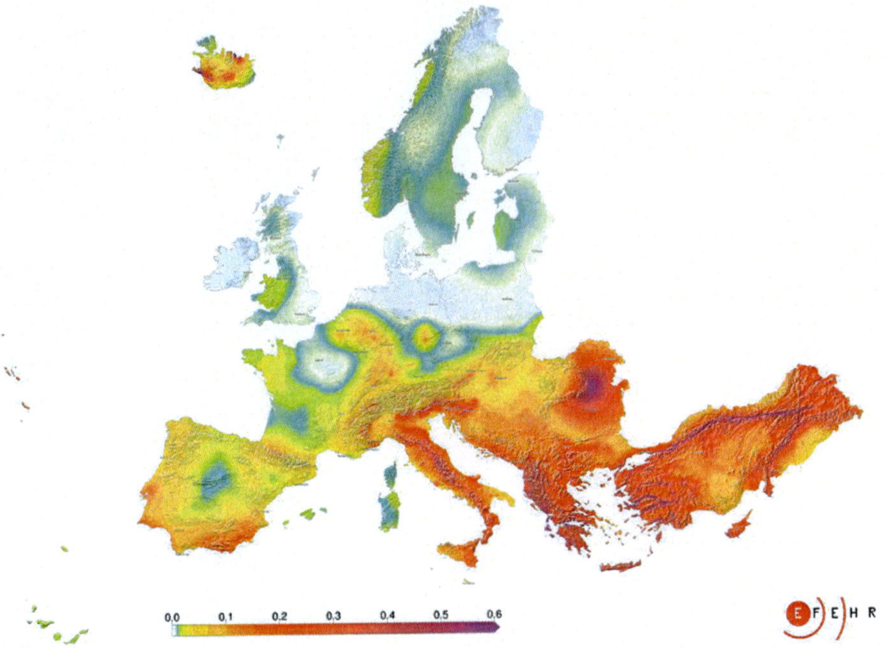

Fig. 5. The hazard map of Europe [6]

While the building stock is gradually aging, since most of it was built during the communist period (more than 40 years ago), many buildings, heritage or not, are now considered vulnerable to strong earthquakes and need urgent interventions. According to an evaluation from 2019, only in the highest seismic risk class (RS1) there are 363 buildings in Bucharest.

Many strengthening interventions, if applied, are made using conventional methods such as shear walls or by jacketing of beams, columns and walls, and this is due to economic reasons. But in some situations, such methods may not offer enough additional strength, or the functionality of the building is not allowed to be interrupted. So other solutions should be found, and recently authorities and engineers applied, even if in few cases, seismic isolation for retrofitting existing heritage buildings and response control systems for new buildings.

For private owners investing in construction of new buildings, the seismic isolation is not yet attractive, because the initial cost is slightly bigger than for a conventional building and since the last big earthquake was more than 45 years ago, the seismic danger seems to be forgotten.

However, academia and research specialists are making efforts to spread the knowledge on the benefits of seismic isolation so the general population would be aware that there are methods and technologies nowadays available that can protect their investment and keep the functionality of the buildings and businesses inside even after major earthquakes.

Events were organized to make the seismic isolation popular among the design and construction engineers, such as the Bucharest seminar on Seismic Isolation, which was funded by the Ministry of Land, Infrastructure and Tourism from Japan and the Japanese Embassy in Bucharest. Within this event, professors, researchers, and professionals from Japan held presentations at the Technical University of Civil Engineering Bucharest, opened to anyone interested in the topic. Moreover, with the support of JSSI, a textbook for students was also published, to introduce the construction engineering principles of the seismic isolation [7].

Due to the long period component of the Vrancea earthquake, seismic isolation applications in Romania need big displacement capacity isolators which can be difficult to be tested in certified laboratories. But since these were steps which were already done before, as will be further presented, more applications are expected within the next years.

This paper will present chronologically a state of the art for the applications of seismic isolation on buildings in Romania.

3 Victor Slavescu Building of the ASE (Economics Studies Academy) in Bucharest

The "Victor Slavescu" building is a national heritage building, being built in 1905 (Fig. 6). The structure is made of clay brick masonry, with concrete slabs, having a timber roof structure covered by metal. The floor over the underground is made of metal beams with masonry arches, a very popular system for the buildings built at the end of the 19th century. The walls' thickness varied along the height, being from 112 cm in the half basement and reaching 56 cm at the second floor. There is also an attic. The foundation is a simple concrete mat foundation with a varying height [8]. The building's dimensions are 55.5 m by 20.87 m in plan and has a height of 22.50 m.

After experiencing two of the strongest earthquakes known in the Romanian seismic history (the 1940 and 1977) and also considering the wear-out of the building, after expert evaluations the building was considered unsafe and strengthening was recommended.

Although initially the strengthening solution involved shear walls, this was totally disrupting the building's functionality, so the owners and engineers looked for other solutions. Thus, the seismic isolation was proposed, to decouple the superstructure from the foundation, and thus to protect it from future big earthquakes.

The final decision on the strengthening solution was based on the fact that the building is an important heritage for both history and architecture and also on the continued functionality during the strengthening works.

A combination of linear and non-linear dynamic analysis was used. Real time-series data obtained from accelerograms recorded by INCERC during the 1977, 1986, and 1990 earthquakes were applied in the analysis. The accelerograms were scaled to a maximum acceleration of PGA = 0.24 g (0.36 g for the isolation system) [8].

Based on an initial modal analysis, the first period of vibration resulted in $T_1 = 0.79$ s. In order to reduce the forces in the structural elements, and also the story drifts, several effective stiffnesses were considered for the isolators. The result was a first period of $T_{isolated} = 2.9$ s.

Fig. 6. "Victor Slavescu" building [8]

The isolation was made using 80 natural rubber bearings (SEP—USA), 700 × 700 × 375 mm, with maximum lateral displacement 600 mm, and 18 dampers (Taylor Devices USA) maximum force 1500 kN and ± 500 mm stroke [8].

Prototype tests were compulsory and were conducted in UCSD in USA on one pair of rubber bearings.

The isolation interface between the building superstructure and the foundation is comprised by an upper support frame, the isolators and dampers, and a lower support frame (Fig. 7).

Fig. 7. The rigid upper and lower concrete frames between which the isolators and the dampers are placed [8]

The displacements at the location of the isolation interface are estimated at 20–25 cm and 30–35 cm for the longitudinal and orthogonal directions, respectively [8]. Reduction stress factors of 3.0–3.2 were obtained through base isolation. The seismic isolation solution was estimated to allow a saving of 8% in structural cost over the conventional strengthening solution [8].

4 Bucharest City Hall

The building was constructed between 1906 and 1911 and it is a national heritage (Fig. 8). The structural system is made of unreinforced brick masonry walls connected with concrete grid slabs (Fig. 9), all resting on a mat foundation of 1.3 m depth. Underneath the foundation, there is a sand pillow which was naturally compacted by flooding during a one year period [9]. The total area of the building is 22,000 m^2, and the dimensions are 89 × 65 m, with a total height of 20.8 m.

Fig. 8. Bucharest city hall (@ Dr. Matsutaro Seki)

Based on an expert evaluation, and its damage it experienced during the past earthquakes, the building was classified as a vulnerable one and it was considered that it needs strengthening. This building was requested by the Ministry to be retrofitted for an MRI of 475 years, equivalent to a 0.36 g PGA, although the code minimum requirement at that time was 0.24 g, thus the final capacity increased even more than as for a new building at that time. Nowadays, the updated code P100-1:2019 increased the minimum PGA value to 0.36 g for Bucharest.

The analysis of the unretrofitted building revealed a first period $T_1 = 0.73$ s.

From the design, it resulted the necessity of large deformation capacity isolators (700 mm), and tests on these could only be made in UC San Diego, because in Pavia, only 500 mm capacity was available. Besides the seismic isolation, the masonry walls in the superstructure were also strengthened and piles were executed with a length of 13 m, because the good foundation soil was found to be between 8.3 m and 15 m from the geotechnical study. The concrete beam floor system was also strengthened, by adding a concrete cover to obtain a rigid diaphragm.

Fig. 9. Bucharest city hall grid slab prepared to be strengthened with a concrete cover

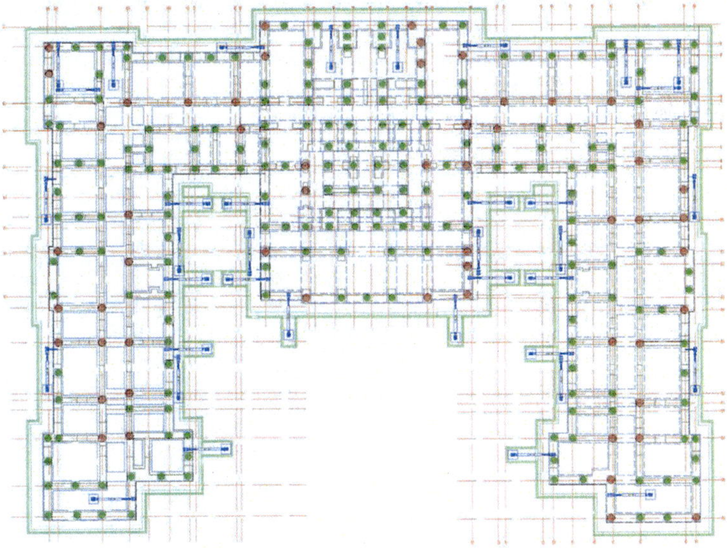

Fig. 10. Bucharest city hall isolation interface plan (Legend: green circle is isolator with a diameter of 1000 mm; red circle is isolator with a diameter of 1050 mm; and blue is viscous damper)

The isolation solution consisted in using 262 isolators of two types (Fig. 10): 206 pieces with a 1000 mm diameter and 56 pieces with a 1050 mmm diameter, but both with a capable displacement of 700 mm and 36 dampers of a max. Displacement of 700 mm and max. Force of 1750 kN. The final values obtained for the periods of the isolated building were $T_{isolated} = 3.3$ s.

In order to accommodate the isolation interface, the masonry walls were cut in the basement, steel boxes were inserted and concreted and thus a rigid R.C. frame was erected to carry the superstructure's loads and transmit it to the isolators (Fig. 11).

Fig. 11. The steel boxes (left) and the rigid frame above the isolation interface (right)

Rubber bearings from Freysinnet were installed, and the contact with the column was done using flat jacks filled with epoxy resin (Fig. 12). The contractor tested the flat jacks to see what happened inside it, and how the epoxy resin spreads (Fig. 12c). The flat jack should have been filled to have a 2.5 cm thickness. The dampers also from Freysinnet were also installed in the building (Fig. 13). Figure 14 shows the isolation gap, which was covered with a concrete slab as a cover for the gap, because the alternative with expansion joints were very expensive.

a b

c

Fig. 12. Rubber bearings installed on steel plated with anchor bolts (a); the flat jack installation after the casting of the concrete mat under the isolators (b); testing the spread of the resin (red color) inside the flat jack (c)

Fig. 13. Damper installed

Fig. 14. The isolation gap for which a concrete cover was made to protect people and objects from falling into the gap

5 Arch of Triumph in Bucharest, 1936

The Arch of Triumph was built in a rush for the military parade after the Independence war, but it deteriorated after some time, so it was reconstructed in 1922 to celebrate Romania's participation in the First World War. Later, it was again reconstructed (1935–1936), this time of stone (Fig. 15) with overall dimensions of 12.00 m × 25.50 m × 25.50 m, crossed by a front 15.20 m height opening [10].

Fig. 15. The Arch of Triumph made of stone (1923) [11]

The foundations are made of reinforced concrete (RC), and the superstructure of RC frames with masonry infills and from the height of + 4.34 m it is entirely made of RC, including the floors.

The building was evaluated in 1994, by the late prof. Radu Petrovici, and considered vulnerable to earthquakes, as its was presenting water infiltrations visible since 1980 and also seismic damages to the stairs and some parts of masonry infills due to the 1977 earthquake. The dimensions of the RC beams and columns lead to the conclusion that they were only designed for gravity loads, thus the frame had no stiffness to lateral loads. The recommendation of strengthening was accompanied by the selection of a non-invasive solution, and the seismic isolation was the most appropriate since the building is also heritage and part of a protected area in Bucharest.

The numerical analysis conducted for the Arch of Triumf used a combination of linear and non-linear dynamics and included data from the accelerograms recorded after the last three major earthquakes in Romania (1977, 1986 and 1990). The accelerograms were scaled at a 0.24 g peak ground acceleration (PGA), with a 0.36 g value for the isolation system). It also revealed a first period $T_1 = 0.3$ s. After analysis and optimization of the isolators' stiffness, the final value for the first period of the isolated building was $T_{isolated} = 3.7$ s, so the stresses would be reduced 3 times.

The application of seismic isolation involved the construction of a superior carrying RC frame, on which the superstructure is resting, and an inferior carrying frame, resting on the foundations. Between them, the isolation interface contains 24 rubber bearings (750 mm diameter) and 8 viscous dampers (Taylor Devices). The absolute displacements at the location of the isolation are estimated at 20–25 cm [10].

Under the existing foundations, micro-piles were executed (Fig. 16), because the soil underneath presented some softer areas and needed strengthening. Then, the masonry in the superstructure was cut to introduce the "supporting steel tables" (Fig. 17) and the superior carrying frame was reinforced (Fig. 18) and concreted (Fig. 19). The isolators (made by SEP, USA) and dampers (made by Taylor Devices, USA) were installed, and the finished isolation interface is shown in Fig. 20. The building is shown in Fig. 21 after the finish application.

Fig. 16. The drawing of the micro piles, retaining wall, upper carrying frame, inferior carrying frame and isolation interface (@ CredoDesign [12])

Fig. 17. The "supporting steel table"

Fig. 18. The reinforcement placement in the superior carrying frame and retaining wall

Fig. 19. The freshly placed concrete in the superior and inferior carrying frames

Fig. 20. The isolation interface and retaining wall (to the right) (@ Credo Design)

Fig. 21. The renovated and strengthened Arch of Triumph

6 Aroneanu Church in Iasi [14]

The structural system of St. Nicolae Aroneanu church (Fig. 22), erected in 1594, consists of brick masonry walls, with a 1.20 m thickness. The foundations are continuous footings made stone masonry with lime mortar, having thicknesses similar to those of the above brick masonry walls. All horizontal structural elements of the church, consisting of domes, arches and pendants, are made of brick masonry [14].

Fig. 22. The Aroneanu church [14]

The in-plan layout of the church is symmetrical with respect to the longitudinal axis and has a rectangular shape, this representing a hint on the fact that the builders were aware of the seismic prone area. Also on the vertical direction, the configuration is well designed, as it has the transition between the walls of the nave and the tower done by two consecutive horizontal bases, the first one being square shaped and the second one being star-shaped [14].

However, in past seismic events, cracks occurred in both vertical and horizontal structural components, demonstrating important losses in terms of structural redundancy [14]. Thus, after an expert evaluation, the only solution was to apply seismic isolation. Friction pendulum isolators were chosen, provided by Maurer (Germany). A horizontal reinforced concrete (R.C.) carrying frame and a raft foundation underneath the existing one were executed, and 48 friction pendulum sliding (FPS) devices were placed between them. Then, ventilation and maintenance exterior culvert (isolation gap) and a R.C slab over the carrying frame [14] (Fig. 23).

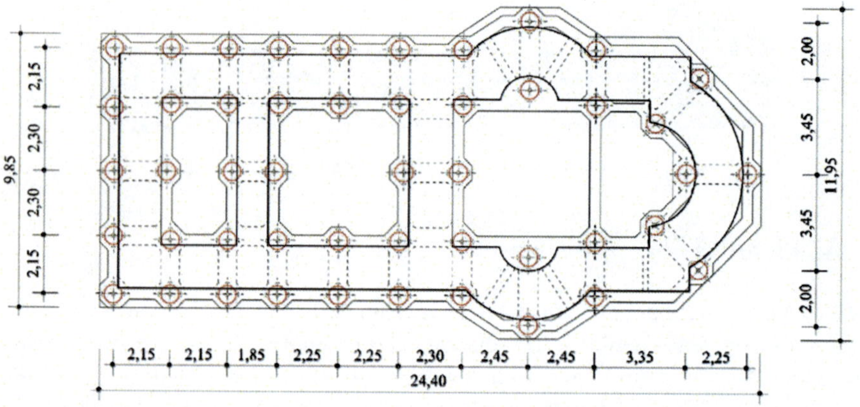

Fig. 23. The position of the friction pendulum isolators in the church [14]

The $T_{isolated}$ was increased to 2.8 s from an initial one $T_1 = 0.23$ s [14]. More details can on this application can be found in [14], both from the analysis and the construction execution point of view.

7 Foisorul de Foc (Fire Tower) Bucharest

The building Foisorul de Foc is nowadays the National Museum of Firefighters, and it represents a national heritage building since 2015, being a representative building from the 19th century. The structure is made of masonry walls, having several parts, developed on the vertical direction. The shape of the building is a perfect circle. The roof is made of steel arches, radially placed, forming a dome and from its center another steel structure rises, the later one having a diameter and height of approx. 3 m.

The built surface is 225 m^2, excluding the steel structure.

The building was built in 1892 to solve two problems for the city administration: for the firefighters to supervise the city and run the alarm in case of fire and to modernize the water supply in the east part of the city.

The structural system consists of 16 pilasters placed tangent to a circle having a 10.8 m diameter. The pilasters have a variable section on the vertical direction (Fig. 24).

Beneath the 20.15 m height, the pilasters are connected with structural walls of 30 cm thickness. All the elements are made of brick masonry with lime mortar, with a granite foundation. The floors are made of steel beams with brick vaults and with a finishing of mosaic marble of 5–10 cm. At the 20.85 m height, there used to be a water reservoir which was disassembled and the replacing structure and also the one above, were made of reinforced concrete, forming a 30 cm thickness cylinder and floor beams.

Fig. 24. The Fire Tower (Foisorul de foc) [13]

After the expert structural evaluation, the building was considered vulnerable to earthquakes due to its weak materials, lack of maintenance in the past 20 years, and also due to water infiltrations at the foundation level. The retrofit solution adopted was seismic isolation together with crack injections in the pilasters and floors, and application of laminated carbon fiber on the damaged elements.

By using 18 rubber dampers with 750 × 750 × 400 mm dimensions, capable of 500 mm lateral displacement, and also with 5 dampers which can develop a 1500 kN force and a 500 mmm displacement, the strengthening solution can provide a stress reduction of almost 3 times on both directions, placing the retrofitted building within the seismic risk class 3–4. The seismic risk class 3 allows some structural degradation, but not significantly affecting the safety of the building. The seismic risk class 4 is corresponding to the new buildings' seismic response.

From the modal analysis, the first period of the building resulted in $T_1 = 0.28$ s. And after the application of the seismic isolation the period is estimate at $T_{isolated} = 3.7$ s.

For this building, too, micro piles were injected underneath the foundation, then the inferior carrying frame was executed and further, the "supporting steel tables" are installed (Figs. 25 and 26). Later, the superior carrying frame will be executed, and then the isolation interface will be equipped with the isolators and dampers, made by Maurer (Germany). In the end, the walls and floors elements will be injected with epoxy and then laminated carbon fiber or steel plates (where the elements are strong enough to allow a strong connection) will be applied to prevent future cracks.

Fig. 25. Cutting of the masonry and concrete foundations

Fig. 26. Installing the "supporting steel tables"

8 Conclusions

Romania has many buildings vulnerable to earthquakes, and many of them are considered important national heritage sites. While for some of them, the surrounding area may not allow the application of seismic isolation, for many of them the method is suitable. But engineers and authorities still need courage to apply seismic isolation, and owners need to understand the benefits of such a method. For this, they all need to be informed. Efforts are being made towards a wider implementation, for existing buildings and for new ones.

Acknowledgments. The authors would like to acknowledge Mrs. Aurora Tarsoaga, the architect who was head of two of the projects presented in this paper (Arch of Triumph and Firetower, both located in Bucharest).

References

1. Clemente, P., Martelli, A.: Seismically isolated buildings in Italy: state-of-the-art review and applications. Soil Dyn. Earthq. Eng. **119**(December 2017), 471–487 (2019)
2. Clemente, P., Martelli, A.: Anti-seismic systems and conditions for their correct use. In: 16th World Conference on Earthquake Engineering, 16WCEE 2017 Santiago Chile, January 9th to 13th 2017, no. 1, pp. 1–28 (2017)
3. Whittaker, A.S., Kumar, M., Kumar, M.: Seismic isolation of nuclear power plants. Nucl. Eng. Technol. **46**(5), 569–580 (2014)
4. Erdik, M., Ülker, Ö., Şadan, B., Tüzün, C.: Seismic isolation code developments and significant applications in Turkey. Soil Dyn. Earthq. Eng. **115**(February), 413–437 (2018)
5. Melkumyan, M.G.: Concepts and features of seismic isolation. In: International Conference on Applied Science, Engineering and Mathematics, no. June, pp. 4–5 (2020)

6. Danciu, L., et al.: The 2020 update of the European Seismic Hazard Model: model overview (2021)
7. Casuta (Dutu), A.: Introduction to the Construction Engineering of Seismic Isolated. MatrixRom (2021)
8. Iordachescu, A., Iordachescu, E.: Seismic Rehabilitation of Victor Slavescu Building—Calea Grivitei—2–2a Through Base Isolation Method, no. figure 1, pp. 514–515 (1977)
9. Iordachescu, A.: Seismic rehabilitation of Bucharest city hall building through base isolation method summary: the Bucharest city hall building was erected between 1906 and 1911. The brick masonry building with concrete slabs has a simple concrete mat foundation 1.30m. In: 15th World Conference on Earthquake Engineering, Lisbon, Portugal (2012)
10. Iordachescu, A.: Seismic rehabilitation of Bucharest arch of triumph through base isolation method. In: Seismic and Energy Renovation for Sustainable Cities (SERC) (2018)
11. Editura Socec & Comp., București, 1923 - România Mare Pitorească - Album cu vederi și porturi populare, pagina 2
12. Tarsoaga, A.: CredoDesign [Online]. Available: https://credodesign.ro/archives/1774#sidr. Accessed: 03 Mar 2020
13. Wikipedia [Online]. Available: https://ro.wikipedia.org/wiki/Foișorul_de_Foc. Accessed: 29 June 2022
14. Lupășteanu, V., Soveja, L., Lupășteanu, R., Chingălată, C.: Installation of a base isolation system made of friction pendulum sliding isolators in a historic masonry orthodox church. Eng. Struct. **188**(March), 369–381 (2019)

Efforts Toward International Harmonization of Seismic Isolation Design Code and Current Status in Japan

Taiki Saito(✉)

Toyohashi University of Technology, Aichi 08544, USA
saito.taiki.bv@tut.jp

Abstract. Structures that use seismic isolation technology have been built since the 1970s. The effectiveness of seismic isolation technology has been verified in many earthquakes. By adopting the seismic isolation technology, the inertial force acting on the building can be dramatically reduced, so that not only the structural safety of the building but also the function of the building can be secured by preventing the fall of the indoor contents. There is no other such excellent seismic technology other than the seismically isolated structure. However, existing design standards of seismically isolated structures such as Eurocode, ASCE, Japanese Design Guide, etc. are not consistent even in their basic philosophy. Also, promotion of good quality seismically isolated structures is needed in all earthquake prone countries. In order to spread the seismic isolation technology more widely and internationally, it is necessary to develop internationally harmonized standards for ensuring the performance of the seismically isolated structure. It should include 1) the design method for the seismically isolated structure to achieve the target performance of the building, 2) the construction and maintenance method for the seismically isolated structure, 3) the manufacturing and testing method for the seismic isolation devices. In 2018, a WG (working group) was established within ISO (International Organization for Standardization) to create an international standard for seismically isolated structures. After active discussions in the committee, a standard has been developed that summarizes the basic concepts that will encompass the seismic isolation standards of various countries.

Keywords: Seismic isolation · International standard · Earthquake design · Building safety

1 Introduction

There are many types of ISO standards, such as "product standards" including emergency exit marks (ISO 7010) and screws (ISO 68), and "management system standards" including quality management systems (ISO 9001) and environmental management systems (ISO 14001). The purpose of international standardization is to provide the same level of quality throughout the world. They also have a significant impact on domestic standards and the activities of the construction industry.

G. P. Cimellaro (Ed.): WCSI 2022, LNCE 309, pp. 47–58, 2023.
https://doi.org/10.1007/978-3-031-21187-4_4

In order to establish a new ISO standard, a new proposal (NP) must be submitted by a national member organization or the secretary of a technical committee (TC)/subcommittee (SC) within ISO, and then adopted by a vote. The working group (WG) is then established under the TC/SC, which in principle works for a period of three years and goes through the stages of working draft (WD), committee draft (CD), draft international standard (DIS), and final draft international standard (FDIS), before the standard is established. At any of these stages, the standard cannot proceed to the next stage without the approval of more than 2/3 of the P-members (Participating Members) of the TC/SC through balloting.

As an earthquake-prone country, Japan has been actively involved in the establishment of ISO standards related to seismic design and loading. For ISO standards on seismic action, a WG with Japan as a convenor was established under TC98 "Bases for the design of structures", and the ISO 3010 "Bases for the design of structures—seismic action on structures" [1] was published in 1988. The ISO standard has been reviewed after enactment, and the third edition was completed in 2017 after a revision process. In 2020, the ISO standard was converted into the JIS (Japanese Industrial Standard) A 3306.

Regarding seismically isolated structures, the ISO standard for elastomeric seismic isolators (ISO 22762 [2]) was published in 2005. The standard was developed by TC45/SC4/WG9 "Elastomeric Isolators" under the chairmanship of Toshio Nishi (Professor Emeritus, Tokyo Institute of Technology). Subsequently, in 2011, a JIS standard was established in accordance with the ISO standard. The domestic secretariat of ISO/TC45 is the Japan Rubber Industry Association.

For the ISO standard on design methods for seismically isolated structures, Japan proposed the establishment of a WG at the 2017 ISO/TC98 meeting (London), and after voting and adoption of a new proposal (NP), the WG was officially established as the ISO/TC98/SC2/WG13 "Design principles of seismically isolated structures" at the 2018 ISO/TC98 meeting (Prague). In 2019, WG13 meetings were held in China (Guangzhou), Russia (St. Petersburg), and the U.S. (Denver, TC/98 meeting). From 2020 onwards, all WG13 meetings were held online due to the COVID-19. As of May 2022, ISO/DIS 23618 "Bases for design of structures—general principles on seismically isolated structures" has been approved and is now in the stage of the final draft of international standard, FDIS [3].

The members of WG13 are nominated from P-member and O (Observing)-member countries of ISO/TC98/SC2. The seven countries participating in WG13 are Japan, China, India, Russia, New Zealand, Poland and Italy. Other countries not belonging to ISO/TC98/SC2 are not allowed to officially comment on the content of the standard, but in the interest of creating an internationally harmonized standard, the author has taken advantage of international conferences on earthquake disaster prevention to introduce the contents of the standard and to hear opinions from experts in other earthquake-prone countries, such as the United States, Chile, and Turkey. In Japan, the draft of the standard has been introduced to engineers and researchers through the Japan Society of Seismic Isolation (JSSI).

The standards related to ISO/TC98 are said to be "Code for Code Writers," and the ISO standard for seismically isolated structures is a summary of basic design principles and does not provide specific numerical values for design. However, the ISO standard covers important issues to be considered when designing seismically isolated structures. Especially, since the quality of the seismic isolation system and its construction and maintenance greatly affect the performance of a seismically isolated structure, the Japanese side strongly insisted that these items be included in the text of the standard.

2 Outline of ISO 23618

(1) Scope

This standard describes the basic principles for the design of seismically isolated structures against earthquake motion. It does not cover bridges and LNG tanks, but some of the basic principles can be referred to, and does not cover seismic isolation systems that reduce vertical responses.

(2) Basic principles of structural planning

It states that a seismically isolated structure has a seismic isolation layer between the upper and lower structures, and that the seismic isolation layer includes isolators that extend the natural period and dampers that increase damping. Then, it describes the basic structural planning for each of the following: seismic isolation layer, upper and lower structures, foundation, seismic isolation device joints, seismic isolation clearance, and nonstructural members and equipment in the seismic isolation layer.

(3) Target performance of the seismically isolated structure

The structure should be able to maintain its function with no damage against earthquakes that may occur during its service life (serviceability limit state). The structure should have limited damage and be repairable, and be able to maintain its function even immediately after the earthquake against the maximum earthquake that may occur at the site (ultimate limit state). The description also states that the structure has sufficient redundancy to protect the occupants against unpredictable natural occurrences and accidents. It also states the target performance of the superstructure, substructure, and the seismic isolation system.

(4) Design seismic force

It describes that the seismic motion and design response spectrum shall conform to the ISO 3010 "Bases for the design of structures—seismic action on structures". The design response spectrum shall be used for the response spectrum analysis using an equivalent linear system, and that the design seismic motion shall be used for the time history response analysis.

(5) Structural analysis

For the response spectrum analysis using an equivalent linear system, the requirements are as follows: the building must be base-isolated, the superstructure must be sufficiently stiff compared to the seismic isolation layer; higher-order vibration modes must be negligible; torsional behavior must be unlikely to occur; the nonlinear restoring force of the seismic isolation system must be expressed by the equivalent restoring force using the equivalent stiffness and the equivalent damping; the upper and lower limits of the restoring force characteristics should be considered in the analysis; the superstructure is designed using the maximum shear force of the isolation system, and no tensile force acts on the isolators. It also describes the design equations for obtaining the equivalent period, equivalent damping factor, maximum response value, and shear force of the superstructure.

In the time history response analysis, the requirements include: the nonlinear restoring force characteristics of the seismic isolation layer shall be explicitly modeled based on the constitutive law of the isolation system; the tensile force acting on the isolation system shall be limited; and the fatigue characteristics of the isolation system against earthquake and wind shall be appropriately considered. It also describes that a sufficient number of earthquake motions should be used in accordance with the ISO 3010, and that the upper and lower limits of the variation of the stiffness and bearing capacity of the seismic isolation system should be considered in the analysis.

(6) Construction management specified in design documents

This section describes the basic items to be checked by the designer for each of the following: construction plan, quality control of the seismic isolation system, temporary construction plan, and construction procedure of the seismic isolation layer.

(7) Maintenance specified in design documents

This section describes the maintenance of the seismic isolation system, performance monitoring, and warning signs for occupants, which should be checked by the designer.

(8) Performance requirement of isolation devices

This section describes the performance information of seismic isolation devices, testing of the seismic isolation devices in accordance with ISO 22762, which should be confirmed by the designer.

(9) Appendix

In addition to the main text of the standard, the following Appendixes are prepared: A. Classification and performance characteristics of isolation devices, B. Construction management of seismically isolated structures, C. Maintenance of seismically isolated structures, D. Wind-resistant design of seismically base isolated buildings, and E. Mid-story seismically isolated buildings.

3 Comparison of Design Codes

In Japan, the Building Standard Law provides two methods for verifying the structural safety of seismically isolated buildings: the response and limit strength calculation and the time history response calculation. The response and limit strength calculation is specified in Notice No. 2009 [6] of the Ministry of Land, Infrastructure, Transport and Tourism (MLIT), Japan, and is a simple response calculation method that uses a one-degree-of-freedom model and the acceleration response spectrum of the earthquake ground motion, assuming the superstructure is rigid. A similar response calculation method is specified as the Simplified Linear Analysis in Eurocode8-10 and the Equivalent Lateral Force Procedure in IBC (ASCE7-16). Table 1 compares the conditions under which the simplified method can be used for the three provisions. In Japan, the height of the building must be less than 60 m and the equivalent period of a seismically isolated building must exceed 2.5 s. In the U.S., the restrictions are more stringent, such as the height must be less than 20 m and the upper limit of the equivalent period is 5.0 s. It is understandable from this example, that it is not easy to provide a specific value in the ISO standard.

Table 1. Limitation of application of the simplified method.

Item	Japan BSL* Notice No. 2009	U.S ASCE7-16	EU Eurocode8–10
Height	< 60 m	< 20 m (65 feet)	Size of plan < 50m
Location of S.I**	Base	Base	Base
Ground condition	1st (stiff) or 2nd (medium)	SA, SB, SC, SD	No rocking foundation
Eq. natural period	2.5 s < Tb	Tb <5 s Tb> 3Ts***	Tb < 3 s Tb > 3Ts***
Eq. damping	No limitation	he < 30%	he < 30%
Distance from fault	No limitation	D > 5 km	D > 15 km for M6.5
Importance factor	None	1.0	1.0, 1.2, 1.4

*BSL: Building Standard Law, **S.I.: Seismic Isolation, ***Ts: period of superstructure

Table 2 compares the target inter-story drift angles for superstructure in case of the simplified method. In Japan, the limit for the inter-story drift angle is less than 1/300 for a horizontal force equivalent to 1.3 times the shear of the seismic isolation layer. In other words, the superstructure of a seismically isolated building is required to maintain its elastic state even during a major earthquake. In the United States, on the other hand, the shear of the seismic isolation layer is reduced by the R value, and the limit of the inter-story deformation angle of the superstructure is 0.015. This can be thought of as allowing a certain amount of damage in the superstructure during a major earthquake. In developing the ISO standard, it was necessary to sort out the differences in target performance by standard and find a harmonious answer.

Table 2. Design criteria of superstructure in the simplified method.

Item	Japan BSL* Notice No.2009	U.S ASCE7-16	Euro Eurocode8-10
Story drift angle	< 1/300	< 0.0015	< 0.01
Lateral force distribution	Ai	Parabolic	1.0 (constant)
Base shear Vs	1.3 Vb*	Vb/R (1 < R < 2)	Vb/q (1 < q < 1.5)

* *Vb* Shear of the seismic isolation layer

Table 3 shows the relationship between the level of design seismic force and the target performance of the building, referred to as the performance matrix. Table 3(a) is the performance matrix for general buildings, and Table 3(b) is the performance matrix for seismically isolated buildings. The level of design seismic force is defined in terms of probability of exceedance or return period. In Japan, the code does not provide a probabilistic definition, but Level 1 (the medium earthquake) is generally considered to have a 50% probability of exceedance in 30 years (44-year return period) and Level 2 (the major earthquake) a 10% probability of exceedance in 50 years (475-year return period). As the target performance of a building, two limit states are considered: the serviceability limit state (SLS) and the ultimate limit state (ULS). In the serviceability limit state, the building is undamaged and remains elastic, while in the ultimate limit state (ULS), the building is damaged but remains safe.

Table 3. (a) Performance matrix of general buildings and (b) Performance matrix of seismically isolated buildings

	Probability of exceedance (Return period)	SLS No damage	ULS Accept damage, protect life
(a). Performance matrix of general buildings			
Level 1	50% in 30 years (44 years)	Japan	
	10% in 10 years (95 years)	Euro	
Level 2	10% in 50 years (475 years)		Japan, Euro (U.S. × 2/3)
Level 3	2% in 50 years (2475 years)		U.S
(b). Performance matrix of seismically isolated buildings			
Level 1	50% in 30 years (44 years)		
	10% in 10 years (95 years)		
Level 2	10% in 50 years (475 years)	Japan	Euro
Level 3	2% in 50 years (2475 years)		(Euro × 1.4*, 1.2**) U.S

*1.4 (Importance factor), **1.2 (Deformation factor)

For general buildings, all regulations target the SLS against Level 1 seismic forces and the ULS against Level 2 seismic forces. In the U.S., Level 3 with a 2% probability of exceedance in 50 years (2475-year return period) is considered for the ULS, but the seismic force is multiplied by 2/3 when used in design, it is considered equivalent to Level 2.

For seismically isolated buildings, as mentioned above, Japan requires that the superstructure of seismically isolated buildings maintain an elastic state against Level 2 seismic forces. This corresponds to the SLS. On the other hand, in the U.S., Level 3 seismic forces are used without reduction, and a certain degree of damage is allowed to the superstructure. This corresponds to the ULS. Eurocode considers the seismic force of Level 2 for the ULS, however, by considering an importance factor of 1.4 and a deformation factor of 1.2, it is considered equivalent to Level 3.

In other words, the target performance of seismically isolated buildings is one level higher than that of ordinary buildings, and is the same for all the standards on the performance matrix. The WG had an active discussion on the description of target performance for seismically isolated buildings in the ISO standard.

4 Design Example

4.1 Benchmark Building

The building is a 7-story condominium building with a RC moment resisting frame structure in the X-direction and a frame structure with shear walls in the Y-direction. The plan size is 27 m × 12.8 m and the building height is 21.4 m. Figure 1 shows the elevation of the building. The isolation gap between the basement of the superstructure and the retaining wall is assumed to be 60 cm.

The information of building height, weight, and horizontal stiffness in each direction are summarized in Table 4.

Lead rubber bearings (LRB) are selected for the isolation devices. Figure 2 shows the arrangement of the devices in the seismic isolation layer at the bottom of the building. The devices are placed under each column, and the total yield force of the seismic isolation layer is set to 4 to 5% of the weight of the superstructure. Based on the axial loads on the devices in Table 5, dimensions and nonlinear characteristics of devices are decided as shown in Table 6.

According to the Notice No. 2009, the design limit deformation, δ_s, at the isolation interface is determined as the minimum value of the limit deformation ${}_m\delta_d$ for all components of the isolation system. The limit deformation ${}_m\delta_d$ for each device is obtained by multiplying the safety factor β ($\beta = 0.8$ for elastomeric isolator) by the ultimate deformation δ_u at 0.3 times the ultimate vertical stress σ_c provided by the manufacturer. Thus,

$$\delta_s = \min({}_m\delta_d), \quad {}_m\delta_d = \beta\,\delta_u \tag{1}$$

Table 7 shows the ultimate deformation of isolation devices in this example. Finally, the design limit deformation in this example is obtained as $\delta_s = 437$ mm.

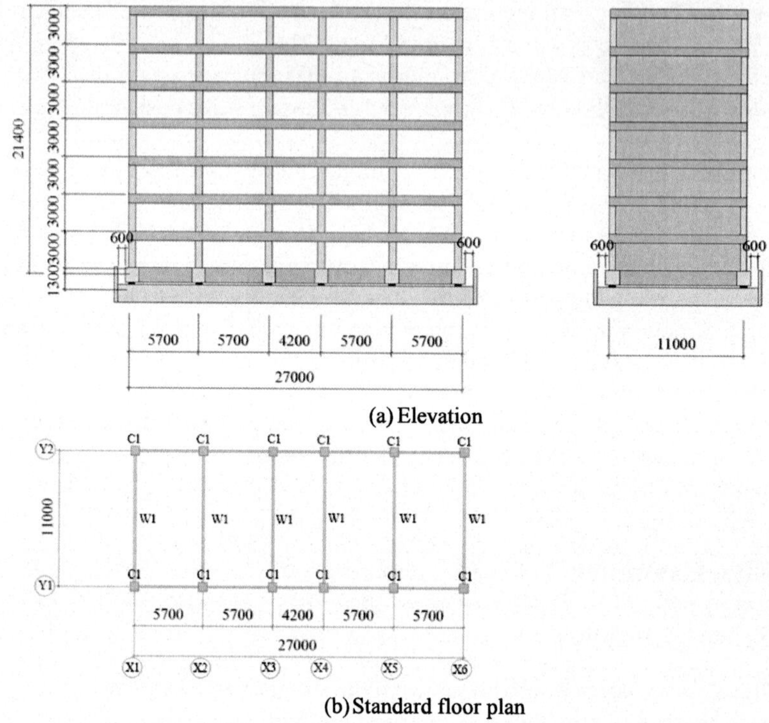

Fig. 1. Elevation and floor plan of the benchmark building

Table 4. Story height, weight, and horizontal stiffness in each direction

Story	Height (mm)	Weight (kN)	Horizontal stiffness (kN/mm)	
			X direction	Y direction
7	3000	4410	951	3243
6	3000	4165	2407	3553
5	3000	4165	1242	5957
4	3000	4165	1336	7950
3	3000	4214	1457	10,183
2	3000	4214	1544	12,966
1	3000	4214	2005	12,814
Isolation layer	1700	5292		

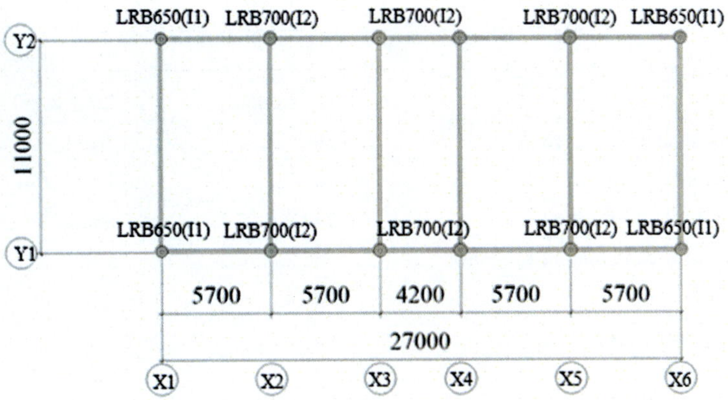

Fig. 2. Layout of isolation devices in the seismic isolation layer

Table 5. Axial loads on the isolation devices (kN)

	X1	X2	X3	X4	X5	X6
Y2	2069	4209	3405	3405	4209	2322
Y1	1874	3171	2566	2566	3171	1874

Table 6. Characteristics of isolation devices

Item	LRB ϕ650	LRB ϕ700
Outer diameter (mm)	650	700
Lead plug diameter (mm)	120	130
Total thickness of rubber (mm)	159.6 4.2 mm × 38 layers	162.0 4.5 mm × 36 layers
Primary shape factor S_1	38.7	38.9
Secondary shape factor S_2	4.1	4.3
Initial stiffness K_1 (kN/mm)	10.695	12.217
Secondary stiffness K_2 (kN/mm)	0.823	0.940
Yield strength (kN)	122.7	140.9
Yield deformation (mm)	11.5	11.5

4.2 Design Seismic Force

The acceleration response spectrum S_0 of 5% damping on the engineering bedrock is given in the Notice No.1446 [5] of MLIT as shown in Fig. 3, which corresponds to an earthquake with approximately a 500-year return period (Level 2).

Table 7. Ultimate deformation of isolation devices

Item	LRB ϕ650	LRB ϕ700
Ultimate vertical stress σ_c (N/mm^2)	30	36
$(1/3)\sigma_c$ (N/mm^2)	10	12
Horizontal limit strain γ (%)	342.27	353.62
Horizontal limit deformation δ_u (mm)	546	573
Safety factor β	0.8	0.8
Limit deformation ${}_m\delta_d$ (mm)	437	459

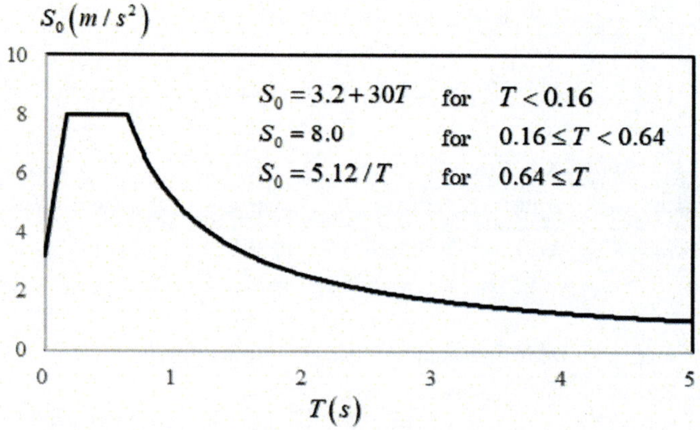

Fig. 3. Acceleration response spectrum S_O (5% damping) at engineering bedrock

The input ground acceleration spectrum is given as follows:

$$S_a = ZG_S S_0 \tag{2}$$

where Z denotes the seismic zone factor (0.7–1.0), G_S is the ground amplification factor as shown in Fig. 4. In this example, $Z = 1.0$ and G_S for the hard soil (Type 1) are adopted.

4.3 Structural Analysis

The response spectrum analysis method for the equivalent linear system (i.e., the response and limit strength calculation in Japan) is used for structural analysis. The effective stiffness, K_e, of the isolation system shall be calculated as the secant stiffness obtained by dividing the peak force by the design maximum displacement, D_M, in the force-deflection behavior of the isolation system.

The effective period of the isolation system, T_e, at the design maximum displacement, D_M, shall be calculated from Eq. (2):

$$T_e = 2\pi\sqrt{M/K_e} \tag{3}$$

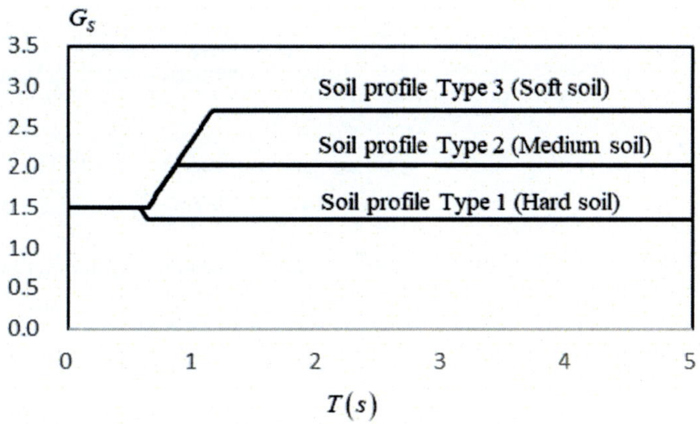

Fig. 4. Ground amplification factor

where, M is the mass of the superstructure. The effective damping ratio of the isolation system, h_d, should be calculated from Eq. (4):

$$h_d = \frac{1}{4\pi}\beta_d\frac{\Delta W}{W}, \quad W = \frac{1}{2}K_e D_M^2 \tag{4}$$

where, β_d is the modification factor of the effective damping of the hysteretic dampers that takes into account the non-stationarity of the seismic response, which is set to 0.8, ΔW is the total energy dissipated in the hysteretic dampers during a full cycle of response at the design maximum displacement D_M, W is the total potential energy in the isolation system at the design maximum displacement, D_M.

The maximum displacement, D_M, of the isolation interface at the center of mass shall be calculated from Eq. (5) by the iterative manner.

$$D_M = \frac{M\ S_a(T_e, h_d)}{K_e} \tag{5}$$

The maximum displacement at the individual element of isolation interface should include additional displacement caused by the torsional movement of isolation interface by an appropriate manner. The Notice No. 2009 provides the maximum displacement as follows:

$$D'_M = 1.1\alpha D_M \tag{6}$$

where α is the safety factor for temperature dependent stiffness changes and property variation in manufacturing of devices (the minimum value is 1.2). Please refer to Ref. [4] for the detail calculation process. In this example, the maximum deformation of the seismic isolation layer is calculated as $D_M = 291$ mm, $D'_M = 384$ mm.

4.4 Verification of the Maximum Displacement

The maximum displacement of the isolation interface must be less than the design limit deformation of the isolation interface. Therefore,

$$D'_M = 384\,\text{mm} < \delta_s = 437\,\text{mm} \quad \ldots\text{O.K.}$$

The isolation gap also needs to be verified. According to the Notice No. 2009, the isolation gap must be larger than either 1.25 times the maximum displacement or 0.20 m plus the maximum displacement. Therefore,

$$1.25D'_M = 480\,\text{mm} < 600\,\text{mm(isolation gap)} \quad \ldots\text{O.K.}$$
$$D'_M + 200 = 584\,\text{mm} < 600\,\text{mm} \quad \ldots\text{O.K.}$$

Please refer to Ref. [4] for the design of superstructure and the time history response analysis.

5 Conclusion

Although there were no major differences among countries in design methods for base isolated buildings, there was a discussion in the WG on the description of target performance, for example, whether damage should be allowed to the superstructure in a major earthquake. Also, some members suggested that more detailed provisions should be made for the design of non-structures and equipment, since it is important to maintain the function of a building after an earthquake. The WG also made progress in preparing Annexes on seismic isolation of bridges and LNG tanks, seismic isolation retrofitting of existing buildings, and other topics not covered in the main text, but the discussion was not sufficiently thorough to include in the standard.

The ISO standard for seismically isolated structures is a compilation of basic design principles, and it is assumed that the specific numerical values required for design should be decided in each country as shown in the design example in this paper. It is expected that the differences in design methods in each country will become clearer based on the design examples compliant with the ISO standard. It is also hoped that this ISO standard will be useful to countries that are trying to create new design codes for seismically isolated structures.

References

1. ISO 3010:2017, Bases for design of structures—Seismic actions on structures (2017). https://www.iso.org/standard/63217.html
2. ISO 22762-1:2018: Elastomeric seismic-protection isolators—Part 1: Test methods (2018). https://www.iso.org/standard/70215.html
3. ISO/FDIS 23618, Bases for design of structures—general principles on seismically isolated structures (2002). https://www.iso.org/standard/76423.html
4. Saito, T., Takano, S.: Design procedures of response spectrum method and time history analysis of seismic isolation buildings in Japan (2010). www.rc.ace.tut.ac.jp/saito/Software/STERA_3D/Sample/SI01/B02_7F_RC_SI_J.pdf
5. Notification of No.1446 of Ministry of Construction, Japan / May 31 (2000)
6. Notification of No.2009 of Ministry of Construction, Japan/October 17 (2000)

Recent Development and Application on Seismic Isolation, Energy Dissipation and Vibration Control in China

Fulin Zhou[1,2,3](✉), Ping Tan[1,2,3], Yangyang Chen[1,2,3], and Yanhui Liu[1,2,3]

[1] Earthquake Engineering and Test Center, Guangzhou University, Guangzhou 510006, China
zhoufl@cae.cn

[2] MOE Key Laboratory of Earthquake Resistance, Earthquake Mitigation and Structural Safety, Guangzhou University, Guangzhou 510006, China

[3] Provincial Key Laboratory of Earthquake Engineering and Applied Technology, Guangzhou University, Guangzhou 510006, China

Abstract. China is a seismic country which 100% of the territory locates in seismic zone. Most of strong earthquakes are over prediction. Most people dead caused by structures collapse. Earthquakes not only cause severe damages for structures, but also cause non-structural elements and inside facilities, cause to stop the city's life, such as hospital, airport, bridge, and power plant, etc. Designers need to use the new techniques to protect the structures, non-structures, and inside facilities. Therefore, the structural isolation, energy dissipation and control are more and more widely used in recent years in China. There are nearly 16000 large or complex structures with isolation, and about 9000 structures with passive energy dissipation or hybrid control in China until 2021. The fields of application include house buildings, schools, hospitals, airports, bridges, power plant, historical or cultural relic protection, industries facilities, retrofit for existed structures, and immersed tunnel under sea or river. China has got some new progress and established relating systems of isolation, energy dissipation and vibration, which include 3D Seismic Isolation system (3DSI), Seismic and Vibration Control system (SVC), immersed Tunnel with Isolation and Energy Dissipation system (TIED), Hybrid TMD Control system for high rise buildings (HTMD), Eddy-Current Damping system (ECD) and Vortex Vibration Control system for bridges (VVC). Paper also introduces the systems of design codes and device standards, and three large scale and advanced testing systems for isolation, energy dissipation and vibration control. Paper also makes discussion for the tendency of development on seismic resistance, seismic isolation, passive and active control technique in the future in China and in the world.

Keywords: Seismic isolation · Energy dissipation · Passive control · Hybrid control

1 Earthquake Tragedy and Lessons in China

A tragedy strong earthquake, Tangshan Earthquake M7.8 happened at 3:15 on 26 July 1976. The epicenter depth was only 13km. The broken faults run through the city. The

G. P. Cimellaro (Ed.): WCSI 2022, LNCE 309, pp. 59–73, 2023.
https://doi.org/10.1007/978-3-031-21187-4_5

whole city become ruin, 240,000 people died, 96% buildings collapsed, including houses, schools, hospitals, office block and any kinds of other buildings (Figs. 1 and 2).

Another tragedy strong earthquake, Wenchuan Earthquake M8.0 happened at 14:28 on 12 May 2008. The epicenter depth was only 17km. The broken faults also run through the city. The whole city become ruin, 90,000 people died, 80% buildings collapsed (Fig. 3).

Fig. 1. Schools in Tangshan 1976

Fig. 2. Hospitals in Tangshan 1976

Fig. 3. Buildings in Wenchuan 2008

In the same area of Wenchuan Earthquake, just after 5 years, another strong Lushan earthquake M7.0 happened in 20 April 2013, over 75% buildings were damaged or collapsed in Lushan County, So call: A Standing Ruin!

But after Lushan Earthquake, A surprise message comes: one of building of Lu Shan County Hospital used Isolation Technique, it is no any damaged, and it performs very well during and after earthquake!

There are 3 buildings in Lu Shan County Hospital, all are reinforced concrete structures with 7 stories and 1 story basement (Fig. 4). One building with seismic isolation, behaved without any damaged for structure and decoration, any falling-down of facilities or equipment inside the building. The persons inside do not have any feeling during earthquake, This Isolation hospital building become unique Rescue Center in whole county after earthquake, Thousands of injured people accept first aid rescue in it (Fig. 6). And the other two hospital buildings with traditional anti-seismic designed, without isolation, persons inside the building feel shaking very severely, building damaged for structure, wall and ceilings, facilities, and equipment seriously fell, the hospital was all break down (Fig. 5).

There are many lessons have been learnt from strong earthquakes in China [1, 2].

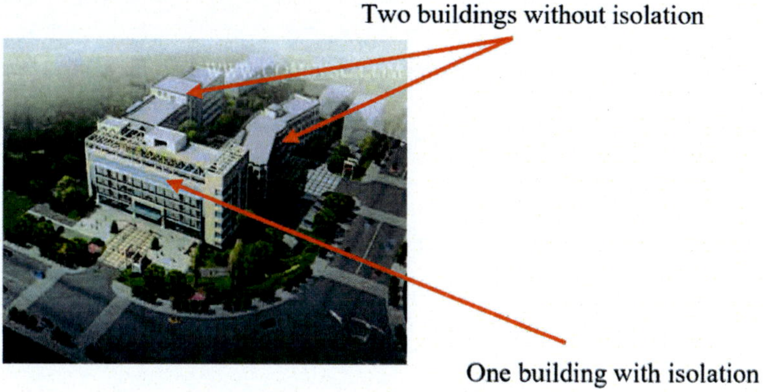

Fig. 4. Lu Shan County Hospital, including three buildings

Fig. 5. Two buildings without isolation in Lu Shan County Hospital, break down

1.1 Most of Strong Earthquakes Are Over Prediction

The designers need to consider strong earthquakes which is possibly over prediction in China (Table 1).

1.2 Most People Dead Caused by Structures Collapse

Depend on the statistics, 90% of people dead caused by building collapse in earthquake. The earthquake not only causes severe damages or collapse of structures, but also causes

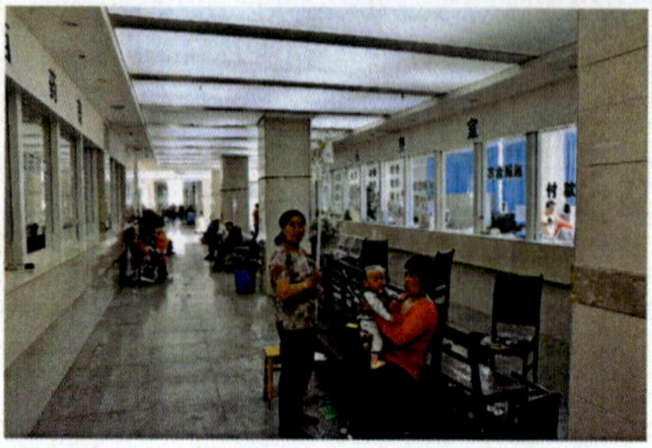

Fig. 6. One building with isolation in Lu Shan County Hospital, everything is perfect

Table 1. Earthquakes severely over prediction in China

Earthquake	Design ground acceleration	Recorded ground acceleration	Times
2008.05.12 Wenchuan earthquake	0.10g	0.96g	10
2013.04.20 Lushan earthquake	0.15g	0.80g	6

severe damages of non-structural elements and inside facilities. It will cause to stop the city life. Like hospital, power plant, and so on.

1.3 Designers Need to Use the New Technique

The isolation, energy dissipation and structural control are effective technique for using. It could protect people life, protect both the building structures and inside facilities in over prediction strong earthquake.

2 Technical Rules for Seismic Isolation and Energy Dissipation in China

Technical rules of seismic isolation and energy dissipation consists three different sets in China [3].

2.1 Technical Specifications

- **Technical specification for seismic isolation with laminated rubber bearing isolators (CECS 2001).** This is the national technical specification for design and construction of buildings and bridges with seismic isolation issued on 2001 in China.
- **Technical specification for Energy Dissipation (CECS 2013).** This is the national technical specification for design and construction of buildings and bridges with energy dissipation issued on 2013 in China.

2.2 Design Codes

- **Seismic design code of buildings with isolation and energy dissipation** (Chapter 12 in national code for seismic design of buildings, **GB 50011–2001 and 2010**). This is a part of national code in China for seismic design of buildings issued on 2001 and 2010 in China,
- **Seismic Design Code for Isolation buildings (GB/T 51408–2021)**. This is the national code for seismic isolation design of buildings structures which is issued on 2021 in China.

2.3 Standards for Isolators and Energy Dissipation Dampers

- **Standard of laminated rubber bearing isolators (GB 20688–2006).** This is the national standard of isolators for laminated rubber bearings issued on 2006 in China.
- **Standard of energy dissipation dampers (JG / T209–2011).** This is the national standard of energy dissipation dampers issued on 2011 in China.

3 Devices and Testing System

3.1 Isolation Devices

There are five kinds of isolation devices have been used in China, include natural rubber bearing, lead rubber bearing, high damping rubber bearing, friction pendulum bearing and 3 direction (3D) isolation system.

3.2 Energy Dissipation Devices

There are five kinds of energy dissipation devices have been used in China, include steel yielding damper, oil damper, BRB bracing damper, smart material (SMA) damper and eddy-current damper (ECD).

3.3 Basic Performance Tests

Tests of mechanical characteristics for isolators are required for every kind of isolator. The general tests include compression tests (capacities, stiffness), compression with shear cycle loading tests (stiffness, damping ratio and maximum horizontal displacement). The results of tests need to reach that, the failure compression stress reach 90 MPa. When horizontal strain is zero, the failure compression stress reach 35 MPa. When horizontal strain is 400%, so, choose designing compression stress $\leq$ 15 MPa. is more enough for safety. The large capability and advanced testing of isolators for some very important project is necessary. China has built one 3D (3 directions) compression and tension with shear cycle loads testing machine for isolators. This new testing machine of isolator possess 3-D load capacities, 10,000ton compression load, 4000ton tension load, 6000ton lateral load, $\pm$ 1.50m horizontal displacement and torsion motion. This large capability and advanced testing system for isolators has been built in Guangzhou University in China (Fig. 7).

3.4 Durability Performance Tests

Tests and investigation of durability for isolator, includes low cycle fatigue failure tests, creep, and ozone aging tests. The results of tests and investigation show that, the maximum thickness of rubber ozone is 5 mm exposed to sunlight and air for 100 years. If make the 10mm cover layer for rubber bearing to isolate the air and sunlight, the rubber bearing inside the cover layer could make sure the working life of rubber bearing to reach over 100 years. In order to explore the regularity of durability for rubber isolator, a long time (50 years) test of full-scale rubber bearing have been doing from 2001 in Guangzhou University in China (Fig. 8).

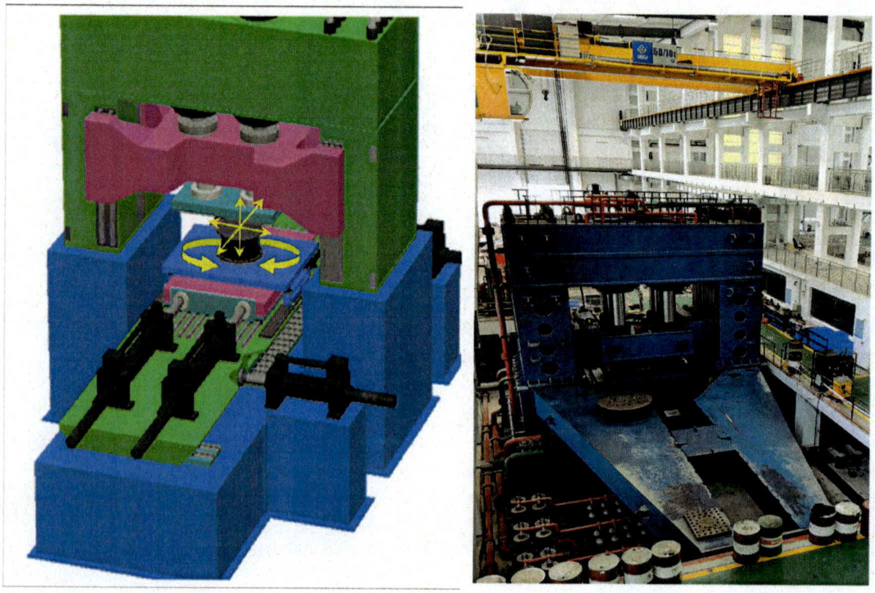

Fig. 7. 3D compression 10,000 ton and tension with shear cycle loads testing machine for isolators

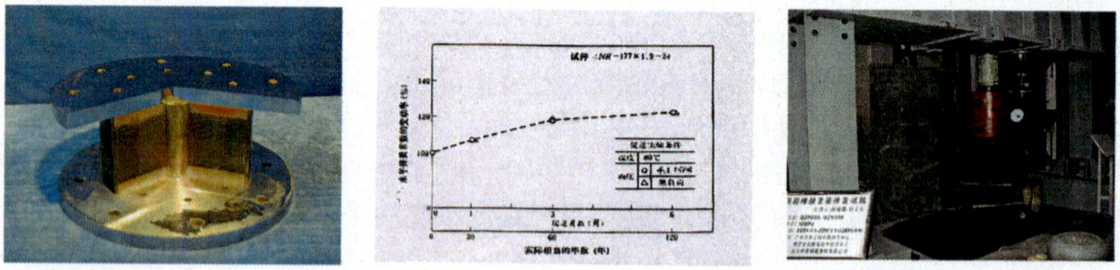

Fig. 8. Creep and ozone aging tests full scale 50 years for rubber bearing (2021–2051)

3.5 Shaking Table Tests

Shaking table tests for large scale structural model, including structure with different location of isolation bearings layers are tested on the shaking table. The testing results show that the isolation effect depends on the ratio of mass of superstructure to the substructure, the base isolation (isolator on the base) is the best, the acceleration of isolation structure could reduce to be 1/4–1/8 comparing with the no-isolation structure. For middle isolation (isolator on the story), the isolation effect also is significant [4].

For some long structures such as bridges, tunnels or long span structures, China has built the multiple shaking tables tests on the long array shaking tables system, which includes three tables with 8 × 10m, 4 × 4m and 4 × 4m, composes 60m long array shaking tables system. This long array shaking tables system could simulate long structures the wave transfer process and multi-point input during shaking table tests. This new shaking table testing system has been built in Guangzhou University in China (Fig. 9).

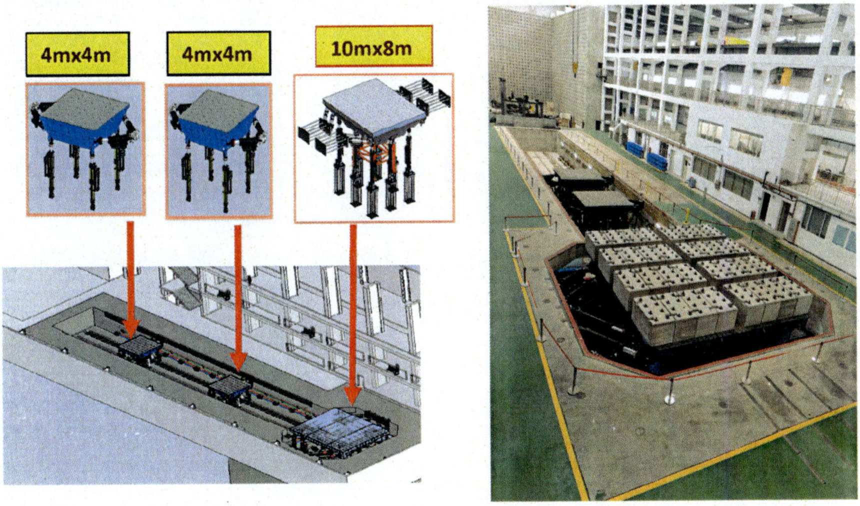

Fig. 9. Multiple shaking tables tests on the long array shaking tables testing system

4 New Progresses and Application On Isolation, Energy Dissipation and Vibration Controls in China

There are nearly 16000 large or complex structures with isolation, and about 9000 structures with passive energy dissipation or hybrid control in China until 2021. The fields of application include house buildings, schools, hospitals, airports, bridges, power plant, historical or cultural relic protection, industries facilities, retrofit for existed structures, and immersed tunnel under sea or river. The most of structural types of buildings are concrete, steel frame and brick structures. It has become a very strong tendency to widely using seismic isolation, energy dissipation and vibration control China now. Some typical projects are illustrated as:

4.1 High Rise House Building with Isolation

The seismic isolation has been used very popular on house buildings with 1–31 stories in China now. The Seismic Isolation Houses buildings group, including 24 high-rise buildings with 31 stories, floor area 980,000 M^2 had been built in Beijing, China, which is the largest area of isolation houses group buildings in the world (Fig. 10). There are many similar isolation houses buildings in different areas in China.

Fig. 10. Seismic Isolation Houses buildings group including 24 high rise buildings, 31 stories, floor area 980.000 M^2

4.2 The School Building with Isolation Experiences the Earthquake

One RC frame 6 stories isolation school building with isolation was built in 2008, then was suffered the Lu Shan Earthquake M7.0 in 20 April 2013. Comparing with the no-isolation building, the acceleration response of isolation buildings was reduced to be 1/6, The teachers told the student children: "When earthquake happen, keep in room, don't run out! this is isolation building, inside is safer than outside!". No any damage for isolation building, no any students were injured in this earthquake (Figs. 11 and 12).

Fig. 11. Seismic isolation school in Wenchuan County

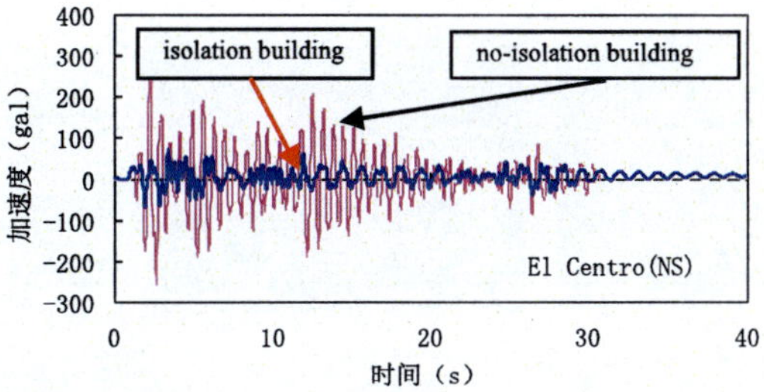

Fig. 12. The records of acceleration response in earthquake

4.3 Isolation House Buildings with 3D Isolators

The Seismically isolated artificial ground which is the largest area with 3D isolation in the world (Fig. 13). There is a large platform (2 stories RC frame) with 1500m wide and 2000m long to cover a railway area in Beijing City. There are 48 isolation RC buildings (7~9 stories RC frame) with floor area 240,000 M^2 built on the top floor of the platforms The rubber bearings layer is located on the top floor of the platform to isolate the seismic motion also to isolate the railway vibration which is 3D isolation.

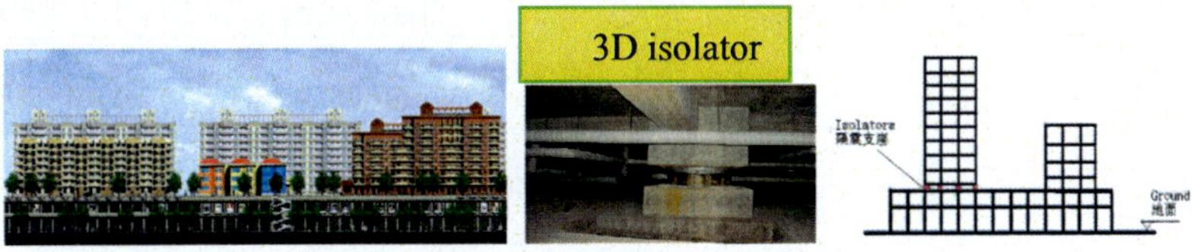

Fig. 13. All 48 isolation buildings with 3D isolators on RC platform

4.4 Seismic Isolation Airports with Large Span or Complex Structures

The floor Area of Isolation Kunming New Airport (2007–2012) is 500,000 M^2. Because the location of Airport is near the seismic faults (10K), it is needed to protect the complex structure of airport, column with curve shape, large wall's glass, large ceilings and the important facilities inside the building during earthquake, only use the isolation! This project uses 1892 Rubber Bearings (ϕ1200 mm) to isolate the ground motion and 108 Oil Dampers to reduce the displacement during earthquake (Fig. 14). This airport terminal was suffered an earthquake M4.5, Chonmin earthquake in 9 March 2015, the acceleration response records has been recorded: terminal floor F3/base = 1/4 (Fig. 15). It is very effect to reduce the structural response for Large Area Airport Terminal.

The Beijing new airport terminal with isolation floor Area 700,000 M^2 (2015–2019) which is the largest floor Area of isolation building in the world at recent. Hainan new airport terminal with isolation floor Area 300,000 M2 (2016–2019) are constructing now. Many new airports with isolation are planning or designing in China now.

Fig. 14. Isolation kunming new airport (2007–2012) terminal

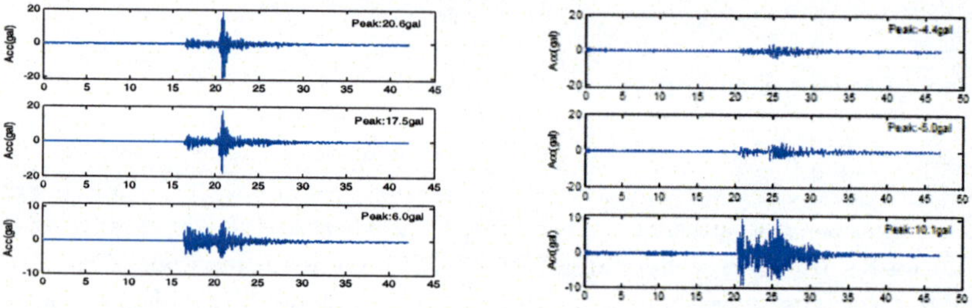

Time history of base

Time history of F3

Fig. 15. Acceleration response records: airport floor F3/base = 1/4.

4.5 Isolation Bridges

This HK-ZHUHAI-MACAO bridge with length 26 km crossing the South China See uses the isolation technique (Fig. 16). It is very effective to avoid the damage cracks at the bottom of piers in the seawater, to move the nonlinear area from the bottom of pier to the isolators on the top of piers, keep the pier on elastic state in earthquake, also to reduce the seismic response to be 1/4 response for traditional bridge.

4.6 Isolation Protection for Historical or Cultural Relic

There are many history relics with thousand years in China. It is very important to protect these history relics, such as statues, tablets, or pictures, etc. in strong earthquakes. Isolation technique is the best option! Putting isolation layer on the base of relics, could reduce the horizontal response of relics to be 1/12 of the ground motion and make sure the safety of relics in strong earthquake (Fig. 17).

4.7 Seismic Retrofit for School and Hospitals

Millions of old buildings of schools, hospital or other lifeline engineering are lack of capacity of seismic resistance in China, which need to retrofit to be satisfied the safety in strong earthquake. Shanxi Province successfully use Isolation to retrofit school buildings, the Government held a National Meeting to extend this demonstrate (Figs. 18 and 19).

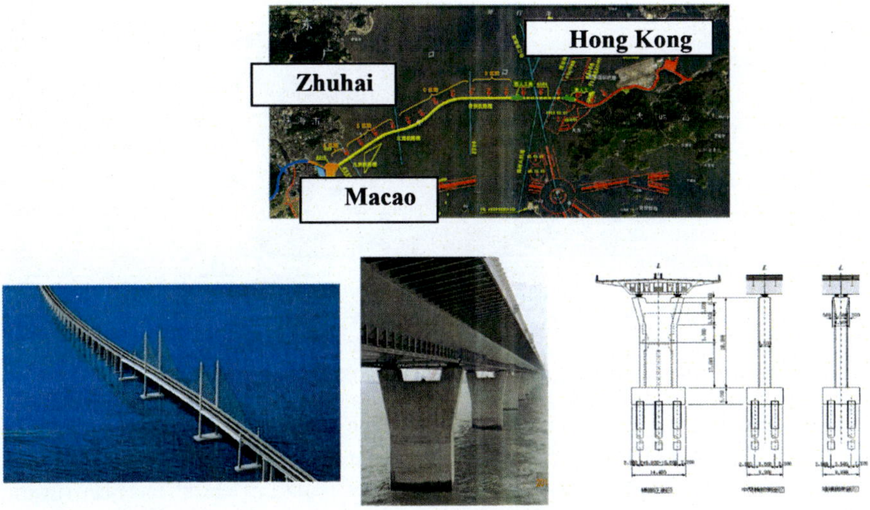

Fig. 16. Seismic Isolation bridge (HK-MACAO-ZHUHAI) 26km

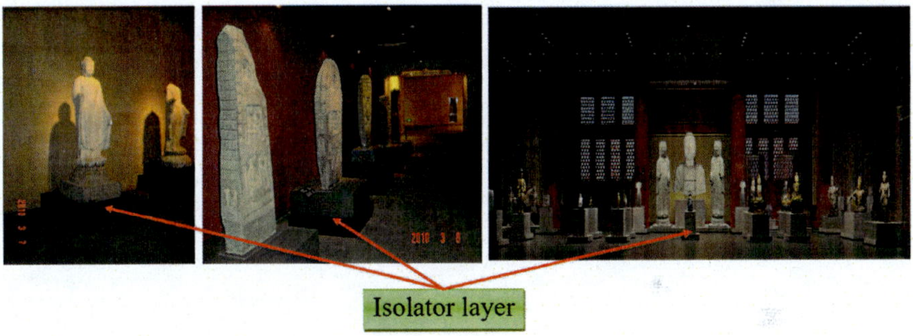

Fig. 17. Isolation protection for history statue and stone tablet (1200 years history)

Fig. 18. Base isolation retrofit in Shanxi Province

Fig. 19. 1st story isolation retrofit in Shanxi Province

4.8 Low Rise House Building with "ELastic Isolation Brick"

There are many houses in rural areas in China. These houses always are low rise, low cost, simple design and simple construction, but most of them are lack of capacity for seismic safety. A new isolation system called "Elastic Isolation Brick" has been invented for this requirement and began to be popularly used in large countryside areas in China (Fig. 20). The cost of this "Elastic Isolation Brick" is only 1/4 of the general rubber bearing, its design is very simple, and its construction is suitable for handmade. The shaking table tests showed that, the low-rise house with "Elastic Isolation Brick" could suffer very strong earthquake with ground acceleration 0.80 g and no any damage, it could become the "safety island" in very strong earthquake.

Fig. 20. Low rise house with "Elastic Isolation Brick"

4.9 Immersed Tunnel with Isolation and Energy Dissipation System (TIED)

There are some tunnels crossing the river or sea, need to control the open width of tunnel joints to avoid the water going to the tunnel to kill the people in the cars or trains inside the tunnel in strong earthquake. One shield tunnel with length 6 km crossing the sea, which locates in seismic area with seismic intensity 0.40g in Southern China (Fig. 22). This tunnel follows the soft design concept to create some soft joints then uses Smart Material (SMA) to control the open width of tunnel joints during earthquake and recover the close state after earthquake, make sure the tunnel becomes watertight tunnel experienced the strong earthquake. Depend on the theory analysis and testing result for this new soft design tunnel system show that, this soft design tunnel could reduce the response about 30–50%, and could make sure the tunnel becomes watertight in very strong earthquakes (Fig. 22).

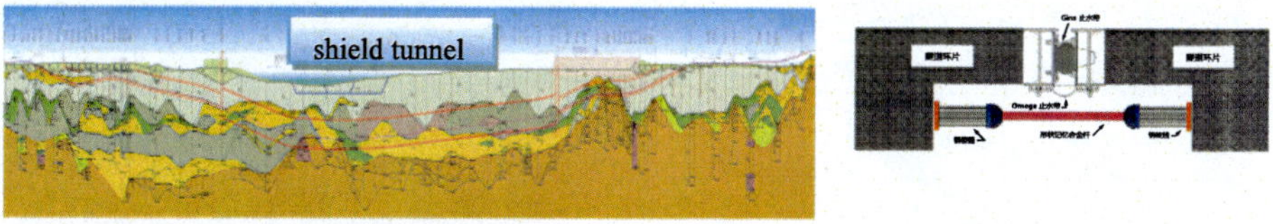

Fig. 21. Shield tunnel with length 6 km crossing the sea in Southern China

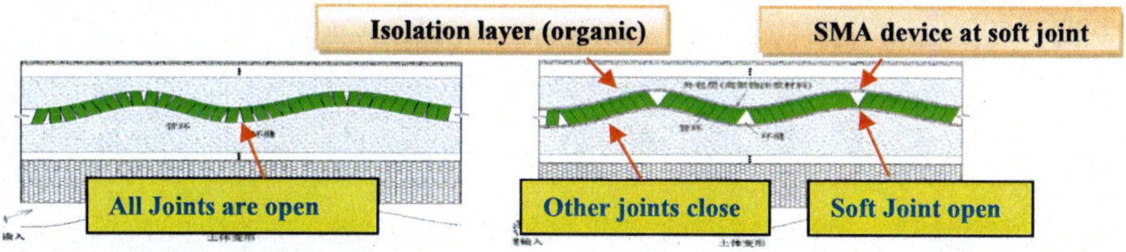

Traditional design tunnel Soft design tunnel

Fig. 22. New soft design concept for tunnel with smart material (SMA)

4.10 Eddy-Current Damper (ECD)

A novel Damper has been developed by Chen [5] The concept is that when the iron plate put in magnetic field will induce large damping. This is a new Generation of Energy Dissipater, which could provide large damping for structures, it is more precise, more durable because no touch between material parts. This new damping device has been used in Shanghai Center tower building with 606m height (Fig. 23). The building with this new system could reduce 30–50% response during strong wind.

Eddy-current damping has many attractive properties and operational performance. By employing the ball screw mechanism, which is capable for producing sufficiently large damping coefficients and apparent mass. This new device has been used on TMD's damper in high-rise buildings, bridges, and other projects. It is more effective, durable, and economical, compared with other types of Dampers above mentioned.

4.11 Hybrid Control (HTMD) for High Tower Structures

Structural control has been used about 30 buildings and bridges. Passive TMD is low cost and stable, but it could not be satisfied for earthquake or wind, Active AMD is effective, but it is too expensive. So, one new control system has been developed, that is called Hybrid Control HTMD (TMD + AMD) which is stable and low cost, and could satisfy the requirement for earthquake or strong wind, so this new system become one of the best options for structures in China [6].

The reasons of using hybrid control (HTMD) for Guangzhou Tower are:

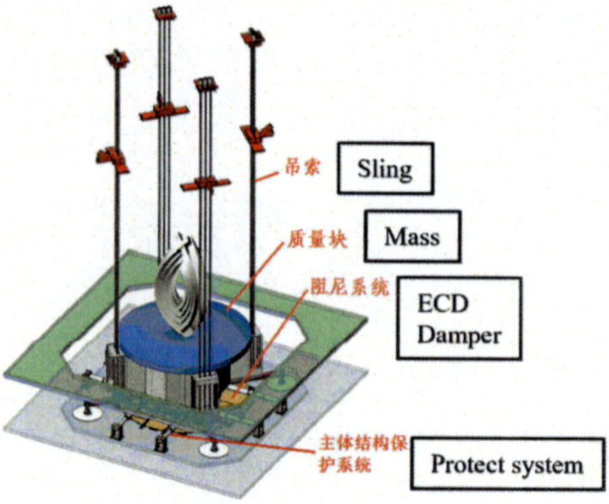

Fig. 23. - Shanghai center building with 606m and Prof. Chen Zeng Qing in the construction site

- Height with 610 M, earthquake and wind load is big problem!
- The structural Plane is ellipse, will torsion in earthquake or wind load
- The slender conformation is not be satisfied in strong wind (Fig. 24).
- TMD is low cost, but can not satisfy the safe requirements. AMD is effective, but too expensive

TMD + AMD = HTMD may be the best balance (Figs. 25, 26).
Using two water tanks as mass of TMD, effectively to reduce the cost of new system.

Fig. 24. Guangzhou tower

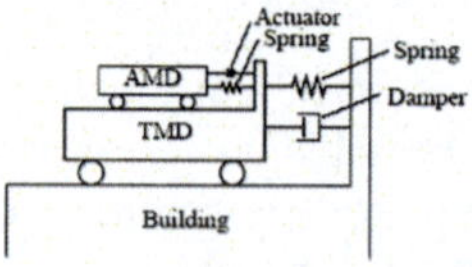

Fig. 25. HYBRID (TMD + AMD)

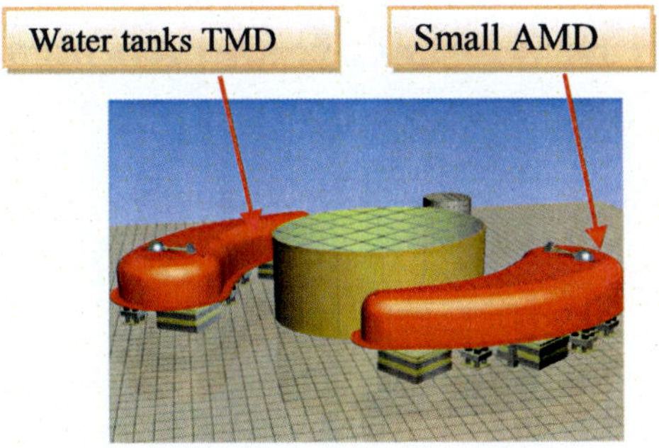

Fig. 26. Using 2 water tanks as mass of TMD

5 Conclusion

The low damage and resilient structures have become the seeking goal for seismic design of buildings structures, and the seismic isolation, energy dissipation and structural control have become one of the best options at present in China!

There are many advantages for seismic isolation, energy dissipation and structural control system is:

- More safe, even in over prediction earthquakes.
- More effective protection for both structures & facilities inside in earthquakes.
- More effective to keep elastic state of supper structures in earthquakes.
- More economical, inexpensive, about add cost 5 ~10% of structural cost.
- More Satisfied for irregular architectural design for anti-seismic or wind load.

So, it could say that, the coming years:

- Traditional anti-seismic structures still is main structure System.
- Seismic isolation, energy dissipation and structural control will be one of main system for anti-seismic structures, it will be more and more widely used in coming years in China!

References

1. Zhou, F.L.: In: Seismic isolation, energy dissipation and structural control system, Beijing, City And Disaster Reduction, vol. 5, (2016)
2. Zhou, F.L.: In: Seismic Control of Structures. Chinese Seismic Publishing House (1997)
3. Zhou F.L., Discussion on Compiling Chinese Design Code of Seismic Isolation and Energy Dissipation, Guangzhou, China, (2014)
4. Huang, X.Y.: In: Earthquake Engineering Research and Test Centre. Technical Reports. Guangzhou University (2003)
5. Chen Z.Q.: In: Dynamics of Structures. China Communications Press (2021)
6. Zhou, F.L., Tan, P., Yan, W.M., Wei, L.S.: Theoretical and experimental research on a new system of semi-active structural control with variable stiffness and damping. Earthq. Eng. Eng. Vib. **1**(01), 130–135 (2002)

Recent Developments and Applications of Seismic Isolation in North America

Ian Aiken(✉) and Cameron Black

SIE, Inc., Berkeley, CA, USA
ida@siecorp.com

Abstract. The number of seismically-isolated building projects continues to grow at a steady, albeit slow rate in the U.S., and applications in Canada are starting to appear. Many projects are architecturally notable, and many also involve increasingly large isolation displacements. The use of isolation for sensitive and critical equipment is becoming more common, and in some cases is extending to state-of-the-art three-dimensional protection. The paper gives an overview of a selection of recent notable projects and discusses recent trends in design and applications.

Keywords: Seismic isolation · Buildings · Industrial · Sensitive equipment · Three-dimensional · USA · Canada

1 Introduction

The first seismically-isolated building in the U.S. was completed in 1985. This was a new, four-story building, the Foothill Communities Law and Justice Center in Rancho Cucamonga, California that utilized high-damping rubber bearings. Soon after, the first seismic isolation retrofit in the U.S. and the world was completed, the Salt Lake City and County Building in Salt Lake City, Utah. These and other early projects initially positioned the U.S. at the forefront of the use of the technology. It is now estimated that there are about 200 seismically-isolated buildings in the U.S., and while not nearly the number that have been completed in various other countries, isolation has continued to be used at a steady, albeit slow, pace with many notable applications and distinctive projects.

The use of seismic isolation for buildings in Canada has taken place much more recently with the first building, the retrofit of an historic school building in Vancouver, B.C., being completed in 2017. Isolation is now being used for a number of notable projects currently underway and is being actively considered by engineers and owners for numerous others.

This paper gives an overview of a selection of recent seismic isolation projects in both countries and provides some observations on trends regarding the use of the technology and applications.

G. P. Cimellaro (Ed.): WCSI 2022, LNCE 309, pp. 74–85, 2023.
https://doi.org/10.1007/978-3-031-21187-4_6

2 Overview of Recent Projects

The following sections briefly describe a selection of projects that have been completed within the last several years or which are currently under construction, in the categories of architecturally notable new buildings, retrofits, essential facilities, data centers, and several categories of non-building applications: equipment, industrial and artwork.

2.1 Architecturally Notable New Buildings

A series of recent building projects in the Los Angeles area have taken advantage of the structural flexibility possible with the use of seismic isolation to achieve dramatic architectural forms. The new home of both the NFL Los Angeles Rams and the Los Angeles Chargers, SoFi Stadium features a sweeping roof structure covering more than 120,000 m^2, supported on 53 pendulum isolation bearings atop massive "blade" columns (Fig. 1.a, .b). The stadium is within meters of the Newport-Inglewood Fault, and the isolation bearings were designed for a maximum displacement of 2.03 m.

Two other dramatic architectural examples are the Academy Museum of Motion Pictures (Fig. 2.a) and the (W)rapper building (Fig. 2.b). The Academy Museum is a spectacular 45.8 m diameter steel and glass orb supported on just eight isolators with a maximum displacement of 750 mm, while the (W)rapper is a unique, 17-story, 73 m tall steel building with no columns or regular framing, instead a helical band exoskeleton supports the structure and achieves entirely column-free interior spaces [1]. The (W)rapper isolation system comprises 18 triple friction pendulum bearings.

Both the Lucas Museum (Fig. 2.c), and the David Geffen Galleries of the Los Angeles County Museum of Art (LACMA) (Fig. 2.d) [2] feature impressive free-form architectural designs, with isolation not only facilitating the architectural and structural forms, but also providing the highest level of protection to the valuable art and contents to be housed in both complexes. All the projects of Fig. 2 have utilized triple pendulum isolation bearings.

2.2 Retrofit Projects

The use of isolation for the retrofit and upgrade of historic buildings in the U.S. has always been a significant sector of application, and in recent years that has expanded into Canada. The first Canadian building isolation project was the retrofit of the Lord Strathcona School in Vancouver (Fig. 3.a). This 1,800 m^2, three-story timber/masonry/stone and masonry building used 30 elastomeric and flat sliding bearings and was completed in 2017 [3]. Currently underway is the retrofit of Centre Block, the main building of the Canadian Parliamentary complex (Fig. 3.b). The project includes the addition of a new, underground visitor welcome center, and will see the transfer of the building onto about 550 isolators [4].

Construction is currently underway on the seismic upgrade of the Salt Lake Temple of The Church of Jesus Christ of Latter-day Saints (Fig. 4.a). The temple, originally constructed over 40 years and completed in 1893 will be supported on 98 triple pendulum bearings with a maximum displacement of 1.52 m (60 in.). The massive, granite walls

(a) Exterior view of stadium

(b) Roof structure with isolators at top of blade columns

Fig. 1. SoFi Stadium, Inglewood, California, USA (photos: SoFi Stadium).

(a) Academy Museum of Motion Pictures
Los Angeles, California, USA
(photo: Renzo Piano Building Workshop)

(b) The (W)rapper
Los Angeles, California, USA
(image: Eric Owen Moss Architects)

(c) Lucas Museum, Los Angeles, California, USA
(image: Lucas Museum of Narrative Art)

(d) David Geffen Galleries, LACMA, Los Angeles, California, USA
(image: Atelier Peter Zumthor & Partner / The Boundary)

Fig. 2. New seismically-isolated buildings, Los Angeles, California, USA.

of the temple required an innovative underpinning and load transfer approach, with the isolators either side of the axis of the walls (Fig. 4.b).

Other historic building retrofits underway, or in the planning stages, include the Oregon Supreme Court in Salem, Oregon (Fig. 5).

(a) Lord Strathcona School
Vancouver, British Columbia, Canada

(b) Centre Block, Canadian Parliament
Ottawa, Ontario, Canada
(photo: Wikipedia)

Fig. 3. Heritage building seismic isolation retrofits, Canada.

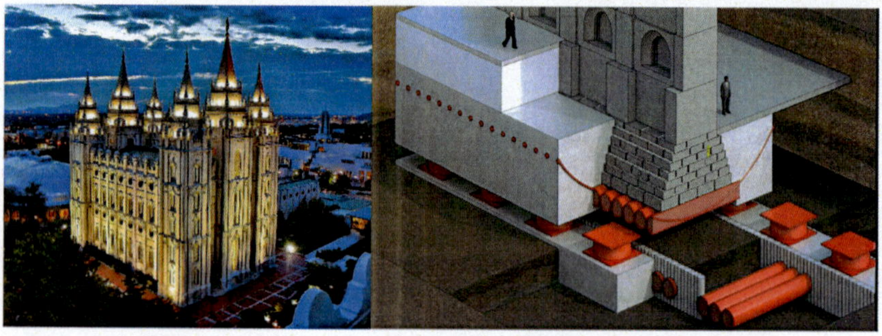

(a) Temple building (b) Load transfer and isolator placement

Fig. 4. Salt Lake temple, the church of Jesus Christ of Latter-day Saints, Salt Lake City, Utah, USA (images: (a) Wikipedia; (b) Intellectual Reserve, Inc.).

2.3 Essential Facilities

Along with the retrofit of historic buildings, isolation has also been consistently used in the U.S. for essential facilities. The following examples in Figs. 6, 7, and 8 show some notable recent examples. The Loma Linda University Medical Center in Southern California is a 100,000 m^2 new hospital building, comprising two patient towers of 16 and 7 stories (Fig. 6.a) [5]. The hospital is located within a few hundred meters of the San Jacinto Fault (see Building "6", Fig. 13) and is exposed to extremely severe near-field shaking. As a result, the isolation system has been designed to accommodate not only large horizontal displacements (1.07 m) using triple pendulum bearings and fluid viscous dampers, but also to provide isolation in the vertical direction (Fig. 6.b). The

Fig. 5. Oregon supreme court, Salem, Oregon, USA (image credit: Wikipedia).

hospital construction is complete, with design details included in the basement to allow the implementation of the vertical isolation system in the coming years.

The other essential facility examples of Figs. 7 and 8 are all in the state of Oregon. Figure 7.a, .b show a new operations center for Oregon's largest energy utility company, and a new headquarters for the Oregon state government Treasury [6]. Both projects have been recently completed and opened in 2021, and both utilize isolation systems with a combination of elastomeric and flat sliding bearings. Along with utility networks, transportation systems are also increasingly regarded as essential facilities, and Fig. 8 shows the currently under-construction main terminal building expansion of Portland International Airport [7]. The roof structure is isolated at the tops of architecturally impressive Y-columns using a total of 68 curved-surface sliding bearings. The use of isolation not only made possible the dramatic architecture but also achieved above-code functionality protection for the facility.

(a) Hospital building general view (b) Isolation system components

Fig. 6. Loma Linda University Medical Center, Loma Linda, California, USA (images: Ref [5]).

(a) Utility company operations center
Tualatin, Oregon, USA

(b) Treasury Resiliency Building
Salem, Oregon, USA

Fig. 7. Essential facilities, Oregon, USA (photos: (a) PGE; (b) GBD architects).

(a) Exterior view, main terminal building

(b) Terminal interior with Y columns

Fig. 8. Terminal Core Roof, Portland International Airport, Portland, Oregon, USA (images: ZGF Architects).

2.4 Data Centers

Seismic isolation has been used for a number of data centers in the last several years. Two recent examples are shown in Fig. 9, a 4-story, 16,000 m^2, center in Silicon Valley, northern California, and a 2-story, 6,000 m^2 project in southern California [8]. Both projects utilize triple pendulum bearings and have been designed for maximum displacements of approximately 800 mm, with the Santa Clara project additionally including fluid viscous dampers to limit the maximum displacement.

2.5 Non-building: Computer and Electrical Equipment

A major new hydroelectric project in the Canadian province of British Columbia, the Site C Clean Energy Project has made extensive use of state-of-the-art isolation systems to provide enhanced seismic performance of both critical electronic equipment and electrical distribution equipment. Figure 10.a shows an example of 3-D seismic isolation platforms used to protect spillway gate control equipment critical to dam safety against 1/10,000 year Maximum Earthquake shaking (a total of 39 platforms of varying sizes are

(a) Data center, Santa Clara, California (b) Data center, Irvine, California

Fig. 9. Seismically-isolated data centers, California, USA (images: (a) NTT/Gensler; (b) TGS/LPA).

used on the project) and Fig. 10.b shows wire rope isolators used to protect high-voltage electrical distribution equipment [9].

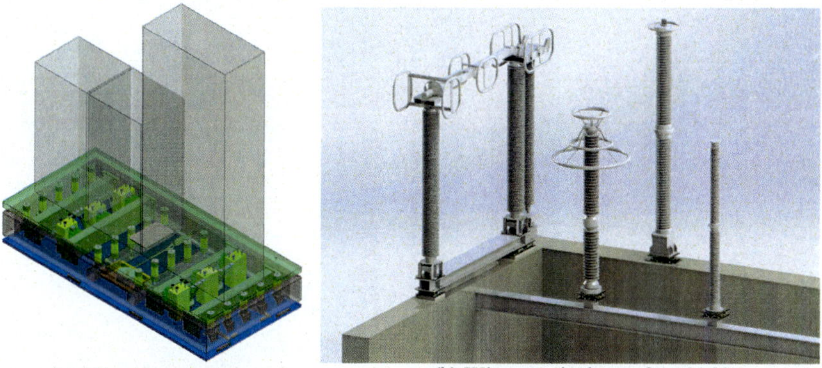

(a) 3D seismic isolation platforms for electrical and electronic control equipment (b) Wire rope isolators for 500 kV electrical equipment

Fig. 10. Equipment seismic isolation, Site C Clean Energy Project Peace River, British Columbia, Canada.

2.6 Non-building: Industrial

The first of a series of modern food and green waste treatment facilities in San Luis Obispo, California has utilized seismic isolation to protect a key component of the facility. The anaerobic digester unit is supported by 52, 800-mm diameter lead-rubber isolators (Fig. 11). The first anaerobic digester unit was completed in 2018, and several more are currently in the planning stages.

2.7 Non-building: Artwork

The final example of seismic isolation is the smallest, and lightest of all of the examples presented. An art object, in the entrance lobby of a hospital in La Jolla, California, is

Fig. 11. Anaerobic Digester Unit, San Luis Obispo, California, USA (photos: Mageba).

isolated using a nonlinear spring-roller system to provide up to 450 mm of movement. A challenging aspect of the design required that there be no mechanical attachments to the sculpture, and this dictated the force-reduction performance requirements for the isolation system (Fig. 12).

Fig. 12. Artwork isolation, La Jolla, California, USA.

3 Trends

Whilst the total number of seismically-isolated buildings in the U.S. over time has not progressed as significantly as in some other countries, many of the applications that have been completed are structurally or architecturally significant, and among those projects that have been completed in recent years a number of general trends can be observed:

- Design displacements for isolation systems have tended to increase over time. This has been due not only to changes (generally increases) in seismic hazard estimations, but also the greater use by designers of longer-period isolation systems. Figure 13 is indicative of this change in maximum design displacement over time, showing the inland region of southern California east of Los Angeles that includes a number of isolated buildings. Building "1" is the first isolated building in the U.S., and Building "6" is the Loma Linda University Medical Center (Fig. 6). It can be seen that design displacements, originally in the range of 380–550 mm have increased to as much as 1.07 m over about 25–30 years. These large displacements present significant design challenges and push the performance limits of isolation devices.
- Many recent projects with large displacement requirements have utilized triple pendulum bearings.
- A number of low-rise (2–3 story) steel-framed buildings have used a combination of elastomeric and flat sliding bearings.
- Fluid viscous dampers continue to be used together with pendulum or elastomeric bearings, to moderate large displacement demands.
- The use of isolation for non-building applications, such as computer systems, sensitive and critical equipment and artwork has increased. These systems have recently included high-performance 3-D capabilities.

Interest in seismic isolation in Canada is growing with more engineers now exploring isolation for both new and retrofit construction, and its use even in areas of relatively low seismicity is proving to be both technically and economically viable, particularly for retrofit applications.

The earliest U.S. isolation projects are now reaching 35 and soon to be 40 years since they were constructed. There has been an ongoing interest in the long-term properties of the isolation bearings for these projects, and several have seen periodic removal and re-testing evaluations. The results of this work will supplement the existing knowledge on the long-term properties of various types of isolation devices. A consideration on several projects has also been the adequacy of the original design displacements and the displacement capacity of the isolators. One such example is Building "2" in Fig. 13 which was originally designed and constructed for a maximum displacement of 355 mm, and after seismic hazard re-assessments following the Northridge earthquake of 1994 was upgraded by the addition of fluid viscous dampers to be able to accommodate the expected higher seismic demand. Recent re-evaluations have been undertaken on a number of the other early projects.

4 Summary

It is nearing 40 years since the design and completion of the first seismically isolated buildings in the U.S. The rate of adoption of the technology has not matched that of some other countries, nonetheless many notable seismic isolation projects have been

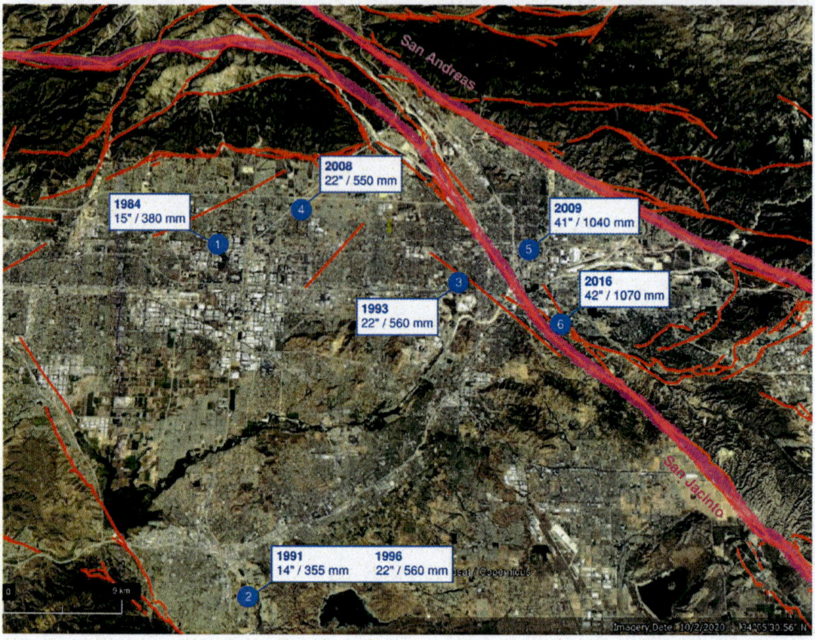

Fig. 13. Maximum design displacements for seismically-isolated buildings in inland southern California, 1984–2016 (image: adapted from Google Earth).

completed and continue to be undertaken, with applications pushing the state-of-art in terms of seismic protection objectives.

The use of seismic isolation for buildings in Canada has started to emerge in the last decade and it is expected that isolation will see increased use in future years, not only in high, but also low-to-moderate seismic regions of the country.

References

1. Darama, H., Nulman, A., et al.: (W)rapper tower: engineering an occupied sculpture. In: Proceedings of 17 World Conference on Earthquake Engineering, Paper No. 2c-0124, IAEE, Sendai, Japan, September (2021)
2. "Earthquakes, Sliders and Art", LACMA UnFramed, 4/29/22. https://unframed.lacma.org/2022/02/16/earthquakes-sliders-and-art
3. Sherstobitoff, J., Latifi, R., et al.: Base isolation seismic upgrade of a heritage school building at Lord Strathcona Elementary School—the first base isolated building in Canada. In: Proceedings of 30th IABSE Symposium—Engineering the Future, IABSE, Vancouver, Canada, September (2017)
4. Bebamzadeh, A., Sherstobitoff, J., et al.: Canada's parliament hill centre block base isolation seismic upgrade: impact of vertical ground motions. In: Proceedings of 12th National Conf. on Earthquake Engineering, EERI, Salt Lake City, USA, June–July (2022)
5. Nielsen, G., Rees, S., et al.: A 3-D seismic isolation system to protect the Loma Linda University Hospital from near-fault earthquakes. In: Proceedings of 17 World Conference on Earthquake Engineering, Paper No. 2g-0295, IAEE, Sendai, Japan, September (2021)

6. Zimmerman, R., Herdich, B.: Resilience-based design of a seismically isolated, continued operations building. In: Proceedings of 12th National Conference on Earthquake Engineering, EERI, Salt Lake City, USA, June–July (2022)
7. Zimmerman, R.: Seismic isolation of the terminal core roof at the Portland International Airport. In: Proceedings of 17th World Conference on Seismic Isolation, Energy Dissipation and Active Vibration Control of Structures, ASSISi, Turin, Italy, September (2022)
8. Richardson, S., Lindorfer, K.: Seismic isolation in the U.S. mission critical sector. In: Proceedings of 17th World Conference on Seismic Isolation, Energy Dissipation and Active Vibration Control of Structures, ASSISi, Turin, Italy, September (2022)
9. Aiken, I., Black, C., et al.: 3D seismic isolation for operational level protection of critical electrical equipment and electronic control equipment for the site C clean energy project. In: Proceedings of 17th World Conference on Seismic Isolation, Energy Dissipation and Active Vibration Control of Structures, ASSISi, Turin, Italy, September (2022)

Seismic Isolation in Chile: An Opportunity for Model Codes Calibration

Ruben Boroschek[1](✉) and Rodrigo Retamales[2]

[1] Universidad de Chile, Santiago, Chile
rborosch@ing.uchile.cl
[2] Seismic Intelligence SpA, Santiago, Chile

Abstract. The application of seismic protection technologies has continued expanding in Chile and other South and Central American countries. This paper presents some selected projects recently designed or built. Most countries in the region have developed local codes or considered the requirements presented in ASCE/SEI 7 for designing seismically isolated structures. Chile is not the exception. Nevertheless, the most recent versions of the Chilean standard, including its current revision, deviates from the US standard because of the recent experience in large earthquakes (M_W 8.8 in 2010, M_W 8.2 in 2014, and M_W 8.3 in 2015), the occurrence of frequent M_W 6.0 or larger (approximately one every month and a half) and economic aspects during construction and maintenance. The main differences include: i) structural system designed for a design basis earthquake; ii) limits to minimum vertical isolation frequency; iii) stricter limits for structure story drifts; iv) requirements for testing 100% production bearings; v) explicit protocol for production bearings testing; vi) explicit requirements for fire protection system; and vii) considerations for minimum safety factors for loads and deformations. In addition, most public hospitals currently under design and construction are incorporating structural health monitoring systems to verify the conditions of the isolation system and the safety of the structure after events with Mercalli Intensity larger than VI. With several seismic events having moment magnitudes larger than 6 every year, it is expected that the Chilean standard becomes empirically validated within the next 10 years.

Keywords: Seismic isolation · Seismic protection · Energy dissipation · Code development

1 Introduction

The use of seismic protection technologies is quickly expanding in Central and South America. From the initial applications in the early 90's to nowadays, several hundred structures have incorporated energy dissipation devices such as viscous, friction, metallic, and tuned mass dampers, and seismic isolation technologies such as high damper rubber, lead rubber, natural rubber, and friction pendulum bearings. Structures such as bridges, health facilities, LNG tanks, wharves, electrical substation equipment, museums, airports, churches/temples, office, educational and residential buildings, hotels, datacenters, control rooms, telescopes, and warehouses have been protected.

G. P. Cimellaro (Ed.): WCSI 2022, LNCE 309, pp. 86–99, 2023.
https://doi.org/10.1007/978-3-031-21187-4_7

Chile has been a pioneering country in the region due to its high seismic activity, characterized by events with moment magnitude greater than 7 and 8 occurring every five and ten years in average, respectively. Only in the last decade, three large events affected the territory. The first of them, the 2010 El Maule M_W 8.8 earthquake, caused approximately 500 fatalities and economical losses close to 33 billion US dollars, according to official figures, corresponding to 15% of the 2010 GDP of the country. The observed seismic performance of structural systems was adequate, because of the application of a strict seismic design code, empirically calibrated, that limits story drifts at diaphragm center of mass (CM) to 0.002 times the story height, and the additional drift at building corners to less than of 0.001 times the story height additional to the CM drift for the service level earthquake. Nevertheless, most economical losses were related to damage to all types of nonstructural components and systems, as described in detail in Miranda et al. [1]. In many cases, nonstructural damage was confounded with structural damage, causing conflicts and legal issues between owners/users and contractors/designers. The other large earthquakes occurred in the decade are the 2014 Iquique M_W 8.2, and the 2015 Canela Baja M_W 8.3. In consequence, authorities, investors, developers, designers, and contractors requested the use of seismic protection technologies to protect function continuity, investment, building contents, and improve users/owner comfort/experience during frequent, moderate, and large magnitude earthquakes.

2 Seismic Environment

Chile is one of the countries with the highest seismic activity in the world, primarily due to the subduction process of the Nazca Plate below the South American continent. This subduction process gives rise to different types of earthquakes, which are classified as inter-plate earthquakes (occurring in the contact zone between the Nazca Plate and South American Plate), intermediate depth and large depth intra-plate earthquakes (occurring within the interior of the Nazca Plate), and shallow intra-plate earthquakes (occurring in the continental crust of the South American Plate). From the late sixteenth century to the present, there has been a high-magnitude ($M_W > 7.5$) earthquake every 5 years on average throughout the Chilean territory. Figure 1 shows the number of seismic events with M_W larger than 6, capable to induce damage, affecting the country in the period 2004–2020. In Fig. 1 it can be observed maximum frequencies correlating to aftershocks following large earthquakes. In average, nine events $M_W > 6$ are observed per year, and in consequence, seismic protection technologies are designed to protect under that scenarios as well, not only during maximum considered earthquakes (MCE).

Another characteristic of the subduction environment is related to the duration of motion. In Fig. 2 it can be observed a typical record registered by RENADIC, Civil Engineering Department of University of Chile. It can be observed that for 20 s the ground accelerations exceed 0.2 g while during 110 and 80 s the accelerations exceed 0.05 and 0.1 g, respectively. This observation has a consequence on the choice of seismic protection technologies, that shall undergo a large number of cycles without losing their energy dissipation capabilities.

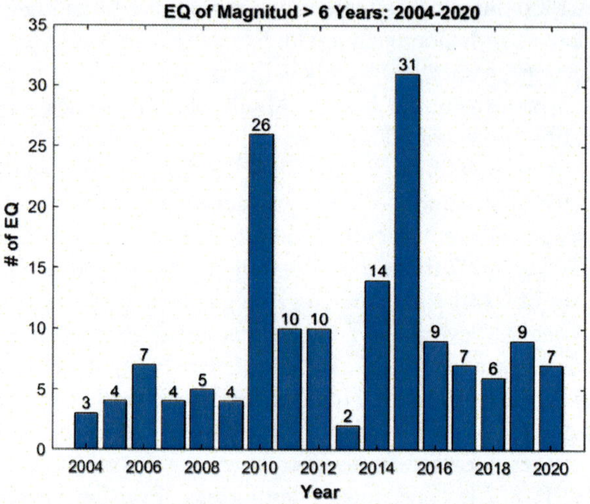

Fig. 1. Number of seismic events with $M_W > 6$.

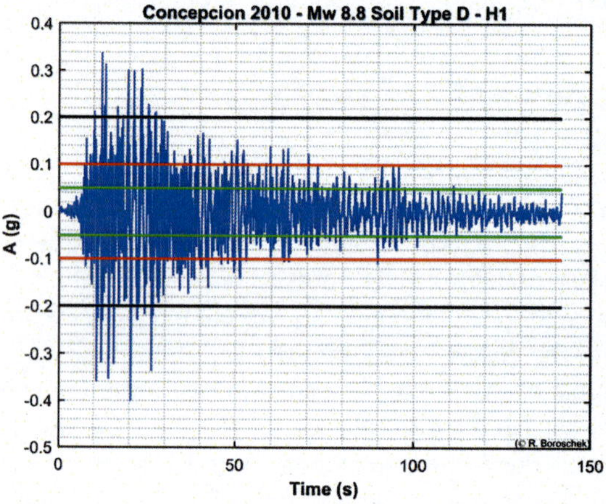

Fig. 2. Typical subduction record (Concepcion Station, El Maule 2010 M_W 8.8 earthquake).

3 First Applications and Current Trends in Chile

3.1 First Applications

The first notable applications of seismic isolation and energy dissipation in Chilean projects include the Andalucia Community Building, developed by Moroni and Sarrazin in 1992, in collaboration with the Ministry of Housing and Urbanism; the Marga-Marga Bridge developed by G+V Engineers in 1996; and the Amolanas bridge. The Andalucia Community Building, shown in Fig. 3, consisted of several low-income dwellings. One of the structures was base isolated and all others were identical but conventional. The isolated building is a 4-story confined masonry structure mounted on 8 high damping rubber bearings (HDRB's). The first level considers reinforced concrete shear walls. The building has a typical floor plan of 10 m × 6 m. The HDRB's are 315 mm in diameter and 320 mm in height. The thickness of the rubber layers is 6.7 mm. Four of these isolators are located in the corners of the building and two at each long side of the plan. The isolators have a vertical load capacity of 350 kN at 200 mm lateral displacement (100% strain). An extensive test series was carried out at the University of California, Berkeley [2]. The design period of the isolated building is 2 s. The base isolated and one of the fixed base buildings were instrumented. The response of the building during the 2010 Maule earthquake (M_W 8.8) was successfully recorded [3], demonstrating the efficiency of elastomeric bearings in reducing roof accelerations from + 1 g (saturated instrument) to 0.22 g when comparing conventional to isolated structure.

Fig. 3. Andalucia community building.

Fig. 4. Marga-Marga bridge.

The second structure incorporating seismic isolation in Chile was the Marga-Marga bridge, shown in Fig. 4 [4]. The bridge, spanning 368 m, consists of 4 steel girders mounted on 36 HDRB's. The height of the tallest reinforced concrete pier is 32 m. Rubber bearing sizes vary among 850 × 500 and 500 × 500 mm^2, depending on the service loads. All isolators are 302 mm in height, with 204 mm of rubber, 16 steel shims 3 mm in thickness, and two 25 mm thick mounting plates. The design displacement is 156 mm. Two prototypes of each type and all production bearings were tested at the University of Chile laboratories. This bridge was fully instrumented by the University of Chile since its construction, using an array of 24 accelerometers. The recorded information demonstrates the reduction in accelerations from the top of the piers to the deck over the isolation system. Further, the recorded data has been widely considered for determining actual dynamic properties and calibrating analysis models used to address the seismic performance of the bridge under several earthquake scenarios.

With regards to energy dissipation technologies, one of the first seismic applications in Chile was the Amolanas viaduct. The length of this steel box girder viaduct is 268 m, and the height of the tallest reinforced concrete pier is close to 100 m. The viaduct has PTFE pot bearings at the abutments and shortest piers and neoprene bearings at the tallest piers. Between the deck and the abutments, the bridge has two viscous dampers in the longitudinal direction, with a capacity 3000 kN each, and a stroke of 200 mm. The bridge was instrumented by the University of Chile with a network of 12 uniaxial accelerometers.

The first application of seismic isolation in a large structure was the La Reina Military Hospital (Fig. 5), developed in 2001 by Misael Astudillo Architects and the Engineering consulting firm Hoehmann Stagno & Associates. The number of beds 330, and the total constructed area 88,000 m^2 approx. The clinical and emergency services are located in a base isolated structure, in order to protect the investment and the functionality of their services. The structural system consists of a moment-resistant reinforced concrete frame with an approximate built area of 50,000 m^2 distributed in 5 levels, including a basement level for parking. The dimensions in plan are 126 m $\times$115 m. The column spacing is 9 m in both directions. The floor height is 5.75 m at the basement level and 4.5 m at the upper levels. The building columns have a typical section of 800 mm $\times$ 800 mm, except in the basement level, where the typical column section is 1100 mm $\times$ 1100 mm. The beams have a typical section of 600 mm $\times$ 900 mm, except in the ceiling of the basement level, where the beams have a 600 mm $\times$ 1100 mm section. The stronger and stiffer system at the basement level was requested for both an elastic response and to control story drifts at this level.

Fig. 5. La Reina Military Hospital.

The hospital is mounted on 164 HDRB's located at the ceiling of the basement level. The isolators are 700 and 900 mm in diameter and are made of 20 rubber layers of 8 mm of thickness and 4 mm thick steel shim plates. Isolators located in the perimeter consider a 150 mm in diameter lead core to increase the energy dissipation capacity of the seismic isolation system. The stiffer isolators are located in the perimeter of the structure in order to reduce torsional effects. The design displacement was 240 mm. The damping provided by the isolation system is close to 12% and the period of the isolated structure is nearly 2.5 s. An interesting conclusion was drawn from this project. The estimated total cost of the base isolation system was between 1 and 2 million US dollars. This value corresponds to 0.9–1.8% of the total cost of the project. Interestingly, the

Fig. 6. Spare friction dampers for transformers electrical substations.

other two bidding offers for the construction of this hospital were 133 and 135 million US dollars, which means that the difference between the winning offer and the following one, nearly 20 million US dollars, was 10–20 times larger than the cost of providing seismic protection to the structure [5].

A couple of additional seismic isolation projects were developed between 2002 and 2010, which included the San Agustin educational building, the San Carlos de Apoquindo Clinic, the Coronel wharf, developed by Sirve S.A.; and the Asociacion Chilena de Seguridad Health Facilities in Santiago and Vina del Mar, designed by CRL Engineers and Patricio Bonelli & Associates, respectively. Among the structures built in the 2002–2010 period that incorporated energy dissipation systems are the Titanium tower (52 stories), designed by Alfonso Larrain Vial Engineers, which included U-shaped metallic dampers, similar to those developed by [6], and the Parque Araucano building (22 stories), that included a Tuned Mass Damper (TMD) system for controlling seismic story drifts. A couple of interesting applications of energy dissipators implemented by the authors during this period are shown in Figs. 6 and 7. Figure 6 shows pre-compressed self-centering friction rings dampers used at the base of high voltage power and current transformers to protect the continuity of operation of the electrical network. As a result of the massive use of these devices in electrical substations, the power outage following the M_w 8.8 Maule earthquake lasted less than 2–3 h in most affected region. Figure 7 shows a glassed pedestrian bridge that joins two buildings at their 8th level. The energy dissipation system, consisting of dogbone shaped high damping rubber dampers, allows for controlling both the impact and excessive separation between the bridge and the buildings through the bumper and tensor behavior of the devices. Additionally, the devices provide damping to the bridge, close to 8% of critical, that allow to control walking induced vibrations. The bridge has exhibited and excellent performance, without any damage to glass, during the 2010 M_w 8.8 and the 2015 M_w 8.3 earthquakes.

Fig. 7. Pedestrian bridge at upper building level.

The above mentioned structures were designed at a time in which there were not codes in Chile for the design of seismically isolated structures or structures incorporating energy dissipating devices. Most of the designs considered as a reference the requirements of the Uniform Building Code of 1997 and the 2000 International Building Code, which become in 2003 the basis for the Chilean code for seismic isolated buildings. Furthermore, the 10–15 structures with seismic protection devices existing in Chile when the 2010 earthquake struck behave as expected, evidencing the abilities of the seismic protection systems to protect the infrastructure and investment, being the basis for current codes described in the following sections.

3.2 Recent Applications

Before year 2010 most existing seismically isolated structures in Chile were buildings of relatively low height, mainly health facilities and bridges. Regarding energy dissipation systems, most applications corresponded to special structures. The use of seismic protection technologies was not massive. Nevertheless, given the excellent response reported during extremely severe earthquakes, the application of seismic protection technologies rapidly expanded to other types of structures such as datacenters, control rooms, museums, laboratories, educational facilities, industrial facilities, wharves, and more importantly, to office and residential buildings. Current trends consider the increased use of lead rubber bearings and natural rubber bearings in replacement of high damping rubber bearings, the isolation of relatively tall buildings (up to 33 levels), and the application of the concept of isolated blocks for large projects, such as hospitals with built areas up to 250,000 m^2.

Regarding the application of damper devices, most of applications consider passive dampers, mainly viscous dampers in buildings facades, and TMD's at building roofs. A few buildings have considered metallic dampers such as TADAS and buckling restrained braces. The incorporation of products mainly manufactured in Japan, Italy, New Zealand, and the United States into the Chilean market has resulted in a reduction of 30% in the cost of isolators and damping devices, contributed significantly to enhance the local testing capabilities, and helped with the technology transfer. The following sections describe a couple of outstanding projects that better demonstrate the main characteristics of current applications of seismic protection technologies.

Torre del Sol Condominium. The building shown in Fig. 8 corresponds to the Torre del Sol condominium, a residential building built in Copiapo in 2011, in a zone with a 100 years seismic gap. This project was developed by Inmobiliaria Santo Domingo Ltda. And built by Constructora ALCORP S.A. The structural design of the building was developed by Patricio Bonelli & Associates while the design of the seismic isolation system, conducted in accordance with the provisions of the Chilean code for isolated structures NCh2745.Of2003 [7], was developed by Ruben Boroschek & Associates (RBA).

The Torre del Sol condominium, with a built surface of 19,000 m^2 approx. And 19 stories (1 level for the isolation system, 2 underground parking levels, 15 residential floors, and 1 level for equipment), was during construction the tallest residential building with seismic isolation in the Americas. The structural system consists of reinforced concrete bearing walls. A typical floor covers 53 m $\times$ 12.9 m while the total height of the

Fig. 8. Torre del Sol Condominium.

building, including the undergrounds, is 49.6 m. The total seismic weight of the structure supported by the isolation system is 155,000 kN. The foundation soil corresponds to a stiff soil with average shear wave velocity in the first 30 m beneath the surface of 598 m/s. For this project, lead rubber bearings were chosen for a number of reasons, among them: these systems have been extensively tested in laboratory and their behavior has been thoroughly demonstrated in real seismic events; the simplicity for predicting and modeling their behavior for the purposes of analysis and design; the stability of their properties when subjected to environmental agents; and the independence of their performance from the loading rate. The number and location of the seismic isolators required was determined largely by the geometry of the building and the decision to prevent uplifts on the bearings. In order to provide stability to the structure and to prevent tension forces on the isolators, 60 cm $\times$ 197 cm beams were considered underneath the parking area on the second underground level. Dead loads were added at the level immediately over the isolation to lower the heigh of the structure center of gravity. Seismic isolators were arranged at the ends of beams and, generally, at the ends of bearing walls. As an alternative to withstand incidental tension forces, the use of cables within the isolators [8] was evaluated. However, this alternative was discarded due to the length of the cables required to prevent yield during the movement of the isolation interface. In order to reduce costs, the 45 seismic isolators used in the structure are all identical. The maximum compressive force on the isolators (considered for design and for testing) is 8200 kN, while the minimum compression load is 0 kN.

Prototypes and all production isolators were subjected to an exhaustive test series combining compression and shear in order to validate the properties assumed during design and to verify the stability of the isolators during extreme seismic loads. The tests were performed at manufacturer's laboratories in McCarran, Nevada. In the design of the isolation system, a series of nonlinear dynamic analyses were carried out taking into consideration 7 seismic records compatible with the demands associated to the maximum considered earthquake defined in the Chilean standard for the seismic zone and ground conditions at the building site. A wide variety of seed records were selected in order to allow for analysis of the different seismic events to which the structure could be exposed during its lifetime. The analysis showed that the use of isolation allows for achieving a reduction of 88% in the average base shear of the structure isolated, when compared to the fixed base structure. The design of the superstructure was made taking into consideration the maximum shear that could be transmitted through the isolation system. In this case, the critical zone of the structure, where the demand of seismic deformations is most concentrated, corresponds to the isolation interface; therefore, the special boundary confinement required by ACI318 was not considered. The demand of displacements at roof level is accommodated by the deformation of the isolators and not by the rotation of concrete walls critical sections.

Critical Datacenter. The second interesting and relatively recent application of seismic isolation is a critical Datacenter nearby Santiago. This building hosts one of the largest high-tech facilities in Chile. The total area of the building is approximately 35,000 m^2, distributed in 2 levels and a mechanical floor. The architectural, structural and isolation system designs, carried out in 2011 by +Arquitectos, Gatica-Jimenez Engineers and Ruben Boroschek and Associates, respectively, considered the stringent requirements of ANSI/TIA-942 [9] for Tier level 4. According to TIA's requirements, the infrastructure shall be fault-tolerant, with electrical power storage and distribution facilities with an expected availability beyond 99.995%. The only way to protect the operation of the building during and after extreme seismic events, and fulfill the ANSI/TIA requirements, is using 115 lead rubber isolators under the structure.

Nunoa Capital Condominium. The Nunoa Capital Building, shown in Fig. 9, whose structure was designed by Rene Lagos Engineers and seismic isolation by RBA, considers 33 levels and a total height of 86 m. This building is the tallest isolated residential building in the Americas since its construction in 2016. This building, with a total constructed area of 42,600 m^2, consists of two identical 29-story towers intended for residential use. Both isolated towers share four underground levels. The seismic-resistant system of the building consists of an eccentric reinforced concrete shear wall core, as well as a perimeter frame and L-shaped walls in the corners. The floor system is comprised of post-tensioned slabs. On top the isolation system, to resist P-Δ effects, lower the building center of mass, preventing tension forces on the isolators, and ease construction, a massive concrete slab 2 m in thickness was considered. The structure is built on top of 24 large diameter rubber isolators, 16 of them are lead rubber bearings (LRB), while the 8 remaining are natural rubber bearings (RB). The seismic isolators are located under the ends of the walls and under the columns. The largest diameter Type C (RB) isolators are 1550 mm in diameter and have a load capacity of more than 40,000 kN. They are located under the most heavily loaded locations at the ends of the walls at elevator shafts. There are 8 Type A LRB isolators, 1150 mm in diameter, with a load capacity greater than 20,000 kN, and 8 Type B isolators, 1350 mm in diameter, with a load capacity of more than 30,000 kN. The stiffest isolators are the Type B isolators and they are located at the perimeter to control the torsion of the structure. Due to the period of the fixed base structure, around 2 s, it was estimated that the period of the isolated structure would be around 5–6 s, and therefore, in accordance with the current Chilean isolation code, a site-specific seismic hazard study was performed. The seismic isolation system allows to achieve an effective vibration period ever 5 s, and about 20% effective damping. Based on the results of the seismic analyses, reductions of shears, absolute accelerations and story drifts by 70–80% are obtained compared to the fixed base building. The isolators are placed on footings connected by beams. The isolated towers and the peripheral structures are separated by a 500 mm isolation gap which is significantly larger than the requirement of the Chilean isolation code, NCh2745.Of2003 [7], in order to minimize the probability of impact between the isolated structure and the adjacent structures.

Fig. 9. Nunoa Capital Condominium.

The Chilean isolation code at the time of design allowed for considering the design base shear as the elastic demand divided by a response modification factor R = 2, but no less than a minimum percentage of the seismic weight that depends on the seismic zone, equal to 5% for this project. Usually, the minimum base shear is between the elastic and the reduced demand, which causes the effective response modification factor R* (the adjusted R factor to achieve the minimum base shear) to be between 1 and 2. Notwithstanding this, in this project, due to the long periods achieved, the elastic demand is less than the minimum. This causes the superstructure to be designed for a seismic demand greater than the elastic demand, which is equivalent to have an effective reduction factor R* less than 1.

The Chilean isolation code for seismic isolation limits the story drifts to be less than 0.25% of the story height for the design basis earthquake (DBE) level. This is different to other codes such as the ASCE/SEI 7 (ASCE/SEI 2010) because it is directly determined from the reduced (by the R* factor) DBE spectrum. In order to control the story drifts, it was necessary to stiffen the building, which was a challenge because a tight architectural plan with almost no space for additional structure. The drifts control was achieved by using 2 outriggers per tower in the uppermost (mechanical) floor, with a height identical to the story height. Even though this location was not optimal, the outrigger's benefits were enough to stiffen the structure and reduce the story drifts below the code limits. The second challenge was to avoid tension forces on the isolators. Although the isolators can resist small tension forces, the design team decided to prevent tension forces on the isolators for the MCE level. This was achieved by using the 2 m in thickness slab resting directly above the isolators. This thick slab stiffens the isolated interface so that the external isolators are coupled with the internal ones, leading to a greater lever arm that resists the overturning moment and reduces the magnitude of the seismic compression/tension forces.

Other Developments. The seismic isolation of relatively tall buildings rapidly expanded among practitioners and investors. An example of that is the Sunset Copiapo condominium (Fig. 10), project developed by Empresas Armas and designed by Patricio Bonelli & Associates and Ruben Boroschek and Associates. This project considers the concept of isolated block to protect three 17-story wall bearing reinforced concrete building, that with a total built surface of 46,000 m^2, share a common isolation slab. The isolated block concept has been further applied for the design of several large hospitals, including the Salvador Hospital, composed of six buildings with a total built area of 130,000 m^2 approx., built on top of a common slab mounted on +550 isolators. Similarly, the Marga Marga and the Quillota-Petorca hospitals, with approximate constructed areas of 80,000 and 70,000 m^2, respectively, use the same concept. In total, approximately 30 hospitals including seismic isolation are currently under construction, design, or bidding process.

Another interesting situation is related to bridge bearings. Current bridge construction practice requests testing in combined compression and shear all neoprene bearings, as normally considered for seismic isolators. Nevertheless, large size neoprene bearings, subject to large loads and displacements, have systematically failed laboratory tests. Aware of this situation, and the impact of failures on construction schedule, many contractors have considered replacing neoprene bearings by lead rubber isolators, with the same effective target stiffness, but with a well-known capacity to accommodate large displacements and loads. The added damping is not considered in design, but it is only considered as a plus or benefit for the bridge seismic performance.

4 Chilean Code for Seismic Isolation Design

4.1 Current Code

The first version of the Chilean standard NCh2745.Of2003: Analysis and Design of Buildings with Seismic Isolation [7] was published in 2003, and basically considered the requirements of Uniform Building Code 1997. Nevertheless, the seismic hazard was defined as a Newmark's spectrum, according to local seismic conditions, mainly controlled by the Nazca and South American plates subduction process. The DBE level earthquake spectrum was defined using both probabilistic analysis and specialist judgement. The DBE peak ground accelerations (PGA's) were determined probabilistically for a probability of exceedance of 10% in 50 years, while the peak ground displacements (PGD's) were determined as the envelopes of all displacement response spectra recorded until 2003. The MCE level earthquake is associated with a probability of exceedance of 10% in 100 years. Given the previous definitions, the ratio between the PGA's for MCE and DBE is 1.2. In addition, the damping reduction factors were determined for Chilean subduction records, obtaining reduction factors larger than the used in the US practice.

In 2013, the code was revised by representatives of academia, government, practitioners, and manufacturers. The update of the standard had the objective of identifying and correcting normative aspects that complicated its practical application, collecting experience from engineers that had applied it, and reconciling its requirements with seismic design standards developed and/or modified after the 2010 Maule earthquake. The new and current standard NCh2745-2013 was published by the National Institute of Normalization (INN in Spanish) in December 2013. In light of the response spectra for the seismic records obtained during the 2010 Maule earthquake, which ratified that the design spectra in the NCh2745.Of2003 standard were conservative for periods close to 3.5 s, it was decided to increase the T_M period from 3.0 to 3.5 s for isolated structures which do not require specific site spectra development.

An important point that was explicitly incorporated in the reviewed code corresponds to the quality control test requirements for the isolator manufacturing process, a decision that used to be left to the discretion of the engineer of record. The updated standard established explicit requirements for testing rubber and frictional pendulum isolators. In particular, and in the case of elastomeric isolators, 100% of production require testing.

For frictional pendulum isolators, production testing was requested, as a minimum, for 15% of the isolators in a project, but no less than 4 units. The difference was based on that for elastomeric bearings the properties of the isolator depend on the characteristics of the rubbers used, which can be affected for several reasons, even climatic, while in the second case, the properties of the isolator depend on the characteristics of the friction between the two surfaces, which were believed at that time easier to reproduce.

The current standard permits that structural elements located above the isolation system in buildings structured using reinforced concrete moment resistant frames could be designed and detailed in conformity with the requirements of chapter 21 of the code ACI-318 [10] for intermediate moment frames instead of special moment frames, but the criteria for the strong column—weak beam shall be met. Similarly, the standard removes the need for boundary elements in walls whenever their disregard is demonstrated via a rational engineering analysis. The structural elements located under the isolation interface in reinforced concrete isolated structures must meet the requirements of ACI-318 for special moment frames, while 80% of the seismic load is not transferred to structural (perimeter) walls. In addition, the code allows for an increase in the response modification factor for designing the structural elements located below the isolation system from 1.0 to 1.5, in consideration of the intrinsic over-resistance of current design procedures. The standard establishes that the response modification factor to be considered for designing the isolated structure does not need to be less than 1.0, so the design demands will not need to be greater than the elastic seismic demands, which is consistent with the requirements of the ASCE/SEI 7-10 standard [11].

In the case of structures with seismic isolation systems based on frictional pendulums, the updated standard established obligatory consideration of time history analysis for designing. This measure is based on the difficulties in predicting the behavior of isolators that base their response on a friction that depends on the vertical load and displacement rate, varying at each second on the isolators. Consequently, current code requires the time history analyses include the vertical components of earthquakes. Complementarily, the code clarifies the procedures for determining the total displacement of structures with this type of isolation system, given that torsion effects were supposed to be controlled given the coincidence of centers of mass and stiffness at any moment in time. The code is further complemented with clarifications regarding the requirements for P-Delta analysis, load combinations to be considered for the design of isolation systems and structural components, inspection and maintenance requirements for isolated structures, gap requirements between isolated and fixed structures and between isolated structures, and requirements for the design of nonstructural components in isolated buildings, among other clarifications.

4.2 Review of Current Code and Other Requirements

Review of Current Code. Current version (2013) of the code is under review. The new revision is expected to be released during 2023. The review process considers correcting the ratio between the MCE and DBE level from 1.2 to 1.4, given the first value is associated to PGA's while the second one is related to PGD's, which control seismic isolation devices. In addition, including the lambda factors in ASCE/SEI 7-16 is under review. Consensus exists in the committee regarding keeping designing the super- and sub-structures for DBE level and the seismic isolation for MCE, deviating from current US practice. Further, consensus exist regarding the need for including in the code a set of equations for designing devices and defining minimum forces and strain safety factors in order to uniformize design criteria considered by practitioners. Additionally, the importance factors (I_P) considered for designing nonstructural components and systems in isolated structures are under review, aiming at rationalizing their design.

In addition to code requirements, practitioners are typically considering bearing pressures limited to 17–18 MPa for the complete load combination. The activation of the isolation system is typically in the range 2–3% of the seismic weight of the structure, to get the system also activated during the frequent and moderate seismic events that periodically affect the country.

Complementarily, the Ministry of Public Works and the Ministry of Health have published their own design criteria for hospitals that consider: i) prohibiting tension/uplift in elastomeric bearings; ii) prohibiting the use of high damping rubber bearings given their aging, scragging and rate dependence issues; iii) requesting performing probabilistic seismic hazard analysis for the project site, no matter soil type or period of the isolated structure, and considering the envelope between code and site specific spectra; iv) requesting the use of the lambda factors in ASCE/SEI 7-16 for design; v) sizing probabilistically the gap around the isolated structure considering probabilities of impact under 10% and 5% for MCE and DBE levels, respectively, and considering a minimum logarithmic standard deviation equal to 0.5, to make independent the gap sizing from the scaling procedure used by designers; vi) requesting the use of passive fire protection systems allowing for inspection (periodic and/or following seismic events or contingencies), capable of deforming at least 10 cycles at the total maximum displacement D_{TM} without exhibiting any type of damage, and limiting the maximum temperature measured by thermocouples located in the surface of the bearing to 140 °C during 2–3 h when subjecting the specimen to the ASTM E119/ISO 834 standard fire test after the cyclic mechanical tests; and vii) requesting the full instrumentation of the isolated structures for their real-time health monitoring.

5 Conclusions

The application of seismic isolation and energy dissipation technologies is rapidly growing in Chile and South and Central American countries. The excellent performance of the protected structures during severe earthquakes, demonstrated throughout the recorded response of instrumented structures, triggered an increased demand for the use of protection devices. The use of protection systems has expanded from demonstration and health facilities projects to office and tall residential buildings. The observation of international experiences and the execution of technology transfer projects have been fundamental for these developments. The effects and benefits of using seismic protection technologies are widely known by authorities, stakeholders, and practitioners. With several seismic events having moment magnitudes larger than 6 every year, it is expected that the updated Chilean standard becomes empirically validated within the next 10 years. Monitoring is the key to continue advancing with the technology.

References

1. Miranda, E., Mosqueda, G., Retamales, R., Peckan, G.: Performance of nonstructural components during the February 27, 2010 Chile Earthquake. Earthq. Spectra **28**(S1), p453-471 (2012)
2. Moroni, M., Sarrazin, M., Boroschek, R.: Experiments on a base-isolated building in Santiago, Chile. Eng. Struct. **20**(8), p720-725 (1998)
3. Boroschek, R., Retamales, R., Aguilar, A.: Seismic response of isolated structures subjected to the M_W 8.8 Chile earthquake of February 27, 2010. In: Proceedings the International Symposium for CISMID 25th Anniversary, Paper N° M-2, Peru (2012)
4. Boroschek, R., Moroni, M., Sarrazin, M.: Dynamic characteristics of a long span seismic isolated bridge. Eng. Struct. **25**, 1479–1490 (2003)
5. Boroschek, R., Retamales, R.: Base isolation in hospitals. In: Proceedings of the Seminar on Design of Health Facilities to Resist Natural Hazards, Barbados, September 30–October 1 (2002)
6. Skinner, R.I., Kelly, J.M., Heine, A.J.: Energy absorption devices for earthquake-resistant structures. In: Proceedings of the 5th World Conference on Earthquake Engineering, Paper N° 370, Rome, Italy (1974)
7. INN: NCh2745.Of2003: Analysis and Design of Structures with Seismic Isolation (in Spanish). National Institute of Normalization of Chile (2003)
8. Kasalanati, A., Constantinou, M.: Testing and modeling of prestressed isolators. J. Struct. Eng. **131**(6), p857-866 (2005)
9. TIA: ANSI/TIA-942-2005: Telecommunications Infrastructure Standard for Data Centers. Telecommunications Industry Association (2005)
10. ACI: ACI 318-08: Building Code Requirements for Structural Concrete and Commentary (2008)
11. ASCE/SEI. ASCE/SEI 7-10: Minimum Design Loads for Buildings and Other Structures. American Society of Civil Engineers (2010)

Optimization as a Tool for Seismic Protection of Structures*

Giuseppe C. Marano(✉), Marco M. Rosso, and J. Melchiorre

Politecnico di Torino, DISEG, Department of Structural, Geotechnical and Building Engineering, Corso Duca Degli Abruzzi, 24, Turin 10128, Italy
giuseppe.marano@polito.it

Abstract. Scientific research within seismic protection techniques is still nowadays an actual and vibrant field. The main goal of earthquake protection is to alleviate the effects of seismic events on civil structures and infrastructures to reduce both human losses and socio-economical impacts in the medium and long-term. Seismic protection techniques are mainly related to the base isolation systems for buildings or bridges, and to energy dissipation-based solutions with the adoption of special devices, e.g. metal hysteretic dampers, viscous dampers, and friction dampers. The optimization procedures play a fundamental role in the dynamic parametric identification of such devices. Current trends in the optimization field are the adoption of computational intelligence meta-heuristic algorithms, which have been inspired by a conjunction of the artificial intelligence community and nature-inspired phenomena. Specifically, the widespread use of meta-heuristic techniques combined with proper seismic protection optimization problem formulations leads to more cost-effective, high-performing, and innovative design solutions. In the present study, a brief review is argued presenting some optimization procedures to accomplish seismic protection tasks.

Keywords: Seismic protection · Viscous dampers · Base isolation devices · Meta-heuristic algorithms · Optimization

1 Introduction

Within the earthquake engineering discipline, seismic or earthquake protection refers to all the procedures implemented to mitigate and diminish the damaging effects of seismic events on civil engineering structures. The main pursued achievement is to reduce as much as possible losses, meant as human losses as a priority and considering, as well, administrative and socio-economical impacts on societies in the medium and long-term [22]. In the existing literature, earthquake

* Supported by Project ADDOPTML.

G. P. Cimellaro (Ed.): WCSI 2022, LNCE 309, pp. 100–113, 2023.
https://doi.org/10.1007/978-3-031-21187-4_8

risk mitigation appears somewhat more widespread terminology, albeit it should properly refer to strengthening interventions only [14]. In order to achieve a better earthquake protection behavior of a structure, the most widespread techniques belong to two main categories: the foremost is denoted as *base isolation* systems, whereas the second class of approaches provides for the installation of special devices acting as *energy absorbers* of the seismic input energy loading [14]. The implementation of base isolators for protecting civil structures against the damaging effects of seismic loadings has become very frequent in recent decades. From the practical point of view, the base isolator devices create a decoupling layer at the top of the foundation level (substructure) which ensures that seismic input energy, and thus the induced damaging effects, of seismic shaking events, do not reach the structural system to be protected (superstructure). To perform that function, isolators permit relative movements between the structural systems and the substructure, i.e. substantially reducing the seismic demand in input to the superstructure, while providing sufficient vertical stiffness to effectively transmit vertical loads in foundations. These devices are commonly composed of thin reinforcing steel plates interspersed between thick rubber pads. The performance of a seismic isolator depends on many factors, such as the rubber typology, the compound, the thickness, and the process of vulcanization of the pads. Additionally, to provide a substantial increase of energy dissipation capacity to base isolators, a core made of lead may also be present [56,57,68]. On the other hand, the second class of energy-based approaches permits dissipating a part of the input energy coming from earthquake events with the advantage of employing devices that can easily be replaced, as necessary, after an event occurrence. The installation of energy adsorbers is extremely convenient both while designing a new structure and even for increasing the seismic protection of existing structures, acting as a retrofitting intervention [2,7,21]. The most widespread devices belongs to three main categories [13]: metallic yielding dampers [11], fluid viscous and viscoelastic dampers and friction-based devices [4,10].

In the following subsection, an introductory discussion related to structural vibration control devices is conducted, differentiating among active, passive, and semi-active techniques. In Sect. 2, the mathematical optimization statements are presented and a brief overview of meta-heuristic algorithms is argued. Finally, in Sect. 3, a brief literature review is presented related to some applications of optimization as a tool for seismic protection.

1.1 Active, Passive and Semi-active Structural Control Devices

Since seismic protection mainly aims to improve or isolate the dynamic behavior of structures, structural control devices have been one of the main subjects of research in the field of earthquake protection engineering for many decades. Several vibration control devices have been proposed in order to reduce vibration due to dynamic loads. Due to their simplicity, efficiency, and reduced costs (compared with other strategies) they are now widely used in many civil engineering applications to reduce vibrations and therefore to increase structural protection against wind, earthquake, and similar events [43,45]. Structural control aims to provide a framework to determine the most effective action to be applied in order to enhance the safety and the dynamic behavior of a structure based on some measuring control response parameters of the structural system itself. Several approaches have been developed to effectively deal with dynamic loads induced by natural hazards, which can be classified as active, passive, hybrid, and semi-active damping strategies.

An *active* structural control system consists of an ensemble of sensors and actuators which can actively modify the response of the structure. Sensors are placed on the structural system to be protected in order to monitor both external excitation levels and the resulting structural response parameters. Based on the information retrieved from the sensors, the control algorithm computes the required control force able to modify the dynamical behavior of the structure and dissipate energy by governing the actuators [31,69]. The actuators usually require a considerable external power source of energy. Moreover, the power supply has to be guaranteed to the actuators to remain functional during seismic events, but this is a challenging task [9]. The control strategy is acknowledged as *feedback control* when only the structural response is measured, whereas it is denoted as *feed-forward control* when only the input excitation is measured. A third scenario is called *feedback-feed-forward* control, and it exploits the measurements of both the response and the excitation quantities [69]. The main advantages of an active control system are: (a) an improved response control of the protected structure; (b) a lower sensitivity to ground shaking behavior; (c) the mitigation of many other dynamic natural hazard input, e.g. wind excitation; (d) the possibility to find the best control strategy considering both computational effort and implementation costs. The latter point may refers to also improve human comfort over structural behavior under some peculiar circumstances [69]. For instance, systems denoted as tuned mass dampers (TMDs) may be also designed as active control systems to minimize the effect of vibrations and also providing immediate damping of vibrations [9]. Active damping strategies of structures rely on the input forces given by electrohydraulic or electromechanical actuators designed to influence several vibration modes of the system. Therefore, active damping is most suited for structures with many degrees of freedom and it is beneficial for a wide range of operating conditions and structures. However, the main drawbacks of active control systems are: (a) high implementation costs; (b) the need for a high-power supply during a natural events such as an earthquake; (c) the possibility of the structure destabilization when the control system is not well designed, or it presents a malfunctioning.

Passive damping strategies do not rely on any actuators and, thus, they do not require any external power source for operation because these devices leverage the motion of the structure to develop the control forces. They are used to increase the energy dissipation capacity of a structure through localized and discrete devices, which may consequently increase also the stiffness and strength of the protected structure. The forces imparted by passive devices are developed in response to the motion of the structure [31]. The passive vibration control techniques for protecting structures dissipate the seismic input energy into both kinetic and strain energy of the passive device, or even heat energy, e.g. in viscoelastic dampers [9]. The performances and efficiecy of passive control devices strongly depends on an optimized design of both the building and the passive device, e.g. the viscoelastic materials added to the building [9].

Hybrid vibration control refers to a combined application of active and passive control systems. A hybrid system incorporates both typologies of devices, this allows the structure to reach the highest levels of performance exploiting the best features of both strategies. Notably, even during a sudden absence of power supply, e.g. during an earthquake event, the hybrid control system remains functional via its mechanical elements [9]. Hybrid techniques are utilized typically in very high-end applications, such as skyscrapers or viaducts, and they incorporate both actuators connected to a high-power source of energy and devices such as isolators.

Control strategies based on *semi-active* devices can be considered as a class of active control systems, denoted also as controllable passive devices in [31], which require less energy than active control systems [9]. These systems combine the best features of both passive and active control systems, e.g. they are not affected by power supply cuts since they can rely on batteries [9,69]. A semi-active system leverages controllable passive devices capable of varying their mechanical properties and characteristic parameters such as damping coefficient and/or stiffness in real-time. The modification of these mechanical properties is based on a control loop that computes the necessary action of the device depending on some structural response control parameters and/or the input excitation level. Since no actuators are adopted in a semi-active system, also in this case as in the passive approach the motion of the structure is used to develop the control forces. Specifically, a noticeable example belonging to this class of vibration control devices is the so-called magnetorheological dampers. These dampers rely on a magnetorheological liquid which is sensible to viscosity increase when the magnetic field intensity from an e electromagnet increase, which strongly affects the damping characteristics [9]. Similarly to active control systems, the available semi-active control strategies are differentiated from what the sensors are demanded to monitor. Therefore, they are referred to as feedback, feed-forward, or feedback-feed-forward if the sensors monitor the response of the system, the excitation of the input, or both, respectively. Semi-active, hybrid, and active control of structures are considered the natural evolution of passive control techniques. They rely on devices capable of processing the data in real time, to develop control actions with the smallest time lag possible. This allows an improved structural behavior and enhanced safety of the structure [31].

2 Optimization Strategies via Meta-Heuristic Algorithms

The main purpose of solving an optimization problem is to find the minimum (or maximum) of certain mathematical functions, namely objective functions (OF) $f(\boldsymbol{x})$. Optimization problems may be stated in two different ways depending on the number of OF involved, i.e. single-objective problems and multi-objective ones. Furthermore, it is possible to differentiate the optimization problems between unconstrained or constrained problems, based on the imposed conditions that must be met by the final solution. Unconstrained optimization problems can be defined as:

$$\min_{\boldsymbol{x}\in\Omega}\{f(\boldsymbol{x})\} \tag{1}$$

meanwhile constrained optimization problems can be defined as:

$$\begin{aligned} &\min_{\boldsymbol{x}\in\Omega}\{f(\boldsymbol{x})\} \\ \text{s.t.}\quad & g_q(\boldsymbol{x}) \leq 0 \quad \forall q = 1,\ldots,n_q \\ & h_r(\boldsymbol{x}) = 0 \quad \forall r = 1,\ldots,n_r \end{aligned} \tag{2}$$

where the vector of the design variables, $\boldsymbol{x} = \{x_1,\ldots,x_j,\ldots,x_n\}^T$, is the vector that contains the parameters to be optimized in order to obtain the best solution for the problem. Thus, the search space appears as a multidimensional domain $\Omega = [x_1^l, x_1^u] \times \cdots \times [x_j^l, x_j^u] \times \cdots \times [x_n^l, x_n^u]$ defined by the Cartesian product (denoted by the $\times$ symbol) of admissible range of values for the design variables involved in the problem. In general, it is possible to define two different categories of constraints: inequality constraints $g_q(\boldsymbol{x})$ and equality constraints $h_r(\boldsymbol{x})$. Being that equality constraint can be converted into two inequality constraints without any loss of generality, it is possible to rewrite the problem considering only the inequality constraints in (2), i.e. $g_p(\boldsymbol{x}) \leq 0$, where $p = 1,\ldots,n_q,n_{q+1},\ldots,n_p$, being $n_p = n_q + 2n_r$.

Over the past two decades, the use of meta-heuristic algorithms and, in particular, evolutionary algorithms for solving structural optimization problems has received much more attention. This is because such problems are often characterized by a discontinuous search space and the objective functions are not always differentiable. For this reason, it is often not possible to adopt gradient-based or quasi-newton methods, since the information about the first and second derivative of the objective function cannot be calculated. On the contrary, meta-heuristic algorithms do not require any first-order (gradient) or second-order (Hessian) information and seem to be more appropriate to handle complex optimization problems. Furthermore, the great majority of meta-heuristic algorithms take inspiration from natural phenomena and provide considerable simple implementations and, for this reason, they are widely used even by practitioners. One of the firstly developed meta-heuristic optimization procedures was the genetic algorithm (GA) proposed by J. Holland around the 1970s [47]. This algorithm still represents one of the most popular population-based evolutionary tools. This method tries to replicate the biological evolutionary process in order to mimic the Darwinian Evolutionary Theory that brings the population of solutions to evolve towards the best optimum. The procedure involves pseudo-random-based operators like crossover, mutation, and selection in order to reproduce the long-term process of evolution in a population with the survival of the fittest individuals [37]. Since the introduction of the genetic algorithm, many variations of it have been developed over the years, such as the Differential Evolution (DE) algorithm [70]. Within meta-heuristic algorithms, it is even possible to classify the various techniques among nature-inspired, swarm-based, physical-inspired phenomena, or statistical-based procedures [38]. A widespread example of a meta-heuristic swarm-based approach is the particle swarm optimization (PSO) algorithm [35,61,63] which is inspired by the behavior of fish schooling or birds flocking in nature. Other famous swarm-based optimization procedures are e.g. the ant colony optimization (ACO) algorithm [24,25], based on the simulation of the behavior of the ant colonies searching for food, or even the artificial bee colony (ABC) [33] and many others. It is worth mentioning some other examples of meta-heuristic algorithms such as the physical-inspired method denoted as simulated annealing (SA) [36], which is a single-solution-based meta-heuristic optimization approach inspired by the annealing in metallurgy, or the charged system search (CSS) algorithm [34]. Furthermore, some authors have developed meta-heuristics on the basis of statistical laws such as the estimation of distribution algorithm (EDA) [53,59].

In the last decades, the meta-heuristic algorithms have been applied in many engineering fields, highlighting their successful capabilities and effectiveness in dealing with real-world constrained problems [8,23,28,41,54,63,72], e.g. dealing with structural design [17,18,20,48,65] and structural optimization tasks [3,15,16,26,27,42,49,71]. In the following section, a brief review of some recent optimization applications for seismic protection is presented.

3 Brief Review of Optimization Applications for Seismic Protection

A well-designed base isolation structural system permits both to reduce the seismic input demand to the superstructure and to reduce the maintenance cost for after seismic events. Furthermore, the performances of passive vibration control devices are extremely sensitive to their dynamic mechanical behavior. Thus, a reliable identification of their mechanical properties play an important key role. In [43], the authors proposed a dynamic identification approach of passive devices by comparing the experimental and the analytical value of the forces experienced by the device under investigation. The authors adopted an optimization procedure for the parametric identification of the passive device based on two evolutionary algorithms, the PSO and the DE. The goal was to minimize the integral difference of the damper's force in time between the experimental testing conducted on a real viscous damper and the analytical simulated response obtained by applying the same time history of displacements of the tested device. The base isolation systems and the superstructure building are usually designed separately, however in [58] the authors proposed a multi-objective optimization approach focused simultaneously on reinforced concrete superstructure elements (beams and columns) and on elastomeric base isolation design parameters, i.e. maximum allowed displacement, rubber type and size. To pursue this task, three objective functions have been considered related to minimization of the top-floor displacement, minimization of the top-floor acceleration, and minimization of the superstructure material cost. In [55], the authors proposed a reliability-based design optimization method with GA for sliding implant-magnetic base isolator bearing. To provide an efficient statement of the optimization problem, one of the main task is an efficient parameter choice. To provide a design parameter reduction, the authors in [55] adopted a variance-based sensitivity analysis. In literature the problem of selecting optimal design parameters is a recurrent issue. In [64], the author proposed an optimization criterion for selecting the design parameters of elasto-plastic dampers based on the dispersion of absolute accelerations of the protected structure during a seismic event. The authors of [66] proposed a method to calculate the optimum design of an original force-limiting floor anchorage system for the seismic protection of reinforced concrete (RC) dual wall-frame buildings. The proposed anchorage system limits the seismic solicitations interposing elasto-plastic links between two structural sub-systems. The optimal solution is obtained by the application of the DE meta-heuristic algorithm and the method is applied to a case study represented by a 12-storey prototype RC dual wall-frame building. The optimization of a passive control strategy that involves the application of a linear dissipative connection in a wall-frame system is presented in [29]. The seismic protection device is designed as a global protection system to protect both the structures. The problem is formulated as a multi-objective optimization problem with conflicting objective functions. The final goal is to use the genetic algorithm in order to minimize the displacements in the frame structure and the shear in the wall.

To test new nonlinear base isolation devices, the current procedure is to analyse the stochastic response and testing experimentally the device with small-scale prototypes [67]. The stochastic response of isolated building is thus derived from a nonstationary random process which usually undergoes to a time-dependent linearization procedure to characterize the time-history of the base isolator nonlinear restoring force. For instance, the authors in [67] adopted a stochastic optimization procedure based on the convergence of different performance indicators, such as base drift or structural acceleration. Moreover, when dealing with uncertainties, it become challenging to provide an effective tuning of base isolation design parameters. Especially due to the need of ensuring both sufficient decoupling properties between substructure and superstructure, possible energy-dissipation features and also enough stiffness to prevent excessice deformations. Thus, the authors in [39] explored an optimization method on a grid-search approach, relying on a probabilistic performance-based optimum seismic design for seismic isolators of a prototype of high-speed railway bridge. A multi-objective optimization process which lead to a Pareto-front optimal solutions has been also analysed in [12]. In this contribution, the earthquake excitation is treated as a non-stationary random process and the desing of frinction-based seismic isolators required a suitable probabilistic characterization of the dynamic response through Monte Carlo simulations.

In [44], a method for the optimal design of Tuned Mass Dampers (TMD) is proposed. The main idea is to define the optimum design of vibration absorbers utilized to reduce undesirable vibrational effects which are originated in linear structures by seismic excitations. The objective function to be minimized is represented by the dimensionless peak of displacement of the protected system with respect to the unprotected one. In [30] the authors proposed a multi-objective optimization of a single tuned mass damper to control vibrations induced in building structures under low-moderate earthquakes. In this case the objective functions consider both the economical and the performance aspects. The optimal design aims to minimize the costs and to maximize the performance in order to control the damage level and the behaviour of components and equipment. In this case the economic cost is considered directly proportional to the mechanical properties of the device while the performance are evaluated as the ratio between the absolute accelerations of the protected structure and the unprotected one. In [19] the optimization of a multi-tuned mass damper inerter (MTMDI) is performed by using the genetic algorithm. The MTMDI system links two adjacent high-rise buildings as an unconventional seismic protection strategy. The authors analysed a real case study of two adjacent high-rise buildings connected by two corridors equipped with the proposed MTMDI system. The optimization problems aimed to minimize the displacement, the inter-story drift and the accelerations on the structure. The problems are treated separately and it is found that the optimally designed MTMDI outperforms both conventional MTMD and single TMDI in acceleration control.

It is worth noting that seismic isolation concepts could also be adopted for protecting equipments, e.g. medical ones in a hospital or some other in some strategic buildings, to avoid damage and remain functional during and after earthquake events. For instance, in [32] the authors provide an optimization design of an equipment isolator denoted as resilient sliding isolation system. This device is composed of a restoring spring and a slider, thus, the design parameters are the period of the system (related to the spring stiffness) and the friction coefficient, respectively.

In conclusion, it is important to keep in mind that the seismic protection techniques described above could benefit from integration with monitoring techniques. In particular, the installation of a comprehensive system that allows monitoring of the structural system with non-destructive techniques (NDT) [40,46,51,52] associated with real-time protection, dampers installation, and damage modeling systems [1,5,6,50,60,62] could be an interesting future development in the field of earthquake engineering. This would further increase the safety and reliability of structures and infrastructure by reducing seismic damage and simultaneously reducing maintenance cost for after seismic events, helping to extend the nominal life of the structures.

4 Conclusions

The brief review reported in the current document shows that the design and optimization of seismic protection devices is still an active and debated research topic. Indeed, besides the optimization algorithms actually implemented, a proper and more efficient optimization problem statement still represents a challenging task. As a matter of fact, the problem statement represents the starting point that tremendously affects the overall effectiveness and engineering soundness of the final result. In particular, a proper choice of the design variables to be included in the design vector, as well as the definition of the objective function to be considered, represents a crucial step in the optimization process. In the case of seismic protection devices, differently than in other fields, the minimization of the cost related to seismic devices themselves rarely constitutes the main objective of the procedure. In general, the optimization problem is pursuing the goal of maximizing the performance of the protection device to be designed. Therefore, the considered objective functions may promote the structural system performances rather than the device's costs themselves. Actually, the production and installation costs of the seismic protection devices are negligible compared with the restoration costs associated with hypothetical seismic-induced severe structural damages. In addition, such devices improve extending the nominal service life of the structure while reducing the maintenance costs. It is worth noting that seismic protection device optimization problems are often multi-objective problems. It is evident that the complexity of the problem cannot be solved by considering a single optimal solution. In general, it is desirable adopting optimization methods that provide designers with the most promising solutions space (e.g. Pareto front) from which the designer may choose the optimal trade-off solution considering conflicting aims. The authors believe that the present brief review may be very helpful to identify the promising future developments within the examined research field. Oversimplified modeling assumptions may lead to design solutions that are quite far from the actual ones which effectively capture all the complex phenomena involved in the seismic protection problems. Therefore, artificial intelligence and, in particular, machine learning techniques are increasingly becoming standard tools because of their capabilities and versatility. These methods should permit to study of more complex problems by reducing the number of simplifications assumptions. In future works, it will be possible to highlight how AI and machine learning techniques can produce significant results when applied to complex earthquake engineering problems. Seismic protection optimization applications with meta-heuristic algorithms represent the first steps in the AI-based direction to reach high-performance and cost-effective protection devices. The upcoming research on seismic protection techniques presents a very promising path in the direction of considering increasingly complex phenomena hitherto neglected, reaching desirable unprecedented levels of efficiency, safety, and reliability never previously achieved.

Acknowledgments. This research was supported by project MSCA-RISE-2020 Marie Skłodowska-Curie Research and Innovation Staff Exchange (RISE) - ADDOPTML (ntua.gr) The authors would like to thank the project ADDOPTML for funding supporting this research.

References

1. Aloisio, A., Alaggio, R., Fragiacomo, M.: Dynamic identification and model updating of full-scale concrete box girders based on the experimental torsional response. Constr. Build. Mater. **264**, 120146 (2020)
2. Aloisio, A., Boggian, F., Tomasi, R.: Design of a novel seismic retrofitting system for RC structures based on asymmetric friction connections and CLT panels. Eng. Struct. **254**, 113807 (2022)
3. Aloisio, A., Pasca, D.P., Battista, L., Rosso, M.M., Cucuzza, R., Marano, G., Alaggio, R.: Indirect assessment of concrete resistance from FE model updating and young's modulus estimation of a multi-span PSC viaduct: Experimental tests and validation. Elsevier Struct. **37**, 686–697 (01 2022). https://doi.org/10.1016/j.istruc.2022.01.045
4. Aloisio, A., Pelliciari, M., Sirotti, S., Boggian, F., Tomasi, R.: Optimization of the structural coupling between RC frames, CLT shear walls and asymmetric friction connections. Bullet. Earthquake Eng. 1–26 (2022)
5. Aloisio, A., Rosso, M.M., Alaggio, R.: Experimental and analytical investigation into the effect of ballasted track on the dynamic response of railway bridges under moving loads. J. Bridge Eng. **27**(10), 04022085 (2022)
6. Aloisio, A., Rosso, M.M., Huang, D., Iqbal, A., Fragiacomo, M., Pei, S.: Nonlinear sdof analytical model of mass-timber building with post-tensioned rocking walls (2022)
7. Aloisio, A., Rosso, M.M., Iqbal, A., Fragiacomo, M.: Hysteresis modeling of timber-based structural systems using a combined data and model-driven approach. Comput. Struct. **269**, 106830 (2022)
8. Asso, R., Cucuzza, R., Rosso, M.M., Masera, D., Marano, G.C.: Bridges monitoring: an application of ai with gaussian processes. In: 14th International Conference on Evolutionary and Deterministic Methods for Design, Optimization and Control. Institute of Structural Analysis and Antiseismic Research National Technical University of Athens (2021). https://doi.org/10.7712/140121.7964.18426
9. Bekdaş, G., Nigdeli, S.M., Kayabekir, A.E.: Introduction and overview: Structural control and tuned mass dampers. In: Optimization of Tuned Mass Dampers, pp. 1–13. Springer (2022)
10. Boggian, F., Tardo, C., Aloisio, A., Marino, E.M., Tomasi, R.: Experimental cyclic response of a novel friction connection for seismic retrofitting of RC buildings with CLT panels. J. Struct. Eng. **148**(5), 04022040 (2022)
11. Bruschi, E., Quaglini, V., Calvi, P.M.: A simplified design procedure for seismic upgrade of frame structures equipped with hysteretic dampers. Eng. Struct. **251**, 113504 (2022)
12. Bucher, C.: Probability-based optimal design of friction-based seismic isolation devices. Struct. Safety **31**(6), 500–507 (2009)
13. Chopra, A.: Dynamics of structures 4th edition (2019)
14. Coburn, A., Spence, R.: Earthquake Protection. Wiley (2003)
15. Cucuzza, R., Costi, C., Rosso, M.M., Domaneschi, M., Marano, G., Masera, D.: Optimal strengthening by steel truss arches in prestressed girder bridges. In: Proceedings of the Institution of Civil Engineers-Bridge Engineering, pp. 1–51 (2022). https://doi.org/10.1680/jbren.21.00056
16. Cucuzza, R., Rosso, M.M., Aloisio, A., Melchiorre, J., Giudice, M.L., Marano, G.C.: Size and shape optimization of a guyed mast structure under wind, ice and seismic loading. Appl. Sci. **12**(10), 4875 (2022)

17. Cucuzza, R., Rosso, M.M., Marano, G.C.: Optimal preliminary design of variable section beams criterion. SN Applied Sciences **3**(8), 1–12 (2021). https://doi.org/10.1007/s42452-021-04702-5
18. De Domenico, D., Qiao, H., Wang, Q., Zhu, Z., Marano, G.: Optimal design and seismic performance of multi-tuned mass damper inerter (MTMDI) applied to adjacent high-rise buildings. Struct. Des. Tall Spec. Build. **29**(14), e1781 (2020). https://doi.org/10.1002/tal.1781
19. De Domenico, D., Qiao, H., Wang, Q., Zhu, Z., Marano, G.: Optimal design and seismic performance of multi-tuned mass damper inerter (MTMDI) applied to adjacent high-rise buildings. Struct. Des. Tall Spec Build. **29**(14), e1781 (2020)
20. De Tommasi, D., Marano, G., Puglisi, G., Trentadue, F.: Morphological optimization of tensegrity-type metamaterials. Compos. Part B: Eng. **115**, 182–187 (2017). https://doi.org/10.1016/j.compositesb.2016.10.017
21. Di Trapani, F., Malavisi, M., Marano, G.C., Sberna, A.P., Greco, R.: Optimal seismic retrofitting of reinforced concrete buildings by steel-jacketing using a genetic algorithm-based framework. Eng. Struct. **219**, 110864 (2020)
22. Di Trapani, F., Sberna, A.P., Marano, G.C.: A genetic algorithm-based framework for seismic retrofitting cost and expected annual loss optimization of non-conforming reinforced concrete frame structures. Comput. Struct. **271**, 106855 (2022)
23. Di Trapani, F., Tomaselli, G., Sberna, A.P., Rosso, M.M., Marano, G.C., Cavaleri, L., Bertagnoli, G.: Dynamic response of infilled frames subject to accidental column losses. In: Pellegrino, C., Faleschini, F., Zanini, M.A., Matos, J.C., Casas, J.R., Strauss, A. (eds.) Proceedings of the 1st Conference of the European Association on Quality Control of Bridges and Structures. pp. 1100–1107. Springer International Publishing, Cham (2022)
24. Dorigo, M.: Optimization, Learning and Natural Algorithms. Ph. D. Thesis, Politecnico di Milano (1992)
25. Dorigo, M., Gambardella, L.M.: Ant colony system: a cooperative learning approach to the traveling salesman problem. IEEE Trans. Evolutionary Comput. **1**(1), 53–66 (1997)
26. Fiore, A., Marano, G.C., Greco, R., Mastromarino, E.: Structural optimization of hollow-section steel trusses by differential evolution algorithm. Int. J. Steel Struct. **16**(2), 411–423 (2016). https://doi.org/10.1007/s13296-016-6013-1
27. Fiore, A., Trentadue, F., Greco, R., Marano, G.C., De Marco, G., Sardone, L., Lagaros, N.D.: Optimización de volumen para arcos empotrados. Hormigón y Acero **71**(292), 71–76 (2020)
28. Greco, R., Marano, G.: Identification of parameters of Maxwell and Kelvin-Voigt generalized models for fluid viscous dampers. JVC/J. Vibr. Cont. **21**(2), 260–274 (2015). https://doi.org/10.1177/1077546313487937
29. Greco, R., Marano, G.C.: Multi-objective optimization of a dissipative connection for seismic protection of wall-frame structures. Soil Dynamics Earthquake Eng. **87**, 151–163 (2016)
30. Greco, R., Marano, G.C., Fiore, A.: Performance-cost optimization of tuned mass damper under low-moderate seismic actions. Struct. Des. Tall Spec. Build. **25**(18), 1103–1122 (2016)
31. Housner, G., Bergman, L.A., Caughey, T.K., Chassiakos, A.G., Claus, R.O., Masri, S.F., Skelton, R.E., Soong, T., Spencer, B., Yao, J.T.: Structural control: past, present, and future. J. Eng. Mech. **123**(9), 897–971 (1997)

32. Iemura, H., Taghikhany, T., Jain, S.K.: Optimum design of resilient sliding isolation system for seismic protection of equipments. Bullet. Earthq. Eng. **5**(1), 85–103 (2007)
33. Karaboga, D., Basturk, B.: A powerful and efficient algorithm for numerical function optimization: artificial bee colony (ABC) algorithm. J. Glob. Optimiz. **39**(3), 459–471 (2007)
34. Kaveh, A., Talatahari, S.: A novel heuristic optimization method: charged system search. Acta Mech. **213**(3), 267–289 (2010)
35. Kennedy, J., Eberhart, R.: Particle swarm optimization. 4, 1942–1948 (1995). www.scopus.com/inward/record.uri?eid=2-s2.0-0029535737&partnerID=40&md5=e6bf04ae50f3268ae545d88ed91d1fc5
36. Kirkpatrick, S., Gelatt, C.D., Jr., Vecchi, M.P.: Optimization by simulated annealing. Science **220**(4598), 671–680 (1983)
37. Lagaros, N.D., Papadrakakis, M., Kokossalakis, G.: Structural optimization using evolutionary algorithms. Comput. Struct. **80**(7), 571–589 (2002). https://doi.org/10.1016/S0045-7949(02)00027-5
38. Lagaros, N.D., Plevris, V., Kallioras, N.A.: The mosaic of metaheuristic algorithms in structural optimization. Arch. Comput. Methods Eng. 1–36 (2022)
39. Li, Y., Conte, J.P.: Probabilistic performance-based optimum design of seismic isolation for a California high-speed rail prototype bridge. Earthq. Eng. Struct. Dynam. **47**(2), 497–514 (2018)
40. Manuello, A., Masera, D., Carpinteri, A.: Ae damage assessment in the bell tower of the turin cathedral. In: Key Engineering Materials, vol. 817, pp. 579–585. Trans Tech Publ (2019)
41. Marano, G., Trentadue, F., Greco, R.: Stochastic optimum design criterion of added viscous dampers for buildings seismic protection. Struct. Eng. Mech. **25**(1), 21–37 (2007). https://doi.org/10.12989/sem.2007.25.1.021
42. Marano, G.C., Trentadue, F., Petrone, F.: Optimal arch shape solution under static vertical loads. Acta Mechanica **225**(3), 679–686 (2013). https://doi.org/10.1007/s00707-013-0985-0
43. Marano, G.C., Greco, R., Quaranta, G., Fiore, A., Avakian, J., Cascella, D.: Parametric identification of nonlinear devices for seismic protection using soft computing techniques. In: Advanced Materials Research, vol. 639, pp. 118–129. Trans Tech Publ (2013)
44. Marano, G.C., Greco, R., Trentadue, F., Chiaia, B.: Constrained reliability-based optimization of linear tuned mass dampers for seismic control. Int. J. Solids Struct. **44**(22–23), 7370–7388 (2007)
45. Marano, G.C., Trentadue, F., Greco, R.: Stochastic optimum design criterion for linear damper devices for seismic protection of buildings. Struct. Multi. Optim. **33**(6), 441–455 (2007)
46. Marasco, G., Rosso, M.M., Aiello, S., Aloisio, A., Cirrincione, G., Chiaia, B., Marano, G.C.: Ground penetrating radar fourier pre-processing for deep learning tunnel defects' automated classification. In: International Conference on Engineering Applications of Neural Networks, pp. 165–176. Springer (2022)
47. Martí, R., Pardalos, P.M., Resende, M.G.C. (eds.): Handbook of Heuristics. Springer, Cham (2018). https://doi.org/10.1007/978-3-319-07124-4
48. Melchiorre, J., Manuello, A., Marmo, F., Adriaenssens, S., Marano, G.: Differential formulation and numerical solution for elastic arches with variable curvature and tapered cross-sections. Eur. J. Mech.-A/Solids **97**, 104757 (2023)

49. Melchiorre, J., Bertetto, A.M., Marano, G.C.: Application of a machine learning algorithm for the structural optimization of circular arches with different cross-sections. J. Appl. Math. Phys. **9**(5), 1159–1170 (2021)
50. Melchiorre, J., Manuello, A., Sardone, L., Marano, G.C.: Damaging configurations in arch structures with variable curvature and tapered cross-section. WCCM-APCOM **2022**, 900 (2022)
51. Niccolini, G., Carpinteri, A., Lacidogna, G., Manuello, A.: Acoustic emission monitoring of the Syracuse Athena temple: scale invariance in the timing of ruptures. Phys. Rev. Lett. **106**(10), 108503 (2011)
52. Niccolini, G., Manuello, A., Marchis, E., Carpinteri, A.: Signal frequency distribution and natural-time analyses from acoustic emission monitoring of an arched structure in the castle of racconigi. Nat. Hazards Earth Syst. Sci. **17**(7), 1025–1032 (2017)
53. Pelikan, M., Hauschild, M.W., Lobo, F.G.: Estimation of distribution algorithms. In: Springer Handbook of Computational Intelligence, pp. 899–928. Springer (2015)
54. Pelliciari, M., Marano, G., Cuoghi, T., Briseghella, B., Lavorato, D., Tarantino, A.: Parameter identification of degrading and pinched hysteretic systems using a modified Bouc-wen model. Struct. Infrast. Eng. **14**(12), 1573–1585 (2018). https://doi.org/10.1080/15732479.2018.1469652
55. Peng, Y., Ma, Y., Huang, T., De Domenico, D.: Reliability-based design optimization of adaptive sliding base isolation system for improving seismic performance of structures. Reliab. Eng. Syst. Safety **205**, 107167 (2021)
56. Quaglini, V., Pettorruso, C., Bruschi, E.: Experimental and numerical assessment of prestressed lead extrusion dampers (2021)
57. Quaglini, V., Pettorruso, C., Bruschi, E.: Design and experimental assessment of a prestressed lead damper with straight shaft for seismic protection of structures. Geosciences **12**(5), 182 (2022)
58. Rizzian, L., Léger, N., Marchi, M.: Multiobjective sizing optimization of seismic-isolated reinforced concrete structures. Proced. Eng. **199**, 372–377 (2017)
59. Rosso, M., Melchiorre, J., Cucuzza, R., Manuello, A., Marano, G.C.: Estimation of distribution algorithm for constrained optimization in structural design. WCCM-APCOM2022
60. Rosso, M.M., Aloisio, A., Cucuzza, R., Marano, G.C., Alaggio, R.: Train-track-bridge interaction analytical model with non-proportional damping: Sensitivity analysis and experimental validation. In: European Workshop on Structural Health Monitoring, pp. 223–232. Springer (2023)
61. Rosso, M.M., Cucuzza, R., Di Trapani, F., Marano, G.C.: Nonpenalty machine learning constraint handling using PSO-SVM for structural optimization. Adv. Civ. Eng. 2021 (2021)
62. Rosso, M.M., Asso, R., Aloisio, A., Di Benedetto, M., Cucuzza, R., Greco, R.: Corrosion effects on the capacity and ductility of concrete half-joint bridges. Available at SSRN 4160407
63. Rosso, M.M., Cucuzza, R., Aloisio, A., Marano, G.C.: Enhanced multi-strategy particle swarm optimization for constrained problems with an evolutionary-strategies-based unfeasible local search operator. Appl. Sci. **12**(5) (2022). DOIurl10.3390/app12052285
64. Rutman, Y.L., Ostrovskaya, N.V.: Damping optimization in seismic isolation systems. In: Key Engineering Materials. vol. 828, pp. 129–135. Trans. Tech. Publ. (2020)

65. Sardone, L., Rosso, M.M., Cucuzza, R., Greco, R., Marano, G.C.: Computational design of comparative models and geometrically constrained optimization of a multi domain variable section beam based on timoshenko model. In: 14th International Conference on Evolutionary and Deterministic Methods for Design, Optimization and Control. Institute of Structural Analysis and Antiseismic Research National Technical University of Athens (2021). https://doi.org/10.7712/140121.7961.18535
66. Scodeggio, A., Quaranta, G., Marano, G.C., Monti, G., Fleischman, R.B.: Optimization of force-limiting seismic devices connecting structural subsystems. Comput. Struct. **162**, 16–27 (2016)
67. Shan, J., Shi, Z., Hu, F., Yu, J., Shi, W.: Stochastic optimal design of novel non-linear base isolation system for seismic-excited building structures. Struct. Cont. Health Monit. **25**(7), e2168 (2018)
68. Shinozuka, M., Chaudhuri, S.R., Mishra, S.K.: Shape-memory-alloy supplemented lead rubber bearing (SMA-IRB) for seismic isolation. Probab. Eng. Mech. **41**, 34–45 (2015)
69. Soong, T., Spencer, B.: Active, semi-active and hybrid control of structures. Bullet. New Zealand Soc. Earthq. Eng. **33**(3), 387–402 (2000)
70. Storn, R., Price, K.: Differential evolution-a simple and efficient heuristic for global optimization over continuous spaces. J. Glob. Optim. **11**(4), 341–359 (1997)
71. Trentadue, F., Fiore, A., Greco, R., de Marco, G., Sardone, L., Marano, G., et al.: Volume optimization of end-clamped arches. In: Proceedings of the International Fib Symposium on Conceptual Design of Struct (Madrid: Torroja Institute), pp. 49–56 (2019)
72. Xue, J., Lavorato, D., Bergami, A., Nuti, C., Briseghella, B., Marano, G., Ji, T., Vanzi, I., Tarantino, A., Santini, S.: Severely damaged reinforced concrete circular columns repaired by turned steel rebar and high-performance concrete jacketing with steel or polymer fibers. Appl. Sci. (Switzerland) **8**(9) (2018). https://doi.org/10.3390/app8091671

Seismicity and Seismic Isolation Design

Near-Fault Earthquake Ground Motion and Seismic Isolation Design

Mustafa Erdik[1(✉)], Bahadır Şadan[2], Cüneyt Tüzün[3], Mine B. Demircioglu-Tumsa[4], Ömer Ülker[4], and Ebru Harmandar[5]

[1] Department of Earthquake Engineering, Boğaziçi University, Istanbul, Turkey
erdik@boun.edu.tr
[2] Department of Civil Engineering, MEF University, Istanbul, Turkey
[3] Department of Civil Engineering, Yaşar University, Izmir, Turkey
[4] Turkish Earthquake Foundation, Istanbul, Turkey
[5] Department of Civil Engineering, Sıtkı Kocman University, Muğla, Turkey

Abstract. Seismic isolation is one of the most reliable passive structural control techniques with adequately established standards for the earthquake protection of structures from earthquakes. However, it has been shown that the seismic isolation systems may not function the best for the near-fault ground motions, since in the proximity of a capable fault, the ground motions are significantly affected by the rupture mechanism and may generate high demands on the isolation system and the structure. In fact, several earthquake resistant design codes state that the seismically isolated structures located at near-fault sites should be designed by considering larger seismic demands than the demand on structures at far-field sites. When the fault ruptures in forward direction to the site most of the seismic energy arrives in coherent long-period ground velocity pulses. The ground-motion prediction equations (GMPEs) typically cannot account for such effects with limited distance metrics and lack adequate data at large magnitudes and near distances. For the reliable earthquake design of the isolated structure in near fault conditions that meets the performance objectives, the 3D design basis ground motion(s) need to be appropriately assessed. Measures in the design of the isolation system, such as modifications in the stiffness and damping characteristics, as well as in the limitation of vertical effects are needed. The behavior of the base-isolated buildings under near-fault (NF) ground motions with fling-step and forward-directivity characteristics are investigated with a rational assessment of design-basis near-fault ground motion, are investigated in a parametric format. The parametric study includes several variables, including the structural system flexibility; number of stories; isolation system characteristic (yield) strength, and the isolation periods related to the post-elastic stiffness. Furthermore, the effect of additional damping by viscous dampers were tested for some selected cases. Important findings observed from the parametric performance results and the overall conclusions of the study are provided.

Keywords: Base-isolated buildings · Fling-step · Forward-directivity · Near-fault ground motions · Pulse period

G. P. Cimellaro (Ed.): WCSI 2022, LNCE 309, pp. 117–152, 2023.
https://doi.org/10.1007/978-3-031-21187-4_9

1 Introduction

It is essential to maintain the functionality of critical structures to form resilient societies after moderate-to-severe ground shaking. Seismic isolation of structures is a mature technique for the design of new and retrofit of existing structures. The earthquake-resistant design of seismically isolated structures can entail sophisticated analysis for the accurate estimation of seismic demands in near-fault regions, and the most accurate estimate of seismic demands of isolated structures can be obtained through nonlinear response history analysis.

Near-fault ground motions have distinguishing characteristics that are related to the faulting mechanism, location of the hypocenter, and direction of rupture propagation relative to the location of the site.

This study investigates the influence of near-fault ground motions on seismically isolated structures by considering their well-known characteristics and provides state-of-the-practice design methodologies for such cases. The near-fault ground motion characterization is reviewed for the proper seismic isolation applications for building-type structures. A set prototype building is exposed to multi-component earthquake ground motions with pulse content in near-fault regions to address problems associated with close proximity to the active faults. Isolation units with a large displacement capacity and sufficient restoring force capacity need to be elaborately selected to accommodate the large displacement demands of these structures.

Earthquakes can be described as a shear dislocation phenomenon that starts to radiate from the focus of the source to the site. Damaging effects of the earthquake emerge when the frequency content of the seismic input motion coincides or is in close range with the fundamental frequency of vibration. Seismic isolation application conceptually aims to decouple structures from the horizontal components of earthquake ground motion by shifting the fundamental period of vibration beyond the damaging range of frequencies. Period shift and the consequential accommodation of displacement demands are generally provided by the installation of elastomeric or curved surface slider bearings to structures. In addition, seismic isolators are used to reduce the acceleration and/or drift response of buildings in order to improve their post-earthquake functionality. Effects of a seismic isolation system on damping and the fundamental period of vibration are shown in Fig. 1.

Prediction of displacement demands in seismically isolated structures is crucial since it has been identified as an important performance indicator. Displacement spectra constitute the main component of displacement-based design (DBD) procedures as per the recommendations developed by Priestly et al. [58]. In some codes, the establishment of displacement spectra is based on the assumption of steady-state behavior. Some standards (e.g., CEN, [20]) provide the characteristic shape of the displacement spectrum on this basis. Apart from the code-based displacement spectra, ground motion prediction equations can be utilized for the computation of displacement spectra. Among others, Bommer and Elnashai [11], Faccioli et al. [31], and Cauzzi et al. [19] have developed empirical relationships for the prediction of spectral displacement amplitudes based on available strong motion data that also contain near-fault data.

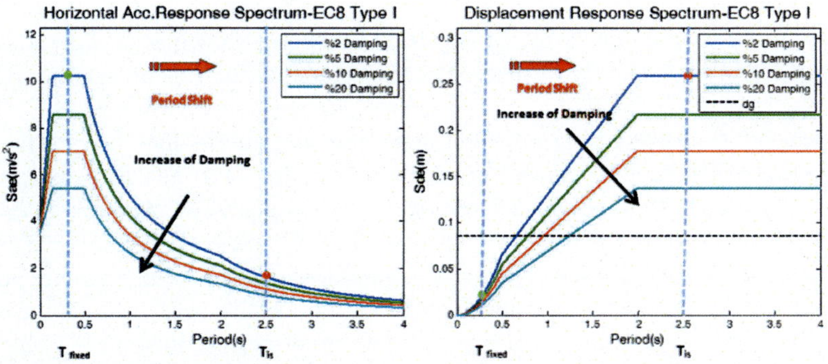

Fig. 1. Effects of seismic isolation on acceleration and displacement spectra.

2 Near-Fault Ground Motions

2.1 Background

Distinctive characteristics of near-fault ground motions have been realized through the dense ground motion arrays deployed in the vicinity of active faults. Near-fault seismic ground motions are frequently characterized by intense velocity and displacement pulses of relatively long duration, which clearly distinguishes them from typical far-field ground motion records.

Near-fault ground motions contain coherent and orientation-dependent effects at intermediate and long periods, such as rupture directivity pulses and fling steps. Extensive damage to some structures in near-fault regions is attributed to these particular features. Earthquakes in close proximity to faults, such as the 1992 Erzincan-Turkey, 1992 Landers-California, 1994 Northridge-California, 1996 Kobe-Japan, 1999 Kocaeli-Turkey, 1999 Chi Chi-Taiwan, and 2011 Christchurch-New Zealand earthquakes, have provided valuable information about the characteristics of this type of ground motions.

2.2 Characteristics of Near-Fault Ground Motions

The large amount of data gathered from site surveys and strong ground motion recordings allowed the earthquake engineering community to understand the near-field effects. It has been recognized that near-source ground motions are dominated by source characteristics and may vary considerably depending on the characteristics of rupture propagation. Post-earthquake observations from major earthquakes indicate that extensive damage to structures is attributed to particular characteristics of near-source records within 0–15 km distance to fault rupture. The main distinctive characteristics of near-source earthquakes can be listed as directivity and fling effects which are associated with long period pulse content and large vertical ground motions.

2.2.1 Directivity Effects

Directivity effects have been studied by numerous researchers as a cause of velocity pulses (e.g., Hall et al. [40]; Somerville and Graves [68]). Rate and duration are the two important factors that influence the large ground displacement demands. The duration of these displacements is closely related to the characteristic slip time on the fault (Heaton [41]).

In near-fault conditions, two distinct displacement patterns may be observed as a consequence of the rupture process (Sommerville et al. [71]). The first displacement pattern associated with double-sided velocity pulses is generally observed in the fault normal component of strike-slip motions. The so-called Fault-Normal Forward Directivity Effect occurs when the fault rupture propagates toward a site with a rupture velocity that is approximately equal to the shear wave velocity. In this case, most of the energy arrives coherently in a single, intense, relatively long-period pulse at the beginning of the record. Near-field locations with dip-slip faulting mechanisms can additionally experience hanging wall effects as well.

The second displacement pattern is linked with the so-called Fault-Parallel Fling Step action that accounts for permanent displacement offsets. Fling Effect (permanent translation) appears in the form of step displacement and one-sided velocity pulse in the strike-parallel direction for strike-slip faults or in the strike-normal direction for dip-slip faults.

Figure 2 (after Mavroedis and Papageorgiu [48]) illustrates the fault normal (forward directivity, ARC Station record) and fault parallel (fling step, SKR Station record) ground motion associated with the 1999 Izmit (Kocaeli) earthquake (Mw7.4) in Turkey. A simple pulse-type ground motion model (black trace) was fitted to the recorded ground motions (gray trace). Time histories of displacement, velocity, and acceleration are illustrated.

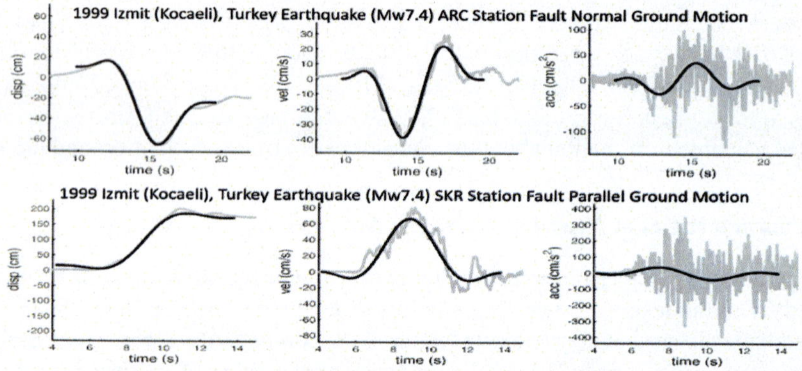

Fig. 2. 1999 Izmit (Mw7.4) earthquake, fault normal and fault parallel ground motion associated with the in Turkey (after Mavroedis and Papageorgiu [48]).

2.2.2 Near-Fault Vertical Ground Motion

The peak value of the vertical component of earthquakes may exceed its horizontal counterparts in the vicinity of the active faults.

Xinle et al. [79] have investigated the characteristics of near-fault vertical ground motions. The vertical to horizontal spectral ratios obtained from 130 sets of selected free-field near-fault ground motion indicates that for a spectral period of less than 0.1 s and greater than about 2 s, the spectral ratios for different pulse period ranges are close to 1. The largest spectral ratio of about 1.3 occurs for pulse periods less than 3 s.

Studies of the vertical/horizontal (V/H) spectral amplitude ratios of ground motion (e.g., Bozorgnia and Campbell [12]; Gülerce et al. [38]) have shown the following important characteristics:

(1) The V/H spectral ratio is sensitive to spectral period, fault distance, local site conditions, and earthquake magnitude and has a distinct peak at short spectral periods (0.05–010 s) that increases with decreasing soil stiffness;
(2) At short spectral periods in near-fault conditions, the (V/H) spectral ratio exceeds unity for soft soil sites since the nonlinearity of site response is stronger in the horizontal component than in the vertical component.
(3) The V/H spectral ratio is generally less than 2/3 at mid-to-long periods since the vertical ground motion attenuates at a higher rate (which increases with decreasing distance from the earthquake source) than horizontal ground motion.

Figure 3 (after Gülerce et al. [38]) illustrates the median V/H ratio for a vertical strike-slip earthquake at a 5 km distance from the fault for (a) rock sites (VS30 = 760 m/s) and (b) soil sites (VS30 = 270 m/s).

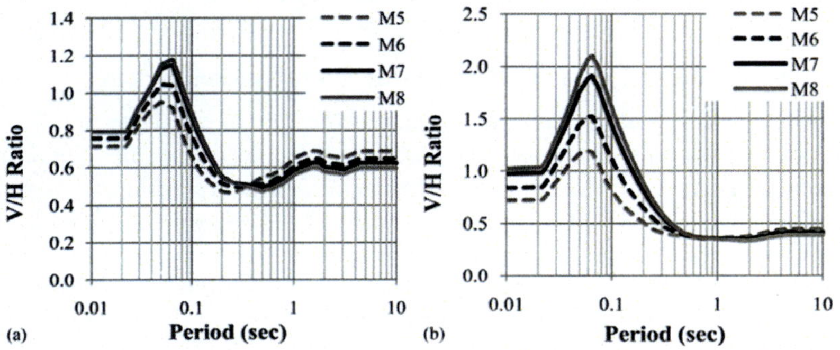

Fig. 3. Median V/H ratio for a vertical strike-slip earthquake at a 5 km distance from the fault for (a) rock sites (NEHRP B/C, $V_{S30} = 760$ m/s) and (b) soil sites (NEHRP D, $V_{S30} = 270$ m/s) (after Gülerce et al. [38]).

The vertical component of the ground motion is described in EN1998-1 (EC8) by an acceleration response spectrum, Sve, denoted as the "vertical elastic response spectrum." The spectrum is anchored to the value of the peak vertical acceleration a_{vg}. For Type-1 earthquakes, the maximum 5% damped spectrum amplitude of the vertical elastic response spectrum is 1.08 times the corresponding horizontal spectral amplitude regardless of earthquake magnitude, distance, and site class.

In ASCE 7-22 [7], Chapter 11.9.2 provides the definition of the MCE_R Vertical Response Spectrum. The vertical spectral amplitude changes as a function of the C_V (Vertical Coefficient) which is tabulated (Table 11.9-1) as a function of the site class and short period horizontal spectral amplitude (S_{MS}). The maximum 5% damped vertical spectral amplitude is 1.58 times the corresponding horizontal spectral amplitude (RotD50), the period ranges of 0.05–01 s for SMS >= 2 and for NEHRP Site Class D.

From Fig. 3, it can be seen that for near-fault conditions, large magnitudes, and soft soil sites, the maximum vertical spectral amplitude can exceed twice the corresponding horizontal spectral amplitude (GeoMean, RotD50) in short periods (0.05–0.1 s). As such, isolated structures in near-fault conditions (especially on the hanging wall side of reverse and oblique faults) are highly vulnerable and require careful consideration of the vertical ground motion excitation for design. Especially, isolation units with relatively low tributary gravity load, and isolation units located below columns that form part of the lateral force-resisting system, can have net uplift or tensile displacements caused by combined large vertical ground motion accelerations and global overturning, thereby inducing high impact forces on the structure and jeopardizing the stability of the isolation units.

2.3 Analysis of the Near-Fault Ground Motion

The current state of knowledge for the quantification of forward directivity effects associated with near-fault ground motion can be incorporated into the analysis and design of structures by utilizing different approaches that involve the frequency domain (response spectra) and time domain (ground motion) modifications.

2.3.1 Response Spectra Associated with Near-Fault Ground Motion

This response spectrum method is used for the specification of response spectra and the selection of time-histories that are compatible with it. This approach relies on the amplification of the spectral ordinates of the conventional design basis spectrum to incorporate the near-fault effects. In this connection, two different spectra may need to be constructed to account for the fault-normal and fault-parallel ground motion components. Empirical spectral amplification factors are determined on the basis of source and distance relationships of the ruptured fault. They are generally independent of site conditions and ground motion intensity.

In the response spectrum method, generally, an empirical model of rupture directivity effects that utilizes spectral amplification factors, such as those originally proposed by Somerville et al. [71], is considered. Somerville's model uses the length ratio ($X = s/L$) and width ratio ($Y = d/W$), respectively, for strike-slip and dip-slip faulting mechanisms. The azimuth and zenith angles are denoted respectively by θ and φ, as illustrated in Fig. 4. In this model, the response spectral ordinates are modified using functions conditional on the directivity of the parameter ($X \cos \theta$ and $Y \cos \phi$) that represent the degree to which the component of fault rupture aligned with the slip direction is towards a site of interest.

The period dependence of the amplitude variations on the average horizontal spectrum, caused by rupture directivity effects shown, are shown in Fig. 5. This figure indicates a transition from coherent to incoherent source radiation and wave propagation conditions at a period of about 0.6 s.

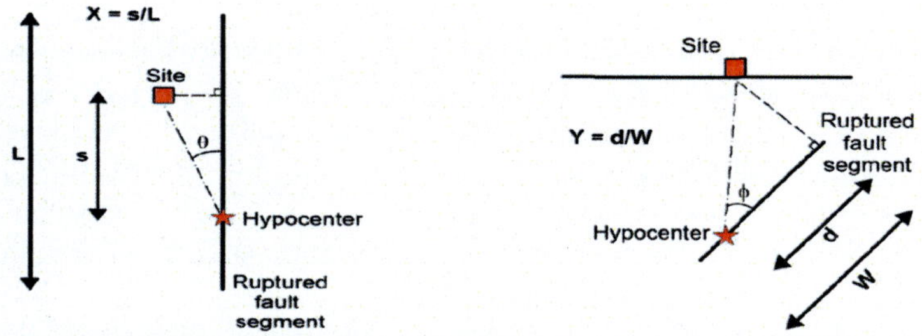

Fig. 4. Directivity parameters for strike-slip (plan view) and dip-slip (vertical section) faults defined by Sommerville et al. [71].

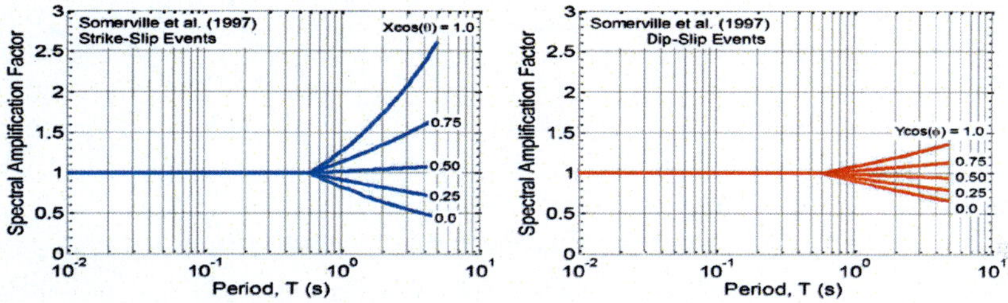

Fig. 5. Empirical model of the response spectral amplitude ratio, showing its dependence on period and on the directivity function (X cos θ for strike-slip; Y cos φ for dip-slip faulting) for strike-slip and dip-slip events (Somerville et al. [71]).

Abrahamson [2] has modified the model to saturate the directivity effect for X cos θ > 0.4 and to reduce directivity effects through the use of tapers for distances > 30 km and M < 6.5. Despite the presence of more recent models developed with far larger NGA-West 1 and NGA-West 2 datasets, the experience has been that in practice, the Somerville et al. [71] model in conjunction with the Abrahamson [2] modification continues to be applied to analyze directivity effects on response spectral ordinates (PEER, [57]).

The Final Report of the NGA-West2 Directivity Working Group (PEER, [56]) and the (PEER, [57]) report entitled "Ground-Motion Directivity Modelling for Seismic Hazard Applications" provide a thorough review and applications of the four directivity models developed on the basis of the NGAWest 2 database and on the numerical simulations of large strike-slip and reverse-slip earthquakes. Models by Shahi and Baker, Spudich and Chiou, Rowshandel, Bayless, and Somerville are covered and elaborated in this report. The first three models are explicitly "narrow-band" such that the spectral amplitude change is concentrated around a pulse period, although the Bayless and Somerville model is broadband for a spectral period T > 0.5 s.

The Shahi and Baker [66] Model is specifically aimed at predicting the characteristics of impulsive ground motions often found at short distances from fault ruptures. Since the directivity pulse amplifies spectral acceleration in a narrow band of periods close to the pulse period, their model consists of a wide-band spectral shape plus a superposed narrow-band spectral shape.

The Spudich and Chiou Model uses of Isochrone Directivity Parameter as the predictor. The Chiou and Spudich model introduce the Direct Point Parameter, based on the Isochrone Directivity Parameter, as the directivity predictor, without presenting empirically derived coefficients.

The Rowshandel Model presented in this report is a major modification of Rowshandel [63]. Specifically, rupture length de-normalization is used, and the direction of rupture and slip both contribute to directivity.

Bayless and Somerville's model is an improved version of the classic Somerville et al. [71] model, updated with new data and a better functional form. Major changes include rupture-length denormalization, a modified dependence on site azimuth, use of azimuth tapers to obviate the need for an excluded zone, and extension of the algorithm to allow directivity calculations for complicated, non-contiguous rupture zones. A set of coefficients is presented that is adequate for simulating directivity for several different GMPEs.

For near-fault ground motions, different target spectra should be used for the fault-normal and fault-parallel components to preserve the near-fault features in the process of spectral matching or scaling of ground motions. The Bayless and Somerville model is the only model in this report that predicts directivity for fault-normal and fault parallel motions as well as azimuthally averaged motion. As stated in PEER [57], this model has been implemented in prior work and has a history of project-based peer review. Since this model has a much straightforward application process for professional applications compared to other models, it will be briefly covered.

Bayless and Somerville Directivity Model

This model is parameterized similarly to the Somerville et al. [71] model (Figs. 4 and 5), except that the dimension of the fault rupturing towards the site is not normalized. The median spectral acceleration from an empirical attenuation relation without directivity effect (Sa) can be modified to obtain the spectral acceleration with directivity effects (Sa_{dir}) by the following Eq. (1).

$$\ln(Sa_{\mathrm{dir}}) = \ln(Sa) + f_{\mathrm{D}} \tag{1}$$

where f_{D} is the directivity effect given by Eq. (2).

$$f_{\mathrm{D}} = (C_0 + C_1 * f_{\mathrm{geom}}) * T_{\mathrm{CD}} * T_{\mathrm{Mw}} * T_{\mathrm{Az}} \tag{2}$$

The directivity effect is quantified as the product of the period and fault-type-dependent constant coefficient, the distance, magnitude, and azimuth tapers, and the geometric directivity predictor (f_{geom}), which correlates the directivity effects with the spatial variation of near-fault ground motions. Referring to Fig. 4, f_{geom} is a function of the fraction of the fault rupture surface that lies between the hypocenter and the site (parameter X or Y) multiplied by the length or width of the fault (parameter L or W), and the angle between the direction of rupture propagation and the direction of waves traveling from the fault to the site (parameter θ or R_{x}/W).

For strike-slip faults:

Geometric Directivity Predictor:

$$f_{\text{geom}}(s, \theta) = \log_e(s) * (0.5 \cos(2\theta) + 0.5) \quad (3)$$

Distance Taper:

$$\begin{aligned} T_{\text{CD}}(R_{\text{rup}}, L) &= 1 && \textit{for} \quad R_{\text{rup}}/L < 0.5 \\ T_{\text{CD}}(R_{\text{rup}}, L) &= 1 - (R_{\text{rup}}/L - 0.5)/0.5 && \textit{for} \quad 0.5 < R_{\text{rup}}/L < 1 \\ T_{\text{CD}}(R_{\text{rup}}, L) &= 0 && \textit{for} \quad R_{\text{rup}}/L > 1.0 \end{aligned} \quad (4)$$

Magnitude Taper:

$$\begin{aligned} T_{\text{Mw}}(M_{\text{W}}) &= 1 && \textit{for} \quad M_{\text{W}} > 6.5 \\ T_{\text{Mw}}(M_{\text{W}}) &= 0 && \textit{for} \quad M_{\text{W}} < 5.0 \\ T_{\text{Mw}}(M_{\text{W}}) &= 1 - (6.5 - M!)/1.5 && \textit{for} \quad 5.0 < M_{\text{W}} < 6.5 \end{aligned} \quad (5)$$

Azimuth Taper:

$$T_{\text{Az}}(Az) = 1 \quad (6)$$

The coefficients C0 and C1 of Eq. (2) for different spectral periods are provided in Table 1 for the average horizontal GeoMean (RotD50), fault-normal (FN), and fault-parallel (FP) acceleration spectra. These coefficients represent the smoothed average coefficients derived from fitting residuals from the four 2008 NGA GMPEs. The equations for the predictor and tapers and the coefficients for the dip-slip fault ruptures are provided in PEER [56], similar to those for the strike-slip fault rupture, and will not be included in this paper.

2.3.2 Time Domain Representation of Near-Fault Ground Motion

There is growing thinking that the response spectrum method is not fully capable of representing the effects of near-fault pulses. The time domain approach is based on the narrow band representation of the forward directivity and fling-step ground motion through its velocity time history. Strong motion recordings can be assumed to justify that a near-fault pulse is a narrow band pulse, and the near-fault ground motions containing forward rupture directivity and fling step effects may be simple enough to be represented by time domain narrow band pulses through velocity time series with its peak ground velocity, predominant pulse period and expected number of cycles. Needless to say, the expected number of significant pulses depends on the physics of the fault rupture (number of asperities) and cannot be estimated a priori via empirical relationships.

The first work on the pulse-like ground motion was employed by Veletsos and Newmark [77] in connection with the analysis of the response of elastoplastic single-degree-of freedom systems subjected to nuclear explosion generated excitation. The normalized response spectra (Pseudo-Velocity/Maximum Ground Velocity) for different damping ratios that would result from a half–cycle displacement pulse (against $2t_1f$, where t_1 is the ¼ of the displacement pulse duration) and a half–cycle velocity pulse (against t_df, where t_d is the velocity pulse duration) are shown in Fig. 6. Regarding the maximum isolation

Table 1. Coefficients of the Bayless and Somerville model for strike-slip ruptures.

Period (s)	Strike-Slip					
	RotD50		FN		FP	
	C_0	C_1	C_0	C_1	C_0	C_1
0.5	0.000	0.000	0.000	0.000	0.000	0.000
0.75	0.000	0.000	−0.080	0.055	0.000	0.000
1	−0.120	0.075	−0.225	0.110	0.015	0.000
1.5	−0.175	0.090	−0.300	0.135	0.030	−0.025
2	−0.210	0.095	−0.325	0.160	0.050	−0.040
3	−0.235	0.099	−0.365	0.185	0.070	−0.045
4	−0.255	0.103	−0.390	0.205	0.080	−0.050
5	−0.275	0.108	−0.410	0.215	0.090	−0.060
7.5	−0.290	0.112	−0.420	0.220	0.100	−0.070
10	−0.300	0.115	−0.425	0.225	0.108	−0.071

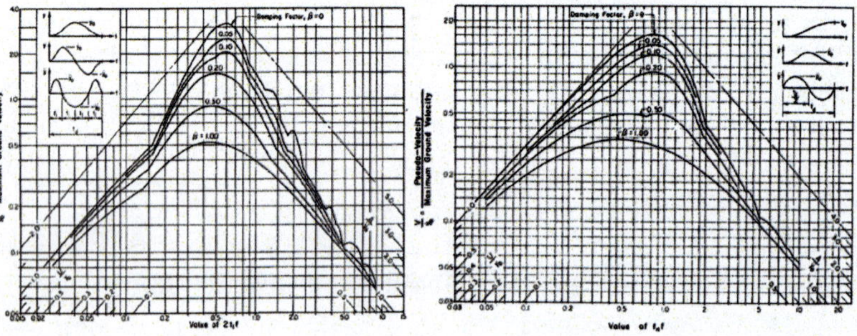

Fig. 6. Normalized response spectra for different damping factors that would result from a half–cycle displacement and half–cycle velocity pulses (Veletsos and Newmark [77]).

system displacement, it can be assessed that, for both cases, the spectral displacements vary between 50 and 80% of the peak ground displacement for damping ratios of about 30% for spectral periods between 2 and 5 s.

Assembly of pulse-like ground motions associated with near-fault ground motion was gathered by many researchers to develop quantitative time-domain identification methods for ground motions containing strong velocity pulses.

Sommerville [70, 71] has obtained the magnitude scaling relationships for the fault-normal forward directivity pulse period (T, Eq. (7)) and the peak velocity (PGV, Eq. (8)) in terms of moment magnitude Mw and closest fault distance (R).

$$logT = -3.1 + 0.5Mw \quad (7)$$

$$logPGV = -1.0 + 0.5Mw - 0.5logR \quad (8)$$

Figure 7 provides a simple representation of a simple full cycle forward directivity velocity pulse together with the associated 5% damped response spectra for different pulse periods and velocity pulse amplitudes. For these plots, different combinations of magnitude (Mw) and fault distance (R) combinations as per Eqs. (7) and (8) are considered.

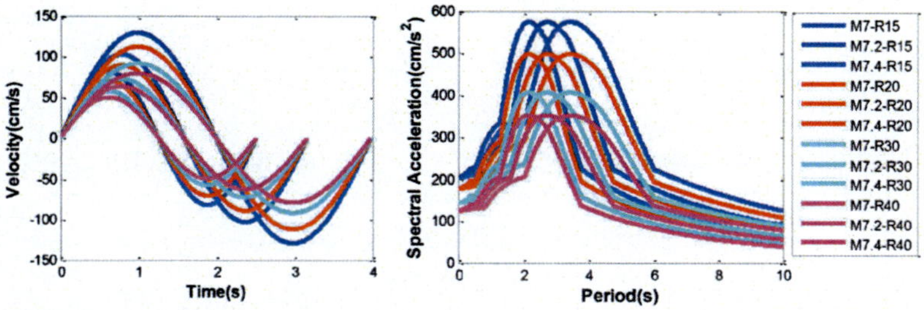

Fig. 7. Full-cycle velocity pulse and the corresponding acceleration spectra for forward rupture directivity (near-fault normal ground motion).

In this connection, Baker [8], Tothong et al. [75], and Shahi and Baker [66] have employed a wavelet transform procedure to identify pulse-type signals in a ground motion time history. Bray and Rodriguez-Marek [15], Rodriguez-Marek and Bray [62], and Bray et al. [14] simplify the characterization of near-fault forward directivity motions by using only the peak ground velocity approximate period of the dominant pulse and the number of cycles of pulse motions of the fault-normal component of forward directivity motions. Mavroeidis and Papageorgiou [48] have proposed an analytical model for the representation of near-field strong ground motions, using the Gabor wavelet defined by the four key parameters of the near-source velocity pulses; namely, the amplitude, prevailing frequency, phase, and the oscillatory character of the signal.

For the representation of near-fault records, one can utilize more complicated pulse models such as those derived based on wavelet functions (e.g., Mavroeidis and Papageorgiou [48]; Mavroeidis et al. [49]; Baker [8]); however, for practical design considerations, their calibration is rather difficult due to the presence of supplementary parameters in the definition of wavelet functions. The normalized displacement spectra of fling-step pulse actions are investigated by Faccioli et al. [31] in connection with the study on the shape of the displacement spectrum. It was shown that the basic features of the displacement spectrum at long periods depend on the shape of the displacement pulse rather than on its derivatives, particularly the acceleration trace.

As such, the use of idealized simple pulses (Fig. 8) offers an advantage in modeling near-fault ground motion for design purposes (Sasani and Bertero [64]; Alavi and Krawinkler [4]; Kalkan and Kunnath [44]; Bhagat et al. [10]). The normalized acceleration, velocity, and displacement of the idealized simple pulse shape and the associated 5 and 30% damped response spectra are provided, respectively, in Figs. 8 and 9.

As it can be assessed, the key parameters involved in the definition of the pulse shapes shown in Fig. 8 are the amplitudes (PGA (A1), PGV, and PGD, which are all related) and the pulse period (Tp). Several researchers (e.g., Alavi and Krawinkler [3]; Mavroeidis

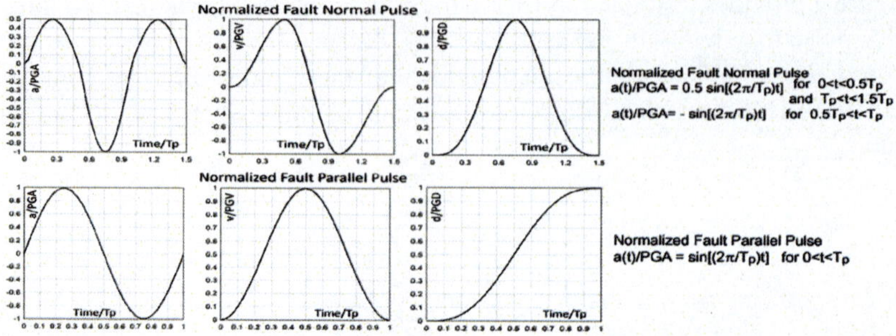

Fig. 8. Idealized sinusoidal fault normal (forward-directivity) and fault parallel (fling-step) pulses, normalized by maximum acceleration, velocity, and displacement.

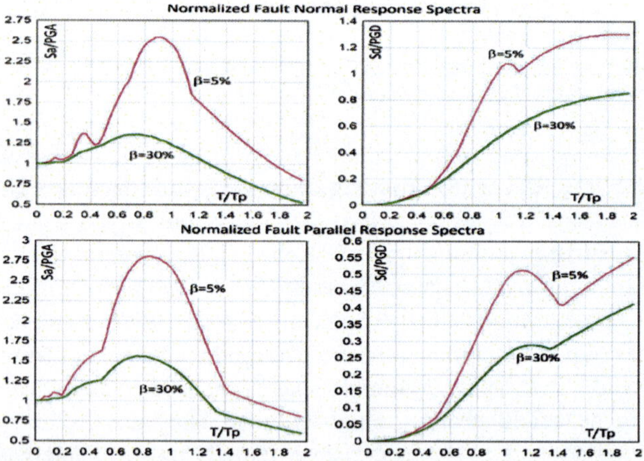

Fig. 9. The normalized 5 and 30% damped acceleration and displacement spectra are associated with the fault normal and fault parallel pulse shapes in Fig. 8.

and Papageorgiou [48]; Bray and Rodriguez-Marek [15]; Somerville [70]; Tang and Zhang [81]; Chen et al. [24]) have studied the relationship between pulse periods and magnitudes, and proposed some empirical models, which are plotted in Fig. 10. It can be assessed that in general, for all site conditions the pulse period has a range between 2 and 5 s for Mw between 7.0 and 7.5, where, the variation of Tp between site conditions is only about 20%. This Tp range almost totally covers the isolated period of vibration range for isolated structures.

Bray and Rodrigues-Marek [15] have thoroughly evaluated the variation of PGV with the earthquake magnitude and site conditions and have shown that the PGV in the near-fault region varies significantly with magnitude and distance, and additionally, the PGV for soil sites is systematically larger than those at rocks sites.

Bray and Rodrigues-Marek [15] compared the proposed relationship for PGV with those developed by Somerville [71] and Alavi and Krawinkler [3]. They have indicated that Somerville [71] and Alavi and Krawinkler [3] propose a much stronger variation of PGV with magnitude compared to Bray and Rodrigues-Marek [15]. This difference

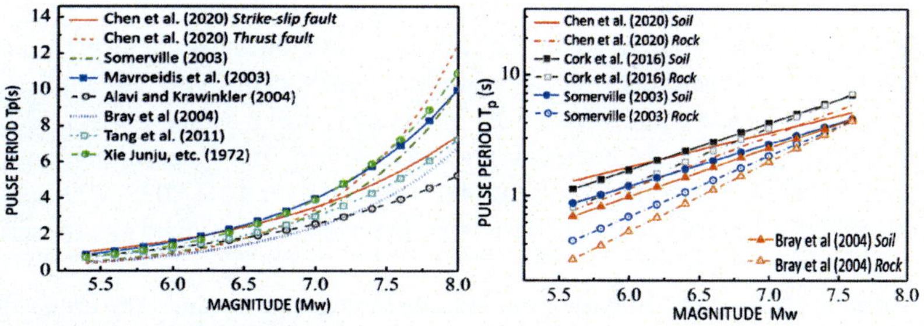

Fig. 10. Relationships between the pulse period (Tp) and the moment magnitude (Mw) (Chen et al. [24]).

essentially stems from the method of magnitude scaling and the amount of data included in the analysis. Chen and Wang [22] have shown that the peak pulse velocity (PGV) is only slightly related to the type of faulting and magnitude, as shown in the left pane of Fig. 11. In the right pane of Fig. 11, the variation of PGV with fault distance for and earthquake with magnitude Mw6 is indicated on the basis of regressions from different authors. It can be seen that Chen and Wang's [22] study yields the lowest values of PGV.

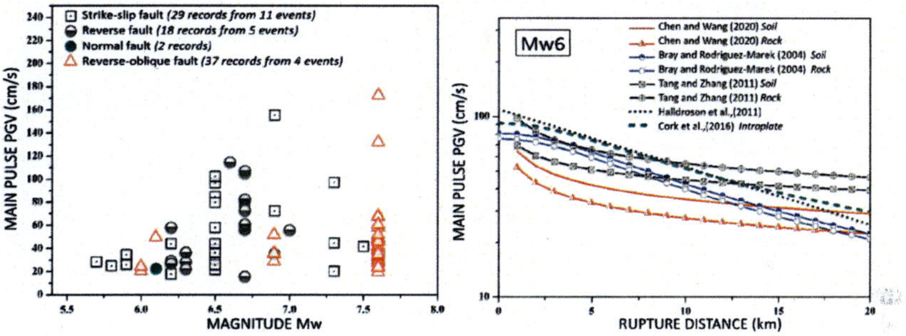

Fig. 11. Variation of PGV with the style of faulting, magnitude, and fault distance.

2.3.3 Ground-Motion Directivity Modelling for Seismic Hazard Applications

Probabilistic (hazard integral and Monte Carlo based) and deterministic procedures can be used for the hazard assessment (Donahue et al. [26]; Gentile and Galasso [34]). Donahue et al. [26], group these procedures under the main titles of A-Posteriori Modification and Incorporation of Directivity into Hazard Integral.

In the A-Posteriori Modification procedure, the results of the classical (directivity-neutral) PSHA are modified to approximately account for the directivity effects. A drawback of this procedure is that the resulting spectrum may not maintain the intrinsic probability level running PSHA with directivity-neutral GMMs and then modify the results in order to approximately account for directivity effects. For this modification, the following alternatives can be considered: (1) Composite Distribution Approach: which involves the deterministic changes in the mean and standard deviation after the hazard deaggragation and the computation of the directivity modified spectral levels associated with each fault; (2) Scenario Event Incorporating Directivity Approach: applies to the conditional mean spectrum (CMS), which is computed for a scenario event identified through disaggregation of the hazard at a specific period, and its modification for the directivity effects. A-Posteriori Modification is adapted directly to UHS, where a single fault dominates the hazard for long periods if the hypocenter is assumed in a conservative manner (Donahue et al. [26]).

In the Incorporation of Directivity into Hazard Integral procedure, the classical hazard integral is extended to incorporate the directivity effects by considering the variability of the location of the hypocenter. This variation is achieved through Modified Moments from Randomized Hypocenter, which modifies the mean and standard deviation of ground motion to approximately account for the effect of variable hypocenter location; Directivity Parameter Randomization, which provides integration over a location-specific distribution of the directivity parameter, indirectly accounting for the variable hypocenter location and; Hypocenter Randomization, which results in integrating directly over alternate hypocenter locations (Donahue et al. [26]).

In addition to several hazard assessment software, the OpenQuake-engine (an open software for seismic hazard and risk assessment) is extended the ordinary hazard integral by adding two additional aleatory variables, hypocenter location and slip angle, to support the near-fault directivity effect (Chen et al. [23]).

3 Performance of Seismically Isolated Buildings Exposed to Near Fault Ground Motion

Extensive studies are conducted for the investigation of the response of conventional (non-isolated) buildings under exposure to near-fault ground motion. As expected, these studies have generally shown that the structural response is maximized in the fundamental period the structure is near the FD pulse period. The ground motions with FD pulses generate high demands that necessitate the dissipation of this input energy with few large displacement excursions, which increases the risk of brittle failure for poorly designed systems.

The presence of FD velocity pulses in the ground motions has been known to impose larger demands on structures than ground motions without velocity pulses (e.g. Almufti et al. [5]; Kitayama and Constantinou [46]; Sasani and Bertero [64]). Using displacement response spectra of idealized and recorded pulse-type ground motions, Sasani and Bertero [64] have shown the efficiency of added damping in controlling the response of multi-degrees of freedom systems acted upon by near-fault ground motion.

The study of Kalkan and Kunnath [45] on the response of steel moment frames under near-fault ground motion excitation demonstrated that the severity of the demands is controlled by the ratio of the pulse to system period and the ground motion records with forward directivity resulted in more instances of higher-mode demand while records with fling-step displacement almost always caused the systems to respond primarily in the fundamental mode. Essentially all of these findings necessitate an explicit consideration of pulse-type ground motions in the design of structures for near-fault conditions. Hall and Ryan [39] investigated the seismic performance of seismically isolated buildings designed per the 1997 Uniform Building Code (UBC). The study indicated that seismically isolated moment-resisting frame buildings under exposure to near-fault ground motions could experience large inter-story drifts that can be controlled with supplemental dampers in the isolation system. Vassiliou et al. [76] and Bao et al. [9] studied the response of isolated structures under pulse-like ground motions and reported the dependence of the structural response on the pulse period. The behavior of base-isolated buildings under near-fault ground motions with fling-step and forward-directivity characteristics is investigated by Bhagat et al. [10] using a 10-story base-isolated building with and without shear walls. Their results indicate lower values of peak inter-story drift ratio for the buildings with shear walls compared to those with moment-resisting frames. The displacement demand at the isolation interface is maximum when the fundamental period of vibration of the building is equal to the pulse period.

Pant and Wijeyewickrema [55] studied the seismic pounding performance of typical four-story base-isolated RC buildings to near-fault ground motions with pulse-like nature. It is found that the peak base displacements under near-fault ground motions are 1.5–1.6 times larger than those under far-fault ground motions., where both ground motion sets are scaled to the same ASCE 7-10 MCER spectrum.

Kitayama and Constantinou [46] have investigated the collapse risk of seismically isolated buildings designed by the procedures of the ASCE/SEI 7-16 standard. The buildings are six-story perimeter frame buildings with special concentrically braced frames located at sites within 5 km of an active fault. The seismic isolation systems are comprised of: (i) triple friction pendulum bearings with high friction coefficients, (ii) triple friction pendulum bearings with the low friction coefficient and viscous dampers, and (iii) triple friction pendulum bearings with low friction coefficients. The paper demonstrated that seismically isolated buildings with either increased isolator displacement capacity or increased superstructure strength compared to the code-based minima could achieve an acceptable collapse risk. Furthermore, the seismic isolation systems that have the best collapse performance for far-field motions are not necessarily the best for near-fault ground motions. The results of Kitayama and Constantinou's [46] study also show that the probabilities of collapse have small differences for the three isolated buildings under exposure to near-field ground motions, while building with the high friction isolation system have the least collapse risk. The behavior of high-damping rubber bearings for base-isolated buildings under earthquake records, under exposure to horizontal impulsive ground motion together with either a pulse-like or a non-pulse-like vertical shaking, were studied by Quaranta et al. [60]. Results show that the maximum displacement of elastomeric bearings is moderately amplified and that significant negative effects on axial capacity, shear strain, and overall stability of elastomeric bearings can occur when the pulse-like vertical excitation is considered in the analysis.

4 Consideration of Near-Fault Ground Motion in Codes

Design codes and their reference standards describe how to design and detail structures in a prescriptive way. Code provisions have historically been developed based on recorded ground motions not sufficiently close to the causative fault. Thus, in general, the current seismic design codes do not explicitly provide an adequate representation of near-fault ground motions. The codes and the earthquake-resistant design practice rely on response spectrum concepts where near-fault characteristics of ground motions are not provided sufficiently (Sommerville [17]; Shahi and Baker [66]; Akkar et al. [82]).

Distinguishing characteristics of near-fault ground motions were not implemented to design codes until the establishment of the 1990 SEAOC Blue Book. Base isolation Subcommittee of the Northern Section of SEAOC has developed N factors, and later, these proposed factors were adopted for the 1997 UBC.

In UBC 1997, the site-source and distance-dependent near-source factors (NA and NV) are introduced on the basis of fault types and fault distances to amplify the elastic design spectrum (i.e., the design base shear). The Na and Nv factors are applied uniformly over respective acceleration and velocity domains. For fault (A-Type) distances less than 2 km, this amplification varies between 50 and 100%. Needless to say, the rationality of constant amplification factors in providing reliable performance levels to near-fault structures are questionable.

For the incorporation of near-fault effects, approaches similar to UBC 1997 are also employed in the Chinese (Chai et al. [21]) and Taiwanese (Li et al. [47]) seismic design codes. The EN Eurocode 8 EN, Part 1 general provisions for near-source effects, does not exist. Part 2 of EC8 ("Bridges") requires, though, the use of site-specific spectra, taking into account near-source effects for bridges located within 10 km of a fault that can produce an earthquake with magnitude => Mw6.5.

New Zealand Standard NZS 1170.5 (2004) uses a near-fault factor (N (T, D)) in the definition of the elastic response spectra to account for near-source effects on the strong motion during the activation of a strike-slip fault: a) forward directivity, resulting in high peak velocities and displacements, and b) polarization of the long-period motions in the near-source region. The near-fault factor must be for sites located within a distance (D) less than 20 km from the major strike-slip faults, for return periods greater than 250 years using the following equations and Table 2 for the maximum values of near-fault factor, Nmax(T).

$$\mathrm{N(T, D) = Nmax(T)\ for\ \ D \leq 2km} \tag{9}$$

$$\mathrm{N(T, D) = 1 + (Nmax(T) - 1) \cdot (20 - D)/18\ for\ \ 2\,km < D \leq 20\,km} \tag{10}$$

$$\mathrm{N(T, D) = 1.0\ for\ \ D > 20\,km} \tag{11}$$

Table 2. Maximum values of near-fault factor, Nmax(T)

Spectral period, T(s)	Nmax(T)
< 1.5	1.0
1.5 < T < 4	0.24T+0.64
4 < T < 5	0.12T+1.12
> 5	1.72

Provisions in ASCE 7-22 [2021] provide guidance on ground-motion selection in consideration of pulse-like ground motions and orientation of the maximum component of ground motion in the FN direction. However, no specific guidance was provided regarding accounting for directivityeffects on intensity measures. In Caltrans (2019), the design spectra for California Bridges are modified for directivity effects by increasing the spectral ordinated at periods larger than 1 s by a factor of 20%. This adjustment factor is applied at locations with a site to rupture plane distance (RRUP) of 15 km or less and linearly tapered to no adjustment at 25 km. The adjustment consists of a 20% increase in spectral values with a corresponding period longer than one second. This increase is linearly tapered to zero at a period of 0.5 s. Since the design spectrum is probabilistically based and includes the influence of multiple faults, the site to rupture plane distance is based on the de-aggregated mean distance for spectral acceleration at a period of 1.0 s.

4.1 ASCE/SEI 7-22 (Near-Fault Ground Motion and Seismic Isolation Design)

In the current ASCE/SEI 7-22 (and also in ASCE/SEI 7-16) standard, the seismic design of seismically isolated structures (Chapter 17) is based on the MCER event only. The response reduction factor can have a maximum value of 2, thereby ensuring that the structure does remain essentially elastic at the MCER level. These standards stipulate that seismically isolated buildings located at near-fault sites should be designed by considering greater seismic demands (than the demand on structures at far-field sites) by using a larger spectral acceleration. For design under near-fault conditions, each pair of horizontal near-fault ground motion components shall be rotated to the fault-normal (FN) and fault-parallel directions (FP) of the causative fault and applied to the building in such orientation.

The Risk-Targeted Maximum Considered Earthquake (MCER) Ground Motion, prepared by the USGS (https://asce7hazardtool.online/; https://doi.org/10.5066/F7NK3C76), incorporates: (1) A target risk of structural collapse equal to 1% in 50 years based upon a generic structural fragility, (2) Factors to adjust from a geometric mean to the maximum response regardless of direction and (2) deterministic upper limits imposed near large, active faults, which are taken as 1.8 times the estimated median response to the characteristic earthquake for the governing fault (1.8 is used to represent the 84th percentile response). A USGS web tool (https://doi.org/10.5066/F7NK3C76) can also be used to determine the MCER (mapped value earthquake hazard) for a specified location.

It should be noted that two metrics are used to characterize ground motion directionality: The RotD50 (essentially equal to the Geometric Mean-GeoMean of the two orthogonal horizontal ground motion components) and the maximum direction RotD100 (MaxDir, orientation independent) values for all orientations (Shahi and Baker [66]). RotD50 is the median value of response spectra of the two horizontal components projected onto all non-redundant azimuths. RotD100 (or maximum direction) is the 100th percentile (the largest possible) value of response spectra of the two horizontal components projected onto all non-redundant azimuths. ASCE-7 defines the design basis of ground motion (MCER) on the basis of the RotD100, which is computed from conversion factors applied to the RotD50-based earthquake hazard maps.

In ASCE/SEI 7-22 provisions, the stipulations and comments regarding the near-fault ground motion, as well as those related to response history analysis, are as follows:

CHAPTER 11 SEISMIC DESIGN CRITERIA
11.4.1 Near-Fault Sites satisfying either of the following conditions shall be classified as a near fault:
• *15 km or less from the surface projection of a known active fault capable of producing Mw7 or larger events, or*
• *10 km or less from the surface projection of a known active fault capable of producing Mw6 or larger events.*
EXCEPTIONS: 1. Faults with an estimated slip rate less than 1 per year shall not be used to determine whether a site is a near-fault site. 2. The surface projection used in the determination of near-fault site classification shall not include portions of the fault at depths of 10 km or greater.
CHAPTER 16 NONLINEAR RESPONSE HISTORY ANALYSIS
16.2.2 Ground Motion Selection
A suite of not less than 11 ground motions shall be selected for each target spectrum. Ground motions shall consist of pairs of orthogonal horizontal ground motion components and, where vertical earthquake effects are considered, a single vertical ground motion component. For near-fault sites, as defined in Section 11.4.1, and other sites where MCER shaking can exhibit directionality and impulsive characteristics, the proportion of ground motions with near-fault and rupture directivity effects shall represent the probability that MCER shaking will exhibit these effects.
16.2.3 Ground Motion Modification
Spectral matching shall not be used for near-fault sites unless the pulse characteristics of the ground motions are retained after the matching process has been completed.
16.2.3.2 Amplitude Scaling
For each horizontal ground motion pair, a maximum-direction spectrum shall be constructed from the two horizontal ground motion components.
16.2.4 Application of Ground Motions to the Structural Model
For near-fault sites, as defined in Section 11.4.1, each pair of horizontal ground motion components representative of a nearby fault source shall be rotated to the fault-normal and fault-parallel directions of the causative fault and applied to the building in such orientation.

The modification of the MCER ground motion obtained from the code-associated seismic hazard map for the near-fault is not explicitly addressed in the ASCE/SEI 7 standards. In ASCE 7-22 [7], Chapter 21 (Site-Specific Ground Motion Procedures for Seismic Design) require that "*21.2.3 Site-Specific MCER Response Spectrum: The site-specific MCER spectral response acceleration at any period, SaM, shall be taken as the lesser of the spectral response accelerations from the probabilistic ground motions of Section 21.2.1 and the deterministic ground motions of Section 21.2.2*". Furthermore, C16.2.2 of ASCE 7-22 [7] (Ground Motion Selection) state that "the maximum direction of response tends to be in the direction normal to the fault strike."

In practical design applications following the ASCE 7-22 [7] Code, the following two methodologies are generally considered.

<u>Methodology 1</u>

(1a) PSHA-based MCER is calculated as per Chapter 21 through PSHA that incorporates directivity effects, summarized in Section 3.2.3 (RotD50) and then converted to (RotD100).
(1b) Develop probabilistic (2%/50) site-specific Fault Normal (FN) spectra through PSHA that incorporate directivity effects, summarized in Section 3.2.3.
(1c) For the probabilistic MCER, use the envelope of (1a) and (1b).
(1d) Compute the 84-percentile deterministic MCER: (RotD50), convert to (RotD100), and apply FN directivity modification.
(1e) Use the lesser of (1c) and (1d) (ASCE 7-22 [7] Ch. 21.2.3) to obtain the design basis MCER.
(1f) The Fault Parallel (FP) ground motion component represents the orthogonal component of horizontal ground motion required to achieve the Average Horizontal ground motion (RotD50 in 1a) when combined with the Fault Normal (FN) spectral ordinates. Noting that the geometric mean combination of the two orthogonal ground motion spectral amplitude components is used to obtain the Average Horizontal ground motion (RotD50), the Fault Parallel (FP) spectral components at each period (T) can be obtained by the FP(T) = [Average Horizontal(T)]2/[FN(T)] relationship.

Methodology 2

(2a) As an alternative to (1a), the ASCE 7-22 [7] code associated MCER (USGS supplied spectra) can be taken equal to the Fault-Normal (FN) ground motion component. The Fault-Normal (FN) ground motion is assumed to be equivalent to the maximum rotated component of ground motion (C16.2.2 of ASCE 7-22 [7]) and, therefore, represents the code definition of site-specific spectra (MCER),
(2b) The Average Horizontal ground motion represents the RotD50 ground motion,
(2c) The Fault Parallel (FP) ground motion component represents the orthogonal component of horizontal ground motion required to achieve the Average Horizontal ground motion when combined with the Fault Normal (FN) spectral ordinates. Noting that the geometric mean combination of the two orthogonal ground motion spectral amplitude components is used to obtain the Average Horizontal ground motion (RotD50), the Fault Parallel (FP) spectral components at each period (T) can be obtained by the

$$\mathrm{FP(T)} = \left[\mathrm{Average\ Horizontal(T)}\right]2/[\mathrm{FN(T)}]\mathrm{relationship.} \tag{12}$$

The vertical ground motion spectrum will be based on the application of the vertical/horizontal (V/H) spectral amplitude ratios on the lesser of the PSHA and DSHA-based horizontal spectral response as per ASCE 7-22 [7] Section 21.2.3. For the application of V/H ratios, the RotD100 spectral metric needs to be converted to the RotD50 spectral metric through the use of conversion factors. The earthquake-resistant design of the Loma Linda Hospital (California) can be provided as a seismic isolation design example based on the 2nd Methodology. The Loma Linda University Medical Center (LLUMC) Replacement Hospital (Fig. 12) is a 17-story high steel frame 3-D seismically isolated 100,000 square meters acute care hospital located 1 km from the M7.9 capable San Jacinto Fault (Nielsen et al. [52]; Nielsen et al. [53]; Nielsen et al. [51]).

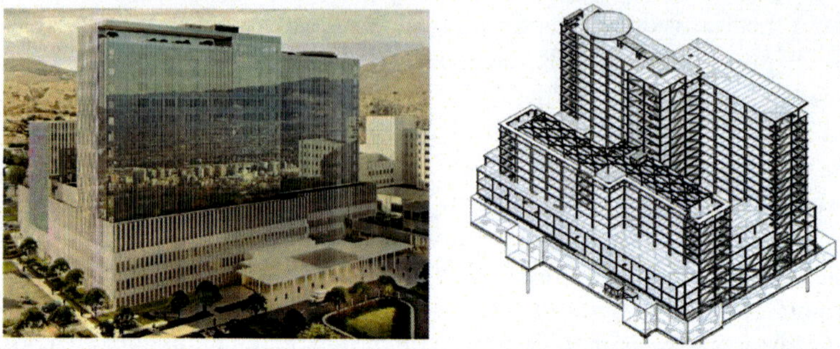

Fig. 12. Picture and an isometric view of the structure LLUMC hospital building.

The design basis response spectra prepared following ASCE 7-16 [6] site-specific MCER specifications are provided in Fig. 13 (Courtesy of A. Dinsick, GeoPentech Inc.).

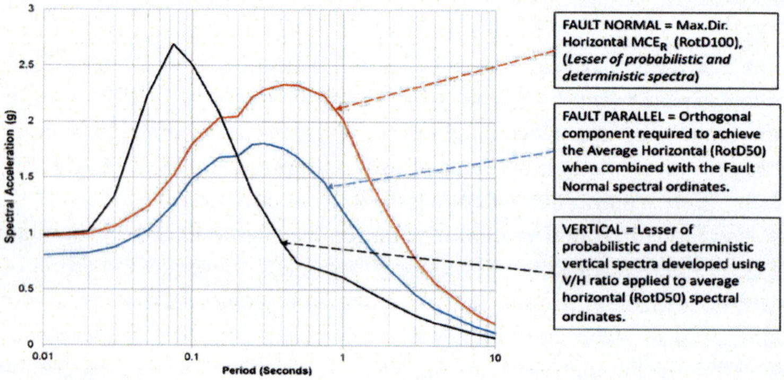

Fig. 13. LLUMC design basis response spectra prepared following ASCE 7-16 [6] site-specific MCER specifications.

The following observations are in order:

- The response spectrum at Maximum Considered Earthquake (MCER) is given for horizontal fault-parallel, horizontal fault-normal, and vertical ground motion directions.
- For use in nonlinear response history analysis and 11 sets of three-dimensional spectrally matched MCER ground motion time histories were generated.
- The ground motions selected included a high proportion of near-fault pulse-like features.
- The fault-normal MCE spectrum is above 1.0 g lateral spectral acceleration at the 2 s period, corresponding to a high spectral displacement demand for a tall building.

- This is exacerbated by the pulse-like nonstationary characteristics of the Project specific ground motions, which act to further increase the displacement demand when applied in the time domain to a nonlinear response history analysis of a yielding lateral system.
- The severe vertical ground motion remains above 1.0 g until approximately 2 Hz, where the vertical frequencies of the floor system and columns are between 4 and 10 Hz.

The lateral base isolation system is comprised of 126 triple friction pendulum bearings (Earthquake Protection Systems) and 104 fluid viscous dampers with 3.6 MN capacity (Taylor Devices). The friction pendulum bearings have a Lateral Displacement Capacity of 1082 mm, Elastic Vertical Load Capacity of 36 MN, and provide an effective period of 4.5 s (Fig. 14). The total equivalent system damping coefficient is 50% of critical damping. The supplemental damping reduced the isolator displacement from 2137 to 1067 mm and greatly reduced the floor accelerations. To reduce the vertical ground motion being transmitted into the building, the vertical isolation system considered comprises an assembly of helical coil springs, fluid viscous dampers, and low-friction sliding shear pins mounted to a suspended concrete-filled steel pedestal.

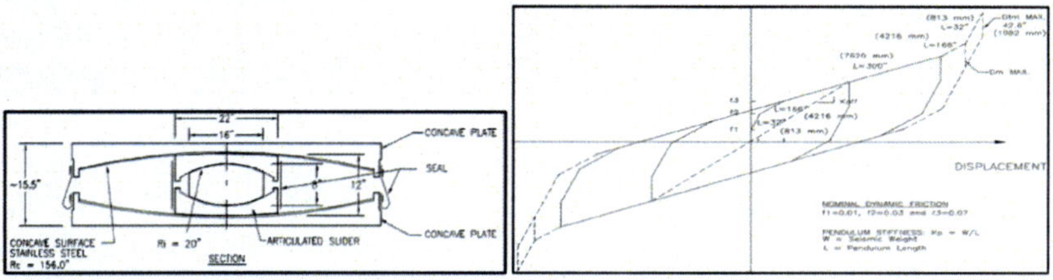

Fig. 14. Triple friction pendulum bearings (Earthquake Protection Systems) used in the isolation system (Courtesy of Dr. A. Mokha)

5 Parametric Study on Near-Fault Parametric Seismic Isolation Design (Earthquake Hazard)

5.1 Introduction

In this chapter, the response of the base-isolated buildings under near-fault (NF) ground motions with fling-step and forward-directivity characteristics are investigated with a rational assessment of design-basis near-fault ground motion, are investigated in a parametric format.

In this connection, we will consider a site with a site class qualified as "engineering bed-rock" with a Vs30 value of 760 m/s, located at a Joyner-Boore distance of Rjb = 2 km from the North Anatolian Fault as illustrated in Fig. 15. As it can be assessed the earthquake hazard at the site is totally controlled by the North Anatolian Fault.

5.2 Near-Fault Ground Motion Spectra

For the assessment of design basis near-fault ground motion spectra, we will consider a probabilistic approach and compute the UHS associated with the 2475-year average return period. The seismic source characterization is illustrated in Fig. 15. The 2475-year UHS for the site-class qualified as engineering bedrock (Vs30 = 760–800 m/s) is provided in Fig. 17 together with ESHM13 (Giardini et al. [35]; Giardini et al. [37]; Woessner et al. [78]) and ESHM2020 (Danciu et al. [25]) spectra for comparison purposes (Fig. 16).

For the incorporation of the near-fault effects on the 2475-year UHS, where a single fault dominates the hazard at long periods, the A-Posteriori Modification Method described in Chapter 2.4 (Donahue et al. [26]) will be considered. For this application, the 2475-year UHS is first subjected to a hazard deaggragation in the long period range, the controlling scenario earthquake is assessed, and the associated deterministic spectrum is developed to proxy the UHS.

The result of deaggragation analysis indicates an earthquake scenario defined by an Mw7.98 magnitude, Rjb = 2.5 km of rupture distance, and epsilon of $\varepsilon = 1.5$. The GMPEs considered in the first approach are also used in this approach to develop the %5 damped spectra. The Fault normal (FN), Fault Parallel (FP), and general direction ground motion spectra that reflect the directivity characteristic are computed using Bayles and Somerville's [56, 73] methodology. The acceleration spectra for FN, FP, and general direction (GeoMean) ground motion are provided in Fig. 18, associated with the scenario earthquake.

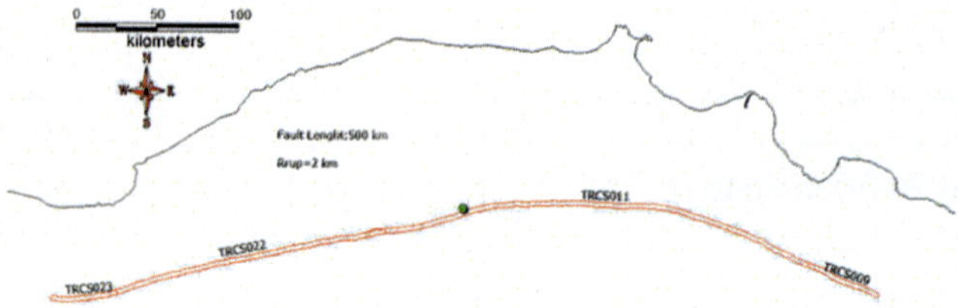

Fig. 15. Location of the site (green dot) at a distance of Rjb = 2 km from the North Anatolian fault.

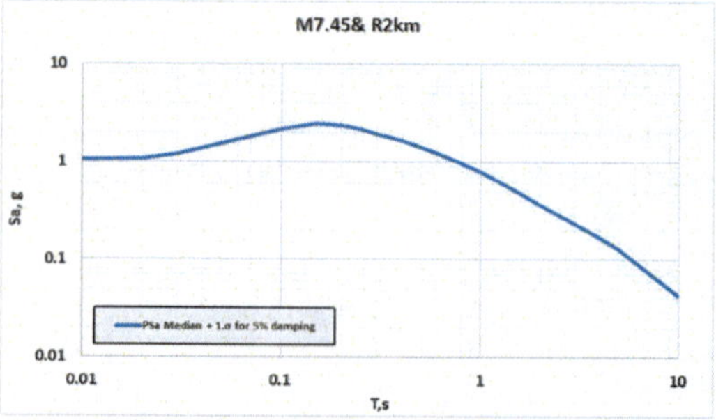

Fig. 16. Deterministic Median + 1 SD spectral acceleration.

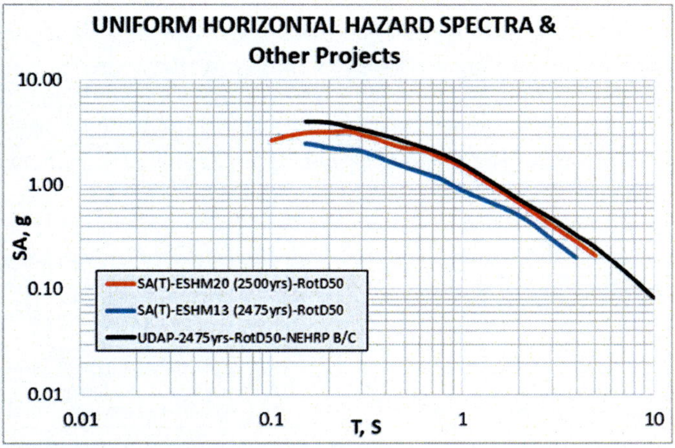

Fig. 17. The resulting 2475-year UHS for the site-class qualified as engineering bedrock together with ESHM13 (Giardini et al. [35]; Giardini et al. [37]; Woessner et al. [78]) and ESHM2020 (Danciu et al. [25]) spectra

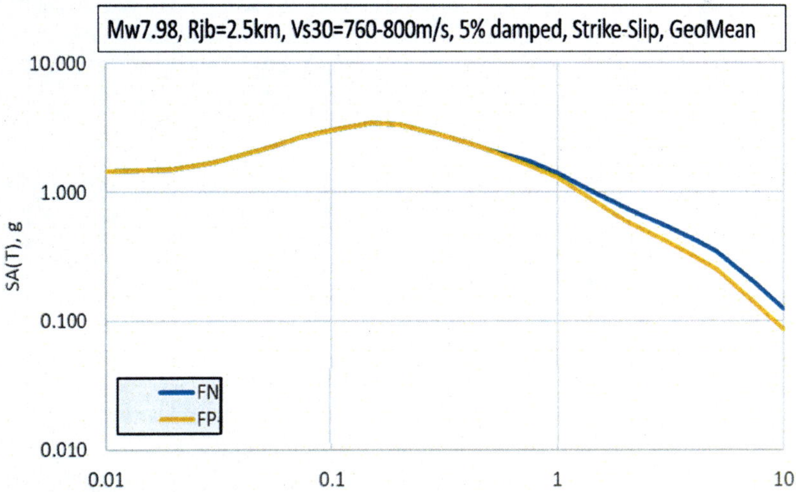

Fig. 18. Acceleration spectra for FN, FP, and general direction (GeoMean) ground motion associated with the scenario earthquake and Median-1.5 SD.

5.3 Near-Fault Ground Motion in Time Domain (Pulse-Based)

As elaborated in Chapter 2.3.2, the NF ground motions with fling-step and forward-directivity characteristics can be identified from the velocity and displacement time histories. The forward directivity effect results in a large amplitude, long period two-sided pulse in the velocity time series of the FN component of the ground motion, whereas the fling step results in a one-sided velocity pulse and permanent displacement at the end of the ground motion record in the FP direction (along the fault plane, in the direction of the slip). The pulse period and pulse amplitude are two major parameters of ground motion records having a velocity pulse. In this study, three pulse periods (Tp = 3, 4, and 5 s) obtained from the relevant references (Bray and Rodrigues-Marek [15]; Alavi and Krawinkler [3]; Sommerville [71]; Bray et al. [14]) are used. We believe that this pulse period range is valid for maximum magnitude in the range between Mw7.0 and Mw7.6.

The algorithm used in the pulse-based near field ground motion approach is shown in Fig. 19, and the acceleration, velocity, and displacement pulses for FN and FP corresponding to the three different pulse periods are presented in Fig. 20.

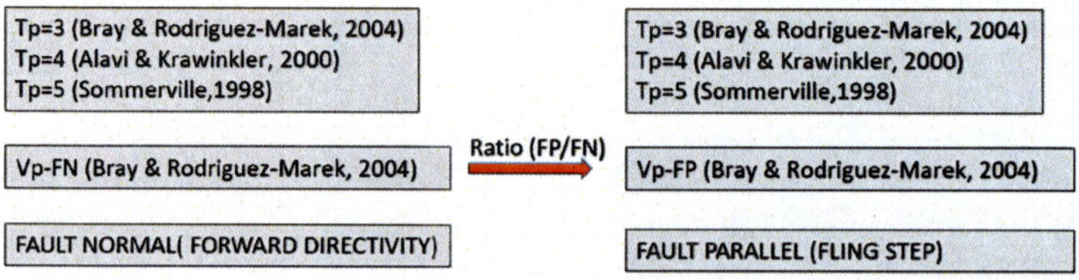

Fig. 19. The algorithm used in the pulse-based near field ground motion approach

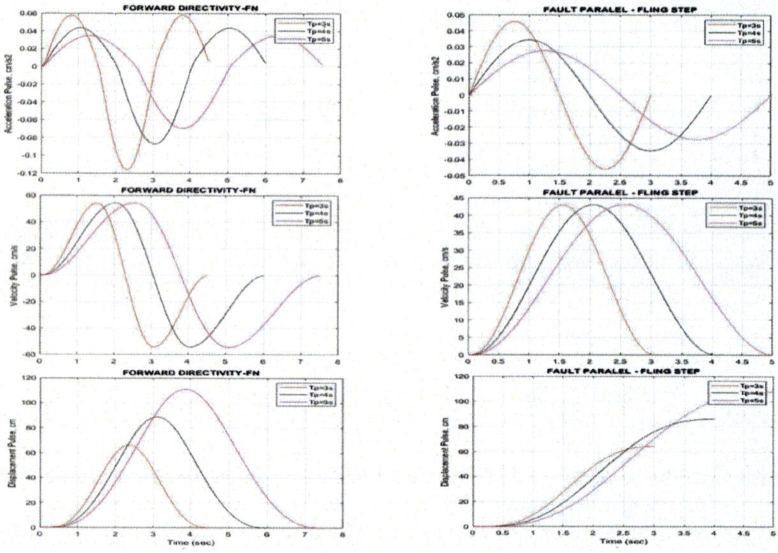

Fig. 20. The acceleration, velocity, and displacement pulses for FN and FP correspond to the three different pulse periods of 3, 4, and 5 s.

5.4 Spectrum Scaled Accelerometric Data

The set ground motions developed in Sect. 5.3 (Fig. 20) are used to develop the scaled accelerometric data compatible with the FN and FP spectra provided in Fig. 21. Figure 22 provides the comparison of the target and scaled ground motion spectra, and the acceleration, velocity and displacement spectra associated with the scaled accelerometric data provided in Fig. 23. As it can be assessed, the PGA and PGV amplitudes of these pulse-type ground motion respectively varies between 150 and 350 cm/s^2 and 110–160 cm/s for the pulse periods of $T_p = 3$, 4 and 5 s. PGA and PGV values in these ranges were observed in the following past strike-slip earthquakes (Mavroeidis and Papageorgiou [48]).

1971 Mw6.6 San Fernando (CA): PGA = 400 cm/s^2, PGV = 130 cm/s, $T_p = 2$ s.
1992 Mw6.6 Erzincan (Turkey): PGA = 180 cm/s^2, PGV = 67 cm/s, $T_p = 2$–3 s.
1999 Mw7.4 Kocaeli (Turkey): PGA = 80 cm/s^2, PGV = 92 cm/s, $T_p = 4$–5 s.
1979 Mw6.5 Imperial Valley (CA): PGA = 160 cm/s^2, PGV = 96 cm/s, $T_p = 3$–4 s.
1989 Mw6.9 Loma Prieta (CA): PGA = 110 cm/s^2, PGV = 60 cm/s, $T_p = 3$ s.
1992 Mw7.2 Landers (CA): PGA = 100 cm/s^2, PGV = 61,000 cm/s, $T_p = 4$ s.

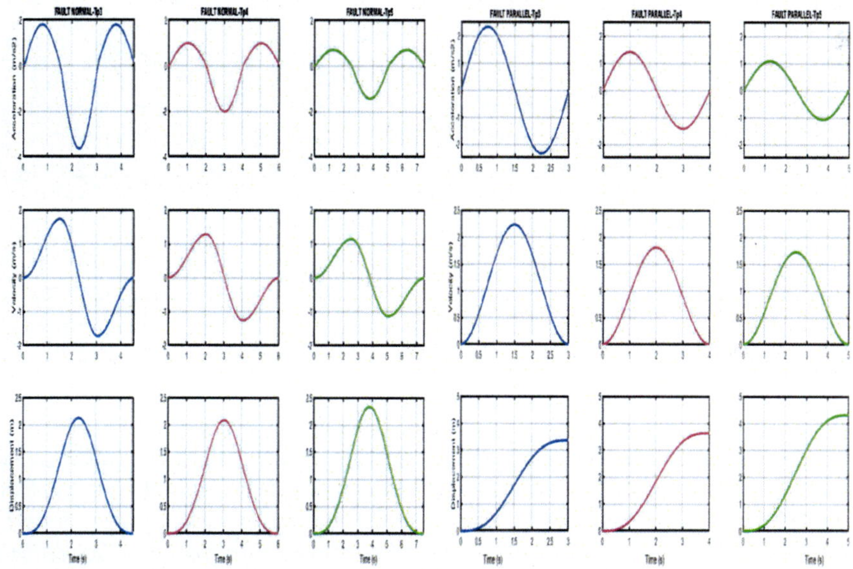

Fig. 21. The FN and FP spectrum compatible scaled accelerometric data.

6 Parametric Study on Near-Fault Parametric Seismic Isolation Design (Structural Analysis)

6.1 Introduction

The near fault performance of seismically isolated buildings, subjected to the fault normal and fault parallel time histories developed in Chapter 5, were investigated through a parametric study.

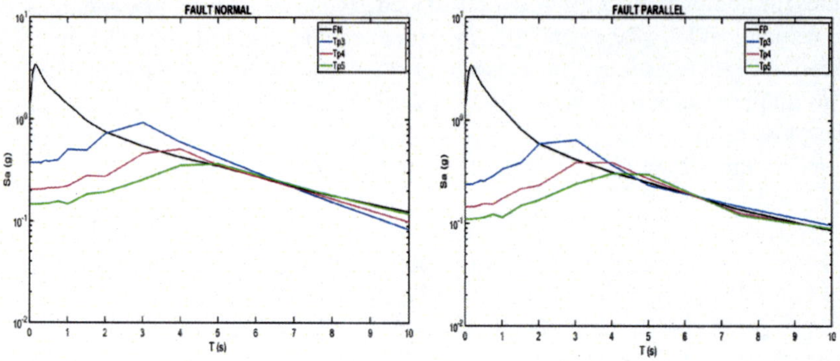

Fig. 22. Comparison of target and scaled-acceleration spectra for FN and FP ground motion.

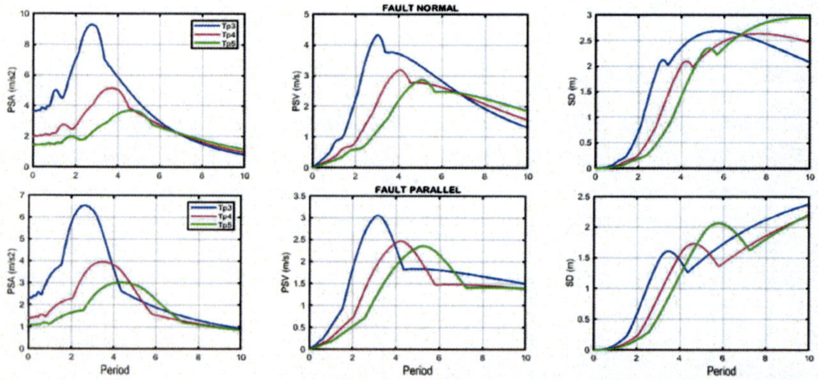

Fig. 23. The acceleration, velocity, and displacement spectra associated with the accelerometric data provided in Fig. 20.

The parametric study includes variables concerning the number of stories, structural system stiffness, and the isolation system characteristics (initial stiffness (K1), yield strength (F_y), and post-elastic stiffness (K2)). The nominal isolation period Ti is defined in terms of a rigid superstructure and the post-elastic stiffness, K2. The effect of supplemental damping in the isolation system by viscous dampers was also tested for some selected cases, and the performance of the structural system with and without the supplemental dampers was compared to demonstrate the need for additional damping systems.

6.2 Structural and Seismic Isolation Modelling

Lumped mass stick models were created in ETABS [https://www.csiamerica.com/products/etabs], representing the buildings with three, five, and eight stories (Fig. 24a). The seismic weight lumped at each story level and at the base is 5000 kN. The total seismic weight of an n-story building makes (n + 1) times 5000 kN (Table 3). The story heights were kept the same. Two different structural systems were considered, a frame system representing a flexural structural system and a shear wall system representing a rigid structure.

Table 3. Structural model properties

Number of stories	3		5		8	
Seismic weight of each story (kN)	5000		5000		5000	
Total seismic weight, W (kN)	20,000		30,000		45,000	
Structural system	Frame	Shear wall	Frame	Shear Wall	Frame	Shear wall
Fixed base period (s)	0.5	0.3	0.75	0.45	1	0.6

A bilinear plastic link element (Fig. 24) is adopted to represent the seismic isolation system, with an initial elastic stiffness (K1), post elastic stiffness (K2), yield force (Fy), and a fixed yield displacement (Dy) of 0.015 m. Initial and post elastic stiffness are derived by the following Eqs. (13) and (14), where T_i is the isolation period, and g is the acceleration of gravity. The yield strength ratio is varied between 4 and 10%, and the Ti is varied between 2.5 and 6 s. For the 3, 5, and 8-story buildings, fixed base first mode vibration periods of 0.5 s and 0.3 s; 0.75 s and 0.45 s; 1.0 s and 0.6 s are respectively considered. For each 3, 5, and 8-story building, the first period refers to a frame and the second period to a shear-wall type building.

$$K1 = \frac{F_y}{D_y} \tag{13}$$

$$K2 = \frac{4\pi^2 W}{gT_i^2} \tag{14}$$

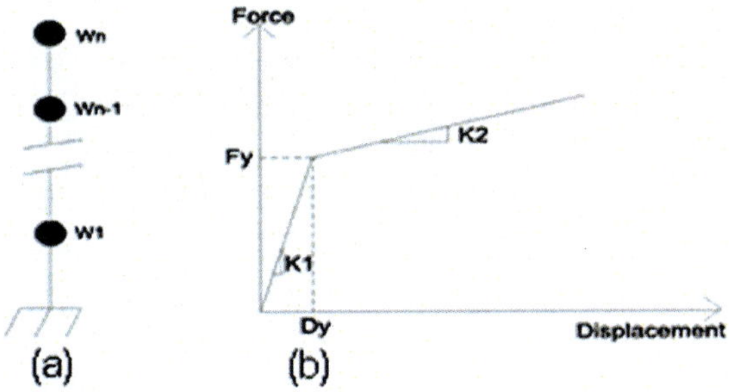

Fig. 24. Stick model representing the building structures (1) and bilinear force-displacement behavior of the seismic isolation system (2).

The models representing the building structure and the isolation system were excited with the three sets of accelerograms in fault normal and fault parallel directions associated with three different pulse periods (Tp = 3, 4, and 5 s), as elaborated in Chapter 5. A direct integration method was employed. An intrinsic modal damping ratio of 2% is considered for the superstructure. A Multilinear plastic link element, following the hysteresis curve given in Fig. 24 (b), was considered to represent the isolation system. The isolation system displacement and base shear were obtained as the vectorial sum of the fault normal and fault parallel components. The mean (average) response associated with the three sets of accelerograms, in fault normal and fault parallel directions, is considered the demand.

6.3 Observation of the Parametric Response

Effect of Superstructure Characteristics on the Seismic Isolation System Response

A parametric study was conducted to investigate the effect of the number of stories (3, 5, and 8) and the superstructure stiffness on the isolation system performance in terms of displacement, base shear, story drifts, and story accelerations while keeping the isolator properties constant (F_y = 5% of W and T_i = 4.0 s). The term "Shear Wall" is used to represent a stiff, while "Frame" is used to represent a flexible structural system. The mean responses were plotted in Fig. 25.

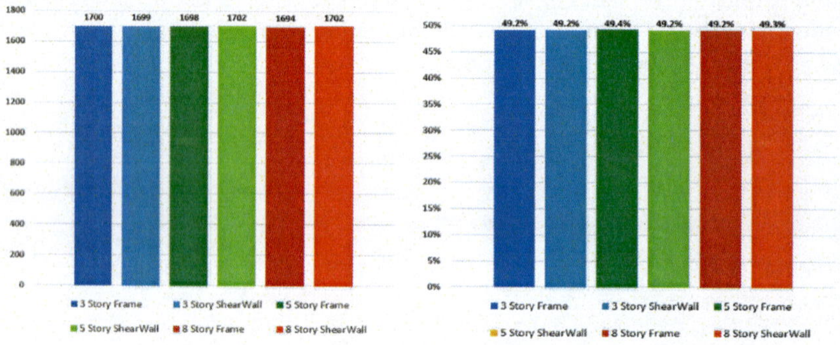

Fig. 25. Mean isolation system displacements and the base shear ratios.

Effect of Seismic Isolation System Yield Force Ratio on Seismic Isolation System Response

A parametric study was conducted to investigate the effect of the seismic isolation system yield force (4, 5, 6, 7, 8, 9, and 10% of the total seismic weight) on the isolation system in terms of displacement, base shear, story drifts, and story accelerations for the five-story frame building. The post elastic stiffness of the seismic isolation system is kept constant, corresponding to an isolation period of T_i = 4.0 s. The mean responses were plotted in Fig. 26.

Effect of Seismic Isolation System Post-Elastic Stiffness on Seismic Isolation System Response

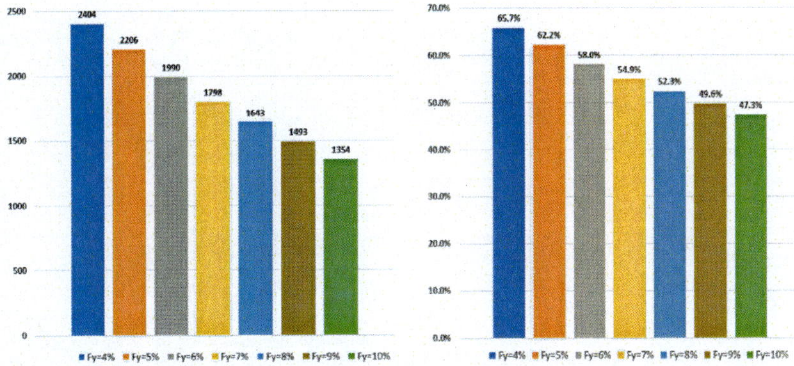

Fig. 26. Mean isolation system displacements and base shear ratios.

A parametric study was conducted to investigate the effect of the seismic isolation system post elastic stiffness (measured as Ti = 2.5, 3.0, 3.5, 4.0, 4.5, 5.0, 5.5, and 6.0 s) on the isolation system performance in terms of displacement, base shear, story drifts, and story accelerations for the five-story frame building. The yield force of the seismic isolation system is kept constant, corresponding to 5% of the total seismic weight. The mean responses were plotted in Fig. 27.

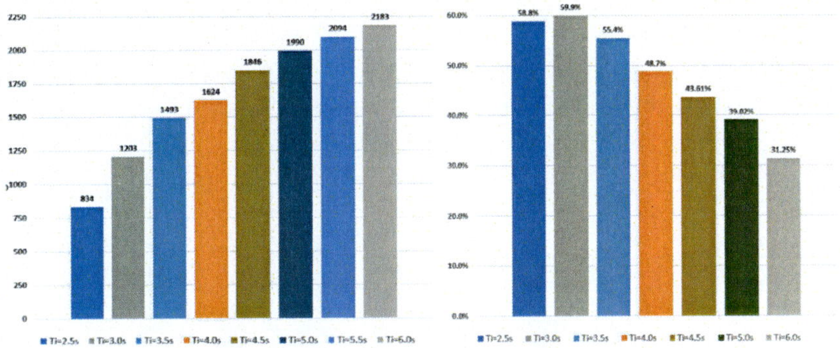

Fig. 27. Mean isolation system displacements and base shear ratios.

Comparison of ELF Analysis and NLTHA Results

In Fig. 28, the ELF and NLTHA results were compared in terms of displacement and base shear for the five-story frame building with constant isolation system properties (Fy = 5% of W and Ti = 4.0 s).

ELF results vary according to the damping modification factors proposed in EC8 [20], TBDY [74], and ASCE 7/22 [7].

The damping modification factor proposed by Priestley et al. [58] for near-fault sites was also included in the comparison (Fig. 28).

Effect of Supplemental Viscous Damping on Seismic Isolation System Response

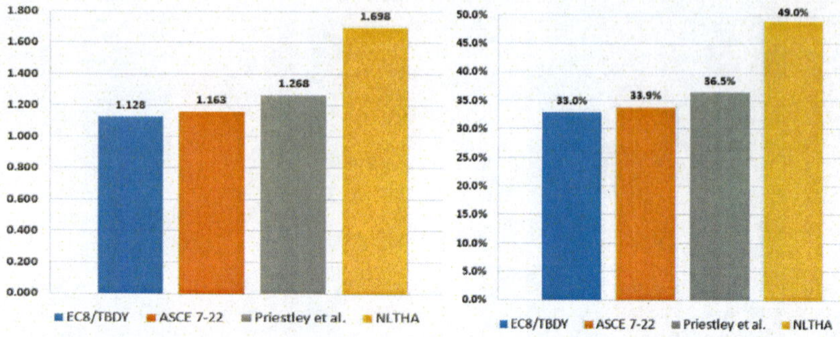

Fig. 28. Mean isolation system displacements and base shear ratios calculated by ELF method and NLTHA.

The seismic isolation system and seismically isolated building performance results in terms of isolation system displacement, base shear, story drifts, and top story acceleration were compared for the five-story flexible building structural system case with constant yield force and varying isolation period with and without supplemental viscous damping at isolation interface. The damping force (F) equation is given in Eq. (15), where V is the ground motion velocity, C is the damping constant taken as 600 kNs/m and α is the damping exponent considered 0.5.

$$F = CV^{\alpha} \tag{15}$$

The effect of supplemental viscous damping on the mean isolation system displacements and the base shear ratios are illustrated respectively in Figs. 29 and 30.

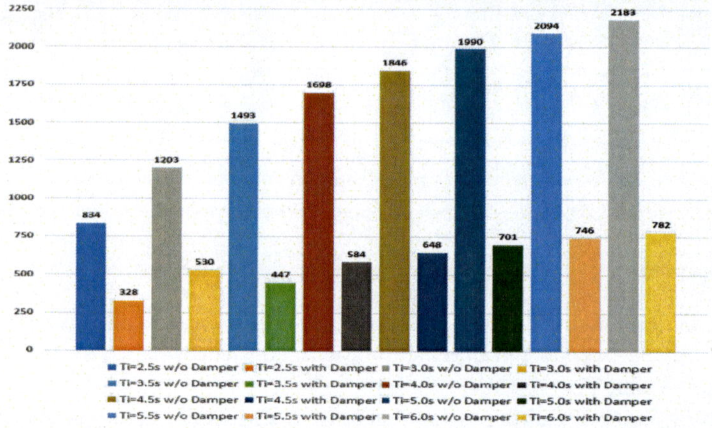

Fig. 29. Effect of supplemental viscous damping on mean isolation system displacements.

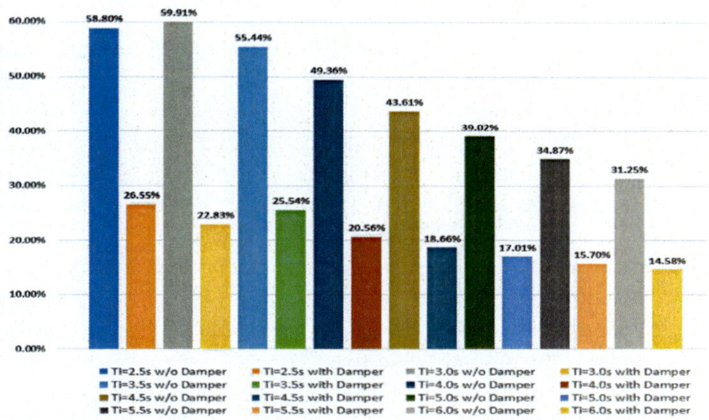

Fig. 30. Effect of supplemental viscous damping on mean base shear ratios.

7 Discussion and Conclusions

For base isolated structures the rational (i.e. state-of-the-art) determination of the long period region of the design basis ground motion is of particular importance due to associated structural safety and the isolation system cost implications. This importance becomes even more crucial for structures exposed to near-fault ground motion effects, since for Mw7-Mw7.5 earthquakes the dominant low frequency pulse period of the near-fault ground motion ($T_p = 3–5$ s) happens to coincide with the typical isolation period of vibration for almost all of the base isolated structures.

In only a few earthquake resistant design codes, the guidelines for the estimation of the near-field ground motion exist (some of which are now outdated) for the modification of the general horizontal direction (RotD50) conventional code-based design spectra. As far as known, the regulations regarding the design of the seismically isolated structures exist only in ASCE SEI 7 codes which stipulate the separate and direction associated application of fault-normal and fault-parallel components of the near-fault design basis ground motion.

This paper presents a review of the current methodologies for the assessment of the near-fault ground motion both in the frequency and time domains, as well as a critical summary of relevant design code regulations. A special study is conducted to find the effect of the near-fault ground motion on the response of a set of base isolated buildings with parameterized characteristics. As it can be observed from the normalized fault-normal (FN) and fault-parallel (FP) near-fault simple pulse-type ground motion spectra presented in Fig. 9, the amplification of the peak ground acceleration (PGA) and the peak ground displacement (PGD), within the normalized period range of $T_p = 0.8–1.2$, vary respectively between 2.2(FN)–2.6(FP) and 1.0(FP)–0.45(FP) for the damping ratio of $\beta = 0.05\%$. These amplifications drop down respectively to 1.2(FN)–1.4(FP) and 0.5(FP)–0.25(FP) for the damping ratio of $\beta = 0.05\%$ (a common value for seismic isolation systems). It should be noted PGA and PGV refer to the long period ($>= 3$ s) of the ground motion and that these amplification levels are applicable to the base of the superstructure for base isolated structures with comparatively rigid superstructure with respect to the isolation system.

The special parametric study investigated the effects of superstructure characteristics, seismic isolation system yield force ratio, post-elastic stiffness and supplemental viscous damping on the seismic isolation system response. Furthermore, a comparison of ELF analysis and NLTHA results is conducted. The major finding of the special parametric study and conclusions drawn from this study are as follows.

- The main conclusion derived from the analysis of different superstructure and different isolation system properties, is that the maximum response is obtained from the fault-normal component of the ground motion.
- The number of stories and the type (frame of shear wall) of the superstructure does not significantly affect the isolation system displacement and the associated base shear ratio.
- As expected, increasing the yield level ratio (thereby increasing the equivalent damping ratio) of the isolation system, results in a reduction in isolation system displacement and the associated base shear ratio.
- As expected, the addition of viscous dampers substantially reduces the isolation system displacements and the associated base shear ratio, regardless of the number of stories and type of the superstructure and yield level ratio.
- The isolation system displacement and base shear ratio computed on the basis of the Effective Lateral Load (ELF) analysis indicate significant under estimation (in the order of 30%) compared to the Nonlinear Time History Analysis (NLTHA).
- Under near-fault conditions and soft soil sites the maximum vertical spectral amplitude can exceed even two times the corresponding general direction (RotD50) horizontal acceleration spectrum amplitude in short periods (0.05–0.1 s). As such, in near-fault conditions, both the isolation system elements and the superstructure are highly vulnerable and require careful consideration of the vertical ground motion excitation for safe design.
- Considering the results of basic design parameters such as isolation system displacement and base shear ratio, it can be deduced that, the design of the seismic isolation system under exposure to near-field ground motion requires the provision of supplemental damping to the system.
- The current seismic design codes do not explicitly provide an adequate representation of near-fault ground motions and only one (or few) of them provide limited guidelines for the design of base isolated structures in near fault conditions.
- In engineering practice, there is no generally accepted methodology for the design of the isolated structures exposed to near-field ground motion. We believe that the methodologies reviewed and developed in this paper will be useful for the design and/or the verification of design of base isolated structures in near fault conditions.

References

1. Abrahamson, N.A., Silva, W.J., Kamai, R.: Summary of the ASK14 Ground-Motion Relation for active crustal regions. Earthq. Spectra (2014). https://doi.org/10.1193/070913EQS198M

2. Abrahamson, N.: Effects of rupture directivity on probabilistic seismic hazard analysis. In: 6th International Conference on Seismic Zonation, Earthquake Engineering Research Institute, CD-2000-01 (2000)
3. Alavi, B., Krawinkler, H.: Effects of near fault ground motions on frame structures. Report TR138, John A. Blume Earthquake Engineering Center. OCLC No. 48748468 (2001)
4. Alavi, B., Krawinkler, H.: Behaviour of moment resisting frame structures subjected to near-fault ground motions. Earthquake Eng. Struct. Dyn. **33**, 687–706 (2004)
5. Almufti, İ., Motamed, R., Grant, D.N., Willford M.: Incorporation of velocity pulses in design ground motions for response history analysis using a probabilistic framework. Earthquake Spectra (2015)
6. ASCE 7-16: Minimum Design Loads for Buildings and Other Structures. American Society of Civil Engineers, USA (2016)
7. ASCE/SEI 7-2022: Minimum Design Loads and Associated Criteria for Buildings and Other Structures. ASCE, NY (2022)
8. Baker, J.W.: Quantitative classification of near-fault ground motions using wavelet analysis. Bull. Seismol. Soc. Am. **97**, 1486–1501 (2007)
9. Bao, Y., Becker, T., Hamaguchi, H.: Failure of double friction pendulum bearings under pulse-type motions. Earthquake Eng. Struct. Dyn. **46**(5), 715–731 (2017)
10. Bhagat, S., Wijeyewickrema, A.C., Subedi, N.: Influence of near-fault ground motions with fling-step and forward-directivity characteristics on seismic response of base-isolated buildings. J. Earthquake Eng. **25**(3), 455–474 (2021). https://doi.org/10.1080/13632469.2018.1520759
11. Bommer, J.J., Elnashai, A.S.: Displacement spectra for seismic design. J Earthquake Eng. **3**(1), 1–32 (1999)
12. Bozorgnia, Y., Campbell, K.W.: Ground motion model for the vertical-to-horizontal (V/H) ratios of PGA, PGV, and response spectra. Earthq. Spectra **32**(2), 951–978 (2016)
13. Bozorgnia, Y., et al.: NGA-West2 research project. Earthq. Spectra (2014). https://doi.org/10.1193/072113EQS209M
14. Bray, B., Rodriguez-Marek, A., Gillie, J.L.: Design ground motions near active faults. Bull. New Zealand Soc. Earthq. Eng. **42**(1) (2009)
15. Bray, J.D., Rodriguez-Marek, A.: Characterization of forward-directivity ground motions in the near-fault region. Soil Dyn. Earthq. Eng. **24**(11), 815–28 (2004). https://doi.org/10.1016/j.soildyn.2004.05.001
16. CALTRANS: CALTRANS SEISMIC DESIGN CRITERIA VERSION 2.0. State of California Department of Transportation (2019)
17. Campbell, K.W., Bozorgnia, Y.: NGA ground motion model for the geometric mean horizontal component of PGA, PGV, PGD and 5% damped linear-elastic response spectra for periods ranging from 0.01 and 10.0 s. Earthq. Spectra **24**, 139–171 (2008)
18. Campbell, K.W., Bozorgnia, Y.: NGA-West2 ground motion model for the average horizontal components of PGA, PGV, and 5% damped linear acceleration response spectra. Earthq. Spectra (2014). https://doi.org/10.1193/062913EQS175M
19. Cauzzi, C., Faccioli, E., Paolucci, R.: A reference model for prediction of long period response spectral ordinates—deliverable D2 (Task 1), Politecnico di Milano (2007)
20. CEN: Eurocode 8: design of structures for earthquake resistance. prEN 1998-1 (Draft No. 6), Comité Européen de Normalisation, Brussels, Belgium (2005)
21. Chai, J., Loh, C., Chen, C.: Consideration of the near-fault effect on seismic design code for sites near the Chelungpu fault. J. Chin. Inst. Eng. **23**(4), 447–454 (2000)
22. Chen, X., Wang, D.: Multi-pulse characteristics of near-fault ground motions. Soil Dyn. Earthq. Eng. **137**, 106275 (2020). https://doi.org/10.1016/j.soildyn.2020.106275(2020)

23. Chen, Y., Pagani, M., Weatherill, G., Cotton, F.: Exploring near-fault directivity effect in probabilistic seismic hazard assessment. In: 16th World Conference on Earthquake, 16WCEE 2017 Santiago Chile, January 9th to 13th 2017 Paper N° 2337 (2017)
24. Chen, Y., Xu, L., Wu, P.: Low-cut filter frequency quantification and its influence on 21 inelastic response spectrum. J. Earthq. Eng. 1–25 (2021). 22 https://doi.org/10.1080/13632469.2021.1927900
25. Danciu, L., Nandan, S., Reyes, C., Basili, R., Weatherill, G., Beauval, C., Rovida, A., Vilanova, S., Sesetyan, K., Bard, P-Y., Cotton, F., Wiemer, S., Giardini, D.: The 2020 update of the European Seismic Hazard Model: Model Overview. EFEHR Technical Report 001, v1.0.0. https://doi.org/10.12686/a15 (2021)
26. Donahue, J.L., Stewart, J.P., Gregor, N., Bozorgnia, Y.: Ground-motion directivity modeling for seismic hazard applications. Review panel: Jonathan D. Bray, Stephen A. Mahin, I. M. Idriss, Robert W. Graves and Tom Shantz. PEER_2019/03 Report (2019)
27. Yang, Y., Zhang, J.: Response spectrum-oriented pulse identification and magnitude scaling of forward directivity pulses in near-fault ground motions. Soil Dyn. Earthq. Eng. **31**(1), 59–76 (2011). https://doi.org/10.1016/j.soildyn.2010.08.006
28. EC-8: Design provisions for earthquake resistance of structure. Eurocode-8, European Committee for Standardization (1998)
29. EC8: Design provisions for earthquake resistance of structures publication ENV-2003-2. Comite European de Normalizations, Brussels (2003)
30. Erdik, M., Sestyan, K., Demircioglu, M., Tuzun, C., Giardini, D., Gulen, L., Akkar, S., Zare, M.: Assessment of seismic hazard in the Middle East and Caucasus: EMME (Earthquake Model of Middle East) project. In: Proceedingd of 15th World Conference on Earthquake Engineering, Lisbon, Portugal (2012)
31. Faccioli, E., Paolucci, R., Rey, J.: Displacement spectra for long periods. Earthq. Spectra **20**, 347–376 (2004)
32. Fu, Q., Menun, C.: Seismic-environment-based simulation of near-fault ground motions. In: Proceedings of the 13th World Conference on Earthquake Engineering, Vancouver, Canada, 1–6 August, p. 15 (2004)
33. Quaranta, G., Angelucci, G., Mollaioli, F.: Near-fault earthquakes with pulse-like horizontal and vertical seismic ground motion components: analysis and effects on elastomeric bearings. Soil Dyn. Earthq. (2022)
34. Gentile, R., Galasso, C.: Hysteretic energy-based state-dependent fragility for ground-motion sequences. Eng. Struct. Dyn. **50**, 1187–1203 (2021)
35. Giardini, D., et al.: Seismic Hazard Harmonization in Europe (SHARE): Online Data Resource (2013). https://doi.org/10.12686/SED-00000001-SHARE
36. Giardini, D., Danciu, L., Erdik, M., Sesetyan, K., Demircioglu, M., Akkar, S., Gülen, L., Zare, M.: Seismic hazard map of the middle east (2016). 10.12686/a1
37. Giardini, D., Woessner, J., Danciu, L.: Mapping Europe's seismic hazard. EOS **95**(29), 261–262 (2014)
38. Gülerce, Z., Kamai, R., Abrahamson, N.A., Silva, W.J.: Ground motion prediction equations for the vertical ground motion component based on the NGA-W2 database. Earthq. Spectra **33**(2), 499–528 (2017)
39. Hall, J.F., Ryan, K.L.R.: Isolated buildings and the 1997 UBC near-source factors. Earthq. Spectra **16**(2) (2000)
40. Hall, J.F., Heaton, T.H., Halling, M.W., Wald, D.J.: Near-source ground motion and its effects on flexible buildings. Earthq. Spectra **11**(4), 569–605 (1995)
41. Heaton, T.H.: Evidence for and implications of self-healing pulses of slip in earthquake rupture. Phys. Earth Planer. lmeriors **64**, 1–20 (1990)
42. Donahue, J.L., et al.: Ground-motion directivity modeling for seismic hazard applications. PEER Report 2019/03 (2019)

43. Hall, J.F., Ryan, K.L.: Isolated buildings and the 1997 UBC near-source factors. Earthq. Spectra 16(2).https://doi.org/10.1193/1.1586118
44. Kalkan, E., Kunnath, S.K.: Effects of fling step and forward directivity on seismic response of buildings. Earthq. Spectra **22**(2), 367–390 (2006)
45. Kalkan, E., Kunnath, S.K.: Effects of fling step and forward directivity on seismic response of buildings. Earthq. Spectra **22**(2), 367–390 (2006)
46. Kitayama, S., Constantinou, M.: Performance evaluation of seismically isolated buildings near active faults. Earthq. Eng. Struct. Dyn. **51**(5)
47. Li, X., Dou, H., Zhu, X.: Response of seismic design code for zones lack of near-fault strong earthquake records. ACTA Seismol. Sinica **20**(4), 447–453 (2007)
48. Mavroeidis, G.P., Papageorgiou, A.S.: A mathematical representation of near-fault ground motions. Bull. Seismol. Soc. Am. **93**, 1099–1131 (2003)
49. Mavroeidis, G.P., Dong, G., Papageorgiou, A.S.: Near-fault ground motions, and the response of elastic and inelastic single-degree-freedom (2004)
50. Vassiliou, M.F., Tsiavos, A., Stojadinović, B.: Dynamics of inelastic base-isolated structures subjected to analytical pulse ground motions. Earthq. Eng. Struct. Dyn. **2013**(42), 2043–2060 (2013)
51. Nielsen, G., Rees, S., Dong, B., Chok, K., Fatemi, E., Zekioglu, A.: Practical implementation of ASCE-41 and NLRHA procedures for the design of the LLUMC replacement hospital. In: 2017 SEAOC Convention Proceedings (2017)
52. Nielsen, G., Sarebanha, A., Biscombe, L., Shao, B., Rees, S.: Development and experimental validation of a 3-D seismic isolation system using Ls-Dyna. In: 17th World Conference on Earthquake Engineering, 17WCEE (2020)
53. Nielsen, G., Rees, S., Zekioglu, A., Sarebanha, A., Biscombe, L., Shao, B., Dong, B.: A 3-D seismic isolation system to protect the Loma Linda University Hospital from near-fault earthquakes. In: 17th World Conference on Earthquake Engineering, 17WCEE (2020)
54. NZS 1170.5 Supp 1:2004. Structural Design Actions Part 5: Earthquake Actions-New Zealand-Commentary. Standards New Zealand (2004)
55. Pant, D.R., Wijeyewickrema, A.C.: Influence of near-fault ground motions on the response of base-isolated reinforced concrete buildings considering seismic pounding. Adv. Struct. Eng. **16**(12), 1973–1988 (2013)
56. PEER 2013 Report: Paul Spudich; Jeffrey R. Bayless; Jack W. Baker; Brian S.J. Chiou; Badie Rowshandel; Shrey K. Shahi; Paul Somerville. Final Report of the NGA-West2 Directivity Working Group, PEER Report 2013-09 (2013)
57. Donahue, J.L., Stewart, J.P., Gregor, N., Bozorgnia, Y.: Ground-motion directivity modeling for seismic hazard applications. PEER Report 2019/03 (2019)
58. Priestley, M.N.: Myths and Fallacies in Earthquake Engineering, Revisited: The Ninth Mallet Milne Lecture, p. 119. IUSS Press, Pavia, Italy (2003)
59. Priestley, M.J.N.: Seismic behaviour of unreinforced masonry walls. Bull. New Zealand Natl. Soc. Earthq. Eng. **18**(2), 191–205 (1985)
60. Quaranta, G., Angelucci, G., Mollaioli, F.: Near-fault earthquakes with pulse-like horizontal and vertical seismic ground motion components: analysis and effects on elastomeric bearings. Soil Dyn. Earthq. Eng. **160** (2022)
61. Gentile, R., Galasso, C.: Accounting for directivity-induced pulse-like ground motions in building portfolio loss assessment. Bull. Earthq. Eng. **19**(15), 6303–6328 (2020). https://doi.org/10.1007/s10518-020-00950-9
62. Rodriguez-Marek, A., Bray, J.D.: Seismic site effects for near-fault forward directivity ground motions. J. Geotech. Geoenviron. Eng. ASCE **132**(12), 1611–1620 (2006)
63. Rowshandel, B.: Directivity correction for the Next Generation Attenuation (NGA) relations. Earthq. Spectra **26**(2), 525–559 (2010). https://doi.org/10.1193/1.3381043

64. Sasani, M., Bertero, V.V.: Importance of severe pulse-type ground motions in performance-based engineering: historical and critical review. In: Proceedings of 12th World Conference on Earthquake Engineering, Paper No. 1302 (2000)
65. Bhagat, S., Wijeyewickrema, A.C., Subedi, N.: Influence of near-fault ground motions with fling-step and forward-directivity characteristics on seismic response of base-isolated buildings. J. Earthq. Eng. **25**(13), 455–447 (2021)
66. Shahi, S., Baker, J.W.: An empirically calibrated framework for including the effects of near-fault directivity in probabilistic seismic hazard analysis. Bull. Seismol. Soc. Am. **101**(2), 742–755 (2011)
67. Michael, S.K., Constantinou, C.: Performance evaluation of seismically isolated buildings near active faults. Earthq. Eng. Struct. Dyn. **51**(5), 1017–1037 (2022)
68. Somerville, P., Graves, R.: Conditions that give rise to unusually large long period ground motions. In: The Structural Design of Tall Buildings, vol. 2(3), pp. 211–232 (1993)
69. Somerville, G.P.: Near Fault Directivity Pulse Model: Development and Application in Scenario and Probabilistic Maps of Pulse Parameters in the San Francisco Bay Area. URS Greiner Woodward Clyde (2004)
70. Somerville, P.G.: Magnitude scaling of the near fault rupture directivity pulse. Phys. Earth Planet. Inter. **137**, 201–212 (2003)
71. Somerville, P.G., Smith, N.F., Graves, R.W., Abrahamson, N.A.: Modification of empirical strong ground motion attenuation relations to include the amplitude and duration effects of rupture directivity. Seismol. Res. Lett. **68**(1), 199–222 (1997)
72. Somerville, P.G., Graves, R.W.: Conditions that gives rise to unusually large long period ground motions. In: The Structural Design of Tall a Buildings, vol. 2, pp. 211–232 (1993)
73. Spudich, P., Bayless, J., Baker, J., Chiou, B.S.J., Rowshandel, B., Shahi, S., Somerville, P.: Final Report of the NGA-West2 Directivity Working Group (2013)
74. TBDY-2018: Türkiye Bina Deprem Yönetmeliği. https://www.resmigazete.gov.tr/eskiler/2018/03/20180318M1.pdf (2018)
75. Tothong, P., Cornell, C.A., Baker, J.W.: Explicit directivity-pulse inclusion in probabilistic seismic hazard analysis. Earthq. Spectra **23**(4), 867–889 (2007)
76. Vassiliou, M.F., Tsiabos, A., Stojabinovic, B.: Dynamics of inelastic base-isolated structures subjected to analytical pulse ground motions. Earthq. Eng. Struct. Dyn. (2013)
77. Veletsos, A.S., Newmark, N.M.: Response spectra for single-degree-of-freedom elastic and inelastic systems. Report No. RTD-TDR-63-3096, Vol. III, Air Force Weapons Lab., Albuquerque, N. Mexico (1964)
78. Woessner, J., et al.: The 2013 European seismic hazard model: key components and results. Bull. Earthq. Eng. **13**(12), 3553–3596 (2015). https://doi.org/10.1007/s10518-015-9795-1
79. Xinle, L., Huijuan, D., Xi, Z.: Engineering characteristics of near fault vertical ground motion and their effect on the seismic response of bridges. Earthq. Eng. Eng. Vib. **6**(4), 345–350 (2007)
80. Bao, Y., Becker, T.C., Hamaguchi, H.: Failure of double friction pendulum bearings under pulse-type motions. Eng. Struct. Dyn. **46**, 715–732 (2017)
81. Tang, Y., Zhang, J.: Response spectrum-oriented pulse identification and magnitude scaling of forward directivity pulses in near-fault ground motions. Soil Dyn. Earthq. Eng. **31**, 59–76 (2011)
82. Akkar, S., Moghimi, S., Arıcı, Y.: A study on major seismological and fault-site parameters affecting near-fault directivity ground-motion demands for strike-slip faulting for their possible inclusion in seismic design codes. Soil Dyn. Earthq. Eng. **104**(2018), 88–105 (2018)

Origin of Seismicity in Italy as a Clue for Seismic Hazard

Carlo Doglioni[1,2](✉)

[1] Istituto Nazionale di Geofisica e Vulcanologia, Roma, Italy
carlo.doglioni@ingv.it
[2] Sapienza University, 00185 Roma, Italy

Abstract. Italian seismicity is generated by the ongoing subduction of the European lithosphere beneath the Alps, and the Adriatic lithosphere beneath the Apennines. The two belts are extremely different due to their opposite polarity relative to the inferred underlying 'eastward' mantle flow. Contractional tectonics is concentrated in low topography areas, whereas extensional tectonics and the larger magnitude seismicity due to normal faulting is preferentially located along the Apennines ridge, where the brittle crustal layer is thicker and the lithostatic load is maximum. Seismicity is the result of dissipation of energy along passive faults but stored mostly in crustal volumes located in the hangingwall of the faults. The 2–5 mm/yr deformation in all Italian tectonic settings prevents the occurrence of great earthquakes (Mw 8) that rather occur in other areas of the world where deformation rates are at least one order of magnitude faster. The maximum event so far recorded in Italy is Mw 7.3, 1693 southeast Sicily. InSAR data nowadays provide a precise definition of the epicentral area of an earthquake, which can be several hundred km^2. The epicentral area is defined as the 'active' domain where the hangingwall is moving along the fault and it is contemporaneously crossed by the seismic waves radiated by the fault plane due to the friction in it. Within the active domain occur the strongest coseismic shaking, both vertical and horizontal. The vertical coseismic motion allows the horizontal shaking to be much more effective.

Keywords: Italian geodynamics · Epicentral areas · Active domain · Passive domain · Vertical motion · Graviquakes · Elastoquakes

1 Geodynamic Setting

Italian geodynamics is shaped by two subduction zones, i.e., the Alps and the Apennines. Their vitality generates the seismicity repeatedly devastating the nation. The two belts pertain to two very different subduction styles. In the Alps the subduction hinge of the downgoing European plate converges relative to the upper Adriatic plate, whereas in the Apennines the Adriatic plate subducts beneath the European plate and the subduction hinge diverges relative to the upper plate (Carminati and Doglioni 2012). This different kinematics determine mostly contraction in the Alps and widely distributed extension in the Apennines (Fig. 1), where the contracting accretionary prism is confined in the

G. P. Cimellaro (Ed.): WCSI 2022, LNCE 309, pp. 153–165, 2023.
https://doi.org/10.1007/978-3-031-21187-4_10

eastern external areas (Po Basin, Western Adriatic, Ionian Sea and Sicily). Contractional seismicity is mostly confined in the low topography frontal ranges of the Alps and Apennines due to the lower lithostatic load, favoring larger differential stress. Unlike the Alps, the extensional seismicity is rather widespread within the Apennines, and it increases in magnitude in highly elevated areas where the brittle thickness is larger and the lithostatic load is maximum (Carminati et al. 2004). There are in average 15.000–20.000 recorded earthquakes per year in Italy, most of them felt only by the INGV seismic network. Between 1900 and 2000 AD, 248 events occurred with Mw $\geq$ 5 and they listed in the INGV Italian Parametric Earthquake Catalogue https://emidius.mi.ingv.it/CPTI15-DBMI15/query_eq/ (Rovida et al. 2022). The extensional area in the hanging wall of the Apennines subduction is where most of the earthquakes occur (Fig. 2). Moving from the Tyrrhenian Sea eastward in the belt, the crustal thickness increases, and the brittle-ductile transition (BDT) deepens. Since the seismogenic layer is controlled by the brittle thickness, and the length of the volume increases with the depth of the BDT (in average 3 times the brittle thickness), the magnitude grows moving from the western side of the Apennines to their main ridge where the brittle crust is thicker (Fig. 3). The volumes adjacent to faults accumulate the energy that will eventually be released by earthquakes, that is elastic energy in strike-slip and contractional settings, whereas is gravitational in extensional tectonic environments. That's why we can distinguish earthquakes into elastoquakes and graviquakes (Doglioni et al. 2015). Based on the involved volumes, we can infer the maximum magnitude that every area in Italy can experience (Fig. 4). Seismic sequence last longer in extensional tectonic settings because the volumes are gravitationally enhanced (Fig. 5). Therefore, also the Gutenberg-Richter b-value differs, as well as the Omori p-value, in the different tectonic settings, as a function of whether the volume moves in favor or against gravity (Fig. 6).

2 Seismic Hazard Remarks

Since the development of InSAR technique and the wealth of new data, the pre- and post-earthquake satellite recording precisely depict the fingerprint of the epicentral areas affected by the coseismic deformation. Moreover, it is noteworthy that the stronger shaking and the highest macroseismic intensity coincide with the area depicted by the SAR images, where the vertical component of the PGA and PGV have been manifested (Fig. 7). During the last decades, the earthquake in Italy have shown PGA much stronger than those expected from the national codes for new constructions. In other words, we should focus our efforts in improving seismic hazard evaluation as we were inside the epicentral area, where the vertical coseismic movements are much larger. When the ground coseismically collapses in the epicentral area of a normal fault-related earthquake, the load of a building decreases and the friction between bricks is attenuated, facilitating the shear imposed by the horizontal shaking, hence producing larger damages (Fig. 8). This area/volume can be defined as the 'active domain', where there occur both the main tectonic movement and the volume is contemporaneously crossed by the seismic waves radiated from the fault surfaces (Fig. 9). It is there, in the future epicentral areas where we should concentrate the seismic hazard assessment; in fact, besides site amplification effects that may occur also outside the epicentral area, any region undergoing tectonic

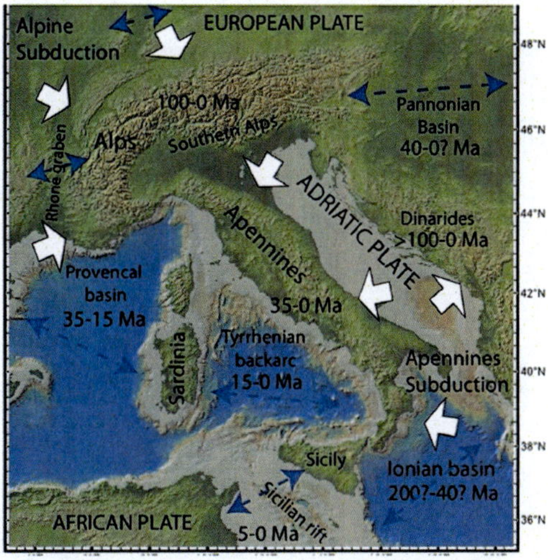

Fig. 1. Italian geodynamics and related seismicity are featured by the subduction of the European plate beneath the Adriatic plate, generating the Alps, and the westward subduction of the Adriatic plate beneath Europe producing the Apennines and the Tyrrhenian backarc basin due to the slab retreat.

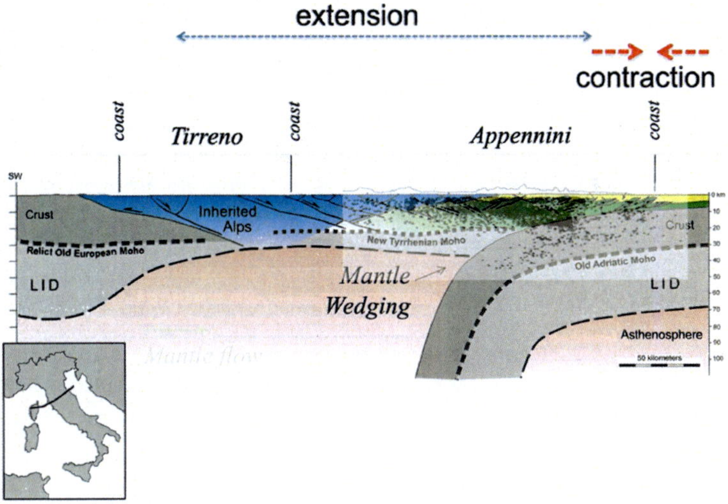

Fig. 2. Cross-section of the central-northern Apennines showing the westward subduction of the Adriatic plate and the distribution of seismicity (thin black dots).

deformation is prone to seismic events. Their magnitude depends on the volume (Fig. 10), whereas their frequency is related to the deformation rate. With a given magnitude, again apart the site effects, the resulting PGA increases as the hypocenter is shallower (Fig. 11).

Seismic hazard maps should be made considering that each area could be an active domain where the shaking will be much higher than the surrounding passive domain.

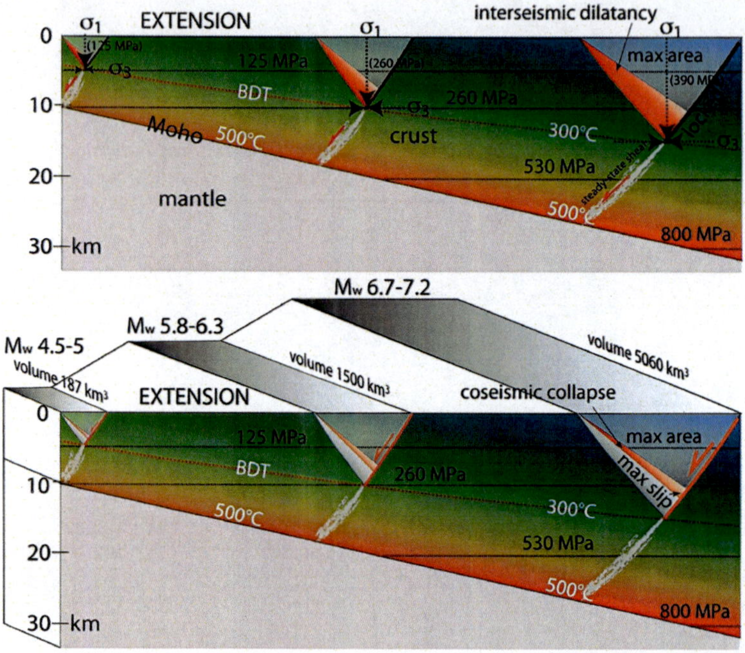

Fig. 3. In extensional tectonic settings, the ratio between the seismogenic layer depth and the hanging wall length appears to be, in average, 3 ± 1. As the thickness increases, the volume enlarges, the fault is longer, and the hanging wall collapse determines a greater fault slip and a consequent higher earthquake magnitude. Note that doubling the seismogenic layer from 5 to 10 km, the volume increases about 1 order of magnitude, as the maximum earthquake size, from M5 to M6. BDT, brittle-ductile transition. The red wedges in the section above suggest the inferred dilated crustal volume permeated by thousands of millimetric microfractures. Therefore, the deeper the BDT, the bigger the volume and magnitude of the earthquake, and the larger the area affected by coseismic subsidence. This explains why seismicity increases in terms of number of earthquakes and their magnitude moving from the Tyrrhenian Sea to the Apennines main ridge (after Bignami et al. 2020; Petricca et al. 2021).

Moreover, new computations of the seismic hazard should include the time dependance of earthquakes (Akinci et al. 2018), the rheological parameters of the lithosphere (Doglioni et al. 2011; Riguzzi et al. 2012, 2013; Petricca et al. 2015), the deformation rates recorded by space geodesy (e.g., Pezzo et al. 2020), the dimension of areas where the vertical component will be concentrated in the active volume (Bignami et al. 2019; Petricca et al. 2021) and seriously include the incompleteness of the catalogs (Scholz 2019).

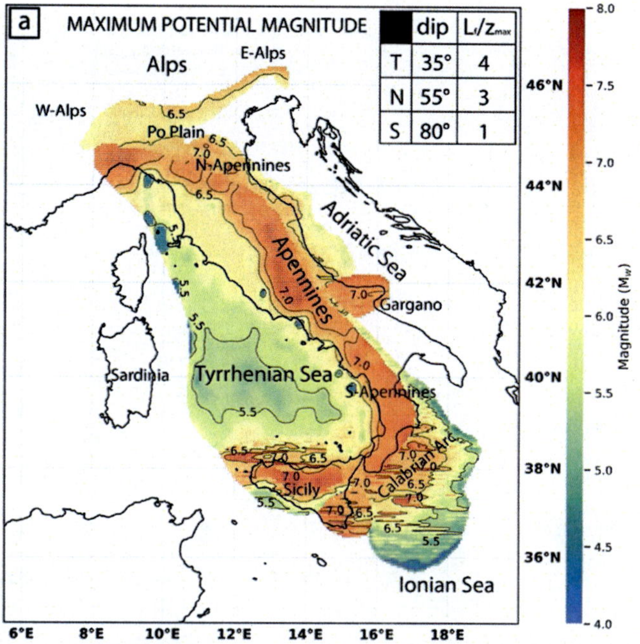

Fig. 4. Map of the maximum potential magnitude (M) predicted for the Italian territory due to normal, thrust and strike-slip faulting (after Petricca et al. 2022). Results assume the following fault dip α and C = Lf/zmax (i.e., fault length—maximum faulting depth ratio): α = 35° and C = 4 (thrust faults); α = 55° and C = 3 (normal faults); α = 80° and C = 1 (strike-slip faults).

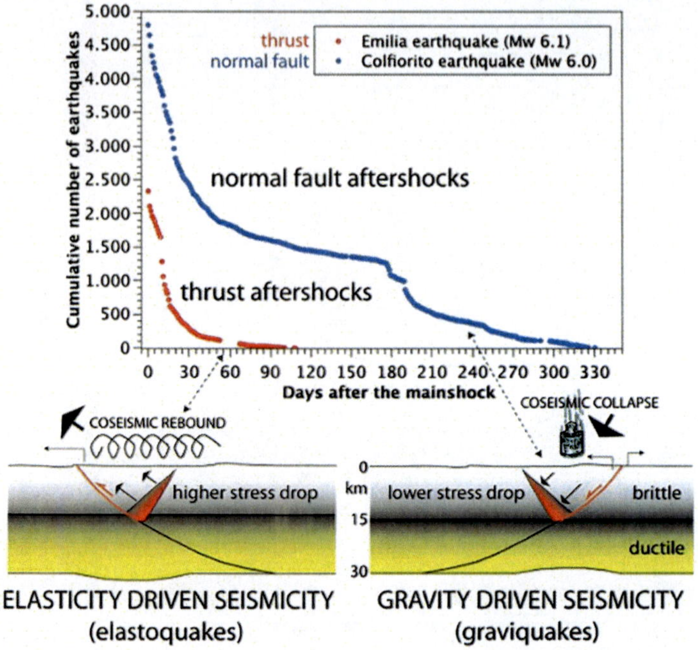

Fig. 5. Aftershocks are more numerous and last longer in extensional tectonic settings because the volumes move in favor of gravity. The contrary occurs in contractional settings because the volume must win gravity (modified after Valerio et al. 2017).

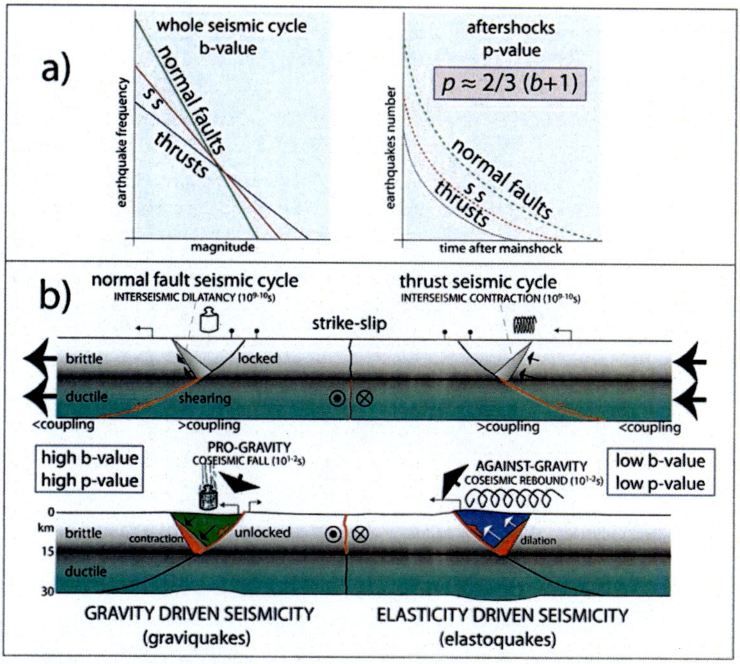

Fig. 6. (a) Left panel, the b-value of the Gutenberg–Richter law differs as a function of the tectonic setting, being higher in extensional tectonic settings. Therefore, normal faults have a steeper alignment, which means more low magnitude events and lower maximum magnitude earthquakes with respect to strike-slip (SS) and thrust faults; right panel, being the p-value of the Omori law correlated to the b-value, it shows different aftershocks evolution in time depending on the tectonic settings. On the base of theoretical considerations and statistical analysis of one hundred seismic sequences, the most reliable relationship between them reads $p \approx 2/3\ (b + 1)$. Larger p-values mean that seismic dynamics moves more rapidly towards stability, which also implies relatively higher probability of secondary ruptures and more numerous aftershocks than the average reference. (b) A geological interpretation of different scaling behaviors can be given in the light of different contributions to energy balance in the long-term rock volumes mobilization. In extensional tectonic settings the seismogenic volumes are smaller, producing lower maximum magnitudes with respect to the other tectonic settings. Contractional settings require more energy to move against gravity and they may involve much larger volumes before reaching the threshold to slip, i.e., producing much larger magnitude events and a lower b-value and p-value (after Zaccagnino et al. 2022).

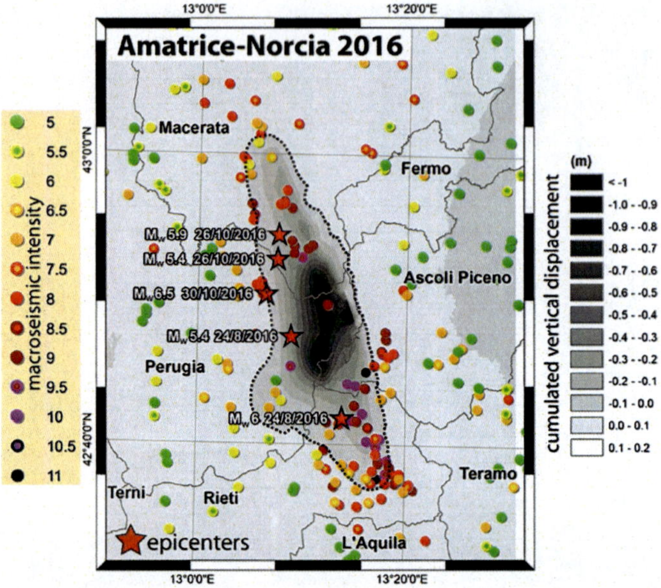

Fig. 7. Coseismic subsidence computed through DInSAR data compared to the macroseismic intensity of the Mw 6 and 6.5 Amatrice-Norcia August 24th and October 30th 2016 Central Italy seismic sequence. Most of the damage occurred in the area that underwent vertical deformation, apart from rare scattered local amplification sites outside the elliptic area (after Al Shawa et al. 2021).

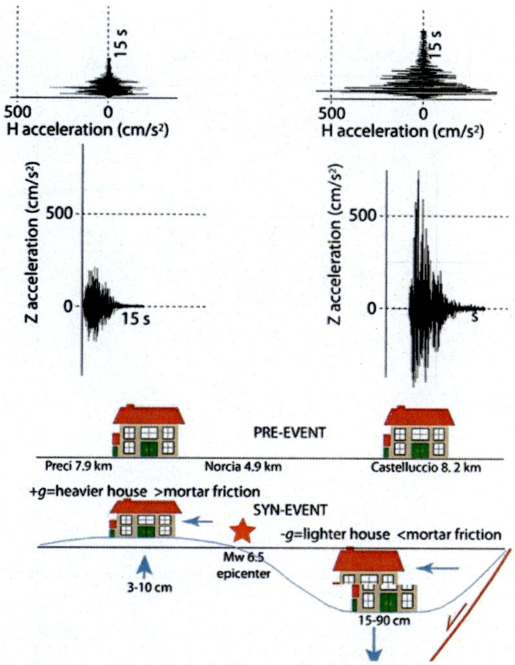

Fig. 8. Higher panels: Ground motion components occurring during the Mw 6.5 October 30th 2016 Norcia mainshock outside (coseismic uplift) and inside the epicentral area (coseismic subsidence). Lower panel: two cases in which the coseismic motion is upward (left) or downward (right) during the increase of horizontal components (after Liberatore et al. 2019). Within the coseismically subsiding area there have been the strongest shaking and the horizontal components acted while the volume and the house were falling down, hence with decreased load acting on the building.

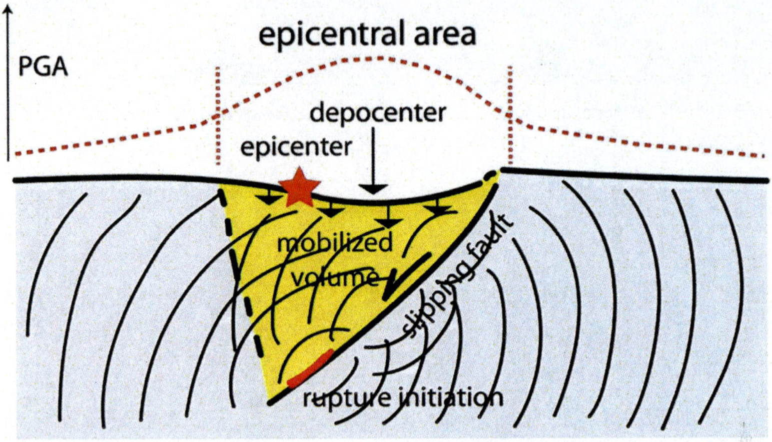

Fig. 9. Cartoon illustrating the overlap of the hanging wall volume undergoing coseismic subsidence and contemporaneously crossed by the seismic waves sourced by the normal fault slip. In the 'active' (yellow) volume where the two phenomena overlap, the peak ground acceleration (PGA) is maximum, and it may be referred to as the epicentral area. Outside the mobilized volume, seismic waves propagate into a 'passive', i.e., static volume. This may explain the higher macroseismic intensity at the surface of the active volume (after Al Shawa et al. 2021).

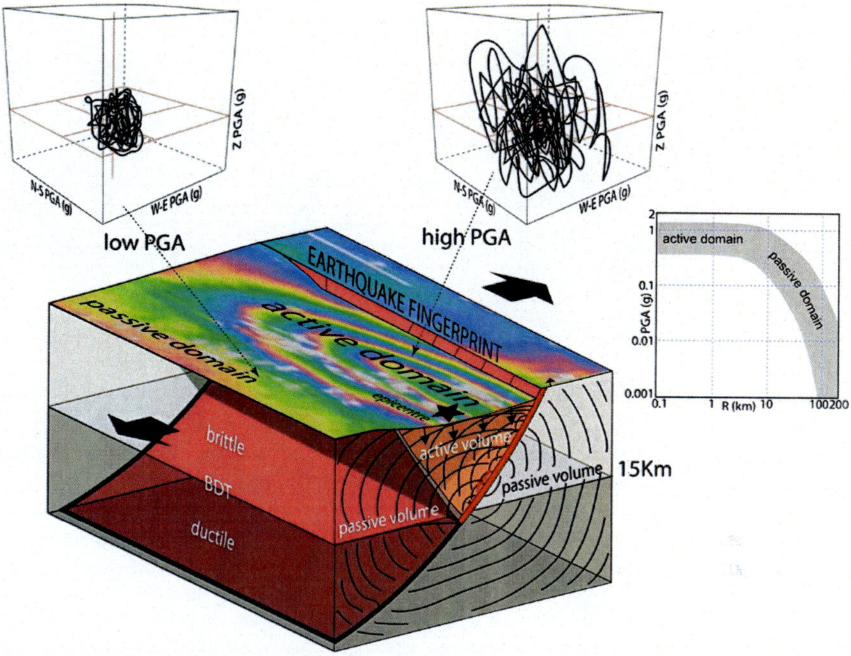

Fig. 10. Sketch illustrating the model. The coseismic ground deformation detected by InSAR data represents the fingerprint of the epicentral area and is located above the volume mobilized in the fault hanging wall in case of extensional setting. The area increases with magnitude. The earthquake epicenter is misplaced relative to the depocenter of the maximum coseismic subsidence where the strongest shaking is recorded. Outside the elongated epicentral area, even if relatively close to the epicenter, the vertical coseismic deformation is drastically reduced as well as the shaking; see the hypothetic trajectory of the acceleration vector inside and outside the epicentral area. Right: diagram of the PGA (g) average data in the gray area from the literature versus distance (R) inferring the active and passive domains (after Petricca et al. 2021).

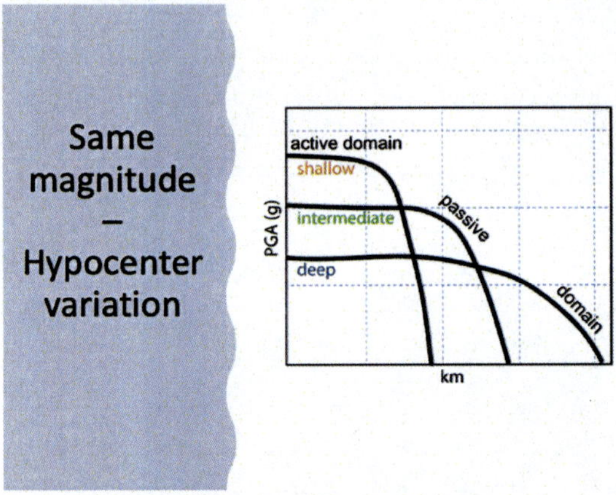

Fig. 11. The highest PGA occur in the epicentral area or active domain, decreasing outside the epicentral area or passive domain. However, assuming the same magnitude, the PGA increases as the rupture occurs at shallower crustal level. Moving deeper, the same energy is partly dissipated by the rocks anelasticity, and the shaking decreases. Although weaker, the active domain becomes wider due to the larger cone of influence.

References

1. Akinci, A., Moschetti, M.P., Taroni, M.: Ensemble smoothed seismicity models for the new Italian probabilistic seismic hazard map. Seismol. Res. Lett. **89**(4), 1277–1287 (2018)
2. Al Shawa, O., Atzori, S., Doglioni, C., Liberatore, D., Sorrentino, L., Tertulliani, A.: Coseismic vertical ground deformation vs. intensity measures: examples from the Apennines. Eng. Geol. **293**, 106323 (2021). https://doi.org/10.1016/j.enggeo.2021.106323
3. Bignami C., Valerio E., Carminati E., Doglioni C., Tizzani P.: Volume unbalance on the 2016 Amatrice—Norcia (central Italy) seismic sequence and insights on normal fault earthquake mechanism. Sci. Reports **9**, 4250 (2019). https://doi.org/10.1038/s41598-019-40958-z
4. Bignami C., Valerio E., Carminati E., Doglioni C., Petricca P., Tizzani P., Lanari R.: Are normal fault earthquakes due to elastic rebound or gravitational collapse? Annals Geophys. **63**, 2, SE213 (2020). https://doi.org/10.4401/ag-8455
5. Carminati, E., Doglioni, C., Barba, S.: Reverse migration of seismicity on thrusts and normal faults. Earth-Sci. Rev. **65**, 195–222 (2004)
6. Carminati, E., Doglioni, C.: Alps vs. Apennines: the paradigm of a tectonically asymmetric Earth. Earth Sci. Rev. **112**, 67–96 (2012). https://doi.org/10.1016/j.earscirev.2012.02.004
7. Doglioni, C., Barba, S., Carminati, E., Riguzzi, F.: Role of the brittle-ductile transition on fault activation. Phys. Earth Planet. Int. **184**, 160–171 (2011)
8. Doglioni, C., Carminati, E., Petricca, P., Riguzzi, F.: Normal fault earthquakes or graviquakes. Sci. Rep. **5**, 12110 (2015). https://doi.org/10.1038/srep12110
9. Liberatore, D., Doglioni, C., Al Shawa, O., Atzori, S., Sorrentino, L.: Effects of coseismic ground vertical motion on masonry constructions damage during the 2016 Amatrice-Norcia (Central Italy) earthquakes. Soil Dyn. Earthq. Eng. **120**(2019), 423–435 (2019). https://doi.org/10.1016/j.soildyn.2019.02.015
10. Petricca, P., Barba, S., Carminati, E., Doglioni, C., Riguzzi, F.: Graviquakes in Italy. Tectonophysics **656**, 202–214 (2015). https://doi.org/10.1016/j.tecto.2015.07.001

11. Petricca, P., Bignami, C., Doglioni, C.: The epicentral fingerprint of earthquakes marks the activated volume. Earth Sci. Rev. **218**, 103667 (2021). https://doi.org/10.1016/j.earscirev.2021.103667
12. Petricca, P., Carminati, E., Doglioni, C.: Estimation of the maximum earthquakes magnitude based on potential brittle volume and strain rate: the Italy test case. Tectonophysics (2022). https://doi.org/10.1016/j.tecto.2022.229405
13. Pezzo, G., Petracchini, L., Devoti, R., Maffucci, R., Anderlini, L., Antoncecchi, I., Billi, A., Carminati, E., Ciccone, F., Cuffaro, M., Livani, M., Palano, M., Petricca, P., Pietrantonio, G., Riguzzi, F., Rossi, G., Sparacino, F., Doglioni, C.: Active fold-thrust belt to foreland transition in northern Adria, Italy, tracked by seismic reflection profiles and GPS offshore data. Tectonics **39**, e2020TC006425 (2020). https://doi.org/10.1029/2020TC006425
14. Riguzzi, F., Crespi, M., Devoti, R., Doglioni, C., Pietrantonio, G., Pisani, A.R.: Geodetic strain rate and earthquake size: new clues for seismic hazard studies. Phys. Earth Planet. Inter. **206–207**, 67–75 (2012)
15. Riguzzi, F., Crespi, M., Devoti, R., Doglioni, C., Pietrantonio, G., Pisani, A.R.: Strain rate relaxation of normal and thrust faults in Italy. Geophys. J. Int. (2013). https://doi.org/10.1093/gji/ggt304
16. Rovida, A., Locati, M., Camassi, R., Lolli, B., Gasperini, P., Antonucci, A.: Italian parametric earthquake catalogue (CPTI15), version 4.0. Istituto Nazionale di Geofisica e Vulcanologia (INGV) (2022). https://doi.org/10.13127/CPTI/CPTI15.4
17. Scholz, C.H.: The mechanics of earthquakes and faulting. Cambridge university press (2019)
18. Valerio, E., Tizzani, P., Carminati, E., Doglioni, C.: Longer aftershocks duration in extensional tectonic settings. Sci. Rep. **7**, 16403 (2017). https://doi.org/10.1038/s41598-017-14550-2
19. Zaccagnino D., Telesca L., Doglioni C.: Scaling properties of seismicity and faulting. Earth Planet. Sci. Lett. **584**, 15 April 2022, 117511 (2022). https://doi.org/10.1016/j.epsl.2022.117511

New Isolation and Energy Dissipation Devices

Seismic Isolation Design for Achieving Resilient Structures

Anoop S. Mokha(✉), Victor Zayas, and Stanley Low

Earthquake Protection Systems, Vallejo, CA, USA
{anoop,victor}@earthquakeprotection.com,
stanley@earthquuakeprotection.com

Abstract. Resiliency is an important consideration while designing critical buildings, bridges, and industrial structures in earthquake prone regions of the world. This is necessary for minimizing post-earthquake disruption to society. Major earthquakes that have occurred every year in the world are a constant reminder that critical structures must remain operational post-earthquake, so that community needs are met. Hospitals need to remain operational in order to treat injured people and save lives. Bridges classified as lifeline structures also need to remain functional so that rescue and recovery operations can be performed. Gas, water and electric facilities also need to remain functional post-earthquake so that basic services are not interrupted. Code provisions (ductility based) for seismic design of structures all over the world have focused primarily on achieving "Collapse Prevention" within acceptable limits, at the expense of inflicting damage to structural, non-structural, architectural elements, and contents. After a major earthquake this results in loss of use and function, as observed in recent Japan, New Zealand and Chile earthquakes. One of the approach to achieve resiliency is through continued functionality design objectives for minimizing damage in structures by absorbing seismic displacement in isolation bearings, maintaining an elastic structure, and minimizing in-structure accelerations and drifts. The paper presents the current pitfalls of code compliant seismic isolation provisions, and a way forward for the right approach for implementing seismic isolation following the "Continued-Functionality Standard (CFS)" to meet the owner's expectation. Examples of resilient isolated structures worldwide designed and constructed at lower costs than conventional ductile structures are presented. Real earthquake performance of CFS designed seismically isolated structures are also presented.

Keywords: Resiliency · Seismic performance · Seismic damage · Continued functionality · Seismic isolation · Collapse prevention

G. P. Cimellaro (Ed.): WCSI 2022, LNCE 309, pp. 169–178, 2023.
https://doi.org/10.1007/978-3-031-21187-4_11

1 Introduction

Before the first development of seismic design codes, earthquake forces were considered an act of God and thus largely ignored. The first recording of an earthquake ground acceleration occurred during the 1940 El Centro earthquake led to the development of seismic design codes that are now practiced all over the world. Engineers thought that designing structures for realistic earthquake forces was not cost-effective. Thus, they adopted a "ductility life-safety" based code approach that allowed structures to get damaged during an earthquake, but largely intended to avoid "collapse" so that human lives would be protected. Over the years this approach was refined with new construction materials and methods. However, till date the basic code objective for earthquake design all over the world has remain unchanged; i.e. "Collapse Prevention". Interestingly, even with new code provisions for seismic protection technologies like seismic isolation and dampers, the basic code objective (Collapse Prevention) remains unchanged. The 2010 Mag. 8.8 earthquake in Chile and the Mag. 6.3 earthquake in New Zealand have demonstrated that Engineers have accomplished the basic objective of "Collapse Prevention" as only 22 buildings collapsed amongst thousands. Earthquake damage of over US$60 billion has nevertheless, left communities devastated.

The Building Code objectives should now change from "Life-Safety" to "Damage Prevention". The world of automotive industry has already started shifting from "Occupant Safety" to "Collision Prevention" with the integration of smart sensors and advanced braking technology. In the US both the Building Code requirements and Building Industry practices are shifting towards resiliency in design and construction to minimize earthquake damage. The American Society of Civil Engineers (ASCE) latest Building Code Standard ASCE 7-22 Section 1.3.3 on Functionality requires Essential Facilities to remain functional and operational after a design level earthquake [1].

2 Code Design of Seismically Isolated Structures

Implementing appropriately seismic isolation technology in building design provides the highest level of safety and damage prevention compared to any available alternate construction technology.

With over 5000 applications which use seismic isolation there are less than 50 isolated structures that have been designed to meet the owner's expectation. The prescriptive structure types specified in building codes, ASCE 7 Chapter 12, were developed for collapse avoidance. Isolated structures designed according to Euro Code 8 [2], with the isolators fabricated according to EN15129 [3], appear to be some of the most dangerous new structures worldwide. The Euro Code 8 required isolator displacement capacity has a 10% probability of being exceeded in 50 years. About 10 earthquakes per year cause ground shaking that exceed this design level. EN15129 prohibits safety retainer rings for pendulum isolators, which essentially will lead to isolator/structure collapse for these stronger earthquakes. Rubber isolators made with poor materials and fabrication are even more dangerous.

The 2011 magnitude 6 Christchurch New Zealand earthquake resulted in damage to the Women's Hospital exceeding 30% of the hospital replacement costs. The hospital was constructed with lead rubber isolators designed for code compliance. The researchers at Canterbury University reported conclusively that "the lead-rubber isolators did not displace", and "the hospital responded essentially as if it was fixed base" [4]. Various degrees of damage have been reported for code designed seismically isolated structures after earthquakes in Japan, New Zealand and United States with resulting loss of facility use and significant repair costs [4, 5]. It is interesting to note that ASCE 7 Chapter 17 allows isolated structures to be constructed with lower lateral strengths as compared to non-isolated structures. When the isolators do not displace during an earthquake, a Chapter 17 compliant isolated structures suffers more damage than a stronger non-isolated ductile structures.

Seismically isolated structures designed for minimum code compliance can not only have damage exceeding 70% of replacement costs, but they also do not comply with the probability of structure failure/collapse limits listed in ASCE 7 Table 1.3-2.

3 Continued Functionality Standard for Resilient Structures

Minimizing seismic damage and maintaining continued functionality and operation has always been desired performance objective of engineers, architects, owners, developers, and planners in earthquake prone regions all over the world. Nowhere this intent is more relevant than in the design of hospitals. Every year many people die when hospitals are not able to function after an earthquake and provide care for the injured persons.

The World Health Organization, and Pan American Health Organization, have published directives that a new hospital needs to be constructed "with a level of protection that best guarantees its capacity to continue functioning" at its "maximum capacity after earthquakes" [6].

The US Resiliency Council [7], REDi Resiliency Rating [8], and the Federal Emergency Management Agency [9] provide guidelines for how to minimize earthquake damage and loss of use. The design criteria specified for the Platinum Resiliency Rating satisfies the Health Ministry requirements for hospitals. Structural engineers have been using these methodologies to advise their clients regarding the expected damage and loss of use for hospitals and other essential structures.

The design criteria to achieve "Continued Functionality" are:

- Elastic structure design, with $R = 1$ for the Design Basis Earthquake
- Seismic story lateral drifts limited to less than 0.003 times the story heights for the Design Basis Earthquake
- Limit the median floor spectra accelerations to less than 0.3 g for each occupied floor, over the frequency range of 0.2–20 Hz, for the Design Basis Earthquake.

These design criteria apply equally for buildings with or without seismic isolators.

4 Achieving Continued Functionality with Seismic Isolation

Installing seismic isolators that are designed for code compliance do not satisfy Continued Functionality design criteria. More importantly they do not satisfy the basic code objective of limiting probability of collapse specified for various structure categories. Building codes have endeavored for decades to eliminate dangerous "soft stories." Isolated structures are extreme soft stories, which can be very dangerous. There are no ASTM or NIST standards for generic materials and fabrication of generic isolators.

Isolated structures designed for minimum code compliance, that install isolators fabricated by un-qualified fabricators, can have FEMA P695 [10] calculated collapse risks 10 times greater than traditional non- isolated structures [11, 12].

The "Continued Functionality Standard" specifies the means and methods to achieve safe isolated structures that retain functionality after earthquakes [13]. Seismic isolator properties and capacities are specified to comply with the "Basic Requirements" of ASCE 7-16, Chapter 1. The ASCE 7 basic "Functionality" requirements specify that "Essential Facilities" have a "reasonable probability to have adequate structural strength and stiffness" that "would not prevent function of the facility immediately following" the design level earthquake. ASCE 7 also specifies "Target Reliability" that limits the "Probability of Failure" to 2.5% for the "Structural Stability" of primary structural components in essential facilities. Compliance with ASCE 7 requires compliance with these "Basic Requirements".

ASCE 7 Sections 1.3.1.1 and 1.3.1.2 require that design loads not exceed the capacity limits specified by a materials standard, such that when applied together with ASCE 7, satisfy the requirements for structure reliability and functionality. The standard specifies the required isolator shear strength and displacement capacities, such that when applied together with the ASCE 7-16, has a reasonable probability of complying with the ASCE 7-16 requirements for reliability and functionality. This standard represent best practices for seismic isolator engineering and manufacture as evolved over 32 years, and validated through isolator performance during extreme MCER events, shake table tests, FEMA P695 and P58 structural analyses, and applications of isolators to more than one hundred million square feet of isolated buildings, bridges, and industrial facilities. Constructing isolated structures without specifying an adequate isolator standard creates serous life safety hazards, and violates ASCE 7 Basic Requirements for material standards, reliability, and functionality.

Specifying the CFS in the construction specifications for new hospitals is the most reliable and economical means of achieving the mandated post-earthquake functionality. Compliance with the criteria in this isolator standard has proven straight forward and cost effective for the construction of 100 million of square feet of important facilities. All hospitals and Structures shown in Figs. 1, 2, 3 and 4 herein were designed for Continued Functionality using Triple Pendulum seismic isolators specially configured to minimize damage to the architectural components, equipment, and contents of the facilities, as well as the structures.

Example seismic damage calculation performed on the Goztepe Hospital in Tukey are shown in Table 1 for four types of structures; Triple Pendulum isolated structure, Single Pendulum isolated structure, Code Compliant isolated structure, and conventional

Fig. 1. Apple headquarters, California, USA.

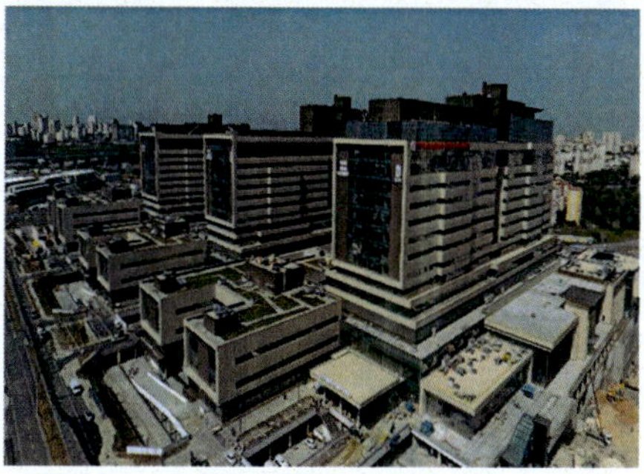

Fig. 2. Ikitelli Hospital, Istanbul, Turkey—10 M. Sq. Ft.

Fig. 3. Loma Linda Hospital, California, USA.

fixed based structure. For the design level earthquake the Triple Pendulum isolated structure results in seismic damage of $< 2\%$, while the Code Compliant isolated structure results in 4 times more seismic damage. Of interest is to note that the ration of damage between Triple and Code Compliant isolated structure increases with decreasing earthquake intensity. Even though the seismic damage calculations are not precise and accurate, they are a valuable tool that allows owners/engineers to make risk and value based decisions, since the relative expected damage between the four structural systems would be of the same order as shown in Table 1.

Fig. 4. Goztepe Hospital, Istanbul, Turkey.

Table 1. FEMA P58 seismic damage (% of replacement cost) estimate for Goztepe Hospital)

Probability of being exceeded in 50 years: EQ recurrence interval	Triple Pendulum bearings (%)	Single Pendulum bearings, same building (%)	Code designed bearings, same building (%)	Same building without seismic isolation bearings (%)
90%: 22 year return	0.08	0.40	1.32	2.89
50%: 75 year return	0.61	1.90	5.10	12.57
10%: 475 year return (DBE)	1.89	4.42	8.18	83.18
2%: 2475 year return (MCE)	3.11	5.17	17.48	98.35

5 Earthquake Performance of Isolated Structures Designed with Continued Functionality Criteria

Ecuador's Bahia Bridge (Fig. 7) maintained full functionality during Ecuador's 2016 M7.8 earthquake and aftershocks. The isolators experienced seismic shear demands more than twice the code maximum considered earthquake, yet the bridge retained full functionality thanks to the isolator factors of safety. In the 24 h after the earthquake, 15,000 vehicles crossed the bridge to evacuate thousands of injured persons from the city, and the patients that were evacuated from the hospital [14]. The Bahia Bridge was used to evacuate hundreds of severely injured persons, and thousands of residents, from the devastated cities of Bahia, Chone, Portoviejo, and Manta, thus saving hundreds of lives.

Texas Instruments' Philippines plant has fragile high precision manufacturing equipment that produce miniscule electronic micro-processors used in medical devices. This Philippines plant produces 60% of Texas Instruments' annual profits. The Triple Pendulum isolators allowed the fragile manufacturing equipment and facility to retain 100% functionality within minutes after the 2019 magnitude 6.1 earthquake [15].

Two large liquid natural gas tanks in Chile maintained full functionality after Chile's 2010 Magnitude 8.8 Earthquake, and a Magnitude 7.1 aftershock epi-centered 3 km from the tanks. For these large liquid natural gas tanks, the leaking of trillions of liters of gas from the concrete tanks would have caused a fire, explosion and major public catastrophe. These tanks supply the gas used to generate most of Chile's electricity, which is essential to hospital functionality.

Acapulco Emergency Response Center (AERC) in Acapulco is a critical communication facility for coordinating emergency responses to natural and man-made disasters. On September 7, 2021 a Magnitude 7.1 earthquake struck the region of Acapulco in Mexico. The epicenter was about 10 km from the town of Acapulco. The Acapulco Emergency Response Center building was approximately 9 km from the epicenter. As expected, the 156 Triple Pendulum isolators protected the AERC building, and maintained uninterrupted function and operation of the response center. Not a single glass panel broke during the Mag. 7.1 earthquake.

Most owners are surprised when relatively new buildings suffer significant earthquake damage. Lawsuits resulting from the 1994 California Northridge earthquake held the structural design engineers responsible for damage losses if they had not warned the owners, and offered options to mitigate the damage. Christchurch New Zealand's 2011 magnitude 6.3 earthquake resulted in 2 building collapses, yet 70% of the downtown buildings were not economically repairable, and thus demolished. Damage to new buildings occurs because the low seismic design strengths specified in structure design standards are sufficient only to reduce the risk of collapse. The CFS provides the construction standard necessary for reducing earthquake damage to acceptable levels.

6 Conclusions

Constructing for post-earthquake functionality according to the CFS is a proven, practical, and alternative to designing and constructing ductile structures for only collapse avoidance. See Figs. 5, 6 and 7.

Reliable post-earthquake functionality of essential facilities is a professional responsibility of structural engineers, a mandate of ASCE 7 Section 1.3.3, Functionality, and a matter of life and death for the public. Installing seismic isolators that are designed for code compliance do not satisfy Continued Functionality design criteria. More importantly they do not satisfy the basic target reliability of limiting collapse.

Structural engineers are society's guardians for minimizing earthquake deaths, loss of use, and economic losses. As a profession we need to go beyond providing designs for minimum cost for minimum code compliance, and protect our clients and society by minimizing earthquake damage and deaths. Implementing Continued Functionality designs with seismic isolation not only achieves owner's expectations, but also satisfies our professional and social responsibilities.

Fig. 5. Elazig Hospital was the nearest hospital to the epi-center of Turkey's 2020 M 6.7 earthquake. The hospital retained 100% functionality with Triple Pendulum seismic isolators.

Fig. 6. Acapulco Emergency Response Center remained fully functional and operational after the Mexico's 2021 M 7.1 earthquake with not a single broken glass. The building was 8 miles from epicenter.

Fig. 7. Ecuador's 2-km Bahia Bridge maintained functionality during Ecuador's magnitude 7.8 earthquake. All of the hospitals in this region were 100% losses, yet the bridge retained 100% functionality.

References

1. ASCE Standard 7 (2022 edition): Minimum Design Loads for Buildings and Other Structures. American Society of Civil Engineers, Washington D.C., USA
2. Eurocode 8: Design of structures for earthquake resistance. European Committee for Standardization (2004)
3. EN15129: Anti-seismic devices—European Standards (2011)
4. Kuang et al.: Christchurch women's hospital: performance analysis of the base isolated system during the series of Canterbury earthquakes 2011–2012. Report CF1297, University of Canterbury, Department of Mechanical Engineering, Christchurch, New Zealand
5. Japan Property Central: 30% of apartments with base isolation systems suffered damage in earthquake. JapanPropertyCentral.com, 30 January 2012 (2012)
6. WHO, World Health Organization: Access to Health Services a Challenge after Chile Earthquake. Media Center, February and March 2010
7. U.S. Resiliency Council: Rating Building Performance in Natural Disasters. California
8. ARUP Engineering Consultants, REDi Rating System: Resilience Based Earthquake Design Initiative for the Next Generation of Buildings, ARUP San Francisco
9. FEMA P58: Seismic Performance Assessments of Buildings. Federal Emergency Management Agency, Washington D.C (2012)
10. FEMA P695: Quantification of Building Seismic Performance Factors. Federal Emergency Management Agency, Washington D.C (2009)
11. Shao, M., Zayas: Member Capacity Factors for Seismic Isolators as Required to Limit Isolated Structure Collapse Risks to Within ASCE 7 Stipulated Structure Collapse Risk Limits. University of California Berkeley, UCB/SEMM-2017/02, Structural Engineering and Structural Mechanics Division
12. Kitayama, S., Constantinou, M.C.: Seismic performance assessment of seismically isolated buildings designed by the procedures of ASCE/SEI 7. MCEER 18-0004, Multidisciplinary Center for Earthquake Engineering Research, Buffalo, NY, USA (2018)
13. UCB CFS: Continued Functionality Standard. University of California Berkeley Civil Engineering, updated 2022 (2021)

14. Morales, Romo, Haro, Sinde, Aroca, Nikolaou, Diaz-Fanaz: A real life resiliency experiment: the Los Caras bridge of Ecuador. In: Proceedings Natural Hazards, Greece
15. Texas Instruments, Post Earthquake Inspection of the Texas Instruments Manufacturing Plant After the 2019 Magnitude 6.1 Philippines Earthquake, EPS Project Report

Mechanical Properties of Thick Rubber Bearings Used in Over-Track Buildings

Ying Zhou[1](✉), Zengde Zhang[1,2], and Michalis F. Vassiliou[2]

[1] College of Civil Engineering, Tongji University, Shanghai 200092, China
yingzhou@tongji.edu.cn
[2] Institute of Structural Engineering, ETH Zürich, 8093 Zürich, Switzerland

Abstract. As the construction of over-track buildings has increased, the need for reducing vertical vibrations caused by subways has emerged. Thick rubber bearings (TRBs) can provide both horizontal and vertical isolation. However, there is still uncertainty about the mechanical properties of TRBs which causes a difficulty in evaluating their performance. In this study, a full-scale TRB with a low first shape factor was designed, manufactured, and then tested under compressive, tensile, and shear loads. Its vertical and horizontal characteristics were identified through experiments. The effects of the vertical pressure, shear strain, and loading frequency on the mechanical behavior of the TRB were explored. Based on the experimentally determined properties, simplified multiple-degree-of-freedom systems were established to evaluate the performance of the TRB in terms of mitigating the vertical vibrations caused by the subway. Under a subway-induced excitation, the acceleration responses of the isolated system using TRBs were substantially reduced. Therefore, TRBs could be used to mitigate vertical vibrations and improve the seismic safety of over-track buildings.

Keywords: Thick rubber bearings · Vertical characteristics · Horizontal characteristics · Over-track buildings · Vibration control

1 Introduction

In order to efficiently utilize the land resources in many metropolises in China, over-track buildings are increasingly being constructed [1, 2]. These buildings are mainly comprised of a large podium used for the parking depot of subways, transfer stories, and superstructures, as shown in Fig. 1. Since the depot usually occupies a large ground area, developing superstructures by transfer stories is a common practice in metropolises [3].

Nevertheless, the structural configuration of these buildings leads to two main problems. First, if the superstructure is fixed to the podium, the lateral stiffness changes abruptly between the superstructure and the podium. Such a problem brings difficulties in the design of transition members. Second, the operation of subways, coming in and out of the podium, causes excessive vibration, which will be transmitted to the superstructure. According to field tests, the vertical frequency of this vibration is normally in the range of 20–50 Hz [4, 5], which is within the limits of disturbing human beings [6].

G. P. Cimellaro (Ed.): WCSI 2022, LNCE 309, pp. 179–189, 2023.
https://doi.org/10.1007/978-3-031-21187-4_12

(a) Photograph of over-track building (b) Section view

Fig. 1. Illustration of over-track building

Occupants in over-track buildings may be affected physically and psychologically [7]. Therefore, there is an urgent need to address these two problems.

Laminated rubber bearings with thick rubber layers, called Thick Rubber Bearings (TRBs), can be used to achieve both vertical and horizontal isolation. Such devices can be employed in the transfer story of over-track buildings to solve the transition problem (horizontal isolation). Meanwhile, TRBs have a low first shape factor (S_1) normally less than 16 to ensure a low working frequency in the vertical direction and therefore achieve vertical isolation [8–11]. For instance, TRBs with a S_1 of 3.20 and 4.56 were designed by Nakamura et al. [10] to mitigate the vibrations in a two-story building adjacent to the railway. This building was measured having a vertical isolated frequency of 7.0 Hz and its effectiveness was validated by field tests [12]. Noted that the vertical design pressure of these TRBs ranged from 1 to 2 MPa [10]. However, mid-rise or high-rise over-track buildings would apply higher vertical pressures. Therefore, TRBs with a higher design pressure should be proposed to allow for their application in over-track buildings.

Additionally, current calculation methods for estimating the mechanical properties of conventional rubber bearings ($S_1 \geq 20$) in different codes may be not applicable to TRBs [9]. To determine their mechanical properties, full-scale tests should be performed. The experimental results can be used to evaluate the isolation effects of TRBs on subway-induced vibrations.

In this research, a TRB with a S_1 of 6.82 was designed and fabricated with a realistic diameter of 800 mm to study its mechanical properties. A series of tests including shear, compression, tension, and stability tests were carried out. Based on the test results, the mechanical properties of this TRB were determined. Finally, simplified multiple-degree-of-freedom systems were studied to evaluate the isolation effect of an over-track building using TRBs.

2 Test Overview of TRB

2.1 Specimen Design

The geometric parameters of the tested TRB are given in Table 1. Because of the increase of S_1 from 4.56 [10] to 6.82, the vertical loading capacity of this bearing was increased.

Its design pressure was determined as 4 MPa, which is 1/4 of the critical buckling compressive load according to ISO 22762-3 [13]. Based on the material tests, the rubber of this TRB specimen has a shear modulus of 0.55 MPa at 100% shear strain.

Table 1. Details of TRB

Property	Symbol (unit)	Value	Property	Symbol (unit)	Value
Outer diameter	d (mm)	800	Thickness of rubber layers	t_R (mm)	27.5
Diameter of hole	d_0 (mm)	50	Number of rubber layers	n	8
Thickness of rubber cover	t (mm)	10	Thickness of total rubber layers	T_R (mm)	220
First shape factor	S_1	6.82	Thickness of steel shims	t_S (mm)	10
Second shape factor	S_2	3.64	Height of bearings	h (mm)	350

2.2 Test Apparatus

The TRB was tested using a single isolator test machine. The machine comprises 4 vertical actuators and 2 horizontal ones, as shown in Fig. 2. The compressive and tensile loading capacities of the machine are 20,000 and 6000 kN, respectively, while the horizontal actuators can reach a stroke of ±500 mm. The vertical displacements were measured with four external linear variable displacement transducers (LVDTs). The mean of four LVDT signals was adopted as the central displacement of the specimen to eliminate the possible effect of rotation of the specimen. The lateral displacement was measured directly from the horizontal actuators.

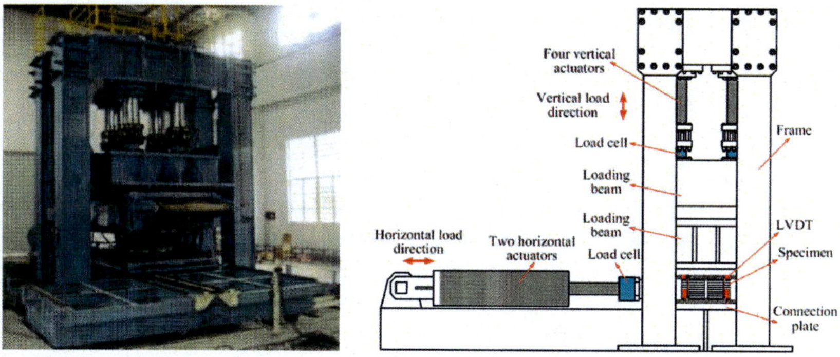

(a) Photograph of test set-up (b) Schematic of test set-up

Fig. 2. Test set-up

Table 2. Summary of testing program

Test type	Load case	Initial pressure σ_0 (MPa)	Uniaxial load amplitude $\Delta\sigma$ (MPa)	Lateral offset γ (%)	Tension strain T (%)	Freq. f (Hz)	No. of cycles
Shear tests	S1	4	n.a	50	n.a	0.004	3
	S2			100			
	S3			150			
	S4	4	n.a	50	n.a	0.01	3
	S5	6	n.a	50	n.a	0.004	3
	S6			100			
	S7			150			
Compression tests	C1	4	1.2	0	n.a	0.004	3
	C2	6	1.8				
	C3	8	2.4				
	C4	4	1.2	50	n.a	0.004	3
	C5			100			
	C6			150			
Tension tests	T1	0	n.a	100	10%	0.004	3
	T2				20%		
Stability tests	ST1	4	n.a	200	n.a	n.a	n.a
	ST2	8					
	ST3	10					
	ST4	12					

2.3 Test Description

In this study, various tests were performed including cyclic shear tests, compression tests, tension tests, and stability tests. Table 2 summarizes the loading conditions of these tests. For shear and compression tests, the loading patterns were based on the provisions of ISO 22762-1 [14]. The shear tests were performed under various vertical loads, lateral offsets, and test frequencies to study their effects on the lateral behavior. Compression tests were carried out under different initial compressive loads and applied lateral offset to explore the variation in the compressive stiffness. According to ISO 22762-3 [13], cyclic tensile tests were performed under lateral offset. Three cycles were performed to examine its tensile properties. In addition, different vertical loads were constantly applied to the TRB, and then this TRB was subjected to a monotonically increasing shear strain of 200% to investigate its stability under a large vertical load.

3 Results of Characteristic Tests

3.1 Horizontal Characteristics

Hysteretic curves of the TRB specimen under vertical pressure of 4 and 6 MPa are shown in Fig. 3 (a) and (b), respectively. According to these loops, the equivalent damping ratio and shear stiffness of each load case can be determined, as shown in the second and third columns of Table 3. It can be found that by increasing the vertical pressure from 4 MPa to 6 MPa, the equivalent damping ratio and shear stiffness under the same shear strain increase by around 40% and 10%, respectively. However, no obvious trends were observed when the amplitude of the shear displacement was changed. In addition, Fig. 3 (c) shows the hysteretic loops of the TRB under different loading frequencies. From this figure and the properties in Table 3, it can be deduced that the exists no rate dependency.

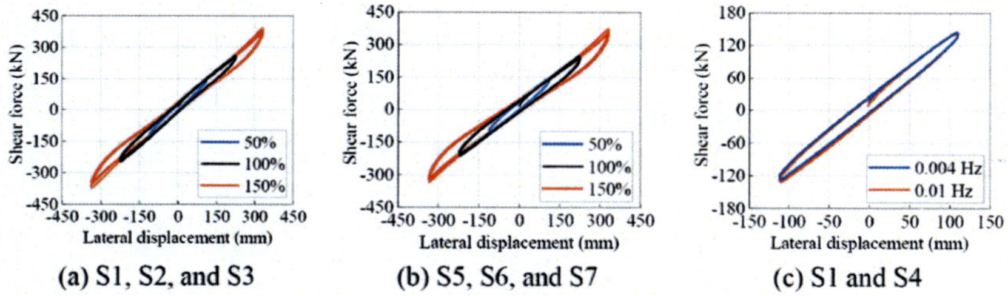

Fig. 3. Hysteretic curves from shear tests

ISO 22762-3 [13] provides a calculation method for the shear stiffness of conventional rubber bearing, K_{h}, which is determined as:

$$K_{\mathrm{h}}=\frac{GA}{T_{\mathrm{R}}} \tag{1}$$

where G is the shear modulus and A is the effective area of the bearings excluding the rubber cover portion. According to Eq. (1), the theoretical shear stiffness of the TRB is calculated and shown in the fourth column of Table 3. A comparison between the experimentally determined results and Eq. (1) shows that the calculation error for the shear stiffness under the design pressure is relatively small (less than 17%), while the error increases with vertical pressure.

3.2 Vertical Characteristics

Compressive stiffness

The experimental curves from the compression tests of the TRB specimen under different vertical pressure and lateral offset are presented in Fig. 4 (a) and (b), respectively. The experimentally determined compressive stiffness is shown in the second column of Table 4. It can be found that the compressive stiffness increases with the applied load,

Table 3. Horizontal properties of TRB under different load cases

Load case	Equivalent damping ratio (%)	Experimental shear stiffness /(kN/mm)	Theoretical shear stiffness /(kN/mm)	Calculation error (%)
S1	5.9	1.2049	1.252	3.9
S2	4.7	1.1074	1.252	13.1
S3	4.3	1.0768	1.252	16.3
S4	6.0	1.2033	1.252	4.0
S5	8.2	1.0913	1.252	14.7
S6	6.6	0.9759	1.252	28.3
S7	5.9	0.9724	1.252	28.8

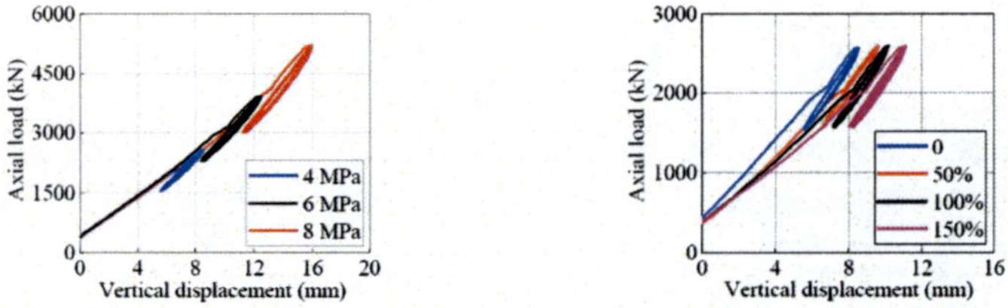

(a) C1, C2, and C3 (b) C1, C4, C5 and C6

Fig. 4. Results from compression tests

and it is almost unchanged under a co-existing shear strain ranging from 50 to 150%. Thus, the influence of vertical load on the compressive stiffness of the TRB is significant.

The theoretical values for the compressive stiffness of rubber bearings without a lateral offset, K_{v0}, can be obtained from the following equation offered in ISO 22762-3 [13]:

$$K_{v0}=\frac{E_c A}{T_R} \tag{2}$$

where E_c is the compression modulus of rubber under the triaxial stress conditions caused by steel shims. To take into account the influence of a lateral displacement, a two-spring model was proposed by Koh and Kelly [15]:

$$K_v=\frac{1}{1+\frac{12}{\pi}\left(\frac{\Delta}{R}\right)^2}K_{v0} \tag{3}$$

where Δ is the applied lateral displacement and R is the radius of bearings. Based on Eqs. (2) and (3), the theoretical stiffness of the specimen under the tested lateral offsets is offered in Table 4. The calculation error for all cases is large ranging from 21.7 to 57.5%, because the above-mentioned methods were proposed for conventional rubber bearings with a high S_1. The compressive stiffness of TRBs with a low S_1 would be significantly underestimated by Eqs. (2) and (3). Therefore, current formulations for the compressive stiffness are not applicable to the TRBs. When using TRBs, their compressive stiffness should be experimentally determined.

Table 4. Compressive stiffness of TRB under different load cases

Load case	Experimental results /(kN/mm)	Theoretical result /(kN/mm)	Calculation error (%)
C1	349.5	273.5	21.7
C2	414.0	273.5	33.9
C3	482.0	273.5	43.3
C4	340.7	250.5	26.5
C5	346.7	200.1	42.2
C6	352.4	149.9	57.5

Tensile stress

ISO 22762-3 [13] and a Japanese guideline [16] provide strict requirements for the tensile properties of rubber bearings: the yield tensile stress of rubber bearings at a co-existing shear strain of 100% is required to be larger than 1.00 MPa [13, 16]. Therefore, tension tests of the TRB specimen subjected to a 100% shear strain were performed under a tensile strain of 10 and 20%.

Figure 5 presents the hysteretic curves from the tension tests, and the yield point of this specimen is illustrated. This point is determined according to ISO 22762-3 [13]. The yield tensile stress was calculated as 1.01 MPa, which is slightly larger than the required value of 1.00 MPa. Thus, the tensile properties of the TRB in this study meet the performance requirements for conventional rubber bearings.

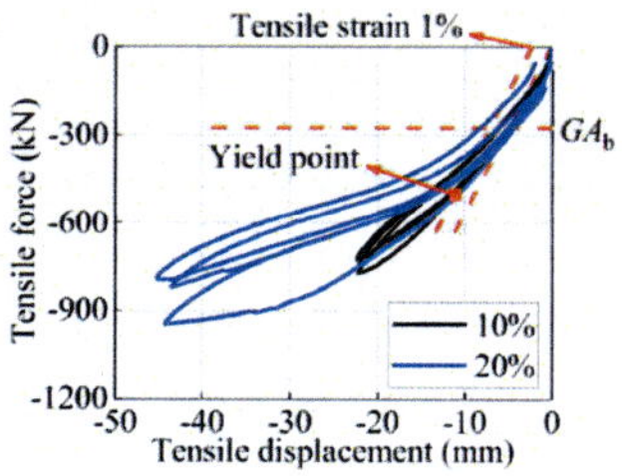

Fig. 5. Hysteretic curves from tension tests

3.3 Stability Characteristics

When performing stability tests, a constant vertical load from 4 to 12 MPa was applied to the TRB. Then, this bearing was subjected to an increasing horizontal displacement. A maximum shear strain of 200% corresponding to a lateral displacement of 440 mm was applied. The test results under four levels of vertical pressure are shown in Fig. 6. With increasing applied vertical pressure, the horizontal stiffness of the bearing decreased. However, the experimental curves were almost linear, and no instability occurred, even under a vertical pressure of 12 MPa, which is three times the design pressure.

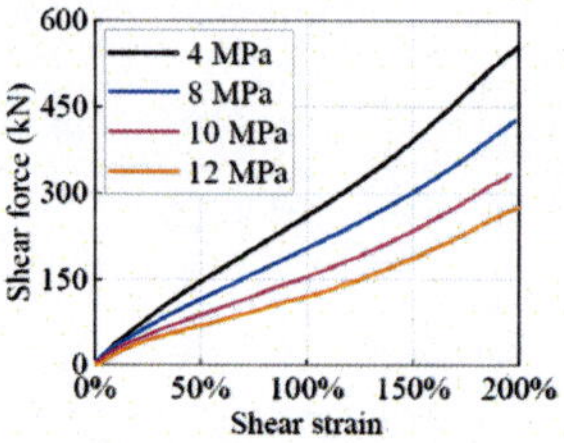

Fig. 6. Stability test curves

4 Effectiveness for Vertical Vibration Control

4.1 Numerical Model of Over-Track Building

In this section, simplified multiple-degree-of-freedom (MDOF) systems were established to verify the isolation effectiveness of TRBs for the vertical vibration caused by the subway. Specifically, a 7-story concrete building designed by Zhou et al. [19] (see Fig. 7 (a)) was selected as a case study. This building would cause a vertical pressure on the bearings close to 4 MPa, which is the design pressure of the TRB of this study. Thus, the examined TRB could be used in this 7-story building. The vertical stiffness of each floor was assumed equal to the vertical stiffness of its columns, i.e. the slabs were assumed as rigid.

To account for the podium in an over-track building, a large mass can be assigned at the bottom of this 7-story building model and is fixed to the ground. Based on this, a simplified MDOF model of the isolated over-track building using TRBs is built in Fig. 7 (b). In parallel, a fixed base over-track building model is developed and shown in Fig. 7 (c).

A vertical acceleration was applied at the base of each model. As an indicative excitation, the acceleration recorded on the first floor in an over-track building was adopted (Fig. 8).

4.2 Numerical Results

The acceleration responses of the first, third, fifth, and seventh stories of the superstructure are illustrated in Figs. 9 (a)–(d), respectively.

The peak acceleration responses of the isolated system using TRBs were reduced by 67.3–77.2%. According to ISO 2631-1 [20], the root mean square acceleration by frequency weighting of each acceleration response, a_{wz}, can be calculated. The maximum

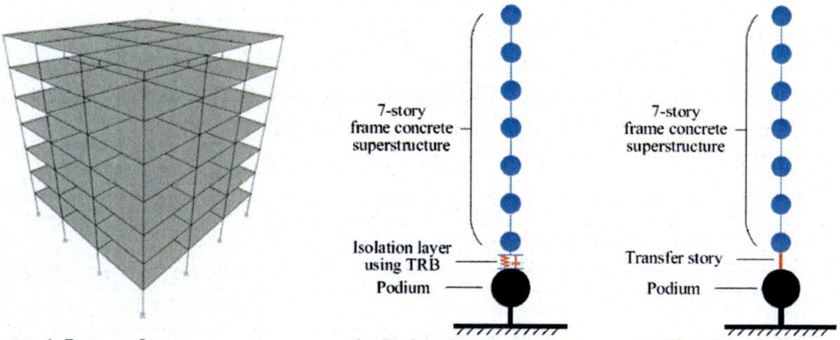

(a) A 7-story frame concrete building model from Ref. [19]
(b) Isolated over-track building model
(c) Fixed over-track building model

Fig. 7. Simplified MODF systems

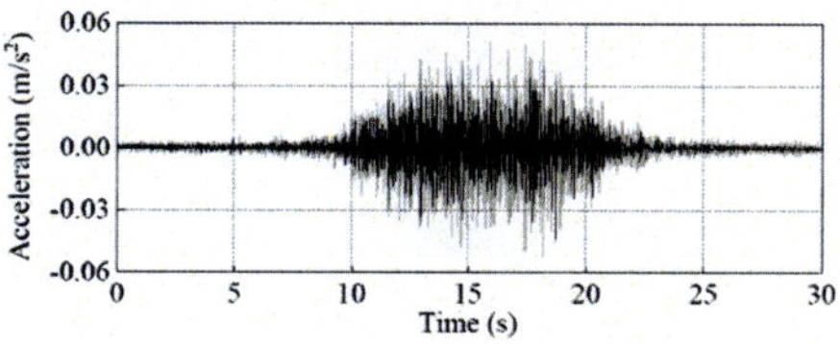

Fig. 8. Subway-induced vibration in vertical direction

value of a_{wz} of the fixed system was 0.0077 m/s^2, while it was significantly reduced to 0.0034 m/s^2 by using TRBs. Hence, the TRBs were numerically verified to be effective for mitigating the vertical subway-induced vibration in over-track buildings.

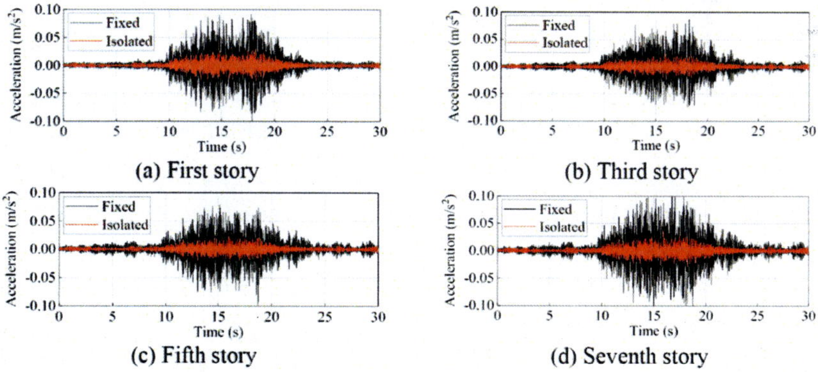

(a) First story
(b) Third story
(c) Fifth story
(d) Seventh story

Fig. 9. Comparison between acceleration responses

5 Conclusions

This research investigated the mechanical properties of a TRB and its isolation effectiveness for subway-induced vibration. A full-scale TRB was designed and tested under compressive, tensile, and shear loads. The vertical and horizontal characteristics were identified and compared to analytical solutions. Based on the experimentally determined properties, the isolation effect of the TRBs was evaluated by simplified MDOF systems. The key findings of this study can be summarized as follows:

- With increasing vertical pressure, the horizontal equivalent damping ratio and compressive stiffness significantly increase.
- Formulations from current codes could be used to predict the shear stiffness of the TRB under design pressure, while they could not provide reliable values for the compressive stiffness with and without applied lateral offset.
- The tensile strength of the TRB meets the performance requirement for conventional rubber bearings. Moreover, the TRB remains stable under a compressive stress three times larger than the design pressure even when subjected to a 200% shear strain.
- Simplified numerical models of fixed and isolated MDOF systems were established to evaluate the effectiveness of TRBs in mitigating vertical vibration. Under a subway-induced excitation, the acceleration responses of the isolated system using TRBs were substantially reduced.

Acknowledgement. The authors gratefully acknowledge the financial support from National Natural Science Foundation of China (Grant No. 51878502) and China Scholarship Council during a visiting study at ETH Zürich (No. 202106260155).

References

1. Zhou, Y., Chen, P., Wang, C., Zhang, L., Lu, L.: Seismic performance evaluation of tall, multitower reinforced concrete buildings with large bottom podiums. Struct. Concr. **19**(6), 1591–1607 (2018)
2. Zhou, Y., Ma, K., Chen, P., Lu, D., Wu, H.: Investigations on train-induced vibration and vibration control of an over-track building using thick-layer rubber bearings. Struct. Des. Tall Spec. Build. **31**(1), e1898 (2022)
3. Shen, B., Lepech, M.D.: Probabilistic design of environmentally sustainable reinforced-concrete transportation infrastructure incorporating maintenance optimization. ASCE J. Infrastruct. Syst. **23**(3), 04016038 (2017)
4. Sanayei, M., Maurya, P., Moore, J.A.: Measurement of building foundation and ground-borne vibrations due to surface trains and subways. Eng. Struct. **53**, 102–111 (2013)
5. Zou, C., Wang, Y., Wang, P., Guo, J.: Measurement of ground and nearby building vibration and noise induced by trains in a metro depot. Sci. Total Environ. **536**, 761–773 (2015)
6. Rasmussen, G.: Human body vibration exposure and its measurement. J. Acoust. Soc. Am. **73**(6), 1–27 (1983)
7. Jik, L.P., Griffin, M.J.: Combined effect of noise and vibration produced by high-speed trains on annoyance in buildings. J. Acoust. Soc. Am. **133**(4), 2126–2135 (2013)
8. Aiken, I.D., Kelly, J.M., Tajirian, F.F.: Mechanics of Low Shape Factor Elastomeric Seismic Isolation Bearings. Earth Engineering Research Center, University of California, College of Engineering (1989)

9. Yabana, S., Matsuda, A.: Mechanical properties of laminated rubber bearings for three-dimensional seismic isolation. In: 12th World Conference on Earthquake Engineering, Auckland, New Zealand (2000)
10. Nakamura, T., Okada, H., Sugiyama, K., Teramura, A.: Development of base isolation system for earthquakes and micro-vibrations using laminated thick rubber bearings: part 1—characteristics tests for laminated thick rubber bearings. Obayashi-Gumi Gijutsu Kenkyusho-Ho **42**, 15–22 (1991)
11. Fukasawa, T., Okamura, S., Yamamoto, T., Kawasaki, N., Somaki, T., Sakurai, Y., et al.: Development on rubber bearings for sodium-cooled fast reactor: part 2—fundamental characteristics of half-scale rubber bearings based on static test. In: Pressure Vessels and Piping Conference. American Society of Mechanical Engineers, V008T08A40 (2015)
12. Teramura, A., Yoshihara, J., Nakamura, M., Ishikawa, R., Yasui, Y., Suzuki, T., et al.: Development of base isolation system for earthquakes and micro-vibrations using laminated thick rubber bearings: part 3—application to a medical facility building subjected to railway-induced vibration. Obayashi-Gumi Gijutsu Kenkyusho-Ho **43**, 7–14 (1991)
13. International Organization for Standardization. ISO 22762-3:2018: Elastomeric seismic-protection isolators—part 3: applications for buildings—specifications, 3rd edn. (2018)
14. International Organization for Standardization. ISO 22762-1:2018: Elastomeric seismic-protection isolators—part 1: test methods, 3rd edn. (2018)
15. Koh, C.G., Kelly, J.M.: A simple mechanical model for elastomeric bearings used in base isolation. Int. J. Mech. Sci. **30**(12), 933–943 (1988)
16. Japan Electric Association. JEAG4614-2013: Technical guidelines on seismic base isolated system for structural safety and design of nuclear power plant. Japan (2013)
17. European Standard. EN 15129: 2009: Anti-seismic devices. CEN, Brussels (2009)
18. Ministry of Housing and Urban-Rural Development of the People's Republic of China. JG/T 118-2018: Rubber isolation bearings for buildings. China (2018)
19. Zhou, Y., Chen, P., Mosqueda, G.: Analytical and numerical investigation of quasi-zero stiffness vertical isolation system. J. Eng. Mech. **145**(6), 04019035 (2019)
20. International Organization for Standardization. ISO2631-1: 1997: Mechanical vibration and shock—evaluation of human exposure to whole-body vibration. Part 1: general requirements. ISO, Geneva (1997)

Bidirectional Model for Shear Behavior of High Damping Rubber Bearings

José A. Gallardo[1,2(✉)], Juan Carlos de la Llera[1,2], José I. Restrepo[3], and Michelle Chen[3]

[1] Department of Structural and Geotechnical Engineering, Pontificia Universidad Católica de Chile, Santiago, Chile
jogallardo@uc.cl
[2] Research Center for Integrated Disaster Risk Management (CIGIDEN), Santiago, Chile
[3] Department of Structural Engineering, University of California, San Diego, San Diego, 9500 Gilman Dr., La Jolla, CA 92093, USA

Abstract. Seismic isolation has become increasingly popular due to the excellent performance observed in buildings during past earthquakes. The seismic response of isolated structures to earthquake input is strongly controlled by the force-deformation constitutive behavior of the isolators. High Damping Rubber Bearings (HDRBs) are one of the most widely manufactured and used isolation systems in practice. Because of the large shear flexibility, the stress-strain constitutive behavior of the elastomeric material controls the overall behavior of the device. Thus, the behavior of HDRBs is highly nonlinear and characterized by the same phenomena as the elastomeric material, which is challenging to model analytically. Consequently, this article describes a simple but sufficiently accurate numerical model for simulating the bi-directional shear behavior of HDRBs under large shear deformations. A brief description of the experimental test data of HDRBs is presented, as well as the importance of including the observed phenomena in the numerical model proposed. The mathematical formulation of the proposed model is summarized, based on a hyperelastic spring and dissipative element connected in parallel. The former considers anisotropic degradation (scragging and Mullins effect), while the latter includes the isotropic hardening phenomenon. The proposed model was validated using experimental results of bi-directional shear tests of cylindrical disks of high damping natural rubber, unidirectional shear cyclic tests, and a bi-directional shear loading history applied to an HDRBs. Since the proposed model is capable of simulating the bi-directional cyclic behavior of HDRBs by considering anisotropic degradation instead of the classical isotropic behavior, which improves the numerical predictions, it can be used in dynamic analyses of buildings isolated with HDRBs.

Keywords: High damping rubber · Base isolation · Nonlinear response · Bidirectional shear · Seismic isolation

G. P. Cimellaro (Ed.): WCSI 2022, LNCE 309, pp. 190–197, 2023.
https://doi.org/10.1007/978-3-031-21187-4_13

1 Introduction

During the last two decades, seismic isolation has increasingly become more popular due to its excellent performance during large earthquakes [1, 2]. The overall dynamic response of a seismically isolated structure depends on the behavior of the seismic isolation devices. Thus, simulating the response of the isolators is critical in predicting of demands on these structures.

Currently, High Damping Rubber Bearings (HDRBs) are probably one of the most widely used isolation systems in practice [3]. These devices are manufactured by superimposing interspersed layers of elastomeric material and steel shims. Because of the large shear flexibility, the stress-strain constitutive behavior of the elastomeric material controls the overall behavior of the device. Thus, the behavior of HDRBs is highly nonlinear, which is challenging to reproduce numerically. Some of the representative phenomena of these devices are shown in Fig. 1, which include a large critical damping ratio (10–20% [4]), stiffening at large strains, anisotropic damage, and isotropic hardening.

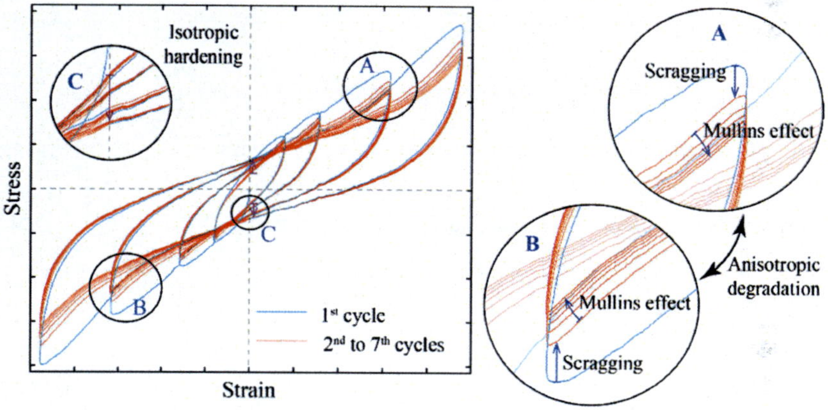

Fig. 1. Phenomena of the shear response of HDRBs

Several numerical models have been proposed to simulate the static and dynamic behavior of HDRBs [5–11]. However, a limited number of them consider bidirectional shear coupling or anisotropic damage. These and the other previously mentioned phenomena are critical in evaluating structural and functional fragilities of essential facilities since seismic isolation is widely used in them.

This work presents a summary of a simple and sufficiently accurate model recently proposed [12] for simulating the bidirectional shear response of the HDRBs. The model considers the bidirectional shear coupling, scragging and Mullins effect (long- and short-term degradation), and isotropic hardening. In the following sections, the basic mathematical formulation is also summarized. Then, the numerical model is validated using different experimental test results and finally, the main conclusions of the work are presented.

2 Mathematical Model

The presented model is an expansion of the proposed in [11] for unidirectional to bidirectional shear. The formulation is based on the decomposition of the force-deformation constitutive curve into a hyperelastic and a dissipative component working in parallel. The degradation (scragging or long-term and Mullins effect or short-term degradation) is considered into the coefficients of the hyperelastic component, and isotropic hardening into the dissipative component. Thus, the shear force vector ($\boldsymbol{F}$) can be expressed as

$$\boldsymbol{F} = \boldsymbol{F}_h + \boldsymbol{F}_d \tag{1}$$

where $\boldsymbol{F}_h$ and $\boldsymbol{F}_d$ are the hyperelastic and the dissipative force components, respectively.

2.1 Hyperelastic Component

The hyperelastic component is derived from the strain energy density proposed in [13]. The formulation considers pure shear in the deformation gradient, and constant stress across the section. Truncating the expression after the third term of the shear component, the force of the hyperelastic component is

$$\boldsymbol{F}_h = \left[a_{10}K_{s1}K_m|\boldsymbol{\gamma}| - a_{20}K_{s2}|\boldsymbol{\gamma}|^3 + a_{30}K_{s3}|\boldsymbol{\gamma}|^5\right]\overline{\boldsymbol{\gamma}} \tag{2}$$

where a_{i0} are the device constants considering the virgin material; K_{si} are the scragging factors; K_m is the Mullins effect factor; $\boldsymbol{\gamma}$ is the strain vector $\left(\boldsymbol{\gamma} = \left[\gamma_x, \gamma_y\right]^T\right)$, and $|\boldsymbol{\gamma}|$ and $\overline{\boldsymbol{\gamma}}$ are the euclidean norm and normalized unit vector of $\boldsymbol{\gamma}$.

2.2 Dissipative Component

The dissipative component represents the energy dissipation capacity of the device. The Bouc-Wen model with isotropic hardening proposed in [14] is considered and neglects the elastic part. Thus, the force of the dissipative component is defined as

$$\boldsymbol{F}_d = f_y \boldsymbol{z} \tag{3}$$

where f_y is the yield force and $\boldsymbol{z}$ is a vector with the dimensionless hysteretic parameters $\left(z = \left[z_x, z_y\right]^T\right)$. Following the procedure described in [15], the dimensionless hysteretic parameters can be written as

$$z_x = (k/f_y)\left[\dot{\gamma}_x - z_x R\right] \tag{4}$$

$$z_y = (k/f_y)\left[\dot{\gamma}_y - z_y R\right] \tag{5}$$

where k is is the initial stiffness multiplied by the total height of the elastomeric material, $\dot{\gamma}$ is the shear strain rate, and R is the coupling factor defined as

$$R = \left[(\beta - \varphi)\left(|\dot{\gamma}_x z_x| + |\dot{\gamma}_y z_y|\right) + (\alpha - \varphi)\left(\dot{\gamma}_x z_x + \dot{\gamma}_y z_y\right)\right]\left(z_x^2 + z_y^2\right)^{\frac{\eta-2}{2}} \tag{6}$$

where β and α are parameters that control the shape of the cycle and satisfy the relation ($\beta + \alpha = 1$); η controls the sharpness of the smooth transition from the elastic to the inelastic zone; and the parameter φ quantifies the isotropic hardening. The latter parameter evolves with the deformation history, which is computed as

$$\varphi = \varphi_{max}\left[1 - \exp\left(-P_{\varphi}\left|\frac{\gamma_{pl}}{\gamma_{yy}}\right|\right)\right] \tag{7}$$

where φ_{max} is the maximum possible value of φ; P_{φ} is the parameter that controls the isotropic hardening rate; γ_{pl} is the plastic strain; and γ_{yy} is the yield strain $\left(f_y/k\right)$.

2.3 Degradation

The parameters K_{si} and K_m of Eq. (2) control the degradation as a consequence of the scragging (long-term degradation) and Mullins effect (short-term degradation). These values are computed as [12]

$$K_{s1} = \exp\left(-C_1\gamma_s^{P_1}\right) \tag{8}$$

$$K_{s2} = K_{s1}\exp\left(-C_2\gamma_s^{P_2}\right) \tag{9}$$

$$K_{s3} = K_{s1}\exp\left(-C_3\gamma_s^{P_3}\right) \tag{10}$$

$$K_m = \exp\left(-C_m\gamma_m^{P_m}\right) \tag{11}$$

where γ_s and γ_m are the considered strain for scragging and Mullins effect, respectively; and the parameters C and P are material properties. The evolution law for parameters γ_s and γ_m can be found in [12]. Figure 2 shows a sketch of the evolution of the parameter K_{s1} as γ_s increases and the effect of the material parameters. The shape of the other degradation parameters is similar to the one shown in Fig. 2.

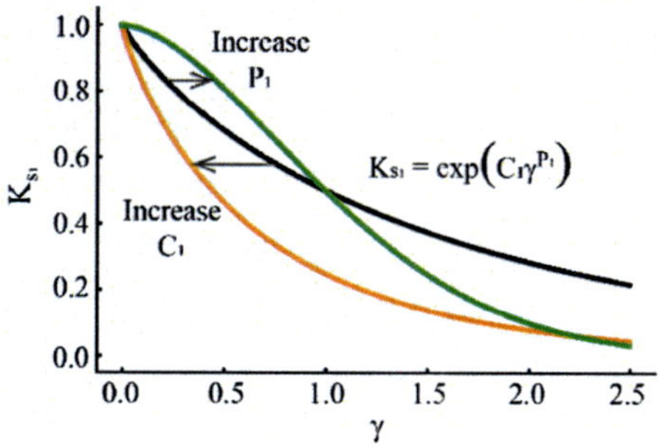

Fig. 2. Sketch of the evolution of parameter K_{s1} as example of the degradation functions

The anisotropic damage of the elastomeric material was considered and implemented using the theoretical base of material directions [16]. Thus, the complete spectrum of strain shear directions is divided into a finite number of spaces, each one characterized by the unit vector of the direction (u_i) and the considered strains for scragging (γ_s) and Mullins effect (γ_m). There is coupling between the evolution of the considered strains of different directions; thus, considering an increase in the considered strain in the i-th direction, the considered strain in the j-th direction also increases. The complete degradation model is presented in [12].

3 Validation

The proposed model was validated using experimental results of: (a) bidirectional shear tests of cylindrical disks using virgin specimens for each test [17]; (b) an HDRB under constant axial load and a unidirectional lateral cyclic load history [18]; and (c) a bidirectional shear test of an HDRB [19].

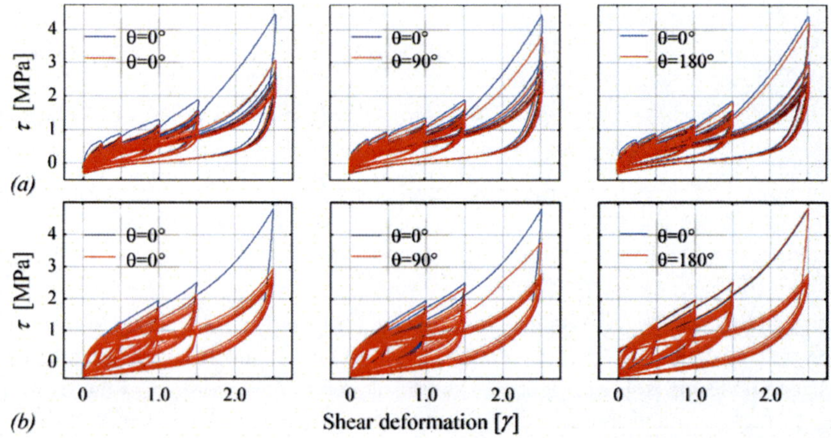

Fig. 3. Double shear tests on rotated directions: a) experimental results (taken from [16]); and b) numerical results

The deformation history of the cylindrical disks consists of six cycles in the reference direction ($\theta = 0°$), then the test stops, and the specimen is rotated ($\theta = 0°$, $90°$, $180°$) to perform the other six deformation cycles. This sequence is performed for five strain amplitudes: 0.25, 0.50, 1.0, 1.50, and 2.50. Figure 3 shows one example of the experimental and numerical test results. The proposed model (Fig. 3b) fits well the experimental test results (Fig. 3a). When the rotation angle is zero, the parameter that controls Mullins effect reboots, which accounts for the recovery of the short-term degradation due to the stop. When the rotation angle equals 90°, the long-term degradation (scragging) has evolved, but not completely, showing anisotropic degradation. When the rotation angle equals 180°, the degradation ocurring at 0° does not affect the response at 180°.

Figure 4 shows the comparison between the numerical and experimental test results of an HDRB under constant axial load and unidirectional lateral cyclic deformation

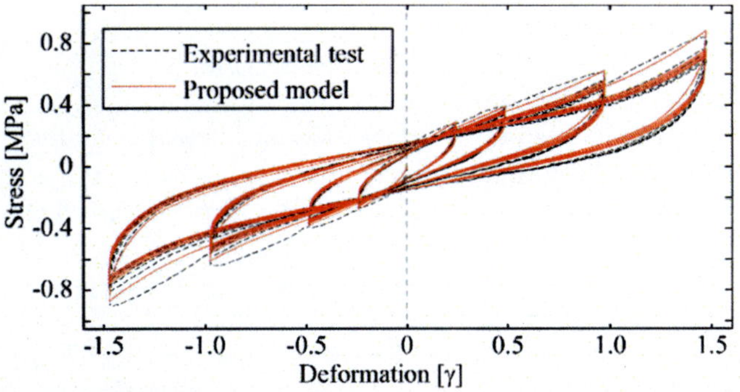

Fig. 4. Stress-strain comparison between numerical (solid red line) and experimental (dashed black line) results of an HDRB

history. The presented model simulates the behavior of the device accurately in the full range of deformations and strain cycles. There is a larger difference when the deformation is negative than when the deformation is positive; this is because the experimental test results are not symmetric, while the numerical result are.

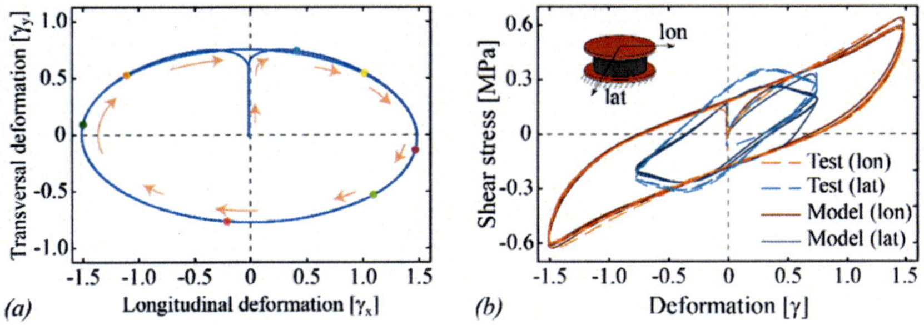

Fig. 5. Comparison of the bidirectional experimental test and the presented model response: a) lateral deformation pattern and b) shear stress versus deformation

Figure 5 shows the results of the bidirectional shear test and numerical results of an HDRB. The load pattern consisted of applying a deformation history with an elliptical orbit in the horizontal plane (see Fig. 5a) and a constant compressive force. The shear deformation response (Fig. 5b) of the numerical model in the longitudinal direction fits the experimental result well while the experimental response in the transverse direction shows a hump protruding in the response in the longitudinal direction. This stiffer response is a consequence of anisotropic damage, while the negative slope after the peak is the effect of bidirectional coupling. The numerical response fits the experimental result reasonably well. The hump due to large anisotropy is not reproduced exactly in magnitude but is reproduced in shape. The different forces at zero deformation in both directions of the experimental results cannot be simulated. This would increase the complexity of the model and make the parameter calibration even more cumbersome.

4 Summary and Conclusion

This paper summarizes a phenomenological model for simulating the shear response of HDRBs. The model has two components connected in parallel: a hyperelastic and a dissipative component. The former includes anisotropic damage (scragging and Mullins effect), while the latter considers isotropic hardening. The main conclusions are:

- The proposed model simulates anisotropic damage reasonably well compared to experimental results. It reproduces the load-direction dependency of the response.
- The model simulates the behavior of HDRBs under lateral cyclic loads accurately in the full range of deformations expected during the lifetime of the devices.
- The bidirectional shear response of HDRBs is simulated reasonably well, including the coupling between the shear directions.
- Due to the evolution laws of the proposed model, it can successfully be used for static and dynamic analyses.

Acknowledgments.
This research has been sponsored by ANID/doctorate scholarship/21201370, Research Center for Integrated Disaster Risk Management (CIGIDEN), ANID/FONDAP/15110017; and FONDECYT project, SImulation Based Earthquake Risk and Resilience of Interdependent Systems and networKs (SIBER-RISK), ANID/FONDECYT/1170836.

References

1. Fujita, T.: Demonstration of effectiveness of seismic isolation in the Hanshin-Awaji earthquake and progress of applications of base-isolated buildings. Technical report. University of Tokyo (1995)
2. De la Llera, J., Mitrani-Reiser, J., Rivera, J., et al.: The 2010 Chile earthquake: a five year reflection. In: Proceedings of the 10th Pacific Conference on Earthquake Engineering, p. 210. AEES (2015)
3. Clemente, P.: Seismic isolation: past, present and the importance of SHM for future. J. Civ. Struct. Heal. Monit. **7**(2), 217–231 (2017)
4. Aghaeidoost, V., Billah, A.: Sensitivity of seismic fragility of base-isolated bridges to lead rubber bearing modeling technique. Struct. Control. Health Monit. **29**(8), e2971 (2022)
5. Kikuchi, M., Aiken, I.: An analytical hysteresis model for elastomeric seismic isolation bearings. Earthq. Eng. Struct. Dyn. **26**(2), 215–231 (1997)
6. Grant, D., Fenves, G., Whittaker, A.: Bidirectional modelling of high-damping rubber bearings. J. Earthq. Eng. **8**(spec01), 161–185 (2004)
7. Yamamoto, M., Minewaki, S., Yoneda, H., Higashino, M.: Nonlinear behavior of high-damping rubber bearings under horizontal bidirectional loading: full-scale tests and analytical modeling. Earthq. Eng. Struct. Dynam. **41**(13), 1845–1860 (2012)
8. Markou, A., Manolis, G.: Mechanical models for shear behavior in high damping rubber bearings. Soil Dyn. Earthq. Eng. **90**, 221–226 (2016)
9. Tubaldi, E., Ragni, L., Dall'Asta, A., Ahmadi, H., Muhr, A.: Stress softening behaviour of HDNR bearings: modelling and influence on the seismic response of isolated structures. Earthq. Eng. Struct. Dyn. **46**(12), 2033–2054 (2017)

10. Miranda, S., Miranda, E., De la Llera, J.: A simplified and versatile element model for elastomeric seismic isolation bearings. Earthq. Spectra **38**(1), 331–357 (2021)
11. Gallardo, J., De la Llera, J.: An advanced numerical model for nonlinear shear behavior of elastomeric bearings with parameter prediction. Under Review (2022)
12. Gallardo, J., De la Llera, J., Restrepo, J., Chen, M.: An advanced numerical model for the multiaxial behavior of high damping rubber bearings. Under Review (2022)
13. Ogden, R.: Nonlinear elastic deformations. Courier Corporations (1997)
14. Karavasilis, T., Kerawala, S., Hale, E.: Hysteretic model for steel energy dissipation devices and evaluation of a minimal-damage seismic design approach for steel buildings. J. Constr. Steel Res. **70**, 358–367 (2012)
15. Harvey, P., Gavin, H.: Truly isotropic biaxial hysteresis with arbitrary knee sharpness. Earthq. Eng. Struct. Dyn. **43**(13), 2051–2057 (2014)
16. Diani, J., Brieu, M., Vacherand, J., Rezgui, A.: Directional model for isotropic and anisotropic hyperelastic rubber-like materials. Mech. Mater. **36**(4), 313–321 (2004)
17. Ragni, L., Tubaldi, E., Dall'Asta, A., Ahmadi, H., Muhr, A.: Biaxial shear behaviour of HDNR with Mullins effect and deformation-induced anisotropy. Eng. Struct. **154**, 78–92 (2018)
18. Miranda, S., De la Llera, J., Miranda, E.: Uncertainty on measurement of elastomeric isolators effective properties. Measurement **180**, 109511 (2021)
19. Chen, M., Restrepo, J., Benzoni, G.: Response of a high damping rubber bearing to multiaxial excitation. J. Test. Eval. **49**(2), 1153–1172 (2019)

Development of an Improved Deformation History Integral Type Hysteresis Model for High-Damping Rubber Bearings

Takahiro Mori(✉), Sadamitsu Takeuchi, and Nobuo Murota

Bridgestone Corporation, Tokyo, Japan
takahiro.mori@bridgestone.com

Abstract. Deformation history integral type hysteresis (DHI) model is a numerical model for seismic isolation bearing which has high non-linearity in restoring force characteristics, such as high-damping rubber bearing (HDR). While the model can represent the complex behavior of HDR including performance change by temperature, aging and manufacturing variation, the scragging effect, which is generally considered in practical design of seismic isolation system, is not taken into account in the conventional model. In this study, the model was improved by defining scragging effect as a function of performance change ratio. In order to confirm the accuracy of the proposed model, time history analysis was carried out and the results of analysis were compared with those of substructure real-time online testing (SROLT) using scaled model of HDR. The analysis results showed good agreement with those of SROLT and the applicability of improved DHI model was verified.

Keywords: Seismic isolation · Scragging · High-damping rubber bearing · Deformation history integral type hysteresis model

1 Introduction

The rubber bearing (RB), made up with layers of alternating rubber and steel plates is the most popular device for providing restoring force and damping characteristics as seismic isolation bearing. In general, RB causes performance fluctuations due to environmental temperature, aging, and repeated loading which should be properly considered when designing the system. They are usually considered as property modification factors [1, 2] in numerical analysis of the isolation system. The shear properties are affected by ambient temperature. As the RB is generally expected to be used for more than 50 years, the aging effect to rubber bearings is also important factor to be considered at design of SI system. The aging factor is usually evaluated by heat accelerated aging test [3] and the change of shear properties after designed service years under nominal temperature at site is assumed, and counted at design of SI system. The shear property of RB changes due to repeated loading. Scragging effect is considered as a part of property change by repeated loading, and defined as the ratio of the shear properties at first cycle

G. P. Cimellaro (Ed.): WCSI 2022, LNCE 309, pp. 198–206, 2023.
https://doi.org/10.1007/978-3-031-21187-4_14

of loading to the nominal shear properties, which is generally defined as average of first three cycles or as properties at third cycle. The stiffness of rubber bearing at first cycle is stiffer than at second and subsequent loading. This phenomena is also called as Mullin's effect [4]. Deformation History Integral (DHI) type hysteresis model is one of numerical models which expresses the non-linear shear stress-strain relationship of RB, especially of HDR [5], that has relatively large nonlinear characteristics in terms of shear properties. This model is applicable for large shear strain deformation in bi-lateral direction and the accuracy of the model were verified by the tests in the past study. Also, the performance change by temperature, manufacturing variation, repeated loading was able to the conventional model by using parameters identified by the test results. However, scragging effect is not considered in the conventional DHI model.

In this study, DHI model is improved to reproduce the shear stress-strain relationship with scragging effect. Using improved DHI model, numerical analysis of single degree of freedom (SDOF) system with HDR under artificial ground motions is conducted. In order to verify the validity of improved model, hybrid simulation method, which combines loading test and numerical simulation by computer, was conducted. The hybrid test method is originally developed for simulation of seismically isolated building, which is called substructure real-time online testing (SROLT) [6].

2 Proposal of Modified DHI Model

Displayed equations are centered and set on a separate line, and numbered sequentially. Equations must be referred in the text as "Eq. (1) defines strain sensitivity". Equations must be referred in the text as "Eq. (1)" for one equation or "Eqs. (1, 2 and 3)" for multiple equations.

Conventional one-dimensional DHI model proposed by [5] is as follows.

$$\tau = W\left(a\gamma + b\gamma^3\right) + \sum_{i=1}^{n}\left\{g_i \int_0^{\Gamma}\left(\gamma'^2 - \frac{2}{3}\gamma\gamma' + 1\right)e^{\frac{\Gamma-\Gamma'}{l_i}}\, d\Gamma'\right\} \tag{1}$$

$$\Gamma = \int_0^t \left|\frac{d\gamma}{dt'}\right| dt' \tag{2}$$

$$W = \theta + (1-\theta)e^{-\frac{\gamma_{\max}}{\beta}} \tag{3}$$

As for details, refer to [5]. In previous study, DHI model which can predict performance change (e.g. temperature dependence, product variation and aging) was developed by defining the material parameters as conversion function of performance change parameters x and y [7]. As for details, refer to [7], 2019. Proposed DHI model considering performance change in previous study, however, cannot represent the property which performance is changing real-time under loading such as repeated loading and scragging effect (i.e. x and y are constant under loading) [7]. However, in case of considering repeated loading or scragging effect, x and y are not constant under loading. Therefore, it was the one of the unsolved matter that the model cannot represent the performance change under repeated loading or scragging effect. On the other hand, authors proposed DHI model considering only repeated loading effect in the past. However, the model cannot represent the scragging effect and the model is in lack of versatility because magnitude of G_{eq} and H_{eq} decrease only exponentially as the function of damage effect and temperature rise.

Under the background, modified DHI model which can represent the performance change considering repeated loading and scragging effect is proposed. In the past studies, performance change by repeated loading is defined as the function of cumulated energy dissipation per unit volume [7, 8], and it is adopted in our study as well. The function form is defined for each rubber type. In this study, performance change by repeated loading and scragging effects is studied for targeting HDR whose G_{eq} and H_{eq} are 0.392MPa and 0.240, respectively (hereafter the HDR is abbreviated as HDR-G4). For the HDR-G4, determined function forms are as follows;

$$\begin{cases} x(\Delta W) = \theta_s + (1-\theta_s)e^{\{(E-\Delta W)/\beta_s\}^{\alpha}}(E > \Delta W) \\ x(\Delta W) = \theta_s + (1-\theta_s)e^{(E-\Delta W)/\beta_s}(E \le \Delta W) \end{cases}, y = 1 \tag{4}$$

$$\Delta W = \int_0^t w\tau \frac{d\gamma}{dt'}dt', \quad \begin{cases} w = \gamma_{\max}/2 \ (\gamma_{\max} < 2) \\ w = 1 \qquad (\gamma_{\max} \ge 2) \end{cases} \tag{5}$$

where, ΔW is cumulated energy dissipation per unit volume with weight w, $\gamma_{\max}$ is experienced maximum shear strain and θ_s, β_s, E and α are material parameters. As can be seen from Eqs. (6, 7), w is added to ΔW in order to predict performance change more accurately. Regarding HDR-G4, it is found that y is not so changed under repeated loading from Eq. (5). Parameters of Eqs. (1–7) are determined to make analysis results correspond with some test results. The test conditions and specifications of the test specimen for parameter identification are shown in Tables 1 and 2, respectively. First of all, a, b, n, g_1–g_n, l_1–l_n, θ and β are determined to make analysis results correspond with 3rd cycle of shear strain dependence test results. It is important point that performance of test specimen include production variation necessarily. In fact, G_{eq} and H_{eq} of strain dependence test specimen are 0.408 MPa and 0.215 respectively. By considering the production variation, material parameters a, b, n, g_1-g_n, l_1-l_n, θ and β at standard state are determined using conversion function [7]. Next, θ_s, β_s, E and α are determine to make analysis results correspond with repeated loading test result. Identified material parameters in Eqs. (1–7) are shown in Table 3. As can be seen from Table 3, as number of plastic term is only one, g_1 and l_1 are simply expressed as g and l. Comparison with strain dependence test is shown in Fig. 1. Comparison with repeated loading test is shown in Figs. 2 and 3. It is found that analysis results show good agreement with test result from Figs. 1, 2 and 3.

3 Specification of SROLT and Time History Analysis for Verification

As for specifications of SROLT, refer to [6]. Although the system cannot carry out test using full-scale specimen, load cell equipped testing machine does not measure inertia force and friction force of linear guide. Therefore, there is no need to correct those effects in this system. Test specimen is just same as Specimen B shown in Table 2. As an aim of this test is verification of proposed DHI model, upper structure of time

Table 1 Test conditions

	Shear strain dependence test	Repeated loading test
Maximum shear strain	0.5–> 1.0–> 1.5–> 2.0–> 2.7	2
Loading wave form	Sine	Sine
Loading frequency [Hz]	0.33	0.33
Number of cycles	3 for each shear strain	50
Compressive stress [MPa]	13	13
Temperature [°C]	Atmospheric temperature	20
Specimen	Specimen A (Full-scale model)	Specimen B (Scaled model)

Table 2 Specifications of test specimen

	Specimen A (Full-scale)	Specimen B (Scaled model)
Shear stiffness [kN/mm]	1.53	0.348
Outer diameter [mm]	1000	225
Inner diameter [mm]	25	0
Rubber thickness [mm]	6.7	1.6
Number of rubber layers	30	28
Total rubber thickness [mm]	201	44.8
Insert plate thickness [mm]	4.4	1.0
First shape factor S_1	36.4	35.2
Second shape factor S_2	4.98	5.02

Table 3 Identified material parameters

	a [MPa]	b [MPa]	g [MPa]	l [–]	θ [–]	β [MPa]	θs [–]	βs [MPa]	E [MPa]	α [–]
Parameters determined for strain dependence test	0.440	0	0.470	0.320	0.350	1.00	–	–	–	–
Parameters for standard state	0.392	0	0.504	0.320	0.350	1.00	0.700	25.0	6.78	1.70

history analysis is assumed one-mass model for simplicity. Horizontal loading is one-directional and degree of freedom is only one. In addition, number of HDR in isolated layer is only one as well. Mass of upper structure is 4390kN and compressive stress

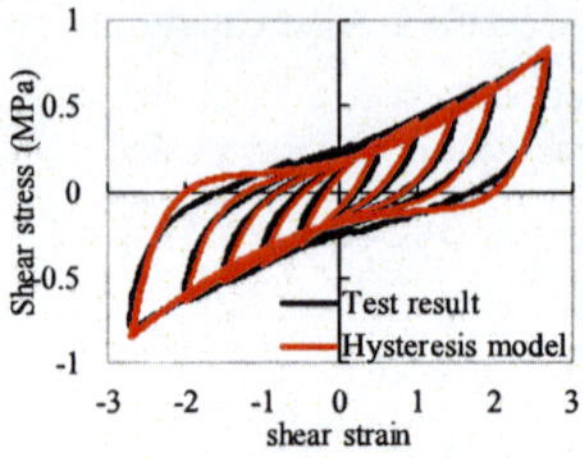

Fig. 1. Comparison with shear strain dependence test (only third cycle)

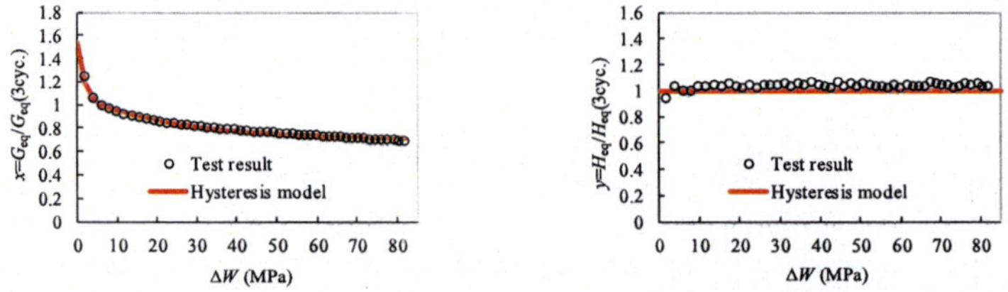

(a) G_{eq}

(b) H_{eq}

Fig. 2. Comparison with repeated loading test (Performance change of G_{eq} and H_{eq})

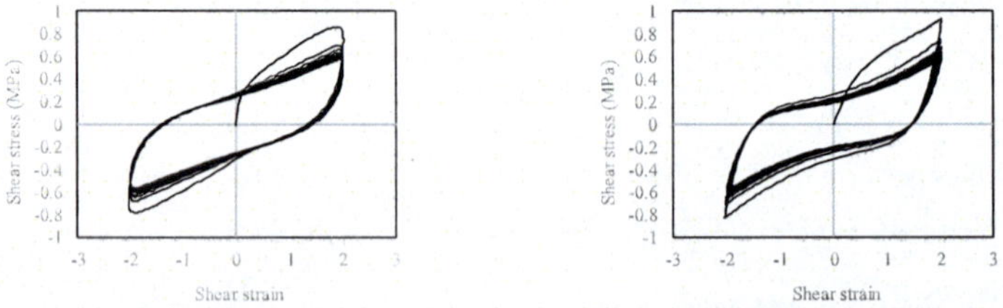

(a) Test result

(b) Hysteresis model

Fig. 3. Comparison with repeated loading test (Hysteresis loop)

for HDR is 5.59MPa. Input earthquake wave is BCJ L2, an artificial wave of Building Center of Japan (Level 2). Assumed virtual HDR specifications in isolated layer is just same as Specimen A shown in Table 2. As mentioned above, test specimen in real is scaled model and assuming virtual HDR specimen is full-scale. Therefore, after shear stress-shear strain relationship is calculated from shear force—shear displacement relationship obtained from real scaled model specimen, shear force-shear displacement relationship of assuming virtual HDR specimen is calculated from the shear stress-shear strain relationship. Shear stress is calculated by dividing shear force by cross section area of rubber. Shear strain is calculated by dividing shear displacement by total rubber thickness. In this study, 3 specimens Model A, B and C were used for verification of proposed DHI model. Test conditions for each specimen is shown in Table 4. For all specimens, benchmark test 1 was carried out first. Regarding Model A, SROLT and benchmark test 2 were carried out under the temperature 20°C after 2-month of interval from the benchmark test 1. Regarding Model B, SROLT and benchmark test

2 were carried out under the temperature 20°C after aging test. Regarding Model C, SROLT and benchmark test 2 were carried out under the temperature 0°C after aging test. Aging condition for Model A and B was carried out by thermal acceleration test and the condition is shown in Table 4.

Table 4 Test conditions

Test No	Case A	Case B	Case C
1	*Benchmark test 1* Maximum shear strain:1, Wave form: sine(0.33Hz, 3cycle), Temperature: 20 °C, Compressive stress: 5.59MPa		
2	**Interval** (Over 2-month)	**Aging by high temperature thermal deterioration** (Equivalent to 105-years)	
3	*SROLT* Input wave: BCJ L2, Compressive stress: 5.59MPa		
	Temperature: 20°C	Temperature: 20 °C	Temperature: 0 °C
4	*Benchmark test 2* Maximum shear strain:1, Wave form: sine(0.33Hz, 3cycle), Compressive stress: 5.59MPa		
	Temperature: 20 °C	Temperature: 20 °C	Temperature: 0 °C

4 Verification of Proposed Model by Comparing SROLT

4.1 Benchmark Test Results

The results of benchmark tests 1 and 2 are shown in Table 5. x" and y" are performance change of G_{eq} and H_{eq} considering temperature and aging effects, respectively. x' and y' are performance change of G_{eq} and H_{eq} considering production variation, respectively. x" and y" are calculated by following equations.

$$x''(*) = \frac{G_{eq}(2,*)}{G_{eq}(1,*)}\frac{G_{eq}(1,A)}{G_{eq}(2,A)} \tag{8}$$

$$y''(*) = \frac{H_{eq}(2,*)}{H_{eq}(1,*)}\frac{H_{eq}(1,A)}{H_{eq}(2,A)} \tag{9}$$

x"(*) and y"(*) are performance change of Case *. $G_{eq}(1,*)$ and $H_{eq}(1,*)$ are test results of benchmark test 1 for Case *. $G_{eq}(2,*)$ and $H_{eq}(2,*)$ are test results of benchmark test 2 for Case *. As can be seen from Eqs. (8, 9) and Table 5, in case of Case A, x" = y" = 1.0 assuming there is no performance change by temperature and aging effects. As can be seen from Table 5, shear modulus is larger in order of Model A, B and C as expected.

Parameter identification result calculated by conversion function [7] and total performance change $x = x'*x"$ and $y = y'*y"$ are shown in Table 6. Parameters in Table 6 were used for time history response analysis.

Table 5 Benchmark test results

	Case A	Case B	Case C
$G_{eq}-1$ (MPa)	0.407	0.408	0.426
$G_{eq}-1$ compared to design value (%)	+3.7	+4.0	+8.7
Performance change of G_{eq} by production variation x'	1.037	1.040	1.087
$G_{eq}-2$ (MPa)	0.376	0.402	0.485
$G_{eq}-2$ compared to design value (%)	−4.0	+2.5	+23.8
Performance change of G_{eq} by temperature and aging effect x''	1.00	1.066	1.231
Total performance change of G_{eq} $x = x'*x''$	1.037	1.108	1.338
$H_{eq}-1$	0.251	0.251	0.239
$H_{eq}-1$ compared to design value (%)	+4.8	+4.6	−0.5
Performance change of H_{eq} by production variation y'	1.048	1.046	0.995
$H_{eq}-2$	0.237	0.219	0.221
$H_{eq}-2$ compared to design value (%)	−1.3	−8.7	−8.0
Performance change of H_{eq} by temperature and aging effect y''	1.00	0.927	0.981
Total performance change of H_{eq} $y = y'*y''$	1.048	0.970	0.977

* $G_{eq}-1$ and $H_{eq}-1$ are test results of benchmark test 1. $G_{eq}-2$ and $H_{eq}-2$ are test results of benchmark test 2

Table 6 Identified material parameters

Model	a [MPa]	b [MPa]	g [MPa]	l [−]	θ [−]	β [MPa]	θs [−]	βs [MPa]	E [MPa]	α [−]
A	0.392	0	0.548	0.320	0.350	1.00	0.700	25.0	6.78	1.70
B	0.445	0	0.542	0.320	0.350	1.00	0.700	25.0	6.78	1.70
C	0.534	0	0.659	0.320	0.350	1.00	0.700	25.0	6.78	1.70

4.2 Comparison Between SROLT and Analysis Results

IN this section, SROLT results and time history response analysis results by various hysteresis model are shown. By comparing analysis with SROLT results, accuracy of proposed DHI model is evaluated. Regarding time history response analysis model, following 3 models were carried out.

1. *Proposed DHI model: DHI model with repeated loading and scragging effects*
2. *Conventional DHI model: DHI model without repeated loading and scragging effects*
3. *Modified bi-linear model* [9]

As for proposed and conventional DHI model, parameter values in Table 6 are used for Eqs. (1–3). Regarding proposed DHI model, parameter values of θ_s, β_s, E and α are specified in Table 6 as well. Regarding parameter values of G_{eq} and H_{eq} at maximum shear strain $\gamma = 1$ for Modified bi-linear model are 0.392MPa*x and 0.24*y, respectively. x and y are shown in Table 5. As earthquake wave BCJ L2 is not so long duration earthquake, accuracy of scragging effect was mainly evaluated in this study.

Comparison of hysteresis loop between SROLT and analysis results are shown in Fig. 4. Maximum shear displacement response of isolated layer and maximum acceleration of upper structure are shown in Fig. 5. As can be seen from Figs. 4 and 5, it has a tendency that proposed model shows the smallest deference from SROLT result. Therefore, it is found that proposed DHI model is more accurate model than other conventional models.

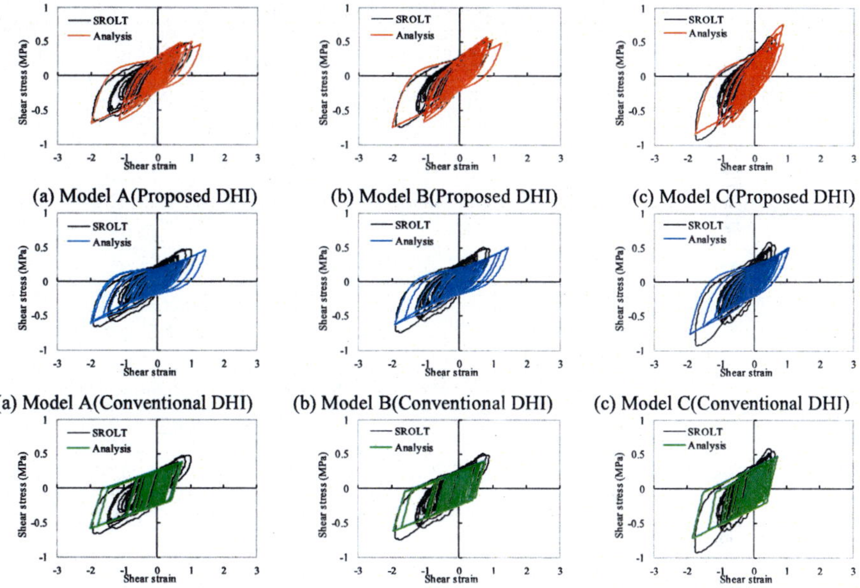

(a) Model A(Proposed DHI) (b) Model B(Proposed DHI) (c) Model C(Proposed DHI)

(a) Model A(Conventional DHI) (b) Model B(Conventional DHI) (c) Model C(Conventional DHI)

Fig. 4. Hysteresis loop of SROLT and time history response analysis

5 Conclusions

The improved DHI model for RB which can take into account of the scragging effect was developed. In this study, HDR is targeted as RB. The parameters of the model were identified by the dynamic loading test using full scale and scaled model HDR. Numerical simulation of SDOF system was conducted under three conditions. They are HDR under nominal condition, under 105 years-aged, and 105 years-aged under 0°C. The amplification factor of shear performance for simulation of the latter two conditions were estimated from the scaled model test. A hybrid simulation method, SROLT, with the test specimen used for parameter estimation, was conducted in parallel under same conditions and ground motions. The results of numerical simulation showed

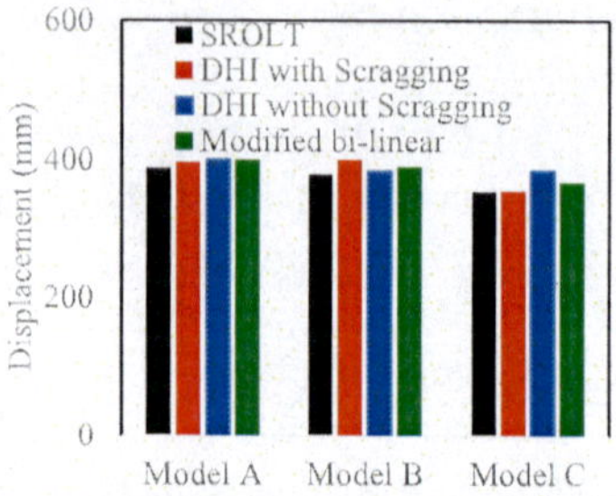

(a) Maximum displacement of isolated layer

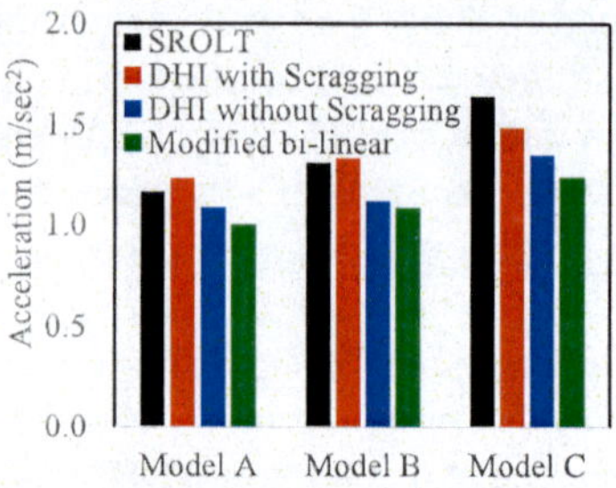

(b) Maximum acceleration of upper structure

Fig. 5. Maximum response results

good agreement with those of SROLT, and the applicability of the improved DHI model was verified.

References

1. McVitty, W.J., Constantinou, M.C.: Property modification factors for seismic isolators: design guidance for buildings, MCEER Technical Report ISSN 1520–295X, (2015)
2. ASCE 7–16 Minimum Design Loads and Associated Criteria for Buildings and Other Structures, American Society of Civil Engineers, Chapter 17 (2017)
3. ISO 22762–1:2018: Elastomeric Seismic Protection Isolators, Part 1: Test methods
4. Mullins, L.: Softening of rubber by deformation. Rubber Chem. Technol. **42**(1), 339–362 (1969)
5. Kato, H., Mori, T., Murota, N., Kikuchi, M.: Analytical model for elastoplastic and creep-like behavior of high-damping rubber bearings. J. Struct. Eng. **141**(9), (2014). https://doi.org/10.1061/(ASCE)ST.1943-541X.000118
6. Takeuchi, S., Mori, T., Sato, D., Inubushi, T., Ougiya, N., Kanda, M.: Development of substructure real-time online testing system for seismically isolated building with a small response delay using existing experimental equipment. J. Technol. Design AIJ **28**(68), 121–126 (2022). https://doi.org/10.3130/aijt.28.121
7. Mori, T., Maruyama, K., Kato, H., Murota, N.: Deformation-history integral type hysteresis model considering performance change for high-damping rubber bearings. In: 16th WCSI (2019)
8. Bridgestone corporation: Variation of characteristic of seismic isolation for long-period earthquake (High-damping rubber bearing), The building Center of Japan, legal conformity appraisals, BCJ evaluations -IB0010–01, (2017)
9. Bridgestone corporation: Variation of characteristic of seismic isolation for long-period earthquake (Lead rubber bearing), The building Center of Japan, legal conformity appraisals, BCJ evaluations -IB0012–01, (2017)
10. Bridgestone corporation: Seismic Isolation Product Line-up (2021). https://www.bridgestone.com/products/diversified/antiseismic_rubber/pdf/leaflet_en_201504.pdf

Application of Isolation in Large Scale Infrastructure in Cold Region in China

Yongfeng Du[1,2(✉)], Chao Zhang[1], and Guanghuan Wang[1]

[1] Institute of Earthquake Protection and Disaster Mitigation, Lanzhou University of Technology, Lanzhou 730050, China
dooyf@lut.edu.cn, zcjg2011@126.com, wanggh715@163.com

[2] International Research Base of Seismic Mitigation and Isolation of Gansu Province, Lanzhou University of Technology, Lanzhou 730050, China

Abstract. This paper presents the application of long, complicated isolated buildings in large scale infrastructure in cold region in China, and carried out quantitative study on the effect of low temperature and shrinkage on long isolated buildings. To investigate how low temperature and shrinkage affect the mechanical behavior of the rubber bearing, a microscopic observation of the effect of low temperature on the rubber isolator is carried out using differential scanning calorimeter (DSC) to determine the scope of freezing low temperature for the isolator to achieve. A mesomechanics model is employed to analyze the effect of micro damage on rubber bearings due to large displacement under low temperature. The dividing of construction elements of isolation layer in several long isolated buildings has been studied using such FEM software tool as Abaqus and SAP2000. The optimum construction scheme of several long isolated buildings has been suggested and verified through machine vision based monitoring in airports and urban complex buildings in typical cold regions in China.

Keywords: Base isolation · Thermal deformation of long isolated structures · Structural health monitoring · Cold region engineering

1 Introduction

Substantial progress has been made in base isolation technology in the world during the past two decades, in that not only the number of applications of base isolation increased rapidly, but also many large scale buildings have been built worldwide. The performance of base isolated structures during several strong earthquakes, *e.g.*, New Zealand on September 27, 2010, Japan on March 11, 2011, Ludian, China on August 3, 2014, Yangbi, China on May 21, 2021, and so on, has shown the effectiveness of base isolation in mitigating earthquake hazard. The examples of successful application of base isolation has promoted further application of base isolation in the high earthquake intensity region. China is among those earthquake prone countries in the world where lots of long, complicated isolated buildings have been built during the past ten years, including several largest isolated airport, largest railway station and largest urban complex in the

G. P. Cimellaro (Ed.): WCSI 2022, LNCE 309, pp. 207–216, 2023.
https://doi.org/10.1007/978-3-031-21187-4_15

world. Some of these large scale isolated buildings are located in cold region in China, and obvious thermo displacement shown up during construction stage, which aroused the attention of isolation community on the life cycle performance and safety of the long isolated buildings in cold region. Figure 1 shows a long isolated airport in south China and the longest isolated medical center in southwest China.

(a) A large isolated airport (b) A large isolated medical center

Fig. 1. Typical examples of long isolated structure in south China.

The mechanical behavior of long isolated buildings has been recognized as an important issue in civil engineering by the authors research team in 2002 after some early applications of isolated buildings. The author's research team made theoretical studies and long term monitoring on the thermal displacement in the isolators of several real isolated long structure from the start of the construction stage, including the longest isolated hospital, a middle school, together with the simulation of Kunming airport during the construction [1].

To understand the mechanism of low temperature on the mechanical behavior of rubber bearing, the effect of low temperature on the microstructure and mechanical properties of rubber materials has been investigated by some scholars since the 1950s. Fuller et al. [2] studied the effect of low temperature crystallization on the mechanical properties of natural rubber bearings, discussed the effect of strain rate on the crystallization ratio, and discussed the effect of additives on the low temperature crystallization temperature and crystallization ratio of rubber. The author's research team has carried out the cyclic test of the isolation rubber bearing under the action of low temperature in China, and the full size laminated rubber bearing has been tested in the low temperature environment of −20 °C, −35 °C, −45 °C, −50 °C. A multi-objective optimal fitting mathematical model for simulating the hysteretic restoring force of rubber isolators using the Bouc-Wen model [3, 4].

This paper presents several factors related to structural properties of long isolated building using both dynamic response during usage stage and thermal displacement in the isolators of the super structure, together with structural monitoring of several large isolated airports and isolated urban complex. A microscopic observation of the effect of low temperature on the rubber isolator is carried out using differential scanning calorimeter (DSC) to determine the scope of freezing low temperature for the isolator to achieve. A mesomechanics model is employed to analyze the effect of micro damage on rubber bearings due to large displacement under low temperature. The dividing of construction elements of isolation layer in a long isolated buildings has been studied using such FEM software tool as Abaqus, Ansis and SAP2000. The optimum construction scheme of several long isolated buildings has been suggested and verified through machine vision based monitoring in airports and urban complex buildings in typical cold regions.

2 Simulation of Low Temperature Deformation

The horizontal stiffness of the isolation layer is small. The designers usually take advantage of the isolation layer to release the non-loaded deformation and temperature stress of the superstructure caused by seasonal temperature changes and concrete shrinkage, and mostly adopt the scheme of reducing or canceling the deformation joints of the superstructure, which is conducive to the integrity of the superstructure. However, the adoption of such a long structural element scheme will accumulate large non-loaded deformation in the isolation bearing and affect the stability and user safety of the rubber isolation bearing, as shown in Fig. 2. The displacement in the isolators caused by temperature differential and shrinkage of concrete in this type of long isolated buildings in cold regions with high temperature fluctuation are calculated using FEM software tool as Abaqus and SAP2000 for 5 different cases as listed in Table 1.

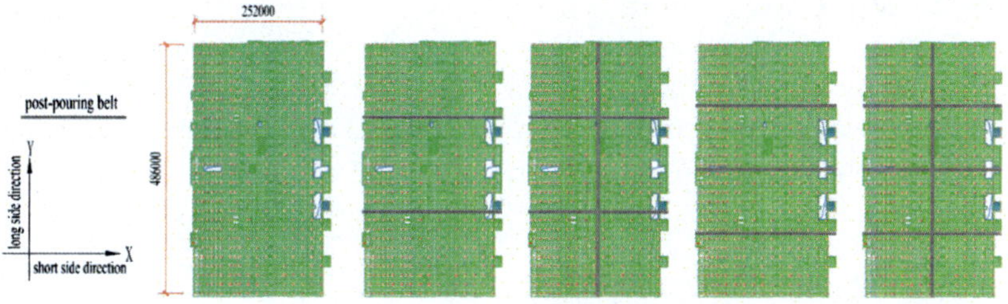

(a) coordinate (b) case 1 (c) case 2 (d) case 3 (e) case 4 (f)case 5

Fig. 2. Layout scheme of post-pouring belt.

Table 1. Layout of post-pouring belts

No. of post-pouring belts	Case 1	Case 2	Case 3	Case 4	Case 5
Long direction	0	2	2	3	3
Short direction	0	0	1	0	1

Figure 3 shows the comparison of the maximum temperature shrinkage deformation of the isolation layer at each observation time point of post-pouring belt with different casting time. It can be seen from Fig. 3 that for controlling temperature and shrinkage deformation of the isolation layer, when post-pouring belt is casted in March of the second year of construction, the maximum temperature and shrinkage deformation in the resultant force direction of isolation layer is smaller at the observation points of the most unfavorable time of temperature shrinkage deformation, and temperature and shrinkage deformation manipulating effect of isolation layer is the best [5].

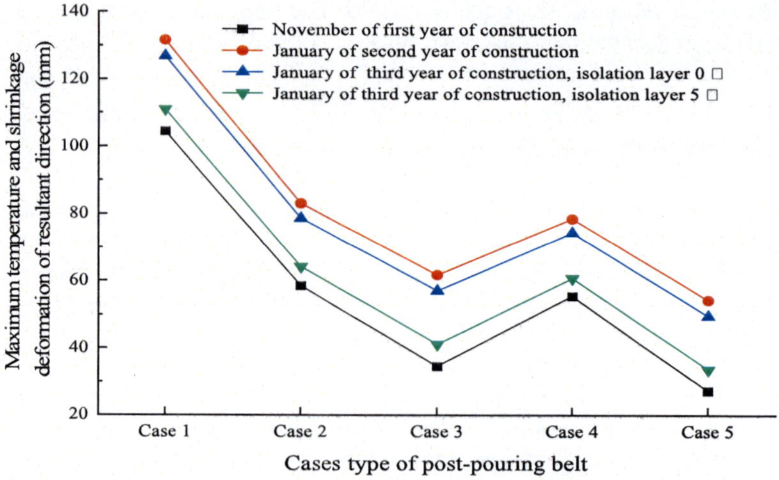

Fig. 3. Maximum thermo and shrinkage deformation for different cases

3 Test of Concrete Shrinkage

To investigate the concrete shrinkage in a long isolated building, over 42 specimens have been worked out for measuring the concrete shrinkage, and the time history of the shrinkage are obtained, as shown in Fig. 4.

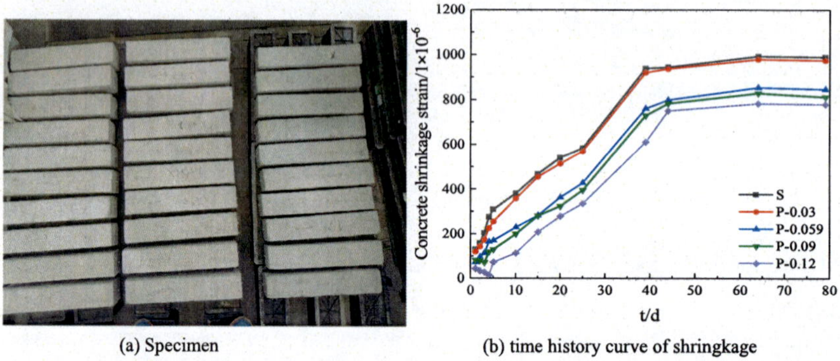

Fig. 4. Test of concrete shrinkage and time history of shrinkage.

Figure 4 shows the shrinkage strain curve of concrete with different expansive agent content. From the figure, the shrinkage trend of each specimen with different expansive agent content is similar, and the shrinkage strain increases rapidly in the early stage, and gradually slows after 40 days. 79 days, S group of concrete shrinkage strain is 990×10^{-6}, expansive agent dosage of 3%, 5.9%, 9%, 12% of concrete shrinkage strain were 976×10^{-6}, 848×10^{-6}, 813×10^{-6}, 781×10^{-6}. The shrinkage strain of concrete decreases with the increase of expansive agent content. Compared with group S, when the expansive agent content is 3%, 5.9%, 9% and 12%, the shrinkage strain of concrete decreases by 1.4%, 14.3%, 17.8% and 21.1%, respectively.

4 Test of Rubber Material and Isolator in Low Temperature

Rubber isolation bearing is a kind of widely used isolation element which is made of rubber and thin steel plate alternately. Studies have shown that crystallization in rubber may occur at low temperature, which will reduce the deformation capacity of rubber, and might be the reason for initial damage under low temperature. One may easily understand, when the rubber isolation bearings work at low temperatures, the rubber will become hard and brittle, and even appear verification situation. The decrease of elasticity of rubber material at low temperature will lead to the decrease of the energy dissipation capacity of rubber isolation bearing, which will affect its ability to isolate the transmission of seismic energy to the superstructure under earthquake action.

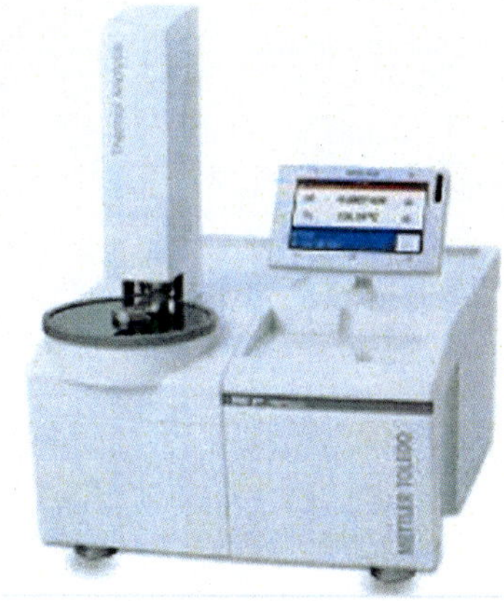

(a) DSC device

(b) Low temperature test device of rubber

Fig. 5. Test facilities for rubber material at low temperature.

Seventy percent of the land area of China is in extremely cold or cold areas. During the construction period, when the isolation layer has not been completely closed from outdoor environment, rubber isolation bearings may work in cold or extremely cold temperature. The issue of rubber isolation bearings in cold environment is extremely concerned by the seismic engineering field and urgently needs to be solved. In order to clarify the applicability of the rubber isolation bearings produced by typical Chinese enterprises at low temperatures, differential scanning calorimetry analysis is adopted to conduct DSC test on the rubber used in the production of rubber isolation bearings in the ultra-long isolation airport studied [6], typical test facilities is shown in Fig. 5. The glass transition temperature T_{gl} of rubber material and the minimum and maximum temperature of rubber crystallization and the scope of crystallization temperature were obtained including the temperature ranging from −50 °C to −60 °C in N_2 atmosphere, which provided basis for setting the temperature of isolator for the subsequent mechanical properties test and numerical simulation analysis of rubber bearing at low temperature. Some rubber specimens, as shown in Fig. 6, were tested for the material decay under natural outdoor temperature in the low temperature environment of up to −12 °C, as shown in Fig. 7.

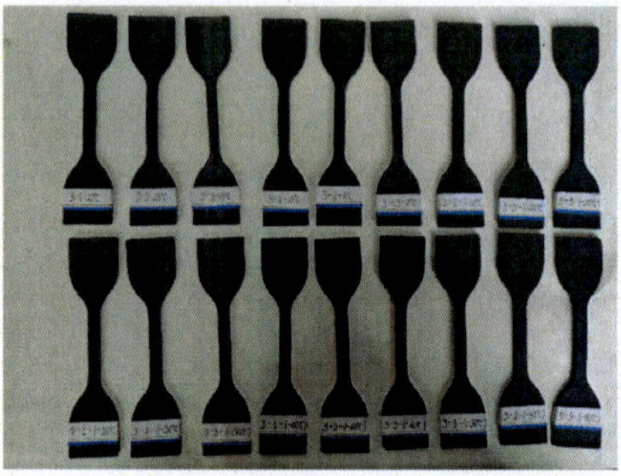

Fig. 6. Low temperature test rubber specimen

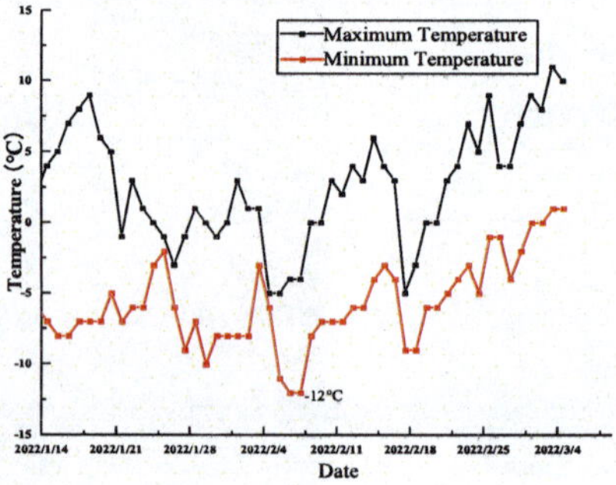

Fig. 7. Environment temperature curve

5 Mesomechanics Model Based Analysis of Micro Damage

One may easily understand, when the rubber isolation bearings work at low temperature at the same time the rubber bearings keep in large deformation state during the construction. Some micro damage may occur in rubber bearings due to the combined action of low temperature and large deformation. The brittleness of the rubber may cause such type of rubber isolation bearings in long isolated structure in cold region become special not only in terms of external force, but also in terms of constitution relationship of material. A mesomechanics model is employed to analyze the effect of micro damage on rubber bearings due to large displacement under low temperature.

Based on a three-dimensional rubber isolator model, the effect of large deformation and low temperature are analyzed considering the initial micro damage in the rubber and in the interface between rubber and steel plate, shown in Fig. 8.

The commercial software tool ANSYS is employed to perform the FEM analysis of rubber isolator, and the elements of SOLID45, SOLID185 are adopted, respectively, to represent steel and rubber shim, and Mooney-Revlin model is used for rubber constitution model.

Fig. 8. Mesomechanics model composite mat.

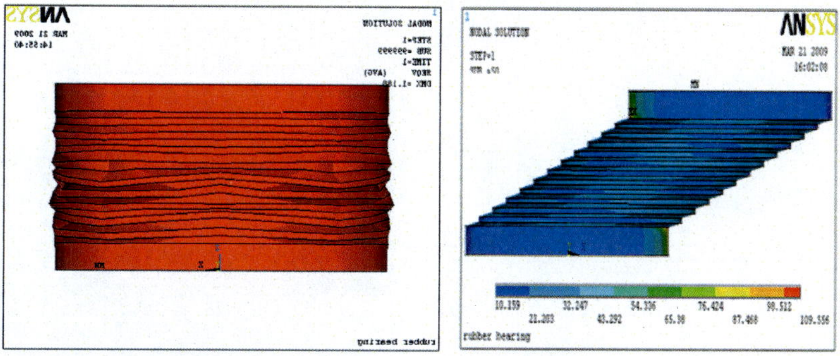

(a) Rubber isolator under compression and shear

(b) Rubber isolator in large deformation

Fig. 9. Distorted rubber isolator and lateral deformation.

The ambient temperature, displacement of rubber isolator and lead core thermal effect on the properties in large deformation state was analyzed using simplified model, preliminary analysis of a real long isolated building are obtained, and compared with the monitored results to derive the model of long isolated buildings in cold regions, and suggest the value of the parameters to represents the rubber isolator for analyzing the long isolated structure, as shown in Fig. 9.

6 Computer Vision Based Monitoring of Isolated Building

For the long isolated buildings in cold regions, the isolation bearing has a large lateral displacement during the construction stage due to combined action of temperature difference and shrinkage. Real-time, accurate monitoring of the deformation of the isolation bearing may provide an effective basis for guiding the construction progress and judging the actual performance of the long isolated structure in cold region. The traditional contact displacement measurement method may be limited by dimension, precision, operation and installation, especially for the isolated buildings in very cold regions, where measurement is rather difficult for manual operation to realize. On the other hand, the computer vision based displacement monitoring of isolation bearing has many advantages, such as long distance, non-contact, high precision, time-saving and labor-saving, multi-point monitoring and so on. The displacement information of isolation bearing of long isolation structure caused by temperature change and other factors such as concrete shrinkage can be obtained in real time. In this method, target tracking is performed on the video image sequence of the structure to get the movement track of the target point in the image, and then the displacement information of the structure is determined by the geometric relationship between the image and the real world.

A displacement monitoring system based on computer vision is established, as shown in Fig. 10. The system uses the network camera as the acquisition equipment, transmits image information wirelessly, and uses the computer equipped with the geometric matching algorithm based on color tracking to obtain the displacement information of the target point, so as to realize the non-contact measurement of the displacement of the isolation bearing in long isolated buildings [7].

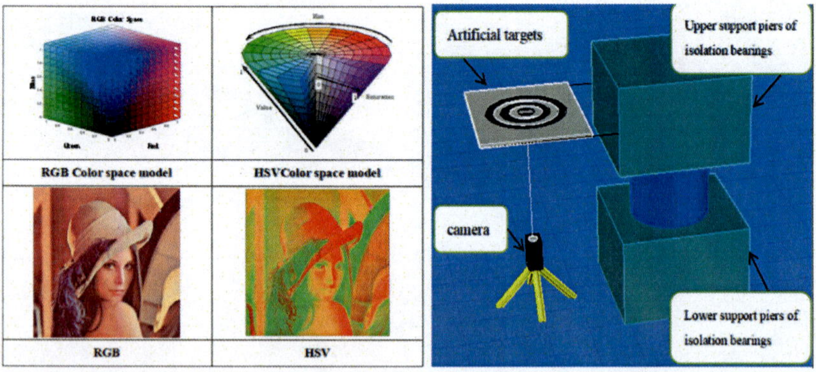

(a) Identification of visional information (b) Monitoring system of isolated buildings

Fig. 10. Computer vision based monitoring system.

The monitoring system is set to collect data periodically every hour [8]. The displacement monitored at the measuring point No.1 from May 16 to June 2, 2021 are selected for plotting displacement curve, as shown in Fig. 11. The position track of point No. 10 within two months was plotted in Fig. 12.

7 Conclusion

Through several numerical examples and experimental study of several long isolated buildings in cold regions in China, one obtains:

(1) Post-pouring belt is an effective scheme for reducing the temperature and concrete shrinkage deformation in the long isolated buildings in cold regions, but the reinforcing bar of the isolation beam and slab should be disconnected.

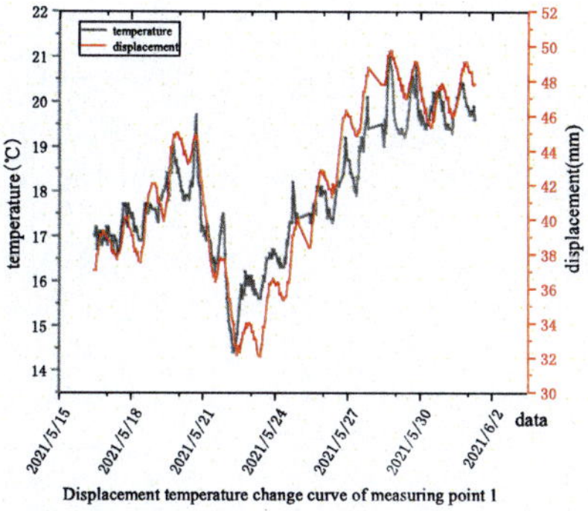

Fig. 11. Displacement curve of point 1

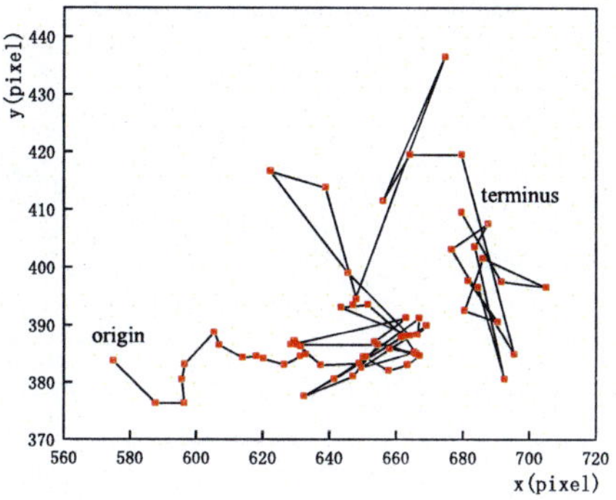

Fig. 12. Position track diagram of point 10

(2) The concrete shrinkage rate may slow down when the concrete casted over certain period, but the shrinkage duration lasts relatively long time for ordinary concrete. Expansive agent is a useful additive for reducing the concrete shrinkage.

(3) Low temperature may reduce the deformation capacity of rubber material, which might be the main reason for initial damage under low temperature.

(4) The computer vision based monitoring is an effective way for measuring the deformation of isolation bearings in the long isolated buildings in cold regions, in that it may measure the displacement in a non-contact way, with relatively high precision, and become easier for unfavorable environment.

Acknowledgements. The authors gratefully acknowledge the financial support provided in part by the Natural Science Foundation of China under Project No. 52178291. This support is gratefully acknowledged.

References

1. Du, Y.F., Zhao, L.J., Zhang, T., Zheng, W.Z., Li, W.R.: Study on construction mechanics of long complicated isolated structures and life-cycle monitoring. Eng. Mech. **32**(07), 1–10+25 (2015)
2. Fuller, K.N.G., Gough, J., Thomas, A.G.: The effect of low-temperature crystallization on the mechanical behavior of rubber. J. Polym. Sci. Part B: Polym. Phys. **42**(11), 2181–2190 (2010)
3. Du, Y.F., Zheng, W.Z., Li, W.R., Li, H., Wang, H.: Study on the dependency of static and dynamic characteristics with environmental temperature for long irregular base-isolated structures. Eng. Mech. **34**(07), 69–78 (2017)
4. Du, Y.F., Shi, C., Zhao, L.J., Li, H.: Optimum scheme of long isolated structures based on life-cycle performance. In: Proceedings 1st International Conference Performance-based and Life-cycle Structural Engineering, pp. 2052–2058. Hong Kong PloyU, Hong Kong, 5–7 Dec (2012)
5. Zhang, C., Wang, G.H., Du, Y.F., et al.: Simulation of Thermo Deformation of Long Isolated Structures based on Probabilistic Temperature Model. To appear in PhD student seminar, Lanzhou University of Technology (2022)
6. Wang, G.H., Zhang, C., Du, Y.F., et al.: Study on the Crystallization Behavior of Natural Rubber and Low Temperature Test Scheme. To appear in PhD student seminar, Lanzhou University of Technology (2022)
7. Du, Y.F., Li, X.X.: Monitoring of Isolator Deformation of Complex Seismic Isolation Structure based on Computer Vision. To appear in Industrial Construction (2022)
8. ISMI Homepage: Field Observation During Construction of Long Isolated Structures (in Chinese). https://www.ismi.lut.edu.cn. Last accessed 20 May 2022

Experimental and Analytical Investigation of Variable Curvature and Friction Pendulum Isolator

Ping Tan[1,2], Jiying Shang[3](✉), Jianping Han[3], Kui Yang[1,2], and Yafei Zhang[4]

[1] School of Civil Engineering, Guangzhou University, Guangzhou 510006, China
[2] Earthquake Engineering Research & Test Center, Guangzhou University, Guangzhou 510405, China
[3] School of Civil Engineering, Lanzhou University of Technology, Lanzhou 730050, Gansu, China
shangjy@lut.edu.cn
[4] GMC Grand-Bay Intelligent Manufacturing and Technology Co. Ltd., Guangzhou 510730, China

Abstract. A novel variable friction pendulum isolator (VFPI) and a novel variable curvature and friction-friction pendulum isolator (VCF-FPI) are developed in this study, which introduces variable mechanical properties into the conventional friction pendulum isolator (FPI). The mechanical behavior and operating principles of the VFPI and VCF-FPI are illustrated. The theoretical constitutive models are subsequently derived based on the mechanical behavior of the VFPI and VCF-FPI. A numerical modeling scheme of VFPI for response-history analysis, composed of existing elements, is proposed, which can be easily achieved in the currently available analysis software. Mechanical property tests for the full-scale VFPI and VCF-FPI are carried out to verify the force-displacement relationships obtained from theoretical analysis and numerical modeling. These investigations conclude that both VFPI and VCF-FPI provide desired changeable stiffness and damping with increasing isolator deformation. The hysteresis loops of VFPI and VCF-FPI are thin in the middle and thick at both ends, which indicates the energy dissipation capability of VFPI and VCF-FPI increases with increasing isolator deformation.

Keywords: Variable curvature and friction-friction pendulum isolator · Variable friction pendulum isolator · Hysteresis loop · Numerical modeling scheme · Mechanical property tests

1 Introduction

Seismic isolation is one of the most efficient methods to upgrade the seismic resistibility of structures. The friction pendulum isolator (FPI) proposed by Zayas is one of the most effective isolation techniques and integrates isolation, energy dissipation, and restoring mechanisms [1–3]. FPI has been demonstrated to be an effective tool for protecting civil engineering structures against earthquake damage through experimental studies

G. P. Cimellaro (Ed.): WCSI 2022, LNCE 309, pp. 217–228, 2023.
https://doi.org/10.1007/978-3-031-21187-4_16

[4, 5] and engineering applications [6, 7]. Therefore, a large body of studies focused on developing more versatile and economic isolation systems based on FPI have been conducted. Hence double concave friction pendulum isolator (DCFPI) [8], triple friction pendulum isolator (TFPI) [9], and quintuple friction pendulum isolator (QFPI) [10] have been come up with in succession. These FPI derivatives, especially TFPI and QFPI, are capable of progressively exhibiting different hysteretic properties at various levels of displacement, which exhibit adaptive stiffness and damping by increasing the isolator deformation. However, the complex mechanism is caused, which isn't beneficial to engineering applications. Another method to achieve adaptive stiffness and damping is implementing the variable friction mechanism. Kong et al. [11], Calvi et al. [12, 13] proposed a variable friction pendulum isolator (VFPI) in which the sliding surface has several parts covered with different friction materials. However, the method of achieving variable friction leads to unstable performance and complicates the machining processes. A variable curvature mechanism is also used to improve the performance of FPI isolation technology. Pranesh et al. [14, 15] first proposed the variable curvature friction pendulum isolator (VCFPI), which is effective for improving the earthquake-proof capability of the structures during low-intensity and high-intensity earthquakes. The type of the variable curvature mechanism can be summarized as follows: the curvature radius of the sliding surface increases with increasing isolator deformation; the curvature radius decreases with increasing isolator deformation. The first type is good for avoiding the low-frequency resonance of the seismically isolated structures under near-fault ground motions but causes large isolator deformation. The second type is good for limiting isolator deformation but decreases the isolation effect.

Given that, the current research team has made more practical attempts to improve conventional FPI. A novel VFPI is proposed, in which the sliding surface is arranged as a series of concentric rings with different roughness [16, 17]. A numerical modeling scheme of the novel VFPI is also proposed, and its accuracy is verified by comparison with results obtained from experimental tests and theoretical analyses. Further, a novel variable curvature and friction-friction pendulum isolator (VCF-FPI) is proposed. The theoretical constitutive model is derived based on the mechanical behavior of the VCF-FPI and verified by mechanical property tests.

2 VFPI Configuration

As Fig. 1a shows, the components of the VFPI are the top plate, internal nested slider, and bottom plate. The upper and lower surfaces of the slider are covered with friction materials. The sliding surface is divided into a series of concentric rings whose roughness increases from the central ring to the outermost ring, as shown in Fig. 1b. The radii of the concentric circles C_1, C_2, and C_i are r_1, r_2, and r_i, respectively. The friction coefficient of the sliding bands C_1, $C_2 \cap C_1$, and $C_i \cap C_{i-1}$ are μ_1, μ_2, and μ_i, respectively, where $\mu_1 < \mu_2 < \ldots < \mu_i$.

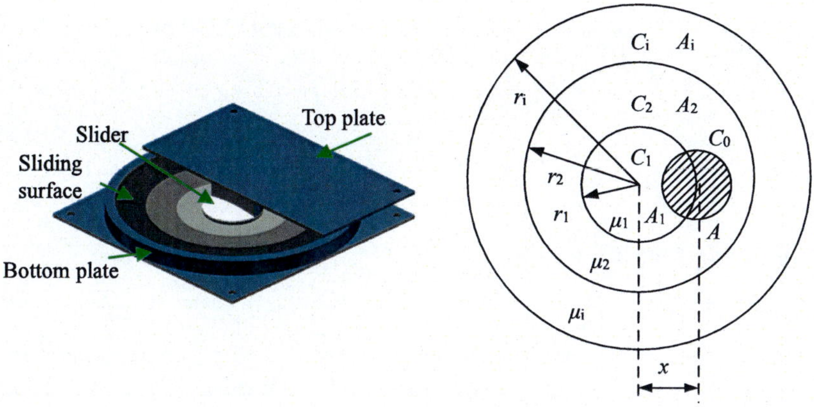

(a) magnified view. (b) schematic plan view of the sliding surface.

Fig. 1. VFPI configuration.

3 Theoretical Constitutive Relationship of VFPI

Sliding will happen when the horizontal force F is greater than the friction force of the central sliding band C_1, f_1. Similar to the traditional FPI, the horizontal force-displacement relationship is given by:

$$F = \frac{W}{R}x + f(x), \tag{1}$$

where W is the vertical load on the isolator; R is the effective radius of curvature of the sliding surface; as shown in Fig. 1b, x is the relative displacement of the slider on the sliding surface; $f(x)$ is the friction force on the sliding interface, which increases with relative displacement x.

The effective friction force increases linearly with relative displacement x [12]. Therefore, the friction force is given by:

$$f(x) = f_1 + tx, \tag{2}$$

where $f_1 = \mu_1 W$; t is the rate of friction force variation from the center to the periphery of the sliding surface and can be expressed as follows:

$$t = \frac{f_i - f_1}{x}, \tag{3}$$

where f_i is the friction force of the ith sliding band. Substituting Eq. (2) into Eq. (1):

$$F = \left(\frac{W}{R} + t\right)x + f_1 \tag{4}$$

Once the sliding is reversed, the force-displacement is expressed by:

$$F = \left(\frac{W}{R} - t\right)x - f_1 \tag{5}$$

Assuming $t = a\%\frac{W}{R}$, Eqs. (4) and (5) can be rearranged as follows:

$$F = (1 + a\%)\frac{W}{R}x + f_1 \tag{6}$$

$$F = (1 - a\%)\frac{W}{R}x - f_1 \tag{7}$$

4 Analysis Model of VFPI

4.1 Modeling Scheme

An analysis model of VFPI for response-history analysis is proposed by connecting the FPI element [18, 19] and damper-friction spring (D-FS) element [18] in parallel. The hysteretic behavior of the FPI element (Fig. 2a) can be represented by the combination of a rigid plastic friction model (Fig. 2b) and a linear spring (Fig. 2c). The hysteretic behavior of the D-FS element (Fig. 3a) can be represented by the combination of a linear hysteretic damping model (Fig. 3b) and a linear spring (Fig. 3c). Similarly, the hysteretic behavior of VFPI (Fig. 4a) can be represented by the combination of a Bow-Tie model (Fig. 4b) [20] and a linear spring (Fig. 4c). On the other hand, the Bow-Tie model (Fig. 4b) can be decomposed into the rigid plastic friction model (Fig. 2b) and the linear hysteretic damping model (Fig. 3b). Therefore, it is reasonable that the hysteretic behavior of VFPI is represented by connecting the FPI element and D-FS element in parallel.

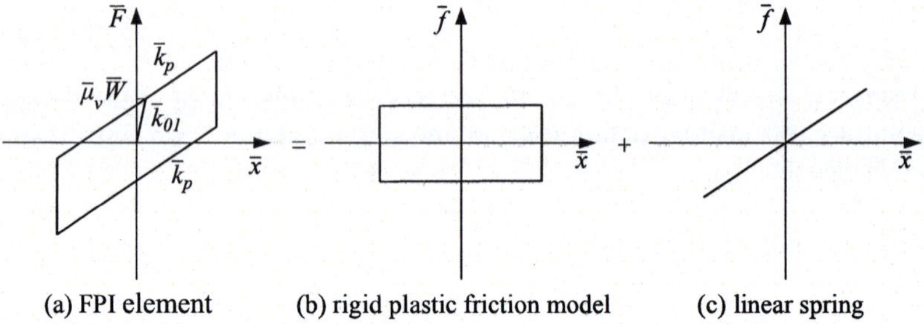

Fig. 2. Decomposition of the hysteresis curve of FPI element.

4.2 Input Parameters

In the analysis model, the modifications made to the input parameters of each element are needed to accurately capture the hysteretic behavior of VFPI. The fundamental principle of the modifications is that the hysteretic curve obtained from the analysis model and actual VFPI have equal elastic stiffness, post-elastic stiffness, and yield strength. The values assigned to the input parameters of the two-element parallel model are summarized in Tables 1 and 2, respectively. In the tables, the overbar notion and standard notion are used to denote parameters associated with the analysis model and the actual behavior, respectively.

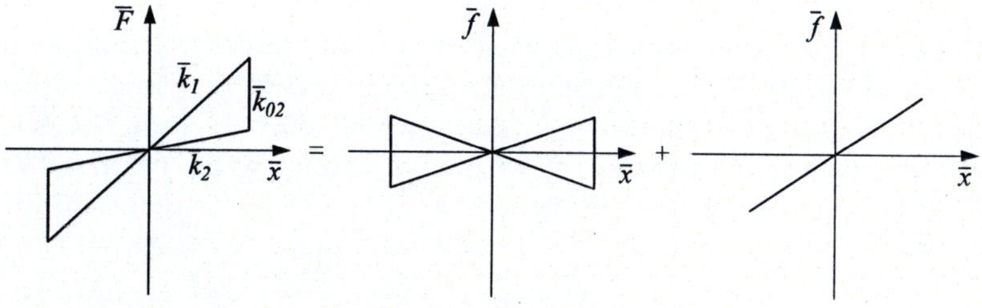

(a) D-FS element (b) linear hysteretic damping model (c) linear spring

Fig. 3. Decomposition of the hysteresis curve of D-FS element.

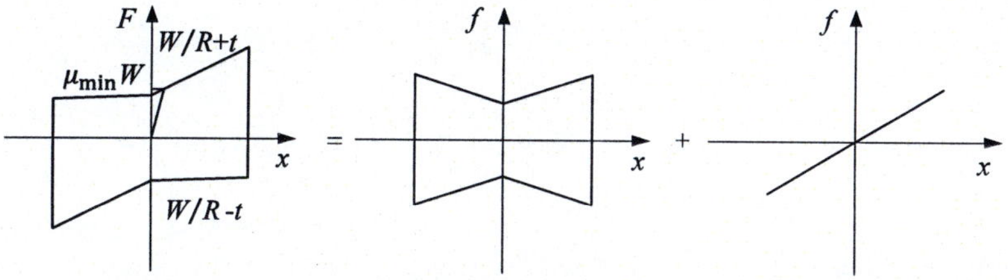

(a) VFPI (b) Bow-Tie model (c) linear spring

Fig. 4. Decomposition of the hysteresis curve of VFPI.

Table 1. Summary of input parameters of the FPI element in the analysis model.

Friction coefficient	Radius of curvature	Elastic stiffness	Vertical stiffness
$\overline{\mu} = 2\mu_1$	$\overline{R} = \frac{R}{2(1-a\%)}$	$\overline{k}_0 = \frac{W}{2d_y}$	$\overline{k}_v = \frac{AE}{2h}$

Note: d_y is the yield displacement of VFPI, assumed to be 0.2 mm; A is the projected area of the slider, h is the height of VFPI; E is a representative modulus assumed to be half of that of the steel

Table 2. Summary of input parameters of the D-FS element in the analysis model.

Loading stiffness	Unloading stiffness	Vertical stiffness
$\overline{k}_1 = (2a\%)\frac{W}{R}$	$\overline{k}_2 = 0$	$\overline{k}_v = \frac{AE}{2h}$

4.3 Validation of the Analysis Model

Mechanical property tests were conducted to validate the proposed VFPI analysis model. A VFPI test isolator is machined, as shown in Fig. 5. A newly developed modified-polyformaldehyde (M-POM) was used as the friction material and covered on the lower surface of the slider. The geometric parameters of the test isolator are listed in Table 3.

Figure 6 shows the experimental hysteresis curves of the VFPI under sinusoidal waves displacement-controlled loading. The displacement-controlled loading cases were carried out at an amplitude of 230 mm and a frequency of 0.138 Hz. By comparison of experimental results, numerical modeling results and theoretical analysis results, the accuracy of the analysis model and analytical algebraic equations are verified. As shown in Fig. 6, the experimental loop is represented by the black solid line, while the numerical analysis loop is represented by the red dashed line, and the theoretical analysis loop is represented by the blue short dashed line. The effective stiffness and effective damping ratio of the hysteretic loops were calculated using Eqs. (8) and (9) to quantify the goodness-of-fit of the results from the experimental tests, numerical analyses, and theoretical analyses. The relative errors of the effective stiffness and effective damping ratio among experimental loops, numerical loops and theoretical analysis loops are within 9% and 5%, respectively.

$$k_e = \frac{F_D}{D}, \tag{8}$$

$$\zeta_e = \frac{E_D}{2\pi k_e D^2}, \tag{9}$$

where D is the isolator deformation, F_{D} is the horizontal force corresponding to D, and E_{D} is the area enclosed by the hysteretic loop.

Table 3. The geometric parameters of the test isolator.

Concentric rings	Curvature radius (m)	Slider diameter (mm)	Ring spacing (mm)	Ring roughness (μm)
Central ring	2	240	200	0.02
Middle ring	2	240	100	0.21
Outermost ring	2	240	100	0.39

Fig. 5. Photograph of the VFPI.

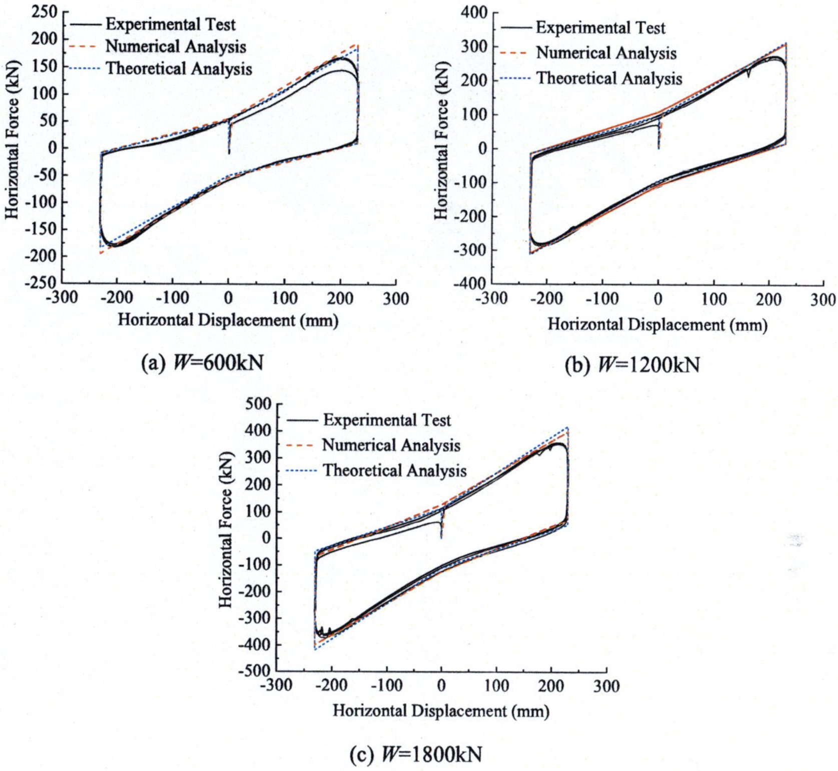

Fig. 6. Comparison of the hysteretic loops from experimental tests, numerical analyses, and theoretical analyses.

5 Configuration and Mechanical Behavior of VCF-FPI

5.1 Configuration

AS shown in Fig. 7, the VCF-FPI is composed of the top and bottom plates, and an internal nested slider. The upper surface of the bottom plate is the sliding surface with several sliding bands with different roughness and curvature radius. These sliding bands are set as a series of concentric rings, as shown in Fig. 8. The roughness of the sliding band increases from the center to the outermost while its curvature radius decreases.

5.2 Theoretical Constitutive Relationship

The parameterized equation of the sliding surface of the VCF-FPI can be expressed in the following equation when the sliding surface has two sliding bands:

$$y(x) = \begin{cases} R_1 - \sqrt{R_1^2 - x^2} & |x| \le d_1 \\ R_1 - \sqrt{R_1^2 - d_1^2} + \sqrt{R_2^2 - d_1^2} - \sqrt{R_2^2 - x^2} & |x| > d_1 \end{cases}, \tag{10}$$

Fig. 7. VCF-FPI configuration.

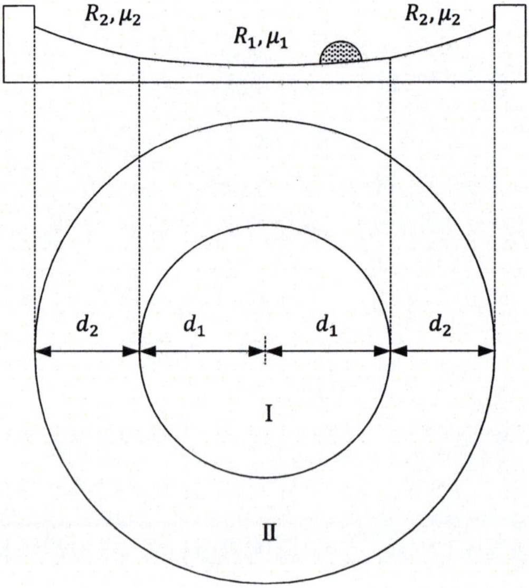

Fig. 8. Cross-section view of the VCF-FPI.

where x is the relative displacement of the slider on the sliding surface. Based on the principle of statical equilibrium, the horizontal force-displacement relationship can be written as:

$$F(x) = F_r(x) + F_f(x), \tag{11}$$

where $F_r(x)$ is the restoring force, $F_f(x)$ is the friction force, $F_r(x)$ and $F_f(x)$ can be expressed as follows:

$$F_r(x) = k_r(x) \cdot x \tag{12}$$

$$F_f(x) = sgn(\dot{x})\mu W, \tag{13}$$

where $k_r(x)$ is post-elastic stiffness of the VCF-FPI, μ is the friction coefficient of the sliding surface, W is the vertical load on the isolator. And the $k_r(x)$ is given by:

$$k_r(x) = Wy'(x)/x \tag{14}$$

Hence, the horizontal force-displacement relationship can be rearranged as follows.
For $|x| \leq d_1$,

$$F(x) = \frac{W}{R_1}x + sgn(\dot{x})\mu_1 W, \tag{15}$$

For $|x| > d_1$,

$$F(x) = \frac{W}{R_2}x + sgn(\dot{x})\mu_2 W, \tag{16}$$

where R_1 and μ_1 are the curvature radius and friction coefficient of the sliding surface in the sliding band I; R_2 and μ_2 are the curvature radius and the friction coefficient of the sliding surface in the sliding band II Considering incomplete contact will happen when the slider crosses the sliding band I, and the slider transition from sliding band I to sliding band I is gradual, Eq. (16) can be rewritten as:

$$F(x) = \frac{W}{R_2}x + sgn(\dot{x})(\alpha\mu_1 + \beta\mu_2)W, \tag{17}$$

where $\alpha, \beta \leq 1$.

5.3 Validation of the Theoretical Model

A VCF-FPI test specimen was designed and processed, and mechanical property tests were conducted to validate the theoretical model of the VCF-FPI. Two sliding bands were divided on the sliding surface with the roughness of $R_a = 0.02$ μm and the curvature radius of $R_1 = 2000$ mm for the central sliding band and with the roughness of $R_a = 0.31$ μm and the curvature radius of $R_1 = 1500$ mm for the outmost sliding band. Figure 9 shows the photo of the VCF-FPI specimen. Tests were carried out with the two applied vertical loads of 1000 kN and 2000 kN, which were kept constant during the application of motion. The sinusoidal displacement-controlled loading cases were carried out at an amplitude of 200 mm and a frequency of 0.095 Hz in the horizontal direction. Figure 10 shows the machine for testing full-scale VCF-FPI.

Figure 11 shows a comparison between the experimental and analytical curves under different vertical loads. The experimental loop is represented by the black solid line, while the theoretical analysis loop is represented by the blue dashed line. Both the experimental loop and theoretical loop are thin in the middle and thick at both ends, which indicates the energy dissipation capability of VCF-FPI increases with increasing isolator deformation. The theoretical algebraic equations accurately reproduce the experimental hysteresis loop when α and β are equal and taken as 0.5. The effective stiffness and effective damping ratio of the hysteretic loops were calculated using Eqs. (8) and (9) to quantify the goodness-of-fit of the results from the experimental tests and theoretical analysis results. The relative errors of the effective stiffness and effective damping ratio between experimental loops and theoretical analysis loops are within 10% and 7%, respectively.

Fig. 9. Photograph of the VCF-FPI.

Fig. 10. Test machine.

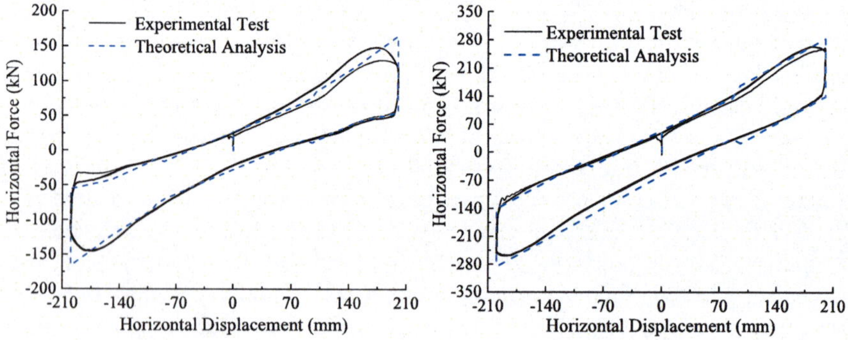

(a) W=1000 kN (b) *W*=2000 kN

Fig. 11. Comparison of the hysteretic loops from experimental tests and theoretical analyses.

6 Conclusions

This paper proposes a novel VFPI and a novel VCF-FPI. The general principles of operation and mechanical behavior of the novel isolators were explained, and their analytic models were presented. Furthermore, the accuracy of the analytical models was validated by carrying out mechanical property tests for full-scale isolators. The following conclusions were drawn from this study:

(1) The VFPI and VCF-FPI are an integration of traditional FPI and variable curvature and friction mechanism, which provide desired changeable stiffness and damping with increasing isolator deformation. The hysteresis loops of VFPI and VCF-FPI are thin in the middle and thick at both ends, which indicates the energy dissipation capability of VCF-FPI increases with increasing isolator deformation.
(2) The theoretical algebraic equations and analysis model proposed for the VFPI can accurately reproduce the experimental hysteresis loop. The relative errors of the effective stiffness and effective damping ratio among experimental loops, numerical loops and theoretical analysis loops are within 9% and 5%, respectively.
(3) The theoretical algebraic equations proposed for VCF-FPI accurately reproduce the experimental hysteresis loop when α and β are equal and taken as 0.5. The relative errors of the effective stiffness and effective damping ratio between experimental loops and theoretical analysis loops are within 10% and 7%, respectively.

Acknowledgments. The authors acknowledge support provided by National key R&D program of China (grant No. 2019YFE0112500), National Natural Science Foundation of China (grant No. 51978185) and Program for Changjiang Scholars and Innovative Research Team in University (grant No. IRT13057).

References

1. Zayas, V.A.: Earthquake protective column support. US Patent 4644714, Feb 24 (1987)
2. Zayas, V.A., Low, S., Mahin, S.: The FPS Earthquake Resisting System: Experimental Report (ucb/eerc-87/01). University of California, Berkeley, CA, Berkeley (1987)
3. Zayas, V.A., Low, S.S., Mahin, S.A.: A simple pendulum technique for achieving seismic isolation. Earthq. Spectra **6**(2), 317–333 (1990)
4. Mokha, A., Constantinou, M.C., Reinhorn, A.M.: Experimental and Analytical Study of Earthquake Response of a Sliding Isolation System with a Spherical Surface. National Center for Earthquake Engineering Research, University at Buffalo, Buffalo (1990)
5. Ren, X., Lu, W., Mosalam, K.M., et al.: Similitude theory for scaled friction pendulum bearings for shaking table experiments. Soil Dyn. Earthq. Eng. **121**, 399–404 (2019)
6. Mokha, A.S., Amin, N., Constantinou, M.C., et al.: Seismic isolation retrofit of large historic building. J. Struct. Eng. **122**(3), 298–308 (1996)
7. Calvi, P.M., Calvi, G.M.: Historical development of friction-based seismic isolation systems. Soil Dyn. Earthq. Eng. **106**, 14–30 (2018)
8. Fenz, D.M., Constantinou, M.C.: Behaviour of the double concave friction pendulum bearing. Earthq. Eng. Struct. Dyn. **35**(11), 1403–1424 (2006)

9. Fenz, D.M., Constantinou, M.C.: Spherical sliding isolation bearings with adaptive behavior: experimental verification. Earthq. Eng. Struct. Dynam. **37**(2), 185–205 (2008)
10. Lee, D., Constantinou, M.C.: Quintuple friction pendulum isolator: behavior, modeling, and validation. Earthq. Spectra **32**(3), 1607–1626 (2016)
11. Kong, D., Fan, F., Zhi, X.: Seismic performance of single-layer lattice shells with VF-FPB. Int. J. Steel Struct. **14**(4), 901–911 (2014). https://doi.org/10.1007/s13296-014-1220-0
12. Calvi, P.M., Ruggiero, D.M.: Numerical modeling of variable friction sliding base isolators. Bull. Earthq. Eng. **14**(2), 549–568 (2016)
13. Calvi, P.M., Moratti, M., Calvi, G.M.: Seismic isolation devices based on sliding between surfaces with variable friction coefficient. Earthq. Spectra **32**(4), 2291–2315 (2016)
14. Pranesh, M., Sinha, R.: VFPI: an isolation device for aseismic design. Earthq. Eng. Struct. Dyn. **29**(5), 603–627 (2000)
15. Murnal, P., Sinha, R.: Earthquake resistant design of structures using the variable frequency pendulum isolator. J. Struct. Eng. **128**(7), 870–880 (2002)
16. Shang, J., Tan, P., Zhang, Y., et al.: Experimental and analytical investigation of variable friction pendulum isolator. Eng. Struct. **243**, 112575 (2021)
17. Shang, J., Tan, P., Han, J., et al.: Performance of seismically isolated buildings with variable friction pendulum bearings under near-fault ground motions. J. Build. Eng. **45**, 103584 (2022)
18. Computers and Structures Inc. Berkeley: SAP2000 V 20. Integrated finite element analysis and design of structures basic analysis reference manual, CA, USA (2017)
19. Fenz, D.M., Constantinou, M.C.: Modeling triple friction pendulum bearings for response-history analysis. Earthq. Spectra **24**(4), 1011–1028 (2008)
20. Weber, F., Distl, J., Meier, L., et al.: Curved surface sliders with friction damping, linear viscous damping, bow tie friction damping, and semiactively controlled properties. Struct. Control. Health Monit. **25**(11), e2257 (2018)

Experimental Investigation on Hysteretic Behavior of a Double Friction Pendulum and Frictional Heating

Esengul Cavdar[1](✉), Gokhan Ozdemir[2], and Ugurcan Ozcamur[2]

[1] ESQUAKE Seismic Isolator Test Laboratory, Eskisehir Technical University, Eskisehir, Turkey
esengulcavdar@eskisehir.edu.tr
[2] Technological Isolation Systems, TIS, Ankara, Turkey
gokhan_ozdemir@eskisehir.edu.tr, ugurcan.ozcamur@tis.com.tr

Abstract. The hysteretic behavior of a sliding seismic isolation system is basically a function of its friction properties. Thus, identification of the coefficient of friction at the sliding interface of the isolator is crucial. It requires displacement controlled dynamic tests to determine friction properties considering loading conditions representative of seismic actions. In these tests, the number of loading cycles is mostly three, and deterioration in friction coefficient, which is mostly due to frictional heating at the sliding interface, is assessed from cycle to cycle. The aim of this experimental study is to investigate the effect of vertical force and loading velocity on friction properties of a full-scale double friction pendulum. During the tests, by means of thermocouples mounted on the isolator, the change in temperature at the sliding interface was monitored too. Isolator tests were conducted at ESQUAKE Seismic Isolator Test Laboratory of Eskişehir Technical University. Results showed that the amplitude of frictional heating is highly dependent on the vertical force and the loading velocity. It is also revealed that the corresponding temperature rise at the sliding interface is not uniform. This is especially more pronounced for high loading velocities.

Keywords: Double friction pendulum · Friction coefficient · Frictional heating · Loading history · Temperature

1 Introduction

Curved surface sliders, or friction pendulum as the trademark name, have been used as a successful and efficient earthquake protection system for buildings, bridges, and viaducts. They are basically composed of concave shaped steel plates and a composite slider. The main features dominating the hysteretic behavior of friction pendulum isolators are the radius of the concave plates and the frictional properties of the slider. Post-yield stiffness of the nonlinear force-deformation relation of the isolator is a function of the radius of the concave shape while friction coefficient of the slider influences the energy dissipation capacity.

G. P. Cimellaro (Ed.): WCSI 2022, LNCE 309, pp. 229–239, 2023.
https://doi.org/10.1007/978-3-031-21187-4_17

Generally, sliding materials used in friction pendulums are made of composites like polytetrafluoroethylene (PTFE) with various fillers and ultra-high-molecular-weight polyethylene (UHMWPE). Several research studies have been conducted to understand the friction properties of such polymer composites under the effect of seismic actions. It is found that the friction coefficient depends on the level of axial load, sliding velocity, and temperature rise at the sliding interface due to repeated cyclic motion [2–8]. For instance, Tyler [2] revealed that the friction coefficient tends to decrease with increasing level of axial load. Dolce et al. [5] investigated the variation in friction coefficient for different sliding velocities. Authors showed that friction coefficient increases with increasing sliding velocity up to 150 mm/s and stays almost constant for larger velocities. Such results indicate that the friction coefficient at the sliding interface of a friction pendulum is not constant. It depends on the loading parameters. And, one of those parameters is the number of loading cycles applied during the tests. Experimental results revealed that there will be a degradation in friction coefficient as the number of loading cycles increases. The reason of the degradation is the temperature rise at the sliding interface [1, 7, 9–12]. It is also reported that temperature dependent degradation of friction coefficient is especially more pronounced for high axial load and sliding velocity. It is to be noted that, any variation in friction properties of friction pendulum may significantly influence the behavior of seismically isolated structures [13]. Thus, effects of axial load, sliding velocity and temperature rise at the sliding interface should rigorously be taken into consideration to idealize the hysteretic behavior of friction pendulums accurately. However, in literature, there are only a few experimental studies that focused on the mutual effects of axial load, sliding velocity and friction degradation due to temperature rise considering full-scale friction pendulums [1, 7, 8, 14]. Most of these studies were conducted with low axial loads (corresponding normal stresses vary in between 15 and 30 MPa) whereas sliding velocities range from 10 mm/s to 400 mm/s. Moreover, very limited number of researches focused on the effect of temperature rise on friction properties of the isolator. Some of these studies also investigated the analytical representation of degradation in friction coefficient due to temperature rise [1, 7, 11, 12]. In such studies, the temperature rise is assumed to be uniform at the sliding interface.

The objective of this paper is to experimentally investigate the temperature rise at the sliding interface of a double friction pendulum (DFP) in order to assess the validity of uniform temperature rise assumption used in analytical models to represent hysteretic behavior of isolators. In the experimental program, several axial loads and sliding velocities were applied to a DFP to mimic different loading conditions and corresponding temperature rises at the sliding interface were monitored by means of thermocouples.

2 Overview of the Test Protocol

A full-scale DFP has been tested at the ESQUAKE Seismic Isolator Test Laboratory of Eskişehir Technical University (see Fig. 1). The main geometrical features of the tested DFP are listed in Table 1. The sliding pad of the DFP is a PTFE-based compound with fillers, developed by the Technological Isolation Systems (TIS).

In order to monitor the temperature at the sliding interface of the DFP, eight thermocouples were embedded into the sliding plate (Fig. 2), with their probes in contact with

Fig. 1. Test setup of ESQUAKE seismic isolator test laboratory.

Table 1. Geometrical features of the tested DFP

Geometric feature	
Diameter of stainless steel (mm)	790
Diameter of sliding pad (mm)	260
Radius of curvature (mm)	2500

the back of the stainless steel. By means of these thermocouples, the temperatures at different points on the stainless steel were continuously recorded during the tests. Then, the recorded force-displacement curves and thermal data were used to assess the degradation in friction coefficient due to temperature rise at the sliding interface for various load combinations.

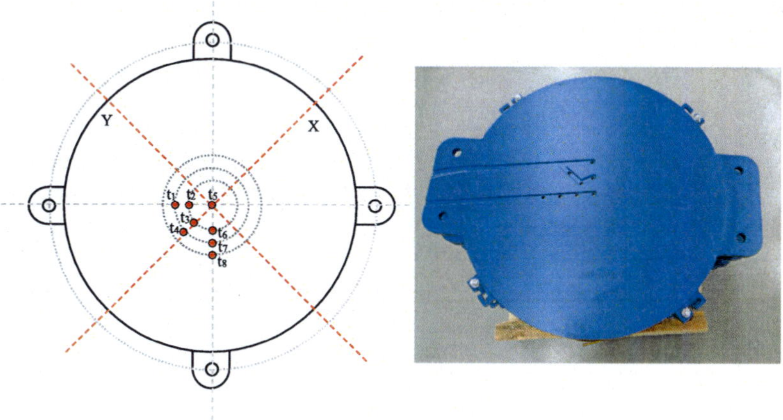

Fig. 2. Configuration of thermocouples.

Test protocol consists of sixteen loading cases. As listed in Table 2, combinations of four different axial loads that corresponds to normal stresses of 20, 40, 60, 80 MPa and four different sliding velocities of 50, 250, 500, and 750 mm/s were considered. In Table 2, d and f are the amplitude and frequency of the cyclic motion, respectively. F_v is the axial load acting on the DFP and p is the corresponding normal stress. V_{max} is the maximum sliding velocity which is computed by (2πfd).

Table 2. Test protocol for the double curve friction pendulum.

Test #	d (mm)	f (Hz)	V_{max} (mm/s)	F_v (kN)	p (MPa)	Cycles (#)
1	300	0.027	50	1062	20	3
2	300	0.133	250	1062	20	3
3	300	0.265	500	1062	20	3
4	300	0.398	750	1062	20	3
5	300	0.027	50	2124	40	3
6	300	0.133	250	2124	40	3
7	300	0.265	500	2124	40	3
8	300	0.398	750	2124	40	3
9	300	0.027	50	3186	60	3
10	300	0.133	250	3186	60	3
11	300	0.265	500	3186	60	3
12	300	0.398	750	3186	60	3
13	300	0.027	50	4247	80	3
14	300	0.133	250	4247	80	3
15	300	0.265	500	4247	80	3
16	300	0.398	750	4247	80	3

3 Results

There are basically four parameters that effect the hysteretic behavior, consequently the friction coefficient, of a friction pendulum. These parameters are (i) breakaway effect, (ii) velocity effect, (iii) cycling effect and (iv) pressure effect. In order to highlight the effects of first three parameters, Fig. 3 is depicted based on the data obtained from test 6 of Table 2. Breakway effect is the phase in which a sudden transition from static to dynamic friction coefficient takes place at the beginning of each return point of the motion. Velocity effect is related to the nature of sliding velocity. Friction coefficient decreases when velocity of motion drops down to zero at each reversal point near the peak isolator displacement. Cycling effect is the gradual reduction of friction coefficient due to temperature rise at the sliding interface. Figure 4 presents the significance of pressure effect which implies that the friction coefficient decreases with increasing axial load. In Fig. 4, the pressure effect is presented for experimental results tests 1, 5, 9 and 13 of Table 2.

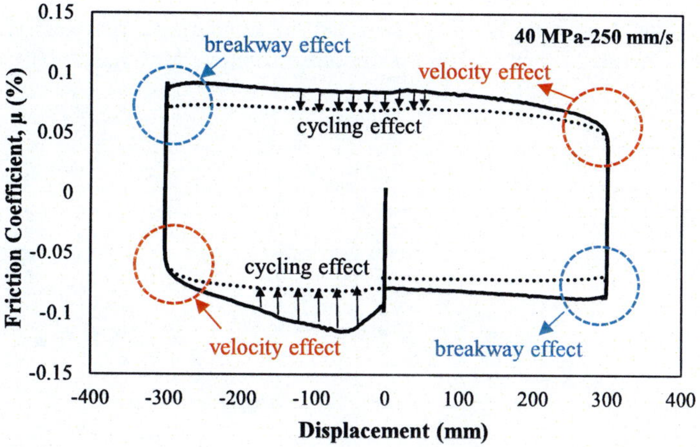

Fig. 3. Identification of parameters that affect the friction coefficient.

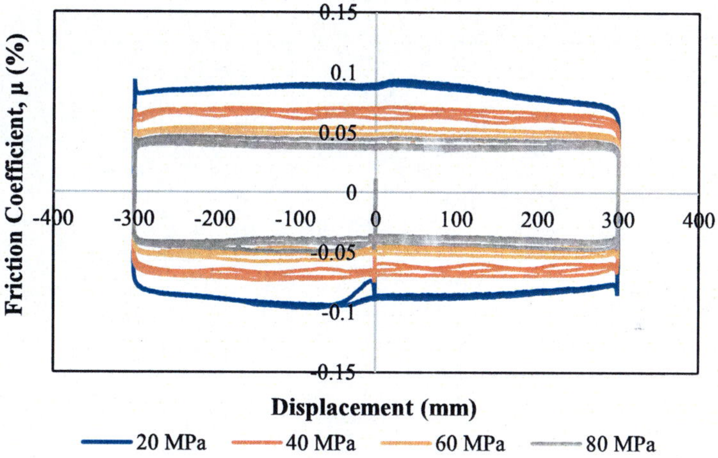

Fig. 4. Investigation of pressure effect on the tested DCFP for tests #1, 5, 9, 13.

In Fig. 4, the variation in friction coefficient is presented for normal stress ranging from 20 MPa to 80 MPa while the sliding velocity is kept constant and equals to 50 mm/s. For 20 MPa normal stress, the friction coefficient is computed as 0.09. It reduces to 0.07, 0.05, and 0.04 for normal stresses of 40, 60 and 80 MPa, respectively. It is clear that the friction coefficient decreases with increasing axial load. Another observation in Fig. 4 is the stability of friction coefficient through the cyclic motion. Friction coefficients computed for each loading cycles are almost indifferent to each other when normal stress is 20 MPa. However, although it is very limited in Fig. 4 because of low sliding velocity, friction coefficient degrades with cyclic motion when normal stress gets larger. Figure 5 presents the temperatures recorded by means of thermocouples in order to assess the relation between the temperature rise at the sliding interface and the degradation in friction coefficient. In Fig. 5, it is shown that the temperature rise gets larger as the normal stress increases. The maximum amounts of temperature rise are 30, 42, 49 and 54 °C for tests 1, 5, 9, and 13, respectively.

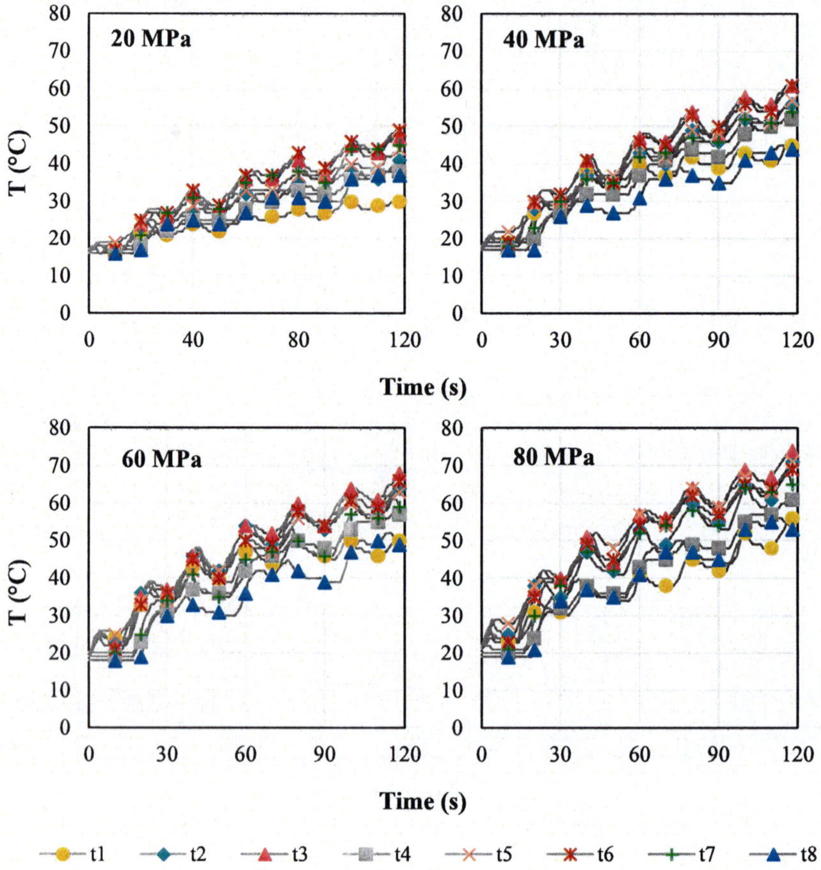

Fig. 5. Temperature rises at the sliding interface recorded in tests # 1, 5, 9, 13

In theory, heat flux (Ω) generated at the sliding interface is given by $\Omega = (\mu \cdot W|v|)/(\pi a^2)$. Here, μ is the friction coefficient, W is the vertical force, v is the sliding velocity and a is the radius of the slider [15]. The generated heat flux is assumed to be equally transferred and distributed to the stainless steel of DFP. Therefore, it is expected to have an amplification in temperature rise proportional to axial load level. However, in accordance with the concept of *thermal control of friction* the corresponding temperature rise is not scaled proportionally with the heat flux [16]. That is why maximum temperature rise is not proportional to amplification in axial load level. Another observation in Fig. 5 is that there is a fluctuation in each recorded data instead of a straight line during the cyclic motion. This behavior is also identified as *flash temperature* by Stachowiak and Batchelor [17].

In an attempt to identify the effect of sliding velocity on friction coefficient, Fig. 6 is presented for the tested DFP. In Fig. 6, the variation of friction coefficient obtained from test # 5, 6, 7 and 8 is given whereas Fig. 7 shows the corresponding temperature rises at the sliding interface. Figure 6 reveals that the degradation in friction coefficient is more pronounced for sliding velocities of 250, 500 and 750 mm/s compared to that of 50 mm/s. And, friction coefficients computed for 250, 500 and 750 mm/s sliding velocities are almost identical and greater than that of 50 mm/s. This indicates that the friction coefficient is not sensitive to change in sliding velocity beyond a certain level. Among the sliding velocities considered in this study, that limit is found to be 250 mm/s. Figure 7 shows that the maximum temperature rises at sliding interface were recorded as 41, 67, 75 and 87 °C by thermocouples in tests 5, 6, 7 and 8, respectively. Thus, as evident in Fig. 6, the degradation of friction coefficient due to the temperature rise at the sliding interface becomes more pronounced at higher sliding velocities.

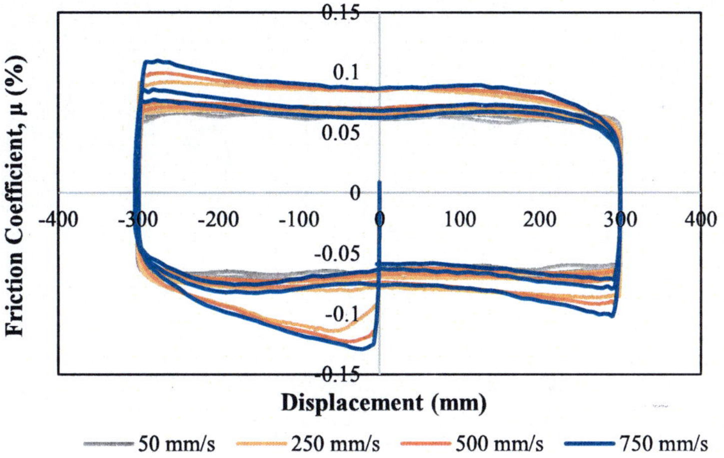

Fig. 6. Variation of friction coefficient with sliding velocity in test protocols of #5, 6, 7, 8.

4 Discussion

In this section, the assumption of having uniform temperature distribution at the sliding interface of a DFP is studied in detail. Accordingly, the recorded temperatures and friction coefficients computed for each cycle of motion are presented in Figs. 8 and 9, respectively.

In Fig. 8, averages of temperatures recorded from all of the eight thermocouples (see Fig. 2) are presented by indicating maximum and minimum values recorded in the same loading. Figure 8 clearly shows that the amount of temperature rise at the sliding interface increases with increasing axial load and sliding velocity. And, among the sliding velocities considered in this study, the assumption of uniform temperature rise at the sliding interface is found to be valid only for 50 mm/s. Regardless of the axial

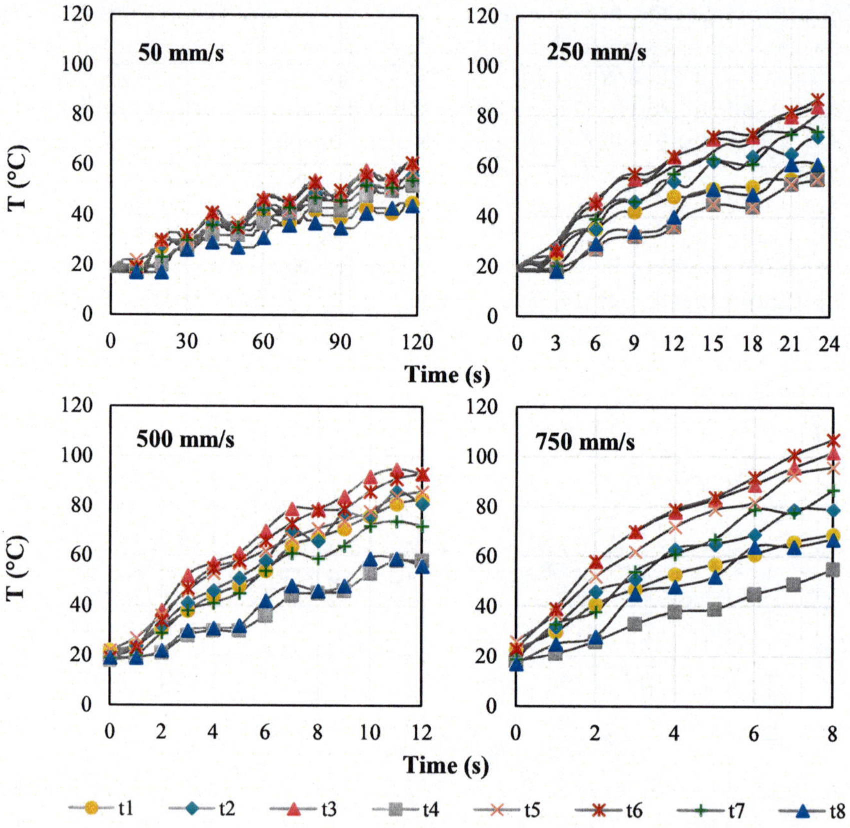

Fig. 7. Temperature rise at sliding interface with increased velocity in test protocols #5, 6, 7, 8.

load level, average temperature ±10 °C gives the range of variation in temperature for 50 mm/s sliding velocity. On the other hand, as the sliding velocity increases, the range of variation in temperature increases, too. For instance, when normal stress is equal to 20 MPa, average temperature ±20 °C defines the range of temperatures for a sliding velocity of 750 mm/s. Furthermore, that range also enlarges with increasing axial load level. The range of recorded temperatures is defined by average temperature ±30 °C when normal stress is 80 MPa. Thus, uniform temperature distribution assumption used in analytical representations used to mimic friction properties of friction pendulums may not be valid for high levels of axial load and sliding velocities.

Figure 9 shows that the friction coefficient is stable at all loading cycles (no degradation) only for 20 MPa normal stress and 50 mm/s sliding velocity (Test 1). Results of tests 5, 9 and 13 (sliding velocity is 50 mm/s but normal stresses are 40, 60 and 80 MPa, respectively) revealed that although it is limited, there is a degradation in friction coefficient from cycle to cycle. In Fig. 9, it is clear that the degradation in friction coefficient becomes more pronounced for loading cases where sliding velocity is greater than 50 mm/s. Such observations are in good agreement with the data presented in Fig. 8. Recorded temperature rises increases with increasing sliding velocity and normal

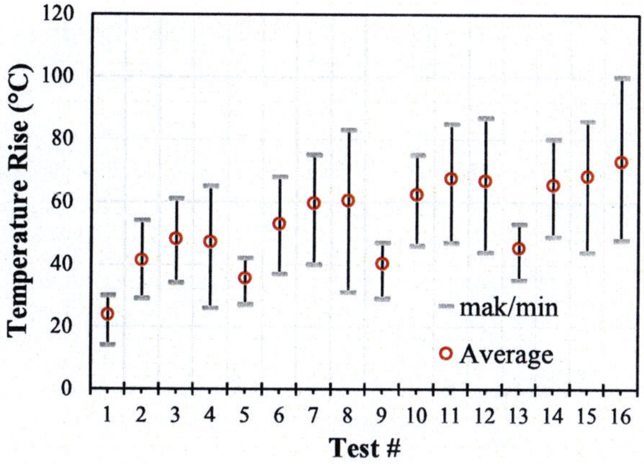

Fig. 8. Variation of temperature rises at sliding interface recorded at all performed tests.

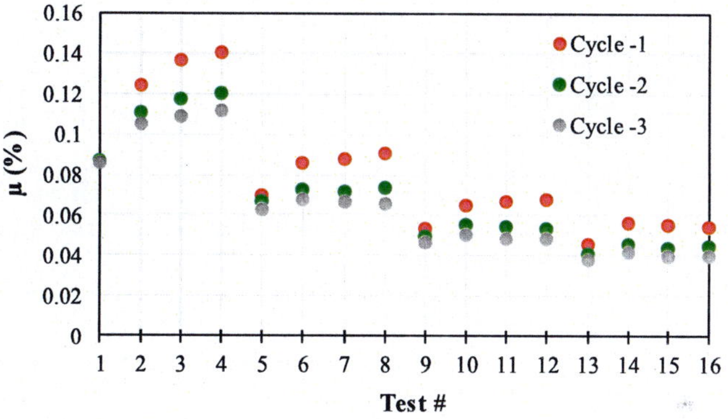

Fig. 9. Friction coefficients computed for each cycle of performed tests.

stress. Thus, the degradation of friction coefficient due to temperature rise becomes more evident.

5 Conclusion

This study was conducted to investigate the temperature rise at the sliding interface of a double friction pendulum (DFP) in order to assess the validity of uniform temperature rise assumption in analytical tools used to model friction properties of isolators. For this purpose, an experimental research in which several axial loads and sliding velocities were applied to a DFP, was carried out. In the tests, temperature at the sliding interface was monitored and corresponding amplifications in temperature during the cyclic motions were recorded. Results revealed the followings:

- Friction coefficient at the sliding interface decreases with increasing normal stress and sliding velocity,

- Friction coefficient increases when sliding velocity increases from 50 to 250 mm/s. However, further increase in sliding velocity does not result in any meaningful variation in friction coefficient,
- Temperature rise at the sliding interface due to cyclic motion gets larger for higher normal stresses and sliding velocities. Thus, the corresponding degradation in friction coefficient becomes more pronounced,
- Recorded temperature data may be assumed to be uniformly distributed at the sliding interface only for very low normal stress and sliding velocity. For larger values of normal stress and sliding velocity, distribution of temperature is not uniform.

It is to be noted that the presented experimental data is based on the behavior of PTFE-based polymer composite used at the sliding interface of the tested DFP.

Acknowledgements. This study was funded by the Disaster and Emergency Management Presidency (AFAD) under grant number UDAP-Ç-21-35 which the authors gratefully acknowledge. Also, the authors want to thank TIS Technological Isolation Systems for their support of the testing campaign.

References

1. De Domenico, D., Ricciardi, G., Benzoni, G.M.: Analytical and finite element investigation on the thermo-mechanical coupled response of friction isolators under bidirectional excitation. Soil Dyn. Earthq. Eng. **106**, 131–147 (2018)
2. Tyler, A.G.: Dynamic tests on PTFE sliding layers under earthquake conditions. Bull. NZ Natl. Soc. Earthq. Eng. **10**, 129–138 (1977)
3. Constantinou, M.C., Caccese, J., Harris, H.G.: Frictional characteristics of teflon-steel interfaces under dynamic conditions. Earthq. Eng. Struct. Dyn. **15**, 751–759 (1987)
4. Mokha, A., Constantinou, M., Reinhorn, A.: Teflon bearings in base isolation. I: Testing. J. Struct. Eng. **116**, 438–454 (1990)
5. Dolce, M., Cardone, D., Croatto, F.: Frictional behavior of steel-PTFE interfaces for seismic isolation. Bull. Earthq. Eng. **3**, 75–99 (2005)
6. Quaglini, V., Dubini, P., Poggi, C.: Experimental assessment of sliding materials for seismic isolation systems. Bull. Earthq. Eng. **10**, 717–740 (2012)
7. Lomiento, G., Bonessio, N., Benzoni, G.M.: Friction model for sliding bearings under seismic excitation. J. Earthq. Eng. **17**, 1162–1191 (2013)
8. Barone, S., Calvi, G.M., Pavese, A.: Experimental dynamic response of spherical friction-based isolation devices. J. Earthq. Eng. **23**, 1465–1484 (2017)
9. Quaglini, V., Bocciarelli, M., Gandelli, E., Dubini, P.: Numerical assessment of frictional heating in sliding bearings for seismic isolation. J. Earthq. Eng. **18**, 1198–1216 (2014)
10. Keikha, H., Amiri, G.G.: Seismic performance assessment of quintuple friction pendulum isolator with a focus on frictional behavior impressionability from velocity and temperature. J. Earthq. Eng. **27**(5), 1256–1286 (2021)
11. Furinghetti, M., Pavese, A., Quaglini, V., Dubini, P.: Experimental investigation of the cyclic response of double curved surface sliders subjected to radial and bidirectional sliding motions. Soil Dyn. Earthq. Eng. **117**, 190–202 (2019)
12. Gandelli, E., Penati, M., Quaglini, V., Lomiento, G., Miglio, E., Benzoni, G.M.: A novel OpenSees element for single curved surface sliding isolators. Soil Dyn. Earthq. Eng. **119**, 433–453 (2019)

13. Castaldo, P., Tubaldi, E.: Influence of FPS bearing properties on the seismic performance of base-isolated structures. Earthq. Eng. Struct. Dyn. **44**, 2817–2836 (2015)
14. Pigouni, A.E., Castellano, M.G., Infanti, S., Colato, G.P.: Full-scale dynamic testing of pendulum isolators (curved surface sliders). Soil Dyn. Earthq. Eng. **130**, 105983 (2020)
15. Constantinou, M.C., Whittaker, A.S., Kalpakidis, Y., Fenz, D.M., Warn, G.P.: Performance of Seismic Isolation Hardware Under Service and Seismic Loading. Technical Report MCEER-07-0012 (2007)
16. Ettles, C.M.: Polymer and elastomer friction in the thermal control regime. ASLE Trans. **108**, 98–104 (1986)
17. Stachowiak, G.W., Batchelor, A.W.: Engineering tribology. Elsevier. Butteworth-Heinemann, Dordrecht (2005)

An Experimental Study on the Effects of Different Pendulum Damper Designs on Structural Behavior

Ersin Aydin[1], Baki Öztürk[2(✉)], Yunus Emre Kebeli[1], and Görkem Gültepe[1]

[1] Nigde Ömer Halisdemir University, Nigde, Turkey
[2] Hacettepe University, Ankara, Turkey
bakiozturk@hacettepe.edu.tr

Abstract. Pendulum dampers, which are the subject of this study, are a type of tuned mass damper (TMD). In general, TMDs are effective since they are designed according to the first mode behavior of the buildings. In this study, the effects of pendulum dampers on the structural behavior are investigated by performing vibration experiments on a 3-storey shear frame model with reduced dimensions. Harmonic loads are applied to the test models with and without pendulums on a one-way shaking table. Firstly, experiments without dampers were performed on the selected 3-storey reduced building model, and structural responses were found. The mass of pendulum dampers is chosen to be around 3% of the total mass of the structural model. One of the main aims of this study is to reveal the effects of the different placements of the dampers on the dynamic behavior. Accordingly, a single pendulum is independently placed on the top floor, the second floor, and finally on the first floor. In addition, experiments are carried out by establishing double damper models and then a triple damper model. By creating seven different damper models, the experiments are repeated under harmonic loads with different frequencies. Tests under a harmonic load equal to the first mode frequency of the undamped model reveal the behavior of the models in the resonant state. Experiments show that the pendulums used are highly effective in the resonance state. Some models with more than one damper have also been shown to be effective at reducing dynamic response.

Keywords: Pendulum dampers · Vibration experiments · Tuned mass dampers · Harmonic loads

1 Introduction

The concept of a tunable mass damper was first introduced by Hermann Frahm in 1909 to eliminate the vibration hazard created by the ship's machinery on the ship's keel. Frahm named this system, which he patented in 1911, as an adjustable vibration damper in those years. This vibration control device, introduced by Frahm, had no natural damping. The instrument was only effective when the natural frequency of the tuned mass and the external impact frequency were very close to each other [1].

G. P. Cimellaro (Ed.): WCSI 2022, LNCE 309, pp. 240–253, 2023.
https://doi.org/10.1007/978-3-031-21187-4_18

Buildings are exposed to many dynamic effects, such as earthquakes, wind, water waves, and construction machinery. What is required from the structure is that it can adapt itself to these loads in a way that is safe, durable, and provides a certain level of comfort. Control systems that can provide this adaptation can be classified as passive, active, semi-active, and base isolation systems. In this study, it is aimed to increase the seismic performance of passively tuned mass damper (TMD) structures added to the structure under the effects of earthquake and wind, and the effects of pendulum dampers, a type of TMD, on the building behavior were investigated by placing them on different floors. The pendulum mass damper, or pendulum vibration damper, which will be examined in this study, can be used as a TMD to suppress unwanted movements of the exposed systems. The pendulum has a low frequency and uses gravity instead of elastic stiffness force, so it is suitable for dampening negative vibrations of bridges or buildings. Passive single TMDs are usually tuned to reduce first mode movement of the structure. It is possible to control other modes with active TMDs. In this study, story displacements and accelerations of the top floor are investigated when pendulum dampers are placed on different floors in a 3-storey building model. The length and total weight of the pendulum were kept constant. Clark [2] showed that a single TMD is generally not very effective in reducing seismic responses for two reasons. First, seismic loads are typically impulsive and rapidly reach their maximum values. Second, earthquake ground motions contain a wide range of frequency components and often cause significant vibration in the foundation and higher modes of a tall building structure. Therefore, a single TMD may fail to reduce the overall response of the structures.

Nagase and Hisatoku [3] developed and installed a new damper to dampen the lateral vibration of buildings in a 37-floor office building, Crystal Tower, Osaka. The tower has a very fine ratio of height to width. The damper can reduce wind-induced movement by 50%. The Crystal Tower is the first tall building in Japan to have this damper, and the damper weight of 540 tons (1190 kips) is large compared to others used elsewhere in the world. The damper is a pendulum type regulated mass damper, and ice thermal storage tanks for air conditioning are used as the moving mass of the damper. Therefore, Crystal Tower's TMD does not need additional bulk weight or space for installation. This article presents some unique features of the damper and shows how simple and practical the damper is. Particular attention is paid to both the connecting joints between the tank and the building and the swaying behavior of the liquid content. Typhoon and earthquake records observed at Crystal Tower have proven that the damper works as expected.

Xu and Igusa [4, 5] proposed a new design concept for TMD to dampen the vibration of a dynamic system called multi-tuned mass dampers (MTMD). The basic configuration of MTMD consists of multiple small oscillators whose natural frequencies are distributed around the natural frequency of the structure's-controlled mode. They studied vibration control using multi-tuned mass dampers (TMDs) with natural frequencies distributed over a frequency range. Yamaguchi and Harnpornchai [6] conducted a parametric study on MTMD-structure systems under harmonic forces. It confirmed the efficiency and offset sensitivity of the MTMD in frequency tuning. One of the important findings in their work is the finding of optimal values in MTMD parameters such as damping ratio and bandwidth of natural frequencies. They based their results on numerical simulations. Kareem and Kline [7] conducted a comprehensive parametric study on the effects of the number of dampers, the damping ratio of individual dampers, mass distribution, and frequency bandwidth. Similarly, Joshi and Jangid [8] showed that multiple TMDs, each tuned to dampen a particular mode, are more effective than a single TMD solution with the same mass ratio. They investigated the optimum parameters of MTMD to suppress the dynamic response of the structure exposed to the base excitation.

Jangid [9] and Bakre and Jangid [10] examined the optimum design issues of TMD parameters. In this research, optimum parameters of multi-tuned mass dampers (MTMD) are investigated by a numerical search technique to suppress the dynamic response of a fundamentally excited and undamped main system. The criterion chosen for optimality is the minimization of the steady-state displacement of the main system under harmonic ground acceleration. The optimum parameters of the MTMD system and the corresponding displacement are obtained for different damping ratios of the main system and different mass ratios of the MTMD system. The explicit formulas for the optimum parameters of MTMD (i.e., damping ratio, bandwidth, and tuning frequency) are then derived using a curve-fitting scheme that can be easily used in engineering applications. The error in the proposed explicit expressions has been investigated and found to be negligible. The effectiveness of the optimally designed MTMD system is also compared to that of the optimum single-adjustment mass damper. It has been observed that the optimally designed MTMD system is more effective for vibration control than a single tuned mass damper. Li [11–13]; Bakre and Jangid [10]; Li and Ni [14]; and Han and Li [15] showed that the scheme of optimum MTMDs is more effective and robust compared to the optimum STMDs of equal mass. The performance of multi-tuned mass dampers to reduce undesirable oscillations in structures under ground acceleration was investigated.

Chen and Wu [16, 17] deployed TMDs according to the maximum acceleration of the building floors in an experimental study and investigated the optimum placement of multiple modulated mass dampers for seismic structures. In addition, these TMDs may still preserve the structural or architectural functions of buildings. A single-set mass damper (STMD) mounted on the upper floor is often used to control the fundamental vibration mode of a building. This indicates that the STMD system is only effective when the dominant frequency of a narrowband earthquake excitation is close to the fundamental frequency of the structure. Therefore, it is not always effective in reducing vibrations caused by various earthquakes. On the other hand, it has been verified that multi-tuned mass dampers (MTMDs) are robust for earthquake ground motions with broad spectrum frequency components and can moderately reduce peak responses even under sudden earthquakes. They reported that MTMDs are more effective than STMD in reducing the acceleration of buildings. Vyas and Bajaj [18] conducted a similar study with pendulum dampers. In their work, it has been shown that the efficiency bandwidth can be increased significantly by using multiple pendulums with different natural frequencies to study the dynamics of autoparametric vibration absorbers.

Li and Qu [19] performed numerical simulations to accurately predict the dynamic properties of MTMD for asymmetrical structures subjected to ground acceleration. In the simulations, the efficacy and robustness of MTMD strategies with different designs were also investigated. In addition, Li [12] reported improved performance of optimum MTMDs compared to optimum STMDs and showed that MTMDs are more robust compared to STMDs. Li and Ni [14] were able to numerically optimize unevenly distributed

MTMDs to minimize the vibration amplitude without placing constraints on the design parameters of the TMDs. This method aims to minimize the root mean square value of the frequency response when the excitation force is harmonic. In this article, Nagarajaiah and Sonmez [20] examine the application of single and multiple semi-active variable stiffness modulated mass dampers STMD/SMTMD for response control of multilayer structures under various excitation types. Casciati and Giuliano [21] evaluated the suitability for wind response reduction and the performance of multiple TMDs used in suspension bridge towers under seismic excitation. They showed improved performance of MTMDs compared to single TMDs. Lin et al. [22] theoretically and experimentally investigated the performance of MTMDs in reducing the seismic response of structures by the motion limitation of MTMDs. Patil and Jangid [23] studied the performance of single and multiple TMDs (MTMDs) when placed at the top of a 76-story wind-loaded benchmark building using sensitivity analysis and minimization of performance criteria.

Gerges and Vickery [24] determined the optimum design formula for planar pendulum type TMDs (PTMDs) under wind and earthquake warnings represented by white noise for optimum design of pendulum regulating mass dampers. Gus'kov et al. [25] analyzed the dynamics of a pendulum vibration absorber. The nonlinear dynamics of a pendulum-type autoparametric vibration damper is considered. The conditions for the existence of periodic motions in the case of monoharmonic excitation are considered. The vibration damping characteristics of the autoparametric damper, and the "classical" (linear) dynamic vibration damper were compared. Lacarbonara and Balleri [26] proposed a passive vibration damping architecture to dampen the transverse vibrations of the struts via tuned pendulum dampers. Linearization of the equations of motion of the mast, which is subjected to a harmonic distributed force, gives rise to the damper and damper transfer functions of the structure. Transfer functions allow investigating the mitigation effects of the pendulum. Using only one pendulum tuned to the frequency of the lowest mode, the effectiveness of the passive vibration potential in reducing the motion and acceleration of the upper section of the mast has been demonstrated.

Lin et al. [27] investigated the vibration control of seismic structures using multi-tuned mass dampers with semi-active friction. Yang et al. [28] studied the design and optimization of multi-tuned mass dampers to suppress machine tool noise. Single and multiple TMD systems are designed to have equal masses; and damping and stiffness values are optimized to improve noise resistance using the minimax numerical optimization algorithm. Moon [29] suggested dividing a single large TMD into multiple smaller TMDs and placing MTMDs vertically along the height of the building to control wind-induced vibrations of the structure. This study revealed that MTMDs can effectively control multiple vibration modes of a high-rise building, and the control efficiency is clearly not less compared to the STMD method. Also, two other distinct values can be appreciated by adopting this method. First, since the mass of each MTMD is much smaller compared to a single large TMD, this method can significantly simplify TMD setup. The reliability of the control system can be significantly increased if some MTMDs do not work properly.

Bandivadekar and Jangid [30] investigated the optimum parameters of multi-tuned mass dampers (MTMD) to reduce the dynamic response of a damped main system subjected to harmonic load. The criterion chosen for optimality is the minimization of the steady-state displacement of the main system under external harmonic excitation. For the same mass ratio and damping in the main system, the optimally designed MTMD system was found to be more effective than the optimum single TMD system. Elias and Matsagar [31–33]; Elias et al. [34] investigated multiple modulated mass dampers (MTMDs) distributed over the height of a 76-story high building for their effectiveness in vibration response control. They also investigated the effectiveness of multimodal control of seismically excited buildings installed with distributed multi-modal mass dampers (d-MTMDs) by comparing the dynamic response with other controllers such as passive friction dampers, semi-active dampers, and single tuned mass dampers. Xiang and Nishitani [35] examined the effectiveness of MTMDs for multimodal control of low-rise buildings with closely spaced frequencies during earthquake ground warnings. Elias et al. [36] investigated the effectiveness of using MTMDs to control the foundation and high vibration modes of a reinforced concrete (RC) chimney under earthquake effects. They performed a parametric study to find the optimum masses and damping ratios for different TMDs. Zuo et al. [37] proposed using multiple TMDs (MTMDs) to control offshore wind turbine tower vibrations under multiple loads (wind, wave, and seismic loads). It has been observed that MTMDs can effectively reduce vibrations and significantly increase the robustness of the wind turbine system. Using MTMDs is more practical than applying a single TMD on the tower top. It also simplifies the installation, maintenance, and replacement of TMDs due to the smaller size and mass of each TMD. Similarly, Xie et al. [38] investigated the application of multiple modulating mass dampers to control the structural loads of bottom-fixed offshore wind turbines, including soil-structure interaction. The use of structural control methods to suppress the structural loads of offshore wind turbines is an economical and viable solution. In this study, the structural responses of a bottom-fixed offshore wind turbine exposed to combined wind and wave excitations were investigated. In addition to the dynamic responses of the wind turbine, the vibration reduction effects of the single tuned mass damper (STMD) and multiple TMD (MTMD) systems were investigated. Hussan et al. [39] investigated the use of a multi-tuned mass damper to reduce multimodal vibration of an offshore wind turbine under seismic excitation.

Deraemaeker and Soltani [40] initiated a study to analytically derive a solution for the optimization of a simple pendulum type TMD connected to a main system considered as undamped and reduced to a single degree of freedom model. Kecik and Warminski [41] investigated the dynamics of a nonlinear semi-active suspension autoparametric pendulum-like system. They performed a vibration analysis of an autoparametric pendulum-like mechanism subjected to harmonic excitation. Roffel et al. [42] studied the dynamic responses of a flexible multi-degree-of-freedom structure coupled with the three-dimensional motions of a nonlinear pendulum tuned mass damper (PTMD). In this article, both planar and three-dimensional motions of PTMD are looked at. The optimal damper parameters obtained using the three-dimensional model are compared with those estimated by the models in which the PTMD motion is linearized in the planar direction.

According to studies in the literature, generally the authors advocated the use of MTMDs and similarly multiple pendulum type TMDs. There are many reasons for this: for practical applications, a smaller mass rather than a large mass facilitates the installation of devices, they can be designed/tuned by increasing the bandwidth for various natural frequencies, they can improve the behavior of the structure in different modes, and they can better handle structural stiffness uncertainties. Considering such systems, it is seen that they have been widely applied to high-rise buildings in the last thirty years. The use of multiple pendulum type TMDs is discussed in this study because it can both improve the structural behavior and reduce the cost. The main purpose of this study is to evaluate and compare the undamped and damped performances of seven models created on a 3-storey structure without changing the total mass.

2 Experimental Study

2.1 The Experimental Setup

The experimental setup used in this study is given in Fig. 1. This assembly consists of a load cell and program, a data loader and program, a shaking table, an accelerometer, three linear displacement meters (LVDT), and a small-scale three-story steel frame identical to the actual structure mounted on the shaking table as shown in figure. The frame is a uniform structure with a weight of 6060 g and a height of 1000 mm, each layer being 330 mm high, with a modulus of elasticity $E = 2 \times 10^5$ MPa and a Poisson's ratio of $\nu = 0.3$. The rectangular floor size of each floor is 350 mm by 305 mm.

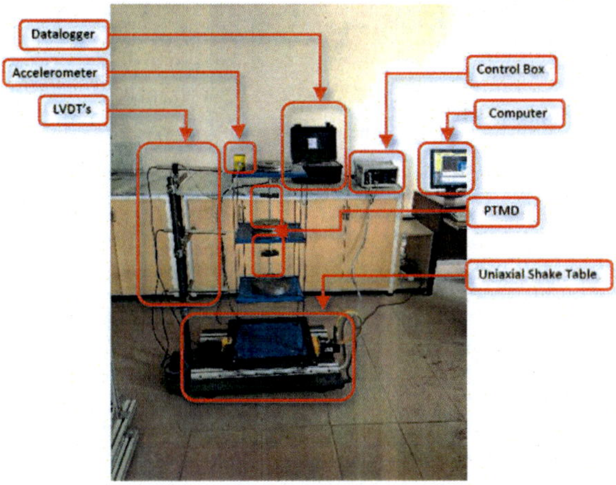

Fig. 1. Experimental setup with PTMD

The weights placed on each floor are different and are given in Table 1. The floors, which consist of a more rigid structure, are supported by four rectangular columns. To ensure a unidirectional movement, the weak sections of the columns are arranged in the shaking direction.

Before the single and multiple pendulums were placed, the weights of these pendulums were added as passive mass to the floor where they would be placed, and the shaking table test of the plain structure was performed. The multiple pendulum type TMD was made for the design purpose, to obtain not only the dynamic parameters of the building,

Table 1. Weights for each floor in a plain structure

Floors	Frame weight (g)	Placed weight (g)	Total floor weights (g)
1st Floor	220	10812	11032
2nd Floor	220	8346	8566
3rd floor	220	6718	6938

but also its response under harmonic load effect for future comparisons. Displacement meters (LVDTs) are installed on each floor and accelerometers are installed on the top floor.

2.2 Loading Procedure

The behavior of the structure in the resonance state has been investigated. The period in the undamped free vibration of the existing structure and its frequency in the resonance state of the period are found. Fourteen different experiments were conducted, and all used a uniform loading procedure. This charge has a harmonic amplitude of 10 mm, a frequency of 1.5 Hz, and a cycle number of 5 (Fig. 2).

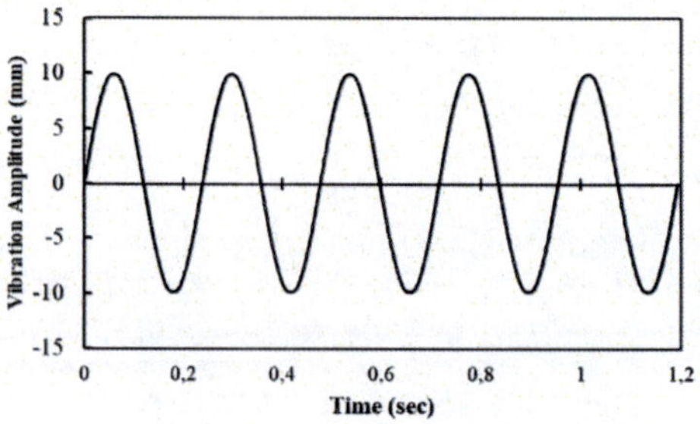

Fig. 2. Loading procedure

2.3 Introducing the Models Used

The total mass of the structure is the same in all models. The total weight of the pendulum is 930 g, and this weight is evenly distributed across the pendulum dampers, regardless of how many pendulum dampers there are. The weights of the pendulum type TMDs designed for the damped system are added to the same floor as a stationary load to create undamped models. All pendulums are equal in length and are 10.5 cm. The nomenclatures representing the models used in the experiments are defined below, and the figures and drawings of the fourteen models examined in this study are given in Table 2 in detail.

M_{01}: There is no damper and the weight of 930 g of damper is passively added to the first floor of the plain construction model.

M_{02}: There is no damper and the weight of 930 g of damper is passively added to the second floor of the plain structure model.
M_{03}: There is no damper and the weight of 930 g of damper is passively added to the third floor of the plain structure model.
M_{012}: There is no damper and the weight of 465 g of damper is passively added to the first and second floors of the plain construction model.
M_{013}: There is no damper and the weight of the damper of 465 g is passively added to the first and third floors of the plain construction model.
M_{023}: There is no damper and the weight of the damper of 465 g is passively added to the second and third floors of the plain construction model.
M_{0123}: There is no damper and the weight of 310 g of damper is passively added to the first, second, and third floors of the plain construction model.
M_1: The weight of the damper placed on the first floor is 930 g.
M_2: The weight of the damper placed on the second floor is 930 g.
M_3: The weight of the damper placed on the third floor is 930 g.
M_{12}: The weight of the damper placed on the first and second floors is 465 g.
M_{13}: The weight of the damper placed on the first and third floors is 465 g.
M_{23}: The weight of the damper placed on the second and third floors is 465 g.
M_{123}: The weight of the damper placed on the first, second and third floors is 310 g.

The models are introduced in Table 2. The figures in the first row are undamped models in which the weights of the pendulum are passively added to the floors, while the figures in the second row are damped models with a pendulum system. θ represents the swing angle of the pendulum type TMD.

2.4 Evaluation of Experiments

When comparing the test results, the accelerometer on the top floor of the model and the values measured by the LVDTs placed on each floor were considered. Time dependent graphs of these data were also drawn within the scope of this study. It aims to obtain the best model because of comparing the graphics among themselves in detail.

When the single PTMD models (M_1, M_2 and M_3) are compared (Fig. 3a-d), M_3, the damped system model placed on the top floor, showed the best performance, as expected. Considering the acceleration and displacement values of the floors, the M_3 model was the best model in this comparison. The M_2 model performed better than the M_1. When the damped M_3 model is used in the structure, when compared to the undamped M_{03} model (Fig. 4a-d), the acceleration and displacement values of the structure are reduced by half. This reduction is a very serious value for a structure.

If the dual PTMD models (M_{12}, M_{13} and M_{23}) are compared among themselves (Fig. 5a-d), the M_{23} model showed the best performance. This model, which has the least acceleration value, has also been the model with the least displacement when viewed in terms of oscillations on each floor. Considering these values, M_{13} performed better than M_{12}. When the damped M_{23} model is used in the structure, when compared to the undamped M_{023} model (Fig. 6a-d), the acceleration and displacement values of the structure decrease, similar to the single pendulum above. In fact, this reduction amount is slightly higher than that of single pendulums.

Table 2. Models with and without dampers

	Floors	Without dampers	With dampers
Placement of damper mass and damper	1st story	M_{01} 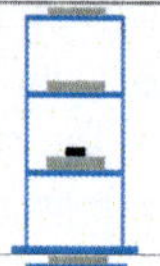	M_1 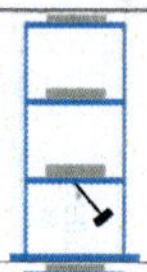
	2nd story	M_{02} 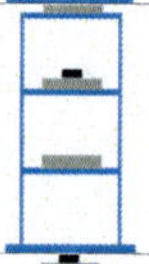	M_2 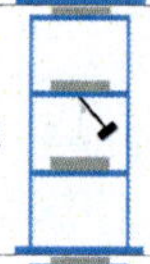
	3rd story	M_{03} 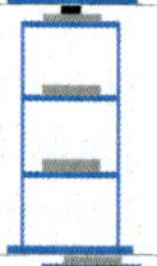	M_3 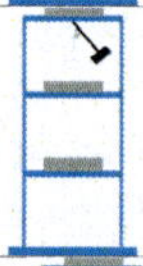
	1st and 2nd stories	M_{012} 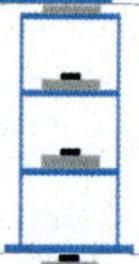	M_{12} 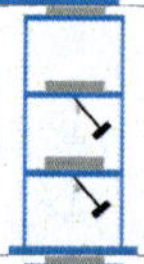
	1st and 3rd stories	M_{013} 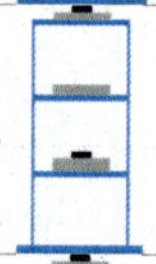	M_{13} 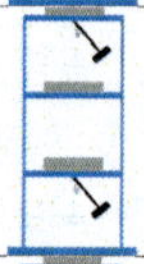
	2nd and 3rd stories	M_{023} 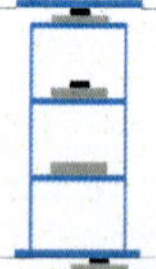	M_{23} 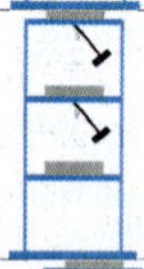
	1st, 2nd and 3rd stories	M_{0123} 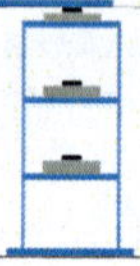	M_{123} 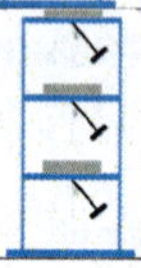

The M_{23} model, which has the best performance among the dual PTMD models, performed better than the M_3 model (Fig. 7a-d), which we can describe as the model of classical TMDs. Acceleration values and displacement values showed that double pendulum type TMDs can give better results than single ones.

The main concern is the behavior of the triple PTMD. Here, the M_{123} model and the M_3 and M_{23} models, which show the best performance among other pendulum types, are compared (Fig. 8a-d). The M_{23} has less acceleration and displacement values and better behavior. In the amplitude and frequency in this study, the triple PTMD model, M_{123} did not outperform the single PTMD model, M_3.

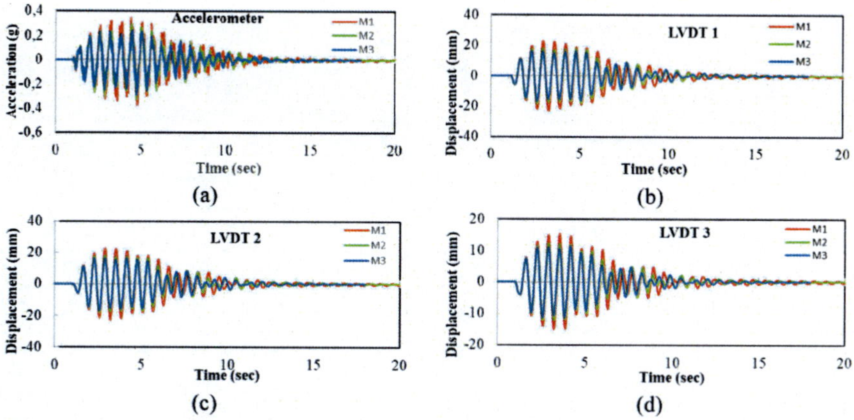

Fig. 3. Comparison of acceleration and displacement measurements of models with a single PTMD; a) Accelerometer measurement b) First story LVDT c) Second story LVDT d) Third story LVDT

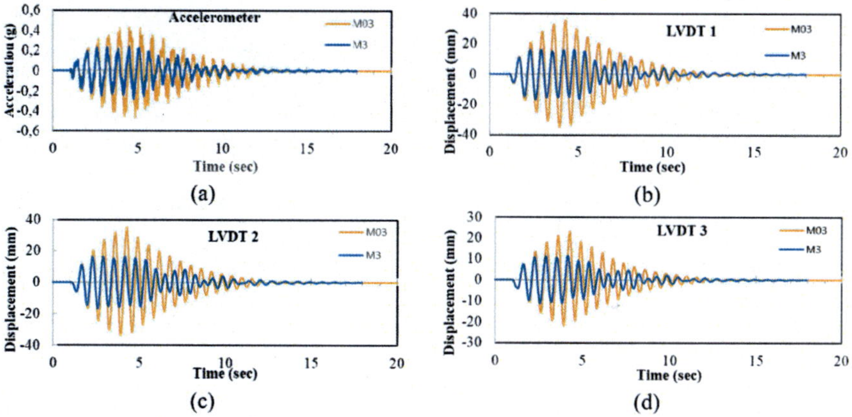

Fig. 4. Comparison of acceleration and displacement measurements of TMD layout models on the third floor; a) Accelerometer measurement b) First story LVDT c) Second story LVDT d) Third story LVDT

3 Results

In this study, the behavior of single, double, and triple PTMDs on a reduced structure model was investigated. The study is based on experiments on a unidirectional shaking table. As an external effect, a harmonic load effect equal to the first natural frequency of the structure is used.

The first parameter studied is the differences between pendulum damped and undamped structures. When the acceleration and displacement values of these values are compared, the vibration damping of PTMD systems is higher than that of undamped structures. Another issue is how multiple PTMDs behave. As can be seen from the graphs

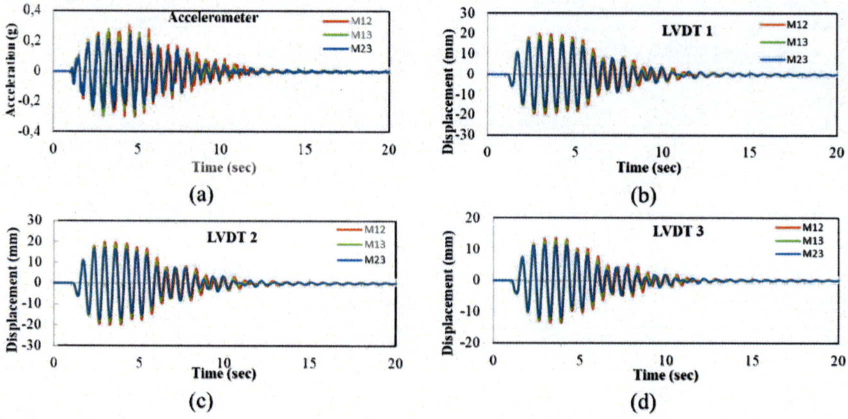

Fig. 5. Comparison of acceleration and displacement measurements of models with dual PTMD; a) Accelerometer measurement b) First story LVDT c) Second story LVDT d) Third story LVDT

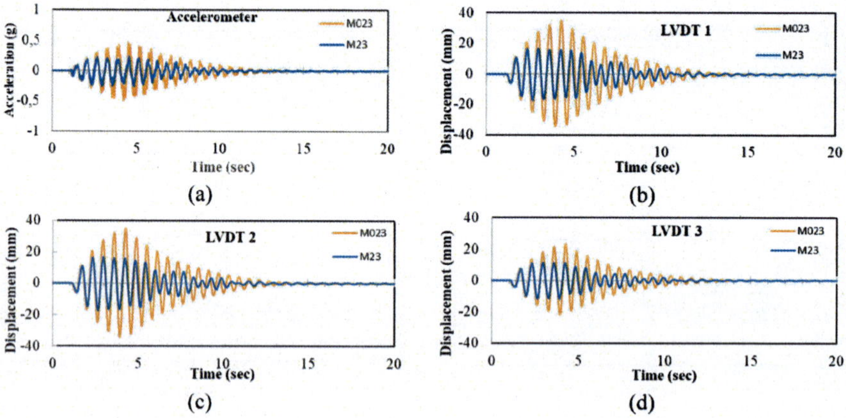

Fig. 6. Comparison of acceleration and displacement measurements of models with PTMD located on the second and third floors; a) Accelerometer measurement b) First story LVDT c) Second story LVDT d) Third story LVDT

above, M_{23}, the dual PTMD model, showed the best performance among single, double, and triple pendulum types. This model has shown us about multiple PTMDs that the different distributions of PTMDs at building layer levels cause significant effects on structural behavior. Considering these values, an optimization problem can also be addressed. In addition to optimization, the behavior of triple and single PTMDs at different frequencies is planned as a continuation of this study in the future. In addition, another subject planned for the future of this study will be to examine how the structure behaves under random earthquake data at different frequencies.

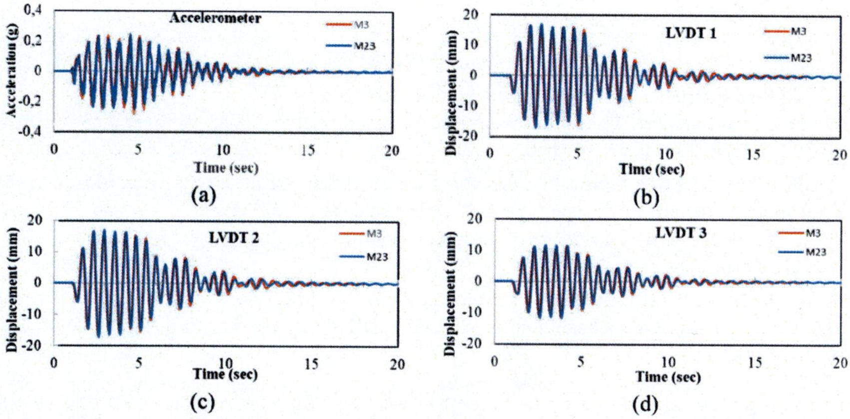

Fig. 7. Comparison of acceleration and displacement measurements of the best single and best dual PTMD models; a) Accelerometer measurement b) First story LVDT c) Second story LVDT d) Third story LVDT

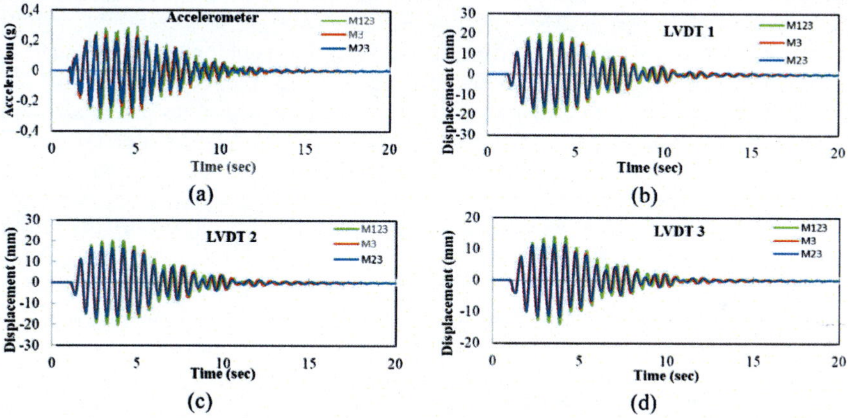

Fig. 8. Comparison of acceleration and displacement measurements of the best single, best dual, and triple PTMD models; a) Accelerometer measurement b) First story LVDT c) Second story LVDT d) Third story LVDT

Among the advantages of the multi PTMD are its simple and practical installation and maintenance. Due to its small size and mass, it can be easily applied to slender high-rise buildings. Here it can be adjusted by changing the natural frequency and length of the pendulum.

References

1. Rana, R., Soong, T.T.: Parametric study and simplified design of tuned mass dampers. Eng. Struct. **20**(3), 193–204 (1998)

2. Clark, A.J.: Multiple passive tuned mass dampers for reducing earthquake induced building motion. In: Proceedings of the 9th World Conference on Earthquake Engineering, vol. 5, pp. 779–784, Aug 1988, Tokyo-Kyoto, Japan (1988)
3. Nagase, T., Hisatoku, T.: Tuned-pendulum mass damper installed in crystal tower. Struct. Des. Tall Build. **1**(1), 35–56 (1992)
4. Xu, K., Igusa, T.: Dynamic characteristics of multiple substructures with closely spaced frequencies. Earthq. Eng. Struct. Dyn. **21**(12), 1059–1070 (1992)
5. Igusa, T., Xu, K.: Vibration control using multiple tuned mass dampers. J. Sound Vib. **175**(4), 491–503 (1994)
6. Yamaguchi, H., Harnpornchai, N.: Fundamental characteristics of multiple tuned mass dampers for suppressing harmonically forced oscillations. Earthq. Eng. Struct. Dyn. **22**(1), 51–62 (1993)
7. Kareem, A., Kline, S.: Performance of multiple mass dampers under random loading. J. Struct. Eng. N. Y. **121**, 348 (1995)
8. Joshi, A.S., Jangid, R.S.: Optimum parameters of multiple tuned mass dampers for base-excited damped systems. J. Sound Vib. **202**(5), 657–667 (1997)
9. Jangid, R.S.: Optimum multiple tuned mass dampers for base-excited undamped system. Earthq. Eng. Struct. Dyn. **28**(9), 1041–1049 (1999)
10. Bakre, S.V., Jangid, R.S.: Optimum multiple tuned mass dampers for base-excited damped main system. Int. J. Struct. Stab. Dyn. **4**(04), 527–542 (2004)
11. Li, C.: Performance of multiple tuned mass dampers for attenuating undesirable oscillations of structures under the ground acceleration. Earthq. Eng. Struct. Dyn. **29**(9), 1405–1421 (2000)
12. Li, C.: Optimum multiple tuned mass dampers for structures under the ground acceleration based on DDMF and ADMF. Earthq. Eng. Struct. Dyn. **31**(4), 897–919 (2002)
13. Li, C.: Evaluation of the lever-type active tuned mass damper for structures. Struct. Control. Health Monit. **11**(4), 259–271 (2004)
14. Li, H.N., Ni, X.L.: Optimization of non-uniformly distributed multiple tuned mass damper. J. Sound Vib. **308**(1–2), 80–97 (2007)
15. Han, B., Li, C.: Characteristics of linearly distributed parameter-based multiple-tuned mass dampers. Struct. Control Health Monit. Official J. Int. Assoc. Struct. Control Monit. Eur. Assoc. Control Struct. **15**(6), 839–856 (2008)
16. Chen, G., Wu, J.: Optimal placement of multiple tuned mass dampers for seismic structures. J. Struct. Eng. **127**(9), 1054–1062 (2001)
17. Chen, G., Wu, J.: Experimental study on multiple tuned mass dampers to reduce seismic responses of a three-storey building structure. Earthq. Eng. Struct. Dyn. **32**(5), 793–810 (2003)
18. Vyas, A., Bajaj, A.K.: Dynamics of autoparametric vibration absorbers using multiple pendulums. J. Sound Vib. **246**(1), 115–135 (2001)
19. Li, C., Qu, W.: Optimum properties of multiple tuned mass dampers for reduction of translational and torsional response of structures subject to ground acceleration. Eng. Struct. **28**(4), 472–494 (2006)
20. Nagarajaiah, S., Sonmez, E.: Structures with semiactive variable stiffness single/multiple tuned mass dampers. J. Struct. Eng. **133**(1), 67–77 (2007)
21. Casciati, F., Giuliano, F.: Performance of multi-TMD in the towers of suspension bridges. J. Vib. Control **15**(6), 821–847 (2009)
22. Lin, C.C., Wang, J.F., Lien, C.H., Chiang, H.W., Lin, C.S.: Optimum design and experimental study of multiple tuned mass dampers with limited stroke. Earthq. Eng. Struct. Dyn. **39**(14), 1631–1651 (2010)
23. Patil, V.B., Jangid, R.S.: Optimum multiple tuned mass dampers for the wind excited benchmark building. J. Civ. Eng. Manag. **17**(4), 540–557 (2011)

24. Gerges, R.R., Vickery, B.J.: Optimum design of pendulum-type tuned mass dampers. Struct. Des. Tall Spec. Build. **14**(4), 353–368 (2005)
25. Gus'kov, A.M., Panovko, G.Y., Van Bin, C.: Analysis of the dynamics of a pendulum vibration absorber. J. Mach. Manuf. Reliab. **37**(4), 321–329 (2008)
26. Lacarbonara, W., Balleri, S.: Vibration mitigation of guyed masts via tuned pendulum dampers. Struct. Eng. Mech. Int. J. **32**(4), 517–529 (2009)
27. Lin, C.C., Lu, L.Y., Lin, G.L., Yang, T.W.: Vibration control of seismic structures using semi-active friction multiple tuned mass dampers. Eng. Struct. **32**(10), 3404–3417 (2010)
28. Yang, Y., Munoa, J., Altintas, Y.: Optimization of multiple tuned mass dampers to suppress machine tool chatter. Int. J. Mach. Tools Manuf. **50**(9), 834–842 (2010)
29. Moon, K.S.: Vertically distributed multiple tuned mass dampers in tall buildings: performance analysis and preliminary design. Struct. Des. Tall Spec. Build. **19**(3), 347–366 (2010)
30. Bandivadekar, T.P., Jangid, R.S.: Optimization of multiple tuned mass dampers for vibration control of system under external excitation. J. Vib. Control **19**(12), 1854–1871 (2013)
31. Elias, S., Matsagar, V.: Wind response control of 76-storey benchmark building with distributed multiple tuned mass dampers. J. Wind Eng. **11**(2), 37–49 (2014)
32. Elias, S., Matsagar, V.: Distributed multiple tuned mass dampers for wind vibration response control of high-rise building. J. Eng. **2014** (2014)
33. Elias, S., Matsagar, V.: Optimum tuned mass damper for wind and earthquake response control of high-rise building. In: Advances in Structural Engineering, pp. 1475–1487. Springer, New Delhi (2015)
34. Elias, S., Matsagar, V., Datta, T.K.: Distributed tuned mass dampers for multi-mode control of benchmark building under seismic excitations. J. Earthq. Eng. **23**(7), 1137–1172 (2019)
35. Xiang, P., Nishitani, A.: Seismic vibration control of building structures with multiple tuned mass damper floors integrated. Earthq. Eng. Struct. Dyn. **43**(6), 909–925 (2014)
36. Elias, S., Matsagar, V., Datta, T.K.: Effectiveness of distributed tuned mass dampers for multi-mode control of chimney under earthquakes. Eng. Struct. **124**, 1–16 (2016)
37. Zuo, H., Bi, K., Hao, H.: Using multiple tuned mass dampers to control offshore wind turbine vibrations under multiple hazards. Eng. Struct. **141**, 303–315 (2017)
38. Xie, S., Jin, X., He, J., Gao, J., Zhang, C., Yan, Y.: Applying multiple tuned mass dampers to control structural loads of bottom-fixed offshore wind turbines with inclusion of soil-structure interaction. Ocean Eng. **205**, 107289 (2020)
39. Hussan, M., Rahman, M.S., Sharmin, F., Kim, D., Do, J.: Multiple tuned mass damper for multi-mode vibration reduction of offshore wind turbine under seismic excitation. Ocean Eng. **160**, 449–460 (2018)
40. Deraemaeker, A., Soltani, P.: A short note on equal peak design for the pendulum tuned mass dampers. Proc. Inst. Mech. Eng. Part K J. Multi-body Dyn. **231**(1), 285–291 (2017)
41. Kecik, K., Warminski, J.: Dynamics of an autoparametric pendulum-like system with a nonlinear semiactive suspension. Math. Probl. Eng. **2011** (2011)
42. Roffel, A.J., Narasimhan, S., Haskett, T.: Performance of pendulum tuned mass dampers in reducing the responses of flexible structures. J. Struct. Eng. **139**(12), 04013019 (2013)

Effect of Over-Stroke Capacity of Curved Surface Sliders on the Collapse Safety of Seismically Isolated Buildings

Antonio Di Cesare and Felice Carlo Ponzo(✉)

School of Engineering, University of Basilicata, Potenza, Italy
felice.ponzo@unibas.it

Abstract. Acceptable probabilities of collapse for seismically isolated buildings could be achieved by an appropriate isolator displacement capacity. This paper investigates the effects of the over-stroke capacity, in addition with the displacement capacity of the isolation system, on the seismic response of a structure isolated with double concave curved surface slider isolators. The case study is a moment resisting steel framed building in which the isolation system assumes different displacement limits. Results from nonlinear static analysis and nonlinear time history analysis demonstrate that over-stroke displacement of the isolators may significantly increase the safety factor in case of earthquakes bigger than the design ones, without drawing on the plastic resources of the structure and changing the construction costs.

Keywords: Base isolation · Double concave curved surface slider isolators · Over-stroke displacement · Pushover analysis · Nonlinear response history analysis

1 Introduction

The Double Concave Curved Surface Slider (DCCSS), obtained combining two Curved Surface Slider (CSS) bearing [1, 2], is one of the most widely used isolator for seismic protection of buildings and bridges. Base isolation system is designed to attain the design displacement at the Maximum Considered Earthquake (MCE).

Recent studies consider the isolation system failure occurring when the actual isolator displacement capacity d_C is exceeded, it can be occur in case of earthquakes stronger than the design one [3].

Results of shake table testing described in [4–7] demonstrated that restraining rim bearings have a strong influence on the structural behavior. A larger collapse margin can be found for configurations without retaining rings, where the inner slider has the possibility of exceeding the nominal geometric displacement capacity. Results show also how the DCCSS behaviour in over-stroke conditions can be affected by the sliding material and the manufacturing of the housing plate border (see Fig. 1. a).

European Standard [8] does not allow for any restraining ring or other mechanical element that serves as end-stroke and that can be damaged in case of a seismic event.

G. P. Cimellaro (Ed.): WCSI 2022, LNCE 309, pp. 254–263, 2023.
https://doi.org/10.1007/978-3-031-21187-4_19

On the other hand, it allows for the use of retaining walls as limit stroke (see Fig. 1. a). Accurate studies to assess the consequences of impacts on the superstructure of seismically isolated buildings under strong ground motions, including pounding against a flexible moat wall, were carried out by Masroor and Mosqueda [9].

Usually, the gap separating the superstructure from the retaining wall is much larger than the displacement capacity of the isolators, and consequently, the inner slider can run on the edge of the sliding surfaces in the over-stroke regime still preserving the ability to support gravity loads and the re-centring capability.

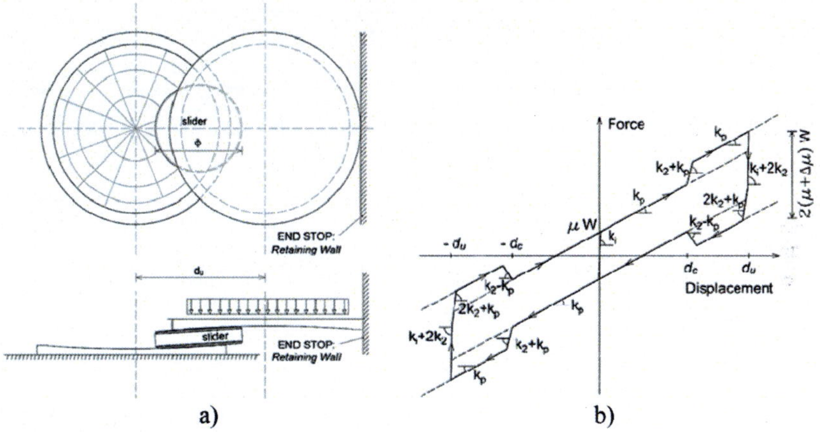

Fig. 1. DCCSS bearing with overstroke displacement: a) overview with retaining wall; b) modified force-displacement cyclic model.

In order to evaluate the effect of the over-stroke capacity on the seismic performance of isolated buildings with strong end-stroke elements, a moment resistant steel frame seismically isolated with modified force-displacement model of DCCSS bearings with over-stroke displacement (see Fig. 1. b) [10, 11] have been analyzed in this paper by means nonlinear static and dynamic analyses varying the ultimate displacement limit of isolators and considering the influence of the seismic stop.

2 Case Study

2.1 Superstructure

The case study is a 2D moment resistant steel frame with 3 bays and 6 story representative of a residential use [3, 12–14]. The steel frame is designed for gravity loads only respecting minimum criteria required by Eurocode [15, 16], while considering low ductility class for structural details. Geometrical dimensions and the cross-sections of structural elements are shown in Fig. 2a. The seismic weights W_i and masses M_i of each floor are shown in Fig. 2b. A lumped plasticity model has been considered for superstructure beam and column elements. Second order effects have not been taken into account. Member capacities were calculated using 275 MPa yield strength for steel materials. Joint panel zones were modeled with rigid elastic elements. Plastic hinges have been concentrated at the end of structural elements considering a distance from the end of nodes corresponding to the element depth [17].

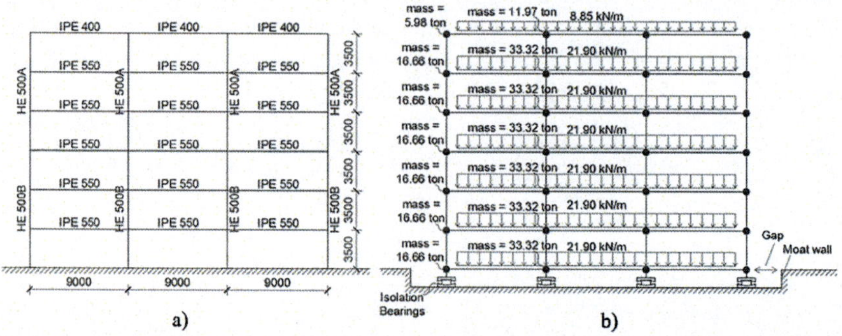

Fig. 2. a) Structural elements of the fixed-base superstructure and building geometry; b) Representation of gravity loads for moment resisting frame of isolated building.

2.2 Isolation Systems

The considered isolated building is shown in Fig. 2b. Elastic beams have been added at the base floor to realize the grid above the isolation system, consisting of four DCCSS isolators placed below the columns. The total seismic weight W of the isolated frame is about 5000kN and the vertical loads on the internal isolators N_{Sd} is 1680kN. The isolation system is designed considering the MCE spectrum shown in Fig. 3a [15]. The 5%-damped elastic spectra of 20 natural seismic records selected to represent the earthquake intensity level at MCE of the isolated structures are reported in Fig. 3b. A reference period (conditioning period) close to the design value of the fundamental period of vibration of $T = 3$ s has been considered. More details about the seismic hazard evaluation and the record selection for nonlinear dynamic analysis are reported in [18].

Main design parameters of the isolation system are reported in Table 1: equivalent radius R_e, the vertical load N_{sd}, friction coefficient μ, initial stiffness k_i, equivalent stiffness k_e, effective period T_e, effective damping ξ_{eff}, design displacement d_{Ed} of the isolation system at the MCE. For the over-stroke stage of motion the mechanical parameters are the increase of stiffness k_2 and friction coefficient $\Delta\mu$.

DCCSS bearing behaviors with and without over-stroke (see Fig. 1b) and with and without end strokes have been considered. *Single FP Bearing* element implemented in OpenSEES [19] has been selected to describe the cyclic behavior of the basic DCCSS isolator (without over-stroke and without end-stop). The friction coefficient is modelled accounting for sliding velocity and contact pressure. The basic model has been modified for the over-stroke displacement by adding three zero-length parallel hinges as described in [7, 11]: two elastic-perfectly plastic gap elements (*elasticPPGap*) and one multi-linear elastic element. Seismic stoppers consisting of a zero-length gap element were added to the model of each isolator.

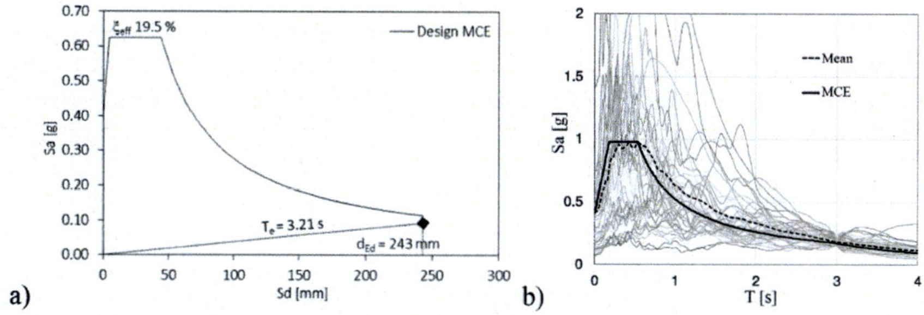

Fig. 3. a) Design MCE spectrum considered for the design of the isolation system; b) Elastic spectra of selected ground motions for base-isolated frame.

Table 1. Isolation system design parameters.

R_e [mm]	N_{sd} [kN]	μ [%]	K_i [kN/m]	K_e [kN/m]	T_e [sec]	ξ_{eff} [%]	d_{Ed} [mm]	k_2 [kN/m]	$\Delta\mu$ [%]
3700	1680	2.9	5000	653	3.2	19.5	243	2500	2%

2.3 Parametric Analysis

A parametric analysis has been performed varying the properties of elements constituting the isolation system. The analysed cases are reported in the Table 2.

In order to take into account the oversizing due to the industrial discretization of the devices the first considered parameter was the ratio between the actual isolator displacement capacity d_c and the design displacement d_{Ed}. The ratio was considered ranging between 1 and 1,5. The overstroke limit has been defined as the ratio between the ultimate displacement d_u and d_c. In that the analyses were carried out varying the ratio between 1 and 2, with 0.25 step.

Moreover, for the case of $d_c/d_{Ed} = 1$, four values of gap element elastic stiffness k_{gap} expressed as proportion of the superstructural horizontal stiffness k_{str} have been considered: (i) rigid ($k_{gap} = \infty$); (ii) stiff ($k_{gap} = 25\%\ k_{str}$); (iii) flexible ($k_{gap} = 10\%\ k_{str}$); (iv) without gap ($k_{gap} = 0$). Plastic deformations of the end-stroke models, as example due to moat wall nonlinearity, were not taken into account. The gap element models are shown in Fig. 4a.

The force-displacement behaviors of four DCCSS model configurations are shown in Fig. 4b for the basic isolation model with displacement capacity $d_c/d_{Ed} = 1.25$ and with: (i) unlimited elasto-plastic behavior (*BI 1.25*); (ii) rigid end-stop (*BI 1.25_Stop*); (iii) over-stroke displacement $d_u/d_c = 2.00$ and unlimited elasto-plastic behavior (*BI 1.25_Over-Stroke*); (iv) with over-stroke displacement $d_u/d_c = 2.00$ and rigid end-stop (*BI 1.25_Over-Stroke_2.00_Stop*).

3 Nonlinear Analyses

The variation of the seismic performances of the isolated structure as function of the before mentioned parameters was evaluated by comparing the results of nonlinear static analysis (Pushover) with those of nonlinear dynamic analysis.

Table 2. Case study of isolation system models and ultimate displacement

d_c/d_{Ed}	d_u/d_c	end-stop	Case ID	d_u *[mm]*
1.00	1.00	w/o	*BI 1.00*	*243.0*
		flexible	*BI 1.00_Flex*	
		Stiff	*BI 1.00_Stiff*	
		rigid	*BI 1.00_Stop*	
	1.25	w/o	*BI 1.00_Over-Stroke 1.25*	*303.8*
		rigid	*BI 1.00_Over-Stroke 1.25_Stop*	
	1.50	w/o	*BI 1.00_Over-Stroke 1.50*	*364.5*
		rigid	*BI 1.00_Over-Stroke 1.50_Stop*	
	1.75	w/o	*BI 1.00_Over-Stroke 1.75*	*425.3*
		rigid	*BI 1.00_Over-Stroke 1.75_Stop*	
	2.00	w/o	*BI 1.00_Over-Stroke 2.00*	*486.0*
		rigid	*BI 1.00_Over-Stroke 2.00_Stop*	
1.25	1.00	w/o	*BI 1.25*	*303.8*
		rigid	*BI 1.25_Stop*	
	1.25	w/o	*BI 1.25_Over-Stroke 1.25*	*379.7*
		rigid	*BI 1.25_Over-Stroke 1.25_Stop*	
	1.50	w/o	*BI 1.25_Over-Stroke 1.50*	*455.6*
		rigid	*BI 1.25_Over-Stroke 1.50_Stop*	
	1.75	w/o	*BI 1.25_Over-Stroke 1.75*	*531.6*
		rigid	*BI 1.25_Over-Stroke 1.75_Stop*	
	2.00	w/o	*BI 1.25_Over-Stroke 2.00*	*607.5*
		rigid	*BI 1.25_Over-Stroke 2.00_Stop*	
1.50	1.00	w/o	*BI 1.50*	*364.5*
		rigid	*BI 1.50_Stop*	
	1.25	w/o	*BI 1.50_Over-Stroke 1.25*	*455.6*
		rigid	*BI 1.50_Over-Stroke 1.25_Stop*	
	1.50	w/o	*BI 1.50_Over-Stroke 1.50*	*546.8*
		rigid	*BI 1.50_Over-Stroke 1.50_Stop*	
	1.75	w/o	*BI 1.50_Over-Stroke 1.75*	*637.9*
		rigid	*BI 1.50_Over-Stroke 1.75_Stop*	
	2.00	w/o	*BI 1.50_Over-Stroke 2.00*	*729.0*
		rigid	*BI 1.50_Over-Stroke 2.00_Stop*	

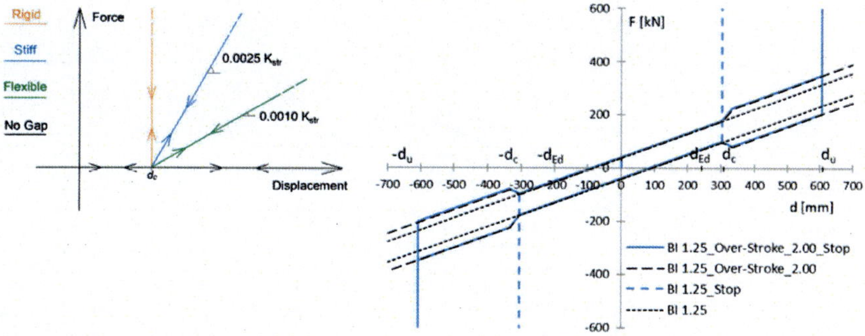

a) b)

Fig. 4. Force-displacement behavior of: a) models with finite stiffness end-stop elements; b) models of isolation bearings with and without overstroke and rigid end-stop

A distribution of lateral forces proportional to the floor masses was used in the pushover analyses. In Fig. 5a the pushover curves of not-isolated and base-isolated building for four values of the gap element elastic stiffness are compared. The red marker represents the structural yield drift limits in terms of global drift ratios of 1%. Pushover curves of base-isolated building models without displacement restraints (black lines) show the basic behaviour of the isolation system, with a first quasi-rigid branch (pre-sliding stage), followed by a second branch accounting for isolators restoring stiffness. The isolated models with displacement restraints highlight a pushover behaviour similar to that of the not-isolated superstructure for displacement greater of the base displacement capacity d_c. Models with flexible displacement restraints show a less sloping curve after the d_c displacement consistent with the gap stiffness K_{gap} but the same maximum base-shear force. Pushover curve of isolators with unlimited stroke displacement (*BI 1.00_1.25_1.50*) shows the basic behavior. Isolators with unlimited over-stroke displacement (*BI 1.00_Overstroke 1.25_1.50_1.75_2.00*) shows the modified behavior of isolators beyond the isolator end-stroke (see Fig. 5b).

In detail, the configuration with geometric displacement capacity $d_c/d_{Ed} = 1.25$ and end-stop at $d_u/d_c = 1$ (*BI 1.25_Stop*), after the first quasi-rigid branch (pre-sliding stage) and the second branch accounting for isolators restoring stiffness, the isolation system stops when the base displacement reaches d_c and superstructure pushover curve is shown again. The system with geometric displacement capacity $d_c/d_{Ed} = 1.25$, and with end-over-stroke at $d_u/d_c = 2$ (*BI 1.25_Over-Stroke 2.00_Stop*) perfectly overlaps the previous one for the first two branches, then, when the geometric capacity displacement is reached, the isolation system enters in the over-stroke stage (see zooming box in Fig. 5b) with a third branch with increased stiffness friction force, and finally, when the end-over-stroke is reached (end of sliding), the superstructure pushover curve is shown again. It is possible to observe that the shape of the first part of the pushover curve of isolated structures can be approximated by an elastoplastic hardening behavior.

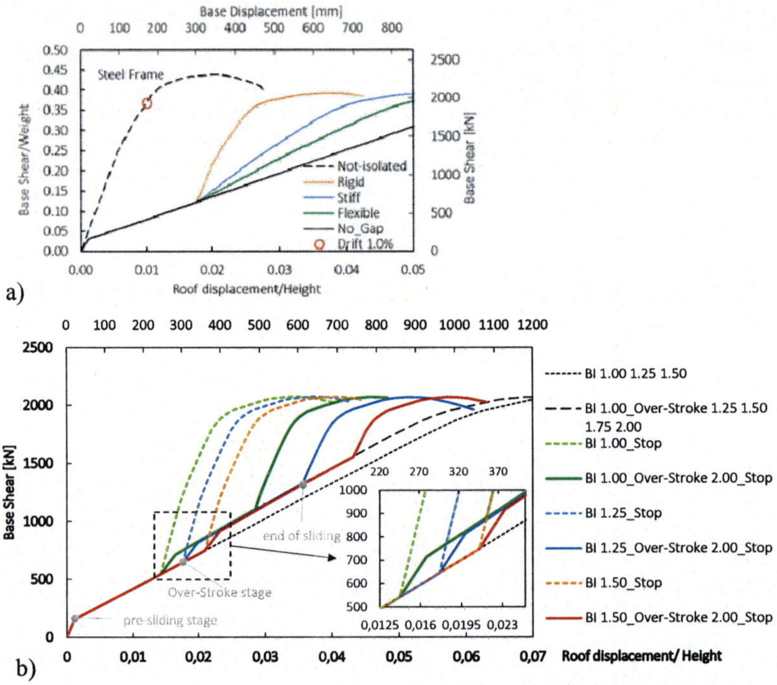

Fig. 5. Pushover curves of: a) not-isolated building and isolated with $d_c/d_{Ed} = d_u/d_c = 1$ and flexible element for end-stop; b) isolated building without and with rigid end-stop.

4 Comparison of Results

In order to adequately reproduce the behaviour of the isolation system with rigid end-stop, the seismic response of base-isolated models has been studied by means of nonlinear time-history analyses considering 20 real ground motions (see Fig. 3b) and compared with pushover curves. In Fig. 6 nonlinear dynamic results are expressed in terms of the maximum superstructure top displacement vs ratio between maximum base shear and total seismic weight (acceleration displacement response spectrum). It is worth pointing out that in the case of impact between the superstructure and the end stop dynamic analyzes provide values of the base shear significantly higher than those obtained from the push-over curves, due to the dynamic effects. The number of earthquakes producing a displacement higher than the ultimate displacement (impact accelerations in Fig. 6) is reported in Table 3; it decreases when the geometric displacement capacity d_c/d_{Ed} and/or the over-stroke displacement d_u/d_c increase.

For the case of isolator without overstroke $d_u/d_c = 1$, non-linear dynamic analyses showed that the displacement capacity is not sufficient to avoid the impact of the superstructure, even when the displacement capacity of the isolator is increased to 50% of design value ($d_c/d_{Ed} = 1.50$). Due to industrial discretization of devices, usual design provides for the adoption of an oversizing of 25% ($d_c/d_{Ed} = 1.25$). In this case with a minimum overstroke ($d_u/d_c = 1.25$) the limit is exceeded just for one earthquake, this

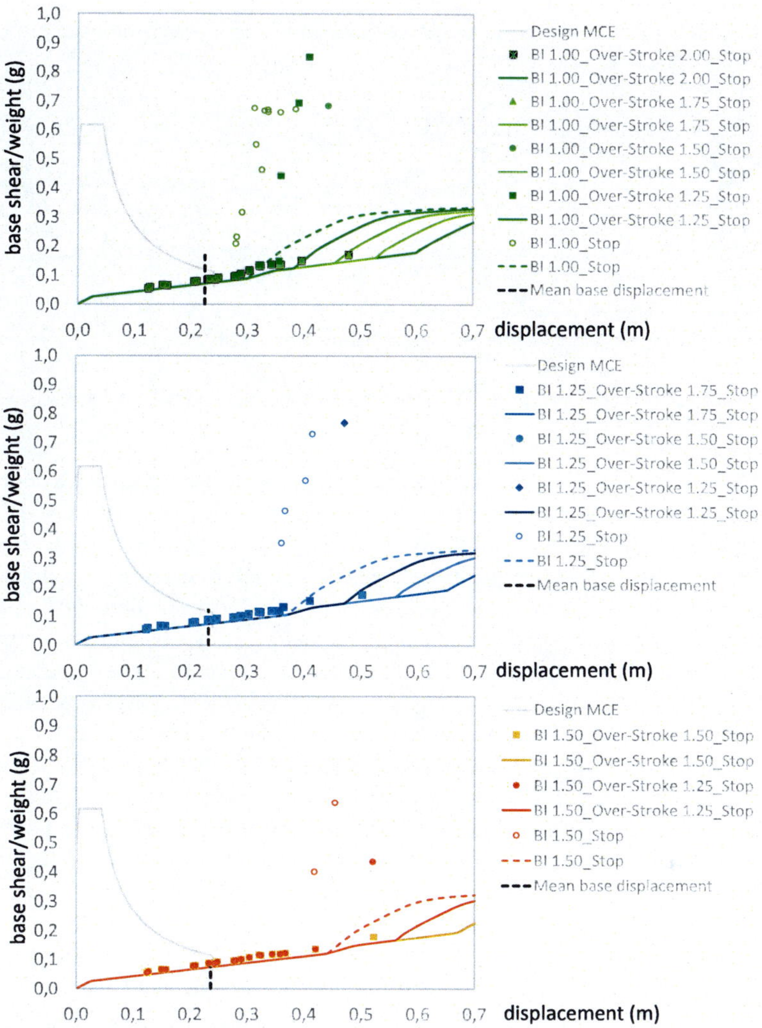

Fig. 6. Nonlinear static versus nonlinear dynamic results in term of maximum top displacement and acceleration at the base level

is mainly due to the variability of the selected ground motions. For the case of usual design (d_c/d_{Ed} = 1.25) with an overstroke of 50% of displacement capacity (d_u/d_c = 1.50), compatible with commercially available devices, the exceedances of ultimate displacement are avoided.

Figure 6 also reports the mean values of maximum base displacements and the design MCE spectrum (see Fig. 3a). For all analyzed cases the mean base displacements are slightly lower than the design displacement d_{Ed}.

Table 3. Number of cases where maximum displacement d_{max} exceeds ultimate displacement d_u

Case study with rigid end stop	$d_c/d_{Ed} = 1.00$	$d_c/d_{Ed} = 1.25$	$d_c/d_{Ed} = 1.50$
$d_u/d_c = 1.00$	11	4	2
$d_u/d_c = 1.25$	3	1	1
$d_u/d_c = 1.50$	1	0	0
$d_u/d_c = 1.75$	0	0	0
$d_u/d_c = 2.00$	0	0	0

5 Conclusions

A case study of a seismically isolated moment resisting steel structure with double concave curved surface sliding bearings was investigated using non-linear static and dynamic analyses. In order to take into account the oversizing due to industrial discretization and the over-stroke displacement capacity of the bearings before reaching the end-stroke, a parametric analysis was performed by varying the displacement limit of the isolation system. Compatibly with commercially available DCCSS devices, the ratio between the actual isolator displacement capacity d_c and the design displacement d_{Ed} and the ratio between the overstroke displacement d_u and d_c were considered ranging between 1 and 1,5 and between 1.25 and 2.00, respectively.

For the case of isolator without overstroke $d_u/d_c = 1$, non-linear dynamic analyses showed that the increase in the displacement capacity of the isolator at $d_c/d_{Ed} = 1.5$ is not sufficient to avoid the effects on the superstructure of the impact at the base against end-stroke stopper. For isolators with limited over-stroke displacement and displacement capacity $d_c/d_{Ed} = d_u/d_c = 1.25$, the elastic behavior of the superstructure is activated just for one earthquake. Impacts against the end-stop are avoided in almost all cases when the over-stroke displacement value is almost 50% of the capacity ($d_u/d_c = 1.5$). Furthermore, the design displacement evaluated by equivalent static analysis is in good agreement with the mean value of maximum base displacement obtained from non-linear dynamic results of all cases. Further studies are currently underway considering different case studies.

Acknowledgements. Authors would like to acknowledge the financial support of RELUIS 2022–2024 project (Work Package no. 15: Contributions to the Italian seismic code for base isolation and energy dissipation systems) funded by the Italian Civil Protection Department.

References

1. Fenz, D.M., Constantinou, M.C.: Behaviour of the double concave friction pendulum bearing. EESD **35**, 1403–1424 (2006)
2. Tsopelas, P., Constantinou, M.C., Kim, Y.S., Okamoto, S.: Experimental study of FPS system in bridge seismic isolation. EESD **25**(1), 65–78 (1996)

3. Kitayama, S., Constantinou, M.C.: Effect of displacement restraint on the collapse performance of seismically isolated buildings. Bull. Earthq. Eng. **17**(5), 2767–2786 (2019). https://doi.org/10.1007/s10518-019-00554-y
4. Sarlis, A.A., Constantinou, M.C.: Model of triple friction pendulum bearing for general geometric and frictional parameters and for uplift conditions. MCEER-13-0010 (2013)
5. Bao, Y., Becker, T.C., Sone, T., Hamaguchi, H.: Experimental study of the effect of restraining rim design on the extreme behavior of pendulum sliding bearings. EESD **47**(4), 906–924 (2018)
6. Furinghetti, M., Yang, T., Calvi, P.M., Pavese, A.: Experimental evaluation of extra-stroke displacement capacity for curved surface slider devices. Soil Dyn. Earthq. Eng. **146** (2021)
7. Di Cesare, A., Ponzo, F.C., Telesca, A.: Improving the earthquake resilience of isolated buildings with double concave curved surface sliders. Eng. Struct. **228**, 111498 (2020)
8. UNI EN. 15129: Anti-seismic devices. European committee for standardization (CEN). Bruxelles, Belgium (2009)
9. Masroor, A., Mosqueda, G.: Experimental simulation of base-isolated buildings pounding against moat wall and effects on superstructure response. EESD **41**(14), 2093–2109 (2012)
10. Ponzo, F.C., Di Cesare, A., Telesca, A., Nigro, D.: Base isolated buildings with curved surface sliders including displacement restraints. In: Proceeding of 14th World Congress in Computational Mechanics (WCCM) ECCOMAS 2020, Virtual Congress, 11–15 Jan 2021
11. Ponzo, F.C., Di Cesare, A., Telesca, A., Pavese, A., Furinghetti, M.: Advanced modelling and risk analysis of RC buildings with sliding isolation systems designed by the Italian Seismic Code. Appl. Sci. **11**, 1938 (2021)
12. McVitty, W.J., Constantinou, M.C.: Property modification factors for seismic isolators: design guidance for buildings. Multidisciplinary Center for Earthquake Engineering Research, Buffalo, NY (2015)
13. ASCE/SEI 7-10: Minimum design loads for buildings and other structures (2013)
14. Bao, Y., Becker, T.C., Sone, T., Hamaguchi, H.: To limit forces or displacements: collapse study of steel frames isolated by sliding bearings with and without restraining rims. SDEE **112**, 203–214 (2018)
15. EN 1998-1. Eurocode 8 (2004) Design of structures for earthquake resistance—Part 1: general rules, seismic actions and rules for buildings. European committee for standardization (CEN), Bruxelles, Belgium
16. NTC: Italian technical code for constructions (in Italian). Ministry of Infrastructures, Italy (2018)
17. Lignos, D.G., Krawinkler, H.: Deterioration modeling of steel components in support of collapse prediction of steel moment frames under earthquake loading. JSE **137**(11), 1291–1302 (2011)
18. Iervolino, I., Spillatura, A., Bazzurro, P.: Seismic reliability of code-conforming Italian buildings. JEE **22**, 5–27 (2018)
19. OpenSEES: Open system for earthquake engineering simulation. PEER Center, University of California, Berkeley (2006). Available at http://opensees.berkeley.edu/. Last Accessed 1 Sept 2020

Crescent Shaped Brace Devices to Strengthen Pinned Beam-Column Connections via Semi-rigid CSB Joints

Michele Palermo, Vittoria Laghi(✉), Stefano Silvestri, Giada Gasparini, and Tomaso Trombetti

Department of Civil, Chemical, Environmental and Materials Engineering, University of Bologna, Bologna, Italy
vittoria.laghi2@unibo.it

Abstract. The present work explores the use of a curved hysteretic steel device, known as Crescent Shaped Brace, to strengthen pinned beam-column connections in order to obtain semi-rigid joints. The kinematics and mechanical behavior of the semi-rigid joint is investigated by introducing a simplified analytical model. The model is then validated through Finite Element analysis. Finally, the possibility of optimizing the joint geometry and mechanical performances using novel metal 3D printing techniques (such as Wire-and-Arc Additive Manufacturing) is briefly presented.

Keywords: Crescent Shaped Braces · Semi-rigid beam-column joints · Moment-rotation response · Wire-and-Arc Additive Manufacturing

1 Introduction

Energy dissipation devices have been developed over the last half century to mitigate earthquake-induced effects on building structures, in order to increase the seismic performances while maintaining reasonable construction costs [1]. The devices are usually classified in three main categories: active, semi-active and passive systems. Metallic (hysteretic) and friction dampers belong to the category of displacement-activated passive supplemental damping systems. Steel dampers take advantage of the hysteretic behavior of the material exceeding its yielding point. Desirable properties of these devices are represented by a stable hysteretic behavior, the ability in sustaining an adequate number of loading-unloading cycles (low-cycle fatigue), long term reliability and low sensitivity to environmental temperature [2].

Since the mid-1970s, a considerable number of hysteretic devices have been developed and tested [3–9]. Among others, in 2009, within the framework of Performance Based Seismic Design [10], a new hysteretic device, known as Crescent Shaped Brace (CSB), has been proposed. Thanks to their particular geometrical and mechanical behavior, CSBs are able to fulfill multiple seismic design objectives in terms of stiffness, strength, ductility and ultimate capacity. So far CSBs have been proposed as enhanced

G. P. Cimellaro (Ed.): WCSI 2022, LNCE 309, pp. 264–272, 2023.
https://doi.org/10.1007/978-3-031-21187-4_21

braces for multi-storey buildings and as connections between frames and RC cores [11–13].

The present work explores the use of CSBs as hysteretic devices to be added to pinned beam-column connections to realize semi-rigid joints. Future developments of this concept see the application of optimization algorithms to improve the performances of the hysteretic connections (from both a geometrical and mechanical perspective), by making use of metal 3D printing production techniques, such as Wire-and-Arc Additive Manufacturing.

2 Crescent Shaped Brace for Semi-rigid Beam-Column Connection

Pinned beam-column connections were typically used in the past for steel industrial buildings designed without current seismic requirements. A possible solution to improve their seismic performances is to strengthen the beam-column pinned joint through replaceable devices with the required stiffness and strength. CSBs can be a possible solution for this purpose, leading to a so-called CSB joint, schematically represented in Fig. 1a.

A CSB made by two straight elements of equal lengths L^*, referred to as "symmetric bilinear CSB" has been widely studied in previous research works [11–14] and is also here considered (Fig. 1c). The angle θ_0 indicates the initial inclination of each straight segment. One end is assumed to be fixed, while the other is supposed free to move along the lateral direction (u indicates the horizontal displacement along the line connecting the two ends). The two end supports do not provide any rotational restraint. The "initial lever arm" d_0 indicates the vertical distance between the knee point of the CSB and the horizontal projection (whose length is referred to as $2L_0$). The normalized initial lever arm is referred to $\xi_0 = d_0/2L_0$ (subscript *0* refers to the undeformed configuration). When subjected to a lateral force F, the two straight elements of the CSB deform due to the interaction of axial force (compression or tension, depending on the direction of F) and bending moment. The angle between the horizontal direction and the chord of one CSB segment in the initial undeformed configuration is indicated with θ_0, while the same angle in the generic deformed configuration is indicated with θ. The corresponding lever arm in the generic deformed configuration is indicated with d (its normalized value is equal to $\xi = d/2L_0$). The vertical displacement of the knee point is indicated with v.

The geometrical and mechanical properties of a "symmetric bilinear CSB" are as follows: L^* is the length of each straight element; L_0 is the projection of L^* in the horizontal plane; h, A and J are the cross-section height, area and moment of inertia (in-plane), respectively; i is the radius of gyration; W_e is the elastic cross-section modulus; E is the material Young's modulus; f_y, ε_y, f_u, ε_u are the yielding strength and strain; and the ultimate strength and strain; r is the hardening ratio.

The kinematic behavior of the CSB can be described considering, at first approximation, a Single Degree Of Freedom (SDOF), for instance the angle θ as follows [12]:

$$
\begin{aligned}
u &= 2L^*(\cos\theta - \cos\theta_0) \\
v &= L^*(\sin\theta_0 - \sin\theta)
\end{aligned}
\tag{1}
$$

The lateral displacement u can be also interpreted as the elongation of the CSB projections, i.e. the displacement along the direction connecting the two ends (see Fig. 1c).

The force displacement response of the equivalent SDOF system is governed by the following set of equilibrium equations [12]:

$$F = \begin{cases} K_L \cdot u & \text{for}\, u \le u_y \\ M_y/d = M_y/\left(d_0 - \frac{u}{2}\frac{(\sin\theta_0 - \sin\theta)}{(\cos\theta - \cos\theta_0)}\right) & \text{for}\, u_y < u \le u_{pl} \\ A \cdot f_u & \text{for}\, u_{pl} < u \le u_u \end{cases} \quad (2)$$

$$K_L = K_\theta \cdot \frac{\Delta\theta}{2 \cdot L^{*2} \cdot \sin\theta(\cos\theta_0 - \cos\theta)} = \frac{3}{2} \cdot \frac{E \cdot J \cdot \Delta\theta}{L^{*3} \cdot \sin\theta(\cos\theta_0 - \cos\theta)} \quad (3)$$

where $u_y = \frac{2}{3} \cdot \frac{d_0^2}{\xi \cdot h \cdot \cos\theta_0} \cdot \varepsilon_y$ is the approximate expression of the displacement leading to the flexural yielding, K_L is the lateral (flexural) stiffness of the system. The maximum lateral force inducing tension in the member is limited by the material ultimate tensile strength f_u, e.g. by the ultimate axial capacity of the straight member F_u (Eq. 2). Note that, in this simplified SDOF mechanical model, the axial flexibility of the CSB has been neglected. The elongation u_{pl} leading to $\theta = 0$ (and corresponding to the axial yielding) is approximately equal to $u_{pl} = 2 \cdot (L^* - L_0)$, while the ultimate elongation u_u can be estimated equal to $u_u = 2 \cdot (L^* - L_0) + 2 \cdot L^* \cdot \widetilde{\varepsilon}_u$, where $\widetilde{\varepsilon}_u$ is the ultimate strain reduced to take into account the prior formation of the plastic flexural hinge. This value needs to be obtained from experimental tests.

The qualitative graphical representations of the *F*-*u* curve of a CSB subjected to tension is shown in Fig. 1c. The *F*-*u* curve is made by: (i) a first elastic range (governed mainly by the flexural stiffness) until the first *flexural yielding* of the knee-point C is achieved at point P_{fy} (or simply P_y); (ii) a pseudo-plastic plateau region, governed by plasticity and flexural stiffness, that ends at the transition point (point P_{gh}) with the sudden increase in stiffness due to significant non-linear *geometrical hardening* effects (reduction of the arm d) determining the engagement of the axial stiffness; (iii) a geometric hardening region that ends at point P_{ay} (or simply P_{pl}) when the brace experiences *axial yielding* while reaching the straight configuration (e.g. the arm d reduces to zero, $\xi = 0$); (iv) a final plastic axial region.

The deformed geometrical configuration of the joint can be described by a single degree of freedom, for instance the angle $\Delta\phi$ representing the relative rotation between the beam and the column (variation between the angle in the deformed configuration and the initial 90° angle in the undeformed configuration). The joint rotation $\Delta\phi$ can be related to the main geometrical parameters describing the deformed configuration of the CSB (namely the elongation along the CSB projection, u, and the rotation angle θ or its

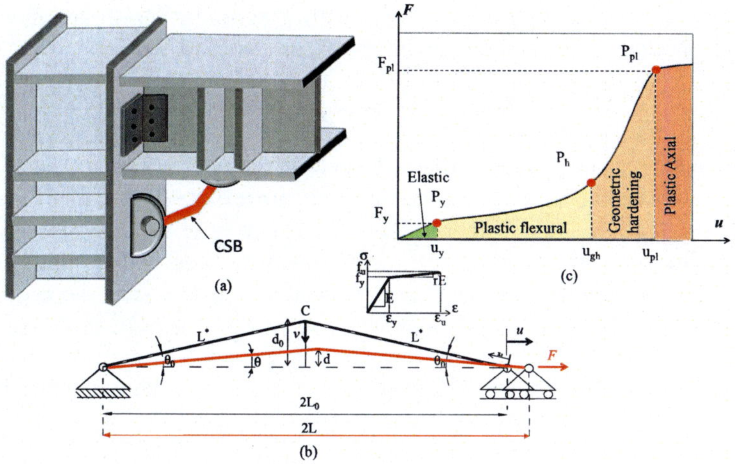

Fig. 1. (a) Schematic representation of the semi-rigid CSB joint. (b) Schematic representation of a CSB under tensile force. (c) Force-displacement behavior in tension.

variation $\Delta\theta = \theta - \theta_0$) through simple analytical kinematic relationships:

$$u = 2L - 2L_0 = 2L^*(\cos\theta - \cos\theta_0) = \sqrt{2}B \cdot \left(\sqrt{1 + \sin\Delta\phi} - 1\right)$$
$$\sqrt{2}B = 2L^* \cos\theta_0$$

$$\sqrt{1 + \sin\Delta\phi} - 1 = \frac{\cos\theta - \cos\theta_0}{\cos\theta_0}$$
$$\sqrt{1 + \sin\Delta\phi} = \frac{\cos\theta}{\cos\theta_0} = \frac{\cos(\theta_0 - \Delta\theta)}{\cos\theta_0} \tag{4}$$

$$\cos\beta = \sqrt{\frac{1 + \sin\Delta\phi}{2}}$$

where β represents the final angle, in the deformed configuration, between the beam and the top straight element of the CSB (see Fig. 2a). B is a geometrical property of the joint (joint arm).

As long as the CSB remains within the elastic field, the rotational stiffness $K_{\theta,\text{joint}}$ of the joint can be estimated according to the following formula:

$$K_{\theta,\text{joint}} = K_{\theta,CSB} \frac{\Delta\theta \cos\theta_0}{\sqrt{1 - \left(1 + \sin\Delta\phi \cdot \cos^2\theta_0\right)}} \frac{\cos\Delta\phi}{\Delta\phi} \sqrt{1 + \sin\Delta\phi} \tag{5}$$

When the joint exhibits a relative rotation $\Delta\phi$, the CSB is elongated of a quantity equal to u and transfers to the end nodes an inclined force equal to F (see Fig. 2a). The joint moment M_{joint} resisting to the joint rotation $\Delta\phi$ can be evaluated as the product of the horizontal component of F ($F_H = F\cos\beta$ with F given by Eq. 2) and the distance between its application point and point O, e.g. $B\cos\Delta\phi$ (as illustrated in Fig. 2a):

$$M_{\text{joint}} = F\cos\beta \cdot B\cos\Delta\phi \quad (6)$$

Figure 2b illustrates the qualitative moment-rotation behavior of two similar semi-rigid connections having the same global geometrical properties (joint arm B) and CSBs of equal cross section, but different values of normalized lever arms ξ_O (equal to 10 and 20%). The comparison shows that, as expected, the connection with $\xi_0 = 10\%$ is characterized by a higher rotational stiffness and yielding bending moment than the connection with $\xi_0 = 20\%$. On the other hand, it exhibits a smaller ductility. The two connections exhibit the same plastic moment, since the CSBs have the same cross-section. P_y and P_{pl} indicate the first yielding and the plastic points of the moment-rotational response of the CSB joint.

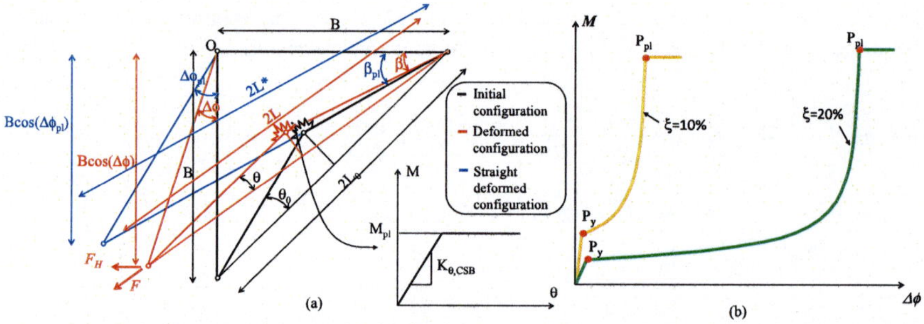

Fig. 2. (a) Schematic representation of the kinematics and mechanical behavior of the semi-rigid CSB joint. (b) Qualitative moment-rotation response of two CSB joints characterized by different normalized lever arms ξ_0.

3 Design Equations

In the preliminary design phase of the semi-rigid CSB joint, it could be useful to have at designers disposal simplified formulas to estimate the two points $P_y = (\Delta\phi_y, M_y)$ and $P_{pl} = (\Delta\phi_{pl}, M_{pl})$ governing the moment-rotation response of the joint. For this aim, the following assumptions and approximations are introduced: $\sin\Delta\phi \cong \Delta\phi$; $B\cos\Delta\phi \cong B$; $\left(\sqrt{1+\Delta\phi} - 1\right)/\Delta\phi \cong 0.5$.

Making use of the introduced approximations, it is possible to estimate the initial joint rotational $K_{\theta,\text{joint},0}$:

$$K_{\theta,\text{joint},0} = K_{\theta,CSB}\frac{\cos^2\theta_0}{2\sin^2\theta_0} \quad (7)$$

Similarly, the rotations corresponding to the CSB elongations at flexural and axial yielding can be estimated through the following analytical relationships:

$$\Delta\phi_{y0} = \frac{4L^* \cdot \cos\theta_0 \cdot u_{y0} - u_{y0}^2}{2L^{*2} \cdot \cos^2\theta_0}$$
$$\Delta\phi_{pl0} = \frac{4L^* \cdot \cos\theta_0 \cdot u_{pl0} - u_{pl0}^2}{2L^{*2} \cdot \cos^2\theta_0} \quad (8)$$

The moment at flexural yielding can be estimated as the product of the initial rotational stiffness and the flexural yielding rotation according to the following relationship:

$$M_{joint,y0} = \frac{K_{\theta,CSB}}{\sin^2\theta_0}\left(\frac{4L^* \cdot \cos\theta_0 \cdot u_{y0} - u_{y0}^2}{4L^{*2}}\right) \quad (9)$$

The (plastic) moment at axial yielding can be estimated as the product of the component of the axial strength of the CSB projected along the horizontal direction (evaluated considering the initial undeformed configuration) and the joint arm B according to the following relationship:

$$M_{joint,pl0} = A_{CSB} \cdot f_u \cdot \cos\beta_{pl0} \cdot B \cdot \cos\Delta\phi_{pl0} \quad (10)$$

4 First Validation Through Finite Element Analysis

A one-storey pinned frame is considered as an applicative example. The frame has a height H equal to 3.00 m and a span width L equal to 6.00 m. The columns are made of HEB240 profiles (moment of inertia around the strong axis equal to $J = 11.26 \cdot 10^7$ mm^4), while the beam is a IPE300 profile (moment of inertia around the strong axis equal to $J = 8.35 \cdot 10^7$ mm^4). The adopted steel grade is S275 with yielding strength $f_y = 275$ MPa. The building is assumed to be located in a high intensity seismic region characterized by a pseudo-acceleration $S_a(T_1) = 1.0$ g corresponding to an earthquake with return period of 475 years. The seismic weight acting on the frame is equal to 180 kN corresponding to a load per unit area of 500 kg/m^2 acting on an influence area of 36 m^2.

A semi-rigid CSB joint is here designed making use of the simplified design equations presented in Sect. 3. A joint arm $B = 500$ mm is considered. It is designed for a target strength ratio ($M_{\text{joint},pl0}/M_{b,pl}$) of around 2.0 and a stiffness ratio ($K_{\text{joint},0}/(EJ_b/L_b)$) of around 0.6. J_b, L_b and $M_{b,pl}$ indicate the moment of inertia, the length and the plastic bending moment of the beam, respectively. A CSB with full square cross-section of 6-cm side and arm $d_0 = 95$ mm corresponding to a normalized arm $\xi_0 = 13.4\%$ allows to satisfy the design requirements.

The SAP2000 [15] Finite Element model of the frame with the semi-rigid CSB joint is shown in Fig. 3a. All structural elements are modelled through elastic beam elements with plastic hinges located at both ends. Each structural element has been modelled by several finite elements to more accurately simulate the mechanical and geometrical non-linearities. Flexural plastic hinges are considered for columns and beams, while uncoupled flexural and axial plastic hinges are considered for the CSBs. The skeleton curves of the adopted plastic hinges are represented in Fig. 3b.

The seismic actions are modelled through an equivalent lateral static force F_{EQ} as shown in Fig. 3a, while the vertical loads in seismic combination are considered through concentrated vertical forces W applied at the top of the two columns.

A non-linear static analysis (pushover) has been carried out to evaluate the moment-rotation response of the joint and the capacity curves in terms of total base shear vs node 2 lateral displacement. The analysis has been conducted in displacement control, by applying an increasing lateral displacement at node 2.

Figure 3c displays the analytical prediction of the moment-rotation response of joint 2, up to the point of complete elongation of the CSB in tension, as compared with the results of the pushover analysis. It can be noted that the analytical estimation is quite accurate within the entire investigated range.

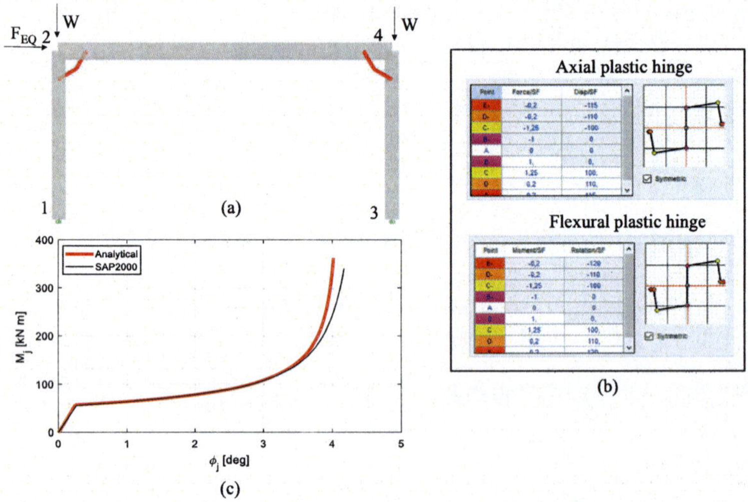

Fig. 3. (a) SAP2000 Finite Element model. (b) Skeleton curves of the flexural and axial plastic hinges. (c) Analytical vs numerical moment-rotation response for the CSB joint in tension.

5 Main Conclusions and Possible Optimizations Using Wire-and-Arc Additive Manufacturing (WAAM) Technology

The present paper introduced the semi-rigid CSB joint to strengthen pinned beam-column connections. An analytical model has been developed to predict the moment-rotation response and its accuracy has been validated through Finite Element analysis.

Future developments see the possibility to further optimize the geometrical and mechanical performances of the CSB joint exploiting the potential of additive manufacturing technologies. A solution to this challenge is offered by Wire-and-Arc Additive Manufacturing (WAAM) [16] technology that is able to fabricate large-scale metal parts with the required structural performances for construction applications. WAAM-produced steel parts can be adequately optimized with high freedom in both shape and thickness. Moreover, the printing process parameters can be engineered to obtain anisotropic material behavior which can be modelled as an orthotropic material both in the elastic [17] and post-yielding stages [18]. All these features can be efficiently used to optimize the joint geometrical shape and the material mechanical properties during the design phase to match the set of target performances.

References

1. Christopoulos, C., Filiatrault, A., Bertero, V.V.: Principles of Passive Supplemental Damping and Seismic Isolation. IUSS Press (2006)
2. Soong, T.T., Spencer, B.F., Jr.: Supplemental energy dissipation: state-of-the-art and state-of-the-practice. Eng. Struct. **24**, 243–259 (2002)
3. Castaldo, P., De Iuliis, M.: Optimal integrated seismic design of structural and viscoelastic bracing-damper systems. Earthq. Eng. Struct. Dyn. **43**, 1809–1827 (2014)
4. Tremblay, R.: Inelastic seismic response of steel bracing members. J. Constr. Steel Res. (2002). https://doi.org/10.1016/S0143-974X(01)00104-3
5. Della Corte, G., D'Aniello, M., Landolfo, R., Mazzolani, F.M.: Review of steel buckling-restrained braces. Steel Constr. **4**, 85–93 (2011)
6. Gray, M.G., Christopoulos, C., Packer, J.A.: Cast steel yielding fuse for concentrically braced frames. In: 9th US National and 10th Canadian Conference on Earthquake Engineering 2010: Including Papers from the 4th International Tsunami Symposium, vol. 6, pp. 4235–4244 (2010)
7. Nobahar, E., Asgarian, B., Mercan, O., Soroushian, S.: A post-tensioned self-centering yielding brace system: development and performance-based seismic analysis. Struct. Infrastruct. Eng. 1–21 (2020)
8. Hsu, H.-L., Halim, H.: Brace performance with steel curved dampers and amplified deformation mechanisms. Eng. Struct. **175**, 628–644 (2018)
9. Jia, L.-J., Ge, H., Maruyama, R., Shinohara, K.: Development of a novel high-performance all-steel fish-bone shaped buckling-restrained brace. Eng. Struct. **138**, 105–119 (2017)
10. Bertero, R.D., Bertero, V.V.: Performance-based seismic engineering: the need for a reliable conceptual comprehensive approach. Earthq. Eng. Struct. Dyn. **31**, 627–652 (2002). https://doi.org/10.1002/eqe.146
11. Palermo, M., Silvestri, S., Gasparini, G., Trombetti, T.: Crescent shaped braces for the seismic design of building structures. Mater. Struct. **48**, 1485–1502 (2015). https://doi.org/10.1617/s11527-014-0249-z
12. Palermo, M., Pieraccini, L., Dib, A., Silvestri, S., Trombetti, T.: Experimental tests on Crescent Shaped Braces hysteretic devices. Eng. Struct. **144**, 185–200 (2017). https://doi.org/10.1016/j.engstruct.2017.04.034
13. Mokhtari, E., Laghi, V., Palermo, M., Silvestri, S.: Quasi-static cyclic tests on a half-scaled two-storey steel frame equipped with Crescent Shaped Braces. Eng. Struct. **232** (2021). https://doi.org/10.1016/j.engstruct.2020.111836
14. Palermo, M., Laghi, V., Gasparini, G., Silvestri, S., Trombetti, T.: Analytical estimation of the key performance points of the tensile force-displacement response of Crescent Shaped Braces. Soil Dyn. Earthq. Eng. **148**, 106839 (2021)
15. SAP2000: (n.d.). www.csiamerica.com/products/sap2000

16. Feucht, T., Lange, J.: 3-D-printing with steel: additive manufacturing of connection elements. In: Advances in Engineering Materials, Structures and Systems: Innovations, Mechanics and Applications—Proceedings of the 7th International Conference on Structural Engineering, Mechanics and Computation 2019, pp. 419–424 (2019). https://doi.org/10.1201/9780429426506-75
17. Laghi, V., et al.: Orthotropic elastic model for wire-and-arc additively manufactured stainless steel. Addit. Manuf. (2020)
18. Hadjipantelis, N., Weber, B., Gardner, L.: Characterisation of the anisotropic response of wire and arc additively manufactured stainless steel. Ce/Papers **4**, 1757–1766 (2021)

Reliability Based Design Optimization of Damped-Outrigger Timber Structure Using Stochastic Spectral Embedding Based Probability Density Evolution Method

Sourav Das and Solomon Tesfamariam(✉)

School of Engineering, The University of British Columbia, Okanagan Campus, Kelowna, BC V1V 1V7, Canada
Solomon.Tesfamariam@ubc.ca

Abstract. This study presents a reliability analysis of stochastic system using the probability density evolution method (PDEM). The PDEM is formulated according to the principle of probability conservation where generalized density evolution equations (GDEEs) are completely decoupled. To estimate the probability density function accurately, a set of representative points of random variables are generated using the GF-discrepancy scheme. A large number of representative points is needed to obtain satisfactory accuracy, which becomes computationally expensive. To reduce the computation burden, a stochastic spectral embedding (SSE) is used as a surrogate model which approximates the original response surface. To illustrate the proposed SSE-based PDEM, two numerical examples are investigated, including the reliability analysis of four-branch problem, and the reliability-based design optimization of a shape memory alloy based damped outrigger tall timber building. Numerical results show that the proposed SSE-based PDEM can estimate failure probability using a very small number of representative points without compromising accuracy compared with Monte Carlo simulation, which leads to a reduction in computational costs.

Keywords: Probability density evolution method · Stochastic spectral embedding · Reliability analysis

1 Introduction

Reliability analysis is an established method to compute probability of failure of structures under structural system and external excitation uncertainties. The reliability analysis methods can be categorized into four sub-groups, i.e.,

G. P. Cimellaro (Ed.): WCSI 2022, LNCE 309, pp. 273–284, 2023.
https://doi.org/10.1007/978-3-031-21187-4_22

analytical-approximation method, numerical-sampling-based method, surrogate-based method, and numerical-integration method [1,2]. This paper focuses on probability density evolution method (PDEM), belongs to the numerical integration method, proposed by Li and Chen [6]. The PDEM is formulated based on the principle of probability conservation. To estimate probability density function (PDF) of the performance function using PDEM, a proper strategy is required to generate the representative points of the random variables present in the system. Generalized F (GF) discrepancy-based method [3] is the most common and popular for estimating PDF using PDEM. To improve the PDF's estimation accuracy, the PDEM requires large number of representative points, which becomes computationally expensive for high-fidelity models. To reduce the computational burden, surrogate model is required which approximates the response surface. Polynomial chaos expansion (PCE) is one of the popular surrogate models in which the output response is expressed in the form of a spectral representation considering the entire domain of random parameters. As a result, it sometimes does not capture the nonlinearity of the system accurately. In this study, stochastic spectral embedding method proposed by Marelli et al. [7] which is an advanced version of PCE is used. This method combines spectral representation and adaptive domain decomposition. The entire domain is divided into a number of subdomains. This discretization occurs adaptively until the residual between original and predicted responses becomes minimal.

The present study sets the following objectives: The reliability analysis of the stochastic system is carried out using first-passage reliability method where probability density function of the performance function is estimated using PDEM; Develop an efficient algorithm, stochastic spectral embedding surrogate based PDEM for reducing the computational cost involved in the traditional PDEM; and asses performance of the proposed method is compared with Monte-Carlo simulation. The outline of the study is structured as follows. Section 2 describes the complete overview of the reliability analysis of the stochastic system using PDEM. Section 3 provides a brief description of stochastic spectral embedding. The numerical applications of the proposed method are presented in Sect. 4. Finally, Sect. 5 presents the concluding remarks.

2 Reliability Estimation of Stochastic System

Consider an N-degrees of freedom structural system that is subjected to external dynamic excitation. The governing equation of motion of the structural system can be written as

$$\mathbf{M}(\boldsymbol{\Theta})\ddot{\mathbf{X}} + \mathbf{C}(\boldsymbol{\Theta})\dot{\mathbf{X}} + \mathbf{f}(\boldsymbol{\Theta}, \mathbf{X}) = \boldsymbol{\Gamma}\eta(\boldsymbol{\Theta}, t) \tag{1}$$

where $\mathbf{M}$, $\mathbf{C}$ are the mass and damping matrix of the structure, respectively; $\mathbf{f}$ is the restoring force vector. The displacement, velocity, and acceleration vectors of the structure relative to the ground are denoted by $\mathbf{X}$, $\dot{\mathbf{X}}$ and $\ddot{\mathbf{X}}$, respectively. The influence vector is represented by $\boldsymbol{\Gamma}$, and the external excitation is represented by $\eta(t)$. In Eq. 1, $\boldsymbol{\Theta}$ is a vector of random parameters associated with the structural properties and external excitation. The governing equation of motion depicted in Eq. 1 can be expressed in state-space form, which is given by

$$\dot{\mathbf{Y}} = \mathbf{A}(\mathbf{Y}, \boldsymbol{\Theta}, t) \tag{2}$$

where $\mathbf{Y}$ is the state vector which is expressed as $\mathbf{Y} = \left[\mathbf{X}^T\, \dot{\mathbf{X}}^T\right]^T = \left[Y_1 \ldots Y_{2N}\right]^T$ and $\mathbf{A} = \left[A_1 \ldots A_{2N}\right]^T$.

2.1 Probability Density Evolution Method

The PDEM proposed by Li and Chen [6] is used for reliability analysis, which is formulated based on the principle of probability conservation. Let $p_{Y_i}(y,t)$ represents the PDF of the i-th component of $\mathbf{Y}(t)$. The change in probability due to the transition of probability through the boundaries y_L and y_R during the time interval $[t, t+\Delta t]$ is expressed as [6]

$$\begin{aligned} \Delta P_J &= \int_{y_L}^{y_R} p_{Y_i}(y, t+\Delta t)dy - \int_{y_L}^{y_R} p_{Y_i}(y,t)dy \\ &= \left[\int_{y_L}^{y_R} \frac{\partial p_{Y_i}(y,t)}{\partial t} dy\right] \Delta t + o(\Delta t) \end{aligned} \tag{3}$$

Similarly, $\mathcal{J}(s,t)$ denotes the probability passing through a point s during unit time. The change of probability within a time interval $[t, t+\Delta t]$ can be written as [6]

$$\begin{aligned} \Delta P_K &= -\mathcal{J}(y_R,t)\Delta t + \mathcal{J}(y_L,t)\Delta t + o(\Delta t) \\ &= -\left[\int_{y_L}^{y_R} \frac{\partial \mathcal{J}(y,t)}{\partial y} dy\right] \Delta t + o(\Delta t) \end{aligned} \tag{4}$$

Using the principle of probability conservation, it is written as

$$\Delta P_J = \Delta P_K \Rightarrow \frac{\partial p_{Y_i}(y,t)}{\partial t} = -\frac{\partial \mathcal{J}(y,t)}{\partial y} \tag{5}$$

where $\mathcal{J}(y,t)$ is the flux of probability whose expression can be found in Li and Chen [6]. Finally,

$$\frac{\partial p_{Y_i\Theta}(y,\theta,t)}{\partial t} + \dot{Y}_i(\theta,t)\frac{\partial p_{Y_i\Theta}(y,\theta,t)}{\partial y} = 0 \tag{6}$$

Equation 6 is also known as the generalized density evolution equation (GDEE). The joint PDF $p_{Y_i\Theta}(y,\theta,t)$ is estimated by solving Eq. 6. The PDF of $Y_i(t)$ can be obtained as $p_{Y_i}(y,t) = \int_{\Omega_\Theta} p_{Y_i\Theta}(y,\theta,t)d\theta$.

2.2 Estimation of First-Passage Reliability Using PDEM

The probability of failure for a first-passage problem can be expressed as

$$P_f = \Pr\{Y(t) \in \boldsymbol{\Omega}_F, \exists t \in [0, T]\} = \Pr\{|Y(t)| > Y_{\mathrm{Thres}}, \exists t \in [0, T]\} \quad (7)$$

where $\Pr\{\cdot\}$ represents the probability operator, the failure domain is denoted by $\boldsymbol{\Omega}_F$, and [0, T] represents the time duration. In Eq. 7, Y_{Thres} denotes the threshold value of $Y(t)$. Also, the probability of failure, P_f in Eq. 7 can be formulated using the principle of equivalent extreme-value event [5] which is given as

$$P_f = \Pr\{Y_{\mathrm{EEV}} \geq Y_{\mathrm{Thres}}\} = \int_{Y_{\mathrm{Thres}}}^{\infty} p_{Y_{\mathrm{EEV}}}(y)dy \quad (8)$$

where $p_{Y_{\mathrm{EEV}}}$ is the PDF of equivalent extreme-value (EEV) of $Y(t)$, expressed as

$$Y_{\mathrm{EEV}} = \max_{t \in [0,T]} (|Y(t)|) \quad (9)$$

The failure probability is estimated by performing one-dimensional integration of $p_{Y_{\mathrm{EEV}}}$, as in Eq. 8. For that reason, our prime concern is to compute $p_{Y_{\mathrm{EEV}}}$ using PDEM. To estimate $p_{Y_{\mathrm{EEV}}}$, a virtual stochastic process is formulated, which is expressed as

$$\mathcal{V}(\tau) = Y_{\mathrm{EEV}} \sin(\omega\tau); \qquad \tau \in [0, 1] \quad (10)$$

where $\omega = 5\pi/2$ [5]. Equation 10 shows that when $\tau = 0$, $\mathcal{V}(\tau) = 0$ and when $\tau = 1$, $\mathcal{V}(\tau) = Y_{\mathrm{EEV}}$. Therefore, it is needed to compute the PDF of $\mathcal{V}(\tau)$ i.e., $p_{\mathcal{V}}(\mathcal{V}, \tau)$ which is obtained by solving GDEE, as in Eq. 6

$$\frac{\partial p_{\mathcal{V}}(\mathcal{V}, \tau)}{\partial \tau} + \dot{\mathcal{V}}(\tau)\frac{\partial p_{\mathcal{V}}(\mathcal{V}, \tau)}{\partial \mathcal{V}} = 0 \quad (11)$$

Once $p_{\mathcal{V}}(\mathcal{V}, \tau)$ is obtained, the PDF of Y_{EEV} is calculated as

$$p_{Y_{\mathrm{EEV}}}(y) = p_{\mathcal{V}}(\mathcal{V}, \tau)|_{\tau=1} \quad (12)$$

Equation 11 is solved using finite difference method where representative points of random variables are generated using GF-discrepancy based method. Readers may refer Chen et al. [3] for details of generation of sampling.

3 Stochastic Spectral Embedding

Stochastic spectral embedding is a novel surrogate model proposed by Marelli et al. [7] that is a combination of spectral representation and adaptive domain decomposition. Consider a computational model $\mathcal{M}(\theta)$ which contains a set of random parameters, $\theta = [\theta_1, \theta_2, \ldots, \theta_n]^T$ with joint probability density function $\theta \sim f_\Theta(\theta)$. Therefore, the output response of the model $\mathcal{M}(\theta)$ can be expressed in the form of spectral representation, which is given by

$$Y = \mathcal{M}(\theta) = \sum_{i=1}^{\infty} a_i \boldsymbol{\Psi}_i(\theta) \quad (13)$$

where a_i is the coefficient of the orthonormal basis function. In practise, the output response can not be decomposed into an infinite number of series as in Eq. 13. Therefore, a truncation is needed, which is expressed as

$$Y = \mathcal{M}(\theta) \approx \hat{\mathcal{M}}(\theta) = \sum_{i=1}^{P} a_i \boldsymbol{\Psi}_i(\theta) \tag{14}$$

where P is the truncation order of the polynomial expansion. The residual is given by

$$\mathcal{R}(\theta) = \mathcal{M}(\theta) - \hat{\mathcal{M}}(\theta) \tag{15}$$

The basis function $\boldsymbol{\Psi}$ has support on the entire domain $\boldsymbol{\Omega}_\Theta$ in polynomial chaos expansion (PCE). Due to that, sometimes PCE is unable to capture the nonlinearity present in the model. The entire domain is divided into smaller subdomains in stochastic spectral embedding (i.e., $\boldsymbol{\Omega}_\Theta^{m,n} \subseteq \boldsymbol{\Omega}_\Theta$, where m is the level of expansion and n is the subdomain index within the level). When $m = 0$, i.e., a single domain is considered, $\boldsymbol{\Omega}_\Theta^{0,1} = \boldsymbol{\Omega}_\Theta$, then residual is the same as for PCE, in Eq. 15. For multi-level of expansion, Eq. 14 is re-written as

$$\hat{\mathcal{M}}(\theta) = \sum_{m=0}^{M} \sum_{n=1}^{N_m} I_{\boldsymbol{\Omega}_\Theta^{m,n}}(\theta) \mathcal{R}^{m,n}(\theta) \tag{16}$$

where the total number of expansion levels and number of subdomains at level $m \in [0, \ldots, M]$ are denoted by M and N_m, respectively. In Eq. 16, I represents the indicator function of subdomain $\boldsymbol{\Omega}_\Theta^{m,n}$. $\mathcal{R}^{m,n}(\theta)$ is the truncated expansion of residual up to level $(m-1)$ which is expressed in following form

$$\mathcal{R}^{m}(\theta) = \mathcal{M}(\theta) - \sum_{l=0}^{m-1} \sum_{n=1}^{N_l} I_{\boldsymbol{\Omega}_\Theta^{l,n}}(\theta) \mathcal{R}^{l,n}(\theta) \tag{17a}$$

$$\mathcal{R}^{l,n}(\theta) = \sum_{i} a_i^{l,n} \boldsymbol{\Psi}_i^{l,n}(\theta) \tag{17b}$$

Readers may refer Marelli et al. [7] for a complete description of stochastic spectral embedding. The flowchart for the reliability analysis using stochastic spectral embedding based PDEM is shown in Fig. 1.

4 Numerical Example

Two different numerical examples are presented to illustrate the reliability analysis using the probability density evolution method. Those are: (1) a four-branch problem in which the probability of failure is estimated from a series system; and (2) reliability-based design optimization of timber outrigger building incorporating shape memory alloy-based damper. To reduce the computational cost, stochastic spectral embedding is used, which approximates the original response surface of an equivalent extreme event. For simulation purposes, `MATLAB` software is used.

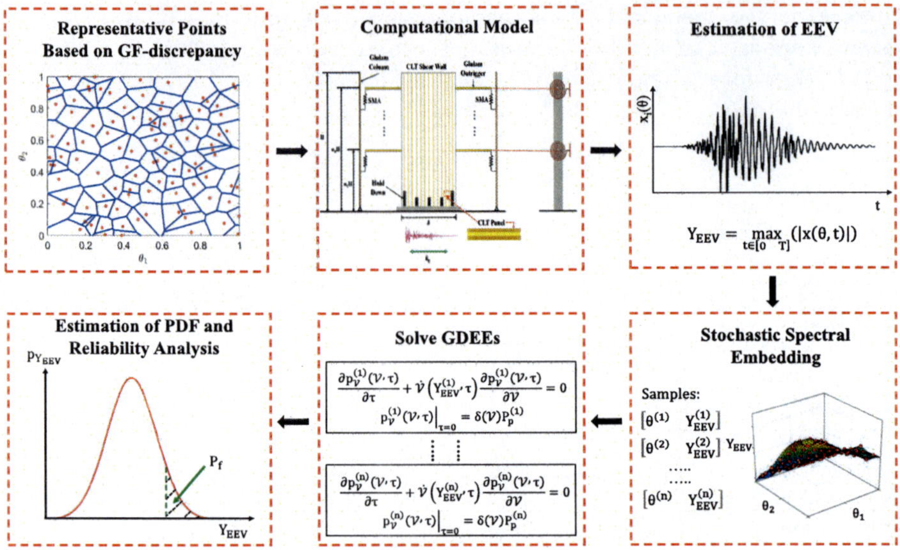

Fig. 1. Flowchart for reliability analysis using PDEM

4.1 Problem 1

As the first example, a four-branch problem is considered whose P_f is given by

$$P_f = \Pr\Big[-0.1(\theta_1-\theta_2)^2+(\theta_1+\theta_2)/\sqrt{2}\geq 3 \cup -0.1(\theta_1-\theta_2)^2 - (\theta_1+\theta_2)/\sqrt{2}\geq 3 \cup (\theta_1-\theta_2)\geq 7/\sqrt{2} \cup (\theta_2-\theta_1)\geq 7/\sqrt{2}\Big] \tag{18}$$

where θ_1 and θ_2 are the random variables which follow the standard Gaussian distribution. According to an equivalent extreme event, the above equation is equivalent to

$$P_f = \Pr\left[Y_{\mathrm{EEV}} \geq 7/\sqrt{2}\right] = \int_{7/\sqrt{2}}^{\infty} p_{Y_{\mathrm{EEV}}}(y)dy \tag{19}$$

where Y_{EEV} is given as

$$Y_{\mathrm{EEV}} = \max\left\{\begin{array}{c} -0.1(\theta_1-\theta_2)^2+(\theta_1+\theta_2)/\sqrt{2}+7/\sqrt{2}-3 \\ -0.1(\theta_1-\theta_2)^2-(\theta_1+\theta_2)/\sqrt{2}+7/\sqrt{2}-3 \\ (\theta_1-\theta_2) \\ (\theta_2-\theta_1) \end{array}\right\} \tag{20}$$

The surrogate model needs to be trained before estimating the probability density function. For this purpose, 100 representative points of random variables are generated using the GF-discrepancy algorithm. Y_{EEV} is calculated for each set of random variables using Eq. 20. With these output responses, the surrogate model is trained. Once training is completed, another set of 1500 representative points are generated, which are passed through the trained surrogate model to obtain the complete response surface.

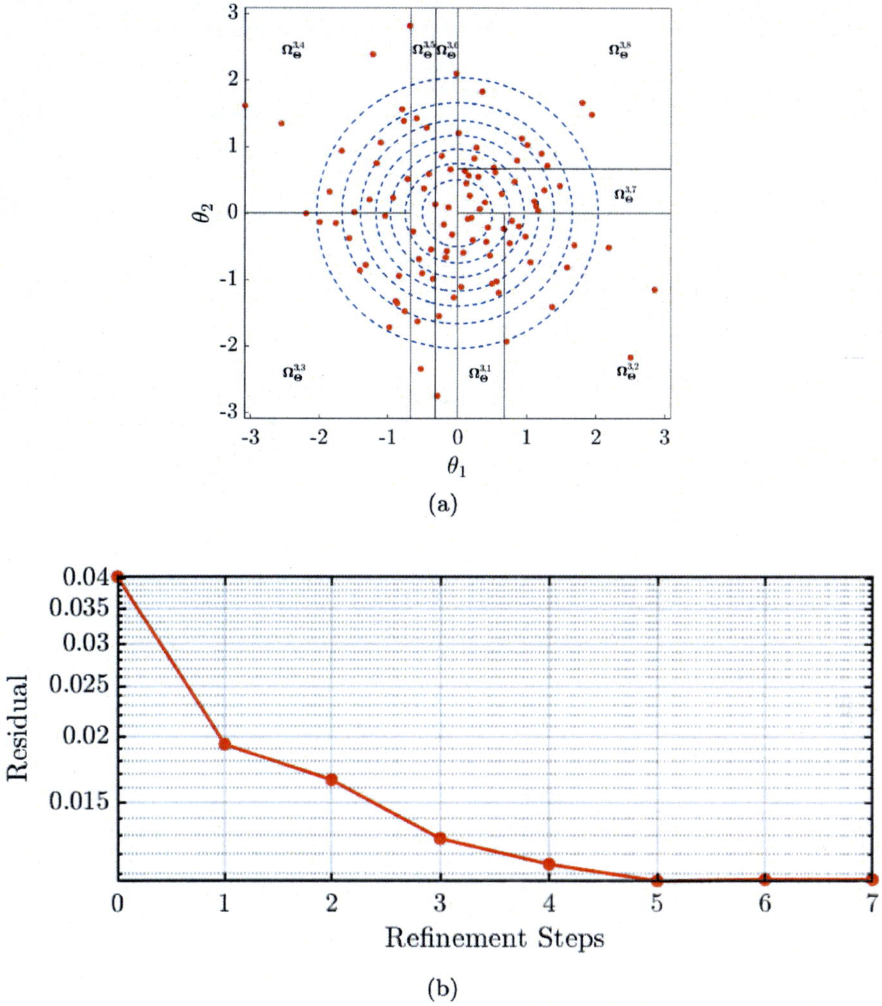

Fig. 2. (a) Subdomains in physical space and (b) degree of accuracy corresponding to increasing number of subdomain

It is seen from Fig. 2 that the stochastic spectral embedding produces a minimum residual when the entire domain of random variables is divided into 8 subdomains, which is seen in Fig. 2(a). Once the surrogate model's accuracy is determined, the virtual stochastic process $\mathcal{V}(\tau)$ is calculated for each value of

Y_{EEV} using Eq. 10. The GDEEs are solved for evaluating the PDF of $\mathcal{V}(\tau)$ using the finite difference method combined with the TVD scheme. Figure 3(a) shows the PDF of $\mathcal{V}(\tau)$. The PDF of Y_{EEV} is obtained by setting τ to 1, as illustrated in Fig. 3(b). Once PDF is obtained, the probability of failure is estimated using Eq. 19. The failure probability is 2.5×10^{-3}. In comparison, a Monte Carlo simulation with 10^6 samples yields a failure probability of 2.43×10^{-3}, which is close to the value obtained using stochastic spectral embedding.

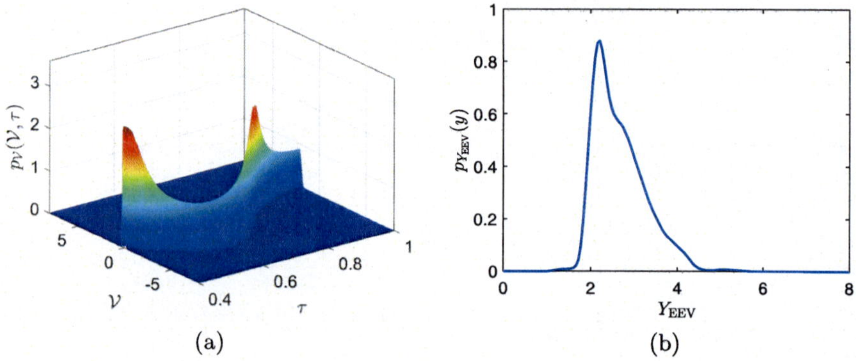

Fig. 3. PDF of (a) virtual stochastic process, $\mathcal{V}(\tau)$ and (b) equivalent extreme event, Y_{EEV}

4.2 Problem 2

As a final example, a shape memory alloy (SMA)-based damped outrigger tall timber building is considered. As shown in Fig. 4, the outrigger is a connecting member between the column and shear wall which helps to reduce structural deformation of the tall building while the structure is subjected to external dynamic loads such as wind, earthquake, etc. The excessive load demand is often seen on the column connected with outrigger, which enhances member size, hence construction cost increases. To reduce load demand, a SMA spring is attached in between the outrigger and the column, which dissipates the energy due to external loads.

With this in view, the initial aim is devoted to the formulation of the governing equation of motion of the SMA-based damped outrigger structure. The core of the structure is modeled as a cantilever beam, and each floor mass is considered as a discrete mass that acts at the junction between the floor and the core of the structure. The outrigger beam is modeled as a torsional spring having constant rotational stiffness, and the moment developed due to the torsional spring is concentrated at the junction of the core of the structure and the outrigger, as shown in Fig. 4. The governing equation of motion is derived using

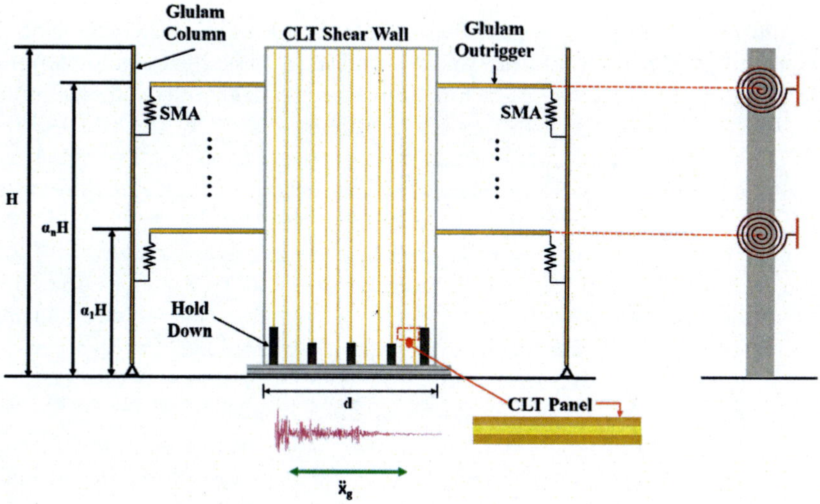

Fig. 4. Schematic representation of SMA-based damped outrigger structure

the Lagrange formulation, which is given by [4]

$$m\frac{\partial^2 u}{\partial t^2} + \frac{\partial^2}{\partial x^2}\left(EI\frac{\partial^2 u}{\partial x^2}\right) - F = 0; \quad 0 < x < H \tag{21a}$$

$$\left[\frac{\partial}{\partial x}\left(EI\frac{\partial^2 u}{\partial x^2}\right)\right]_{x=0} = 0 \tag{21b}$$

$$\left[EI\frac{\partial^2 u}{\partial x^2} + 2\sum_{j=1}^{n} K_{r_j}\frac{\partial u(\alpha_j H, t)}{\partial x} - \sum_{k=1}^{N}\hat{M}_k\frac{\partial u(H_k, t)}{\partial t}\right]_{x=H} = 0 \tag{21c}$$

where u is displacement of the structure which is a function of time and position i.e., $u(x,t) = \Phi(x)q(t) = \sum_i \phi_i(x)q_i(t)$. ϕ_i and q_i denote the ith mode-shape of the system and displacement of the ith mode shape, respectively. H is the total height of the building; $EI(x)$ and $m(x)$ are the continuous beam's flexural stiffness and the mass per unit height, respectively; the jth equivalent rotational stiffness due to outrigger is denoted by K_{r_j}; n is the total number of outriggers considered in the structure. It is noted that the governing equation of motion in Eq. 21 is a partial differential equation. Equation 21 is converted into an ordinary differential equation which is given by [4]

$$\begin{aligned}[M][\ddot{q}(t)] + [C][\dot{q}(t)] + [K][q(t)] &= I_g m H \ddot{x}_g(t) \\ &- 2\left\{F_{SMA}\left(\frac{d}{2}\right) + K_r\right\}[\bar{\phi}\,\bar{\phi}^T][q(t)]\end{aligned} \tag{22}$$

where $\ddot{x}_g$ represents the ground acceleration, I_g the influence vector, and F_{SMA} the force induced by SMA spring. The details of the derivation of the equation of motion and load-deformation hysteresis of the SMA spring are found in [4].

To illustrate this problem, a 20-storey timber building is considered. The architectural plan of the building can be found in [4]. The beams and columns of the building are designed using glulam, and the core shear wall is cross-laminated timber (CLT). The material properties of glulam and CLT, the cross-sectional details of the beam, column, and shear wall are taken from [4]. Two SMA-based outriggers are considered. The objective of this study is to estimate the optimal positions of outriggers and tuning parameters of SMA spring by minimizing the probability of failure of the system. One of the design parameters is the normalised outrigger location, α, which ranges from 0 to 1 (Fig. 4). Other design parameters associated with SMA hysteresis parameters [4] include the initial austenite stiffness (k_a), the maximum design displacement ($x_{s,max}$), and the ductility ratio (μ_s) of SMA. Also, the initial austenite stiffness is estimated from the normalized transformation strength of SMA, which is $F_0 = k_a x_{ys}/(mH)$, and ranges $F_0 \in [0.1, 0.5]$, where x_{ys} is the yield displacement of SMA. The mean value of the other parameters, $x_{s,max}$ and μ_s are taken as 0.2 m and 20, respectively. The coefficient of variation of these two variables is assumed to be 20%, which follows a uniform distribution. The bounds are applied to the design parameters to avoid numerical instability during optimization. For this reason, the bounds are taken as $x_{s,max} \in [0.13, 0.27]$ and $\mu_s \in [13, 27]$, respectively. This study also takes into account the material uncertainties, which are the modulus of elasticity of CLT in longitudinal and transverse directions, i.e., E_1 and E_2, and the modulus of elasticity of glulam beam and column (E_g). Therefore, it is seen that there are two sets of random variables, i.e., $\theta = [\alpha_1, \alpha_2, F_0, \mu_s, x_{s,max}]$ which are design variables and $\epsilon = [E_1, E_2, E_g]$. The variables relating to CLT and glulam material properties, namely E_1, E_2, and E_g, are assumed to have Gaussian distributions with mean values of 11.7, 9 and 12.4 ($\times 10^9$ Pa), respectively, and c.o.v is 10%. The objective function is taken as the maximum value of the inter-storey drift ratio of the floors. To compute that, an ensemble of ground motions are considered, which are selected based on the conditional mean spectrum-based record selection method for the Vancouver site (Canada) at a time period of 2.0 s. The probability of failure of the system is defined as follows

$$P_f = \Pr\left[Y_{\mathrm{EEV}} \geq Y_{\mathrm{Thres}}\right] = \int_{Y_{\mathrm{Thres}}}^{\infty} p_{Y_{\mathrm{EEV}}}(y)dy \tag{23}$$

where Y_{EEV} is given as

$$Y_{\mathrm{EEV}} = \frac{1}{N_g} \sum_{j=1}^{N_g} \max_{1 \leq i \leq 20} \left\{ \max_{t \in [0,T]} |\mathrm{IDR}_{i,j}(\theta, \epsilon, t)| \right\} \tag{24}$$

where N_g denotes the total number of ground acceleration time histories and IDR the inter-storey drift ratio. In Eq. 23, Y_{Thres} is the allowable inter-storey drift ratio, which is taken as 2.5% according to NBC 2015 [8]. The procedure to evaluate the PDF of Y_{EEV} is same as before. Once PDF is obtained corresponding to each value of θ, the probability of failure is estimated using Eq. 23. The obtained value of failure probability corresponding to each set of design variables,

θ are used to carry out the optimization problem. The objective function is defined as follows

$$\begin{aligned} \theta^* &= \underset{\theta}{\operatorname{argmin}} \left[P_f(\theta) \right] \\ \text{s.t. } \theta_{ll} &\leq \theta \leq \theta_{ul} \end{aligned} \tag{25}$$

Figure 5 shows the PDF of Y_{EEV} and the response surface of the probability of failure corresponding to design variables, θ. The optimal value is taken where the failure probability is the minimum. The optimal locations of the outriggers from the ground level are 21 m and 57 m, respectively. The other obtained optimal design parameters are $[F_0, \mu_s, x_{s,max}] = [0.27, 22.15, 0.18]$, respectively.

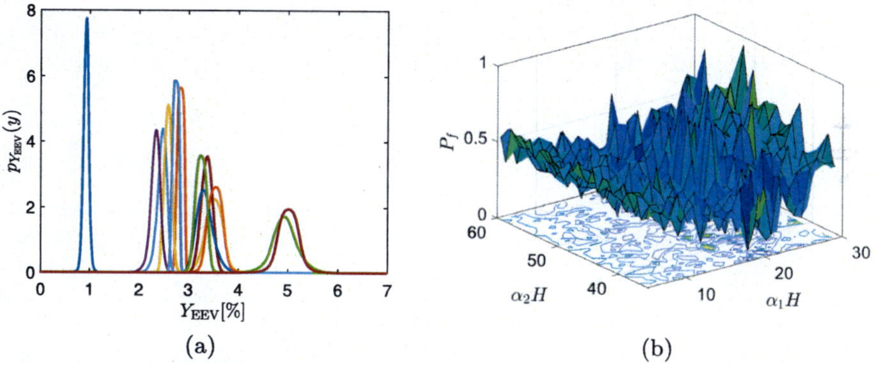

Fig. 5. (a) PDF of equivalent extreme event and (b) probability of failure surface corresponding to design parameters

5 Conclusions

The numerical investigations presented in this study are focused on reliability analysis of stochastic systems using the probability density evolution method (PDEM). The PDEM is computationally expensive for high-fidelity models as it requires a large number of sample points. To reduce the computational burden, stochastic spectral embedding is used as a surrogate model. The major observations from this study are listed below:

- PDEM enables to estimate the joint probability density function of the equivalent-extreme event of an MDOF system by solving GDEEs efficiently.
- When compared to Monte Carlo simulation, surrogate assisted PDEM can accurately estimate the probability of failure while using fewer samples of design variables.

This study clearly indicates the efficiency of the proposed surrogate model based PDEM for the time-independent reliability analysis of the stochastic system. This method can be extended to time-varying system such as deteriorating system, which the authors wish to address in their future works.

Acknowledgements. This study was funded by Natural Sciences and Engineering Research Council of Canada under the Discovery Grant programs (RGPIN-2019-05584).

References

1. Au, S.K., Beck, J.L.: Estimation of small failure probabilities in high dimensions by subset simulation. Probabilistic Eng. Mech. **16**(4), 263–277 (2001)
2. Bourinet, J.M., Deheeger, F., Lemaire, M.: Assessing small failure probabilities by combined subset simulation and support vector machines. Struct. Saf. **33**(6), 343–353 (2011)
3. Chen, J., Yang, J., Li, J.: A GF-discrepancy for point selection in stochastic seismic response analysis of structures with uncertain parameters. Struct. Saf. **59**, 20–31 (2016)
4. Das, S., Tesfamariam, S.: Multiobjective design optimization of multi-outrigger tall-timber building: using SMA-based damper and Lagrangian model. J. Build. Eng. **51**, 104358 (2022)
5. Li, J., Chen, J.B., Fan, W.L.: The equivalent extreme-value event and evaluation of the structural system reliability. Struct. Saf. **29**(2), 112–131 (2007)
6. Li, J., Chen, J.: The principle of preservation of probability and the generalized density evolution equation. Struct. Saf. **30**(1), 65–77 (2008)
7. Marelli, S., Wagner, P.R., Lataniotis, C., Sudret, B.: Stochastic spectral embedding. Int. J. Uncertain. Quantif. **11**(2) (2021)
8. NBCC: National Building Code of Canada 2015. National Research Council of Canada, Ottawa (2015)

Research on the Development of the Three-Dimension Seismic Isolation System for Lightweight Buildings

Yinglu Wang(✉) and Osamu Takahashi

Tokyo University of Science, Katsushika-ku, Tokyo, Japan
wangyinglu241@gmail.com

Abstract. In the seismic isolation structure, the seismic isolation device is designed to have low horizontal rigidity and high vertical rigidity for obtaining a long horizontal period and stable vertical support capacity. However, due to this, the seismic isolation structure cannot be expected to have a seismic isolation effect in the vertical direction. For protecting human and physical safety from earthquakes, it is necessary to consider the vertical component of the earthquake. In this research, after confirming the mechanical performance of the existing seismic isolation device, a three-dimensional seismic isolation system will be constructed by combining the horizontal seismic isolation device and the vertical seismic isolation device. The purpose is to analyse the seismic response at the building level using the constructed three-dimensional seismic isolation system and propose a three-dimensional seismic isolation system which can provide the better seismic isolation effect.

Keywords: Base-seismic isolation · Dimensional combined isolation bearing · Seismic response

1 Introduction

In the 1970s, with the development of computers and structural analysis methods, the seismic behavior of buildings could be estimated to a certain extent, ushering in the era of low-rise buildings. Seismic isolation has emerged.

In seismic isolation structures, seismic isolators are designed to have low horizontal stiffness and high vertical stiffness in order to obtain long horizontal periods and stable vertical support capacity. However, this does not allow the seismic isolation structure to be effective in the vertical direction. The Northridge earthquake occurred in January 1994 in northwestern Los Angeles, U.S.A., and attracted much attention in earthquake engineering when it recorded vertical accelerations of more than 1G. In the Great Hanshin-Awaji Earthquake, the support plates of buildings were subjected to vertical tensile forces, causing localized uplift and destruction. According to various earthquake damage studies to date, the vertical component of earthquakes must also be considered in order to protect human and physical safety from earthquakes. Therefore, it is considered that seismic isolation structures should have a vertical seismic isolation function.

G. P. Cimellaro (Ed.): WCSI 2022, LNCE 309, pp. 285–295, 2023.
https://doi.org/10.1007/978-3-031-21187-4_23

In this research, after confirming the mechanical performance of existing seismic isolation devices, a three-dimensional seismic isolation system is constructed by combining horizontal and vertical seismic isolation devices. The purpose of this research is to propose a three-dimensional seismic isolation system that can demonstrate a higher seismic isolation effect by analyzing the seismic response at the building level using the constructed three-dimensional seismic isolation system.

2 Vibration Testing and Traffic Vibration Observation

2.1 Test Purpose

The objective is to understand the periodic characteristics and damping characteristics of the existing upper and lower seismic isolators by conducting vibration tests, and to build an analytical model.

2.2 Test Specimen Overview and Test Method

The existing upper and lower seismic isolation devices verified in this test consisted of spring pads and rubberized coil springs (Figs. 1 and 2).

The spring pads were placed at the four corners of the 20-ft container, and constant microtremors and traffic vibration were observed (Fig. 3). The dead weight of a 20-ft container is approximately 2.2–2.4 tons.

The rubberized coil springs were divided into two versions, one with rubber and the other without, and each was placed at the four corners of the weight, which was the maximum load, and vibration tests were conducted using a vibration machine (Fig. 4).

2.3 Test Result

From the overlaid graph of the time history waveforms of the traffic vibration observation results (Fig. 5), it was found that the acceleration observed at the container floor was 2.56 times greater than that observed at ground level. From this, it can be said that the use of the spring pad at the building level is neither a vibration-isolation effect nor a possible seismic isolation device. The analytical model of the spring pad is not constructed in the analytical modeling.

In the results of the vibration test using the vibration machine, the time history waveform shows that the response acceleration of the weight is reduced by more than 50% from the input acceleration. In addition, by checking the Fourier spectrum of the input waveform, it can be seen that the frequency of excitation is around 10 Hz. Therefore, it can be said that the rubberized coil spring is demonstrating its vibration-isolation effect (Fig. 6).

3 Vibration Testing and Traffic Vibration Observation

There, analytical models of seismic isolation sliders used for horizontal isolation measures and bellows-type air springs and rubberized coil springs used for vertical isolation devices are constructed using Midas iGen finite element analysis software.

Fig. 1. Spring pad rubber

Fig. 2. Coil spring with rubber

Fig. 3. Traffic vibration observation

Fig. 4. Vibration machine vibrations

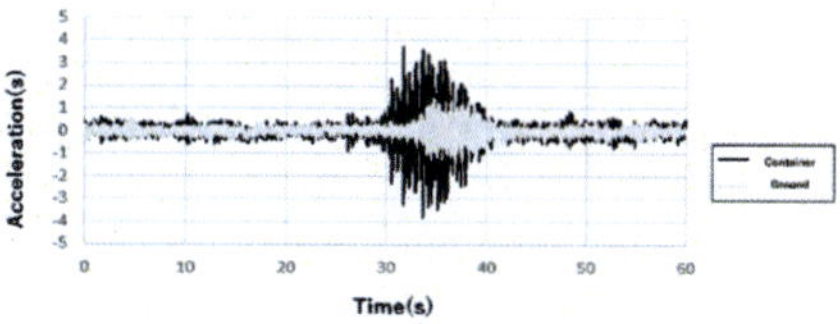

Fig. 5. Traffic vibration observation results

3.1 Modeling of the Specimen

For the rubberized coil spring, an analytical model is proposed using the results obtained from the tests. The analytical model for the rubberized coil spring consists of vertical stiffness and vertical damping. The vertical stiffness and damping coefficients are calculated based on the constant vibration waveform and free vibration waveform observed from the test. The transfer function was calculated from the constant vibration waveform, and the natural period of the two versions with and without rubber was 0.153 s

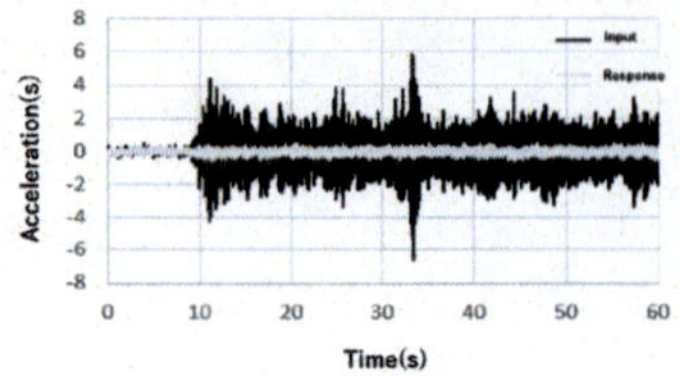

Fig. 6. Result of vibration machine vibration

and 0.248 s, respectively, and the vertical stiffness of the device alone was calculated to be 12.4 and 153.4 N/mm. The damping constant of the coil spring with rubber was calculated to be 0.102 using Eq. (1) from the free vibration waveform. The test specimen reproduction model is shown in Fig. 9.

$$h = \left(\frac{lnd}{2\pi}\right)/\sqrt{1+\left(\frac{lnd}{2\pi}\right)^2} \tag{1}$$

3.2 Reconstruction of Isolation Slider and Bellow-Type Air Spring Analysis Model

In the literature [1] analytical models for seismic isolation sliders and bellows-type air springs were constructed using SNAP analysis software. However, in this study, we would like to construct a three-dimensional seismic isolation system analysis model in combination with a rubberized coil spring, so it is necessary to reconstruct the analysis model using Midas iGen finite element analysis software.

For the analytical model of the bellows-type air spring, the specimen analytical model is reconstructed using the horizontal stiffness of 7.65 N/mm, vertical stiffness of 794 N/mm, and vertical damping constant of 0.02 according to Ref. [2] (Fig. 7).

For the analytical model of the seismic isolation slider, the specimen analytical model is reconstructed using the horizontal stiffness of 29.8 N/mm, friction coefficient of 0.195, and horizontal damping constant of 0.02 from literature (Figs. 8 and 9).

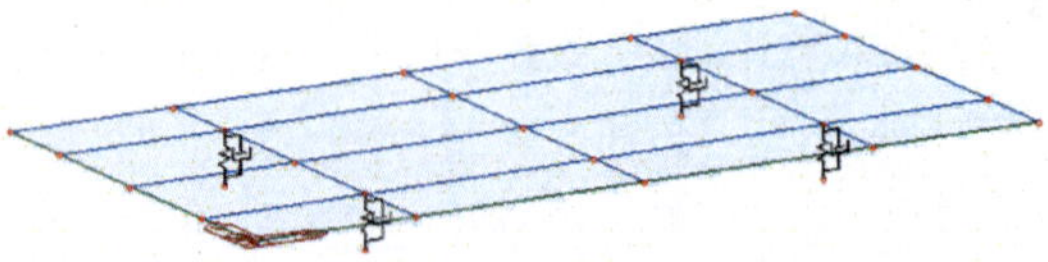

Fig. 7. Reproduction model of bellows-type air spring test body

3.3 Validation of Analytical Models

Response analysis will be performed using the specimen analysis model constructed in the previous section.

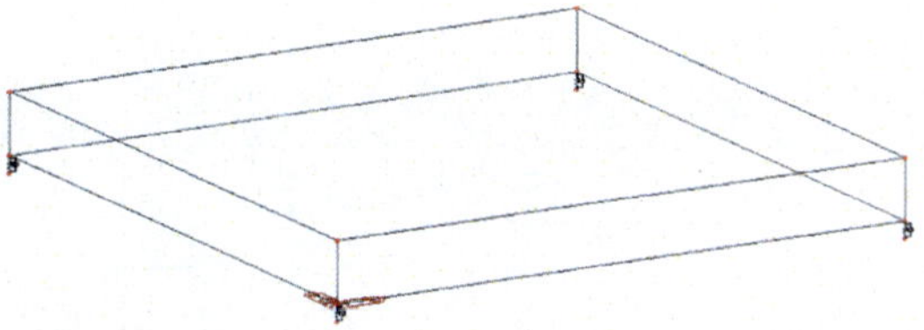

Fig. 8. Reproduction model of seismic isolation slider test body

Fig. 9. Reproduction model of coil spring test body

For the validation of the bellows-type air spring and seismic isolation slider analysis models, time response analysis was performed with the input seismic waves used in the tests according to Refs. [1, 2], and the results of the analysis and the results from the tests were summarized. The time response acceleration graphs (Figs. 10 and 11) showed that the waveforms could generally be followed.

To verify the validity of the rubberized coil spring analytical model, a time response analysis was performed using excitation waves from the vibration machine excitation. It was confirmed that the analytical results generally agreed with the experimental results (Fig. 12).

From the above, the validity of the analytical model was confirmed.

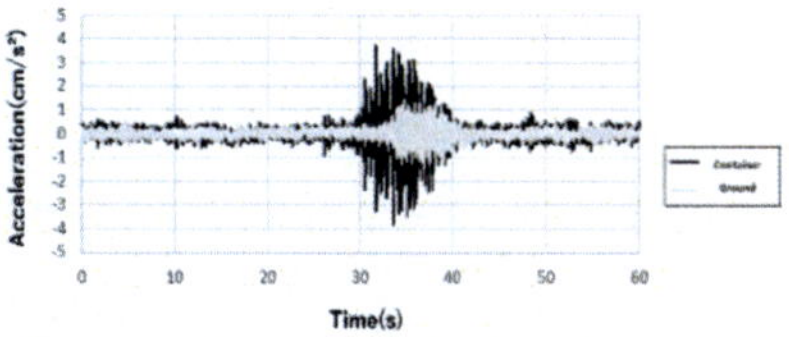

Fig. 10. Traffic vibration observation results

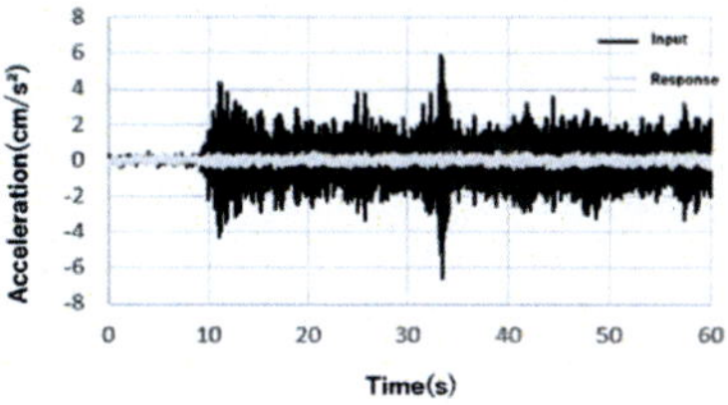

Fig. 11. Vibration machine vibration results

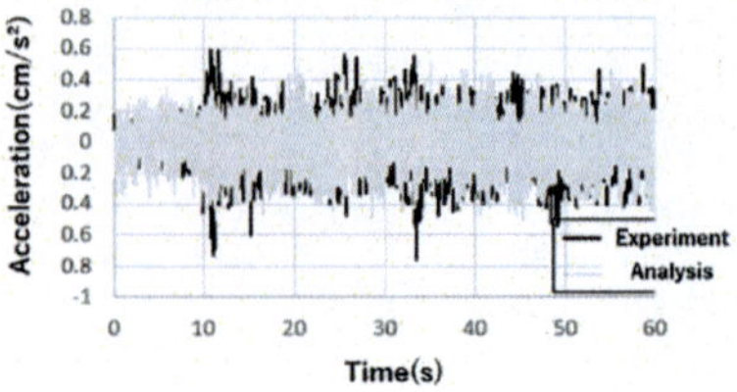

Fig. 12. Coil spring vibration machine time history response acceleration Z

4 Vibration Testing and Traffic Vibration Observation

4.1 Superstructure Overview

The building to be analysed is a two-story container house employing 20-ft and 40-ft containers. The plan view of the two-story container house employing 20-ft containers is 6.058 m in the X-direction, 2.348 m in the Y-direction, and 6.29 m in the Z-direction. The plan shape is 12.192 m in the X-direction, 2.348 m in the Y-direction, and 6.29 m in the Z-direction (Figs. 13, 14 and 15).

Fig. 13. Container house perspective view

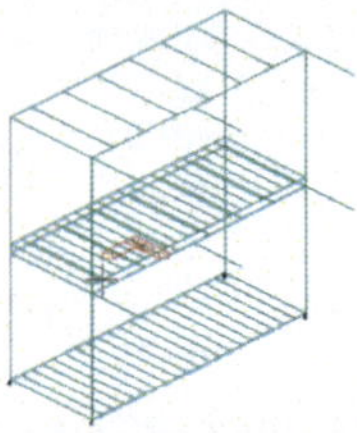

Fig. 14. 20 ft building model

4.2 Design of Seismic Isolation Layer

For the seismic isolation layer, the analytical model in the previous section is considered as one unit, and a combination of two seismic isolation devices (Table 1) is proposed to form a three-dimensional seismic isolation system. In addition, the seismic isolation layers are designed by setting the natural period and damping constants in the horizontal and vertical directions to the targeted values (Table 2). The seismic isolation layers are

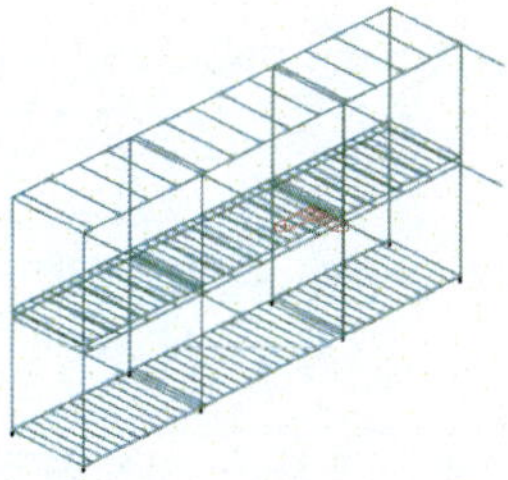

Fig. 15. 40 ft building model

placed at eight corners for the 40-ft container house and at four corners for the 20-ft container house (Figs. 16 and 17).

The horizontal seismic isolators of the 3D seismic isolation system are selected based on the reduction rate of horizontal response acceleration. Figures 18 and 19 show that the reduction rate of response acceleration of bellows-type air springs is greater than that of seismic isolation sliders, so the horizontal isolation device of the three-dimensional seismic isolation system should be a bellows-type air spring.

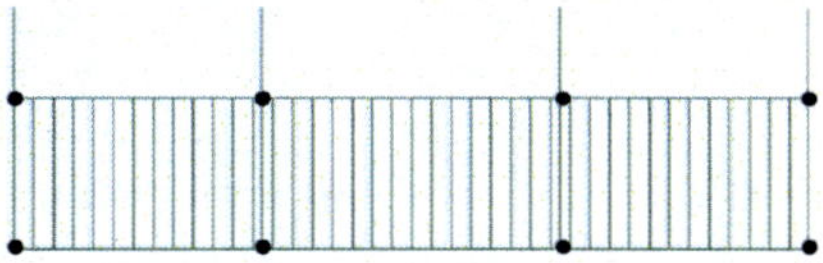

Fig. 16. 40 ft container house seismic isolator arrangement

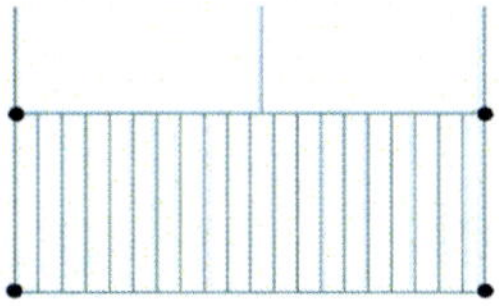

Fig. 17. 20 ft container house seismic isolator arrangement

4.3 Response Analysis Overview

In order to confirm the seismic performance of the building, a time history response analysis using seismic waves assumed for the site will be conducted. As criteria for determining seismic resistance, we confirm that the maximum response inter-story deformation angle of the superstructure is less than 1/200 in the horizontal direction, that the seismic isolation layer has a design limit displacement of 50 cm or less, and that the response acceleration reduction rate is approximately 40% in the vertical direction.

The input seismic waves are assumed to be three observed waves and three KOKUJI waves, assuming that seismic motions occur extremely rarely (Table 3).

Table 1. Seismic isolation layer design

Seismic isolation layer design	
Coil spring rubber + Seismic isolation slider	Coil spring rubber + Bellows type air spring

Table 2. Seismic isolation system design

Design	Inherent cycle	Rigidity [kN/cm]		Decay constant
	[s]	20 ft	40 ft	–
Vertical	1.80	0.22	0.27	0.15
	1.60	0.28	0.34	
	1.40	0.36	0.44	
	1.20	0.49	0.60	
	1.00	0.71	0.87	
	0.80	1.11	1.35	
Horizontal	4.00	0.04	0.05	0.15

Table 3. Input seismic wave parameters

Wave	X-direction	Z-direction	Duration
	[cm/s^2]	[cm/s^2]	[s]
El Centro	510.00	284.18	53.74
Hachinohe	333.79	166.00	50.98
Taft	486.31	327.71	54.28
KOKUJI-Kobe	216.86	168.74	120.00
KOKUJI-Hachinohe	274.75	186.76	234.00
KOKUJI-Random	219.36	193.85	120.00

4.4 Response Analysis Result

For the horizontal analysis results, the maximum response acceleration for each seismic wave is shown in Figs. 20 and 21. The results show that the response acceleration of the superstructure is within 150 gal for all seismic waves. It is still reduced to about 30% for observed seismic waves and 50% for KOKUJI waves with respect to the input acceleration.

For the analytical results in the vertical direction, the maximum response acceleration for each seismic wave with a natural period of 1.8 s is shown in Figs. 22 and 23. The analysis results confirm that the response acceleration in the vertical direction of the superstructure is low for each seismic wave, and that the seismic isolation effect of the upper and lower base isolation system is demonstrated. In addition, the observed seismic wave is reduced to about 30% of the input acceleration, and the KOKUJI wave is reduced to about 50% of the input acceleration. The design target of 40% reduction has been met.

For the horizontal analysis results, the response values for the maximum displacement at each seismic wave are shown in Figs. 24 and 25. It can be seen that the maximum displacement is about 15.94 cm at 2F, and the maximum displacement of the seismic isolation layer is 15.6 cm, which does not exceed the clearance of 50 cm.

The analysis results confirm that the maximum interlaminar deformation angle of the superstructure is 1/2837, which is within 1/200.

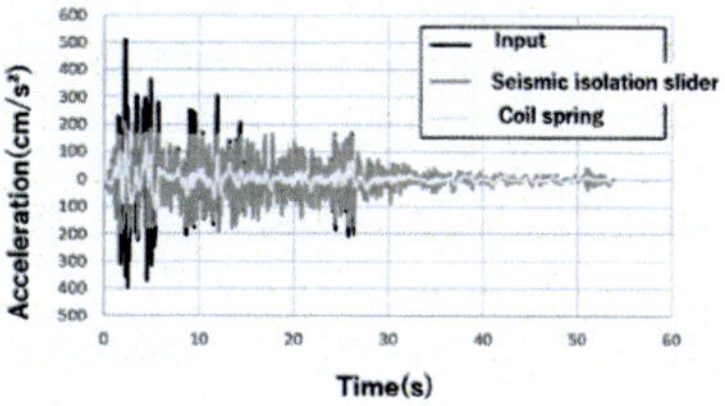

Fig. 18. 40 ft container house ELCENTRO response acceleration X

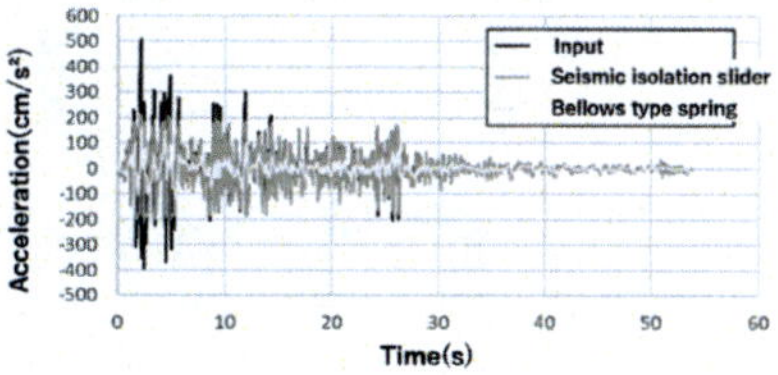

Fig. 19. 20 ft container house ELCENTRO response acceleration X

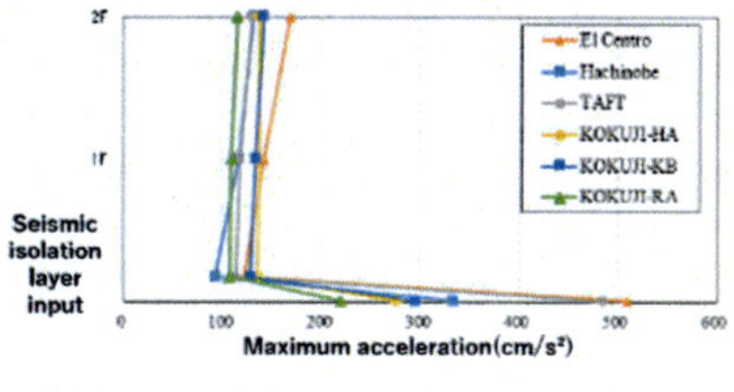

Fig. 20. 20 ft maximum response acceleration X

5 Summary

The results of the vibration test using a vibration machine confirmed the vibration-isolation effect of the rubberized coil spring.

The horizontal natural period of the seismic isolation layer was set to 4 s, which was confirmed to have a higher seismic isolation effect.

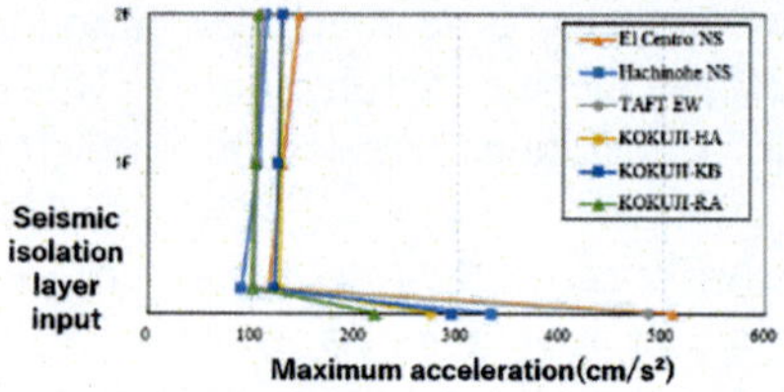

Fig. 21. 40 ft maximum response acceleration X

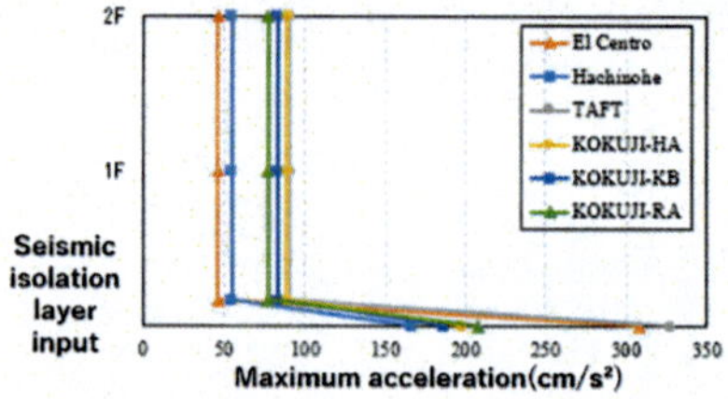

Fig. 22. 20 ft maximum acceleration Z

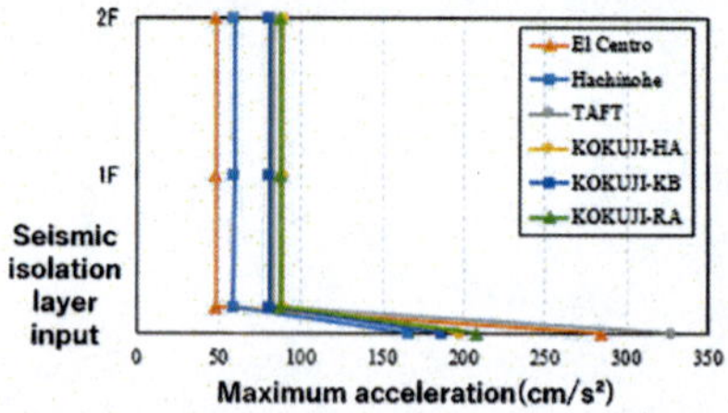

Fig. 23. 40 ft maximum acceleration Z

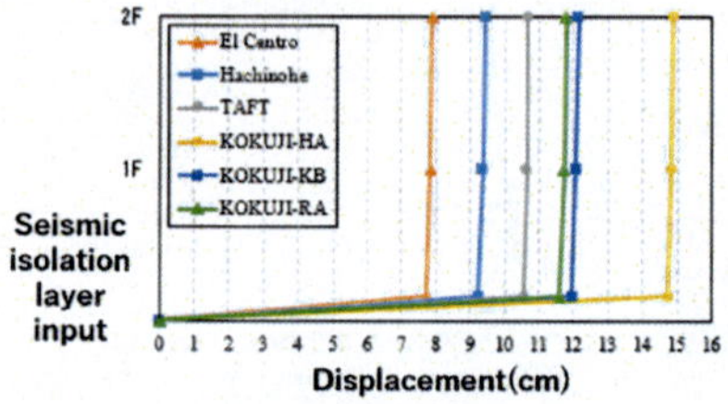

Fig. 24. 20 ft maximum displacement X

When the horizontal natural period was set to 4 s, the bellows-type air spring was confirmed to have a higher vibration-isolation effect.

When the natural period in the vertical direction of the isolation layer is set to 1.4 s or more, the bellows-type air spring has a higher seismic isolation effect.

The clearance of the seismic isolation layer in the vertical direction needs to be further investigated for the three-dimensional seismic isolation system proposed in this study.

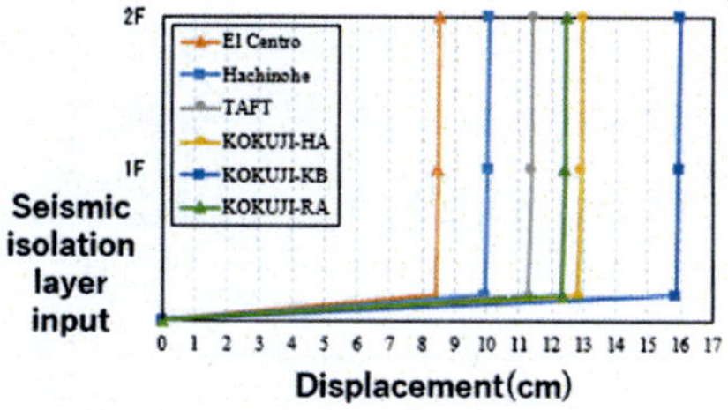

Fig. 25. 40 ft maximum displacement X

References

1. Chujo, A., et al.: Research on development of 3-D seismic isolation device using air levitation system, p. 2. Master's thesis, Department of Architecture, Graduate School of Engineering, Tokyo University of Science, 2020
2. Liang, S., et al.: Research on the development of seismic isolation slider, p. 2. Master's thesis, Department of Architecture, Graduate School of Engineering, Tokyo University of Science, 2021
3. Shibata, A.: Latest Seismic Structure Analysis, 3rd edn. (2003)
4. Architectural Institute of Japan: Seismic Isolation Structure Design Guidelines, 4th edn. (2013)

Numerical Assessment of Ultra-Low Cycle Fatigue Performance of Buckling Restrained Aluminum Shear Yielding Dampers

Deepak Yadav and Dipti Ranjan Sahoo(✉)

Indian Institute of Technology Delhi, New Delhi 110016, India
drsahoo@iitd.ac.in

Abstract. Long period seismic ground motions are known to cause large amplitude cyclic deformation in the energy dissipation devices installed in high-rise structures. Low cycle fatigue damage due to the large amplitude cyclic deformation in energy dissipation devices is of great concern. In this study, ultra-low cycle fatigue performance of buckling restrained aluminum shear yielding dampers (BR-AlSYDs) is assessed through numerical investigation on finite element program ABAQUS. Material properties of annealed Al-6063 alloy shear yielding devices are calibrated using ABAQUS platform against the cyclic experiment results. These calibrated material properties are adopted to design and investigate the BR-AlSYDs for their low cycle fatigue performance numerically. Two geometrical profiles of energy dissipator core plates are adopted in BR-AlSYD specimens which are hourglass and rectangular shapes. It is observed that the hourglass shaped energy dissipator core plates show better low cycle fatigue performance than rectangular shaped energy dissipator core plates. Further, fatigue performance parameters of Coffin-Manson relation are calibrated for BR-AlSYDs and accordingly Miner's rule is proposed to assess the fatigue performance of these dampers for any random seismic excitation.

Keywords: Ultra low cycle fatigue · Shear yielding device · Aluminum · Buckling restrained · Passive energy dissipation device

1 Introduction

Passive energy dissipation devices are widely researched and used for seismic vibration control of structures. Among these, metallic energy dissipation devices are very common due to the material availability and their great energy dissipation potential through nonlinear deformation capacity. Shear yielding devices (SYDs) are the type of metallic energy dissipation devices which are used for energy dissipation through shear yielding of an energy dissipator core plate. These devices are typically installed in frame systems over a chevron brace assembly and below the beams. However, core plates of the SYDs are prone to shear buckling which eventually compromises the energy dissipation capacity of these devices. To overcome this critical issue, restrainers can be used to restrain the shear buckling of the core plate. These SYDs in which shear buckling of the core

G. P. Cimellaro (Ed.): WCSI 2022, LNCE 309, pp. 296–303, 2023.
https://doi.org/10.1007/978-3-031-21187-4_24

plates is restrained are known as buckling restrained shear yielding devices (BR-SYDs). In the present study, core plates of the BR-SYDs are made of the annealed Al-6063 alloy aluminum which has high ductility and thus the devices are named as buckling restrained aluminum shear yielding devices (BR-AlSYDs).

Long period seismic ground motions usually affect the high-rise structures severely and cause large inter-story drifts [1]. This large inter-story drift is the reason for large amplitude cyclic deformations in the BR-AlSYDs installed in the frame system. Low cycle fatigue damage in the BR-AlSYDs is of great concern and importance which has not been investigated extensively in previous studies. Present study is focused on the low cycle fatigue performance assessment of the BR-AlSYDs. Further, fatigue damage law based on Coffin-Manson relation has been calibrated and presented.

2 Numerical Investigation of BR-AlSYDs

Numerical simulations for the low cycle fatigue performance of buckling restrained aluminum shear yielding devices (BR-AlSYDs) were conducted on a finite element analysis software ABAQUS. Modelling technique, details of specimens and loading history for numerical investigation of BR-AlSYDs are discussed in following sections.

2.1 Modelling Technique

The numerical model of BR-AlSYD created in ABAQUS is shown in Fig. 1. MPC constraint is used to attach the bottom connection zone of the core plate of BR-AlSYDs to a fixed reference point. The top connection zone of the core plate is also tied to another reference point using MPC constraint. The reference point attached to the BR-AlSYD's top connection zone is subjected to a constant amplitude cyclic displacement time history. The core plates of the BR-AlSYDs are modelled using C3D8R elements with hourglass control. C3D8R is a general-purpose linear brick element with a single integration point with reduced integration technique. This represents the eight node-element, each of which has three degrees of freedom. In the past, solid elements were employed to model shear yielding devices and BRBs [2, 3]. C3D8R elements have also been used to model the restrainers. The core plates are meshed using hex mesh control. Restrainers are modelled using coarse mesh. Since, the objective of restrainers is to prevent out-of-plane buckling of core dissipator plates, they are modelled as thick plates with suitable boundary conditions.

The interaction between core plate and restrainers is assigned as hard contact in the normal direction and frictionless contact in the tangential direction. This restrainer mechanism avoids the shear buckling of the core plates.

2.2 Details of the Specimens

The BR-AlSYD specimens in this numerical investigation were designed using annealed Al-6063 aluminum alloy plates. A total of seven specimens are numerically investigated under the effect of constant amplitude displacement controlled cyclic loading. Six specimens with hourglass shaped core plates and one specimen with a rectangular shaped

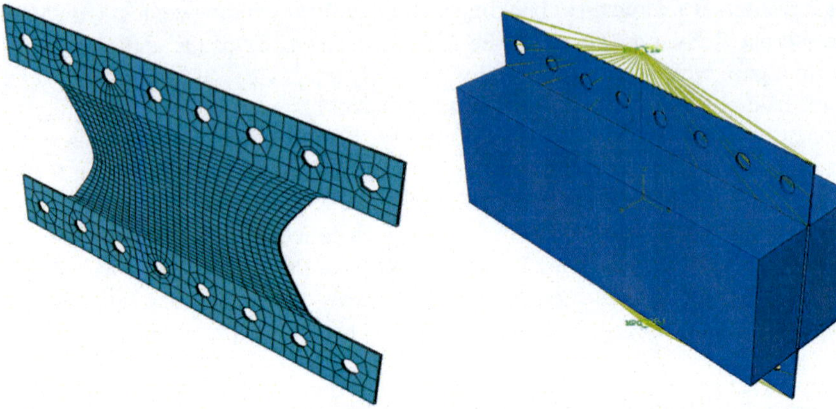

(a) (b)

Fig. 1. Numerical model of BR-AlSYDs: (a) Core plate, and (b) the device assembly.

core plate were created using annealed Al-6063 alloy plates. Core plates are confined between the built-up restrainers in all these specimens to prevent the plate buckling. Hourglass shaped core plates have been compared to the rectangular shaped core plate in terms of low cycle fatigue performance. Core plate dimensions, such as the thickness of the plate (t_w), length of the plate (l_w), depth of the plate (d_w) and equivalent length of the hourglass shaped plate (l_{eq}) for numerical investigations are listed in Table 1. Table 1 also includes the loading protocol used for each specimen.

Table 1. Geometric details of the specimens.

Specimens	l_w (mm)	l_{eq} (mm)	d_w (mm)	t_n (mm)	Imposed loading history (%)	Core plate shape
BR-AlSYD-1	250	290	175	3	6	Hourglass
BR-AlSYD-2	250	290	175	3	8	Hourglass
BR-AlSYD-3	250	290	175	3	10	Hourglass
BR-AlSYD-4	250	290	175	3	12	Hourglass
BR-AlSYD-5	250	290	175	3	14	Hourglass
BR-AlSYD-6	250	290	175	3	16	Hourglass
BR-AlSYD-7	250	–	175	3	6	Rectangular

Under constant amplitude cyclic loads, BR-AlSYD specimens have been investigated. According to ATC 24 [4], these specimens are subjected to a displacement history that represents the low cycle fatigue loading. As illustrated in Fig. 2, cyclic loads are indicative of the first three cycles at shear strain values of 0.0025, followed by constant amplitude cycles at the desired shear strain levels. The ratio of the in-plane horizontal displacement to the depth of the core plate, d_w, of the BR-AlSYDs is termed as shear strain.

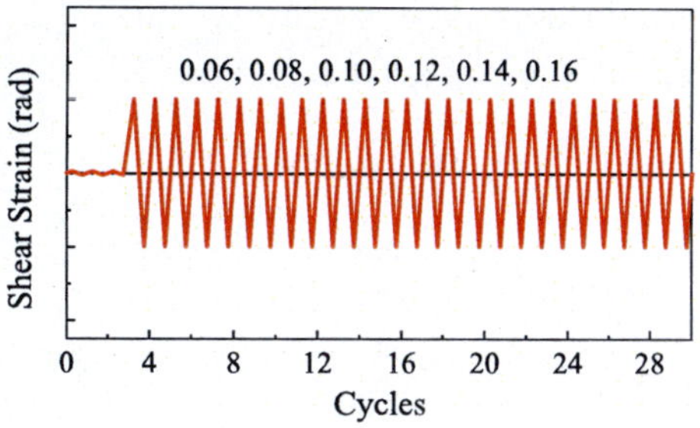

Fig. 2. Imposed constant amplitude cyclic loading history.

2.3 Material Properties

To calibrate the material mechanical properties of annealed Al-6063 alloy SYDs under cyclic loads, experiments conducted by Jain et al. [5] have been considered. Detailed procedure to calibrate the material mechanical properties is presented elsewhere [6]. Validated material mechanical properties are shown in Table 2. These material parameters have been considered for low cycle fatigue analysis of BR-AlSYDs in the present study.

Table 2. Material cyclic hardening parameters.

Aluminum alloy	E (MPa)	σ_y (MPa)	Q_∞ (MPa)	b	C (MPa)	Υ
Al-6063	18700	33.5	21.7	5.9	2750	85

Notes: E = Modulus of elasticity, σ_y = Yield stress, Q_∞ = Limiting isotropic hardening component, b = Rate of change in yield surface, C = Kinematic hardening modulus, Υ = Rate of recovery.

3 Low Cycle Fatigue Performance

3.1 Fatigue Damage Law

Low cycle fatigue damage of BR-AlSYDs can be evaluated using the Coffin-Manson relation given below as Eq. (1).

$$N_f = A^{-1}(\delta\varepsilon)^{-a} \tag{1}$$

where, N_f is number of required cycles for fatigue failure, $\delta\varepsilon$ is the strain range, A and a are the fatigue performance parameters.

Fatigue performance parameters can be calibrated against the experimental or numerical analysis results. Miner's rule states that the damage (*D*) can be summed up linearly for each cycle in the devices under fatigue loading. To evaluate the fatigue seismic performance of BR-AlSYDs for random cyclic loading, Miner's rule can be expressed as given below in Eq. (2).

$$D = A\Sigma(\delta\varepsilon)^a \tag{2}$$

3.2 Analysis Results

Figure 3 presents the stress contours and fatigue failure in BR-AlSYDs having hourglass and rectangular shaped core plates. All BR-AlSYD specimens with hourglass shaped core plates failed in the similar way. In these specimens, the fatigue failure began around one-third of the depth of the core plate and then spread in the core plate. Fatigue failure began near the corners of BR-AlSYD with the rectangular shaped core plate, and subsequently fracture propagated along the connection zones.

The numerical results also showed that BR-AlSYDs with hourglass shaped core plates had a more uniform stress distribution in the middle region of the plate than BR-AlSYDs with rectangular shaped core plate. As a result, the hourglass shaped core plates can be taken as an optimal shape for BR-AlSYD specimens for better fatigue resistance. As shown in Fig. 3, the middle region of the core plate exhibiting the stress range in red color was fully utilized in the BR-AlSYD-7 specimen, while the remaining part of the core plate was only partially utilized in dissipating the energy. Figure 3 also shows the hourglass shaped core plates were almost fully utilized in dissipating the energy.

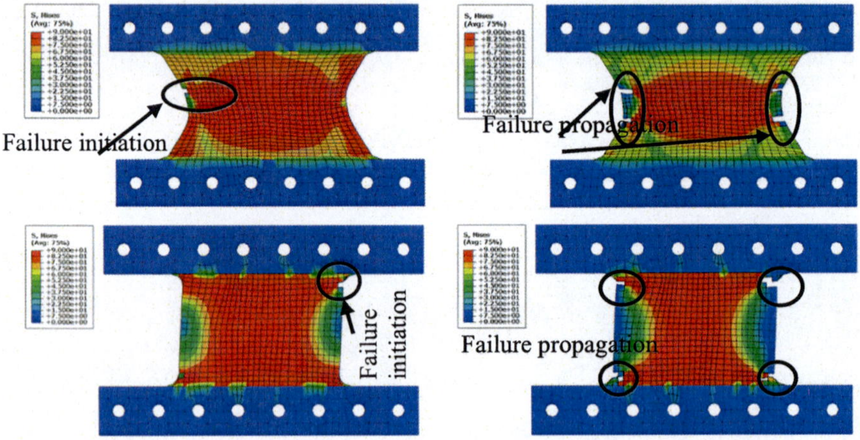

Fig. 3. Stress contours and low cycle fatigue failure initiation in BR-AlSYDs.

Figure 4 shows the hysteresis curves and backbone curves of BR-AlSYDs. BR-AlSYD specimens yield at 33 MPa approximately and then exhibits strain hardening. Peak shear stress observed in the specimens was approximately 40 MPa. Low cycle fatigue performance of hourglass and rectangular shaped core plates can be compared from backbone curves in Fig. 4. BR-AlSYD-1 and BR-AlSYD-7 have been subjected to 6% constant cyclic shear strain loading. The BR-AlSYD-1 specimen with the hourglass shaped core plate performs better than the BR-AlSYD-7 specimen with the rectangular shaped core plate. This is mostly owing to early fracture initiation due to stress concentration near the corners of rectangular shaped core plates.

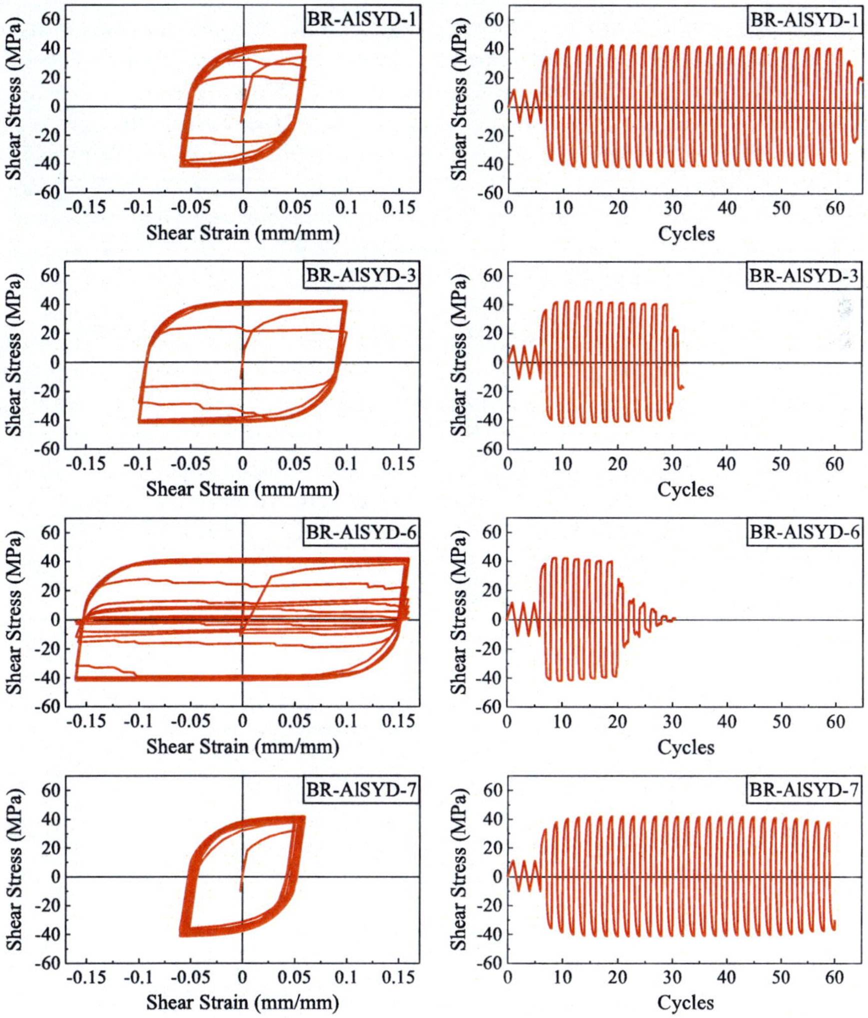

Fig. 4. Hysteresis and backbone curves of BR-AlSYDs.

3.3 Proposed Fatigue Damage Law for BR-AlSYDs

As analysis progressed, fatigue damage in the core plates of BR-AlSYDs started accumulating. When the D reached unity, the elements were deleted from the numerical models and shear fatigue failure was assumed to have occurred. Prior to fatigue failure, the BR-AlSYD-1 specimen having hourglass shaped core plate lasted sixty-two cycles of 6% constant amplitude shear strain time history. The BR-AlSYD-7 specimen having rectangular shaped core plate required fifty-four cycles for fatigue failure when subjected to 6% constant amplitude cyclic shear strain. Furthermore, fatigue failure was observed in BR-AlSYD-2, BR-AlSYD-3, BR-AlSYD-4, BR-AlSYD-5, and BR-AlSYD-6 specimens after thirty-eight, thirty, twenty-five, twenty-two and twenty cycles, respectively. These specimens were subjected to constant amplitude shear strain time histories of 8%, 10%, 12%, 14%, and 16%, respectively.

Equation (3) is the calibrated Coffin-Manson relation for BR-AlSYDs. R-Square of 0.977 shows the goodness of the curve fit to obtain the fatigue performance parameters A and a. As a result, Miner's rule can be expressed as Eq. (4) which can be used to assess the extent of fatigue damage in BR-AlSYDs.

$$N_f = (0.241)^{-1}(\delta\varepsilon)^{-1.259} \tag{3}$$

$$D = 0.241\,\Sigma(\delta\varepsilon)^{1.259} \tag{4}$$

A and a are the fatigue performance parameters calibrated based on the strain range, $\delta\varepsilon$, for BR-AlSYDs having hourglass shaped core plate as 0.241 and 1.259, respectively.

Figure 5 shows the plot of the number of cycles required for fatigue failure vs. constant amplitudes of shear strain for BR-AlSYDs with hourglass shaped core plates using numerical analysis results and fatigue damage law.

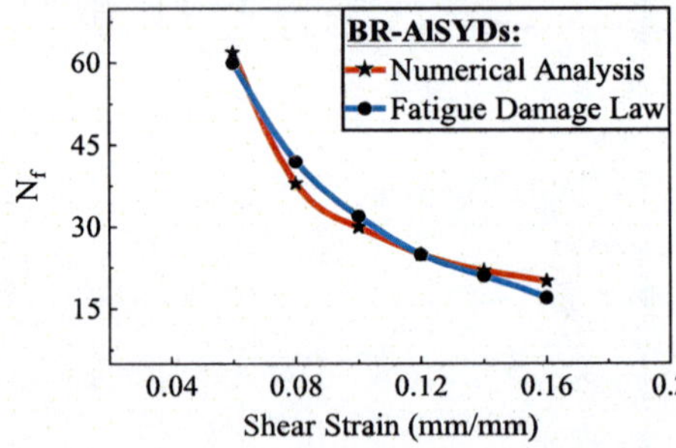

Fig. 5. Number of cycles to fatigue failure from numerical analysis and fatigue damage law.

4 Conclusions

In this study, the low cycle fatigue behavior of the BR-AlSYDs has been investigated numerically on ABAQUS. The core plates of BR-AlSYD specimens are designed using annealed Al-6063 aluminum alloy for numerical analysis. It is observed that the hourglass shaped core plates show better distribution of the stresses as compared to the rectangular shaped core plates. The fatigue performance parameters for the Coffin-Manson relation are calibrated based on the numerical results and can be used to evaluate the extent of fatigue damage under random seismic excitations using the Miner's rule.

References

1. Inoue, Y., et al.: Fatigue-resistant Fe-Mn-Si-based alloy seismic dampers to counteract long-period ground motion. Jpn. Archit. Rev. **4**(1), 76–87 (2021)
2. Dusicka, P., Itani, A.M., Buckle, I.G.: Finite element investigation of steel built-up shear links subjected to inelastic deformations. Earthq. Eng. Eng. Vib. **3**(2), 195–203 (2004)
3. Pandikkadavath, M.S., Sahoo, D.R.: Cyclic testing of short-length buckling-restrained braces with detachable casings. Earthq. Struct. **10**(3), 699–716 (2016)
4. ATC 24: Guidelines for Cyclic Seismic Testing of Components of Steel Structures (1992)
5. Jain, S., Rai, D.C., Sahoo, D.R.: Postyield cyclic buckling criteria for aluminum shear panels. J. Appl. Mech. **75**(2), 021015 (2008)
6. Yadav, D., Sahoo, D.R.: Validation of hysteretic behavior and prediction of energy dissipation potential of aluminium shear yielding devices. Int. J. Mech. Sci. **194**, 106204 (2021)

Validation of a New Equation to Estimate the Yield Strength of Lead Rubber Bearings

Masoud Pourmasoud[1](✉), Alan R. L. Park[2], Iman Hajirasouliha[3], James Lim[1], and Amirmahmoud Behzadi[4]

[1] Department of Civil and Environmental Engineering, University of Auckland, Auckland, New Zealand
mpou570@aucklanduni.ac.nz
[2] Low Damage Design Company, Hamilton, New Zealand
[3] Department of Civil Engineering, The University of Sheffield, Sheffield, UK
[4] Department of Civil Engineering, Shomal University, Amol, Iran

Abstract. The design of seismic isolators as specified by building code requirements is well defined and well established. However, there is still some refinement required for these design practices, especially in terms of estimating the yield strength and confinement of lead cores. The Lead Core is a key component in a Lead Rubber Bearing (LRB) that provides energy absorption resulting in the damping of a unit/system. Historically, designers have used a standard value for lead yield (7–10 MPa) depending on individual preferences. However, based on prototype test results, it has been shown that lead yielding is variable and strongly influenced by a number of physical parameters. This piece of research has focused on validating a newly developed equation to estimate the yield strength of lead rubber bearing regarding the confinement category of the lead cores. The yield strength results from prototype tests were compared with the predicted yield strengths from the proposed equation. The output trends present an appropriate match between the prototype test and the proposed equation so that the difference for most cases falls within ±5%.

Keywords: Seismic isolation · Lead core · Yield strength · Confinement category

1 Introduction

Seismic isolation systems have become commonplace worldwide due to their effectiveness in providing a low damage design solution for a variety of structures including – but not restricted to – hospitals, schools, nuclear power plants, control towers, and retrofitting of existing buildings [6, 10, 19, 21]. Code-compliant buildings are generally designed for a specific earthquake tolerance although more significant events are possible, which can escalate the accelerations imparted on the building and affect the content's performance [8, 9, 20]. An optimized seismic isolation system should diminish floor accelerations and drifts to less than the damage thresholds which are specified by

G. P. Cimellaro (Ed.): WCSI 2022, LNCE 309, pp. 304–313, 2023.
https://doi.org/10.1007/978-3-031-21187-4_25

seismic codes [14]. It is particularly important for the case of seismic isolators that are subjected to high peak ground motions in near field zones [1, 3, 13]. It should be noted that the vertical component of ground motions in a near field zone can also increase considerably the structural axial loads as well as horizontal responses [18].

Seismic isolated structures are generally designed according to the specifications that are assumed for isolator bearings rather than their accurate characteristics [4, 11, 17]. Seismic design codes such as ASCE 07-22 simplify the complexity of seismic isolator behaviour to facilitate an easier designing process [7, 9]. The adequacy of a seismic isolator's specifications is assessed during three distinctive stages [4]. The first stage is to determine the components specification of the isolator, where the resilience of each component of isolator bearings shall comply with relevant codes under recommended actions and be approved by the seismic isolation designer. In second stage, the isolator units are designed under the load combinations and lateral displacements to pass the prototype/production tests criteria [5]. The final stage is to assess the performance of the group of isolators to meet the design targets and structural expectations, using key performance parameters such as the maximum base shear and displacement. In addition, the durability and maintenance of isolators need to be addressed during the building lifetime.

In a conventional lead rubber bearing, the lead core is the key component that provides damping. Lead plugs can provide high damping capacity due to their relatively low yield strength of about 10 MPa and a perfect elastic-plastic behavior [17]. Although 10 MPa is known as a nominal number for lead yield strength (σ_y), parameters such as the magnitude of applied axial load, dimension of lead core, lateral shear strain of bearing, and the adopted cyclic loads (degradation) can also considerably affect the lead yield strength [12]. Kelly proposed 7–8.5 MPa as the pragmatic yield level of the lead core depending on the axial load and level of confinement, with 10.5 MPa as the theoretical yield level [11]. Furthermore, it was reported that more shim plates and thinner rubber layers restrain the lead core from bulging into the rubber layers, which help enhance yield strength along with utilizing confining plates at the top and bottom of the lead core. The multilinear behavior of a lead rubber bearing tends to be bilinear, providing appropriate confinement for the lead core, so a yield strength closer to the nominal value will be achieved.

Pourmasoud et al. investigated the influential parameters on the lead core confinement and the lead yield strength of lead rubber bearings regarding the prototype tests' results of more than 300 tests [15, 16]. For each test, the characteristic strength was calculated from data extracted from the third hysteresis loops and then divided to the characteristic strength under zero axial load (Q_0) to achieve the normalised characteristic strength (Q/Q_0). The applied axial loads were also normalised by the buckling load of each bearing (P/P_{cr}) to obtain the characteristic strength trend versus the applied axial load.

Figure 1 shows the trends extracted from all the selected tests. Generally, three different trends are recognised. The green line represents Low Confinement (LC) and low yield strength. This characteristic strength, even under high axial loads ($P/P_{cr} > 0.5$), is approximately 20% more than the characteristic strength under zero axial load ($Q/Q_0 = 1.2$). It should be noted that for all the case studies, the axial loads under seismic load combinations were more than 50% of the buckling load. The orange dots show the High Confinement (HC) trend with normalised characteristic strength of more than 1.6. The yield strength of this category is quite close to the theoretical yield strength (10 MPa). The outcomes between the two aforementioned categories (green and orange) are designated Medium Confinement (MC) and their normalised characteristic strength values lie between 1.2 and 1.6.

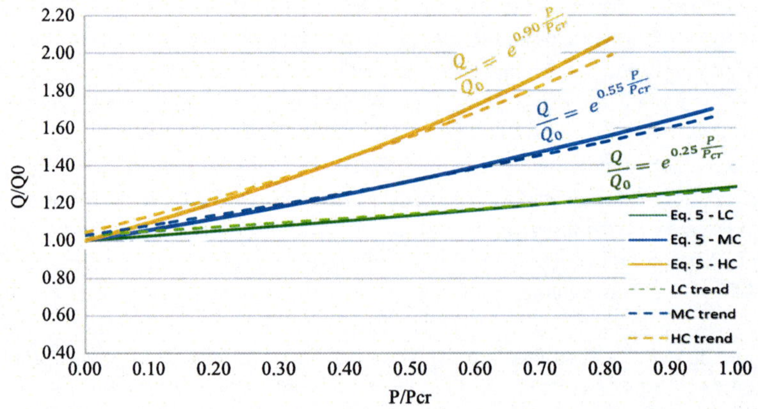

Fig. 1. Confinement categories and their effect on the normalized characteristic strength

Based on the results presented in Fig. 1, for each confinement category the characteristic strength of a lead rubber bearing is achievable through Eq. 1.

$$\frac{Q}{Q_0} = e^{\alpha \frac{P}{P_{cr}}} \tag{1}$$

where α is the confinement exponent that is 0.25 for low confinement, 0.55 for medium confinement, and 0.90 for high confinement. Figure 1 compares the trends regarding Eq. 1 (dashed lines) with the data obtained for each confinement category. The proposed confinement exponents in Eq. 1 were obtained to provide the best agreement with the presented results using a regression statistical optimization method.

2 Proposed Equation to Estimate a Lead Core's Yield Strength

Figure 1 illustrates the general trend of (Q/Q_0) versus the influential factor of (P/P_{cr}). However, to obtain specific values, the confinement conditions as well as the yield/character strengths under zero axial loads (Q_0, σ_{y0}) shall be also taken into account. This section addresses the conditions that can be used to define confinement levels. The two main influential factors are the slenderness ratio (H_l/D_l) and the area ratio (A_l/A_b), where H is the lead core height, D_l is the lead core diameter, A_l is the cross area of the lead core, and A_b is the bonded area of the isolator. In addition, the slenderness ratio over the area ratio is defined as Influential Factor Ratio (IFR) according to Eq. 2.

$$IFR = \frac{(H_l/D_l)}{(A_l/A_b)} \tag{2}$$

Figure 2 illustrates the trend for σ_{y0} versus IFR when an accurate yield strength is known. This diagram is extracted from testing isolators under zero axial load and obtaining the corresponding yield strength of the lead core based on the confinement condition. The diagram has a steep slope for IFRs less than one but changes to a smoother slope with increasing IFR, which indicates that the axial load influence is more significant for fewer IFRs. In other words, better confinement results in less variation of yield strength regardless of the axial load. To obtain the yield strength under zero axial load, Eq. 3 can be employed:

$$\sigma_{y0} = 0.35\ln(IFR) + 6 \tag{3}$$

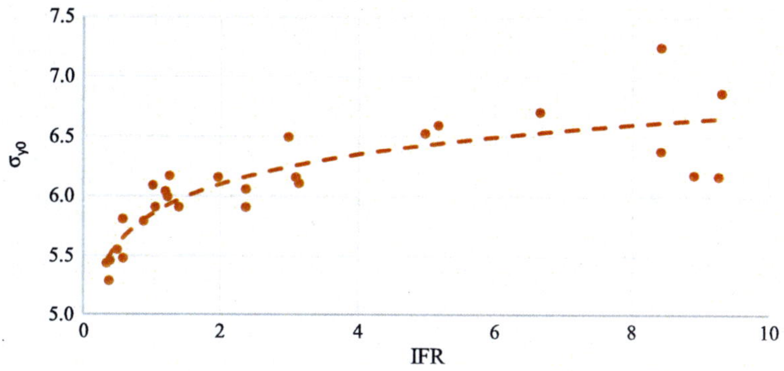

Fig. 2. The trend of yield strength versus IFR based on Eq. 3.

Combination of Eqs. 2 and 3 results in the following equation to calculate the yield strength of lead rubber bearings as a function of their configuration, the imposed axial loads, and other influential factors.

$$\sigma_y = [0.35\ln(IFR) + 6]e^{\alpha \frac{P}{P_{cr}}} \tag{4}$$

Equation 4 provides a practical perspective regarding the actual yield strength of the lead core before running the prototype tests. This equation can help to narrow the gap between upper/lower bounds of λ_{test} and λ_{spec} mentioned in [2], which is beneficial for the design of super-structures as discussed before.

3 Validation

This section validates the accuracy of the developed equation to estimate the yield strength of a lead rubber bearing. To this end, seven lead rubber bearings subjected to various axial loads and lateral displacements were employed. Table 1 shows the physical dimensions and applied axial loads. Table 2 shows the applied axial loads and lateral displacements. The isolators were tested regarding the displacement increments mentioned in [2, Section 17.8.2.2]. However, the bearings' specifications need to be fall within the allowable ranges under D_M and $0.67D_M$.

Table 3 demonstrates a sample tests' results from isolator 7. The energy dissipated and its equivalent yield strength from prototype tests ($\sigma_{y\text{-result}}$) is reported. Also, the yield strength regarding the proposed equation for the lead cores ($\sigma_{y\text{-equation}}$) were obtained and compared with the test results. Both approaches represent a range of yield strength varies between 5.0 and 7.8, depends on the applied axial load and the lateral displacement.

Table 1. Isolators' mechanical specification and confinement category.

Isolator	D_b	D_l	H_l	$P_{cr\text{-static}}$ (kN)	$\frac{H_l}{D_l}$	$\frac{A_L}{A_b}$ (%)	IFR	Confinement category	α
1	870	100	330	20703	3.3	1.45	2.27	MC	0.5
2	720	160	330	5154	2.06	5.5	0.37	LC	0.25
3	770	170	330	8029	1.94	5.4	0.36	LC	0.25
4	1020	170	330	25278	1.94	3.0	0.64	LC	0.25
5	1020	100	373	34797	3.73	1.04	3.6	HC	0.9
6	1220	120	373	71183	3.11	1.03	3.1	HC	0.9
7	770	85	326	13201	3.84	1.3	2.8	MC	0.5

Table 2. Isolators' axial loads and lateral displacements.

Isolator	Gravity load (kN)	Max seismic load (kN)	Min seismic load (kN)	D_M	0.67*D_M	Shear strain (%) at D_M
1	3012	6349	530	390	261	150
2	720	2356	−171	264	177	102
3	1847	4074	−228	264	177	102
4	5141	14977	−343	264	177	102
5	5450	11278	500	341	228	126
6	10053	21932	500	341	228	126
7	1629	2667	530	248	166	102

It is notable that the yield strength from both approaches for each test are reasonably close so that under the maximum load and maximum displacement (PT-9) the $\sigma_{y\text{-result}}$ and $\sigma_{y\text{-equation}}$ are achieved 7.85 MPa and 7.58 MPa, respectively. The average of total yield strength ratios shows that the predicted $\sigma_{y\text{-equation}}$ have been 4.0% more than $\sigma_{y\text{-result}}$. It worth mentioning that PT-12 shows a maximum difference between the test and the equation which might be due to the cyclic loading effect on the lead core strength as the isolator had experienced 33 loading cycles before PT-12.

Figure 3 presents the yield strengths from each test carried out on three isolators as the samples. The comparison between the results shows similar trend between the tests and the proposed equation which provide designers estimating the yield strength

Table 3. Tests' results and yield strength comparison between prototype test and proposed equation – Isolator 7.

Isolator 7								
Test	P (kN)	Disp (mm)	Ah (kN.m)	σ_y *result (MPa)*	P_{cr} (disp)	$(P/P_{cr})_{disp}$	σ_y *equation (MPa)*	$\frac{\sigma y-\text{equation}}{\sigma y-\text{result}}$
PT-4	1629	166	23.5	*6.24*	9412.03	0.17	*6.94*	1.11
PT-5a	1629	248	41.9	*7.44*	7602.64	0.21	*7.08*	0.95
PT-5b	1445	257	42.9	*7.35*	7408.21	0.20	*7.02*	0.95
PT-5c	1912	238	40.1	*7.42*	7819.75	0.24	*7.19*	0.97
PT-5d	2374	223	37.9	*7.49*	8147.44	0.29	*7.36*	0.98
PT-8	2667	166	25.8	*6.85*	9412.03	0.28	*7.33*	1.07
PT-9	2667	248	44.2	*7.85*	7602.64	0.35	*7.58*	0.97
PT-12	530	166	19	*5.04*	9412.03	0.06	*6.55*	1.30
PT-13	530	248	34.8	*6.18*	7602.64	0.07	*6.59*	1.07
							Average	*1.04*

of the lead rubber bearing at any level of shear strain and axial load. For majority of tests, the yield strength from the tests have been fallen with ±5% of the yield strength extracted from the test. In addition, it shall be noted that, there is no guaranty that the tests' results from two identical isolator give rise to the same numbers. It is because, the rubber is made from a natural material which its specification is not absolute and has an influential impact on the confinement of the lead core. So, even in practice a partial difference between the results of two identical isolators are quite likely.

Table 4 and Fig. 4 summarize the yield strength results under the maximum displacements and maximum seismic loads which usually are the target during the design and isolators' prototype tests. The yield strength ratios ($\frac{\sigma y-\text{equation}}{\sigma y-\text{result}}$) show the accuracy of the predicted yield strength in comparison with the tests results. The fifth case study shows the tests' yield strength relatively less than the equation outcome. It is because this test was carried out after 36 cycles of loading, which caused a partial rounding in the circumference of the lead core and a correspondingly decrease in the lead area. Generally, the test results show that the lead cores' yield strength have varied between 6 MPa and 10 MPa depends on their confinement conditions and proposed equation has predicted that with about 3% difference in average.

4 Summary and Conclusion

Lead core plays an important role in the performance of Lead Rubber Bearings (LRB) as it provides the damping effect of the system. Underestimation of the lead core's yield strength causes insufficient damping and extra travel on the isolation level, while overestimation produces a stiff system that transmits unfavorable base shear to the superstructure. The yield strength of the lead core is suggested to be between 6 and 10 MPa in

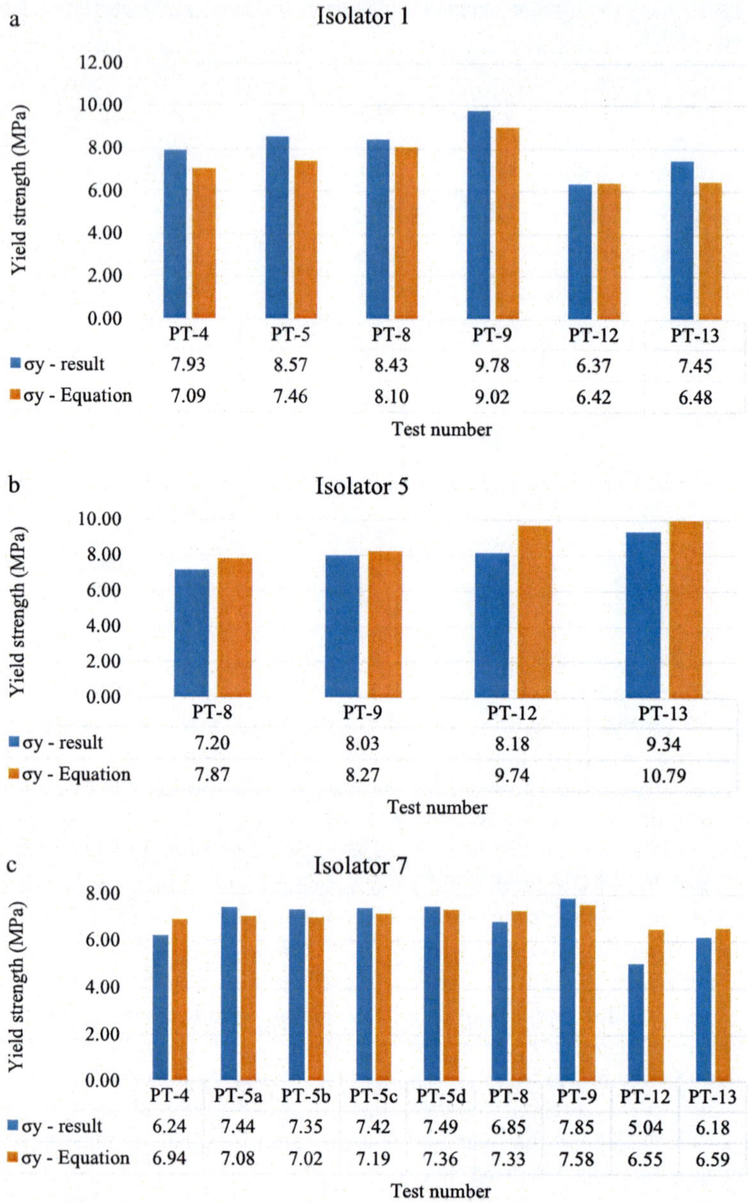

Fig. 3. Yield strength comparison between the prototype tests and the proposed equation. (a) Isolator 1; (b) Isolator 5; (c) Isolator 7.

current codes/guidelines depending on loading conditions. Yield strength can be calculated from the hysteresis loops that are obtained from prototype tests. However, there can be incompatibilities between the design targets and test results. This study investigated

Table 4. Yield strength comparison under maximum seismic loads and maximum displacements.

Isolator	Pressure (MPa)	Shear strain (%)	$(P/P_{cr})_{disp}$	σ_y *result (MPa)*	σ_y *equation (MPa)*	$\frac{\sigma y-\text{equation}}{\sigma y-\text{result}}$ (%)
1	11.73	118.18	0.72	*9.78*	*9.02*	92.28
2	6.49	80.00	0.88	*6.42*	*7.05*	109.73
3	9.73	80.00	0.92	*6.85*	*7.10*	103.70
4	19.86	80.00	0.90	*7.29*	*7.31*	100.36
5	14.95	91.42	0.57	*9.34*	*10.79*	115.50
6	20.06	91.42	0.48	*9.59*	*9.87*	102.91
7	5.67	68.40	0.29	*7.49*	*7.36*	98.32
					Average	*103.3*

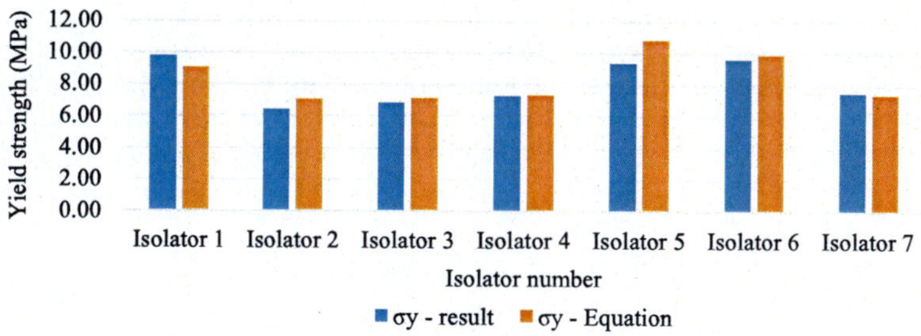

Fig. 4. Yield strength comparison under maximum seismic loads and maximum displacements.

more than 300 tests to develop a new approach to estimate the yield strength of lead cores with high accuracy prior to prototype testing.

The following conclusions were made based on the new approach:

- Lead rubber bearings are energy dissipative even under zero/tension and shear actions. While the hysteresis loops of isolators under tension/zero axial loadings are not as inflated as an isolator under heavy axial loading (particularly those with $P/P_{cr} > 0.5$), they still provide a reasonable amount of damping. Currently this effect is neglected in the design process of lead rubber bearings.
- A variety of influential factors, including applied axial load, slenderness of the lead core, and thickness of the rubber layers, confer different levels of confinement for the lead core. Three confinement categories were recognised based on the trend of the normalised characteristic strengths versus the normalised axial loads. The developed Influential Factor Ratio (IFR) shows that IFRs less than 1.0 represent low confinement, IFRs of approximately 2.0 indicate medium confinement, and when IFR exceeds 7.0, high confinement is achieved.

- The yield strength of the lead core directly depends on the level of confinement, the isolator travel amplitude, and the applied axial load. These effects are generally ignored in the code specified design criteria. To address this issue, a new design equation was developed to estimate lead core yield strength by taking into account all the key influential factors.
- Accuracy improvement of the yield strength calculation narrows the gap between the upper bound and lower bounds offered by [2] and is beneficial for more efficient design of super-structures.

Acknowledgements. Special thanks to those who have assisted in this research, in particular, the technical support from Low Damage Design Co.

References

1. Alhan, C., Öncü-Davas, S.: Performance limits of seismically isolated buildings under near-field earthquakes. Eng. Struct. **116**, 83–94 (2016)
2. ASCE: Minimum Design Loads and Associated Criteria for Buildings and Other Structures. ASCE/SEI 7-22. ASCE, Reston, VA (2022)
3. Bhagat, S., Wijeyewickrema, A.C., Subedi, N.: Influence of near-fault ground motions with fling-step and forward-directivity characteristics on seismic response of base-isolated buildings. J. Earthq. Eng. 1–20 (2018)
4. BS EN 1337-3: Structural Bearings - Part 3: Elastomeric Bearings. BSI Standard Publication (2005)
5. BS EN 15129: European Standard on Anti-Seismic Devices. BSI Standard Publication (2018)
6. Clemente, P., Martelli, A.: Recent applications of seismic isolation in Italy. In: 16th World Conference on Seismic Isolation, Energy Dissipation and Active Vibration Control of Structures. Russian Federation, St. Petersburg (2019)
7. Constantinou, M.C., Kalpakidis, I., Filiatrault, A., Ecker Lay, R.A.: LRFD-Based Analysis and Design Procedures for Bridge Bearings and Seismic Isolators. MCEER 11-0004. Univ. at Buffalo, Buffalo, NY (2011)
8. FEMA: Multi-Hazard Loss Estimation Methodology - Earthquake Model. HAZUS-MH MR4 Technical Manual. Federal Emergency Management Agency (FEMA), Washington, DC (2003)
9. FEMA: Building the Performance You Need. FEMA P-58-7. FEMA, Washington, DC (2018)
10. Jampole, E., Swensen, S., Miranda, E., Deierlein, G.G.: Parametric study of seismic isolation properties for light-frame houses. J. Struct. Eng. **146**(10), 04020207 (2020)
11. Kelly, T.E.: Base Isolation Design of Structures: Design Guidelines. Holmes Consulting Group Ltd (2001)
12. McVitty, W.J., Constantinou, M.C.: Property Modification Factors for Seismic Isolators: Design Guidance for Buildings. MCEER-15-0005. Univ. at Buffalo, Buffalo, NY (2015)
13. Özuygur, A.R., Farsangi, E.N.: Influence of pulse-like near-fault ground motions on the base-isolated buildings with LRB devices. Pract. Period. Struct. Des. Constr. **26**(4), 04021027 (2021). https://doi.org/10.1061/(ASCE)SC.1943-5576.0000603
14. Pal, S., Hassan, A., Singh, D.: Optimization of base isolation parameters using genetic algorithm. J. Stat. Manag. Syst. **22**(7), 1207–1222 (2019). https://doi.org/10.1080/09720510.2019.1614338

15. Pourmasoud, M.M., Lim, J.B.P., Hajirasouliha, I., McCrum, D.: Multi-directional base isolation system for coupled horizontal and vertical seismic excitations. J. Earthq. Eng. **26** (2020). https://doi.org/10.1080/13632469.2020.1713925
16. Pourmasoud, M.M., Park, A.R.L., Hajirasouliha, I., Lim, J.B.P.: Investigation of influential factors on yield strength of lead rubber bearings. In: NZSEE 2021, Annual Conference (2021)
17. Skinner, R.I., Robinson, W.H., McVerry, G.H.: An Introduction to Seismic Isolation. Wiley (1993)
18. Wei, B., Zuo, C., He, X., Jiang, L., Wang, T.: Effects of vertical ground motions on seismic vulnerabilities of a continuous track-bridge system of high-speed railway. J. Soil Dyn. Earthq. Eng. **115**, 281–290 (2018)
19. Whittaker, D.: Recent developments in New Zealand in seismic isolation, energy dissipation and vibration control of structures. In: 16th World Conference on Seismic Isolation, Energy Dissipation and Active Vibration Control of Structures. Russian Federation, St Petersburg (2019)
20. Whittaker, D., Parker, W., Pettinga, D., Pierta, D., McVerry, G., Sidwell, G.: Guideline for the Design of Seismic Isolation Systems for Buildings. New Zealand Society for Earthquake Engineering (2019)
21. Xie, L., Wang, X., Zeng, D., Jia, J., Liu, Q.: Resilience-based retrofitting of adjacent reinforced concrete frame-shear wall buildings integrated into a common isolation system. J. Perform. Constr. Facil. **36**(1), 04021100 (2022)

International Standards on Seismically Isolated Structures

Seismic Isolation Design Comparison of Japan, China, USA and Eurocode

Demin Feng[1(✉)], Taiki Saito[2], Honglei Wu[3], and Wenguang Liu[4]

[1] Fujita Corporation, Tokyo, Japan
feng@fujita.co.jp
[2] Toyohashi University of Technology, Toyohashi, Japan
[3] Tongji Architectural Design (Group) Co., LTD, Shanghai, China
[4] Shanghai University, Shanghai, China

Abstract. More and more seismically isolated buildings have been constructed in the world. One of the reasons for the dramatic increase is the stipulation of design standards. In October 2021, "ISO/FDIS 23618-2022 Bases for design of structures - General principles for seismically isolated structures" was approved.

In this paper, seismic isolation design and devices testing standards of Japan, China, the United States of America and Eurocode, are compared using the same building model. For seismic loads, design response spectra were calculated and compared. The locations of the buildings were assumed to be Tokyo, Beijing, San Francisco and Reggio Calabria, Italy, respectively, in order to take into account the seismicity.

The building model is a 7-story RC building supported by lead rubber bearings. In each analysis case, the weight, height, super-structure and ground conditions of the building were all the same. In Japan, China and Eurocode, the same LRB devices were used. However, the seismic load in USA was quite large, larger size devices were used. Both the equivalent linearization method and the time history response analysis method were used for the response analysis of the building. The response analysis results such as the response displacement of the seismic isolation system, the shear force coefficients of the super-structure, and the story drift were used as comparison parameters.

Keywords: Seismic isolation · Design code · Equivalent linear method · Time history analysis · ISO 23618

1 Introduction

In October 2021, "ISO/FDIS 23618-2022 Bases for design of structures - General principles for seismically isolated structures" was approved. In this paper, detailed design procedures of Japan (MLIT Notification No. 2009), China (GB/T 51408-2021), USA (ASCE 7-16) and Eurocode (EC8) are compared to propose a common design procedure.

Seismic isolation design codes are introduced first. Seismic load, analysis methods, major load combinations and device testing methods are compared. A 7-story RC building model is used to demonstrate the design procedures. Analysis results of equivalent linear method and time history analysis method are summarized at last.

G. P. Cimellaro (Ed.): WCSI 2022, LNCE 309, pp. 317–325, 2023.
https://doi.org/10.1007/978-3-031-21187-4_26

2 Seismic Isolation Design Codes

The seismic load, analysis methods and test methods for seismic isolation devices of Japan, China, the United States of America and the Eurocode are summarized in Table 1.

Table 1. Concept of the seismic isolation design codes.

Contents		Japan	China	USA	EC8
Seismic load	Hor. spectrum	Defined			
	Ver. spectrum	0.3 g	0.2–0.4 g	defined	
Analysis method	Lateral force		○		○
	ELM	○		○	○
	Spectrum		○	○	○
	THA	○	○	○	○
Major load combinations	Gravity	D + L	1.3D + 1.5L	1.2D + 1.6L 1.4D	1.35D + 1.5L
	Seismic	D + L ± E	1.0D + 0.5L + $1.0E_h$ + $0.4E_V$	1.2D + 0.5L + 1.0E	1.0D + 0.3L ± 1.0E
			0.9D−$1.0E_h$−$0.5E_V$	0.9D−1.0E	
Devices testing method	Prototype	Factory	○	○	Factory
	Production	○	○	○	○

D: dead load; L: live load; $E_{h, V}$: horizontal or vertical seismic load

In each Code, the horizontal design spectrum is defined as the seismic load. The vertical design spectrum is defined in the USA and Eurocodes, while in Japanese and Chinese Codes, the vertical response coefficient is defined. For the analysis method, the static method (lateral force or equivalent linearization method) and the dynamic method (response spectrum analysis or time history analysis method) are defined.

The limit state design method is adopted in China, USA and EC8, while the allowable stress design method is adopted in Japanese code. The load combinations usually give impacts to the design of isolation devices. In Chinese and USA Codes, the vertical compression and tension load combinations are different, where the tension load design may be critical than other Codes. There are many mechanical tension-resistant devices applied in China and USA, while there are few applications in Japan.

The quality control of the isolation devices is very important. Strict prototype test method is stipulated in all Codes. In Japanese code and EN 15129, factory surveillance is conducted periodically. For the production control, 100% products are tested in Japan and USA, while a percentage is allowed in China and EN 15129.

ULS seismic load and response of the super-structure defined in the Codes are summarized in Table 2. In Japanese Code, seismic loads corresponding 500-year return period are used. Drift value is limited to 1/300 in ELM corresponding with 1st stage design of seismic design and about 1/150 or less in THA. In Chinese Code, 475-year load is used for the design of the super-structure and substructure, while 2475-year load is used to check both the super-structure and isolation devices. In addition, a maximum 10,000-year load is also defined to check both the super-structure and isolation devices. In USA, from ASCE 7-16, a MCE event having 1% probability of building collapse in 50 years is used to design both the super-structure and isolation system. The EC8 uses 475-year load and there is no requirement for drift response of the super-structure. In USA and EC8, response modification coefficient is adopted so that the model of the super-structure is usually treated as elastic. Except Chinese Code, the bound properties of the isolation system are introduced to obtain the response results. The response deformation of rubber bearings is usually limited to about 250% shear strain in practice.

Table 2. ULS seismic load and response of the super-structure defined in the codes.

Contents		Japan	China	USA	EC8
Return period (year)	ULS	500[a]	475 1600–2475	2475[a]	475
Model of super-structure		Non-linear	Non-linear	Linear (response mod. coef. R_I)	Linear (behaviour factor q)
bound properties		○		○	○
Deformation of RB (%)	ULS	267[b]	min(300, 0.55D)	250[b]	250[b]
Drift (RC frame)	ULS	1/150–1/300[b]	1/400 1/100	1/67	

a: estimated; b: in practice

3 Design Examples

3.1 Analysis Model

The 7-story RC building model presented by Saito et al. (2011) is used to demonstrate the design procedures. The super-structure is slightly modified as shown in Feng (2022), where the fundamental periods of the fixed-base model are Tx = 0.564, 0.190, 0.107 s (frame direction) and Ty = 0.238, 0.105, 0.087 s (shear wall direction), respectively.

LRB devices are selected as they have appropriate restoring force and damping capacity. Diameters from 650 to 750 mm were used in Japan, China and EC8 design model. Diameter of 900 mm was used in USA design model due to the large value of the MCE_R seismic load. The nominal design properties of the isolation system are summarized in Table 3. In USA design model, large amount of dampers was needed to resist the deformation of the isolation system.

Table 3. Nominal design properties of the isolation system.

Item	Symbol	Unit	Japan, China, EC8	USA
Mass	M	Ton	3,555	
Yielding load of lead plug	Q_d	kN	1,092	2,780
Ratio	Qd/W	%	3.1	8.0
Initial stiffness	K_1	kN/m	137,806	199,068
Post-elastic stiffness	K_2	kN/m	10,600	15,313
Vertical stiffness	K_V	kN/mm	34,502	49,536

3.2 Seismic Load

To consider the seismic region coefficients, the target construction sites are assumed to be in Tokyo, Beijing, San Fransisco and Reggio Calabria respectively. A fixed soil profile is assumed in all cases, where the average shear wave velocity within the top 30 m is about 209 m/s (Feng 2006). Typically, seismically isolated buildings should be located on relatively stiff ground. In the Japanese code, an iterative procedure was used to calculate the site amplification coefficient, rather than using the amplification coefficients defined in the Code.

The 5% code acceleration spectra and pseudo velocity spectra were calculated to compare the seismic load level at each location shown in Fig. 1. Typically, a seismically isolated building usually has about 20% critical damping in ULS earthquakes shown in Fig. 2. USA code gave the largest spectra, almost 1.5 times larger than Japan. The pseudo velocity spectra increased with the period in Chinese code while the pseudo velocity spectra remained constant or decreased in other Codes.

3.3 Analysis Results

ELM. The analysis results by the equivalent linear method (ELM) are summarized in Table 4. Although an equivalent linear analysis based on a single-degree-of-freedom (SDOF) system is defined in all codes, there are different limitations on the applicability of the method as summarized by Feng (2006, 2022). Bound properties of the isolation system were determined considering variability and uncertainty in production, temperature and ageing properties *etc.* In Chinese code, response results by both 475 and 2475-year seismic load were calculated without considering boundary properties. An equivalent mass of 85% is introduced in Chinese code.

The convergence procedure shown by Feng (2006) was used to obtain the response results. The shear strain of LRB and the equivalent damping ratio obtained by LBDP

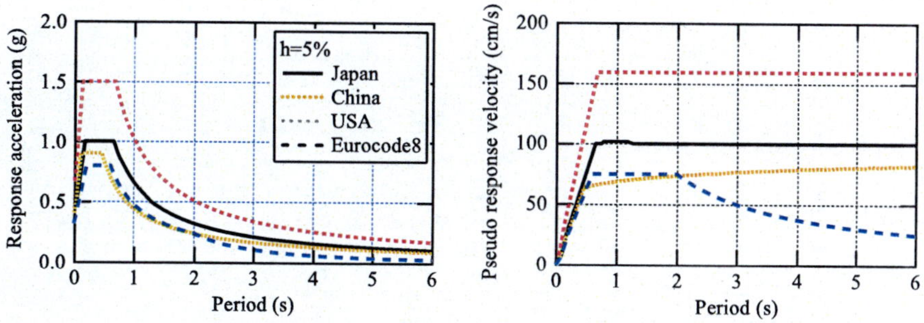

Fig. 1. 5% damping acceleration (left) and pseudo velocity (right) response spectra.

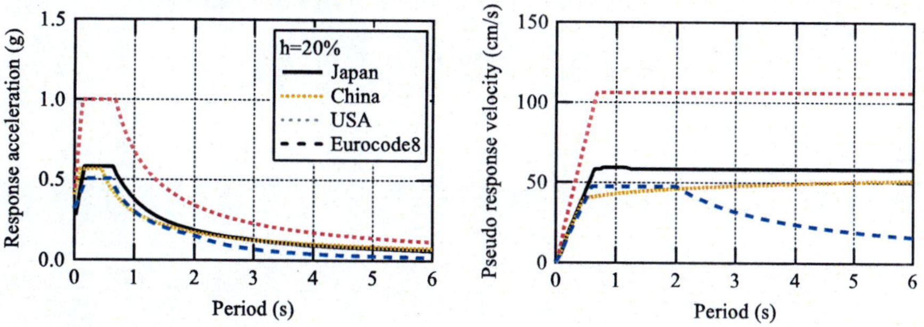

Fig. 2. 20% damping acceleration (left) and pseudo velocity (right) response spectra.

were 278%, 11% in Japanese Code and 270%, 14% in USA Code, respectively. The base shear force coefficient of the super-structure obtained by UBDP was 0.149 in Japanese Code and 0.194 in USA Code. The vertical distributions of the shear force are shown in Fig. 3 comparing with the response results based on THA. In Japanese, USA and EC8 Code, the vertical response was obtained as 0.3 g, $0.2S_{MS} = 0.3$ g and $a_{vg}(3\tau) = 0.75$ g respectively. The moat/seismic gap was obtained considering eccentricity of the isolation system and safety factor *etc.*, which was 0.688 m in Japanese Code and 0.633m in USA Code. In EC8, combination of two horizontal direction was considered to obtain the response results. Due to the limitation of the paper, the restrictive requirements check of ELM can be found in Feng (2022).

THA. Time history analysis method (THA) is mostly used in all codes. In Japanese code, the maximum response values from six pairs ground motions were adopted, while in the other codes, the average response values from ten pairs ground motions were adopted. All ground motions were compatible with the 5% design spectra shown in Fig. 1. In USA and EC8 Code, combination of two horizontal direction was considered. In Japanese, USA and EC8, vertical spectra were calculated, which were used in the vertical response analysis based on the response spectrum analysis (RSA).

A 3D frame model was used in this study for both RSA and THA. In the horizontal time history response analysis, the super-structure was assumed as non-linear in Japanese and Chinese Codes, while elastic in USA and EC8 Codes. LRB devices were idealized

Table 4. Response results by ELM (equivalent linear method).

Item	Symbol	Unit	Japan		China		USA		EC8	
			UBDP	LBDP	475 yr	2475 yr	UBDP	LBDP	UBDP	LBDP
Effective mass	M	Ton	3,555		3,022		3,555		3,555	
Initial stiffness	K_1	kN/m	173,084	120,305	137,806		282,417	175,533	184,887	114,643
Post-elastic stiffness	K_2	kN/m	13,314	9,254	10,600		21,724	13,503	14,222	8,819
Yielding load of lead	Q_d	kN	1,411	891	1,092		4,688	2,157	1,575	810
Response disp. of IS	δ_r	m	0.283	0.444	0.080	0.268	0.310	0.541	0.133	0.140
Response disp. THA				0.378		0.194		0.270		0.144
Shear strain		%		278		167		270		88
Equivalent period	T_e	s	2.769	3.531	2.215	2.851	1.952	2.833	2.322	3.101
Eq. Damping ratio	ξ		0.168	0.111	0.320	0.171	0.246	0.142	0.269	0.238
Spec. Reduction coef	F_h		0.640	0.793	0.550	0.658	0.628	0.755	0.560	0.589
Response at top of sub-structure	V_b	kN	5,179		1,928	3,934	11,424		3,624	
Response at bottom of super-structure		kN					10,722			
Response mod. Coef	R_I, q						$R_I = 1.875$		$q = 1$	
Design base shear force	$V_s = V_{st}/R_I$	kN	5,179		1,926	3,934	5,719		3,624	
Shear coef. of IS	C_{r0}		0.149		0.055	0.113	0.328		0.104	
Shear coef. of SS	C_{r1}		**0.153**		**0.064**	**0.130**	**0.194**		**0.104**	

(continued)

Table 4. *(continued)*

Item	Symbol	Unit	Japan		China		USA		EC8	
			UBDP	LBDP	475 yr	2475 yr	UBDP	LBDP	UBDP	LBDP
Vertical response		g	0.3		–	–	0.3		0.75	
Eccentricity coef				1.1				1.17		1.012
Safety factor				(+ 0.2 m)		1.2				1.2
Seismic gap		(m)		**0.688**		**0.322**		**0.633**		**0.170**

as a bilinear model. In the horizontal response analysis, the structural damping was set to Rayleigh type, the damping ratio of the seismic isolation system was set to 0, and the damping ratio corresponding to the 1st and 2nd natural periods of the super-structure was set to 3%. The THA analysis tool used was SERA3D Ver10.8 by Saito (2021). RSA was conducted for Chinese Code by PKPM which is the 1st commercial program with the convergency procedure. In the vertical response analysis, RSA was used. Since the vertical rigidity of the seismic isolation system is hard, the vibration mode of the beam was remarkable. In the vertical response analysis, the structural damping was set to Rayleigh type too. The damping ratios corresponding to the 1st and 2nd vertical natural periods were both set to 3%. ETABS V18 was used as the analysis tool.

In Fig. 3, response results based on THA compared with ELM, RSA and seismic design are shown together. Deformation of the isolation system are shown in Table 4. Shear force coefficient comparing with the seismic design and drift comparing with the Code limitation are shown in Fig. 3.

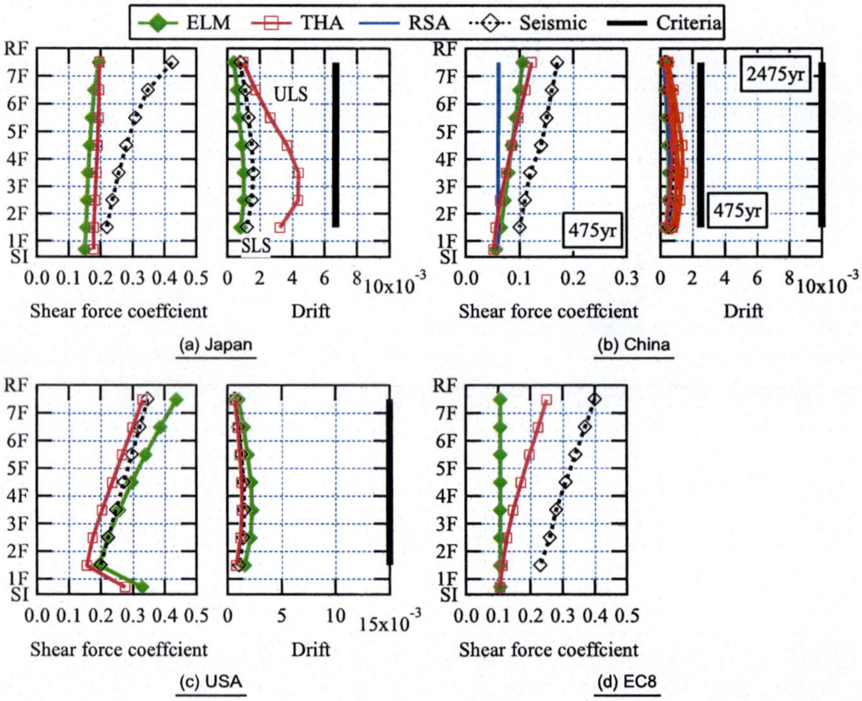

Fig. 3. Response results based on THA compared with ELM, RSA and seismic design.

The main findings are shown below.

- In the results of Japanese Code, the drift of the seismic isolation design by ULS was larger than the drift of seismic design by SLS.
- In Japan, ELM and THA can be selected independently. In practice, the ratio between ELM and THA is about 2:8. In this design case, shear force due to KOBE NS phase

(near filed ground motion) were the maximum and larger than ELM. The deformation of isolation system based on ELM was larger.

- In China, the 475-year seismic load is used to design the super-structure based on RSA. Since the drift of the super-structure due to 2475-year seismic load has to be checked, THA is used together with the spectrum method. The maximum response values between RSA and THA are taken as design load.
- In USA, the response results based on THA has a limitation from the ELM. The shear force based on ELM was slightly larger than the seismic design, which is only case in this study. Different response reduction coefficients ($R_I = 1.875$ for SI, $R = 5$ for normal seismic design) were main reason.

4 Conclusions

Detailed design procedures of seismically isolated buildings were compared to propose a common one based on Japanese, Chinese, USA and EC8 Codes. The seismic load, analysis methods and test methods for seismic isolation devices were summarized first. A 7-story RC building model was used to demonstrate the design procedures. Response results based on the equivalent linear method and time history analysis method were compared.

References

ASCE: Minimum Design Loads for Buildings and Other Structures, SEI/ASCE 7-16. American Society of Civil Engineers, Reston, VA (2016)

BCJ: The Building Standard Law of Japan on CD-ROM (2016)

European Standard: EN 15129:2018, Anti-seismic devices (2018)

European Standard: Eurocode 8: Design of structures for earthquake resistance—Part 1: General rules, seismic actions and rules for buildings (2004)

Feng, D., et al.: A Comparative study of seismic isolation codes worldwide, 1st ECEES (2006)

Feng, D., et al.: Certification system of seismic isolation devices in Japan, 16WCSI (2019)

Feng, D.: Seismically isolated buildings, Appendix A, pp. 330–371. Ohmusha (2022) (in Japanese)

Ministry of Construction, China: Standard for seismic isolation design of building GB/T 51408-2021 (2021) (in Chinese)

Saito, T.: STERA 3D (Structural earthquake response analysis 3D). rc.ace.tut.ac.jp/saito/software.html (2021)

Saito, T., Hiramatsu, M., and Kani, H.: Design procedure of seismically isolated buildings in Japan, 12WCSI (2011)

Statistical Analysis of Rubber Compounds Material Tests for Seismic Isolation Bearings and Code Provisions Comparison

Fabio Micozzi[1], Andrea Dall'Asta[1], Laura Gioiella[1], Laura Ragni[2(✉)], and Virginio Quaglini[3]

[1] University of Camerino, Ascoli Piceno, Italy
fabio.micozzi@unicam.it
[2] Università Politecnica Delle Marche, Ancona, Italy
laura.ragni@univpm.it
[3] Politecnico Di Milano, Milan, Italy

Abstract. Rubber compounds are extensively used for manufacturing rubber bearings for seismic isolation. Current seismic codes prescribe the use of design properties of seismic isolators derived from experimental evaluations to account for the effects of strain rate representative of the design seismic condition, cyclic degradation, frequency and environmental conditions (especially the temperature variation). Moreover, ageing phenomena, which such devices undergo during their service life, should be taken into account. Such tests should be carried out on devices, even though the European Standard on seismic devices (EN 15129) allows for both high and low damping rubber bearings the use of data from material tests performed on the elastomer used in their manufacture, (although extrapolation the results of materials tests to the scale of isolation devices is not always straightforward). In this study a set of experimental data obtained from type tests performed by different manufacturers on both low and high damping compounds with nominal stiffness ranging from 0.4 MPa to 1.3 MPa is statistically analysed. The effects of shear deformation, cyclic degradation, ageing, frequency and temperature on equivalent shear stiffness and equivalent damping are shown, providing also a view of the intra-supplier variability and inter-supplier variability. As general modification factors, also known as λ-factors, are provided by the codes in case of unavailability of experimental data, a comparison between the codes values and the obtained experimental results is shown.

Keywords: Upper-lower bound analysis · Rubber bearing · Seismic isolation

1 Introduction

Current seismic codes prescribe the use of design properties of seismic isolators determined from standardized tests performed under specified conditions to account for the change of mechanical parameters during the lifetime of the structure due to different effects. With reference to elastomeric devices, the main effects are the variation of

G. P. Cimellaro (Ed.): WCSI 2022, LNCE 309, pp. 326–336, 2023.
https://doi.org/10.1007/978-3-031-21187-4_27

mechanical properties due to the loading history (especially maximum shear deformation and number of cycles), ageing and environmental conditions (mainly temperature variation). Moreover, there are other source of variability (such as the production variability or the frequency content of the earthquake) also influencing the final response of the isolation system. Some sources of variability can be taken into account by using accurate models [1, 2] or simplified models in combination with modification parameters, to account for the full range of variation. The latter approach was introduced first by Constantinou et al. [3] and then implemented into several seismic codes, such as the AASHTO guide specifications, the European code on bridges (EN 1998—part 2 [4]) and the European Standard on seismic devices (EN 15129 [5]). This approach is based on property modification factors (also called λ-factors) that represent the ratio between the maximum or minimum value of mechanical parameters with respect to nominal values, defined as the unaged 3th cycle properties, according to [5] at the reference strain amplitude, frequency and temperature. Such properties are the equivalent linear shear stiffness G and equivalent viscous damping coefficient ξ or the post elastic stiffness K_p and the force at zero displacement F_o, for the bilinear models. All the above-mentioned codes suggest default values of λ-factors, but they also specify that it is preferred to derive these values directly from tests carried out on the devices during the qualification process by the manufacturers. However, in the European context, for elastomeric isolators made of High Damping Rubber (HDR) and Low Damping Rubber (LDR) the EN 15129 [5] allows to perform material tests on the elastomer utilized for the bearing manufacturing instead of full-scale or reduced-scale tests on the isolation bearings to define the variation of the parameters related to strain amplitude, frequency, temperature, repeated cycle and ageing. This study statistically analyses a dataset provided by the Politecnico di Milano composed by 18 rubber compounds (7 LDR and 11 HDR) belonging to 5 different manufacturers, with a nominal shear stiffness (G_O) in the range 0.4–1.3 MPa and a nominal damping (ξ_O) in the range 0.034–0.167. The data refer to the results of Type Testing of elastomer for isolators according to the EN 15129 clause 8.2.2 [5], consequently they are already averaged results of a set of 3 samples for each test [6].

In particular, in the first part of the paper the statistical results are showed for each test, focusing on the results in terms of variability of the ratio between the equivalent linear shear stiffness G and the equivalent viscous damping coefficient ξ measured in the test and the nominal or reference values by highlighting median, interquartile and maximum-minimum value. Results have been analysed separately for HDR and LDR specimens and for the different manufacturers. Moreover, by grouping all the test data at the nominal conditions, also the production variability has been analysed.

The hypothesis that results of materials tests can be automatically extended to isolation devices is not always appropriate or straightforward because the device dimension and the vulcanization process used to fabricate the isolator can influence significantly the final cyclic behaviour of the bearing as well as the ageing phenomena. In particular, for some effects the variability increases, when moving to a full-scale device, with respect to the results obtained through material test, i.e. the loading history effect or the production variability. On the other hand, the temperature or ageing effects measured by material tests could be overestimated with respect to full-scale bearing. Moreover, other source of variability known in literature (i.e. [7]) are only depictable from bearing tests. However, the trend associated to the rubber compound is usually related to the final full-scale bearing behaviour. For this reason, in the second part of the paper, the obtained results are compared with λ-factors provided by [5], with comments on how such modification factors are currently provided.

2 Test Results

Data analysed in this study includes cyclic tests carried out on elastomeric specimens at different strain amplitudes, frequencies, temperatures, ageing conditions and number of cycle. Results are analysed in terms of ratio between experimental values (G and ξ) and nominal values (G_0 and ξ_0). According to the standard, experimental values are measured at the 3rd cycle of each test (unless the code specifies a different cycle), while the nominal value is conventionally defined as the property evaluated in reference conditions: shear deformation $\gamma = 1$, temperature T = 23 °C and frequency f = 0.5 Hz.

The statistical outcomes are presented through boxplot, where the box shows the 25th–75th percentiles of the samples whereas the internal line is the sample median. The extended segments instead show the minimum-maximum values range, except for the data considered outliers (values more than 1.5 interquartile range away from the box limits). Where the data are too much to show as boxplots, a simpler median line with all the data in the background have been used to improve the readability of the figures.

2.1 Effect of Strain Amplitude

The strain amplitude influences the mechanical behaviour of the rubber [8, 9], as shown in Fig. 1, where the variation of stiffness and damping ratios are related to the maximum shear strain. The effect on the stiffness for shear strains lower than the reference one ($\gamma = 1$) is higher for HDR than for LDR. Both of them, instead, show median values close to one up to $\gamma = 2.5$, suggesting an almost constant behaviour for deformation larger than $\gamma = 1$, which is the common range of the design strains. For what concerns damping, the variability is almost linear for HDR with a median value ranging from 1.3 up to 0.75 by increasing shear strain. It is worth to observe that, although LDR has a similar trend, the damping ratio increases towards values higher than 1 for values of γ higher than 2. In [5] there is no limit for the amplitude dependence of the mechanical parameters, but these figures show that the variability between different compound is reduced especially for $\gamma > 1$.

2.2 Effect of Repeated Cycles

The results of G and ξ measured at each cycle, with respect to the values recorded at the 2nd cycle, are reported in Fig. 2 (a, b). All the compounds are in compliance with the limit of 0.7 prescribed in [5] for the ratios recorded between the 2nd and the 10th cycle. Moreover, it is worth to note that all the values are higher than 0.85, with a median even higher than 0.9, suggesting that the cyclic degradation of rubber characteristics have been reduced during time by the manufacturers and it is currently a marginal source of

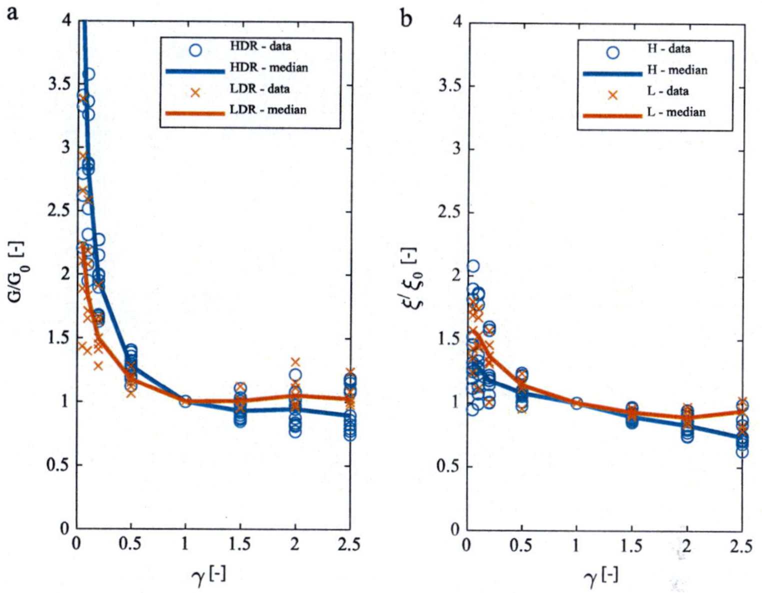

Fig. 1. Effect of deformation amplitude for (a) shear modulus; (b) damping ratio.

variability of the material. Also, the limit of 0.6 for the ratio between the 1st and the 10th cycle for G (Fig. 2c) is largely satisfied (the lower value is 0.71, while the median value is 0.83). There are no limits prescribed by the code for the variation of the damping ratio between the 1st and the 10th cycle, nevertheless the related results are reported in Fig. 2d.

2.3 Effect of Frequency

The effect of frequency on the response (again in terms of G and ξ, normalized to unity with respect to the results achieved at 0.5 Hz) is limited by the code [5] within a range of 0.8–1.2 for frequencies spanning from 0.1 to 2 Hz. Figure 3 (a, b) shows that all the values are in compliance with the code limits. Regarding the shear modulus ratio for HDR, the values are close to 1 for a frequency of 0.1 Hz, i.e. negligible effect, while the values are slightly higher for a frequency of 2 Hz (with a median value around 1.13). The effect of frequency for LDR is instead always negligible, as expected.

2.4 Effect of Ageing

The ageing effect on the rubber compound, according to [5], shall be estimated to be less than 20% over the expected service life of the isolator. The tests results are reported in Fig. 4 (a, b). The code allows to consider this condition satisfied for the bearings if the elastomer material satisfies the requirements under the standard ageing conditions. In the Authors' opinion, even if the accelerated aging test performed on rubber material is substantially different respect to the real ageing process of bearings, it is expected that the real ageing effect should be lower than the one tested on material. For example external surface of the elastomeric bearings are normally fabricated with a layer of cover rubber

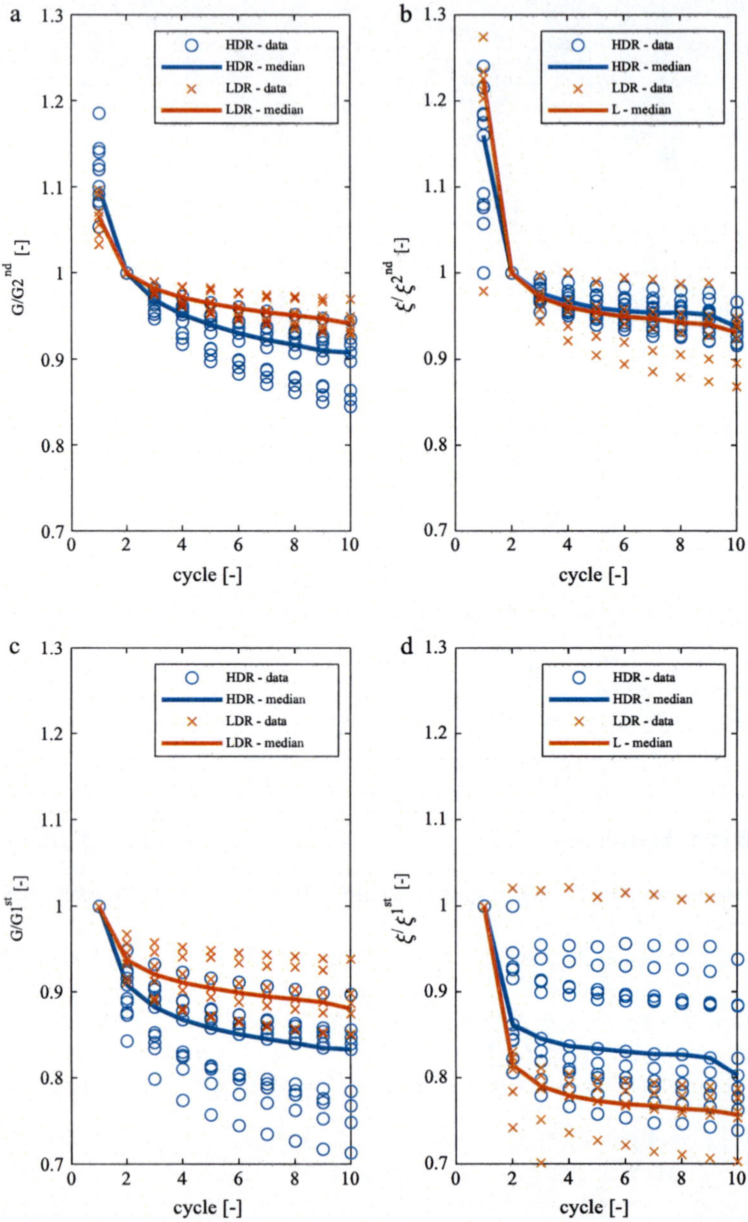

Fig. 2. Cyclic effect for (a, c) shear modulus; (b, d) damping ratio with respect to the (a, b) second and (c, d) first cycle.

that includes anti-oxidants to protect the core from significant infiltration by oxygen and ozone [10, 11]. Moreover, this external surface reduces its porosity with ageing, leading to a less and less oxidation from the surface to a critical depth [12]. Thus, material tests provide an upper (and conservative) limit for the real ageing effect [13].

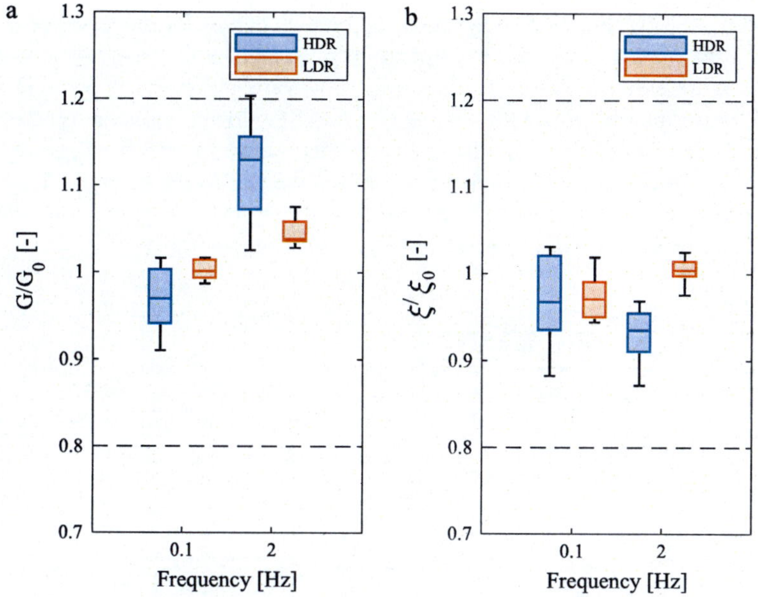

Fig. 3. Frequency effects for (a) shear modulus; (b) damping ratio.

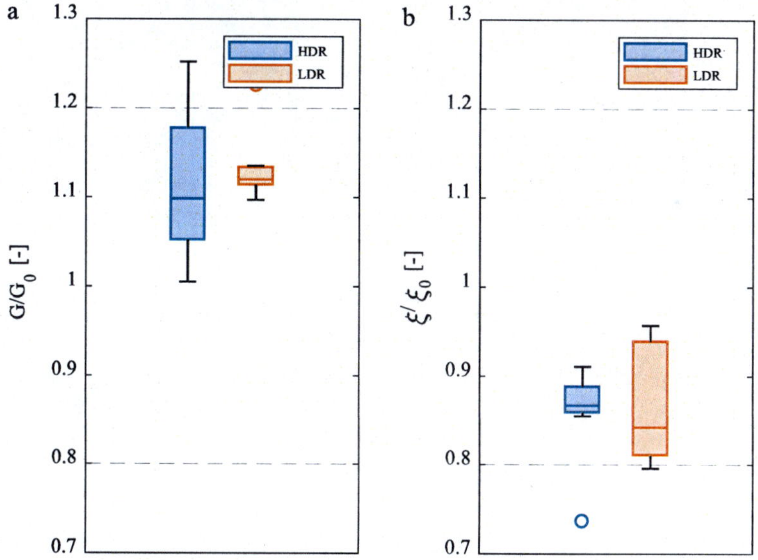

Fig. 4. Ageing effects for (a) shear modulus; (b) damping ratio.

2.5 Effect of Temperature

As for the ageing effect, also the tests for the temperature effect on rubber bearings, according to [5], may be substituted by material tests. The values at the lowest temperature shall not differ by more than + 80% or −20% from the corresponding values at 23 °C, and the values at the highest temperature shall not differ by more than ± 20%

from those at 23 °C. Figure 5 shows data and median curves for the tests available in a range of −40 + 50 °C, even if the temperature suggested by the code are between −20 + 40 °C. With reference to Fig. 5, the results shown are not always in compliance with the code prescriptions, especially for the *G* ratio at the lower temperature; the median curve for HDR, indeed, is inside the code limit only down to −15 °C. Looking at each value, most of the data are inside the code limit in the temperature range −10 + 40 °C, that is actually enough for most of the applications (except for bearings used in bridges in very cold regions, which can experience very low temperatures).

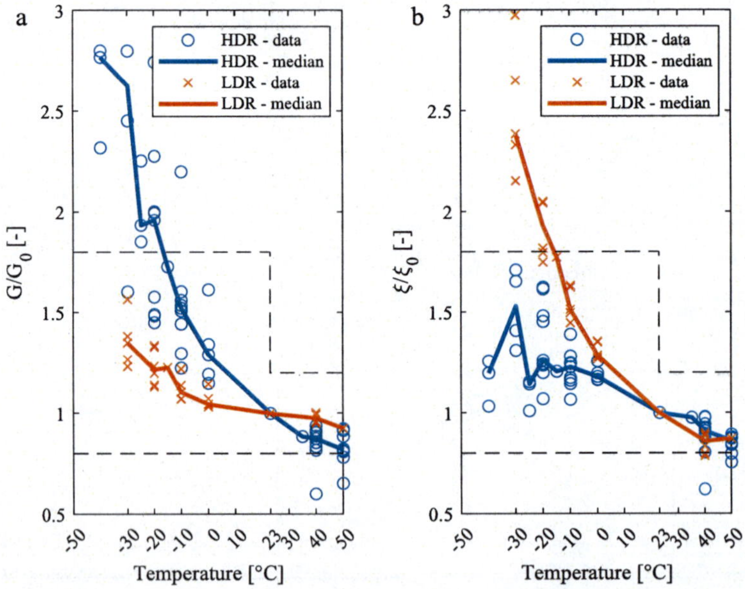

Fig. 5. Temperature effect for (a) shear modulus; (b) damping ratio.

2.6 Production Variability

As shown in the previous chapter, all the results are shown in terms of ratio to the nominal value in reference conditions ($\gamma = 1$, temperature T = 23 °C and frequency f = 0.5 Hz). Consequently, for each test a set of values refers to the nominal condition. Grouping this data, a larger set can be collected to properly analyse the production variability on each rubber compound. Figure 6 shows the variability of the ratio between the measured values of *G* and ξ with respect to the nominal ones declared by the manufacturers, for both the HDR and LDR compounds. All data of the shear stiffness (Fig. 6a) are in the range prescribed by the code (0.8–1.2). Moreover, most of the data (25th–75th percentiles) are close to 1, especially for the damping ratio, suggesting that usually the suppliers have a good control of the production process and real values match the nominal ones. The large number of outliers for the LDR damping is justified as the damping for this compound is not controlled (usually it is around 0.05). This is in compliance with the results on test performed also on bearings [14] where a statistical analysis has been performed to define the overall production variability and also the one within and between batch variability.

To have an insight on these data, showing them for each manufacturer, it is possible to make the same evaluation but enabling the analysis of intra-manufacturer variability and inter-manufacturer variability. The variation range is often smaller respect the whole data, as shown in Fig. 6c, d, confirming that the individual manufacturer has a specific compound, that is able to control.

3 Code Comparison of λ-Factors

Starting from the λ-factors provided by the EN 15129 (annex J) [5] and the EN 1998—part 2 (annex J) [4] in terms of K_p and F_0 it is possible to define the equivalent values in terms of G_0 and ξ_0 through simple formulas reported in ISO 22762 [15]. In this way it is possible to compare directly the λ-factors provided by the code and the mean value obtained by the data previously shown. Moreover, to deal with the variability of the data and the low number of samples, the 90% confidence interval is also reported (using the Student's t-distribution). The comparison is reported in Fig. 7, for G (a, c) and ξ (b, d) related to LDR (a, b) and HDR (c, d). In each tile the values for ageing and two on three temperature (0 and −10 °C) considered by the code (for HDR with $\xi_0 \leq 15\%$) are reported. The −30 °C temperature is not considered in this comparison as most of the data for that temperature are not in compliance with the limit prescribed by EN 15129 (see Subsect. 2.5) and therefore the associated λ-factor is not available for the design. It is worth to note that in [5] HDRs compounds are divided in two categories based on the nominal ξ_0. In the comparison only the λ-factors related to $\xi_0 \leq 15\%$ are reported as only one compound (with a damping of 16.7%) out of 11 exceeds this limit. Regarding LDR (Fig. 7 a, b), the values suggested by the code are in agreement with the data obtained by the tests regarding the shear stiffness, while the λ-factors obtained for the damping are different. However, the damping of LDR is negligible and usually not considered in the design process. Moreover, the underestimation of the damping for low temperatures should be considered on the safety side and so as an extra source of reliability of the system.

On the contrary, for HDR (Fig. 7 c, d) the tests show an underestimation of the temperature effects on the shear stiffness. As mentioned before [5] allows to extend the material tests at the bearing scale but, given the data obtained, specific full-scale tests for low temperature are strongly recommended in the case of seismic isolation in cold region with a design temperature lower than 0 °C. The ageing λ-factor for HDR damping provided by the code is 1, while the data show a reduction of the damping capacity around 0.85. This effect, even if somehow intuitive is not recognized by the code and should be better investigated. On the contrary, as a higher damping could be considered a positive effect on the seismic response, the λ-factor for temperature effects on ξ may be acceptable as quite close to the ones provided by the code and on the safety side.

The implementation of a direct λ-factor format in terms of G and ξ is highly recommended because the equivalent linear parameters are commonly used by the designer to define the isolation system and it is possible to obtain a good linear approximation of the seismic response even for HDR [16]. On the contrary, bilinear models are no more commonly used because new models, also implemented in commercial software [17] can be easily modified using the equivalent linear λ-factors.

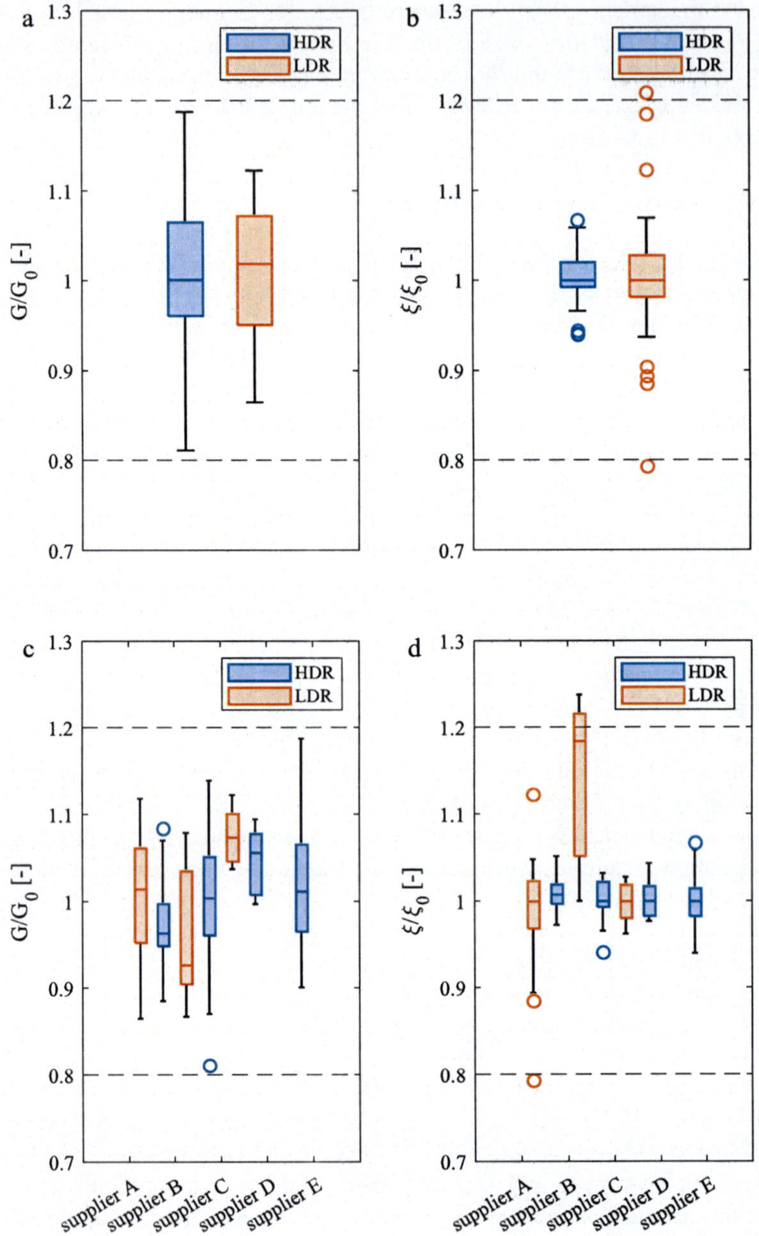

Fig. 6. Production variability for (a, c) shear modulus; (b, d) damping ratio for (a, b) overall data and (c, d) each supplier.

In general, it is important for the structural engineering which design the base-isolation system, that λ-factors should be provided by the manufactures and tuned based on their current production. This is import for all the effects but essential for some effects, such as the low temperature or the repeated cycling effects. To this purpose,

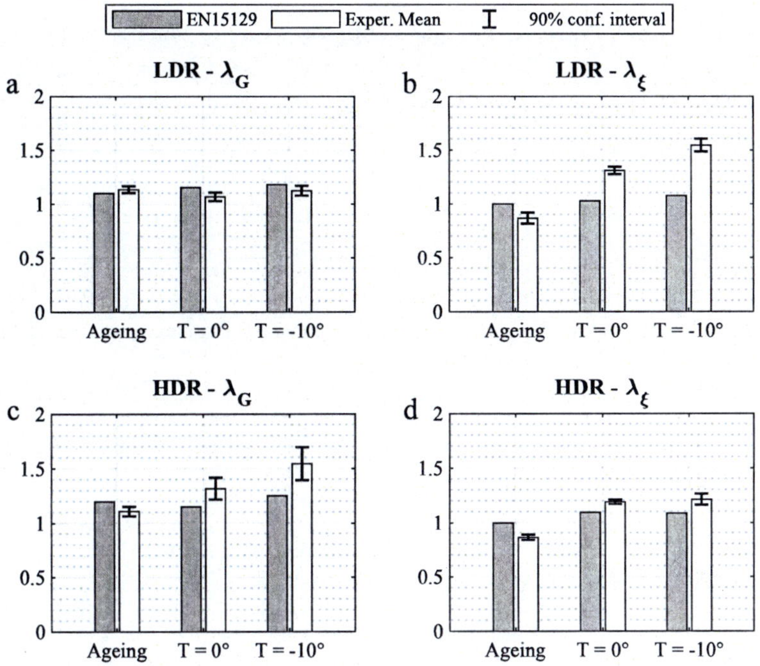

Fig. 7. λ-factors comparison between code values and data.

the manufacturers should provide material (when significant) or bearing tests properly carried out during the qualification procedure. Also the codes (in particular the EN15129 [5] in the European contest) should be improved by requiring bearing tests (in addition to material tests) to properly define λ-factors.

4 Conclusion

This paper describes a statistical analysis of the material tests carried out according to EN 15129 [5] on a set of different rubber compounds from five different manufacturers. Based on the outcomes of the present study, the following conclusions can be drawn::

- Among the different sources of variability, temperature and ageing are quite important, while frequency and cyclic degradation seems to be of lower importance or even negligible;
- Production variability is overall inside the code limit of 0.8–1.2 and even lower for each individual manufacturer, however material tests do not allow to assess the effect of the production process of the elastomeric bearing, which could introduce important sources of variability. Therefore a comparison with tests performed on isolation devices is required to extend this result to the bearing scale and use it by the designers;
- Also for the others effects comparison between material and device tests should be performed in order to validate the results shown in this paper;

- Current λ-factors provided by the code EN 15129 [5] are related to the bilinear model while values in terms of equivalent linear parameter (stiffness and damping) should be also provided directly by the manufacturers.

References

1. Mazza, F.: Effects of the long-term behaviour of isolation devices on the seismic response of base-isolated buildings. Struct. Control. Health Monit. **26**(4), e2331 (2019). https://doi.org/10.1002/stc.2331
2. Ragni, L., Micozzi, F., Tubaldi, E., Dall'Asta, A.: Behaviour of structures isolated by HDNR bearings at design and service conditions. J. Earthq. Eng. (2022). https://doi.org/10.1080/13632469.2020.1776792
3. Constantinou, M., Tsopelas, P., Kasalanati, A., Wolff, E.: Property modification factors for seismic isolation bearings. Technical Report MCEER-99-0012. MCEER, NY (1999)
4. CEN EN 1998-2: Eurocode 8: design of structures for earthquake resistance—Part 2: Bridges. Brussels, Belgium (2005)
5. CEN EN 15129: Antiseismic Devices. Brussels, Belgium (2009)
6. Ferroni, D., Vazzana, G., Cuminetti, D., Quaglini, V., Dubini, P., Poggi, C.: Certification of anti-seismic devices according to the European Standard EN 15129:2009: Tasks for manufacturers and notified bodies. 15 WCEE, Lisboa (2012)
7. Quaglini, V., Dubini, P., Vazzana, G.: Experimental assessment of high damping rubber under combined compression and shear. J. Eng. Mater. Technol. **138**(1), 011002 (2015). https://doi.org/10.1115/1.4031427
8. Payne, A.R.: The dynamic properties of carbon black-loaded natural rubber vulcanizates. Part I. J. Appl. Polym. Sci. **6**(19), 57–63 (1962)
9. Tubaldi, E., Ragni, L., Dall'Asta, A., Ahmadi, H., Muhr, A.: Stress softening behaviour of HDNR bearings: modelling and influence on the seismic response of isolated structures. Earthq. Eng. Struct. Dyn. **46**(12), 2033–2054 (2017)
10. Constantinou, M.C., Whittaker, A.S., Kalpakidis, Y., Fenz, D.M., Warn, G.P.: Performance of seismic isolation hardware under service and seismic loading: MCEER-07-0012 (2007). http://mceer.buffalo.edu/publications/catalog/reports/Performance-of-Seismic-Isolation-Hardware-Under-Service-and-Seismic-Loading-MCEER-07-0012.html
11. CEN EN 1337-3: Structural bearings—Part 3: Elastomeric bearings (2005)
12. Itoh, Y., Gu, H.S.: Prediction of aging characteristics in natural rubber bearings used in bridges. J. Bridge Eng. **14**(2), 122–128 (2009). https://doi.org/10.1061/(ASCE)1084-0702(2009)14:2(122)
13. Paramashanti, K.Y., Itoh, Y., Kito, S., Muratani, K.: Experimental investigation of aging effect on damping ratio of high damping rubber bearing. J. Struct. Eng. **57A**, 769–779 (2011)
14. Micozzi, F., Ragni, L., Dall'Asta, A.: Statistical modelling of HDNR bearing properties variability for the seismic response of isolated structures. In: Proceedings 6th European Conference on Computational Mechanics (ECCM 6), 7th European Conference on Computational Fluid Dynamics (ECFD 7) 11–15 June 2018, vol. 12. Glasgow, UK (2018). https://congress.cimne.com/eccm_ecfd2018/admin/files/filePaper/p1754.pdf
15. ISO 22762-3: Elastomeric seismic-protection isolators. Part 3: applications for buildings—specifications. International Organization for Standardization (2018)
16. Dall'Asta, A., Ragni, L.: Dynamic systems with high damping rubber: nonlinear behaviour and linear approximation. Earthq. Eng. Struct. Dyn. **37**(13), 1511–1526 (2008). https://doi.org/10.1002/eqe.825
17. Masaki, N., Mori, T., Murota, N., Kasai, K.: Validation of hysteresis model of deformation-history integral type for high damping rubber bearings. In: Proceedings of the 16th World Conference on Earthquake Engineering, paper 4583. Santiago, Chile (2017)

Code Provisions About λ-Factors of HDRBs for the Upper and Lower Bound Analyses: Hystorical Review

Laura Ragni[1](✉), M. Gabriella Castellano[2], Andrea Dall'Asta[3], Laura Gioiella[3], Samuele Infanti[2], and Fabio Micozzi[3]

[1] Università Politecnica Delle Marche, Ancona, Italy
laura.ragni@univpm.it
[2] FIPMEC S.R.L, Padova, Italy
[3] University of Camerino, Camerino, Italy

Abstract. Most of the current seismic codes on seismic isolation prescribe upper and lower bound analyses to account for the variability of mechanical properties of seismic isolators. This procedure should be based on experimental data provided by manufacturers during qualification tests. When such data are not available, codes provide the so-called property modification factors or λ-factors for different kind of devices, giving maximum upper and lower bound coefficients to apply to nominal properties of bearings. Different sources of variability are considered, from the variability due to the bearing production to environmental or behavioural effects (such as temperature, contamination, ageing or repeated cycles). However, the origin of these values is not straightforward: some of them are based on wide experimental data, whereas others are based on few and old indications given by technical literature. The aim of this paper is to illustrate in detail such factors, especially for HDRBs (High Damping Rubber Bearings). In particular, it is shown that in the European context, both the EN 1998—part 2 (on bridges) and the EN 15129 (on anti-seismic devices), provide values deriving (with some transcription inconsistencies) from the second version of the AASHTO guide specification, which in turn proposes values derived from the MCEER-99-0012 report written in 1999 and never updated. The aim of this paper is to illustrate in detail such factors and to make the reader aware that most of factors derive from limited and/or old data, especially for HDRBs (High Damping Rubber Bearings).

Keywords: Seismic isolation · High damping rubber bearings · Elastomeric bearings · Property modification factors · Upper and lower bound analyses

1 Introduction

The design of seismically isolated structures and the prediction of their dynamic response are currently based on nominal properties of isolation bearings. However, in order to increase their reliability current seismic codes, such as EN 1998—part 2 [1], EN15129 [2], AASTHO [3] and ASCE 7-16 [4], suggest to account for some effects related to

G. P. Cimellaro (Ed.): WCSI 2022, LNCE 309, pp. 337–348, 2023.
https://doi.org/10.1007/978-3-031-21187-4_28

changes in the mechanical properties of the bearings, such as those due to temperature variation, ageing or loading history. All the codes use simplified approaches, based on upper and lower bound coefficients to apply to nominal properties, in order to perform upper and lower bound analyses. They also agree that values of such coefficients should be based on experimental data provided by the specific manufacturer during the qualification tests of the devices. Nevertheless, in case of unavailability of such data, codes provide the so-called property modification factors or λ-factors for different kind of devices that can be applied to nominal properties of bearings to obtain the bounding values. Nowadays, both in the national and international contest, few manufacturers make such information available directly from the product line-up catalogue, thus during the design phase (occurring before the qualification procedure) these data are often unknown. For this reason, the values of the modification factors proposed by the codes are often the only reference for structural engineers. Regarding the European context, both [1, 2] suggest values of the factors derived from the second edition of the AASHTO (dated 1999) [5] that have never been modified or updated in the subsequent editions of AASHTO: dated 2010 [6] and 2014 [3].

The aim of this paper is to illustrate in detail such factors and to make the reader aware that most of them derive from limited and/or old data that do not reflect the actual production of seismic devices, especially for HDRBs (High Damping Rubber Bearings). Consequently, such default values of λ-factors should be eliminated and replaced by updated values or directly by experimental values provided by the specific manufacturer.

2 Seismic European Standards: EN 1998 and EN 15129

Some suggestions about the variability of isolation bearings properties are given in EN1998-part 2 [1], dedicated to bridges. Sect. 7.5.2.4, indeed, specifies that nominal Design Properties (DP) of isolator units shall be evaluated according to EN15129 [2]. However, since nominal properties of isolator units (hence those of the isolating system) may vary due to environmental or behavioral effects (such as temperature, contamination, aging or repeated cycles), two sets of design properties of the isolating system, should be used in the analyses: Upper Bound Design Properties (UBDPs) and Lower Bound Design Properties (LBDPs). Thus, two analyses shall be performed: one using the UBDPs that generally leads to the maximum forces and the other using the LBDPs, which usually provides the maximum displacements. This variability should be established in accordance with Annex J of EN1998-part 2 [1], unless different values of the properties are derived by performing appropriate tests. Recommended values of λ-factors have to be applied to the following design properties: the post elastic stiffness (K_p) and the force at zero displacement (F_o). For elastomeric isolators recommended values for aging and temperature are reported in Tables 1 and 2, respectively, while unitary values are suggested for contamination and cumulative travel. Four isolation devices are considered: LDRBs (Low Damping Rubber Bearings), two kinds of HDRBs (HDRBs[1] with equivalent damping coefficient $\xi \leq 0.15$ and equivalent shear modulus at 100% of strain G > 0.5 MPa and HDRBs[2] with equivalent damping coefficient $\xi > 0.15$ and equivalent shear modulus at 100% of strain G > 0.5 MPa) and LDBs (Lead Rubber Bearings).

Table 1. λ-factors provided for aging by EN1998—part 2 [1]

Component	$\lambda_{max, f1}$	
	K_p	F_0
LDRBs	1.1	1.1
HDRBs[1]	1.2	1.2
HDRBs[2]	1.3	1.3
LRBs	–	1.0

[1] with equivalent damping coefficient $\xi \leq 0.15$ and equivalent shear modulus at 100% of strain G > 0.5 MPa
[2] with equivalent damping coefficient $\xi > 0.15$ and equivalent shear modulus at 100% of strain G > 0.5 MPa

Table 2. λ-factors provided for temperature by EN1998—part 2 [1]

Component	$\lambda_{max, f2}$					
Design temperature (°C)	K_p			F_0		
	LDRBs	HDRBs[1]	HDRBs[2]	LDRBs	HDRBs[1]	HDRBs[2]
20	1.0	1.0	1.0	1.0	1.0	1.0
0	1.3	1.3	1.3	1.1	1.1	1.2
−10	1.4	1.4	1.4	1.1	1.2	1.4
−20	1.5	2.0	2.5	1.3	1.4	2.0

[1] with equivalent damping coefficient $\xi \leq 0.15$ and equivalent shear modulus at 100% of strain G > 0.5 MPa
[2] with equivalent damping coefficient $\xi > 0.15$ and equivalent shear modulus at 100% of strain G > 0.5 MPa

Moreover, in order to account for the reduced probability of simultaneous occurrence of the maximum adverse effects of all factors, combination factors (ψ_{fi}) should be assumed in accordance with Table 3. The effective UBDP used in the design should be estimated as follows:

$$\text{UBDP} = max\ \text{DP}_{\text{nom}} \cdot \lambda_{\text{U,f1}} \cdot \lambda_{\text{U,f2}} \ldots \lambda_{\text{U,f5}} \tag{1a}$$

with modification factors $\lambda_{\text{U,f1}} = 1 + (\lambda_{\text{max,fi}} - 1)\psi_{fi}$. For the effective LBDP (and relevant modification factors $\lambda_{\text{L,fi}}$) a similar format should be used, in conjunction with $\lambda_{\text{min,fi}}$. However for the commonly used elastomeric and friction bearings, it may be assumed in general that $\lambda_{\text{min,fi}} = 1$and therefore:

$$\text{LBDP} = min\ \text{DP}_{\text{nom}} \tag{1b}$$

The European code on anti-seismic devices EN 15129 [2] suggests a similar approach, but it accounts also for the variation within the production supply. More in detail,

Table 3. Combination factors provided by EN1998—part 2 [1]

Importance class	ψfi
III	0.90
II	0.70
I	0.60

a production variability equal to ± 20% is considered (unless a lower variability has been agreed for the factory production control tests). Moreover, this code clearly specifies that UBDPs and LBDPs shall be determined from the type tests and, for elastomers bearings, only by considering the following variations: temperature (between upper and lower temperature, reported as T_U and T_L) and ageing. However, when such experimental values are not available, the Annex J of [2] provides reference factors that may be used for the estimation of the UBDPs and LBDPs. It is also worth to observe that these factors totally coincide with values reported in the Annex J of [1] (Tables 1 and 2).

In combining these three effects a combination factor of 0.7 is suggested, for the production variability and temperature variation, whereas a factor of 1.0 for the ageing variation. In any case, it is specified that for all elastomeric bearings the ratio between the upper and lower bound design property values shall be less than 1.8.

As described in detail in the next section, it is emphasized that the above illustrated values for λ-factors currently suggested by European codes have been taken from the second edition of AASTHO [5] which derives its data from the MCEER-99-0012 report [7]. This version of the AASTHO code is dated back 1999 and λ-factors have been never updated in its next editions of 2010 [6] and 2014 [3].

3 AASHTO Guide Specification for Isolated Building

3.1 Second Edition of AASTHO (1999)

With the overall objective of improving the reliability and longevity of base-isolated structures, the second edition of AASTHO [5] provided for the first time upper and lower bound values of properties of seismic isolation bearings to be used in the analysis and design of seismic isolated structures, particularly with reference to bridges. The bounding values of properties are determined by using property modification factors or λ-factors, which are proposed for several effects including ageing, contamination, travel, temperature and scragging for sliding and/or elastomeric isolators based on experimental results and on the understanding of the behaviour of seismic isolation bearings at the code drafting time. Essentially, these values are based on the MCEER-99-0012 report [7], written a few months earlier. It is worth to note that in this last document background information that led to the definition of λ-factor values for sliding bearings are widely presented, differently those related to elastomeric devices are only briefly reported. The Authors of the report themselves state that "*given the rather recent origin of the seismic isolation technology, there is a complete lack of data on aged isolation bearings. In the course of the research several laboratory and in field data are collected for sliding bearings, while very limited data are collected for elastomeric bearings*". Moreover, in section "5.6 System property modification factors for elastomeric bearings" the Authors also recognize that for elastomeric bearings λ-factors strongly depend on the elastomeric compound and that a variety of compounds is used for their manufacturing and experimental data are scarce. Thus, values presented in their document and assumed in [5] are primary based on a "*proposal of the Authors*". In the following subsections λ-factors for elastomeric systems that have to be applied to both the post-elastic stiffness (K_d) and the characteristic strength (Q_d) are reported and commented for each considered effect.

In details, for what concerns the aging effect, values of λ-factor are assumed (for both K_d and Q_d) equal to 1.1 for LDRBs, 1.2 for HDRBs with small differences (smaller than 25%) between scragged and unscragged properties and 1.3 for HDRBs with large differences (larger than 25%) between scragged and unscragged properties (Table 4). It can be observed that the values suggested by the European codes (Table 1) are in agreement with those proposed in Table 4 of [5], by assuming that HDRBs[1] and HDRBs[2] correspond to HDRBs with small and large differences between scragged and unscragged properties, respectively. However, the Authors of [7] affirm in the report itself that the first two values are based on limited data and the last one is based on the engineering judgement. More in details, the Authors cite some papers dealing with old NRBs not provided with sufficient anti-oxidant, differently from modern bearings. The most relevant information considered are those reported by Clark et al. [8], referring to some tests carried out on LDRBs and HDRBs in 1983 in conjunction with the construction of the Foothill Communities Law and Justice Center in California and to those carried out after 12 years of service on a pair of HDRBs removed from the building. Although, there is uncertainty in the interpretation of the results (because original tests and re-tests were done with different machine configurations and probably different rates of loading), the Authors conclude that the effective stiffness increase is about the 25%. On the basis of this study they conclude that, except for improperly cured bearings, the increase of mechanical properties due to aging is expected to be modest: in the order of 10% for LDRBs (or NRBs) and 20% for HDRBs (with small differences between scragged and unscragged properties) over a period of approximately 30 years. For what concerns the distinction between elastomeric isolators with small or large difference between unscragged and scragged properties, the Authors' opinion is that such large difference is typical of very high damping elastomeric isolators characterized by incomplete curing of the elastomer, which implies continuation of chemical processes in the rubber and therefore a potential for aging. According to this opinion, they advanced the proposal of increasing up to 30% the aging effect (over the same period of approximately 30 years) for HDRBs with large differences between scragged and unscragged properties. Finally, they explicitly criticize results coming from accelerated aging of rubber specimens that cannot provide, in their opinion, any substantial evidence of the long-term properties of in-situ elastomeric bearings.

For what concerns the temperature effect, values of λ-factor assumed for K_d and Q_d in [7] and reported in [5] (Table 5), are based on the results of Nakano et al. [9] and Kim et al. [10]. Data refer to LDRBs and HDRBs with effective damping less than 15% and with moderate difference between unscragged and scragged properties, whereas no data are available for HDRBs with effective damping larger than 15%. Thus, as outlined

Table 4. λ-factors provided for aging by AASTHO 1999 [5]

Component	$\lambda_{max,\,f1}$	
	K_d	Q_d
LDRB	1.1	1.1
HDRBs[a]	1.2	1.2
HDRBs[b]	1.3	1.3
LRB	–	1.0

[a] Small differences between scragged and unscragged properties
[b] Large differences between scragged and unscragged properties

by the Authors themselves, the difference between values of these two kinds of HDRBs (see Table 5) "*has been arbitrarily assumed*" by the task group writing the code. A further remark of the Authors refers to the duration of exposure to low temperatures. Unfortunately, all the studies analysed by the Authors on the effect of low temperature on the properties of elastomeric seismic isolation bearings neglect this aspect: some of them even do not report the duration of exposure or report an approximate exposure time. This is certainly a source of uncertainty affecting these data.

Table 5. λ-factors provided for temperature by AASTHO 1999 [5]

Component	$\lambda_{max,\,f2}$					
Design temperature (°C)	Q_p			K_d		
	LDRB	HDRB[a]	HDRBs[b]	LDRB	HDRB[a]	HDRB[b]
20	1.0	1.0	1.0	1.0	1.0	1.0
0	1.3	1.3	1.3	1.1	1.1	1.2
−10	1.4	1.4	1.4	1.1	1.2	1.4
−30	1.5	2.0	2.5	1.3	1.4	2.0

[a] Small differences between scragged and unscragged properties
[b] Large differences between scragged and unscragged properties

In this case values suggested by the European codes (Table 2) would be in agreement with those proposed in Table 5 (by assuming that HDRBs[1] and HDRBs[2] correspond to HDRBs with small and large differences between scragged and unscragged properties) only if values referring to K_p and F_0 in Table 2 were inverted. This suggests that a transcription error was made when writing the annex J of EN 1998-part 2 [1] and of EN15129 [2], also because the British version of these codes, BS EN 1998-2 [11] and BS EN15129 [12], are in accordance with Table 5.

Regarding the loading history effect, it is generally recognized that LDRBs do not exhibit much difference between unscragged and scragged properties. Differently, the additives used in the rubber compound to enhance the damping capacity of HDRBs and the curing process are responsible for a significant loading history effect and thus for the bigger difference between the unscragged and scragged properties. It is also know that rubber bearings recover their virgin properties and that significant recovery occurs within a short period of time. Thus, Authors of [7] specify that if the nominal properties are based on scragged condition (i.e. stable hysteretic cycles), λ-factors may be used to estimate the unscragged (i.e. virgin) properties of the isolation bearings. However, at the drafting time of the report of [7], few published results were available in a form that can be suitable for the calibration of appropriate values of λ-factors. In such technical report the Authors have observed that in testing small scale HDRBs with effective damping less than about 0.15, the λ-factor for both the characteristic strength and the post-yield stiffness is of the order of 1.2 (or even less), as reported in Kasalanati et al. [13]. Moreover, they declare that improperly cured bearings may exhibit much higher damping and much higher ratio between unscragged and scragged properties, as also confirmed by data reported in Cho and Retamal [14], relevant to bearings with effective damping in the range 0.15–0.20. Based on these limited data, the task group writing the [5] decided to propose the factors showed in Table 6.

It should be noted that these values of properties modification factors for scragging have not been adopted by the European codes, also because they define as nominal properties the 3rd cycle data, thus coefficients reported in Table 6 (referring to full scragged properties, i.e. almost 10th cycle data) cannot be applied. However, alternative values have not been given, thus, if experimental data provided by manufacturers working in the European context are unavailable, no indication exists regarding on how to account for the scragging effect.

Table 6. λ-factors provided for scragging by AASTHO 1999 [5]

Component	$\lambda_{max,\, fl}$	
	Q_d	K_d
LDRB	1.0	1.0
HDRBs with $\xi < 0.15$	1.2	1.2
HDRBs with $\xi > 0.15$	1.5	1.8

[a] Small differences between scragged and unscragged properties
[b] Large differences between scragged and unscragged properties

Finally, this AASTHO edition [5] specify that the probability of occurrence of joint events and the implication on the structural analysis should be based on a statistical analysis of the variations in mechanical properties with time. However, a simple procedure is described in [7] and implemented in the code. It is based on the concept of system property modification factor for adjusting maximum values of λ-factors. Such a procedure uses an adjustment factor (a) as follow:

$$\max(\lambda) = 1 + (\lambda_{\max} - 1)a \tag{2a}$$

$$\min(\lambda) = 1 + (\lambda_{\min} - 1)a \tag{2b}$$

where max (λ) and min (λ) are the adjusted maximum and minimum system property modification factors, respectively. Based on engineering judgment and the desire to employ a conservative approach for mission-critical bridges, the following adjustment factors are proposed: 1.0 for critical bridges, 0.75 for essential bridges and 0.66 for all other bridges.

3.2 Updated Editions of AASHTO (2010 and 2014)

In 2007 a new technical report, the MCEER-07-0012 report [15], has been developed with the aim of updating the previous document [7], which first proposed the analysis procedure to account for the effects of changing mechanical properties then implemented in specifications for seismic isolation design [1–3, 5, 16]. The scope of the new report is to further understand the effects of heat, loading and aging on seismic isolators and dampers to enable a better prediction of response over the lifetime of seismically protected structures. More in detail, section 12 of the aforementioned document focuses on system property modification factors (λ-factors) and clearly specifies two important issues: (i) the approach based on λ-factors is a simplified one (the most probable maximum and minimum values of mechanical properties should be established by statistical analysis of the distribution of the properties), (ii) the maximum and minimum values of mechanical properties presented in the document are default values and all manufacturers are encouraged to develop specific values for all system property modification factors. As concern default values, it is worth to note that for friction bearings the proposed λ-values are based on a wide experimental campaign carried out at the University of Buffalo and at Queen's University in Canada and/or on the technical literature analysis. Differently, for elastomeric bearings only few data available in the technical literature are collected and discussed (except for some data about LDRBs deriving from tests carried out at the Buffalo University).

For the temperature effect the same references of the old report [5] are analysed, except for one paper, i.e. Yakut and Yura [17] published after the 1999, investigating the low temperature performance of LDRBs as a function of temperature and duration of exposure for four materials. Based on these studies, λ-values collected in Table 7 are proposed, where it is specified that data for HDRBs are for elastomers described as HDRB[a] in Table 5. However, the task group in charge of updating the AASHTO guide specification did not change values of the previous edition. Thus, in both the 3th [6] and 4th [3] editions of AASTHO values of Table 5 are maintained.

For the effect of scragging, which notably affects HDRBs, in addition to the references already commented in the old report [5], two further studies are analysed, i.e. Thompson et al. [18] and Morgan et al. [19]. The first one reports an experimental campaign carried out on 23 moderate-scale rubber bearings with effective damping in the range between 0.05–0.17. The study establishes that the scragging effect increases as the shear modulus of rubber decreases and the effective damping increases. The second paper extends the first one and presents results from tests performed on 45 moderate-scale rubber bearings fabricated by six manufacturers using a total of 12 compounds. Outcomes show that the percentage reduction in the effective shear modulus from the first to third cycle strongly depends on the formulation of the rubber compound, the vulcanization profile used to fabricate the isolator and the imposed strain history. Data

Table 7. λ-factors provided for temperature by Constantinou et al. [15]

Component	$\lambda_{max, f2}$			
Design temperature (°C)	Q_P		K_d	
	LDRB, LRB	HDRB	LDRB, LRB	HDRB
20	1.0	1.0	1.0	1.0
0	1.2	1.2	1.1	1.2
−10	1.4	1.4	1.1	1.4
−30	1.8	2.3	1.3	2.0

reported in these two works suggest that the greater is the volume of filler (maximized by high damping and low shear modulus), the greater is the possible scragging effect. Moreover, another important outcome is that all rubber bearings recover their virgin properties and the time to fully recover them depends on the rubber compound, the extent of its curing and the ambient temperature. Finally, experimental cruciform tests analysed indicate that scragging along one axis significantly influences response on the perpendicular axis, but the degree of interaction appears to be compound dependent. Based on these two documents, values reported in Table 8 are proposed in [15] as defaults ones to apply to full scragged properties (assumed as nominal properties).

Table 8. λ-factors provided for scragging by Constantinou et al. [15] for HDRBs

Effective shear modulus at 100% of shear strain (MPa)	Effective damping at 100% of shear strain	$\lambda_{max, f1}$	
G ≤ 0.45	ξ < 0.10	Q_d	K_d
G > 0.45	ξ > 0.10	1.0	1.0
G ≤ 0.45	ξ < 0.10	1.2	1.2
G > 0.45	ξ > 0.10	1.5	1.8

Differently, for low damping and low modulus elastomer, usually used in lead-rubber bearings, little-to-no scragging is assumed. Alternatively, it is specified in the report that λ-factor can be experimentally defined (at the appropriate conditions of axial load and shear strain as the ratio between the 1st (virgin) to the full scragged or 3rd cycle properties (coherently with the nominal properties assumed).

Practically, based on [18, 19]. Authors of [15] definitely advice against the use of HDRBs because their load history dependent behaviour due to the scragging effect is too much compound-dependent and thus difficult to predict. Also for this reason, values of λ-factors have not been updated for this effect in [3, 6]. Value of Table 6 are maintained and considered as default values (to use in the case of absence of experimental data of the specific manufacturer), despite they date back 20 years ago and thus do not refer to more recent rubber compounds.

In the new report [15] data of [18, 19] are discussed also for the aging effect, in addition to the information already reported and commented in [7]. In particular, based on these papers, Authors [15] argue that HDRBs are generally not significantly affected by ozone and oxygen, because they are fabricated with a layer of cover rubber that includes anti-oxidants to protect the core of the bearing from significant infiltration by oxygen and ozone. Age-related stiffening or hardening is rather due to continued vulcanization of the elastomer that can lead to an increase in the effective shear modulus with time. The percentage of increase in effective modulus will vary depending on several factors including the completeness of the initial curing and the temperature. The ongoing vulcanization of the rubber matrix occurs more rapidly in the first few years after the rubber is compounded, while it slows later on, over time. For this reason, the effective shear modulus will likely reach a limit value, which can only be evaluated, for a given compound and vulcanization profile by long term aging studies. Since high damping can be achieved by incomplete curing, HDRBs will generally exhibit large differences between the virgin and scragged properties and a larger aging effect. However, it is recognized that data on the aging characteristics of elastomeric bearings with effective damping exceeding 15% do not exist. Despite this consideration, λ-factors of Table 4 have been confirmed, including the value of 1.3 for HDRBs with large differences between scragged and unscragged properties.

Finally, also for this effect data on LRBs show an insignificant change in the characteristic strength and some minor increase in the effective stiffness of the bearings because the characteristic strength of LRBs is dictated by the yield strength of the lead core that will not change with time.

4 Conclusion

In this paper, an historical review of the approach proposed by current seismic codes to account for the property variability of isolation bearings is presented. The simplified approach is based on the use of the so called λ-factors or property modification factors to perform upper and lower bounds analyses. In particular, the origin of values provided by the informative annex of Eurocodes is illustrated and commented by pointing out that they are based on old and limited experimental data or on the “engineering judgment”. For example, with reference to HDRBS, currently several manufactures are able to produce HDRBs with low modulus, adequate dissipation capacity and complete initial curing, thus avoiding large scragging or ageing effects.

For this reason, it is strongly suggested to review these reference values in accordance to the philosophy proposed by more recent international codes, where the variability range should be should be agreed between the structural engineer and the manufacturer and only experimentally evaluated by the manufacturing.

To conclude, from one side structural engineers and researchers on this field are encouraged to perform numerical analyses aimed at evaluating the impact of these variations on the response of base-isolated structures. On the other hand, all manufactures should produce and make available experimental data about the property variability range of their own products. At this regard, investigations about the correspondence between material tests and the accelerated aging procedure to reflect the real variability of the bearings' properties in service condition should be carried out, in order to avoid too much challenging experimental tests for the producers. These efforts will make possible the use of high quality HDRBs also for strategic structures, without any uncertainty of their behaviour during all the lifetime of the base-isolated structure.

References

1. EN 1998: Eurocode 8: design of structures for earthquake resistance—Part 2: Bridges, European Committee for Standardization, Brussels, Belgium (2006)
2. EN15129: Antiseismic Devices. European Committee for Standardization, Brussels, Belgium (2009)
3. AASTHO: Guide Specification for Seismic Isolation Design, 4th edn. America Association of State Highway and Transportation Officials, Washington, DC (2014)
4. ASCE 7-16: Minimum Design Loads and Associated Criteria for Buildings and Other Structures. American Society of Civil Engineers (2016)
5. AASTHO: Guide Specification for Seismic Isolation Design, 2th edn. America Association of State Highway and Transportation Officials, Washington, DC (1999)
6. AASTHO: Guide Specification for Seismic Isolation Design, 3th edn. America Association of State Highway and Transportation Officials, Washington, DC (2010)
7. Constantinou, M.C., Tsopelas, P., Kasalanati, A., Wolff, E.D.: Property Modification Factors for Seismic Isolation Bearings. Technical Report MCEER-99-0012, Multidisciplinary Center for Earthquake Engineering Research, State University of New York at Buffalo, Buffalo, NY (1999)
8. Clark, P.W., Kelly, J.M., Aiken, I.D.: Aging studies of high-damping rubber and lead-rubber seismic isolators. In: Proceedings of the 4th U.S.-Japan Workshop on Earthquake Protective Systems for Bridges, pp. 75–89. Published by the Public Works Research Institute, Ministry of Construction, Japan as Technical Memorandum No. 3480 (1996)
9. Nakano, O., Nishi, H., Shirono, T., Kumagai, K.: Temperature-dependence of base-isolation bearings. In: Proceedings of the 2nd U.S.-Japan Workshop on Earthquake Protective Systems for Bridges, Technical Memorandum No. 3196. Public Works Research Institute, Tsukuba Science City, Japan (1993)
10. Kim, D.K., Mander, J.B., Chen, S.S.: Temperature and strain effects on the seismic performance of elastomeric and lead-rubber bearings. In: Proceedings of the 4th World Congress on Joint Sealing and Bearing Systems for Concrete Structures, Vol. 1, pp. 309–322. American Concrete Institute, Farmington Hills, Michigan, Publication SP-164 (1996)
11. BS EN 1998-2: Eurocode 8: Design of Structures for Earthquake Resistance Part 2: Bridges (2012)
12. BS EN 15129: Anti-Seismic Devices (2018)
13. Kasalanati, A., Constantinou, M.C.: Experimental Study of Bridge Elastomeric and Other Isolation and Energy Dissipation Systems with Emphasis on Uplift Prevention and High Velocity Near Source Seismic Excitation. Technical Report MCEER-99-0004, Multidisciplinary Center for Earthquake Engineering Research, State University of New York at Buffalo, Buffalo, NY (1999)
14. Cho, D.M., Retamal, E.: The Los Angeles County emergency operations center on high damping rubber bearings to withstand an earthquake bigger than the big one. In: Proceedings ATC-17-1 Seminar on Seismic Isolation, Passive Energy Dissipation, and Active Control, March, San Francisco, CA, Applied Technology Council, Redwood City, CA (1993)

15. Constantinou, M.C., Whittaker, A.S., Kalpakidis, Y., Fenz, D.M., Warn, G.P.: Performance of Seismic Isolation Hardware Under Service and Seismic Loading. Technical Report MCEER-07-0012, Multidisciplinary Center for Earthquake Engineering Research, State University of New York at Buffalo, Buffalo, NY (2007)
16. ASCE 7-05: Minimum Design Loads and Associated Criteria for Buildings and Other Structures. American Society of Civil Engineers (2005)
17. Yakut, A., Yura, J.A.: Parameters influencing performance of elastomeric bearings at low temperatures. J. Struct. Eng. ASCE **128**(8), 986–994 (2002)
18. Thompson, A.C.T., Whittaker, A.S., Fenves, G.L., Mahin, S.A.: Property modification factors for elastomeric seismic isolation bearings. In: Proceedings, 12th World Congress in Earthquake Engineering. New Zealand (2000)
19. Morgan, T., Whittaker, A.S., Thompson, A.: Cyclic behavior of high-damping rubber bearings. In: Proceedings, 5th World Congress on Joints, Bearings and Seismic Systems for Concrete Structures. American Concrete Institute, Rome, Italy (2001)

Design of Base-Isolated Building as Per Indian Code Provisions and Practices

Vasant Matsagar[1](✉) and Ratish Jain[2]

[1] Indian Institute of Technology (IIT) Delhi, Hauz Khas, New Delhi 110 016, India
matsagar@civil.iitd.ac.in

[2] Resistoflex Dynamics Private Limited, 7, Sector 2, Noida, Uttar Pradesh 201 301, India

Abstract. Detailed design of a base-isolated building as per Indian code provisions and practices is made and described here to facilitate corroboration of the design with other international codes and standards. Upon reviewing the latest developments and applications of the base isolation technology in India, the code development is presented. Particularly, the code provisions and practices in India have been elaborated in light of the state-of-the-art and state-of-the-practice. Then, by taking an example building, base isolation system has been designed and provided to achieve higher seismic performance. Assessment of the design outcomes is made to understand the implications of the code recommendation, especially from techno-economics viewpoints.

Keywords: Base-isolated building · Performance-based earthquake engineering · Indian seismic code provisions · Passive vibration control · Benefit-cost analysis

1 Introduction

Base isolation is one of the few ways to mitigate storey drift and acceleration, the main causes of structural damage [1, 2]. Base isolation separates a superstructure from its foundation through the use of sliding bearings, elastomeric bearings or a combination of the two [3]. Together, these bearings act as a layer with low lateral resistance and reduce earthquake-induced deformations by concentrating deformations to the isolation layer, resulting in increasing building's natural period. In effect, the transmissibility and pseudo-acceleration decreases, removing the fundamental trade-off between decreasing storey-drift at the expense of increasing acceleration and vice versa [4].

It is well known that better performance and cost effectiveness of base isolation systems have been widely discussed and extensively researched in earthquake-prone regions such as, New Zealand, Japan, United States, Italy etc. [5, 6]. This is primarily because of the advancement in seismic code provisions in such regions that helps to implement this concept in the field. However, very few studies are available in the context of Indian subcontinent to assist both decision makers and structural designers to have a persuasive opinion on whether to adopt base-isolated buildings or conventional fixed-base buildings for improved seismic resistance. Recent addition of base isolation

G. P. Cimellaro (Ed.): WCSI 2022, LNCE 309, pp. 349–360, 2023.
https://doi.org/10.1007/978-3-031-21187-4_29

guidelines in Indian seismic code IS 1893: 2020 (Part 1 and Part 6) [7, 8] has allowed the designers to understand the advantages of base isolation systems, particularly to mitigate the effects of earthquake induced displacements and forces, in addition to giving guidelines to design base-isolated structures. The present study demonstrates the detailed design of a real-life reinforced concrete (RC) building with base isolation system based on Indian code provisions to understand the higher performance and cost-benefits (both in short- and long-term) with respect to fixed-base case.

2 Numerical Modeling and Design

2.1 Fixed-Base Building

The present study considers a seven storey (B + G + 5) reinforced concrete (RC) moment-resisting frame (MRF) building (see Fig. 1a), representing an office building in National Capital Region (NCR) of Delhi, India for analysis and design. The RC building has a floor plan of 27.77 m × 10.05 m as shown in Fig. 1b. The elevation of the building in Fig. 1c shows that the total height of the frame structure is 26.44 m, having basement height of 3.9 m and storey height of 3 m. This RC building is designed with: (a) M25 (28-day compressive strength of 25 MPa) and M30 (28-day compressive strength of 30 MPa) grade concrete for beams and columns, respectively, and (b) Fe500D (yield strength of 500 MPa) grade reinforcement steel [9]. The RC frame building is modeled in ETABS® (version 18.1.1), a finite element (FE)-based programme, with appropriate behavior of structural members and boundary conditions.

The site-specific soil report of the NCR region shows a bearing capacity of 357.94 kN/m^2, and raft foundation is designed (having a uniform thickness of 800 mm) based on the behavior of local site and prevailing conditions. The raft is modeled in SAFE® software as per the IS 456: 2000 [9] code provisions, and important parameters such as, soil pressure, punching shear, and settlement are checked to verify the design through quasi-static and dynamic analyses. The outcomes of pressure contour show that the maximum pressure below the raft is 210 kN/m^2, which is less than the allowable bearing capacity of the base soil. Moreover, the shear stress developed due to external load is less than the shear resistance of the foundation, resulting all the design factors for punching shear less than 1. These results confirm the safe design of the raft foundation.

Additional aspects which require attention during structural modeling are that: (i) end moments are to be released for the beams to allow the transfer of load from secondary to primary beams, (ii) beam-column connections should not be over-designed, (iii) floor slabs are modeled as thin shell elements, and (iv) area stiffness properties of the shell elements are suitably modified to ensure that beams do not act integral with the floor slab.

Static analysis of the RC frame building is performed under the basic load combinations of dead load (DL) and live load (LL). This analysis helps in assessing member forces and member deformations under quasi-static load cases, and these analysis parameters are essential for subsequent design of the frame members. Further, Eigen value (free vibration) analysis is performed and the frequencies for the first three modes are found out to be 0.66 Hz (i.e., fundamental mode along y-direction), 0.8 Hz (i.e., fundamental mode along x-direction), and 0.93 Hz (i.e., first torsional mode).

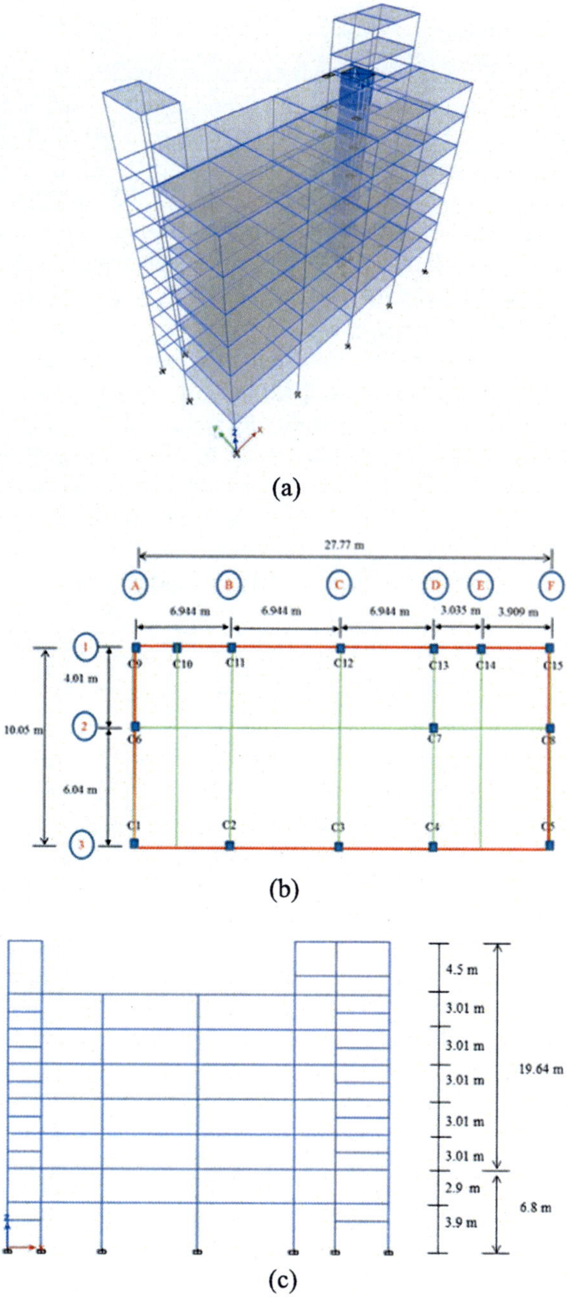

Fig. 1. (a) The 7-storey (B + G + 5) RC moment-resisting frame building with its (b) floor plan and (c) elevation

2.2 Base-Isolated Building

For the considered RC frame building, base-isolators are decided to be placed at stilt level (see Fig. 2a) to isolate the superstructure from ground level. The rationale for this choice is to reduce the transfer of base shear to the superstructure during seismic events. In this study, double-curvature friction pendulum bearings (FPBs) are considered at the stilt level of each column (layout shown in Fig. 2b) and the FPBs are designed in accordance with the provisions of UBC 1997 [10] in absence of Indian standard then at design stage of the structure. It may be noted that the NCR region is in seismic Zone IV in accordance with the IS 1893: 2016 (Part 1) [7], which is equivalent to the seismic Zone III of the UBC 1997 [10].

Additional parameters considered for the analysis and design of base-isolators are: (a) zone factor (Z) 0.3 (seismic Zone III, UBC 1997, Table 16-I), (b) soil profile as stiff (UBC 1997, Table 16-J), (c) seismic source of Type A (UBC 1997, Table 16-U), (d) near-source factor as 1.0 (UBC 1997, Table 16-S), (e) maximum considered earthquake (MCE) response coefficient, M_m=1.5 (UBC 1997, Table A-16-D). Seismic design code procedures are followed to ensure that the FPBs will safely support the maximum gravity and live loads during the entire life of the structure, and also provide a period shift and hysteric damping during one or more design basis earthquake(s) - DBE.

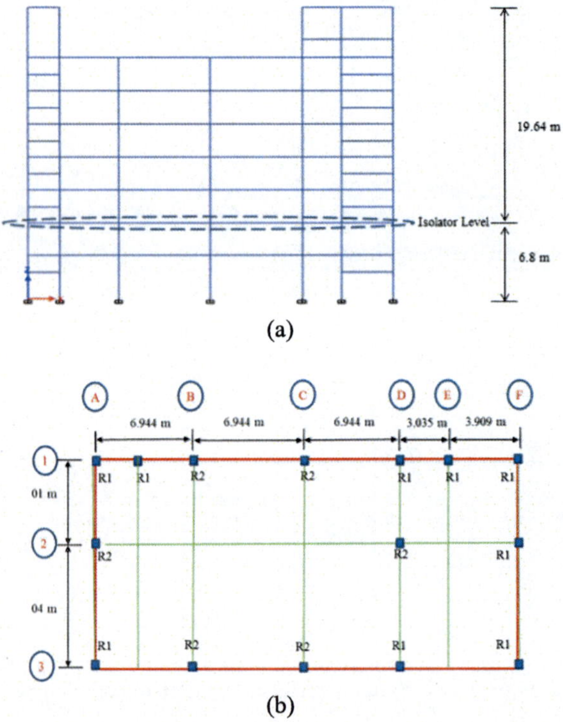

Fig. 2. (a) Elevation view of the base-isolated building with the (b) location of isolators at stilt level

2.3 Design of Isolators

In the RC building, the fundamental period of the base-isolated structure at MCE (T_M) and DBE (T_D) conditions as 4 s and 3.5 s, respectively, which corresponds to 3.25 and 2.85 times the fundamental period of the structure in fixed-base. The lateral stiffnesses of the bearings corresponding to T_M and T_D are evaluated as $K_M = 6222.21$ kN/m and $K_D = 8126.97$ N/m, respectively [8]. Then, the design displacements D_M and D_D are estimated as follows [8]:

$$D_M = \frac{\left(\frac{g}{4\pi^2}\right)}{B_M} C_{VM} T_M = 530 \text{ mm} \tag{1}$$

$$D_D = \frac{\left(\frac{g}{4\pi^2}\right)}{B_D} C_{VD} T_D = 460 \text{ mm} \tag{2}$$

In Eqs. (1) and (2), the notations B_M and B_D denote the damping coefficients having value 1.35 (i.e., effective damping ratio of 25%), and g is gravitational acceleration. In addition, the values of seismic coefficients C_{VM} and C_{VD} are taken from UBC 1997 (Table 16-R) for the given seismic zone and soil profile type. Further, torsional effect is included to estimate the lateral stiffness and displacement of the bearings. Additional eccentricity provision by the Indian standard (IS) is 5% of the plan dimension perpendicular to the loading. Also, an additional +10% variation is assumed for K_M and K_D to evaluate the maximum value of effective stiffness as 6844.43 kN/mm and 8939.66 kN/mm, respectively.

The design of friction pendulum bearings (FPBs) in this case included calculation of the radius of curvature (R) of the isolators in MCE (R_M) and DBE (R_D) conditions, which are estimated as follows:

$$R_M = T_M^2 \frac{g}{4\pi^2} = 3973.32 \text{ mm} \tag{3}$$

$$R_D = T_D^2 \frac{g}{4\pi^2} = 3040.80 \text{ mm} \tag{4}$$

For the considered RC frame structure two types of FPBs are considered to accommodate the superstructure loads (W) up to 1300 kN (isolator R_1) and up to 2650 kN (isolator R_2). The coefficient of static friction (μ) for the FPBs is considered as 0.02 [3, 4]. The effective stiffness (K_{eff}) of both isolator R_1 and R_2 is evaluated through bilinear idealization as follows [8, 11]:

$$K_{eff} = \frac{W}{R} + \frac{\mu W}{D} \tag{5}$$

In Eq. (5), the effective stiffnesses for isolator R_1 and R_2 are estimated to be 600 kN/m and 1300 kN/m, respectively; D denoting design isolation displacement. Further, the effective damping (β_{eff}) provided by the base isolation systems, i.e., isolator R_1 and R_2 are expressed as follows:

$$\beta_{eff} = \frac{2}{\pi} \frac{\mu}{\mu + \frac{D}{R}} \tag{6}$$

In Eq. (6), the effective damping for isolator R_1 and R_2 are estimated to be 18% and 18%, respectively. Based on the design results, distribution of isolator R_1 and R_2 below the columns at stilt level is shown in Fig. 2(b). Furthermore, the outcomes of Eigen value analysis shows that the first three global modes are 3.31 s (i.e., fundamental mode along y-direction), 3.24 s (i.e., fundamental mode along x-direction), and 0.612 s (i.e., first torsional mode). This confirms that FPBs at the base of 1st storey level elongates the fundamental period of the RC building in the fixed-base.

3 Analysis and Performance Assessment

3.1 Response Spectrum Analysis

Response spectrum analysis is performed in the present study (along x- and y-directions) in ETABS® to assess the better performance of the FPB-isolated RC frame as compared to the fixed-base case. Seismic load pattern along x- and y-directions for the response spectrum analysis is calculated as per the provisions of the IS 1893: 2016 [7], and is summarized as follows.

FPB-isolated condition (load pattern EQ X - seismic) [IS 1893: 2016]

i. Direction and eccentricity
   Direction = X + Eccentricity Y
   Eccentricity ratio = 5% for all diaphragms
ii. Structural period
   Fundamental period along X direction, $T = 3.245$ s
iii. Factors and coefficients
   Seismic zone factor, Z (IS 1893: 2016, Table 3) = 0.24
   Response reduction factor, R (IS 1893: 2016, Table 9) = 3.0
   Importance factor, I (IS 1893: 2016, Table 8) = 1.0
   Soil type (IS 1893: 2016, Table 2) = B
iv. Seismic response
   Spectral acceleration coefficient (IS 1893: 2016, Sect. 6.4.2), $\frac{S_a}{g} = 0.419$.
v. Equivalent lateral force
   Seismic coefficient (IS 1893: 2016, Sect. 6.4.2), $A_h = \frac{ZIS_a}{2Rg}$.
   Base shear, $V_b = A_h \times W$

The values of base shear for the RC frame building in fixed and isolated conditions are shown in Table 1. Then, the distribution of base shear as lateral force along the height of the structure is evaluated as per the provisions of the IS 1893: 2016 [7, 8] and shown in Table 2 for both fixed and isolated conditions.

3.2 Performance Assessment

The results from the response spectrum analysis (RSA) have demonstrated in terms of structural responses such as, storey shear, absolute storey acceleration, relative storey displacement, and inter-storey drift. The profile of peak storey shear along the height

Table 1. Estimated base shear

Base condition	Direction	T (s)	W (kN)	V_b (kN)
Fixed	X + Ecc. Y	1.259	29350.028	760.805
	Y + Ecc. X	1.522	29350.028	629.594
Isolated	X + Ecc. Y	3.245	21462.971	359.818
	Y + Ecc. X	3.307	21462.971	353.057

Note Ecc. denotes eccentricity

Table 2. Distribution of base shear along the height of the structure (IS 1893: 2016 [7, 8])

Floor	Height (m)	Fixed (lateral force)		Isolated (lateral force)	
		X-dir (kN)	Y-dir (kN)	X-dir (kN)	Y-dir (kN)
Roof	26.44	61.529	50.918	42.519	41.720
Terrace	21.85	267.394	221.278	158.876	155.891
4	18.84	164.268	135.938	84.019	82.441
3	15.83	117.560	97.285	47.908	47.008
2	12.82	76.672	63.449	21.173	20.775
1	9.81	45.129	37.346	5.321	5.221
Isolator level	6.8	21.144	17.498	0	0
Basement	3.9	7.104	5.879	–	–
Base	0	0	0	–	–

Note dir represents direction

of the structure is plotted in Fig. 3 for fixed and base-isolated (both at DBE and MCE) conditions. This figure shows that the installation of FPBs in the RC frame reduces storey shear considerably as compared to the fixed-base case. The variation in peak floor acceleration at different stories along x- and y-directions are displayed in Figs. 4a and 4b, respectively. It is quite evident from Fig. 4 that there is a significant reduction in the peak floor acceleration for the base-isolated case as compared to the fixed-base case. Furthermore, the peak floor displacement and inter-storey drift profiles along the height of structure are presented in Figs. 5 and 6, respectively. The results confirm that FPBs installed at the stilt level of each column produce noticeable reduction in the inter-storey drift at higher storeys of the building. Thus, it can be summarized from the outcomes of the RSA that reduction in storey acceleration and inter-storey drift for FPBs-isolated condition may have beneficial impact on the performance of the structure and the in-house non-structural components as well.

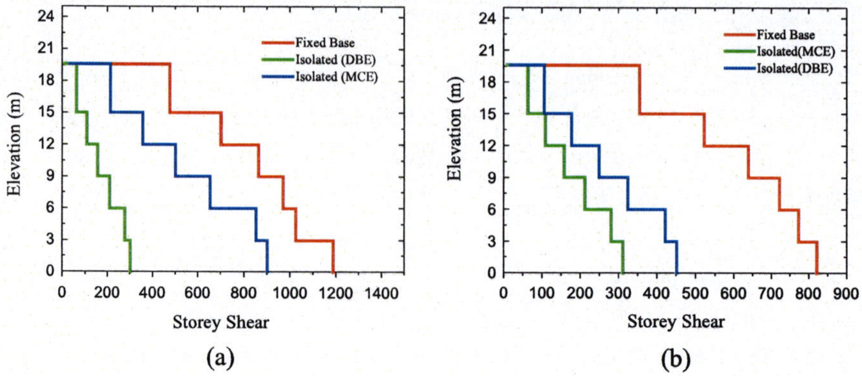

Fig. 3. Variation in peak storey shear (kN) along the height for fixed and isolated conditions: (a) x-direction and (b) y-direction

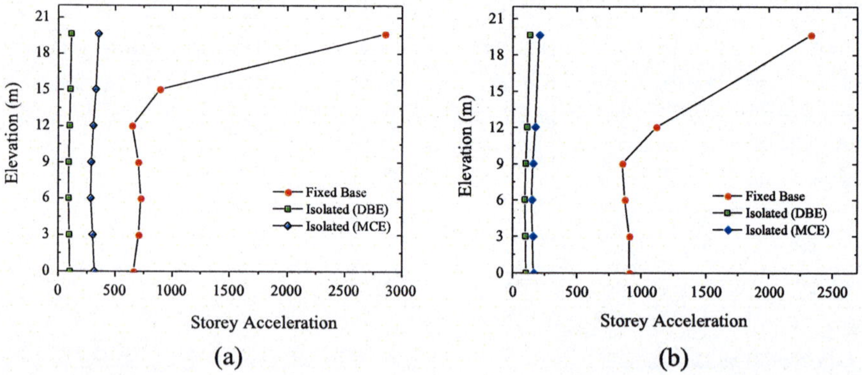

Fig. 4. Variation in peak acceleration (mm/s^2) along the height for fixed and isolated conditions: (a) x-direction and (b) y-direction

4 Benefit-Cost Analysis

Upon conducting numerical modeling and seismic analysis of the fixed-base and base-isolated RC frame buildings, detailing and cost analysis for the RC frame under both the conditions are carried out. In this study, the cost analysis of the structure is performed in the STAAD® Advanced Concrete Design (RCDC) software, and the foundation in SAFE® software. It may also be noted that cost evaluation of the building is limited to structural members such as, beams, columns, slab, walls and foundation. Further, benefit-cost analysis is performed using Federal Emergency Management Agency (FEMA) Benefit-Cost Analysis (BCA) tool [12].

A detailed cost analysis of construction materials is carried out and based on which the final cost for each structural component is assessed. The cost so determined is then compared and a conclusive summary is presented with regards to cost escalation or reduction upon introduction of the base isolation systems in this project. Also, saving of material cost due to non-ductile detailing of the superstructure above isolator level is

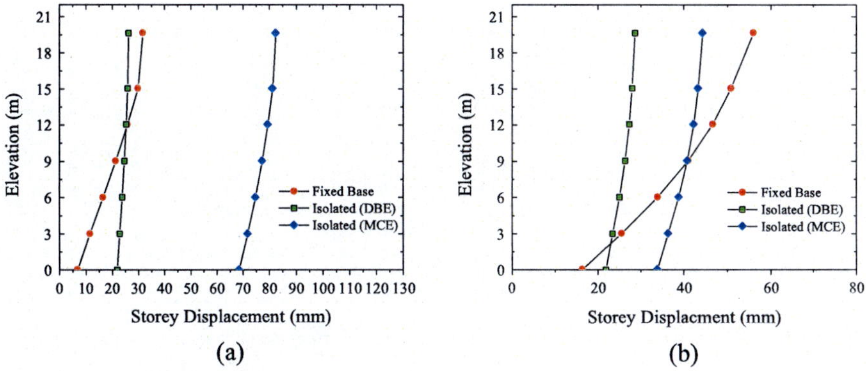

Fig. 5. Variation of peak displacement along the height for fixed and isolated conditions: (a) x-direction and (b) y-direction

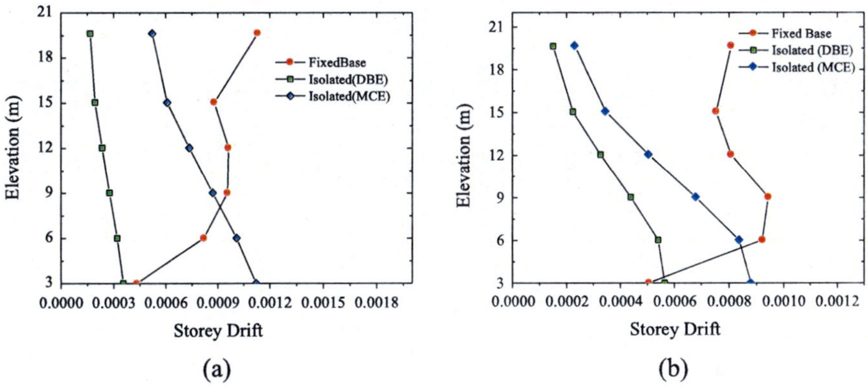

Fig. 6. Variation in peak inter-storey drift along the height for fixed and isolated conditions: (a) x-direction and (b) y-direction

estimated. Based on the RSA results, the RCDC software assisted in ductile as well as non-ductile detailing. In this study, the general prevailing rates of construction materials in Indian Rupees (Rs.) considered for the benefit-cost analysis are given in Table 3. Total cost incurred for the construction of each of the building components for the fixed-base and base-isolated conditions are shown in Fig. 7. It is important to observe that the savings in cost of materials for the base-isolated case is primarily due to reduced ductility demand and structural deformations. Figures 8a and 8b present the contribution of each member and their overall share in the total cost estimation, and also assist in determining this material cost savings for the base-isolated case.

5 Conclusions

A series of quasi-static and dynamic analyses are performed to comprehend the improved performance and cost-benefits of a real-life base-isolated building in India. The stiffness

Table 3. General prevailing rates of used construction materials

S. No.	Construction materials	Rate (Rs.)
1	Reinforcement	50 per kg
2	Concrete (M30)	4000 per m^3
3	Concrete (M25)	3800 per m^3

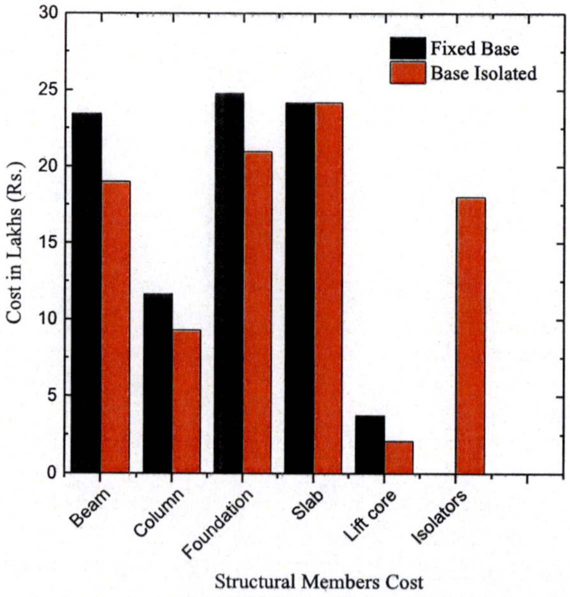

Fig. 7. Cost of structural members both in fixed and isolated conditions

parameters of friction pendulum bearings in this case are designed and analyzed to maximize the seismic performance of the structure. The installed friction pendulum bearings demonstrate positive impact by increasing the fundamental period of the structure and thereby reducing the force transferred to the superstructure during strong ground shaking. Also, an attempt has been made to give an elaborate idea about the cost-benefits of the base-isolated buildings vis-é-vis fixed-base buildings. Moreover, this study outlines the relevant issues concerning the practical financial benefits of base-isolated structure. The parameters of the building and the site conditions were deliberately selected in such a way that the application and financial benefits of base-isolated structure can be conveyed to designers and decision makers in the easiest possible way.

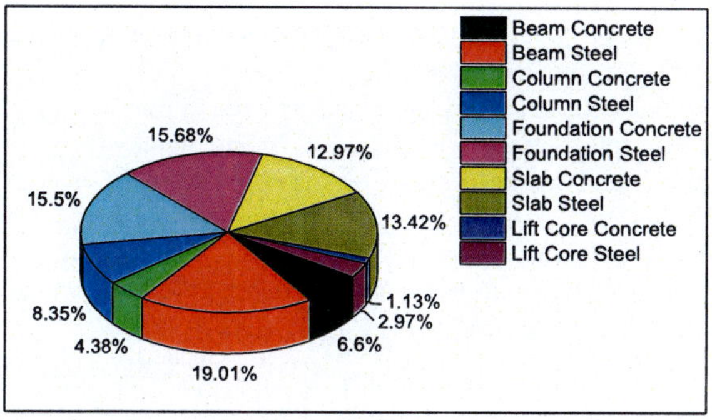

(a)

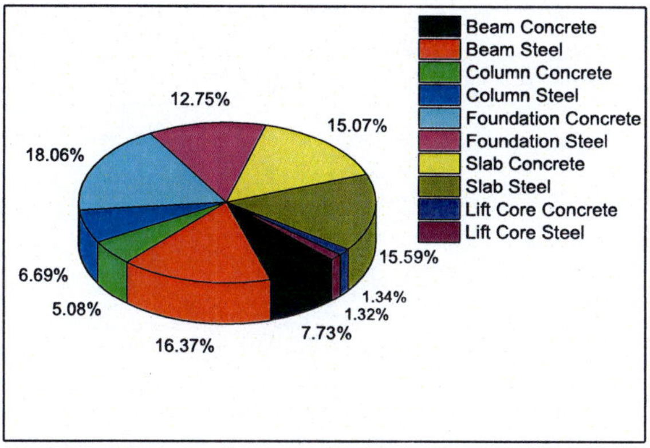

(b)

Fig. 8. Material wise cost estimation for (a) fixed-base structure, and (b) base-isolate structure

References

1. Buckle, I.G., Mayes, R.L.: Seismic isolation: history, application and performance - a world view. Earthq. Spectra **6**(2), 161–201 (1990)
2. Mayes, R.L.: Impediments to the implementation of seismic isolation. Earthq. Spectra **6**(2), 283–296 (1990)
3. Madhekar, S., Matsagar, V.: Passive Vibration Control of Structures. CRC Press, Taylor & Francis Group, UK (2022)
4. Panda, J., Ray-Chaudhuri, S.: A new optimization approach to enhance seismic performance of lead rubber bearing-isolated steel moment-resisting frames under extreme events. Curr. Sci. **122**(1), 77–86 (2022)
5. Goulet, C.A., Haselton, C.B., Mitrani-Reiser, J., Beck, J.L., Deierlein, G.G., Porter, K.A.: Evaluation of the seismic performance of a code-conforming reinforced-concrete frame building-from seismic hazard to collapse safety and economic losses. Earthquake Eng. Struct. Dynam. **36**, 1973–1997 (2007)
6. Vinci, G., Serino, G., Pampanin, S.: Multi-criteria cost-benefit analysis for base-isolated buildings. In: NZSEE Conference, New Zealand Society for Earthquake Engineering (NZSEE), pp. 10–12 (2015)

7. Indian Standard IS 1893 (Part 1): Criteria for Earthquake Resistant Design of Structures. Bureau of Indian Standards (BIS) (2016)
8. Indian Standard IS 1893 (Part 6): Indian Standard Criteria for Earthquake Resistant Design of Structures: Base Isolated Buildings. Bureau of Indian Standards (BIS) (2020)
9. Indian Standard IS 456: Plain and Reinforced Concrete - Code of Practice. Bureau of Indian Standards (BIS) (2000)
10. Uniform Building Code (UBC): Structural Design Requirements. Vol. 1 (1997)
11. Matsagar, V.A., Jangid, R.S.: Influence of isolator characteristics on the response of base-isolated structures. Eng. Struct. **26**(12), 1735–1749 (2004)
12. Federal Emergency Management Agency (FEMA): Benefit-Cost Analysis Tool. Version 5.2.1. (2015)

Complex Modal Shapes Superposition Response Spectrum Based Design Method for Seismically Isolated Structures in China

Ping Tan[1,2(✉)], Kui Yang[1,2], Jiajun Tan[1,2], and Wenzhi Zheng[1]

[1] School of Civil Engineering, Guangzhou University, Guangzhou 510006, China
ptan@gzhu.edu.cn

[2] Earthquake Engineering Research and Test Center, Guangzhou University, Guangzhou 510006, China

Abstract. Seismic isolation technology has become a practical strategy for seismic-resistant design. Several studies have also been proposed to promote the applications of seismic isolation technology in structural engineering and the corresponding design procedures. This study introduces the newly issued Building Seismic Isolation Design Standards of China, in which the framework of an integrated direct design method(IDM) for seismically seismically isolated structures is proposed based on the concept of the Complex-mode-superposition response spectrum. The proposed method is different from the traditional two-stage seismic reduction coefficient design method. The new standard design spectrum for the seismically isolated building is given in the issued Standard. The dynamic history analysis method, the IDM, and the divisional design method(DDM) are employed to conduct the seismic isolation design using the seismically isolated benchmark model. The critical design parameters and analysis results from the China standard are reported and evaluated to facilitate the comprehensive comparisons of the seismic isolation design codes from various countries.

Keywords: Building · Seismic isolation design standards · Benchmark model · CCQC · Design parameters · Design method

1 Introduction

Seismic isolation is a technology that separates the structure from the detrimental effect of ground motions through isolators with flexibility and energy dissipation capability [1]. In recent decades, seismic isolation has received significant attention as a very effective seismic-resistant technology, and several theoretical and experimental studies have been conducted [2–4]. Besides, the seismic isolation standards for structures in many countries, such as China, Japan, the United States, and European countries, have been promulgated. Notably, the seismic isolation Code for seismic design of buildings in China (GB50011–2010) has been enacted [5]. These designed structures based on the Code have successfully experienced the Wenchuan earthquake in 2008, the Yushu earthquake in 2010, and the Lushan earthquake in 2013 [6–8]. So far, there are more

G. P. Cimellaro (Ed.): WCSI 2022, LNCE 309, pp. 361–376, 2023.
https://doi.org/10.1007/978-3-031-21187-4_30

than 10,000 seismically isolated buildings in China, and the number of them is still increasing. The higher requirements for seismic isolation design method are needed to accommodate the new development stage.

The DDM specified in the Code for seismic design of buildings requires that the seismic-reduced factor is first obtained by calculating the shear force and bending moment of the seismically isolated and seismically resistant structures, respectively. The seismic design of the superstructure is carried out under reduced frequent earthquakes. The design procedure is too cumbersome, and it is challenging to consider seismically isolated and resistant structures when selecting seismic records. In addition, the seismically isolated structure is a typical non-proportional damping structure due to the different energy dissipation characteristics between the superstructure and the seismic isolation interface. Thus, the traditional modal superposition response spectrum method for response analysis needs to be improved.

To consider the non-proportional damping, several studies have focused on the complex modal theory. Zhou [9] proposed the Complex Mode Superposition method and complex complete quadratic combination (CCQC) formula. Chen and Tan [10, 11] derived a recipe for the truncation error of complex mode and proposed a static correction method to improve the error. Besides, a way to determine a reasonable number of modes was proposed. By optimizing the Complex Mode Superposition method, the complex-mode-superposition response spectrum method considering the influence of higher-order modes was proposed.

Further, Chen and Tan [12] have investigated the mode superposition response spectrum method and complex-mode-superposition response spectrum method for seismic response analysis using a Benchmark model. Results show that the larger the damping ratio of the seismic isolation interface is, the larger the error of forced decoupling by the mode superposition response spectrum method shall be. Meanwhile, the energy dissipation capability of the superstructure is overestimated by the mode superposition response spectrum method, which leads to unrealistic results. These researches forcefully promoted the development of the seismic isolation design method in China. In September 2021, the first national standard dedicated to guiding the seismic isolation design, Building Seismic Isolated Design Standards (GB/T 51408–2021) [13], was officially promulgated, in which the IDM is precisely based on the theory of the complex-mode-superposition response spectrum method and CCQC.

Based on the seismic codes and seismic isolation standards in China, this paper mainly introduces two methods: divisional design and integrated design, including their essential parts. To facilitate the mutual complementation of the seismic isolation design methods with those of other countries, the implementation process, design considerations, and calculation indexes of the two design methods are analyzed based on the benchmark model.

2 Divisional Design Method for Seismically Isolated Structure (DDM)

The DDM proposed in the code for seismic design of buildings (GB50010–2001) is mainly used in seismic isolation design. The isolated structure is divided into superstructure, isolation layer, and substructure. The primary purpose is to obtain the seismic-reduced factor through response history analysis, establish the relationship between isolation design and seismic design through the seismic-reduced factor, and transform the isolation design into the traditional seismic design. The basic design process is shown in Fig. 1.

2.1 Superstructure Design

The maximum value of the horizontal seismic impact coefficient of the superstructure after seismic isolation is calculated as follows:

$$\alpha_{\max 1} = \beta \alpha_{\max} / \psi \tag{1}$$

where $\alpha_{\max 1}$ and $\alpha_{\max}$ are the maximum values of the horizontal seismic impact coefficients for seismic isolation and non-seismic isolation, respectively, and ψ is the adjustment coefficient, which is generally taken as 0.8. $\alpha_{\max}$ is mainly defined by the response spectrum shown in Fig. 2.

2.2 Isolation Layer Design

The design of the isolation layer is mainly to consider whether the performance of the isolation bearing meets the specification requirements. It includes explicitly eccentricity of isolation layer, the limit value of tensile and compressive stress, limit displacement of isolation bearing, and wind resistance bearing capacity.

3 Integrated Design Method of Isolated Structure (IDM)

To fully consider the non-proportional damping characteristics of the seismically isolated structures and avoid the unreasonable bearing parameters, response spectrum, and seismic wave selection in the DDM. Based on the concept of equivalent linearization, the complex mode theory, and the CCQC, IDM considering the non-proportional damping characteristics of seismically isolated structures, is proposed in the standard for seismic isolation design of buildings. The traditional "DDM" is replaced by the "IDM," and the existing response spectrum is improved based on referring to the current response spectrum regulations of Japan, North America, and Europe. On the issue of seismic fortification targets, the original 'no damage under frequent earthquakes, repairable under fortification earthquakes and not collapsed in rare earthquakes" was upgraded to "not bad in fortification earthquakes, repairable in rare earthquakes and not collapsed in very rare earthquakes," and performance-based design was introduced.

The IDM is different from the DDM in that the isolation layer, superstructure, and substructure are analyzed and designed as a whole. The core complex mode theory

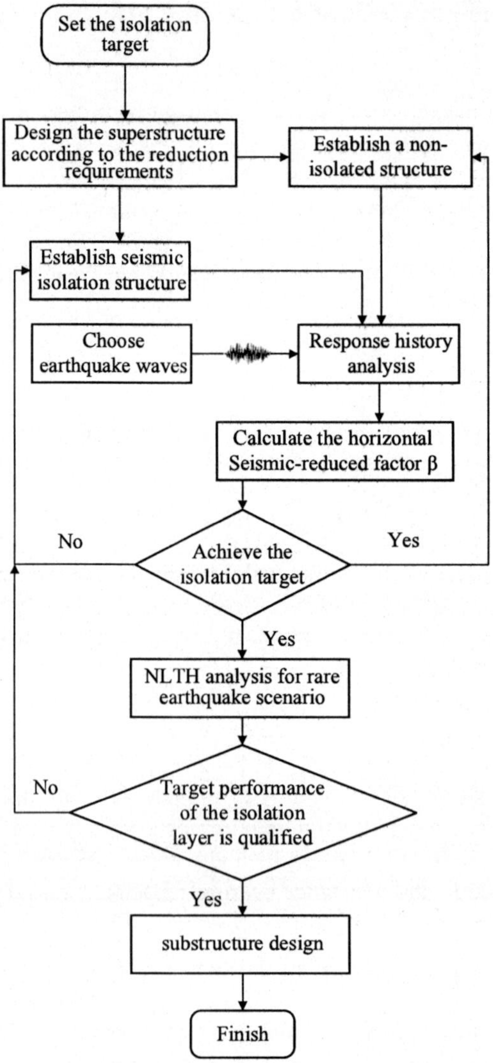

Fig. 1. DDM for seismically isolated structures

can avoid the error caused by a non-proportional damping system such as a forced decoupling isolation structure. CCQC can well combine it with the response spectrum to facilitate the design. In addition, the internal force distribution characteristics of the isolated system can be fully considered in the overall layout, making the procedure more reasonable and accurate. The fundamental principle of the design procedure is to consider the fortification earthquake scenario and estimate the structural response for the rare earthquake scenario. The analysis method used for the design, particularly for the fortification earthquakes scenario is the mode-superposition response spectrum method. The method used in estimating the structural response for the rare earthquake scenario

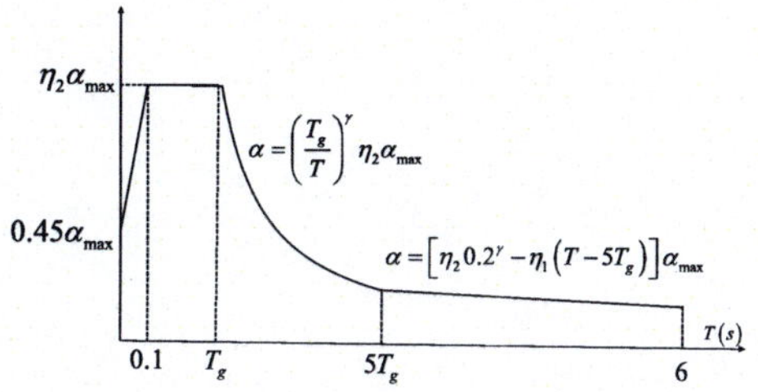

Fig. 2. Seismic influence coefficient curve of seismic code [5]

is the time-history analysis. The design flow is shown in Fig. 3. The specific steps of this method are as follows:

(1) Firstly, the structural performance objectives are determined;
(2) Establish the finite element model, select the bearing according to the vertical load, preliminarily determine the bearing layout scheme, and establish the isolation analysis model;
(3) The equivalent linearization of the isolation bearing is carried out by the iterative method, and the equivalent parameters of the bearing used for fortification earthquakes design are obtained. The principle of iterative calculation is shown in Fig. 4. Where k_y is the post-yield stiffness of the lead-core rubber bearing, Q_d is the yield force, D_i and D_y are the initial displacements and yield displacement of the i-th bearing, respectively;
(4) Through the integrated design analysis, the internal force of the structure is obtained. By calculating the story drift, whether the structure meets the requirements of moderate seismic elasticity is judged, and the reinforcement calculation is carried out;
(5) The modified response spectrum is taken as the target spectrum to select the seismic records of the isolated structure directly and check whether the bearing and story drift meet the standard requirements under rare earthquakes;
(6) Compile the isolation design report.

It is worth noting that the response spectrum in step (5) above is different from that used in the DDM. For such long-period systems, The straight-line descent section of the seismic design response spectrum by referring to the current response spectrum regulations of Japan, North America, and Europe has been removed from the Chinese Code and directly extended from the curve descent of the third section to the 6S section. A new isolation design response spectrum is proposed, which aligns more with the seismic dynamic response characteristics of such long-period structures. The response spectrum is shown in Fig. 5. The response spectrum is mainly modified by the iterative process's damping ratio of the isolation layer.

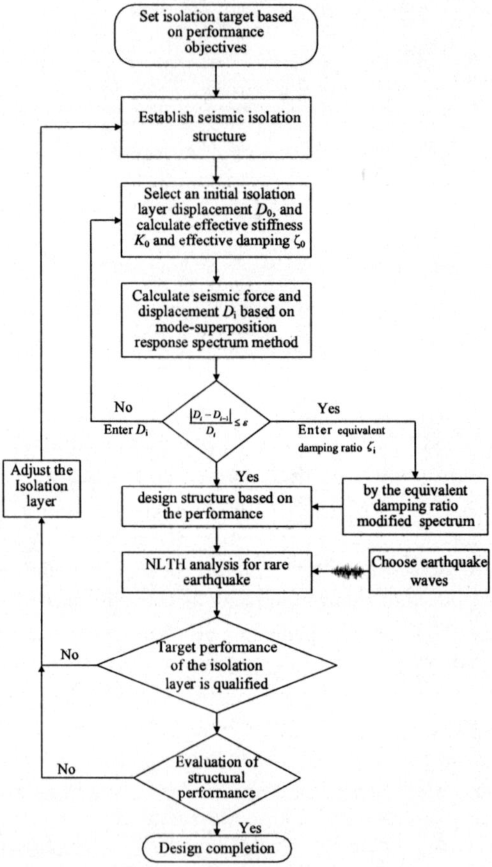

Fig. 3. IDM for seismically isolated structures

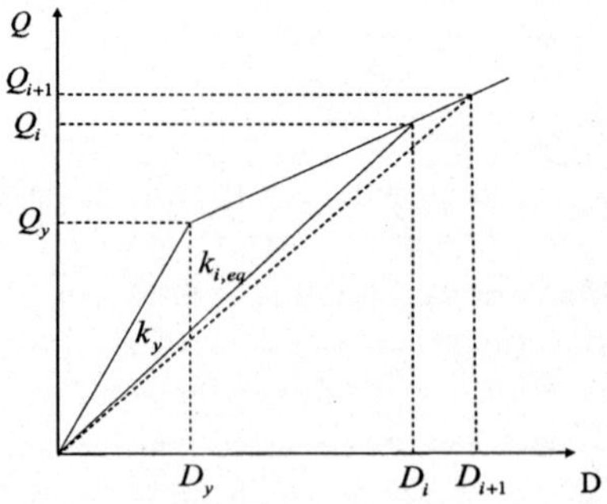

Fig. 4. Principle of equivalent linear iteration

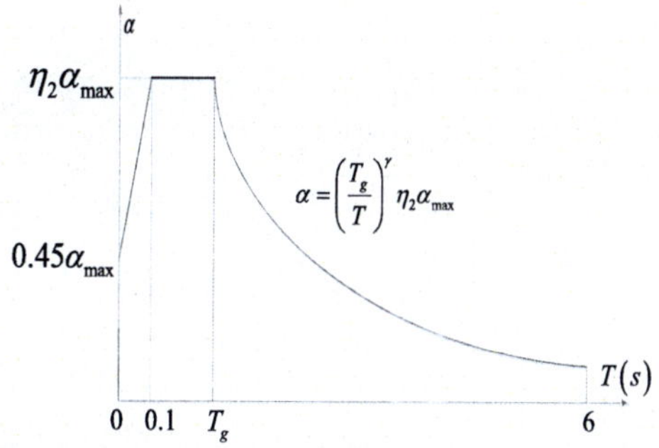

Fig. 5. Seismic impact coefficient curve in isolation standard [13]

3.1 Basic Theory of Complex Mode Decomposition Response Spectrum

The equation of motion after equivalent linearization can be expressed as follows:

$$\mathbf{M}\ddot{\mathbf{y}} + \mathbf{C}_{eq}\dot{\mathbf{y}} + \mathbf{K}_{eq}\dot{\mathbf{y}} = -\mathbf{M}\mathbf{r}\ddot{x}_g(t) \tag{2}$$

where, $\mathbf{C}_{eq}$ denotes the non-classical damping matrix and $\mathbf{K}_{eq}$ denotes the equivalent linearized stiffness matrix.

In order to decouple the above equations, it is necessary to convert them into 2n first-order differential equations by using the Foss transformation [16], as follows:

$$\mathbf{A}\dot{\mathbf{v}} + \mathbf{B}\mathbf{v} = -\mathbf{A}\tau\ddot{x}_g(t) \tag{3}$$

In the formula, $\mathbf{A} = \begin{bmatrix} \mathbf{0} & \mathbf{M} \\ \mathbf{M} & \mathbf{C}_{eq} \end{bmatrix}$, $\mathbf{B} = \begin{bmatrix} -\mathbf{M} & \\ & \mathbf{K}_{eq} \end{bmatrix}$, $\mathbf{v} = \begin{Bmatrix} \dot{\mathbf{y}} \\ \mathbf{y} \end{Bmatrix}$, $\tau = \begin{bmatrix} \mathbf{r} \\ \mathbf{0} \end{bmatrix}$

Matrices **A** and **B** are symmetric non-positive definite matrices. Usually, the eigenvalue λ_i corresponding to Eq. (2) appears as a conjugate pair, and the eigenvector $\boldsymbol{\psi}_i$ has the form $\boldsymbol{\psi}_i = \left[\lambda_i \boldsymbol{\Phi}_i^{\mathbf{T}} \boldsymbol{\Phi}_i^{\mathbf{T}}\right]^{\mathbf{T}}$, where Φ_i is the i-th order vibration deformation mode (vibration mode). Referring to the eigenvalue expression of the classical damping system, here λ_i can be expressed as $\lambda_i = -\xi_i\omega_i + \mathrm{i}\omega_i\sqrt{1-\xi_i^2}$, $\mathrm{i} = \sqrt{-1}$. ω_i and ξ_i are the i-th natural frequency and mode damping ratio, as follows:

$$\begin{aligned} \omega_i &= \sqrt{[\mathrm{Re}(\lambda_i)]^2 + [\mathrm{Im}(\lambda_i)]^2} \\ \xi_i &= -\frac{\mathrm{Re}(\lambda_i)}{\mathrm{Im}(\lambda_i)} \end{aligned} \tag{4}$$

where Re(·) and Im(·) denote taking the real and imaginary parts of the complex numbers, respectively.

Using the complex mode superposition method, the displacement vector can be expanded as:

$$\mathbf{y}(t) = \sum_{i=1}^{n} \left[p_i\dot{q}_i(t) + \varphi_i q_i(t)\right] \tag{5}$$

where, $\varphi_i = -2\text{Re}\left(\overline{\lambda}_i \mathbf{\Phi}_i \eta_i\right)$ and $\eta_i = \mathbf{\Phi}_i^{\mathbf{T}}\mathbf{Mr}/\left(2\mathbf{\Phi}_i^{\mathbf{T}}\mathbf{M}\Phi_i + \mathbf{\Phi}_i^{\mathbf{T}}\mathbf{C}\Phi_i\right)$ is the participation coefficient of modal shapes; $\dot{q}_i(t)$ and $q_i(t)$ are the velocity and displacement responses of the single degree of freedom system under the action of $\ddot{x}_g$, which are determined by the following equations:

$$R(t) = \sum_{i=1}^{n} \left[\alpha_i \dot{q}_i(t) + \beta_i q_i(t)\right] \tag{6}$$

where, $\alpha_i = \nu^{\mathbf{T}} \mathbf{\rho_i}$, $\beta_i = \mathbf{\nu}^{\mathbf{T}} \varphi_{\mathbf{i}}$.

Assuming that the time history $\ddot{x}_g$ of earthquake acceleration is a Gaussian stationary process with zero mean value, the peak response coefficient of each mode response is equal to the total peak response coefficient. Reference uses the contour integral method to deduce that the maximum value of $R(t)$ is expressed as the combination of peak response of each vibration mode, as follows:

$$R_{\max}^2 = \sum_{i=1}^{n}\sum_{j=1}^{n} \begin{bmatrix} \alpha_i\alpha_j\rho_{ij}^{vv} V_i V_j + (\alpha_j\beta_i - \alpha_i\beta_j) \\ \rho_{ij}^{vd} V_i D_j + \beta_i\beta_j\rho_{ij}^{dd} D_i D_j \end{bmatrix} \tag{7}$$

where, ρ_{ij}^{vv}, ρ_{ij}^{vd} and ρ_{ij}^{dd} are the mode shape correlation coefficients of mode shape velocity and velocity, velocity and displacement, and displacement and displacement, respectively. V_i and D_i are the received spectral value and displacement spectral value of mode I, respectively.

The expression of mode shape correlation coefficient is as follows ($r = \omega_i / \omega_j$):

$$\rho_{ij}^{vv} = \frac{8\sqrt{\xi_i\xi_j}\left(\xi_i + r\xi_j\right)r^{3/2}}{\left(1 - r^2\right)^2 + 4\xi_i\xi_j r\left(1 + r^2\right) + 4\left(\xi_i^2 + \xi_j^2\right)r^2} \tag{8}$$

$$\rho_{ij}^{vd} = \frac{8\sqrt{\xi_i\xi_j}\left(1 - r^2\right)r^{3/2}}{\left(1 - r^2\right)^2 + 4\xi_i\xi_j r\left(1 + r^2\right) + 4\left(\xi_i^2 + \xi_j^2\right)r^2} \tag{9}$$

$$\rho_{ij}^{dd} = \frac{8\sqrt{\xi_i\xi_j}\left(r\xi_i + \xi_j\right)r^{3/2}}{\left(1 - r^2\right)^2 + 4\xi_i\xi_j r\left(1 + r^2\right) + 4\left(\xi_i^2 + \xi_j^2\right)r^2} \tag{10}$$

4 Example Analysis

4.1 Overview of Target Structure

The benchmark model is selected to conduct the proposed design mothod. The standard fortification class is 8 degree, the design peak acceleration is 0.2 g, and the seismic design grouping is group II with a characteristic period of 0.4 s. The lead-core rubber bearing (LRB) and natural rubber bearing (LNR) are selected. To ensure the correctness of the model parameters, a hybrid model of solid plus simplified type is used in this paper.The model display and the plan layout of isolation bearings are shown in Fig. 6.

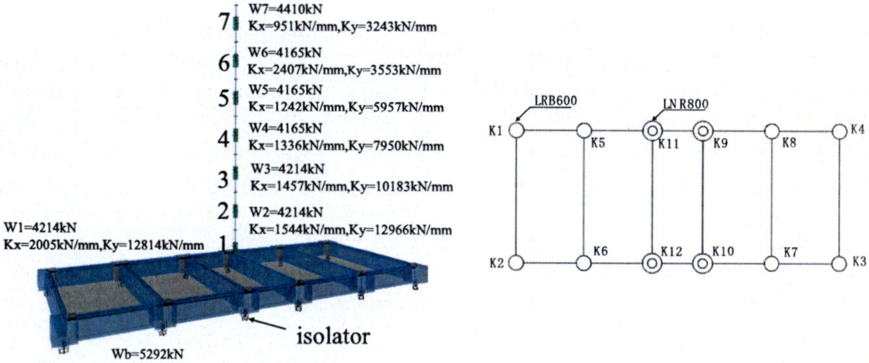

Fig. 6. Finite element model of seismically isolated structures and layout of bearings

4.2 Determination of Equivalent Parameters of Isolator

In this paper, two design methods are adopted for seismic isolation, and the parameters of the initial selection of bearing for the DDM are shown in Table 1. The equivalent stiffness and equivalent damping ratio of the bearing are taken as 100 and 250% of the shear deformation corresponding to the fortification earthquakes and rare earthquakes, respectively, which the bearing manufacturer generally provides. However, in practice, the bearing displacement is often in error with the actual value, which may lead to a large or small corresponding seismic-reduced factor, affecting the design of the superstructure.

Table 1. Isolatior parameters for DDM

Isolation device	LRB600	LNR800
Diameter(mm)	600	800
Yielding strength:Q_y(kN)	90.4	/
Equivalent stiffness:k_{eq}(kN/mm)	1.681(100%) 1.081(250%)	1.197(100%)
Yeilding stiffness:k_d(kN/mm)	0.928(100%) 0.779(250%)	/
Ratio of post yield stiffness α	0.1	/
Limit design deformation:u_h(mm)	330	440
Equivalent damping ratio ξ_{eq}	28.5%	5%

The difference between the IDM and the DDM is that the equivalent parameters (k_{eq}, ξ_{eq}) are determined by iterative calculation based on the fortification earthquakes analysis. An ideal linear model is slected for LNR800 without providing additional damping. The LRB600 adopts a double-fold line. The iterative results are shown in Table 2. The bearing displacement in the table is the maximum displacement of LRB600 in the iterative process. It can be seen from the table that under the fortification earthquakes, the equivalent damping ratio of the structure is 26.72%, and the equivalent stiffness of LRB600 under the movement of the fortification earthquakes is 2.3 kN/mm. In the DDM under a fortification earthquakes, the equivalent stiffness of LRB600 is taken as 1.681 kN/mm, which leads to the small seismic-reduced factor, which is non-conservative for the structural design.

Table 2. Iteration procedure

Series Number	D/mm	T_i(s)	k_{eqi} kN/mm	ξ^l_{eqi}(structure)	Error (%)
8	65.8	2.472	2.300	0.2672	0.1

T_i: isolation period in the *i*th step, k_{eqi}: equivalent stiffness in the *i*-th step, and ξ^l_{eqi}: damping ratio of the isolation layer in the *i*-th step

4.3 Preliminary Inspection of Isolation Layer Performance

The seismic code and isolation standard requires the preliminary checking of the rationality of the isolator placement. The wind resistance capability of the isolation layer, the vertical compressive stress of the bearing under the expected gravity load value, and the isolation layer's eccentricity must be checked.

(1) Wind resistance checking calculation of isolation layer

Under the design wind load, the isolated structure should not produce horizontal displacement. That is, the design value of wind load on the isolated structure should be less than the sum of yield force shear strength of each bearing of the isolated structure, and the design value of wind load is calculated by Eq. (5). Table 3 shows the wind load checking results of the isolated structure, and it can be seen that the design value of wind load is less than the yield force of the seismic isolation layer, which meets the requirements.

Table 3 Wind load checking calculation of isolated structure

Wind direction	Standard value (kN)	Design value (kN)	Yield force of isolation layer (kN)	YES/NO
X	188.22	263.51	723.2	YES
Y	451.03	631.44	723.2	YES

(2) Bearing vertical stress calculation

The vertical compressive stress of the isolator under the expected value of gravity load should not exceed the required reference surface pressure limit value. In this paper, the maximum compressive stress of the seismic isolation bearing is 11.23 MPa under gravity load, which is by the code requirements.

(3) Eccentricity test

The code requires that the center of stiffness of the seismic isolation layer should coincide with the center of mass, and the eccentricity should not be greater than 3% under the action of fortification earthquakes. The eccentricity of the isolated structure in the x-direction and y-direction is 1.0% and 2.8%, respectively, less than the 3% specified in the code.

4.4 Seismic Records

Figure 7 shows the seismic wave response and target response spectra selected by the DDM and IDM. The DDM requires the average seismic wave response spectra to be statistically consistent with the normative response spectra. The target response spectrum of the integrated design is "three-stage," which is more suitable for the long-period structure and only needs to take into account the main period of the isolated structure, so it can be seen that the selected seismic waves and the target response spectrum match well under large damping.

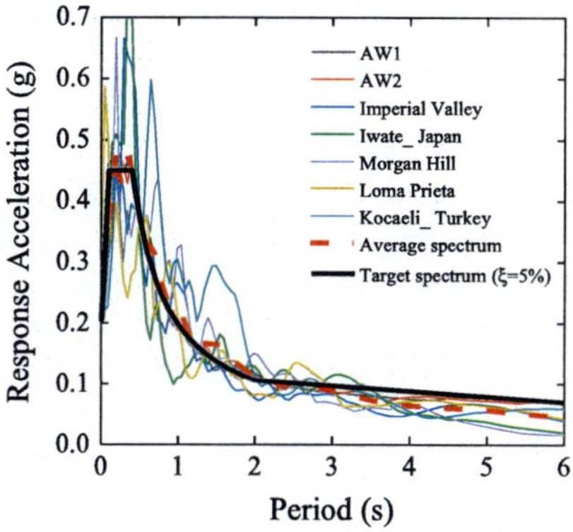

(a) Based on the DDM provisions

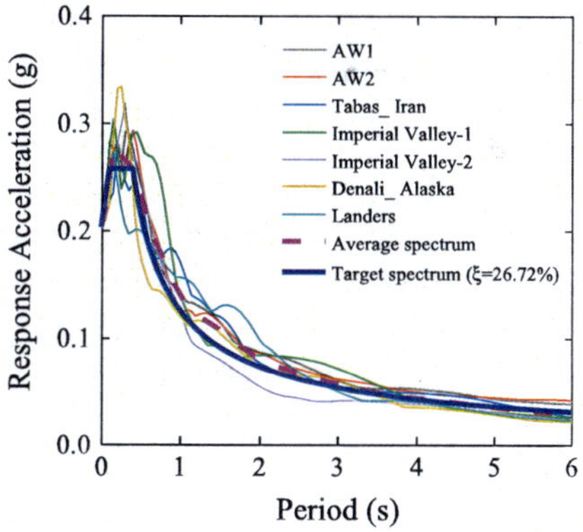

(b) Based on the IDM provisions

Fig. 7. Selection of seismic wave and comparison with the target response spectrum

4.5 Isolation Efficiency Evaluation

Figure 8A shows the seismic-reduced factor β obtained by DDM, and its value is the average floor shear ratio before and after isolation. Figure 8b shows the floor shear ratio before and after isolation obtained using IDM response spectrum analysis. The two design methods have effectively reduced the seismic force. This paper's maximum seismic-reduced factor of the DDM is 0.355 in the X direction, as shown in Fig. 8a. It can be seen that it meets the seismic code requirements for degree reduction design. The maximum shear ratio of the IDM is also in the X direction, and the maximum value is 0.246, as shown in Fig. 8b, which also meets the requirement that the bottom shear ratio before and after isolation is less than 0.5 in the isolation standard. The maximum shear ratio of the two is not the same, and the latter is more efficient. The maximum shear ratio in the IDM is mainly affected by the selected seismic wave and the equivalent parameters of the bearing.

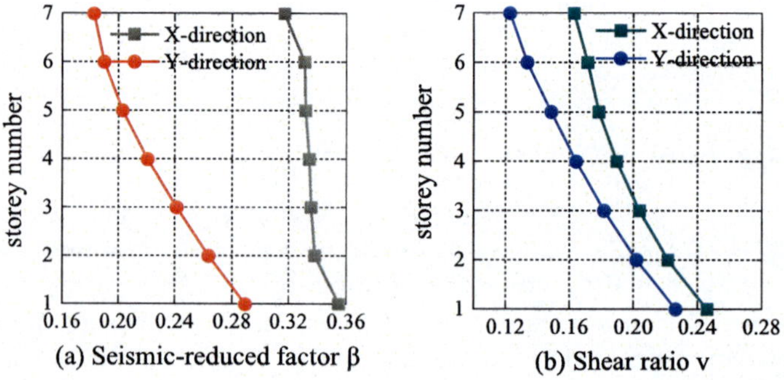

Fig. 8. Evaluation diagram of damping effect of different isolation design method

4.6 Superstructure Design of Isolated Structure

The IDM is to take the response spectrum modified by the equivalent damping ratio after iterative calculation as the design response spectrum and directly carry out the elastic design of seismically isolated structures under fortification earthquakes. The traditional DDM determines the reduction of one degree or half a degree according to the code requirements after obtaining the maximum seismic-reduced factor β and then carries out the frequent earthquakes elastic design for the non-isolated structure. The maximum earthquake influence coefficient $\alpha_{\max}$ is obtained from Eq. (1). The fortification intensity of this project is 8 (0.2 g), which can be designed according to 7 (0.1 g) according to the above Sect. 4.5.

Figure 9 shows the design target response spectra for the IDM and DDM. The maximum seismic shadow acceleration for the IDM is 0.257 g. For the DDM, the maximum seismic acceleration is 0.072 g. According to the one-degree design, the maximum seismic-reduced factor β ratio of the seismic isolation is between 0.27 and 0.4. In the following, the floor shear force is obtained by the structural design based on the mode decomposition and complex mode decomposition response spectrum methods corresponding to the DDM and IDM. Secondly, the flat floor stiffness is obtained by model analysis. Finally, the floor shear force is divided by the floor stiffness to get the floor displacement, and the story drift is calculated.

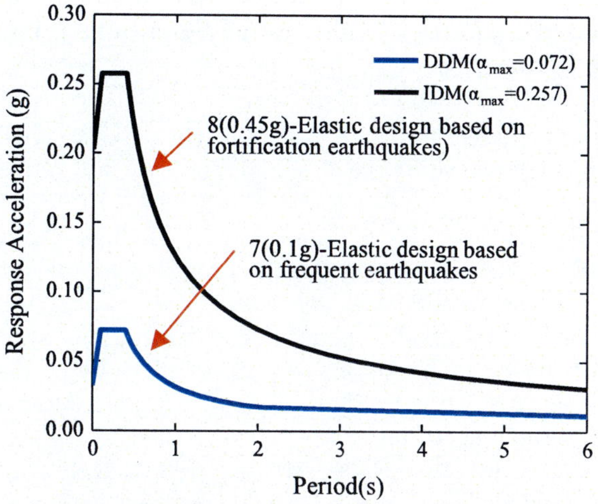

Fig. 9. Superstructure design response spectrum

Figure 10a shows the story drift after adopting the once-degree reduction design for the superstructure after adopting the DDM. The seismic code requires the frame-shear wall structure to be designed with peak story drift within 1/800. The IDM requires the superstructure to be designed with the modified response spectrum as the design spectrum for fortification earthquakes elasticity with the story drift within 1/500. It can be seen from Fig. 10b that the design of the superstructure according to both design methods meets the requirements of the respective codes, and the DDM is much smaller than the code requirements, which can further optimize the superstructure to improve the economy.

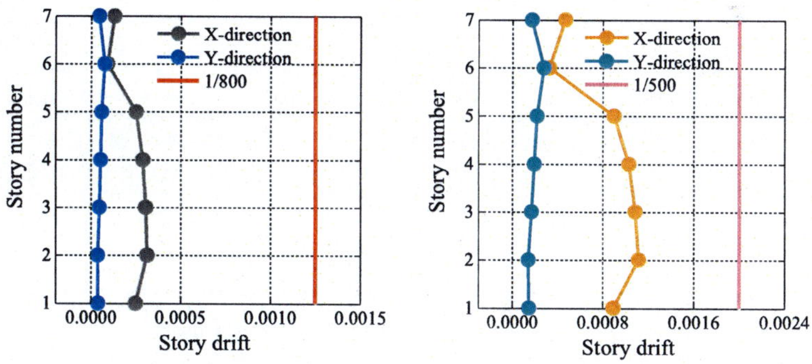

(a) Based on frequent earthquakes seismic elastic design

(b) Based on fortification earthquakes seismic elastic design

Fig. 10. Story drift of each directionis

5 Nonlinear Time History Analysis of Seismically Isolated Structures

Nonlinear analysis is required for DDM and IDM. The purposes are to check the performance of the bearing and verify the elastic-plastic story drift of the structure under rare earthquakes.

For rare earthquakes, according to the regulations of China, the maximum tensile stress limit of rubber isolation bearing should not appear tensile stress; when a few isolation bearings appear tensile stress, the tensile stress should not be greater than 1 Mpa. The maximum horizontal displacement of the seismic isolation layer under a large earthquake should not be greater than the smaller of 0.55 times the effective diameter of the seismic isolation bearing and three times the thickness of the rubber layer of the seismic isolation bearing.

Since the seismic waves selected by the two design methods are different, the seismic wave analysis results in Fig. 7a are recorded as NO1, and the seismic wave analysis results in Fig. 7b are recorded as NO2 for comparative analysis. In this paper, the maximum compressive stress of the bearing is 12.355 mpa, and there is no tension bearing. The maximum compressive stress of the seismic wave selected by the two design methods is basically the same under rare earthquakes, and both meet the code's requirements. However, the maximum displacement of the bearing under the action of the NO2 seismic wave is 163.5 mm, and the maximum displacement of the No1 seismic wave is 120.2 mm. Although the displacement also meets the code requirements according to the DDM, it is not conservative. Therefore, selecting the seismic wave according to the method in the IDM is recommended.

Figure 11 gives the average responses of the story drift of the seismically isolated structure under rare earthquakes. The figure shows that the story drift is much smaller than the code requirement. The results are very similar, which indicates that the displacement is still mainly concentrated in the isolation layer during large earthquakes, and the relative displacement of the upper floors is smaller. The upper part can basically remain elastic.

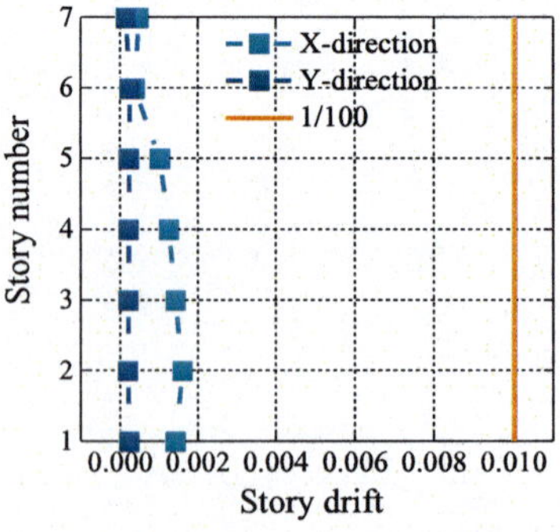

(a) Based on seismic code requirements

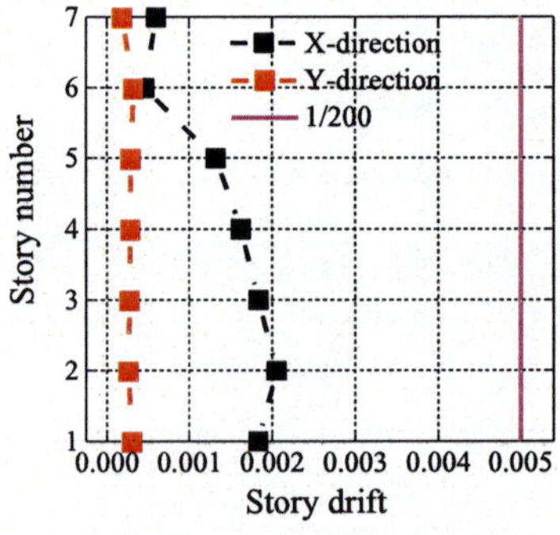

(b) Based on isolation standard

Fig. 11 Story drift of each directionis is based on a great earthquake

6 Conclusions

Both DDM and IDM methods play an important role in promoting the application of seismic isolation technology in China. The IDM is a development of the DDM to overcome its shortcomings and further expand the seismic isolation technology to nuclear power plants, historical buildings, masonry buildings, and village buildings.

1. Using the DDM, the structural story drift is much smaller than the code requirements, so optimizing the design that not structural dimensions are recommended to improve the economy.
2. The acceleration reduction rate of the isolated structure using artificial wave response spectrum is significantly greater than that of the code response spectrum. The artificial wave generated by a 5% damping ratio will cause the seismic response to result smaller when applied to the isolated structure. It is more reasonable to select the seismic wave using the damping ratio of the isolation layer.
3. The equivalent parameters corresponding to 100% and 250% shear deformation cannot be simply taken as the bearing parameters. The bearing parameters obtained by iteration under the design response spectrum should be used.

Acknowledgments. The authors acknowledge support provided by National key R&D program of China (grant No. 2019YFE0112500), National Natural Science Foundation of China (grant No. 51978185) and Program for Changjiang Scholars and Innovative Research Team in University (grant No. IRT13057).

References

1. Kelly, J.M.: Aseismic base isolation: review and bibliography[J]. Soil Dyn. Earthq. Eng. **5**(4), 202–216 (1986)
2. Naeim, F., Kelly, JM.: Design of Seismic Isolated Structures: From Theory to Practice. John Wiley & Sons (1999)
3. Ismail, M., Rodellar, J., Ikhouane, F.: An innovative isolation device for aseismic design[J]. Eng. Struct. **32**(4), 1168–1183 (2010)
4. Skinner, R.I., Robinson, W.H., Mcverry, G.H.: An introduction to seismic isolation. Wiley, Chichester (1993)
5. GB50011-2010. Code for seismic design of buildings. Beijing, China: China Construction Industry Press. (in Chinese) (2010)
6. Kawashima, K., Takahashi, Y., Ge, H., et al.: Reconnaissance report on damage of bridges in 2008 Wenchuan, China, earthquake[J]. J. Earthquake Eng. **13**(7), 965–996 (2009)
7. Gong, M., Lin, S., Sun, J., et al.: Seismic intensity map and typical structural damage of 2010 Ms 7.1 Yushu earthquake in China. Nat. Hazards **77**(2), 847–866 (2015)
8. Xing, J.H., Lu, M., Li, H.W., et al.: The seismic damage investigation and phenomenon analysis of space grid structures in Lushan Ms 7.0 earthquake. Appl. Mech. Mater. **501**, 1535–1541 (2014)
9. Zhou, X., Yu, R.: CCQC method for seismic response of non-classically damped linear system based on code response spectra. Eng. Mech. (02), 10–17+9 (2006) (in Chinese)

10. Chen, H., Tan, P., et al.: The truncation error and improvement of complex mode superposition method. J. Vib. Eng. **30**(04), 556–563 (2017). (in Chinese)
11. Chen, H.: The dynamic response analysis method of non-classically damped systems. Guangzhou University (2017) (in Chinese)
12. Chen, H., Tan, P.: Complex modal shapes superposition response spectrum approach based on vibration isolation structure benchmark model. J. Vib. Shock **36**(23), 157–163 (2017). (in Chinese)
13. GB/T51408-2021. Standard for Seismic Isolation Design of Building. Beijing, China: China Planning Press (2021) (in Chinese)
14. Foss, K.A.: Co-ordinates which uncouple the equations of motion of damped linear dynamic systems (1958)

Hidden Pitfalls in Double Curved Surface Sliders (DCSS)

Renzo Medeot(✉)

Engineering Consultant, Rome, Italy
medeot@medeot.eu

Abstract. Dr. Victor Zayas invented and patented the seismic isolator that is known under the Trade Name "Friction Pendulum Bearing®" in 1985, which together with Lead Rubber Bearing revolutionized modern seismic hardware. Prior to the expiration of Dr. Zayas's patent, Prof. Chong-Shien Tsai filed a patent in Taiwan (R.O.C.) in 1991, which represents an interesting evolution thereof. In all the seven (7) versions illustrated in the Prof. Tsai's Patent a hinged slider separates the two concave surfaces. At the turn of the last century, a further version of Double Curved Surface Slider (DCSS) appeared in Japan, in which the intermediate slider is solid, in that the hinge has been removed. Being constructively much simpler, it has met with success with all seismic hardware manufacturers. However, this simplified version has produced serious anomalies. These consist of high eccentricities that cause significant deformations at the edges of the sliding material (liner) with its consequent wearing, as well as strong vibrations transmitted to the structure and even the phenomenon of "banging". The paper elaborates the mathematical models of the Kinematics of the two types of Double Curved Surface Sliders, identifies the critical parameters and quantifies the values thereof that produce the observed adverse phenomena. Finally, the paper reports the list of factors, the concomitant presence of which suggests not adopting the simplified solution.

Keywords: Double Curved Surface Sliders (DCSS) · Vibrations · Stick-slip

1 Introduction

In recent years vibration phenomena induced by the bearing system have been observed in bridge structures where seismic isolators of the DCSS type had been adopted. This unwanted effect can be explained by the non-synchronous movements in the two curved surfaces that Professor Michael C. Constantinou of State University of New York at Buffalo, NY, studied in detail as early as 2004. For his theoretical and experimental study, Prof Constantinou used four devices described as follows: "The bearing consists of two facing concave surfaces. The upper and lower concave surfaces have radii of curvature R_1 and R_2, respectively, that might not be equal.

The coefficients of friction on these two sliding surfaces are μ_1 and μ_2, respectively. An articulated slider separates the two concave surfaces."

G. P. Cimellaro (Ed.): WCSI 2022, LNCE 309, pp. 377–387, 2023.
https://doi.org/10.1007/978-3-031-21187-4_31

In truth, the vibrations, sometimes accompanied by sound phenomena (banging), have been observed only in cases where the slider is solid, that is, without the articulation (hinge).

Prof. Constantinou in his report had correctly pointed out that "The articulation is necessary for proper distribution of pressure on the sliding surfaces and to accommodate differential movements along the top and bottom sliding surfaces".

The purpose of this paper is to examine the effects of replacing the articulated slider with a solid slider.

2 Principles of Operation of Double Curved Surface Sliders

Dr. Victor Zayas invented and patented the seismic isolator that is known under the Trade Name "Friction Pendulum Bearing®" in 1985. Prior to the expiration of Dr. Zayas's patent, Prof. Chong-Shien Tsai filed a patent in Taiwan (R.O.C.) on July 10th, 1991, which represents an interesting evolution and innovation of Friction Pendulum Bearing®.

The title given by Prof. Tsai to his Patent is very generic, namely: "*Transfor-mation of Seismic Energy*", which is applicable to all dissipative seismic devices existing on the market except Shock Transmitters.

Prof. Tsai has never extended the validity of his patent outside Taiwan (R.O.C.) and does not appear to have presented it in international events (Conferences, Symposia etc.). This explains why it has remained perfectly unknown for so many years outside his country.

It is not known if its use in Japan at the turn of the century is a "re-invention" of Prof Tsai's patent or a legitimate violation of the same.

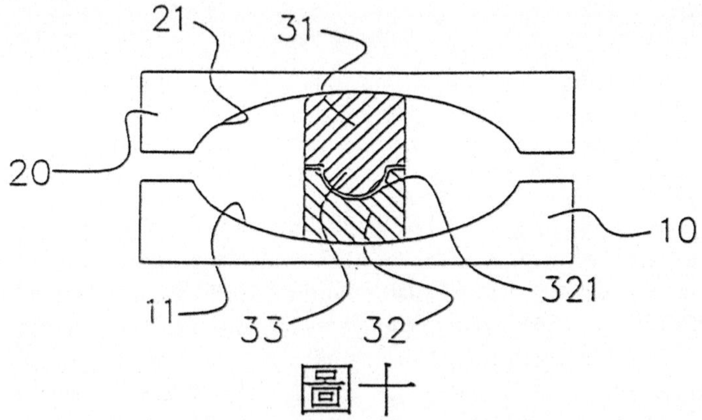

Fig. 1. Original sketch of Prof. Tsai's double curved surface isolator

Figure 1 schematically represents the device of Prof. Tsai's Patent, which consists. of two facing concave surfaces separated by a hinged slider. In all the seven (7) versions illustrated in the Prof. Tsai's Patent the hinge (articulation) is present.

In principle, the hinge is not necessary for the functioning of the DCSS, but it is needed for the proper pressure distribution on the sliding surfaces.

Nonetheless, it should be noted that the hinge in the intermediate element is not only intended to accommodate the rotations induced by the structure (e.g. deformation of the girders in the presence of live loads on bridges), but also for its correct functioning in both service and seismic conditions, as we shall see later on.

Its successful extensive use in seismically isolated buildings (where the rotations impressed by the structure are almost absent) led to the misinterpretation of the function performed by the hinge and to the mistaken conclusion that it is always legitimate to remove it from the intermediate slider.

This fact can cause unwanted effects in some applications (eg. bridge structures).

3 Statics of Double Curved Surface Slider

For the purpose of this report, the case with hinged intermediate element repr-esented in Fig. 2 is considered first, where the radii of curvature R of the two primary sliding spherical surfaces are equal, r is the radius of the secondary surface and t is the thickness of the concave lower intermediate element.

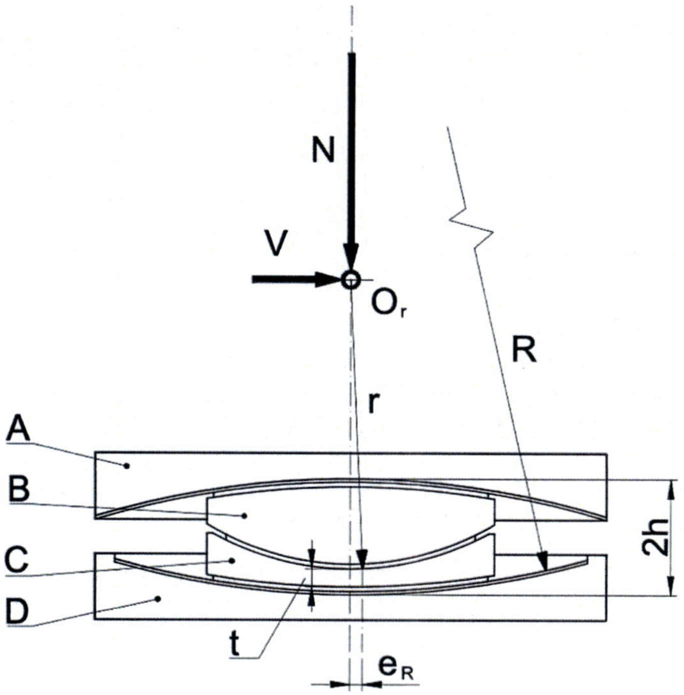

Fig. 2. DCSS with hinged intermediate slider

The subscripts "r" and "R" refer to the parameters relating to the spherical hinge (secondary surface) or the primary surfaces of radii r and R respectively.

If we imagine impressing a slow horizontal movement to the upper concave element A of the isolator, the resultant of the vertical load N and of the horizontal friction reaction force V must pass through the center O_r of the spherical convex element B, so it is concluded that the eccentricity e_r of the resulting load on the primary surface is equal to (see Fig. 3):

$$e_r = \frac{V}{N}(r + t) = \mu_{r,s1}(r + t) \quad (1)$$

where $\mu_{r,s1}$ the static coefficient of friction at the beginning of the movement and t is the minimum thickness of the lower concave element C.

With reference to Fig. 3 below relating to the case of a device with a solid intermediate slider and repeating the same calculation procedure, it is found that the eccentricity is equal to:

$$e_R = \mu_{R,s1} \cdot R \quad (2)$$

Combining Eqs. (1) and (2) we obtain the ratio of the eccentricities of the transmitted load N for the two cases examined:

$$\frac{e_R}{e_r} = \frac{\mu_{R,s1} \cdot R}{\mu_{r,s1} \cdot (r + t)} \quad (3)$$

Considering that:

- The coefficient of friction of the main sliding surfaces $\mu_{R,s1}$ is much larger than that of the hinge $\mu_{r,s1}$ (the latter is lubricated).
- R is much larger than (r + t)

we can draw a first conclusion that the elimination of the hinge from the intermediate slider entails a substantial increase in the eccentricity of the vertical load (by a factor from 6 to 10) whenever the device is subjected to a displacement of any origin, including those caused by rotations.

A "Frictional Resistance Test" according to EN 15129 was conducted on a DCSS with solid intermediated slider having the following characteristics:

- Vertical Load capacity N = 33000 kN.
- Radius of curvature R = 2200 mm.

At the beginning of the movement the measured Horizontal Load was:

- V = 3630 kN

Therefore the static coefficient of friction is equal to $\mu_{R,s1} = 0.110$.

The dynamic coefficient of friction $\mu_{dyn} = 0.055$ was also measured.

Thus the eccentricity of the vertical load N at the beginning of each impressed movement is equal to:

$$e_R = \mu_{R,s1} \cdot R = 0,110 \times 2200 = 242\,\text{mm} \quad (4)$$

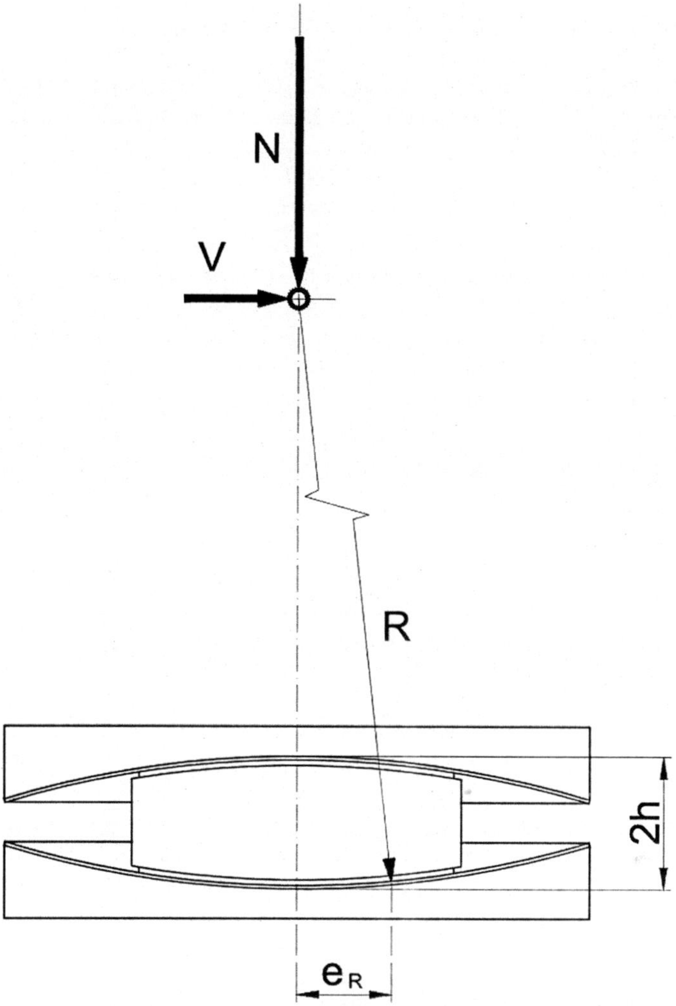

Fig. 3. DCSS with solid intermediate slide

The large values of the eccentricity in this type of seismic isolators results in a pronounced non-uniformity in the distribution of the contact pressures inside the surface of the sliding liner, which can induce permanent deformation in the liner, as indeed was observed during the testing campaign.

We remind that the idea of Prof. Tsai was that of accommodating the impressed translations by means of the two primary curved surfaces, and the impressed rotations by means of the hinge of the intermediate slider.

The idea of giving the second task to the two primary surfaces is subject to compliance with certain conditions, as we will see below.

4 Kinematics of Double Curved Surface Slider

For simplicity of treatment of the mathematical model, the Kinematics of Double Curved Surface Slider will first be examined for the case of the DCSS with hinged intermediate element, also because for this type of device there are some experim-ental results available.

The presence of the hinge allows to develop a logical path to describe the kinem-atics of this device that would be possible with the solid intermediate element only if one knew the physical law of the reaction of the sliding liner to deformation by compression.

For the purpose of this report, the DCSS shown in Fig. 4 is considered, which has a cylindrical symmetry about the central vertical axis and a specular (mirror-like) symmetry with respect to the horizontal plane passing through the pivot point.

Therefore the radii of curvature R of the two main sliding spherical surfaces are equal, as well as the distances h between the pivot point and the intercept of the vertical symmetry axis and the main curved surfaces.

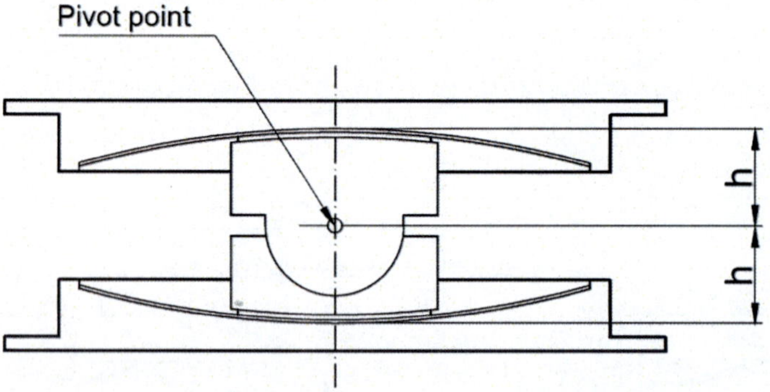

Fig. 4. Structure of the device used for the kinematics of DCSS

Even if identical sliding materials are used, it would not be conservative and realistic to consider perfectly identical the coefficients of friction of the two main sliding surfaces.

In service conditions, the DCSS acts as a structural bearing and therefore shall comply with the requirements of the European Norm EN 1337: *Structural Bearings.*

Said Norm gives a typical curve μ *vs* d of a generic polymeric sliding material at the transition from initial static $\mu_{s,1}$ to dynamic coefficient of friction μ_{dyn} due to the sudden break-away of Van der Waals forces (see Fig. 5).

The transition between the initial static coefficient $\mu_{s,1}$ and the dynamic one μ_{dyn} takes about 0.2 s.

With reference to Fig. 6, if we impress a horizontal movement to the upper element A, it is extremely unlikely that the initial static coefficient of friction $\mu_{s,1}$ at the first movement is identical in both the two primary sliding surfaces.

If we assume that the static coefficient of friction of the lower primary surface is smaller, then the sliding movement will take place on the latter and will continue over time, as the dynamic coefficient of friction μ_{dyn} that develops on said surface is much lower (about the half) than the initial static one $\mu_{s,1}$ present on the other primary surface.

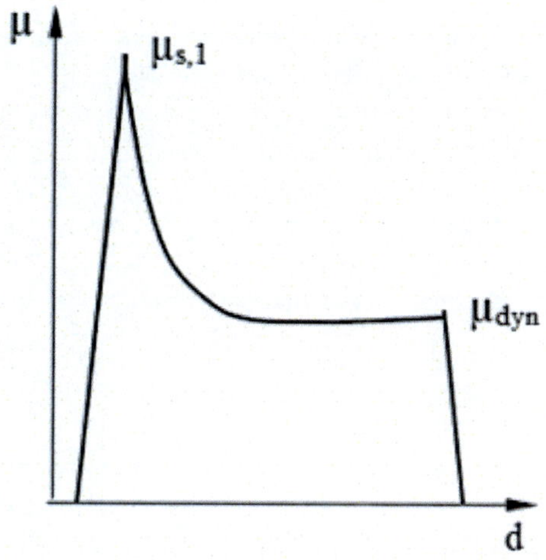

Fig. 5. Characteristic curve μ versus d of a polymeric sliding material after a pre-loading

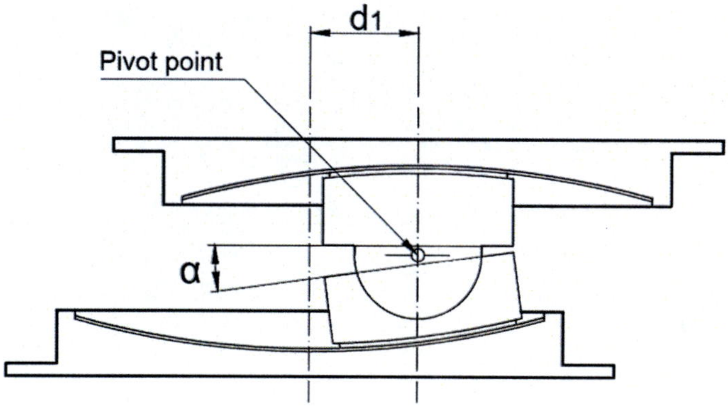

Fig. 6. Displacement d_1 with sliding occurring on the lower primary surface only

With reference to the symbols of Fig. 2, the movement of the upper surface (element B) will start only when element C enters an area of the lower primary surface with an average slope α such that the horizontal component of the friction force compensates for the force resulting from the difference $\Delta\mu$ between initial static $\mu_{s,1}$ and dynamic μ_{dyn}. Coefficients of friction.

With a simple force balance it is shown that this will happen at a displacement d_1 equal to:

$$d_1 = \Delta\mu(R - h) \tag{5}$$

We conclude that, during the impressed horizontal movements, relative rotations are generated in the DCSS between the two components B and C of the hinged intermediate element, which are easily accommodated by the hinge.

The transition between the first phase (when just one surface slides) to the second phase (when both surfaces slide together) is smooth.

The above has been demonstrated in the above mentioned experimental campaign conducted by Professor Michael Constantinou.

In the above studies on the phenomenon of non-concomitant displacements in Double Curved Surface Sliders, a seismic isolator with hinged intermediate slider was used.

The results are shown in Fig. 7, where the smooth transition between the two phases is clearly visible with the change of the slope (stiffness) in the experimental loop.

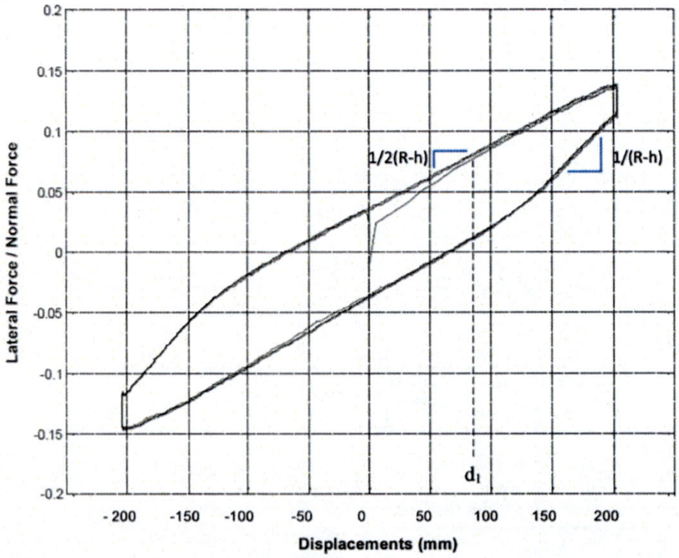

Fig. 7. Loop Force versus displacement recorded by Prof Constantinou

It should be noted that the phenomenon of non-concomitant displacements does not depend on the presence or absence of a hinge in the intermediate element.

Indeed, the studies of Prof. Constantinou were conducted on a hinged DCSS and a non-concomitant displacement equal to $d_I = 86$ mm has been observed (see Fig. 7).

However in this case no damages or permanent deformations in the sliding material were reported.

If the intermediate slider is solid, the relative rotation shall be accommodated by the deformation of the protruding parts of the sliding material.

On this matter, the European Norm EN 1337, in the sub-clause *6.8 Design verification for sliding surfaces* requires the following:

"*6.8 Design verification for sliding surfaces*

Deformation of sliding materials shall not be used to accommodate rotations."

Let's consider now the case of a solid intermediate slider with the same assumptions made for the previous case.

In the first phase the movement proceeds, but with a marked concentration of the pressure on the edge in the direction of the movement due to the high eccentricity of vertical load, as described in Chap. 3.- Statics.

In this phase the intermediate slider is dragged by the upper element A and is wedged between the two curved surfaces.

The rotation of the intermediate element can only be accommodated by the deformation of the protruding part of polymeric sliding elements (Fig. 8).

Fig. 8. Permanent deformation of the encased polymeric sliding materialdue to excess of contact pressure

When the deformation capacity of the latter runs out, the temporary block of the seismic isolator occurs and this means that in both curved surfaces we have a static coefficient of friction.

From this point on, the thermal displacement of the bridge deck is accommodated by the axial deformation of the latter (with large stiffness) and the flexural deformation of the piers (with lesser stiffness, due to their flexibility in the longitudinal direction).

This blocking phase can last a few minutes and the associated displacement can reach several mm, depending on the bridge span length and thermal gradient. When the horizontal force deriving from the deformation of the piers exceeds the frictional forces of the two sliding elements, the seismic isolator is released with a rapid ejection of the intermediate solid slider by the horizontal component of the vertical load in the opposite direction to that of the thermal movement.

Following the release, the seismic isolator is dragged by the pier and the bridge deck into a position close to that congruent with the ambient temperature.

The rapid ejection of the solid slider produces vibrations or even a "banging", which is particularly intense in the case of steel bridge decks.

Banging can be an unacceptable disturbance in urban areas.

However, the most worrying pitfall of these phenomena is the wear of sliding elements.

Not many experts know that the wear Δh of the sliding element depends on the product (pressure x velocity) according to a bilinear law similar to the one shown in Fig. 9.

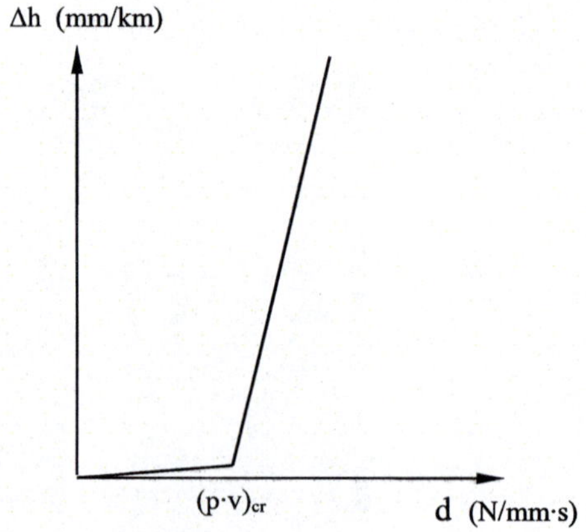

Fig. 9. The bilinear law of wearing versus (p · v)

For low (p · v) values, such as those produced by movements of thermal origin, the wearing rate (mm/km) is negligible.

When the critical value $(p \cdot v)_{cr}$ is exceeded, the wearing rate increases rapidly.

The $(p \cdot v)_{cr}$ value and the slopes of the two branches of the characteristic curve depend on the nature of the polymeric sliding material.

This is precisely what happens in the presence of vibrations induced by traffic, but above all by the phenomenon described in this paper.

From some experimental findings it appears that the latter occurs at values much higher than $(p \cdot v)_{cr}$.

The use of DCSS in bridges is too recent to have any valid statistical data.

Furthermore, the attitude of the Authorities, which are not inclined to disseminate information on a subject that could create alarm for users of road transport systems, does not help.

5 Conclusions

1. The version of DCSSs with solid intermediate slider presents some drawbacks, in that:
   - its use is subject to some restrictions for large-span bridges.
   - it causes important eccentricities of the vertical load (large stress concentration at the edges).
   - It produces parasitic displacements under heavy traffic.

- It is subjected to magnified stick-slip effect.

2. Its use is discouraged in the case of concomitance of two or more of the following adverse factors:
   - bridge decks with long spans → (large rotations and displacements).
   - steel bridge decks (poor thermal inertia) → (high rate of thermal displacements).
   - bridges erected in areas with large temperature range → high rate of thermal displacements.
   - use of sliding materials with high coefficient of friction → (large $\Delta\mu$ → large accumulation of potential elastic energy).
   - long natural period of the isolated structure → (means large R → large eccentricity → high stresses at the edges of sliding elements).

References

1. Constantinou, M.: Friction pendulum double concave bearing. Technical Report—State University of New York at Buffalo, NY 29 Oct 2004
2. Medeot, R. Zordan, T.: The new performance-oriented European standard on anti-seismic devices. In: 16th World Conference on Earthquake, 16WCEE 2017 Santiago de Chile, Jan 2017

Full Scale Dynamic Tests on Concave Curved Surface Sliders: Comparison of Time History and Cyclic Sinusoidal Tests

A. E. Pigouni[1], M. G. Castellano[2(✉)], N. Hima[3,4], and S. Infanti[5]

[1] Testing Laboratory, FIP MEC Srl, 35030 Selvazzano Dentro, Italy
[2] Academy Unit, FIP MEC Srl, 35030 Selvazzano Dentro, Italy
maria.gabriella.castellano@fipmec.it
[3] DICAM, University of Trento, 38123 Trento, Italy
[4] FIP MEC Srl, 35030 Selvazzano Dentro, Italy
[5] Director of Testing Laboratory, FIP MEC Srl, 35030 Selvazzano Dentro, Italy

Abstract. Full scale dynamic testing of seismic isolation devices in combined compression and shear is nowadays required by all international standards on seismically isolated structures and/or on anti-seismic devices, for prototype tests, and sometimes for production control (or acceptance) tests. The Type tests aim at understanding the effects on the isolator's hysteretic behavior of different actions, such as vertical loads, amplitude and velocity of the horizontal displacements, and at introducing such effects in proper modelling of seismically isolated structures. The dynamic tests required by the standards are usually cyclic sinusoidal tests under different vertical loads, with different amplitude, frequency and number of cycles, thus resulting very demanding, in terms of global energy dissipation and cumulated displacement. Testing protocols for production control are usually simpler. As a contribution to future revision of the testing protocols in the standards, this paper presents a comparison between cyclic loading tests and Time History (TH) tests performed under the design vertical load on a pendulum isolator characterized by ±450mm maximum displacement capacity. The input of TH tests is the displacement TH resulting as output of non-linear dynamic analyses on the structure in which the isolator are installed. A total of 9 time-history tests were performed: 3 couples of records of Turkish earthquakes, scaled to the MCE spectrum, and 3 generated earthquakes compatible with the same spectrum. The results are compared with that of cyclic sinusoidal tests similar to those required by standards, in terms of energy dissipation and cumulated displacement.

Keywords: Seismic isolation · Dynamic testing · Curved surface slider · Time history

1 Introduction

1.1 Testing Protocols Requirements

Prototype tests (or Type tests) and quality control (or factory production control) tests on seismic isolation devices are required by all current Standards and regulations, such as the

G. P. Cimellaro (Ed.): WCSI 2022, LNCE 309, pp. 388–402, 2023.
https://doi.org/10.1007/978-3-031-21187-4_32

European Standards EN15129:2009 on Anti-seismic devices [1], American Association of State Highway and Transportation Officials AASHTO 2010 [2], Uniform Building Code (UBC) 1997 [3], ASCE/SEI 7–05 [4], NCh2745:2013 [5]. The prototype series of tests intend to verify the device's performance characteristics under a wide range of seismic and non-seismic load combinations, as well as its possible variations during all the life of the structure. Principal design properties, namely force vs displacement hysteretic loops or effective characteristics (effective stiffness and equivalent viscous damping values) are required to be determined, through a series of dynamic tests. The quality control tests are conceived in order to verify the quality and constancy of production with reference to the prototype behavior and/or to the nominal characteristics. In all the above-mentioned standards, the Prototype tests should be performed on minimum two full-scale specimens of each type, and size identical or similar to that used in the design.

Full-scale type test procedure for Curved Surface Sliders (CSS), as per EN 15129:2009 definition, or pendulum isolators, requires the assessment of at least the following characteristics: (a) vertical load bearing capacity; (b) horizontal restoring stiffness; (c) friction coefficient (or energy dissipation capability); and (d) horizontal displacement capacity. Characteristic (b) and (c) are the most important parameters that define the horizontal hysteretic cycle and its possible variation under different conditions. A large series of dynamic tests are required by all standards in order to verify the effects of vertical load, displacement amplitude, frequency, repeated cycling (and associated temperature increase) on these two parameters. Often the shape of the signal to be applied is indicated as well. The specified frequency when associated to the design displacement provides the range of velocity necessary to perform the tests. The peak velocity often is very high, that is why in the last 20 years many new testing rigs have been designed and built [6].

In all the Standards, the loading history and the number of cycles to be applied is described. In all cases type testing begins with one or a set of static tests in order to verify the resistance of the isolator under service load conditions. The EN15129:2009 requires prior to any test, the isolator to be subjected to a pre-loading equal to the design load and a load bearing capacity test by applying a vertical load equal to two times the design load. The Wind test is included in all Standards: usually 20 fully reversed cycles are performed at the non-seismic displacement applying the design axial load. EN15129:2009 includes a Frictional resistance test under service conditions.

After the static tests, the dynamic sliding isolation tests follow. All Standards require the performance of a series of dynamic tests by applying three fully reversed cycles, at various amplitude values expressed as percentage of the design displacement or the maximum displacement, under a certain vertical load, in most cases equal to the quasi-permanent vertical load (i.e. average load during earthquake). Three displacement increments according to the EN15129 and four in NCh2745, ASCE/SEI 7–05 and AASHTO 2010.

The EN15129, ASCE/SEI 7–05 and NCh2745 require two seismic tests, where three fully reversed cycles are performed with an amplitude equal to the design seismic displacement and the applied vertical load equal to the maximum and minimum seismic load, respectively. In the NCh2745 these tests are performed at different amplitude (expressed as percentage of the maximum displacement), instead of at maximum displacement only.

The EN15129 would require a seismic performance verification performed with the simultaneous application of a sinusoidal displacement input waveform in two perpendicular directions (bidirectional test), obtaining a "clover leaf" path. However, since few testing rigs are capable of performing a multidirectional test, both EN15129 [1] and AASHTO 2010 [2] require instead to perform a test that can be completed after a rotation of 90 ° of the bearing in order to involve a displacement path perpendicular to the one verified with previous tests.

Finally, AASHTO 2010, ASCE/SEI 7–05, NCh2745 require two stability tests (static tests) applying a vertical load equal to the maximum and minimum seismic load, respectively.

In all Standards from nine up to twelve fully reversed cycles are performed at maximum displacement and maximum velocity. Minimum three values of vertical loads are applied: quasi-permanent load, maximum and minimum seismic load.

Regarding the factory production control tests (acceptance tests) in every Standard different percentage of production tests is required. EN15129:2009 requires testing 5% of the production, ASCE 7–05 and AASHTO 2010 100% of the production, NCh2745 15%. Of course, the number of tests required for the factory production control are much less than for type testing. The EN15129:2009, AASHTO 2010 and NCh2745 require a vertical load bearing capacity test at maximum load. All Standards ask for a combined compression and shear test by subjecting the isolator to three fully reversed cycles of loading at the design displacement. Furthermore, NCh2745 asks for a second sliding test with a velocity not less than 150 mm/s and smaller displacement (25 mm). The ASCE 7–05 is quite incomplete concerning the specifications for factory production control tests, compared to the other standards.

The aim of the research described in this paper is to give a contribution for future revision of testing protocols, both for type tests and for factory production control tests.

2 Experimental Campaign

2.1 The Double Curved Surface SLIDER'S Characteristics

The Double Concave Curved Surface Slider (DCCSS) used for this experimental campaign is identical to those designed and manufactured by FIP MEC for the seismic isolation system for an ammonia tank located in Samsun, Turkey [7]. It is characterized by an equivalent radius of curvature of 4.5 m, maximum seismic vertical load of 3100 kN, design vertical load 2110 kN, maximum horizontal displacement ±450 mm and a maximum design velocity of 530 mm/s. It is equipped with an Ultra-High Molecular Weight Poly-Ethylene (UHMW-PE) sliding material. Figure 1 shows the device during the prototype testing performed, whereas Table 1 lists the type tests required for these isolators according to the European Standard EN15129:2009. Actually, the testing protocol in this case included additional tests required by the Client, but the complete results of the type tests are not within the scope of this paper. However, it is worth noting here that the complete type test procedure according to the EN15129 for this project

Table 1. Type test matrix according to EN15129:2009

Test Name		Ampl. [m]	Max Velocity [m/s]	Frequency [Hz]	Load Shape	Vertical Load [kN]	No. Cycles	Total displacement [mm]
Pre-Test		–	–	–	Constant	2210	–	–
Load Bearing Capacity		–	–	–	Constant	4420	–	–
Frictional Resistance		±0.006	0.0001	0.00417	Triangular	2210	1	24
Sliding Isolation Tests	Service	±0.025	0.005	0.032	Sine	2210	20	2000
	Benchmark	±0.359	0.05	0.022	Sine	2210	3	4308
	Dynamic 1	±0.090	0.530	0.940	Sine	2210	3	1080
	Dynamic 2	±0.180	0.530	0.470	Sine	2210	3	2160
	Dynamic 3	±0.359	0.530	0.235	Sine	2210	3	4308
	Seismic	±0.359	0.530	0.235	Sine	3098	3	4308
	Seismic	±0.359	0.530	0.235	Sine	1322	3	4308
	Bi-directional	±0.359	0.53	0.235	Sine	2210	3	4308
	Property verification	±0.359	0.53	0.235	Sine	2210	3	4308
Post-Test		–	–	–	Constant	2210	–	–

consists in applying a total displacement of 6195 mm, excluding service and benchmark tests, i.e. considering only the dynamic tests at maximum velocity (530 mm/s).

2.2 Dynamic Test Facility

The tests were carried out at FIP MEC's Test Laboratory of Padua (Italy) using the Biaxial Dynamic Test Facility (Fig. 1a). It's a two-degree-of-freedom system able to accommodate full-scale isolation devices of different sizes and capable of applying a wide range of vertical and horizontal loads, displacements and velocity. The performance characteristics are: static maximum vertical load of 30000 kN and 20000 kN dynamically; horizontal actuator capacity of 3000 kN; ±500 mm total horizontal stroke and ±1570 mm/s maximum velocity [6].

It consists of steel vertical test frame coupled to a horizontal frame designed to withstand the vertical and lateral load imposed to the isolator specimen. The specimen is installed between the loading platen and the bottom moving platen. The relative horizontal displacement is imposed to the isolator by moving the bottom platen connected to the horizontal actuator. Considering that the most of the required dynamic tests never exceed 5 full continuous cycles, the available power system can provide for both vertical and horizontal actuators consumption. The general arrangement of the biaxial test facility of FIP MEC new laboratory is presented in [6]. Figure 1b shows a double curved surface slider during testing.

The measurements of the displacements and the horizontal loads are carried out using a displacement transducer and a load cell of 500 kN and 3000 kN; the vertical load is measured by means of load cells, connected to the amplifier control system actuator whose output signals are acquired through a measuring amplifier. This amplifier, connected via PC with data acquisition and processing software DIA Dem (National Instruments), provides the final signal then processed in the graphs.

The inertial effects were evaluated considering the average acceleration measured by means of two accelerometers arranged according to the horizontal movement axes, and then extracted from the signal measured with the load cell arranged to the same axis.

(a) (b)

Fig. 1. Biaxial test system - FIP MEC Dynamic Testing Laboratory **(a)** DCSS under testing **(b)**

2.3 Test Procedure

The DCCSS was subjected to a series of dynamic tests, inspired at the current standards and guidelines, imposing a sinusoidal input waveform with an amplitude equal to 20% (D1), 40% (D2), 60% (D3), 80% (D4) and 100% (D5) of the design seismic displacement (d_{bd} = 359 mm). The frequency was maintained constant at $f = 0.235$ Hz, the vertical load applied was equal to 2210 kN, and the velocity was chosen in relation to the displacement d_x, according to formula (1):

$$v_{Ed} = 2 \cdot \pi \cdot f_x \tag{1}$$

The vertical load applied in all tests was equal to the quasi-permanent load N_{Sd} (average load during earthquake). Table 2 presents the test matrix of the dynamic tests performed with sinusoidal signal input.

According to the EN15129:2009, the Benchmark test was performed at the beginning of the test procedure, applying the quasi-permanent load N_{Sd} and imposing an amplitude equal to the design seismic displacement d_{bd} at a peak sliding velocity of v = 50 mm/s.

In each test three (3) sinusoidal cycles were performed, imposing also an entrance and exit cycle of smaller amplitude, controlling in this way the initial acceleration which otherwise is very high (Fig. 2d). In the elaboration of the experimental data presented later on, these entrance and exit cycles are excluded.

Table 2. Experimental campaign test matrix

Test Name	Label	Main DOF	Ampl. [mm]	No. Cycles
Benchmark	P1	Long	±359	3
Dynamic 1	D1	Long	±090	3
Dynamic 2	D2	Long	±180	3
Dynamic 3	D3	Long	±269	3
Dynamic 4	D4	Long	±359	3
Dynamic 5	D5	Long	±450	3

The dynamic friction coefficient for the sinusoidal tests is calculated from the energy dissipation, when computed for 3 cycles, as follows in Eq. (2):

$$\mu_{dyn,3} = \frac{1}{3}\sum_{i=1}^{3} \frac{A_{h,i}}{4 \cdot N_s \cdot d_x} \tag{2}$$

where $A_{h,i}$ is the area enclosed within the hysteresis loop in the i-cycle; N_s is the value of vertical axial load under which the isolator was tested; d_x is the value of the peak horizontal displacement achieved during the test.

After the dynamic tests (Table 2) the device was subjected to three sets of displacement time-histories corresponding to non-linear time-history analyses carried out on the isolated structure with different input time-histories, namely three near-fault natural earthquakes occurred in Turkey (Erzincan 1992, Kocaeli 1999 and Duzce 1999), each for two horizontal components (Fig. 2a, b, c). After these 6 tests, other three displacement time-histories were imposed, named MCE1, MCE2, MCE3, corresponding to non-linear time-history analyses of the structure subjected to 3 generated acceleration time histories, compatible with the design spectrum at Maximum Considered Earthquake (MCE). Such displacement time-histories are characterized by peak displacements up to 355 mm, reaching peak velocities higher than 900 mm/s. The vertical load was kept constant in all tests, as in the sinusoidal tests, for comparison reasons.

The average dynamic friction coefficient for time history tests was calculated using Eq. (3):

$$\mu_{dyn,TH} = \frac{A_{h,TH}}{N_s \cdot d_{tot}} \tag{3}$$

where $A_{h,TH}$ is the area enclosed within the hysteresis loop of the time history test, d_{tot} is the cumulated displacement and N_s is the value of vertical axial load under which the isolator was tested.

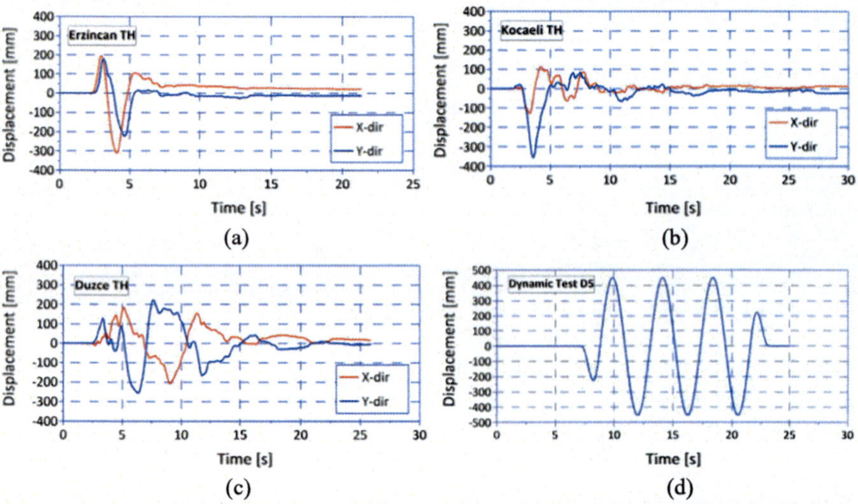

Fig. 2. Displacement Time Histories applied to the isolator **a** Erzincan, **b** Kocaeli, **c** Duzce and **d** sinusoidal signal of dynamic test D5

In the following, the results obtained from the dynamic cyclic tests are presented and compared with those from the earthquake time-history analyses and the MCEs. The time histories tests are compared with the cyclic sinusoidal tests mainly in terms of cumulated displacement and total energy dissipation obtained.

3 Experimental Results

3.1 Dynamic Tests

The hysteretic responses of the dynamic tests performed with sinusoidal input waveform are presented in Fig. 3. The average friction coefficient of the five tests was 9.1% for the 1st cycle and 8.2% the average of the three cycles. Although the total displacement after the five dynamic sinusoidal tests was 16190 mm and the total energy dissipated was 2854 kJ, the isolator's behaviour is very stable (Fig. 3f). Despite the large difference amongst the 5 tests in terms of peak velocity, no significant difference is observed.

The hysteretic loops of the natural and artificial time history tests are given in Fig. 4. The loops of the three near-fault natural earthquakes in both directions exhibit a more irregular behaviour than the artificial time-histories, due to the impulsive and instantaneous peak displacements up to 355 mm (Kocaeli Y-dir), reaching peak velocities up to 774 mm/s with peak acceleration up to 0.66g (Erzincan X-dir). Although during the artificial time history tests the device exhibit also very high accelerations (about 1.0 g) and displacements, the irregularity of the hysteretic loops, in particular in the restoring stiffness, seems to be attributed mainly to the impulsive movements of the natural time-histories. It is worth noting that Pavese et al. [8], who have carried out bi-directional time history tests based on scaled 1994 Northridge earthquake, attribute the irregularity of the hysteretic loops mainly to the continuous variation of the frictional force direction associated to the bi-directionality of the tests. However, the results shown in Fig. 4 a, b, c, d, e, and f, show similar irregularity even in unidirectional time history tests. As noted in [8], the vertical load variation during the tests is also an important factor that contribute to this irregularity, especially at peak accelerations. In effect, it is well known that it is not easy at all to maintain the vertical load constant during these high velocity and high displacement tests on CSS, because of the vertical movement associated to horizontal movement due to the sliding on curved surfaces.

In order to reduce the effects of the fluctuations of vertical load on the shape of the hysteretic loops, the horizontal forces measured in both sinusoidal and seismic TH tests were normalized. The vertical load variation during the first three sinusoidal tests (D1, D2 and D3) was less than 8%, whereas in tests D4 and D5 it reached a value of 18% due to the larger displacements. For this reason, the horizontal force was normalized with respect to the average vertical load, in order to reduce the variability of the horizontal force with respect to the vertical load. The comparison of the loops as acquired and normalized shows the effectiveness of such procedure (Fig. 5a). The same procedure was applied to the results of the natural and artificial time history tests. However, in this case the shape change in the hysteretic loop is almost negligible (Fig. 5b), due to the very small vertical load variation during most of the earthquake simulations. The maximum vertical load variation is achieved only during the cycle of the maximum displacement.

Table 3 presents the cumulated displacement and energy dissipation for all tests. The total displacement cumulated throughout the sinusoidal tests (from D1 to D5) was 16193 mm with a total cumulated energy dissipated equal to 2854 kJ. In the 6 natural seismic time-history tests the total displacement cumulated was 8939 mm and the total cumulated energy dissipated equal to 1851 kJ. Comparing these numbers, it is evident that the total displacement applied to the device in the 5 cyclic tests similar to those required

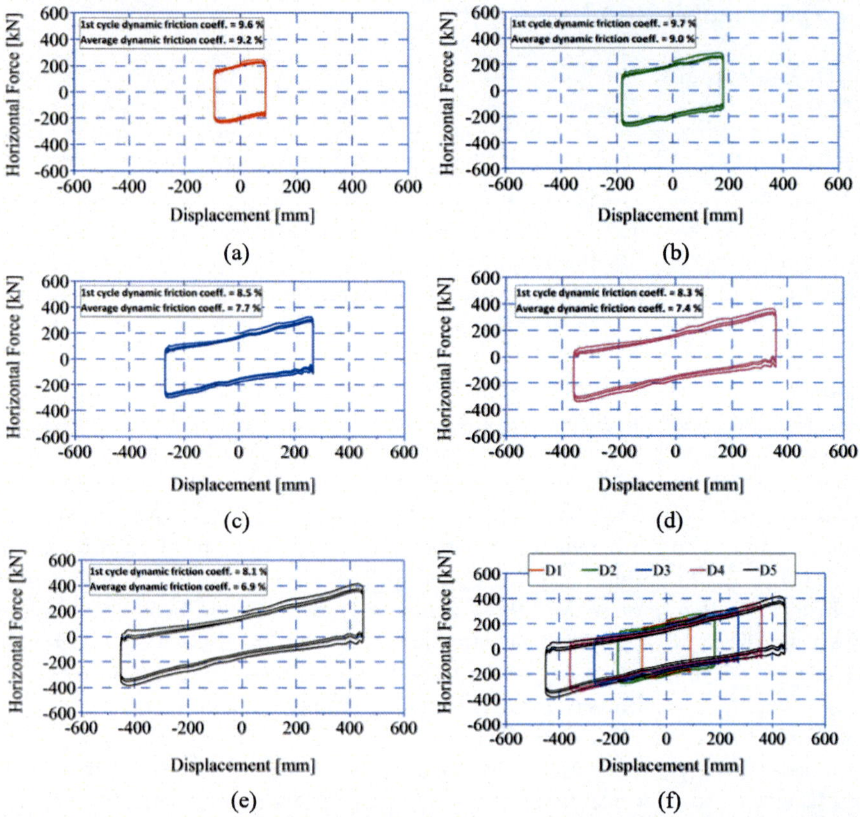

Fig. 3. Force vs. displacement graphs obtained in dynamic cyclic tests: **a** D1, **b** D2, **c** D3, **d** D4, **e** D5, **f** comparison of all the dynamic cyclic tests

by the Standards is almost two times greater than that applied in the six components of the 3 natural earthquakes together. Of course, this is mainly due to the near fault nature of the natural earthquakes considered, that gives only one cycle at large displacement, and few cycles with much smaller displacement. It can be observed that in the 3 artificial time-histories the total cumulated displacement is 9800 mm, much larger than in natural earthquakes, and the total energy dissipated was 1861 kJ.

Considering the type tests performed at maximum displacement and maximum velocity, according to EN 15129 (Table 1), the total cumulated displacement was 21540 mm, much greater (from 7 to 12 times) than any of the three natural earthquakes, considering the linear combination of vectors of the two directions X-Y (Table 3), and about 3 times greater than the total cumulated displacement of the three natural earthquakes (X-Y combined).

Comparing the single tests, it is clear that in the sinusoidal tests D4 and D5 the device exhibits larger values of total displacement and therefore of energy dissipation than in any other test. The natural THs show a total displacement that ranges from about 960 mm up to 2183 mm (Duzce Y-dir), with energy dissipations ranging between about 210 kJ

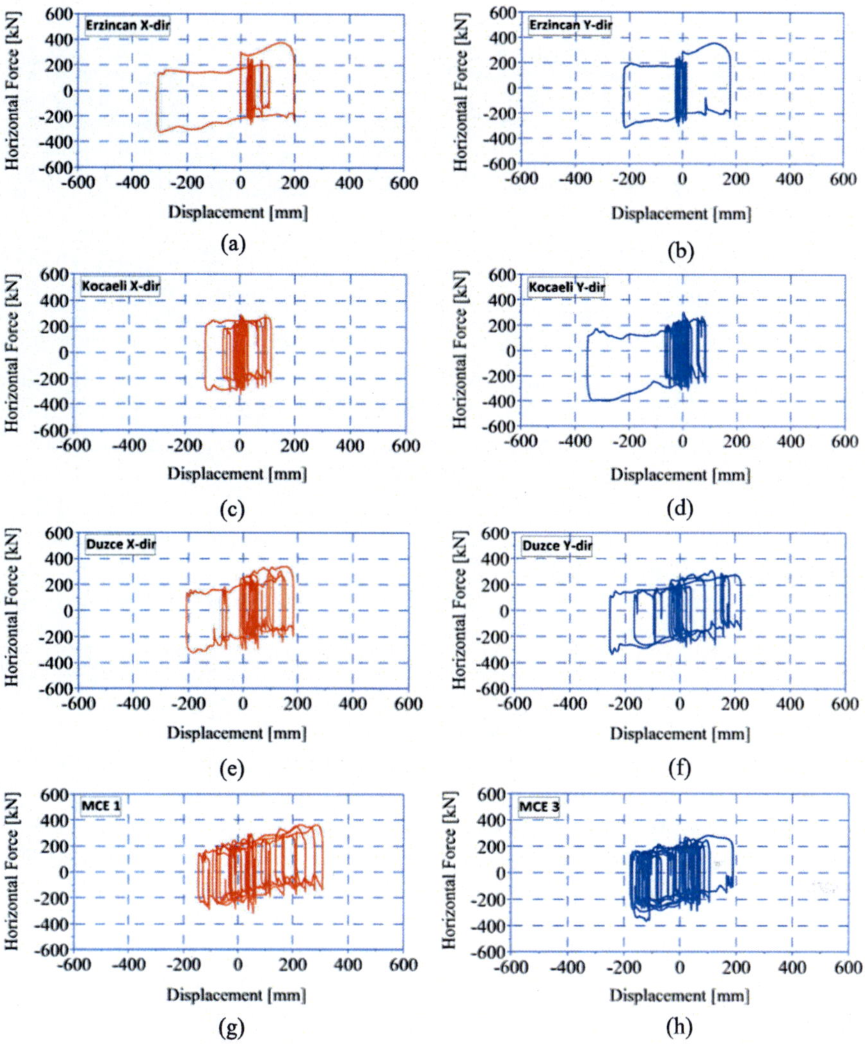

Fig. 4. Force vs. displacement graphs obtained in time history tests

and 433 kJ. Each MCE test accumulated a total displacement of more than 3000 mm with total energy dissipation up to 681 kJ. This confirms that artificial spectrum-compatible earthquakes are in general more severe than natural earthquakes.

It is clear that the accumulated displacement is very important to quantify the severity of a test in terms of possible wearing of the sliding material. From the data in Table 3 it is also evident a correlation between the total cumulated displacement in each test and the total dissipated energy. To check this correlation, Fig. 6 presents total energy dissipation vs. total displacement for all the tests. Despite the quite different inputs imposed, and the strong differences amongst the tests in terms of maximum displacement and maximum velocity reached (Table 3) the linear correlation for cyclic sinusoidal tests and TH tests are

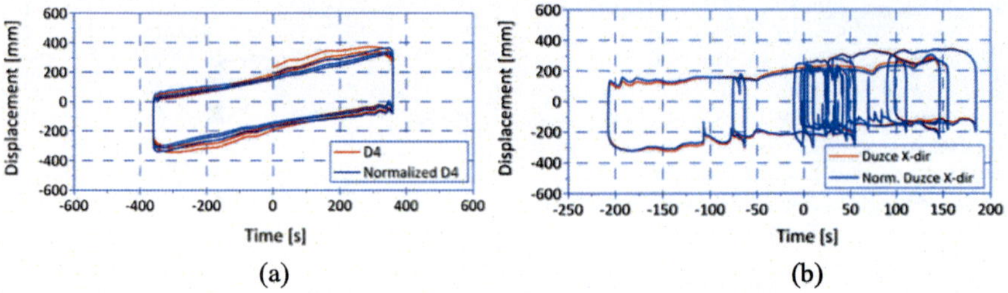

Fig. 5. Normalization comparison graphs. **a** Test D4 and **b** Seismic TH Duzce X-dir

quite similar. This demonstrates that the device behaviour in terms of energy dissipation capacity is not significantly dependent on the shape of the displacement input (sinusoidal or earthquake simulation), but mainly on the total cumulated displacement. Thus, the cyclic dynamic tests required by the Standards can predict the behaviour of the device when subjected to an earthquake. It is also evident that the dynamic tests at different amplitudes can cover a wide range of possible earthquakes in terms of total displacement performance and energy dissipation.

In Fig. 7 the two histograms present the results of the total displacement (Fig. 7a) and total energy dissipation (Fig. 7b).

It is clear that the natural and artificial time-histories applied are covered by the dynamic test D4 which is the test required by all Standards, performed at the design displacement. Although the total displacement of the D4 test is greater from 16% (MCE 3) up to 58% (Erzincan), the total energy dissipation (Fig. 7b) the device can exhibit during an earthquake can be as the D4 test (e.g.: Duzce only 15% less energy dissipated than during the D4 test).

Finally, Fig. 8 presents the comparison of the hysteretic loops between dynamic tests and TH tests. The selection of the tests to compare was based on the similarity of the maximum displacement reached. The behavior of the device is almost the same with perfect overlapping in many cases. Although the device underwent 14 dynamic and seismic tests its response is very stable in terms of restoring stiffness and friction coefficient. Comparing D2 with MCE 3 (Fig. 8 b), which are respectively the 2nd test and the last test performed, it can be noticed the similar behavior; the friction coefficient of the MC3 results only 3% lower than the D2 test. The same performance was observed by comparing D2 with Duzce X-dir. Again, the friction coefficient reduction was only 3.6%, with perfect overlapping of the hysteretic loops (Fig. 8 d).

Table 3. Total displacement and energy dissipation test results

Test name	Max. Velocity Reached mm/s	Max. Displacement [mm]	Total Displacement [mm]	Combined Total Displacement [mm]	Total Energy Dissipation [kJ]
Dynamic 1	133	90	1086	–	217.7
Dynamic 2	266	180	2163	–	422.9
Dynamic 3	397	269	3231	–	569.5
Dynamic 4	530	359	4311	–	743.6
Dynamic 5	664	450	5402	–	899.8
Seismic-Erzincan X	774	307	1275	1795	275.5
Seismic-Erzincan Y	574	222	957		212.4
Seismic-Kocaeli X	620	127	1332	2366	278.3
Seismic-Kocaeli X	568	355	1605		331.0
Seismic-Duzce X	424	207	1587	2973	321.1
Seismic-Duzce Y	675	257	2183		432.6
MCE 1	923	308	3184	–	605.2
MCE 2	778	314	3004	–	574.2
MCE 3	725	188	3612	–	681.4

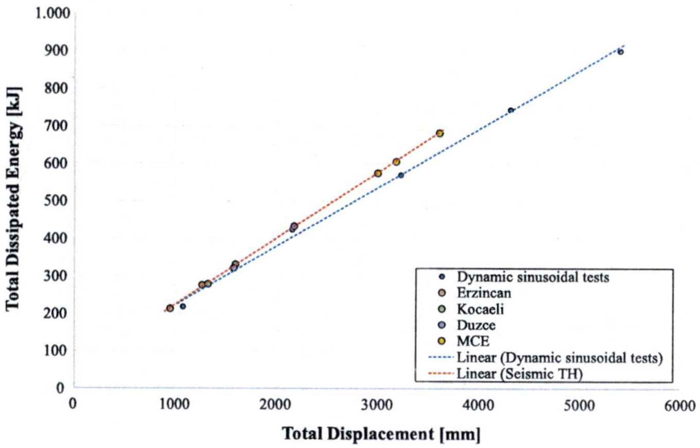

Fig. 6. Total energy dissipation vs. total displacement test results

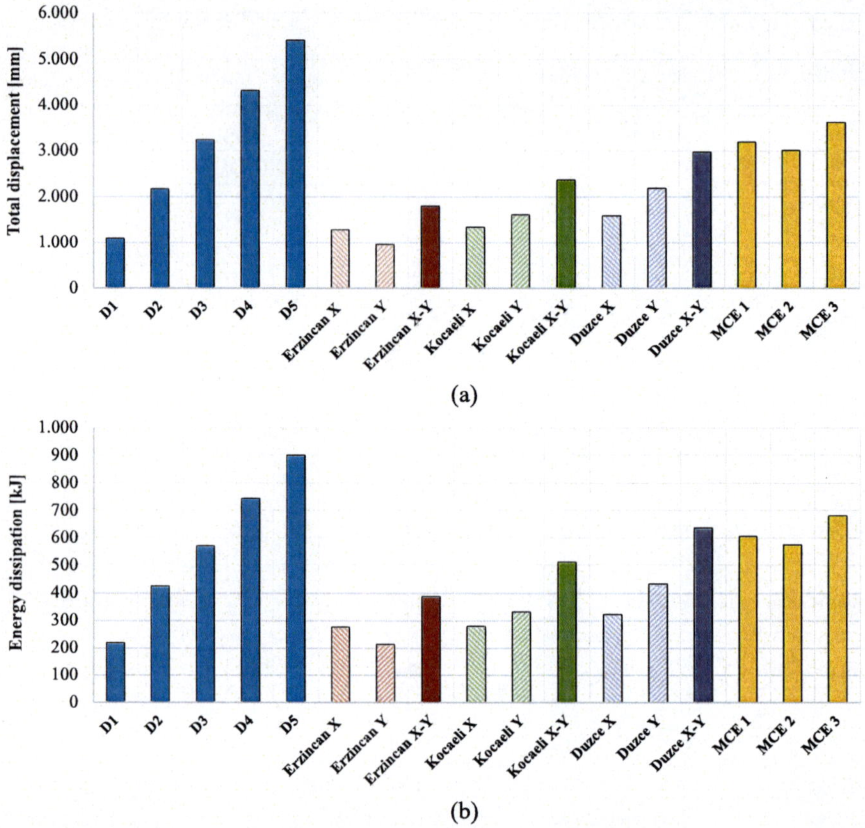

Fig. 7. Sinusoidal and TH test histograms **a** Total displacement accumulated and **b** total energy dissipation.

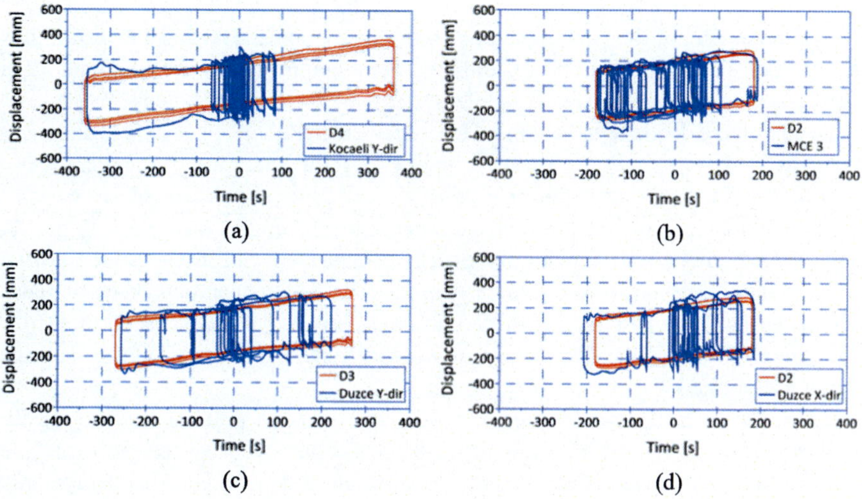

Fig. 8. Comparison between sinusoidal tests and TH tests with similar maximum displacement

4 Conclusions

The prototype testing (Type test) of the Curved Surface Sliders seismic isolation devices is required in all international standards. Type testing is composed by a series of cyclic sinusoidal tests under different vertical loads, with different amplitude, frequency and number of cycles thus resulting very demanding, in terms of global energy dissipation and cumulated displacement.

As a contribution to future revision of the testing protocols in the standards, the present paper showed a comparison between cyclic loading tests and Time History (TH) tests performed under the design vertical load on a double concave curved surface slider characterized by design vertical load 2110 kN, design horizontal displacement ±359 mm, maximum horizontal displacement ±450 mm and a maximum design velocity of 530 mm/s. The device was subjected to five sinusoidal dynamic tests inspired at the current standards and guidelines and a total of 9 time-history tests. The displacement TH applied are the output of non linear analyses of the isolated structure when subjected to 3 couples of records of Turkish earthquakes (Erzincan, Kocaeli and Duzce), scaled to the site specific MCE spectrum, and 3 generated earthquakes compatible with the same spectrum.

The time histories tests were compared with the cyclic sinusoidal tests mainly in terms of cumulated displacement and total energy dissipation obtained.

Initially, the behavior of the device was analyzed considering the hysteretic loops obtained in each test. In the hysteretic response of the dynamic tests, despite the large difference amongst the 5 tests in the maximum velocity reached, no significant difference was observed. The loops of the three near-fault natural earthquakes in both directions exhibited a more irregular behaviour than the artificial time-histories due to the impulsive and instantaneous peak displacements and accelerations. The impulsive movements and the vertical load variations seem to be the main reasons of the irregularity of the response.

The comparison of the hysteretic loops between dynamic tests and TH tests showed that, although the device in this experimental campaign underwent 14 dynamic and seismic tests, its response was very stable in terms of friction coefficient.

The cumulated displacement applied to the device in the 5 cyclic tests was almost two times greater than that applied in the six components of the 3 natural earthquakes together, not surprising due to the near fault nature of those earthquake. It was observed that in the 3 artificial time-histories the total cumulated displacement was much larger than in natural earthquakes with much more energy dissipation confirming that the artificial spectrum-compatible earthquakes are in general more severe than natural earthquakes.

Considering the type tests performed at maximum displacement, according to EN 15129, the total cumulated displacement was much greater than any of the natural or artificial TH applied.

Despite the quite different inputs imposed, and the strong differences amongst the tests in terms of maximum displacement and maximum velocity reached, the linear correlation for cyclic sinusoidal tests and TH tests was quite similar. This demonstrated that the device behaviour in terms of energy dissipation capacity is not significantly dependent on the shape of the displacement input (sinusoidal or earthquake simulation), but mainly on the total cumulated displacement. Thus, the cyclic dynamic tests required by the Standards can predict the behaviour of the device when subjected to an earthquake. It was also evident that the dynamic tests at different amplitudes can cover a wide range of possible earthquakes in terms of total displacement performance and energy dissipation. However, future experimental research with input related to other earthquake records - in particular far field earthquakes - would be useful.

References

1. European Standard EN 15129:2009: E Anti-seismic devices CEN, Bruxelles, Nov 2009
2. American Association of State Highway and Transportation Officials AASHTO 2010: Guide Specifications for Seismic Isolation Design, 3rd edn, July 2010
3. Uniform Building Code (UBC): Earthquake Regulations for Seismic—Isolated Structures, Division IV (1997)
4. Structural Engineering Institute ASCE/SEI 7-05: Seismic Design Requirements for Seismically Isolated Structures, Chap. 17
5. Norma Chilena, NCh2745:2013: Análisis Edificios Con Aislacion Sísmica
6. Infanti, S., De Toni, S., Pigouni, A.E.: Dynamic testing protocols for seismic protection devices: new challenges for test facilities. In: 14th World Conference on Seismic Isolation, 9–11 Sept, San Diego USA (2015)
7. Castellano, M.G., Marcolin, L.: Seismic isolation of an ammonia tank in Turkey through pendulum isolators. In: 16th World Conference on Earthquake, 9–13 Jan 2017, Santiago Chile (2017)
8. Pavese, A., Furinghetti, M., Casarotti, C.: Experimental assessment of the cyclic response of friction-based isolators under bidirectional motions. In: Soil Dynamics and Earthquake Engineering, Nov 2018, vol. 114, pp. 1–11 (2018)

Seismic Isolation and Energy Dissipation in Bridges and Viaducts

Seismic Isolation of Bridges: Practice-Oriented Considerations

M. Molinari, D. Pastorelli(✉), M. Cademartori, and S. Dellacasagrande

Rina Consulting SPA, Via A. Cecchi 6, 16129 Genoa, Italy
{marco.molinari,daniele.pastorelli,marcello.cademartori,
simone.dellacasagrande}@rina.org

Abstract. Seismic isolation is an effective method to control seismic forces, seismic displacements and to govern modal shapes on a structure or a part of it. Seismic isolation devices are widely used in the design of new bridges and retrofit of existing structures due to their effectiveness in the reduction of seismic forces and the relatively simple methods of installation. This work aims to provide practical considerations to be included when dealing with the design of new bridges equipped with seismic isolator devices and in particular the paper focuses on the study of seismic isolator stiffness variation and inclusion of soil-structure interaction. The purposes of the paper are a) to set a fast design procedure for the design of seismically isolated bridges b) to investigate how the stiffness of an equivalent system composed of foundations, piers and bearings can be affected by varying the stiffness of the individual components. Both these objectives will be described following a sort of flow chart containing the different main steps to be included for the design of seismically isolated bridges. This study can provide practical oriented considerations to the designers dealing with seismic design of isolated bridges and can help to understand how stiffness variations of the single components might affect the global behavior of the structure and thus the sizing of the structural elements.

Keywords: Seismic isolation · Bridges · Fast design procedure · SSI

1 Basic Concept of Seismic Isolation of Bridges

1.1 Introduction

Seismic isolation techniques have increased their usage during the last years, especially in high intensity earthquake prone zones.

Bridges are, in most cases strategic structures which need to be protected from damages ensuring functionality after a seismic event and seismic isolation could be used to concentrate damages in few elements, easy to check and replace.

The main reason to prefer seismic isolation over other seismic design choices is to take advantage of the period elongation, by means of a decrease in stiffness, in order to reach convenient values in terms of acceleration (i.e., forces) demand for the structure

G. P. Cimellaro (Ed.): WCSI 2022, LNCE 309, pp. 405–416, 2023.
https://doi.org/10.1007/978-3-031-21187-4_33

or a part of it [1]. With this approach, it is possible to protecting a part of the structure, assuring an elastic response.

Since bridges are usually simple structures regarding the expected global behavior and most of the mass is concentrated at the deck level, seismic isolation could be particularly convenient to apply corrections to the global response by means of the modification of stiffness distribution.

A bridge could be irregular from the point of view of the longitudinal distribution of mass (e.g. in case of a variation of the deck width along the longitudinal axis) or stiffness (e.g. for the presence of piers with different transversal sections or different heights), hence the main goal of the above mentioned modification is to have an isolated system with optimized motion capacity in order to avoid torsion and with respect to isolation devices and expansion joints displacement capabilities.

In addition, seismic isolation is also effective to increase the energy dissipation capacity of the system or to regularize the seismic behavior of the structure.

At the same time, within the design process of an isolated structure such a bridge, Soil-Structure Interaction (SSI) is an important issue to be considered and it may have greater or lesser extent in sizing phase of the structural elements.

This work aims to show the effect of the variation of substructures (i.e., piers and foundation piles) stiffness on an isolated bridge, in order to understand its influence on the global behavior.

2 Simplified Preliminary Design of Seismic Isolation System

The present research work aims at providing a simplified preliminary design procedure for a fast initial design of the isolation system of a bridge, according to the most usual typologies of devices adopted in common practice for these structures. Similar procedures have been already developed and are available in literature, especially for buildings and other civil structures [2].

The procedure here presented is based on a simplified schematization of the bridge in sub-components as represented in Fig. 1: each support of the bridge is treated as a system of springs (characterized by their values of stiffness) and masses. The approach followed is that of the "sub-component method" presented in Priestley et al., 1996 [3].

The springs accounted in this approach, placed in series, from the bottom to the top of the support, are the following:

- Resultant spring of the foundation piles (the foundation piles are seen as springs in parallel), calculated from geotechnical assessment of pile-soil interaction.
- Spring of the pier (stiffness calculated from the Young Modulus E, the Moment of Inertia J_{pier} of the cross section and the height of the pier h_{pier}), considering both longitudinal and transversal inertia.
- Resultant spring of the isolation system (the isolation devices are seen as springs in parallel).

A preliminary knowledge of deck geometries, position of piers and abutments, expected mass of the deck in the seismic condition, mass and inertia of deck, piers

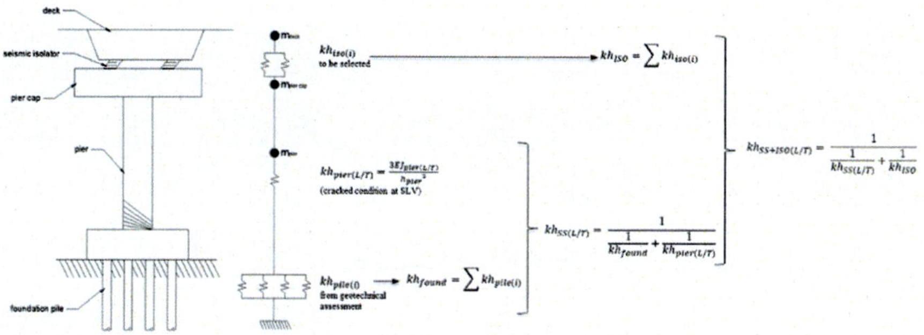

Fig. 1. Simplified spring-mass system (foundation piles, pier and isolation devices).

and pier caps is necessary to set up the procedure. In parallel, it is important to know the foundation layout, hence a preliminary evaluation of the soil springs due to SSI.

Since abutments are assumed extremely rigid with respect to the piers, masses and inertia of the abutments and foundation plinth is neglected.

Figure 2 shows a flow chart of the simplified procedure here presented.

A first evaluation of the period of piers, taken as single elements (i.e., without isolation system and SSI), is performed with the purpose to assess the influence of concrete cracking in the positioning of the first mode of the piers in the pseudo-acceleration response spectrum. In this step it is also possible to determine the position of the center of mass of the bridge, CoM.

The second step is to repeat the same evaluation, considering the SSI effect (i.e., springs of the foundation system) to assess its influence in the determination of the substructures natural periods, and subsequentially on the seismic action these can receive due to their own mass and stiffness. Dynamic stiffness of pile group can be calculated with analytic procedure available in literature, e.g., the one proposed by Novak et al. [4].

Finally, the stiffness of the isolation devices is chosen with the aim to meet several requirements:

- the period of the structure must be close to 2 s uncoupling of the motion between the deck and the substructures;
- the center of stiffness of the bridge, CoS, must be as close as possible to its center of mass, CoM, to avoid torsional modes of the deck, reducing the displacements of the deck and to make them compatible with commercial expansion joints and isolation devices [6].

It is worth to note that the simplified procedure here described can be more easily applied to Elastomeric Damping Rubber Bearings [5], since the assumption of linear-elastic behavior of the force-displacement curve is more applicable than for Friction Pendulum [7].

Additionally, design equivalent viscous damping of the isolation devices does not represent a key parameter in the selection of the types of isolation devices and plan layout.

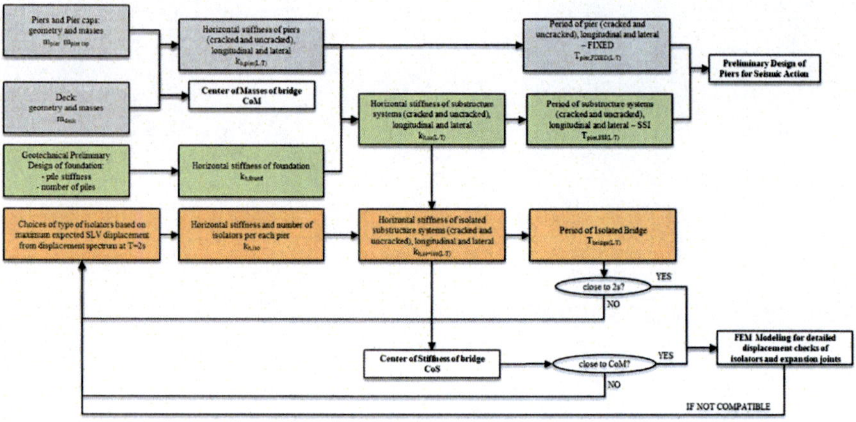

Fig. 2. Flow chart of the procedure.

On the other hand, damping represents a very important parameter to limit actions and displacement on substructures and foundations.

3 Sensitivity Analysis on the Effect of Geometry Variations and Soil-Structure Interaction on the Seismic Response of Isolated Bridges

This chapter reports the sensitivity analyses carried out with the purpose of providing practical considerations potentially useful in the design of the isolated bridges, especially when the design flow requires quick assumptions necessary to go in parallel with the design phases.

3.1 Description of the Case Studies

The bridge of the case study here presented is composed by a steel-concrete composite deck with a main bridge and two access ramps. The bridge is in a very high seismic area and presents a total length of 330 m with typical span between 35 and 50 m. The main deck is supported by 7 reinforced concrete piers and 2 abutments, 1 pier and 2 additional abutments support the deck of the access ramps. The deck is a continuous structure with transversal expansion joints at the abutments only. All the substructures are founded on foundation piles.

Figure 3 shows a longitudinal and a plan view of the bridge.

Starting from the layout of the bridge shown above, four different cases have been analyzed with the purpose of analyzing the effect of SSI and bridge geometries on the design of the isolation system and its effect on the substructures and foundation.

The cases showed in Fig. 4 have been examined:

- Case A: Rock and squat piers (pier height between 6.5 m and 9.5 m).
- Case B: Non-competent soil and squat piers (pier height between 6.5 m and 9.5 m).

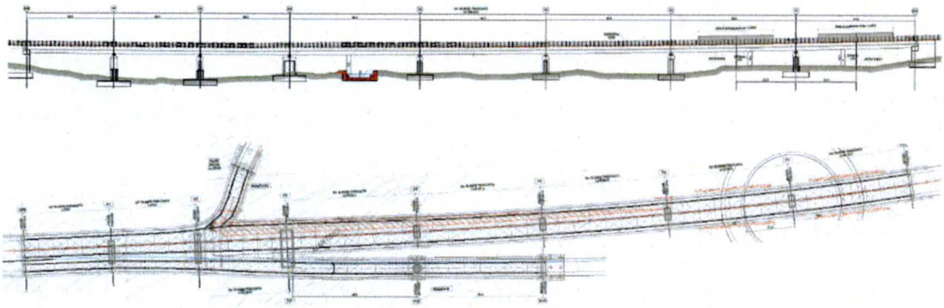

Fig. 3. Case study bridge longitudinal and plan view.

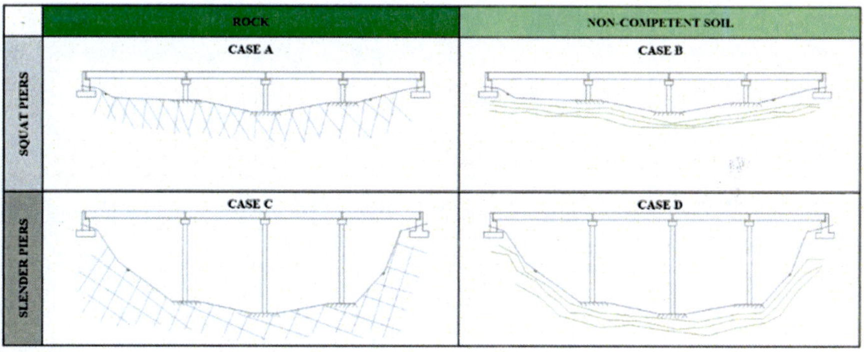

Fig. 4. Simplified and qualitative sketches of the four cases assessed in the sensitivity analysis.

- Case C: Rock and slender piers (pier height between 16 m and 24 m).
- Case D: Non-competent soil and slender piers (pier height between 16 m and 24 m).

For each of these cases, the simplified procedure previously described has been applied.

First, the isolation system has been selected to design an isolated bridge having a longitudinal and transversal period of about 2.0 s and the center of stiffness as close to center of masses as possible. For all the cases, the cracked stiffness for the piers has been considered with a reduction of the gross inertia equal to 50%, considering a SLV event.

3.2 Seismicity of the Site

The bridge is located in a high seismic area in Italy. The Acceleration Response Spectrum at the site is showed in Fig. 5 (a) for the SLV event. Peak ground acceleration is 0.46 g, and the range of periods of the plateau is included within 0.19 s and 0.55 s. The Displacement Spectrum represented in Fig. 5 (b) shows the expected displacement at the site for the SLV event.

Due to high seismicity of the site, the expected displacement of the deck at natural period of 2 s is around 300 mm. This parameter shall be considered as a first reference to select appropriate isolation devices.

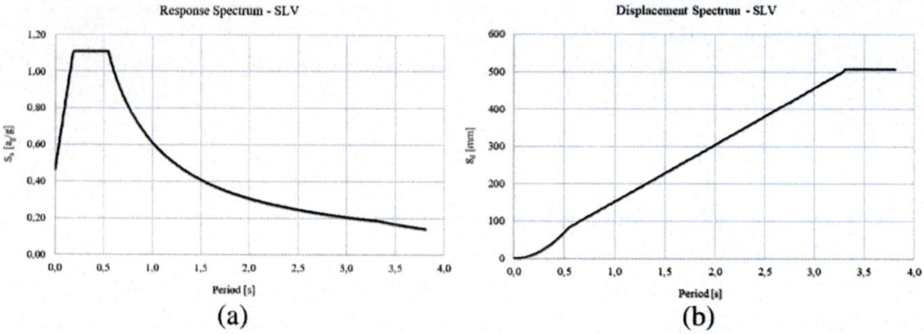

Fig. 5. Response spectrum (a) and displacement spectrum (b) at the site for SLV event.

For comparison purposes only, the same response spectra are adopted in the study for all the four cases.

3.3 Substructures Dynamic Properties

Per each of the 7 piers of the bridge, Fig. 6 shows the longitudinal and transversal first natural periods, considering the fixed base hypothesis and the soil-structure interaction effects.

Light red and red columns represent the longitudinal natural periods in fixed based and SSI condition respectively. Light blue and blue columns represent the transversal natural periods in fixed based and SSI condition respectively. The grey strip represents the periods of the plateau of the SLV seismic spectrum at the site previously showed in Fig. 5.

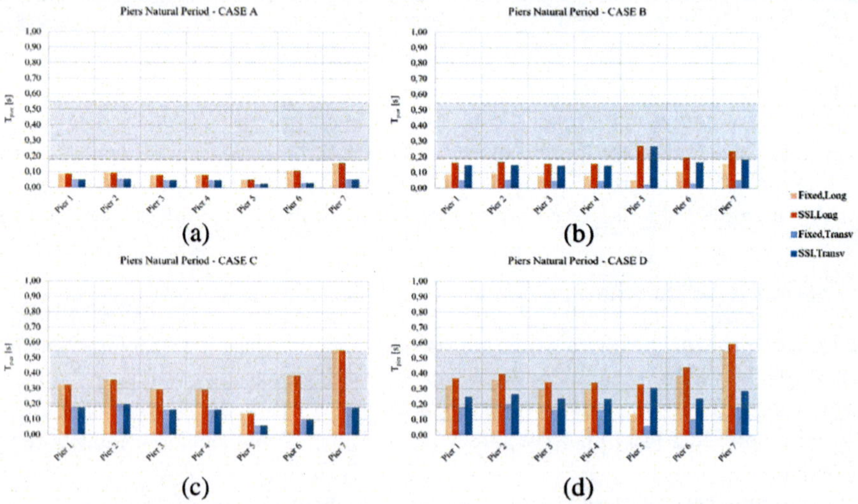

Fig. 6. Longitudinal and transversal natural period of piers – Fixed based vs. SSI effects: (a) Case A; (b) Case B; (c) Case C; (d) Case D.

For both the cases of squat and slender piers in good soil condition (cases A and C respectively), the effect of SSI in modifying substructures stiffness and natural period can be neglected. On the contrary, in case of non-competent soil (cases B and D), SSI cannot be ignored since this effect can provide very different values of natural periods of the piers, especially when dealing with squat piers or in the transversal direction of the bridge (typically more rigid than longitudinal for higher inertia of the pier). This is in general agreement with the indications found in literature (see for example Priestley et al. 1996 [1], Priestley et al. 2007 [3]).

In addition, the same figure shows that for all the 4 cases, both longitudinal and transversal natural periods of the substructures are far from the estimated natural periods foreseen for the isolated deck (around 2.0 s). The more the substructures are stiff, the more the uncoupling degrees between deck and substructure is expected.

3.4 Isolation System

As previously described in Chapter 2, the selection of the isolators (types of devices and layout) is one of the main goals of the proposed procedure. For this reason, different isolators, and relevant plan positioning can be expected for each of the four cases. In general, for all the 4 cases, Elastomeric Damping Rubber Bearings have been foreseen.

The main characteristics of the isolation devices adopted in the four cases are summarized in Table 1. For all the cases, isolators with displacement capacity of 350 mm have been selected in order to satisfy the expected displacement demand at the design period of 2.0 s. An important aspect to highlight is that the definition of the isolator stiffness in the different locations is driven by the necessity to accomplish the two goals: the uncoupling criteria and the reduction of the eccentricity between CoM.

Table 1. Elastomeric damping rubber bearings adopted in the four case studies

	Type A	Type B	Type C
Nominal horizontal stiffness kh_{ISO} (kN/mm)	3.20	1.90	4.20
Displacement capacity (mm)	350	350	350
Equivalent viscous damping coefficient ξ (%)	15	15	15
Dynamic Shear Modulus G for $\gamma = 1$ (MPa)	1.4	1.4	1.4
Maximum ULS vertical load (kN)	6900	7400	12000

3.5 Uncoupling Degree

Table 2 summarizes the main properties of the isolated deck in the four cases analyzed. In detail the table reports the first longitudinal and transversal natural period of the isolated deck in the four cases, as well as the longitudinal and transversal distances between center of masses and center of stiffnesses of the bridge.

Table 2. Natural periods of isolated deck and CoM – CoS distances for the analyzed cases

	Case A	Case B	Case C	Case D
Longitudinal Period T_L (s)	1.80	1.82	2.06	2.08
Transversal Period T_T (s)	1.79	1.81	1.98	2.01
CoM – CoS Longitudinal distance (m)	−0.4	−0.4	−1.0	−1.0
CoM – CoS Transversal distance (m)	−5.1	−5.0	0.6	0.8

Per each of the 7 piers of the bridge, Fig. 7 shows the effect of uncoupling deck and substructures with seismic isolation systems. The yellow columns represent the horizontal stiffness of isolators only, placed on top of each pier. Red and blue columns represent the longitudinal and transversal horizontal stiffness of the substructure in the isolated configuration considering isolation, flexural inertia of the piers and SSI effect.

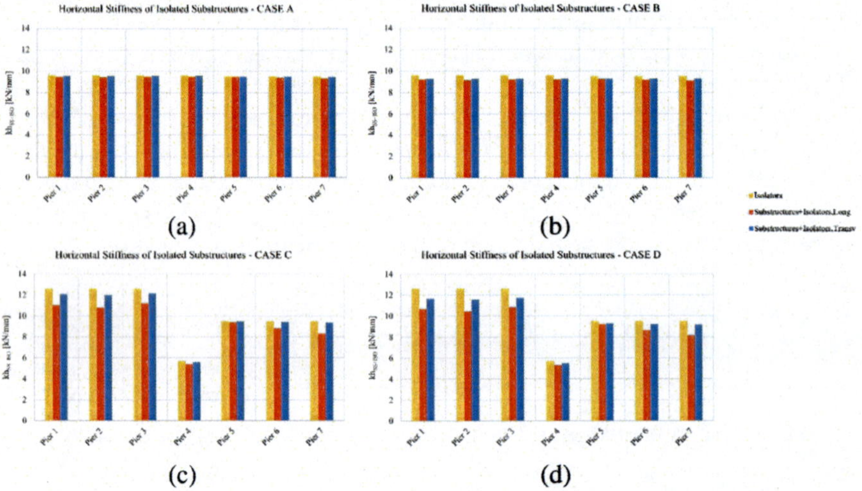

Fig. 7. Horizontal Stiffness of Substructures vs. Horizontal Stiffness of Isolators: (a) Case A; (b) Case B; (c) Case C; (d) Case D.

For the analyzed scenarios, independently of the soil type, the isolator type and layout are mainly governed by the stiffness of the piers, and it can be argued that SSI does not represent a key factor to define the most appropriate isolation system. Indeed, despite very different soil conditions, Cases A and B present the same isolator layout and the same natural period of the isolated deck in both longitudinal and transversal direction. The same situation occurs for cases C and D, which present same isolators configuration and very similar natural period for both longitudinal and transversal directions. This result shows an important beneficial effect of the bridge isolation: given the uncoupling between deck and substructure, the seismic behavior of seismically isolated bridges is not greatly influenced by SSI effect.

In addition, Fig. 7 suggests another important consideration. Uncoupling is almost complete when dealing with squat piers for both rock and non-competent soil (in cases A and B the red and blue columns are almost overlapped with yellow ones). On the other hand, uncoupling tends to reduce when dealing with slender piers: in cases C and D the overall horizontal stiffness of each substructure starts to decrease, especially in longitudinal direction, where piers are more flexible (red columns show the tendency to low from the yellow ones).

3.6 Actions on Bridge Foundations

This section has the goal of showing the seismic actions transferred from the deck and the substructures to the foundations. Figure 8 reports the total forces transferred from the substructure to foundations for both longitudinal and transverse direction.

Masses due to foundation plinths and abutments are not included. The first column represents the weight transferred to ground, composed by two terms: the light grey is the deck component, while the dark grey is the substructure component. The second and the third columns of each diagram represent the total longitudinal and transversal base shear due to seismic SLV event respectively.

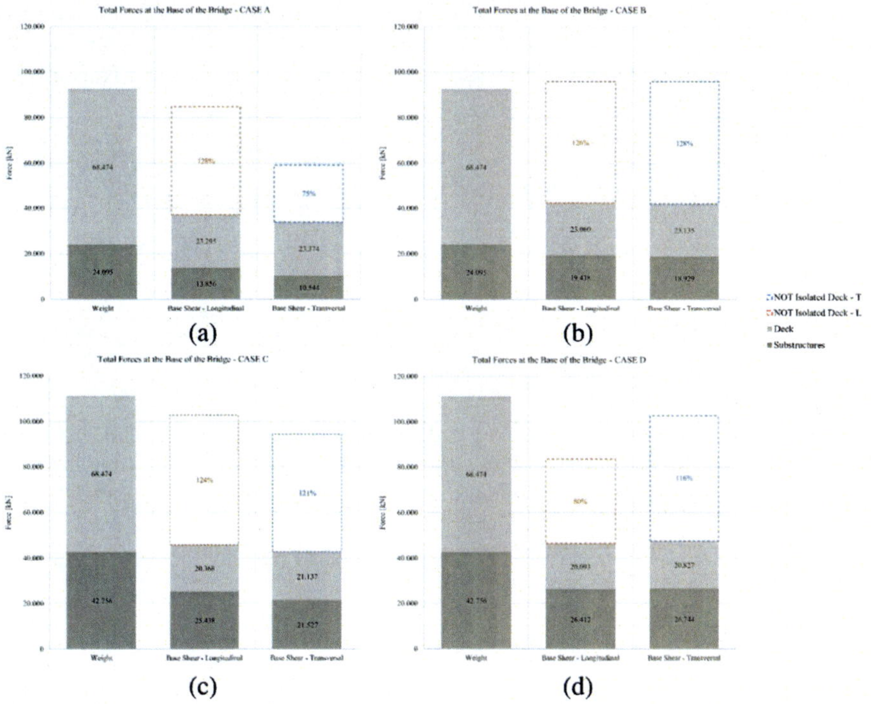

Fig. 8. Total forces at the base of the bridge – Deck and Substructures contributes: (a) Case A; (b) Case B; (c) Case C; (d) Case D.

The red and the blue dashed lines show the values of the longitudinal and transversal base shear in case of non-isolated deck.

From this figure it is worth to highlight some significant considerations.

The effect of isolation system is always beneficial for the 4 cases analyzed also when uncoupling is not total (cases C and D): indeed, the reduction of the base shear with respect to the non-isolated condition is significant.

As second consideration, the contribution of SSI is relevant when dealing with non-competent soil to avoid underestimation of base shear. Indeed, in both cases of squat and slender piers, the base shear due to the deck part is approximatively the same, while the SSI provide an increment of the portion of base shear due to substructures.

3.7 Validation of the Simplified Procedure

The methodology here described has been tested for several bridges with different geometries and number of piers showing a very good agreement with the results of the more refined Finite Element Analyses.

As an example, the FE model here presented includes all the elements of the bridge (i.e., deck, isolators, pier caps, piers, foundation plinths and foundation piles) used in the design of the Case B. The main deck has been modeled with beam elements, considering the effective mass, weight, and inertia of beams, diaphragms and slab. The isolators have been modeled as elastic link placed on top of each pier cap. The piers have been modeled as beam elements. Finally, foundation plinths have been modeled as shells, connected to the foundation piles, modeled as springs.

Table 3. Comparison of Natural Period - FE model vs. simplified procedure

	FE model	Simplified procedure (Case B)
Longitudinal period T_L (s)	1.89	1.82
Transversal period T_T (s)	2.00	1.81

The model has been subjected to several sanity checks and the same has been used to perform the verification of all the element of the bridges, under different static and seismic load conditions. The figures here reported (Fig. 9) shows the first natural modes of the deck resulting from modal analysis.

Table 3 show compares the values of natural period of the simplified procedure with those resultants from the FE analysis.

Considering the different levels of complexity of the two models, a good match is shown with differences between 4% and 10%.

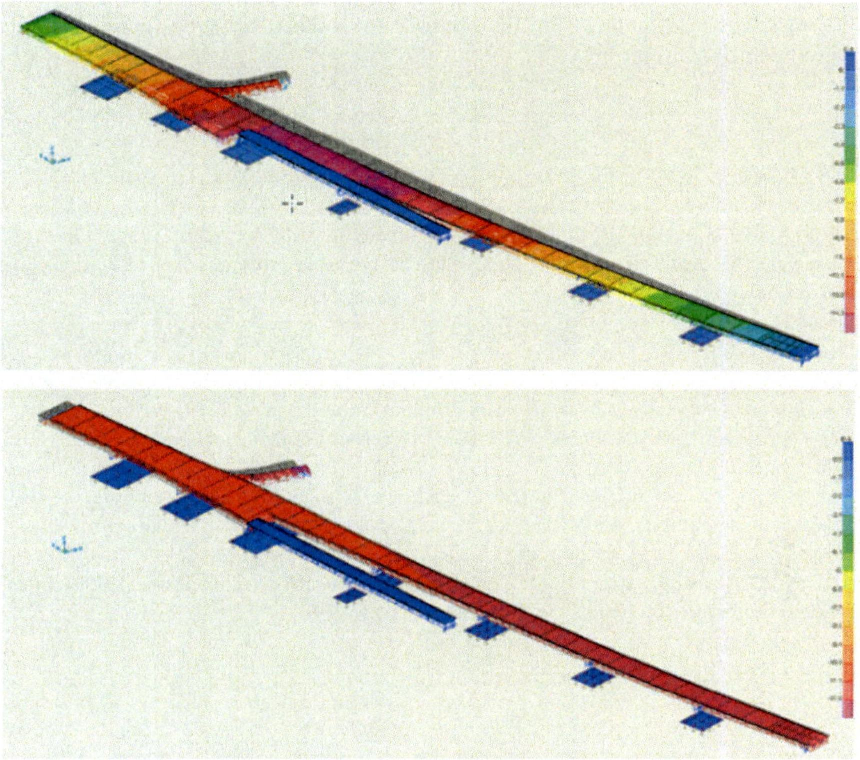

Fig. 9. First modal shapes from FE model: (a) Long $T_L = 1.89$ s; (b) Transv $T_T = 2.00$ s.

4 Conclusions and Practice Oriented Considerations

Within the design process of an isolated bridge, it is important to be aware of the following aspects, in order to correctly steer the design approach:

1. The simplified procedure allows to define the choice of the isolation layout (i.e. the distribution of the isolator's stiffness on the different locations) also in very complex systems prior to set a FE analysis. This can be a quite important time-saving operation in terms of modelling and can be a useful simplified check of the FEM results at the same time.
2. Increasing the stiffness of the piers, the SSI effect increases its importance: the neglection of the SSI effect can lead to an underestimation of the seismic action acting on the foundations.
3. The SSI, however, does not affect the choice of the isolation devices, nor the global fundamental period of the isolated structure. The choice of the devices is indeed greatly influenced by deck's geometry and mass and pier's layout.
4. SSI does not affect the type of devices, but it is important to include SSI in seismic design of bridges, especially for soft soil and stiff structures, because the contribution to the base shear can be significant. Indeed, higher modes effect might be much

affected by SSI with the results that the base shear due to the sub structure can varies significantly considering SSI.

References

1. Priestley, M.J.N., Seible, F., Calvi, G.M.: Seismic Design and Retrofit of bridges (1996)
2. Furinghetti, M.: Definition and validation of fast design procedures for seismic isolation system. MDPI (2022)
3. Priestley, M.J.N., Calvi, G.M., Kowalsky, M.J.: Displacement-Based Seismic Design of Structures (2007)
4. Novak, M., Sheta, M.: Dynamic Response of Piles and Pile Groups. Faculty of Engineering Science. The University of West Ontario, London, Ontario, Canada (1974)
5. Christopoulos, C., Filiatrault, A.: Principles of Passive Supplemental Damping and Seismic Isolation. IUSS Press (2006)
6. Cademartori, M., Miano, A., Fiorillo, A., Aliotta, A., Di Ludovico, M., Prota, A., Dellacasagrande, S., Fatnassi, A., Pastorelli, D.: Functional acceptance tests for the seismic isolation devices of the new San Giorgio bridge in Genoa. COMPDYN (2021)
7. Calvi, G.M., Pietra, D., Moratti, M.: Criteri per la progettazione di dispositivi di isolamento a pendolo scorrevole. Progettazione Sismica numero **3** (2010)

Effects of Ice and Water and Contamination on Friction Pendulum Bearings

R. Grijalva Alvarado(✉) and K. Ryan

University of Nevada - Reno, Reno, NV 89557, USA
rgrijalva@nevada.unr.edu, klryan@unr.edu

Abstract. Water and ice contamination in Friction Pendulum System (FPS) bearings has been observed on several seismically isolated bridges in Alaska. However, there have been few studies on how water and ice contamination affect FPS bearings frictional properties. In this study, two formerly in-service double pendulum, a new double pendulum and a new single pendulum bearing were tested. Each bearing was subjected to displacement controlled loading using a single bearing test assembly that supports the bearing between the earthquake simulator and a stabilizing A-shaped steel frame that loads the bearing axially and restrains the bearings top plate from horizontal movement. An idealized bilinear loop was fitted to the data points of each test, excluding the first and last cycle. Uneven sliding between both sliding surfaces was observed on tests on double pendulum bearings subjected to water, ice or soil contamination. Sliding restricted to one surface results in approximately twice the post yield stiffness compared to balanced sliding on two surfaces. Also, lower coefficient of friction was observed during wet tests compared to dry tests. Furthermore, ice resistance was calculated by comparing the first cycle on frozen tests with the idealized bilinear loops of the subsequent cycles. Ice resistance is hypothesized to be a function of ice volume and doesn't seem to have a big impact on heavily loaded bearings.

Keywords: FPS · Seismic isolation · Ice · Water · Single pendulum · Double pendulum

1 Introduction

Friction Pendulum System (FPS) bearings are seismic isolation devices widely used around the world to protect bridges and buildings. The original FPS bearing is the Single Pendulum bearing (SPB) [1], which consists of an articulated slider on a spherical sliding surface (Fig. 1(a)). Later, the Double Pendulum bearing (DPB) was invented, which is a more compact version of the SPB. A DPB consists of an articulated slider between two spherical sliding surfaces above and below (Fig. 1(b)). The slider displacement on the sliding surface(s) absorbs earthquake demands to isolate the superstructure. Furthermore, energy is dissipated as a result of friction between the slider and the sliding surface(s).

Water, soil and other types of contamination has been observed in FPS bearings in Alaskan bridges, despite the seals around the bearings. Similar contamination problems

G. P. Cimellaro (Ed.): WCSI 2022, LNCE 309, pp. 417–425, 2023.
https://doi.org/10.1007/978-3-031-21187-4_34

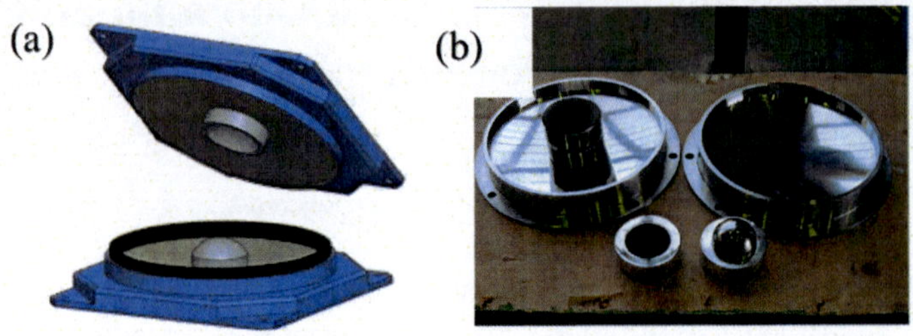

Fig. 1. Friction Pendulum System (a) SPB [2] (b) DPB [3]

have been observed by Washington DOT [4] and in Tappan Zee bridge in New York [5]. Furthermore, in northern Alaska regions, air temperature is commonly below freezing, and water trapped inside the bearing is believed to be frozen during much of the year.

There are several documented studies on the effects of axial pressure, sliding velocity, temperature, sustained axial load, long-term traffic loading, corrosion, lubrication and others on frictional properties of FPS bearings [6]. However, there are no known studies on the effects of water or ice contamination on the frictional properties of FPS bearings. Ice contamination is of concern to Alaska as several of Alaskan isolated bridges are located in cold northern regions.

In this study, one SPB and three DPB were tested. The SPB was a new bearing representative of Robertson River Bridge abutment bearings (Robertson). Two of the DPBs were formerly in-service abutment bearings from Susitna River Bridge (Aged Susitna 1 and 2) while the third DPB was an identical new bearing (New Susitna). A single bearing test assembly was designed that uses one of University of Nevada Reno the shake tables to impose cyclic motions to the bearings. The objective of this study is to characterize the four bearings tested under water, ice and soil contamination relative to clean dry conditions. In ice contaminated tests, the bearings were characterized after the ice was broken. Furthermore, ice resistance is interpreted from the measured bearing shear force.

2 Experimental Methodology

The experimental setup is shown in Fig. 2. The shake table was used to impose cyclic loading to the bearings. A steel frame located on top of the bearing to restrain the top plate from horizontal movement. Furthermore, the steel frame transmitted axial load from a hydraulic ram to the top of the bearing. On the opposite end, the steel frame was pin connected to two rigid steel columns, which allowed the steel frame to rotate to accommodate the vertical movement caused by the bearing horizontal movement. Figures 2(b) and 2(c) show elevation views for both types of bearings; an additional adapter plate was used in between Robertson bearing and the steel frame to accommodate the difference in height between the two bearing types. String pots were used to measure bottom plate and inner slider (DPBs only) displacement relative to the steel frame, in both horizontal directions. A load cell placed below the bearing measured axial and shear force in the bearings. An insulation box was designed to freeze water inside the bearings. With the bearing axially loaded, the inner bearing cavity was filled with water using a funnel and hose. Then the insulation box was wrapped around the bearing, filled with dry ice, and left to freeze overnight, with dry ice refills every six hours.

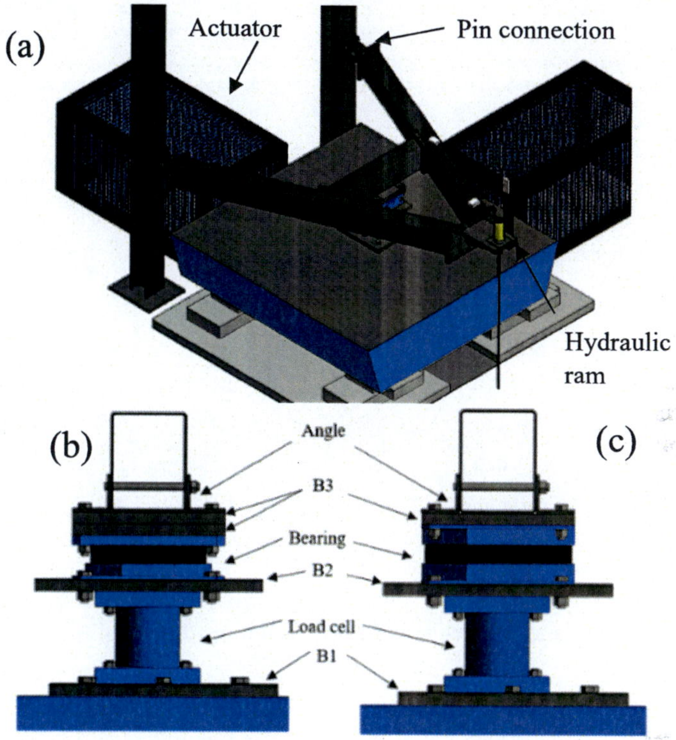

Fig. 2. Experimental setup (a) 3D rendering (b) SPB setup elevation (c) DPB setup elevation.

2.1 Test Specimens

The test bearings for this study were supplied by Earthquake Protection Systems (EPS). Figure 3 shows the cross section drawings of the bearings tested. The SPB is shown in Fig. 3(a) while the DPB is shown in Fig. 3(b). Bearings design properties were calculated using AASHTO 2014 Guide Specifications for Seismic Isolation [7]. Furthermore, bearings design parameters are shown in Table 1.

2.2 Loading Protocol

The shake table was used to impose displacement controlled loading to the bearings. Loading protocols included constant amplitude and increasing amplitude sine waves (ramping sine). The latter one was only applied to DPBs to eventually initiate sliding on both sliding surfaces. Target axial forces were chosen based on what could be easily

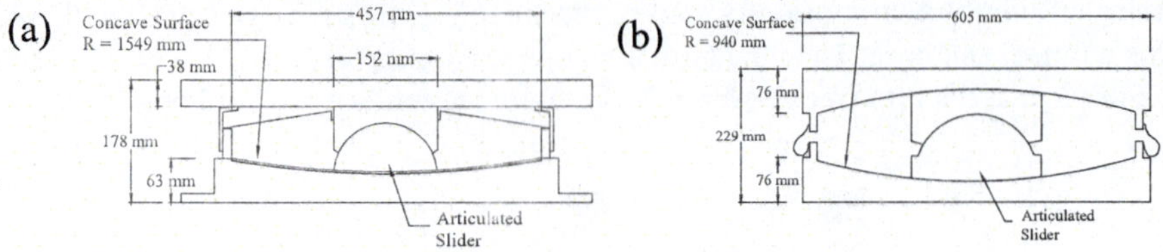

Fig. 3. Test specimen (a) SPB cross section (d) DPB cross section.

Table 1. Bearing design parameters.

Spectral parameters				Bearing response			
	SPB	DPB			SPB	DPB	
S_S	0.4164	1.025	g	μ	0.10	0.075	–
S_1	0.2302	0.371	g	T	2.50	2.75	s
Soil type	D	B	–	W	80.07	444.82	kN
PGA	0.175	0.486	g	K_{eff}	0.129	0.589	kN/mm
S_{DS}	0.612	1.025	g	T_{eff}	1.578	1.743	s
S_{D1}	0.447	0.362	g	D	103	94	mm

achieved with the hydraulic rams available in the laboratory. Furthermore, maximum amplitude, axial load and frequency were varied in constant amplitude sine waves, as shown in Table 2.

Table 2. Loading protocol variations.

	SPB	DPB
Displacement (mm)	103, 26, 52, 77 and 129	94, 24, 47, 71, 118, 203 (ramping sine) and 254
Frequency (hz)	0.63, 1.06, 0.80, 0.69 and 0.59	0.57, 0.96, 0.73, 0.63, 0.54, 0.29, 1.14, 0.48 (ramping sine) and 0.45
Axial load (kN)	111.21 and 66.72	444.82 and 244.65

2.3 Bearing Characterization Procedure

Figure 4 shows an example of axial load time history, absolute hysteresis loop and normalized hysteresis loop (shear force normalized by axial force) for a test on SPB (Robertson) at 66.72 kN target axial load and 103 mm peak displacement. Variation in axial load was observed in each cycle (Fig. 4(a)). Hence, normalized hysteresis loops were used to characterize the bearings to mitigate this effect. To characterize the bearing

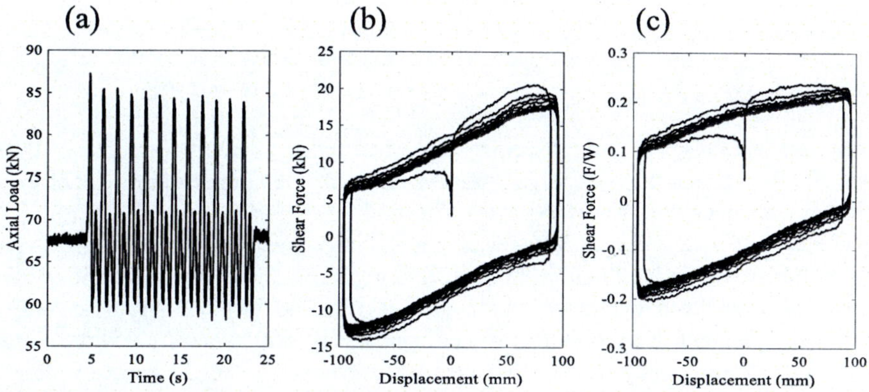

Fig. 4. SPB at 66.72 kN target axial load and 103 mm displacement (a) Axial load history (b) Hysteresis loop (c) Normalized hysteresis loop.

response, each cycle was isolated. The first and last cycles were omitted to avoid any influence of static friction or decrease in shear force at the last cycle.

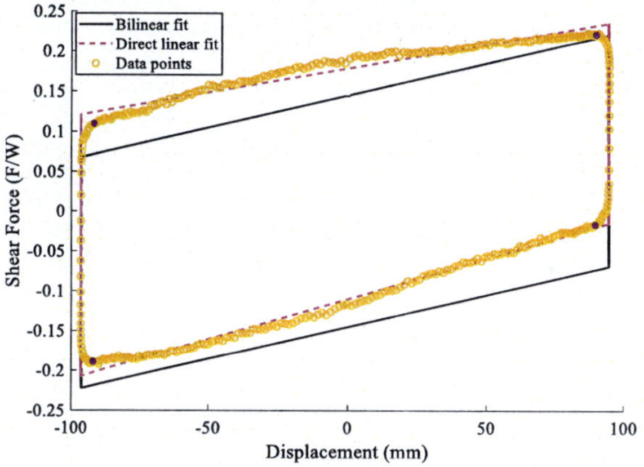

Fig. 5. Idealized bilinear loop fit.

Figure 5 shows an example of the characterization process using the second cycle of one of the dry tests on the SPB at 66.72 kN axial load and 103 mm absolute maximum displacement in X direction. A linear curve was fitted to the top and bottom boundaries of the normalized hysteresis loop (dashed lines in Fig. 5) for the given cycle. Then, the slopes of both linear curves were averaged to compute a single the post-yield stiffness. The energy dissipated per cycle (EDPC) was then calculated as the area under the normalized hysteresis loop. Then, the average coefficient of friction (μ) (Y intercept) was calculated by dividing EDPC by 2 $\times$ total displacement. The solid black lines in Fig. 5 show the idealized bilinear loop after following the process described above.

3 Results

3.1 Single Surface Sliding

During the test program, several factors were observed to affect the response of DPBs under the cyclic loads imposed. Figure 6 shows three examples of displacement histories of the independent sliding surfaces observed on the DPB. In Fig. 6(a) for a test under dry conditions, both sliding surfaces were equally engaged, as assumed in design. This is hereafter referred to as "double surface sliding". In Fig. 6(b) (wet test), both sliding surfaces were engaged, but the movement gradually shifted from bottom surface only to equal sliding on both surfaces. On the other hand, in Fig. 6(c) (wet test) only the bottom sliding surface was engaged, which is hereafter referred to as "single surface sliding".

The displacement contribution of the dominant sliding surface relative to the total displacement, in percentage (%SST), was calculated to determine the relationship between type of sliding (e.g. single or double surface sliding) and pendulum length. The %SST was calculated for each cycle (except for first and last) in each test and related to the average pendulum length calculated using the procedure described in Sect. 2.3.

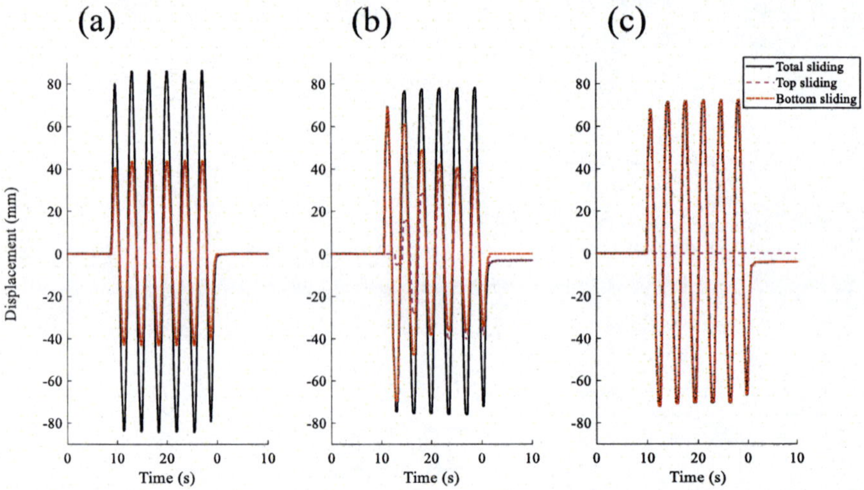

Fig. 6. Displacement history (a) Perfect double surface sliding (b) Uneven double surface sliding (c) Pure single surface sliding.

Figure 7 shows a scatter plot relating %SST and pendulum length for each cycle of every dry, wet and frozen test conducted on Aged Susitna 1. Each point in Fig. 7 represents a cycle. Top and bottom solid lines in Fig. 7 represent the theoretical pendulum length values for double surface sliding and single surface sliding, respectively. %SST = 50 represents double surface sliding and %SST = 100 represents single surface sliding.

Furthermore, Fig. 7 shows that when %SST is close to 100 (sliding dominated by one surface) the pendulum length gets closer to the theoretical pendulum length for single surface sliding (940 mm), and when %SST is close to 50 (both sliding surfaces equally engaged) the pendulum length gets closer to the theoretical pendulum length for double surface sliding (1880 mm).

3.2 Effects of Contamination on Frictional Properties

Next, characterized friction coefficient μ vs cycle over all the tests is shown in Fig. 8, with different subplots in each bearing. Each point represents a cycle and all points connected with lines represent the cycles from the same test. Y-direction tests have been aligned with their reciprocal X-direction test and dry, wet and frozen tests with the same loading protocol are aligned. Data for frozen tests represents the free sliding phase on one surface of a DPB or after ice breakaway has occurred. In general, observed μ in dry tests was higher than the theoretical value for all bearings. Furthermore, on wet tests, a decrease in μ with respect to dry tests was observed on all DPBs, as shown in Figs. 8(b), 8(c) and 8(d). Data is inconclusive as to how ice contamination affects μ. In contrast to the literature ([6, 8]), observed μ was higher on the new DPB than the aged DPB. However, manufacturing processes have changed since the formerly in-service bearings were installed.

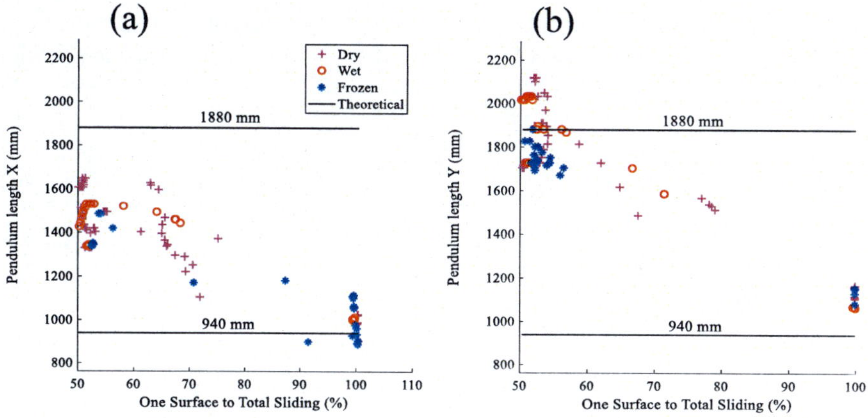

Fig. 7. One surface sliding relative to total displacement related to pendulum length Aged Susitna 1 (a) X test (b) Y test.

3.3 Ice Strength

Ice breakaway strength or additional resistance caused by ice adhesion to the slider and sliding surface was quantified. To do so, normalized hysteresis loops were isolated and characterized as described earlier. To interpret ice breakaway strength, the average characterized bilinear loop was subtracted from the first cycle and then multiplied by the instantaneous axial load of the first cycle, resulting in an absolute value of ice strength. Figure 9 shows the measured ice strength for several tests conducted on Robertson bearing. The blue dotted line shows the ice strength for a test conducted with a 32 mm

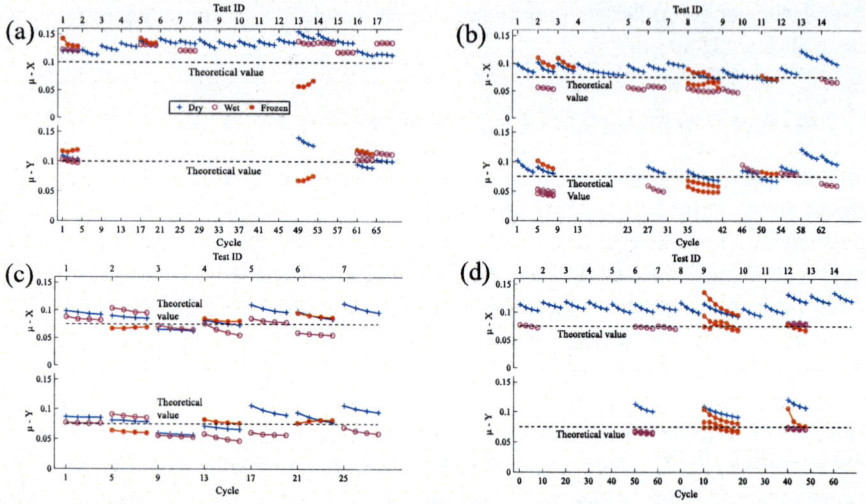

Fig. 8. A μ per cycle (a) Robertson (b) Aged Susitna 1 (c) Aged Susitna 2 (d) New Susitna.

ice layer and the ice strength for this test is smaller than the rest of the tests which were conducted using a 64 mm ice layer. Hence, as the ice volume increases, ice strength increases. Furthermore, ice strength does not seem to increase with axial load, as the highest value of ice strength was achieved at the lowest axial load (dashed-dot line). However, the data is limited to make a strong conclusion about the relationship between axial load and ice strength.

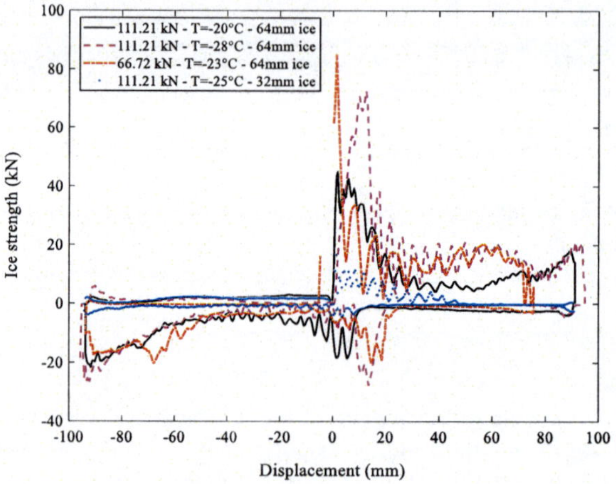

Fig. 9. Ice strength Robertson bearing.

4 Conclusions

Uncontaminated FPS bearings, as well as water and ice contaminated FPS bearings, were tested using constant and increasing amplitude sine waves. The bearings were then characterized and the properties of contaminated and uncontaminated bearings were compared. Additional resistance due to ice contamination was estimated. The test program showed that water, ice or soil contamination can induce uneven sliding between both sliding surfaces or even single surface sliding, increasing post-yield stiffness and, as a consequence, the shear force demands on bridge piers. Water contamination tends to reduce the coefficient of friction, hence reducing the energy dissipation of the system and thus increasing displacement demands on the device. Ice strength depends mainly on ice volume and temperature, but not on axial load.

References

1. Zayas, V., Low, S.S., Mahin, S.A.: The FPS earthquake resisting system, experimental report. Report No. UCB/EERC-87/01, Earthquake Engineering Research Center, Univ. of California, Berkeley, Calif. (1987)
2. Gillich, G.R., Iancu, V., Gillich, N., Korka, Z.I., Chioncel, C.P., Hatiegan, C.: Decoupling the structure from the ground motion during earthquakes by employing friction pendulums. IOP Conf. Ser. Mater. Sci. Eng. **294**, 012025 (2018). https://doi.org/10.1088/1757-899x/294/1/012025
3. Morgan, T.A., Mahin, S.A.: The Use of Base Isolation Systems to Achieve Complex Seismic Performance Objectives. Pacific Earthquake Engineering Research Center (2011)
4. WSDOT Bridge and Structures: SR522 Seismic Isolation Bearings (2015)
5. Tappan Zee Constructors: FPT Bearing QC Inspection Report: Proof Testing of Affected Bearings (2017)
6. Constantinou, M.C., Whittaker, A.S., Kalpakidis, Y., Fenz, D.M., Warn, G.P.: Performance of seismic isolation hardware under service and seismic loading. Technical Rep. MCEER-07-0012. Multidisciplinary Center for Earthqueake Engineering Research, State Univ. of New York at Buffalo, NY (2015)
7. American Association of State Highway and Transportation Officials: Guide Specifications for Seismic Isolation Design (2014)
8. McVitty, W.J., Constantinou, M.C.: Property modification factors for seismic isolators: design guidance for buildings. MCEER Report 15-0005 (2015)

Performance Assessment of a Bridge Seismically Isolated with Lead Rubber Bearings at Low Temperature

Volkan Karuk, Cansu Yasar, Esengul Cavdar, and Gokhan Ozdemir(✉)

ESQUAKE Seismic Isolator Test Laboratory, Eskisehir Technical University, Eskisehir, Turkey
gokhan_ozdemir@eskisehir.edu.tr

Abstract. This study investigates the variation in structural response of a bridge seismically isolated with lead rubber bearings (LRBs), when subjected to ground motion excitations at low ambient temperature. For this purpose, first, a full-scale LRB was conditioned at −20 °C, −10 °C, 0 °C and 20 °C temperatures for 24 h and tested under dynamic conditions in order to determine the change in its characteristic strength and post-yield stiffness. Recorded force-displacement curves were then utilized to represent the deteriorating hysteretic behavior of the LRB in case of different temperatures. Seismically isolated bridge was idealized by a single-degree-of-freedom model and analyses were performed by OpenSees structural analysis program. Nonlinear response history analyses (NRHA) were conducted considering the strength deterioration of the LRB due to the lead core heating during cyclic motion. In the analyses, eleven pairs of ground motion records, selected and scaled to represent a target spectrum, were used. Response quantities assessed in the analyses were maximum isolator displacement and maximum isolator force for the isolation system and maximum acceleration for the bridge deck. It is revealed that as the ambient temperature decreases, the amount of amplification in maximum isolator forces and accelerations can be up to 30%.

Keywords: Seismic isolation · Lead rubber bearing · Low temperature · Lead core heating · Bridge

1 Introduction

Seismic isolation is an innovative technology that protects structures against probable damages of earthquakes. The philosophy of this technique is to lengthen the fundamental period of the structure by means of structural elements with low horizontal stiffness between the superstructure and substructure. There are various types of isolation systems and the lead rubber bearing (LRB) is one of the most widely used isolators. LRB is composed of alternate layers of rubber and steel shims with a lead core at the center. While the steel shims provide vertical rigidity, rubber is responsible for the lateral flexibility. The lead core at the center provides additional damping to the isolator. The widespread use of the LRBs in several structures such as bridges, hospitals, data centers leads to research

G. P. Cimellaro (Ed.): WCSI 2022, LNCE 309, pp. 426–437, 2023.
https://doi.org/10.1007/978-3-031-21187-4_35

studies focused on the performance of the LRBs under different loading conditions. One of these conditions is low ambient temperature.

LRB behavior is dominated by the properties of the isolator components, rubber and lead core. Numerous studies conducted with rubber samples reveal that rigidity of the rubber increase under low temperature (Roeder et al. 1987; Yakut and Yura 2002(a, b); Cardone et al. 2011). On the other hand, the limited experimental research carried on LRB showed that both characteristic strength and post yield stiffness of the isolator increases due to reduction in ambient temperature (Robinson 1982; Constantinou et al. 2007; Park et al. 2012). Furthermore, they exhibited that the amount of increment in characteristic strength is larger than the post yield stiffness (Zhang and Li 2020).

In addition to experimental studies, in literature, there are also analytical studies that investigated the seismic performance of bridges isolated with LRB at low temperature. For instance, Billah and Todorov (2019) investigated the seismic response of an isolated bridge with LRB through nonlinear response history analyses (NRHA). They used non-deteriorating bilinear force-displacement curves to idealize hysteretic behavior of LRBs in NHRAs. In their study, the performance of the isolated bridge was evaluated under two different temperature conditions, reflecting the temperatures observed in Canada, namely summer (+ 25 °C) and winter (– 30 °C). Similarly, Deng et al. (2019) examined the detrimental effects of low temperature on the seismic responses of bridge equipped with LRB. In this sense, LRB hysteretic curves were obtained for temperatures ranging from – 30 °C to 40 °C. LRBs were represented by non-deteriorating force-displacement curves. On the other hand, Wang et al. (2019) carried out an analytical study with two LRB mathematical models, which can be classified as small and medium size. Different from the above studies, mathematical model of LRB was obtained with the deteriorating model proposed by Kalpakidis and Constantinou (2009a, b). Dynamic analyses under ground motion excitations were repeated for LRB properties calculated for temperatures between – 40 °C and 40 °C.

The literature review showed that in most of the studies, the idealization of LRBs rely on the empirical relations or modification factors obtained for small-to-medium size isolators. However, in highly seismic regions, the size of the LRB increases due to high displacement demands. Thus, there is a need to conduct experimental studies to present test results of a large size LRB tested at large shear strains and dynamic conditions after conditioned at low temperature. In addition, analyses generally were performed with mathematical models that neglect the deterioration in strength of LRBs due to temperature rise in the lead core under cyclic motion. In the light of these observations, the aim of this study is to determine the variation in seismic performance of a bridge seismically isolated with LRBs considering the change in mechanical properties of LRB when exposed to low temperature. For this purpose, a large size LRB was tested dynamically after conditioned at temperatures of – 20 °C, –10 °C, 0 °C and 20 °C. Recorded force-displacement curves were used to idealize deteriorating hysteretic response of LRBs. Then, NRHAs were conducted with idealized single-degree-of-freedom bridge model under the effect of eleven pairs of ground motion records. Maximum isolator displacements (MIDs), maximum isolator forces (MIFs) and maximum accelerations (MAs) were the response quantities used to evaluate the variation in seismic performance of the LRB isolated bridge when exposed to low temperature.

2 Test Specimen, Loading Protocol and Test Set-Up

The bearing tested in this study is an LRB which has a rubber diameter of 850 mm. The lead core diameter of the isolator is 220 mm. The overall height of the isolator is 339 mm including top and bottom plates together with the end shim at the top. There are 18 layers of rubber with 10 mm individual thickness and 17 layers of steel shim of 3 mm thickness. Figure 1a and 1b present the picture and geometric features of the tested LRB, respectively.

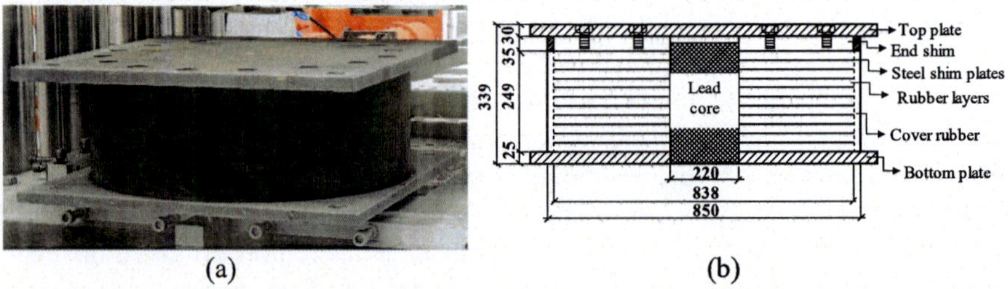

Fig. 1. (a) View and (b) geometric features of the tested LRB (units are in mm).

A series of tests were conducted to determine the hysteretic properties of LRB at low ambient temperatures. During the tests, three fully reversed loading cycles were subjected to the LRB conditioned at −20 °C, −10 °C, 0 °C and 20 °C with an exposure time of 24 h. The amplitude of the sinusoidal motion is equal to 198 mm that corresponds to 100% shear strain as given in Fig. 2a. All displacement-controlled tests were performed with a loading frequency of 0.5 Hz (peak loading velocity of 622 mm/s). The vertical force acting on the bearing during the cyclic motion was 3177 kN that corresponds to a normal stress of 5.8 MPa. The loading protocol followed in the experimental program is presented in Table 1. Tests were conducted at ESQUAKE Seismic Isolator Test Facility of Eskişehir Technical University. The test set-up in ESQUAKE shown in Fig. 2b is capable of applying dynamic excitations in both vertical and horizontal directions. Thus, vertical force applied to the specimen can be kept at a certain level during the horizontal motion. Table 2 presents the maximum loading capacities of test apparatus in ESQUAKE. In compliance with the test program given in Table 1, desired low temperatures (0, −10 and −20 °C) to which LRB was exposed were provided by the environmental chamber placed next to the test set-up at ESQUAKE (Fig. 2c). During the low temperature tests, first, environmental chamber was set to the proper temperature and then LRB was conditioned in this chamber for 24 h. Afterwards, the LRB was taken from the thermal chamber and mounted to the test set-up. Then, cyclic motion was run as shown in Fig. 2d. The total time spent for installation and testing of the isolator is approximately 10 min. After each test, there was a dwell period of 24 h in the laboratory conditions to prevent the LRB from cumulative thermal effects.

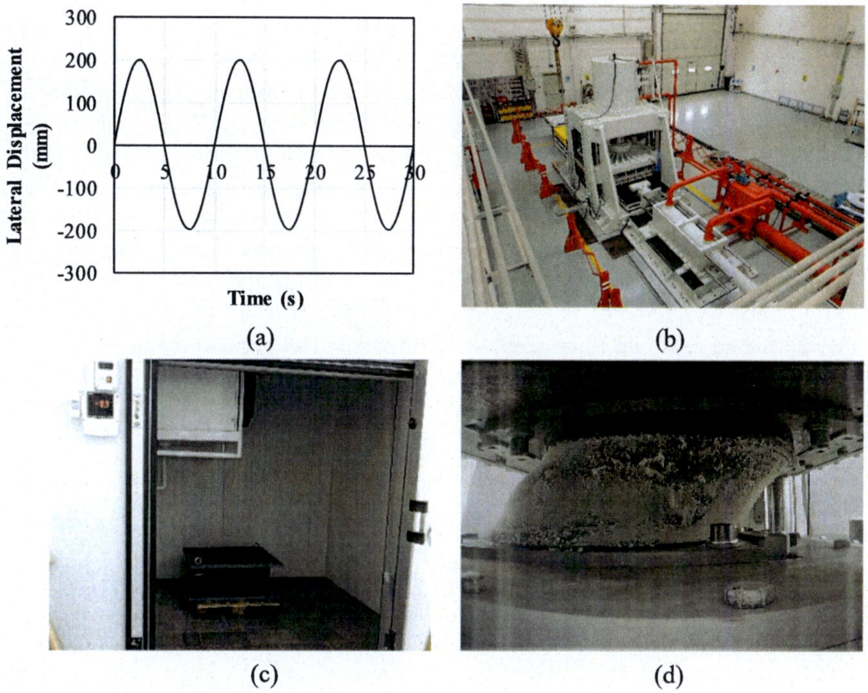

Fig. 2. (a) Applied loading history to LRB (b) seismic isolator test set-up (c) environmental chamber (d) a view taken during cyclic motion.

Table 1. Applied test protocol.

Test No.	Temperature (°C)	Exposure time (h)	Shear strain (%)	Frequency (Hz)	Cycle number
1	20	-	100	0.5	3
2	0	24	100	0.5	3
3	−10	24	100	0.5	3
4	−20	24	100	0.5	3

Table 2. Loading capacities of ESQUAKE test set-up

Maximum vertical load	20.000 kN
Maximum horizontal load	2.000 kN
Maximum horizontal stroke	600 mm
Maximum velocity	1000 mm/s

3 Experimental Results

During the tests, force-displacement curves of LRB were recorded for temperatures of −20 °C, −10 °C, 0 °C and 20 °C (Fig. 3a). Mechanical properties, namely characteristic strength (Q) and post-elastic stiffness (k_2) of LRB at each temperature were quantified using recorded force-displacement curves. In order to calculate Q and k_2, the least square method (LSM), which is one of the regressions analyses, was used to obtain the best-fit line. The principle of this method is to find the best-fit to a data by minimizing the sum of squared difference between the recorded value and the fitted value. The definition of Q (Q_{pos} and Q_{neg}) and k_2 ($k_{2(lower)}$ and $k_{2(upper)}$) values according to LSM is given in Fig. 3b for an example force-displacement curve. The ultimate values of Q and k_2 for each cycle of motion are computed as ($Q_{pos} + Q_{neg}$)/2 and ($k_{2(lower)}$ and $k_{2(upper)}$)/2, respectively. Table 3 gives the mechanical properties calculated at −20 °C, −10 °C, 0 °C and 20 °C for each loading cycle.

Table 3. Calculated Q and k_2 values for each loading cycle.

	Cycle	Temperature			
		20 °C	0 °C	−10 °C	−20 °C
Q (kN)	1st	156	215	277	307
Q (kN)	2nd	131	177	212	253
Q (kN)	3rd	122	158	188	230
k_2 (kN/m)	1st	2000	2695	3106	3122
k_2 (kN/m)	2nd	1865	2496	2792	2866
k_2 (kN/m)	3rd	1808	2382	2652	2721

4 Mathematical Model

4.1 Modeling of Temperature-Dependent Behavior of Lead Rubber Bearings

The hysteretic behavior of LRBs is generally modeled by a generic non-deteriorating force-deformation relation. However, experimental studies performed with LRB reveal that there is a reduction in strength of the isolator at each cycle, as can be observed in Fig. 3a. Kalpakidis and Constantinou (2009a) showed that the main reason for the deteriorating behavior of LRB is the temperature rise in the lead core from cycle-to-cycle. The authors proposed a mathematical model that calculates the deterioration in strength of LRB as a function of the lead core temperature. Accordingly, the instantaneous temperature of the lead core is used to update the strength of the isolator via reducing the initial yield stress of the lead (σ_{YL0}), instantly. The relation between lead core temperature (T_L) and initial yield stress of the lead (σ_{YL0}) is defined by the Eq. (1). Here, E_2

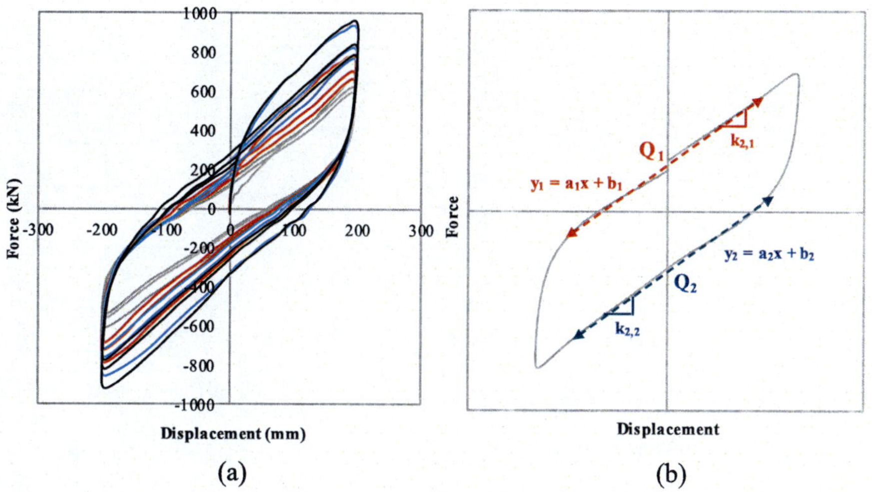

Fig. 3. (a) Comparison of force displacement curves of LRB recorded at different temperatures. (b) Application of LSM method.

is a constant, which is given as 0.0069/°C. More information about the deteriorating mathematical model of LRB can be found in Kalpakidis and Constantinou (2009a).

$$\sigma_{YL}(T_L) = \sigma_{YL0} \exp(-E_2 T_L) \tag{1}$$

The validity of the deteriorating model proposed by Kalpakidis and Constantinou (2009a) in representing the hysteretic behavior of LRBs is demonstrated for low ambient temperature in Cavdar et al. (2022). It is seen that the deteriorating material model is highly accurate to simulate the actual behavior of LRBs under low temperatures as well.

4.2 Modeling of Seismically Isolated Bridge

The structural model investigated within the scope of this study is a continuous, three-span, cast-in-place concrete box girder structure with a 30° skew. The total length of the bridge is 97.5 m with lengths of 30.5, 36.5 and 30.5 m for each span. The intermediate bents are composed of two circular columns with a cap beam. The isolation system consists of eight LRBs, two isolators at each abutment. Figure 4a and b schematically shows bridge bent elevation and analytical representation of the bent, respectively. Dicleli and Buddaram (2007) concluded that the seismic response of a typical seismically isolated bridge can be acquired by a single degree of freedom (SDOF) model. In this modeling approach, the piers and deck of the bridge are considered as rigid. As seen in Fig. 4b, bridge analytical model is represented by an LRB supporting a tributary bridge superstructure with infinite in-plane rigidity. The total weight of the bridge above the isolation level is 24956 kN.

In the present study, bidirectional bilinear model implemented in OpenSees (2001) were used to simulate LRBs hysteretic behavior. The validity of this mathematical model

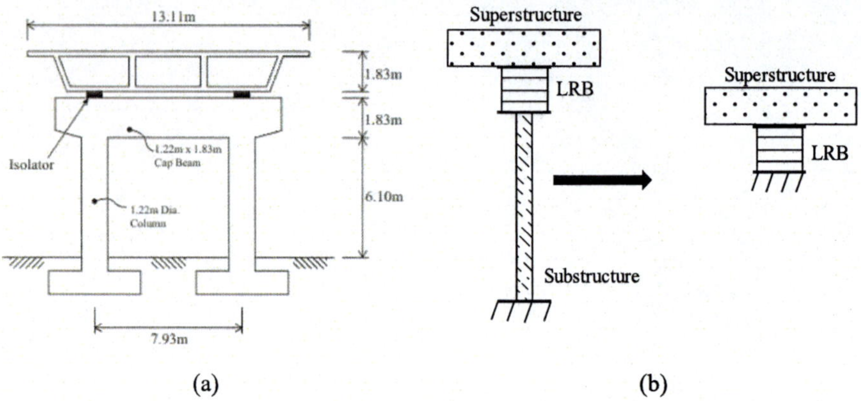

Fig. 4. (a) Bent geometry of the seismically isolated bridge (b) analytical model.

based on the equations derived by Park et al. (1986) was verified in Mokha et al. (1993). Related expressions are defined in Eqs. (2)–(4).

$$\begin{Bmatrix} F_x \\ F_y \end{Bmatrix} = c_d \begin{Bmatrix} \dot{U}_x \\ \dot{U}_y \end{Bmatrix} + K \begin{Bmatrix} U_x \\ U_y \end{Bmatrix} + (\sigma_{YL}(T_L)A_L) \begin{Bmatrix} Z_x \\ Z_y \end{Bmatrix} \quad (2)$$

$$Y \begin{Bmatrix} \dot{Z}_x \\ \dot{Z}_y \end{Bmatrix} = (A[I] - B[\Omega]) \begin{Bmatrix} \dot{U}_x \\ \dot{U}_y \end{Bmatrix} \quad (3)$$

$$[\Omega] = \begin{Bmatrix} Z_x^2\left[\mathrm{sgn}\left(\dot{U}_x Z_x\right) + 1\right] & Z_x Z_y\left[\mathrm{sgn}\left(\left(\dot{U}_y Z_y\right) + 1\right]\right. \\ Z_x Z_y\left[\mathrm{sgn}\left(\left(\dot{U}_x Z_x\right) + 1\right]\right. & Z_y^2\left[\mathrm{sgn}\left(\dot{U}_y Z_y\right) + 1\right] \end{Bmatrix} \quad (4)$$

In the above equations, F_x and F_y are the isolator forces and U_x and U_y are the isolator displacements in both horizontal directions. Y and K are the yield displacement and post-yield stiffness of the bilinear force-deformation relation of isolators, respectively. Z_x and Z_y are hysteretic dimensionless quantities that account for the interaction of hysteretic seismic isolator forces in orthogonal horizontal directions. C_d represents the energy dissipation of the rubber and A_L is the cross-sectional area of the lead core. A and B values should satisfy the relation of A = 2B (Nagarajaiah vd., 1989) and were chosen as A = 1 and B = 0.5 in order to assure that Z_x and Z_y values vary between + 1 and − 1. Finally, [I] is the unit matrix, sgn stands for the signum function and overdot refers to differentiation with respect to time.

5 Dynamic Analyses

5.1 Selection of Ground Motions for Dynamic Analyses

Eleven pairs of ground motion records were considered in the analyses. The selection criteria recommended by TEC (2018) and ASCE/SEI 7-16 (2017) is the similarity in terms of the magnitude of ground motion, distance to the fault rupture, fault rupture mechanism and soil type. Ground motion records were downloaded from the Pasific Earthquake Engineering Research (PEER) Center strong ground motion database. Table 4 gives the characteristics of considered ground motions. In Table 4, M_w is the magnitude of the ground motion; R is the closest distance to fault rupture; PGA, PGV, and PGD stand for peak ground acceleration, peak ground velocity, and peak ground displacement, respectively. 5% damped response spectra of ground motions listed in Tables 4 are given in Fig. 5a.

Table 4. Characteristics of selected ground motions.

#	EQ Name	Station	M_w	Rrup (km)	Soil class	PGA (g)	PGV (cm/s)	PGD (cm)
1	Imperial valley-06	Delta	6.5	22.0	D	0.26	26.9	17.2
2	Taiwan smart1(45)	Smart1 O08	7.3	54.8	D	0.16	25.6	9.7
3	Superstition hills-02	Westmorland fire sta	6.5	13.0	D	0.21	30.5	20.3
4	Northridge 01	Canyon Country - W Lost Cany	6.7	12.4	D	0.40	45.2	13.4
5	Kobe, Japan	Fukushima	6.9	17.9	D	0.19	34.2	14.7
6	Dinar Turkey	Dinar	6.4	3.4	D	0.28	35.0	8.3
7	Kocaeli Turkey	Izmit	7.5	7.2	C	0.19	33.0	19.0
8	Chi-Chi Taiwan	CHY036	7.6	16.0	D	0.25	43.7	30.0
9	Chi-Chi Taiwan	TCU059	7.6	17.1	D	0.16	53.3	49.0
10	Niigata Japan	NIGH11	6.6	8.9	D	0.51	48.2	14.1
11	El Mayor Cucapah	Tamaulipas	7.2	26.6	D	0.21	41.1	37.8

5.2 Scaling of Ground Motions

The followed procedure for scaling of ground motion records is composed of two complimentary phases. In the first stage, sum of the weighted squared errors between the geometric mean of the two horizontal components and the target spectral values in the relevant period range is minimized. The second step of scaling was performed to assure

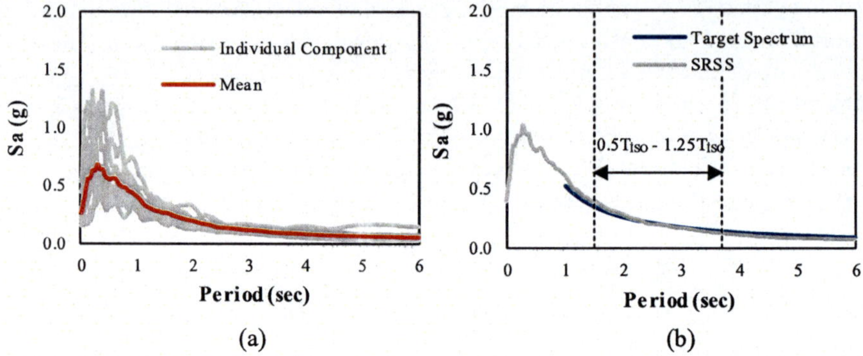

Fig. 5. (a) Geometric mean spectrum of ground motion records (b) scaled average spectrum and target spectrum.

the requirements of TEC (2018). According to this, in the $0.5T_{iso}$–$1.25T_{iso}$ period range defined for seismic isolated structures, the average of square-root-of-sum-of-squares (SRSS) spectra from all ground motion pairs shall not fall below 1.3 times the corresponding ordinate of the target response spectrum. The final scale factor to be applied to the ground motion is defined as the product of two factors obtained in both phases. Figure 5b shows the scaled average SRSS spectrum of the considered ground motions together with the 5% damped design spectra for DD-2 according to TEC (2018).

6 Analytical Results

6.1 Maximum Isolator Displacements

In this section, variation of MID for the analyzed seismic-isolated bridge due to change in ambient temperature is presented. MID is calculated as maximum of $((D_x)^2 + (D_y)^2)^{1/2}$ where D_x and D_y are the isolator displacements in horizontal x and y directions, respectively. Figure 6a shows comparison of the maximum, minimum and average MID values acquired from all ground motion records at each ambient temperature. Besides, Fig. 6b presents the average amplification in MIDs at low temperatures (0 °C, −10 °C, −20 °C) in comparison to the ones obtained at the reference temperature (20 °C). Figure 6 reveals that as the ambient temperature decreases, MID also decreases significantly. The amount of the amplification in MID is about 35% for both −20 °C and −10 °C. On the other hand, it is calculated as 20% for 0 °C.

6.2 Maximum Isolator Forces

In this section, the variation of MIF in the analyzed seismic-isolated bridge due to the change in ambient temperature is presented. MIF is defined as maximum isolator force in any direction, max (F_x, F_y). F_x and F_y are the isolator forces in horizontal x and y directions, respectively. According to Fig. 7a, as a result of the increase in stiffness and strength in the LRB exposed to low temperature, the shear force transferred to

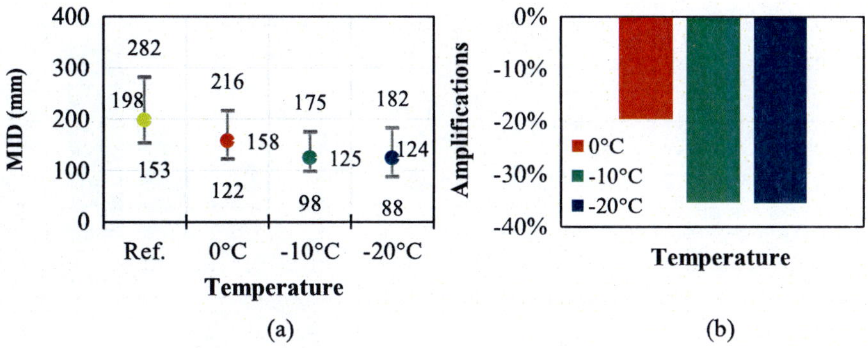

Fig. 6. (a) Comparison of MIDs at different temperatures (b) variation of MID.

the substructure from the isolator increases. To indicate the variation more clearly, the amount of amplification in maximum shear force is presented in Fig. 7b. Accordingly, amplifications in average shear forces are 31%, 20% and 13% for −20 °C, −10 °C and 0 °C, respectively.

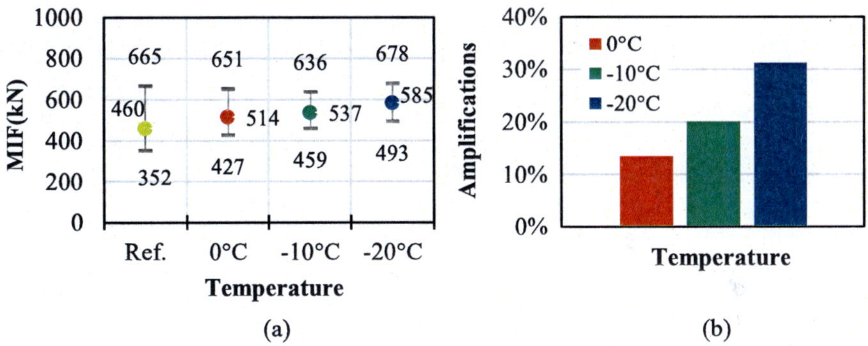

Fig. 7. (a) Comparison of MIF values at different temperatures (b) variation of MIF.

6.3 Maximum Accelerations

In this section, the response of the analyzed model is studied in terms of maximum acceleration defined as the maximum of $((a_x)^2 + (a_y)^2)^{1/2}$ where a_x and a_y are the accelerations in x and y directions, respectively. The maximum, minimum and average accelerations obtained for 11 ground motions are depicted in Fig. 8a whereas Fig. 8b presents the amount of variation in maximum accelerations at low temperatures in comparison to the reference temperature. Figure 8 clearly shows that maximum acceleration tends to increase with decreasing temperature. The amounts of amplification are in between 11% and 28%.

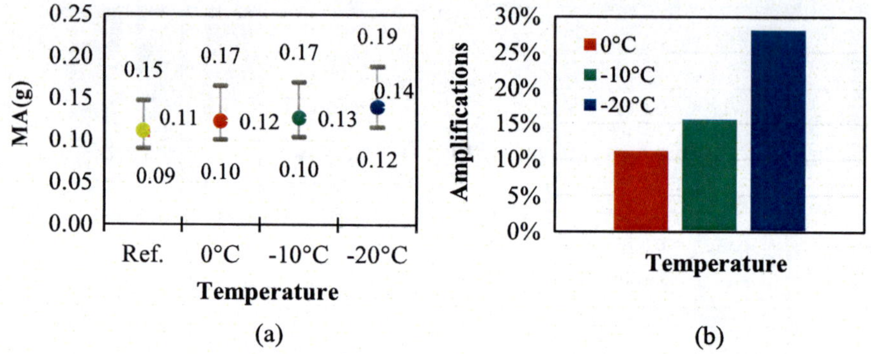

Fig. 8. (a) Comparison of MA values at different temperatures (b) variation of MA.

7 Conclusions

This study focused on the response of a bridge isolated with LRB when exposed to low ambient temperatures under bidirectional excitations of ground motions. First, tests were conducted with an 850 mm diameter LRB at temperatures of 20 °C, 0 °C, −10 °C and −20 °C. In the tests, the loading frequency and the amplitude of shear strain were 0.5 Hz and 100%, respectively. Once the corresponding force-displacement curves were recorded, they were idealized to be used in dynamic analyses. Experimental and analytical results of this study revealed the following conclusions:

1. There is an increase in both characteristic strength and post-elastic stiffness with decreasing temperature. The increase in characteristic strength is more pronounced than post-elastic stiffness.
2. MIFs and MAs tend to increase with decreasing temperature. Among the temperatures considered in the study, −20 °C is the most critical case with almost 30% amplification in both MIF and MA.

Acknowledgments. This study was funded by the Turkish Scientific and Technical Research Council (TÜBİTAK) under grant number 119M573 which the authors gratefully acknowledge. Second author of the manuscript has a scholarship from 100/2000 program of the Council of Higher Education (YÖK).

References

ASCE-7/16: Minimum design loads and associated criteria for buildings and other structures. American Society of Civil Engineers (2017)

Billah, M., Todorov, B.: Effects of subfreezing temperature on the seismic response of lead rubber bearing isolated bridge. Soil Dyn. Earthq. Eng. **126**(105814), 1–13 (2019)

Cardone, D., Gesualdi, G., Nigro, D.: Effects of air temperature on the cyclic behavior of elastomeric seismic isolators. Bull. Earthq. Eng. **9**(4), 1227–1255 (2011)

Cavdar, E., Ozdemir, G., Karuk, V.: Modification in response of a bridge seismically isolated with lead rubber bearings exposed to low temperature. IMO Tech. J. **33**(5), 1–24 (2022)

Constantinou, M.C., Whittaker, A.S., Kalpakidis, Y., Fenz, D.M., Warn, G.P.: Performance of Seismic Isolation Hardware Under Service and Seismic Loading. Technical report, NY: MCEER=07-2012, Buffalo (2007)

Deng, P., Gan, Z., Hayashikawa, T., Matsumoto, T.: Seismic response of highway viaducts equipped with lead-rubber bearings under low temperature. Eng. Struct. **209**(110008), 1–17 (2019)

Dicleli, M., Buddaram, S.: Equivalent linear analysis of seismic-isolated bridges subjected to near fault ground motions with forward rupture directivity effect. Eng. Struct. **29**(1), 21–32 (2007)

Kalpakidis, I.V., Constantinou, M.C.: Effects of heating on the behavior of lead-rubber bearings. I: theory. J. Struct. Eng. **135**(12), 1440–1449 (2009a)

Kalpakidis, I.V., Constantinou, M.C.: Effects of heating on the behavior of lead-rubber bearings. II: verification of theory. J. Struct. Eng. **135**(12), 1450–1461 (2009b)

Mokha, A.S., Constantinou, M.C., Reinhorn, A.M.: Verification of friction model of Teflon bearings under triaxial load. J. Struct. Eng. (ASCE) **119**(1), 240–261 (1993)

Nagarajaiah, S., Reinhorn, A.M., Constantinou, M.C.: Nonlinear dynamic analysis of three-dimensional base isolated structures (3D-BASIS). Technical Report NCEER-89-0019, National Center for Earthquake Engineering Research, State University of New York at Buffalo (1989)

OpenSees: Open System for Earthquake Engineering Simulation; Version: 2.1.0. University of California, Pacific Earthquake Engineering Research Center, Berkeley, California (2001)

Park, Y.J., Wen, Y.K., Ang, A.H.: Random vibration of hysteretic systems under bi-directional ground motions. Earthq. Eng. Struct. Dynam. **14**(4), 543–557 (1986)

Park, J.Y., Jang, K.S., Lee, H.P., Lee, Y.H., Kim, H.: Experimental study on the temperature dependency of full-scale low hardness lead rubber bearing. J. Comput. Struct. Eng. **25**(6), 533–540 (2012)

Pacific Earthquake Engineering Research (PEER) Center: Strong motion data base (2012). Available from https://ngawest2.berkeley.edu/

Robinson, W.H.: Lead rubber hysteretic bearings suitable for protecting structures during earthquake. Earthq. Eng. Struct. Dynam. **10**(4), 593–604 (1982)

Roeder, C.W., Stanton, J.F., Taylor A.W.: Performance of Elastomeric Bearings (No. 298). National Cooperative Highway Research Program, Transportation Research Board, Washington, DC (1987)

TEC: Turkish Earthquake Code: Specifications for Building Design Under Earthquake Effects. Ankara, Turkey (2018)

Yakut, A., Yura, J.A.: Evaluation of low-temperature test methods for elastomeric bridge bearings. J. Bridg. Eng. **7**(1), 50–56 (2002)

Yakut, A., Yura, J.A.: Parameters influencing performance of elastomeric bearings at low temperatures. J. Struct. Eng. **128**(8), 986–994 (2002b)

Zhang, R., Li, A.: Experimental study on temperature dependence of mechanical properties of scaled high-performance rubber bearings. Compos. B Eng. **190**(107932), 1–8 (2020)

Wang, H., Zheng, W.Z., Li, J., Gao, Y.Q.: Effects of temperature and lead core heating on response of seismically isolated bridges under near-fault excitations. Adv. Struct. Eng. **22**(14), 2966–2981 (2019)

Seismic Design and Performance Assessment of the Post-tensioned Bridge Piers

Yu Shen[1](✉), Fabio Freddi[2], Jianzhong Li[1], and Yongxing Li[3]

[1] State Key Laboratory of Disaster Reduction in Civil Engineering, Tongji University, Shanghai 200092, China
yushen@tongji.edu.cn

[2] Department of Civil, Environmental & Geomatic Engineering, University College of London, London WC1E 6BT, UK

[3] Engineering (Bridge, Tunnel & Subgrade) Design Research Institute, China Railway Design Corporation, Tianjin 300308, China

Abstract. Post-tensioned (PT) rocking piers have emerged as a promising solution for bridge substructures, significantly increasing their seismic performance and resilience. These piers are characterized by unbonded PT tendon and energy dissipation (ED) components providing, respectively, self-centering and dissipating capacities. Despite the potential advantages, their practical applications are still limited due to the lack of design standards and guidelines. In this context, developing a simple design procedure represents an essential step in making the use of such systems more accessible to the bridge design community. The present study introduces a design procedure for PT bridge piers by defining a performance index λ related to both initial PT force level and ED device amount. The design process is based on the improvement of a conventional monolithic bridge pier (*i.e.,* similar dimensions and steel amount with superior self-centering behavior and load resistance). For illustration purposes, the design procedure is applied to a case study bridge pier, and its seismic performance is compared with the corresponding conventional monolithic pier. The results show that a PT bridge pier that satisfies the design objectives of residual drift and lateral force (*i.e.,* superior self-centering behavior and load resistance) can be conceived and designed through the proposed simple design approach.

Keywords: Bridge engineering · Unbonded post-tensioned piers · Performance index · Seismic · Design approach

1 Introduction

In the last few decades, the bridge construction sector has been characterized by the increasing use of precast bridge construction technologies. Among others, unbonded post-tensioned (PT) rocking piers are characterized by low-damage and self-centering responses representing promising solutions in high seismicity regions [1–3]. However, the designs and analysis of this novel technology require advanced tools and sophisticated procedures, which constrain its accessibility to bridge designers.

G. P. Cimellaro (Ed.): WCSI 2022, LNCE 309, pp. 438–446, 2023.
https://doi.org/10.1007/978-3-031-21187-4_36

In this context, developing a simple design method to obtain the main design parameters of unbonded PT pier is essential, which is the focus of this study. The present study is based on a performance index λ, which allows for simplifying the design procedure by providing the key design parameters of the unbonded PT pier. The feasibility and effectiveness of the proposed approach are assessed by considering a case study bridge and comparing the hysteretic response of the innovative bridge piers and an equivalent conventional monolithic pier.

2 Post-tensioned (PT) Bridge Pier System

Figure 1a shows a typical PT bridge pier system, featuring an unbonded PT tendon at the center of the column's cross-section and the energy dissipation (ED) elements/mild steel bars at the interface between the column base and footing. The system would exhibit the rocking behavior at the base interface under the seismic events. During the rocking motion, the PT and ED components are respectively expected to experience the elastic and plastic deformations, which are related to the re-centering and ED capacities, combing to produce the flag-shaped hysteretic behavior with small residual deformation, as shown in Fig. 1b.

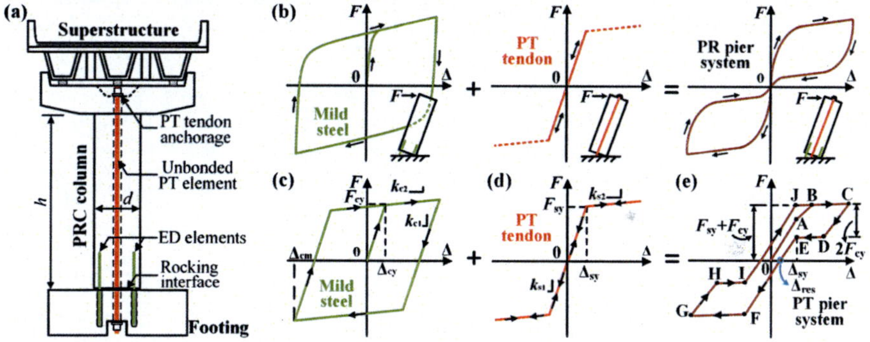

Fig. 1. Typical unbonded PT pier system: (a) sketch representation; (b) expected force-drift relationship; (c) bilinear elastic-plastic model; (d) bilinear elastic model; and (e) idealized flag-shaped hysteretic model ($\Delta_{sy} < \Delta_{cy}$).

2.1 Hysteretic Characteristics

A satisfactory flag-shaped hysteretic behavior (*i.e.*, sufficient ED and acceptable residual deformation) of the unbonded PT pier can be achieved by tuning the initial force of the PT bar and the amount of mild steel bars [3]. To facilitate the interpretation of the contribution of either component to the flag-shaped hysteretic behavior, an idealized bilinear elastic behavior (Fig. 1c) and a bilinear elastic-plastic behavior (Fig. 1d) are respectively used to model the properties of the ED and PT bars. For the bilinear elastic-plastic behavior of ED bars, shown in Fig. 1c, k_{c1} and k_{c2} are respectively the initial and post-yield stiffness; and F_{cy} and Δ_{cy} are respectively the yield force and drift. For the bilinear elastic behavior of the PT bar, shown in Fig. 1d, k_{s1} and k_{s2} are respectively the initial stiffness and post-yield stiffness; and F_{sy} and Δ_{sy} are respectively the yield force and drift.

Based on the combination of the two above-mentioned simplified bilinear models, the multi-linear flag-shaped model representing the hysteretic behavior of the unbonded PT pier is obtained, as shown in Fig. 1e. In the model, the force resistance and drift capacity of the unbonded PT pier can be represented through the superposition of the behavior of two fundamental bilinear models. It should be noted that Δ_{sy} is assumed to be smaller than Δ_{cy} according to the material properties [4]. Due to $k_{c2} \ll k_{c1}$ and $k_{s2} \ll k_{s1}$, the lateral force at Point E in Fig. 1e can be approximately expressed as $F_{sy}-F_{cy}$ (*i.e.*, $F_{sy} + F_{cy}-2F_{cy}$). Thus, the residual drift of the unbonded PT pier Δ_{res} can be approximately calculated by the following Eq. (1):

$$\Delta_{res} \approx \left(F_{cy} - F_{sy}\right)/(k_{s1} + k_{c2}) + \Delta_{sy} \tag{1}$$

or alternatively by the following Eq. (2):

$$\Delta_{res} \approx \left(k_{c1}\Delta_{cy} + k_{c2}\Delta_{sy}\right)/(k_{s1} + k_{c2}) \tag{2}$$

again, considering $k_{c2} \ll k_{c1}$ and $k_{s2} \ll k_{s1}$, the Eqs. (1) and (2) can be approximately rewritten as follow:

$$\Delta_{res} \approx F_{cy}/k_{s1} = F_{cy}/F_{sy} \times \Delta_{sy} \tag{3}$$

Thus, the Δ_{res} is mainly related to the three parameters F_{cy}, F_{sy}, and Δ_{sy}.

2.2 Performance Index λ

The proposed design procedure is based on the dimensionless index λ, defined by the following Eq. (4):

$$\lambda = F_{sy}/F_{cy} \tag{4}$$

Combining Eqs. (3) and (4) gives the Δ_{res} as:

$$\Delta_{res} = \Delta_{sy}/\lambda \tag{5}$$

A typical PT pier designed and previously investigated by Li [5] is selected for case study purposes and to investigate the effect of index λ on the hysteretic response of the unbonded PT pier. The pier clearance height h and sectional depth d (Fig. 1a) are 8.3 m and 2.1 m, respectively. The pier has 23 #38 longitudinal ED bars (corresponding to a reinforcement ratio ρ_E of 0.75%), and the PT bar is initially stressed with 749 MPa (corresponding to a PT ratio η_P of 0.15). More key design parameters of the pier are listed in Table 1. According to Sect. 2.1, the hysteretic behavior of this pier is decomposed into a bilinear elastic-plastic curve and a bilinear elastic curve, and relevant parameter values are provided in Table 2.

Table 1. Key design parameters of the unbonded PT pier studied by Li [5].

Reinforcement ratio (%)		Axial force ratio	
ED bars ρ_E	PT bar ρ_P	Initial PT force η_P	Gravity load η_G
0.75	0.44	0.15	0.10

Table 2. Parameters of two simplified bilinear models.

Bilinear elastic-plastic model			Bilinear elastic model		
F_{cy} (kN)	Δ_{cy} (m)	k_{c2} (kN/m)	F_{sy} (kN)	Δ_{sy} (m)	k_{s2} (kN/m)
1085	0.025	1294	1259	0.014	1411

A parametric analysis with varying λ values is carried out on the unboned PT pier. Figure 2a shows the variation of the residual drift by varying λ when the lateral target drift is 2%. The analysis shows a high sensitivity of the residual drift with respect to λ for values within the interval 0.6 < λ < 1.0. Conversely, the residual drift is less sensitive to the variability of λ for values larger than 1.0 or smaller than 0.6. To enable a more explicit interpretation of the response within the high sensitivity region (*i.e.,* 0.6 < λ < 1.0), Fig. 2b compares three lateral force-drift curves with λ = 0.2, 0.8, and 1.6. It can be observed that the residual drift depends on the location of Point *E*. When the ordinate of the Point *E* is larger than 0 (λ = 1.6), the residual drift is small; conversely, the residual drift is considerable when the Point *E* is located below the zero-ordinate axis (λ = 0.2 and 0.8). In addition, the extent of the highly sensitive λ values is related to the slope of the line that intersects with the zero-ordinate axis. Figure 2b shows that for the hysteretic curve with λ = 0.8 (within the sensitive region), the intersection between the hysteretic curve and the zero-ordinate axis is located on the Line *DE*, whose slope is relatively small (*i.e.,* $k_{c2} + k_{s2}$). Therefore, a small variation in λ would result in a substantial change in residual drift.

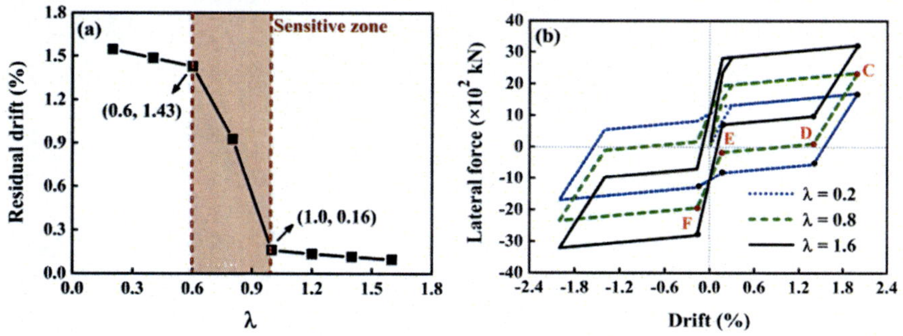

Fig. 2. The effect of λ on the hysteretic responses of the unbonded PT bridge pier: (a) residual drift; and (b) lateral force-drift behavior.

3 Design Procedure

The present section proposes a simple design procedure for the unbonded PT pier based on the λ index and an equivalent monolithic pier. The objective of the design is to obtain a PT pier with similar dimensions and steel amount to the monolithic pier, characterized by comparable load resistance but with limited residual drifts. The procedure includes four primary steps illustrated in the flowchart shown in Fig. 3.

3.1 Step 1: Preliminary Preparations

Based on the reference monolithic pier, the basic design parameters, such as configurations, targeted load resistance (*i.e.*, the ultimate force of monolithic pier $F_{\mathrm{m,R}}$), and material properties, are determined. Also, a total axial force ratio η_{tot} ($= \eta_P + \eta_G$) of the PT pier is selected, with a recommendation of 0.15–0.20.

3.2 Step 2: Pure Rocking Analysis

The η_P is determined by η_{tot} - η_G; and the reinforcement ratio ρ_P of the PT bar is calculated by the following Eq. (6):

$$\rho_P = (\eta_P \times f_c)/\left(0.5\varepsilon_{py} \times E_{pt}\right) \tag{6}$$

where ε_{py} and E_{pt} are the yield strain and elastic modulus of the PT bar; and f_{c} is the concrete strength of the pier. Note that $0.5\varepsilon_{py}$ indicates the PT initial strain (*i.e.*, the initial strain in the PT bar is set as half of its yield strain [2]). In addition, the simplified analysis method based on interface section analysis [4] is conducted on a pure rocking pier (*i.e.*, without ED components) to obtain the F_{sy}.

3.3 Step 3: Determination of Energy Dissipation (ED) Amount

Selecting the index λ (initial recommendation of 1.0 or a larger one based on the analysis results in Sect. 2.2) and the F_{cy} is determined accordingly. Then, the reinforcement ratio of ED bars ρ_E is calculated according to the kinematics of rigid bodies (contact length c at the rocking interface $= 0.25d$ at the target drift [6]) and the assumption of yielding of all ED bars at the target drift. It should be noted that λ is preliminarily selected and would be adjusted if the check conditions were not satisfied in the next step.

3.4 Step 4: Re-centering and Force Capacities Checks

Once the ρ_P and ρ_E are calculated, a trial arrangement of PT tendons (bars or strands) and energy-dissipation elements should be made, and the Δ_{res} and load resistance $F_{\mathrm{m,H}}$ of PT pier can be successively checked. Note that the residual ratio allowable limitation Λ is set as 1% [7], and the $F_{\mathrm{m,H}}$ should be in the range of [0.8, 1.2]$F_{\mathrm{m,R}}$. If these conditions are not satisfied, successive iterations are required.

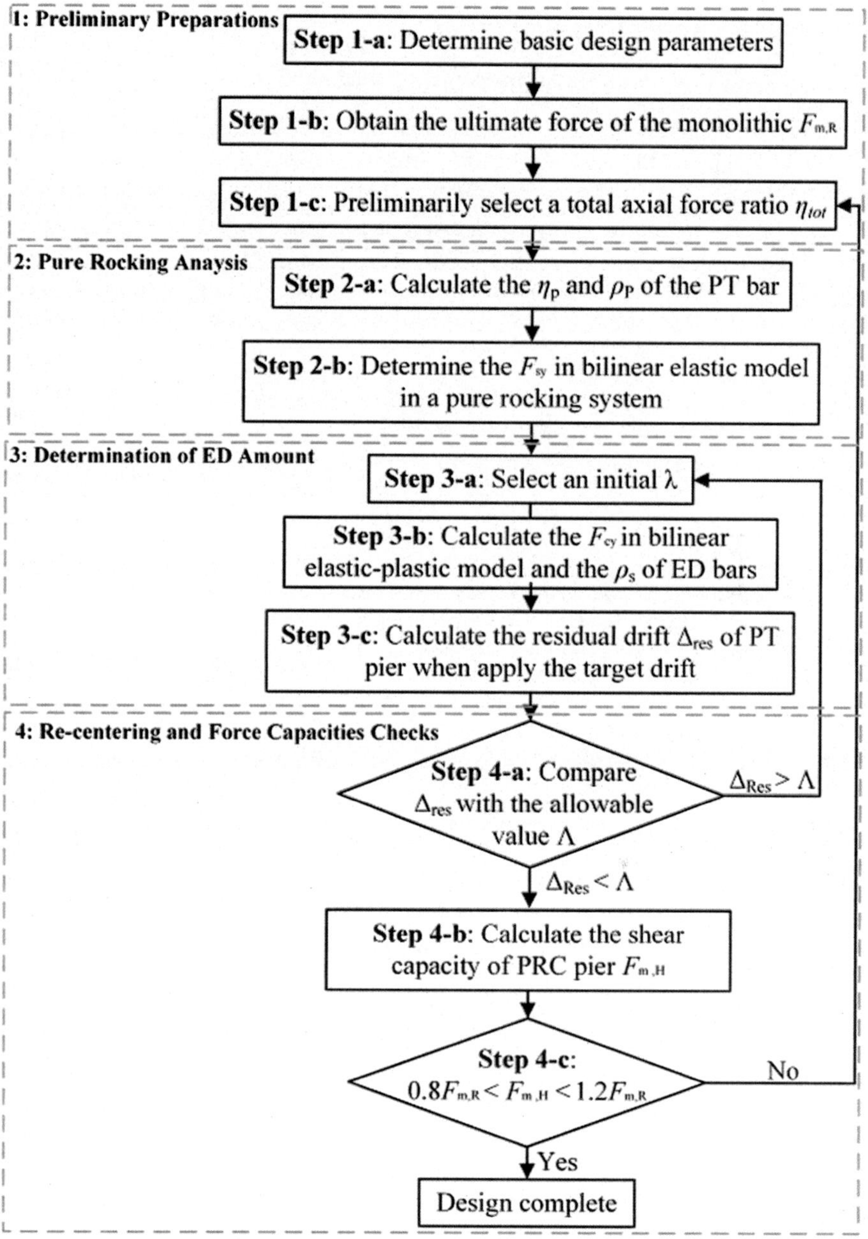

Fig. 3. Flowchart illustrating the design procedure.

4 Design Example and Validation

4.1 Case Structure - The Monolithic Bridge Pier

The typical urban viaduct bridge, shown in Fig. 4, is selected as case study structure to illustrate and validate the design procedure of the unbonded PT pier. The bridge span has a total length of 3 × 35 m and a width of 12 m. As depicted in Fig. 4a, the fixed bearing is only installed at Pier 3 (P3), while the bearing supports at other piers are based on expansion joints. Thus, Pier 3 is redesigned to include the PT system. This pier is characterized by longitudinal and lateral reinforcement ratios of 1.31% and 0.61%, respectively. The axial force caused by the superstructure gravity load G is 7730 kN, corresponding to a η_G of 0.09.

4.2 Design of the PT Bridge Pier and Validation

The corresponding unbonded PT pier is designed based on the design procedure presented in Sect. 3. The target drift is assumed equal to 3%, and Table 3 provides the assumed material properties (f_y: yield stress; f_c: compressive stress; E: Young's modulus; and ε_u: ultimate strain). Table 4 reports the main design parameters for the monolithic per and the unbonded PT pier. The reinforcement ratios between the monolithic and PT piers (1.31 and 1.20%) are close, and their force resistances at target drift are within 20% difference (3537 and 4192 kN).

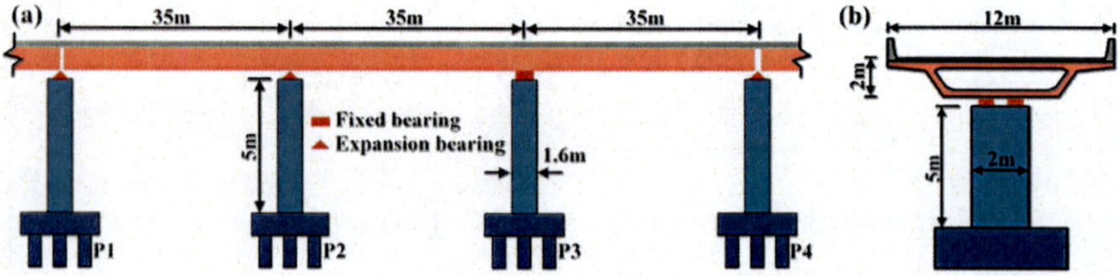

Fig. 4. Schematic diagram of the case bridge structure: (a) front view; and (b) lateral view.

Table 3. Material properties.

Material	f_y (MPa)	f_c (MPa)	E (MPa)	ε_u
Concrete	–	26.8	32.5	0.006
ED bars	400	–	200	0.14
PT bars	1080	–	195	–

Figure 5 compares the hysteretic behavior of the monolithic and designed PT pier. The hysteretic curve of the monolithic pier is plump with considerable residual deformations during unloading (Fig. 5a), while a distinct flag-shaped hysteretic curve with minimal residual drift (<1%) is obtained by the PT pier (Fig. 5b). The validation of the design procedure is also confirmed by the residual drift and the equivalent damping ratio ξ_{eq} at each applied drift (Fig. 6).

Table 4. Parameters of the case monolithic pier and designed PT pier.

Pier type	ρ_E (%)	ρ_P (%)	η_P (%)	F_u (kN)
Monolithic	1.31	–	–	3537
PRC	0.90	0.30	5.99	4192

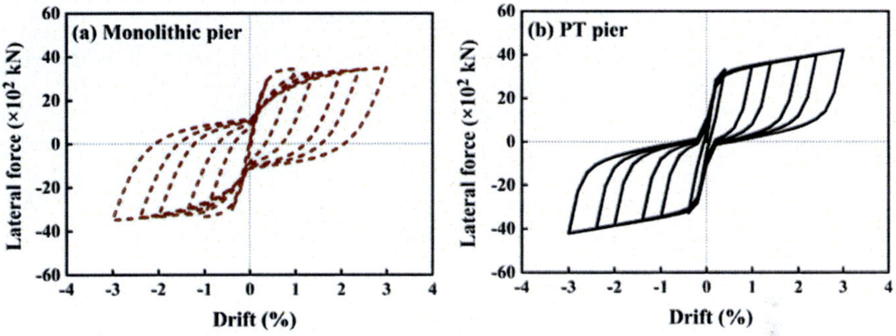

Fig. 5. The hysteretic behavior of: (a) monolithic pier; and (b) PT pier.

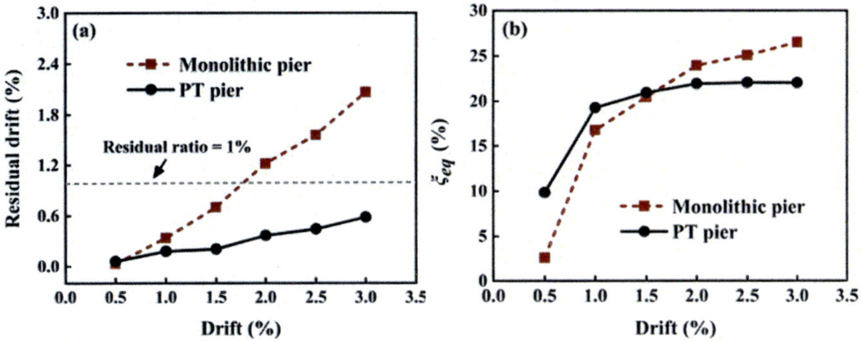

Fig. 6. Cyclic responses comparison between the monolithic and PT piers: (a) residual drift; and (b) equivalent damping ratio ξ_{eq}.

5 Conclusions and Furter Work

The present paper introduces a dimensionless index λ accounting for both self-centering and energy dissipation capacity of the unbonded PT pier. A simple design procedure is presented and validated via a case study bridge where a monolithic pier has been redesigned into an unbonded PT pier. The comparison of the hysteretic behavior between the two solutions highlights the effectiveness of the design method.

References

1. Palermo, A., Pampanin, S., Marriott, D.: Design, modeling, and experimental response of seismic resistant bridge piers with posttensioned dissipating connections. J. Struct. Eng. **133**(11), 1648–1661 (2007)
2. Roh, H., Ou, Y.C., Kim, J., Kim, W.: Effect of yielding level and post-yielding stiffness ratio of ED bars on seismic performance of PT rocking bridge piers. Eng. Struct. **81**, 454–463 (2014)
3. Shen, Y., Liu, X., Li, Y., Li, J.: Cyclic tests of precast post-tensioned concrete filled steel tubular (PCFT) columns with internal energy-dissipating bars. Eng. Struct. **229**, 111651 (2021)
4. Li, Y., Li, J., Shen, Y., Xu, W.: Cyclic behavior and simplified design method of hybrid rocking columns with external energy-dissipators. J. Earthq. Tsunami **14**(6), 2050026 (2020)
5. Li, Y.: Research on the seismic performance of hybrid rocking columns. Ph.D. Dissertation, Tongji University, China (2021)
6. Shen, Y., Freddi, F., Li, J.: Experimental and numerical investigations of the seismic behavior of socket and hybrid connections for PCFT bridge columns. Eng. Struct. **253**, 113833 (2022)
7. JRA: Design specifications of highway bridges-Part 5: seismic design. Japan Road Association, Tokyo, Japan (2016)

Seismic Performance of Isolated Bridges Under Extreme Shaking

Claudio Sepulveda, Ricardo Bustamante, and Gilberto Mosqueda(✉)

University of California San Diego, La Jolla, San Diego, CA 92093, USA
{csepulve,ribustam}@eng.ucsd.edu, gmosqueda@ucsd.edu

Abstract. Seismic isolation is a proven strategy for protecting critical infrastructures from the damaging effects of design-level earthquake shaking. For seismically isolated highway bridges, the expected seismic performance is to provide limited service under a safety evaluation-level ground shaking with minimal to moderate damage. The behavior under extreme shaking, corresponding to a large return period seismic hazard, is not well understood and could induce significant damage. In these rare events, the seismic isolation system can be subjected to displacement demands beyond its design capacity, resulting in failure of the bearings or exceeding the clearance and pounding against the abutment backwalls. To better assess the seismic performance of a prototype highway bridge subjected to earthquakes beyond design considerations, this study examines the bridge lateral displacement, the transfer of forces to the substructure, and potential failure modes of seismically isolated bridges. Advanced modeling approaches are considered to capture bearing characteristics such as hardening for large strain, and a pounding macro-elements is implemented. Results show that for beyond design shaking, the bearings can reach the maximum shear strain capacity, significant residual deformation of the abutment can result from pounding, and the columns can experience moderate damage. The progression of damage is identified and interpreted to assess downtime.

Keywords: Bridge · Seismic isolation · Extreme shaking

1 Introduction

Seismic isolation has proved to be an effective alternative for the design of civil infrastructure requiring high seismic performance. Seismically isolated bridges have been studied by several authors [1–3], showing the effectiveness of this technology through numerical simulation and experiments. Seismic codes such as Caltrans Seismic Design Criteria (Caltrans SDC) [4] provide recommendations for modeling and analyzing highway bridges in California including the use of seismic isolation. Design codes and prediction of seismic hazards have evolved over time, with notable differences among them. For instance, ASCE 7-22 [5] for buildings uses a seismic hazard associated with 2475 years return period (probability of exceedance of 2% in 50 years), while Caltrans SDC uses 975 years (probability of exceedance of 5% in 50 years). This difference in the

G. P. Cimellaro (Ed.): WCSI 2022, LNCE 309, pp. 447–454, 2023.
https://doi.org/10.1007/978-3-031-21187-4_37

seismic hazard design level raises concerns over the expected performance of infrastructure at the regional scale. This study examines the seismic performance of seismically isolated bridges considering shaking beyond design considerations. Maximum response and possible failure modes are predicted using advanced models to capture expected behavior.

2 Seismic Hazard

This study considers ground motions scaled to the two levels of seismic hazard corresponding to 975- and 2475-year return periods. The Unified Hazard Tool [6] was used to estimate a Uniform Hazard Spectrum (UHS) assuming a location for a bridge in Downtown Los Angeles on soil with a shear wave velocity equal to 360 m/s.

The nonlinear time-history analysis of the bridge is performed using the RotD50 [6] of thirty pairs of ground motions scaled using a target period of 2.5 s. This value is representative of effective periods between 975 and 2475 years. The records are selected using the probability of exceedance of 2% in 50 years spectral ordinate and scaled using the Conditional Spectrum methodology [7]. For the probability of exceedance of 10% in 50 years seismic hazard, the same set is downscaled by a factor of 1.4 to bring the records to the corresponding spectral ordinate. Thus, the same set of motions is used for both seismic hazards to enable a performance comparison.

3 Modeling

3.1 Prototype of Seismically Isolated Bridge

The prototype bridge was developed based on an example provided by the Federal Highway Administration Seismic Design [7] and later modified by Constantinou et al. [8] to include seismic isolation. The three-span bridge has a cast-in-place concrete box girder deck with a length of 30 m between abutments and bents and 37 m between bents in the center span. The abutment gap distance with the deck is equal to 53 cm. There is no shear key in the transverse direction that limits displacements. Bents are composed of two 6 m tall columns joined by a cap beam. Each bent abutment has two lead rubber bearings to create the isolated layer, giving an effective period of 2.13 s with effective damping of 27% associated with a design displacement of 23 cm.

3.2 Nonlinear Model

An OpenSees [9] model was developed to simulate the nonlinear behavior of the bridge. For the deck and cap beam, linear elastic elements were used considering the gross section of the elements. Three displacement-based nonlinear beam elements were used for each bent column, considering one element for each potential plastic hinge zone and another for the middle part of the column. The fiber-based section of the column elements includes unconfined concrete for the cover zone and confined concrete for the core. The footing stiffness is modeled with columns supported by linear springs for all 6 DOFs. The model proposed by Marquez et al. [10] was used for the lead rubber bearings since it captures the horizontal force resistance of rubber bearings under large strains, including strain hardening and heating effects [11]. The impact force-indentation behavior between the deck and the abutment back wall is represented by a *Hertzdamp* model [12] implemented in OpenSees by Hughes and Mosqueda [13]. The abutment backfill soil was modeled using the hyperbolic force-displacement model by Wilson and Elgamal [14]. Finally, for the nonlinear characterization of the abutment backwall, an Abaqus [15] model was generated to obtain the force-deformation behavior. Volumetric 8-node elements with a Concrete Damaged Plasticity material model was used to simulate concrete behavior, while truss elements considering a plastic model with isotropic hardening were defined to simulate the rebar contribution. A monotonic analysis was carried out to obtain a force-deformation curve of the abutment being pushed by a rectangular elastic block representing the deck cross-section. Figure 1 shows a scheme of the base-isolated bridge model used.

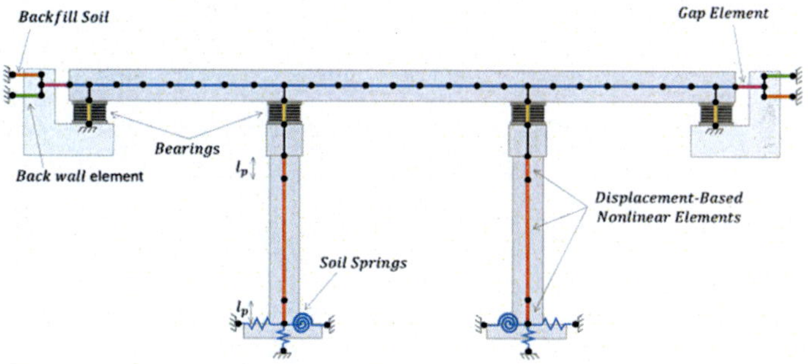

Fig. 1. Scheme of the nonlinear model of the isolated bridge

4 Results

The seismic performance of the isolated bridge is evaluated through time history analysis with the sets of motions scaled to the two hazard levels. The element response is presented in detail first for a typical record, followed by a summary of peak response for all records.

4.1 Local Element Response

Figure 2 shows the displacement orbit of the deck for both seismic hazard levels. The dashed line represents the abutment gap distance indicating that the deck pounds against the east abutment for the larger seismic hazard.

The response of the bearings at the abutment is shown in Fig. 3 for both seismic hazards in the two orthogonal horizontal directions. For the 2475 year return period, a maximum shear strain of 300% is observed, with strain hardening effects starting around 250% of shear strain deformation [10]. The peak at about −150% shear strain in the transverse direction is due to bi-directional coupling and needs to be further investigated in the model.

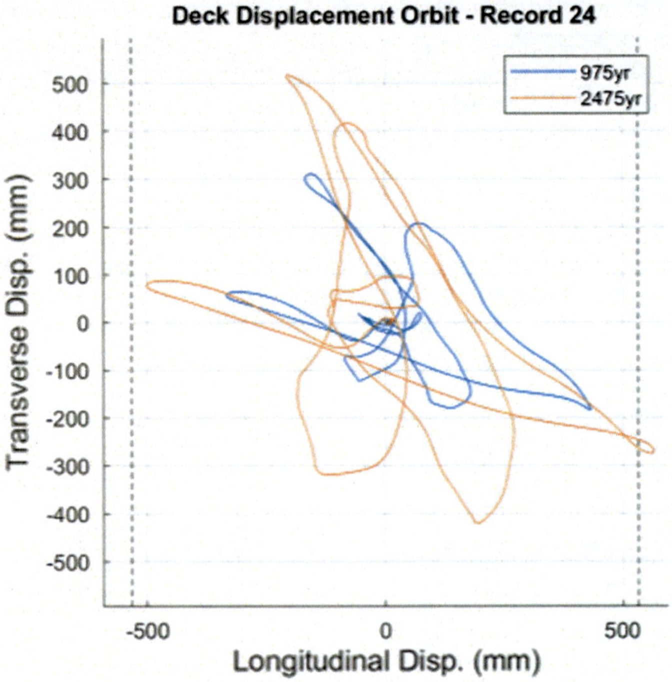

Fig. 2. Orbital deck displacement for both seismic hazards

For 2475 years return period, the abutment response due to an impact is shown in Fig. 4a disaggregating the result in terms of the backwall and backfill soil contributions. The maximum deformation reached by the abutment is about 30 mm. The backwall reaches the maximum force capacity then fails by shear. After this point, the backfill soil contribution is negligible, but continues to increase with increasing deformation.

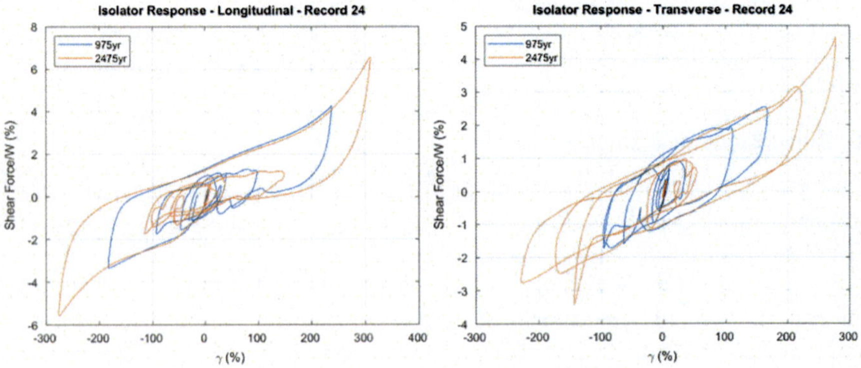

Fig. 3. Bearing response for both seismic hazards

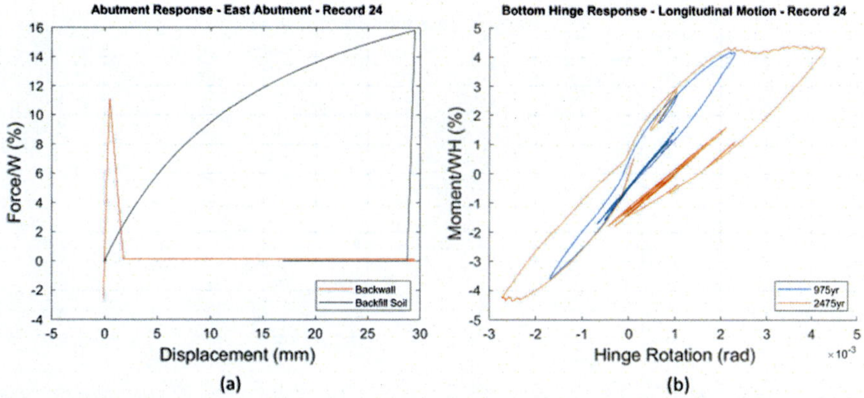

Fig. 4. (a) East abutment response for 2475 years return period due to the deck impact (b) moment-rotation of column bottom hinge in the longitudinal direction for both seismic hazards

With respect to the bridge structural system, the most severe damage is expected at the bottom hinge in the longitudinal direction. The hysteresis for the south-east column bottom hinge is shown in Fig. 4b for both seismic hazards. The moment shown is normalized by the deck weight multiplied by the column height. The column hinge shows plastic deformation after a rotation of about 2e-3 rad, after which the moment remains relatively constant. This yield rotation θ_y is exceeded in both directions (positive and negative) for the beyond-design shaking, while for the design-level, it is slightly exceeded in only one direction.

4.2 Bridge System Response

The overall performance of the seismically isolated bridge is analyzed by examining the maximum response for all the records. Figure 5 shows the maximum horizontal displacement for both ground motion levels. For the following plots, the records are sorted based on the maximum orbit displacement for 2475-year motions. For 2475 year scenario, the maximum values are about twice the design-level ground motions.

The maximum rotation of the plastic hinge at the column base is shown for all records in Fig. 6a. Values are presented separately in the longitudinal and transverse direction given the different deformation behavior of the bent. In the longitudinal direction, column hinges exceed θ_y = 2e-3 rad. For just a few ground motions for 975 year case, while for 2475 year case, half of the ground motions exceed such threshold. However, none of the hinges showed strength degradation, therefore no severe damage to the columns is expected if the column is well confined [16]. Figure 6b shows the maximum bent displacement for longitudinal direction normalized by the maximum longitudinal deck displacement. For low demands on the column rotation, the bent contribution is high because the isolators have a high initial stiffness. At large seismic intensities, resulting in larger column rotation, the bent contribution remains around 25% as the isolators take much of the deformation. However, with increasing demands and as the hinge rotation exceed 8e-3 rad the bent contribution starts to increase reducing the effectiveness of

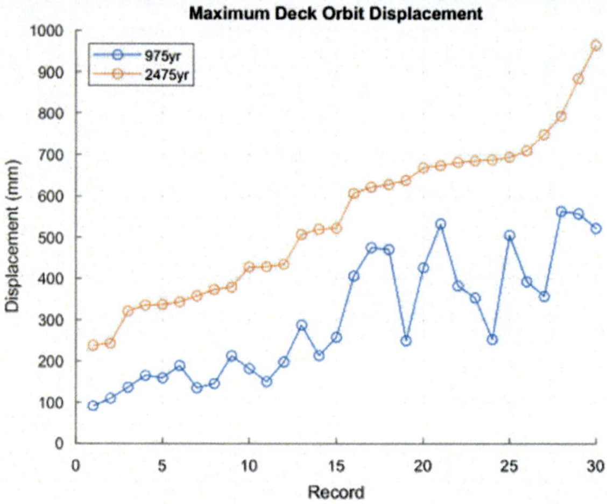

Fig. 5. Maximum deck displacement for both seismic hazards

the isolation system. As the bearings resistance continues to increase, larger forces are transferred to the columns and increase the nonlinear demands. While this behavior is expected [8] it has not been studied in detailed nonlinear analysis.

For beyond design level, 11 out of the 30 ground motions produced pounding, while for design level, no impacts are recorded. According to Wilson and Elgamal [17], the maximum load capacity of the soil can be reached at a deformation of about 0.027–0.03 times the wall height. For this bridge prototype, such a threshold is equivalent to about 76 mm, which is not reached in any of the ground motions. Beyond this value, the soil starts to lose strength.

The maximum shear strain for design level is about 300%, a value for which hardening is barely manifested and a bearing failure is unlikely according to tests for similar bearing sizes and axial stresses [18]. For beyond design shaking, four ground motions produced shear strains higher than 400%, a value for which bearing shear failure may occur [19]. One of the ground motions resulted in strain larger than 500%, which is likely to result in failure of the bearing and the models are no longer reliable.

5 Conclusions

This study assesses the performance of a seismically isolated bridge under design level and beyond design level ground shaking. For design level shaking, the isolated bridge shows good performance, with very limited nonlinear behavior at the columns and maximum shear strain in lead rubber bearings of about 300%.

For beyond design shaking, 11 of 30 ground motion pairs resulted in abutment pounding. Although several impacts indicated the potential for abutment deformations, those are below the strength degradation limit and not likely to fully develop a failure wedge. The columns moment-rotation behavior shows significant yielding but without strength degradation that was incorporated in the model. Regarding shear strain at the

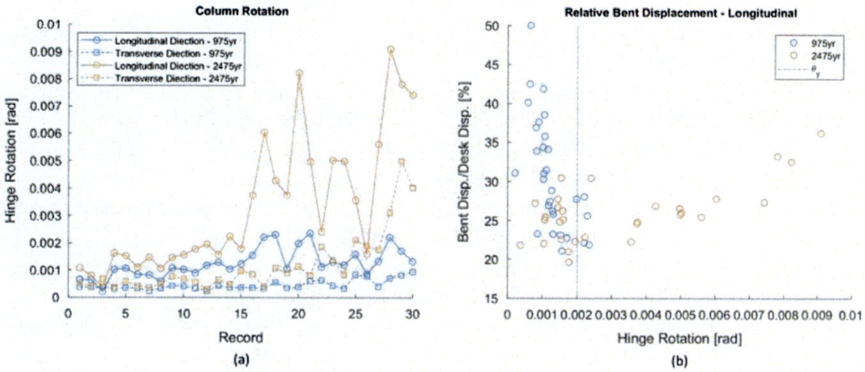

Fig. 6. Maximum rotation at (a) bottom hinge of column (b) relative bent displacement

bearings, 4 out of 30 ground motions produced a deformation greater than 400%, which raises concerns regarding a potential limit for bearing failure.

The results show that the failure mode for a seismically isolated bridge for beyond design shaking is likely bearing failure in the transverse direction. Shear keys were not considered in the study, but it will be examined in future work.

Acknowledgements. The authors would like to acknowledge the generous support of PEER - Pacific Earthquake Engineering Research Center. This research was funded under PBE – M7. The opinions, findings, conclusions or recommendations expressed in this publication are those of the authors and do not necessarily reflect the views of the study sponsors, Pacific Earthquake Engineering Research Center (PEER), or the Regents of the University of California.

References

1. Tsopelas, P., Constantinou, M.C., Kim, Y.S., Okamoto, S.: Experimental study of fps system in bridge seismic isolation. Earthq. Eng. Struct. Dyn. **2**(5), 5–18 (1996)
2. Warn, G.P., Whittaker, A.S.: Performance estimates in seismically isolated bridge structures. Eng. Struct. **26**(9), 1261–1278 (2004)
3. Buckle, I.G., Constantinou, M.C., Dicleli, M., Ghasemi, H.: Seismic Isolation of Highway Bridges. Multidiciplinary Center for Earthquake Engineering Research, University at Buffalo (2006)
4. CALTRANS: CALTRANS Seismic Design Criteria Version 2.0 State of California Department of Transportation (2019)
5. ASCE 7-22: Minimum Design Loads and Associated Criteria for Buildings and Other Structures. American Society of Civil Engineers (2022)
6. Boore, D.M.: Orientation-independent, nongeometric-mean measures of seismic intensity from two horizontal components of motion. Bull. Seismol. Soc. Am. **100**(4), 1830–1835 (2010)
7. Mast, R., Marsh, L., Spry, C., Johnson, S., Griebenow, R., Guarre, J., Wilson, W.: Seismic Design of Bridges. Design Example No. 4: Three-Span Continuous CIP Concrete Bridge. 1st edn. National Technical Reports Library—NTIS, United States (1996)

8. Constantinou, M.C., Kalpakidis, I., Filiatrault, A., Lay, R.A.E.: MCEER: Earthquake Engineering to Extreme Events (2011)
9. McKenna, F., Scott, M.H., Fenves, G.L.: Nonlinear finite-element analysis software architecture using object composition. J. Comput. Civ. Eng. **24**(1), 95–107 (2009)
10. Marquez, J.F., Mosqueda, G., Kim, M.K.: Modeling of lead rubber bearings under large cyclic material strains. J. Struct. Eng. **147**(11) (2021)
11. Kalpakidis, I.V., Constantinou, M.C.: Effects of heating on the behavior of lead-rubber bearings. I: theory. J. Struct. Eng. **135**(12), 1440–1449 (2009)
12. Lankarani, H.M., Nikravesh, P.E.: A contact force model with hysteresis damping for impact analysis of multibody systems. J. Mech. Des. **112**(3), 369–376 (1990)
13. Hughes, P.J., Mosqueda, G.: Evaluation of uniaxial contact models for moat wall pounding simulations. Earthq. Eng. Struct. Dyn. **49**(12), 1197–1215 (2020)
14. Wilson, P., Elgamal, A.: Full-scale shake table investigation of bridge abutment lateral earth pressure. Bull. New Zeal. Soc. Earthq. Eng. **42**(1), 39–46 (2009)
15. Smith, M.: ABAQUS/Standard User's Manual, Version 6.9. Dassault Systèmes Simulia Corp, United States (2009)
16. Priestley, M.J.N., Seible, F., Calvi, G.M.: Seismic Design and Retrofit of Bridges, 1st edn. John Wiley & Sons, United States (1996)
17. Wilson, P., Elgamal, A.: Large-scale passive earth pressure load-displacement tests and numerical simulation. J. Geotech. Geoenviron. Eng. **136**(12), 1634–1643 (2010)
18. Feng, D., Chen, C., Liu, W., Tanaka, K.: A performance test study on Chinese g4 lead rubber bearings. In: 13th World Conference on Earthquake Engineering, Vancouver, Canada (2004)
19. Japan Nuclear Energy Safety Organization, Proposal of Technical Review Guidelines for Structures with Seismic Isolation (2013)

The Role of an Advanced Quality System for the Control of Performance of Lead Rubber Bearings: The Case of Puente Industrial De Biobio

M. Sartori[1](✉), C. Galbiati[1], G. Bresaola[1], S. Barone[1], I. Zivanovic[2], and I. Alende[2]

[1] Freyssinet Product Company Italia, Milano, Italy
mauro.sartori@freyssinet.com
[2] Freyssinet, Rueli-Malmeison, France

Abstract. The 6 km toll road "Via Puente Industrial" represents one of the major public infrastructure under development in Chile, linking the Municipality of Hualpén to San Pedro de la Paz [1]. The total investment of the project, about 160 million dollars, includes the construction of the main section "Puente Industrial", that is the fourth viaduct over the Biobio River in the Conception region, which also incorporates various complementary works at intersections at different levels and railway overpass. The bridge will have a total length of 2.5 km, 2 abutments and 55 piers. Due to the high seismicity of the site, the structure will be seismically isolated with Lead Rubber Bearings (LRB). For this project Freyssinet designed, produced and finally tested a total of 784 ISOSISM® LRB. The high number of devices to produce, together with the very demanding performance and testing requirements, asking to test full scale all the 784 isolators with a maximum variation of 10% on the average performance, represented a real challenge. Freyssinet successfully achieved it, going even beyond the requirement providing an overall variation close to half of the required value, thanks to strict quality and production processes, allowing to control all the steps from characterization of incoming materials, up to vulcanization and test of the isolators. This paper aims at analyzing the performance characteristics of the lead rubber bearings provided for this important project with special focus on the quality control processes that Freyssinet carries on during the full production cycle. This was a mandatory gate to guarantee the performances beyond the requirements of the applicable standard.

Keywords: Seismic isolation · Lead rubber bearings · Full-scale testing · Quality process

1 Introduction

1.1 General Overview of the Project

The toll road "Via Puente Industrial" represents one of the key infrastructure projects under development in Chile. The conception of the project dates back to the 70s, being

G. P. Cimellaro (Ed.): WCSI 2022, LNCE 309, pp. 455–464, 2023.
https://doi.org/10.1007/978-3-031-21187-4_38

considered since then as a public infrastructure work necessary for the regional urban development.

In the recent years, the province of Concepción has increased its levels of vehicular congestion due to the flow increase of trucks and private cars traveling between the municipality of Talcahuano, Hualpén, Concepción and San Pedro de la Paz and to the neighboring ports. Given this, the local and government authorities made the decision to improve the road infrastructure to mitigate traffic, improve circulation to industrial areas, decongest Route 160 and Av. Pedro Aguirre Cerda, ensure logistic routes in the region and reduce the risk of accidents.

The total investment of the project is about 160 million dollars and includes the construction of "Puente Industrial", the fourth viaduct over the Biobio River in the Concepción region of the country, and it incorporates various complementary works at different levels and railway overpass.

Figure 1 shows the geographical area interested by the economic investment, with indication of the new Puente Industrial, under construction at the time of writing this paper.

Fig. 1. Map view of Chile's region interested by the infrastructure investment.

1.2 Characteristics of Puente Industrial

Puente Industrial is a very long infrastructure that connects both banks of the Biobio River and crosses the Los Batros estuary too. It has a total length of 2.520 m, of which 2.094 m straight and 427 m curved, joining the municipalities of San Pedro de la Paz and Hualpén. A rendering of the bridge is shown in Fig. 2.

The superstructure is made up of 56 simply supported decks with a span of 45 m each. The infrastructure consists of 2 abutments and 55 piers, each one composed of deep foundations, specifically 4 pier-piles with 2.5 m diameter.

The bridge is located in a high seismic area, therefore it will be protected from earthquake thanks to the seismic isolation approach. Seismic isolation is in fact a response modification technique that reduces the effects of earthquakes on bridges and other structures.

Fig. 2. Rendering of Puente Industrial.

The bridge will be protected against earthquake thanks to Freyssinet's ISOSISM® anti-seismic devices. The following sections will provide an overview of the technical performances of the isolators, the production/quality process and the prototype and quality control tests performed at ISOLAB®, the advanced dynamic testing facility of Freyssinet Group.

2 FREYssinet's Anti-seismic Devices for Earthquake Protection of the New Puente Industrial

2.1 The Isolation System

For the isolation system of Puente Industrial, the use of anti-seismic devices type Lead Rubber Bearings (LRB) was envisaged, in order to combine horizontal flexibility, to reduces seismic forces in the substructures, and energy dissipation capacity, to limit relative displacements between the superstructure above the isolators and the substructure below.

For this project, Freyssinet designed, produced and tested 784 ISOSISM® LRB, of which 14 (type LRB-N 800 × 246.5) for the abutments and 770 (type LRB-N 650 × 204) for the piers, due to different performance requests.

The devices are able to guarantee high level of dissipation (up to 25% equivalent viscous damping) for the expected total design earthquake and displacement capacity up to more than 55 cm.

Table 1 gives a summary of the main characteristics of the isolators.

As previously anticipated, the implementation of the seismic isolation system by means of ISOSISM® LRBs has determined an important decoupling of the superstructure motion (i.e. deck) from that of the substructure (i.e. piers and foundations). In particular, the increase in the vibration period of the bridge to about 2 s was achieved thanks to the lateral flexibility of the elastomeric component, while the dissipative capacity of the lead core was exploited to reduce the relative displacements of the devices but

Table 1. Characteristics and performances of ISOSISM® LRB.

Parameter		ISOSISM® LRB-N 800x246.5	ISOSISM® LRB-N 650x204
Location		Abutment	Piers
Quantity		14	770
Rubber diameter	D_r [mm]	800	650
Total rubber height	T_r [mm]	246.5	204
Total height	T [mm]	460	370
External plate width/length	B [mm]	850	700
Weight	W [kg]	1152	609
Dead load	D [kN]	1007	1007
Live load	L [kN]	695	695
Service load	N_{SLS} [kN]	1701	1701
Max seismic load	P_{max} [kN]	550	550
Total design displacement	TDD [mm]	±286	±229
Maximum displacement	d_{max} [mm]	±555	±400
Effective stiffness at TDD	K_{eff}(TDD) [kN/mm]	2.24	1.97
Effective damping at TDD	ξ_{eff}(TDD) [-]	23%	25%
Energy dissipated at TDD	EDC(TDD) [kNm]	265	163

also the stresses transmitted in the substructure. The combined effect of high horizontal flexibility and energy dissipation has made it possible to optimize the design of piers and foundation piles, reducing structural sections as well as longitudinal and transversal steel reinforcement.

Figure 3 shows a 3D view of ISOSISM® LRB isolator.

2.2 Quality Process for High Level Production

The LRBs (Lead Rubber Bearings) supplied for this project are made of alternating layers of a natural rubber compound and steel sheets bonded together by vulcanization. A central lead core completes the device providing the desired amount of energy dissipation.

Achieving a maximum variation of ±10% on the average effective stiffness represented a real challenge for the production.

The starting point to succeed, is even before the production begins, by checking the incoming raw materials. In other words, only high-quality raw materials, rubber, lead and steel, can provide a high quality vulcanized device.

On this type of devices, performances are influenced by two main contributions: the cured rubber and the internal lead core. Lead core behavior is straight forward and not

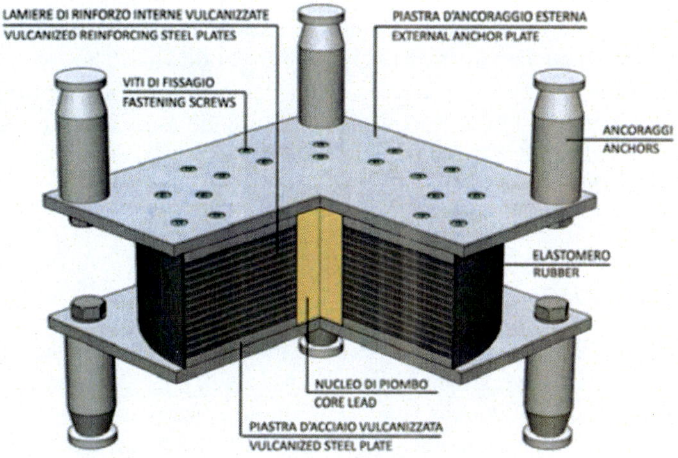

Fig. 3. 3D view of ISOSISM® LRB

examined in this article while the cured rubber behavior is naturally subjected to wider production variability.

The production of such a high number of devices required multiple rubber lots, knowing that a lot is defined as a sequence of continuously manufactured batches while a batch consists in an individual mix or blend of mixes of the same composition. Moreover, to meet the client delivery schedule on site, the production was split in multiple sessions over several months.

To respect the project specifications and ensure a tight tolerance on the full-scale testing, every lot of compound was monitored at the source (milling factory) with different tests such as rheometric curves, physical-mechanical characterization tests, ozone and aging tests etc.

Once each rubber lot was certified to meet the project and Freyssinet internal specifications (which are more demanding to ensure the highest quality standards), the material was sent to Freyssinet production site where a new set of tests were performed on each batch. This allowed the production department to characterize each batch through this double check, in order to guarantee an uniform performance throughout all the mass production.

The manufacturing process of an elastomeric isolator involves the separate treatment of two different materials: rubber and steel. Steel parts quality controls aim to check the correct application of the bonding agent. Although these steps are fundamental to the final outcome and ultimate resistance of the devices, they are not directly responsible for constancy of performance in term of stiffness and energy dissipated.

Finally, rubber and steel parts are bonded together through vulcanization which consists in applying heat and pressure to the assembled elements in a mold for a determined period.

Vulcanization is one of the, if not the, most important steps of the production process, especially when such tight tolerance on production variability is required. Temperature and vulcanization timing are highly dependent on the rubber type and device geometry however this have been mastered by Freyssinet through its experience in thousands of elastomeric devices supplied in hundreds of projects worldwide. The definition of these parameters is hence the key for the success of this last but crucial step of the production. In addition to a good design of this phase, to ensure the same result on each device, vulcanization temperature was monitored on each device by means of a thermocouple inserted in the mold, to immediately detect and manage any unexpected difference in the vulcanization process.

2.3 Testing of the Isolators

Following the strict quality control described in the previous section, the isolators have been tested full-scale to check their performances.

Before starting the production, two units of each size have been subjected to a Prototype Test protocol in accordance with AASHTO "Guide Specification for Seismic Isolation Design". This very severe protocol aims at characterizing the tested isolators in order to check the performances in term of effective stiffness K_{eff} and energy dissipation (EDC), comparing them with the design values and to prove a stable behavior of the Lead Rubber isolators, both in term of performances and bonding. The test protocol for the two types of isolators supplied for the new Puente Industrial are shown in Table 2.

The prototype tests were performed in November 2021 at ISOLAB, the Freyssinet dynamic testing facility in Montebello della Battaglia, Pavia, Italy. ISOLAB is a laboratory that Freyssinet developed specifically for dynamic tests of seismic protection devices where the products are tested according to the main international standards and according to the highest calibration requirements like ASTM E4 and ISO 7500 as well as quality system. The laboratory, with its 30 years of experience in the field, in addition to prototype and production quality test, is also dedicated to the R&D that Freyssinet group constantly carries on to be always in first line to provide innovative reliable solutions for the seismic protection of structures and infrastructures.

The prototype tests of the ISOSISM® LRB-N 800x246.5 to be installed on the abutments were performed on the ISOLAB's 70 MN testing machine able to test devices up to 70.000 kN static vertical load and up to 20.000 kN static horizontal load. The testing bench can also develop a vertical dynamic force up to 18.000 kN and a horizontal dynamic force up to 5.000 kN with 1000 mm maximum stroke and 1000 mm/s peak velocity. Figure 4a shows the 70 MN machine.

The prototype tests of the ISOSISM® LRB-N 650 × 204 to be installed on the piers were tested on the ISOLAB's 50 MN testing machine able to test isolators or bearings up to 50.000 kN of vertical load and up to 20.000 kN of horizontal load. Figure 4b shows the 50 MN machine.

The resulting force-displacement loops from where it is possible to determine the effective stiffness (K_{eff}) and the energy dissipated (EDC) are shown in Fig. 5. The difference of the resulting effective stiffness with the corresponding design values is very limited being it less than 2%.

The mass production is still ongoing at the moment of writing of this paper. Anyway, the full quantity for the abutments and more than 50% of the isolators for the piers were produced and successfully tested.

(a) (b)

Fig. 4. Test benches at ISOLAB: a) 70 MN; b) 50 MN

AASHTO Guide specification for seismic isolation design requires for the effective stiffness to not exceed the 20% of variation with respect to the design value for any device tested and that the average of all the devices does not exceed the 10% of the design value. The high level of quality process implemented allowed to achieve an excellent result in term of constancy of performances, with a variation on the single device not greater than 15% and an average value not greater than 6%. Figure 6 shows the results in term of stiffness obtained on the full number of devices tested, highlighting the importance of a strict quality control along the production phases.

3 Conclusions

It is well known that design and erection of big structures requires high level of expertise and competences, as well as manufacturing level from design stage to production and installation.

The paper presented an interesting case study where Freyssinet designed, produced and finally tested of a total of 784 ISOSISM® LRB, of which 14 isolators for the abutments (type ISOSISM® LRB-N 800 × 246.5) and n° 770 for the piers (type ISOSISM® LRB-N 650 × 204). The high number of devices to produce, together with the very demanding performance and testing requirements, asking to test full scale all the 784 isolators, represented a real challenge. Freyssinet successfully achieved it, going even beyond the requirements providing an overall variation close to half of the required value. This was possible only because of strict quality and production processes together acting to keep under control all the steps from characterization of incoming material up to the vulcanization of the full isolator and finally to the testing of the full mass production. In the paper, the performance characteristics of the lead rubber bearings have been presented together with the main test results. Special focus to the quality process has been given, that allowed to finally get the exceptional results here presented.

Table 2. Prototype test protocol for ISOSISM® LRB 650 × 204.

Test #	Test name	label	dof	Despl.	Forma de carga	Max vel	freq	carga vert	ciclos
				[mm]		[mm/s]	[Hz]	[kN]	[#]
1	Termico §GSID 13.2.2.1	TH	long	± 40	triangular	–	–	1007	3
2	Viento y frenado: ensayo pre-sismico §GSID 13.2.2.2	WB1	long	± 30	sine	–	–	1007	20
3	Sismico §GSID 13.2.2.3	S1	long	± 222	sine	10	0.007	1007	3
4	Sismico §GSID 13.2.2.3	S2	long	± 56	sine	10	0.029	1007	3
5	Sismico §GSID 13.2.2.3	S3	long	± 111	sine	10	0.014	1007	3
6	Sismico §GSID 13.2.2.3	S4	long	± 167	sine	10	0.010	1007	3
7	Sismico §GSID 13.2.2.3	S5	long	± 222	sine	10	0.007	1007	3
8	Sismico §GSID 13.2.2.3	S6	long	± 278	sine	10	0.006	1007	3
9	Viento y frenado: ensayo post-sismico §GSID 13.2.2.4	WB2	long	± 30	sine	–	–	1007	3

(*continued*)

Table 2. (*continued*)

Test #	Test name	label	dof	Despl.	Forma de carga	Max vel	freq	carga vert	ciclos
				[mm]		[mm/s]	[Hz]	[kN]	[#]
10	Verification de desempeno sismico §GSID 13.2.2.5	SP	long	± 222	sine	10	0.007	1007	3
11	Estabilidad §GSID 13.2.2.6	ST1	long	± 333	triangular	2	0.002	2500	1
12	Estabilidad §GSID 13.2.2.6	ST2	long	± 333	triangular	2	0.002	260*	1

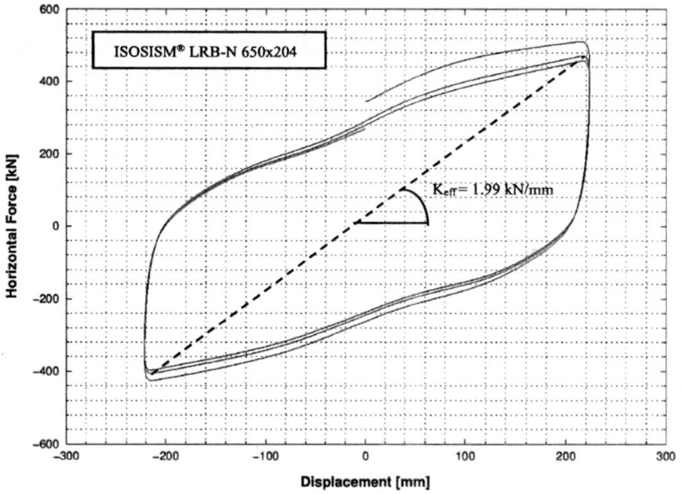

Fig. 5. Force-displacement loop for ISOSISM® LRB-N 650 × 204

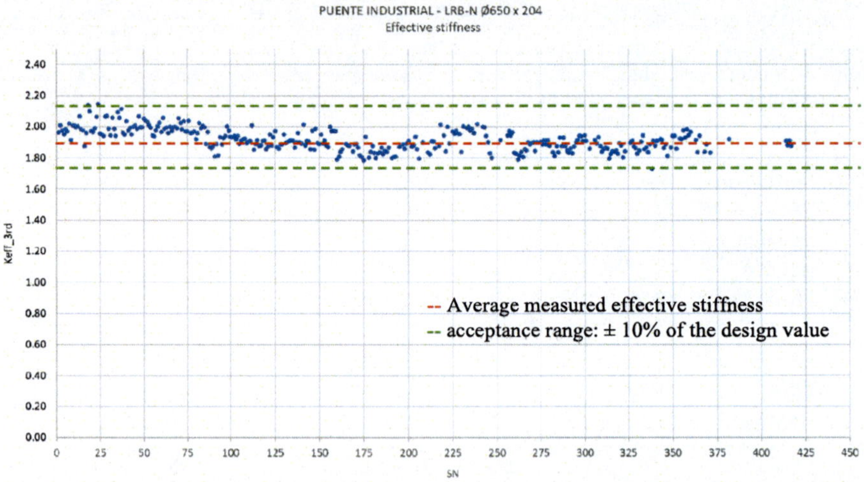

Fig. 6. Effective stiffness from quality control tests for ISOSISM® LRB-N 650 × 204

Reference

1. Sartori, M., Bresaola, G., Turkdonmez, B., Alende, I., Zivanovic, I.: Lead rubber bearings: a solution for the seismic protection of strategic buildings. 17WCEE. Sendai (2020)

The Bridges on Çanakkale Highway: A Huge Application of the Seismic Protection Technology in Europe

S. Barone[1(✉)], M. Sartori[2], C. Galbiati[3], M. Ambor[4], G. Bresaola[5], and I. Zivanovic[6]

[1] Sales Area Manager, Freyssinet Product Company Italia, Milano, Italy
stefano.barone@freyssinet.com
[2] Tech. Department Manager, Freyssinet Product Company, Milano, Italy
mauro.sartori@freyssinet.com
[3] Tech. Department Coordinator, Freyssinet Product Company, Milano, Italy
claudio.galbiati@freyssinet.com
[4] Tech. Department Manager, Atlante, Milano, Italy
michal.ambor@freyssinet.com
[5] Project Engineer, Freyssinet Product Company, Milano, Italy
giorgio.bresaola@freyssinet.com
[6] Tech. Department Deputy Director, Freyssinet, Milano, Italy
ivica.zivanovic@freyssinet.com

Abstract. Çanakkale-1915 is the longest mid-span suspension bridge in the world and it is part of a mega-scale investment in transportation and infrastructure in Turkey. The bridge across the Dardanelle (Çanakkale) Strait carries a new highway connecting Europe and Asia and it will represent a new alternative to the Bosporus passage. The approach bridges are seismically isolated with high-friction curved surface sliders to accommodate the high displacement demand while dissipating the seismic energy transmitted to the piers and foundations which otherwise would be very large and costly due to the extreme seismic excitation foreseen for that region. This paper gives a complete overview of the seismic isolation system of both Asian and European approach bridges. More than 70 isolators were designed and manufactured by Freyssinet. In addition, two viaducts of the approaching highway will be presented, both equipped with viscous dampers to dissipate a huge quantity of energy and designed and tested for extremely high performances. These structures represent an application of mega anti-seismic devices, with displacement capacity up to more than one meter and design velocity above one meter per second.

Keywords: Seismic isolation · Curved surface sliders · High friction coefficient · Viscous dampers

1 Introduction

1.1 General Overview of the Project

1915Çanakkale Bridge and Malkara-Çanakkale Motorway Project is a mega-scale investment in transportation and infrastructure in Turkey. It comprises 88 km of

G. P. Cimellaro (Ed.): WCSI 2022, LNCE 309, pp. 465–477, 2023.
https://doi.org/10.1007/978-3-031-21187-4_39

motorway including the 1915Çanakkale Bridge, and 12 km of access roads streching between Malkara and Çanakkale. The project constitutes a segment of the 324-km-long Kinali-Tekirdağ-Çanakkale-Savaştepe Motorway Project.

Besides the 1915Çanakkale Bridge, the project also includes the construction of 2 approach viaducts, 2 concrete viaducts, 43 overpasses, 40 underpasses, 12 bridges, 228 culverts, 12 junctions, 4 motorway service areas, 2 operation and maintenance centers and 5 toll plazas.

Figure 1 shows the geographical area interested by the economic investment.

Fig. 1. Geographic location of the infrastructure.

The construction of 1915Çanakkale Bridge and Malkara-Çanakkale Motorway started on 18th of March 2017 and ended with the opening to traffic on 18 March 2022. Figure 2 shows a real photo of the bridge.

Fig. 2. 1915Çanakkale suspension bridge with its approaches viaducts.

1915Çanakkale is the longest mid-span suspension bridge in the world with its 2023 m long mid span and 318 high towers. Two approaches give access to the bridge from the two sides for a total length of 4608 m. The deck carries six lanes (three in each direction) of motorway, together with two walkways on each side for maintenance.

Not only the bridge will cut down the crossing of the Çanakkale Strait to six minutes but the Project itself will be making an enormous contribution to the country's economy. The project ranks first amoung Europe's large-scale PPP investments with a value of nearly 3 bilion euros and crown the 100th anniversary of Turkish Republic.

In addition to the exceptional figures of the motorway and of the main bridge over the Dardanelles, Turkey is one of the most seismic countries in the world, with Anatolian fault crossing its territory. The fault is 1500 km long and its north branch reaches Çanakkale. Very strong earthquakes of the Turkey history with more than 8 events above mangnitude 7, make each construction project a challenge, that only estraordinary solutions are suitable to cope with.

Freyssinet, world leader in seismic protection technologies, was able to achieve the stringent and high performance standards requested for the earthquake protection of the infrastructure. In particular, the Group was involved in the supply of 70 curved surface sliders with controlled friction for the seismic isolation of the two approaches viaducts and 48 fluid viscous dampers for the two concrete viaducts.

This paper aims to describe the exceptional performances of the supplied seismic protection devices which required the execution of tests at the CALTRANS testing facility, the most powerful laboratory of the world, located at the University of California (San Diego, USA).

2 Seismic Protection of Canakkale Highway with Freyssinet Technology

2.1 ISOSISM® PS—Curved Surface Sliders

The approach viaducts to the Canakkale bridge were seismically isolated through the installation of ISOSISM® PS, i.e. the curved surface sliders supplied by Freyssinet. Different types were designed according to loads and displacements at the different supports of the Asian and European approach. Tables 1 and 2 give a summary of the overall characteristics of the sliding isolators.

Table 1. Characteristics of ISOSISM® PS for the Asian approach viaduct.

ISOSISM®	PS 17000.1700	PS 28000.1220	PS 28000.1300	PS 15000.1360
Position	SPA	PA 1 5 7 8	PA 2 3 4 9 10	AA
Number of devices	4	16	20	4
Service vertical load N_{Sd} [kN]	6700	1417	13640	5400
Max static vertical load N_{ULS} [kN]	17000	28000	28000	15000
Max seismic vertical load N_{Ed} [kN]	12890	23925	23840	8545
Static displacement d_s [mm]	± 254	± 287	± 256	± 192
Design displacement d_{bd} [mm]	± 635	± 420	± 450	± 510
Maximum displacement d_{max} [mm]	± 850	± 610	± 650	± 680
Radius of curvature R_{eq} [mm]	3700	3700	3700	3700
Dynamic friction coefficent $\mu(N_{Sd})$	5.4%	4.4%	4.6%	5.9%

ISOSISM® PS devices are equipped with a special high-performance sliding material, named ISOGLIDE®, that governs the energy dissipated through the friction coefficient developed between the stainless steel surface and the sliding material during an earthquake movement. ISOGLIDE® is a frictional material tested in accordance with European Standard EN 15129 and covered by ETA. It is characterized by excellent wear (up to 10 km) and temperature resistance (up to 90 °C) and high compressive strength (up to 180 MPa).

Table 3 reports the main global dimensions of the devices. All the isolators are CE-marked and protected with EN ISO 12944–2 C5-H anticorrosion system.

Table 2. Characteristics of ISOSISM® PS for the European approach viaduct.

ISOSISM®	PS 15000.1680	PS 31500.1630	PS 31500.1540	PS 16000.2180
Position	AE	PE 1	PE 2 3 4 5	SPE
Number of devices	4	4	16	4
Service vertical load N_{Sd} [kN]	6000	14250	14050	6550
Max static vertical load N_{ULS} [kN]	15000	31500	31500	16000
Max seismic vertical load N_{Ed} [kN]	10795	20870	26900	13155
Static displacement d_s [mm]	$\pm$ 208	$\pm$ 211	$\pm$ 222	$\pm$ 245
Design displacement d_{bd} [mm]	$\pm$ 570	$\pm$ 510	$\pm$ 525	$\pm$ 760
Maximum displacement d_{max} [mm]	$\pm$ 840	$\pm$ 815	$\pm$ 770	$\pm$ 1090
Radius of curvature R_{eq} [mm]	3700	3700	3700	3700
Dynamic friction coefficent $\mu(N_{Sd})$	5.4%	4.8%	4.9%	5.5%

Table 3. Global dimensions of the devices.

ISOSISM®	Sliding plate diameter [mm]	Total height [mm]
PS 17000.1700	1760	360
PS 28000.1220	1580	275
PS 28000.1300	1615	295
PS 15000.1360	1365	240
PS 15000.1680	1645	320
PS 31500.1630	1890	380
PS 31500.1540	1860	375
PS 16000.2180	2140	535

Figure 3 shows the technical drawing of ISOSISM® PS 16000.2180, the device with higher displacement capacity ($\pm$ 1090 mm).

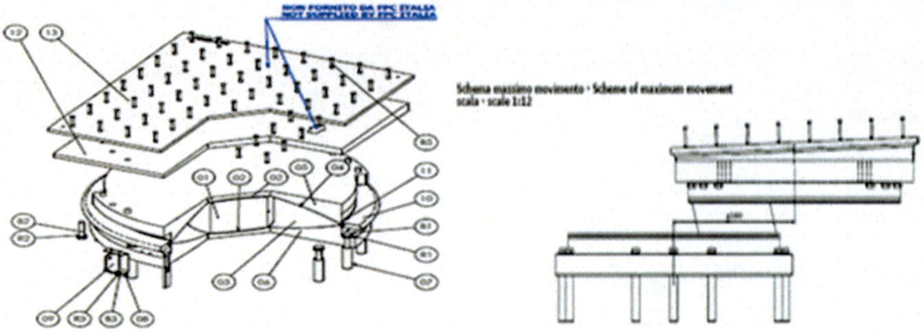

Fig. 3. ISOSISM® PS 16000.2180: 3D view (left) and maximum movement scheme (right).

2.2 ISOSISM® FD—Fluid Viscous Dampers

V01 and V02 represent a brilliant example of how Freyssinet Group is able to offer a complete package from design to construction of civil structures, always proposing avant-garde solutions to meet technical and economical needs related to the structure itself. Freyssinet studied and presented an alternative technical project with the goal to improve the structural seismic response. In fact, being a high seismic risk area, the technical concept of the viaducts had to obtain the best compromise between the strength of the structure and its ability to dissipate the expected strong seismic energy, while guaranteeing the correct daily functionality in static conditions. The most efficient solution to mitigate the seismic risk was to differentiate the dynamic behavior of the viaducts in the two main direction. In longitudinal direction, a set of fluid viscous dampers type ISOSISM® FD were provided at the abutments, with very high mechanical and hydraulic performances, in order to dissipate a huge quantity of seismic energy and to avoid to overload horizontally the fixed piers of the viaduct. In transversal direction the transmission of the seismic forces from the deck to the piers is entrusted to the concrete shear keys with energy dissipation in 4-shaft piers plastic hinges. Figure 4 shows the V06 viaduct under construction while Fig. 5 the layout of the devices for both V01 and V06 viaducts.

A total of 48 ISOSISM® FD were supplied with the technical characteristics shown in Table 4.

Figure 6 shows ISOSISM® FD 4000.925 at the end of the assembly phase.

In addition to the viscous dampers, Freyssinet supplied free and guided spherical bearings type TETRON SB, equipped with ISOGLIDE® sliding material. The characteristics of the bearings are not discussed here for sake of brevity.

3 Testing

3.1 Experimental Response of ISOSISM® PS

Type Test (TT) and Factory Production Control Test (FPCT) were performed on the sliding isolators according to EN 15129 requirements. The devices were tested at the University of California in San Diego (USA). Figure 7 shows ISOSISM® PS 31500.1540 under

Fig. 4. V06 viaduct under construction.

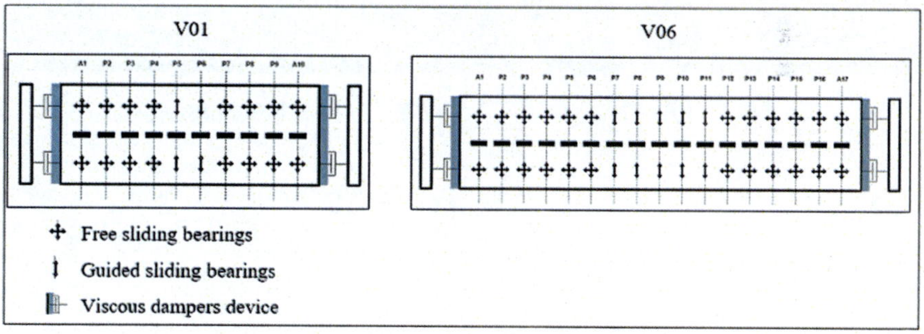

Fig. 5. Devices layout for V01 and V06 viaducts.

Table 4. Charasteristics of ISOSISM® FD.

ISOSISM®	FD 4000.1170 (prototype)	FD 4000.620	FD 4000.925
Position	–	V01	V06
Number of devices	1	20	28
Max force F_{max} [kN]	5080	4090	4225
Max displacement d_{max} [mm]	585	310	462
Max velocity v_{max} [mm/s]	1800	790	1060
Energy dissipation per cycle EDC [kJ]	5599	4080	4390
Max internal fluid flow, f [l/min]	14700	4800	6300
Pin-pin length in def. Position L [mm]	4980	3200	4200

Fig. 6. ISOSISM® FD 4000.925 after the assembly process.

the CALTRAN Testing system, during the Type Test qualification. Table 5 shows the complete Type Test protocol performed on ISOSISM® PS 17000.1700 and ISOSISM® PS 31500.1540.

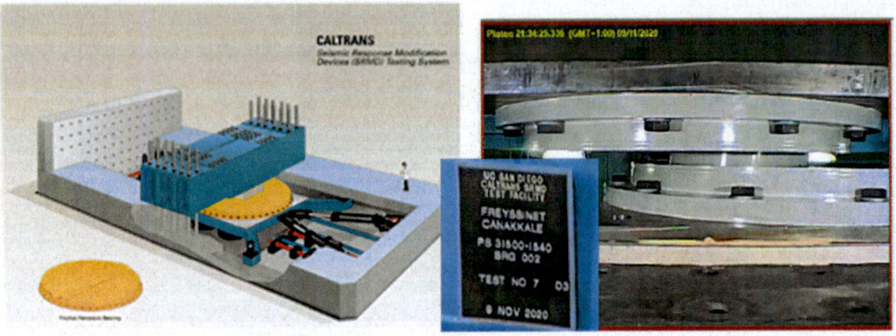

Fig. 7. ISOSISM® PS 31500.1540 tested with CALTRAN press.

Figure 8 shows the clover leaf displacement input, (to match the total sliding path of Dynamic 3 test run) and the force-displacement response of the isolator. In the same Fig. 8, the equivalent acceleration input impressed to the device is also shown to demonstrate the performance of the isolator under a real acceleration input.

TT were successfully completed on all devices, including additional tests performed on ISOSISM® PS 31500.1540 (Fig. 8). Factory Production Control Tests (FPCT) were performed at ISOLAB, the innovative testing facility of Freyssinet Group, located in

Table 5. Type Test protocol according to EN 15129.

Test	dof	n	PS 17000.1700			PS 31500.1540		
			d_{max} [mm]	v_{max} [mm/s]	N [kN]	d_{max} [mm]	v_{max} [mm/s]	N [kN]
Frictional resistance	hor	1	–	0.1	6700	–	0.1	14050
Service	hor	20	± 254	5	6700	± 222	5	14050
Benchmark	hor	3	± 635	50	6700	± 525	50	14050
Dynamic 1	hor	3	± 159	1186	6700	± 131	992	14050
Dynamic 2	hor	3	± 318	1186	6700	± 263	992	14050
Dynamic 3	hor	3	± 635	1186	6700	± 525	992	14050
Seismic 1	hor	3	± 635	1186	12890	± 525	992	26898
Seismic 2	hor	3	± 635	1186	1032	± 525	992	4064
Bidirectional	hor	3	± 635	1186	6700	± 525	992	14050
Bidirectional clover leaf (B1.1)	hor-long	–	–	–	–	± 525	992	14050
	hor-lat		–	–		± 262	496	
Bidirectional time history (B1.2)	hor-long	–	–	–	–	± 221	685	14050
	hor-lat		–	–		± 144	418	
Property verification	hor	3	± 635	1186	6700	± 525	992	14050
Load bearing capacity	vert	1	–	–	13400	–	–	28100

Montebello della Battaglia, Pavia (Italy). Figure 9 shows the 70 MN dynamic press, the most powerful of the Lab, which allows to perform static tests up to 70 MN and dynamic tests with horizontal force up to 3000 kN, 1000 mm stroke and peak velocity up to 850 mm/s.

3.2 Experimental Response of ISOSISM® FD

The prototype tests were first performed at SISMALAB testing lab in Crispiano (Italy), with max testing velocity up to 500 mm/s, for a preliminary check of the damper's performances. Subsequently, the devices were tested at UCSD testing lab in San Diego (USA), were the official EN 15129 testing protocol (Table 6) was performed by the European Notified Body. Figure 10 shows the damper during the qualification test at UCSD lab.

The testing campaign was successfully completed, satisfying all the acceptance criteria of EN 15129 and the dampers were CE marked with a velocity up to 1.8 m/s. Figure 11 shows the experimental response of the dampers (force-velocity and force-displacement response).

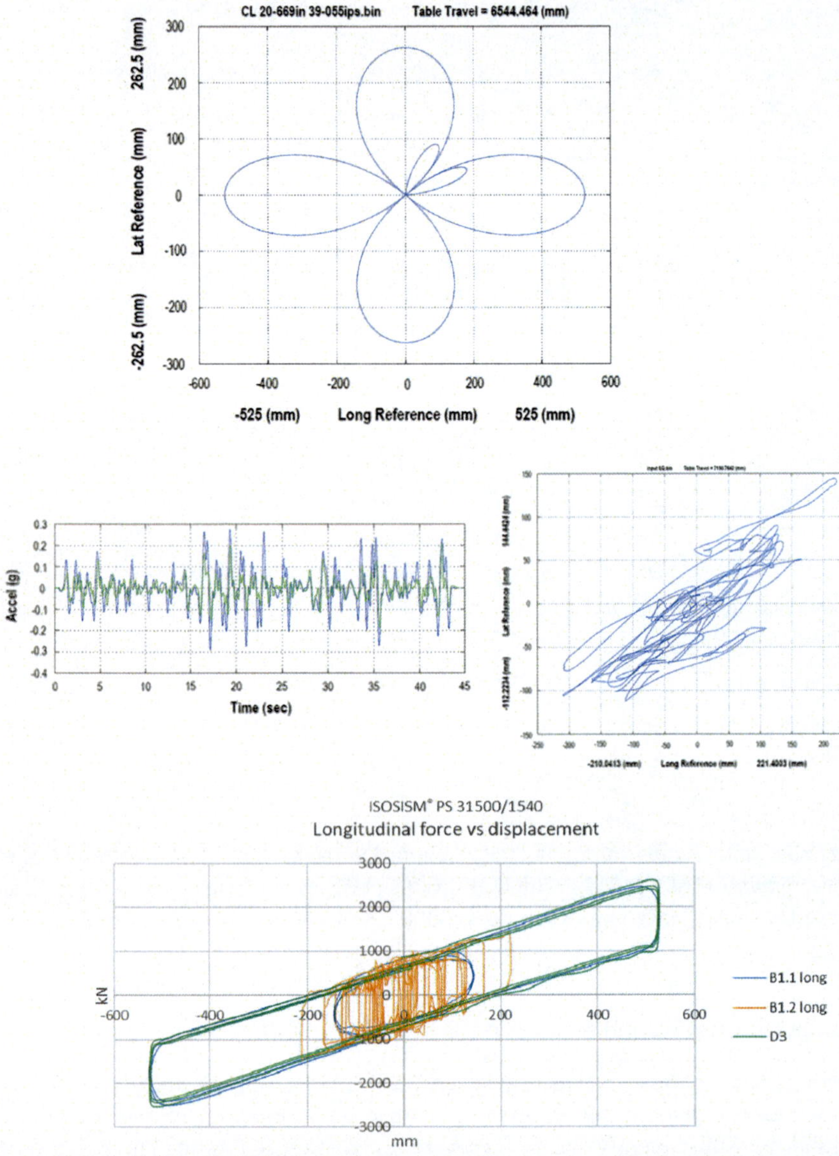

Fig. 8. ISOSISM® PS 31500.1540 testing: clover leaf input (top), acceleration/displacement time history input (center), force-displacement response (bottom)

4 Conclusions

1915Çanakkale Bridge and Malkara-Çanakkale Motorway Project represent a mega-scale investment in transportation and infrastructure in Turkey and includes, among the various civil works, longest mid-span suspension bridge in the world.

Fig. 9. ISOLAB 70 MN testing system.

Table 6. ISOSISM® FD 4000.1170 TT protocol & ISOSISM® FD 4000.925 FPC protocol.

Test	n	FD 4000.1170—TT		FD 4000.925—FPC	
		d_{max} [mm]	v_{max} [mm/s]	d_{max} [mm]	v_{max} [mm/s]
Seal wear	1000	± 10	–	–	–
Low velocity	1	± 10	0.01	± 10	0.01
Consitutive law [1%]	3	± 480	18	± 323	11
Consitutive law [25%]	3	± 480	450	± 323	265
Consitutive law [50%]	3	± 480	900	± 323	530
Consitutive law [75%]	3	± 480	1350	± 323	795
Consitutive law [100%]	3	± 480	1800	± 323	1060
Damping efficiency	3 + 3	± 480	1800	± 323	1060

The region is characterized by one of the highest seismicity level in the world therefore, in order to ensure high safety standard and optimize overall costs of the project, it was required to protect the main infrastructures from earthquakes implementing modern seismic protection technologies.

In particular, Freyssinet, world leader in earthquake protection systems, was responsible for the supply of curved surface sliders type ISOSISM® PS for the seismic isolation of 1915Çanakkale approach bridges and fluid viscous dampers type ISOSISM® FD for the seismic protection of V01 and V06 viaducts.

Fig. 10. Qualification of ISOSISM® FD 4000.1170 at UCSD lab (USA).

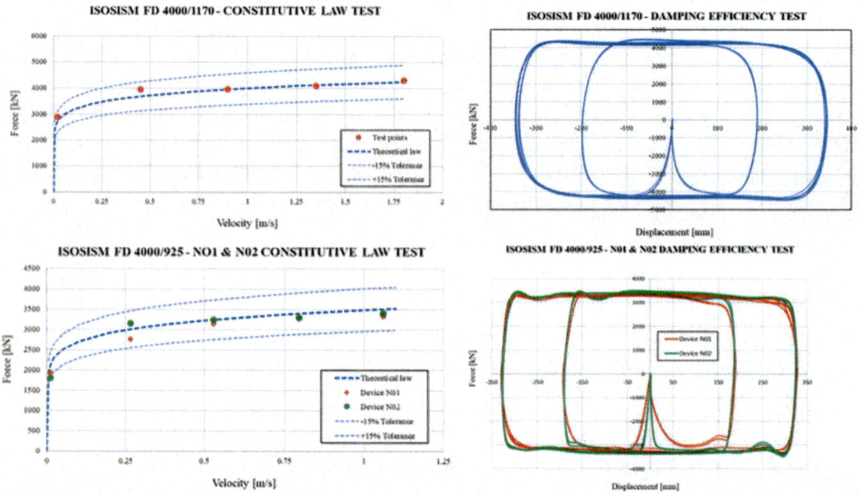

Fig. 11. Experimental response of the dampers: ISOSISM® FD 4000.1170 (top) and ISOSISM® FD 4000.925 (bottom)

The exceptional seismicity of the region required the use of devices with extraordinary performances which Freyssinet was capable to supply thanks to its advanced technical, production and testing capabilities.

References

1. Barone, S., Calvi, G.M., Pavese, A.: Experimental dynamic response of spherical friction-based isolation devices. J. Earthquake Eng. **9**, 1465–1484 (2019)
2. Barone, S., Pavese, A., Calvi, G.M.: Characterization of frictionla properties of different sliding materials for curved surface sliders. In: 7th International Conference on Advances in Experimental Structural Engineering (2017)
3. Sartori, M., Barone, S., Galbiati, C., Turkdonmez, B., Zivanovic, I.: Alibekoy and Kagithane viaducts: advanced seismic protection solutions in high seismicity region. In: 16th World Conference on Seismic Isolation, Energy Dissipation and Active Vibration Control of Structures (2019)

Fluid Viscous Dampers for the 1915 Çanakkale Bridge in Turkey

A. E. Pigouni[1], R. Borella[2], S. Infanti[3], and M. G. Castellano[4(✉)]

[1] Testing Laboratory, FIP MEC Srl, 35030 Selvazzano Dentro, Italy
[2] Engineering and Innovation Unit, FIP MEC Srl, 35030 Selvazzano Dentro, Italy
[3] Director of Testing Laboratory, FIP MEC Srl, 35030 Selvazzano Dentro, Italy
[4] Academy Unit, FIP MEC Srl, 35030 Selvazzano Dentro, Italy
maria.gabriella.castellano@fipmec.it

Abstract. The 1915 Çanakkale Bridge is a recently completed suspension bridge, characterized by a main span of 2023 m which is the world's record. The bridge restraint system is quite complex in order to guarantee safety under traffic, wind and seismic actions. It comprises 8 fluid viscous dampers, 4 large bearings designed to withstand compression load as well as tension load and horizontal displacement, 8 lateral elastomeric bearings for transversal restraint and 8 hydraulic end-stop devices. This paper focuses on the eight special fluid viscous dampers that give large energy dissipation capacity to the bridge under earthquake. They are characterized by a load capacity of 5000 kN, a stroke of ± 1.25 m and a maximum design velocity of 1 m/s. The paper describes the dynamic and static tests carried out on two full scale viscous dampers according to the test protocol that includes the European Standard EN 15129:2009 type tests and additional contract specifications.

Keywords: Testing · Fluid viscous damper · Bridge

1 Introduction

The 1915 Çanakkale Bridge is a suspension bridge with 2023 m long main span and two 770 m long side spans; it is the longest suspension bridge in the world by the length of central span. The bridge is part of the Kivali-Balikesir highway connecting Eastern Thrace and Anatolia avoiding passing through the Istanbul junction. The project site is located in the area of the North Anatolian Fault, that forms the tectonic boundary between the Eurasian Plate and the Anatolian block of the African Plate. The site area could be therefore subjected to strong seismic events.

2 Seismic Protection System

Special structural devices have been implemented to the suspended deck and side spans of the bridge, in order to control the bridge response to the strongest design dynamic

G. P. Cimellaro (Ed.): WCSI 2022, LNCE 309, pp. 478–487, 2023.
https://doi.org/10.1007/978-3-031-21187-4_40

action, i.e. earthquake with 2475 years return period, and at the same time ensure the functionality of the bridge under service condition. As shown in Fig. 1, the bridge deck is supported at the side span piers by four free sliding spherical bearings with anti-lifting capacity, carrying vertical load up to 22,000 kN and at the same time tension loads up to 6800 kN, allowing horizontal movements up to ± 1525 mm along the longitudinal axis of the bridge. Eight lateral elastomeric bearings, arranged at the abutments and at the towers, act as elastic buffers for the bridge deck for all service and seismic transversal loads, up to 26,000 kN.

The restraining system includes eight special hydraulic end-stop devices installed between the suspended deck and the tower legs, in order to limit the non-seismic longitudinal movement of the deck to ± 900 mm and protect the expansion joints from possible damage. They are axial devices with a piston-cylinder system able to resist a static maximum load of 13,500 kN; with a diameter greater than 1 m, they are among the largest hydraulic devices ever built for bridge application.

In longitudinal direction, eight special Fluid Viscous Dampers (FVD) were installed between the suspended deck and the tower legs, providing the longitudinal restraint to the structure, when necessary, as well as an effective damping system in case of seismic actions. The Fluid viscous dampers are axial devices whose force is proportional to velocity through the relationship $F = c\ v^{\alpha}$, the exponent α usually ranging from 0.01 ÷ 0.15 (highly non-linear devices) to 0.5 for seismic applications, and to 1 (linear devices) for wind applications. The highly non-linear FVDs are often preferred due to their almost constant reaction within a wide velocity range, thus higher energy dissipation during the relatively short duration of the earthquake, in comparison with FVDs with higher α exponent, and consequently lower displacement at a given level of maximum load transmitted to the structure [1–4].

The FVDs (referred as OT(P)500/2500) are characterized by a load capacity of 5000 kN, a stroke of ± 1.25 m and a maximum design velocity of 1 m/s. In addition, the dampers are capable of accommodating the thermal contraction/expansion movements of the bridge as well as any other low velocity displacements without appreciable reaction, whilst at the same time ensuring an almost rigid connection between the deck and the piers when high velocity movements are applied to the structure, generating forces lower than 5000 kN; i.e., the devices act as temporary connection devices (or shock transmission units) for forces up to 5000 kN. When the dynamic input, e.g. a strong earthquake, generates forces higher than 5000 kN, the devices act as viscous dampers. Thus, the devices are contemporarily shock transmission units and viscous dampers, depending on the intensity of the dynamic action.

It is worth noting that, in order to ensure an even load distribution on the structural elements of the tower, a flow rate regulator hydraulically connects the two dampers installed at each tower. The final dimensions of the devices, reaching 10.35 m in length at their maximum extended configuration, make these devices among the longest viscous dampers ever built in the world.

All fluid viscous dampers were designed, constructed and tested in FIP MEC in Italy. This paper focuses on the full-scale Type Tests performed on two (2) FVDs, carried out at FIP MEC Laboratories. The tests were performed according to the European Standards EN15129:2009 and the contract specifications in order to confirm the characteristics

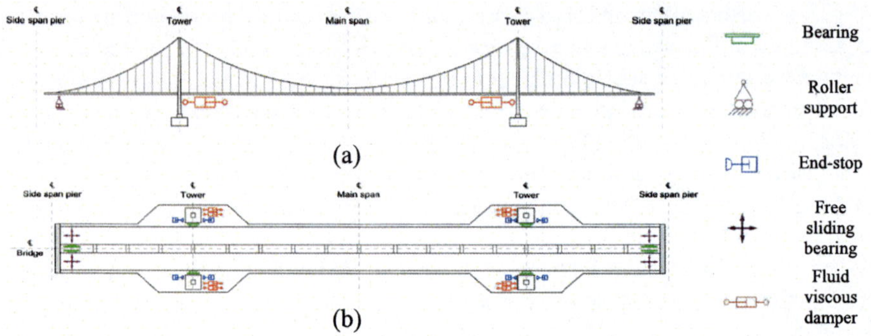

Fig. 1. Bridge bearing and seismic protection system layout: a) longitudinal restraining system under earthquake; b) restraining system under all loads.

of the devices and the main design assumptions of the seismic dissipation system of the bridge. In the following paragraphs the testing program, procedure and results are presented.

3 Full-Scale Tests on Fluid Viscous Dampers

3.1 Experimental Set up

The FIP MEC's Dynamic Test Facility satisfied the huge demand of hydraulic power required for this test campaign [5]. The test rig comprises a 10 m long steel frame support structure housing a servo-hydraulic actuator. The device under testing is connected to the fixed part of the rig and to the hydraulic actuator piston on the other side. Externally to the frame, the actuator—with a capacity of 11000 kN and a total stroke of ± 300 mm—is fixed with its piston aligned with the center-line of the frame. The damper's clevis plates are pinned to the actuator piston at one side and to the end plate on the opposite side (Figs. 2 and 3). The device was tested with its center line, laid horizontally in its centered condition.

The command and control functions are achieved by using electronic MTS and Bosch-Rexroth units connected with various measurement transducers. The applied load was monitored by means of pressure transducers reading directly the pressure acting on the actuators. The imposed displacements were measured by means of two SSI transducers (total stroke 600 mm); the first connected to the piston actuator and the second between the actuating rod and the damper's cylinder. The second transducer eliminates the effects of gaps and elasticity within the system.

All tests were conducted monitoring in real time the load (pressure), displacement and temperature, using a MGC HBM measuring amplifier connected to a PC equipped with a DIA-Dem data acquisition and processing software (National Instruments). All tests were conducted at room temperature in the range of variation permitted (23 ± 5 °C).

3.2 Experimental Program

The type testing was carried out according to the European Standard EN15129:2009 [6] (clause 7, Velocity Dependent Devices) and the contract specification, including:

Fig. 2. FIP MEC Test facilities: 11,500 kN Axial Test System for viscous damper testing. Fluid viscous damper under testing.

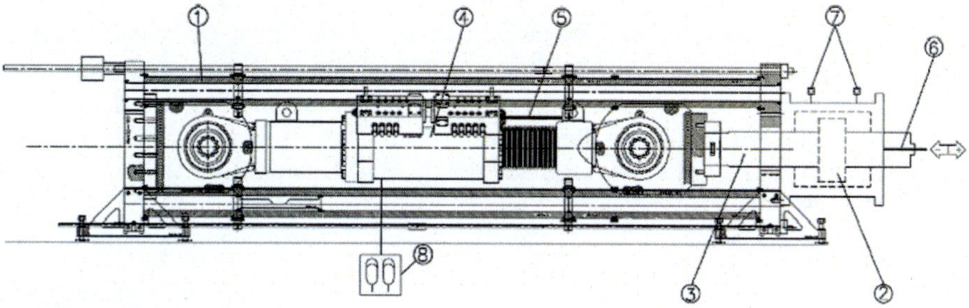

1. Test rig frame
2. Servo control actuator
3. Actuator piston
4. Damper test arrangement
5. Dispalcement transducer on device
6. Displacement transducer on actuator
7. Pressure transducers
8. Damper's accumulators

Fig. 3. Test set-up scheme.

(a) Static Stiffness Test;
(b) Static Pressure Test;
(c) Low velocity test;
(d) Three dynamic load tests: Impulsive load test, dynamic loading test, constitutive law test;
(e) Two Fatigue, Wear and durability tests: Seal wear test and Damping efficiency test.

Due to the huge amount of power required to test the damper (greater than 6.25 MW), the constitutive law test and damping efficiency test were carried out at the maximum design force F_{max} and at an equivalent reduced velocity. The real on-site condition (i.e. $v_{max} = 1.0$ m/s) was simulated by excluding 50% of the relief valves of the main circuit in order to ensure that the flow passing through the single operating valve during the tests is equal to the maximum flow rate per each valve expected during the seismic event.

4 Test Procedure and Results

4.1 Static Stiffness Test

The static stiffness test was carried out in order to measure the compressibility of the oil that fills the cylinder body of the damper, when loaded up to a force of 4500 kN. The valve system was kept closed and the device was loaded up to the target load. The deformation measured was calculated 2.3 and 2.5% of the total stroke, for the two devices (maximum allowable value was 4% of the total stroke). Figure 4 shows the force versus displacement graph measured during this test on one damper.

4.2 Static Pressure Test

Before the installation of the accumulators, the devices were imposed to the static pressure test in order to assess their ability to withstand for more than 120s an internal pressure of 125% of the pressure corresponding to the maximum nominal force. The pressure read during the test and the final examination of the hydraulic system showed no visible leakage or signs of physical deterioration of the units.

4.3 Low Velocity Test

The aim of this test is to measure the reaction of the dampers when subjected to thermal bridge movements. One fully reversed cycle at a constant velocity of 0.02 mm/s and 25 mm amplitude was implemented and the maximum reaction force was about 120 kN, i.e. less than 10% of the nominal design load (500 kN), as required by the European Standards and the client's specifications (Fig. 5).

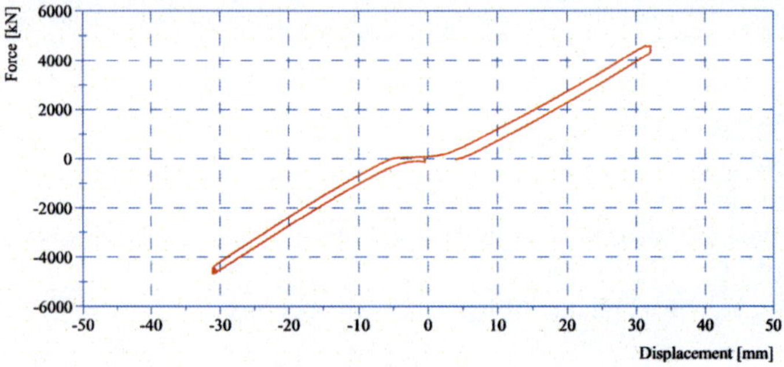

Fig. 4. Static stiffness test. Force versus displacement.

4.4 Impulsive Load Test

This test aims at verifying the non-seismic performance of the device, in particular at checking the activation velocity when subjected to its design load applied as an impulse.

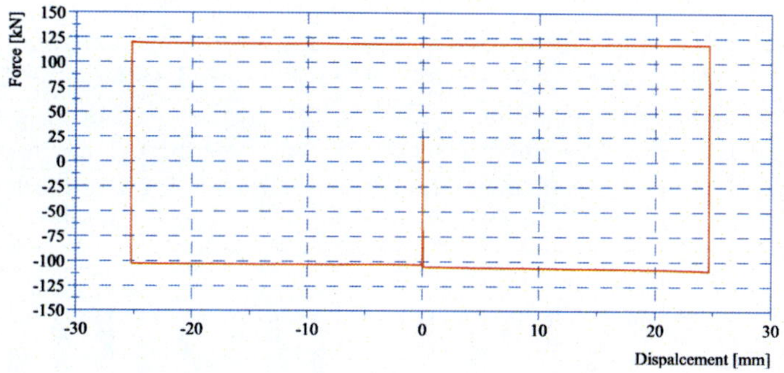

Fig. 5. Low velocity test. Force versus displacement.

The test was performed under force control by imposing the load (about 4500 kN) and recording the resulting velocity. The load was maintained constant for approximately 5 s. and then reversed to reach the same load in the opposite direction. This procedure was repeated for five settings of the flow control rate valve, ranging from 0.02 to 4.0 mm/s (5 valve positions). Additionally, for each of the 5 velocities (valve positions), 3 tests were performed at different stroke positions of the piston, i.e. central and ± 150 mm off-axis.

Figure 6 shows that both dampers at the same valve position and in all 3 piston positions, reached almost the same velocity. This ensures a uniform distribution of pressures among the two dampers of each pylon and thus assures an equilibrated distribution of loads on the structural elements.

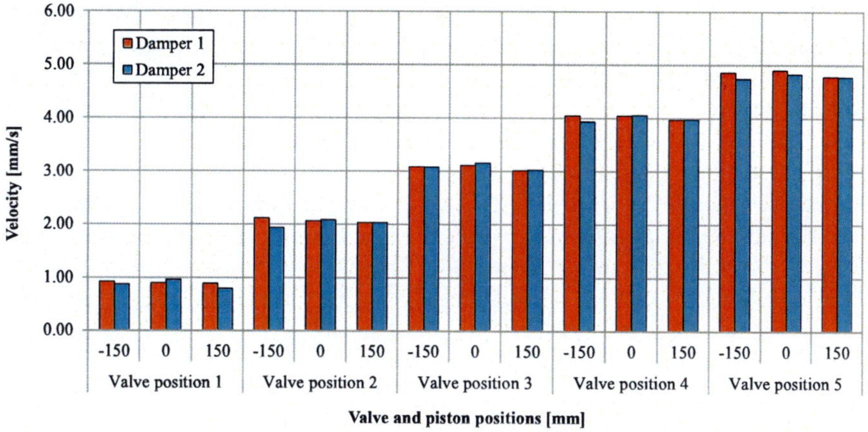

Fig. 6. Impulsive load test—velocity results for all valve and piston positions, for both dampers.

4.5 Dynamic Loading Test

The dynamic load test aims at evaluating the damping reaction stability of the devices by applying 10 sinusoidal cycles at ± 200 mm amplitude and 10 mm/s velocity. The load vs. displacement was recorded during the test and the Energy Dissipation per Cycle (EDC) was calculated. The results show a variance less than 15% of the EDC at the 2nd cycle (reference cycle according to EN 15129) for all the 9 following cycles. As it is clear from Fig. 7, a very stable reaction was achieved through the cycling and a negligible EDC variation was obtained, ranging from 0.3% to 2.2% for both dampers.

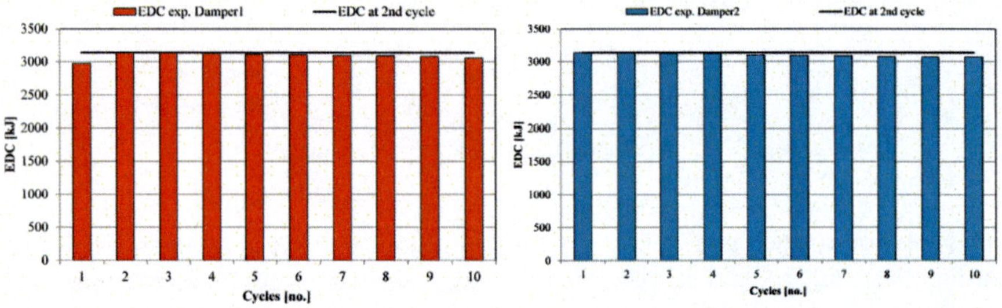

(a) (b)

Fig. 7. Energy dissipation results compared with the 2nd cycle (reference cycle). a) Damper 1 and b) Damper 2.

4.6 Constitutive Law Test

The scope of the constitutive law test is to determine the damper's force vs. velocity curve, i.e. the parameters C and α which define the constitutive law. Five series of 3 sinusoidal cycles were carried out, imposing the test displacement at amplitude equal to $\pm$ 200 mm, with frequencies corresponding to velocities 25, 50, 75 and 100% of the maximum velocity.

For each velocity, the experimental force values were calculated as the average of the positive and negative intercepts with the force axis of the second hysteresis loop cycle. The Energy Dissipation per Cycle (EDC) calculated varies from 0.4 to 2.9% between the second and the third cycle for each velocity test, ensuring a variance less than 15%, as required by the client's specifications (Fig. 8).

Figure 9a shows that both dampers provided a very low velocity exponent, i.e. a force almost constant with velocity. Figure 9b gives the force vs. displacement curves obtained at all velocities, for one of the dampers. From the hysteresis loops it is evident a very stable behavior at any velocity. The hysteretic loop shows the compressibility of the fluid as well, quite evident due to the very large stroke of the device ($\pm$1.25 m).

4.7 Damping Efficiency Test

The test aims at evaluating the energy dissipating capacity of the devices and their reaction stability. Five (5) sinusoidal cycles were imposed at a frequency producing the maximum velocity (500 mm/s) and an amplitude equal to the test design displacement $\pm$ 200 mm, at ambient temperature. Both devices showed a stable behavior with a very small variation of the maximum force. Figure 10 shows the force vs. displacement loops

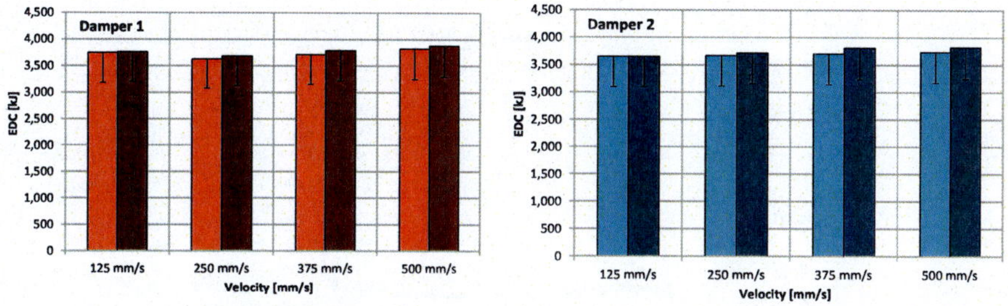

(a) (b)

Fig. 8. Constitutive Law test results. Energy dissipation variation between 2nd and 3rd cycle. a) Damper 1 and b) Damper 2.

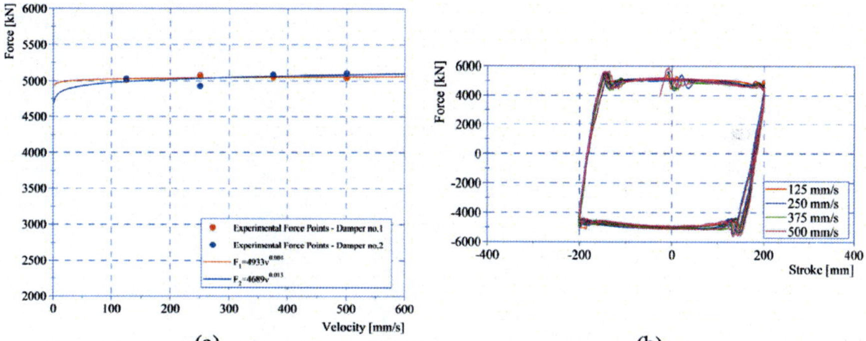

(a) (b)

Fig. 9. Constitutive Law test results. a) Force versus velocity graph and b) Force versus displacement curves at all velocities (damper 1).

obtained for both dampers tested at the maximum design velocity of 500 mm/s (1.0 m/s simulation velocities). It is clear that both devices maintained a steady output force reaction at maximum velocity in all five cycles.

The EDC variation per cycle in comparison to the first cycle was calculated for both devices ranging from 0.6 to 3%. Furthermore, higher energy dissipation was achieved (up to 35%) compared to the minimum allowable value ($EDC_{min} = 2907$ kJ), corresponding to the minimum reaction force of 5000 kN (Fig. 11).

4.8 Seal Wear Test

The scope of the seal wear test is to ensure that the sealing system works over the assumed design life of the device without leakage of the internal fluid. Both dampers were tested for 10000 cycles at force equal to 10% of the nominal reaction. The test was completed successfully with no leakage of the internal fluid.

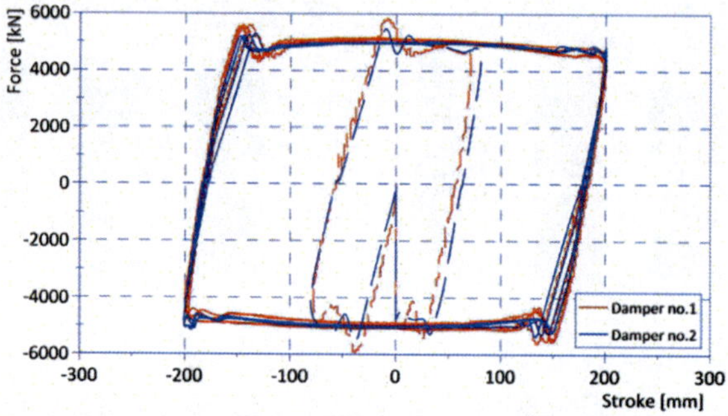

Fig. 10. Damping efficiency test. Hysteretic loops

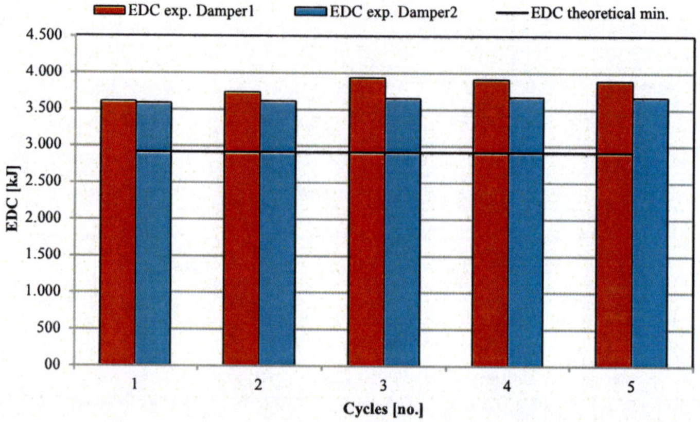

Fig. 11. Damping efficiency test. EDC compared with the minimum allowable EDCmin

5 Conclusions

The full-scale dynamic and static tests of two Fluid Viscous dampers were performed in the FIP MEC Test Laboratories, in order to confirm their characteristics and the main design assumptions of the seismic dissipation system used for the 1915 Çanakkale Bridge in Turkey. The tests were performed according to the European Standards EN 15129:2009 and some additional tests were carried out as per client's specifications.

The dampers, nominated as OT(P) 500/2500, are designed with a load capacity of 5000 kN, a stroke of ± 1.25 m and a maximum design velocity of 1 m/s. Both devices demonstrated a very stable and constant behavior under dynamic conditions at any velocity up to the maximum test velocity of 500 mm/s (simulation of 1.0 m/s). A steady output force reaction was obtained with the negligible maximum variation of 5% within the entire velocity range. The Energy Dissipation per Cycle was steady even when subjected to a high number of cycles and/or full design velocity (max. Variation 2.9%). The hydraulic system of both dampers performed satisfactorily with no leakage or signs of physical deterioration in all the tests. The behavior of the devices as shock transmission units was also checked with success.

The dampers have been installed and the bridge is in operation.

Acknowledgements. The authors wish to express their sincere gratitude to all laboratory staff and technicians, as well as to the technical department staff who designed the devices, for the active and fruitful contribution to this project.

References

1. Infanti, S., et al.: Seismic protection of Seohae bridge approaches through viscous dampers. In: IABSE Conference on Cable-Supported bridges. Korea, Seoul 12–14 June (2001)
2. Infanti, S., Papanikolas, P., Theodossopoulos, G.: Rion-Antirion Bridge: Full-scale testing of seismic devices. In: Proceedings of fib Symposium Concrete Structures in Seismic Regions, Athens, Greece (2003)
3. Infanti, S., et al.: Seismic protection of the Guadalfeo bridge by viscous dampers. In: Proceedings of First European Conference on Earthquake Engineering and Seismology. Geneva, Switzerland, September 3–8 (2006)
4. Infanti, S., and Castellano, G.M.: Sheikh Zayed Bridge Seismic Protection System. In: Proceedings of 10th World Conference on Seismic Isolation, Energy dissipation and Vibrations Control of Structures. Istanbul, Turkey, May 27–30 (2007)
5. Infanti, S., et al.: Dynamic testing protocols for seismic protection devices: new challenges for test facilities. In: 14th World on seismic isolation, Energy Dissipation and Active Vibration Control of Structures. USA, San Diego 9–11 September (2001)
6. European Standard EN 15129:2009: E Anti-seismic devices, CEN, Bruxelles, November (2009)

Seismic Isolation and Energy Dissipation in High-Risk Plants, Existing and Cultural Heritage Structures

Application of the Gradient Based Optimization to the Structural Systems with the Supplemental Damping Devices

Suat Gündemir(✉)

Department of Civil Engineer, Istanbul, Turkey
suatgundemir@gmail.com

Abstract. Even though there are so many studies upon optimization of supplemental damping devices integrated to structural systems, the vast majority of them remain only as a subject of the research articles of the Journals and cannot find a place itself in common practice due to lack of the regulations about the issue in codes and standards. In this paper, the issue is dealt with the most fundamental inputs and outputs of a general problem. All efforts are made to show how the procedure could frequently be used in common practice and what the attained benefits could be from the design and application point of view. The analysis of the structural systems in which the "seismic response control systems" have been adapted requires considering the special numerical analysis techniques which provide "solutions in frequency domain". By this means, not only the "dynamic response characteristics" of the structure shall be determined but also the variation of this response with the change of the "design parameters forming the structural control systems" shall be able to be investigated from the optimization point of view. Since conventional analysis tools and methods are not capable of providing such information, advanced numerical analysis method based on the "state-space approach" is considered along with the "gradient based optimization algorithm" in the proposed study. In this paper, viscous fluid dampers are examined as a supplemental damping devices in the analysis of a multi degree of freedom structure. Dampers are modelled with the Maxwell model in which the spring and the dashpot are placed in series. For a given brace stiffness, viscous damping coefficients and the distribution of the damping devices between stories are optimized by a procedure to reach a "target performance level". The optimization is achieved by minimizing an "objective function" which rates the structural responses of the controlled system with the uncontrolled one. Generally, mean square responses of the inter-story drift, floor acceleration and base shear are among the structural response quantities considered for this procedure.

Keywords: Seismic response control systems · Supplemental damping devices · Signals and systems · State-space approach · Numerical optimization · Gradient based optimization algorithm

G. P. Cimellaro (Ed.): WCSI 2022, LNCE 309, pp. 491–503, 2023.
https://doi.org/10.1007/978-3-031-21187-4_41

1 Introduction

1.1 Flow Diagram of the Analysis

The flow diagram of the analysis is based on the general principles of the signals and systems. The central concept in signal and system analysis is that of the transformation of a signal. The signals, which are functions of one or more independent variables, contain information about the behavior or nature of some phenomenon, whereas the systems respond to particular signals by producing other signals or some desired behavior [3].

In this regard, a linear time invariant system [6] accompanied with the optimization tool are presented below in a simple block diagram. A block diagram consists of blocks that represent different parts of a system and signal lines that define the relationship between the blocks (Figs. 1 and 2).

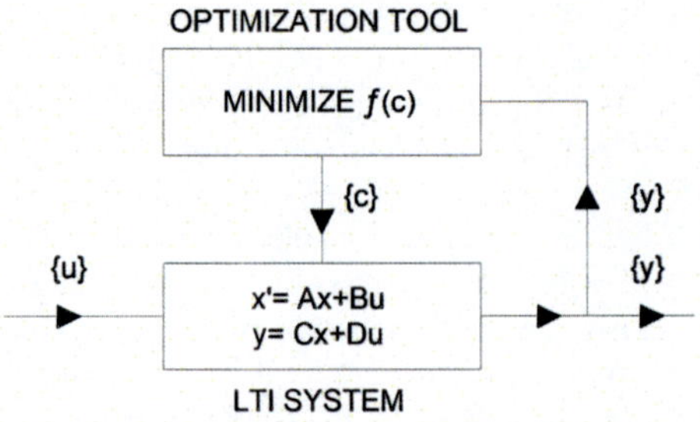

Fig. 1. Flow diagram of the analysis

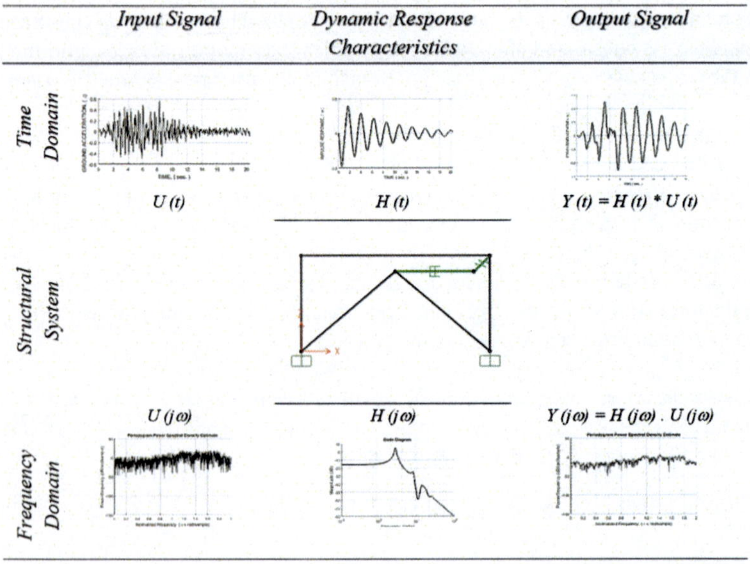

Fig. 2. Extended representation of linear time invariant systems for structural systems

1.2 Seismic Response Control Systems

Seismic response control systems are divided into three categories defined as (a) seismic isolation systems, (b) energy dissipation devices and (c) dynamic vibration absorbers. When properly designed, these devices enhance the performance of the structure by modifying their dynamic response characteristics. General illustrations of these devices integrated with the buildings are also presented in the figure below.

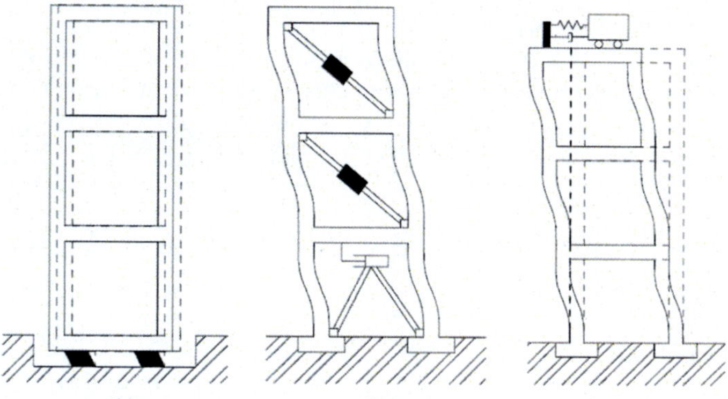

Fig. 3. Seismic response control systems, (a) seismic isolation systems, (b) energy dissipation devices, (c) dynamic vibration absorbers, [4]

Each of these control devices has different design parameters and so the optimization quantities. Since Maxwell model based viscous damper is adopted in this study, integration of this damping device with the structural system is shown in the figure below. As seen from the figure, damper represented by the Maxwell model have two design parameters consisting of a spring (chevron-type brace) and a dashpot (viscous damper) placed in series (Fig. 4).

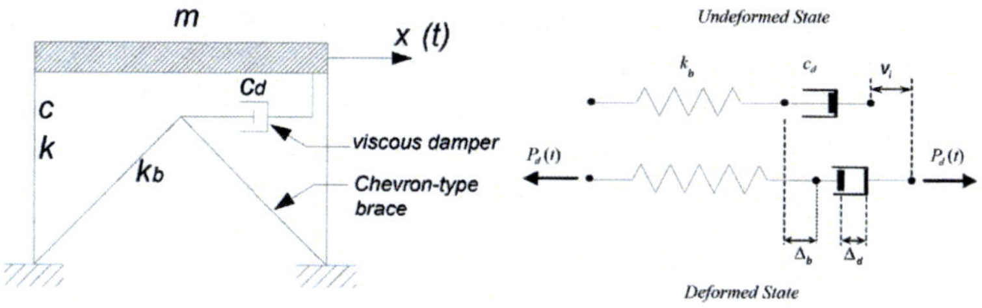

Fig. 4. Maxwell model based viscous damper and integration to the structural system

The optimal brace stiffness and damping coefficients can be determined through an iterative process. The reduction in response and hence improved performance of the structure can be achieved by a strategic combination of the brace stiffness and viscous damper coefficients, instead of relying on damper coefficients alone as conventionally been done [1].

Several researchers [1, 4] noted that increased brace flexibility tends to reduce the effectiveness of the viscous dampers. So, it was suggested that a brace stiffness that is ten times the story stiffness can be considered essentially rigid without compromising the damping effectiveness significantly.

Since similar recommendations are given in different sources like [5], chevron-type brace stiffness is considered as ten times the story stiffness in the current case study as well. Therefore, viscous damping coefficients and the distribution of the damping devices between stories are considered as significant design parameters in the optimization.

2 State Space Formulation of the Mdof Systems with Multiple Maxwell Models

2.1 Analytical Model

The dynamic response of a multistory building with **n** degrees of freedom installed with viscous dampers at the top of Chevron-type braces is governed by the following equation,

$$\mathbf{M\ddot{x}} + \mathbf{C\ddot{x}} + \mathbf{Kx} + \sum_{\mathbf{i=1}}^{\mathbf{m}} \boldsymbol{r_{d,i}.f}_{\mathbf{d,i}} = -\mathbf{Me\ddot{x}}_{\boldsymbol{g}}\mathbf{(t)} \tag{1}$$

where **M**, **C** and **K** are mass, inherent damping and stiffness matrices of the structure, respectively. The n-dimensional vector **e** is the influence vector that couples the structural degrees of freedom to ground motion direction and $\boldsymbol{f_{di}}$ is internal force of the $\boldsymbol{i_{th}}$ damper. The number of possible locations for installation of control devices is assumed to be $\boldsymbol{m}$. The effect of damper force on the structural degrees of freedom is described through the n-dimensional $\boldsymbol{r_{di}}$ influence vectors.

The auxiliary differential equations governing the behavior of each Maxwell elements are given as follows with a little modification than that of [2].

$$f_{d,i} = c_{d,i}\dot{\Delta}_{d,i} = k_{b,i}.\Delta_{b,i} \tag{2}$$

$$\Delta_{b,i} = v_i - \Delta_{d,i} \tag{3}$$

$$c_{d,i}.\dot{\Delta}_{d,i} + k_{b,i}.\Delta_{d,i} = k_{b,i}.v_i \tag{4}$$

$$v_i(t) = r^T_{d,i}.x(t) \tag{5}$$

$$\mathrm{C_d\dot{\Delta}_d + K_b\Delta_d = K_bDx} \tag{6}$$

where $\mathbf{K_b}$ and $\mathbf{C_d}$ are diagonal stiffness and damping matrices of Maxwell elements and matrix $\mathbf{D}$ transforms the structural displacements to the inter-story drifts $\mathbf{v}$.

$$\mathrm{D} = \begin{bmatrix} r_{d,1}^T \\ \vdots \\ r_{d,m}^T \end{bmatrix}_{\mathrm{mxn}} \tag{7}$$

$$\mathrm{C_d}\dot{\Delta}_\mathrm{d} + \mathrm{K_b}\Delta_\mathrm{d} = \mathrm{K_b}\mathrm{v} \tag{8}$$

Differentiating of Eq. (8) with respect to time yields,

$$\Gamma\ddot{\Delta}_\mathbf{d} + \dot{\Delta}_\mathbf{d} = \dot{\mathbf{v}} \tag{9}$$

$$\Gamma = \mathrm{K_b^{-1}C_d} \tag{10}$$

Substituting story displacement vector $\mathbf{x}$ in Eq. (1) with the inter-story drift vector $\mathbf{v}$, the equation of motion can be rewritten as,

$$\mathbf{MD^{-1}\ddot{v} + CD^{-1}\dot{v} + KD^{-1}v + LC_d\dot{\Delta}_d = -Me\ddot{x}_g(t)} \tag{11}$$

$$\mathrm{L} = \begin{bmatrix} r_{d,1} & \dots & r_{d,m} \end{bmatrix}_{\mathrm{nxm}} \tag{12}$$

Transforming the simultaneous Eq. (9) and Eq. (11) into the state space representation gives the following equation that is used for dynamic analysis of MDOF passively controlled structures,

$$\dot{\mathbf{z}}(\mathbf{t}) = \mathbf{Az(t)} + \mathbf{E}\ddot{\mathrm{x}}_\mathrm{g}(\mathrm{t}) \tag{13}$$

where $\mathbf{A}$ and $\mathbf{E}$ are system and input matrices respectively and the state vector $\mathbf{z}$ are defined as follows,

$$\mathbf{A} = \begin{bmatrix} \mathbf{0} & \mathbf{I} & \mathbf{0} \\ -\hat{\mathbf{M}}^{-1}\hat{\mathbf{K}} & -\hat{\mathbf{M}}^{-1}\hat{\mathrm{C}} & -\hat{\mathbf{M}}^{-1}\mathbf{LC_d} \\ \mathbf{0} & \Gamma^{-1} & -\Gamma^{-1} \end{bmatrix}_{(2\mathrm{n}+\mathrm{m})\,\mathrm{x}\,(2\mathrm{n}+\mathrm{m})}$$

$$\mathbf{E} = \begin{Bmatrix} \mathbf{0} \\ -\hat{\mathbf{M}}^{-1}\mathbf{Me} \\ \mathbf{0} \end{Bmatrix}_{(2\mathrm{n}+\mathrm{m})\,\mathrm{x}\,1} \quad \mathbf{z(t)} = \begin{Bmatrix} \mathbf{v}\,(\mathbf{t}) \\ \dot{\mathbf{v}}\,(t) \\ \dot{\Delta}_\mathbf{d}(\mathbf{t}) \end{Bmatrix}_{(2\mathrm{n}+\mathrm{m})\,\mathrm{x}\,1} \tag{14}$$

$$\hat{\mathrm{M}} = \mathrm{MD^{-1}};\ \hat{\mathrm{C}} = \mathrm{CD^{-1}};\ \hat{\mathrm{K}} = \mathrm{KD^{-1}} \tag{15}$$

2.2 Performance Criteria

The objective is to optimize the viscous damping coefficients and the distribution of the damping devices between stories that will minimize an objective function which measures the reduction in some response quantity (or quantities) of design interest. The response quantities of the design interest could be the mean square responses of the inter-story drifts, floor accelerations, base shear and etc. Accordingly:

Mean square story drift response is defined as,

$$\sigma_{d,i}^2 = \frac{1}{2\pi}\int_{-\infty}^{\infty} |V(j\omega)|^2 d\omega; \quad i = 1, 2, 3, \ldots, n \tag{16}$$

Mean square floor acceleration response is defined as,

$$\sigma_{a,i}^2 = \frac{1}{2\pi}\int_{-\infty}^{\infty} \left|\ddot{X}(j\omega)\right|^2 d\omega; \; i = 1, 2, 3, \ldots, n \tag{17}$$

$$\ddot{X}(j\omega) = D^{-1}.\ddot{V}(j\omega) \tag{18}$$

using the relations above, the objective functions are specified as follows for drift and acceleration based performance criteria:

for drift based performance criteria,

$$f_d(\mathrm{c}) = \frac{\sum_{\mathrm{i=1}}^{\mathrm{n}}(\sigma_{\mathrm{d,i}}^2)_{\mathrm{controlled}}}{\sum_{\mathrm{i=1}}^{\mathrm{n}}(\sigma_{\mathrm{d,i}}^2)_{\mathrm{uncontrolled}}} \tag{19}$$

for acceleration based performance criteria,

$$f_a(\mathrm{c}) = \frac{\sum_{\mathrm{i=1}}^{\mathrm{n}}(\sigma_{\mathrm{a,i}}^2)_{\mathrm{controlled}}}{\sum_{\mathrm{i=1}}^{\mathrm{n}}(\sigma_{\mathrm{a,i}}^2)_{\mathrm{uncontrolled}}} \tag{20}$$

In this study, the objective function corresponding to drift based performance criteria is evaluated.

2.3 Optimal Search Approach

In the analysis, gradient based optimization technique is adopted. In optimization, gradient method is an algorithm to solve the problems of the form,

$$\begin{aligned} &\text{minimize} \quad f(x) \\ &\text{subject to} \quad g_j(x) = \sum_{i=1}^{m} a_{ji} x_i - b_j \geq 0; j = 1, 2, 3 \ldots\ldots, n_g \end{aligned} \tag{21}$$

with the search directions defined by the gradient of the function at the current point without disregarding the constraint functions. When the general form of the constrained optimization function is rearranged for the current case study [5].

$$\begin{aligned} &\text{minimize} \quad f(c) \\ &\text{subject to} \quad g(c) = \sum_{i=1}^{m} c_i - C_T = 0 \quad ; i = 1, 2, 3 \ldots\ldots, m \\ &\qquad\qquad 0 \leq c_i \leq c_i^u \qquad\qquad ; i = 1, 2, 3 \ldots\ldots, m \end{aligned} \tag{22}$$

where ***f(c)*** is the objective function, **c** is the vector containing design variables $\mathbf{c_i}$, $\mathbf{c_i}^{\mathbf{u}}$ is the upper bound value of the damping coefficient for the $\boldsymbol{i_{th}}$ damper, $\mathbf{C_T}$ is the total amount of damping coefficient value to be distributed in the structure and **m** is the number of the viscous dampers.

Although there are two design parameters in a Maxwell damper - damping coefficient for the viscous damper and the stiffness coefficient for the chevron-type brace - here only the damping coefficient values are chosen as independent variables. This assumes that for a given damper, the stiffness coefficient is just a constant factor times the damping coefficient value.

3 Numerical Case Study

3.1 Structural Model and the Initial Design Parameters of the Maxwell Model

In order to demonstrate the application of the above formulation, a numerical case study is carried out. A 10-story shear frame has been considered for the installation of viscous dampers to reduce its response and improve its seismic performance. The story masses are assumed concentrated at each floor level and there is one degree of freedom associated with each floor mass. Accordingly, prepared structural model, story mass, stiffness, initial design parameters of the Maxwell model and all other assumptions are shown in the Table 1.

Table 1. Structural model and the initial design parameters of the maxwell model

Structural Model	Story	Mass	Stiffness	$C_T = 5{,}000$	K_b
	n	ton	kN/m	kN.s/m	kN/m
	(1)	(2)	(3)	(4)	(5)
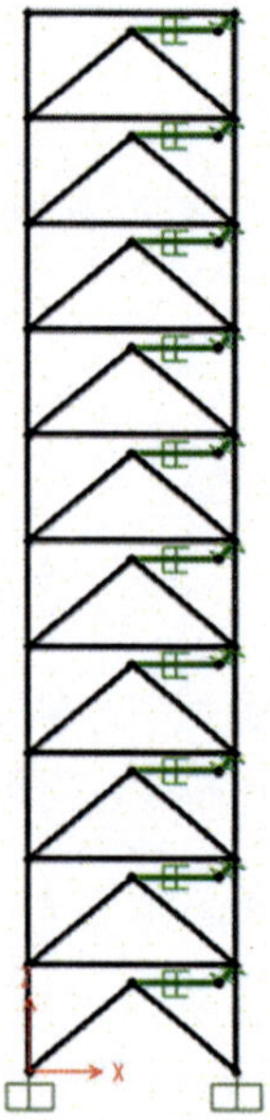	10	100.0	33,628.0	500.0	336,280.0
	9	100.0	33,628.0	500.0	336,280.0
	8	100.0	33,628.0	500.0	336,280.0
	7	100.0	33,628.0	500.0	336,280.0
	6	100.0	33,628.0	500.0	336,280.0
	5	100.0	33,628.0	500.0	336,280.0
	4	100.0	33,628.0	500.0	336,280.0
	3	100.0	33,628.0	500.0	336,280.0
	2	100.0	33,628.0	500.0	336,280.0
	1	100.0	33,628.0	500.0	336,280.0

Note For classical Rayleigh damping, $C = \mu M + \lambda K$, where $\mu = 0.157$, $\lambda = 0.0119$. Lower and upper bond of the damping coefficients for dampers are assumed as 0.0 and 1,500 kN.s/m, respectively

3.2 Seismic Ground Excitation

In the analysis, both white noise spectrum and the seismic acceleration record of Gölcük-1999 earthquake are studied as seismic ground excitations. Acceleration time series and corresponding spectral density of Gölcük-1999 earthquake are also presented in the Fig. 5.

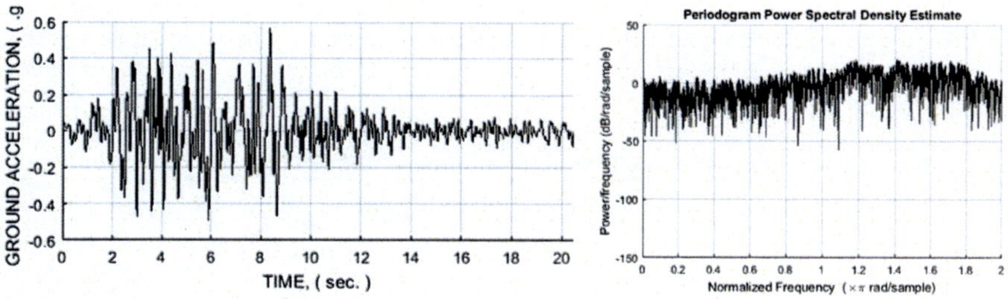

(a) (b)

Fig. 5. Acc. Time Series (a) & Corresponding Spectral Density (b) of Gölcük-1999 Earthquake

3.3 Optimal Search Result

Optimal search results are determined separately for specified seismic ground excitations and results are presented in the figures below. It is realized that search algorithm yields a similar optimal viscous damper coefficients and a same distribution of the damping devices between the stories, under each considered ground excitations. As seen from the plots, the search algorithm redistributes initial damping coefficients of the last six floors into the first four floors and finally drives them to zero damping capacity. Besides, about 12.0% and 10.0% improvements over initial uniform distribution of damping are achieved for each ground excitations respectively, which can be attributed to optimization process (Fig. 6).

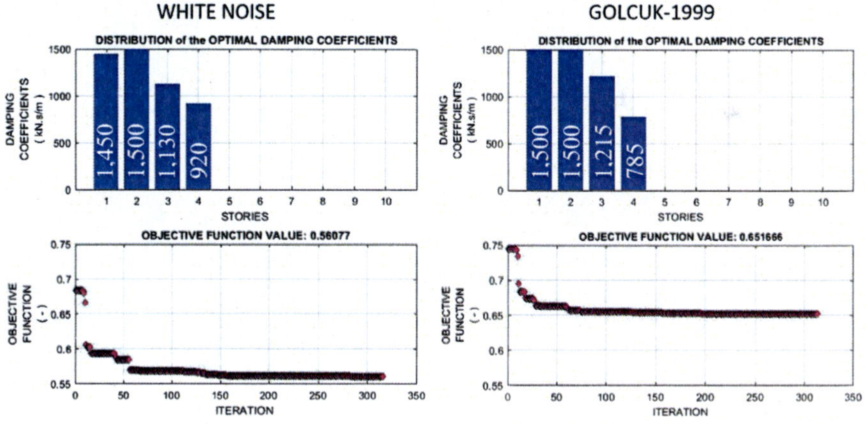

(a) (b)

Fig. 6. Optimal search results for considered seismic ground excitations

Other than, a comparison of the drift response reductions along the stories are also presented in the figure below.

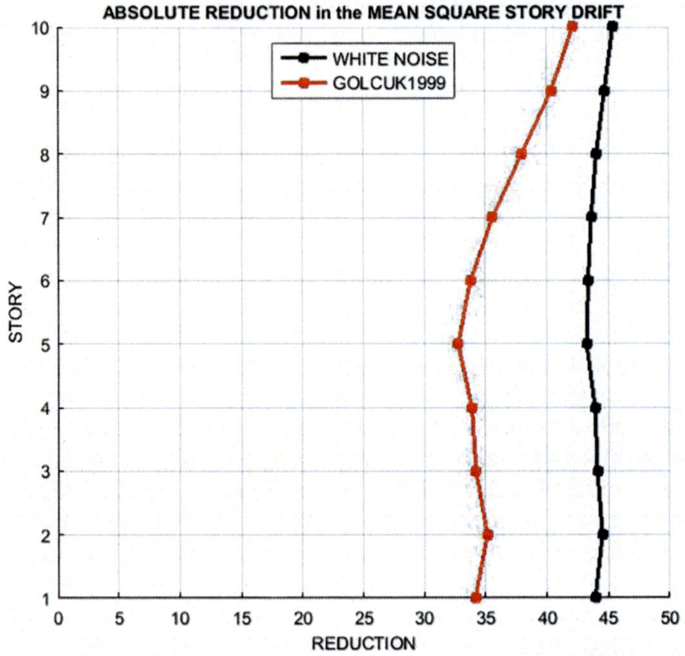

Fig. 7. Comparison of the drift response reductions along the stories

3.4 Dynamic Response Characteristics

Dynamic response characteristics of the structure are also determined for controlled and uncontrolled cases. Damper parameters corresponding to optimal point are considered in the controlled case analysis. The attained findings are presented in the plots and tables given below for comparison purpose. Consequently, (a) pole-zero map, (b) bode diagram of the inter-story drift responses and (c) a table of the frequency and damping values for each structural mode are given as follows (Fig. 8 and Table 2).

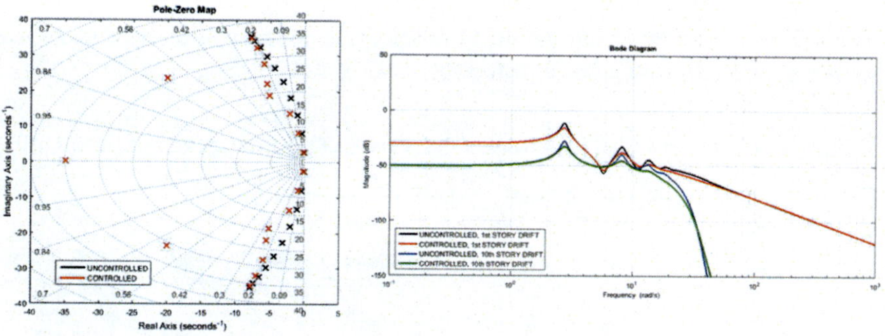

(a) (b)

Fig. 8. Dynamic response characteristics, pole-zero map (a), bode diagram of the inter-story drift responses (b)

Table 2. Table of the frequency and damping values for each structural mode

	Uncontrolled			Controlled		
Mode	Poles	Frequency	Damping	Poles	Frequency	Damping
ω	–	rad/s	–	–	rad/s	–
(1)	(2)	(3)	(4)	(5)	(6)	(7)
1	$-0.123 \pm 2.74i$	2.74	0.0449	$-0.220 \pm 2.74i$	2.75	0.0802
2	$-0.475 \pm 8.15i$	8.16	0.0582	$-0.945 \pm 8.16i$	8.21	0.1150
3	$-1.150 \pm 13.4i$	13.40	0.0856	$-2.210 \pm 13.7i$	13.80	0.1590
4	$-2.080 \pm 18.2i$	18.30	0.1130	$-5.090 \pm 18.8i$	19.50	0.2620
5	$-3.190 \pm 22.6i$	22.90	0.1390	$-5.500 \pm 21.9i$	22.60	0.2430
6	$-4.380 \pm 26.5i$	26.90	0.1630	$-5.870 \pm 27.6i$	28.20	0.2080
7	$-5.540 \pm 29.8i$	30.30	0.1830	$-20.00 \pm 23.6i$	30.90	0.6460
8	$-6.580 \pm 32.4i$	33.00	0.1990	$-6.950 \pm 32.0i$	32.80	0.2120
9	$-7.390 \pm 34.3i$	35.00	0.2110	$-34.90 \pm 0.00i$	34.90	1.0000
10	$-7.900 \pm 35.4i$	36.30	0.2180	$-7.790 \pm 34.8i$	35.70	0.2180

3.5 Comparison of the Responses from the Optimal Search and a Commercial Software

10-story shear frame is reanalyzed by a well-known but non-iterative commercial software, considering the damper parameters corresponding to final optimal point. Structural responses from the paper formulation and the commercial software are compared for the check of the proposed procedure. The compared plots representing the 1st story drift responses are presented in the figure below. Here, there are benefits to say that while Fig. 9-a shows linear proportionality in drift response reductions, Fig. 7 gives the rate of the mean square response reduction, numerically. From the figures below, it is understood that responses from the paper formulation shows an excellent match with the responses from the commercial software.

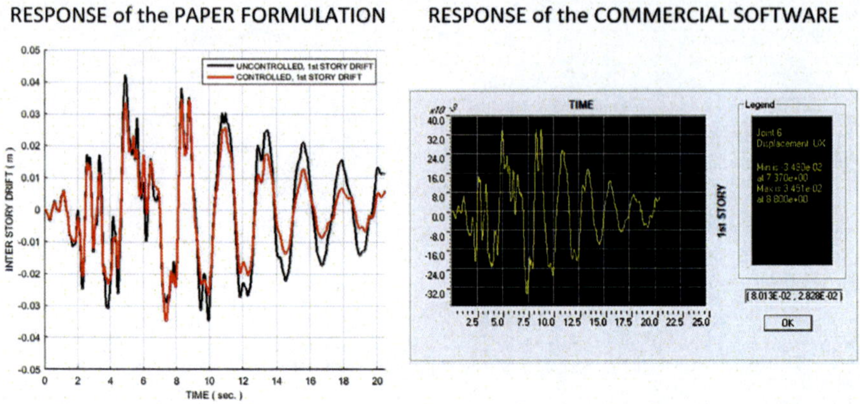

(a) (b)

Fig. 9. Comparison of the responses from the optimal search and a commercial software

4 Conclusions

In this paper, an approach for calculating the optimum design parameters for viscous fluid dampers represented by Maxwell model is presented. For a given brace stiffness and by providing an initial guess of the damper coefficients, iterative search procedure is carried out for determining the optimum damper coefficients until the target reduction in structural response is achieved. Even though only the damper coefficients are assumed to be independent variables in the current study, the brace stiffness can be included as another independent variables if required.

In the analysis, both white noise spectrum and an acceleration record from a true-life seismic event are investigated. It is realized that search algorithm yields a similar optimal viscous damper coefficients and a same distribution of the damping devices between the stories, under each considered ground excitations. Thus, it is concluded that about 10.0% more additional improvements over initial uniform distribution of damping could be achieved by the proposed optimization process.

Appendix A

Other projects of the author can be seen in the following links:

1. Advanced Excel Application for "Site Specific Seismic Ground Response Analysis" with fully dynamic algorithm.

   https://www.linkedin.com/feed/update/urn:li:activity:6782672730585088000/

2. Advanced Excel Application for the Structural Analysis of the Finite Element Space Frames with fully dynamic algorithm.

https://www.linkedin.com/posts/suatgundemir_finiteelementspaceframe-directstiffnessmethod-activity-6818866593632796673-4gmV.

References

1. Chen, Y., Chai, Y.: Seismic design of structures with supplemental Maxwell model-based brace-damper systems (2008)
2. Jamshidi, M., Ahmadizadeh, M. (2012). Design of passive structural control systems using viscous fluid dampers considering brace flexibility (2012)
3. Oppenheim, A., & Willsky, A. (tarih yok). Signals and Systems. Boston University
4. Singh, M., Moreschi, L.: Seismic analysis and design with Maxwell dampers (2003)
5. Verma, N.P.: Viscous dampers for optimal reduction in seismic response (2001)
6. Williams, R.L., II., Lawrence, D.A.: Linear State-Space Control Systems. John Wiley and Sons Inc, Ohio (2007)

3-D Seismic Isolation for Operational Level Protection of Critical Electrical and Electronic Control Equipment for the Site C Clean Energy Project

Ian Aiken[1](✉), Cameron Black[1], Blair Lawrence[2], Apollo Zhang[3], Nicole Cheang[3], Mohammed Mohammed[4], and Patrice Mclean[5]

[1] SIE Inc., Berkeley, CA, USA
{ida,black}@siecorp.com
[2] Klohn Crippen Berger Ltd., Vancouver, BC, Canada
blawrence@klohn.com
[3] BC Hydro, Vancouver, BC, Canada
{apollo.zhang,nicole.cheang}@bchydro.com
[4] Dynamic Isolation Systems Inc., Reno, NV, USA
mmohammed@dis-inc.com
[5] SNC-Lavalin Inc., Montreal, QC, Canada
patrice.mclean@snclavalin.com

Abstract. The Site C Clean Energy Project is a new, 1,100 MW hydroelectric generating station on the Peace River in northeastern British Columbia, Canada. Seismic isolation has been adopted to protect more than 150 pieces of equipment deemed critical for dam safety and to meet the project's operational requirements. Three-dimensional seismic isolation platforms are being used to protect the critical equipment, with a total of 39 platforms throughout the spillway headworks structure. The design of the isolation systems followed the framework of IEEE 693-2018, Annex W, with nonlinear analyses and shake table testing of the equipment and the isolation platforms to confirm their performance. Seismic isolation is also being used to protect seismically-vulnerable electrical equipment at the interface of the powerhouse generator step-up transformers and the 500 kV transmission system. The paper will describe the design and testing of the isolation systems being implemented on the project and key aspects of their benefits and effectiveness.

Keywords: Seismic isolation · 3-D isolation · Wire rope isolator · Three-dimensional · Electrical equipment · Hydroelectric

1 Introduction

The Site C Clean Energy Project is a new, 1,100 MW hydroelectric generating station on the Peace River in northeastern British Columbia, Canada. The main features of the

G. P. Cimellaro (Ed.): WCSI 2022, LNCE 309, pp. 504–515, 2023.
https://doi.org/10.1007/978-3-031-21187-4_42

project are a 1,050 m long, 60 m high, earthfill dam and an 800 m roller-compacted concrete buttress that includes the spillways and the six-turbine, generating station (Fig. 1). Two new 500 kV transmission lines will connect the Site C facilities to an existing substation and the provincial grid.

Although the project is not located in a region of high seismicity, several important project safety and economic design requirements ultimately led to the adoption of innovative seismic isolation solutions for key aspects of the project.

Three-dimensional seismic isolation platforms are being used to protect more than 100 pieces of critical dam safety equipment and to ensure full operation following an earthquake. Seismic isolation using wire rope devices is also being used to protect 45 pieces of seismically-vulnerable electrical equipment at the interface of the powerhouse generator step-up transformers and the 500 kV distribution system. The paper will describe the design and testing of the isolation systems being implemented on the project and key aspects of their benefits and effectiveness.

Fig. 1. Powerhouse, intakes and spillway construction, February 2022.

2 Project Performance Criteria

The project considered two key levels of seismic hazard: the Design Basis Earthquake (DBE) and the Maximum Design Earthquake (MDE), with annual exceedance probabilities of 1/2475 and 1/10,000, respectively. The project design criteria required that flows past the dam shall be safely managed after an MDE, leading to the requirement that all the primary and back-up power supplies and electronic control equipment required to operate the spillway gates be fully operational after an MDE event.

A second important criterion, but associated with the generating capability of the facility, rather than dam safety, applied to much of the seismic design of the powerhouse and intake. The project owner required that the facility be able to generate electricity within 72 h of a DBE event. This requirement was not intended to stipulate immediate post-earthquake functionality, rather that it could be met by repair and/or replacement of damaged components or equipment, so long as this could be reasonably achieved within the three-day time-frame.

The above performance requirements resulted in two different implementations of seismic isolation to protect different types of equipment on the project, and these applications will be described in the paper.

3 3-D Isolation for Critical Gate Control Equipment

Detailed three-dimensional, time-history analyses of the dam civil structures were performed for the project. These analyses incorporated site geologic variability, consideration of an upper, nominal and lower bound foundation stiffness and earthquake directionality effects. In total, five sets of three component ground motions—scaled to the appropriate hazard level—were used in the analyses of the dam and its associated structures.

One of the key products of the spillway and intake/powerhouse structure analyses were Floor Design Spectra (FDS) that were used extensively for the subsequent design of components and equipment. FDS were calculated at numerous locations, termed points of interest (PI) throughout the structures. Figure 2 shows the PI's in the spillway headworks (labeled as PI-xx in the figure).

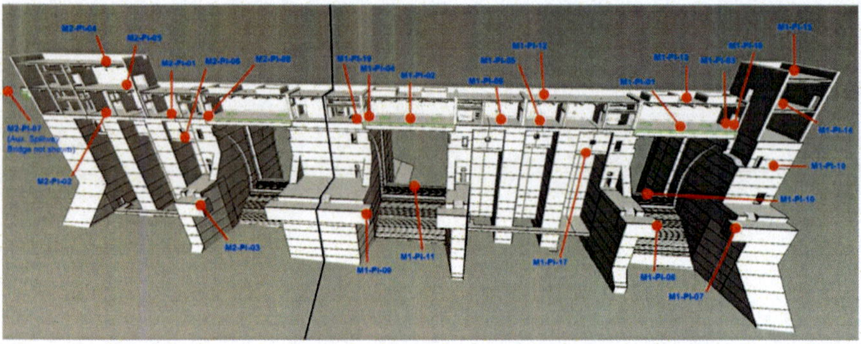

Fig. 2. Schematic view of the spillway and headworks structure.

Whilst a general seismic qualification level of IEEE 693-2005 High RRS was established for all the project electrical equipment, this level of seismic demand was significantly inadequate when compared to the FDS at the locations of the critical gate control equipment in the spillway headworks. In the spillway headworks there is substantial structural amplification of seismic shaking in the short period range. Figure 3 shows the 2%-damped response spectra for one PI compared to the IEEE 693–2005 High RRS. Faced with the equipment design challenge of such high seismic demands, three-dimensional seismic isolation was concluded as the only practical means of achieving the required level of equipment protection and performance.

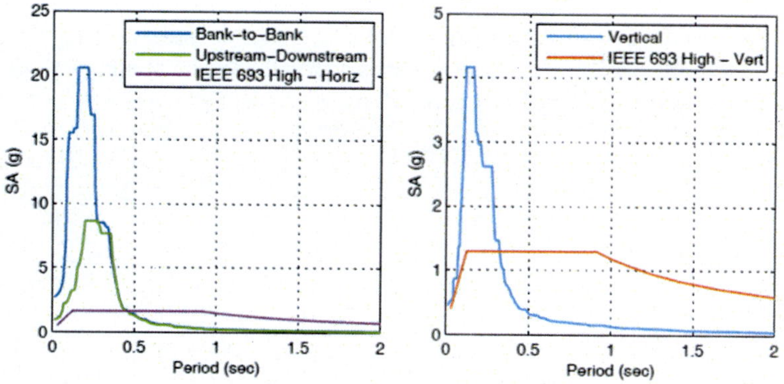

Fig. 3. FDS for a PI in the spillway headworks, horizontal (left) and vertical (right).

A two-step approach was taken to achieve the seismic protection of the critical components required to be immediately available and fully operational after an MDE event for spillway and low-level outlet gates. The two-step process involved: a. qualifying the critical equipment to IEEE Std 693-2005 High RRS level by shake table testing of the equipment, and b. providing additional protection in the form of 3-D seismic isolation platforms to ensure that the shaking experienced by the equipment on the platforms is reduced to a level not more than the IEEE Std 693-2005 High level. One exception to the shake table testing requirement was for the station service transformers which were qualified by analysis per the requirements of IEEE Std 693-2005.

IEEE Std 693-2005 does not provide detailed recommendations for equipment with seismic protective systems or for equipment on isolation platforms. The above approach is essentially that prescribed for Category 3 equipment in IEEE Std 693-2018, Annex W [1]. Category 3 equipment is equipment with protective systems which is required to be shake-table tested but for several reasons outlined in IEEE Std 693-2018, cannot be tested together with the protective system [2]. For the Site C Clean Energy Project, the design and manufacturing of the equipment and isolation platforms occurred simultaneously and therefore, shake table testing of the actual equipment on the platforms was not possible.

3.1 Isolation Platform Seismic Loading

The spillway includes three wire rope hoist operated radial gates and six hydraulically operated low level outlet gates. All of these gates have complete operational redundancy with primary and back-up power supplies and electronic control equipment. As such, depending on the specific equipment, there are numerous units of the same equipment, and therefore isolation platforms, throughout the spillway headworks. The spatial variability of the platforms necessitated an enveloping approach to the relevant point of interest FDS for the platform design process. Figure 4 shows examples of envelope design spectra for one particular isolation platform. The figure shows the enveloping of all FDS for the upper bound and lower bound foundation stiffness assumptions, shown left and right, respectively. Sets of earthquake time histories were then generated to match the envelope spectra, using original project time histories from the spillway headworks as seed motions, and finally, these time histories were used for the nonlinear time-history verification analyses and the performance validation shake table tests.

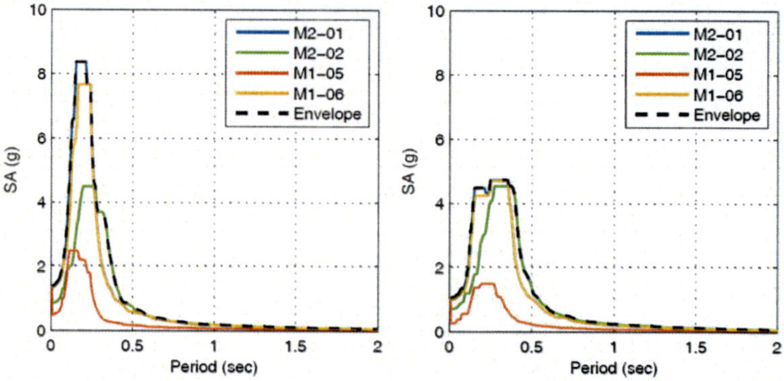

Fig. 4. Enveloped spectra for upper and lower bound foundation stiffnesses (left and right).

3.2 Isolation Platform Designs

The general design requirements for the isolation platforms were simple:

a. the total height of the platforms should not exceed 400 mm,
b. the total horizontal MDE seismic movement of the platforms should not exceed 150 mm, and
c. under MDE level FDS excitation the platforms attenuate both the horizontal and vertical seismic input to the supported equipment to less than the IEEE Std 693–2005 High RRS demand.

Criterion (a) was associated with vertical clearances within rooms, and particularly to overhead cable trays; (b) was related to functional and operational layouts of equipment within rooms, and (c), the most important, linked the seismic protection provided by the platforms to the qualification level of the equipment on top.

The 3-D isolation platforms are all based on a common design concept and followed the same general design process. The horizontal and vertical isolation functions are fully de-coupled and comprise different functional mechanisms. Horizontally, the platform is free to move in any direction on very low friction cross-linear bearings with a rolling resistance about 0.1% to 0.3% of the supported weight. The horizontal restoring force and damping is provided by a proprietary Multi-Directional Spring Unit which has a flag-type hysteresis that provides a bilinear loading stiffness and about 13–15% equivalent viscous damping [3–5]. The vertical isolation system operates on top of the horizontal system and comprises linear vertical springs, linear viscous dampers and linear guides to eliminate rocking deformations and ensure vertical-only translational movement. Figure 5 shows

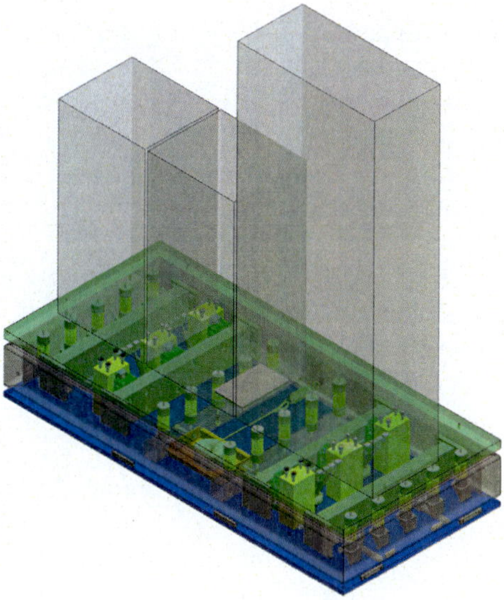

Fig. 5. Typical seismic isolation platform, isometric view

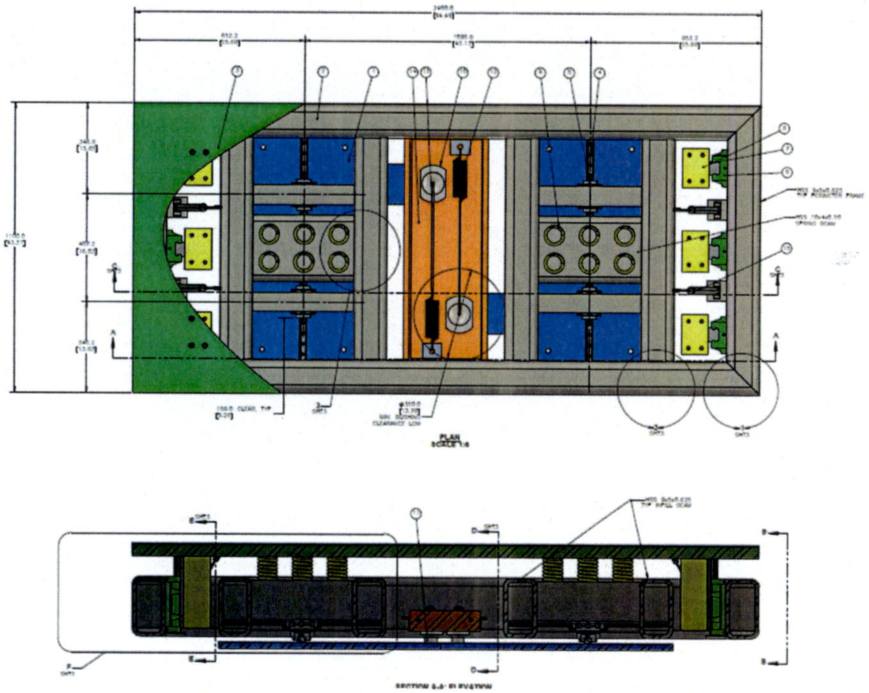

Fig. 6. Typical seismic isolation platform, cross-sectional details

a typical isometric view of a platform and Fig. 6 typical cross-sections with the key functional components.

The platform designs were initially developed using equivalent linear and response spectrum methods. Subsequently, detailed nonlinear time-history verification analyses were performed on a subset of the designs, selected to be representative of the entire range of all the platform designs.

There are a total of 39 3-D seismic isolation platforms on the project, comprising 15 different designs, to protect the gate control and associated power supply equipment. The locations of the platforms in the spillway headworks can be seen in the plan view of Fig. 7, where the platforms are indicated by the red rectangles. The platforms range in size from the smallest at only 0.7 × 1.2 m in plan, supporting a single item of control equipment of 380 kg, up to the largest at 2.75 × 12.0 m in plan which supports 30,391 kg of transformer and switchgear. The quantity of each design varies from just one unit for each of ten platforms, up to nine units for the most commonly-used design.

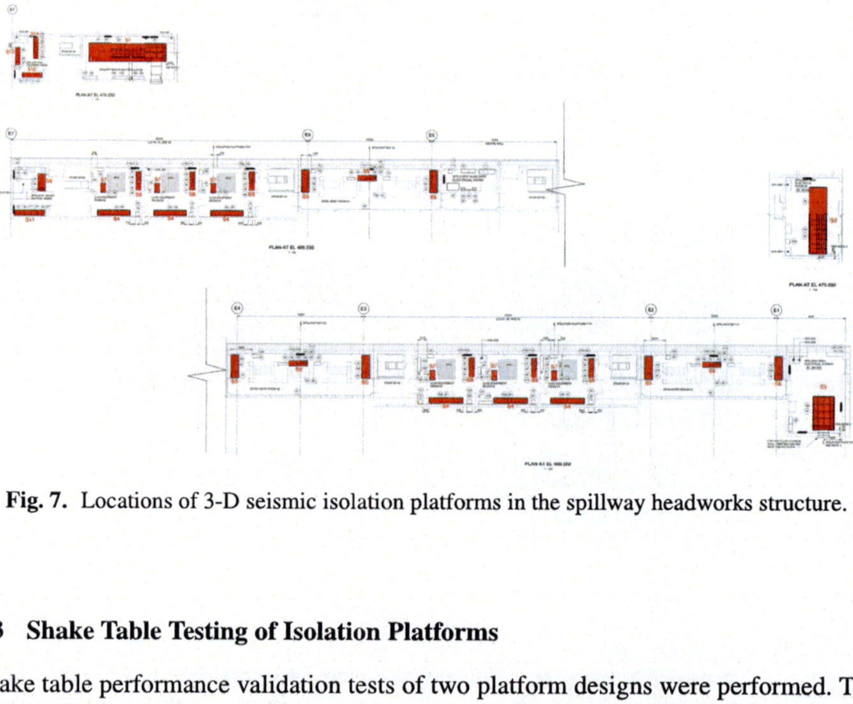

Fig. 7. Locations of 3-D seismic isolation platforms in the spillway headworks structure.

3.3 Shake Table Testing of Isolation Platforms

Shake table performance validation tests of two platform designs were performed. The smallest platform, denoted S07, and a second design, denoted S15, were selected for the shake table tests. S07, because of its small size, was the most mechanically challenging of all the platform designs, and S15 was selected to be representative of the mid-range of platform sizes. It is noted that the largest platform designs could not be tested because of the lack of shaking tables of the appropriate size (to accommodate the platforms) and three-dimensional performance capabilities. The weight of the supported equipment is an important factor in the horizontal and vertical dynamic behavior of the isolation platforms. The parallel design and manufacturing of the equipment and the isolation platforms did not allow the use of the actual equipment for the platform shake table tests, instead equipment mock-ups that accurately represented the weight, stiffness, and center of mass characteristics of the equipment were included in the tests. The shake table test results confirmed that the platform designs effectively reduced both the horizontal and vertical equipment demands to below the IEEE Std 693 High RRS levels for the MDE level of shaking (Fig. 8). Additional beyond-MDE level tests were undertaken, up to 150% of the MDE input, to confirm that the platforms did not reach their displacement capacity limits even for 50% higher inputs.

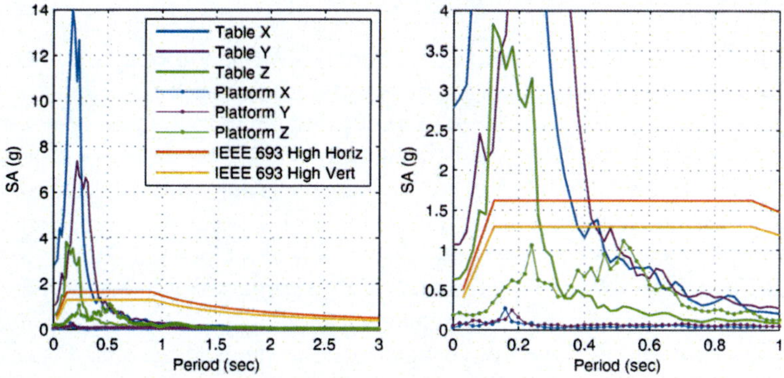

Fig. 8. Comparison of horizontal and vertical spectra below and on top of platform S15 (Table and Platform), respectively and IEEE 693-2005 High RRS. Right side is enlarged view to more clearly show the top of platform responses.

3.4 Shake Table Testing of Equipment

All the equipment to be installed on the 3-D isolation platforms, with the exception of the station service transformers, was separately qualified to the IEEE Std 693-2005 High RRS level via shake table testing. Two shake table test programs were undertaken by the owner to qualify a total of 12 different items of equipment. IEEE Std 693-2005 allows for qualification of equipment by group, with a successful qualification test of the most seismically vulnerable piece of equipment in the group serving as the basis for qualification of all other equipment items in the group. Most of the equipment supported on the isolation platforms are cabinets or rack-mounted items, which allowed for straight-forward grouping criteria to be applied.

The project required that all other equipment not installed on the isolation platforms also be qualified to IEEE Std 693-2005 High, with most specifications requiring that the qualification be by shake table testing. These qualifications were performed by the equipment manufacturers themselves either specifically for the Site C Clean Energy Project or previously as is common practice in the electrical industry.

3.5 Production Testing Protective System Devices

In addition to the shake table performance validation tests of the platforms, component testing of the various isolation system devices was performed as part of the platform manufacturing process. This included stiffness verification tests of the springs for both the horizontal and vertical isolation mechanisms, and cyclic dynamic tests of the fluid viscous dampers in the vertical isolation system.

4 Isolation of 500 kV Equipment

The use of seismic isolation to protect high-voltage distribution equipment at the interface to the transmission grid utilized a quite different approach from that for the gate control equipment and was undertaken to meet a different performance goal. Large and seismically-vulnerable equipment associated with nine 500 kV generator step-up transformers is mounted on tall reinforced concrete fire walls that surround each transformer (Fig. 9). Due to the height of the walls, this equipment is subjected to significantly amplified seismic shaking. Each transformer includes a Motor-Operated Disconnect switch (MOD), a Capacitive Voltage Transformer (CVT), a Surge Arrester (SA) and three insulator posts. The PI spectra at these equipment locations on the fire walls, in some cases, substantially exceeded the levels of shaking to which the equipment had been seismically qualified so it was highly unlikely that the equipment would sustain the PI level of shaking without failure (see Fig. 10 which shows the PI spectra for the MOD and CVT, on the left and right, respectively). Whilst the project design basis did not explicitly prohibit equipment damage, the expectation of the resumption of electricity generation within 72 h of a DBE event meant that a repair-or-replace approach was not feasible considering both the short time-frame and the potentially large amount of equipment damage.

Seismic isolation was therefore selected to provide a higher level of protection to the 500 kV equipment to meet the project performance objective. The details of the equipment, and their locations atop the fire walls (and for some, on beam elements) presented significant constraints and challenges for isolation. Wire Rope Isolators (WRIs) were ultimately selected, which while providing some flexibility in the horizontal and vertical directions, offer the primary benefit of rocking mode flexibility which achieves substantial reductions in seismic bending demands on taller, slender equipment (Fig. 11.a) [6].

A detailed nonlinear finite element study was undertaken to evaluate the response of the MOD, CVT, SA and insulators supported on WRIs. The equipment was modelled with expected stiffness and mass properties while each of the WRIs was modelled using two elements in parallel, a piece-wise linear force-deformation backbone curve in combination with a nonlinear damper to represent the physical behavior of the WRI shown in Fig. 11.b.

To determine the seismic response at each location of the equipment, a detailed finite element analysis of the fire walls had previously been completed. The nonlinear time history analysis of the equipment on the WRI platforms made use of the five, three-component time histories which defined the wall motion at each location.

The analysis determined the maximum bending moments at the base of each component for the case when the equipment was mounted on WRIs and when it was fixed directly to the top of the wall or beam. In all cases, the mean maximum bending moments for the equipment mounted on WRIs were below the manufacturer's published rating. This was not the case for the equipment fixed directly to the top of the walls or cross beam (i.e., without WRI isolation). In several cases the response atop the WRIs was five to ten times less than for the case when the equipment was rigidly mounted. The analyses also evaluated the displacement at the top of the equipment, which is important for the bus work interconnects. It was observed that even when flexibility was added to the base of the equipment in the form of a WRI platform, the deformation at the top of the equipment was less than for the case when the equipment was fixed at its base. In some instances, the isolated equipment displacements were less than half of those for the fixed-base condition.

Fig. 9. 500 kV equipment on top of GSU transformer fire walls

Tests of the WRI devices were undertaken to establish the detailed properties used in the nonlinear design analyses, and production tests of all WRI devices was performed prior to equipment installation.

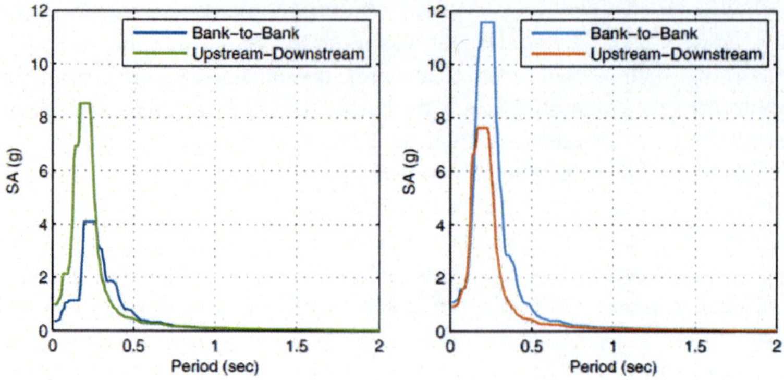

Fig. 10. PI spectra at the location of the MOD (left) and CVT (right)

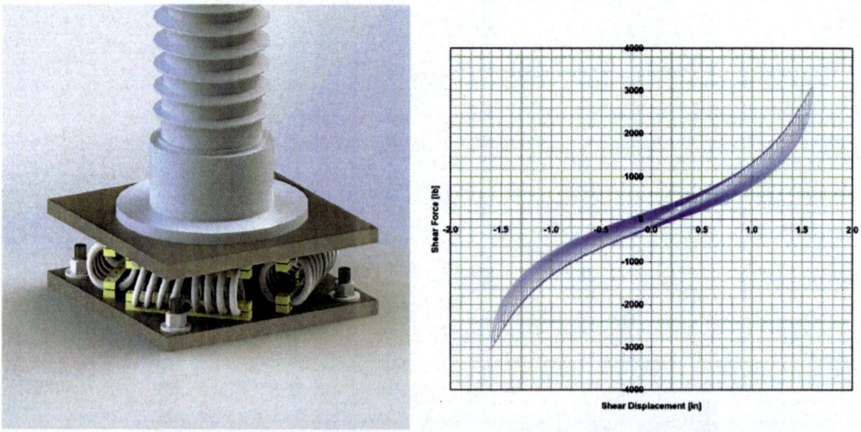

(a) Wire rope isolators under insulator

(b) Wire rope isolator hysteresis loop (50 cycles, 0.5 Hz)

Fig. 11. Wire rope isolator installation configuration, and shear hysteretic behavior

5 Conclusions

The Site C Clean Energy project is using two types of seismic isolation to achieve enhanced seismic protection of a wide range of critical equipment. High-performance 3-D isolation platforms are used to achieve post-MDE functionality protection of gate control equipment critical to reservoir water level control and dam safety, and wire rope isolation is used to protect 500 kV electrical equipment at the interface to the regional grid, ensuring post-DBE event generating capability. The implementation of the isolation systems utilized nonlinear analyses, shake table systems testing and spring, damper and wire rope device testing to confirm properties and behavioral characteristics.

Acknowledgements. Many others at the authors' companies have participated in and contributed to the design and implementation of seismic isolation for the Site C equipment. Equipment qualification shake table testing was performed at the University of British Columbia, Vancouver, Canada, and 3-D seismic isolation platform shake table performance validation testing at the University of Nevada, Reno, USA. The valuable work of all others involved and these testing laboratories is gratefully acknowledged.

References

1. Aiken, I., Black, C. et al.: New IEEE 693 guidelines for electrical substation equipment with protective systems. Paper No. 2f-0029, Proc. 17th World Conference on Earthquake Engineering, Sendai, Japan (2020)
2. IEEE Power & Engineering Society: IEEE recommended practice for seismic design of substations. IEEE Std 693-2018. New York (2019)
3. Cui, S., Bruneau, M., Kasalanati, A.: Behavior of bidirectional spring unit in isolated floor systems. J. Struct. Eng. **136**(8). American Society of Civil Engineers, August (2010)
4. Cui, S., Bruneau, M., Kasalanati, A.: Seismic response case study of isolated floor system having special biaxial spring units. J. Struct. Eng. American Society of Civil Engineers, July (2016)
5. Kasalanati, A., Ng, T. et al.: Seismic isolation of sensitive equipment. In: Paper No. 2676, Proc. 16th World Conference on Earthquake Engineering, Santiago, Chile (2017)
6. Alessandri, S., Giannini, R., Paolacci, F., Malena, M.: Seismic retrofitting of an HV circuit breaker using base isolation with wire ropes. Part 1: Preliminary tests and analyses. Eng. Struct. (2015). https://doi.org/10.1016/j.engstruct.2015.03.032

A Comparative Study on Isolator Modeling Approaches

Eser Çabuk[1](✉), Uğurhan Akyüz[1], Fatih Sütçü[2], and Murota Nobuo[3]

[1] Middle East Technical University, Ankara, Turkey
{eser.cabuk,han}@metu.edu.tr
[2] İstanbul Technical University, İstanbul, Turkey
fatih.sutcu@itu.edu.tr
[3] Bridgestone Corporation, Tokyo, Japan
nobuo.murota@bridgestone.com

Abstract. Due to the complexity in the development of highly nonlinear models, engineers tend to use more practical ones such as smoothed or sharp bilinear models. It has been shown that differences in hysteresis characteristics may lead to variance in performance parameters, resulting in possible overdesign or underdesign. In this study, to assess the effect of the modelling approach on seismic performance, high damping rubber bearing was designed and evaluated in terms of structural performance for a selected hospital building. Commonly used bilinear and highly nonlinear isolator hysteresis models from the literature were adopted, and structural models were created for 2475 years return period earthquake level. Three-dimensional nonlinear time history analyses were conducted on the 10-story building model under a set of 11 ground motions. For each model, structural performance is evaluated and compared in terms of maximum isolator displacements, base shear reactions, story accelerations, and inter-story drifts. The results show that when bilinear models are used instead of more accurate nonlinear models, there is significant variation in the superstructural response, especially for sharply bilinear models for the considered building.

Keywords: Seismic isolation · Nonlinear model · Bilinear model · High damping rubber bearing

1 Introduction

In the analytical modeling stage, researchers and engineers tend to use different mathematical models suggested by design codes or the literature to define the hysteretic behavior of isolators for nonlinear time history analysis (NTHA). These include bilinear models (sharp and smoothed) and highly nonlinear models. The bilinear models are more practical and can easily be implemented in the structural analysis by following the design code guidelines. In contrast, nonlinear models which accurately capture the actual behavior are more complex to implement and require better comprehension of nonlinear dynamic isolator characteristics. These models were used in both research and design projects. However, previous studies [1–3] have shown that characteristics

G. P. Cimellaro (Ed.): WCSI 2022, LNCE 309, pp. 516–523, 2023.
https://doi.org/10.1007/978-3-031-21187-4_43

of the hysteresis loops might introduce significant variations in structural performance parameters.

The aim of this study is to assess the effect of the isolator modeling approaches on earthquake performance of 10-story seismic-isolated reinforced concrete hospital building, with plan irregularity, under a set of 11 ground motions. The structural system consists primarily of beam-column frame elements. In addition, shear walls are present at the core regions, which significantly affects the isolation system design and evaluation. High-damping rubber type isolators were designed for the building due to its highly nonlinear response under earthquake excitation, especially at higher shear strain levels.

2 Analysis Model

2.1 Building Information

The building is one of the 10-story T-shaped blocks of a seismic-isolated reinforced concrete (RC) hospital which was built in a high seismic zone in Turkey. The structural system consists mostly of column and beam (frame) elements and shear walls supporting the core (elevators and stairs) regions to limit the lateral displacements on the superstructure. The plan geometry is 115 m in the X direction and 75 m in the orthogonal Y direction, and the total height is 40.9 m.

The pedestal sizes are 1250 mm × 1250 mm below the isolation layer. The column dimensions are 900 mm × 900 mm for the first floor and 800 mm × 800 mm for the upper floors. Moreover, beam dimensions of 800 mm × 600 mm were used throughout the structure. In addition, a limited number of columns and beams of different sizes were also used in the structural system. The shear wall thickness is 300 mm. The seismic weight is evaluated to be 500,000 kN. Cracked section stiffnesses were used for all members for effective stiffness. 3-D structural model (Fig. 1a) was created and analyzed using the commercial structural analysis software ETABS [5], and the isolation layout is given in Fig. 1b.

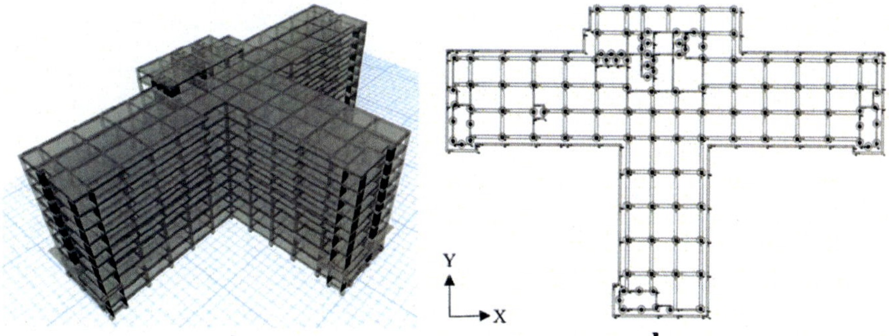

a **b**

Fig. 1 a. 3-D structural model. **b.** Isolation system layout.

2.2 Selected Ground Motions and Nonlinear Time History Analysis

The building is located in a high seismic region in Turkey. The shear wave velocity at 30 m depth of the soil, Vs_{30}, was retrieved as 350 m/s. Maximum Considered Earthquake (MCE) level elastic design spectrum was constructed according to the Turkish Seismic Code [4]. In order to conduct nonlinear time history analyses, eleven ground motions were selected from the PEER NGA-West2 Ground Motion Database [11]. The characteristics of the ground motions and their scale factors for MCE level are shown in Table 1. A comparison of acceleration spectrums for each ground motion to the target spectrum is shown in Fig. 2 for MCE level seismic intensity.

Two orthogonal components of each earthquake ground motion define each dynamic loading condition. Also, the second batch of loading cases was created by interchanging the directions of the components. Therefore, a total of 22 loading conditions were created. The most critical cases for each pair were selected and the average of 11 selected critical conditions was used to evaluate the results. Fast nonlinear analysis [5] which was implemented in ETABS, was chosen as the analysis method due to its computational efficiency.

Table 1. Characteristics of selected ground motions and scale factors.

Result ID	Reqord seq.	Tp-pulse	Event	Year	Magnitude	Vs30	Scale factor
1	187	–	Imperial valley-06	1979	6.53	348.7	3.892
2	316	4.389	Westmorland	1981	5.9	348.7	2.030
3	549	–	Chalfant valley-02	1986	6.19	303.5	2.826
4	558	–	Chalfant valley-02	1986	6.19	316.2	1.724
5	821	–	Erzican_Turkey	1992	6.69	352.1	1.125
6	850	–	Landers	1992	7.28	359.0	3.530
7	1116	–	Kobe_Japan	1995	6.9	256.0	2.487
8	1158	–	Kocaeli_Turkey	1999	7.51	281.9	1.281
9	1605	–	Duzce_Turkey	1999	7.14	281.9	0.947
10	2752	–	Chi-Chi_Taiwan-04	1999	6.2	258.9	2.814
11	6893	–	Darfield_New Zealand	2010	7	344	1.517

2.3 Isolator Design and Nonlinear Models

An isolation system design for a target period of approximately 3 s and maximum displacement of 0.38 m was designed. The selected maximum displacement corresponds to 190% shear strain for the HDRB isolators having 0.2 m total rubber height. The lower bound for the reduction of the isolator properties was assumed as 0.9 for this study. Four

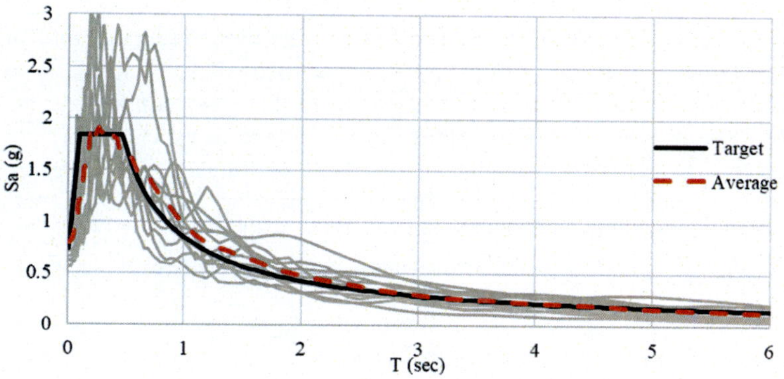

Fig. 2. Target spectrum and average spectrum of 11 ground motions.

types of Bridgestone HDRB isolators were selected according to their corresponding column axial loads under the seismic conditions. The one-directional bilinear envelope force-displacement relationship of the four types is given in Fig. 3.

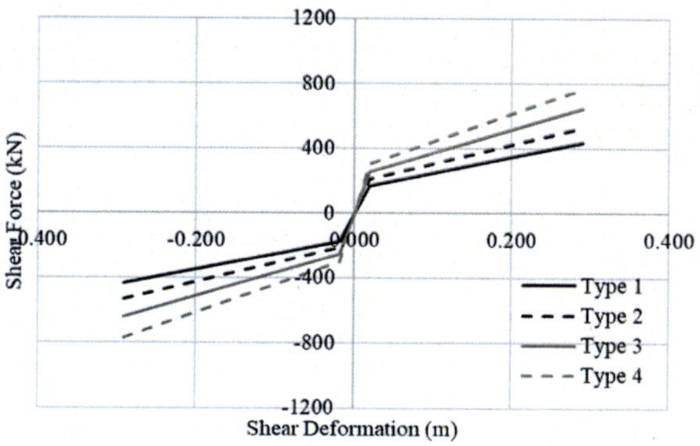

Fig. 3. Target bilinear force-displacement relationship of selected HDRB.

An illustration of the force-displacement relationship of the sharp bi-linear (BL) model is shown in Fig. 4a with the dotted line. Three parameters are associated with this model: the initial elastic stiffness (K_1), the characteristic strength (Q_d) and the post-yield stiffness (K_2). A secondary parameter, yield displacement, is optionally neglected in most cases due to a small magnitude of elastic deformation of the isolator. A sharp and sudden corner is provided between stiffness transitions. After the maximum load in each cycle is reached, the unloading curve follows a parallel line of the same length with the previously loaded segments in the opposite direction until it reconnects the envelope curve. The sharp bilinear model, implemented in the structural analysis programs as a general hysteretic element, does not account for the bilateral excitation of the ground motion components (CSI Analysis Reference Manual [5]).

Another bilinear model, called as smoothed bilinear (or Bouc-Wen, BW) model, was developed by Bouc [6], Wen [7], and Park et al.[8]. This analytical model can be applied to a wide range of hysteretic systems for both elastomeric and sliding isolators. In Fig. 4a, with a solid line, hysteresis is defined by three parameters: elastic stiffness, post-yield stiffness ratio to elastic stiffness, and yield force. Unlike the sharp bilinear model, the transition between initial and post-yield stiffness is smoothed by a circular yield surface function, and the two lateral degrees of freedom are coupled. Therefore, the model can capture nonlinear characteristics of isolators under biaxial lateral excitation. The vertical degree of freedom is linearly elastic and uncoupled from the lateral directions. This model is implemented in ETABS software as Rubber Isolator link element [6].

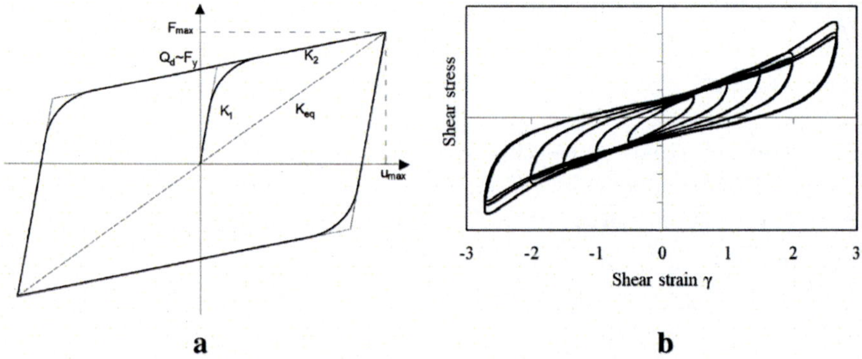

Fig. 4 a. Sharp and smoothed bilinear model. **b.** DHI model conceptual behavior.

A nonlinear time-independent elastoplastic model (Deformation History Integral Type, DHI [9, 10]) was constructed for HDRB isolators. The DHI model can accurately capture the highly nonlinear behavior of rubber bearings under biaxial and uniaxial excitations. A conceptual shear stress-strain behavior is given in Fig. 4b. The hysteresis characteristics highly depend on the shear strain and represents stiffness increase when the shear strain exceeds 200%. This model is also implemented in ETABS, named as High Damping Rubber Isolator Link [5]. The axial behavior is linearly elastic and independent from the two coupled shear directions. The following parameters are required to define the DHI model: isolator cross-sectional area, isolator effective height, added elastic stiffness, hysteretic parameters that affect the hysteretic behavior, damage parameters that define the damage function, and elastic stiffness degradation (resistance ratio and control strain).

3 Analyses Results

Time history displacement response in both lateral directions was obtained for each MCE level ground motion with lower bound parameters. The horizontal displacements were directionally combined at each time step. The average results of the eleven ground motions were used to determine the maximum displacement. Additional displacements due to torsion were neglected.

Average maximum combined displacements for the DHI, smoothed BW, and sharp BL models were gathered as 36.3, 36.2, and 35.0 cm, respectively (Table 2). Excessive deformations are observed in the ground motions RSN316 (Westmorland, 1981) and RSN821 (Erzincan, 1992), which have pulse-like behavior and near-fault effects in the records. In these types of cases, the BL model yielded the maximum deformations. When the ground motions are evaluated separately, the displacement response could be underestimated or overestimated with bilinear models. The system's deformation response is highly dependent on the seismic input characteristics.

Table 2. Maximum displacement results for each model.

Eq ID	DHI model	BW model	BL model
RSN0187	41.51	41.47	45.09
RSN0316	61.32	61.22	64.79
RSN0549	25.68	25.63	20.23
RSN0558	20.02	19.99	18.05
RSN0821	47.44	47.35	55.21
RSN0850	20.45	20.41	20.80
RSN1116	31.25	31.21	26.50
RSN1158	42.63	42.55	38.09
RSN1605	44.58	44.51	40.35
RSN2752	33.02	32.94	35.52
RSN6893	31.23	31.20	20.44
Dmax(ave)	36.28	36.22	35.01

The superstructure response in terms of story force, story acceleration and inter-story drift in X and Y directions are shown from Figs. 5, 6 and 7, respectively. The isolation level is located at Floor 0, and there is a basement floor (Floor-1) below the isolation interface. In other words, the first floor above the isolator is Floor 1. Among all performance parameters, the overestimation of the response by the sharp BL model is significant. The smoothed BW model showed more compatible results with the DHI model but resulted in underestimation or overestimation of the response.

4 Conclusions

In this study, performance evaluation of a 10 story seismic-isolated hospital building with irregular T-plan shape is conducted using various nonlinear isolator models. For HDRB type isolators, the deformation history integral type (DHI) model, the smoothed Bouc-Wen (BW) bilinear model, and the sharp bilinear (BL) model were considered. The superstructure and isolation system response for various mathematical models is evaluated and compared in terms of maximum isolator displacements, structural forces, floor accelerations, and inter-story drifts. Based on the results, the following conclusions can be made about the investigated building:

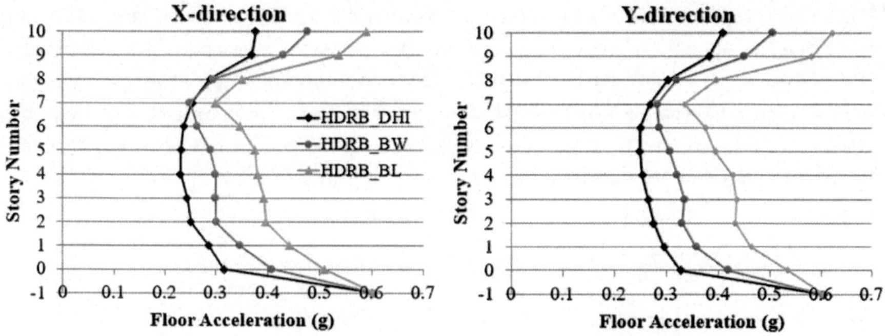

Fig. 5. Floor accelerations in X (left) and Y (right) direction.

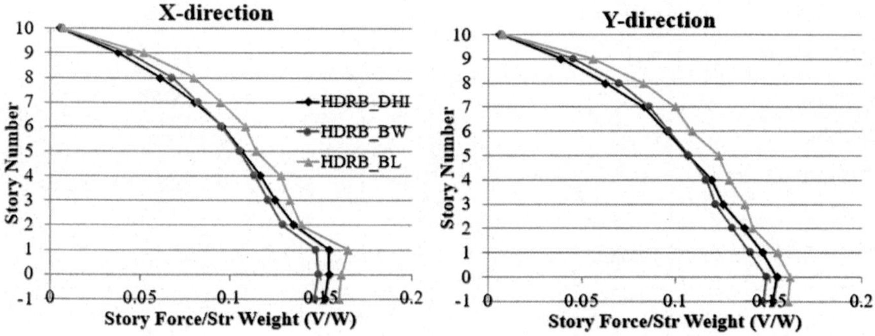

Fig. 6. Structural forces in X (left) and Y (right) direction.

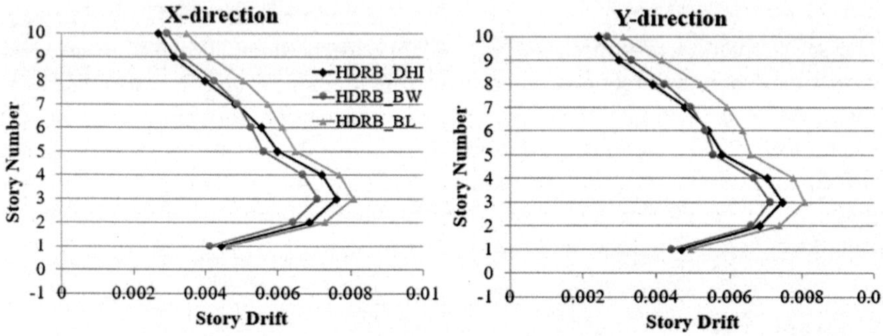

Fig. 7. Story drifts in X (left) and Y (right) direction.

The bilinear models can overestimate or underestimate the maximum isolation system displacements compared to the more sophisticated DHI model. The result displacements are also highly dependent on the ground motion characteristics. In addition, using idealized bilinear models resulted in a considerable increase in the superstructure response. The sharp bilinear model's sharp transitions could result in an overestimated response. The smoothed bilinear model yields a superstructural response that is more compatible with the sophisticated models.

Although the smoothed bilinear model predicted relatively more accurate results for this building, both idealized bilinear models overestimated the superstructural response. This might result in an exceedance of performance criteria, leading to overdesign. Most design guidelines do not include recommendations for highly nonlinear sophisticated models. Engineers should carefully assess the model's capabilities and possible outcoming structural response before implementing in analyses.

References

1. Çabuk, E.: Seismic performance comparison of base-isolated hospital building with various isolator modeling approaches. MS thesis, Middle East Technical University (2021)
2. Mavronicola, E., Komodromos, P.: On the response of base-isolated buildings using bilinear models for LRBs subjected to pulse-like ground motions: sharp vs. smooth behavior. Earthq. Struct. **7**(6), 1223–1240 (2014)
3. Ray, T., Sarlis, A.A., Reinhorn, A.M., Constantinou, M.C.: Hysteretic models for sliding bearings with varying frictional force. Earthq. Eng. Struct. Dyn. (2013)
4. The Disaster and Emergency Management Presidency of Turkey: Turkish Building Seismic Code (2018)
5. Computers and Structures Inc.: CSI Analysis Reference Manual (2017)
6. Bouc, R.: Modele mathematique d'hysteresis. Acustica **24**(1), 16–25 (1971)
7. Wen, Y.K.: Method for random vibration of hysteretic systems. J. Eng. Mech. Div. (1976)
8. Park, Y.J., Wen, Y.K., Ang, A.H.S.: Random vibration of hysteretic systems under bi-directional ground motions. Earthq. Eng. Struct. Dyn. **14**(4) (1986)
9. Kato, H., Mori, T., Murota, N., Kikuchi, M.: Analytical model for elastoplastic and creep-like behavior of high-damping rubber bearings. J. Struct. Eng. (2015)
10. Masaki, N., Mori, T., Murota, N., Kasai, K.: Validation of Hysteresis Model of Deformation-History Integral Type for High Damping Rubber Bearings (2017)
11. Pacific Earthquake Engineering Research Center: PEER Ground Motion Database. University of California, Berkeley

Seismic Base Isolation of a Strategic Historic Masonry Building

Tobia Zordan[1](✉), Alessandra Romano[2], and Fulvio Parisi[3]

[1] BOLINA Ingegneria, CEN-TC 340, 30174 Venice, Italy
zordan@bolinaingegneria.com
[2] BOLINA Ingegneria, 30174 Venice, Italy
[3] Dipartimento di Strutture per l'Ingegneria e l'Architettura, Università Federico II, 80125 Naples, Italy

Abstract. Historic masonry buildings are often heavily damaged by earthquakes, resulting in huge losses of cultural heritage. Several studies have shown the effectiveness of seismic vulnerability mitigation strategies based on base isolation systems (BISs), which allow damage to masonry buildings and their contents to be significantly reduced or even avoided. This paper presents preparatory activities and base isolation project for seismic retrofitting of an existing strategic masonry building with historical-architectural value, which is located in Perugia (Italy) and hosts the headquarters of Umbria Region, Perugia public prosecutor's office and some offices of the Italian Ministry of Infrastructure and Transportation. The building consists of five floor levels and has an L-shaped plan with size equal to approximately 1600 m^2. The structure was built in the early 1950s using different materials and construction technologies. The BIS design started from a historical investigation on the original design, fabrication and modifications of the masonry building, based on archival data and available literature. An extensive investigation was carried out through on-site experimental tests on both subsoil and masonry structure, as well as geometric surveys with laser scanner technology, allowing an excellent knowledge of the building (including the development of a building information model) and the calibration of two three-dimensional structural models based on both finite element method (FEM) and equivalent frame method (EFM). The structural characteristics, historical value and public function of the building delineate a case study with considerable importance in the field of seismic isolation of existing masonry buildings.

Keywords: Seismic base isolation · Historic building · Masonry · BIM · FE analysis · Construction

1 Introduction

The World's largest Global Economy sector, at least for most industrialized Countries, resides within the field of buildings and infrastructures which are meant to ease and protect human life, represent the welfare and heritage of Peoples and host a great part of the investments of both individuals and groups. A recent increased frequency of non-extreme

G. P. Cimellaro (Ed.): WCSI 2022, LNCE 309, pp. 524–531, 2023.
https://doi.org/10.1007/978-3-031-21187-4_44

and extreme events such as earthquakes and other environmental extreme conditions (like tsunamis, storms, droughts, floods, landslides etc.) is recorded worldwide.

As a consequence, any strategies dealing with risk assessment and management (with all related uncertain scenarios), proved to be effective in protecting the economic value of the built environment and able to keep the function of buildings and infrastructures uninterrupted in the aftermath of a major event, receives great interest from different and large number of stakeholders.

Anti-seismic design of structure and infrastructures has been largely introduced in Standards worldwide based on the concept of ductility.

This approach and the idea of allowing a structural system to enter its post-elastic field to dissipate energy, implicitly accept the concept of damage. This approach, and the acceptance of the occurrence of structural damage (and therefore also of consequent non-structural damage), even if shared and commonly accepted on the basis of cost-benefit evaluations by Legislators and Standardization Bodies, have proved to be unsatisfactory and below expectation if not under a merely structural point of view, certainly, and often, from a Social and Economical point of view. A damaged structure needs to be assessed and possibly, if required, repaired or retrofitted. This evidence is the cause of enormous costs. A suitable example is represented by the management of the aftermath of an earthquake, when large parts of the inhabitants of the hit areas are moved away from their houses pending on checks on the damaged structures.

Base isolation concept, namely the idea of de-coupling the sub-structure from the superstructure applying a cut and inserting specific devices able to lower the seismic demand on the superstructure (shifting the fundamental vibration period of the structure in the response spectrum towards higher values associated to lower levels of acceleration) is an ancient technique based on self-construction (Fig. 13) and empirical trial and error practice. Examples are known rom the past in many earthquake prone regions. In more recent times, starting from the early 1970s, base isolation has started become a standard for new construction in areas with medium and high seismicity and specific industrial products are constantly provided, certified and installed according to the specifications of specifically oriented product Standards (i.e. European Standard UNI EN 15129 Anti-seismic devices, 2018) (Fig. 1).

(a) (b) (c)

Fig. 1. Seismic base isolation techniques: general layout (a) and historic evidence based on self-construction (b), (c).

Base isolation is known to be an effective tool in providing uninterrupted functionality for structures and infrastructures hit by a severe earthquake maintaining the structure within its elastic limit state and fully protecting its overall value and the value of the contents. The possibility to provide also existing structures and infrastructures (and specifically buildings originally built with a fix base) with base isolation, has become in recent years a reality. After the earthquake that hit the city if L'Aquila in central Italy in 2009, several cases of the application of this technique have taken place in Italy, mainly boosted by economic incentives introduced by the Italian Government for the re-construction, in order to promote an increased and diffuse seismic resilience of the National built environment. One technique consists in applying a cut at a certain level of the existing structure and to insert the anti-seismic devices one by one.

2 Seismic Base Isolation of Palazzo Partigiani in Perugia, Italy

The present work illustrates the process of knowledge and the structural analyses and design finalized to the introduction of a seismic base isolation system on a strategic masonry building of historical and architectural interest, located in Perugia, Italy. The building hosts the headquarters of the Umbria Region, the Public City Tribunal and a displacement of the National Ministry of Infrastructure and Transport. The building, originally built in the early 50s, is characterized by an L-shaped plan with a floor area equal to about 1600 m^2; it develops in elevation on 5 levels and is characterized by a mic concrete and masonry bearing structure (Fig. 2).

2.1 On-Site Assessment

The structural design for the base isolation of the building began with a historical investigation of the origins of the building based on archival data and on the examination of the available bibliography.

The survey on construction materials, together with an extensive campaign of investigations on geometry of foundations and on the superstructure, has made possible to obtain an exhaustive knowledge of the typological and mechanical characteristics of the materials and construction details. The survey and investigation phase, besides all, showed how the building in its seventy years of life, has undergone two main interventions; the first one with the superelevation of the building with the creation of a further storey; the second one, aimed at an energy requalification.

An extensive campaign of non-destructive investigations was also carried out to clearly identify the reinforced concrete part of the structure and the masonry one Termography georadar and pacometer survey were extensively used for this purpose (Fig. 3).

Specimens were extracted and tested to determine the mechanical properties of concrete. Leeb rebound hardness test were provided to identify the type of reinforcement steel (Fig. 4).

Single and double flat jack testing method were used to determine the acting stresses (single flat jack) or to evaluate the mechanical parameters of the masonry (double) (Fig. 5).

Furthermore, pull-out tests on concrete were correlated to sample tests in order to obtain a general mapping of the actual properties of structural concrete (Fig. 6).

Fig. 2. The building at the end of construction in 1950 and today.

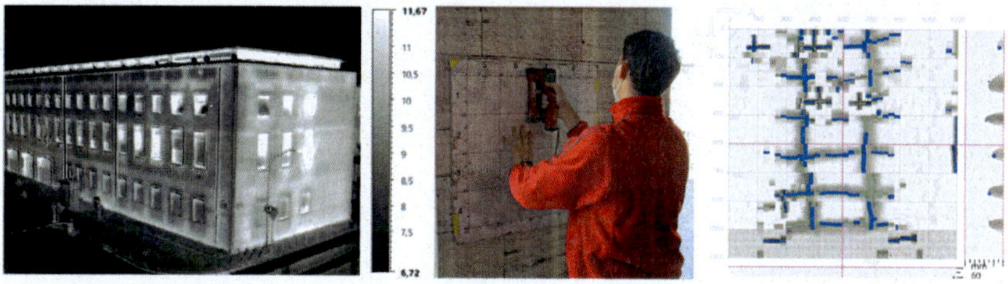

Fig. 3. Termography, georadar and pacometer test.

Fig. 4. Leeb rebound hardness test and concrete samples.

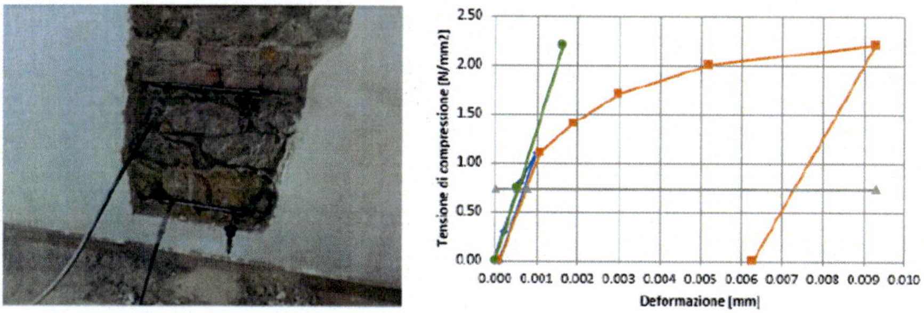

Fig. 5. Double flat jack testing method applied on masonry structures.

Extensive endoscopy testing offered the chance to assess directly shape and appearance of elements otherwise inaccessible to direct observation, making it possible to carry out useful evaluations and to determine the exact sequence of layers in different parts of the structure (Fig. 7). The endoscopes gave also evidence of the state of conservation of materials and structural elements. Finally geotechnical samples were obtained from boreholes and the geometry and consistence of the foundation system was derived (Fig. 8).

Fig. 6. Pull-out tests.

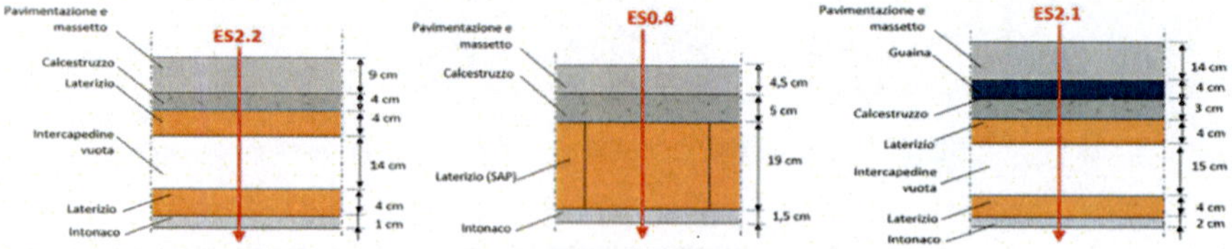

Fig. 7. Endoscopy with direct assessment of different layers and degree of conservation of materials.

Fig. 8. Geotechnical samples obtained from boreholes and foundation assessment.

2.2 3D Modelling and Analysis

The analysis of the building has been performed through the implementation of its structure on two different softwares, in order to better understand the structural response. In particular, for the study of the masonry superstructure, a macro-element model was implemented using "3Muri", while for design and verification of the base isolation system reference was made to a finite element model in MasterSap environment (Fig. 9). Experimentally assessed mechanical properties of materials were used to define the material properties used for the analyses. The two structural models have been calibrated in terms of dynamic response comparing the natural periods the mode shapes and the associated participating masses. Analyses carried out on the original fixed base structure shown remarkable flaws in terms of structural capacity with reference to seismic forces. A target isolation period was then defined in order to design the final intervention (Fig. 10). The isolation system combining different types of devices, was able to minimize the eccentricity between the centers of mass and stiffness of the structure (Fig. 11).

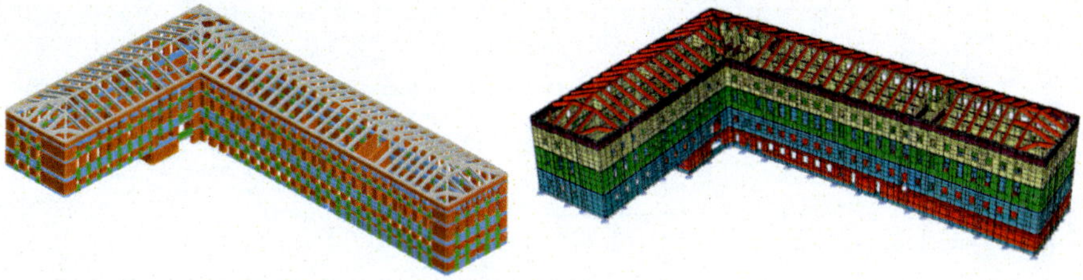

Fig. 9. The macro-element model used to check the mixed masonry and concrete superstructure (left) and the FE model (right) used to design the base isolation system.

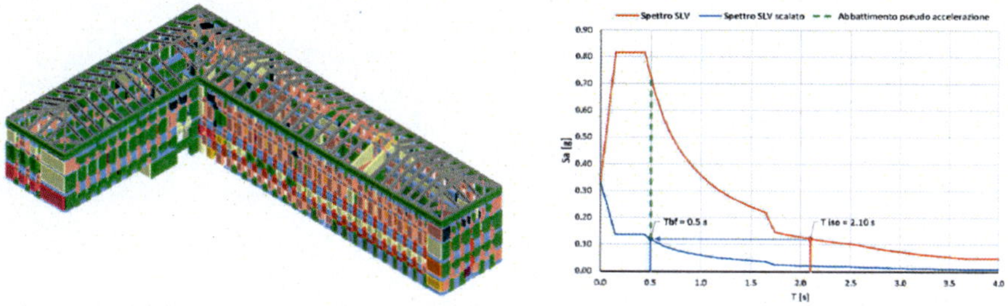

Fig. 10. Insufficient capacity towards seismic action shown by the macro-element model (left) and the identification of the base isolation target period (right).

Provided an almost perfect decoupling of modes in the 2 main x and y directions, as represented in Fig. 12, a maximum displacement of 185 mm was found at the level of devices, while the superstructure was found have an almost negligible interstorey drift (Fig. 12) and could provide a suitable capacity in the final design base isolated layout (Fig. 13).

2.3 Construction Phases

The sequence of phases aiming to introduce base isolation at the level of the building foundations is presented in Fig. 14.

3 Conclusions

The present paper has presented the introduction of seismic base isolation on a public building of historic value. The seismic retrofitting of the entire structure was achieved limiting the interventions and the foundation level without interfering with the superstructure and leaving the building functions almost uninterrupted during work execution.

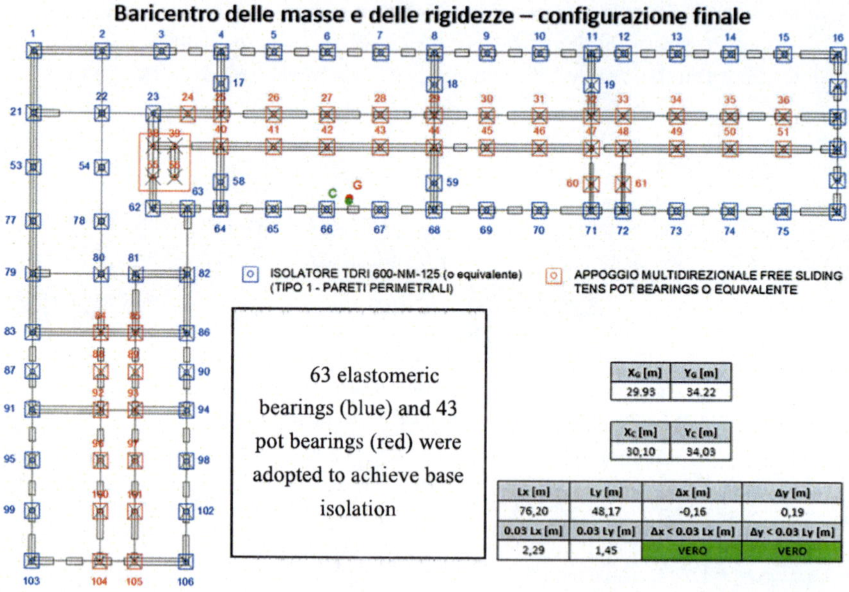

Fig. 11. Anti-seismic devices layout.

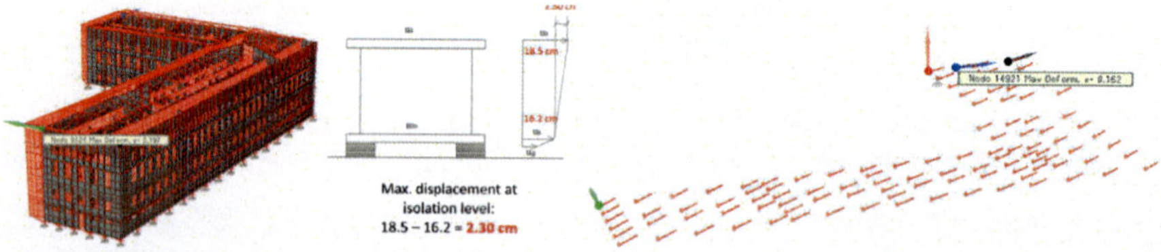

Fig. 12. Maximum theoretical displacements at isolators calculated at the limit state of collapse. Despite the L-shaped plan of the building an almost perfect decoupling of shape modes in the 2 main building direction is displayed.

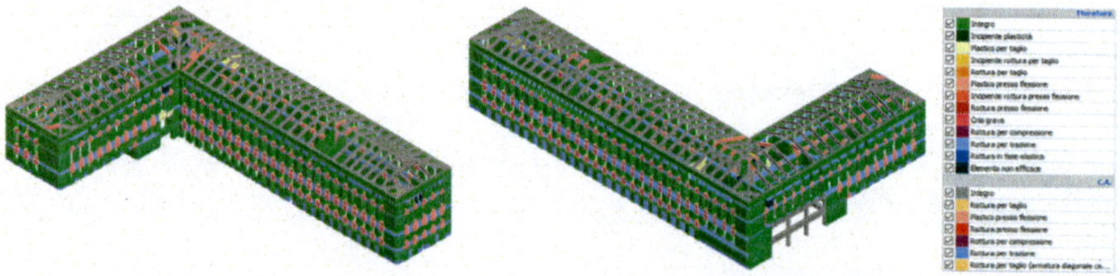

Fig. 13. Map of structural checks on the base isolated building superstructure presents suitable results in terms of capacity highlighting the effectiveness of the reduced demand on the superstructure which was unable to withstand horizontal seismic forces in its original layout (fix base).

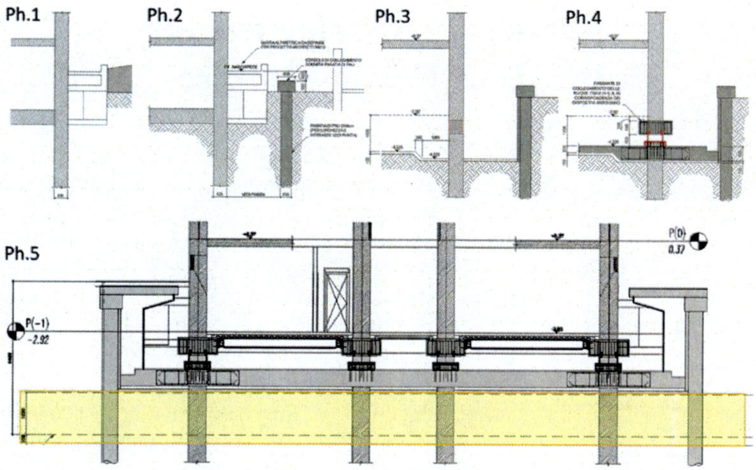

Fig. 14. Construction phases in typical cross section. Phase 1: original site condition. Phase 2: realization of a diaphragm wall. Phase 3: Masonry segments cut at isolation level. Phase 4: realization of new confining concrete girders at the base of existing masonry walls, removal of existing masonry (decoupling between substructure and superstructure) and final positioning of seismic isolators. Phase 5: Activation of base isolation system and realization of finishes.

Acknowledgements. Authors would like to acknowledge Eng. Francesco Renò MSc, for the contribution given to this work during his graduation thesis.

References

1. Frangopol, D.M., Kallen, M.J., Van Noortwijk, J.M.: Probabilistic models for life-cycle performance of deteriorating structures: review and future directions. Prog. Struct. Mat. Eng. **6**(4), 197–212 (2004)
2. Dickson, E., Baker, J., Hoornweg, D., Tiwari, A.: Urban Risk Assessments—Understanding Disaster and Climate Risk in Cities. Urban Development Series, The World Bank, Washington DC (2012)
3. Stojanović, M., Mitković, P., Mitković, M.: The scenario method in urban planning. Arch. Civil Eng. **12**(1), 81–95 (2014)
4. Olsen, J.: Adapting Infrastructure and Civil Engineering Practice to a Changing Climate. ASCE American Society of Civil Engineers, Reston-Virginia (2015)
5. DMRH—Disaster Management Reference Handbook. Center for Excellence in Disaster Management and Humanitarian Assistance, Havana Cuba (2017)
6. Medeot, R., Zordan, T.: EN15129 the new performance-oriented European Standard on anti-seismic devices. In: 16th World Conference on Earthquake Engineering, 16WCEE 2017, ID3820, Santiago, Chile (2017)
7. Polese, M., Gaetani d'Aragona, M., Di Ludovico, M., Prota, A.: Sustainable selective mitigation interventions towards effective earthquake risk reduction at the community scale. Sustain. J. **10**, 2894 (2018)
8. CEN: EN 15129. Anti-Seismic Devices, European Committee for Standardization, Brussels, Belgium ((2009))
9. Zordan, T., Medeot, R.: Revision of EN15129. In: IALCCE 2020—The 7th International Symposium on Life-Cycle Civil Engineering, Shanghai, China (2020)

Multi-storey Building Retrofit by ADAS-Equipped Braces

Terenzi Gloria[1](✉), Sorace Stefano[2], Melani Damiano[3], and Fuso Elena[2]

[1] Department of Civil and Environmental Engineering, University of Florence, 50139 Florence, Italy
gloria.terenzi@unifi.it
[2] Polytechnic Department of Engineering and Architecture, University of Udine, 33100 Udine, Italy
{stefano.sorace,elena.fuso}@uniud.it
[3] Department of Civil and Environmental Engineering, University of Florence, 50139 Florence, Italy
damiano.melani@stud.unifi.it

Abstract. An alternative passive energy-based retrofit design is proposed for a reinforced concrete building, seismically retrofitted in 2013 and damaged by the 2016 Central Italy earthquake. An evaluation analysis carried out by referring to the pre-2013 conditions shows unsafe stress states in most structural members and severe damage in the masonry partitions and perimeter infills. The alternative retrofit strategy of the building, which consists in the incorporation of a dissipative bracing system equipped with pressurized fluid viscous spring-dampers, allows attaining a substantial seismic performance improvement. This is assessed by a safe response of all columns and beams, and at most a very slight damage in the masonry panels, in post-intervention conditions.

Keywords: ADAS dampers · Energy-based design · Seismic assessment · Seismic retrofit · Frame buildings · Reinforced concrete structures

1 Introduction

Incorporation of dissipative bracing systems is an emerging seismic retrofit strategy for frame structures. Among the several types of devices currently adopted as passive protection elements, Added Damping and Stiffness (ADAS) steel dissipaters have a well-established tradition [1–3]. This is a consequence of their plain working principle, based on the elastic-plastic behaviour of the constituting plates, as well as of their relatively easy installation. In spite of this, the design of ADAS dampers is not simple, because it requires a proper balance between the addition of energy dissipation and horizontal translational stiffness. The former allows remarkably reducing the plastic demand on the structural members. The latter is always beneficial in terms of lateral displacements but, as long as the dampers respond elastically, it causes a significant increase in base shear, and thus in the stress states of the frame elements and the foundations.

G. P. Cimellaro (Ed.): WCSI 2022, LNCE 309, pp. 532–545, 2023.
https://doi.org/10.1007/978-3-031-21187-4_45

In view of this, an energy-based pre-sizing criterion of dampers was recently proposed by the first two authors [4, 5], aimed at achieving a joined control of stress states and displacements. According with this criterion, the total number of plates constituting ADAS devices is tentatively fixed by: a. pre-estimating the energy dissipation capable of providing a target drop of base shear in retrofitted conditions; b. computing the energy dissipation by assuming a maximum displacement of the dissipaters calibrated on a target reduction of storey drifts.

In this paper a new energy-based pre-sizing procedure is formulated, where the energy dissipation capacity of the dampers is evaluated by expressly taking into account the reduction of the fundamental vibration period of the structure caused by the bracing system-related increase of horizontal stiffness.

This new sizing criterion is demonstratively applied to a representative case study, i.e. a 6-storey residential building with reinforced concrete structure situated in the municipality of L'Aquila, Abruzzo, Italy. Based on the results of a seismic assessment analysis initially carried out, three different installations of the ADAS dissipaters in plan and along the height of the building are comparatively examined, with the aim of providing practical suggestions for the best placement of the protective system for target performance levels.

2 Case-Study Building

The case study building is located in L'Aquila, Abruzzo, Italy, and was designed according with the 1986 edition of the Italian Seismic Standards. It has a (21.05 × 11.45) m^2 sized rectangular plan (Fig. 1, referred to the lower three storeys), and is articulated in six above-ground storeys, with inter-storey heights equal to 3.4 m (first storey) and 3.04 m (remaining ones).

The structure is constituted by a reinforced concrete frame skeleton. Columns have the following cross sections: 600 × 300 mm^2 (highlighted in red in Fig. 1), 550 × 300 mm^2 (blue), 500 × 300 mm^2 (green) and 400 × 300 mm^2 (black) on the three lower storeys, and 300 × 300 mm^2, on the three upper storeys. Perimeter beams have out-of-depth sections sized 300 × 500 mm^2, and internal beams have in-depth sections sized 800 × 300 mm^2 (2–5 alignments in Fig. 1) and 600 × 300 mm^2 (B and C alignments), at all floors.

A view of the finite element model of the structure, generated by SAP2000NL software [6], is displayed in Fig. 2. A modal analysis carried out by this model highlighted two main translational modes in current state, along *X* and *Y* axis in plan, with vibration periods equal to 1.21 s and 1.18 s, and effective masses of 84% and 82.7%, respectively.

Based on the data extracted from the available technical documentation, the following mechanical properties were assumed for the materials: mean cubic compressive strength of concrete equal to 25 MPa; yield stress of the reinforcing steel bars equal to 430 MPa. These were assumed as reference values both for the finite element and stress check analyses.

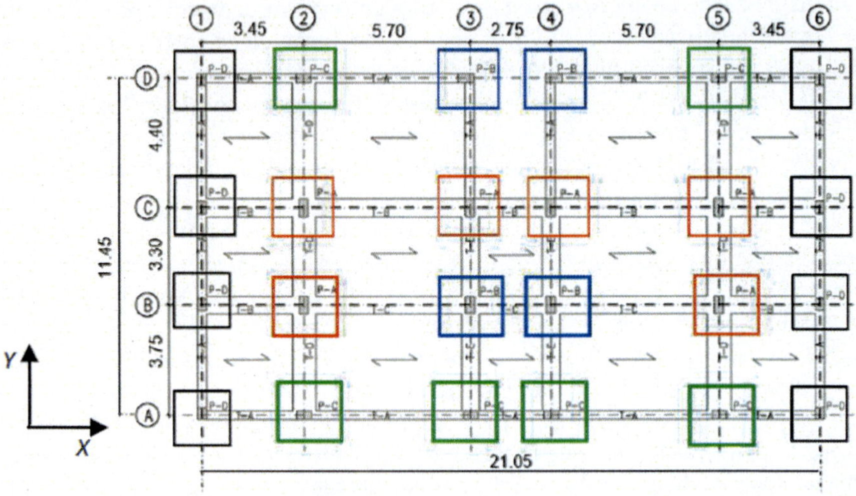

Fig. 1. Structural plan of the building.

3 Seismic Assessment of the Building

The seismic assessment study was developed via time-history analysis by using in input a set of seven accelerograms for the two axes in plan and the vertical axis, generated from the pseudo-acceleration response spectra plotted in Fig. 3. Therein, the three spectral graphs for the horizontal and vertical components are referred to the Serviceability Design Earthquake (SDE), Basic Design Earthquake (BDE) and Maximum Considered Earthquake (MCE) hazard levels, for C-type soil and T1-type topographic category. The corresponding peak ground acceleration values are: 0.156 g (SDE), 0,347 g (BDE), and 0.407 g (MCE), for the horizontal components; 0.045 g (SDE), 0.180 g (BDE) and 0.261 g (MCE), for the vertical one.

The results of the analysis in current conditions are synthesized in Tables 1 and 2. Table 1 reports the maximum storey shears along the two axes, V_x and V_y, and their ratios, ρ_{sx} and ρ_{sy}, to relevant strength values, V_{xR} and V_{yR}. Table 2 lists the maximum inter-storey drifts, u_x and u_y, and their ratios, ρ_{ux} and ρ_{uy}, to the Immediate Occupancy drift performance limit for buildings with masonry infills and partitions, Id_{IO}, fixed at 0.5% of the inter-storey height h by the Italian Technical Standards [7].

As shown in Table 1, ρ_s ratios greater than 1 are found for the two lower storeys at the SDE and for the four lower storeys at the BDE, reaching values of about 3 for the latter. ρ_s values below 1 come out only for the top storey at the MCE, the maximum values for which are equal to 3.66.

4 Dissipative Bracing-Based Retrofit—Sizing Criterion and Application to Different Design Solutions

The seismic assessment study was developed via time-history analysis by using in input a set of seven accelerograms for the two axes in plan and the vertical axis.

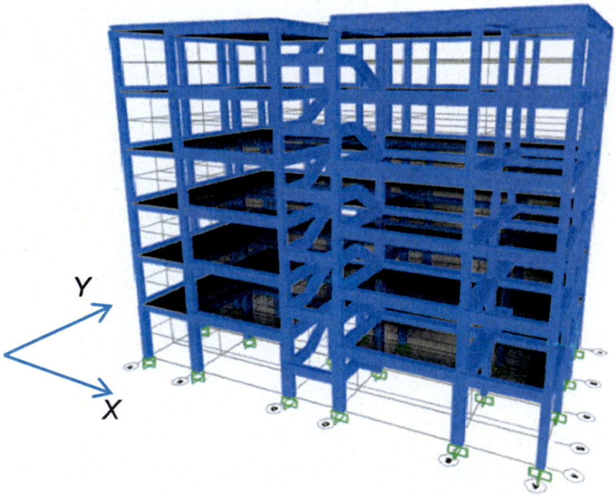

Fig. 2. View of the finite element model of the structure.

As mentioned in the Introduction, the retrofit hypothesis proposed herein consists in the installation of a dissipative bracing system incorporating ADAS steel dampers with triangular shape (also named T-ADAS in the literature) as protective devices. The geometry of a plate and the schematic hysteretic cycle of a damper are traced out in Fig. 4.

The preliminary sizing procedure of the dampers proposed in this study is based on a quick estimation of the protective system energy dissipation capacity by which a pre-established reduction of lateral displacements and base shear is attained. For frame structures with a substantially regular geometry in plan and elevation, this evaluation is carried out in terms of spectral quantities, by referring to the periods of the first translational modes along the two coordinate axes in plan (which a predominant portion of modal masses is associated with, in the above-mentioned hypothesis of structural regularity). Both in current and retrofitted conditions, these periods are assumed to be included in the wide period interval T_C-T_D of the response spectra that corresponds to the constant pseudo-velocity branch, where the fundamental vibration periods of low through mid-to-high rise reinforced concrete frame structures are always situated.

For each one of the two axes in plan, named T_{IN} the fundamental period in current state, the procedure starts by fixing a tentative reduction, $\Delta S_{D\text{-}}$, of the spectral displacement S_D computed for T_{IN}. Then, named $S_{V,cost}$ the value of the horizontal branch of the pseudo-velocity spectrum, the period in retrofitted conditions, T_{FIN}, given by the difference between T_{IN} and the period reduction ΔT caused by $\Delta S_{D\text{-}}$, can be evaluated as follows:

$$T_{FIN} = T_{IN} - \Delta T = T_{IN} - \frac{2\pi \Delta S_D}{S_V} \tag{1}$$

As illustrated in Fig. 5, a rise (ΔS_{A+}) in the initial pseudo-acceleration ordinate comes out when passing from T_{IN} to T_{FIN}, as a consequence of the stiffening effect of

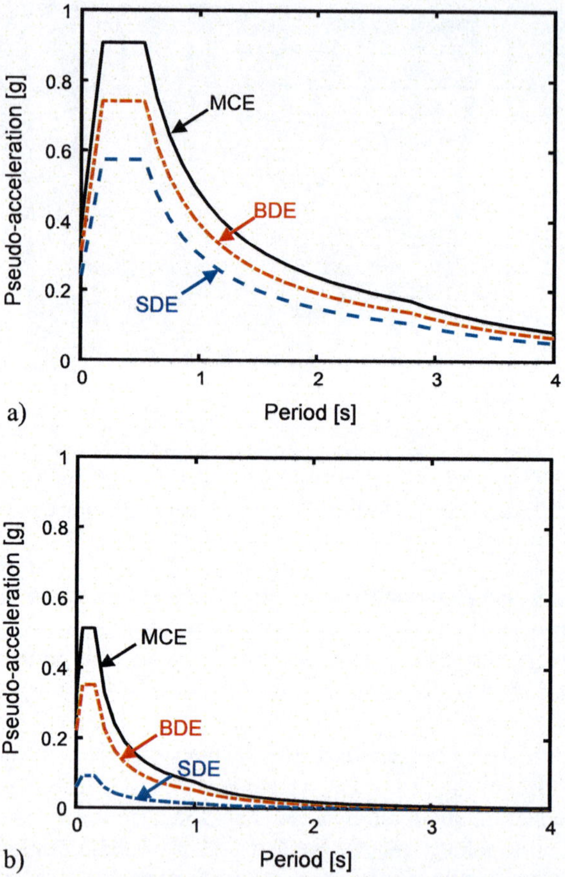

Fig. 3. Horizontal (a) and vertical (b) pseudo-acceleration elastic response spectra for L'Aquila—SDE, BDE and MCE hazard levels.

the dissipative bracing system. The supplemental damping contribution of the latter must guarantee a drop in the spectral ordinate, $\Delta S_{A\text{-},diss}$, significantly exceeding ΔS_{A+} and capable of reaching the target $S_A(T_{FIN})$ value in retrofitted conditions. The corresponding reduction in base shear, ΔV, is obtained by multiplying $\Delta S_{A\text{-},diss}$ by the seismic mass M of the building.

Hence, the tentative energy dissipation capacity to be assigned to the protective system is computed as:

$$E_D = 4\Delta V \Delta S_D \quad (2)$$

Based on the values calculated by Eq. (2) for the two axes in plan, $E_{d,x}$, $E_{d,y}$, the corresponding total numbers of plates, $n_{p,x}$, $n_{p,y}$, are evaluated by dividing $E_{d,x}$, $E_{d,y}$ by the maximum energy that each plate can dissipate in a cycle bounded by the maximum estimated displacement, $E_{d,pl}$, multiplied by 11 to take into account also the energy dissipated in the remaining cycles, characterized by smaller amplitudes [4, 8]. In the

Table 1. Current conditions: maximum storey shear and ρ_S values.

Storey	S1	S2	S3	S4	S5	S6
V_{xR} (kN)	1996	1814	1667	1696	1696	1696
V_{yR} (kN)	2337	1912	1813	1696	1696	1696
SDE						
V_x (kN)	2772	1846	1485	1157	827	378
V_y (kN)	2571	1520	1379	1949	752	374
ρ_{sx}	1.39	1.02	0.89	0.68	0.42	0.22
ρ_{sy}	1.10	0.79	0.76	0.62	0.44	0.22
BDE						
V_x (kN)	5967	4245	3155	2419	1638	725
V_y (kN)	6193	3654	3288	2531	1664	767
ρ_{sx}	2.99	2.34	1.89	1.43	0.96	0.43
ρ_{sy}	2.65	1.91	1.81	1.49	0.98	0.45
MCE						
V_x (kN)	7315	5259	3934	3002	2053	590
V_y (kN)	7383	4403	3857	2993	2074	975
ρ_{sx}	3.66	2.89	2.36	1.77	1.21	0.52
ρ_{sy}	3.16	2.30	2.12	1.76	1.22	0.57

$E_{d,pl}$ calculation the maximum plate displacement, d_{max}, is tentatively put as equal to ID_{IO}, in order to obtain a substantial constraint of drifts up to the highest normative earthquake level considered in the analysis (BDE or MCE).

The diagonal braces are sized by evaluating the increase in lateral stiffness of the structure, ΔK, related to the ΔT transition from T_{IN} and T_{FIN} induced by the incorporation of the protective system, as follows:

$$\Delta K = \frac{4\pi^2 M}{\Delta T^2} \tag{3}$$

For the case study application, by referring to the nomenclature in Fig. 4, the geometric sizes selected for the plates are: $H = 150$ mm, $t = 15$ mm, and $B = 75$ mm. The constituting steel is type S275, with yield stress and tensile strength equal to $f_{yk} = 275$ N/mm^2 and $f_{tk} = 430$ N/mm^2, respectively. By assuming a target $S_D(T_{FIN})$ value of the building equal to 0.5% of its total height, i.e. $S_D(T_{FIN}) = 93$ mm, by applying relations (1–3), the following demands in terms of plate numbers are computed for the two references axes:

X axis—$T_{IN,x} = 1.21$ s, $S_{D,x}(T_{IN,x}) = 159.4$ mm, $S_{V,x} = 665.7$ mm/s, $\Delta S_{D\text{-},x} = (159.4–93)$ mm $= 66.4$ mm, $T_{FIN,x} = 0{,}583$ s, $S_{A,x}(T_{FIN,x}) = 0.723$ g, $M = 1675.4$ kN/g, $E_{d,x} = 1632.6$ KNm, $E_{d,plate,x} = 3.58$ KNm;

Table 2. Current conditions: maximum inter-storey drift and ρ_u values.

Storey	S1	S2	S3	S4	S5	S6
u_R (mm)	17.0	15.3	15.3	15.3	15.3	15.3
SDE						
u_x (mm)	13.4	14.4	13.7	11.7	7.7	3.6
u_y (mm)	9.8	11.3	11.3	10.3	7.0	3.7
ρ_{ux}	0.78	0.94	0.89	0.76	0.50	0.23
ρ_{uy}	0.57	0.74	0.74	0.67	0.46	0.24
BDE						
u_x (mm)	28.1	36.9	41.8	35.5	24.8	11.2
u_y (mm)	23.6	32.5	33.6	27.5	12.1	9.3
ρ_{ux}	1.65	2.41	2.73	2.32	1.62	0.73
ρ_{uy}	1.39	2.12	2.19	1.79	0.79	0.61
MCE						
u_x (mm)	34.2	48.0	47.5	41.2	31.4	14.6
u_y (mm)	31.3	44.6	50.9	40.5	22.1	17.7
ρ_{ux}	2.01	3.14	3.01	2.69	2.05	0.95
ρ_{uy}	1.84	2.91	3.32	2.65	1.44	1.15

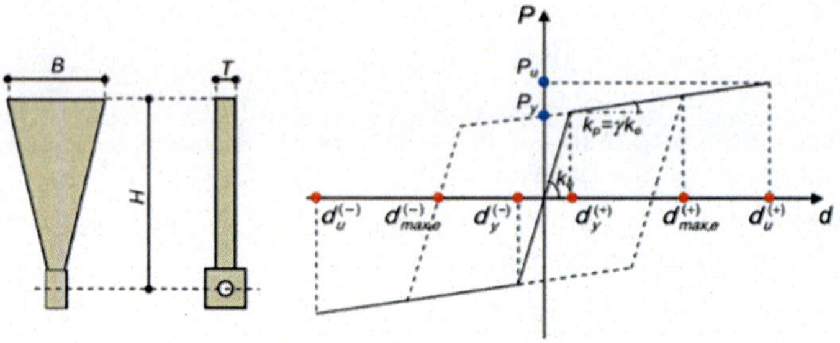

Fig. 4. Geometry of a triangular plate and hysteretic cycle of a device.

Y axis—$T_{IN,y}$ = 1.18 s; $S_{D,y}(T_{IN,y})$ = 155.4 mm, $S_{V,y}$ = 665.7 mm/s, $\Delta S_{D\text{-},y}$ = (155.4–93) mm = 62.4 mm, $T_{FIN,y}$ = 0,591 s, $S_{A,y}(T_{FIN,y})$ = 0.72 g, $E_{d,y}$ = 1481 KNm; $E_{d,plate,y}$ = 3.58 KNm.

The ratios of $E_{d,x}$ to $E_{d,plate,x}$, and $E_{d,y}$ to $E_{d,plate,y}$ result in the following numbers of plates in *X* and *Y*: n_{px} = 455 and n_{py} = 414. Thus, the total number of plates for the building, $n_{p,tot}$, is equal to 869.

Based on this preliminary sizing, three different installation solutions—named RS1, RS2, RS3 in the following—were designed by varying the distribution in plan and elevation of the dissipative braces, as described below. In all cases, no dampers are

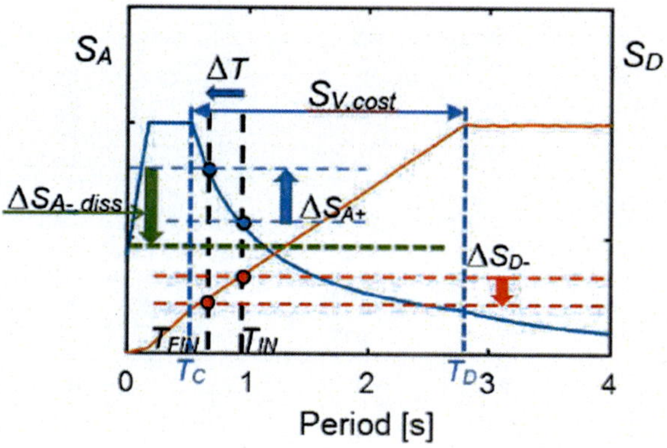

Fig. 5. Quantities involved in the sizing procedure.

installed on the upper floor, because of the elastic response of relevant structural members in current state up to the BDE. By complying with Eq. (3), a circular tubular section with a 150 mm diameter, 4.6 mm thick in *X* and 3 mm thick in *Y*, was selected for the supporting steel braces.

RS1 retrofit solution. 12 plate-dampers placed in 6 bays along *X*, and 8 in *Y*, giving rise to $n_{p,x} = 360$, $n_{p,y} = 480$, and $n_{p,tot} = 840$. This choice is aimed at minimizing the architectural impact of the intervention in *X*. Figure 6 shows the positions in plan of the dissipative braces along *X* (evidenced in yellow), and *Y* (in red).

RS2 retrofit solution. 12 plate-dampers placed in 8 bays both along *X* and *Y*, with $n_{p,x} = n_{p,y} = 480$, and $n_{p,tot} = 960$, nearly coinciding with the tentative plate numbers calculated by the sizing procedure.

RS3 retrofit solution. 12 plate-dampers on the first storey, 14 on the second and third, 10 on the fourth and 8 on the fifth in 8 bays for each direction, resulting in $n_{p,x} = n_{p,y} = 464$, and $n_{p,tot} = 928$. This distribution is approximately proportional to the first modal shapes of the structure in *X* and *Y*. Figure 7 shows the positions in plan of the dissipative braces along *X* (in yellow), and *Y* (in red) mutually adopted for the RS2 and RS3 solutions.

4.1 RS1—Results

The two fundamental translational modes in *X* and *Y* are kept, with modified periods of 0.67 s and 0.64 s, and effective masses of 73.5% and 80.7%, respectively. The mass reductions are due to a little transfer to the torsional mode contributions caused by the asymmetric installation of the bracing system in plan (along 6 bays in *X*, and 8 in *Y*, as stated above). The results of the time-history analyses are recapitulated in Tables 3 and 4. A generalized improvement of seismic performance is observed, except for the first storey, where the energy dissipation benefits do not compensate for the response increase caused by the stiffening effects of the protective system. Moreover, the even greater benefits obtained in *X*, in spite of the lower number of plates assumed for this direction, are as a consequence of the greater amount of energy dissipated, as compared to *Y*.

Table 3. Retrofitted conditions—RS1: maximum storey shear and ρ_s values.

Storey	S1	S2	S3	S4	S5	S6
V_{xR} (kN)	1996	1814	1667	1696	1696	1696
V_{yR} (kN)	2337	1912	1813	1696	1696	1696
SDE						
V_x (kN)	1975	1085	764	552	342	98
V_y (kN)	2278	1106	877	585	380	141
ρ_{sx}	0.98	0.59	0.46	0.32	0.20	0.06
ρ_{sy}	0.97	0.58	0.48	0.34	0.22	0.08
BDE						
V_x (kN)	4689	2841	1979	1448	958	223
V_y (kN)	5331	3236	2578	1725	1128	402
ρ_{sx}	2.35	1.56	1.18	0.85	0.56	0.13
ρ_{sy}	2.28	1.69	1.42	1.01	0.66	0.23
MCE						
V_x (kN)	6326	3844	2357	1917	1233	275
V_y (kN)	6614	3797	3022	2018	1317	456
ρ_{sx}	3.17	2.12	1.41	1.13	0.73	0.16
ρ_{sy}	2.96	1.98	1.66	1.19	0.77	0.27

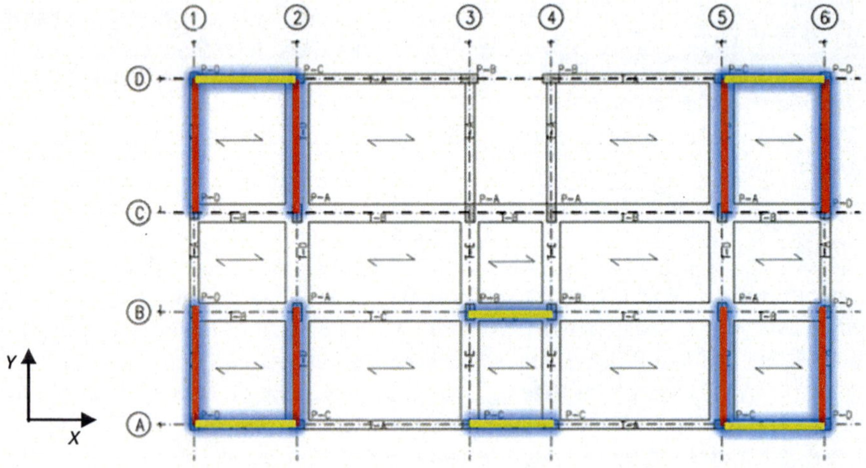

Fig. 6. RS1 retrofit solution: positions of the dissipative bracing system alignments in plan.

Table 4. Retrofitted conditions—RS1: maximum inter-storey drift and ρ_u values.

Storey	S1	S2	S3	S4	S5	S6
u_R (mm)	17.0	15.3	15.3	15.3	15.3	15.3
SDE						
u_x (mm)	8.8	8.6	7.4	5.5	3.6	1.6
u_y (mm)	9.6	9.5	7.9	6.5	4.8	2.3
ρ_{ux}	0.52	0.56	0.48	0.36	0.23	0.10
ρ_{uy}	0.56	0.62	0.52	0.42	0.31	0.15
BDE						
u_x (mm)	23.9	23.4	20.6	15.8	10.3	4.3
u_y (mm)	26.6	26.3	22.1	18.6	13.5	6.5
ρ_{ux}	1.40	1.53	1.35	1.03	0.67	0.28
ρ_{uy}	1.56	1.72	1.44	1.22	0.88	0.42
MCE						
u_x (mm)	29.5	28.9	25.6	19.5	12.9	5.2
u_y (mm)	31.5	31.3	26.2	21.9	15.9	7.6
ρ_{ux}	1.73	1.88	1.67	1.27	0.84	0.33
ρ_{uy}	1.85	2.04	1.71	1.43	1.04	0.49

4.2 RS2—Results

The two fundamental translational modes in *X* and *Y* have periods of 0.622 s and 0.619 s in *X* and *Y*, and effective masses of 72.2% and 76.5%, respectively, highlighting small reductions as compared to the RS1 solution, but virtually equal values of periods and closer values of masses for the two axes.

As highlighted in Tables 5 and 6, the time-history response in terms of stress states is virtually coincident with the RS1 one in *Y*, whereas it is slightly decreased (SDE), increased (BDE), or very similar (MCE), in *X*. At the same time, drifts are lowered up to about 10% in *Y* and 15% in *X*.

4.3 RS3—Results

The two fundamental translational modes have periods of 0.621 s and 0.618 s in *X* and *Y*, and effective masses of 75.3% and 77.1%, respectively, i.e., practically coincident periods with the RS2 solution, and an about 3% increased mass in *X* and less than 1% in *Y*, as compared to it.

As illustrated by Tables 7 and 8, the first mode-proportional distribution of dampers along the height provides, with a 3.3% smaller number of plates than for RS2, an additional regularization of the demand on the various storeys due to the highest dissipation offered by the protective system on the second and third ones, in terms both of stress states and drifts.

Table 5. Retrofitted conditions—RS2: maximum storey shear and ρ_s values.

Storey	S1	S2	S3	S4	S5	S6
V_{xR} (kN)	1996	1814	1667	1696	1696	1696
V_{yR} (kN)	2337	1912	1813	1696	1696	1696
SDE						
V_x (kN)	1711	1020	691	490	316	62
V_y (kN)	2262	1106	849	588	384	140
ρ_{sx}	0.86	0.56	0.41	0.29	0.19	0.04
ρ_{sy}	0.97	0.58	0.47	0.35	0.23	0.08
BDE						
V_x (kN)	5039	3064	2171	1572	1012	164
V_y (kN)	5259	3233	2581	1730	1129	407
ρ_{sx}	2.52	1.68	1.30	0.93	0.59	0.09
ρ_{sy}	2.25	1.69	1.42	1.02	0.66	0.24
MCE						
V_x (kN)	6224	3757	2662	1984	1281	192
V_y (kN)	6676	3850	3068	2049	1333	449
ρ_{sx}	3.12	2.07	1.59	1.17	0.75	0.11
ρ_{sy}	2.86	2.01	1.69	1.2	0.78	0.26

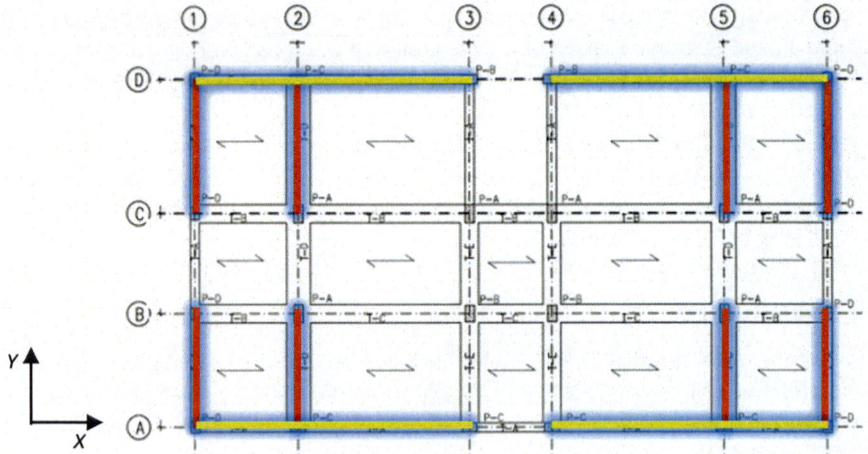

Fig. 7. RS2 and RS3 retrofit solutions: positions of the dissipative bracing system alignments in plan.

Table 6. Retrofitted conditions—RS2: maximum inter-storey drift and ρ_u values.

Storey	S1	S2	S3	S4	S5	S6
u_R (mm)	17.0	15.3	15.3	15.3	15.3	15.3
SDE						
u_x (mm)	7.4	7.1	6.1	4.4	2.7	0.9
u_y (mm)	8.7	8.5	6.1	4.4	2.7	0.9
ρ_{ux}	0.43	0.46	0.39	0.28	0.17	0.06
ρ_{uy}	0.51	0.55	0.47	0.37	0.26	0.14
BDE						
u_x (mm)	21.7	21.5	18.9	14.0	8.5	2.5
u_y (mm)	25.6	24.9	21.0	17.0	12.1	6.6
ρ_{ux}	1.27	1.4	1.23	0.91	0.55	0.16
ρ_{uy}	1.50	1.63	1.37	1.11	0.79	0.43
MCE						
u_x (mm)	26.8	26.4	23.2	17.6	10.9	3.1
u_y (mm)	30.5	29.6	25.0	20.3	14.3	7.8
ρ_{ux}	1.57	1.72	1.51	1.15	0.71	0.2
ρ_{uy}	1.79	1.93	1.63	1.32	0.93	0.51

5 Conclusions

The energy-based sizing procedure for ADAS dampers proposed in this paper, based on simple spectral relations for the original structure, schematized like a SDOF system in the hypothesis of a substantial regularity in plan and elevation, was demonstratively applied to the seismic retrofit of a multi-storey reinforced concrete residential building.

Starting from the estimated minimum number of plates needed to reach the target performance, three different solutions for the installation of the protective system were designed and comparatively evaluated.

Although the assessed performance was similar for the three hypotheses, both in terms of stress states and inter-storey drifts, the analyses highlighted that the most effective distribution of plates is obtained by conceiving it as proportional to the lateral deformation of the original structure.

This is underlined by the RS3 retrofit solution, which complies with this concept. Indeed, although including a smaller total number of plates as compared to RS2, characterized by a uniform distribution of plates in elevation, RS3 guarantees the best combined performance, as it constrains further the response of the second and third storey, where the highest drifts resulted in original conditions. At the same time, very slight differences are observed between RS2 and the other uniform-distribution intervention, RS1, in spite of the fact that the latter incorporates a 25% smaller number of plates in X, as a consequence of the greater energy dissipated by the ADAS dampers in this direction.

Table 7. Retrofitted conditions—RS3: maximum storey shear and ρ_s values.

Storey	S1	S2	S3	S4	S5	S6
V_{xR} (kN)	1996	1814	1667	1696	1696	1696
V_{yR} (kN)	2337	1912	1813	1696	1696	1696
SDE						
V_x (kN)	1636	938	638	521	351	52
V_y (kN)	2028	1032	818	581	385	121
ρ_{sx}	0.81	0.52	0.38	0.34	0.21	0.03
ρ_{sy}	0.87	0.54	0.45	0.34	0.22	0.07
BDE						
V_x (kN)	5058	2948	2086	1659	1102	136
V_y (kN)	5222	3021	2422	1722	1140	354
ρ_{sx}	2.53	1.62	1.25	0.98	0.65	0.20
ρ_{sy}	2.23	1.58	1.33	1.01	0.67	0.21
MCE						
V_x (kN)	6222	3601	2534	2055	1406	162
V_y (kN)	6578	3570	2852	2015	1323	384
ρ_{sx}	3.12	1.98	1.52	1.21	0.83	0.09
ρ_{sy}	2.81	1.86	1.57	1.18	0.78	0.22

Table 8. Retrofitted conditions—RS3: maximum inter-storey drift and ρ_u values.

Storey	S1	S2	S3	S4	S5	S6
u_R (mm)	17.0	15.3	15.3	15.3	15.3	15.3
SDE						
u_x (mm)	7.3	7.0	5.8	4.7	3.2	1.0
u_y (mm)	8.6	8.2	6.9	5.8	4.2	2.2
ρ_{ux}	0.43	0.46	0.38	0.31	0.21	0.06
ρ_{uy}	0.50	0.53	0.45	0.38	0.27	0.14
BDE						
u_x (mm)	21.7	20.8	18.3	14.6	9.3	2.6
u_y (mm)	25.4	23.9	20.2	17.0	12.2	6.4
ρ_{ux}	1.27	1.36	1.19	0.95	0.60	0.17
ρ_{uy}	1.49	1.56	1.32	1.10	0.79	0.42
MCE						
u_x (mm)	26.6	25.4	22.3	17.9	11.9	3.1
u_y (mm)	30.6	28.7	24.3	20.4	14.7	7.6
ρ_{ux}	1.56	1.66	1.46	1.17	0.77	0.20
ρ_{uy}	1.80	1.87	1.59	1.33	0.96	0.49

Acknowledgements. Financial support from ReLUIS-DPC Project 2022–2023, Project-WP15: Normative contribution for Isolation and Dissipation, is gratefully acknowledged.

References

1. Aiken, I.D., Nims, D.K., Whittaker, A.S., Kelly, J.M.: Testing of passive energy dissipation systems. Earthq. Spectra **9**(3), 335–370 (1993)
2. Hanson, R.D., Soong, T.T.: Seismic design with supplemental energy dissipation devices. Publication MNO-8 EERI–Earthquake Engineering Research Center: Oakland, CA (2001)
3. Teruna, D.R., Majid, T.A., Budiono, B.: Experimental study of hysteretic steel damper for energy dissipation capacity. Adv. Civil Eng., Article ID 631726, 12 p (2015). https://doi.org/10.1155/2015/631726
4. Sorace, S., Terenzi, G., Mori, C.: Passive energy dissipation-based retrofit strategies for R/C frame water towers. Eng. Struct. **106**, 385–398 (2016)
5. Terenzi, G.: Energy-based design criterion of dissipative bracing systems for seismic retrofit of framed structures. Appl. Sci. **8**(2), 1–22 (2018)
6. CSI, 2021: SAP2000NL. Theoretical and users' manual, Release 22.03. Computers & Structures Inc., Berkeley, CA, USA (2021)
7. Italian Council of Public Works: NTC 2018 technical standards on constructions [in Italian] Rome, Italy (2018)
8. Gandelli, E., De Domenico, D., Quaglini, V.: Cyclic engagement of hysteretic steel dampers in braced buildings: a parametric investigation. Bull. Earthq. Eng. **19**(12), 5219–5251 (2021). https://doi.org/10.1007/s10518-021-01156-3

The Effect of Nonlinear Response of the Primary System in Nonconventional TMDs

Francesco Esposito, Diana Faiella, Mario Argenziano, Giuseppe Brandonisio, and Elena Mele(✉)

Department of Structures for Engineering and Architecture, University of Naples Federico II, Piazzale Tecchio 80, 80125 Naples, Italy
{francesco.esposito8,diana.faiella,mario.argenziano, giuseppe.brandonisio,elenmele}@unina.it

Abstract. Tuned Mass Dampers (TMD) have been widely used in the passive vibration control of engineering structures, both under wind and seismic loads. Under wind loads, a proper frequency tuning ensures a significant response reduction of the primary structure. While under wind loads the structure is likely to remain in the elastic range of behaviour, when large earthquakes occur it can experience inelastic deformations and damage, with a consequent de-tuning effect. In this paper, nonconventional TMDs with large mass ratios, realized through Interstory Isolation Systems (IIS), are considered and the influence of the inelastic behaviour of the primary system on the overall structure response is analysed. In detail, IIS is employed for retrofitting existing masonry buildings through a vertical extension, isolated at its base and built up on the roof of the existing structure. Pivot-type hysteretic models are used to grasp the mechanical behaviour of the masonry structure. Based on a case-study building, time history analyses on 3D Finite Element models are implemented. The results of nonlinear time history analyses show that in the building with the vertical extension both displacements and hysteretic energy dissipated by the masonry piers, are lower than in the as-is building, confirming that, although a de-tuning effect can occur, the inelastic engagement of the original structure is significantly reduced.

Keywords: Nonconventional TMD · Masonry structure · Masonry nonlinearities · Steel vertical addition · Seismic retrofit

1 Introduction

Tuned Mass Dampers (TMD) with large mass are particularly suitable for reducing the seismic response of buildings [1, 2] and can be realized by converting an upper portion of the building into the TMD itself [3, 4]. For this purpose, the upper portion should be connected to the lower part by means of a seismic isolation layer, designed for activating the desired dynamic interaction between the two structural portions, i.e. for tuning the relevant fundamental frequencies. Such systems are appointed as Intermediate (or Inter-story or Mid-story) Isolation System (IIS) or nonconventional TMDs.

G. P. Cimellaro (Ed.): WCSI 2022, LNCE 309, pp. 546–556, 2023.
https://doi.org/10.1007/978-3-031-21187-4_46

The IIS technique is very interesting when the needs of seismic improvement and vertical extension should be contemporarily satisfied, as happens for the built heritage of several European cities located in seismic zones. In this case, a vertical extension, isolated at its base, can be built on the roof of the existing building. Actual applications of this approach mainly refer to reinforced concrete buildings [5, 6]. However, isolated vertical extensions seem particularly effective for masonry buildings. In fact, other than exploiting the mass damping effect exerted by the vertical extension on the existing building, two additional benefits can be obtained: (1) additions of several extra-floors thanks to the masonry's overstrength for bearing an increase of gravity load, since historical masonry structures usually have compression strength demand to capacity ratio quite low, between one-fifth and one-tenth [7, 8]; (2) stabilizing effect of the additional compression that reduces the wall tensile stresses due to bending and shear effects of lateral loads.

This approach has been already proposed by the authors in [9, 10], by examining some existing masonry buildings through spectrum-compatible time history analyses in which a linear behaviour is assumed for the masonry structure. However, when large earthquakes occur, the overall structure can experience inelastic behaviour. In such a case, the effectiveness of TMDs attached to inelastic primary structures is affected by both the intensity and duration of the dynamic excitation, and can be strongly reduced by the de-tuning effect due to the elongation of the fundamental period of the damaged structure [11]. Furthermore, for assessing the seismic behaviour of the primary structure not only the peak displacement response but also the accumulated damage, i.e. the hysteretic energy absorption [12–14], should be accounted for.

In this paper a masonry aggregate extended by means of a steel isolated structure is assumed as case study. Optimal dynamic properties of several nonconventional TMDs are selected by means of linear parametric analyses performed through response spectrum analyses on 1D lumped mass models. Then, nonlinear dynamic analyses are carried out on 3D Finite Element models. The hysteretic behavior of the masonry structure is represented through a Pivot-type model [15]. The TMD effectiveness in reducing the seismic response, the plastic engagement of masonry elements, and the overall damage are evaluated by comparing the extended configurations to the original structure in terms of displacement, walls' inelastic demand, accumulated hysteretic energy.

2 The Case Study

2.1 Brief Description of the Case Study

The case study is an existing aggregate of two-story masonry buildings extended through a one-story steel building, isolated at its base, and built up on the top of the existing structure (Fig. 1). To create a unique level of the isolation layer, the height difference between the buildings of the aggregate is compensated through a steel structure, added on the roof of the building on the left (Fig. 1a) prior to build the additional story. The structure, located in Pozzuoli (Naples, Italy), was built in the 16th century, and is utilized for residential and tertiary occupancy; the new steel addition is utilized for tertiary occupancy. Front view, floor plans and dimensions of the case study are also provided in Fig. 1.

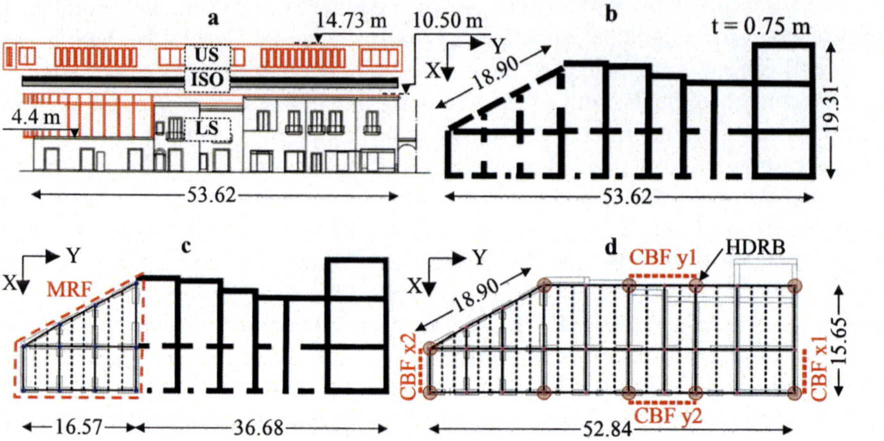

Fig. 1. Case study building: (a) front view; (b) first floor plan; (c) second floor plan, (d) schematic representation of the isolation layer.

The mechanical properties assumed for the tuff masonry material are: weight 16 kN/m^3, Young modulus 1080 MPa, shear modulus 360 MPa, compression strength 3.0 MPa [16]. The filling structure is a spatial Moment Resisting Frame (MRF), designed to not alter the dynamics of the masonry structure; the superstructure is a Concentric Braced Frame (CBF) with two perimeter braced bays for each direction. These steel structures are made of steel grade S275 (f_{yk} = 275 MPa, f_{yd} = 262 MPa); the cross-sections adopted are listed in Table 1. Then, ten high damping rubber bearings (HDRB) are placed at intersections of orthogonal masonry walls.

Table 1. Cross sections of the steel structures.

	MRF		CBF		
	Columns	Beams	Columns	Beams	Diagonals
GL	CHS 219.1 × 16	IPE 240/360	CHS 219.1 × 16	IPE 240/360 RHS 900 × 450 × 4	
LL	CHS 219.1 × 16	IPE 360	CHS 219.1 × 16	IPE 360	HEB 160

Key: GL = Gravity Load; LL = Lateral Load; CHS = Circular Hollow Section; RHS = Circular Hollow Section. The dimensions are reported in [mm].

The case study is a nonconventional TMD or intermediate isolation system (IIS), in which the structure can be ideally subdivided into three parts (Fig. 1a): the lower structure (LS), i.e. the masonry aggregate plus the filling MRF; the isolation system (ISO); the upper structure (US), i.e. the CBF. Two design configurations are adopted for the isolated vertical addition, in which only the properties of the isolation system vary (Fig. 1d). They are selected based on wide parametric analyses reported in a previous paper [10], where the dynamic properties (mass and stiffness) of the isolated vertical addition and the fundamental period of the existing building are assumed as design variables. The main results of the parametric analysis referring to the case study are briefly recalled in the following Subsect. 2.2.

2.2 Choice of the Design Solutions Through Simplified Parametric Analysis

The design configurations of the nonconventional TMDs are selected based on the results of the parametric analyses mentioned above, i.e. response spectrum analyses (RSA) carried out on lumped mass (3DOF) models. The 3DOF IIS model (Fig. 2a) represents the masonry building extended through the steel structure equipped with seismic isolation system at its base. The mass, stiffness, and damping constants of each DOF are respectively indicated with *m*, *k*, and *c*; the subscripts *LS*, *ISO*, and *US* refer to the DOF of the lower structure, isolation system, and upper structure. The SDOF LS model is also considered for comparison purpose, with properties derived according to the N2 method [17] applied to the masonry structure (see Sect. 2.3). The elastic response spectrum is prescribed by the Italian seismic design code [16] for the case study site (Pozzuoli: $a_g =$ 0.162 g, $F_0 = 2.347$, $T_C{}^* = 0.333$ s, $S = 1.47$, $\xi = 0.05$, Fig. 2b). The design parameters considered in the parametric analyses are: (1) the mass ratio $\mu = (m_{US} + m_{ISO})/m_{LS}$, (2) the stiffness ratio $K = k_{US}/k_{LS}$, (3) the isolation period T_{ISO}. The damping constants c_{LS}, c_{ISO}, and c_{US}, are obtained from the values of the damping ratios of the lower structure, isolation system, and upper structure, i.e.: $\xi_{LS} = 0.05$, $\xi_{ISO} = 0.15$, $\xi_{US} = 0.02$. It is worth observing that this 3DOF model is a system characterized by non-proportional damping and complex-valued natural modes [18]. Therefore, the eigenvalue complex problem is solved in the state space for extracting the system vibration characteristics; then, RSA is carried out by considering complex modal superposition methods (e.g. [19]).

The results of the parametric RSA are provided in Fig. 2c, in particular with reference to $\mu = 0.25$ and $K = 1.25$, and as a function of the isolation period T_{ISO}. The plot provides the trend of displacement in non-dimensional form, i.e. ratio between the displacement of the first DOF in the 3DOF IIS model (i.e. the top displacement in the masonry structure) and the SDOF LS counterpart: $\overline{d} = d_{1,3DOF\ IIS}/d_{1,3DOF\ LS}$.

The two red points in the plot identify two selected design configurations, respectively appointed as IIS1 and IIS2 and characterized by isolation period of 0.5 s and 5.0 s. The values of the response parameter ratios for the two solutions are smaller or nearly equal to one: IIS1: $\overline{d} = 0.692$; IIS2: $\overline{d} = 0.983$. The solution of conventional vertical addition (LS + US), without the isolation system, is also considered for comparative purpose.

2.3 3D Finite Element Models

Different three-dimensional Finite Element (FE) models are considered: the primary system equipped with the nonconventional TMD structure, 3D IIS, Fig. 3a (i.e. the masonry structure extended through isolated CBF); the primary system, 3D LS, Fig. 3b (i.e. the masonry structure as standalone); the aggregate with the conventional vertical extension, LS + US, Fig. 3c (i.e. the masonry structure with the steel CBF structure without any isolation system at its base). The original structure is composed by masonry piers modelled as frame elements that contribute to the global response only with their

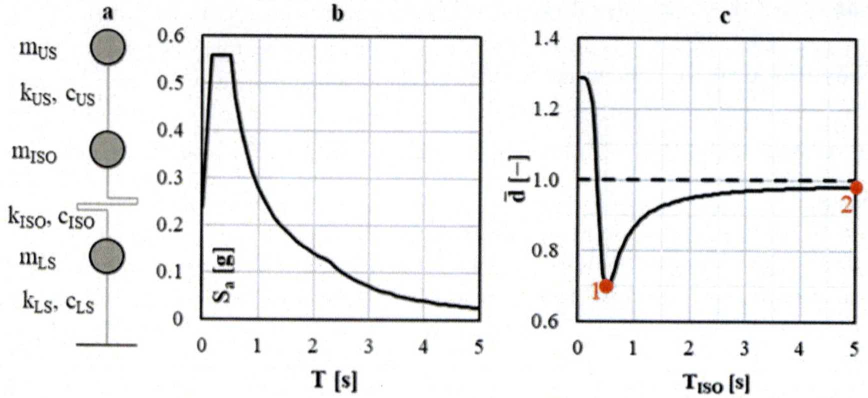

Fig. 2. (a) 3DOF IIS model; (b) (Life Safety Limit State) LLS Elastic Acceleration Response Spectrum of the site of Pozzuoli (Italy); (c) displacement ratio as a function of the isolation period T_{ISO}.

in-plane behavior along each principal direction. The isolation layer is composed by isolator devices modelled as linear springs. At each level, all the nodes are connected through a rigid floor diaphragm. The mass is concentrated in the master joint of each level. The damping ratios assigned to the lower structure, the isolation layer, and the upper structure are respectively equal to 0.05, 0.15, and 0.02.

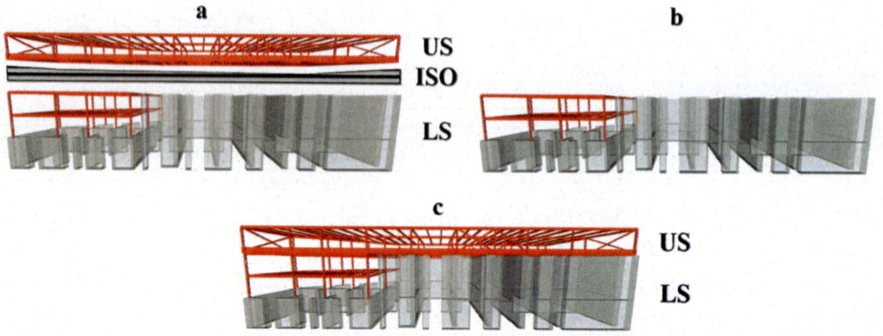

Fig. 3. 3D Finite Element models: (a) nonconventional TMD (IIS); (b) primary structure (LS); (c) conventional vertical addition (LS + US).

2.4 Assessment of Masonry Building Capacity

The masonry building capacity is evaluated by carrying out push over analyses (Fig. 4) through the software SAP2000 [20]. For this purpose, a bilinear model is defined for both shear and bending behaviour of the masonry walls (Fig. 4a). Concerning the bending behaviour, two plastic hinges are assigned at both ends of the piers; for the shear behaviour, one plastic hinge is assigned at the middle of each pier. The plastic hinge properties are derived by applying the capacity models suggested by the Italian seismic code [16] for both yielding and ultimate limit states. Two types of loading pattern, triangular and uniform, are considered along both principal axes of the building plan. As collapse condition, a peak strength reduction of 20% is assumed. As observed in Fig. 4b, the curves obtained from the analyses with triangular loading pattern show the lowest capacity of the masonry structure. Further, the failure of piers aligned along Y direction is due to the attainment of the ultimate bending capacity (Fig. 4c), while the piers along X (not depicted in figure for the sake of brevity) fail for the attainment of both ultimate bending and shear capacities. The dynamic properties of the reduced-order 1D SDOF LS, are derived according to the N2 method [17] (Fig. 2a): equivalent mass $m_{LS} = 2200.64$ kNs^2/m, tangent stiffness $k_{LS} = 639883$ kN/m, period $T_{LS} = 0.368$ s.

3 Effect of Masonry Nonlinearities

As already mentioned in Sect. 2.2, IIS is characterized by non-proportional damping. To account for the full damping matrix, and to consider coupling between modes of vibration, direct-integration time history analyses (THA) are here adopted and developed on the 3D IIS models by means of SAP2000 [20]. THA are also performed on the 3D LS and LS + US models. A set of seven pairs of acceleration records compatible with the reference spectrum for the site of Pozzuoli (Fig. 2b) is adopted as seismic input and defined by means of the software REXEL v 3.5 ($PGA = 0.1$–0.3 g in the period range 0.15–2 s) [21].

For assessing the effect of the masonry nonlinearities, both linear and nonlinear THA analyses are carried out, in which the masonry structure is considered either elastic or inelastic, while the isolated vertical addition is always assumed elastic. The hysteretic behaviour of the masonry structure is firstly described in Sect. 3.1, while the seismic behavior of the as-is and design configurations is discussed in Sect. 3.2. For the sake of brevity, in the following two sub-sections the results of the analyses are only provided for the weakest Y direction, by considering for each model the average of the peak values of the response parameters obtained from the seven input waves (in the Y direction).

3.1 Hysteretic Behaviour of the Masonry Structure

The Pivot-type model [15] is assumed as hysteretic model for the plastic hinges of the masonry walls (Fig. 4a). The shape of the hysteresis loop and the stiffness and strength degradation are controlled by the mechanical parameters α_1, α_2, β_1, β_2, and η. In this paper, the values assigned to the parameters are 0.45 for α_1, α_2, β_1, β_2, and zero for η [22–24] (Fig. 5a). The average values of the maximum plastic rotation demand to capacity ratios (DCR), calculated for the seven acceleration records in the walls at the base floor of the 3D LS model, are depicted in Fig. 5b. Analogously to what observed in Fig. 4c, in which the piers collapse for the attainment of the ultimate bending capacity in Y direction, the most stressed piers show values of plastic rotation DCR in the range of 0.15–0.20.

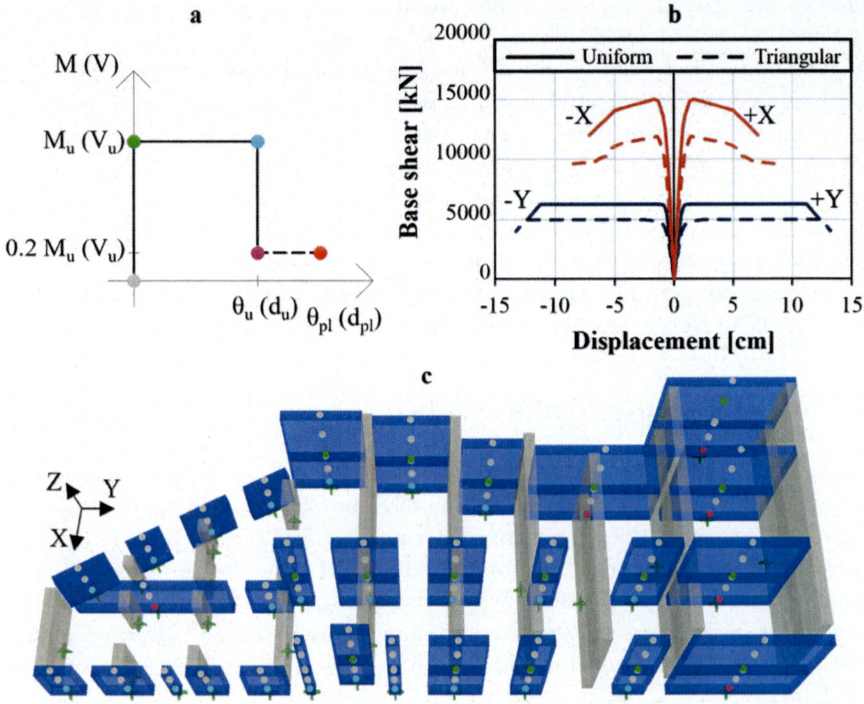

Fig. 4. (a) Plastic hinge: moment-plastic rotation and force-displacement responses; (b) 3D LS model: push over curves; (c) 3D LS model: plastic hinges in the walls (Y direction).

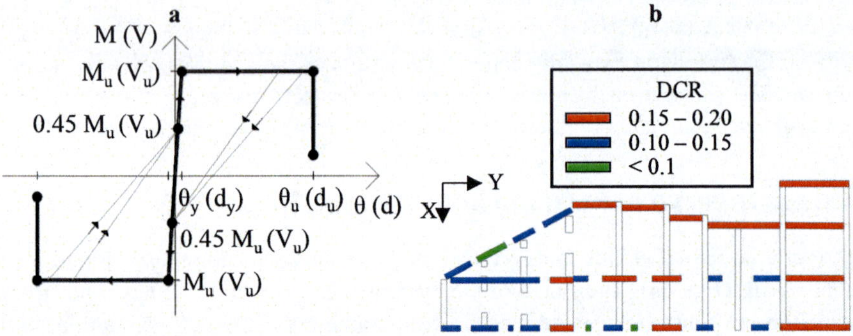

Fig. 5. (a) Pivot-type model; (b) DCR in terms of chord rotation for the piers at the base floor of the 3D LS model.

3.2 Seismic Assessment of the Different Configurations

For the different design configurations, i.e. LS + US, IIS1, IIS2, the plot in Fig. 6a provides the average value of the peak displacements at the top of the LS as a function of the fundamental period T_1 of the structural complex. Both linear (L) and nonlinear (NL) responses of the THA for the 3D models are considered and normalized to the linear counterpart of the 3D LS model. The curve of the displacement ratio $\overline{d}$ as a function of T_1, obtained from the RSA performed on the 3DOF IIS models (Sect. 2.2), is also depicted in the figure for comparative purpose. The blue markers, indicating the LS top-displacement values derived from linear analyses, clearly follow the trend of the 3DOF RSA curve. The comparison between the value obtained for the solution IIS1 at 0.5 s from linear THA (blue triangle marker) and the RSA curve suggests a reduction of the displacement response slightly larger than the one predicted. Therefore, the adoption of simplified models for the choice of the design configuration of the vertical addition seems on the safe side when the existing building works in the elastic field. When the masonry nonlinearities are considered, the effectiveness of both solutions IIS1 and IIS2 decreases, with the former still showing a reduction of the displacement response, while the latter exhibits an increase of response.

The effect of the damage accumulation in the existing masonry structure, both in the as-is and design configurations, is evaluated by means of a hysteretic energy absorption ratio, i.e. $\rho_E = E_h/E_{i,}$, with E_h hysteretic energy and E_i input energy, under the different ground motions. The Fig. 6b provides the average ratio $\rho_{E,ave}$ calculated for the various models, which shows almost equal values for the LS and LS + US models (0.51 and 0.48), and smaller values for the IIS models (0.16 and 0.41 for IIS1 and IIS2, respectively), with minimum value obtained, as expected, for the IIS1 solution.

For comparing the plastic rotations in the different configurations, the Fig. 7 depicts the ratio between DCRs (demand to capacity ratios); in particular, DCR calculated for the masonry walls in the LS + US and IIS models are normalized to the DCRs in the reference LS model (Fig. 5b). The figure highlights that the minimum engagement of the masonry structure is obtained by adopting the IIS1 design solution (Fig. 7b); instead, the plastic rotation demand to capacity ratios in the LS + US model are larger than in the LS model (Fig. 7a), while in the IIS2 model, the DCRs are almost equal to the counterparts in the LS model (Fig. 7c).

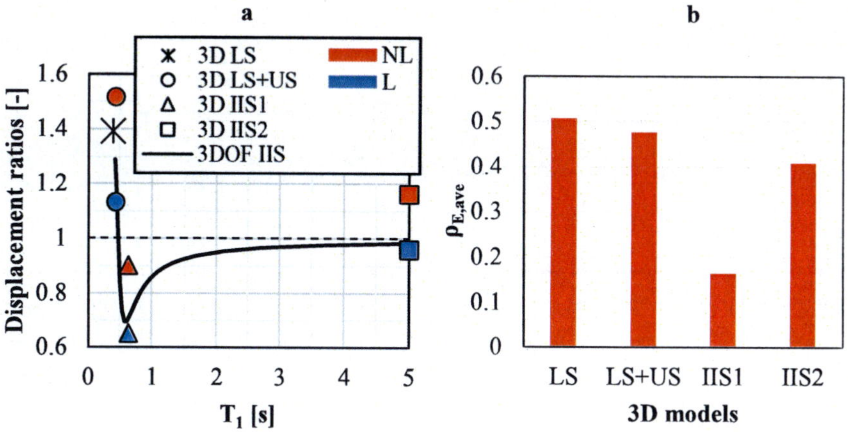

Fig. 6. (a) Displacement ratios vs. fundamental period T_1; (b) average cumulated hysteretic energy ratio $\rho_{E,ave}$.

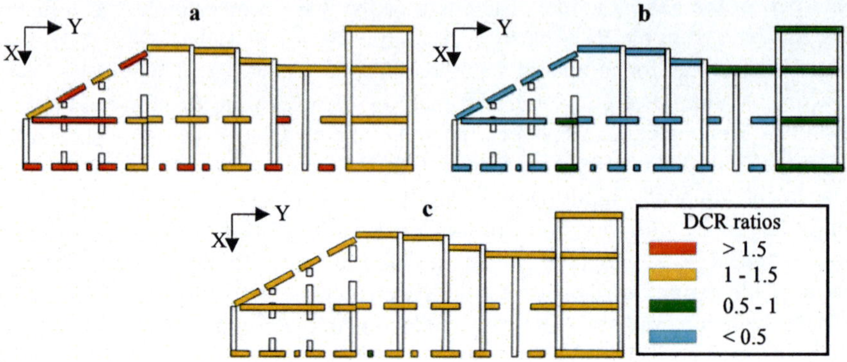

Fig. 7. Demand to capacity ratios in terms of chord rotation of the piers at the base floor in the 3D models: (a) LS + US; (b) IIS1; (c) IIS2.

4 Conclusions

This paper deals with the effect of nonlinear response in intermediate isolation systems (IIS) working as nonconventional TMDs. A two-story masonry aggregate located in Pozzuoli (Naples, Italy), superelevated by means of a one-story isolated vertical addition in structural steelwork, is considered as case study. The optimal dynamic properties of the isolated vertical addition are firstly identified through linear parametric analyses, by carrying out complex response spectrum analyses on lumped mass models. Then, the seismic behavior of the IIS configurations is assessed through 3D Finite Element analyses.

From the results of the linear parametric analyses, remarkable reductions of the displacement response of the existing masonry structure can be obtained by designing isolation systems with low-medium isolation periods thanks to the tuning effect. When the masonry nonlinearities are considered, a de-tuning effect is observed and the effectiveness of the isolated extension as nonconventional TMD decreases. However, reductions in terms of top displacement and cumulated hysteretic energy in the existing building are still obtained, particularly for low-medium isolation periods. Instead, in the case of conventional vertical extension (without the isolation system) the seismic inelastic engagement in the original masonry structure increases. The proposed approach should be suggested when building extension and seismic improvement are contemporarily required for existing masonry structures. Despite the actual behavior of masonry structures depends on several factors and can exhibit significant differences from one case to another, the IIS solutions seem particularly effective for improving the seismic response of masonry buildings that work in the elastic range or in moderately inelastic range.

References

1. Rana, R., Soong, T.T.: Parametric study and simplified design of tuned mass dampers. Eng. Struct. **20**(3), 193–204 (1998)
2. De Angelis, M., Perno, S., Reggio, A.: Dynamic response and optimal design of structures with large mass ratio TMD. Earth. Eng. Struct. Dynam. **41**(1), 41–60 (2012)
3. Mele, E., Faiella, D., Argenziano, M.: Innovative mass-damping-based approaches for seismic design of tall buildings. In: COMPDYN 2021, the 8th ECCOMAS Thematic Conference on Computational Methods in Structural Dynamics and Earthquake Engineering. Streamed from Athens, Greece, 27–30 June (2021)
4. Mele, E., Faiella, D.: Innovative mass-damping approaches for sustainable seismic design of tall buildings. In: Sustainable high-rise buildings: design, technology and innovation. Al-Kodmany K, Du P, Ali M, editors. The Institution of Engineering and Technology (The IET) (2022). Book DOI: https://doi.org/10.1049/PBBE003E, Chapter DOI: https://doi.org/10.149/PBBE003E_ch11
5. Dutta, A., Sumnicht, J.F., Mayes, R.L., Hamburger, R.O.: An innovative application of base isolation technology. In: Proceedings of 18th Analysis and Computation Speciality Conference-structures Congress, 2008, pp. 841–854, (2009)
6. Tsuneki, Y., Torii, S., Murakami, K., Sueoka, T.: Middle-story isolated structural system of high-rise building. J. Disaster Res. **4**(3), 229–238 (2009)
7. DM: Norme Tecniche per la progettazione, esecuzione e collaudo degli edifici in muratura e per il loro consolidamento. Gazzetta Ufficiale della Repubblica Italiana **1987**, 285 (1987). (in Italian)
8. Heyman, J.: The stone skeleton. Int. J. Solids Struct. **2**(2), 249–256 (1966)
9. Faiella, D., Calderoni, B., Mele, E.: Seismic retrofit of existing masonry buildings through inter-story isolation system: a case study and general design Criteria. J. Earthq. Eng., 1–37 (2020)
10. Argenziano, M., Faiella, D., Bruni, F., De Angelis, C., Fraldi, M., Mele, E.: Upwards—vertical extensions of masonry built heritage for sustainable and antifragile urban densification. J. Build. Eng., **44**, 102885 (2021)
11. Domizio, M., Garrido, H., Ambrosini, D.: Single and multiple TMD optimization to control seismic response of nonlinear structures. Eng. Structures **252**, 113667 (2022)
12. Lukkunaprasit, P., Wanitkorkul, A.: Inelastic buildings with tuned mass dampers under moderate ground motions from distant earthquakes. Earthq. Eng. Struct. Dyn. **30**(4), 537–551 (2001)
13. Sgobba, S., Marano, G.C.: Optimum design of linear tuned mass dampers for structures with nonlinear behaviour. Mech. Syst. Signal Process. **24**, 1739–1755 (2010)
14. Quaranta, G., Mollaioli, F., Monti, G.: Effectiveness of design procedures for linear TMD installed on inelastic structures under pulse-like ground motion. Earthq. Struct., **10**(1), 239–260 (2016)
15. Dowell, R.K., Seible, F., Wilson, E.L.: Pivot hysteresis model for reinforced concrete members. ACI Struct. J. **95**(5), 607–617 (1998)
16. NTC: Consiglio Superiore Lavori Pubblici, Aggiornamento delle Norme Tecniche per le Costruzioni. Gazzetta Ufficiale della Repubblica Italiana **2018**, 42 (2018). (in Italian)
17. Fajfar, P., Gaspersic, P.: The N2 method for the seismic damage analysis for RC buildings. J Earthquake Engng. Struct. Dyn. **25**(1), 31–46 (1996)
18. Veletsos, A.S., Ventura, C.E.: Modal analysis of non-classically damped linear systems. Earthq. Eng. Struct. Dyn. **14**(2), 217–243 (1986)
19. Sinha, R., Igusa, T.: CQC and SRSS methods for non-classically damped structures. Earthq. Eng. Struct. Dyn. **24**(4), 615–619 (2012)

20. Computers and Structures, Inc.: SAP2000 linear and nonlinear static and dynamic analysis and design of three-dimensional structures (2009)
21. Iervolino, I., Galasso, C., Cosenza, E.: REXEL: computer aided record selection for code-based seismic structural analysis. J. Earthq. Eng. **8**(2), 339–362 (2010)
22. Magenes, G., Bolognini, D., Braggio, C.: Simplified methods for non-linear seismic analysis of masonry buildings. CNR—National Group for Seismic Protection, Rome (2000) (in Italian)
23. Pasticier, L., Amadio, C., Fragiacomo, M.: Non-linear seismic analysis and vulnerability evaluation of a masonry building by means of the SAP2000 vol 10 code. Earthquake Engng. Struct. Dyn. **37**, 467–485 (2008)
24. Ozturk, B.: Seismic behavior of two monumental buildings in historical Cappadocia region of Turkey. Bull. Earthq. Eng. **15**(7), 3103–3123 (2017). https://doi.org/10.1007/s10518-016-0082-6

Isolated Artificial Ground for the Seismic Safety in the Urban Reconstruction of Castelluccio di Norcia

Marco Mezzi(✉) and Alessandro Fulco

University of Perugia, 06125 Perugia, Italy
marco.mezzi@unipg.it

Abstract. The effects of seismic attacks, especially in ancient or historical district of hit towns, depend not only on the intensity of the seismic action, but most of all on the quality of materials and construction technique. The negative consequences of traditional design and construction approaches appeared in all their negative evidence in many areas of Central Italy affected by recent seismic sequences of medium intensity (M5-M6) in 2016–2017 where entire villages were destroyed. Now the reconstruction should solve the problem to rebuild with safety but preserving the historical aspect of buildings and landscape. This paper presents a particular application of the known technique of seismic isolation for the reconstruction with integral seismic protection of the worldwide known village of Castelluccio di Norcia in Umbria (Central Italy). The adoption of a seismic isolation system at city scale involves the construction of a large floating slab, having the dimensions of the entire compartment, stepped due to the site orography, and supported by seismic isolators, above which to construct buildings that will present the aesthetic and constructive characteristics of the collapsed original ones. The solution allows a correct interpretation of the objective to rebuild "as it was, where it was", safeguarding the landscape, prolonging the lifetime, saving the expected cost.

Keywords: Base isolation · Artificial ground · Ground isolation

1 Introduction

On the occasion of seismic events that hit areas Italy, even in recent years, there have been extensive damage to buildings, in many cases with the almost total destruction of entire inhabited centers or entire neighborhoods. In 2009, on the occasion of the L'Aquila earthquake ($M = 5.8$), the town of Onna was largely destroyed [1]. The main events of August 24, 2016 ($M = 6.0$) and October 26 and 30, 2016 ($M = 5.9$ and 6.1) of the seismic sequence of the 2016–2017 Central Italy earthquake resulted in the total destruction of the small towns of Arquata del Tronto, Pescara del Tronto, Castelluccio di Norcia. Even the large historic center of Amatrice, significantly damaged in the first strong earthquake, was finally razed to the ground on October 26, 2016 [2].

G. P. Cimellaro (Ed.): WCSI 2022, LNCE 309, pp. 557–574, 2023.
https://doi.org/10.1007/978-3-031-21187-4_47

The buildings that form the agglomerations we are talking about were characterized by a high seismic vulnerability [3] due to well-known multiple causes: low quality of the materials and construction methods, generally being spontaneous building; lack or absence of earthquake-resistant details; decay of materials and the damage associated with the age of the buildings and the attacks suffered over the lifetime; absence or inadequacy or harmfulness of maintenance works.

The inhabited centers in question are typically made up of historic buildings, even if for the most part lacking of specific architectural values, and, above all, they are characterized by having a historical and landscape value in their entirety. The historical and landscape relevance, associated with the set of buildings that constitute them and their urban structure, is then accompanied by a social and cultural significance as the population recognizes the urban structure as a whole, as well as the individual components that make it up, even if individually they may not have an architectural relevance, as a fundamental element of its aggregation and compactness, as well as an identifying sign of its own identity. The historical, landscape, social, cultural and identity values are also perceived outside the community and in many cases these villages take on an identifying and symbolic meaning of general recognition.

On the occasion of catastrophic earthquakes in which these centers are heavily damaged or completely destroyed, the typical problems concerning the principles that must guide the reconstruction arise again. The basic question always concerns the choice between the option of rebuilding "where it was, as it was" and that of rebuilding instead with different structural solutions or even rebuilding in different sites, relocating the old towns.

Except in rare cases, and precisely for the reasons listed, we have always proceeded to rebuild "where it was" and in ways not very dissimilar to "as it was". Unfortunately, the use of materials and construction equipment, improved but not dissimilar to the original ones, means that the seismic vulnerability of the reconstructed settlements is not significantly improved and such as to avoid major damage in subsequent earthquakes. Therefore, every time a strong earthquake occurs, the same problems arise: suspension of social life and activities, temporary relocation of the inhabitants, implementation of new important works on the buildings.

Following the 2016–2017 Central Italy earthquake, to overcome the contradictions of a traditional reconstruction without hypothesizing relocations or distortions of the settlements, proposals were formulated [4–6] to rebuild the destroyed villages according to the principle of "where it was, as it was", creating new constructions on seismically isolated platforms having the dimensions of the entire inhabited center. Indeed the proposals were based on the application of the seismic isolation technique to soil (ground isolation or artificial ground).

This paper presents the results of a feasibility study that was the base for the reconstruction of the entire town of Castelluccio [7] on a single seismically isolated platform, evaluating the main aspects of technical-economic feasibility:

- earthquake-resistant performance of the solution;
- construction technologies;
- compatibility of the urban fabric;
- cost/benefit analysis.

2 The Site of Castelluccio di Norcia

An emblematic situation of what has been exposed so far, both with reference to the characteristics of the site and to the present condition of destruction, is represented by Castelluccio di Norcia, a hamlet of the municipality of Norcia. Despite its small size, the village of Castelluccio has a series of landscape, agro-food, tourist and cultural factors that synergistically determine a symbolic importance with a significance and notoriety that goes beyond the borders of Italy.

The area of the municipality of Norcia suffered significant earthquakes in 1979–1980, in 1989, in 1996–1997 with more or less significant reconstruction works. The high seismic vulnerability of the town of Castelluccio resulted from the studies carried out after the earthquakes of 1996–1997 [3]. On the occasion of the earthquakes of October 26 and 30, 2016, the historic center of Castelluccio di Norcia was practically razed to the ground as shown in the images in Fig. 1. Now a urban plan for the reconstruction has been designed (Fig. 2a).

(a) (b)

Fig. 1. Aerial views of the damaged post-earthquake conditions of Castelluccio di Norcia: (a) from W; (b) from S.

3 Solution Based on Ground Isolation

The term "ground isolation" defines a solution that provides for the seismic isolation of large platforms above which various constructions are built [8, 9]. Referring to the post-seismic reconstruction proposals based on the application of the "ground isolation" previously formulated [4, 5] and to the realizations referred to in the aforementioned works, a collaboration activity was started between the Department of Civil and Environmental Engineering of University of Perugia, the Special Office for Reconstruction of Umbria Region, and the Municipality of Norcia aimed at defining a seismic isolation system on a large platform for the reconstruction of Castelluccio di Norcia.

The use of ground isolation for post-earthquake reconstruction can only be planned in cases in which the destruction of a portion or an entire inhabited center is complete, and therefore a total reconstruction of the entire village or neighborhood should be provided. Within the historic center of Castelluccio, an area of approximately 6000 m^2

(a) (b)

Fig. 2. (a) Extract of the reconstruction plan with aggregate reference number; (b) in plan location of the area to be seismically isolated.

was identified to be rebuilt on an isolated platform. The area (Fig. 2b) has an irregular in-plan shape with overall dimensions of 90 × 80 m and is characterized by average slopes equal to 25% in the NS direction and 15% in the EW direction. Before the seismic events of 2016, the area included 18 masonry aggregates of 1, 2 and 3 stories, which covered about 4150 m^2 of area of the grounds, with 59 residential units with a total floor area of about 12880 m^2 and a volume of about 38650 m^3. Among the buildings there were also the oratory church of the Sacramento, completely destroyed, and the church of Santa Maria dell'Assunta, of which only a portion of the apse remains.

The total mass of the buildings involved in the project is equal to—assuming a unit mass of 1.3 t/m^2 (masonry buildings)—approximately 17300 t. Since the building land area is 4150 m^2, the unit mass of buildings to be insulated is 4.10 t/m^2. To this contribution must be added the mass relating to the "remaining" area, required for the restoration of the urban aspect, that is estimated equal to 11500 t: the unit mass of the "remaining" area to be insulated, equal to 2120 m^2, is equal to 5.40 t/m^2.

4 Feasibility Conditions

The fundamental aspects that are the subject of the evaluation of the technical feasibility of the project are the following [4, 10–12]:

- definition of the heights of buildings and plates;
- optimization of excavations;
- consolidations and terracing;
- foundation and substructure;
- isolation system;
- isolated plate;
- basements for elevation compensation;
- plants and sub-services;
- restoration of the urban fabric.

4.1 Establishment of the Elevations

The trend of the altitudes, with very significant slopes, represents one of the most problematic and characterizing aspects of the project. Figure 3a shows, as an example, the development of the elevations along one of the transversal streets of the village.

In order to achieve a reconstruction spatially equivalent to the pre-earthquake situation, the aspect of the elevations must be considered as a priority. The base quotas and the elevation development of the buildings must be maintained or restored, both for the principle of restitution "where it was" and for avoiding incompatibility in the solutions of continuity between the isolated portion and the fixed base boundary. The isolated plate, which acts as the foundation of the buildings, is stepped and organized on a limited number of staggered elevations defined in order to satisfy two criteria:

- to envelop the base quotas of buildings (Fig. 3b) with a lowering from 0.70 m (minimum) to 4.00 m (maximum);
- to limit the elevation differences of the steps below 6.00 m.

The resulting stepped plate has 11 reference elevations (Fig. 3c) which cover the maximum difference in height of 25 m between the lowest and the highest area.

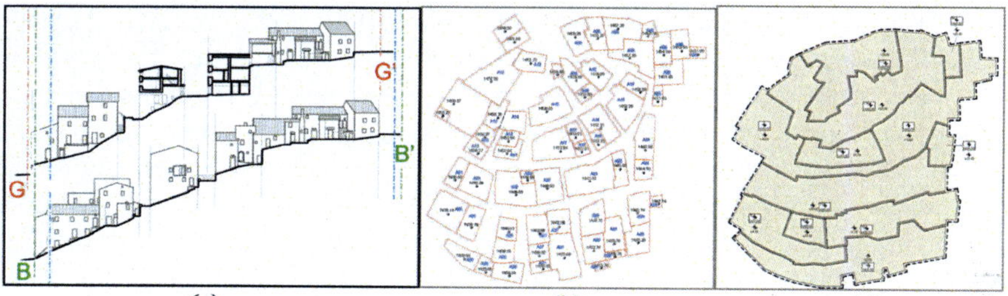

(a) (b) (c)

Fig. 3. (a) typical development of buildings along the streets; (b) quotas of buildings' base; (c) elevations of the stepped isolated plate.

Below the plate, the isolation interface and the foundation structure are provided. Figure 4a shows a constructive section of the solution. The base quotas of the excavation follow the same trend of the plate but are deepened by 1.50 m to cover the height of the isolation system and foundation structures. Similarly, the vertical retaining walls are set back by about 0.80 m with respect to the vertical faces of the stepped isolated plate. Ultimately, there is an increase in excavation volume of approximately 8000 m^3 compared to that of a conventional fixed-base solutions.

A total number of 301 devices is provided for the seismic isolation of the plate. Their positioning (Fig. 4b) derives from: (i) the arrangement at a distance of about 5–6 m, compatible with the vertical bearing capacity of the isolated plate, (ii) the modulation on the shapes of the horizontal shelves, (iii) the optimization of the dynamic response. The isolated plate has a height of 0.70 m which derives from a pre-dimensioning based on the assumption of a permanent load of 40 kN/m^2 and an accidental load of 8 kN/m^2, derived from the estimates of the provided superstructure (buildings and fillings). In order to satisfy the strength check, a total reinforcement percentage equal to 0.8% is provided.

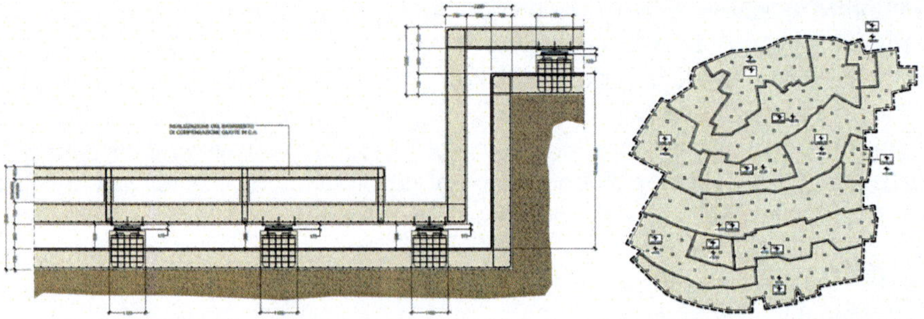

Fig. 4. (a) Section of the isolated plate and sub-structures; (b) in plane location of isolators.

The isolators are arranged on pedestals having a section of 1.15 × 1.15 m set on a foundation consisting of a grid of 0.70 m thick r.c. beams. The space between the lower edge of the isolated plate and the upper edge of the foundation is limited to a height of 0.80 m, allowing for inspection operations also using self-moving automated systems.

The equipment serving the buildings are installed in the space between the foundation and the isolated plate and are collected in the tunnels created in correspondence with the roads. The plants network include: water supply, sewer, electricity, optical fiber and telephone, gas. As regards the water (white/black) leaving the houses, pipes for the collection of the descendants coming from each house are provided to converge to the lowest point of the plate where there will be flexible pipes descending towards the "fixed" ground below the isolated compartment.

The urban fabric is reconstituted in a way that substantially corresponds to that pre-existing at the 2016 seismic event (Fig. 5a) according to the concept "where it was, how it was", except for minor variations provided by the urban Recovery Plan.

Above the isolated plate, both the road network and the buildings will be reconstituted. The correct positioning in elevation is got through rigid r.c. boxed bases, located under the buildings and roads, which ensure the compensation of the differences between the quotas of the isolated plate and those of roads and buildings. These volumes do not have a functional definition but could be used as basements or for public use. The achievement of the design road quotas can be locally obtained through the filling of landfill material. Figure 5b shows a schematic cross section of the project aiming at the restoration of the pre-existing "urban" share (highlighted in green).

5 Isolation System

Given the importance of the project, the design has been carried out ensuring to the isolated system a performance level higher than the standard one, considering that this can be achieved without significant cost increases. It has been considered a reference

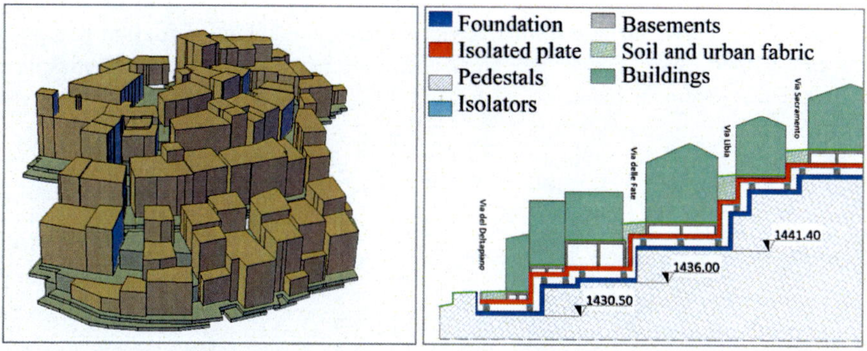

(a) (b)

Fig. 5. (a) 3D simulation of rebuilt compartment; (b) typical section of restored urban fabric.

period $V_r = 200$ years which corresponds to a nominal life of 200 years and class of use II ($C_u = 1.0$ for ordinary buildings), or to a nominal life of 100 years and class of use IV ($C_u = 2.0$ for strategic buildings). A design seismic action corresponding to a return period Tr = 2475 years is therefore considered.

On the basis of the available information, the subsoil category is B (semi-rigid soil) and the topographical condition is T3 (slope > 15%). Ultimately, the elastic acceleration spectrum for the collapse limit state (SLC) is characterized by the following values of the seismic parameters [13]:

- bedrock acceleration $a_g = 0.441$ g;
- amplification coefficient $F_O = 2.45$;
- characteristic period $T_C{}^* = 0.497$ s;
- soil amplification coefficient S = 1.20;
- horizontal ground acceleration PGA = 0.529 g.

Table 1 reports the characteristics of the various components that contribute to the seismic mass to be isolated: area A, load per unit area G, total weight P. The total isolated seismic mass is approximately equal to Miso = 45000 t. The average load acting on the isolated plate is Q = 70 kN/m2.

Table 1. Load contribution of the isolated components.

Component	A (m^2)	G (kN/m^2)	P (kN)
Plate shelves	6270	17.5	109,725
Plate walls	2672	17.5	46,760
Buildings	4169	40	168,959
Non built areas	2120	55	113,227
Total			438,671

The use of curved surface sliding devices with a coefficient of friction μ equal to 2.5% is assumed. This type of device is characterized by a bi-linear (elastic-plastic like) force-displacement behavior. The effective stiffness is equal to the sum of two contributions.

$$K = N(1/R + \mu/\,d) \tag{1}$$

where.

N is the axial load,

R is the radius of curvature,

μ is the coefficient of friction,

d is the maximum displacement.

Since the stiffness is proportional to the axial load and therefore to the mass that determines it, the isolation system is characterized by a substantial coincidence of the centers of mass and stiffness and therefore by the absence of significant torsional effects. Actually, this condition is not fully attained because also the friction coefficient is dependent on the axial load. The eccentricity between the center of mass and the stiffness is anyhow reduced to values lower than 0.2% of the platform dimensions.

From the preliminary assessments carried out with the aim of having an oscillation period of the isolated system in the interval 2.50–3.50 s, the adoption of market devices with the following characteristics appears appropriate:

D_g = 880 mm, diameter;

H = 173 mm, total height;

N_{Ed} = 6000 kN, maximum vertical load in the presence of horizontal displacements;

K_d = 591 kN / m, effective lateral stiffness for mean axial force;

K_v = 10000000 kN / m, vertical stiffness;

d_E = 350 mm, maximum displacement (collapse limit state).

In order to optimize the response of the isolated compartment by mitigating the torsion effects, the possible additional installation of dissipating devices, strategically located and working in parallel and synergistically with the isolator system, has been hypothesized. The adoption of magneto-rheological (MR) dissipating devices is hypothesized which, through an intelligent system for regulating their stiffness, allow to control the response with an instantaneous re-centering.

6 Isolation System Design

For the purposes of evaluating the seismic response of the system, a numerical model was built in which the stepped plate is reproduced with two-dimensional elements (Fig. 6) and the devices with linear elements. The superstructure (constructions and infrastructures) was considered in terms of loads acting on the plate with the values already reported in Table 2. Linear dynamic analyses [13] were carried out to determine the response of the isolated platform and to carry out checks on the isolators.

Table 3 shows the periods of oscillation and the participating masses in the longitudinal X (EW) and transverse Y (NS) direction of the first three modes. The two main translational modes have oscillation period values within the target range and are characterized by nearly unitary participant mass rates.

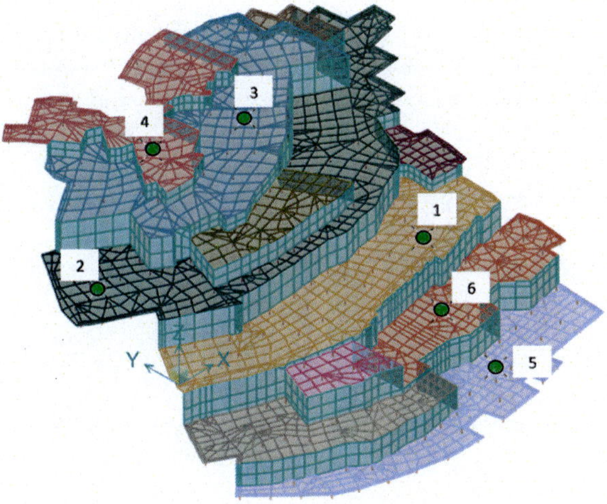

Fig. 6. Modello di calcolo della piastra isolata.

Table 2. Results of modal analysis.

Mode	Period (*s*)	Participating mass X-dir (%)	Participating mass Y-dir (%)
1	3.16	98.0	0.02
2	3.16	0.01	99.0
3	3.16	0.01	0.01

The maximum values of displacement and axial force on the isolators at the collapse limit state (SLC) are equal to 335 mm and 4201 kN, respectively, both compatible with the performance of the hypothesized devices [14]. The maximum horizontal force value is 360 kN. None of the isolators exhibit tensile axial force. The 1.15×1.15 m section pillars of the substructure meet the SLV checks with reinforcement percentages less than 1.00%.

As regards the other categories of loads, in the examined situation the effects of the wind are not relevant [15], while the deformations associated with thermal variations are evaluated in the order of 10 mm in consideration of the thermally insulated conditions of the substructure. The effects of temperature and shrinkage under construction can be mitigated and compensated by means of suitable technological-constructive solutions.

7 Simulation of the Response Under Actual Earthquakes

In order to analyze the actual seismic response of the isolated compartment, non-linear dynamic analyses were carried out using the accelerograms recorded in the destructive seismic event of October 30, 2016 as input. The values of the peak accelerations for the three components corresponding to the X (EW), Y (NS) and Z (VERT) directions of the model are equal to 0.420 g, 0.634 g, and 0.801 g, respectively. Figure 7 shows the elastic acceleration response spectra of the two horizontal components of the event, as well as, as a comparison, the SLC site spectrum provided by code. It can be observed how, in the field of the fundamental periods of ordinary masonry and r/c structures (between 0.1 s and 0.5 s), the values of the spectral accelerations double those of the SLC site spectrum, reaching values of 2.00 g. On the contrary, in the range of oscillation periods typical of seismically isolated systems (>3.00 s), the spectral accelerations of the records are lower than those predicted by the standard spectrum, with values around 0.10–0.15 g. Therefore, considering the event of October 30, 2016, the reduction in the response spectral accelerations of the isolated base systems compared to the fixed base ones is about 15 times.

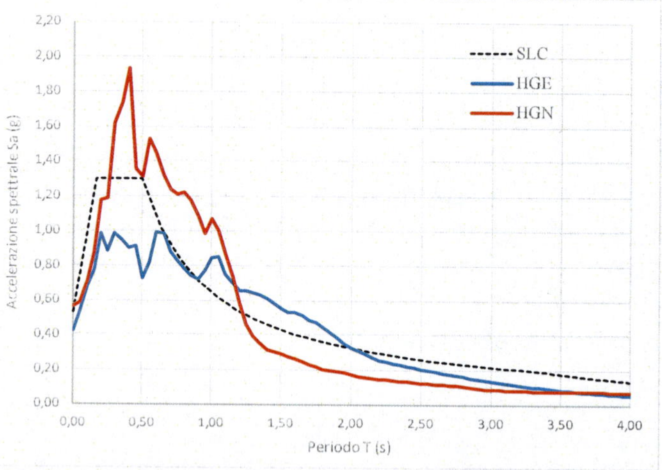

Fig. 7. Elastic response spectra of recorded accelerograms of October 30, 2016 and site code spectrum.

The nonlinear behavior of the isolators was reproduced in the numerical model by means of an elastic-plastic law characterized by the following parameters:

$K_1 = 60$ MN/m, stiffness of the initial elastic branch;

$F_1 = 60$ kN, force and displacement at the elastic limit;

$d_1 = 1$ mm, displacement at the elastic limit;

$K_2 = 386$ kN/m, stiffness of the plastic branch, considering the average axial load.

The values derive from the geometric and physical characteristics of the hypothesized devices:

$R = 3700$ mm, radius of curvature;

$\mu = 4\%$, coefficient of friction (taking into account actual vertical loads).

Table 4 shows, for the event of October 30, 2016 the longitudinal and transversal displacement components, s_x and s_y, the resultant displacement, s_{res}, the acceleration components, u_x and u_y, and the acceleration reduction factors, $\alpha_x = PGA_x/u_x$ and $\alpha_y = PGA_y/u_y$. The values are calculated in some significant points, highlighted in Fig. 6, where the sample buildings are located.

The maximum displacement of the isolators is 274 mm, lower than those resulting from linear analyses with site spectra. The maximum accelerations evaluated on the isolated plate vary from 0.15 g and 0.20 g, that is the seismic isolation determines a reduction in acceleration at the base of the buildings varying between 2.4 and 3.6 times that of the fixed base condition.

Table 3. Displacements, accelerations and reduction factors from nonlinear dynamic analysis (records of Oct 30, 2016).

Bldg.	s_x (mm)	s_y (mm)	s_{res} (mm)	u_x (mm)	u_y (mm)	α_x	α_y
1	197	189	273.00	0.145	0.178	2.89	3.56
2	197	190	273.70	0.151	0.177	2.79	3.58
3	197	190	273.70	0.159	0.176	2.65	3.61
4	197	190	273.70	0.168	0.196	2.51	3.24
5	199	189	274.45	0.176	0.198	2.39	3.20
6	198	189	273.72	0.170	0.214	2.47	2.96

To evaluate the seismic response of the superstructures, the numerical model was integrated with linear elements constrained to the plate and characterized by periods of oscillation in the range of those of the real buildings (0.10–0.40 s). The response was calculated for both the isolated and fixed base condition by applying the recorded accelerograms of October 30, 2016.

Table 5 shows, for various elements simulating the superstructures, the following parameters: the number of floors assumed; the fundamental period of oscillation, T_{FB}; the maximum values of the top displacements with respect to the base in the two conditions of base isolation, $\delta_{x,BI}$ and $\delta_{y,BI}$, and fixed base, $\delta_{x,FB}$ and $\delta_{y,FB}$. The isolated solution determines a reduction in displacements, compared to the fixed base solution, varying from 3 to 13 times (on average 8 times), with an average interstory drift approximately equal to 0.1%. The displacement values calculated for the fixed base condition exceed the limits of the linear behavior of the structures, but also those of the plastic range, with a ductility demand overwhelming that available for the typical constructions: they therefore identify conditions of severe collapse of the superstructures, like those actually occurred on the occasion of the actual event.

Figure 8 shows, for the isolated and fixed base configurations, respectively, the acceleration diagrams at the base and at the top of the element reproducing a superstructures characterized by an oscillation period equal to 0.20 s, in the most significant time window. In the case of isolated plate, the amplifications of the accelerations from the base to the top are very low, going from values equal to about 0.10 g at the base up to values of about 0.20 g at the top. On the other hand, in case of fixed base systems acceleration values go from about 0.5 g at the base to 2.0 g at the top. The isolated solution therefore determines a reduction of floor accelerations by about 10 times.

Table 4. Characteristics and response of superstructures (records of Oct 30, 2016).

Bldg.	Stories (no)	T_{FB} (s)	$\delta_{x,BI}$ (mm)	$\delta_{y,BI}$ (mm)	$\delta_{x,FB}$ (mm)	$\delta_{y,FB}$ (mm)
1	3	0.3	1.66	3.77	48.75	41.96
2	4	0.4	4.75	14.09	56.99	99.77
3	1	0.1	0.14	0.24	2.4	2.58
4	2	0.2	0.6	1.76	15.79	22.68
5	1	0.1	0.46	0.78	2.4	2.47
6	2	0.2	0.19	4.31	15.92	20.65

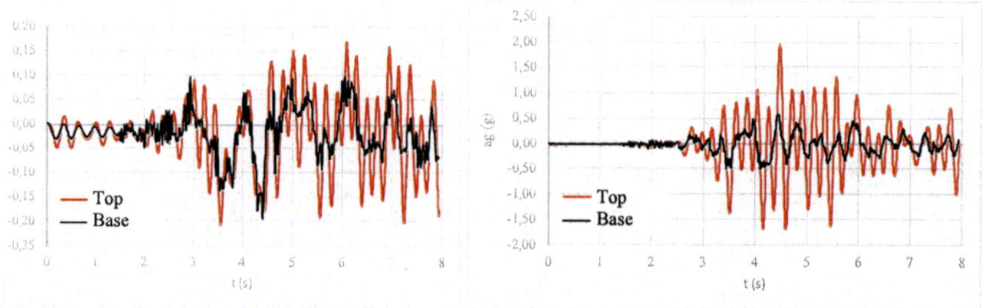

Fig. 8. Floor accelerations at the base and at the top of buildings: (a) base-isolated solution; (b) fixed-base solution.

The resulting displacement and acceleration values of the buildings in the isolated solution are even below the limits corresponding to the operational limit state of of the constructions.

As regards the response of the isolating devices, Fig. 9 shows, as an example, the force-displacement diagrams of the components in the two directions of one isolator. The maximum displacement reaches the value of 200 mm for both directions and the maximum force, exerted by the device and transmitted to the substructure, is in the order of 250 kN, i.e. approximately 1.5 times lower than the values resulting from linear dynamic analyses which can therefore be considered prudential.

8 Economic Aspects

Aiming at performing economic comparisons, the direct and indirect costs related to the reconstruction of the superstructures as well as the additional costs associated with the adoption of the isolated solution have been analyzed. In this context, prompt assessments of the expected consequences over time horizons of 10, 50 and 100 years have been carried out.

First of all, the cost necessary for the restoration of the considered sector has been assessed, concerning both the reconstruction (direct costs) and the emergency and post-event management (indirect costs). The former were assessed in accordance with the

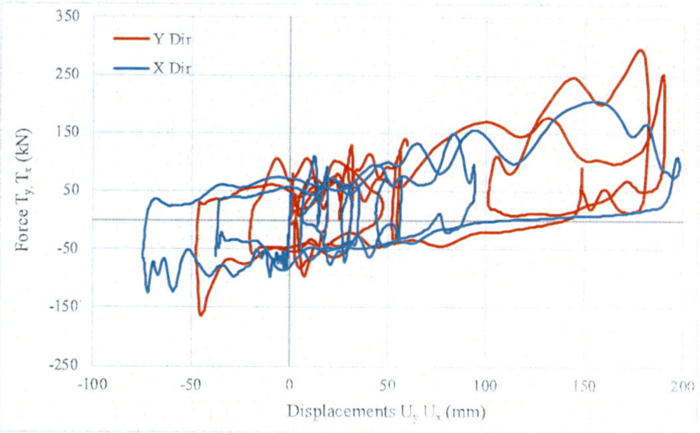

Fig. 9. Response force-displacement diagrams of the two components of a typical isolator.

ordinances of the Commissioner for reconstruction, considering a unit cost, including the various envisaged increases, of 1740 €/m^2 for a total area of approximately 13000 m^2. Adding the costs related to non-residential reconstruction (public, cultural, social infrastructures), estimated by the Commissioner at € 14 million, it results an estimate of direct costs of 36 M€. As validation of the assumed costs, the funding provided by the government for the entire Umbria region and equal to 3.7 G€ were taken into consideration. It was therefore estimated, proportionally, that resources of € 80–120 million (including indirect costs) are allocated to the area affected by the intervention in question.

To integrate the sector with a "ground-isolation" system, a total cost, additional to that already foreseen for the superstructures, equal to 5 M€ was estimated on the basis of a prompt calculation of the works and related costs: 0.50 M€ for excavation; 0.60 M€ for substructure and foundation; 0.70 M€ for the devices; 2.70 M€ for the stepped isolated plate; 0.20 M€ for the basements; 0.20 M€ for the repost. Therefore, the economic impact of the sector isolation system is equal to 15% of the direct costs and to 4–6% of the total costs.

It is also necessary to consider that an extreme seismic event entails further costs connected to the social, cultural and organizational consequences, such as the depopulation, the critical issues of resettlement and social reintegration, the deficit in production activities, the compensation of investments, the state debt. Moreover, last but not least, the direct consequences on human beings (dead and injured) must be taken into consideration. All these aspects are not evaluated in this work but assume a decisive importance in the choices and strategies for managing extreme events.

To define the convenience criteria in the context of an optimal reconstruction strategy, it is appropriate and necessary to move from economic estimates to performance evaluations [16, 17].

The conventional reconstruction project strategy, in line with current codes, would provide for the achievement of the life safety limit state for a seismic intensity with an exceeding probability of 10% in 50 years. Exceeding this limit state would result in costs substantially of the same order as those of the event that occurred and for which the reconstruction is carried out, but lower damage is to be expected even for events of lesser intensity.

Once L levels of consequences have been defined, it is possible to estimate the expected cost in N years, C_{EXP}, as Query ID="Q1" Text="Equation are not sequential"

$$C_{EXP}^{(N)} = \left[\sum_{l=1}^{L-1} C_{EG,l} \cdot \left(P_l^{(N)} - P_{l+1}^{(N)} \right) + C_{EG,L} \cdot P_L^{(N)} \right] \tag{2}$$

where

$$P_l^{(N)} = 1 - e^{N/T_{R,l}} \cong 1 - \left(1 - \frac{1}{T_{R,l}} \right)^N \tag{3}$$

is the probability of occurrence, in N years, of an event of seismic intensity for which l-th level consequences are reached;

L is the number of considered scenarios;

$T_{R,1}$ is the return period of the seismic intensity corresponding to the l-th scenario;

$C_{EG,l}$ = total costs corresponding to the l-th level of consequences.

The economic consequences must include both the direct costs associated with repairing the damage and the indirect costs. As regards indirect costs, the following aspects must be considered: (a) displacement of the inhabitants; (b) interruption of activities; (c) emergency management. At this stage, an amount of indirect costs has been assumed, simplified and as a precaution, approximately double that of direct costs. A preliminary estimate of the expected consequences was carried out, with reference to the damage levels provided for by the Commissioner's ordinances (Ord. 19 and subsequent) for private and public reconstruction, i.e. L0, L1, L2, L3-L4. The parametric direct cost of reconstruction, $C_{U,DIR}$, was attributed to each of the four levels in accordance with the aforementioned ordinances and therefore the overall direct costs of C_{DIR}. The direct costs of the public works, $C_{DIR,OP}$, were estimated with the same proportions for all the levels, being known the amount allocated for public works corresponding to the L3-L4 state observed in Castelluccio di Norcia. The total direct cost, $C_{DIR,tot}$, is then given by the sum of the quantities provided for private and public works. Also the total equivalent parametric cost $C_{U,DIR,tot}$ can be computed. Indirect costs, C_{IND}, were estimated with the previous proportions for all the levels, considering for level L3-L4 an amount approximately double that of direct costs. This assumption is congruent with the actual allocations foreseen by the reconstruction plan defined by the Commissioner. Finally, it is possible to obtain the overall C_{EG} costs potentially achievable for each level of damage.

$$C_{EG} = C_{DIR,tot} + C_{IND} \tag{4}$$

The following table reports the values of the parameter described above considering the reconstruction area equal to 13000 m^2.

The four damage levels have been associated with the performance levels (limit state) provided by the code [13] for the design of buildings, therefore the operational (SLO), damage (SLD), life-safety (SLV) and collapse (SLC) limit state have been considered

Table 5. Direct, indirect, total costs for the four considered damage levels.

Damage level	$C_{U,DIR}$ (€/m^2)	C_{DIR} (M€)	$C_{DIR,OP}$ (M€)	$C_{U,DIR,tot}$ (€/m^2)	$C_{DIR,tot}$ (M€)	C_{IND} (M€)	C_{EG} (M€)
L0	300	3.900	2.333	479.49	6.233	0.390	6.623
L1	1.000	13.000	7.778	1.59829	20.778	6.500	27.277
L2	1.375	17.875	10.694	2.19765	28.569	26.813	55.381
L3-L4	1.800	23.400	14.000	2.87692	37.400	46.800	84.200

respectively corresponding to the damage level L0, L1, L2, L3-L4. The return periods $T_{R,l}$ of the seismic intensity corresponding to the limit states for a conventional reference life of 50 years, can be then attributed to each of the damage levels and the exceeding probabilities of the damage level $P_{l(10)}$, $P_{l(30)}$, $P_{l(100)}$ for three significant time horizons of 10, 30, 100 can be defined. Table 6 reports.

Table 6. Probability of exceeding in the 3 significant time horizons: 10, 30, 100 years.

Damage level	l (damage level index)	$T_{R,l}$ (years)	$P_{l(10)}$	$P_{l(50)}$	$P_{l(100)}$
L0	1	30	0.288	0.816	0.966
L1	2	50	0.183	0.636	0.867
L2	3	475	0.021	0.100	0.190
L3-L4	4	975	0.010	0.050	0.098

The expected costs probabilistically obtained through the Eq. (1) for the three time horizons considered of 10, 50 and 100 years are equal to 6563 M€, 22,792 M€ and 32,466 M€, respectively. It is therefore obtained that the expected consequences over the next 50 years for the urban sector in question conventionally rebuilt, are equal to approximately 23 M€. This cost is about 5 times higher than the immediate cost increase associated with seismic isolation and does not take into account the consequences on people. Considering instead a time horizon of 100 years, the total cost associated with the expected consequences would be more than 6 times higher than that associated with seismic isolation.

The expected cost for the sector equipped with "ground isolation" is instead practically absent, in fact, as illustrated in this work, even an extreme seismic event, of an intensity corresponding to or greater than that of SLC, such as the one recorded in 2016, would not produce significant damage on the sector that would remain below the operational limit state. Moreover, in addition to mitigating (canceling) the economic consequences, there would be an almost total reduction of the socio-cultural consequences that are not strictly economically assessable and above all of the consequences on people.

9 Urban Dimension and Fruition of the Infrastructure

Another important element, not related to the structural aspects and seismic protection, is given by the infrastructural nature and the urban dimension of the work, and is represented by the possibility of making a part of it visible and open to visitors with the creation of a multifunctional space accessible from the outside. Accessibility to the infrastructure is guaranteed by an annular path (also designed for maintenance and monitoring) and through the creation of multi-purpose vaulted hypogeal spaces.

On a strictly functional level, it is envisaged a museum use aimed at illustrating the main characteristics of the seismic isolation intervention and telling the millennial history of the territories of the plateau of Castelluccio di Norcia. The architectural form of the rooms is characterized by a double arch structural system resting on a reticular mesh of seismic isolators. The size of the rooms, to be checked in the subsequent design phases, is approximately 650 square meters.

10 Conclusions

The work carried out shows the feasibility of the reconstruction of an entire village or section of an urban center affected by destructive seismic events, using the seismic isolation of the whole section according to the technology defined "ground isolation" or "artificial ground". In particular, the work illustrates the study concerning the application to the town of Castelluccio di Norcia, practically destroyed in the earthquake of 30 October 2016, for which the reconstruction project is currently being carried out according to this isolated solution under the guidelines established by the Commissioner for the reconstruction.

Due to the dramatic reduction of the seismic response allowed by the isolation technology, the objective of rebuilding "how it was, where it was" can be fully implemented, even applying techniques that can be traced back to traditional ones and by building constructions with "almost zero" expected damage, even for extreme expected earthquakes characterized by return periods of 1000–2000 years.

The economic evaluations of the expected costs show that the extra costs associated with the use of the ground isolation technique are 3–6 times lower than the expected costs over the reference life for buildings rebuilt according to conventional criteria. Moreover there are the enormous benefits associated with the absence of direct consequences on people and social, cultural, organizational consequences, all components whose values cannot be, and are not, evaluated economically.

References

1. EERI: The Mw 6.3 Abruzzo, Italy, Earthquake of April 6, EERI Special Earthquake Report (2009)
2. Mazzoni, S. Fulco, A., Castori, G., Galasso, C., Calvi, P., Dreyer, R., Fisher, E., Sorrentino, L., Wilson, J., Penna, A., Magenes, G.: 2016–17 Central Italy Earthquake Sequence: Seismic Retrofit Policy and Effectiveness. Earth. Spect. **34**(4) (2018)
3. Valluzzi, M.R., Munari, M., Modena, C., Cardani, G., Binda, L.: Analisi di vulnerabilità sismica degli aggregati storici: il caso di Castelluccio di Norcia. Convegno ANIDIS. Pisa (2007)
4. Fulco, A., Comodini, F., Mezzi, M.: Isolamento sismico a grande scala per la salvaguardia del tessuto urbano nella ricostruzione post-sisma. Convegno ANIDIS. Ascoli Piceno (IT) (2019)
5. Mezzi, M: Meta-isolation of entire urban sectors for a resilient post-earthquake reconstruction preserving the pre-existing landscapes. In: 44th Conf. on Our World in Concrete & Structures. Singapore (2019)
6. Mezzi, M., Fulco, A.: Seismic isolation of whole city quarters for an effective and preservative reconstruction. In: Procs. EURO-MED-SEC-3-ISEC-Holistic Overview of Structural Design and Construction. Limassol, Cyprus (2020)
7. Fagotti, G., et al.: The integrated project for the post-earthquake reconstruction of Castelluccio di Norcia: procedures, techniques, implementation. XIX ANIDIS, Torino (2022)

8. Martelli, A., et al.: New activities performed in Italy on innovative anti-seismic techniques for civil and industrial structures. Proc. ASME-PVP Conf. Boston, USA, PVP-Vol. **387**, 311–326 (1999)
9. Zhou, F. L., Yang, Z., Liu, W.G., Tan, P.: New seismic isolation system for irregular structure with the largest isolation building area in the world. In: 13th World Conference on Earthquake Engineering, Vancouver, Canada. August (2004)
10. Comodini, F., Mezzi, M.: Performance comparison of isolated, dissipated and fixed-based steel buildings. In: 13th Int. Conf. on Civil, Structural and Environmental Engin. Computing, Chania, Crete, Greece. CIVIL-COMP PROCEEDINGS, Stirlingshire:B.H.V. Topping and Y. Tsompanakis, Civil-Comp Press, ISBN: 978-1-905088-47-8, ISSN: 1759-3433 (2011). https://doi.org/10.4203/ccp.96.71
11. Mezzi, M., Comodini, F., Rossi, L.: Base isolation option for the full seismic protection of an existing masonry school building. In: 13th Int. Conf. on Civil, Structural and Environmental Eng. Computing, Chania, Crete, Greece. Civil-Comp Proceedings, Stirlingshire: B.H.V. Topping, Civil-Comp Press, ISBN: 978-1-905088-47-8, ISSN:1759-3433. https://doi.org/10.4203/ccp.96.72
12. Mezzi, M., Petrella, P.: Experiences from the reconstruction after the 2009 L'Aquila Earthquake. Damage, Vulnerability, Retrofitting, ISEC-7 New Developments in Structural Engineering and Construction. Honolulu, USA (2013)
13. NTC2018: Norme Tecniche per le Costruzioni D.M. 17 gennaio 2018, Consiglio Superiore dei Lavori Pubblici, GU Repubblica Italiana, Roma (2018)
14. EN15129. Anti-seismic devices. CEN. Brussels (2009)
15. Ubertini, F., Comodini, F., Fulco, A., Mezzi, M.: A simplified parametric study on occupant comfort conditions in base isolated buildings under wind loading. Advances in Civil Engineering, Hindawi Publishing Corporation, Article ID 3524975, 13 pages (2017)
16. Comodini, F., Fulco, A., Mezzi, M., Petrella, P.: Fast risk assessment of losses in r/c buildings. Procs. 16WCEE. Santiago, Chile (2017)
17. Fulco, A., Mezzi, M., Comodini, F.: Practical procedure to assess the expected consequences of earthquakes on buildings. Procs. 17thWCEE. Sendai. Japan (2020)
18. Edited by Vacanas, Y., Danezis, C., M., Yazdani, S., and Singh, A

Numerical Application of Viscoelastic Devices for Improving the Out-Of-Plane Behaviour of a Historic Masonry Building

Nuno Mendes(✉), Elesban Nochebuena-Mora, and Paulo B. Lourenço

Institute for Sustainability and Innovation in Structural Engineering (ISISE), University of Minho, Campus de Azurém, 4800-058 Guimarães, Portugal
nunomendes@civil.uminho.pt

Abstract. It is well known the low capacity of unreinforced masonry (URM) buildings to withstand seismic excitations. Thus, many reinforcement techniques (tie rods, buttresses, collar beams, etc.) have been developed, aiming at improving their dynamic response. Such techniques increase the strength and sometimes the ductility of the structural members. However, in the last two decades, vibration control techniques have been tested and applied as an alternative to reduce the seismic response of URM buildings. These techniques, based on performance approaches, rely on the dissipation of kinematic energy by placing passive devices able to add damping into the structure. Among them, viscoelastic (VE) devices can supply self-centring capacity and dissipate energy through the shear deformation of viscoelastic layers. In contrast with other dampers, VE devices can have linear behaviour which simplifies their design and application to URM structures. This paper presents the results of a parametric analysis comparing the dynamic response of a URM building strengthened with a traditional technique and with VE devices. A FEM model was generated using a prototype based on viceregal buildings (18th century) located in Mexico. The response was evaluated in terms of displacements and energy.

Keywords: Energy dissipation · Viscoelastic dampers · Historical buildings · Unreinforced masonry · Seismic performance

1 Introduction

Passive vibration control systems are devices capable of improving the seismic response of buildings by dissipating energy. The additional supplied damping reduces the excessive deformations and, consequently, less ductility or inelastic behaviour of the structure is required to withstand strong earthquakes [1]. Viscoelastic (VE) dampers dissipate energy by the shearing of elastomeric layers bonded to plates. If properly designed, VE dampers can behave in the linear range (opposite to yielding or friction devices), which simplifies the analysis of buildings. Despite the extensive research on the characterization of VE dampers and their application to framed structures, for example [2–4],

G. P. Cimellaro (Ed.): WCSI 2022, LNCE 309, pp. 575–585, 2023.
https://doi.org/10.1007/978-3-031-21187-4_48

any attempt has been made to study their feasibility for the retrofit of existing unreinforced masonry buildings [5]. Therefore, this work explores the application of VE dampers in this type of structure and presents the preliminary results of a set of numerical simulations. A FEM model of an unreinforced masonry building was generated using Abaqus/Explicit [6]. The model represents a partial prototype of a characteristic viceregal dwelling from Mexico. The explicit solver was adopted since it can analyse dynamic problems with high nonlinear material response (quasi-brittle behaviour), which causes convergence difficulties for implicit methods. Although explicit solvers are conditionally stable, Abaqus/Explicit automatically computes a conservative time increment to guarantee the stability of the solution [7].

2 Numerical Model

The numerical model represents the front gallery and a part of the lateral rooms of a viceregal dwelling (18th century) located in Mexico. Walls and foundation were built with rubble unreinforced masonry and lime mortar. The thickness of the façade and exterior lateral walls is 0.90 m, whereas the thickness of the inner walls varies from 0.80 m (ground floor) to 0.60 m (upper floor). Floors and roof are composed of a large set of timber joists (section 0.15 cm^2 × 0.20 cm^2) spacing 30 cm, placed perpendicular to the façade and supported in pockets in the walls (no rigid connection) (Fig. 1).

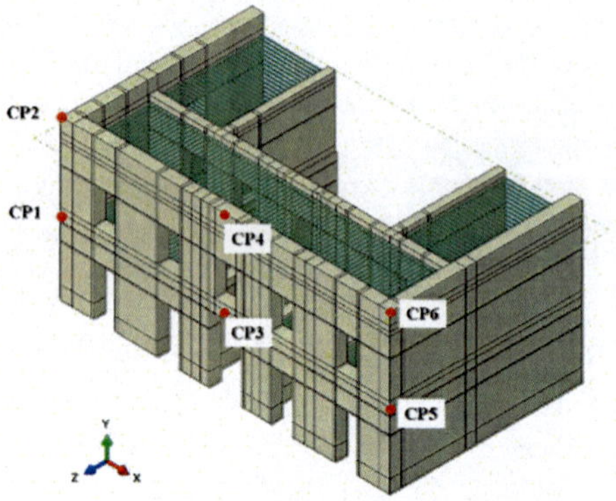

Fig. 1. Geometry of the numerical model.

2.1 Mesh

Walls and foundation are composed of 3D-solid linear hexahedral elements (C3D8R) with reduced integration to decrease the computational time and avoid the shear locking effect. Four elements through the thickness of the walls were used to capture possible bending behaviour [7, 8]. To minimize the onset and propagation of the hourglass effect (anomalous distortion presented in reduced integration elements), Abaqus introduces artificial energy to counteract deformation [7]. Recommended values for artificial energy range between 5 and 10% of the internal energy [9, 10]. The joists were simulated using linear beam elements (B21) with additional mass considering the dead and live loads. Connections between walls and joists are represented by translator elements with nonlinear properties (high stiffness for compression and low stiffness under tension). VE dampers were simulated using axial connections with a linear stiffness and a linear damping coefficient. To avoid stress concentration, the axial connections were attached to steel plates (0.30 m^2 × 0.30 m^2) anchored to the walls.

2.2 Mechanical Properties

Density, Young's modulus and compression strength for masonry were obtained from Chávez et al. [11]. Regarding the tension properties, the strength was assumed 10% of the compression strength. Foundation and walls have the same properties since both were built with the same procedure. Timber for joists and lintels is assumed to be linear during all the analysis since no damage in these elements is expected. Density and Young's modulus for timber were taken from Guzmán [12]. Table 1 presents the mechanical properties of the materials.

Table 1. Mechanical properties.

	Density (kg/m^3)	Young's modulus (N/m^2)	Poisson's ratio	Compression strength (N/m^2)	Tension strength (N/m^2)	Tension fracture energy (N/m)
Rubble masonry	1794	1.43e+09	0.2	1.90e+06	1.90e+05	50
Timber	610	8.8e+09	0.3			

The concrete damage plasticity (CDP) model was adopted to describe the nonlinear behaviour of the masonry under cyclic loads. This model is provided in Abaqus and has been also used to simulate the response of unreinforced masonry buildings for nonlinear static and dynamic analyses [13, 14]. The stress-strain compression and tension curves (Fig. 2) were calculated based on the values in Table 1. Two damage parameters in the model describe the tensile cracking and compressive crushing, which were calculated using an approximation expressed as $d = 1 - \sigma/\sigma_{max}$, where σ is the stress after the maximum strength σ_{max}.

CDP model requires other five parameters to describe the yield surface. These parameters are: the dilatancy angle ψ, for which a value of 10° was adopted; the ratio between the biaxial and uniaxial compressive strength f_{b0}/f_{c0}, assumed equal to 1.16; the parameter K, set to 2/3, as recommended by Abaqus [7]; the viscoplastic regularization, assumed as zero; he eccentricity, considered as 0.1.

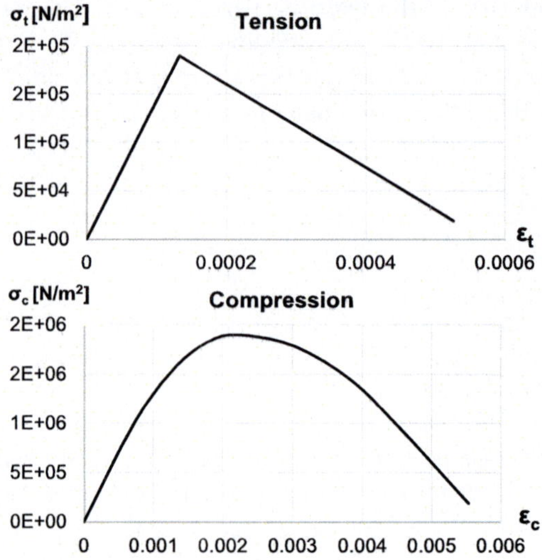

Fig. 2. Tensile and compression uniaxial nonlinear curves calculated for the nonlinear behaviour of the masonry.

3 Design of the Viscoelastic Dampers

The behaviour of VE materials can be described using shear storage modulus G', corresponding to the elastic component, and the shear loss modulus G'', related to the viscous component, which controls the energy dissipation. The relationship between both parameters is the loss factor $\eta = G''/G'$. The properties are sensitive to frequency, temperature, and strain. Some VE materials present a linear behaviour up to 150% of shear strain [15]. The equations relating the moduli to an equivalent stiffness k and equivalent damping coefficient c are:

$$k = \frac{nG'A}{t}, \quad c = \frac{nG''A}{\omega t} \tag{1}$$

where n is the number of layers, A is the shearing area of one layer, t is the thickness of the layer, and ω is the excitation frequency (rad/s) [16]. Aiken and Kelly [15] proposed a simplified procedure to approximately estimate the volume of material and dimensions required to achieve a certain damping ratio. The thickness t of a single layer is selected based on an operating strain. Then, the number of dampers and the number of layers per damper determine the shear area A. Finally, values for k and c can be calculated with Eq. (1) for a given G' and G''.

4 Seismic Load

The explicit solver is more efficient when the dynamic actions are introduced as velocities instead of accelerations or displacements [7]. North-South component of the Emilia Romagna's earthquake was used for the analyses. Figure 3 depicts the signal recorded in the seismic station in Mirandola (Italy), on May 29th, 2012 [17], that was applied only in one direction to cause the out-of-plane collapse of the façade and rear wall.

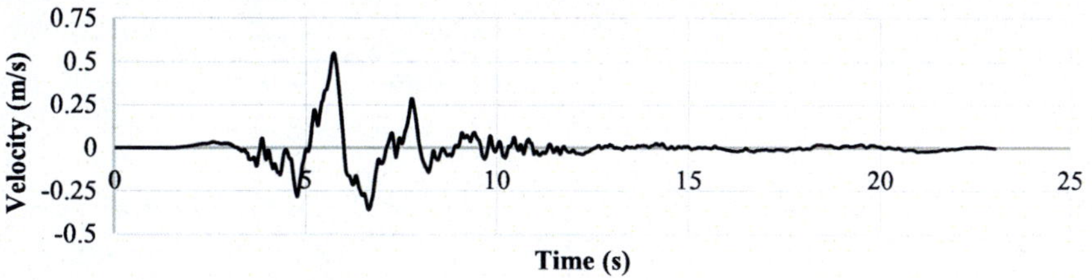

Fig. 3. Velocities of the North-South component of the Emilia Romagna's earthquake.

5 Reinforcement Proposals Using Dampers

Several solutions for the reinforcement with dampers were proposed by changing the number and distribution of dampers. To calculate the properties of the dampers, G' and G'' were selected from the datasheet of the 3M polymer 112 for an ambient temperature of 20 °C. For the first proposal, dampers were placed at the connection between the façade and all the joists of the floor and roof, while the other end of the joists was rigidly fixed to the rear wall. The dynamic analysis with 100% of the seismic load demonstrated the inefficiency of the solution. Other strategies with steel ties, which are embedded in the orthogonal walls and attached to the façade with dampers, also had an inadequate performance. Figure 4 shows other two tested configurations with diagonal steel ties. In both cases, dampers are located only at the level of the roof, whereas the ground floor is reinforced with traditional ties (no dampers).

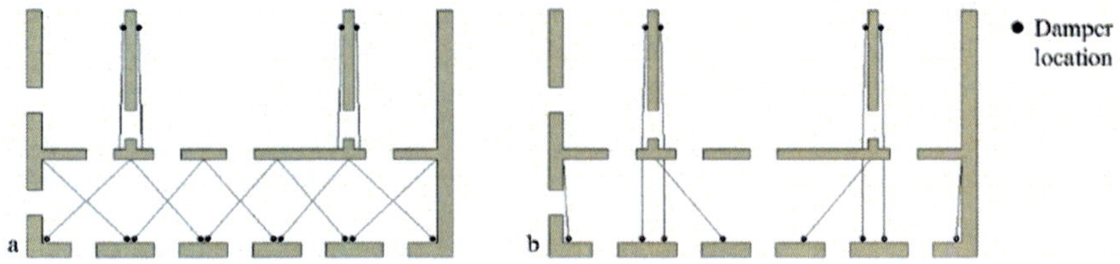

Fig. 4. Damper location at the level of the roof in a) Model-C and b) Model-D.

Table 2 presents the estimates for k and c for both proposals, additionally, the values were reduced by a factor of 0.5 and 0.25 for a sensitivity analysis.

6 Results

To assess the effectiveness of the dampers, the results were compared with other two conditions corresponding to the unreinforced building (Model-UR) and reinforced with

Table 2. Equivalent stiffness and damping coefficient of the VE dampers.

	Model-C100	Model-C50	Model-C25	Model-D100	Model-D50	Model-D25
k	2.70E+07	1.35E+07	6.74E+06	4.49E+07	2.25E+07	1.12E+07
c	1.61E+06	8.03E+05	4.02E+05	2.67E+06	1.34E+06	6.68E+05

steel tie rods with the same configuration as Model-C and Model-D. Model-UR was subjected to an incremental analysis showing severe damage for an amplitude of 75% of the seismic load. The façade completely detached from the perpendicular walls and collapsed out-of-plane around second 6.2 of the analysis (Fig. 5). Artificial energy significantly increased after this point, which indicates that subsequent data may not be reliable.

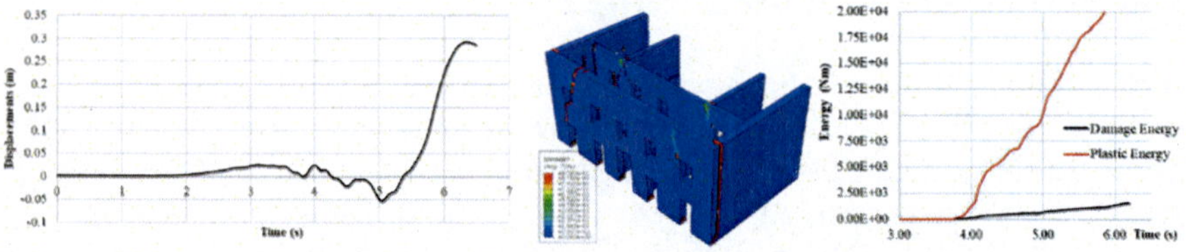

a b c

Fig. 5. a) Time-history displacement of control point CP4, b) damage pattern at second 6.2 (cracks in red), and c) damage and plastic energies.

The analyses with the rest of Model-C and Model-D were performed applying 100% of the seismic load, assuming that the reinforcement can enhance the seismic response of the building. In the following, the results are presented in terms of maximum displacement of control points at the façade (see Fig. 1), damage distribution, and comparison of damage and plastic energy. It is noted that for the current application, VE dampers only work under tension, since the ties avoid the outward movement of the façade. Whereas for the inward displacement, ties buckle and are not able to resist compression forces, causing the VE material to remain unreformable and thus, no energy is dissipated.

6.1 Model-C

The results of the four models are compared at instant 5.9 s. This time was set because Model-C-Ties presented an increment of artificial energy higher than 10% after this point due to the large deformation of damaged elements. In this model, the façade is still detached from the rest of the building. Orthogonal walls collapsed before the façade since the inertia forces transmitted by the ties produced shear cracks along both levels of the walls. Additionally, high shear stress was developed at the base. Models with dampers exhibited a better response given that part of the energy was dissipated by the devices. As a result, the plastic energy and energy dissipated by damage diminished considerably as shown in Fig. 6. Models with dampers are more flexible and, thus, displacements are larger as depicted in Fig. 7. However, they performed better in decreasing the damage. Despite Model-C50 and Model-C25 had similar behaviour in terms of energy, it can be seen in Fig. 8 that the damage pattern is different. Model-C25 presented a more distributed damage along the façade and its foundation.

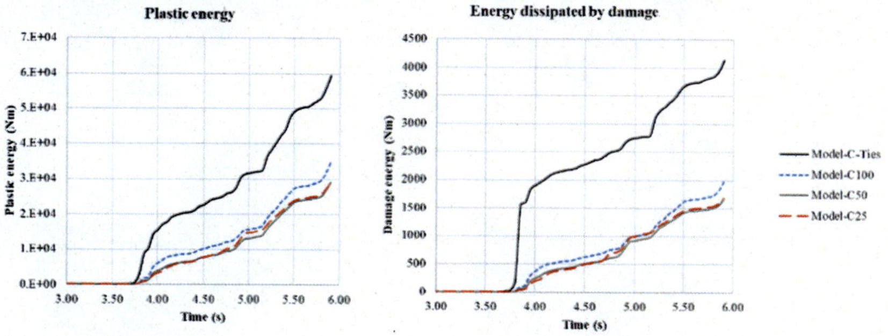

Fig. 6. Comparison between damage energy and plastic energy for Model-C with different properties of the dampers.

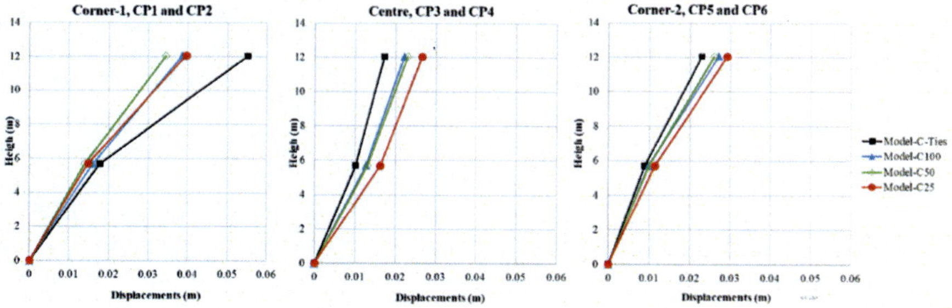

Fig. 7. Maximum displacements of the control points of the façade up to second 5.9 (Model-C).

6.2 Model-D

The results for Model-D are compared at time 6.05 s, when the artificial energy of Model-D100 increased considerably. In this case, the model reinforced with ties had a better capacity to reduce the damage than Model-D100. The total force (sum of elastic and viscous forces) produced by the dampers in Model-D100 is too high, which caused local damage. However, Model-D-Ties also presented severe cracks later in the analysis. There was an improvement when the properties of the dampers were reduced to 50 and 25%. The plastic and damage energies decreased, mainly in Model-D25 (Fig. 9). When comparing the results with the previous solution, Model-D25 had a similar performance that Model-C25 but with less number of dampers. It should be noted that the properties are different even if the same material is used, i.e., the size and volume are also different. Models with reduced properties of the dampers exhibited larger displacements, with permanent deformation of the facade between 1.5 (corner 2) and 5 cm (corner 1), see Fig. 10.

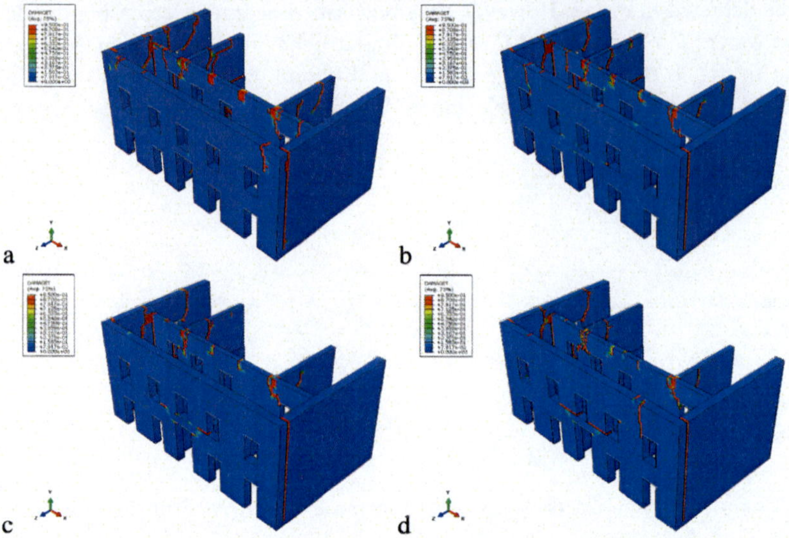

Fig. 8. Damage pattern at second 5.9 (cracks in red) a) Model-C-Ties, b) Model-C100, c) Model-C50, and d) Model-C25.

However, Model-D50 and Model-D25 were able to reduce the damage considerably, whereas the model with ties and full properties caused the collapse of some orthogonal walls. The damage pattern of Model-D50 and Model-D25 are similar (Fig. 11), but the latter can dissipate more energy with less material.

In general, comparing all the models in terms of energy, the damage and plastic energies of the unreinforced model are lower than Model-C and Model-D. This means that cracking is concentrated in specific areas (at the corners) instead of a more distributed damage. However, this also implies that the façade of the unreinforced model collapses under smaller intensities of the earthquake.

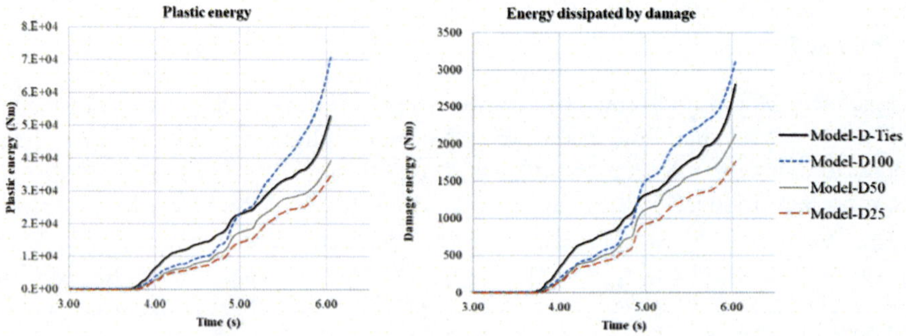

Fig. 9. Comparison between damage energy and plastic energy for Model-D with different properties of the dampers.

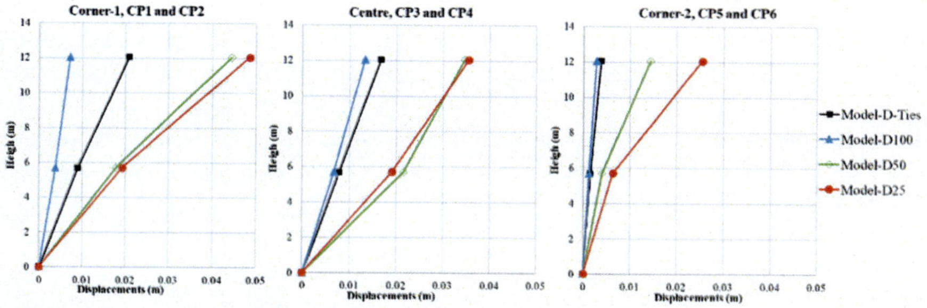

Fig. 10. Maximum displacements of the control points of the façade up to second 6.05 (Model-D).

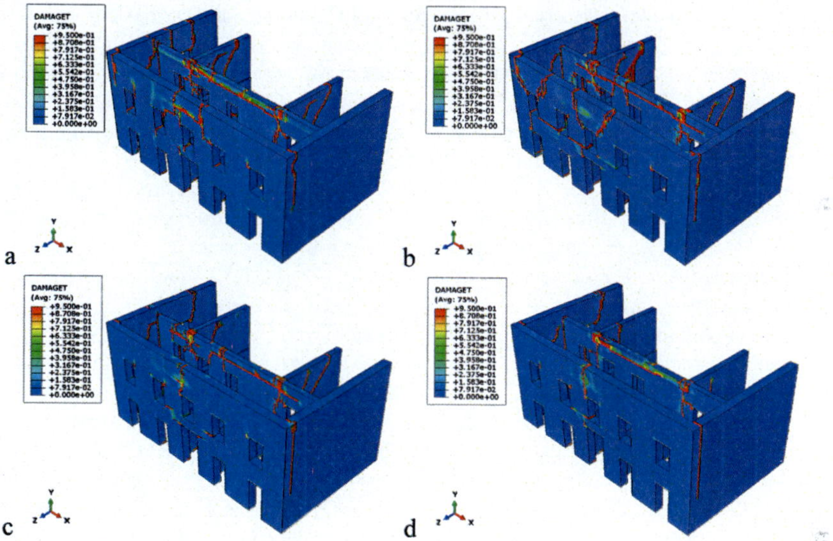

Fig. 11. Damage pattern at second 6.05 (cracks in red) a) Model-D-Ties, b) Model-D100, c) Model-D50, d) Model-D25.

7 Conclusions

One of the biggest challenges for the application of dampers to unreinforced masonry buildings arises from the low ductility of the structural system and the material itself. Masonry cracks under small displacements can occur due to its poor tensile strength. The parametric analysis showed that dampers calculated with the method proposed by Aiken and Kelly are too stiff and can cause damage. A reduction in properties seems to be better for unreinforced masonry. A proper combination of properties (stiffness and viscosity) should be chosen for more deformable dampers even if the displacements are larger than those allowed by rigid connections. The analyses showed that traditional techniques can be improved when dampers are added to the system. Despite the building becoming more flexible, the seismic response improved. That is, a certain level of damage can be accepted for strong earthquakes. Finally, the solutions herein presented must be optimized and other configurations should be studied.

Acknowledgements. This work was supported by the Mexican National Council for Science and Technology (CONACYT) and the Foundation for Science and Technology (FCT) within the scope of project VISCO-D (EXPL/ECI-EGC/0940/2021).

References

1. Soong, T.T., Lai, M.L.: Correlation of experimental results with predictions of viscoelastic damping for a model structure. In: Proceedings of Damping. San Diego, California (1991)
2. Ranaei, O., Aghakouchak, A.A.: Experimental and numerical study on developed elastomeric layers based on natural and butyl matrix rubbers for viscoelastic dampers. Mech. Time-Depend. Mater. **26**(1), 211–233 (2021). https://doi.org/10.1007/s11043-021-09484-2
3. Samali, B., Kwok, K.C.S.: Use of viscoelastic dampers in reducing wind-and earthquake-induced motion of building structures (1995)
4. Shu, Z., Ning, B., Li, S., Li, Z., Gan, Z., Xie, Y.: Experimental and numerical investigations of replaceable moment-resisting viscoelastic damper for steel frames. J. Constr. Steel Res. **170** (2020). https://doi.org/10.1016/j.jcsr.2020.106100
5. Nochebuena-Mora, E., Mendes, N., Lourenço, P.B., Covas, J.A.: Vibration control systems: a review of their application to historical unreinforced masonry buildings. J. Build. Eng. **44** (2021). https://doi.org/10.1016/j.jobe.2021.103333
6. Dassault Systemes Simulia Corp: Abaqus/CAE 2019 (2018)
7. Dassault Systèmes: Abaqus Analysis User's Guide (2016)
8. Mostafawi, H.: Experimental and Numerical Analysis of the Shear Ring (2014)
9. Storheim, M., Notaro, G., Johansen, A., Amdahl, J.: Comparison of ABAQUS and LS-DYNA in simulations of ship collisions. In: Proceedings of the ICCGS. Ulsan (2016)
10. Zheng, G., Nie, H., Chen, J., Chen, C., Lee, H.P.: Dynamic analysis of lunar lander during soft landing using explicit finite element method. Acta Astronaut. **148**, 69–81 (2018). https://doi.org/10.1016/j.actaastro.2018.04.014
11. Chávez, M.M., Durán, D., Peña, F., García, N.: Caracterización mecánica de muestras de mampostería extraídas de templos conventuales mexicanos del siglo XVI. In: Construction Pathology, Rehabilitation Technology and Heritage Management (2020)
12. Guzmán, J.A.S.: Fichas técnicas sobre características tecnológicas y usos de maderas comercializadas en México (2007)
13. Castellazzi, G., D'Altri, A.M., de Miranda, S., Chiozzi, A., Tralli, A.: Numerical insights on the seismic behavior of a non-isolated historical masonry tower. Bull. Earthq. Eng. **16**(2), 933–961 (2017). https://doi.org/10.1007/s10518-017-0231-6
14. Tarque, N., Camata, G., Spacone, E., Varum, H., Blondet, M.: Nonlinear dynamic analysis of a full scale unreinforced adobe model. Earthq. Spectra **30**, 1643–1661 (2014). https://doi.org/10.1193/022512EQS053M
15. Aiken, I.D., Kelly, J.M.: Earthquake Simulator Testing and Analytical Studies of Two Energy-Absorbing Systems for Multistory Structures. Berkeley (1990)
16. Ge, T., Huang, X.H., Guo, Y.Q., He, Z.F., Hu, Z.W.: Investigation of mechanical and damping performances of cylindrical viscoelastic dampers in wide frequency range. Actuators **10** (2021). https://doi.org/10.3390/act10040071

17. Moretti, M., Abruzzese, L., Augliera, P., Azzara, R.: Terremoto in Emilia Romagna: le attività del Pronto Intervento Sismico durante il primo mese di emergenza. Modalità e tempistica. Quaderni di Geofisica (2013)

Resilience and Sustainability: New Challenges and Future Technologies for the Protection of Structures and Infrastructures

State-of-the-Art of Resilience Using Bibliometric Analysis

Melissa De Iuliis[1] (✉), Alessandro Cardoni[2], and Gian Paolo Cimellaro[2]

[1] Sapienza University of Rome, Via Eudossiana 18, 00184 Rome, Italy
melissa.deiuliis@uniroma1.it

[2] Politecnico di Torino, Corso Duca degli Abruzzi, 24, 10129 Torino, Italy
{alessandro.cardoni,gianpaolo.cimellaro}@polito.it

Abstract. The interest in the concept of Resilience in the scientific community has been growing consistently over the past few years to study the functionality and behavior of systems against natural and man-made hazards. This is reflected by the number of journal articles that can be accessed in the Web of Science database. In this paper, a bibliometric and visualization method is applied to explore the status of resilience research in civil engineering applications by analyzing the journal papers published from 1996 to 2020. The bibliometric analysis aims at consolidating the state of the art by identifying influential journals, most cited articles, the geographic distribution of resilience publications including the research institutions by country, the author keywords distribution, and the co-authorship status. The concept of resilience is investigated through eight subject categories identified by the authors in the literature: Recovery time strategies and Downtime, Critical infrastructures, Probabilistic approaches, Fuzzy logic approaches, Structural health monitoring, Health Care facilities, Emergency management and Decision-making, Community and Urban Resilience. Results show that resilience research has increased rapidly since its introduction, most notably in the last seven years. In terms of the geographical region of the studies, most of them have been carried out in the USA, the United Kingdom, China, and Italy. Finally, based on the author keywords analysis, it is possible to observe that recovery strategies, critical infrastructures, vulnerability, and community resilience have attracted prominent attention during the past decade.

Keywords: Resilience · Review · Bibliometric analysis · Bibliometric visualization · Civil engineering

1 Introduction

The increased frequency of natural and man-made disasters worldwide has attracted deep attention to the concept of resilience in recent years. Resilience firstly emerged in academic literature in the 1970s from an ecological study on the functionality responses of the interacting population [1]. Over the last years, the topic of resilience has spread into several fields such as ecology, management, economics, social, and engineering.

G. P. Cimellaro (Ed.): WCSI 2022, LNCE 309, pp. 589–598, 2023.
https://doi.org/10.1007/978-3-031-21187-4_49

Hence, resilience itself is a multidisciplinary and broad concept and its definition is highly variable depending on the subject area.

In engineering, the concept of resilience is relatively new compared to other fields [2]. Resilience engineering is related to the capacity of a system to absorb environmental shocks and maintain its functionality and it is defined as the ability to "withstand stress, survive, adapt and bounce back from a crisis or disaster and rapidly move on" [3]. Bruneau et al. [4] defined resilience as "the ability of organizations and communities to mitigate hazards, contain the effects of disasters and carry out recovery strategies to mitigate the effects of further earthquakes". The resilience of a system depends on its performance level, which ranges between 0% (e.g., no service is available) and 100% (e.g., high level of the service). When a disaster occurs at time t_0 causing damages to the system, its functionality drops until it is restored to its initial state over the recovery period ($t_1 - t_0$).

Numerous applications and frameworks for evaluating resilience in the civil engineering field are available in the literature. Particularly, these studies address the resilience evaluation of systems subjected to natural hazards. Most of the studies focus on the resilience quantification aimed at planning for mitigation, recovery strategies of a physical system (e.g., lifelines). For instance, Liu et al. [5] developed a framework to combine dynamic modeling with resilience analysis. The framework has been applied to two interconnected critical infrastructures for a numerical evaluation of the resilience in terms of design, operation, and control parameter values for given failure scenarios. A conceptual model in which the concept of resilience is correlated to the concept of recovery is presented by [6]. In their work, the relationships between a community's business, lifeline networks, and neighborhood are related.

Most authors have been developing frameworks to quantify resilience at the community level. A quantitative framework for assessing resilience at the community level is the PEOPLES framework [7]. PEOPLES framework, which improves the resilience research at the Multidisciplinary Center of Earthquake Engineering Research (MCEER), consists of seven dimensions: Population, Environment, Organized government services, Physical infrastructures, Lifestyle, Economic, and Social capital [8, 9]. Recently, probabilistic, and fuzzy approaches have been introduced to estimate the recovery time and the resilience of buildings and infrastructures after a seismic event [10–13]. Moreover, the concept of resilience is also studied in the structural health monitoring research area [14–17]. Zhang et al. [18] described a strategy for enhancing structural resilience through the implementation of structural health monitoring (SHM) technologies into civil infrastructures.

The great variety of applications and methodologies that can be found in the literature shows the necessity of systematic analysis to indicate how resilience research has evolved over time and become an emerging topic in the scientific community. Several literature reviews have been conducted on resilience, recently. For instance, Annarelli and Nonino [19] reviewed studies on organizational resilience emphasizing strategic and operational management areas. A bibliographic review of papers on resilience related to industrial ecology was performed by Meerow et al. [20]. In their work, the design of resilient industrial ecosystems is discussed. A year later, a critical review of the qualitative and quantitative resilience frameworks from a built-environment engineering perspective was provided by Cerè et al. [21]. Furthermore, Fraccascia et al. [22] reviewed the state of the art on the resilience of complex systems in different research areas through bibliometric tools.

The main purpose of this paper is to identify trends in resilience research in the context of civil engineering through a bibliometric and visualization analysis based on quantitative analysis of journal papers published on Web of Science (WoS). The remainder of this paper is structured as it follows: Sect. 2 introduces the data source and bibliometric methodology. Section 3 illustrates the results and their classification, including historical analysis, the geographical distribution of the research works, the author keyword analysis among the paper reviewed. Finally, Sect. 4 summarizes the findings, identifies the gaps in literature and directions for future research.

2 Literature Review Methodology

The study is based on bibliometric research conducted in March 2021 to investigate the impact and importance of resilience in the field of civil engineering. The expected results of conducting the bibliometric analysis proposed in this work can be replicated by implementing the steps presented in Fig. 1.

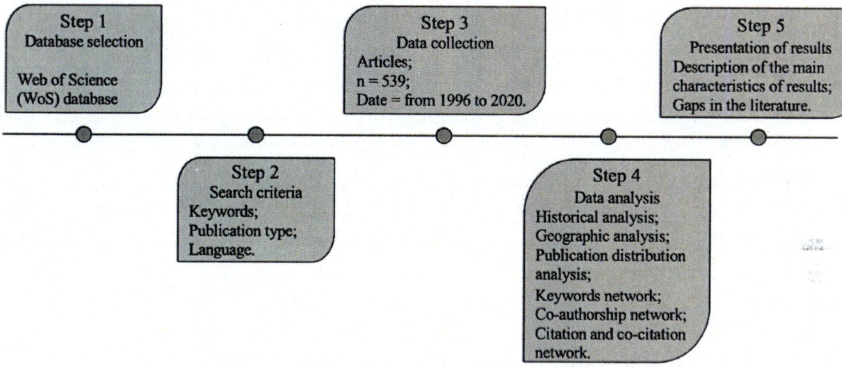

Fig. 1. Steps of the bibliometric analysis for mapping the state of the art

Five steps are followed: (1) choosing the databases to cover relevant literature in the field of study; (2) defining inclusion and exclusion criteria and selecting the studies based on the criteria; (3); collecting and extracting the bibliometric data for the subsequent analyses, (4) performing quantitative data analysis through the construction of visualization bibliometric networks, and (5) presentation and discussion of the results. The data was retrieved from WoS, a widely used database. WoS is an academic citation indexing and search service of Thomson Reuters' Web of Knowledge that covers journals, conference papers, websites, and patents [23]. Once the scientific platform to be used has been established, the search criteria of the publications need to be settled. Generally, the main search criteria applied in the bibliometric analysis are specific terms to the field of study, area of knowledge, publication type, and language. To identify resilience in civil engineering applications, keyword searches within the titles, keywords, and abstracts of the different research outputs have been applied. No limit of the year was applied to the search criterion. However, the oldest article, according to the keywords, was from 1996. Thus, research articles were extracted from a period of 1996–2020. Only journal articles that have been cited and published in English were selected for further analysis. Furthermore, those papers having titles and abstracts where the word resilience does not appear among the keywords were omitted. A very strict and generally limiting way of categorizing publications was applied in this work to accurately reflect the dissemination of resilience in civil engineering applications in academic research. First, postprocessing of the literature data was necessary to exclude papers that were insufficiently relevant (e.g., papers where resilience was not the main topic but only mentioned as a suggestion for future works, papers where the word resilience was among keywords, but they did not really address the topic). All the papers including titles and abstracts were read and analyzed and the selection was refined by removing non-resilience-related works. After these steps, 539 papers were selected as constituting the final database.

The full record and cited references include bibliometric parameters such as publication information on the title, author(s), year of publication, abstract, source title (i.e., journal name), affiliation, author's keywords, language, and the number of citations (if any). The exported records were used for different analyses: (i) historical analysis was conducted for understanding the temporal trend of resilience research in civil engineering; (ii) geographical distribution analysis was carried out to analyze the research output in different countries and leading institutions or universities taking into account the authors' affiliations as the criterion to locate the place of their origin; (iii) publication distribution analysis was investigated to highlight the most popular journal, the most cited records along with the corresponding authors in the realm of resilience research, and finally (iii) keywords analysis was performed for the author's keywords to understand the nature, linkages, and concepts used by the author(s). The analyzed database was imported into VOS-viewer 1.6.14 software to build visualization networks of major co-occurrence keywords to create easily readable and comprehensive keywords in figures [24–26].

3 Results and Discussion

3.1 Historical Analysis

Results showed an increasing number of studies on the concept of resilience applied to civil engineering applications. The analyzed publications started in 1996 to study resilience in civil engineering applications. Figure 2 shows the annual number of the most productive papers with a steady increase after 2011 and was contributed by the major industrialized countries (USA, UK, Italy, China, etc.). According to the extracted database, the first study which used the concept of resilience in civil engineering was published by Fox and Suidan [27]. Most of the research on resilience within civil engineering was published during 2013–2018, i.e., 60%. This means a growing knowledge accumulation and research interest of the scientific community in resilience discourse.

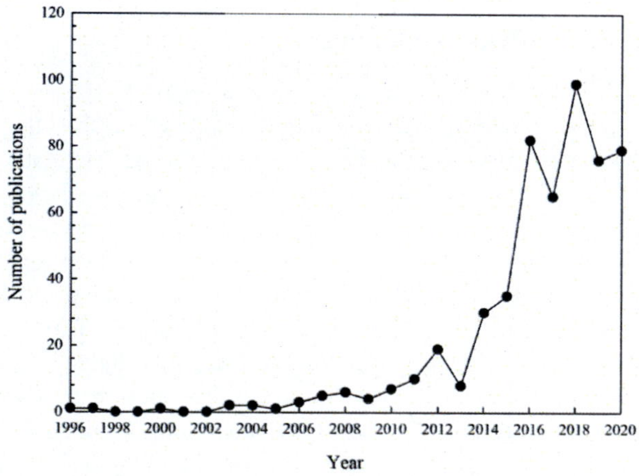

Fig. 2. Most productive papers on resilience in civil engineering applications (1996–2020)

3.2 Number of Publications by Countries and Institutes

The 539 papers analyzed were published in 52 different countries worldwide (Fig. 3). A mapping approach is used to represent the country's performance in civil engineering resilience where the authors' affiliation in the records was used as the criterion to locate the place of their origin. The most productive country in this regard is the USA with 307 cited publications (i.e., 57%) followed by UK, Italy, China, and Australia with 63, 56, 54, and 27 cited publications, respectively. Few to no authors come from Central Asia, the Middle East, and Western Africa. Therefore, the mapping approach indicates the dominance of western researchers. This can be justified by considering that researchers in these countries have resilience-related publications in non-English languages, which this study does not capture.

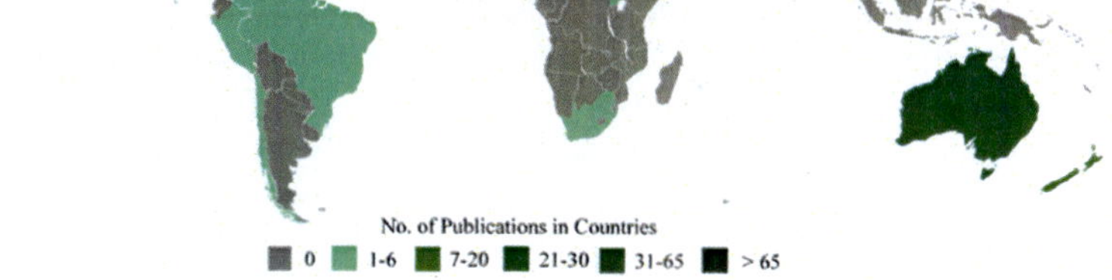

Fig. 3. Distribution of publications by country

3.3 Number of Publications by Journals

This section aims at analyzing the most popular journals in the realm of resilience research in the civil engineering field, the highly cited articles on the topic, and the authorship. The analysis examines the 539 academic articles published in 94 journals and authored by 1431 researchers. Table 1 lists the top 5 journals in which the analyzed publications appear. The top journal with 38 cited papers in resilience in civil engineering is *Natural Hazards Review,* published by the American Society of Civil Engineers (ASCE). This is followed by the *Journal of Infrastructure Systems* (34 articles) and the *Journal of Structural Engineering* (33 articles). Furthermore, the actual number of citations that each cited resilience article has attracted was investigated. The top 5 journal articles with the most citations and their authors, publication year, and the title of the journals are presented in Table 2. The most prominent contributor in the area is Professor M. Bruneau who is the author in four of the top 5 cited journal papers. Similarly, Professors G.P. Cimellaro (Italy) and A. Reinhorn (USA) have also achieved excellent recognition with their names appearing as authors and co-authors in the top 5 paper list.

Table 1. Top 5 journals with the largest number of publications (1996–2020)

Rank	Journal	Year of first publication	No. of articles
1	Natural Hazards Review	1986	38
2	Journal of Infrastructure Systems	1995	34
3	Journal of Structural Engineering	1955	33
4	Engineering Structures	1970	26
5	Earthquake Spectra	1984	25

3.4 Author Keywords Analysis

Keyword analysis reveals the exact context of resilience studies in the civil engineering area. VOSviewer software was used to construct a visualization map to picture the linkages and co-occurrence of chosen keywords by studied authors [28]. In the map, each keyword is represented by a circle whose diameter represents the number of links with other keywords. Hence, a larger circle means more links with other keywords. The thickness of the line between two or more circles represents the frequency of co-occurrence of the keywords together. Furthermore, the distance between two nodes reflects the strength of the relation between two keywords.

Figure 4 shows the major keywords used in resilience studies in the civil engineering area. The figure represents a minimum of 5 times the co-occurrence of a keyword in extracted records. Only 61 out of a total of 1453 authors' keywords met the threshold covering the studied interval, which equals to 4.13%. VOSviewer divided the keywords of resilience-related publications into 6 clusters. Common keywords revealed are "community resilience", "sustainability", "recovery", "climate change", and "earthquake".

Table 2. Top 5 articles with most citations (1996–2020)

Rank	Title	Year	Author(s)	No. of citations	Journal
1	A framework to quantitatively assess and enhance the seismic resilience of communities	2003	Bruneau, M; Chang, SE; Eguchi, RT; et al.	1490	Earthquake Spectra
2	Urban hazard mitigation: creating resilient cities	2003	Godschalk, David R.	534	Natural Hazards Review
3	Framework for analytical quantification of disaster resilience	2010	Cimellaro, Gian Paolo; Reinhorn, Andrei M; Bruneau, Michel	479	Engineering Structures
4	Multiobjective genetic algorithms for design of water distribution networks	2004	Prasad, TD; Park, NS	323	Journal of Water Resources Planning and Management
5	A three-stage resilience analysis framework for urban infrastructure systems	2012	Ouyang, Min; Duenas-Osorio, Leonardo; Min, Xing	306	Structural Safety

Table 3 lists the 5 most frequently used author keywords along with their occurrences in the analyzed publications and total link strength with other keywords. The keyword "resilience" has the highest frequency of 147. Other keywords with a high frequency include "community resilience" (37), "sustainability" (24), and "recovery" (22). The total link strength of a node is the sum of link strengths of such node over all the other nodes and it can be used to indicate the relationships among two nodes [29]. For example, the node "resilience" presents thicker lines with "community resilience" (50), "recovery" (58), and "sustainability" (36). The relationship between "resilience" and "recovery" (link strength of 58) implies the close integration of recovery programs within the resilience estimation.

4 Conclusions

This paper proposes a bibliometric literature review of the 539 publications retrieved from WoS regarding resilience in civil engineering applications. The outcomes of the study are investigated with a temporal analysis aiming at identifying intellectual communities, leading researchers, key knowledge of resilience in civil engineering applications. The following conclusions were drawn from this study:

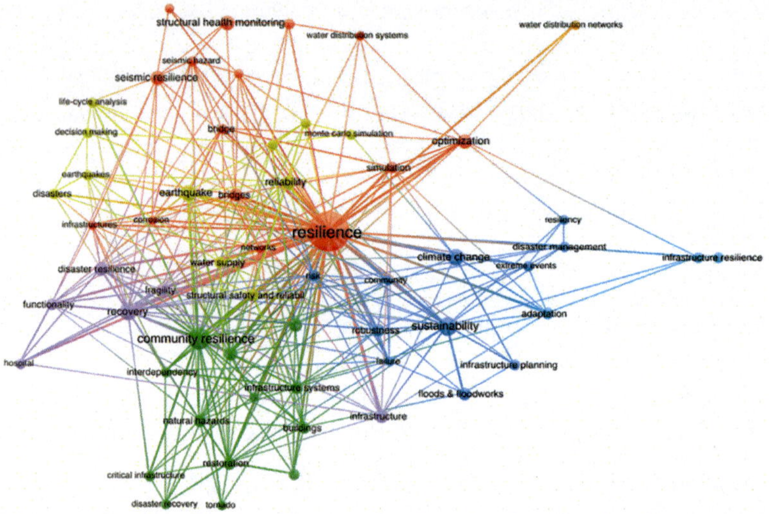

Fig. 4. Map of civil engineering resilience of author's keywords (5 co-occurrence criteria with 61 keywords)

Table 3. Top 5 most frequency of author keywords (1996–2020)

Rank	Author keywords	Occurrences	Total link strength
1	Resilience	147	220
2	Community resilience	37	50
3	Sustainability	24	36
4	Recovery	22	58
5	Earthquake	18	31

1. Publications in resilience applied to civil engineering have significantly increased during the last 7 years, meaning that there is a huge number of journals, institutions, and countries involved in resilience research.
2. The research output is distributed worldwide. However, a clear disparity has been observed among developed and developing countries of the world. Industrialized countries such as the USA, China, UK, and Italy predominated in terms of article count and citation scores.
3. By analyzing the distribution of author keywords, it could be concluded that the research on community resilience, sustainability, recovery, and infrastructure attracted more attention, which are probably the most representative topics to the frontiers of resilience. Furthermore, many author keywords emerged after the year 2013, showing a growing interest in resilience research probably due to numerous natural and man-made hazards.

Compared to the more traditional literature reviews, the bibliometric method can be generalized and used as a tool to map discipline knowledge. A periodically bibliometric analysis should be conducted to further improve the resilience knowledge map provided in this study, as the field continues to evolve.

References

1. Holling, C.S.: Resilience and stability of ecological systems. Annu. Rev. Ecol. Syst. **4**(1), 1–23 (1973)
2. Hosseini, S., Barker, K., Ramirez-Marquez, J.E.: A review of definitions and measures of system resilience. Reliab. Eng. Syst. Saf. **145**, 47–61 (2016)
3. Wagner, I., Breil, P.: The role of ecohydrology in creating more resilient cities. Ecohydrol. Hydrobiol. **13**(2), 113–134 (2013)
4. Bruneau, M., et al.: A framework to quantitatively assess and enhance the seismic resilience of communities. Earthq. Spectra **19**(4), 733–752 (2003)
5. Liu, J.J., Reed, M., Girard, T.A.: Advancing resilience: an integrative, multi-system model of resilience. Person. Individ. Differ. **111**, 111–118 (2017)
6. Miles, S.B., Chang, S.E.: Modeling community recovery from earthquakes. Earthq. Spectra **22**(2), 439–458 (2006)
7. Cimellaro, G.P., et al.: PEOPLES: a framework for evaluating resilience. J. Struct. Eng. **142**(10), 04016063 (2016)
8. Renschler, C.S., et al.: A Framework for Defining and Measuring Resilience at the Community Scale: The PEOPLES Resilience Framework. MCEER Buffalo, NY (2010)
9. De Iuliis, M., Kammouh, O., Cimellaro, G.P.: Measuring and improving community resilience: a fuzzy logic approach (2022). arXiv preprint arXiv:2204.04335
10. De Iuliis, M., et al.: Resilience of the built environment: a methodology to estimate the downtime of building structures using fuzzy logic. In: Resilient Structures and Infrastructure, pp. 47–76. Springer (2019)
11. De Iuliis, M., et al.: Downtime estimation of building structures using fuzzy logic. Int. J. Disaster Risk Reduct. **34**, 196–208 (2019)
12. Hosseini, S., Barker, K.: A Bayesian network model for resilience-based supplier selection. Int. J. Prod. Econ. **180**, 68–87 (2016)
13. Yodo, N., Wang, P.: Resilience modeling and quantification for engineered systems using Bayesian networks. J. Mech. Des. **138**(3) (2016)
14. Gattulli, V., Potenza, F., Piccirillo, G.: Multiple tests for dynamic identification of a reinforced concrete multi-span arch bridge. Buildings **12**(6), 833 (2022)
15. Gattulli, V., et al.: Dynamical models of a suspension bridge driven by vibration data. Smart Struct. Syst. Int. J. **27**(2), 139–156 (2021)
16. Gattulli, V., Lofrano, E., Paolone, A., Potenza, F.: Measured properties of structural damping in railway bridges. J. Civ. Struct. Heal. Monit. **9**(5), 639–653 (2019). https://doi.org/10.1007/s13349-019-00358-3
17. Domaneschi, M., et al.: Laboratory investigation of digital image correlation techniques for structural assessment. In: Bridge Maintenance, Safety, Management, Life-Cycle Sustainability and Innovations, pp. 3260–3266. CRC Press (2021)
18. Zhang, Y., Ayyub, B., Huang, H.: Enhancing civil infrastructure resilience with structural health monitoring. In: Resilience Engineering for Urban Tunnels, pp. 3–12 (2018)
19. Annarelli, A., Nonino, F.: Strategic and operational management of organizational resilience: current state of research and future directions. Omega **62**, 1–18 (2016)

20. Meerow, S., Newell, J.P., Stults, M.: Defining urban resilience: a review. Landsc. Urban Plan. **147**, 38–49 (2016)
21. Cerè, G., Rezgui, Y., Zhao, W.: Critical review of existing built environment resilience frameworks: directions for future research. Int. J. Disaster Risk Reduct. **25**, 173–189 (2017)
22. Fraccascia, L., Giannoccaro, I., Albino, V.: Resilience of complex systems: state of the art and directions for future research. Complexity **2018** (2018)
23. Reuters, T.: Social sciences citation index. In: Web of Knowledge. Thomson Reuters (2012)
24. Van Nunen, K., et al.: Bibliometric analysis of safety culture research. Saf. Sci. **108**, 248–258 (2018)
25. Cancino, C., et al.: Forty years of computers and industrial engineering: a bibliometric analysis. Comput. Ind. Eng. **113**, 614–629 (2017)
26. Gall, M., Nguyen, K.H., Cutter, S.L.: Integrated research on disaster risk: is it really integrated? Int. J. Disaster Risk Reduct. **12**, 255–267 (2015)
27. Fox, P., Suidan, M.T.: Shock and transient loading on anaerobic reactor coupled with adsorber. J. Environ. Eng. **122**(1), 18–24 (1996)
28. Van Eck, N.J., Waltman, L.: VOSviewer manual. Leiden: Univeristeit Leiden **1**(1), 1–53 (2013)
29. Pinto, M., Pulgarín, A., Escalona, M.I.: Viewing information literacy concepts: a comparison of two branches of knowledge. Scientometrics **98**(3), 2311–2329 (2013). https://doi.org/10.1007/s11192-013-1166-6

Observed Seismic Behaviour of Base Isolation Systems in Italy

Giacomo Buffarini[1], Paolo Clemente[1], Ugo Ianniruberto[2], Chiara Ormando[2](✉), and Federico Scafati[3]

[1] ENEA, Casaccia Research Centre, 00123 Rome, Italy
[2] Department of Civil and Computer Science Engineering, University of Tor Vergata, 00133 Rome, Italy
chiara.ormando@uniroma2.it
[3] Department of Civil, Construction-Architectural and Environmental Engineering, University of L'Aquila, 67100 L'Aquila, Italy

Abstract. The observed behaviour of seismically isolated structures in different countries allowed testifying the suitability of this technique to protect structures against strong earthquakes. In Italy, up to now, base isolated buildings have been subjected only to low energy earthquakes. This paper describes and compares the observed behaviours under a low energy earthquake of three buildings with different isolation systems. In the first one, seismically isolated by means of HDRBs, the structure showed resonance frequencies higher than the design ones but the decoupling was always guaranteed also thanks to the high stiffness of the superstructure. In the second one, seismically isolated with CSSs, the onset of motion was governed by the friction, which was much higher than the design one. Finally, in a building seismically isolated with HDRBs and SDs, the onset of motion was governed by the SDs and the successive motion by the HDRBs.

Keywords: Seismic isolation · Observed seismic behaviour · Low energy earthquake · Rubber bearings · Flat surface sliders · Curved surface sliders

1 Introduction

The number of structures protected by seismic isolation increased remarkably all over the world in the last decades, so a complete list of them cannot be done. Nowadays seismic isolation is a technology that is ready for a wide application and is used both to build new constructions and to retrofit the existing ones [1-6]. The excellent behavior exhibited by seismically isolated buildings during strong earthquakes encouraged the adoption of base isolation as the best system to protect buildings against seismic events. Actually, the extensive adoption of the new anti-seismic systems started after strong earthquakes in several countries [7].

For example, the number of seismic isolated buildings in Japan increased very much after the 1995 Hyogo-ken Nanbu earthquake ($M = 7.3$), when two seismically isolated buildings in the epicentre area, near Kobe (among these the communication minister in

G. P. Cimellaro (Ed.): WCSI 2022, LNCE 309, pp. 599–610, 2023.
https://doi.org/10.1007/978-3-031-21187-4_50

Sanda City, about 30 km far from the epicentre), reported no damage during the quake. Later, a reinforced base isolated concrete building in Ojiya City supported very well the 2004 Mid Niigata earthquake ($M = 6.8$). All the isolated buildings, more than 100, which were affected by the 2011 Tohoku earthquake, behaved quite well, even though they had been designed to withstand less severe earthquakes. Among these, the 4-storey National Western Art Le Corbusier Museum in Tokyo, retrofitted in 1999 by inserting high damping natural rubber bearings in a sub-foundation [8]. After these events and other unmentioned earthquakes in which all seismic isolated buildings behaved very well, the number of applications in Japan increased significantly.

In China, base isolation became very popular when two concrete seismic isolated buildings and a 6-storey masonry building showed an excellent behaviour during the 2008 Wenchuan earthquake ($M_W = 7.9$). Furthermore, during the Lushan earthquake ($M_W = 7.0$) of April 20th, 2013, two primary school buildings, one close to the other, one conventionally founded, and the second base isolated, suffered very different seismic accelerations [9, 10]. The same occurred for the two buildings of the county hospital (seven floors above ground and one basement), one with conventional foundations and one with base isolation. The first was damaged to partitions, roof and contained equipment, and was unusable after the earthquake; on the contrary, the seismically isolated building remained fully undamaged and operational, allowing to heal thousands of injured people.

Finally, in California the USC hospital in Los Angeles showed very good behaviour during the 1994 Northridge Earthquake. It was about 30 km far from the epicentre [11-13].

While the behaviour shown by seismically isolated buildings under severe earthquakes was always good, the behaviour under low energy earthquakes has not been investigated in detail. This is true for both high damping rubber bearings (HDRB) systems, which under low intensity dynamic actions exhibit vibrations of lower amplitudes but characterized by higher frequencies, due to the more rigid behaviour of the rubber at lower strains [14, 15], and curved surface sliders (CSS) systems, where the friction could be too high for the onset of motion [16]. This occurred, for example, for the Jovine School in San Giuliano di Puglia, Italy [17], seismically isolated by means of HDRBs and SDs, whose accelerometric network recorded some seismic events of low energy at the site but no evidence of the isolation participation was found. If the seismic isolation doesn't work as assumed in the design, the seismic effects in the superstructure could be higher than those evaluated during the design phase [18]. The onset of motion should be particularly analysed.

It is worth noting that the isolation devices must have low horizontal stiffness, which allows the relative displacements between the upper and lower faces in case of seismic events, but also an adequate stiffness in order to avoid vibrations under low horizontal actions, such as wind, traffic or ambient vibrations.

This paper describes and compares the observed behaviour, under a low energy earthquake, of three different structures with different isolation systems. The first structure is seismically isolated by means of an isolation system made of HDBRs only, the second one by means of an isolation system composed of CSSs and the last one by an isolation system with HDRBs and SDs. The behaviour of these isolation systems was analysed under the same quake that occurred on October 30th, 2016, at Norcia, Italy. Due to the distance from the epicentre, the effects at the site were quite low.

2 Main Features of the Isolation Devices

Base isolation systems are usually designed to face the design earthquake at the site and the equivalent characteristics, in terms of stiffness and damping, are used for the seismic check in the linear analysis [19, 20]. Under very low intensity earthquakes, and also under ambient vibrations, the isolation system could not be put in action and the higher modes of the superstructure could be activated. As a result, some amplification of the response in terms of acceleration could occur. In order to analyse in detail this aspect the case of HDRB systems and that of CSS systems must be distinguished.

The force-displacement curves relative to a cycle test of HDRBs show that their behaviour varies strongly from one cycle to the other, depending on the maximum displacement reached (Fig. 1). Therefore, also in the linear analysis, assuming for each cycle the equivalent values relative to the maximum displacement, both shear modulus and damping ratio vary very much as a function of the shear strain γ. As a result, the values for very low shear strain could be quite greater than the equivalent values, usually evaluated for the shear strain $\gamma = 1$.

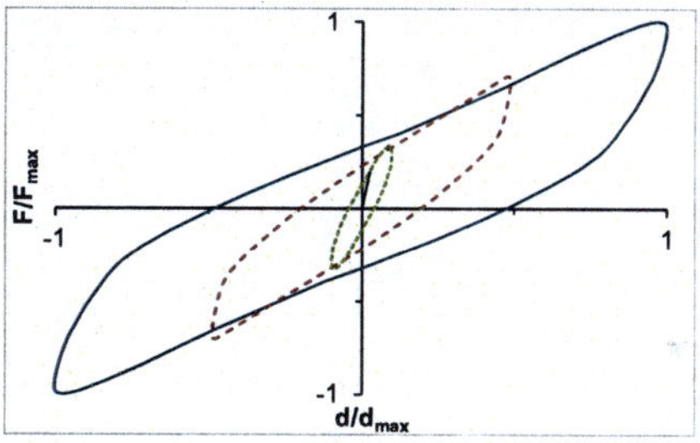

Fig. 1. Typical force-displacement diagram for a HDRB.

The behaviour of CSSs is strongly influenced by the static frictions at the onset of motion (μ_0) and at the maximum displacement (μ_1) when the velocity sign changes (Fig. 2) [21, 22]. Furthermore, the friction μ during the dynamic phase showed significant variations, even though it is assumed to be constant. In any case, it is $\mu < \mu_1 < \mu_0$.

Finally, the behaviour of SDs can be modelled with a rigid-perfectly plastic law, which can be seen as the limit case of Fig. 2, when $R \to \infty$ and so $k \to 0$.

3 Case Studies

The following seismically isolated buildings, provided with monitoring systems, are considered in this paper:

- The Operative Centre in the Civil Protection Centre of Umbria Region at Foligno, instrumented by means of 12 accelerometers [23].

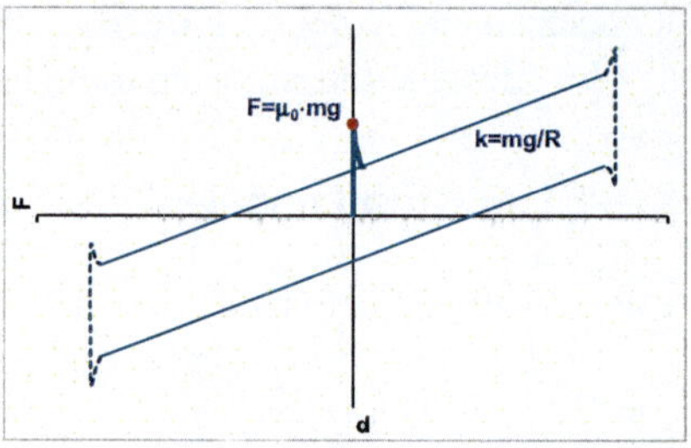

Fig. 2. Typical force-displacement diagram for a CSS.

- The C4 building of the C.A.S.E. project in L'Aquila, built for temporarily hosting the homeless residents after the April 6th, 2009, earthquake, instrumented with 10 accelerometers [24].
- The Forest Ranger Building in the Civil Protection Centre of Umbria Region at Foligno, instrumented by means of 12 accelerometers [25].

In order to verify the seismic behaviour of these buildings, the seismic event that caused the maximum effects at the sites was considered. This is also the earthquake of the seismic sequence with the maximum magnitude. It occurred on October 30th, 2016 ($M_w = 6.5$), at a distance of 36 km from the two buildings at Foligno and 62 km from the building at L'Aquila.

3.1 The Civil Protection Centre at Foligno, Italy

The building that hosts the Operative Centre of the Umbria Civil Protection in Foligno has a very interesting architectural design with a hemispherical shape. It has four storeys above the ground level and an underground storey. The superstructure, formed of ten flying buttresses, is supported by ten HDRB isolators, located along the perimeter under the flying buttresses (Fig. 3). The foundations are composed of reinforced concrete plinths, each of them supported by four piles and are connected to each other by means of beams. The HDRB devices (diameter = 1000 mm, total thickness of the rubber layers $t_e = 240$ mm, shear modulus $G = 0.4$ N/mm^2, damping $\xi = 15\%$) have an equivalent horizontal stiffness $K_e = 1310$ N/mm. The design fundamental frequency of the isolated structure was equal to 0.38 Hz (fundamental period = 2.66 s) [26, 27].

A permanent seismic monitoring network was installed in the building. It is composed of twelve accelerometers, deployed in the building as shown in Fig. 3.

The analysis of the time histories recorded at the basement, at the first floor just above the isolation system and at the top (Fig. 4) pointed out that no amplification occurred from the basement to the top. The frequency content changed significantly (Fig. 5) because of the variability of the isolator stiffness. It is worth noting that the value of the fundamental frequency (about 1.0 Hz) is quite far from the first resonance frequencies of the superstructure considered fixed at its base. Furthermore, the time frequency analysis pointed out that the resonance frequency diminishes during the strong phase and recovers the original value at the end of the record. This occurrence also justifies the presence of more peaks around 1.0 Hz in the Fourier spectra.

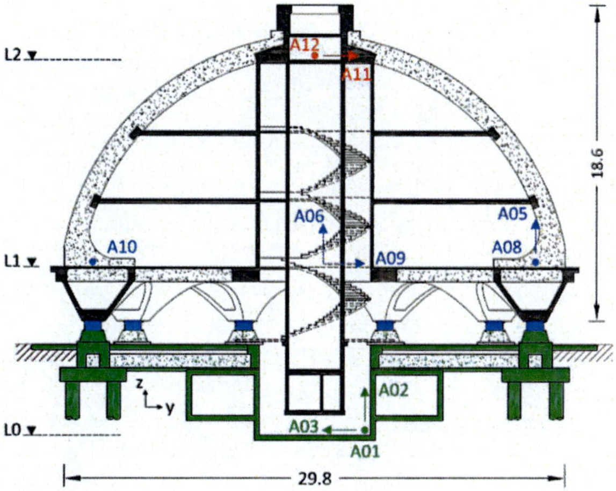

Fig. 3. Scheme of the building and deployment of the sensors.

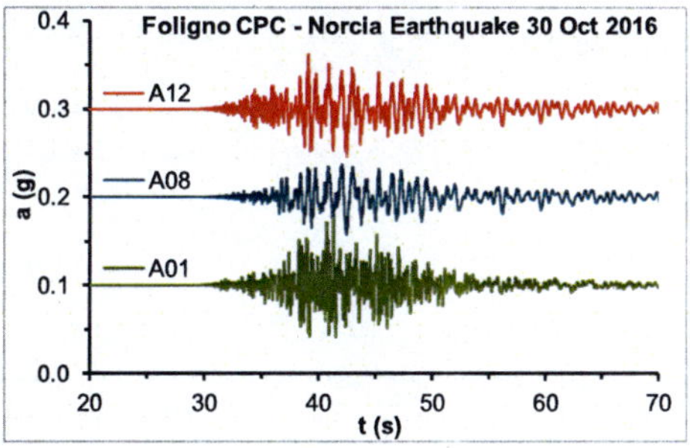

Fig. 4. Time histories recorded during the earthquake of 30 October 2016, at L0 (A01), L1 (A08) and L2 (A12) in *x* direction.

The relative particle motion between the gravity centre of L1 and L0 is plotted in Fig. 6. It can be considered representative of any isolators due to the translational characteristic of the movement. The maximum relative displacement was about 10 mm: no preferential direction can be seen.

3.2 Residence Building in L'Aquila

The superstructure of the C4 building of the C.A.S.E. Project is composed of a three story building in wood X-LAM panels raising up from a reinforced concrete slab, 21 × 57 m in plan and 0.50 m thick, supported by 40 single curved surface sliders [28]. These isolation devices are placed at the top of 40 circular hollow cross-section steel columns, with a diameter of 0.8 m. They are 2.7 m tall and rise up from a reinforced concrete foundation plate, which has the same size of the slab (Fig. 7).

The building is equipped with a seismic monitoring system (Fig. 7), composed of a three-axial accelerometer located at the foundation slab (L0: A08, A09 and A10), three

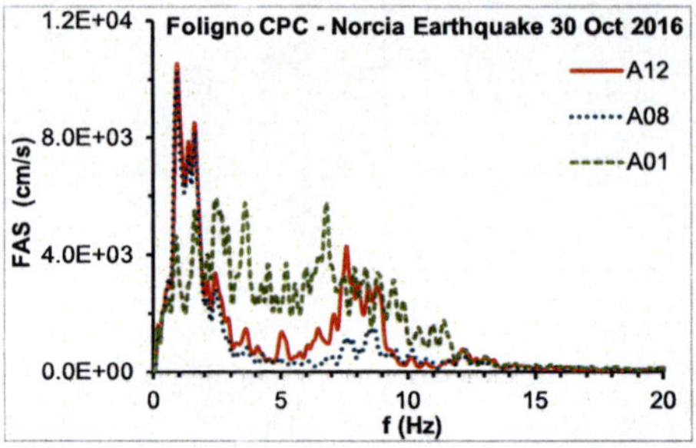

Fig. 5. Fourier spectra, at L0 (A01), L1 (A08) and L2 (A12) in x direction.

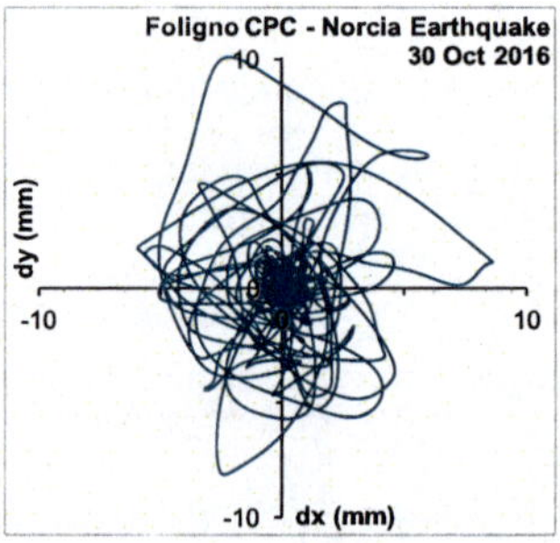

Fig. 6. Particle motion between the gravity centres of L1 and L0.

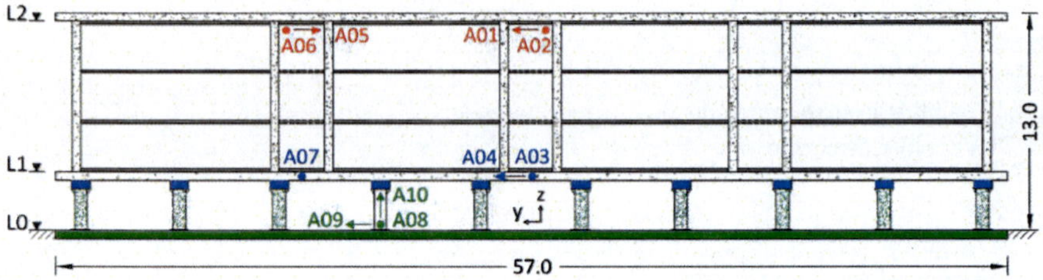

Fig. 7. Vertical section of the C4 building in L'Aquila and accelerometric network.

accelerometers at the base slab of the superstructure, just above the isolation interface (L1: A03, A07 and A04) and four accelerometers at the top of the building (L2: A02, A06, A01 and A05).

The recordings obtained during the considered event are plotted in Fig. 8. The amplification from the basement to the top is apparent and the absence of a suitable filtering is confirmed by the Fourier spectra (Fig. 9). In Fig. 10a the drift between the relative particle motion between the first floor and the basement is plotted for a corner isolator. A relative displacement of 5 mm can be observed in x direction, while the displacement in y direction is negligible. In Fig. 10b the inertial forces due to the earthquake are plotted and compared to the maximum friction forces evaluated considering the actual vertical loads on the structure. It can be seen that the force required to put in motion the isolation system is overtaken only in a few instants so that the building behaves like a fixed base structure for most of the seismic event due to the high static friction in the isolation devices.

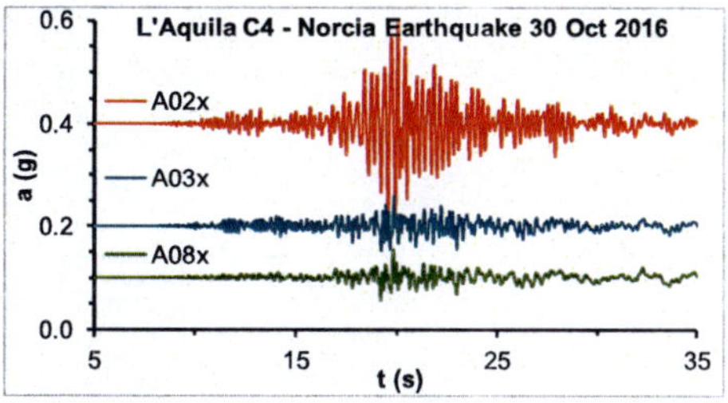

Fig. 8. C4 building: accelerometric recordings obtained during the Norcia Earthquake at L0, L1 and L2 (elaboration of data of DPC-PCM, Italy).

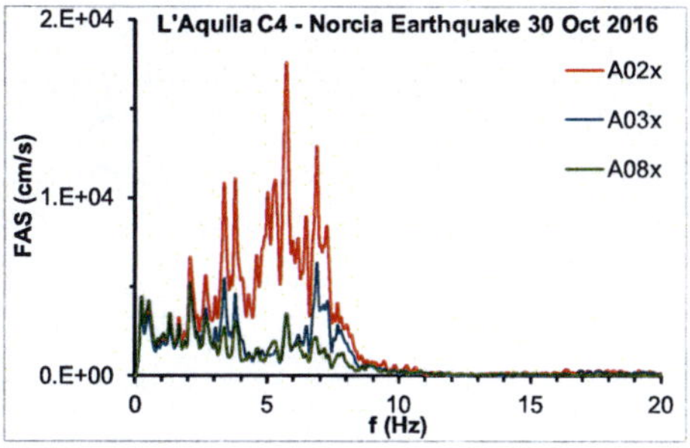

Fig. 9. C4: Fourier spectra at L0, L1 and L2 (elaboration of data of DPC-PCM, Italy).

3.3 The Forest Ranger Building at Foligno

The Forest Ranger building of the Umbria Civil Protection Centre in Foligno is a seismically isolated reinforced concrete building (Fig. 11). It has an underground level and two floors above the ground. The maximum dimensions in plan are about 16 × 31 m, in x and y direction, respectively. The inter-floor heights are 3.14, 4.14, and 3.34 m, for the underground, first and second level, respectively.

The isolation system is composed by 12 HDRBs of diameter 700 mm (shear modulus $G = 0.4$ N/mm^2, damping $\xi = 15\%$, equivalent horizontal stiffness = 541 N/mm), located along the perimeter of the main rectangular portion; 4 flat slider devices (SD) with a lubricated steel-PTFE (polytetrafluoroethylene) interface, having a nominal friction factor of 1.0%, located at the internal column of the main rectangular portion; 4 HDRBs of diameter 550 mm (shear modulus $G = 0.4$ N/mm^2, damping $\xi = 15\%$, equivalent horizontal stiffness = 317 N/mm), located at the columns external to the main portion.

The Forest Ranger building has been instrumented by means of twelve accelerometers, deployed in the building as follows (Fig. 11): three accelerometers (A01, A02 and

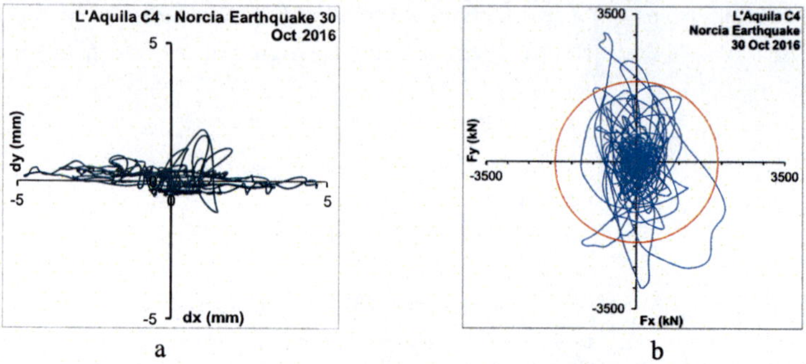

Fig. 10. C4 building: (a) particle motion at a corner isolator; (b) Inertial forces (blu line) compared to the maximum friction forces (red circle).

A03) were on the basement, under the isolation system; six accelerometers (A04, …, A09) were on the first floor, just above the isolation system; three accelerometers (A10, A11 and A12) were at the top of the building. In Fig. 12 the recordings obtained in the transversal direction are shown. It is apparent the absence of amplification to the top and also a change in the frequency content, confirmed also by the Fourier spectra (Fig. 13), in which the amplification at a frequency around 1 Hz is apparent. In Fig. 14 the relative particle motion between the first floor and the basement is plotted for a corner isolator.

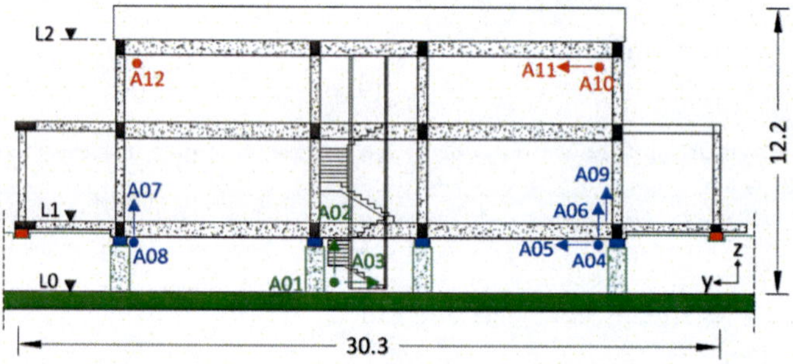

Fig. 11. The Forest Ranger Building in the Civil Protection Centre of Umbria Region at Foligno, Italy, and sensor deployment.

4 Conclusions

Base isolated buildings are characterised by low resonance frequencies when subject to strong earthquakes, but exhibit vibrations of lower amplitudes with higher frequencies under low intensity dynamic actions. In this paper, three different case studies have been presented:

- A building seismically isolated by means of HDRBs only that, under a low energy earthquake at the site, showed a fundamental frequency much higher than that assumed

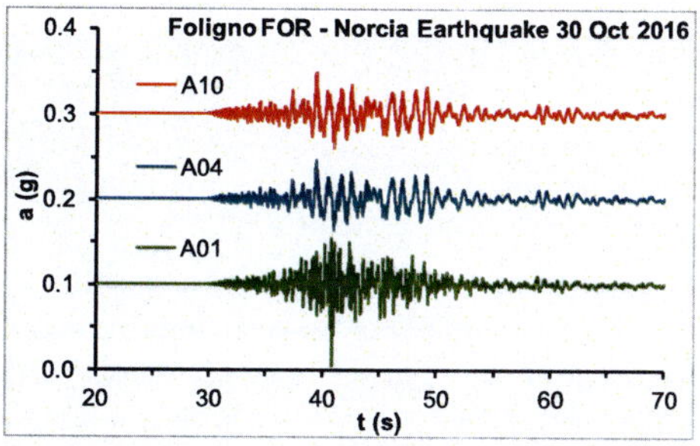

Fig. 12. Forest Ranger Building at Foligno: accelerometric recordings obtained during the Norcia Earthquake at (a) L0, (b) L1 and (c) L2.

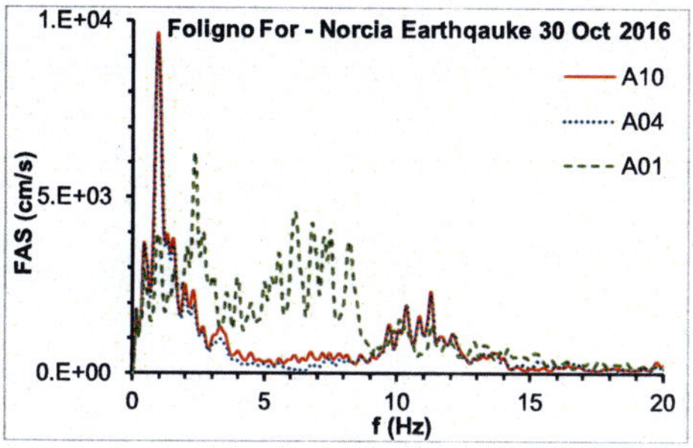

Fig. 13. Forest Ranger Building at Foligno: Fourier spectra at (a) L0, (b) L1 and (c) L2.

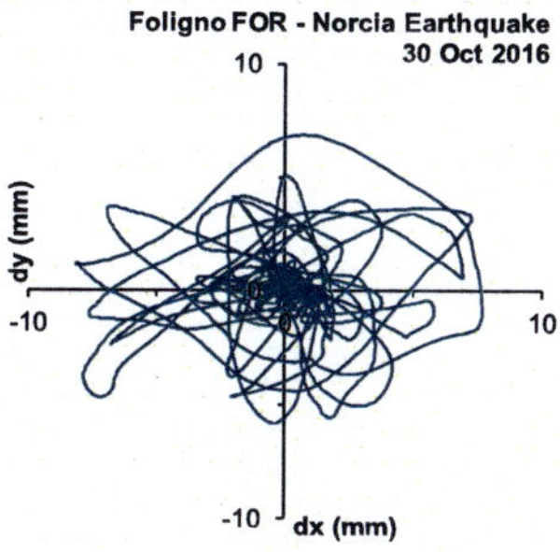

Fig. 14. Relative particle motion at a corner isolator.

in the design but not closed to the fundamental frequencies of the superstructure. This occurrence guaranteed a suitable decoupling of motion between the superstructure and the soil-foundation system.

- A building seismically isolated by means of CSSs, in which the seismic behaviour is governed by the static friction. If this has been designed for the ULS vertical loads, the actual friction coefficient could be quite greater under the vertical seismic loads.
- A building seismically isolated by means of HDRBs + SDs. In this case, the onset of motion is governed by the static friction. After that, the stiffness of the rubber

determines the effective resonance frequencies, which could become close to that of the superstructure.

These aspects must be accounted for when designing seismically isolated buildings.

Acknowledgements. The monitoring of the buildings at the Civil Protection Centre at Foligno is part of a research project organized by ENEA in collaboration with the Umbria Region. The monitoring of the buildings of the C.A.S.E. project is part of the Seismic Observatory of Structures project, managed by the Italian Department of the Civil Protection (http://www.protezionecivile.gov.it/jcms/it/osservatorio.wp).

References

1. Clemente, P., Martelli, A.: Recent applications of seismic isolation in Italy. In: Proceedings of the 16th World Conference on Seismic Isolation, Energy Dissipation and Active Vibration Control of Structures (16WCSI, St. Petersburg, July 1–6), Keynote lecture, pp. 21–36. (2019). https://doi.org/10.37153/2686-7974-2019-16-21-36
2. Clemente, P., Martelli, A.: Seismically isolated buildings in Italy: state-of-the-art review and applications. Soil Dyn. Earthq. Eng. **119**, 471–487 (2019). https://doi.org/10.1016/j.soildyn.2017.12.029
3. Clemente, P., De Stefano, A.: Application of seismic isolation in the retrofit of historical buildings. In: Brebbia, C.A., Maugeri, M. (eds.) ERES 2011 (Sept. 7–9, Chianciano, Italy), WIT Press, Southampton, UK 120, pp. 1746–4498. (2011). https://doi.org/10.2495/ERES110041
4. De Stefano, A., Clemente, P., Invernizzi, S., Matta, E., Quattrone, A.: Innovative technique for the base isolation of existing buildings. In: Papadrakakis, M., Papadopoulos, V., Plevris, V. (eds.) Computational Methods in Structural Dynamics and Earthquake Engineering (COMPDYN 2015, Crete Island, Greece, 25–27 May 2015) 1, 377–387, NTUA, Athens, Greece, (2015). ISBN: 978-960-99994-7-2
5. Bongiovanni, G., Buffarini, G., Clemente, P., Saitta, F., Tripepi, C.: Retrofit of existing buildings with seismic isolation: design issues and applications. In: Proceedings of the 16th World Conference on Seismic Isolation, Energy Dissipation and Active Vibration Control of Structures (16WCSI, St. Petersburg, July 1–6), pp. 111–122. (2019). https://doi.org/10.37153/2686-7974-2019-16-111-122
6. Saitta, F., Clemente, P., Buffarini, G., Bongiovanni, G.: Vulnerability analysis and seismic retrofit of a strategic building. J. Perform. Construct. Facilities **31**(2), 04016085 (2017). https://doi.org/10.1061/(ASCE)CF.1943-5509.0000948
7. Clemente, P.: Seismic isolation: past, present and the importance of SHM for the future. J. Civ. Struct. Heal. Monit. **7**(2), 217–231 (2017). https://doi.org/10.1007/s13349-017-0219-6
8. Matsuda, K., Kasai, K., Yamagiwa, H., Sato, D.: Responses of base-isolated buildings in Tokyo during the 2011 Great East Japan Earthquake. In: Proceedings 15th World Conf. on Earthquake Engineering (15WCEE, Lisbon, 24–28 Sep) (2012)
9. Zhou, F.L., Tan, P., Heisa, W.L.H., Xian, X.L.: Lu Shan earthquake M7.0 on 2013.4.20 and recent development on seismic isolation, energy dissipation and structural control in China. In: Proceedings 13th World Conf. on Seismic Isolation, Energy Dissipation and Active Vibration Control of Structures (13WCSI, Sendai, Sept 24–26), JSSI, Tokyo (2013)
10. Zhou, F.L., Tan, P., Heisha, W., Huan, X.Y.: Progress and application of seismic isolation, energy dissipation and control in civil and industrial structures and design codes in China. In: Proceedings 14WCSI (Sep 9–11 2015, San Diego, CA USA) (2015)

11. Çelebi, M.: Successful performance of a base-isolated hospital buildings during the 17 January 1995 Northridge earthquake. The Struct. Design of Tall Build. **5**(2), 95–109 (1996). https://doi.org/10.1002/(SICI)1099-1794(199606)5:2<95::AID-TAL71>3.0.CO;2-7
12. Nagarajaiah, S., Sun, X.: Response of base-isolated USC hospital building in Northridge earthquake. J. Struct. Eng. **126**(10), 1177–1186 (2000). https://doi.org/10.1061/(ASCE)0733-9445(2000)126:10(1177)
13. Nagarajaiah, S., Sun, X.: Base-Isolated FCC building: impact response in Northridge earthquake. J. Struct. Eng. **127**(9), 1063–1075 (2001). https://doi.org/10.1061/(ASCE)0733-9445(2001)127:9(1063)
14. Clemente, P., Bongiovanni, G., Buffarini, G., Saitta, F., Castellano, M.G., Scafati, F.: Effectiveness of HDRB isolation systems under low energy earthquakes. Soil Dynamics & Earthquake Engineering 118, 207–220, Elsevier Science Ltd, (2019). https://doi.org/10.1016/j.soildyn.2018.12.018
15. Clemente, P., Bongiovanni, G., Buffarini, G., Saitta, F., Scafati, F.: Monitored seismic behavior of base isolated buildings in Italy. In: Limongelli, M., Çelebi, M. (eds), Seismic Structural Health Monitoring, pp. 115–137. Springer (2019). https://doi.org/10.1007/978-3-030-13976-6_5
16. Ormando, C., Clemente, P., Ianniruberto, U., Scafati, F.: Onset of motion of curved surface sliders used in seismic isolation systems. Pract. Period. Struct. Des. Constr. **26**(3), 04021015 (2021). https://doi.org/10.1061/(ASCE)SC.1943-5576.0000580
17. Bongiovanni, G., Buffarini, G., Clemente, P., Saitta, F., De Sortis, A., Nicoletti, M., Rossi, G.: Behaviour of the base isolated Jovine School in San Giuliano di Puglia, Italy, under the December 20th 2013 earthquake. In: Proceedings of the 14th World Conference on Seismic Isolation, Energy Dissipation and Active Vibration Control of Structures (14WCSI, San Diego, Sep 9–11 2015) (2015)
18. Clemente, P., Bongiovanni, G., Benzoni, G.: Monitoring of seismic isolated buildings: state of the art and results under high and low energy inputs. In: Proceedings of the New Zealand Society for Earthquake Annual Conference and 15th World Conference on Seismic Isolation, Energy Dissipation and Active Vibration Control of Structures (NZSEE2017 and 15WCSI, Wellington, Apr 27–29), Invited lecture. New Zealand Society for Earthquake Engineering NZSEE (2017)
19. Clemente, P., Buffarini, G.: Base isolation: design and optimization criteria. Seismic Isolation and Protect. Syst. **1**(1), 17–40 (2010), Mathematical Science Publisher. https://doi.org/10.2140/siaps.2010.1.17
20. Tripepi, C., Clemente, P.: Graphic procedure for the optimum design of elastomeric isolators. Practice Periodical on Struct. Design Construct. **26**(1), 04020058–1_9 (2021). https://doi.org/10.1061/(ASCE)SC.1943-5576.0000547
21. Lomiento, G., Bonessio, N., Benzoni, G.: Friction model for sliding bearings under seismic excitation. J. Earthquake Eng. **17**(8), 1162–1191 (2013). https://doi.org/10.1080/13632469.2013.814611
22. Saitta, F., Clemente, P., Bongiovanni, G., Buffarini, G., Salvatori, A., Grossi, C.: Base isolation of buildings with curved surface sliders: basic design criteria and critical issues. In: Advances in Civil Engineering, vol. 2018, **1569683**, Hindawi, pp. 14. (2018). https://doi.org/10.1155/2018/1569683
23. Clemente, P., Di Cicco, A., Saitta, F., Salvatori, A.: Seismic behaviour of base isolated civil protection operative centre in Foligno, Italy. J. Perform. Construct. Facilities **35**(4), 04021027 (2021). https://doi.org/10.1061/(ASCE)CF.1943-5509.0001589
24. Scafati, F., Ormando, C., Clemente, P., Bongiovanni, G.: Observed behavior of buildings seismically isolated with CSSs under a low energy earthquake. J. Civ. Struct. Heal. Monit. **12**(2), 225–243 (2022). https://doi.org/10.1007/s13349-021-00539-z

25. Salvatori, A., Bongiovanni, G., Clemente, P., Ormando, C., Saitta, F., Scafati, F.: Observed seismic behaviour of a HDRB and SD isolation system under far fault earthquakes. Infrastructures **7**(2), 13 (2022). https://doi.org/10.3390/infrastructures7020013
26. Salvatori, A., Di Cicco, A., Clemente, P.: Monitoring of base isolated building subjected to far fault earthquake. In: Rainieri, C., Fabbrocino, G., Caterino, N., Ceroni, F., Notarangelo, M.A. (eds.) CSHM 2021. LNCE, vol. 156, pp. 311–338. Springer, Cham (2021). https://doi.org/10.1007/978-3-030-74258-4_21
27. Salvatori, A., Di Cicco, A., Clemente, P.: Seismic monitoring of buildings with base isolation. In: Papadrakakis, M., Fragiadakis, M. (eds) Computational Methods in Structural Dynamics and Earthquake Engineering (COMPDYN 2019, Crete Island, Greece, 24–26 June 2019), ID 19221, NTUA, Athens, (2019). https://2019.compdyn.org/proceedings/
28. Turino, R. (ed.): L'Aquila, il progetto C.A.S.E.. IUSS Press, Pavia (2010)

Versatile Aseismic Isolation Based on Practical Applications of Advanced Materials for Sustainable Resilience Against Earthquakes

Mohammad Noori[1(✉)], Jian Zhang[2], Eltahry Elghandour[1], Donatello Cardone[3], and Peyman Narjabadifam[4]

[1] Department of Mechanical Engineering, California Polytechnic State University, San Luis Obispo, CA 93405, USA
mnoori@calpoly.edu
[2] Department of Civil & Environmental Engineering, University of California, Los Angeles, CA 90095, USA
[3] School of Engineering, University of Basilicata, 85100 Potenza, Basilicata, Italy
[4] Department of Civil Engineering, Faculty of Engineering, University of Bonab, 5551395133 Bonab, Iran

Abstract. Effective protection against earthquakes has always challenged humanity. Aseismic isolation (AI) has provided an excellent solution and isolation systems (ISs) have improved seismic safety of the built environment. However, this improving impact of AI on seismic resilience is not appropriate to its capabilities and further attempts are required. Advanced materials provide great opportunities to make AI more effective in sustainable seismic resilience. Shape memory alloys (SMAs) are known as the most favorable advanced materials for this purpose. SMA-based superelasticity-assisted slider (SSS) brings the advantages of SMAs in the practice of AI by innovatively combining them with the technically preferred sliding ISs. SSS is pioneered to seismically protect a broad variety of structures using a single platform. This IS can be implemented both traditionally by its construction-industry-friendly structure and in the industrialized style by means of isolation units. It is also facilitated by SSS to utilize other advanced materials such as nanomaterials and metamaterials. All these are discussed in this paper. The versatility of SSS is demonstrated by explanatory technical drawings and roles of alternative configurations that result in various hysteretic behaviors are investigated. Possible design strategies are studied for a typical office building and a sensitive medical equipment. Applications of other advanced materials are suggested and seismic performances of SSS are compared with those of currently used ISs in earthquake protection of a typical reinforced concrete building. It is shown that high seismic performances can be obtained by the practical applications of the advanced materials in the AI technology to obtain sustainability in seismic resilience.

Keywords: Aseismic isolation · Advanced materials · Versatility · Sustainability · Seismic resilience · Broad variety of structures · Built environment

G. P. Cimellaro (Ed.): WCSI 2022, LNCE 309, pp. 611–622, 2023.
https://doi.org/10.1007/978-3-031-21187-4_51

1 Introduction

Human casualties and economic losses experienced by recent earthquakes indicate the public safety risks of seismically vulnerable structures in the built environment and the need for further advancements in the field of structural earthquake engineering (SEE). Figure 1 represents some examples of the various forms of the seismic damage experienced in the current century.

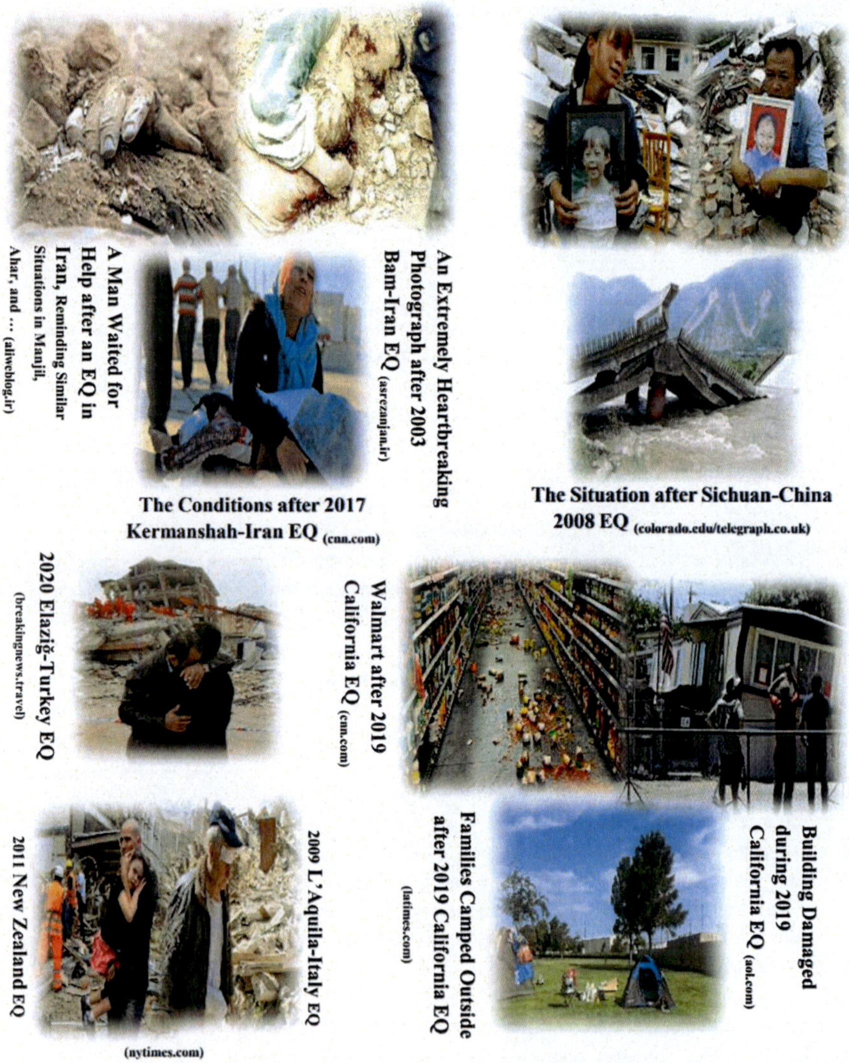

Fig. 1. Some examples of the recent devastating seismic disasters, indicating that structural engineering profession requires more than that provided by the current anti-seismic technologies.

From the perspective of structural engineering, the nonstructural damage experienced in a hypermarket (Walmart in Fig. 1) and the damaged confidence on the seismic resilience of structures forcing people to camp outside because of an earthquake in a country like US in the twenty-first century are enough to understand the situation and there is no need to discuss the structural damage or the heartbreaking experiences that no one could stay indifferent to them.

While the achievements of humankind in developing SEE technologies have been remarkable and helpful in reducing the tragic consequences of earthquakes, the brief discussion provided above based on Fig. 1 shows that SEE technologies need some advancements to improve sustainable seismic resilience in the built environment that will consequently reduce the public safety risks of earthquakes. It would be rational and wise to apply the required advancements on the technologies that have been successful in protecting structures against earthquakes. Statistics show that aseismic isolation (AI) is well accepted as the most effective technique in protecting structures against earthquakes [1–3] but still requires improving advancements to be appropriately effective in sustainable seismic resilience. Advanced materials can be used to make the required advancements practically available. Shape memory alloys (SMAs) are a class of advanced materials suited to this purpose [4–7]. Austenitic SMAs are preferred to martensitic SMAs because of their favorable characteristics such as superelasticity with large strain plateaus, adequate energy dissipation capacity through flag-shaped hysteretic loops, and high fatigue and corrosion resistance [8–11]. SMA-based AI can be realized in many ways and different isolation systems (ISs) can be used. Among the most promising ISs, rubber bearings have widely been implemented in practice [12–14], and sliding isolators have attracted a high attention due to their practical advantages including the further elongation of the natural period, the insensitivity to the frequency content of excitation, the lower transmission of seismic accelerations into the superstructure, and the more robust stability due to the separate functions of carrying weight and providing isolation [15–17]. Flat sliding bearings (FSBs) are the simplest sliders but possess higher degrees of the advantages of sliding isolators [18]. Principally, an FSB requires a proper restoring mechanism. Restoring mechanism can typically be provided by combining the FSBs with laminated rubber bearings (LRBs). The friction pendulum system (FPS) and its variants are the other engineering solutions provided up to date. FPS has become a well-known practical IS due to its technical advantages. There are, however, some limitations with FPS and its variants [19, 20]. SMA-based recentering is a relatively modern alternative solution for the restoring problem of sliding ISs [21–24]. Various SMA-based sliding ISs and SMA-based devices that can be used in sliding ISs or provide ideas for this purpose have been proposed by different researchers [25–31], generally suffering from various defects that mainly include impracticality and limited usefulness. SMA-based superelasticity-assisted slider (SSS) has recently been proposed [32–35] to overcome some of these shortcomings by its construction-industry-friendly structure and other advantageous features such as the versatility provided by its alternative configurations. This paper deals with the versatility of SSS obtained by the pioneered utilization of SMAs and discusses the opportunities provided to further advance AI technology by utilizing other advanced materials to further improve sustainable seismic resilience in the built environment. Section 2 introduces SSS, Sect. 3 discusses the versatility of this modern IS, and Sect. 4 provides concluding remarks through comparing SSS with currently used ISs.

2 Practical Applications of Advanced Materials in SSS

SSS materializes advanced AI idea by practically combining superelasticity with sliding phenomenon. Superelasticity in SSS is innovatively provided by SMA wire ropes (or cables) arranged in alternative novel and conventional configurations with using specifically engineered connections. For the purpose of sliding isolation, flat sliding bearings (FSBs), which possess higher degrees of the advantages of sliding isolators – as already emphasized in Sect. 1, are used. The high-tech construction-industry-friendly structure of SSS is demonstrated in Fig. 2.

As seen in Fig. 2, the pioneered structure of SSS makes it possible to practically exploit the advantages of both sliding isolation and superelasticity by utilizing austenitic SMA cables, which can be fabricated off-site or on-site. The possibility for benefitting from printed SMA cables [36, 37] is an additional advantage provided by SSS. With SSS, it is also facilitated to utilize other advanced materials to further advance AI performances. Nanomaterials, for example, can be used to improve the sliding isolation effect by reducing the coefficient of friction and wear [38–40] in the FSBs that can be composed of mirror-polished stainless steel (SUS) or galvanized steel sheets against polytetrafluoroethylene (PTFE), polyethyleneterephtalate (PET), high-density polyethylene (HDPE), or ultrahigh molecular weight polyethylene (UHMWPE) pads. Metallic or nonmetallic 3d/4d-printed metamaterial or periodic material pads [41], biomimetic architected elements [42], or combinations thereof can similarly be used to enhance the vertical isolation effect that can alternatively be provided by utilization of some mechanisms or added elements of traditional or advanced materials such as steel or SMA coil springs [43], super high damping rubber pads [44], viscous or steel dampers [45], or telescoping piers [46], as indicated by Noori and Narjabadifam [34].

3 On the Versatility of SSS

The versatility of SSS is basically because of its pioneered structure introduced in Sect. 2. This versatility is a consequence of the provided alternative isolation unit (IU)-based industrialized and IU-less traditional styles of implementation (already demonstrated in Fig. 2), the various hysteretic behaviors of the IS that are provided by its alternative configurations and their geometric variants (as detailed in Fig. 3), the advantages of SMA cables over the other similar structural elements (investigated based on Fig. 4), and the modularity of the IS (discussed at the end of this section).

Figure 3 shows how SSS provides various hysteretic behaviors because of different amounts of geometric nonlinearity included in its alternative configurations (SSS-v/d/h/o/l/u/c, respectively obtained by vertical/diagonal/horizontal/O-shaped/L-shaped/U-shaped/C-shaped arrangements of the SMA cables) and their geometric variants that can be obtained by changing the inclination angle (α) in SSS-d or the ratio between the lengths of the vertical and horizontal arms of the SMA cables (L_v/L_h) in SSS-o/l/u/c. Each type of hysteresis (relaxation chair hysteresis, cleaver type hysteresis,

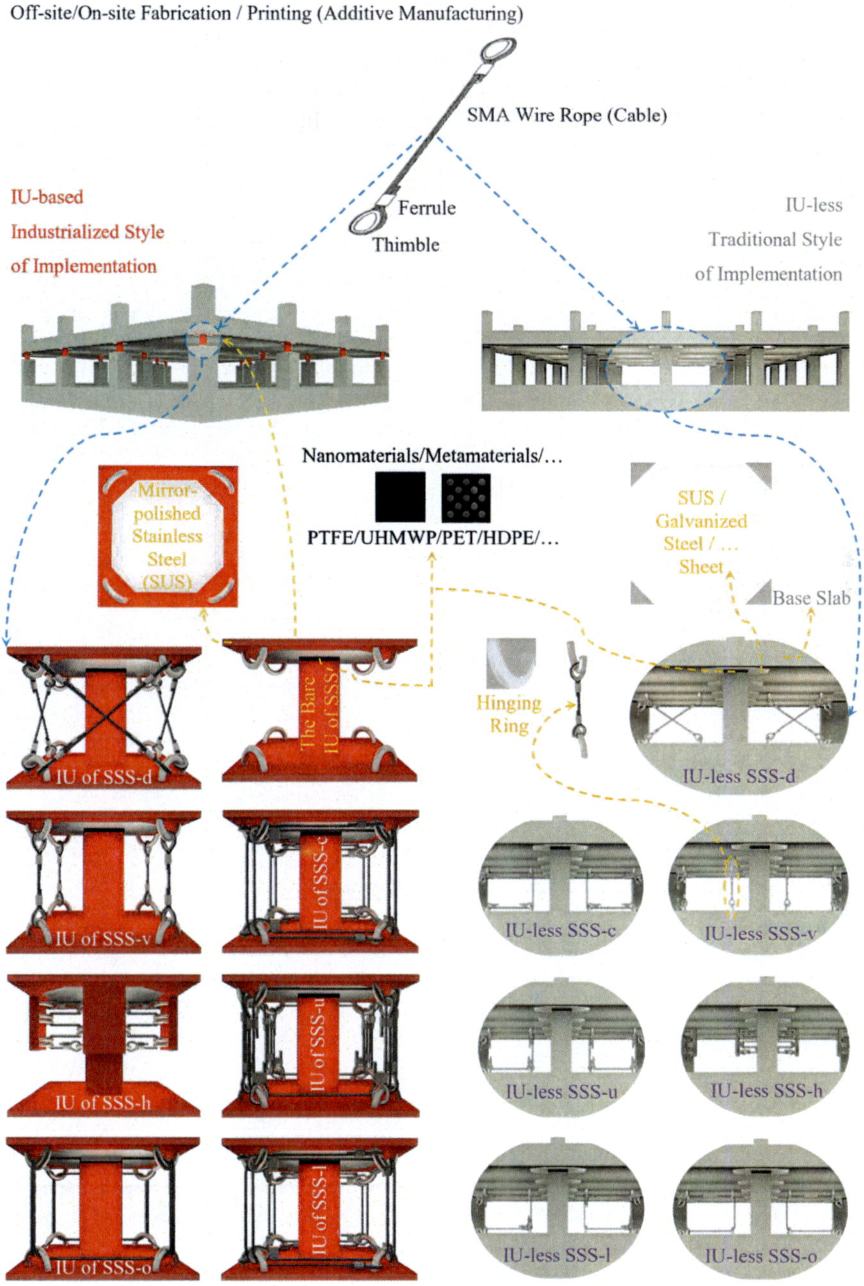

Fig. 2. Practical applications of advanced materials and technologies in SSS.

inclined bow-tie hysteresis, pure flag-shaped hysteresis, and modified flag-shaped hysteresis) can be used to satisfy the required performance objectives in different projects.

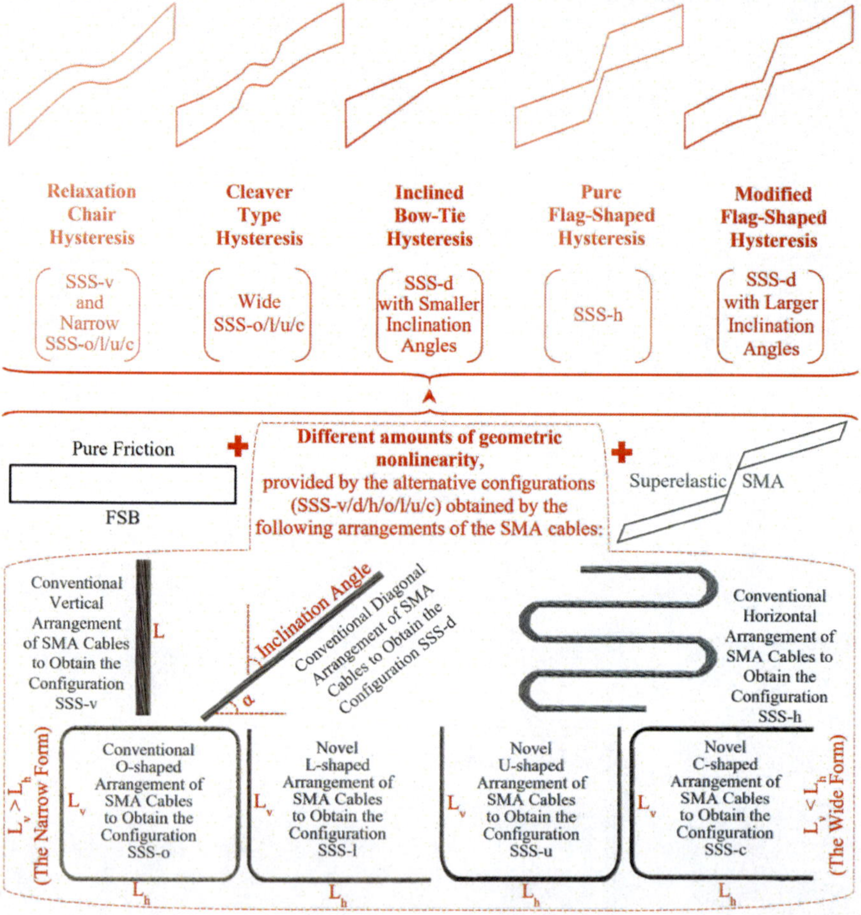

Fig. 3. Various hysteretic behaviors of SSS provided by different amounts of geometric nonlinearity included in its alternative configurations (SSS-v/d/h/o/l/u/c, respectively obtained by vertical/diagonal/horizontal/O-shaped/L-shaped/U-shaped/C-shaped arrangements of the SMA cables) and their geometric variants.

This behavior-based versatility helps SSS seismically protect a variety of structures ranging from buildings and bridges to equipment (medical/industrial) and art objects.

Regarding the versatility resulted from the advantages of SMA cables over the other structural elements, it is because various cross-section layouts of cables can be used to design SSS for different projects. The cross-section layouts mostly suited for structural engineering applications are represented in Fig. 4, and they can also be used in SSS.

It is well known that utilization of cables instead of wires, wire bundles or bars (used in the prior art) provides specific technical advantages with regard to integrity and redundancy. All these advantages are also guaranteed by SSS but its versatility is because of the possibility for using various cross-section layouts. Table 1 typically represents how this versatility of SSS provides alternative design options for the aseismic isolation

Fig. 4. The most suitable cross-section layouts of cables to be used in SSS.

of two example cases representing first the large-scale structures with an office building and second the small-scale structures with a sensitive medical equipment. The office building has 20 columns and the total weight of the medical equipment is 10 tons. Both the cases are designed with a target displacement of 30 cm in the IS. A typical lubricated SUS-PTFE surface is assumed for the FSB component of the system in both the cases. The SMA component considered in the first case (for the IS of the office building) is a commercial NiTi cable with a 7 × 7 cross-section layout. In the second case (for the IS of the medical equipment that is assumed to be composed of four IUs), however, the SMA cable has a 1 × 3 cross-section layout with the same alloy considered for the office building.

As far as the roles of the modularity of SSS in its versatility are concerned, the most important role of this advantageous feature is the separation of isolation and recentering mechanisms, which makes it possible to practically design the system in different projects. Also, it is because of the modularity of the IS that the IU-less style of implementation shown in Fig. 2 becomes possible. Moreover, the critical recentering elements of the system can easily be replaced in a case of overloading. In summary, the modularity of SSS adds to its versatility that can help this IS and the AI technology to be appropriately effective in sustainable seismic resilience.

4 Concluding Remarks Based on a Comparison Between SSS and Currently Used Isolation Systems

To be widely used in the practice of structural engineering, SSS needs also to prove that its unique versatility is accompanied by competitiveness in terms of seismic performances. This section, in this paper, summarizes the results obtained from a case study carried out for this purpose. In this case study, seismic performances of SSS are compared to those of FPS and high-damping LRB (HRB) that are known as the most famous currently used ISs. Fixed-base performances of the building are used as the reference in all the comparisons. The case study building is a four-story moment resisting reinforced concrete residential building and SSS is represented by its vertical arrangement of the cables to obtain the meaningful comparison including all the other configurations that result in better performances due to the hysteretic characteristics shown in Fig. 3. The structural model of the building was subjected to seven natural, natural/artificial, and artificial accelerograms suggested by Italian seismic engineering network of laboratories at national universities for the studies of isolators. The accelerograms were scaled to match (on average) the relevant spectrum of EC8, which is the type I for the soil type C.

Table 1. The versatility of SSS that is because the possibility for using various cross-section layouts by the SMA cables.

The structure	Design options
A typical 3-story office building located in a high-seismicity region (large-scale structure)	SSS-v, with 0.82-m-long 7 × 7 cables made up of 1.75-mm-diameter wires
	SSS-d_{75}, with 1.58-m-long 7 × 7 cables made up of 1.54-mm-d wires
	SSS-d_{45}, with 3.36-m-long 7 × 7 cables made up of 1.36-mm-d wires
	SSS-h, with 4.61-m-long 7 × 7 cables made up of 1.63-mm-d wires
	SSS-c_n, with $L_v = 0.7m$ and $L_h = 0.125m$ 7 × 7 cables made up of 1.13-mm-d wires
	SSS-c_{w1}, with $L_v = 0.25m$ and $L_h = 1m$ 7 × 7 cables made up of 0.75-mm-d wires
	SSS-c_{w2}, with $L_v = 0.1m$ and $L_h = 1.62m$ 7 × 7 cables made up of 0.67-mm-d wires
A sensitive medical equipment (small-scale structure)	SSS-v, with 0.82-m-long 1 × 3 cables made up of 2.2-mm-d wires
	SSS-d_{75}, with 1.58-m-long 1 × 3 cables made up of 1.88-mm-d wires
	SSS-d_{45}, with 3.36-m-long 1 × 3 cables made up of 1.73-mm-d wires
	SSS-h, with 4.61-m-long 1 × 3 cables made up of 2-mm-d wires
	SSS-c_n, with $L_v = 0.7m$ and $L_h = 0.125m$ 1 × 3 cables made up of 1.38-mm-d wires
	SSS-c_{w1}, with $L_v = 0.25m$ and $L_h = 1m$ 1 × 3 cables made up of 1.02-mm-d wires
	SSS-c_{w2}, with $L_v = 0.1m$ and $L_h = 1.62m$ 1 × 3 cables made up of 0.95-mm-d wires

Figure 5 shows the force-displacement behaviors of the ISs under the case study building, together with the story displacements and accelerations. The story displacements and accelerations are the average responses over the seven accelerograms, but the force-displacement behaviors represent the responses of the ISs subjected to one of the accelerograms. The promising competitiveness of SSS is basically proven by comparing the force-displacement diagrams. By comparing the story displacements and accelerations, however, it can be concluded that seismic performances of structures can further be improved by SSS that is capable of better controlling the responses by the more effective hysteretic behaviors shown in Fig. 3. The effectiveness of AI in terms of controlling the

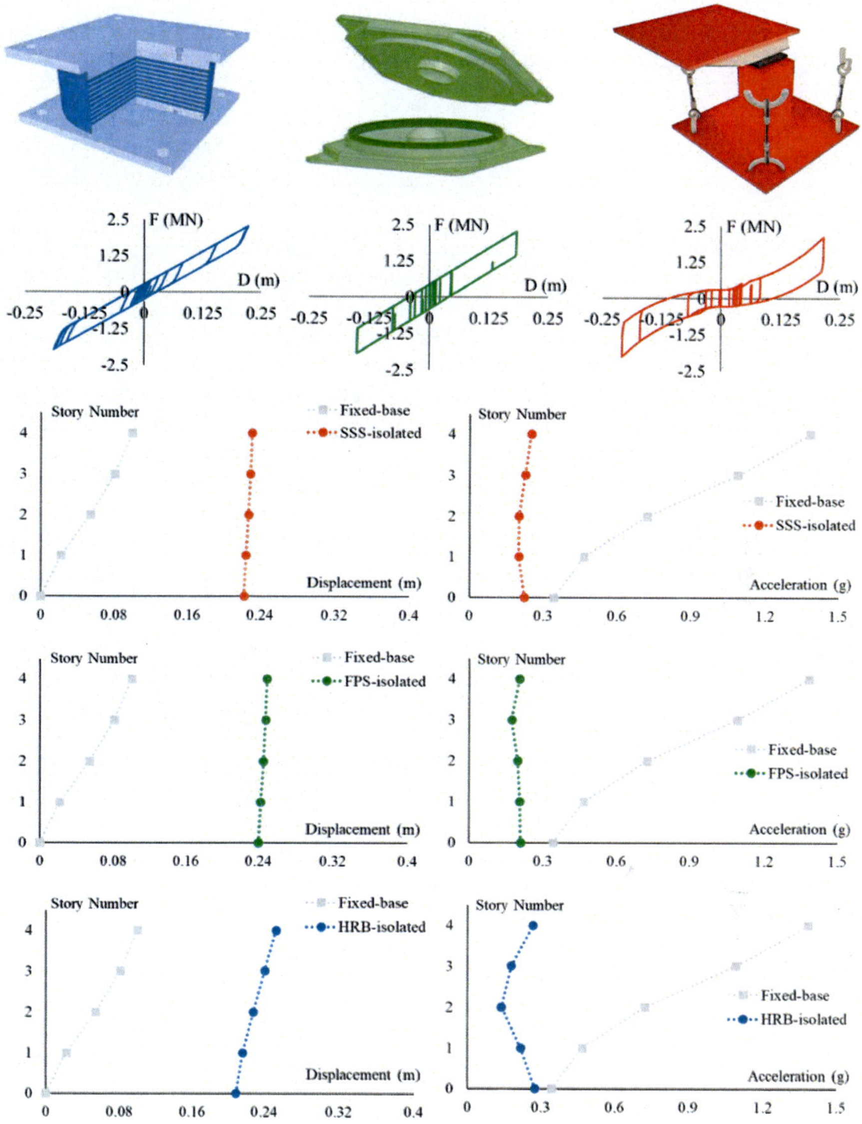

Fig. 5. The competitiveness of SSS.

structural responses will indeed be increased by using the cleaver type hysteresis, the inclined bow-tie hysteresis, and the pure/modified flag-shaped type hysteresis provided by SSS.

Acknowledgements. The authors would like to thank Professor Ertugrul Taciroglu from the University of California at Los Angeles, Professor Dipanjan Basu from the University of Waterloo, Professor Tao Wang from the Key Laboratory of Earthquake Engineering and Engineering Vibration at the Institute of Engineering Mechanics of China Earthquake Administration, and Engineers

Mehrdad Kiani and Rasa Eradat from the Sharestan Tarh Tabriz Consultants for all their advice and help during the research and in the preparation of this paper.

References

1. Martelli, A., Clemente, P., De Stefano, A., Forni, M., Salvatori, A.: Recent development and application of seismic isolation and energy dissipation and conditions for their correct use. In: Ansal, A. (ed.), pp. 449–488. Springer, Switzerland (2014)
2. Takagi, J., Wada, A.: Recent earthquakes and the need for a new philosophy for earthquake-resistant design. Soil Dyn. Earthq. Eng. **119**, 499–507 (2019)
3. Wallemacq, P., House, R., McClean, D.: Economic Losses, Poverty and Disasters 1998–2017, pp. 1–31. United Nations Office for Disaster Risk Reduction (UNDRR)/Center for Research on the Epidemiology of Disasters (CRED), Geneva/Brussels, Switzerland/Belgium (2018)
4. Desroches, R., Smith, B.: Shape memory alloys in seismic resistant design and retrofit: a critical review of their potential and limitations. J. Earthquake Eng. **8**, 415–429 (2004)
5. Dolce, M., Cardone, D.: Mechanical behavior of shape memory alloys for seismic applications: part 2-austenite NiTi wires subjected to tension. Int. J. Mech. Sci. **43**, 2657–2677 (2001)
6. Noori, M., Narjabadifam, P.: Innovative civil engineering applications of smart materials for smart sustainable urbanization. J. Civ. Eng. Urbanism **9**, 24–35 (2019)
7. Van Humbeeck, J.: Non-medical applications of shape memory alloys. Mater. Sci. Eng., A **273–275**, 134–148 (1999)
8. Alvandi, S., Ghassemieh, M.: Application of shape memory alloys in seismic isolation: a review. Civ. Eng. Infrastruct. J. **47**, 153–171 (2014)
9. Motavalli, M., Czaderski, C., Bergamini, A., Janke, L.: Shape memory alloys for civil engineering structures—On the way from vision to reality. Arch. Civ. Eng. Environ. **2**, 81–94 (2009)
10. Sherif, M.M., Ozbulut, O.E.: Tensile and superelastic fatigue characterization of NiTi shape memory cables. Smart Mater. Struct. **27**, 015007 (2018)
11. Song, G., Ma, N., Li, H.N.: Applications of shape memory alloys in civil structures. Eng. Struct. **28**, 1266–1274 (2006)
12. Feng, Q., Kong, Q., Jiang, J., Liang, Y., Song, G.: Detection of interfacial debonding in a rubber–steel-layered structure using active sensing enabled by embedded piezoceramic transducers. Sensors **17**, 2001 (2017)
13. Narjabadifam, P., Tiong, P.L.Y., Mousavi-Alanjagh, R.: Effects of inherent structural characteristics on seismic performances of aseismically base-isolated buildings. Iran. J. Sci. Technol.—Trans. Civ. Eng. **44**, 1385–1401 (2020b)
14. Nakamura, Y., Okada, K.: Review on seismic isolation and response control methods of buildings in Japan. Geoenviron. Disasters **6**(1), 1 (2019). https://doi.org/10.1186/s40677-019-0123-y
15. Fallah, N., Zamiri, G.: Multi-objective optimal design of sliding base isolation using genetic algorithm. Sci. Iranica **20**, 87–96 (2013)
16. Higashino, M., Hamaguchi, H., Minewaki, S., Aizawa, S.: Basic characteristics and durability of low-friction sliding bearings for base isolation. Earthquake Eng. Eng. Seismol. **4**, 95–105 (2003)
17. Panchal, V., Jangid, R.: Seismic response of structures with variable friction pendulum system. J. Earthquake Eng. **13**, 193–216 (2009)
18. Constantinou, M., Mokha, A., Reinhorn, A.: Teflon bearings in base isolation (II-modeling). J. Struct. Eng. **116**, 455–474 (1990)

19. Calafell, R.L., Roschke, P.N., De la Llera, J.C.: Optimized friction pendulum and precast-prestressed pile to base-isolate a Chilean masonry house. Bull. Earthq. Eng. **8**, 1019–1036 (2010)
20. Calvi, G.M., Calvi, P.M., Moratti, M.: Seismic isolation of buildings using devices based on sliding between surfaces with variable friction coefficient. Innovative Infrastruct. Solutions **2**(1), 1–18 (2017). https://doi.org/10.1007/s41062-017-0081-8
21. Casciati, S.: SMA-based devices: insight across recent proposals toward civil engineering applications. Smart Struct. Syst. **24**, 111–125 (2019)
22. Desroches, R., McCormick, J., Delemont, M.: Cyclic properties of superelastic shape memory alloy wires and bars. J. Struct. Eng. **130**, 38–46 (2004)
23. Narjabadifam, P., Watandust, A.S.: Latest trends on the application of shape memory alloy cables (wire ropes) in structural earthquake engineering. Int. J. Adv. Eng. Res. Appl. **4**, 9–21 (2018)
24. Wilde, K., Gardoni, P., Fujino, Y.: Base isolation system with shape memory alloy device for elevated highway bridges. Eng. Struct. **22**, 222–2298 (2000)
25. Attanasi, G., Auricchio, F., Fenves, G.L.: Feasibility assessment of an innovative isolation bearing system with shape memory alloys. J. Earthquake Eng. **13**(sup1), 18–39 (2009)
26. Cardone, D., Narjabadifam, P., Nigro, D.: Shaking table tests of the smart restorable sliding base isolation system (SRSBIS). J. Earthquake Eng. **15**, 1157–1177 (2011)
27. Casciati, F., Faravelli, L., Hamdaoui, K.: Performance of a base isolator with shape memory alloy bars. Earthq. Eng. Eng. Vib. **6**, 401–408 (2007)
28. Logiadis, I., Logiadis, E., Logiadis, G.: Antiseismic connector of limited vibration for seismic isolation of an structure. US Patent—5669189 (1997)
29. Ozbulut, O.E., Silwal, B.: Performance assessment of buildings isolated with S-FBI system under near-fault earthquakes. Smart Struct. Syst. **17**, 709–724 (2016)
30. Zheng, W.Z., Wang, H., Li, J., Shen, H.J.: Performance evaluation of bridges isolated with SMA-based friction pendulum system at low temperatures. Soil Dyn. Earthq. Eng. **125**, 105734 (2019)
31. Zhang, Z.-X., Ping, Y., He, X.: Self-centering shape memory alloy-viscoelastic hybrid braces for seismic resilience. Materials **15**, 2349 (2022)
32. Narjabadifam, P.: Shape memory alloy (SMA)-based Superelasticity-assisted Slider (SSS). In: 7th International Conference on Seismology and Earthquake Engineering. Tehran, Iran (2015)
33. Narjabadifam, P., Noori, M., Cardone, D., Eradat, R., Kiani, M.: Shape memory alloy (SMA)-based Superelasticity-assisted slider (SSS): an engineering solution for practical aseismic isolation with advanced materials. Smart Struct. Syst. **26**, 89–102 (2020a)
34. Noori, M., Narjabadifam, P.: Earthquake protection systems, methods and apparatus using shape memory alloy (SMA)-based superelasticity-assisted slider (SSS). US Patent—11313145 (2022)
35. Narjabadifam, P., Noori, M., Taciroglu, E., Zhang, J., Khoshnevis, B., Cardone, D., Basu, D., Wang, T., Elghandour, E., Noroozinejad Farsangi, E., Lotfi, R., Chavoshi, M., Sattarian, D., Stirnimann, O.F.: Sustainable earthquake resilience with the versatile shape memory alloy (SMA)-based superelasticity-assisted slider. Sens. **22**, 6876 (2022)
36. Alagha, A.N., Hussain, S., Zaki, W.: Additive manufacturing of shape memory alloys: a review with emphasis on powder bed systems. Mater. Des. **204**, 109654 (2021)
37. Xue, L., et al.: Laser powder bed fusion of defect-free NiTi shape memory alloy parts with superior tensile superelasticity. Acta Mater. **229**, 117781 (2022)
38. Spear, J.C., Ewers, B.W., Batteas, J.D.: 2D-nanomaterials for controlling friction and wear at interfaces. Nano Today **10**, 301–314 (2015)
39. Fuadi, Z., et al.: Effect of graphene nanoplatelets on tribological properties of bacterial cellulose/polyolester oil bio-lubricant. Front. Mech. Eng. **8**, 810847 (2022)

40. Li, P., et al.: Role of nanoparticles in achieving macroscale superlubricity of graphene/nano-SiO_2 particle composites. Friction **10**, 1305–1316 (2022)
41. Al Rifaie, M., Abdulhadi, H., Mian, A.: Advances in mechanical metamaterials for vibration isolation: a review. Adv. Mech. Eng. **14**, 1–20 (2022)
42. Fraternali, F., Singh, N., Amendola, A., Benzoni, G., Milton, G.W.: A biomimetic sliding–stretching approach to seismic isolation. Nonlinear Dyn. **106**(4), 3147–3159 (2021). https://doi.org/10.1007/s11071-021-06980-5
43. Zhou, Z., Wong, J., Mahin, S.: Potentiality of using vertical and three-dimensional isolation systems in nuclear structures. Nucl. Eng. Technol. **48**, 1237–1251 (2016)
44. Pourmasoud, M.M., Lim, J.B.P., Hajirasouliha, I., McCrum, D.: Multi-directional base isolation system for coupled horizontal and vertical seismic excitations. J. Earthquake Eng. **26**, 1145–1170 (2022)
45. Lee, D., Constantinou, M.C.: Combined horizontal–vertical seismic isolation system for high-voltage–power transformers: development, testing and validation. Bull. Earthq. Eng. **16**(9), 4273–4296 (2018). https://doi.org/10.1007/s10518-018-0311-2
46. Kammula, V., Erochko, J., Kwon, O.-S., Christopoulos, C.: Application of hybrid-simulation to fragility assessment of the telescoping self-centering energy dissipative bracing system. Earthquake Eng. Struct. Dynam. **43**, 811–830 (2014)

Seismic Isolation of the Terminal Core Roof at the Portland International Airport

Reid B. Zimmerman(✉) and Christopher Pitt

KPFF Consulting Engineers, Portland, OR 97204, USA
reid.zimmerman@kpff.com

Abstract. The Terminal Core Redevelopment project is an approximately 1.6 billion USD project to functionally expand and seismically upgrade the existing main terminal at the Portland International Airport in Portland, Oregon, USA. At the center of the project is the seismic isolation of a new approximately 350,000 square foot (32,500 m^2) mass timber roof. A total of 68 double-concave friction pendulum isolators with displacement capacities ranging between 16 and 22.5 in. (406–572 mm) sit on top of each tine of 34, approximately 53-foot (16.2-m) tall concrete-filled steel Y-Columns on which the mass timber roof is supported. Performance-based seismic design for both a Risk-Targeted Maximum Considered Earthquake (approximately equal to the 2500 year return period event) and a deterministic event on the Cascadia Subduction Zone was pursued. Extensive nonlinear response history analyses considered multiple cases of foundation stiffness and seismic isolator coefficients of friction as well as explicit, time-step-by-time-step assessment of isolator-induced P-Delta and V-h moments. Full-scale, dynamic testing of prototype seismic isolators was performed at EUCENTRE. Quasi-static testing of 100% of production isolators was conducted at either EUCENTRE or SISMALAB prior to shipping to the project site.

Keywords: Seismic isolation · Performance-based design · Dynamic testing

1 Project Description

The Terminal Core Redevelopment (TCORE) project is an approximately 1.6 billion USD project to functionally expand and seismically upgrade the existing main terminal at the Portland International Airport in Portland, Oregon, USA. At the center of the project is the seismic isolation of a new approximately 350,000 square foot (32,500 m^2) mass timber roof designed by ZGF Architects and KPFF Consulting Engineers and owned by the Port of Portland. See Fig. 1. The existing ticketing, baggage claim, security, and concessions areas are within multiple, seismically separate but functionally homogeneous buildings which create the existing terminal core lower level (deplaning), main level (enplaning), and roof. These existing buildings are seismically vulnerable, both due to ground shaking and to liquefaction-induced settlement. The new mass timber roof is being erected over the existing terminal core buildings – completely covering them and expanding the terminal core to the west – while the airport is fully open and functioning.

G. P. Cimellaro (Ed.): WCSI 2022, LNCE 309, pp. 623–631, 2023.
https://doi.org/10.1007/978-3-031-21187-4_52

To accomplish this, the roof is supported by thirty-four, approximately 53-foot (16.2-m) tall concrete-filled steel Y-Columns spaced at approximately 100 ft (30.5 m) on center. See Fig. 2. Approximately half of the Y-Columns pass through new holes cut in the existing buildings while the other half are connected to a new structure which creates the enplaning level extension to the west. Double-concave friction pendulum seismic isolators with equal coefficients of friction and radii of the top and bottom surfaces were supplied by Maurer of Munich, Germany and are placed at the top of each tine of each Y-Column. See Fig. 2. The seismic isolators protect both the mass timber roof above and the architecturally exposed, geometrically complex Y-Columns below by serving as a force-limiter in the seismic load path.

Fig. 1. Rendering of the TCORE project viewed from the west (courtesy of ZGF Architects).

Fig. 2. Rendering of the TCORE project viewed from inside on enplaning level (courtesy of ZGF Architects). Seismic isolators occur at the top of each tine of Y-Columns.

2 Seismic Analysis and Design

2.1 Governing Code

The TCORE project was permitted as a Risk Category III structure under Section 104.10 *Alternate materials, design and methods of construction and equipment* of the 2019 Oregon Structural Specialty Code (OSSC) [1]. The 2019 OSSC is based on the 2018 International Building Code (IBC) [2]. Performance-based seismic design was pursued, adopting criteria from ASCE 7-16 Chap. 16 *Nonlinear Response History Analysis* and Chap. 17 *Seismic Design Requirements for Seismically Isolated Structures* [3].

2.2 Seismic Hazard Levels and Bounding Analysis Cases

To demonstrate code-equivalent performance, seismic analysis and design was performed for the Risk-Targeted Maximum Considered Earthquake (MCE_R) which is approximately equal to the 2500 year return period event. In addition to this probabilistically-derived hazard, the structure was voluntarily analyzed and designed for a deterministic seismic hazard corresponding to a mean plus one standard deviation estimate of a Mw 9.0 event (deterministic $M9 + 1\sigma$) on the Cascadia Subduction Zone. The Cascadia Subduction Zone stretches from northern California, United States to Vancouver Island, Canada and is approximately 60 miles (100 km) to the west of the project site. This deterministic hazard scenario was selected to reduce structural damage and improve building repairability, motivated by the Port of Portland's goal of regaining 30% passenger capacity for the Portland International Airport within three months of a Mw 9.0 earthquake on the Cascadia Subduction Zone. A suite of eleven ground motions represented the MCE_R hazard; a suite of seven ground motions represented the deterministic $M9 + 1\sigma$ hazard.

In order to develop a design for the structure which was robust to uncertainties in pile and seismic isolator properties, multiple bounding analyses were performed. Table 1 indicates the combination of seismic hazard, pile stiffness, and seismic isolator friction considered which resulted in six distinct analysis models. Expected, lower-bound, and upper-bound pile stiffness were provided by the geotechnical engineer of record, GRI, and affected the base of Y-Column stiffness as described in Sect. 2.4 Analysis Results. Seismic isolator friction was varied in accordance with ASCE 7-16 Chap. 17 *Seismic Design Requirements for Seismically Isolated Structures* [3] using property modification (λ) factors coordinated with the seismic isolator manufacturer, Maurer. While it may have been possible to justify reducing the number of combinations (e.g., only pairing lower-bound pile stiffness with lower-bound seismic isolator friction), it was judged that uncertainty in pile stiffness and seismic isolator friction were essentially uncorrelated (i.e., lower-bound pile stiffness was just as likely to occur for lower-bound seismic isolator friction as for upper-bound seismic isolator friction). Therefore, all six combinations were considered. Certain combinations control certain engineering demand parameters (e.g., seismic isolator displacement is controlled by lower-bound seismic isolator friction while Y-Column forces are controlled by upper-bound seismic isolator friction).

Table 1. Six combinations of seismic hazard levels and bounding analysis cases considered

Seismic Hazard Level	Pile Stiffness	Seismic Isolator Friction
MCE_R (code required)	Lower-bound	Lower-bound
	Upper-bound	
	Lower-bound	Upper-bound
	Upper-bound	
Deterministic M9 + 1σ (voluntary)	Expected	Lower-bound
		Upper-bound

2.3 Analytical Modeling

The seismic analyses of the TCORE lateral force-resisting system were performed in PERFORM-3D [4]. PERFORM-3D was chosen to facilitate consideration of multiple sources of inelasticity in the structure including the seismic isolators, buckling-restrained braces, and steel moment frame beams in flexure as well as P-Delta (approximate large displacement) effects. Figure 3 shows the modeling of Y-Columns with micropiles or pipe piles at their base depending on location, the enplaning level below the roof in the center foreground laterally supported by buckling-restrained braced frames, and the roof composed of glulam beams and steel plate girders and a mass timber diaphragm. The seismic isolators are modeled at the tops of each of the two tines of the Y-Columns. Difficult to see in Fig. 3 but also present in the PERFORM-3D model are the two north and south three-story mechanical mezzanines which penetrate through but are structurally separated from the seismically isolated roof but connected at the enplaning level. They rely upon steel special moment frames in the north-south direction and a full-story-height truss moment frame in the east-west direction with a buckling-restrained brace as the yielding diagonal in the trusses. A single-story central mezzanine above the enplaning level but below the seismically isolated roof is also present and laterally supported by steel special moment frames.

The double-concave friction pendulum isolators in PERFORM-3D [4] were modeled by two single friction pendulum elements in series. Since PERFORM-3D's isolator element can only distribute isolator-induced P-Delta moment to one element end, this modeling approach was necessary to adequately characterize, on a time-step-by-time-step basis, the P-Delta and V-h moments generated in a double-concave friction pendulum isolator. One single friction pendulum isolator element was modeled such that it distributed the isolator-induced P-Delta and V-h moments down (i.e., to the Y-column below) and the other was modeled such that it distributed the isolator-induced P-Delta and V-h moments up (i.e., to the steel plate girder and glulam beams above). The two isolator elements resisted equal shear and shared the total relative horizontal displacement equally as a result of the series system modeling approach. While it would have been possible to simplify isolator-induced P-Delta and V-h moments during the analysis – including them only as a post-processing step – the interaction of these moments with the building response was judged to be important enough to consider explicitly. This is especially true because varying axial forces in the seismic isolators occur due to vertical load redistribution between the two tines of the same Y-Column as the column laterally deflects (i.e., one tine rises and the other falls during lateral deflection in the plane of the Y).

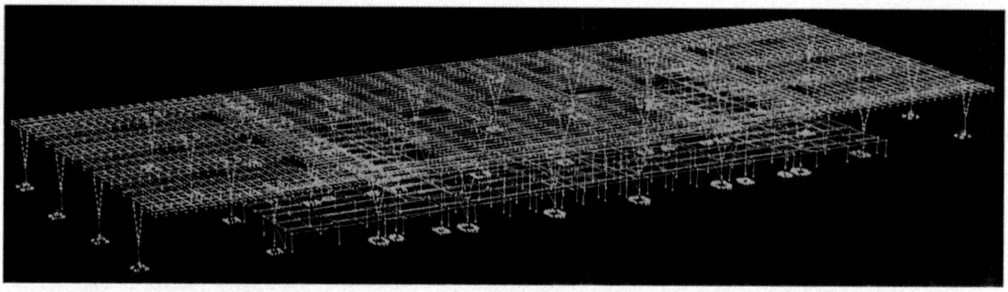

Fig. 3. PERFORM-3D model for the TCORE project viewed from the northwest.

2.4 Analysis Results

In more typical seismic isolation configurations where a relatively stiff superstructure is placed on top of seismic isolators (i.e., base isolation), the total seismic displacement demand concentrates in the isolation plane. For the TCORE project where the seismic isolators are placed on top of relatively flexible Y-Columns more akin to a seismic isolation configuration used for a bridge, the seismic displacement demand is shared between the Y-Column lateral deflection and the seismic isolators. While the seismic displacement demand at the roof is approximately equal at each isolator location (i.e., minimal torsion), the isolator displacements differ by location as a result of the varying support conditions at each Y-Column. Most notably, Y-Columns can be separated into two general groups: (1) Y-Columns which are back-stayed at the enplaning level, and (2) Y-Columns which cantilever their full height.

Y-Columns which are back-stayed at the enplaning level are the west-central columns where a new enplaning floor is constructed as part of the project. This enplaning level is laterally supported by buckling-restrained braced frames which are much stiffer than the Y-Columns. However, since the braced frames and Y-Columns are laterally connected at the enplaning level, the floor diaphragm serves to enforce displacement compatibility between the two systems, effectively converting the Y-Columns from cantilevers to propped cantilevers. Doing so reduces the top of Y-Column seismic displacement and, since the Y-Column and isolator displacements must sum to the roof displacement, the isolator displacement demands increase at these locations. Furthermore, the enplaning level responds at its own, much lower period than the seismically isolated roof. Since the top of Y-Column seismic displacement is partially "driven" by the response of the enplaning level, the Y-Columns can, for certain points in time for certain ground motions, displace in the opposite direction of the roof displacement. Therefore, isolator displacement demands at Y-Columns which are back-stayed at the enplaning level can exceed the roof displacement demand.

Y-Columns which cantilever their full height occur at all other locations except where the new enplaning floor is constructed. These Y-Columns either penetrate through, but are not connected to, the existing enplaning floor or are constructed outside the footprint of the enplaning floor; the latter condition occurs for the Y-Columns on the northernmost and southernmost grids. Unlike the Y-Columns which are back-stayed at the enplaning level, the Y-Columns which cantilever their full height help share the roof displacement demand with the seismic isolators. The amount of roof displacement demand shared by these Y-Columns varies by location and is dependent on the pilecap stiffness at the Y-Column base. Notably, many of the micropile groups at the Y-Column bases are complex and asymmetric in order to miss existing pile caps at the surface and existing piles at depth.

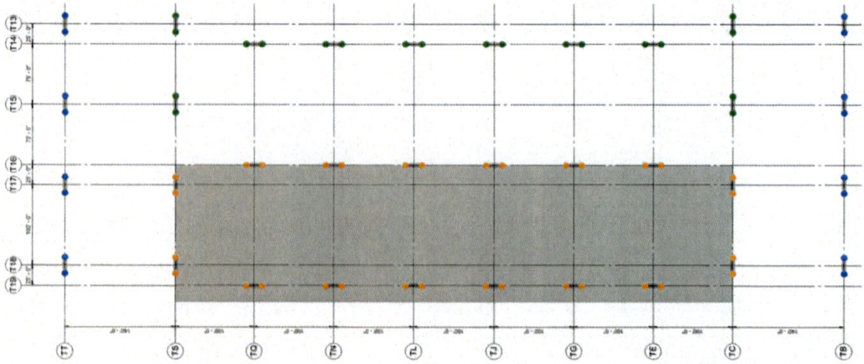

Fig. 4. Isolator displacement capacities by location. Green, blue, and orange indicate 16 in. (406 mm), 19 in. (483 mm), and 22.5 in. (572 mm) isolator displacement capacities, respectively. Grey area indicates extent of new enplaning level below. North is to the left.

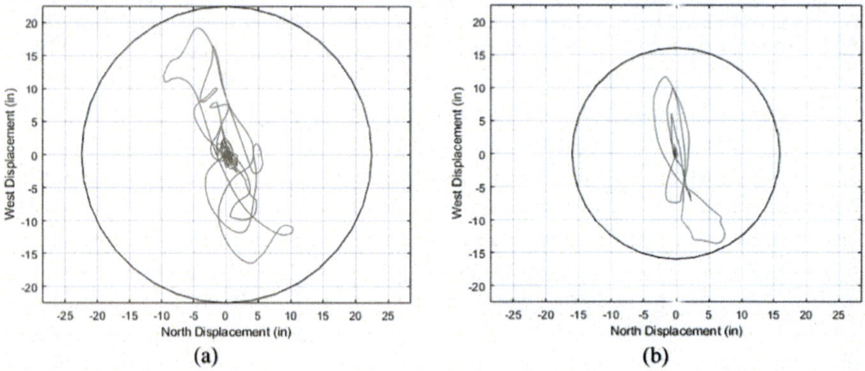

Fig. 5. Isolator displacement traces for suite maximum MCE_R for (**a**) representative location where Y-Column is back-stayed at the enplaning level with 22.5 in. (572 mm) displacement capacity, and (**b**) representative location where Y-Column cantilevers its full height with 16 in. (406 mm) displacement capacity.

As a result of the varying support conditions at each Y-Column as described above, isolator displacement capacities differ by location as shown in Fig. 4. Figure 5 further shows isolator displacement traces for a representative location for each Y-Column support condition. Isolator displacement capacities were determined as the larger of (a) 1.5× suite mean for the MCE_R hazard, (b) suite maximum for the MCE_R hazard, and (c) suite mean for the M9 + 1σ hazard. The 1.5 factor on suite mean for the MCE_R hazard, in combination with a restraining ring in the seismic isolator design which stops the inner sliding "puck" at the isolator displacement capacity, was applied based on engineering judgment to limit the likelihood of isolator overstroke and loss of gravity-carrying capacity. Seismic isolator displacement demand was measured as the peak resultant displacement (i.e., peak displacement at any angle in plan) at any point in time for each ground motion before performing the suite mean or maximum operation to establish isolator displacement capacity.

3 Seismic Isolator Testing

3.1 Prototype Testing

Full-scale, dynamic prototype testing of the seismic isolators was performed at the European Centre for Training and Research in Earthquake Engineering (EUCENTRE) in Pavia, Italy. Figure 6 shows one of the prototype seismic isolators during testing at the instant of time when it is near its full displacement capacity. The purpose of the prototype testing was to validate and refine the testing property modification (λ_{test}) factors (i.e., the factors which address uncertainty of the coefficient friction primarily due to axial stress, heating, and rate of loading). A test matrix was developed and implemented in accordance with the testing requirements of ASCE 7-16 Chap. 17 *Seismic Design Requirements for Seismically Isolated Structures* [3] that included multiple vertical load levels, peak lateral displacements (and corresponding peak velocities for sinusoidal loading), and cycles.

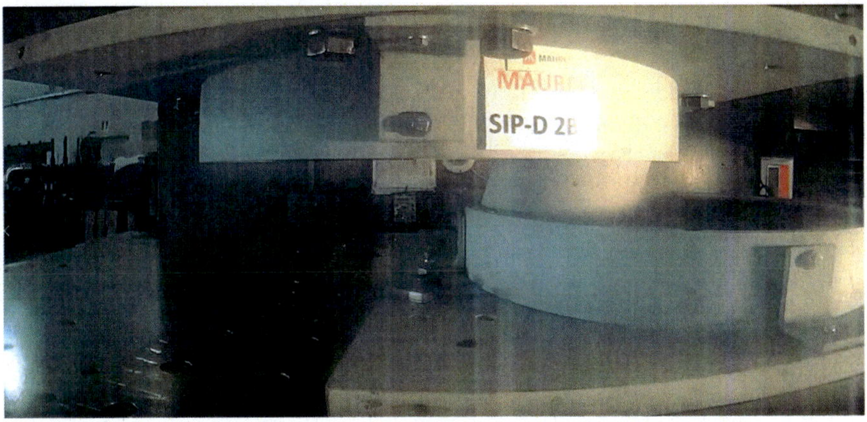

Fig. 6. Prototype isolator during testing at instant of time near its full displacement capacity.

Before and after the prototype test sequence, the prototype isolators were subjected to the quasi-static production test sequence. Half of the prototype isolators were then shipped to SISMALAB in Crispiano, Italy and subjected to the identical production test sequence just completed at EUCENTRE. This was a unique project requirement in order to cross-check results from EUCENTRE with those from SISMALAB since, as discussed in Sect. 3.2 Production Testing, production isolator testing was split between the two laboratories. This requirement was necessary to ensure that production testing results at SISMALAB could be anchored back to the prototype isolator performance established by EUCENTRE testing.

Prototype seismic isolators were refurbished after prototype testing by replacing the sliding material which occurs on the top and bottom of the inner "puck" as well as, where required, repolishing the stainless steel liner on the two concave outer surfaces. The refurbished prototype isolators then joined the group of production isolators that were ultimately production tested and used on the project.

3.2 Production Testing

Quasi-static production testing was performed on 100% of all seismic isolators used on the project in accordance with ASCE 7-16 Chap. 17 *Seismic Design Requirements for Seismically Isolated Structures* [3]. Forty production isolators were tested at SISMALAB and the remaining 28 were tested at EUCENTRE. Although production testing was performed quasi-statically whereas prototype testing was performed dynamically, the ASCE 7-16 Chap. 17 requirement to perform the production test sequence on prototype isolators before the prototype test sequence, as described in Sect. 3.1 Prototype Testing, allowed production testing results to be anchored back to prototype isolator performance. Specification property modification (λ_{spec}) factors to represent manufacturing variability from the design coefficient of friction were set as 0.85 and 1.15 for the average over all isolators of a common type and 0.8 and 1.2 for any individual isolator.

4 Conclusion

A unique configuration of double-concave friction pendulum seismic isolators on top of relatively flexible Y-Columns to support a new mass timber roof achieved the Port of Portland's goal for a safer, more resilient, and functionally enhanced terminal core at the Portland International Airport. Extensive analysis, physical testing, and engineering validation was performed to ensure that the novel design was robust to uncertainties in seismic hazard, structural properties, and analytical modeling approaches. When completed in 2024, the Terminal Core Redevelopment project will provide approximately twenty million travelers each year the opportunity to appreciate seismic isolation in a visible, architecturally celebrated position.

References

1. Oregon Structural Specialty Code (OSSC): International Code Council (2019)
2. International Building Code (IBC): International Code Council (2018)
3. Minimum Design Loads and Associated Criteria for Buildings and Other Structures, ASCE 7-16: American Society of Civil Engineers, Structural Engineering Institute, Reston, Virginia, USA (2017)
4. User Guide, PERFORM-3D, Nonlinear Analysis and Performance Assessment for 3D Structures, Version 7, Computers and Structures Inc. (2018)

Coupling of Structural Additions for the Mitigation of Seismic Response in Existing Buildings

Diana Faiella, Mario Argenziano, Francesco Esposito, and Elena Mele(✉)

Department of Structures for Engineering and Architecture, University of Naples Federico II, Piazzale Tecchio 80, 80125 Naples, Italy
elenmele@unina.it

Abstract. In the present paper, additions in structural steelwork are utilized for giving new life to old buildings in regions characterized by medium/high seismic hazard. Two models are here proposed, i.e.: vertical addition for masonry buildings and lateral addition for r.c. buildings. For the model of vertical addition, the connection between the masonry and steel structures is realized by means of an Intermediate Isolation System (IIS). For the model of lateral addition, an exoskeleton (EXO) is connected to the existing building by means of rigid or flexible and dissipative link. Two buildings, representative of the heterogeneous Italian building stock, are selected as case studies. Parametric analyses are firstly performed on lumped mass models to explore the feasibility and effectiveness of the IIS and EXO systems in reducing the seismic response of the case studies; then, once selected the design configurations of the new additions, more refined 3D FE models are adopted for the detailed analysis of the two solutions.

Keywords: Vertical addition · Lateral addition · Structural steelwork · Intermediate isolation system · Exoskeleton · Seismic retrofit

1 Introduction

A sustainable model of urban growth should be based on the density increase and use/reuse of existing building stock, with no additional land consumption and saving the carbon burden that already exists in the built environment. In order to make feasible such strategy in regions of medium/high seismic hazard, a common approach is proposed, based on well-known principles of structural dynamics. Two models, both based on the concept of addition, are proposed: vertical addition for masonry buildings (Fig. 1a), and lateral addition (exoskeletons) for r.c. buildings (Fig. 1b). The design parameters are the dynamic properties of the structural addition and of the connection between the existing and new structural parts.

In the case of vertical extension (rooftop addition) the connection is realized by means of a seismic isolation system, located on the roof of the existing building and at the base of the new structure [1, 2], thus giving rise to an Intermediate Isolation System (IIS) for the structural complex; the isolation system should be designed for converting

G. P. Cimellaro (Ed.): WCSI 2022, LNCE 309, pp. 632–642, 2023.
https://doi.org/10.1007/978-3-031-21187-4_53

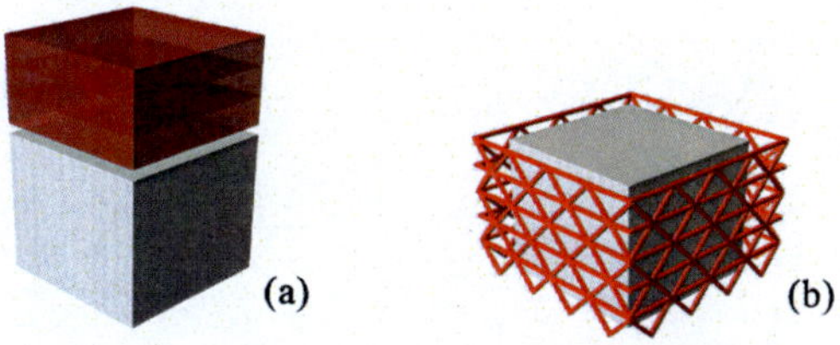

Fig. 1. (a) Vertical addition, (b) lateral addition.

the vertical addition into a huge mass damper, with remarkable reduction of the seismic demand on the structural complex, even with respect to the standalone existing building.

In the case of lateral addition, exoskeletons (EXO) made by steel diagrid, hexagrid, or nonconventional structural patterns, are added to wrap the existing building along the whole perimeter [3–5]. This exoskeleton can also support ventilated or double skin façades, as well as satisfy solar shading/lighting requirements and integrate energy retrofit interventions [5, 6]. The connection between the existing building and the exoskeleton can be either "total" (rigid), by coupling in parallel the two structures, or "partial" (flexible and dissipative) [4, 5, 7, 8].

In this context, two case studies are selected for exploring the feasibility and effectiveness of steel additions for the seismic retrofit of existing buildings.

2 Simplified Models

IIS and EXO systems can be preliminary described through simplified lumped mass models. In the three-degree-of-freedom IIS model (3DOF IIS, Fig. 2a), m_i, k_i, c_i are the mass, stiffness and damping constant of the *i-th* DOF, with $i = 1, 2, 3$, i.e.: the lower structure (LS), the isolation system (ISO), and the upper structure (US), respectively. In the two-DOF model of the existing building (EB) connected to the exoskeleton (2DOF EB + EXO, Fig. 2b), m_1, k_1, c_1 and m_2, k_2, c_2 refer to the properties of the existing building and the exoskeleton, respectively. Instead, k_3 and c_3 are the stiffness and damping constant of the connection, which can be purely elastic ($c_3 = 0$), purely viscous ($k_3 = 0$), viscoelastic ($k_3 \neq 0$, $c_3 \neq 0$), by assuming either a Kelvin-Voigt (K-V) or a Maxwell (M) model [4, 7, 8].

3 Case Studies

Two Italian buildings are selected as case studies (Fig. 3). After a brief description of the buildings, their capacity is assessed by means of push over analyses [9], developed on the relevant 3D FE models through the software SAP2000 [10]. Considering the contribution of the fundamental mode in the weakest (Y) direction, the dynamic properties of the reduced-order SDOF, i.e. the LS in the 3DOF IIS model (Fig. 2a) or the EB in the 2DOF EB + EXO model (Fig. 2b), are derived.

The *first case study* is a 16th century aggregate of four masonry buildings (Fig. 3a), located in Pozzuoli (Naples, Italy) [2], with residential and tertiary occupancy. It falls

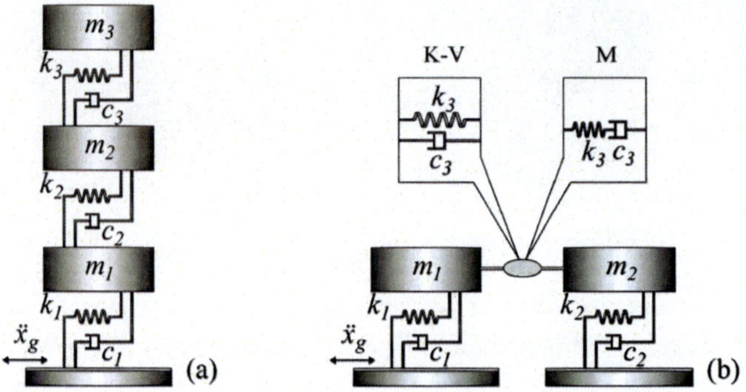

Fig. 2. Simplified lumped mass models: (a) 3DOF IIS; (b) 2DOF EB + EXO.

within a plan (P.U.A. [11]), recently issued for the town, aimed at reconstructing some rooftop volumes, demolished in the 1980s. Plan dimensions and heights of the buildings are provided in Fig. 3a. The mechanical properties assumed for the tuff masonry material are: average weight of 16 kN/m^3, Young modulus of 1080 MPa, shear modulus of 360 MPa, compression strength of 3.0 MPa [9]. The dynamic properties of the SDOF LS are: $m_1 = 2201$ kNs^2/m, $k_1 = 639883$ kN/m, $T_1 = 0.37$ s.

The *second case study* (Fig. 3b) is a four-story residential building assumed to be located in L'Aquila (Italy), built in 1972 [5, 12]. Lacking some structural data of the existing building, a simulated design has been carried out for gravity loads according to the Italian building code in force at the time of construction [13]; then the assessment under seismic loads is carried out according to the current NTC 2018 [9]. The plan dimensions, height, and sections are provided in Fig. 3b. The building materials are steel FeB44k and concrete C 25/30. The dynamic properties of the SDOF EB are: $m_1 = 575$ kNs^2/m, $k_1 = 27420$ kN/m, $T_1 = 0.91$ s.

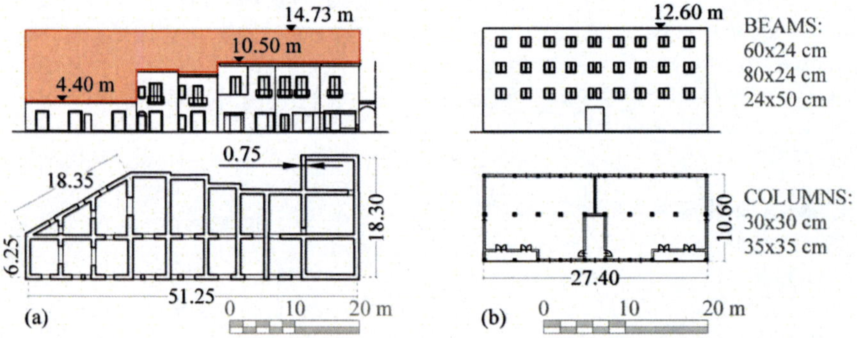

Fig. 3. (a) First case study, (b) second case study.

4 Design Solutions

The solutions adopted for the seismic retrofit of the two case studies are realized by means of additions in structural steelworks (S275, $f_{yk} = 275$ MPa, $f_{yd} = 261$ MPa). The design solutions are firstly described; parametric analyses are then performed on the lumped mass models introduced in Sect. 2, to explore the seismic behavior of such systems with response spectrum analyses; finally, spectrum-compatible time history analyses are performed on the relevant 3D FE models [10]. The design acceleration response spectrum is the elastic spectrum at the Life Safety Limit State (SLV, 475 years of return period) [9]. The set of spectrum-compatible acceleration records is defined with the software REXEL v 3.5 (PGA = 0.1 – 0.3 g in the period range 0.15 – 2 s) [14].

4.1 Vertical Addition

The vertical addition is defined according to the P.U.A. of Pozzuoli [11] in terms of floor area and stories to rebuild (Fig. 3a). A "filling" steel structure has been preliminarily added on the left side of the aggregate in order to regularize the top edges of the buildings and to provide the floor level for the isolation system; then, a single-story steel structure is realized on the top of the isolation devices. The idea is to design the isolation layer for converting the new isolated addition into a mass damper able to reduce the seismic demand with respect to the as-is configuration [2].

Parametric analysis. In the 3DOF IIS model (Fig. 2a) the design parameters are the mass ratio $\mu = (m_2 + m_3)/m_1 = 0.25$; the stiffness ratio $K = k_3/k_1 \in [0.1, 2]$; the isolation ratio $I = T_2/T_3 \in [0.1, 50]$. The design spectrum is depicted in Fig. 4a for the site of Pozzuoli. The damping constants c_1, c_2, c_3, are obtained from the values of the equivalent viscous damping ratios usually adopted for the corresponding structure types, i.e.: $\xi_1 = 0.05$ (masonry), $\xi_2 = 0.15$ (high damping rubber), $\xi_3 = 0.02$ (steel). The adoption of different values of damping ratios within the same structure leads to a system characterized by non-proportional damping and complex-valued natural modes [15]. The response spectrum analyses are carried out by considering the complex modal superposition method described by Sinha and Igusa [16]. The results are provided in Fig. 4b in terms of displacement ratio $\overline{d}$, i.e. the first-DOF displacement of the 3DOF IIS model normalized to the SDOF LS counterpart, by varying both the parameter K and the isolation period T_2.

From Fig. 4b it can be observed that the ratio $\overline{d}$ does not significantly varies by varying the parameter K, as also observed in [2]. Instead, $\overline{d}$ assumes values larger than one for very small isolation periods, values smaller than one in the range of small-medium isolation periods (e.g. for $T_2 \in [0.4, 2]$ s, $\overline{d} \in [0.69, 0.95]$), and values that tend to one in the range of large isolation periods (e.g. $T_2 > 2$ s).

When $\overline{d} > 1$ the upper structure is rigidly connected to the lower structure and works as a conventional vertical addition [1]. When $\overline{d} < 1$ the isolated vertical addition and the lower structure dynamically interact and the former works as a mass damper for the latter (*mass damping effect* [1]). When $\overline{d}$ tends to one, a perfect disconnection occurs between the isolated vertical addition and the lower structure, with the former working as a base isolated structure without any dynamic interaction with the existing building (*isolation effect* [1]).

The design solution is defined by choosing the minimum value of the red curve passing through $K = 1.25$, i.e.: $\overline{d} = 0.69$, $T_2 = 0.5$ s, $I = 4$.

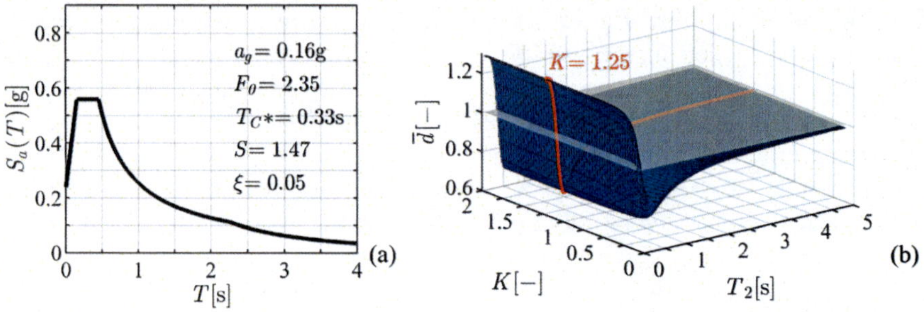

Fig. 4. (a) Elastic acceleration response spectrum (SLV, [9]), (b) $\overline{d}$ vs. K and T_2.

Choice of the design configuration. The steel vertical addition is a single-story Concentric Braced Frame (CBF), with two and four braced bays along the X and Y direction, respectively. The cross sections adopted for the structural members are: columns CHS 219.1 × 16; beams IPE 240 and 360; diagonals HEB 120 and 160; beam grillage on the roof of the existing building, RHS 900 × 450 × 40 mm, HEB 700 and 900, IPE 400 and 550. The isolation system consists of 10 high damping rubber bearings ($\xi_{eq} = 0.15$) placed at the intersections of the orthogonal walls of the lower structure (Fig. 5a). The steel structure, introduced to fill the height gap between the left building and the remaining aggregate, is a two-story Moment.

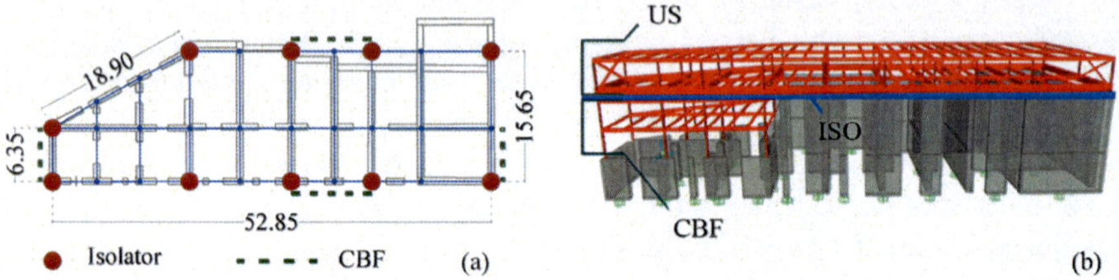

Fig. 5. (a) Isolation system, (b) 3D IIS model.

Resisting Frame (MRF) designed for gravity loads only, with beams IPE 240 and 360 and columns CHS 219.1 × 16.

3D models and analyses. The 3D FE model of the building with intermediate isolation IIS (Fig. 5b) accounts for the nonlinear behaviour of the masonry structure, while the isolated vertical addition is considered elastic. The Pivot-type model [17] is assumed as hysteretic model for the plastic hinges of the masonry walls. The shape of the hysteresis loop and the stiffness and strength degradation are controlled by the mechanical parameters α_1, α_2, β_1, β_2, and η. The values here assigned to the parameters are 0.45 for α_1, α_2, β_1, β_2, and zero for η [18–20]. The isolation system is modelled with linear springs. As mentioned above, the IIS system is characterized by non-proportional damping. Therefore, nonlinear direct-integration time history analyses are developed in SAP2000 accounting for the full damping matrix [10]. The same analyses are also performed for the LS model.

When the masonry nonlinearities are accounted for, the effectiveness of the isolated vertical extension as tuned mass damper should be assessed by considering both the peak-displacement response and the effect of the damage accumulation in the existing building [21–24]. For this aim, Fig. 6a shows the peak displacement profiles for the seven input waves in the retrofitted configuration, and the average response for both the IIS and LS models. Instead, Fig. 6b provides a hysteretic energy absorption ratio $\rho_{E,LS(\mathrm{IIS})}$ between the hysteretic energy, $E_{h,LS(\mathrm{IIS})}$, and the input energy, $E_{i,LS(\mathrm{IIS})}$, of the two configurations, i.e.: $\rho_{E,LS} = E_{h,LS}/E_{i,LS}$ and $\rho_{E,IIS} = E_{h,IIS}/E_{i,IIS}$, under the different ground motions. The average values of the hysteretic energy absorption ratios are also provided in Fig. 6b. The results of the analyses show that the displacements of the LS decrease in the IIS configuration, as well as the hysteretic energy absorption ratios with an average value of 0.16 with respect to the average value of 0.51 of the LS configuration. This implies a smaller inelastic engagement of the existing building in the IIS configuration.

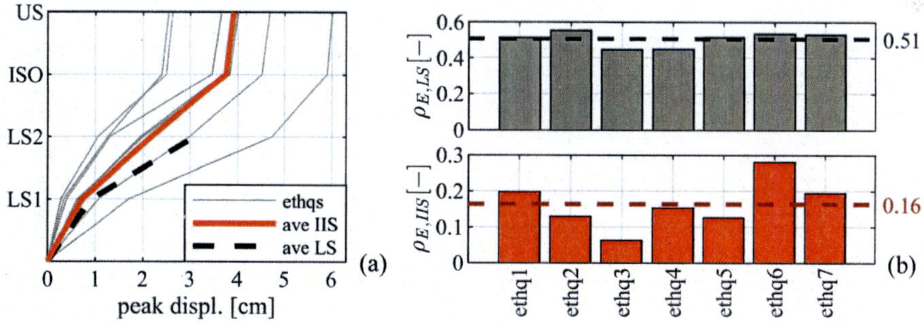

Fig. 6. (a) peak story displacements, (b) cumulated hysteretic energy ratios.

4.2 Lateral Addition

The seismic retrofit of the r.c. building is achieved by means of a steel diagrid exoskeleton connected to the existing structure by either a rigid connection or a viscoelastic connection. The diagrid is configured as a narrow grid of diagonal members that guarantees high stiffness and strength against lateral actions [25]. The idea is to preserve the elastic behaviour of the existing building, by transferring a great part of the seismic demand from the existing building to the diagrid in the case of rigid connection, or by reducing the seismic demand on the existing building thanks to the dissipation offered by the viscoelastic connection [12].

Parametric analysis. For the 2DOF EB + EXO model (Fig. 2b) two types of connection are considered, i.e. the rigid connection and the viscoelastic connection (K-V model). For the rigid connection, the design parameters are the mass ratio $\mu = m_2/m_1 = 0.05$ and the stiffness ratio $k_E = k_2/k_1 \in [3, 15]$. For the viscoelastic connection, the design parameters are the mass ratio $\mu = 0.05$; the stiffness ratios $k_E \in \{3, 5, 10, 15\}$ and $k_L = k_3/k_1 \in [1, 10]$; the damping ratio $c_L = c_3/c_1 \in [1, 10]$. The spectrum is depicted in Fig. 7a for the site of L'Aquila. A constant damping ratio is assumed equal to 0.05 for the rigid connection. When the damping of the connection is considered by means of the parameter c_L, its different value with respect to the counterparts of the existing building and the exoskeleton leads to a system characterized by non-proportional damping. As in the case of IIS systems (Sect. 4.1), the response spectrum analyses are carried out by considering the complex modal superposition method described by Sinha and Igusa [16]. The results are provided in terms of displacement ratios, namely: $\overline{d} = d_{2DOF\ EB+EXO}{}^{(EB)}/d_{SDOF\ EB}$ and $\overline{d}_y = d_{2DOF\ EB+EXO}{}^{(EB)}/d_{y,SDOF\ EB}$. In particular, $d_{SDOF\ EB}$ and $d_{y,SDOF\ EB}$ are the displacement and the displacement at the elastic limit in the SDOF EB model, respectively; $d_{2DOF\ EB+EXO}{}^{(EB)}$ is the displacement of the existing building in the 2DOF EB + EXO model. The displacement ratios are provided in Fig. 7b for the case of rigid connection, by varying the parameter k_E, and in Fig. 8 for the case of viscoelastic connection, by varying k_L and c_L for each value of k_E. By increasing the stiffness of the exoskeleton for the case of rigid connection (Fig. 7b), both the ratios $\overline{d}$ and $\overline{d}_y$ decrease. The ratio $\overline{d}$ is smaller than one, therefore the demand on the existing building is drastically reduced; the ratios $\overline{d}_y$ is equal to, or smaller than, one for $k_E > 9$ implying that the r.c. elements remain in the elastic field. For the case of viscoelastic connection (Fig. 8), by increasing both k_L and c_L the ratios $\overline{d}$ and $\overline{d}_y$ decrease. The ratio $\overline{d}$ is always smaller than one by varying the parameter k_E, as observed for the case of rigid connection (Fig. 7b); instead, the ratio $\overline{d}_y$ is equal to, or smaller than, one for large values of the design parameters, i.e.: $k_E = 5$, k_L and c_L equal to 10; $k_E = 10$, k_L and c_L in the range 7–10; $k_E = 15$, k_L and c_L in the range 5–10.

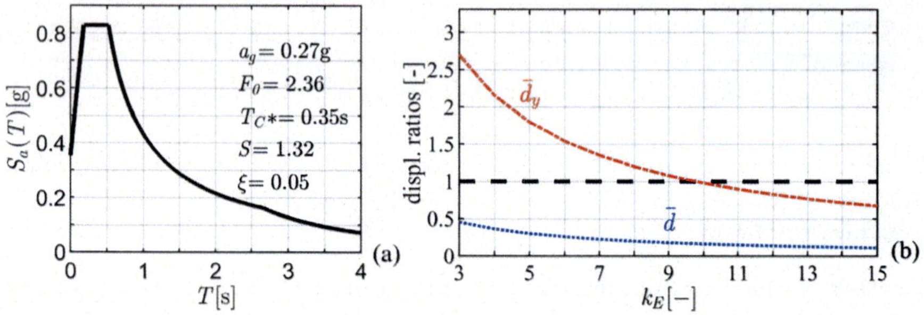

Fig. 7. (a) Elastic acceleration response spectrum (SLV, [9]), (b) displ. ratios vs. k_E.

The design solution is selected by considering the rigid connection only (Fig. 7b): $k_E = 10$, $\overline{d} = 0.17$, $\overline{d}_y = 0.98$.

Choice of the design configuration. The geometric characteristics of the triangular module are provided in Fig. 9a in terms of: diagonal angle (θ) and length (L_d), module height (h), number of diagonals on each façade (n), number of modules along the perimeter (n_k) and elevation (n_m). The diagrid is designed by adopting the procedure provided in Ref. [25] for tall buildings. As observed in Ref. [25], thanks to the high rigidity of the triangular module, particularly for this low-rise building, the local

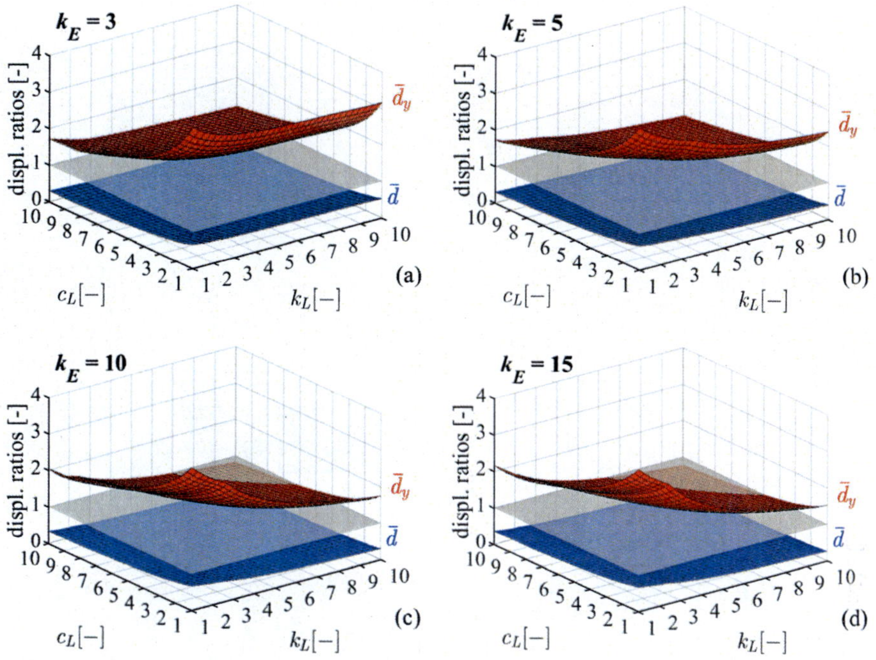

Fig. 8. Displ. Ratios vs. ratios k_L and c_L: (a) $k_E = 3$, (b) $k_E = 5$, (c) $k_E = 10$, (d) $k_E = 15$.

strength requirements (members sizing for strength and stability) prevail on global stiffness requirements. Hence, the members cross sections are derived to obtain $\overline{d}_y \leq 1$; then, the corresponding stiffness ratio is calculated, resulting in $k_E = 12$, which, being larger than 9, guarantees that the r.c. building is preserved from any damage. The profiles adopted for diagonal members are: SHS 140 mm × 8, 6.3 mm (X direction); SHS 180 × 20, 12.5, 8, 6.3 mm (Y direction).

3D models and analyses. In the 3D FE model of the retrofitted building rigidly connected to the diagrid (3D EB + EXO, Fig. 9b) both the existing building and the diagrid are considered elastic, checking *a posteriori* that the Demand to Capacity Ratio is smaller than one (DCR < 1). In the 3D FE model of the existing building, the structure is assumed inelastic, by adopting the Takeda model [26] as hysteretic model for the plastic hinges at the ends of columns and beams.

This configuration is characterized by almost the same mass and larger overall stiffness with respect to the standalone existing building ($T_{EB} = 0.94$ s, $T_{EB+EXO} = 0.27$ s). The results, provided in terms of peak story displacements (Fig. 10a) and displacement ratios $\overline{d}$ and $\overline{d}_y$ (Fig. 10b), show a drastic reduction of the engagement of the existing building, which remains in the elastic field.

5 Conclusions

This paper deals with the seismic retrofit of old buildings in seismic areas. Two design solutions are developed by means of additions in structural steelworks. For the first case,

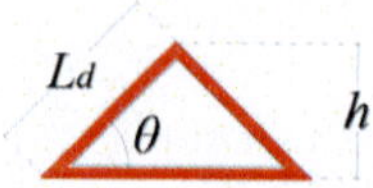

Dir.	θ [°]	L_d[m]	h[m]	n[-]	n_k[-]	n_m[-]
X	48	2.12	1.58	10	40	8
Y	47	2.14	1.58	4	40	8

(a)

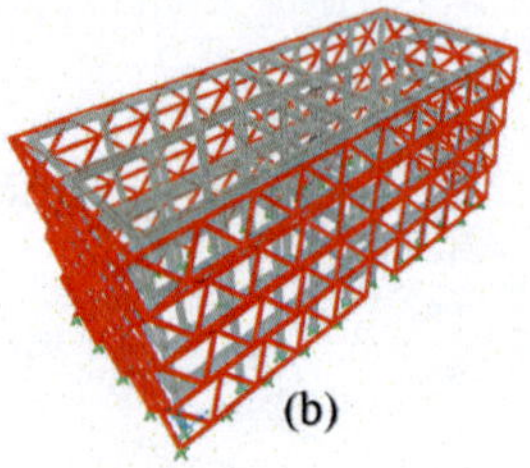

Fig. 9. (a) Geometrical characteristics of the diagrid triangular module, (b) 3D EB + EXO.

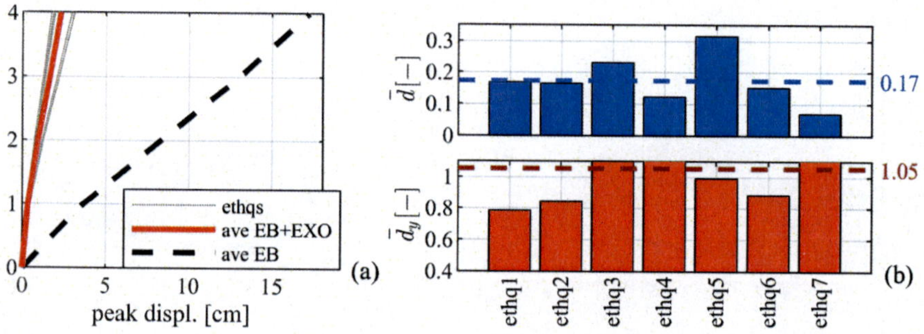

Fig. 10. (a) Peak story displacements, (b) displ. ratios.

a base isolated vertical extension is mounted on the roof of a masonry aggregate, thus realizing an intermediate isolation system. For the second case, a diagrid exoskeleton is rigidly connected to an existing r.c. building. Parametric analyses are firstly developed on the lumped mass models of the IIS and EXO systems, by means of response spectrum analyses to define the design configurations for the new additions. Then, spectrum-compatible time history analyses are carried out on the relevant 3D FE models of the design solutions. From the results of the dynamic analyses, it emerges that: (i) in the IIS solution, by carefully designing the isolation system, the new structure can be converted into a giant mass damper able to reduce the seismic response of the structural complex; (ii) in the EXO solution, the high rigidity of the diagrid, though increasing the global seismic response of the new configuration, preserves the elastic behaviour of the existing building.

The strategy suggested in this paper for the seismic retrofit of existing buildings, based on steel additions and proper design of the link between the old and new structural parts, appears effective and particularly feasible for a quick site assembly, with almost no disruptions of the activities that are hosted in the existing structures.

References

1. Faiella, D., Calderoni, B., Mele, E.: Seismic retrofit of existing masonry buildings through inter-story isolation system: a case study and general design criteria. J. Earthq. Eng. 1–37 (2020)
2. Argenziano, M., Faiella, D., Bruni, F., De Angelis, C., Fraldi, M., Mele, E.: Upwards-vertical extensions of masonry built heritage for sustainable and antifragile urban densification. J. Build. Eng. **44**, 1–17 (2021)
3. Mele, E., Fraldi, M., Montuori, G.M., Perrella, G., Della Vista, V.: Hexagrid-Voronoi transition in structural patterns for tall buildings. Frattura ed Integrità Strutturale **47**, 186–208 (2019)
4. Reggio, A., Restuccia, L., Martelli, L., Ferro, G.A.: Seismic performance of exoskeleton structures. Eng. Struct. **198**, 1–11 (2019)
5. Passoni, C., Guo, J., Christopoulos, C., Marini, A., Riva, P.: Design of dissipative and elastic high-strength exoskeleton solutions for sustainable seismic upgrades of existing RC buildings. Eng. Struct. **221**, 1–23 (2020)
6. Menna, C., Del Vecchio, C., Di Ludovico, M., Mauro, G.M., Ascione, F., Prota, A.: Conceptual design of integrated seismic and energy retrofit interventions. J. Build. Eng. **38**, 1–16 (2021)
7. Gattulli, V., Potenza, F., Lepidi, M.: Damping performance of two simple oscillators coupled by a visco-elastic connection. J. Sound Vib. **332**, 6934–6948 (2013)
8. Luongo, A.: Free vibrations and sensitivity analysis of a defective two degree-of freedom system. AIAA J. **33**(1), 120–127 (1995)
9. NTC: Consiglio Superiore Lavori Pubblici, Aggiornamento delle Norme Tecniche per le Costruzioni. Gazzetta Ufficiale della Repubblica Italiana **42**, 274–277 (2018) (in Italian)
10. Computers and Structures, Inc.: SAP2000 (2009)
11. P.U.A.: Piano Urbanistico Attuativo del Comune di Pozzuoli. Ricomposizione volumetrica degli edifici parzialmente demoliti c/o crollati a seguito ai fenomeni bradisismici in coerenza con il P.R.G. art. 26 della L.R. n°16 del 22/12/2004 I° Stralcio. Delibera n. 84 del 24/06/2015 (2015) (in Italian)
12. Faiella, D., Faella, V., Alaio, E., Mele, E.: Adeguamento sismico di edifici esistenti in c.a. tramite esoscheletro diagrid. Costruzioni Metalliche Sept-Oct 2019, 9–23 (2019) (in Italian)
13. DM 30 maggio: Norme tecniche alle quali devono uniformarsi le costruzioni in conglomerato cementizio, normale e precompresso ed a struttura metallica (1972) (in Italian)
14. Iervolino, I., Galasso, C., Cosenza, E.: REXEL: computer aided record selection for code-based seismic structural analysis. Bull. Earthq. Eng. **8**, 339–362 (2010)
15. Veletsos, A.S., Ventura, C.E.: Modal analysis of non-classically damped linear systems. Earthq. Eng. Struct. Dyn. **14**(2), 217–243 (1986)
16. Sinha, R., Igusa, T.: CQC and SRSS methods for non-classically damped structures. Earthq. Eng. Struct. Dyn. **24**(4), 615–619 (1995)
17. Dowell, R.K., Seible, F., Wilson, E.L.: Pivot hysteresis model for reinforced concrete members. ACI Struct. J. **95**(5), 607–617 (1998)
18. Magenes, G., Bolognini, D., Braggio, C.: Simplified methods for non-linear seismic analysis of masonry buildings. CNR—National Group for Seismic Protection, Rome (2000) (in Italian)
19. Pasticier, L., Amadio, C., Fragiacomo, M.: Non-linear seismic analysis and vulnerability evaluation of a masonry building by means of the SAP2000 vol 10 code. Earthquake Eng. Struct. Dyn. **37**, 467–485 (2008)
20. Ozturk, B.: Seismic behavior of two monumental buildings in historical Cappadocia region of Turkey. Bull. Earthq. Eng. **15**(7), 3103–3123 (2017). https://doi.org/10.1007/s10518-016-0082-6
21. Domizio, M., Garrido, H., Ambrosini, D.: Single and multiple TMD optimization to control seismic response of nonlinear structures. Eng. Struct. **252**, 113667 (2022)

22. Lukkunaprasit, P., Wanitkorkul, A.: Inelastic buildings with tuned mass dampers under moderate ground motions from distant earthquakes. Earthq. Eng. Struct. Dyn. **30**(4), 537–551 (2001)
23. Sgobba, S., Marano, G.C.: Optimum design of linear tuned mass dampers for structures with nonlinear behaviour. Mech. Syst. Signal Process. **24**, 1739–1755 (2010)
24. Quaranta, G., Mollaioli, F., Monti, G.: Effectiveness of design procedures for linear TMD installed on inelastic structures under pulse-like ground motion. Earthq. Struct. **10**(1), 239–260 (2016)
25. Montuori, G.M., Mele, E., Brandonisio, G., De Luca, A.: Design criteria for diagrid tall buildings: stiffness vs. strength. Struct. Des. Tall Spec. Build. **23**(17), 1294–1314 (2014)
26. Takeda, T., Sozen, M., Nielsen, N.: Reinforced concrete response to simulate earthquakes. J. Struct. Div. **96**(12), 19–26 (1970)

Verification of Actual Displacement Scaled Displacement-Control System with High-Static-Low-Dynamic Stiffness and Rotational Inertia for Seismic Isolation

Jun Iba[1(✉)], Koichi Watanabe[2], Kou Miyamoto[2], Ken Ishii[3], and Masaru Kikuchic[3]

[1] Graduate School of Engineering, Hokkaido University, Sapporo, Japan
iba.jun.j9@elms.hokudai.ac.jp
[2] Shimizu Corporation, Tokyo, Japan
[3] Faculty of Engineering, Hokkaido University, Sapporo, Japan

Abstract. Amid increasing concerns about the occurrence of mega earthquakes, recent studies have shown that the deformation of conventional seismic isolators may exceed their expected design range. To address this issue, we have been developing a displacement-control system for seismic isolation equipped with high-static-low-dynamic stiffness and rotational inertia. In this paper, we report the results of large shaking-table experiments involving a comparison of the response of a seismic-isolation model of our displacement control system we developed to that of a time-response analysis model we constructed on the basis of our model.

Keywords: Seismic isolation · Displacement control system · HSLDS · Rotational inertia · Shaking table test

1 Introduction

Seismic isolation reduces the acceleration of structures by extending their natural periods. However, there is a trade-off relationship between response acceleration and displacement. Therefore, when a long-period earthquake occurs, a conventional seismically isolated building may resonate, resulting in response displacement that exceeds the expected design range. If the displacement of the seismically isolated structure exceeds the clearance of the seismic-isolation layer, ultimate phenomena, such as a collision with a retaining wall, are likely to occur. In fact, a long-period pulse earthquake with a large amplitude occurred in Kumamoto, Japan in 2016 [1], and some of the seismically isolated buildings were observed to have response displacements that exceeded the design assumptions. Therefore, displacement control for such buildings is important, especially for long-period earthquake waves. Supplemental dampers are used to control displacement, but dampers increase the floor acceleration and interstory drift at the cost of controlling displacement.

To solve this problem, we propose the displacement control system (DCS). Figure 1 shows the diagram of this system. The DCS combines high-static-low-dynamic stiffness

G. P. Cimellaro (Ed.): WCSI 2022, LNCE 309, pp. 643–651, 2023.
https://doi.org/10.1007/978-3-031-21187-4_54

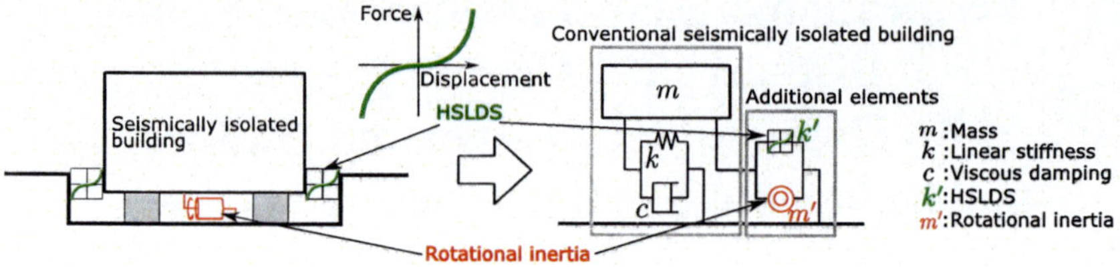

Fig. 1. Diagram of DCS.

(HSLDS) [2] and rotational inertia [3] in conventional seismically isolated buildings. HSLDS increases as the deformation of the isolation layer increase. It keeps the deformation of the seismic-isolation layer within a clearance between seismically isolated buildings and retaining wall. However, when it increases, the natural period of the seismically isolated building becomes short, causing large absolute acceleration. To suppress the absolute acceleration, the DCS is equipped with rotational inertia. Rotational inertia adds a virtual mass to the installed structure and reduces acceleration. The virtual mass is not the actual mass and does not increase the seismic forces.

The purpose of this study was to investigate the performance of the DCS on actual-displacement scale through shaking-table experiments and confirm the validity of the time-history analysis (THA) for the DCS. In the experiments, the frequency-resonance curve was obtained to understand the basic characteristics of the DCS. Seismic excitation was also conducted to verify the displacement-suppression effect of the DCS. The experimental responses were then compared with the responses of the THA to confirm THA consistency for a seismic-isolation model of our displacement control system we developed.

2 Testing DCS Through Shaking-Table Experiments

2.1 Configuration of DCS Seismic-Isolation Model

Figure 2 shows the four configurations of the DCS seismic-isolation model we constructed, where m is a mass, k is linear stiffness, c is damping, k' is nonlinear stiffness, i.e., HSLDS, and m' is virtual mass of the rotational inertia. The S model is a conventional seismically isolated structure, R model is the S model with rotational inertia, N model is the S model with HSLDS, and NR model is the S model with rotational inertia and HSLDS.

Table 1 lists the parameters of the S model, and Fig. 3 shows the DCS model of the actual-displacement scale. Vertical support and horizontal flexibility of the mass were achieved using a low-friction linear roller bearing system. The model was designed to have a movable range of 74 cm and natural period of about 3 s, referring to a previous study [4]. The stiffness of the linear spring is determined from the relationship of the mass of the model and natural period of the S model. The four linear springs were used to pull the model equally from both sides in the moving direction to reproduce linear stiffness. The stiffness of the linear spring in Table 1 was calculated from the force-displacement relationship by conducting dynamic excitation on the model. The friction force generated by the linear roller bearing was also measured by conducting static excitation on the model.

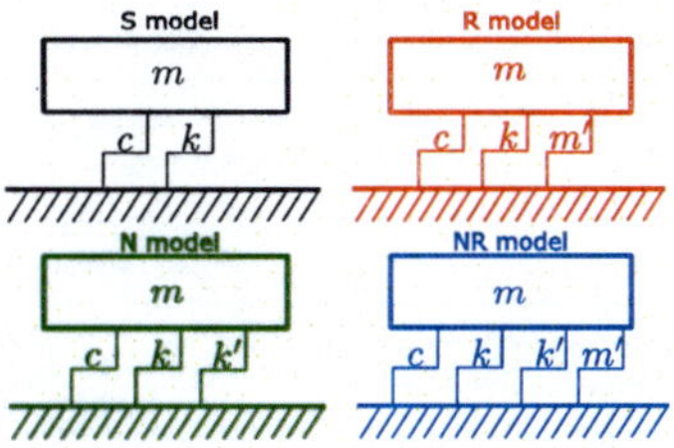

Fig. 2. Four configurations of the DCS seismic-isolation model

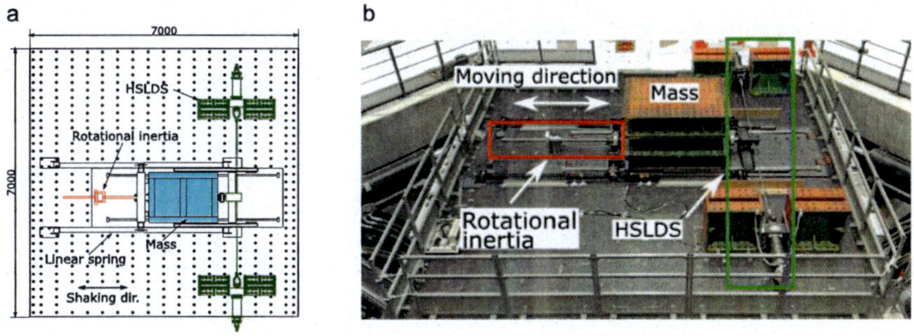

Fig. 3. (a) Design and (b) photograph of DCS seismic-isolation model

Table 1. Parameters of S model

Parameters	Value
Movable area of model	± 740 mm
Mass	4038 kg
Linear stiffness	13597 N/m
Natural period of S model	3.4 s
Friction	100N

2.2 Ground Motions

We conducted sweep and seismic excitations for the S, R, N, NR models to investigate their basic properties and behaviour. Figure 4a shows an example of sweep excitation, and Fig. 4b shows the pseudo-response spectrum for input seismic motions. Table 2 summarizes the information for the input waves.

Sweep excitations were conducted to obtain resonance curves. The response of the nonlinear models (N and NR model) shows hysteresis characteristics in which the response differs depending on the direction of change in the excitation frequency. Therefore, two excitation patterns were used to derive the resonance curves: one in which the excitation frequency changes from low to high (Sweep up) and the other in which it changes from high to low (Sweep down).

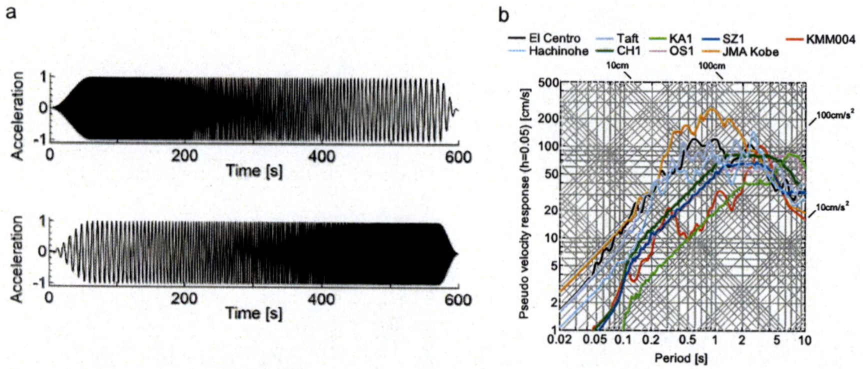

Fig. 4. Input waves; (a) Example of sweep excitation (frequency 0.1 ↔ 1.2) and (b) pseudo-response spectrum for input seismic motion

Table 2. (a) Sweep excitation and (b) input wave magnification and peak ground acceleration

a

Model	Acceleration [cm/s^2]	Frequency [Hz]
S	10	0.1 ↔ 1.2
R	15	0.1 ↔ 0.4
N	20	0.1 ↔ 1.2
NR	30	0.1 ↔ 1.2

b

Wave	Input magnification	PGA [cm/s2]
El Centro NS	1.49	509
Hachinohe NS	1.47	340
Taft EW	2.82	496
CH1	0.5	132
KA1	1.0	46
OS1	0.5	131
SZ1	0.4	113
JMA Kobe NS	1.0	818
KMM004 EW	0.3	104

We selected nine seismic waves with different characteristics. El Centro NS, Hachinohe NS, and Taft EW are seismic waves customarily used in Japan for building response evaluation, CH1, KA1, OS1, and SZ1 [5] have long-period characteristics with long duration, JMA Kobe NS is a short-period pulsed seismic wave, and KMM004 EW is a long-period pulsed seismic wave. The input magnification was determined so that the maximum response displacement of the S model would not exceed the range of motion of the model.

2.3 Test Results and Discussion

Figure 5 shows the resonance curves of the S, R, N, and NR models obtained from the sweep excitation. The frequency ratio on the horizontal axis is the excitation frequency (ω) divided by the natural frequency of the S model (ω_0). The hysteresis characteristics were confirmed for the linear models (S and R) as well as the nonlinear models (N and NR). This is because, the friction generated by the guide rails of the sliding parts of the model affected the response. Another reason is that the sweep speed of the sweep wave was not sufficiently slow. Turbulent responses were confirmed in the resonance curves for each model configuration after the maximum response of each excitation (for example, the response of the S model near the frequency ratio of 1.2 for Sweep up). This turbulent response is explained by the theory that free vibrations occur when the response changes drastically then disappear as the frequency ratio changes. When the N and NR models were subjected to Sweep up excitation, the resonance curves increased with the increase in the frequency ratio then dropped off suddenly. When Sweep down was excited, the response displacement increased suddenly at a specific frequency ratio. This phenomenon of rapid increase or decrease in response to a slight change in the excitation frequency in a nonlinear system is called the jump phenomenon. The results indicate that there is a jump phenomenon for full-scale displacement models with HSLDS.

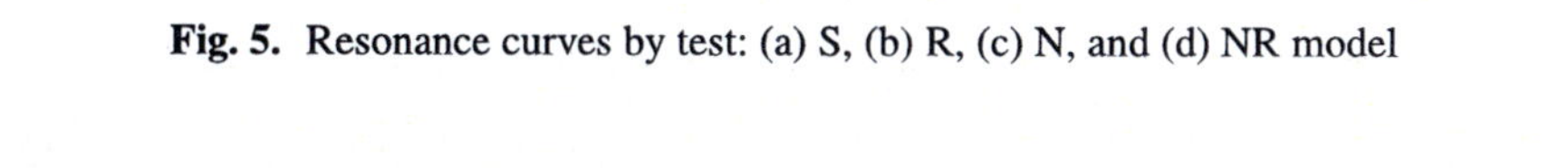

Fig. 5. Resonance curves by test: (a) S, (b) R, (c) N, and (d) NR model

Figure 6a shows the maximum response displacements of the S, R, N, and NR models for each seismic wave, and Fig. 6b shows the maximum response acceleration. The response acceleration is filtered using a moving average method with a window width of 0.5 s. Figure 6a shows that the N model reduced the maximum response displacement for many waves compared with the S model. This difference was especially large for long-period seismic waves such as Hachinohe NS, OS1, CH1, SZ1, and KMM004 EW. Figure 7a shows that the natural period of the N model became shorter than that of the

S model around 155 s. This phenomenon is consistent with the hypothesis that HSLDS increased as the response displacement increased, and the natural period of the model decreased. The reduction in response displacement in the N model is in agreement with the theory that resonance is prevented by varying the natural period.

As shown in Fig. 6a and b, the R and NR models showed almost the same maximum response. Figure 7b also shows that the time-history responses of the R and NR model follow the same trajectory. This phenomenon is due to the virtual mass of rotational inertia being too large. The HSLDS had little effect on the response in the NR model because the response was small. As shown in Fig. 6b, the NR model had a larger maximum response acceleration than the S model for short-period waves such as El Centro NS, Taft EW, and JMA Kobe NS. Saito et al. [6] reported that for high-frequency excitation, the R model with a virtual mass equivalent to the mass of the model has half the response acceleration of the peak ground acceleration (PGA). Since the R and NR models' maximum response acceleration for a short-period wave is about half that of the PGA, setting a larger virtual mass causes the maximum acceleration to increase.

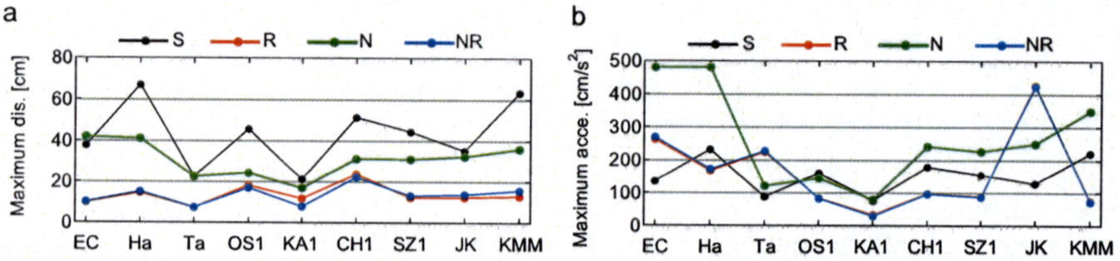

Fig. 6. Maximum response for seismic excitation: (a) Maximum displacement, (b) maximum acceleration

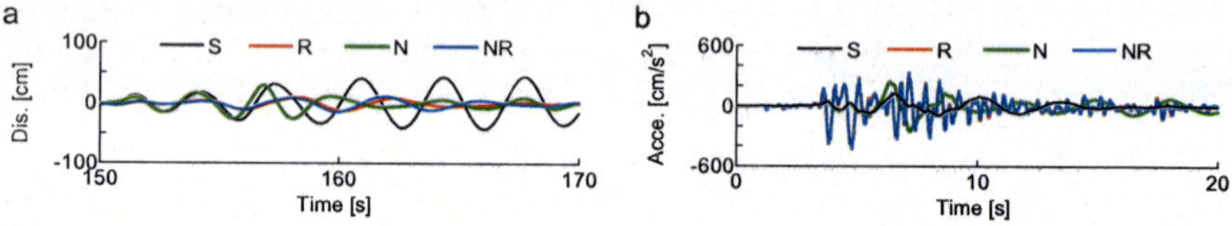

Fig. 7. Time-history response for seismic excitation: (a) Displacement for SZ1, and (b) acceleration for JMA Kobe NS

3 Validity of Time-History Analysis for Displacement Control System

This section discusses the validity of the THA for the DCS by comparing the test and THA results.

3.1 Analytical Model

Table 3 shows the parameters of the analytical model. The analytical model was constructed on the basis of the DCS seismic-isolation model, as illustrated in Fig. 8 The mechanical characteristics of the analytical model reflect the stiffness of the linear spring and HSLDS measured with the test. The friction force reflects the values measured by applying static force to the seismic-isolation model. The virtual mass of rotational inertia and viscous damping were determined by combining the time-response waveforms from the test and THA results.

3.2 Comparison of Test and Time-History Analysis Results

Figure 9 shows the resonance curves of the four models obtained by the sweep excitation in the test and THA. The test and THA results are consistent. Figure 10 shows the time-history response for seismic excitation. Comparison of the test and THA response shows that THA for the DCS is reliable.

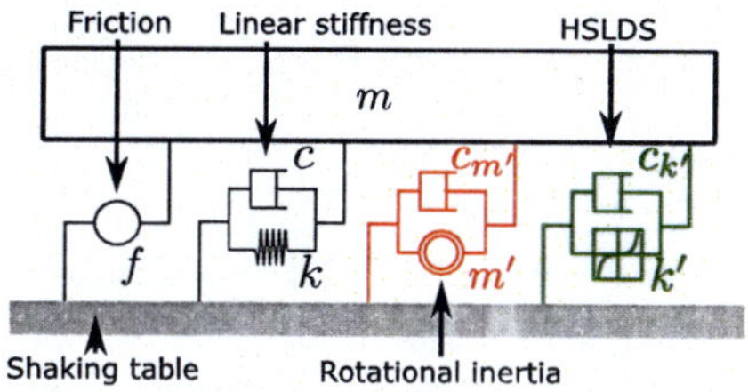

Fig. 8. Analysis model

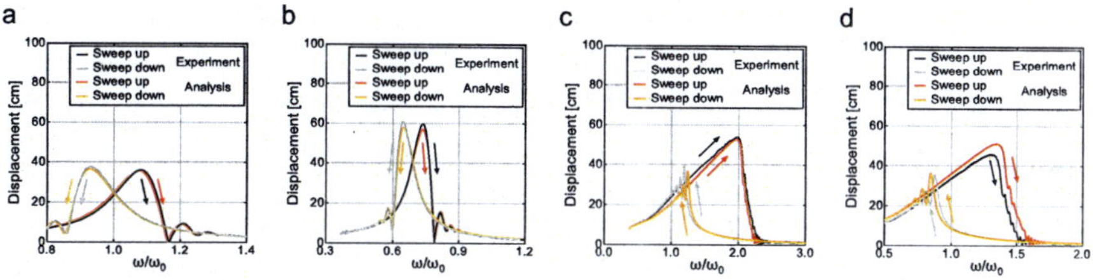

Fig. 9. Comparison of test and analysis results for resonance curves: (a) S model, (b) R model, (c) N model, and (d)NR model

4 Conclusions

We conducted shaking-table experiments to determine the behavior of the DCS on full-scale displacement. The DCS seismic-isolation model can reproduce its S, R, N, and NR model configurations that indicate a conventional seismically isolated construction, the S model with rotational inertia, S model with HSLDS, and S with rotational inertia and HSLDS. From the excitation tests, resonance curves and time-history responses for the S, R, N, and NR models were obtained. The resonance curves of the N and NR models show the jump phenomenon for full-scale displacement models with HSLDS. From the results of seismic wave excitation, HSLDS was confirmed effective for displacement

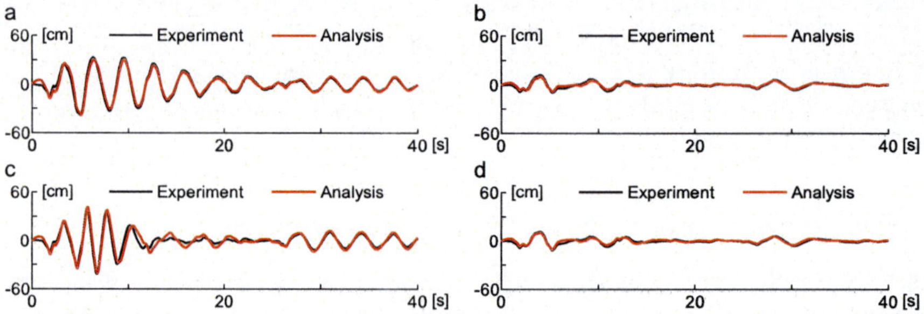

Fig. 10. Comparison of test and analysis results for seismic excitation: (a) S model, (b) R model, (c) N model, and (d) NR model

Table 3. Parameters for analysis model

Parameter	Value
Mass, m	3950 kg
Linear stiffness, k	13597 N/m
Nonlinear stiffness, k'	182947 N/m^3
Equivalent mass of rotational inertia, m'	4300 kg
Friction, f	100 N
Viscous damping coefficient, c	248 N s/m
Viscous damping coefficient of rotational inertia, c'_m	260 N s/m
Viscous damping coefficient of hardening device, c'_k	300 N s/m

suppression against long-period seismic waves. It was also confirmed that in some cases the maximum acceleration increases for short-period seismic waves when the virtual mass is set to a large value. The reliability of THA for DCS was also confirmed on full-scale displacement to compare with the test and THA results.

Acknowledgements. This work was supported by JSPS KAKENHI Grant Number JP19H00788. We thank Dr. Frank McKenna of the University of California, Berkeley who assisted in the implementation of the numerical analysis model in the OpenSees program.

References

1. Morita, K., Takayama, M.: Behavior of seismically isolated buildings during the 2016 Kumamoto earthquakes. In: Proceedings of the NZSEE Conference 2017, Wellington, New Zealand, pp. 27–29. April (2017)
2. Carrella, A., Brennan, M.J., Waters, T.P., Shin, K.: On the design of high-static-low-dynamic stiffness isolator using linear mechanical springs and magnet. J. Sound. Vib. **315**, 712–720 (2008). https://doi.org/10.1016/j.jsv.2008.01.046

3. Isoda, K., Hanzawa, T., Tamura, K.: A study on response characteristics of a SDOF model with rotating inertia mass dampers (in Japan). J. Struct. Construct. Eng. AIJ **74**(642), 1469–1476 (2009). https://doi.org/10.3130/aijs.74.1469
4. Tanaka, Y., Fukuwa, N., Tobita, J., Mori, M.: Development and analysis of database for base-isolated buildings in Japan (in Japanese). J. Technol. Design AIJ **17**(35), 79–84 (2011). https://doi.org/10.3130/aijt.17.79
5. Ministry of Land, Infrastructure, Transport and Tourism. Countermeasures against Long-period Ground Motions Caused by a Large Earthquake along the Nankai Trough in High-rise Buildings, etc., https://www.mlit.go.jp/jutakukentiku/build/jutakukentiku_house_fr_000080.html (Accessed 17 June 2022)
6. Saito, K., Yogo, K., Sugimura, Y., Nakaminami, S., Park, K.: Application of rotary inertia to displace ment reduction for vibration control system. In: Proceedings of the Thirteenth World Conference on Earthquake Engineering, Vancouver, B.C., Canada (2004)

Seismic Protection of Non-structural Elements and Equipment, Statues and Art Objects

Improved Structural Serviceability and Seismic Protection by Adaptive Isolators and Dampers

Peter Huber[1](✉) and Felix Weber[2]

[1] MAURER SE, Munich, Germany
p.huber@maurer.eu
[2] Maurer Switzerland GmbH, Zurich, Switzerland

Abstract. In order to enhance the serviceability of civil structures during seismic action base isolators and dampers need to perform at high efficiency independent of the ground shaking level. However, base isolators and dampers are commonly designed to the shaking level of the Design Basis Earthquake (DBE) whereby their efficiencies are poor for earthquakes with lower peak ground accelerations (PGAs). This shortcoming can only be improved if the friction of curved surface sliders is not constant and if hysteretic dampers can produce hysteretic behaviour at various yielding levels. This article presents two new types of curved surface sliders and hysteretic dampers which operate at high efficiency over a wide range of PGAs. This makes it possible to enhance structural serviceability for weak and frequent earthquakes and ensure structural safety for strong motion earthquakes. The dynamic behaviours of the new adaptive curved surface slider and hysteretic damper types are presented. The isolation performance computed by non-linear time history analysis confirms their improved functioning.

Keywords: Adaptive · Curved surface slider · Hysteretic damper · Seismic

1 Introduction

Depending on the fundamental time period T_1 of civil structures they are protected against seismic excitation either by base isolators (T_1 in the plateau and first part of constant velocity range of response spectra) or inter-story hysteretic dampers (T_1 in the second part of constant velocity range of response spectra). Base isolators such as Curved Surface Sliders (CSSs) decouple the structure from the shaking ground by their low coupling stiffness due to their large curvature and add damping to the structure by nonlinear friction damping. Due to the nonlinearity of friction damping CSS can be optimized to one shaking level only, i.e. to one selected Peak Ground Acceleration (PGA) only [1], commonly the PGA of the Design Basis Earthquake (DBE). The behaviour of Hysteretic Dampers (HDs) can be described by the Maxwell model, i.e. HDs produce a stiffness force when operated in their pre-yield region due to small inter-story relative motions while they exert a hysteretic damping force with low second branch stiffness when inter-story displacements are sufficient big to operate HDs in their post-yield region [2]. Hence, the structural isolation by CSS and the structural damping by HDs at

G. P. Cimellaro (Ed.): WCSI 2022, LNCE 309, pp. 655–664, 2023.
https://doi.org/10.1007/978-3-031-21187-4_55

low ground shaking levels due to weak but frequent earthquakes may be insufficient to maximally protect non-structural elements and sensitive equipment which are crucial to ensure the serviceability of vulnerable structures such as hospitals, schools, government buildings, fire and police stations etc. In order to overcome these drawbacks of CSSs and HDs the following new adaptive types of CSSs and HDs are developed:

- Sliding Isolation Pendulum of type Adaptive "SIP®-Adaptive"
- Modular Adaptive Shear Hysteretic Damper "SHARK®-ADAptive"

Their dynamic behaviours are described by force displacement curves and the enhanced structural isolation is demonstrated by non-linear time history analysis.

2 SIP®-Adaptive

2.1 Concept

The SIP®-Adaptive is a CSS with two primary sliding surfaces and an articulated slider that decouples the relative motions on both primary sliding surfaces (Figs. 1a and b, 2).

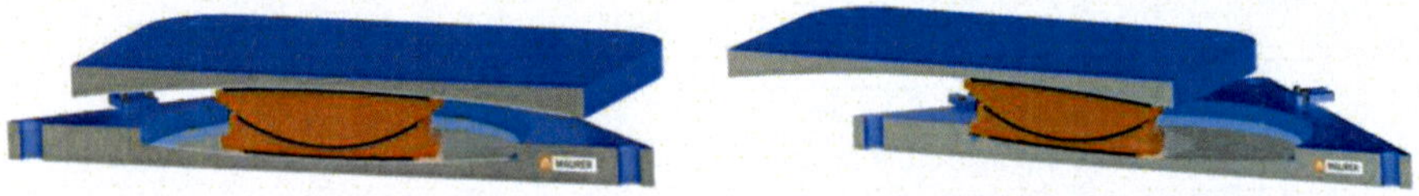

a b

Fig. 1. **a** SIP®-Adaptive in center position **b.** SIP®-Adaptive fully deflected position

This basic feature allows designing the two primary sliding surfaces with different friction coefficients and different effective radii. The bottom sliding surface is designed with lower friction coefficient to active relative motion at low PGAs (below DBE). The primary sliding surface is constrained by a recess to limit relative motion (Fig. 2). Sliding surface 2 is designed with greater friction to provide the optimum damping at DBE and above without end stopper recess. The effective radii are selected to produce stiffening and increased damping behaviours at the shaking level of the Maximum Considered Earthquake (MCE).

2.2 Optimum Design by Dynamic Simulation

The parameters of the SIP®-Adaptive, i.e. its friction coefficients, effective radii and displacement capacities of both primary sliding surfaces, are model-based optimized. For this the following model of the SIP®-Adaptive with superstructure is used (Fig. 3).

The superstructure with total mass m_s is modelled as linear multi-degree-of-freedom (DOF) system with as many DOFs as storeys. The equations of motion of the upper

Fig. 2. Bottom concave plate with recess and articulated slider of SIP®-Adaptive

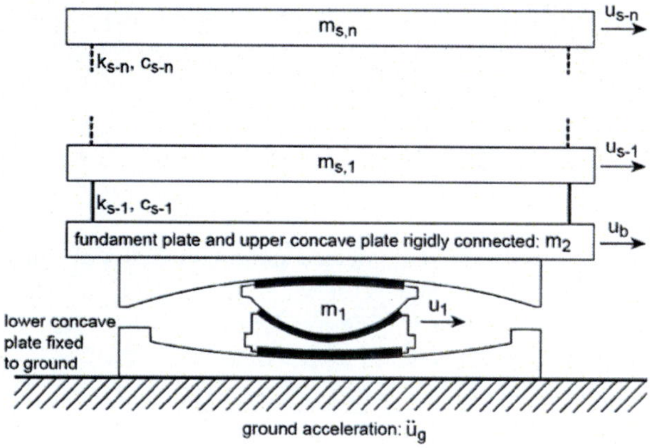

Fig. 3. Sketch of structure (multi-DOF) with SIP®-Adaptive

bearing plate with mass m_2 (including the fundament plate of the structure) and the slider mass with m_1 become

$$m_2\ddot{u}_2 + f_2 = c_s(\dot{u}_{s1} - \dot{u}_2) + k_s(u_{s1} - u_2) - m_2\ddot{u}_g \quad (1)$$

$$m_1\ddot{u}_1 + f_1 = f_2 - m_1\ddot{u}_g \quad (2)$$

where the forces f_2 and f_1 describe the longitudinal forces on surfaces 2 and 1 that are given by the sum of the friction force given by the friction coefficients μ_2 and μ_1, the restoring stiffness force due to the effective radii R_{eff-2} and R_{eff-1} and the linear stiffness force (recess stiffness k_{recess}) of the recess of sliding surface 1 that is only activated at the displacement capacity d_1

$$f_2 = \mathrm{sgn}(\dot{u}_2 - \dot{u}_1)\mu_2 g(m_s + m_2) + (u_2 - u_1)\frac{g(m_s + m_2)}{R_{eff-2}} \quad (3)$$

$$f_1 = \mathrm{sgn}(\dot{u}_1)\mu_1 g(m_s + m_2 + m_1) + u_1\frac{g(m_s + m_2 + m_1)}{R_{eff-1}} + f_{recess} \quad (4)$$

$$f_{recess} = \begin{cases} 0 & : |u_1| \le d_1 \\ k_{recess}\,(|u_1| - d_1)\,\mathrm{sign}(u_1) & : |u_1| > d_1 \end{cases} \quad (5)$$

The variables used in Eqs. (1–5) are:

- $\dot{u}_{s1}, u_{s1}$: velocity, displacement of story 1 relative to ground
- $\ddot{u}_g$: ground acceleration
- $\ddot{u}_2, \dot{u}_2, u_2$: absolute acceleration, velocity and displacement relative to ground of the mass m_2 (top bearing plate and fundament plate)
- $\ddot{u}_1, \dot{u}_1, u_1$: absolute acceleration, velocity and displacement relative to ground of slider mass m_1

The displacement capacity of sliding surface 2 is not constrained whereby the total displacement capacity of the SIP®-Adaptive is not limited. The coupled, stiff and non-linear differential Eqs. (1-5) are solved in the time domain using the Matlab®-solver ode15s(stiff/NDF) with maximum relative tolerance of 1e-3 and variable step size with upper bound of 1e-5 s.

2.3 Isolation Performance

Absolute structural acceleration (~ base shear) and total isolator displacement are computed by non-linear time history analysis for two different designs of the SIP®-Adaptive and compared to the conventional CSS with friction damping (CSS-FD) that is optimized at the PGA of the DBE (Figs. 4a,b, 5a, b). As theoretical benchmark the fully linear isolator with optimized linear viscous damping (CSS-VD) is computed as well. The following observations can be made:

- On the one hand, the SIP®-Adaptive reduces peak absolute structural acceleration within the entire PGA-range due to the fact that damping is smaller at lower shaking levels and damping is greater at higher shaking levels and, on the other hand, the total isolator displacement at MCE is smaller than for the CSS-FD due to the increasing stiffness and friction behaviours of the SIP®-Adaptive at shaking levels of the MCE.
- Due to the two friction values of the sliding surfaces 1 and 2 it is possible to tune the minimum base shear at lowest shaking levels by μ_1 (e.g. used as wind locking force) independently of the performance at DBE.

3 Modular Shear Hysteretic Damper SHARK®-Adaptive

3.1 Concept

Conventional shear hysteretic dampers (HD) are characterized by one yielding force which leads to the drawback of far suboptimal energy dissipation at low ground shaking levels because then the HD might not or only partially be operated in its yielding regime. Consequently, the damping provided to the structure is far below the maximum possible at PGAs significantly below the PGA of the DBE whereby the proper functioning of sensitive equipment within vulnerable buildings, e.g. hospitals, is not granted. To overcome this problem and thereby enhance significantly structural serviceability the Modular Shear Hysteretic Damper SHARK®-Adaptive has been developed.

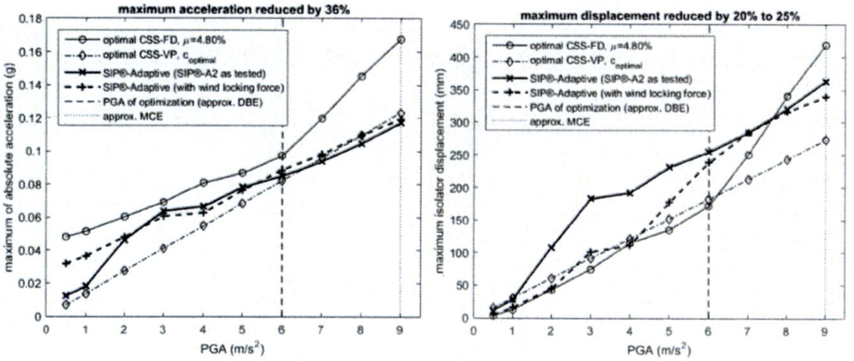

a b

Fig. 4. **a** Absolute structural acceleration **b**. Displacement depending on PGA leve

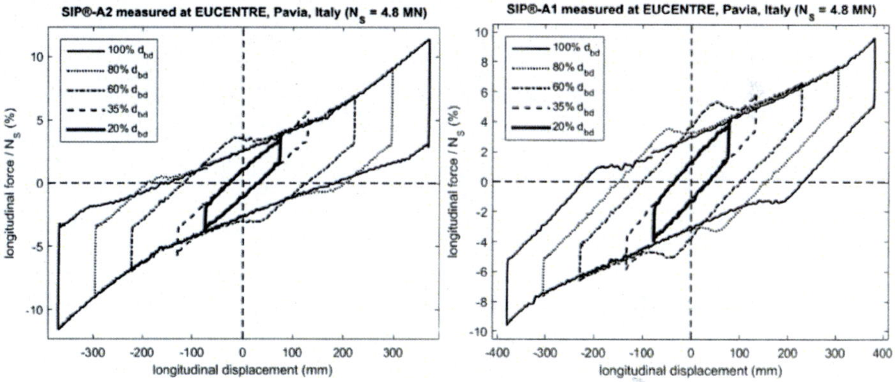

a b

Fig. 5. **a** SIP®-A2 test plot **b**. SIP®-A1 test plot

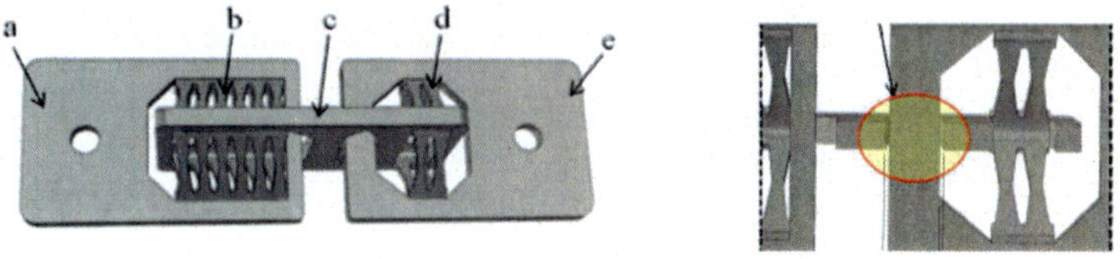

a b

Fig. 6 **a** Sketch of entire SHARK®-Adaptive **b.** Close-up of gap element

The basic working behaviour of this 2-stage hysteretic damper is as follows (a, e: main damper connecting plates):

- One smaller (d) and one bigger (b) shear hysteretic element are coupled in series by a connecting plate (c) with gap element (Fig. 6a and b).

- At low PGAs (≤50% of DBE) the smaller hysteretic element is operated in its post-yield regime while yielding in the bigger element is not activated (Fig. 7, small inner force displacement loop).
- At medium PGAs (~DBE) the smaller and the bigger hysteretic elements are activated (Fig. 7, smaller and partly greater force displacement loops).
- At high PGAs (~MCE) the energy dissipation is dominated by the bigger hysteretic loop which reduces maximum structural displacements.

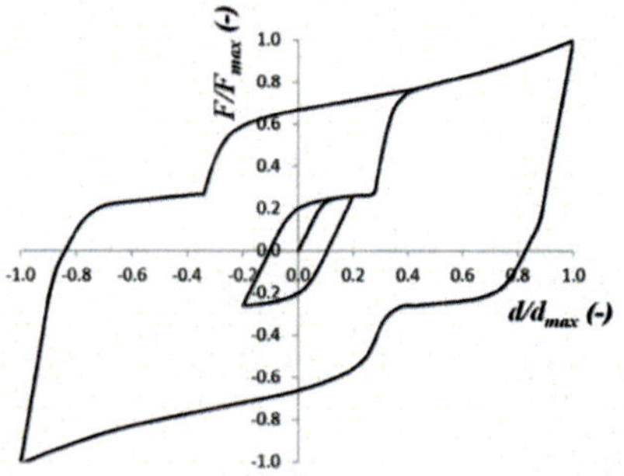

Fig. 7. Force displacement plot of SHARK®-Adaptive computed by FEM

3.2 Non-linear Time History Analysis of Structure with SHARK®-Adaptive

Non-linear time history analysis is performed for the 3-storey reinforced concrete building Hospital Lamezia Terme ($T_1 = 0.7$ s) that is located in a high-seismic area in southern Italy [3]. The structure is modelled as linear shear frame and the common HD and the SHARK®-Adaptive, respectively, are modelled by the nonlinear hysteretic damper model. Seven site-specific accelerograms are used as ground excitation for each of the following ground shaking levels (according to Italian code):

- FDE (weak frequent earthquakes): 7 accelerograms with PGA = 0.17 m/s^2
- DBE: 7 accelerograms with PGA = 0.45 m/s^2
- MCE: 7 accelerograms with PGA = 0.50 m/s^2

The dynamic simulations are performed in Matlab®/Simulink® adopting the same solver with same settings as used for the simulations of the SIP®-Adaptive.

The pushover analysis of the structure shows that plastic deformation in the columns occurs when the total deformation over all three stories is more than 0.45% of entire structure height, i.e. 60 mm. The common design target for HD devices is based on the approach [3] to avoid plastic deformation in the columns at DBE.

The design of the SHARK®-Adaptive dampers targets at achieving the following performances [3]:

1. FDE: Reduced peak floor accelerations (PFA) to maximally protect sensitive equipment of the structure and ensure serviceability; inter-story drift $\leq 0.5\%$ to protect drift-sensitive non-structural elements (infills); this requirement is fulfilled here by the requirement on the total drift $\leq 0.45\%$.
2. DBE: Ensure total structural deformation $\leq 0.45\%$ as for the HD.
3. MCE: Reduced maximum column deformations to increase structural safety at maximum ground shaking level.

The pre-yield and post-yield stiffness coefficients of the SHARK®-Adaptive are assumed to be equal to the values of the HD. The SHARK®-Adaptive design parameters are therefore:

- yielding force f_a of the smaller hysteretic element,
- size of gap element of the smaller hysteretic element, and
- yielding force f_b of the bigger hysteretic element.

Two solutions are derived by dynamic simulation fulfilling the design targets:

- SHARK®-Adaptive -1: $f_{a-i} = 0.5 \cdot f_{HD-i}$, $gap = [6.1; 5.8; 8.7]$ mm, $f_{b-i} = 1.32 \cdot f_{HD-i}$
- SHARK®-Adaptive -2: $f_{a-i} = 0.6 \cdot f_{HD-i}$, $gap = [5.9; 5.5; 8.5]$ mm, $f_{b-i} = 1.24 \cdot f_{HD-i}$

where f_{HD-i} denotes the yielding force of the common HD of storey i.

The mean PFA and mean total structural displacement resulting from 7 accelerogram are given in Table 1 for the ground shaking levels FDE, DBE and MCE. The comparison of the data reveals:

1. **Significantly reduced PFAs for weak earthquakes (FDE):** PFAs are reduced in average by -29.2% by SHARK®-Adaptive-1 and −26.5% by SHARK®-Adaptive-2.
2. **Approximately same PFA at DBE:** If the design of the SHARK®-Adaptive aims at obtaining a reasonable PFA reduction at FDE (SHARK®-Adaptive-2➔ −6.5%) then the same average value of PFA at DBE can be achieved as for the HD. If the SHARK®-Adaptive design aims at further reducing PFAs (SHARK®-Adaptive-1➔−29.2%) then the average PFA at DBE is marginal higher (+11.3%).
3. **Reduced total structural displacement at MCE:** The SHARK®-Adaptive reduces the maximum displacement over all storeys at MCE due to the bigger hysteretic element which enhances structural safety against unexpected too large plastic deformations.

3.3 Proof of Concept

One shear hysteretic element of the SHARK®-Adaptive was tested at Bundeswehr University Munich (Fig. 8). Besides the common force displacement measurements also strain measurements were made using a 3-dimensional scan optical system (Fig. 9). The measurement data were used to further improve the FE-modelling to be able to extrapolate the FE-model for future design purposes (Fig. 10).

Table 1. Mean PFA and mean total structural displacement resulting from 7 FDE, DBE and MCE accelerograms

	Ground shaking level	Maximum displacement over all storeys (mm)	PFA storey 1 (g)	PFA storey 2 (g)	PFA storey 3 (g)
No HD	FDE*	Far greater than 60	**0.257**	**0.295**	**0.374**
	DBE*	Far greater than 60	0.731	0.933	1.390
	MCE*	Far greater than 60	0.831	1.020	1.477
Common HD	FDE	21.4	**0.305**	**0.372**	**0.444**
	DBE	60	0.674	0.662	0.603
	MCE*	**75**	0.761	0.754	0.681
SHARK®-Adaptive—1	FDE	20.3	**0.249**	**0.288**	**0.257**
	DBE	60	0.782	0.666	0.711
	MCE*	**72**	0.787	0.720	0.766
SHARK®-Adaptive—2	FDE	21.8	**0.245**	**0.291**	**0.288**
	DBE	60	0.653	0.639	0.686
	MCE*	**71**	0.690	0.689	0.727

* Computed with linear 3-DOF model of structure

Fig. 8. Testing one single hysteretic element of the SHARK®-Adaptive at Bundeswehr University Munich: left under max. Elongation, right under max. compression

4 Summary

The new base isolator SIP®-Adaptive and the new hysteretic damper SHARK®-Adaptive are presented. Both devices target at improving the seismic protection of structures and their sensitive equipment for frequent earthquakes with low to medium PGAs to enhance the serviceability of vulnerable structures including their content such as hospitals, schools, government buildings, fire and police stations etc. The performance results of the SIP®-Adaptive demonstrate that this base isolator can reduce base shear within a large PGA range and—at the same time—the maximum isolator displacement at MCE. The results of the SHARK®-Adaptive point out that PFAs at weak and frequent earthquakes can be significantly reduced while guaranteeing approximately the same PFAs at DBE and reduced total structural displacements at MCE.

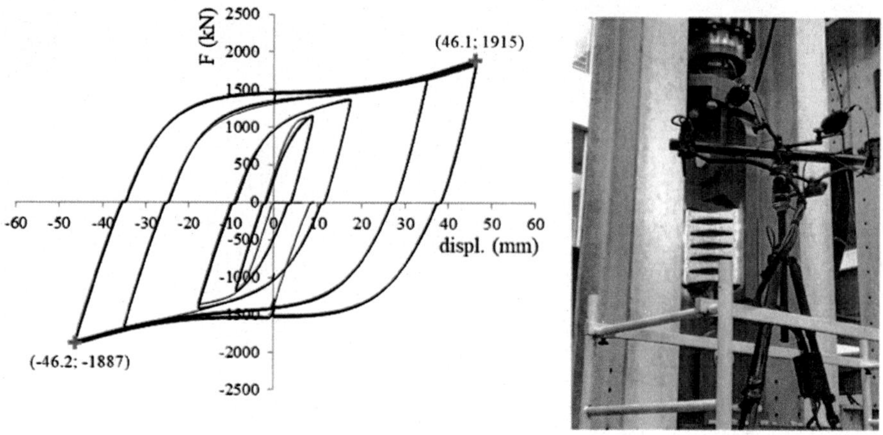

Fig. 9. **a** Measured force displacement plots **b.** 3-D scan with optical system

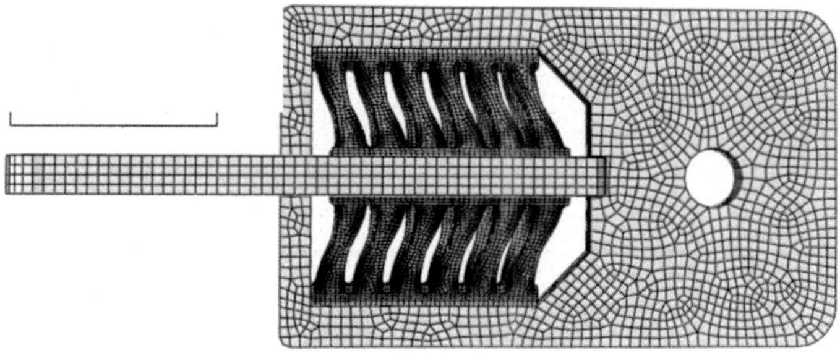

Fig. 10. FE-modelling for stress analysis and optimization

References

1. Weber, F., et al.: Curved surface sliders with friction damping, linear viscous damping, bow tie friction damping and semi-actively controlled properties. Struct. Control Health Monit (2018). https://doi.org/10.1002/stc.2257
2. Takeuchi, T. et al.: Buckling restrained braces and damping steel structures. US patent No. US20050055968A1 (1999)
3. Gandelli, E. et al.: Seismic retrofit of hospitals by means of hysteretic braces: influence on acceleration-sensitive non-structural components. Front. Built Environ. (2019). https://doi.org/10.3389/fbuil.2019.00100

Preliminary Results in the Design and Testing of Earthquake-Proof Glass-Aluminium Partition Walls

Rocco Ditommaso[1], Fabrizio Scozzese[2], Antonello Mossucca[1], Gianluca Auletta[1], Antonio Di Cesare[1], Domenico Nigro[1], Alessandro Zona[2](✉), Felice Carlo Ponzo[1], and Andrea Dall'Asta[2]

[1] Scuola di Ingegneria, Università degli Studi della Basilicata, Viale dell'Ateneo Lucano 10, 85100 Potenza, Italy

[2] Scuola di Architettura e Design, Università degli Studi di Camerino, Viale della Rimembranza 3, 63100 Ascoli Piceno, Italy

alessandro.zona@unicam.it

Abstract. Recent years have seen an increasing demand in the use of glass partition walls in office buildings, with major concerns regarding their behaviour and safety in case of strong seismic events. This study illustrates the development of an innovative solution for aluminium-glass partition walls that can resist, without any damage, horizontal accelerations as those measured during strong earthquakes. The proposed solution is based on a highly dissipative interface rubber between the aluminium frame and the glass plate, the latter constituting the largest portion of the mass of the partition wall. During a seismic event the glass plate moves relatively to the aluminium frames, thus, activating a dissipative mechanism that is basically a scaled-down version of what is seen in rubber bearings used in seismic isolation. In this article the initial structural concept is presented, preliminary numerical analyses illustrated, the preparation of preliminary small-scale prototypes followed by full-scale prototypes, and the execution of shake-table tests to simulate extreme seismic events are discussed. It is shown that very encouraging structural performances can be achieved in a product that is economically comparable to currently non-earthquake proof solutions, without problems or limitations in its every-day use.

Keywords: Glass partition walls · Experimental shaking table test · Energy dissipation · Base isolation

1 Introduction

This work was developed within the research project "S.A.F.E. Sustainable Design of Anti-Seismic Furniture as Smart Life-Saving Systems during an Earthquake" funded by the Italian Minister of University and Research, having as general aim the development and design of innovative school and office furniture able to protect people in those buildings that cannot guarantee sufficient level of seismic safety, as it is the case

G. P. Cimellaro (Ed.): WCSI 2022, LNCE 309, pp. 665–674, 2023.
https://doi.org/10.1007/978-3-031-21187-4_56

of existing structures built before the entrance in force of seismic regulations. Three types of furniture system were the object of the S.A.F.E. Project, namely school desks, self-supporting shelving units, and glass-aluminium partition walls. The present paper focuses on the latter product, while details about the other two typologies can be found in a dedicated contribution.

During a vibratory event such those induced by earthquakes, internal and external walls of a building are subjected to accelerations that might compromise their in-plane and out-of-plane integrity and stability, with potential risk of damages and injuries even in structures properly designed to withstand seismic accelerations. As highlighted by the recent literature, non-structural elements can modify the dynamic behaviour of a buildings both in ambient vibration condition and during strong motion earthquake (e.g. [1, 3–7, 10–13, 15, 16]). Especially during seismic events it is necessary to ensure that non-structural elements, e.g., partitions walls and curtain walls, do not negatively interfere with the designed resisting mechanisms of the building, i.e., non-structural elements must not adversely modify the structural behaviour but should rather provide help, to some extent, to favourably improve the dynamic response of the building. Conventional partition walls, as the most of non-structural components, are not designed with proper attention to seismic (or other dynamic) actions but they are only verified to satisfy the requirements of functionality and acoustic comfort. One of the few exceptions found in the literature is represented by the work of Petrone et al. [10] concerning the realization of internal partitions that meet the requirements of seismic resistance.

In this work, an innovative solution for glass-aluminium partition walls is proposed, conceived to withstand from light to strong vibratory events, e.g., earthquakes and vibrations from operating machines, without experiencing any form of damage or out-of-plane overturning, thus, creating a safer space for people living the environments in which they are installed (life-saving functions). The seismic-resistant property is obtained by exploiting a passive seismic protection concept (e.g., [8, 9, 14]), inspired by seismic base isolation systems, and scaled to the dimensions of the wall. The seismic dissipation function is assigned to specific elements, identified by means of elastomeric PADs with viscoelastic properties, able to realize a dissipative coupling between each glass plate and the perimetral aluminium frame. The product presented herein is covered by the European Patent n. 21425061.5/EP21425061 AREF: SAFE22021 (2021) titled “Controlled deformability partition wall with vibration-induced energy dissipation”.

In the following of the paper, some preliminary results retrieved from the experimental campaign carried out at the Seismic Laboratory of the University of Basilicata to investigate the performance of the sub-components of the proposed glass-aluminium partition wall are presented. Moreover, some details on the numerical modelling strategy are given and an overview of the main outcomes obtained during shaking table tests on both small-scale and full-scale prototypes is provided.

2 Components and Scaled Model Tests

Several tests were performed on the adopted materials, assembled components, and partition wall prototypes. The experimental campaign was divided into the following phases:

1. Experimental tests for the mechanical characterization of the new structural components of the proposed partition wall;
2. Quasi-static cyclic tests on a small vibrating table for the characterization of the proposed wall system equipped with the new structural components;
3. Dynamic seismic simulation tests on a large vibrating table for the analysis of the behaviour of a three-dimensional prototype of a partition wall module.

Fig. 1. Test equipment used for experimental characterization of SAFE partition wall and its new components: a) test facilities managed by the Instron Shenck® actuation system; (b) large vibrating shaking table operated by the MTS® system.

The experimental tests on assembled components were carried out using a small shaking table characterized by a 40 kN maximum horizontal force and ±150 mm maximum stroke, while global tests on the whole system were carried out on a unidirectional 4.00 × 3.00 m shaking table characterized by a 500 kN maximum force and a ±250 mm maximum stroke.

3 Preliminary Characterization of Single Components and Materials

Static and dynamic tests were performed to characterize the mechanical characteristics of materials and components constituting the prototype system. The test equipment shown in Fig. 1a, able to control force, displacement, and temperature respectively, was used to perform several tests on the double-sided PE foam tape characterized by highly conformable PE-foam, high immediate bonding strength, even on rough or uneven surfaces, high bonding strength at low mounting pressure and high bonding strength on low surface energy substrates. The tape is characterized by a 4.0 N/cm nominal initial value of adhesion and 200% break elongation. The double-sided PE foam tape is used to connect the aluminium frame to structural and non-structural elements of the buildings.

It must guarantee sufficient levels of adhesion even under dynamic loads. Preliminary analyses of data acquired during the tests carried out considering different input shape, load velocity, and surface typology, provided the results in Table 1.

Table 1. Table captions should be placed above the tables.

N. Test	Input	d	Max velocity	f	Surface	Adhesion	σ_{med}	F_{max}	τ_{med}
		[mm]	[mm/sec]	[Hz]		[hours]	[MPa]	[kN]	[MPa]
1	RAMP	+ 10	0.33	0.0083	Virgin	24	0.00	3.46	0.35
2	RAMP	+ 10	0.33	0.0083	Cleaned with res	0	0.00	2.83	0.28
3	SIN	± 10	16	1.0	Cleaned without res	0	0.00	4.16	0.42
4	SIN	± 10	16	1.0	Cleaned with res	0	0.00	3.87	0.39
5	RAMP	+ 10	0.33	0.0083	Cleaned with res	0	0.90	3.44	0.34
6	SIN	± 10	16	1.0	Cleaned with res	0	0.90	4.18	0.42
7	RAMP	+ 10	0.33	0.0083	Cleaned with res	0	1.80	2.94	0.29
8	SIN	± 10	16	1.0	Cleaned with res	0	1.80	3.83	0.38

The second investigated material was the methacrylate joint, used to couple glass sheet, as represented in the following figure.

During the experimental tests performed on the shaking table shown of Fig. 2, methacrylate exhibits a shear behaviour depending on velocity and displacement amplitude. Particularly, as depicted in Fig. 3, by increasing the velocity of load application the equivalent viscous damping factor increases from around 23 to 30% and shear force from around 0.6 to 1.3kN.

The latest investigated material is the dissipative rubber PAD that represents the most innovative element of the proposed partition wall system. Particularly, the tests were performed considering several rubber compounds and applying the shear force on a PADs couple by means the apparatus shown in the Fig. 4. For each PAD couple 12 quasi-static cyclic tests were carried out, reproducing as many series as possible of cyclic displacements of sinusoidal shape with increasing amplitude from a minimum of ±2.5 mm to a maximum of ±17.5 mm. Experimental tests shown that the PADs break occurred when amplitude reached a displacement equal to about ±15 mm. Test 12 was carried out just with the aim of evaluating the friction of the PADs after rupture. In Fig. 4 the test apparatus is depicted while Fig. 5 shows preliminary results retrieved from the test.

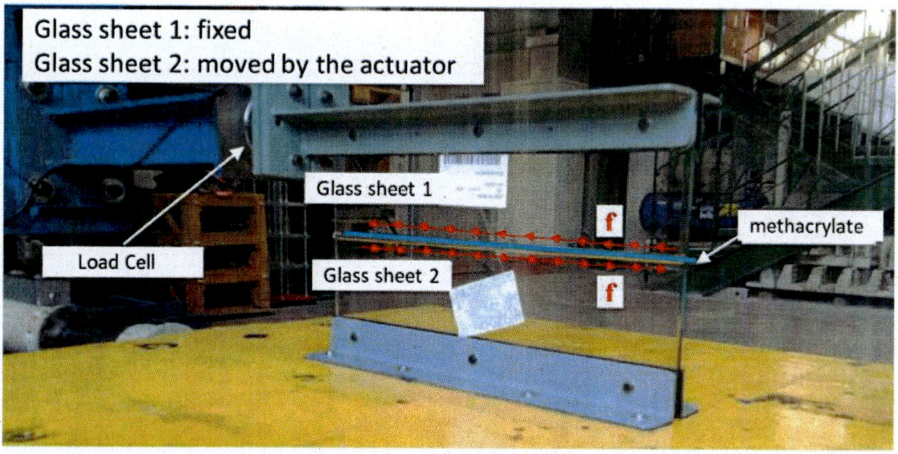

Fig. 2. Test apparatus for the mechanical characterization of the methacrylate joint—shear test

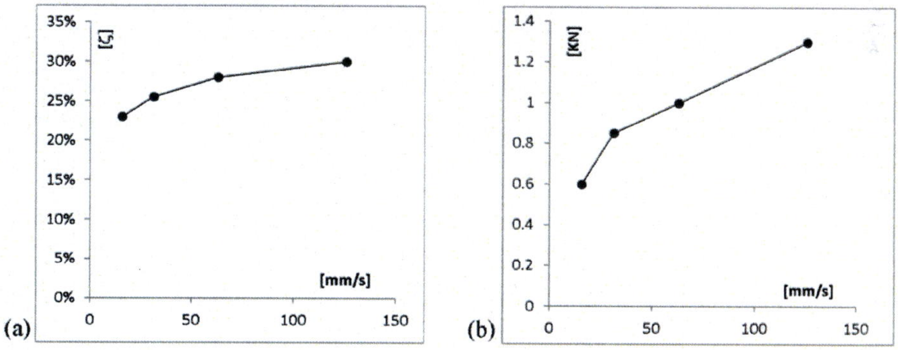

Fig. 3. Representation of (a) Equivalent viscous damping factor and (b) Shear Force as a function of velocity

Figure 6 shows the test apparatus used to characterize the mechanical behaviour of the assembled glass-aluminium partition wall. Several displacement transducers and one load cell were used to acquire data related to the mechanical behaviour of the scaled partition wall under cyclic displacement.

Two main configurations were considered for test. In Configuration 1, the pair of low friction sliding feet on which the single slab rests were positioned between the pair of PADs while, in Configuration 2, feet were located externally to the PADs.

Figure 7a shows the overall experimental cyclic behaviour of the partition walls for increasing amplitude of the applied cyclic base displacement. In the Fig. 7b and c the degrading trend of the system force (FH) and stiffness due to the damage of the PAD is schematised. Fig. 7d shows the comparison between Test 3 and Test 4 realized considering the same displacement amplitude and a different input frequency, 1 and 3Hz respectively. The results show a negligible dependence of maximum reaction force, stiffness, and dissipative capacity of the system from the input test frequency.

Fig. 4. Test apparatus for PADs characterization

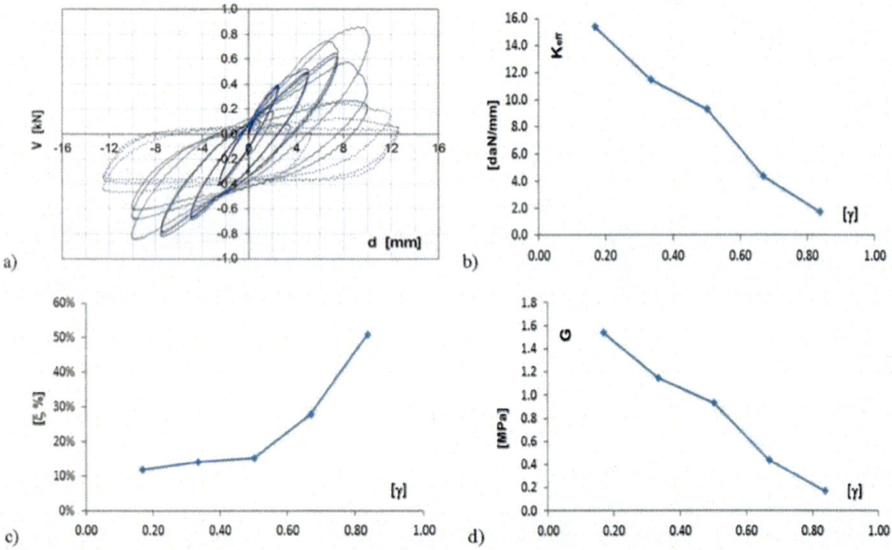

Fig. 5. Experimental results for PADs pair 18A-20G—Config. C1: a) horizontal displacement (γ) vs shear force; b) horizontal deformation (γ) vs shear stiffness (K_{eff}); b) horizontal deformation (γ) vs equivalent damping (ξ); horizontal deformation (γ) versus G Modulus.

3.1 Numerical Model Calibration

A numerical model of the partition wall was developed in SAP2000 V23 [2] and results from the dynamic tests performed on the scaled partition wall were used to calibrate the main model parameters. The geometry of the tested samples is well reproduced by the

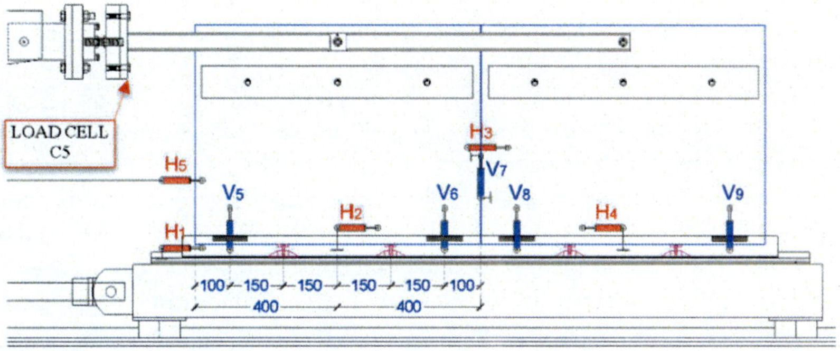

Fig. 6. Sensor set-up for monitoring the behaviour of the assembled glass-aluminium SAFE partition wall

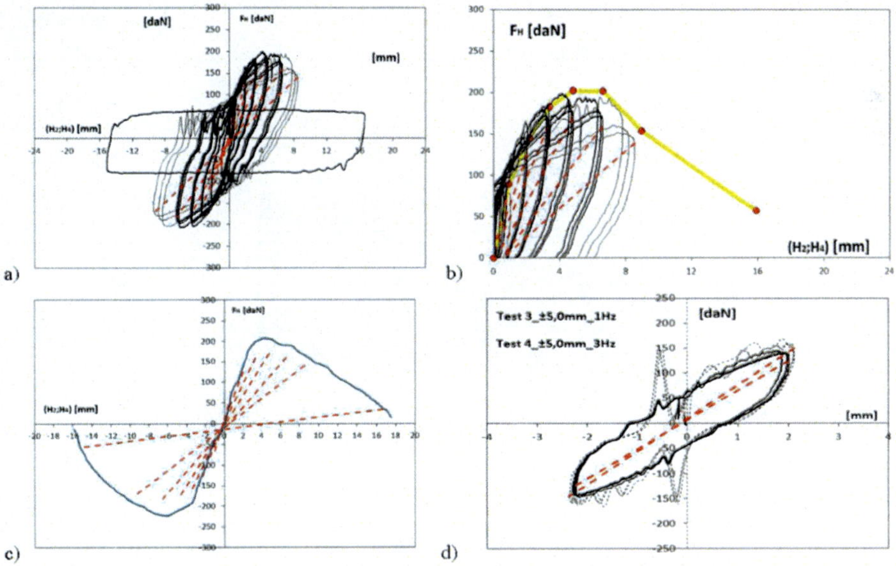

Fig. 7. a) Force-displacement cyclic behaviour at increasing displacement; b) maximum force envelope detail; c) maximum and minimum force envelope; d) Comparison of the prototype cyclic behaviour for different input frequency

finite element model (Fig. 8), glass panels are modelled by 10 mm thick shell elements, additional concentrated masses of about 50 kg per panel are added, and rigid links are adopted to simulate the upper horizontal contrast rod.

For what concerns the main wall sub-elements, the following modelling strategy was pursued:

- rubber pads at the wall base are modelled as viscoelastic links, with equivalent properties (stiffness $k_{eff}(\gamma)$ and viscous coefficient $ceq(\gamma)$) calibrated, at proper shear strain levels γ, on the basis of the results of the mechanical laboratory characterization previously carried out on the sub-components;

- similarly, methacrylate joint response is simulated through a set of viscoelastic links connecting the adjacent panels and activated by their relative sliding motion (properties derived from experimental tests);
- sliding supports are inserted at the base of each panel and modelled through link elements with elastoplastic behaviour along the horizontal direction (accounting for the low friction property) and having unilateral constraint property along the vertical direction, i.e., a gap element link is adopted to allows contact in compression only, with possibility of traction detachment of panels from the base supports.

The model, as detailed above, was subjected to controlled displacement time-history analyses by using harmonic sinusoidal input consistent with the adopted experimental test protocol, i.e., 3 cycles at increasing amplitudes of 5 mm, 10 mm, and 12.5 mm with frequency 1 Hz. Experimental and numerical results showed a good agreement; thus, the validated model was extended to simulate the response of real-scale shake table tests discussed later.

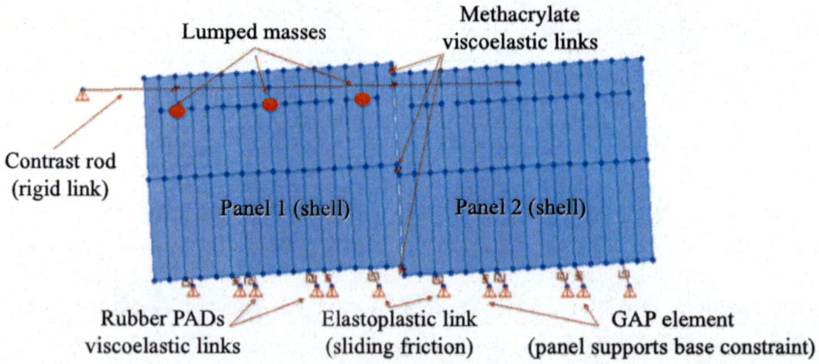

Fig. 8. Numerical model calibrated based on the results from the dynamic tests performed on the scaled partition wall

4 Full Scale Tests and Discussion on Preliminary Results

Experimental tests were performed on a full-scale prototype of the proposed glass-aluminium partition wall.

A typical "C" configuration, representative of a relevant number of real applications made by three side walls, schematically shown in Fig. 9, one of which is equipped with a door (wall 1), was considered. The in-plan geometrical dimensions are $L = 3540$ mm (length at the finished long walls, measured outside-outside on the vertical uprights), $B = 3000$ mm (transversal dimension, measured axis-axis long walls), $H = 2700$ mm (total finished height).

The sequence of shaking table tests performed at the Seismic Laboratory of the University of Basilicata was planned gradually, increasing the PGA from 50 to 125% of

Fig. 9. Test setup installation on the 3D full-scale prototype, shake table and metallic contrast frame

the target design intensity provided in the NTC 2018 for L'Aquila city and soil class "C". During the experimental campaign the relative vertical displacement between glass plates reached a maximum value around 1.5mm, allowing for an additional seismic energy dissipation due to the methacrylate hysteresis contribute. It is also worth to highlight that during strong motion tests the maximum relative displacement never reached the limit value of 25 mm allowed by the seismic joint between aluminium frame and glass panels. The acquisition system was composed by 10 accelerometers, 27 displacement transducers, and 1 load cell. The shaking table tests confirmed the high performance of the proposed glass-aluminium partition walls with no hints of damages also during the strongest motions that were tested. Data analyses are in progress and more detail will be provided in future publications.

Acknowledgements. This study was made within the MIUR-PON project "S.A.F.E. Sustainable Design of Anti-Seismic Furniture as Smart Life-Saving Systems during an Earthquake" coordinated by the University of Camerino, Principal Investigator Professor Lucia Pietroni. The Authors are grateful to the PI and her industrial design research team (Jacopo Mascitti, Davide Paciotti, Daniele Galloppo, and Alessandro Di Stefano) at the University of Camerino which are relevant cocreators of the partition wall here presented. The Authors acknowledge the constructive interactions with the industrial partner (Styloffice) in the design stage and in developing the tested prototypes.

References

1. Aiello, C., Caterino, N., Maddaloni, G., Bonati, A., Franco, A., Occhiuzzi, A.: Experimental and numerical investigation of cyclic response of a glass curtain wall for seismic performance assessment. Constr. Build. Mater. **187**, 596–609 (2018)
2. CSI: SAP 2000 V23 Integrated finite element analysis and design of structures basic analysis reference manual. Computers and Structures Inc, Berkeley, California, USA (2022)

3. Ditommaso, R., Iacovino, C., Auletta, G., Parolai, S., Ponzo, F.C.: Damage detection and localization on real structures subjected to strong motion earthquakes using the curvature evolution method: the Navelli (Italy) case study. Appl. Sci. **11** (2021)
4. Ditommaso, R., Ponzo, F.C.: Automatic evaluation of the fundamental frequency variations and related damping factor of reinforced concrete framed structures using the short time impulse response function (STIRF). Eng. Struct. **82**, 104–112 (2015)
5. Ditommaso, R., Ponzo, F.C., Auletta, G., Iacovino, C.: Numerical investigation on the effects of non-structural components on the elastic fundamental period of buildings. In: Proceedings of the 6th International Conference on Structural Engineering, Mechanics and Computation, SEMC 2016, Cape Town, 5 September 2016 through 7 September 2016. Code 179629 pp. 266−270. (2016)
6. Gerardi, V., Ditommaso, R., Auletta, G., Ponzo, F.C.: Reinforced concrete framed structures: numerical validation of two physical models capable to consider the stiffness contribution of infill panels on framed structures in operative conditions. Ingegneria Sismica **35**, 1–21 (2018)
7. Kazantzi, A.K., Miranda, E., Vamvatsikos, D.: Strength-reduction factors for the design of light nonstructural elements in buildings. Earthquake Eng. Struct. Dynam. **49**, 1329–1343 (2020)
8. Micozzi, F., Scozzese, F., Ragni, L., Dall'Asta, A.: Seismic reliability of base isolated systems: sensitivity to design choices. Eng. Struct. **256**, 114056 (2022)
9. Pavia, A., Scozzese, F., Petrucci, E., Zona, A.: Seismic upgrading of a historical masonry bell tower through an internal dissipative steel structure. Buildings **11**(1), 24 (2021)
10. Petrone, C., Magliulo, G., Manfredi, G.: Shake table tests on standard and innovative temporary partition walls. Earthquake Eng. Struct. Dynam. **46**, 1599–1624 (2017)
11. Picozzi, M., et al.: Real time monitoring of structures in task force missions: the example of the Mw = 6.3 Central Italy Earthquake, April 6, 2009. Nat. Hazards **52**, 253–256 (2010)
12. Ponzo, F.C., Ditommaso, R., Auletta, G., Iacovino, C., Mossucca, A., Nigro, A., Nigro, D.: Localization of damage occurred on framed structures: analysis of the geometric characteristics of the fundamental mode shape. Ingegneria Sismica **31** (2014)
13. Ponzo, F.C., Ditommaso, R., Auletta, G., Mossucca, A.: A fast method for structural health monitoring of Italian strategic reinforced concrete buildings. Bull. Earthq. Eng. **8**, 1421–1434 (2010)
14. Scozzese, F., Gioiella, L., Dall'Asta, A., Ragni, L., Tubaldi, E.: Influence of viscous dampers ultimate capacity on the seismic reliability of building structures. Struct. Saf. **91**, 102096 (2021)
15. Scozzese, F., Terracciano, G., Zona, A., Della Corte, G., Dall'Asta, A., Landolfo, R.: Analysis of seismic non-structural damage in single-storey industrial steel buildings. Soil Dynam. Earthquake Eng. **114**, 505−519 (2018)
16. Steneker, P., Filiatrault, A., Wiebe, L., Konstantinidis, D.: Integrated structural-nonstructural performance-based seismic design and retrofit optimization of buildings. J. Struct. Eng. **146**(8), 04020141 (2020)

Preliminary Results in the Design and Testing of Earthquake-Resistant School Furniture

Laura Gioiella[1], Fabio Micozzi[1], Alessandro Zona[1](✉), Andrea Dall'Asta[1], Martina Sciomenta[2], Gabriele Tamagnone[3], and Massimo Fragiacomo[2]

[1] Scuola di Architettura e Design, Università degli Studi di Camerino, Viale della Rimembranza 3, 63100 Ascoli Piceno, Italy
{laura.gioiella,fabio.micozzi,alessandro.zona, andrea.dallasta}@unicam.it

[2] Dipartimento di Ingegneria Civile, Edile-Architettura e Ambientale, Università degli Studi dell'Aquila, Monteluco di Roio, 67100 L'Aquila, Italy
{martina.sciomenta,massimo.fragiacomo}@univaq.it

[3] New Model Institute for Technology and Engineering, Blackfriars Street, Hereford HR4 9HS, UK
gabriele.tamagnone@nmite.ac.uk

Abstract. This study illustrates the development of innovative solutions for school furniture designed to protect children and teachers in school buildings that cannot guarantee sufficient level of safety in case of seismic events. Attention is given to school desks and shelving units, the former having as objective the protection against falling debris from the ceilings, the latter the protection against damages and overturning of partition walls. In the past, school desks were the focus of seismic designs resulting in expensive and heavy solutions that, for these reasons, have not found practical application. Shelving units, on the other hand, did not receive special attention in their seismic design, other than a stronger connection to the walls. However, partition walls play a very important role during a seismic event due to the different interaction with the structure that can lead to anticipated damages and danger for people, even for moderate seismic events. In this article briefly presents the initial structural concept, preliminary numerical analyses, the preparation of full-scale prototypes, and the execution of experimental tests to simulate extreme loading conditions following structural collapse. It is concluded that very encouraging structural performances against extreme loading conditions and impacts can be achieved using products that are economically affordable with no constraints in their every-day use.

Keywords: Life-saving furniture system · School desks · Self-supporting shelving unit · Earthquake · Design

1 Introduction

Schools should be safe places where children are hosted during their training years to develop their abilities and skills and to receive an education that make them prepared to become our future. Even though these aspects are of paramount importance

G. P. Cimellaro (Ed.): WCSI 2022, LNCE 309, pp. 675–684, 2023.
https://doi.org/10.1007/978-3-031-21187-4_57

and widespread recognized, schools are often hosted in old buildings, realized without modern safety standards, lacking resilience and robustness towards multiple natural hazards. For example, Italy, known to be an area prone to natural hazards such as earthquakes or hydrogeological phenomena, is characterized by a school building stock where more than 60% of the facilities were designed without modern seismic standards. It is frequent that retrofit interventions are promoted only after the occurrence of natural hazards, therefore, operating as reaction rather than as prevention. In this context, school furniture could become obstacles and barriers that aggravate the danger conditions or could be a possibility of salvation in case of building damages or collapse. The latter condition provides the idea of a life-saving furniture system for schools. The system is designed to protect children and teachers in a classroom within school buildings that cannot guarantee sufficient level of safety in case of seismic damages. This paper focuses on school desks and self-supporting shelving units by presenting their initial structural concepts, preliminary numerical analyses, the preparation of full-scale prototypes, and the execution of experimental tests to simulate extreme loading conditions following structural collapse.

2 Objectives and Structural Concepts

2.1 School Desk

School desks are simple objects that can be made resistant to falling masses in the case of seismic damages through an oversizing of all its components, as can be found in many examples in the technical literature. However, the oversizing path produces school desks that are impractical in everyday use being very heavy to organise in the configurations required by a given class, and commonly very expensive, thus, unaffordable by public administrations managing schools. To overcome these limitations, in the study here presented a different path was followed: instead of using oversized elements arranged into conventional desk shapes, elements commonly used for regular school desk constitute an innovative desk shape able to emphasize structural performance. Specifically, the proposed desk (Fig. 1) has an external frame providing direct support to the tabletop and an inner frame with similar geometry is connected to the outer frame at the four corners and at the basis, along the shorter sides of the desk. The inner frame provides support to the outer frame, collaborates with a bending progressive damage in case of strong overloading (being the larger inclination of the legs of the inner frame properly optimized for this task), improves overall stability, and provides a backing protection in case of damage/failure in the upper part. The resulting desk is much lighter as compared to existing earthquake-proof desks, more economical being based on materials, sections, dimensions of elements and manufacturing processes already adopted in common desks, does not have moving components that could harm users, has a bending-dominated damage in case of significant overloading that exploits the ductility of steel to dissipate impact energy while preserving the volume of the protection area.

2.2 Self-supporting Shelving Unit

Standard shelving units are installed next to a wall and there attached through mechanical fasteners. Alternatively, they can be set up through base connections with the floor

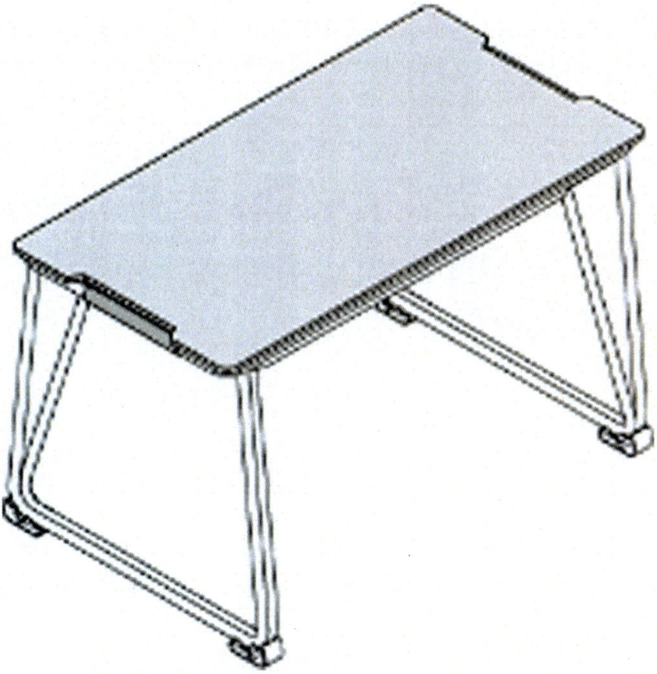

Fig. 1. Proposed dual-frame school desk.

and top connections with the ceiling, with adjustment devices for upright height, thus, defining a self-supporting structure. The review of the state-of-the-art shows that classic bookshelves did not receive special attention in their seismic design, other than a stronger connection to the classroom walls, as also underlined in [1–5] and in international standards related to the reduction of the risks of non-structural earthquake damage [6] as well as in California [7] and Japanese schools [8]. An important drawback of these solutions is that, if the partition walls do not meet the seismic requirements, the equipped walls anchored to them undergo the same damages of the partition walls or could be even a worsening factor. Similarly, shelving units with base connection at the floor and top connection at the ceiling that do not have elements capable of stabilizing their structural behaviour under seismic actions, cannot resist to the forces generated by the overturning of the nearby walls not designed for seismic safety. The partition walls, indeed, play a very important role during a seismic event due to the different contribution in terms of stiffness with the frames and their mutual interaction that can lead to downtime even in the case of no structural damages.

The proposed self-supporting shelving unit (Fig. 2) tries to overcome these drawbacks by combining both a functional and a protection aspect. The former is exploited in the operational conditions, i.e., absence of adverse events, when the shelving unit works as a bookshelf where interactive whiteboard, teaching material, and audio-visual projection systems can be stored (Fig. 2a). The latter, instead, is exploited in case of seismic events (Fig. 2b) and it aims at guaranteeing the safety of the nearby individuals pursuing two goals: contrasting the overturning of the infill partition wall located behind the shelving unit and protecting against falling objects in the case of damages of the ceilings.

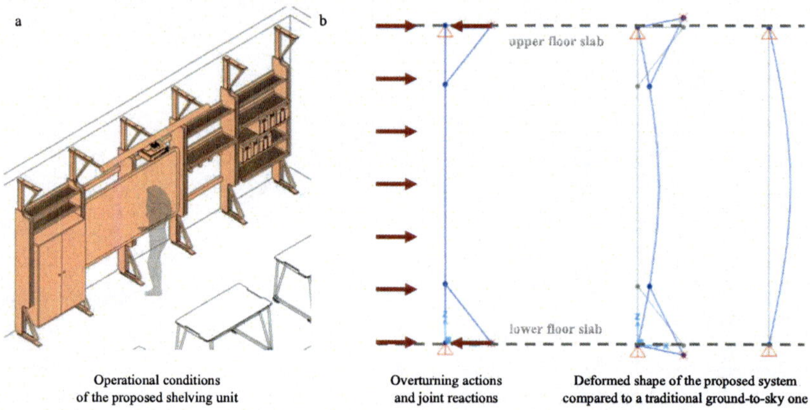

Fig. 2. Working conditions of the proposed shelving unit: (a) operational ones; (b) in case of earthquake.

Figure 2b shows the working mechanism of the shelving unit in contrasting the overturning, highlighting the seismic behaviour of its principal element: the telescopic steel upright. It is composed of three elements, a central tubular hollow rectangular profile sized to work mainly for bending and axial forces and two symmetric stiffening triangular profiles, subjected to axial force and hosting the central profile. The triangular profiles can slide over the central one, achieving the desired height configuring a ground-to-sky system and then fixed together with the central profile by means of self-tapping screws for metallic carpentry. Thanks to this geometric configuration, the triangular profiles can develop a static scheme equivalent to a bridge with inclined legs, which reduces the free length of inflection of the central profile and simultaneously empowers the performance of the restraints, both at the base and on top, thanks to the compressive action developed. This way, the overturning occurs only in case of the breaking through the upper and lower floors to which the profiles are connected, or for the overcoming of the resistance of the system itself. This latter aspect is the substantial difference with the common ground-to-sky furnishing systems, in which the foot and head restraints are usually not designed to withstand horizontal seismic actions. Furthermore, a certain pre-compression is given to the uprights during the setting up that increases the efficiency of a layer of elastomeric material, located at the interface between the metal support plates at each restraint and the floors, in regularizing the contact surfaces and increasing the friction against sliding.

Figure 3 describes the structural frame constituted by horizontal crossbeams and vertical uprights, so that it can withstand the stress induced by the possible overturning of a partition wall during an earthquake. Each module of the shelving unit spans 1.20 m and is suitable for the insertion of a shelter at its top that contrasts and contains any falls of debris due to damage to the floor slabs, thus creating a life-saving area for people below.

This protection area allows to achieve the second objective of the life-saving shelving unit. It is also worth to observe that the proposed system is realized with prefabricated elements, which are easy to assemble and transport.

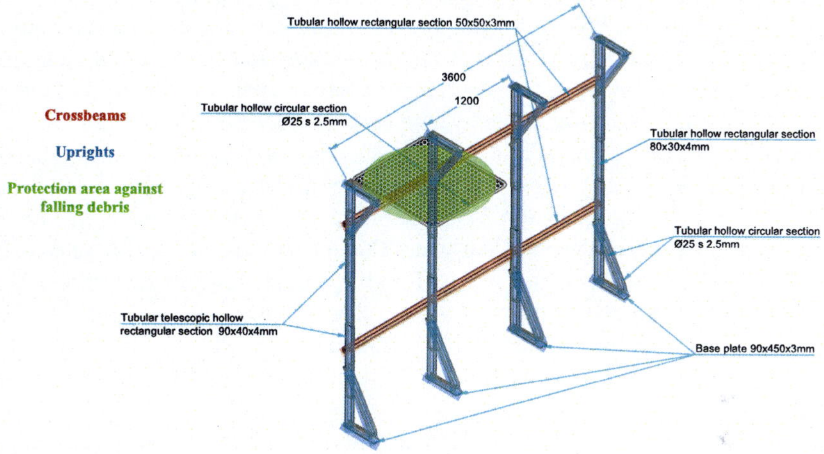

Fig. 3. Layout of the proposed self-supporting shelving unit.

3 Preliminary Numerical Analyses

3.1 School Desk

The shape of the school desk was optimized considering vertical distributed and concentrated loads, both symmetric and non-symmetric, through a finite element model implemented in the SAP 2000 software (Fig. 4). The most critical loading condition was identified being a concentrated force at one corner with global stability being the most demanding verification. However, it was clear for the very beginning that this preliminary numerical study could only provide initial indications, while deeper insight could only be based on impact tests.

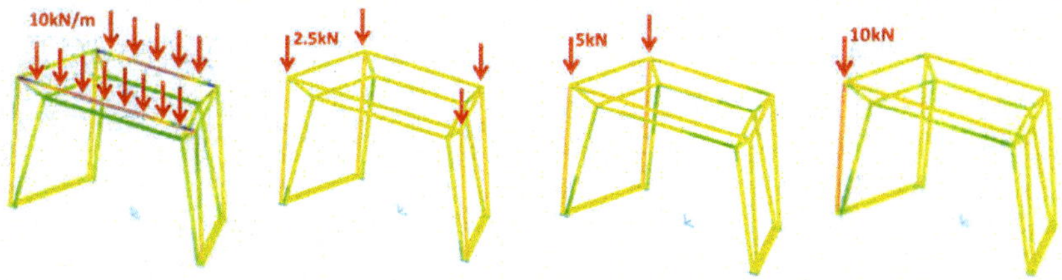

Fig. 4. Examples of preliminary numerical simulations for the design of the school desk.

3.2 Self-supporting Shelving Unit

The shelving unit was designed to sustain the overturning action produced by a typical infill partition wall, used within reinforced concrete (RC) frame structures realized between 1960–1990, consisting of a 2 cm thick layer of plaster that covers each of the two faces of 8 cm thick hollow bricks. For the estimation of both the dimension and the maximum acceleration suffered by the infill wall, the Don Giussani school in Ascoli Piceno, a city of moderate seismic hazard in the nearby of the seismic crater of the 2016 Central Italy earthquakes, was taken as case study. With reference to the infills under consideration (length 7.60 m, height 3.0 m), a shelving unit was dimensioned having a total length of 3.60 m and a depth of 0.45 m, encompassing 4 uprights able to span a height between 2.70 and 3.0 m and 2 crossbeams able to distribute among the uprights the overturning action of the partition wall. The seismic action representative of the demand for the dimensioning and check of infills partition towards overturning action and consequently to the shelving unit was defined according to [9] $F_a = S_a W_a / q_a$ where $S_a = 1.96$ is the maximum acceleration in g experienced by the non-structural component during an earthquake having a probability of occurrence of 10% in 50 years (i.e., a return period of 475 years); $W_a = 36$ kN is the weight of the non-structural component and $q_a = 1$ is the behaviour factor. Accordingly, the seismic design action results in $F_a = 71.5$ kN.

Two Finite Element Models (FEMs) of a single upright related to the maximum height of 3.0 m were implemented in the SAP2000 software (Fig. 5). The models differ for the number of connection points between the central tubular profile and the two telescopic stiffening profiles. Once known the stresses on the structural elements, their dimensions were checked: central profile 80 × 30 × 4 mm, telescopic tubular profiles 90 × 40 × 4 mm, circular stiffening profiles Φ25 mm s = 2.5 mm, base and top plates 90 × 450 × 3 mm.

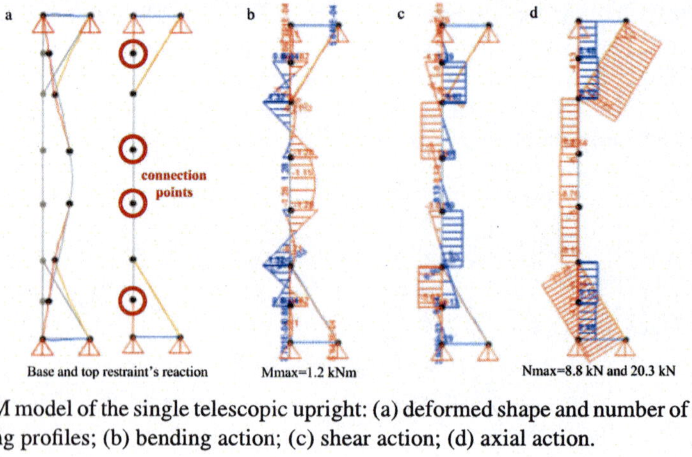

Fig. 5. FEM model of the single telescopic upright: (a) deformed shape and number of connection points among profiles; (b) bending action; (c) shear action; (d) axial action.

4 Full-Scale Prototypes and Experimental Test Results

4.1 School Desk

The experimental tests of the proposed school desk involved both static loading under concentrated and distributed loading areas (symmetric and non-symmetric) as well as impact tests where masses made of bricks or portion of traditional floors were suddenly released (Fig. 6). The tested prototypes were realized by the two industrial partners involved in the project. Many results and indications were obtained from the tests. Given space limitations in this article, a brief selection of some impact test is reported.

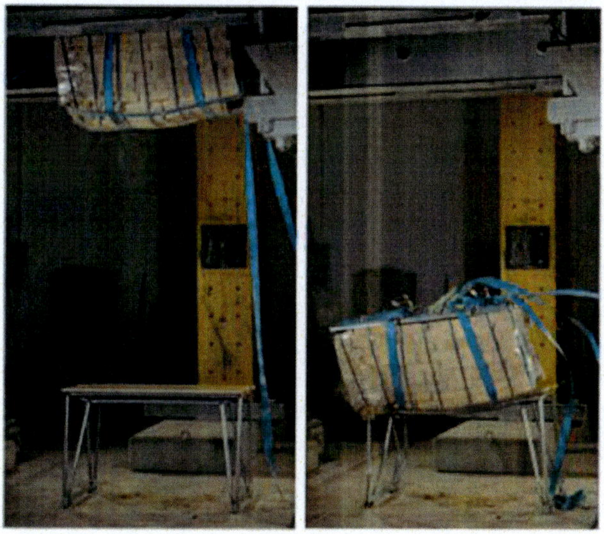

Fig. 6. Two moments of an impact test on one desk prototype.

The main aim of impact tests was highlighting the effect of seismic events on school desks on the chance that some floor collapses occur. In the field of civil engineering, impact tests are uncommon, and these carried out were particularly challenging due to several reasons: (i) furniture are usually considerate as non-structural components, so it is not required to test their load bearing capacity against impact dynamic loads and (ii) full scale impact tests on the whole desk were carried on, otherwise from other engineering fields that test apart sub-components and reduced-size specimens. These latter conditions required to establish a novel tests methodology for the tests setting-up and execution. Tests were performed by releasing a load instantaneously from a height of 2 m, using a suitable manual quick release device. The amount and type of load was deeply analysed to guarantee the same features of real floors in terms of weight, stiffness, and strength. Two different scenarios were accounted for the existing floor: (i) a floor built with ancient techniques as brick vaults and domes, (ii) a traditional reinforced concrete and hollow tiles mixed floor. To reproduce the flexibility of vaults, a series of brick stacks were packed. Bricks were simply overlapped and wrapped with clamped bands and plastic film stripes to simulate their poor bonding during the collapse. A steel plate was settled on the top of the stack to allow its hooking and lifting, further belts were used to increase the holding. The area of the stack was assessed to be equal to the tabletop, while the height varied based on the two mass amounts: respectively 450 and 600 kg. The effect of a traditional reinforced concrete and hollow tiles floor was simulated by releasing a portion of this floor having as base dimensions 1.0 × 1.2 m and a height of 0.24 m. The weight of these floor portions was 300 kg. In addition, to understand the effect of lighter but concentrated loads, additional tests were carried out by making a smaller (0.36 m × 0.12 m) brick stack weighting 36 kg.

The reliability of earthquake-proof desks as well as those of traditional ones was tested. The earthquake-proof desks gave appreciable experimental evidence: (i) the tabletops were never damaged during the impact, regardless of the amount and type of load; (ii) the horizontal and inner vertical steel frames only deflect without collapsing; (iii) the outer vertical frames were damaged only if the 600 kg mass was not centred otherwise it bent outwardly in the lower part with a ductile mechanism. After the impact with 300 kg mass, the traditional desks overturned by losing the load bearing capacity on one leg; the tabletop collapsed after the impact with the smaller mass of 36 kg.

4.2 Self-supporting Shelving Unit

The two industrial partners involved in the project of the life-saving furniture system realized 4 uprights each; neither of them used the cross-sections of the structural members designed because of supplier availability. The result are two versions of the shelving unit, one with heavier profiles and one with lighter profiles, which represent a sort of Upper Bound Design Properties (UBDPs) and Lower Bound Design Properties (LBDPs). Consequently, the tests performed on these prototypes provided the opportunity to evaluate the influence of the different stiffness on the activation of the inclined leg static scheme expected, on the ultimate capacity of the members and on the efficiency of the different restraints tested. These latter spans from double-sided adhesive tape to mechanical fasteners for concrete surfaces, i.e., increasing the maximum force capacity.

The static experimental tests performed on the Laboratory of the University of L'Aquila aims to shed light on: (i) the efficiency of the different solutions for the base restraints in contrasting the sliding; (ii) the efficiency of the top restraint with different values of prestress, when the upright is installed below a concrete joist of the floor or below the brick element for lightening the floor; (iii) verify the effective activation of the inclined leg mechanism and characterizing which element of the telescopic uprights leads to the failure of the system.

Figure 7 shows the deformed shape and the results achieved during one of the tests executed on a prototype describing the LBDPs of the telescopic uprights.

5 Conclusions

This paper described the objectives and the structural concepts of a school desk and of a self-supporting shelving unit designed within a system of new life-saving furniture thought for buildings that are not able to ensure the safety of persons in case of natural

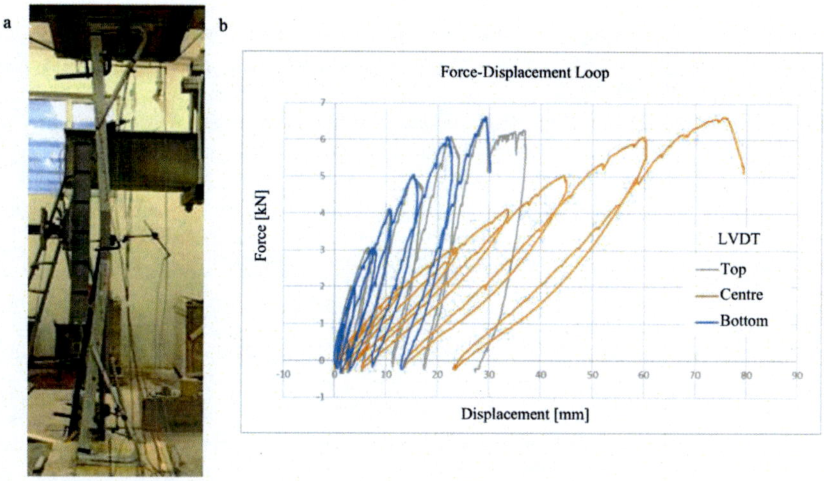

Fig. 7. Static test performed on the LBDPs prototype: (a) deformed shape at the end of the test; (b) force-displacement loop recorded by three displacement transducers installed along the telescopic upright.

hazards like earthquakes. A wide experimental campaign was conducted with the aim of verify the efficiency of the furniture and to characterize their ultimate capacities towards different load and restraints conditions. It is shown that very encouraging structural performances against extreme loading conditions and impacts can be achieved using products that can be economically affordable with no constraints in their every-day use.

Acknowledgements. The research was carried out within the activities of the project "S.A.F.E. Sustainable Design of Anti-Seismic Furniture as Smart Life-Saving Systems during an Earthquake" funded by the Italian Ministry of University and Research within the framework PON 2015–2018 in the Specialization Area of "Design, Creativity and Made in Italy", coordinated by the University of Camerino, Principal Investigator Professor Lucia Pietroni. The Authors are grateful to the PI and her industrial design research team (Jacopo Mascitti, Davide Paciotti, Daniele Galloppo, and Alessandro Di Stefano) at the University of Camerino, which are relevant cocreators of the furniture here presented. The Authors acknowledge the constructive interactions with the industrial partners (Camillo Sirianni and Vastarredo) in the design stage and in developing the tested prototypes. The Authors are also grateful to the technicians of the Laboratory of the University of L'Aquila, Edoardo Ciuffetelli and Alfredo Peditto, for their valuable support during the tests.

References

1. Filiatrault, A.: Seismic evaluation of modular office furniture system. Earthq. Spectra **7**(54), 524–541 (1991)
2. Filiatrault, A., Kuan, S., Tremblay, R.: Shake table testing of bookcase—partition wall systems. Can. J. Civ. Eng. **31**, 664–676 (2004)

3. Chen, M.C., et al.: Full-scale structural and nonstructural building system performance during earthquakes: part I—specimen description, test protocol, and structural response. Earthq. Spectra **32**(2), 737–770 (2016)
4. Di Sarno, L., Magliulo, G., D'Angela, D., Cosenza, E.: Experimental assessment of the seismic performance of hospital cabinets using shake table testing. Earthq. Eng. Struct. Dynam. 1–21 (2018)
5. Pürgstaller, A., Gallo, P.Q., Pampanin, S., Bergmeister, K.: Seismic demands on nonstructural components anchored to concrete accounting for structure-fastener-nonstructural interaction (SFNI). Earthq. Eng. Struct. Dynam. **49**, 589–606 (2020)
6. FEMA: Reducing the risks of nonstructural earthquake damage—a practical guide. FEMA **E-74** (2012)
7. California Emergency Management Agency: Guide and Checklist for Nonstructural Earthquake Hazards in California Schools (2011)
8. Educational Facilities Research Center: Case studies of seismic nonstructural retrofitting in school facilities. National Institute for Educational Policy Research, Study Group Report on the "Investigative Research for Promoting the Seismic Retrofitting of School Facilities", Japan (2005)
9. Ministry of Infrastructure and Transport: NTC 2018. Aggiornamento delle "Norme Tecniche per le costruzioni" DM 17/01/2018 (2018) (in Italian)

Experimental Seismic Response Characterisation of Brackets for Use in Ventilated Façade Systems

Simone Peloso, Emanuele Brunesi(✉), and Elisa Rizzo Parisi

European Centre for Training and Research in Earthquake Engineering, EUCENTRE, Via Ferrata 1, 27100 Pavia, Italy

{simone.peloso,emanuele.brunesi,elisa.rizzoparisi}@eucentre.it

Abstract. Ventilated façades are complex non-structural elements composing the building system, the latter being intended as a unique system integrating both structural skeleton/components and non-structural items, owing for instance to issues of seismic performance and loss. In light of this, experimental response characterisation is of paramount importance, leading indeed to data gathering for key components or sub-systems, which in turn could inform numerical modelling efforts, with a view to design methodology definition. Thus, this paper presents testing efforts meant to characterise the cyclic behaviour of brackets – connecting the panel elements in the façade system of interest – for both in-plane and out-of-plane earthquake-induced actions, assuming prescriptions by FEMA461 for the displacement-controlled loading protocol that relies upon monotonic test counterparts for reliable definition of ultimate target displacement. Test outcomes are presented with emphasis placed on stiffness and dissipation capacity for the case-study bracket typology.

Keywords: Façade systems · Component testing · In-plane testing · Out-of-plane testing · Response characterisation

1 Introduction

Non-structural elements (NSEs) are not designed to be part of the load-bearing system of a building although they are subjected to the same dynamic environment as the supporting structure during an earthquake, with moderate and large seismic events in urban regions of countries with well-established seismic design provisions that however have repeatedly shown the importance and vulnerabilities of NSEs [1–6], demonstrating also how high earthquake-induced losses can be. These elements, defined as the items in a building that are not part of the building's structural load-bearing system, include architectural elements, building utility systems and building contents, with ventilated façade systems being an interesting example because of not only inertia effects but also those related to building distortions and separation as well as non-structural interaction [7].

G. P. Cimellaro (Ed.): WCSI 2022, LNCE 309, pp. 685–695, 2023.
https://doi.org/10.1007/978-3-031-21187-4_58

With this in mind, experimental seismic response characterisation of ventilated façade systems and components is of paramount importance, as façades are complex NSEs integrated in the building system, the latter being intended as a unique system that combines, together, both structural skeleton/components and non-structural items. Tests were thus carried out, in pseudo-static cyclic fashion, to investigate the behaviour of brackets connecting panel elements for both in-plane (IP) and out-of-plane (OoP) seismic actions, with particular emphasis posed on the former case. In more details, whilst the OoP capacity of the brackets – and hence that of the façade – can somehow be analysed on the basis of wind-loading considerations, the resistance of the brackets in the other horizontal direction, namely IP, should be carefully and accurately assessed since façade systems are both acceleration- and drift-sensitive. Product standards available nowadays for ventilated façades do not require the evaluation of this parameter, nor are methods for its determination specified in current codes and standards, implying that research is needed on this front/aspect.

2 Review of EADs and Relevant Standards

With the aim of identifying the best approach for the execution of the tests and therefore for the characterisation of the brackets to be tested, a review of relevant EADs – namely European Assessment Documents – and standards is given hereinafter. This is not only the first and perhaps most relevant step for the identification of behaviour and response mechanisms of the façade systems but also provides guidance for the design of the experimental methods for the products to test and eventually qualify.

Additionally, even if the pace of research on NSEs and their seismic behaviour/performance is increasing sharply in the last years [8], the focus is so broad that it is still almost unrealistic to find ad hoc specifications for all different kinds of NSE typologies and archetypes (as well as components). Thus, it is necessary to understand how to deal with elements that are not yet fully characterised and/or covered by specifications, which had also been developed taking into account technical and scientific knowledge at the time of their issue (or earlier).

Especially from a seismic point of view, which is not exhaustively covered to date, the actions (i.e. inertia forces) are generated by the element of the kit having the greatest mass, which is the panel in most cases. Its characteristics and the type of action can affect in different ways the behaviour of the connection to the structure. On the one hand, therefore, it is difficult to generalise the behaviour of all parts that make up the connection without studying a wide range of case studies. On the other hand, studying the behaviour of the façade system as a whole could be crucial. In this respect, it is certainly worth recalling that accelerations and inter-storey drifts are dual and mutual, implying that façade panels, their connections and the supporting frame will possibly deform during an earthquake (in which case they do so being driven by issues of displacement compatibility). As such, inter-storey displacements play a pivotal role not only because the drift is an easy-to-evaluate indicator of the level of damage to the façade but also owing to deformation-induced/triggered effects, all which can be significant or not depending on type of panel and interaction with the primary structure. Interactions between actions on a system as well as between the components of a system shall thus be evaluated carefully and on a case-by-case basis.

One of the most relevant standard is EAD 090062-00-0404 [9], which is applicable to kits for mechanically fixed external wall claddings, or external wall claddings mechanically fixed as referred to as therein, composed by (i) cladding elements, (ii) cladding fixings, (iii) sub-frame components, (iv) thermal insulation products, and (v) ancillary components. Although the effects of seismic actions are not covered by EAD 090062-00-0404 [9], this EAD sets the working life of cladding kit (i.e. 25 years or shorter depending on the environmental conditions to which the kit is subjected) and identifies methods and criteria for performance assessment, out of which the most significant are (i) for wind load resistance, (ii) for mechanical resistance and (iii) for bracket resistance specifically.

The test method for the assessment of wind load resistance shall be in line with what indicated in Annex E of the EAD 090062-00-0404 [9], which provides two test types: wind suction and pressure. In both cases, the principle is to establish the effects on the assembled cladding kit. The mechanically weakest case, at least, shall be tested with uniformly distributed loads applied in subsequent steps on the surface of the assembled cladding kit until failure, the latter being determined by breaks or permanent deflection of any cladding element, fixing, profile or bracket, or by falling of detached component or by failure or detachment of the kit sub-frame.

For what concerns mechanical resistance, key items are identified as follows: (i) cladding element, (ii) connections between the cladding element and fixing, (iii) cladding fixing, and (iv) sub-frame components. More in detail, bracket resistance – as per Sect. 2.2.12.16 of EAD 090062-00-0404 [9] – shall be tested according to the method described in Annex L of this EAD. Their fixing to the sub-frame has to be evaluated under shear and tension loads, namely the weight load for vertical load and the wind load for horizontal load, respectively. The test specimen shall consist of one brackets (asymmetric layout) or two brackets, in which case testing can be undertaken by means of two brackets in opposition on both sides of the profile. The weakest bracket configuration shall, at least, be tested (for example if several length of wings are available), and the profile defined for the system shall be used in the test. Whenever this is not possible, a square or rectangular section steel tube of 1.5 mm minimum thickness may be considered. Also the profile-bracket fixing must correspond to that used in the system and the worst fixing position shall be assumed. The bracket-substrate fixing must be chosen according to the type of substrate. When no manufacturer specifications are available for the fixings, bolts of at least 6 mm diameter shall be used in conjunction with washers. These anchors shall not represent a weak point of the test specimen. The maximum specified distance from the profile of the fixing anchor on the support shall be considered.

For the test, attention should be paid to the measurement of load and displacement so that the load-displacement capacity curve of the tested system could be obtained. The test procedure for both vertical and horizontal loads relies upon a semi-cyclic test protocol: in each cycle an increasing load is applied, at a constant load or displacement rate, to the profile and then returned to zero. Constant load/displacement rate is considered to avoid mobilising a dynamic failure of the specimen. For load-controlled tests, EAD 090062-00-0404 [9] suggests increasing steps of 100 N for vertical load tests and 200 N for horizontal load tests, both with a constant rate <5000 N/min. For displacement-controlled tests, increasing steps of 0.25 mm, 0.5 mm or 1.0 mm are suggested, depending on the bracket behaviour under study, with a constant rate ≤5 mm/min.

EAD 090034-00-0404 [10] is applicable to kits composed of sub-frame and fixings for fastening cladding and external wall elements, and includes the following items: (i) skin element fixing (for some bracket families), (ii) sub-frame metallic vertical profiles, (iii) sub-frame metallic skin element fixings, and (iv) ancillary components. For what concerns the assessment of bracket resistance, criteria are given in Sects. 2.2.11 and 2.2.12, with Annex H that specifies how to assess the load bearing capacity and the wind resistance. The test procedure provided is the same as the one described in EAD 090062-00-0404 [9].

AAMA 501.4-18 [11] and AAMA 501.6-18 [12] could also be named, along with EN 13830:2015+A1:2020 [13]. The former two are American standards applicable to curtain walls subjected to seismic inter-storey drift, the first of which defines a static test methodology to determine the failure modes for exterior wall systems and allows correlating the design approaches with the horizontal displacement performance/capacity. Thus, it allows evaluation of the performance of elements, identifying failure modes and any possible need for changes in their design. Even if the systems covered by AAMA 501.4-18 [11] are other than those of interest herein, it is worth noting that similar experimental methods could be adopted for the testing of entire façade portions whereby performance of a system as a whole is to be evaluated. AAMA 501.6-18 [12] defines a dynamic test methodology, intended as the dual of the previously introduced static method of testing. This methodology may be used to determine the failure condition of the exterior wall systems with an in-plane dynamic crescendo displacement. An approach of this type implies that all parts of the test specimen shall be full size and of the same actual materials of the building wherein the curtain wall is installed. The same consideration can be made for connections and fixings, which must correspond to those used in the system. Analogously, EN 13830:2015+A1:2020 [13] includes a systematic framework of requirements, test methods, and compliance criteria for curtain walling kits, with emphasis on the safety in use and the serviceability.

3 Definition of Testing Procedure for Brackets

In this section, the main aspects regarding the testing procedure are presented in detail. Firstly, a brief overview regarding the rationale behind the design of the test setup and the definition of the assumed, displacement-controlled loading protocol is proposed to highlight the primary assumptions that were considered for the execution of the test.

The chosen test set includes fifteen different specimens, combination of three sizes, i.e. small, medium and large, and five lengths, as summarised in Table 1.

As can be inferred from above matrix (see Table 1), this implies a total of fifteen specimens, which are two of a kind when considering the direction of testing, namely IP and OoP. More specifically, each of those fifteen brackets were tested twice, in plane and out of plane, one bracket at a time in accordance with clause L.3.2 of EAD 090062-00-0404 [9]. In other words, a decision was made to opt for an asymmetric layout consisting

Table 1. Sizes and lengths of the test set.

Size	Length (mm)
Small (S)	40–80–140–220–300
Medium (M)	40–80–140–220–300
Large (L)	40–80–140–220–300

of a single bracket – instead of undertaking tests on a pair of brackets in opposition on both sides of the profile – so that interaction one another could be avoided. On the other hand, interaction with the profile was considered, by testing brackets with a 60 × 40, 1.8 mm-thick aluminium profile. Both EAD 090062-00-0404 [9] and EAD 090034-00-0404 [10] allow the use of a square or rectangular steel section whenever no profile is (yet) identified for a system. Alternatively, use can be made of the actual profile as permitted by both EADs. Out of the two permitted solutions, however, the latter option was favoured in this case, and was opted for in order (i) to reproduce the actual interaction between bracket and profile, and (ii) to identify the weakest link in the bracket-screwed-connection-profile chain, all with their own possible failure mechanisms. This means that the lower resistance (and the corresponding mechanism or interacting mechanisms) could be evaluated accordingly.

The bracket-profile connection was chosen to be along the longer side of the profile (i.e. the 60 mm one) and was made with two S-AD 01 S 5.5 × 16 self-drilling screws. Additionally, the contact length between the bracket and the profile was set equal to 20 mm, as this configuration was deemed the most critical.

For each bracket per type and length, a monotonic test was performed first, both for initial characterisation of the specimen as well as for calibration of displacement target for the successive cyclic tests. Note that, for OoP testing, the brackets were monotonically tested separately, in compression and tension, because the specimen behaviour was considered very different – in the two directions – in terms of failure mode. Indeed, the target displacement for the compression test was expected to be driven by issues of instability, especially for the medium-long brackets. On the contrary, the likeliest critical point for determining the target displacement in tension was deemed to reflect issues of plasticity of the test specimen. Pseudo-static cyclic tests at increasing displacement amplitudes were carried out right after, considering five specimens per size and length. In what follows, a summary table is provided with number of specimens per test (see Table 2).

3.1 Test Setup

As can be inferred from Fig. 1, tests were performed by means of an electromechanical MTS machine with a load cell of 10 kN capacity (see Fig. 1a). The test setup was designed in such way that the same plates could be used for both IP and OoP tests. It consisted of four main steel pieces/components:

Table 2. Number of specimen per test type.

Test type	Number of specimen per size and length	Total number of specimens per test	Total number of specimens
Monotonic – IP	1	15	195
Monotonic – OoP	2* (1T and 1C)	30	
Cyclic – IP	5	75	
Cyclic – OoP	5	75	

- a base plate for connection to the bottom part of the machine, wherein brackets are also fixed for OoP tests (A);
- a vertical plate with stiffeners for the fixing of brackets for IP tests (B);
- a bottom piece consisting of horizontal and vertical plates with stiffener, which was used for fixing the two rails (above the horizontal plate) and the profile (on the vertical plate) for the IP tests (C);
- an upper horizontal plate for connection to the mobile part of the machine, wherein the aluminium profile is fixed for the OoP tests and for fixing the guides for the IP tests (D).

Three-dimensional views of the test setup for IP and OoP testing are shown in Fig. 1b, c, each of which contains labels to facilitate the identification of each single component, as per the above reported description.

Note that the base plate, which featured holes in different positions so as to move the vertical plate with stiffeners (B) depending on the bracket length for the IP tests, was assumed to rest on the test machine and was fixed to it by means of four M8 bolts. In the OoP test case, the brackets are fixed directly to the base plate as can be seen in Fig. 1c.

Additionally, it is worth noting that the vertical plate (B) featured a hole for fixing of the bracket during IP tests. Within its thickness, it had two holes for connection to the lateral stiffeners, which were also bolted to the base plate (A). The other elements were designed to be fixed to the mobile part of the machine and consisted of one piece for the OoP tests (i.e. D) and two pieces for the IP tests (i.e. C and D). In the latter case, component C and component D were connected together by two linear guides that allow for displacement in the direction perpendicular to that of the test so as to avoid axial tension being mobilised in the specimen.

3.2 Loading Protocol

Initial monotonic tests, with a constant displacement rate of 5 mm/min in accordance with EAD 090062-00-0404 [9] and EAD 090034-00-0404 [10], accompanied cyclic tests for calibration of ultimate displacement. For the IP monotonic tests, the brackets were loaded pushing them downwards, meaning that the edge of their bottom/bolted part made contact with the stiffened vertical plate of the setup. Conversely, for the OoP monotonic tests, a decision was made to load the brackets separately in both directions

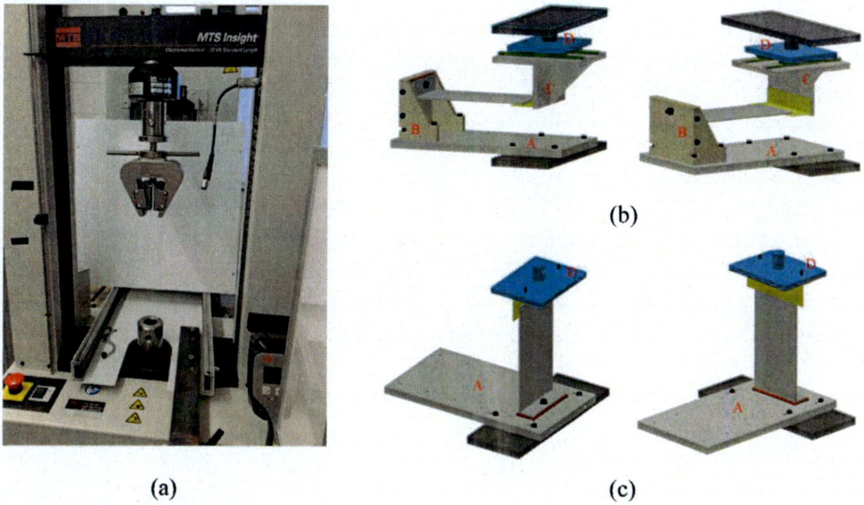

Fig. 1. Test rig: (a) testing machine; (b) schematics of test setup for IP testing; (c) schematics of test setup for OoP testing.

to determine the different failure mode corresponding to each case (i.e. tension and compression).

For cyclic pseudo-static tests, FEMA461 [14] was taken as reference for the loading protocol (see for instance Filiatrault et al. [15]); starting from the identified target displacement, the amplitude of each cycle was calculated as a percentage of it. The number of chosen steps or displacement amplitudes was ten and for each of them two cycles were performed in displacement control. Figure 2 shows conceptual representation of FEMA461 [14] cyclic loading protocol as well as an example with the first amplitude a_1, or the targeted smallest deformation amplitude Δ_0, taken as 0.048 times Δ_m, which is the targeted maximum deformation amplitude (as per monotonic test).

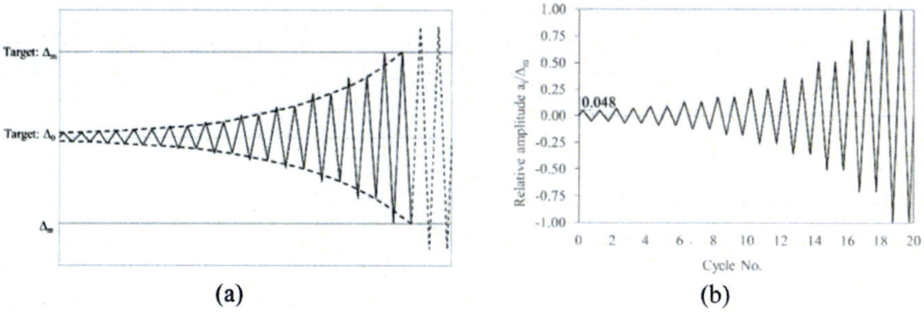

Fig. 2. FEMA461 cyclic loading protocol: (a) conceptual representation; (b) example with $a_1 = 0.048\Delta_m$.

Note that a decision was taken to consider symmetrical displacement histories for IP tests, while for OoP testing they were simply made asymmetrical as a different behaviour in the two directions was expected.

4 Test Results

Figure 3 presents the normalised force-displacement response curves of a mid-length/mid-size bracket-profile for IP and OoP directions of testing, whereas Figs. 4 and 5 show trends obtained in terms of normalised secant stiffness and normalised dissipated energy for the former and latter testing cases, respectively.

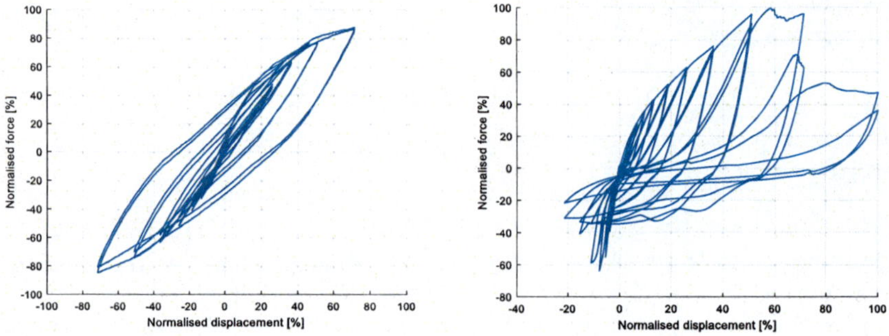

(a) (b)

Fig. 3. Normalised force-displacement response: (a) IP testing; (b) OoP testing.

As can be seen from Fig. 3a, the IP response is approximately symmetric and the hysteresis loops are hardened in character, with a fairly stable unloading branch, the stiffness of which resembles the loading one. As far as the OoP behaviour is concerned instead, the hysteretic force-displacement response curve is highly asymmetric, as expected, due to the fact that different damage mechanisms took place in tension and compression (i.e. yielding/plasticisation and buckling, respectively), with the displacement capacity corresponding to incipient buckling being approximately one fifth of the tensile displacement level at which the maximum force is recorded. Peak compressive resistance-to-peak tensile resistance ratio is approximately two thirds. Moreover, after tensile and compressive force peaks are reached, the resistance drops to approximately half the peak for both tension and compression (when referring to the ultimate displacement amplitude imposed before bringing the test to a halt).

The abovementioned trends are reflected by normalised stiffness decay curves (see Figs. 4a and 5a), with an almost linear decrement for IP testing that is opposed to a pair of curves having two different slopes for the OoP test case. Lastly, it is worthwhile to note that, somehow expectedly, the – normalised – dissipated energy increases as the imposed displacement increases (Figs. 4b and 5b), with small-deformation cycles associated with the IP response that show minor energy dissipation capacity, as can be gathered from Fig. 4b.

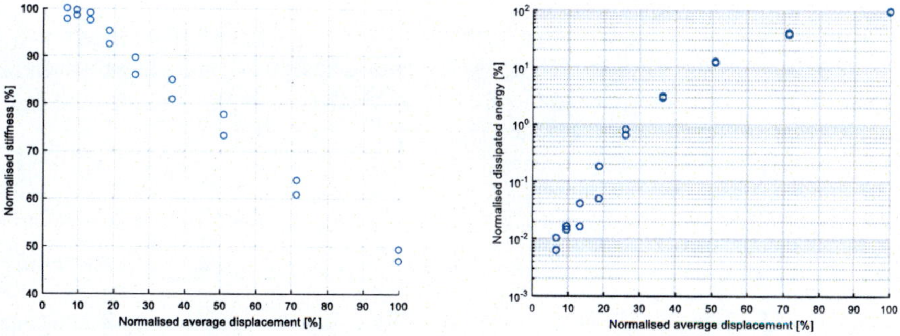

(a) (b)

Fig. 4. Experimental characterisation – IP testing: (a) normalised stiffness; (b) normalised dissipated energy.

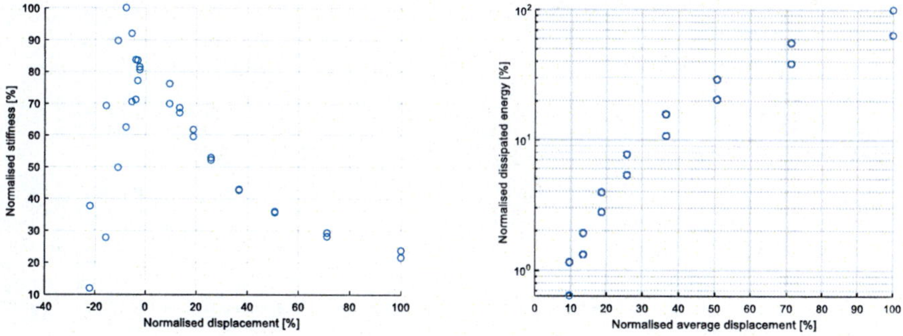

(a) (b)

Fig. 5. Experimental characterisation – OoP testing: (a) normalised stiffness; (b) normalised dissipated energy.

5 Conclusions

The piece of research described in this paper has delved into pivotal aspects related to experimental testing for seismic response characterisation of brackets for use in ventilated façade systems. Tests, carried out in pseudo-static cyclic fashion, were undertaken assuming the displacement-controlled loading protocol recommended in FEMA461 [14] for both in-plane and out-of-plane directions. The study allowed the following conclusions to be drawn:

- The out-of-plane behaviour was primarily driven by issues of yielding/plasticisation and buckling in tension and compression, respectively, whilst the in-plane response was flexural-dominated, with stable and fairly symmetric hysteresis loops that are hardened in character.
- Secant stiffness reduction and energy dissipation capacity were quantified, and were proven to be characterised by stable trends with the imposed displacement, the latter being taken as a measure of damage for the considered façade components.

- Test data collected during the experimental testing campaign described herein may be valuable input information to feed numerical and/or analytical modelling towards definition of reliable seismic design methodology, which is currently missing in technical literature as well as design codes and standards.

References

1. Filiatrault, A., Uang, C.M., Folz, B., Christopoulos, C., Gatto, K.: Reconnaissance report of the February 28, 2001 Nisqually (Seattle-Olympia) earthquake. In: Structural Systems Research Project Report No. SSRP-2000/15, 62 pp. Department of Structural Engineering, University of California, San Diego, La Jolla, CA, USA (2001)
2. Chock, G., et al.: Compilation of observations of the October 15, 2006, Kiholo Bay (Mw 6.7) and Mahukona (Mw 6.0) earthquakes, Hawaii, 53 pp. Earthquake Engineering Research Institute, Oakland (2006)
3. Gupta, A., McDonald, B.M.: Performance of building structures during the October 15, 2006 Hawaii earthquake. In: Proceedings of the 14th World Conference on Earthquake Engineering, 8 pp. Beijing, China (2008)
4. Ricci, P., De Luca, F., Verderame, G.M.: 6th April 2009 L'Aquila earthquake, Italy: reinforced concrete building performance. Bull. Earthq. Eng. **9**(1), 285–305 (2011)
5. Belleri, A., Brunesi, E., Nascimbene, R., Pagani, M., Riva, P.: Seismic performance of precast industrial facilities following major earthquakes in the Italian territory. J. Perform. Constr. Facil. **29**(5), 04014135 (2015)
6. Perrone, D., Calvi, P.M., Nascimbene, R., Fischer, E.C., Magliulo, G.: Seismic performance of non-structural elements during the 2016 central Italy earthquake. Bull. Earthq. Eng. **17**(10), 5655–5677 (2018). https://doi.org/10.1007/s10518-018-0361-5
7. Rodriguez, D., Perrone, D., Filiatrault, A., Brunesi, E.: A probabilistic strong floor motion duration model for seismic performance assessment of non-structural building elements. Earthq. Eng. Struct. Dynam. **50**(15), 4161–4179 (2021)
8. Filiatrault, A., Sullivan, T.: Performance-based seismic design of nonstructural building components: the next frontier of earthquake engineering. Earthq. Eng. Eng. Vib. **13**(1), 17–46 (2014). https://doi.org/10.1007/s11803-014-0238-9
9. EAD 090062-00-0404: Kits for External Wall Claddings Mechanically Fixed, 81 pp. European Organisation for Technical Assessment (EOTA), Brussels, Belgium (2018)
10. EAD 090034-00-0404: Kit Composed by Subframe and Fixings for Fastening Cladding and External Wall Elements, 37 pp. European Organisation for Technical Assessment (EOTA), Brussels, Belgium (2016)
11. AAMA 501.4-18: Recommended Static Test Method for Evaluating Window Wall, Curtain Wall and Storefront Systems Subjected to Seismic and Wind-Induced Inter-Story Drift, 11 pp. American Architectural Manufacturers Association (AAMA), Schaumburg, IL, USA (2018)
12. AAMA 501.6-18: Recommended Dynamic Test Method for Determining the Seismic Drift Causing Glass Fallout from Window Wall, Curtain Wall and Storefront Systems, 15 pp. American Architectural Manufacturers Association (AAMA), Schaumburg, IL, USA (2018)
13. EN 13830:2015+A1:2020: Curtain Walling—Product Standard, 94 pp. European Committee for Standardisation (CEN), Brussels, Belgium (2020)
14. FEMA461: Interim Testing Protocols for Determining the Seismic Performance Characteristics of Structural and Nonstructural Components, 138 pp. Federal Emergency Management Agency (FEMA). Washington, DC, USA (2007)

15. Filiatrault, A., Perrone, D., Brunesi, E., Beiter, C., Piccinin, R.: Effect of cyclic loading protocols on the experimental seismic performance evaluation of suspended piping restraint installations. Int. J. Press. Vessels Pip. **166**, 61–71 (2018)

Shake Table Tests on a New Passively Controlled System with Pulley Amplification Mechanisms for Suspended Ceilings

Ryo Majima[1], Tomonari Haruyama[1], Shigeki Sakai[2], Yasuo Yamasaki[3], and Taiki Saito[1](✉)

[1] Department of Architecture and Civil Engineering, Toyohashi University of Technology (TUT), Aichi 441-8580, Japan
{majima.ryo.bk,haruyama.tomonari.be,saito.taiki.bv}@tut.jp
[2] Technical Research Institute, Hazama Ando Corporation, Ibaraki 305-0822, Japan
sakai.shigeki@ad-hzm.co.jp
[3] Technical Research Institute, Nishimatsu Construction Co., Ltd., Kanagawa 243-0303, Japan
yasuo_yamasaki@nishimatsu.co.jp

Abstract. Passive response control systems have been adopted as a common practice to mitigate the seismic response of structures. Meantime, suspended ceiling systems (SCS) have frequently suffered severe damage among nonstructural elements during earthquakes in the last decade. This paper presents the shake table test results on a new passive control system developed for indirect-SCSs employing a damped cable system equipped with pulley tackle amplification mechanisms, named "Pulley Damper Ceiling System (PDCS)". In this study, the seismic performance of the SCSs with and without the presence of the proposed system is investigated based on a series of full-scale shake table tests. In particular, one-direction shake table tests on a total of six SCSs, a non-damped SCS, a steel-braced-SCS, and four SCSs with different PDCS configurations, are primarily tested to evaluate the fundamental response reduction effect of the PDCS. Additionally, two-directions shake table tests on the SCSs with PDCS are carried out to confirm the seismic response reduction efficiency and dynamic behavior of the PDCS under bidirectional earthquake motions. The experimental observations showed that the PDCS successfully amplified the damper displacements and velocities. Besides, the PDCS sufficiently mitigates both displacement and acceleration response by comparing to the conventional and braced SCSs.

Keywords: Suspended ceiling system · Passive control system · Damped cable system · Pulley tackle mechanism · Shake table test

1 Introduction

Several energy dissipation devices (i.e., viscous, viscoelastic, and hysteretic dampers) have been successfully developed to eliminate or limit the earthquake-induced damage to buildings [1]. However, the installation of dampers either with diagonal or chevron

G. P. Cimellaro (Ed.): WCSI 2022, LNCE 309, pp. 696–706, 2023.
https://doi.org/10.1007/978-3-031-21187-4_59

auxiliary braces only can deform the dampers that are less than or equal to the story drift where devices are attached. Both configurations are usually appropriate for flexible buildings, although stiff structures respond to an earthquake with small story drifts. On the basis of these considerations, various attempts to integrate amplification mechanisms into the damper with the purpose of increasing the actual deformation that the damper receives have been proposed by many researchers. The representatives of amplification mechanisms are toggle system [2, 3], lever-arm system [4, 5], and seesaw system [6, 7]. They are used to magnify the displacement of the structure and deliver to the damper based on the geometric mechanisms. Another example is the pulley tackle mechanism. The concept to incorporate the pulley system with a passive damper was firstly proposed by Saito [8] and the displacement amplification effect was investigated by some research groups through full-scale experimental tests [8, 9] and a shake table test using a scale model [10].

Meantime, only a little attention has been paid to the damage to non-structural building components including suspended ceilings, exterior walls, fire extinguishing equipment, and piping equipment. When considering non-structural elements, the suspended ceiling system (hereinafter referred to as SCS) is one of the most frequently used in a large-scale space building and is fragile during earthquakes. The vulnerability of SCSs in past earthquakes has been extensively documented. For example, the falling of ceiling panels in indoor sports facilities, gymnasiums, shopping malls, and airport terminals, was previously reported during the 2011 Tohoku Earthquake [11] and the 2016 Kumamoto Earthquake [12]. Besides, the collapse of SCS has been reported in various countries, for example, in the 1994 Northbridge earthquake in the United States [13], the 2016 Gyeongju Earthquake, and the 2017 Pohang earthquake in Korea [14].

In August 2013, the Ministry of Land, Infrastructure, Transport and Tourism (MLIT) published a new technical bulletin on the suspended ceilings system (Notice No. 771) [15]. According to this technical bulletin, the installation of steel rigid braces to SCS is required, if the SCS has an area of more than 200-m^2 and the mass per unit area is over 2 kg. The effectiveness of the braced-SCS has been investigated through a shake table test of a full-scale school gym at the E-Defense earthquake simulator in Japan [16]. However, Lee et al. [17] indicated that the difference in each ceiling panel behavior including the relative displacement causes the panel to fall out, thus, the arrangement of steel brace must be considered carefully. An additional concern about the rigid-SCS is when subjected to the strong vertical excitation [18]. Because the rigid elements constrain the ceiling movement, the acceleration difference between the fixed ceiling grid and the unfixed part may make the ceiling unstable and drop off.

To overcome those problems, and maintain the functionality of buildings even after suffering from an earthquake, the authors of this paper proposed an innovative passive control system for SCSs (pulley damper ceiling system, PDCS) aiming to reduce both displacement and acceleration response of SCSs, and presented the derivation of constitutive equation [19]. In this study, one and two directions full-scale shake table tests were conducted to confirm the feasibility of the proposed damping system and evaluate the seismic response reduction effect. Besides, based on the experimental results, the dynamic behaviors of the SCS including the ceiling torsion and energy absorption of the proposed damping system were investigated.

2 Mechanism of the Pulley Damper Ceiling System

Figure 1 illustrates the system configuration and the arrangement of the PDCS. The PDCS consists of two pulley tackle mechanisms, one passive energy dissipating device, and an SCS. A continuous wire is reciprocally stretched n times between the Pulley A fixed at the primary structure and the Pulley B embedded on the SCS. The basic operating principle of this proposed system is that, if the horizontal displacement of the SCS is "D" due to lateral forces, then the amount of damper deformation "d" is amplified by the number of wire thread "n" between Pulley A and B. In addition, because the wire cable is diagonally stretched with the angle "θ", it is necessary to consider the influence of the wire stretching angle. Therefore, the damper displacement can be expressed as Eq. (1). The pulley tackle mechanism simply allows the damper to absorb more energy in proportion to the number of wire loops between each pulley group, even with a small damper capacity.

$$d = D \times n \times \cos\theta \tag{1}$$

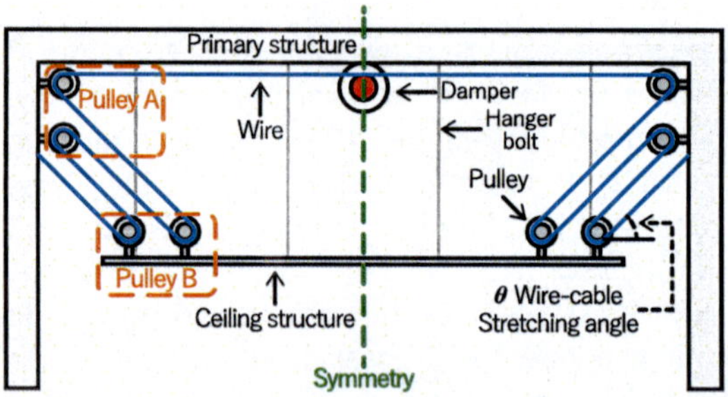

Fig. 1. Concept illustration of the proposed pulley damper ceiling system (PDCS).

3 Shake Table Tests

3.1 Shake Table Test Program

From 2020 to 2022, to evaluate and qualify the response control performance of the proposed ceiling system, shake table tests were carried out using the earthquake simulator at the structural engineering laboratory of the Hazama Ando Technical Research Institute and Nishimatsu Technical Research Institute, in Japan. The information on the outline of the shake table program is summarized in Table 1. The details of test specimens and input motions are described in the following subsections.

This experimental program consists of two phases. The first phase is to confirm the maximum response reduction effect by comparing with and without the presence of the damping system under unidirectional excitation and to find the optimal system configurations. The second phase is to investigate the dynamic behavior including the torsion of the ceiling system under bidirectional excitations.

Table 1. Information on the shake table test conditions.

Test name	Phase I	Phase II
	Unidirectional test	Bidirectional test
Laboratory location	Hazama Ando Technical Research Institute	Nishimatsu Technical Research Institute
Shake table specifications	Table size: 4m × 4 m Maximum payload: 784 kN Peak acc. (x, y, z): 1G, 3G, 1G	Table size: 5.5m × 5.5 m Maximum payload: 637 kN Peak acc. (x, y, z): 2G, 2G, 2G
Year	2020 and 2021	2022
Number of the damping systems	2 for X-direction	2 for each X and Y direction
Number of the test cases	6 cases	2 cases
Input motion	White noise wave, Sinusoidal wave, Simulated earthquake wave	
Excitation direction	X-direction	X-Y directions

3.2 Shake Table Test Cases

In this study, a total of four models for unidirectional and bidirectional shake table tests were prepared. Table 1 presents the specimens list with its model illustrations.

In Phase I, six cases: an unreinforced SCS (Case C), a reinforced SCS with steel braces (Case CB), two damped-SCSs with a wire but without a viscous damper (Case CW), and two damped SCSs with both a wire and a viscous damper (Case CWD), were primarily tested. Additionally, in the case of SCS with PDCS, for each model, two different configurations of wire stretching angle (0-degree and 60-degree) were applied. Here, in terms of prevention of the ceiling panel torsion during this unidirectional test, V-braces were placed in the Y-direction for all the test cases.

As the continuing experiment, in Phase II, two cases: a non-damped SCS (Case C), and a damped-SCS (Case CWD) with both a wire and a damper, were performed. The major difference between Phase I and Phase II is the arrangement of the damping system. As the system design is illustrated in Fig. 2, while the two DPCSs were installed only in the X-direction during the Phase I (red lines in Fig. 2 [a]), four damping systems, two for each X and Y direction (red and blue lines in Fig. 2 [b]) were placed (Table 2).

3.3 System Components

The damping system in this experiment is composed of a stainless wire cable, pulleys, and a rotational viscous damper. A stainless wire-cable consists of 7 × 19 strands with 2.75 kN braking force, and 2.0 mm in nominal diameter. Pulleys which consist of three sheaves with a 30 mm internal diameter with a ball bearing inside were applied as a cable slider to smoothly transfer the ceiling displacement to the damper.

Figure 3 (a) presents the 3D perspective of the newly developed rotational nonlinear viscous damper that produce a damping force from the hydraulic resistance of the silicon

Table 2. Specimens list of the shake table test.

Conventional model		Wire model	Damper model
(SCS)	(SCS + Brace)	(SCS + Wire)	(SCS + Wire + Damper)
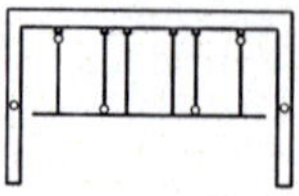 Case C	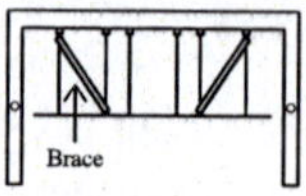 Case CB	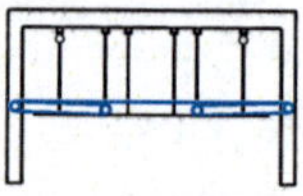 Case CW0	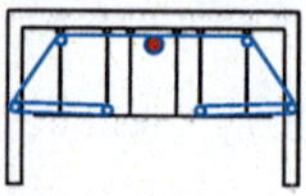 Case CWD0
		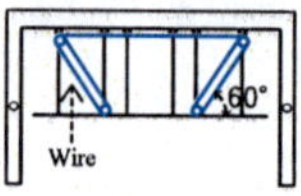 Case CW60	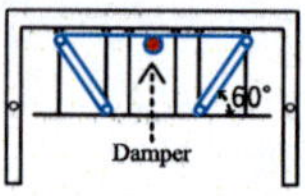 Case CWD60

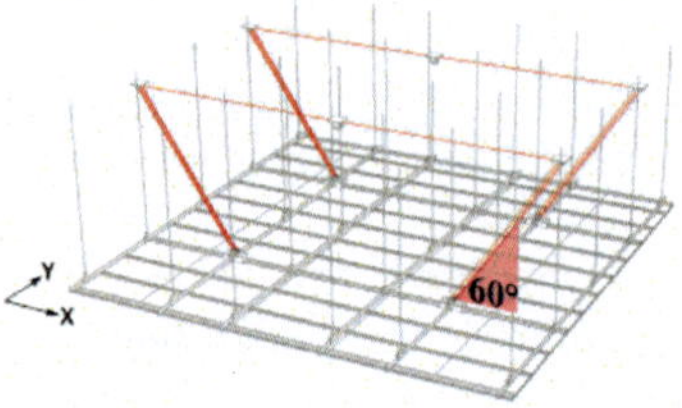

(a) Unidirectional excitation test

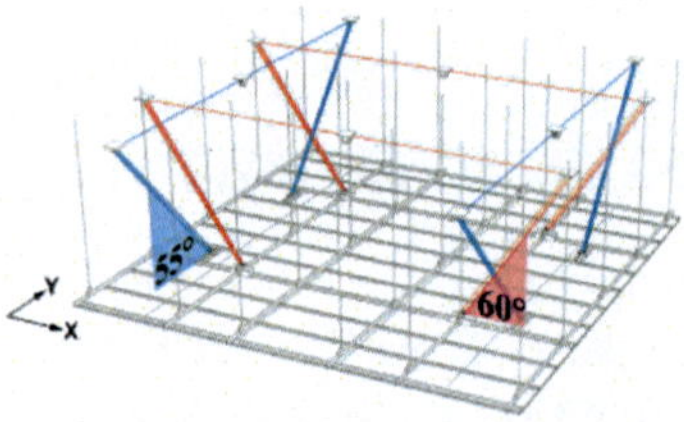

(b) Bidirectional excitation test

Fig. 2. Pulley damper system arrangements and wire stretching angles.

when the shaft rotates. This energy dissipation device has two rotary dampers of 60 mN m-rated torque. To follow up on the large amplified wire stroke, the wire was wrapped around the spool several times. The force-velocity hysteresis loop is shown in Fig. 3 (b).

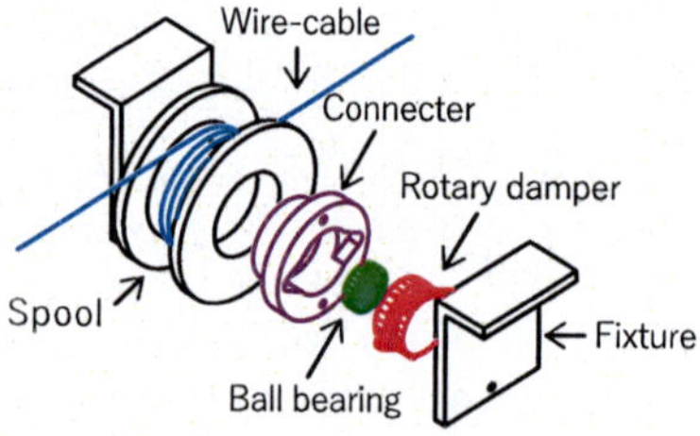

(a) 3D view of the damper

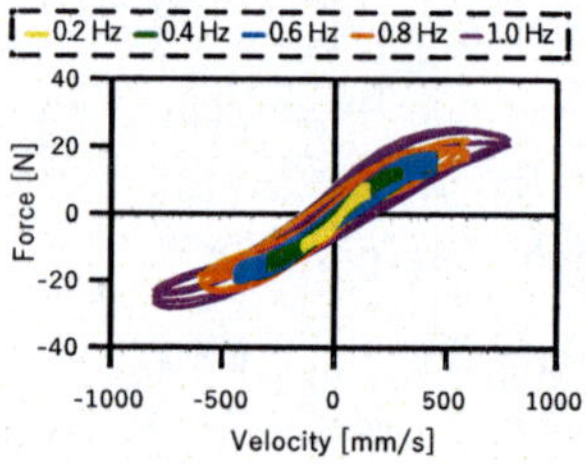

(b) Sinusoidal excitations test results

Fig. 3. Pulley damper system arrangements and wire stretching angles.

3.4 Specimen Details

Figure 4 (a) shows the specimen overview used in the shake table test. To hang SCSs and install the proposed damping system, an overall 5800 mm (length) × 3500 mm (width) × 3200 mm (height) steel rigid test frame was firstly designed and prepared. It is then, indirect-SCS (Fig. 4 [b]) where continuous ceiling boards are fixed on the bottom of the carrying runner structure was constructed. The ceiling structure, 4020 mm (length) × 3020 mm (width) in area size, with a weight of 200 kg, was hung 1500 mm from the test frame by 30 hanger bolts of 9 mm diameter.

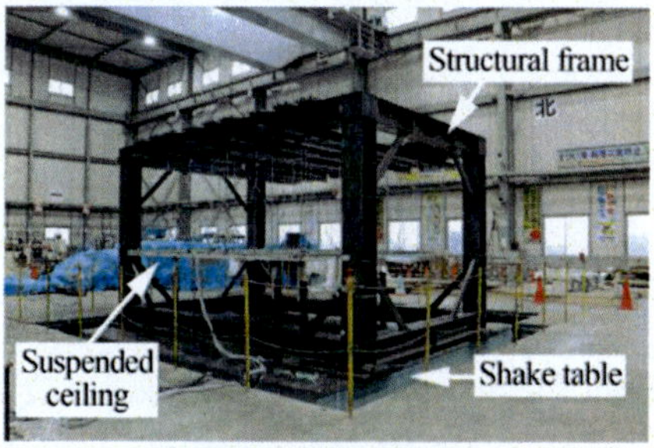

(a) Specimen overview

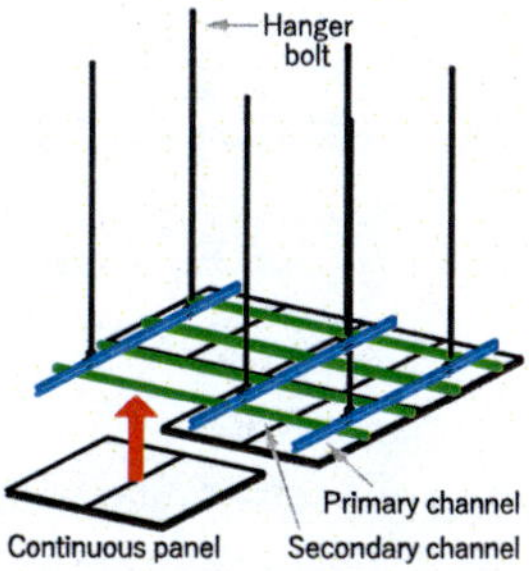

(b) The detail of the SCS structure

Fig. 4. Specimen view in Phase II.

3.5 Input Motions and Measurement Plan

A white noise wave test with a random phase with a maximum of 0.1 m/s/s including the frequency range from 0.1 Hz to 50 Hz was initially used to observe the dynamic properties in the elastic range. Then incremental earthquake excitation tests using simulated earthquake ground motions with random phases generated based on the design spectrum specified in the Japanese Building Standards Law were performed. Figure 5 shows the simulated earthquake wave and spectrum curve. In the case of the bidirectional test, the input intensity ratio of X-direction and Y-direction was 3 to 2. Besides, harmonic excitation using a sinusoidal wave having the period of Case C (1.37 s) was carried out.

To monitor the dynamic behavior of the structure, accelerations at 9 locations (4 at structural frame, 4 on ceiling panel, and 1 on shake table surface) and displacements at 4 points (2 in parallel for each X and Y direction) were installed. Also, 2 draw-wire displacement sensors and 4 force measurement equipment (consisting of strain gages and copper plate) were inserted into the wire in every damping system. A major concern in DCS is the cable slackness especially when the direction of the inter-drift changes from forward to backward [20, 21]. Thus, a pre-tension force of 50 N was introduced to the wire which is larger than the damping force of the viscous damper (about 30 N) and the total friction force (about 6 N) was mainly produced around pulley-wire contacts, to make sure that the cable is not being slack during the earthquake excitations.

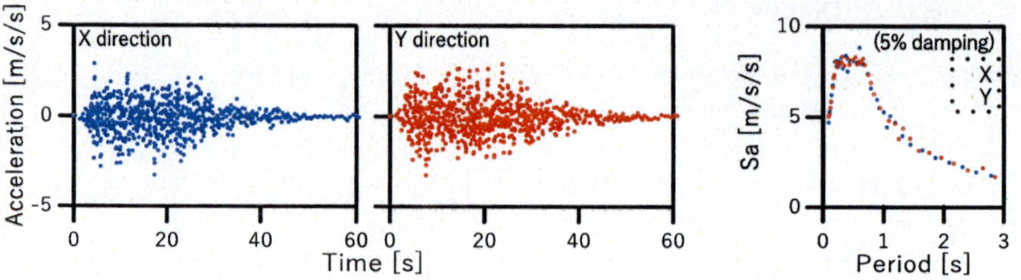

(a) Acceleration time history waves (b) Spectral acceleration

Fig. 5. Simulated earthquake motion used in shake table tests.

4 Shake Table Tests Results

4.1 Unidirectional Excitations

Figures 6 (a) and 7 (a) present the maximum displacement and acceleration response on the ceiling panel surface under each earthquake wave intensity, respectively. Also, the response amplification ratio of the displacement and acceleration to Case C under major peak ground acceleration (PGA) are summarized in Figs. 6 (b) and 7 (b).

As Fig. 6 shows, Case CB, in which the SCS was rigidly fixed to the floor above by the steel braces, performed the largest displacement response reduction at an average of 3% to the unreinforced Case C. However, the acceleration response of Case CB was amplified by more than 200% of Case C. Meanwhile, Case CWD effectively mitigated both displacement and acceleration to less than half of Case C. The difference in response reduction between CWD0 and Case CWD60 is considered to be caused by the changes in horizontal stiffness given by the stretched wire.

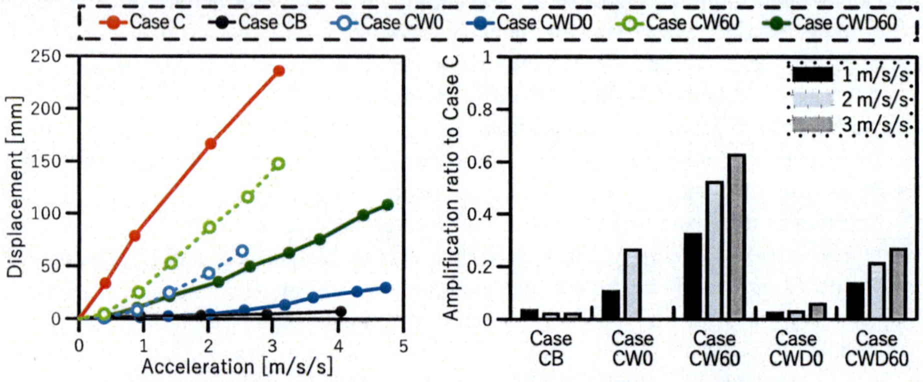

(a) Maximum response (b) Response amplification

Fig. 6. Results of displacement response at the ceiling panel surface.

Figure 8 shows the displacement and force relation of the ceiling panel and damping system under sinusoidal wave excitation. Considering the friction force generated around

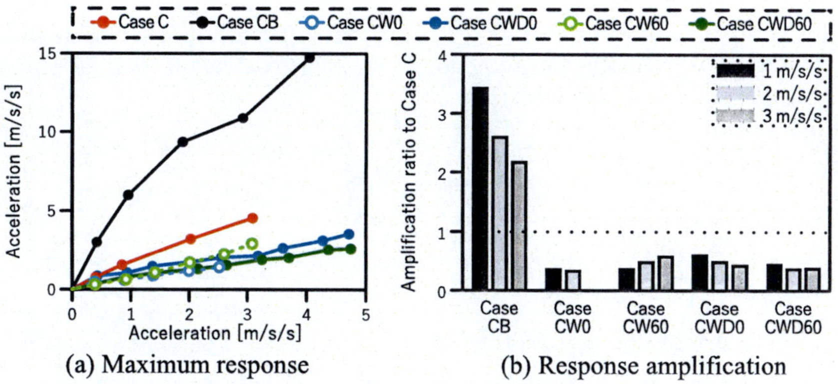

Fig. 7. Results of acceleration response at the ceiling panel surface.

pulleys, the total energy dissipation by the damping system becomes the sum of viscous damper and friction (the red loop in Fig. 8). Here, because the input energy area and the dissipated energy area size are equal, the proposed system absorbed most of the input energy.

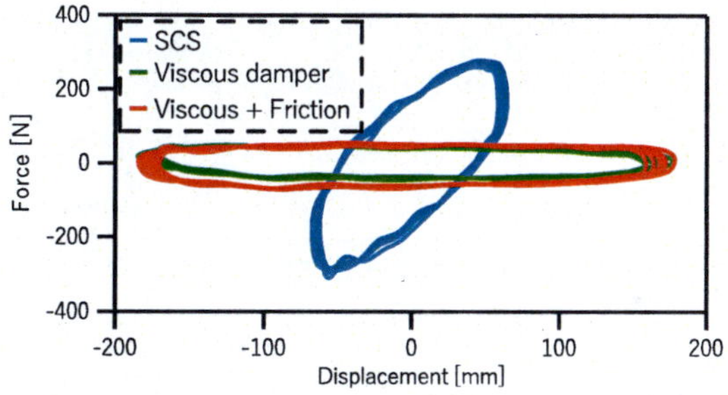

Fig. 8. Hysteresis loop of the ceiling and damping system under a sinusoidal wave.

4.2 Bidirectional Excitations

Figure 9 (a, b) compares the displacement time history record at X1 and X2 points in parallel (Fig. 9 [e]), and Fig. 9 (c, d) shows the rotational angle of the ceiling panel, under PGA 2 m/s/s of simulated earthquake wave.

From the results of X-direction excitation (Fig. 9 [a, c]), the maximum distance error between X1 and X2 points was nearly zero, and the rotation of the ceiling was not overserved during the experiment. Also, for XY directions excitation (Fig. 9 [b, d]), the maximum distance difference was less than 4mm (equal to 0.015 rad). Additionally, from the comparison of the X1 records between the X-direction and XY-directions tests under the same intensity earthquake, the wave shape and period matched very well, and the ceiling rotation was very small. Therefore, it is confirmed that the influence of two-directions excitations on the proposed damping system in response reduction effect is very limited. Also, during the excitations, wire cable coming off from the pulley or breaking of wire was not observed (Fig. 10).

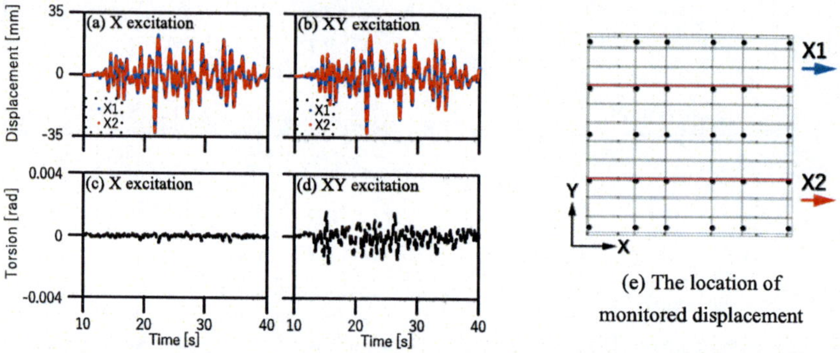

Fig. 9. Displacement time histories and torsion under 2 m/s/s acceleration.

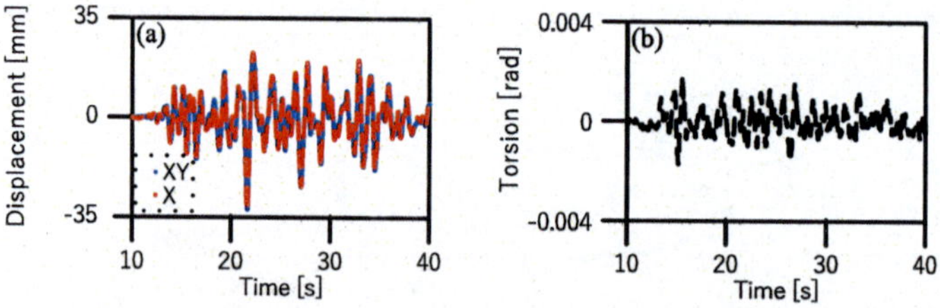

Fig. 10. Displacement record comparison under one and two-direction excitations.

5 Conclusions

This paper presented the experimental investigation of the new passive control system for the suspended ceiling system, namely, Pulley Damper Ceiling System (PDCS). To evaluate and understand the dynamic behavior of the proposed damping system, a series of unidirectional and bidirectional shake table tests using various types of full-scale suspended ceiling systems were carried out.

Six types of specimens: an unreinforced SCS (Case C), a reinforced SCS with steel braces (Case CB), and 4 damped SCSs by applying different PDCS configurations such as with and without the presence of viscous damper and wire stretching angle, were primarily tested to evaluate the fundamental response reduction effect of the PDCS. Unidirectional tests using simulated earthquake ground motions demonstrated that the proposed system is capable to reduce of both displacement and acceleration ceiling responses to less than 50% of Case C.

It is then, two-directions shake table tests on the SCSs with PDCS were performed to confirm the seismic response reduction efficiency and dynamic behavior of the PDCS under bidirectional earthquake motions. The experimental results showed that the proposed system is able to control and provide a successful response reduction effect even under bidirectional excitations without the torsion of the ceiling structure.

Acknowledgments. Authors gratefully thank the practice committee of the development of the ceiling vibration control system: M. Uchikawa (Sato Kogyo Co., Ltd.), E. Nishimura (Toda Corp.), Y. Yamasaki (Nishimatsu Construction Co., Ltd.), H. Ryujin (Maeda Co., Ltd.), and R. Doi (Kumagai Gumi Co., Ltd.).

References

1. Kasai, K., Fu, Y., Watanabe, A.: Passive control systems for seismic damage mitigation. J. Struct. Eng. **124**, 501–512 (1998)
2. Constantinou, B.M.C., Tsopelas, P., Hammel, W., et al.: Toggle-brace-damper seismic energy dissipation systems. J. Struct. Eng. **127**, 105–112 (2001)
3. Feng, H., Zhou, F., Ge, H., et al.: Energy dissipation enhancement through multi-toggle brace damper systems for mitigating dynamic responses of structures. Structures **33**, 2487–2499 (2021)
4. Mosquera, J.S.B., Almazán, J.L., Tapia, N.F.: Amplification system for concentrated and distributed energy dissipation devices. Earthq. Eng. Struct. Dyn. **45**, 935–956 (2016)
5. Zhang, Z., Bi, K., Hao, H., et al.: Development of a novel deformation-amplified shape memory alloy-friction damper for mitigating seismic responses of RC frame buildings. Eng. Struct. **216**, 110751 (2020)
6. Kang, J.-D., Tagawa, H.: Experimental evaluation of dynamic characteristics of seesaw energy dissipation system for vibration control of structures. Earthq. Eng. Struct. Dyn. **43**, 1889–1895 (2014)
7. Katsimpini, P.S., Papagiannopoulos, G.A., Askouni, P.K., et al.: Seismic response of low-rise 3-D steel structures equipped with the seesaw system. Soil Dyn. Earthq. Eng. **128**, 105877 (2020)
8. Saito, T., Maegawa, T., Denno, S., et al.: New seismic response control system using block and tackle. In: 16th World Conference on Earthquake Engineering (2017)
9. Jung, I.Y., Ryu, J., Oh, J., et al.: Experimental investigation on displacement amplification mechanism of steel wire rope-pulley damping systems with viscous damper. Eng. Struct. **248**, 113206 (2021)
10. Majima, R., Hayashi, K., Saito, T.: Development of new passive vibration control system in coupled structures with block and tackle. Soil Dyn. Earthq. Eng. **159**, 107319 (2022)
11. Motosaka, M., Mitsuji, K.: Building damage during the 2011 off the Pacific coast of Tohoku Earthquake. Soils Found. **52**, 929–944 (2012)
12. Nakaso, Y., Kawaguchi, K.: Investigation on Large Enclosures Damaged during the "2016 Kumamoto Earthquake" (Japanese). Seisan Kenkyu **68**, 421–426 (2016)
13. Reitherman, R., Sabol, T.: Nonstructural damage. Earthq. Spectra **11**, 453–514 (1995)
14. Jun, S.C., Lee, C.H., Bae, C.J., et al.: Shake-table seismic performance evaluation of direct- and indirect-hung suspended ceiling systems. J. Earthq. Eng. (2020)
15. Ministry of Land, Infrastructure: Transport and Tourism, Determining a Safe Structural Method for Specific Ceilings and Specific Ceiling Structural Strengths (Japanese), vol. 221 (2013)
16. Sasaki, T., Aoi, A., Tagawa, H., et al.: Design and structural response of full-scale steel gymnasium specimen (Japanese). J. Struct. Constr. Eng. **82**, 831–841 (2017)

17. Lee, J.-S., Jung, D.-I., Lee, D.-Y., et al.: Performance evaluation of rigid braced indirect suspended ceiling with steel panels. Appl. Sci. **11**, 1986 (2021)
18. Soroushian, S., Maragakis, E.M., Ryan, K.L., et al.: Seismic simulation of an integrated ceiling-partition wall-piping system at E-defense. II: evaluation of nonstructural damage and fragilities. J. Struct. Eng. **142**, 04015131–04015131 (2016)
19. Majima, R., Sakai, S., Saito, T.: Development of vibration control structure on suspended ceiling using pulley mechanism. Appl. Sci. **12**, 3069 (2022)
20. Mousavi, S.A., Zahrai, S.M.: Slack free connections to improve seismic behavior of tension-only braces: an experimental and analytical study. Eng. Struct. **136**, 54–67 (2017)
21. Pekcan, G., Mander, J.B., Chen, S.S.: Experiments on steel MRF building with supplemental tendon system. J. Struct. Eng. **126**, 437–444 (2000)

Advanced Digital Video Analyses to Estimate the Dynamic Behavior for Proper Design of a Base-Isolation System of the Sarcophagus of the Spouses at the National Etruscan Museum in Rome: Preliminary Results

Vincenzo Fioriti[1], Antonino Cataldo[1], Ivan Roselli[1], Alessandro Colucci[1], Paolo Clemente[1], Miriam Lamonaca[2], and Luigi Sorrentino[3](✉)

[1] ENEA, Casaccia Research Centre, Rome, Italy
[2] National Etruscan Museum of Villa Giulia in Rome, Rome, Italy
[3] Department of Structural and Geotechnical Engineering, Sapienza University of Rome, Rome, Italy
luigi.sorrentino@uniroma1.it

Abstract. A preliminary characterization of the input vibrations and of the dynamic behavior of the Sarcophagus of the Spouses at the National Etruscan Museum in Rome by advanced digital videos analyses is illustrated. This activity focuses on the development of a protection system for museum assets in order to reduce potential damage due to traffic and earthquakes. The vibrations induced by roads, tramways and an underground train line were measured by seismometers and analyzed to characterize the base excitation to the Sarcophagus. Given the high vulnerability of the Sarcophagus, no instrumentation could be placed on it, so that conventional contact sensors could not be used. Consequently, non-contact video-based techniques were considered and, in particular, high-speed and high-resolution video camera footage was recorded. The video was analyzed by advanced algorithms derived from the motion magnification method, which is a video signal elaboration method able to enormously amplify the tiniest movements of the acquired objects. A preliminary estimation of the main dynamic characteristics of the Sarcophagus was performed. The associated frequencies will be used to calibrate the numerical models and to optimize the design of the base-isolation system. The used digital video analysis technique is also discussed in terms of its many advantages and its application criticalities.

Keywords: Artifact · Dynamic identification · Motion magnification · Non-contact technique · Traffic-induced vibrations

1 Introduction

Whereas earthquakes have always happened in certain region of the Earth, the build-up of motor and rail traffic in many countries and especially in large cities has introduced

G. P. Cimellaro (Ed.): WCSI 2022, LNCE 309, pp. 707–716, 2023.
https://doi.org/10.1007/978-3-031-21187-4_60

a new source of vibrations for infrastructures, buildings and their content. At least the vibrations generated by rail traffic, being it long-distance railways, subways or lighter tramways, are bound to increase as population in cities increases and rapid mass transportation becomes the only viable solution for high-density areas. According to the "United Nations Global Sustainable Transport Conference" [1], rapid mass transportation is still ineffective, unsafe, not sustainable and inaccessible to a large share of the urban population worldwide, and should be encouraged as envisioned for instance by the "Strategies for a new mobility policy in Italy" [2]. Railway traffic-induced vibrations are studied since the late 1970s [3], and successive detailed studies were carried out on different types, such as columns [4, 5], archaelogical sites [6, 7] and other heritage constructions [8, 9]. Technical standards set vibration measurement procedures and peak velocity thresholds to avoid damage to buildings [10–13]. However, these limits might be unconservative for architectural and cultural heritage, especially for those artifacts already damaged during their substantially long life [14].

An example of such an artifact is the Sarcophagus of the Spouses at the National Etruscan Museum of Villa Giulia in Rome, a masterpiece of the 6th century B.C. found in Cerveteri, about 35 km north-west of Rome. The statue is divided into four parts, is hollow and is made of ceramic, was shattered in about 400 fragments when recovered and suffers from the continuous vibrations induced by an underground railway and a surface tramway. The sarcophagus will be protected from these vibrations as an application of the research and development project MONALISA (MONitoraggio Attivo e Isolamento da vibrazioni e Sismi di oggetti d'Arte = Active monitoring and isolation from vibrations and earthquakes of art objects), funded by Lazio Regional Government and Italian Ministry of University and Research.

In order to properly design the solution for the isolation of the statue from traffic-induced vibrations, it is necessary to adequately evaluate the input induced by both railway and tramway, as well as the dynamics of the sarcophagus. To pursue the first goal a set of velocimeters was deployed in the museum and recorded the floor motion. The second goal is more challenging because conservation concerns ruled out the application of any sensor on the statue surface. Therefore, a non-contact technique was adopted, relying on the magnification of the motion recorded by a high-velocity, high-resolution digital camera. The preliminary interpretations of these experimental campaigns are presented hereinafter.

2 Input Vibrations Generated by Traffic

2.1 Procedure Description

The vibration data were recorded on April 26, 2022 according to the ISO 4866 standard [7], which collects measurements procedures (type of instrumentation, location, data processing and analysis, etc.) accepted by international and national regulations.

Vibration data were acquired by digital recorders equipped with triaxial velocimeters provided with a GPS antenna for time synchronization [15]. Four sensors were located at the four corners of the sarcophagus display case (Fig. 1a). Two-hour-long time-histories were recorded at each position at a sampling frequency of 200 Hz. During vibration acquisition, the time and type of vehicle passage was also noted. The underground railway line is located under the museum, while the surface tramway is located in Viale delle Belle Arti, adjacent to the museum (Fig. 1b).

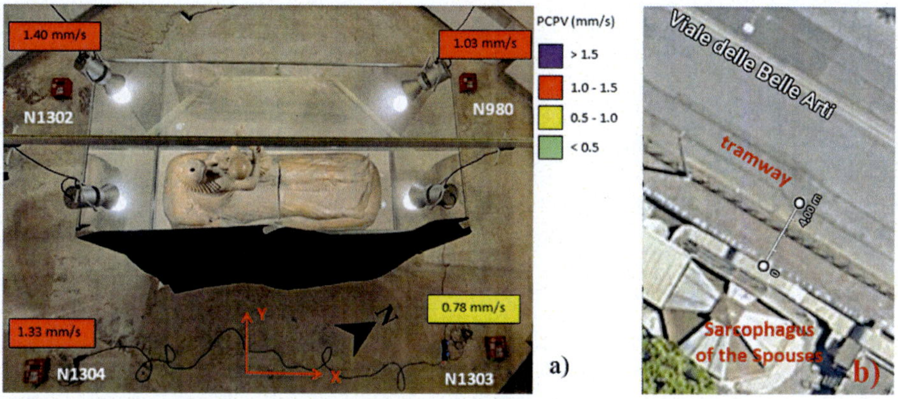

Fig. 1. (a) Room of the Sarcophagus of the Spouses: Peak Component Particle Velocity by the four recorded seismographs and (b) evaluation of distance from nearest tramway.

2.2 Railway and Tramway Vibrations

During an acquisition period of two hours, twenty-seven passages were recorded, among which twelve tramway passages and fifteen train passages. Considering that the acquisition was carried out since 10:20 a.m. to 12:20 a.m., it can be hypothesized that the frequency of passages might be even higher during rush hour (around 8:00 a.m.). This high frequency of passages may play a role in the potential damaging effect of traffic-induced vibration. In fact, in order to assess how the intensity of traffic vibration affects the fatigue resistance of exposed object, the international standards generally consider the following parameters [11, 12]:

- the duration of vibration intensity, classified as short-term (occasionally or transient events) or long-term (permanent or continuous events) vibrations;
- the peak velocity, calculated as Peak Particle Velocity (PPV) or Peak Component Particle Velocity (PCPV).

The analysis of the recorded time histories (Fig. 2a) revealed that seven events of remarkable intensity occurred in just twelve minutes. The peak velocity distribution, evaluated as PCPV (Fig. 2b), shows two classes of vibration intensity: at PCPV lower than 0.4 mm/s and in the range 0.6–1.6 mm/s. The first class is related to the railway, whereas the second class is induced by the tramway.

The highest values recorded near the Sarcophagus are reported in Fig. 1a. Although the obtained PCPV values are lower than typical limits prescribed by international standards for archaeological structures (1.5 mm/s for the Swiss standard [12] and 2.5 mm/s for the German one [11]), they still highlights a substantial excitation capable of eliciting a dynamic response of the sarcophagus. This response will be studied in the following section by means of a motion magnification analysis.

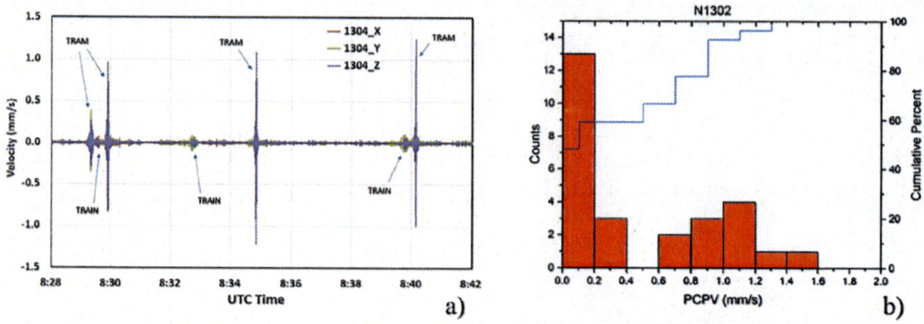

Fig. 2. (a) Recorded velocity time-history at location N1304 with tram and train passages indicated; (b) PCPV distribution and cumulative analysis at N1302.

3 Magnification of the Motion of the Sarcophagus

3.1 Procedure Description

The video-based vibration monitoring techniques offer a unique property: each pixel of a frame is a "virtual" sensor, meaning that each pixel can provide information and, therefore, a very high sensor density on the object surface, also allowing the selection of the points to be analyzed at any time after the video recording. In addition, all these analyses can be perform without any contact with the investigated object surface. The motion magnification (MM) algorithm falls into this category of methodologies [16]. It looks at tiny movements embedded in a conventional digital video amplifying the pixel time series, while preserving the general topology of the frames. There have been many attempts to solve this problem, and finally a group of MIT researches succeeded [17, 18], providing a viable MM algorithm at a reasonable computational cost. The algorithm acts like a microscope for motion in video sequences, affecting only some groups of pixels and unveiling motions hardly visible with the naked eye. The motion magnification uses the spatial resolution of the video-camera to extract physical properties from images to make inferences about the dynamic behavior of the object, and thus is a contactless method offering simultaneous, quasi-continuous, measurements. If we compare the constraints of traditional accelerometers and video techniques [19], the first are affected by contact problems, sparse discrete points, sensitivity to temperature, aggressive chemical products, measurement of absolute values. Most of all, practically all standard sensors, are contact sensor, including laser based ones, since the beam hits the surface of the object. Clearly, when the object is a valuable archaeological relic, every alteration must be avoided. On the other hand, in-door videos are affected by moving objects, sensitivity to alternate current lighting, reflections, smoke, Intrinsic limits of camera sensitivity, 2D only monitoring (in case of a single camera), quantitative errors in calculations, camera displacements, and by the presence of large shifting of the object. Sadly, all of these issues are present in the MM analysis of the sarcophagus.

Moreover, the statue is enclosed in a glass case whose light reflections and shadows affect heavily the MM elaborations. In addition, this is probably the first time that MM methodology was applied to a sculpture, therefore no standard reference exists. Nevertheless, given the extreme vulnerability of the statue, non-contact video-based techniques seem to be an appropriate investigation approach. From the magnified videos a preliminary estimate of the main vibration frequencies of the sarcophagus could be proposed. These frequencies and their association to various parts of the statue will be used to calibrate the numerical models and to optimize the design of the base-isolation system.

The first step of the procedure is the video recording. The camera resolution and the frame rate are the main parameters for the MM elaboration. The higher they are, the better, but as usual this conflicts with economic costs and calculation time. We used a commercial tablet at 28 frame per second (fps) and a xiQ USB3 superspeed camera Ximea, endowed with a 9–90 mm objective F1.8 Kowa Optronics, at 130 fps and a resolution of 632 × 508 pixels. Videos from the tablet were used for the low frequency analysis in the range 0.5–14 Hz, while the xiQ camera was used in the range 0.5–60 Hz. During the first two campaigns of January 31, 2022 (since 9:00 a.m. to 1:00 p.m. local time, 9 °C, 87% humidity) and April 26, 2022 (since 9:00 a.m. to 1:30 p.m. local time, 18 °C, 62% humidity) several videos from different point of views and different lighting conditions were shot. This redundancy was of the outmost importance, because it directly affects the pixel intensity value.

The videos were at least 20 s long, allowing a minimum frequency resolution of 0.05 Hz. Many light sources are present in museum, appropriately placed to enhance the artistic details of the statue. However, the magnification procedure favors a uniform, constant, direct illumination of the scene, otherwise noise is introduced and amplified, along with the useful information [13, 20, 21].

During the second campaign this problem was reduced by turning off some lights and setting a black cloth at the back of the statue (relative to the camera) to reduce reflections. Moreover, the number and the direction of light sources must be taken into account; ideally, one wants only a powerful source perpendicular to the object, in order to avoid shadows. Many non-perpendicular sources produce spurious peaks on the frequency spectrum (Fig. 3). Another source of illumination disturbance is caused by the 50 Hz pulsation of the electric power grid, that reverberates on the video recording, even if imperceptibly to the naked eye. Lighting problems are reported for battery powered sources that are not reliable as the color temperature and intensity of the light fluctuate [22].

The magnification algorithm requires the specification of some parameters. Among them the amplification and the magnification frequency range. The amplification is bounded above by the noise; if too large, the noise will exceed the actual signal.

The frequency range of the MM analysis can be limited. From one hand, the magnification is more precise when the range is narrow. On the other hand, when the frequencies of interest are unknown *ex-ante*, a wide frequency range must be taken into consideration.

Once the video was magnified, a region of interest (ROI) was selected and a MM signal was extracted averaging the pixel values in the ROI [23]. Then, the MM signal extracted from the ROI can be analyzed by usual processing tools for modal or statistical analysis. It is important to note that disturbances of any kind reverberate over the signal and, as a consequence, on the subsequent calculations. Hence, determining the peaks relates to disturbances as done in Fig. 4 reduces their relevance.

Fig. 3. High frequency camera experimental set-up. Lighting sources and related glass reflections can be observed.

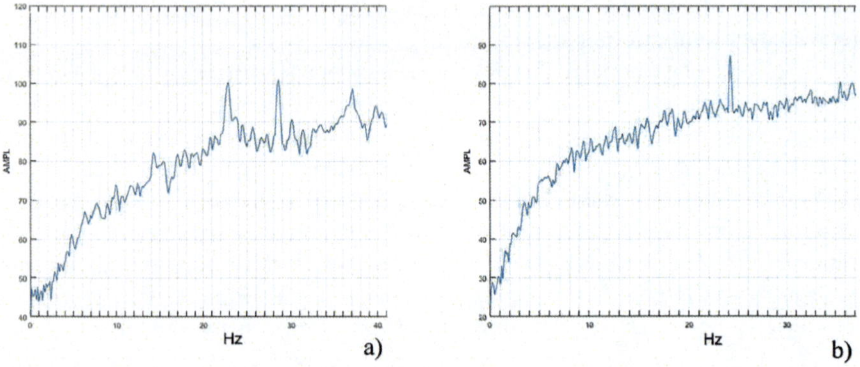

Fig. 4. Effect of illumination conditions on the frequency response of the bust of the statue: when the corridor lighting is on (a) and when it is off (b).

3.2 Vibrations of Parts of the Sarcophagus

Here and in the following, we show some very preliminary results of the analysis in the frequency domain. Because many issues such as the role of noise, disturbances and of the technical limitations of the cameras have to be established more clearly, these analyses should be considered as not conclusive.

The extremities of the artifact, hands and feet, are of particular interest, because sensitive to vibrations (especially the feet), according to informal reports. Frequencies between 0 and 13.3 Hz are considered reasonably precise, outside this range the Nyquist theorem does not guarantee enough precision for a sampling rate of 28 Hz. In fact, using a 28 fps video-camera, the maximum frequency allowed would be 14 Hz, frequencies above this threshold will introduce spurious peaks because of the aliasing. However, given the heavy presence of noise, as a rule of thumb it would be better at least a 5% larger sampling frequency. Finally, the duration T of the video is inversely correlated to the frequency resolution in the frequency domain: the larger T, the better the resolution Δf. Unfortunately, a large T means a much longer elaboration time and, most of all, T should be restricted to the duration of the perturbation caused by the tram/train passage, that are known to last 20–30 s [13].

Three main groups of frequency ranges seemed to give a more relevant response: 8–9 Hz, 10–11 Hz, 12–13 Hz. With respect to the whole picture, the feet ROI of the woman and of the man show a prevalence of the frequencies about 12.2 Hz and 12.9 Hz, confirming that the feet have a peculiar response to excitations. The shape of the ROI indicates that the motion is according to the z axis. The same analysis was repeated several times to identify peaks coming from spurious sources, such as people talking or passing by, vibrations of the camera basement, accidental bumps in the glass case. The woman's head ROI (Fig. 5) is characterized by a strong response along the z axis, and a rich frequency content, but it is not completely clear, given the current state of investigations, if some peaks are due to reflections or to spurious disturbances.

3.3 Vibrations Related to the Wooden-Covered Support

The Sarcophagus is placed on a wooden-covered support with four short steel legs atop a low metal pedestal which lies directly on the ground floor of the museum, which is at the same level of the adjacent street with the tramway (Fig. 6). Through the metal base pedestal, vibrations are transmitted to the sarcophagus by this steel support, thus it is very important to investigate their response to excitations. Analyses were carried out from the same video to ensure the same conditions, moreover the area of the ROI is small, therefore the lighting cannot vary too much over the surface.

Figure 6 shows clearly the presence of noise especially in the ROI a and c, along the z axis. This disturbance is likely due to the tramway, but the underground railway may well have an impact. Hence, further investigations is needed to distinguish the effect of tramway from railway.

Moreover, it seems that noise decreases from ROIa to ROIc. From these considerations the usage of noise reduction techniques could be appropriate, even if filtering will yield less precise results.

Also, the different frequency content between axes z and x is evident: along the z axis the dynamic response is much more complicated and noisy, as well as larger in intensity. Because some perturbation directions are more dangerous for the statue than others, this finding is significant to design a protection device.

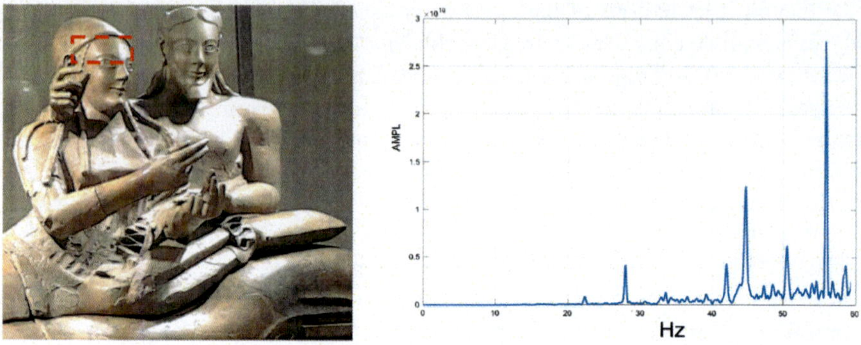

Fig. 5. High frequency estimation of the woman's head ROI (red dotted area). The subplot shows the response in the vertical direction: some peaks are due to disturbances and will be identified by comparison with other MM tests (data sampling frequency is 120 Hz).

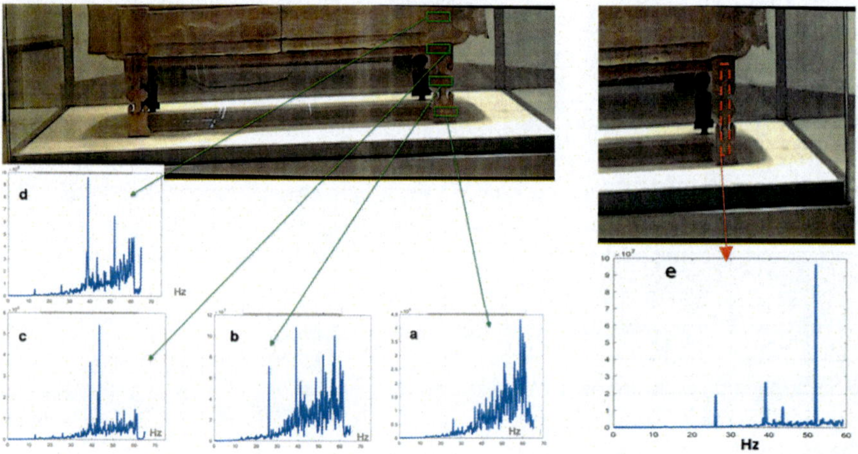

Fig. 6. High frequency estimation of right wooden-covered support along the z axis (a–d) and along the x axis (e): the amplitude (arbitrary units) of subplot b is one order of magnitude smaller than a, c, d (data sampling frequency is 120 Hz).

4 Conclusions

In the present paper the preliminary results of an experimental in-situ study carried out on the Sarcophagus of the Spouses at the National Etruscan Museum in Rome was illustrated. In fact, a proper estimate of the dynamic behavior of the statue will be essential for an effective design of its base-isolation system. The results of the analyses, performed by means of the magnified motion methodology and resorting to a non-contact technique, revealed a preliminary frequency of vibration of the feet of the figures at about 12.2–12.9 Hz. Such indications are intended to be a first step for investigating other portions of the sarcophagus along multiple axes.

Unfortunately, the site is characterized by high levels of environmental noise and disturbances that affect the video recordings. In the future careful filter processing will be considered to test the robustness of identified frequencies.

Acknowledgements. This work was partially carried out under the MONALISA (MONitoraggio Attivo e Isolamento da vibrazioni e Sismi di oggetti d'Arte = Active monitoring and isolation from vibrations and earthquakes of art objects) research project funded by Lazio Regional Government and Italian Ministry of University and Research, project id. 305-2020-35576. The opinions expressed in this publication are those of the authors and are not necessarily endorsed by the funding bodies.

References

1. UN-Secretary-General: United Nations Global Sustainable Transport Conference. Ash-gabat (2016)
2. MEF: Strategie per una nuova politica della mobilità in Italia. Ministero Eco-nomia e Finanza, Roma (2019)
3. Verhas, H.P.: Prediction of the propagation of train-induced ground vibration. J. Sound Vib. **66**(3), 371–376 (1979)
4. Bongiovanni, G., Buffarini, G., Clemente, P., Saitta, F.: Time and frequency domain analyses in the experimental dynamic behaviour of the Marcus Aurelius column. Int. J. Archit. Herit. **15**(1), 64–78 (2021). https://doi.org/10.1080/15583058.2019.1706785
5. Bongiovanni, G., Buffarini, G., Clemente, P., Rinaldis, D., Saitta, F.: Experimental vibration analyses of a historic tower structure. J. Civ. Struct. Health Monit. **7**(5), 601–613 (2017). https://doi.org/10.1007/s13349-017-0245-4
6. Puzzilli, L., Bongiovanni, G., Clemente, P., Di Fiore, V., Verrubbi, V.: Effects of anthropic and ambient vibrations on archaeological sites: the case of the circus maximus in Rome. Geosciences **11**, 463 (2021). https://doi.org/10.3390/geosciences11110463
7. Bongiovanni, G., Buffarini, G., Clemente, P., Rinaldis, D., Saitta, F.: Dynamic characteristics of the Amphitheatrum Flavium northern wall from traffic-induced vibrations. Ann. Geophys. **60**(4), S0439 (2017). https://doi.org/10.4401/ag-7178
8. Clemente, P.: Extending the life-span of cultural heritage structures. J. Civ. Struct. Health Monit. **8**(2), 171–179 (2018). https://doi.org/10.1007/s13349-018-0278-3
9. De Stefano, A., Matta, E., Clemente, P.: Structural health monitoring of historical heritage in Italy: some relevant experiences. J. Civ. Struct. Health Monit. **6**(1), 83–106 (2016). https://doi.org/10.1007/s13349-016-0184-y
10. UNI 9916: Criteria for the Measurement of Vibrations and the Assessment of Their Effects on Buildings; Italian code; UNI, Milan, Italy (2004)
11. DIN 4150-3: Erschütterungen im Bauwesen - Einwirkungen auf bauliche Anlagen. Deutsche Institut für Normung, Berlin (1999)
12. SN 640312a: Vibrations—Vibration Effects in Buildings. Swiss Standards, Switzerland (2013)
13. ISO 4866: Mechanical vibration and shock vibration of buildings. Guidelines for the measurement of vibrations and evaluation of their effects on buildings (2010)
14. Clemente, P., Rinaldis, D.: Protection of a monumental building against traffic-induced vibrations. Soil Dyn. Earthq. Eng. **17**(5), 289–296 (1998)
15. Roselli, I., et al.: Urban transport vibrations and cultural heritage sites in Rome: the cases of the temple of Minerva Medica and of the Catacomb of Priscilla. WIT Trans. Ecol. Environ. **223**, 335–343 (2017)

16. Zona, A.: Vision-based vibration monitoring of structures and infrastructures: an overview of recent applications. Infrastructures **6**(1), 4 (2021)
17. Yu-Wu, H. et al.: Eulerian Video Magnification for Revealing Subtle Changes in the World (2016). https://people.csail.mit.edu/mrub/papers/vidmag.pdf
18. Wadhwa, N., Wu, H.Y., Davis, A., Rubinstein, M., Shih, E., Mysore, G.J., Chen, J.G., Buyukozturk, O., Guttag, J.V., Freeman, W.T., Durand, F.: Eulerian video magnification and analysis. Commun. ACM **60**(1), 87–95 (2017)
19. Smieja, M., Mamala, J., Praznowski, K., Ciepliński, T., Szumilas, L.: Motion magnification of vibration image in estimation of technical object condition—review. Sensors **21**, 6572 (2021). https://doi.org/10.3390/s21196572
20. Forliti, S., Fioriti, V., Roselli, I., Tatì, A., Colucci A.: Motion magnification analysis for monitoring cultural heritage buildings and archeological sites. In: IMEKO TC-4 International Conference on Metrology for Archaeology and Cultural Heritage, Trento (2020)
21. Fioriti, V., Roselli, I., Tatì, A., Romano, R., De Canio, G.: Motion magnification analysis for structural monitoring of ancient constructions. Measurement **129**, 375–380 (2018)
22. Pavelka, K., Khan, A.: University of Texas (2016). https://www.nps.gov/articles/000/using-eularian-video-magnification-evm-to-study-the-effect-of-fluctuations-in-relative-humidity-on-cultural-heritage-materials.htm
23. Fioriti, V.A., Roselli, I., De Canio, G.: Modal identification from motion magnification of ancient monuments supported by blind source separation algorithms. In: Papadrakakis, F. (eds.) COMPDYN '19, 7th ECCOMAS Thematic Conference on Computational Methods in Structural Dynamics and Earthquake Engineering Crete, Greece, 24–26 June 2019 (2019)

Seismic Protection of the Goddess of Morgantina Statue Through an Innovative Base-Isolation Device: Validation by Shake-Table Tests

Elena Alberti[1], Giovanni Ausenda[2], Maria Chiara Castino[3], Francesco Lo Iacono[1], and Giacomo Navarra[1(✉)]

[1] Facoltà di Ingegneria ed Architettura, Università degli Studi di Enna Kore, Enna, Italy
giacomo.navarra@unikore.it
[2] Kyneprox S.r.l., via dell'Artigianato, Vigano di Caggiano (MI), Italy
[3] Somma S.r.l., via Carlo Mirabello 12a, Rome, Italy

Abstract. Monuments and historic structures have been extensively damaged by major seismic events that occurred in the Mediterranean area in the last decades. A large research activity has been devoted to the mitigation of seismic effects in historic buildings, but also the museums contents deserves the attention of scholars due to their inestimable values and cultural significance. In particular, statue-like objects revealed a poor dynamic behaviour during past earthquakes and are prone to overturn if seismic excitations overcame an acceleration threshold. In scientific literature a lot of analytical research are present in non-linear dynamic field but experimental works in this area still remain limited. In this research work, the results of an extensive test campaign, aiming at characterizing the dynamical behaviour and assessing the seismic protection level of an innovative bidirectional base-isolation device is presented and commented. The isolator consists in a series of multiple articulated quadrilateral mechanism elements, patented as KSJ (Kinematic Steel Joint), that allows the free movements of a rigid base in both horizontal directions, providing pendulum-like behaviour and re-centring capabilities. Shake-table tests on a full-scale replica of the statue of the Goddess of Morgantina have been carried out at the L.E.D.A. Research Institute at the Kore University of Enna, with varying load conditions, acceleration time-histories and intensities.

Keywords: Shake-table testing · Seismic protection · Experimental dynamics

G. P. Cimellaro (Ed.): WCSI 2022, LNCE 309, pp. 717–725, 2023.
https://doi.org/10.1007/978-3-031-21187-4_61

1 Introduction

Recently, some major seismic events that struck Mediterranean area, caused extensive damages to monuments and historic structures, especially in Turkey, Greece and Italy. Also museums contents, artefacts and statues have been affected by earthquakes [1] and their seismic safeguard deserves the attention of scholars due to the high cultural significance and the inestimable value of most of them. Unfortunately, in seismic codes museum contents are considered as non-structural elements [2] and no specific provisions are indicated.

Starting from 1984, seismic protection of art objects has been systematically studied at Jean Paul Getty Museum in Malibu (California, USA) by investigations and practical approaches [3]. Museum objects, statues and displays are generally modelled as rigid bodies and analysed with the non-linear dynamics tools. Since the pioneering work by Housner [4], this subject attracted the attention of several researcher through, in most of cases, analytical studies [5–8], whereas the experimental works in this area remain limited. In particular, amphoras or statue-like items objects, due to the height of their centres of gravity respect to the width of their bases, during earthquakes generally experiences strong non-linear dynamic effects that may lead to their overturning. In these cases base isolation devices revealed to be very effective in seismic protection.

The features of an isolation system are: (i) low lateral stiffness to decouple the motion of the isolated item from the ground; (ii) significant damping level to dissipate seismic energy and to limit the displacement demand; and (iii) self-centre capability to return at the rest position as the dynamic load ends [9].

This paper presents a research conducted in the framework of the Research Project *eWAS - An Early-Warning System for Cultural Heritage*, funded by the MUR (Italian Ministry of University and Research), whose final goal is the implementation of a monitoring system integrated with smart-wireless and ICT technologies to evaluate the natural hazards undergone by the monumental and historical heritage of some areas of the Sicily. An extensive study to characterise the dynamic performance of an innovative bidirectional base-isolation device is presented. The isolator consists in a series of multiple articulated quadrilateral mechanism elements, patented as KSJ (Kinematic Steel Joint), that allows the free movements of a rigid base in both horizontal directions, providing pendulum-like behaviour, decoupling from the ground motion and re-centering capabilities.

The Goddess of Morgantina statue (depicted in Fig. 1-a) has been selected as case-study, due to its significance and its peculiar story. In fact, this 2.24 m tall statue, was carved in the 5th century BC in Sicily by a direct disciple of Phidias, operating in Magna Graecia. It was illegally excavated from the archaeological site of Morgantina (Sicily) in late 1970s s and dissected into three parts to be illegally exported. The statue was sold in Switzerland in 1986 and then acquired and exposed by the Paul Getty Museum in Malibu for several years. After a long legal dispute between Italy and the United States governments, the statue was finally returned to Italy on 2011. When exposed in California, the statue was equipped with a base-isolating device, designed at Paul Getty Museum, which has been also maintained in its present placement.

A full scale replica of the statue has been constructed (see Fig. 1-b) and shaking table tests have been carried out at L.E.D.A. (*Laboratory of Earthquake engineering and Dynamic Analysis*) Research Institute of the "Kore" University of Enna, Italy. It has been proven that the allowable stroke of the device and the damping characteristics are adequate to strongly reduce the acceleration suffered by the statue.

The paper is organized as follows: in Sect. 2 the base isolation device is described and its dynamical characterization is presented, whereas in Sect. 3 the results of seismic shaking table tests are illustrated and, finally, some conclusions are drawn.

(a) (b)

Fig. 1. a) The statue of the Goddess of Morgantina, in its present placement; b) the full scale replica with the isolation device on the shaking tables at L.E.D.A.

2 Characterization of the Isolation Device

Seismic isolation devices have lately received an increasing interest due to their ability to reduce the accelerations transmitted to the isolated objects masses and, then, their potential damage. Roughly, two categories of isolators currently available on the market can be defined: elastomeric and friction pendulum bearings. The cost for these device and their maintenance are issues that have limited the general use of the base-isolation technology. In order to overcome these difficulties and to extend the applicability of the base-isolation also to the protection of non-structural elements or cultural heritage items, Kyneprox S.r.l. has designed and patented a new isolator device which exploits the beneficial effects of a double articulated quadrilateral with crossing rods, called *Kinematic Steel Joint* (KSJ). Since KSJ elements are made of cut and folded steel sheets, this device presents production and maintenance costs lower than other similar base isolation device [10].

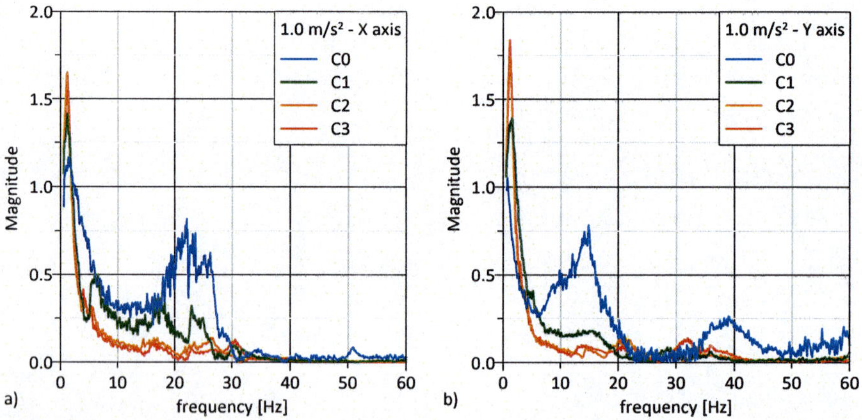

Fig. 2. FRFs during random noise test at 1.0 m/s^2 RMS: a) X direction; b) Y direction.

KSJ elements can be assembled in different configurations depending on the shape and the size of the protected item. In the present case, KSJ elements are deployed to realize the device illustrated in Fig. 1-b, consisting of a rigid base fixed on the top of two double-rail stacked systems of KSJ elements to allow horizontal bidirectional relative displacements with respect to the ground. The prototype has been designed to support and protect a mass of 900 kg, similar to that of the statue of the Goddess of Morgantina.

The behaviour of the device is similar to a pendulum-like isolator, since it produces vertical upward displacements when it is subjected to horizontal movements from the central rest position. In this way, the object to be protected is forced to follow a trajectory that lies on of a double-curved concave surface. The device can resist the overturning moment since sliding bearings prevent uplift but cannot isolate the vertical component of seismic excitation, as occurs in most isolation systems [9].

The restoring forces and the self-centring capabilities largely depend on the slope of the trajectories, which have been identified in a previous work [11]. It has been shown that the device allows for horizontal displacements of ±0.150 m in each direction, giving rise to upward vertical displacements up to about 15 mm. Moreover, the trajectories are contained in a surface which can be approximated by an elliptical paraboloid. The equivalent oscillation periods are thus variable along the stroke and their values have been estimated to be close to 2.5 s, typical values of base-isolation devices. Furthermore, no significant variations have been observed in both direction and in the range of interest of co-ordinates.

Several dynamic shaking table broadband random noise tests have been conducted to measure the Frequency Response Function (FRF) of the device. More information about the research facility and the test sequence may be found in [11,12]. Test were performed for each horizontal axis and for four different load configurations, realized by fixing heavy steel plates to the upper rigid base of the device. The considered load level are C0, with no added masses, C1 and C2 levels with 320 kg and 760 kg added masses, respectively, and the highest load level C3 in which the same mass of the full-scale replica of the Goddess of Morgantina statue (about 900 kg) is applied. For each tests, FRFs of signals from an accelerometer on the shaking table (R1, input) and one on the rigid base (A1, output) have been computed. For brevity's sake Fig. 2 reports only the FRFs amplitude plots for each load level, for each direction and for the input intensities of 1.0 m/s^2 RMS.

Results show the ability of the device to strongly reduce the horizontal accelerations regardless of the mass load. Specifically, the results obtained for load level C0 have a marginal importance with respect to the normal operating configurations of the device. As the masses increase, the dynamic behaviour of the device does not depend on the load conditions. For frequencies higher 5 Hz a significant reduction in the transmission of vibrations is observed, whereas a narrow magnification frequency range between 1.0 Hz and 1.15 Hz is present.

3 Assessment of the Seismic Protection Level

To assess the seismic protection level of the device, the following four natural historical ground motion acceleration time-histories have been selected: (1) Kobe (16/01/1995, $M_w = 6.9$, recorded at Takatori, Japan); (2) Irpinia (23/11/1980, $M_w = 6.9$, rec. at Sturno, Italy); (3) Norcia (30/10/2016, $M_w = 6.5$, rec. at Norcia, Italy); and (4) L'Aquila (06/04/2009, $M_w = 6.1$, rec. at L'Aquila V. Aterno, Italy). These signals have been selected to have different characteristics in terms of frequency content, impulsive behaviour, duration and intensity. It worths to say that only horizontal components have been considered.

In the following, the results of two sets of shaking table tests are presented. During the first set a parametric analysis on the device loaded with rigid steel plates, accordingly to the load levels already defined, has been conducted, whereas the second set includes the assessment of the seismic protection level on the full-scale replica of the Goddess of Morgantina statue.

The experimental results of the first set of seismic tests are reported in terms of peak acceleration, RS plots, stroke values and residual displacements. Figure 3 compares the peak accelerations recorded at shaking table and on the device for all seismic tests. It seems that the isolator is able to clip the accelerations transmitted by the device to a certain level, about 1 m/s^2, which does not depend on the load level or on the imposed seismic wave. This effect is probably related to the overcoming of a friction threshold that more in-depth studies are investigating.

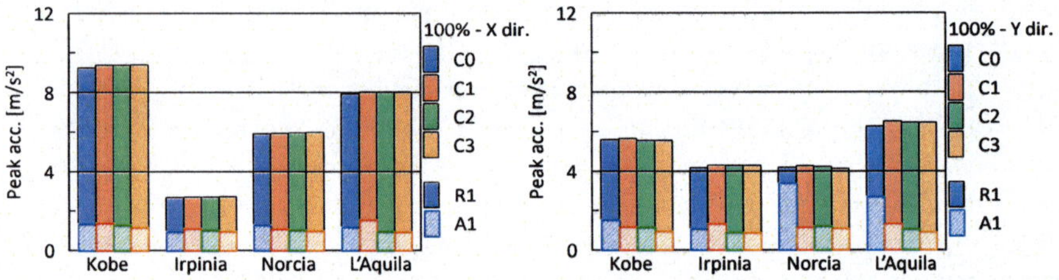

Fig. 3. Comparison in terms of peak accelerations.

Figure 4 reports the comparison between the RS of the input and the RS of the accelerations recorded on the rigid base for Norcia and L'Aquila signals, for all load levels previously defined and for both horizontal directions. RS have been computed for 5% damping level. Similar results have been obtained also for Kobe and Irpinia signals [11]. The RS curves prove that the isolation device effectively reduce the accelerations transmitted to the protected item and that the seismic protection degree is independent of the load mass.
During seismic tests, also residual displacements of the device and the peak rail strokes have been annotated. Maximum residual displacements are about 10 mm and 20 mm in X and Y direction, respectively, thus proving the self re-centring ability of the isolated base as the ground motion ends. The design adequateness of the device is also confirmed by the fact that, for all the tests conducted, the strokes of the device rails were lower than the allowable ones.

Basing on an accurate 3D laser-scanner survey and 3D-printed form-work, a full-scale replica of the Goddess of Morgantina statue has been constructed. A fibre-reinforced compensated shrinkage cementitious mortar has been selected for its highly fluidity and for its unit mass similar to that of the real statue, so that the mass and inertial properties of the actual statue and of the replica are very close each other. The second set of seismic tests has been performed with the statue mounted on the isolating device. The same four natural historical ground motion acceleration time-histories have been applied with intensities ranging from 20% to 120% of the actual one.

Figure 5 illustrates the trends of peak accelerations at shaking table and on the rigid base with the increasing input intensity. Also the acceleration ratio, chosen as a measure of the protection level, is plotted for both directions and with reference to Kobe and L'Aquila signals. Similar results have been obtained for the other seismic time histories but have not reported here for brevity's sake.

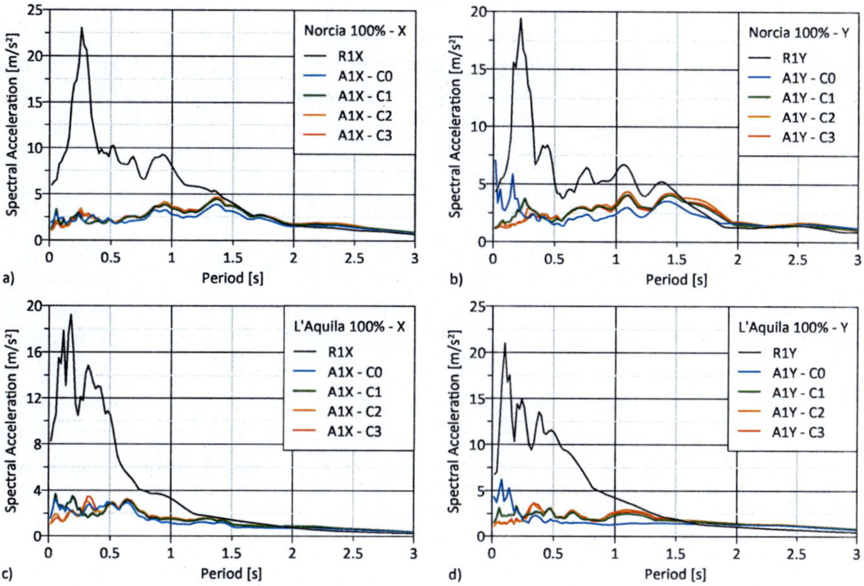

Fig. 4. RS during seismic tests: a) and b) Norcia; c) and d) L'Aquila signals.

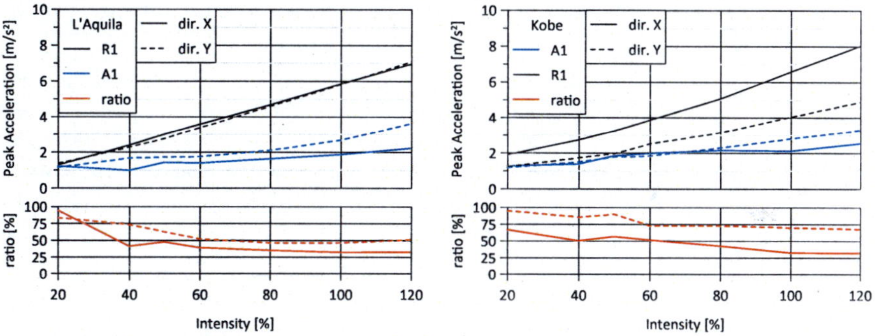

Fig. 5. Peak accelerations and protection level during L'Aquila and Kobe seismic tests.

Results shown that the device is able to strongly reduce the acceleration transmitted to the statue and that it works better in X direction rather than in Y direction. Moreover, for lower intensities, a worse seismic protection is obtained, probably since the sliding starts when a threshold, due to friction forces, is overcame. For higher intensities, as the device operates by clipping the accelerations transmitted to the statue, the acceleration ratio decreases up to levels of about 30% and 50% in X and Y directions, respectively.

This difference of behaviour is due to the presence of the statue by means of two concurring effects. In fact, the X rails, being located under the Y rails, are less affected by the overturning moments deriving from a payload with a high centre of gravity. Moreover, the connection between the statue and the rigid base is made by two threaded bars located inside the statue and placed side by side, so that the stronger axis of the connection coincides to the X direction.

4 Conclusions

Shaking table tests have been carried out at L.E.D.A. Research Institute of Kore University of Enna on an innovative bi-directional isolating device based on the patented *Kinematic Steel Joint* system in order to characterize its dynamic behaviour and to assess the ability of the device in reducing the seismic acceleration transmitted to statue-like art objects.

The device kinematic characteristics have been identified by analysing the displacement trajectories of the device rigid base and equivalent vibration periods have been computed. Shaking table random noise tests allowed for the dynamic characterization of the isolation in four increasing load configurations. The device proved to significantly attenuate the horizontal components of acceleration regardless the variation of mass loads.

Then, the horizontal components of four historical earthquakes have been applied to show that the device drastically reduces the accelerations transmission to the protected item for all loading configurations, both in terms of Response Spectra and in peak accelerations. Moreover, the re-centring property of the device is assessed by measuring residual displacements.

Finally, to achieve realistic results, a full-scale replica of the Goddess of Morgantina statue, chosen as case-study, has been constructed, placed on the isolation device and subjected to seismic tests with increasing intensities. The results show that, even if the presence of the statue affects the performance of the device, a very good protection level is achieved. Further studies are regarding the development of a theoretical model of the isolation device in order to have a deep insight on the experimental results and for design purposes.

Acknowledgements. This work has been carried out in the framework of the research project "eWAS An Early WArning System for Cultural Heritage" (ARS01_00926), funded by Italian Ministry for University and Research, whose financial support is gratefully acknowledged.

References

1. Parisi, F., Augenti, N.: Earthquake damages to cultural heritage constructions and simplified assessment of artworks. Eng. Fail. Anal. **34**, 735–760 (2013)
2. Spyrakos, C.C., Maniatakis, C.A., Taflampas, I.M.: Application of predictive models to assess failure of museum artifacts under seismic loads. J. Cult. Herit. **23**, 11–21 (2017)
3. Agbabian, M.S., Ginell, W.S., Masri, S.F., Nigbor, R.L.: Evaluation of earthquake damage mitigation methods for museum objects. Stud. Conserv. **36**(2), 111–120 (1991)

4. Housner, G.W.: The behaviour of inverted pendulum structures during earthquakes. Bull. Seismol. Soc. Am. **53**, 403–417 (1963)
5. Shenton, H.W., Jones, N.P.: Base excitation of rigid bodies. I: Formulation. J. Eng. Mech. **117**(10), 2286–2306 (1991)
6. Spanos, P.D., Koh, A.S.: Rocking of rigid blocks due to harmonic shaking. J. Eng. Mech. **110**(11), 1627–1642 (1984)
7. Palmeri, A., Makris, N.: Response of rigid structures rocking on viscoelastic foundation. Earthq. Eng. Struct. Dynam. **37**, 1039–1063 (2008)
8. Caliò, I., Marletta, M.: Passive control of the seismic rocking response of art objects. Eng. Struct. **25**(8), 1009–1018 (2003)
9. Christopoulos, C., Filiatrault, A.: Principles of Passive Supplemental Damping and Seismic Isolation. IUSS Press, Pavia, Italy (2006)
10. Guerrini, G., Ausenda, G., Graziotti, F., Penna, A.: An innovative seismic isolation device based on multiple articulated quadrilateral mechanisms: analytical study and shake-table test. In: Proceedings of XVIII ANIDIS Conference, Ascoli Piceno, Italy (2019)
11. Castelli, F., Lentini, V., Lo Iacono, F., Navarra, G.: Shake-table tests for the dynamic characterisation of an innovative isolator for seismic protection of statues. In: Geotechnical Engineering for the Preservation of Monuments and Historic Sites III, pp. 389–400. CRC Press, London (2022)
12. Lo Iacono, F., Navarra, G., Oliva, M., Cascone, D.: Experimental investigation of the dynamic performances of the L.E.D.A. shaking tables system. In: Proceedings of the 23rd AIMeTA Conference 2017, vol. 5, Salerno, Italy, pp. 897–915 (2017)

Base-Isolation of a Rocking Object on a Rocking Pedestal: Response to Pulse-Type Ground Motion

G. Destro Bisol[1](✉), M. J. DeJong[2], D. Liberatore[1], and L. Sorrentino[1]

[1] Department of Structural and Geotechnical Engineering, Sapienza University of Rome, Via Antonio Gramsci 53, 00197 Rome, Italy
giacomo.destrobisol@uniroma1.it

[2] Department of Civil and Environmental Engineering, University of California, Berkeley, 760 Davis Hall, Berkeley, CA 94720, USA

Abstract. The rocking response of rigid blocks, such as statues, storage systems, and slender structures in general, was documented both in the field and in the laboratory. These objects/structures must be protected from earthquakes, as their value may be inestimable in some cases. The use of base isolation devices for this structural typology was already studied and successfully applied to real cases. Previous research has focused on the isolation of a single rigid block system, although there are circumstances in which the use of a single block is insufficient to define the response. Configurations in which one block is stacked on top of another, such as ancient Greek and Roman structures composed of large heavy members or statues placed on a pedestal, are examples. In this paper the rocking response of two stacked rigid blocks placed on a seismic isolation device is investigated. The dynamics of this multi-degree of freedom system is briefly summarized, together with the conditions associated to an impact or a change in the rocking configuration. Many parameters govern the response of the system and to understand their influence on the dynamic behavior, analysis of the response to a pulse-type ground motion is carried out. In general, the seismic isolation device is shown to be effective for a wide range of scenarios. However, the study also highlights how, for large period pulses or short period devices, the response of the isolated system can be amplified compared to the scenario where the blocks are placed on a rigid foundation. This phenomenon is due to the amplification of the acceleration of the isolation device for specific ratios between the period of the pulse and that of the isolator; the need of a careful design of the device is evident.

Keywords: Multibody dynamics · Overturning stability of art objects

1 Introduction

When subject to earthquakes, some objects, or structures, such as statues, obelisks, storage systems and transformers, show a dynamic behavior that can be modelled considering the structure as a rigid block. There are also particular cases in which one block is placed over the other and the model of a single rigid body is not, by itself, able to adequately

G. P. Cimellaro (Ed.): WCSI 2022, LNCE 309, pp. 726–732, 2023.
https://doi.org/10.1007/978-3-031-21187-4_62

define the response. Such a configuration, formed by stacked blocks, can be thought of as the model of ancient Greek and Roman structures composed of large heavy members and of a piece of machinery or statue placed on a pedestal [1]. These artefacts may have significant historical and cultural value, such as statues and museum pieces, and it is therefore essential to take measures to preserve them. Past earthquakes demonstrated the vulnerability of these objects to horizontal forces, highlighting the need to protect them against seismic events; base isolation devices can be used for this purpose.

In the literature, a wide number of analytical studies on the dynamics of rigid blocks is available, starting from the pioneering work of Housner [2]. The study of the two-stacked-rigid-block system free to rock on a rigid foundation, named "fixed base system" from now on, was explored by Spanos et al. [1]. The authors conclude that the analytical formulation of this nonlinear problem is challenging. Its complexity is associated with the transition from one rocking mode (or pattern) to another, each one being governed by a set of highly nonlinear equations. Furthermore, the transition from one pattern to another can happen either through impact, occurring between the bodies or between the lower body and the foundation, or by an instantaneous separation of the two blocks.

The rocking response of rigid objects placed on a seismic isolation device is a subject that has also received attention, mainly from the need to protect slender art objects. These studies primarily focus on the seismic protection of relatively small size blocks, such as art objects or human-size statues, and they generally conclude that seismic isolation protects such objects from overturning. Caliò and Marletta [3] show that the system may effectively be applied to safeguard art objects during earthquakes. Vassiliou and Makris [4] investigate in depth up to what size of free-standing objects the application of seismic isolation is beneficial, and deduce that large free-standing structures may not benefit from the base isolation technique. Recently, the focus moved on the study of innovative constitutive laws for the isolation system [5].

Hereinafter, the response to sine-pulse excitation, of the system formed by two stacked rigid blocks placed on a seismic isolation device, is investigated. Although an extensive experimental campaign was conducted by Fragiadakis et al. [6], a numerical model capable of capturing the response of this kind of system was not, to date, developed. For this three degrees-of-freedom (DOFs) system, the equations of motion will be presented, for all the possible patterns, using a Lagrangian approach. After deriving all the relations that govern the rocking pattern transition, a code in MatLab [7] was developed. This analytical model will then be used to understand the effect of the base isolation device on the two-stacked-block system, when subject to sine-pulse ground motion.

2 Formulation

Procedures that rely on energy principles or analytical mechanics are useful for the formulation of the equations of motion of complicated systems [8]. In the model formed by two stacked rigid blocks standing on a base isolation device (named from now on "isolated system"), the number of variables and all the different patterns in which the system can move make the use of variational methods suitable for deriving the equations of motion. If the base isolation behavior is assumed to be linear, the protection device contributes one DOF. The system is characterized by five patterns (Fig. 1). For each pattern, the analytical equation of motion is formulated with a Lagrangian approach [9].

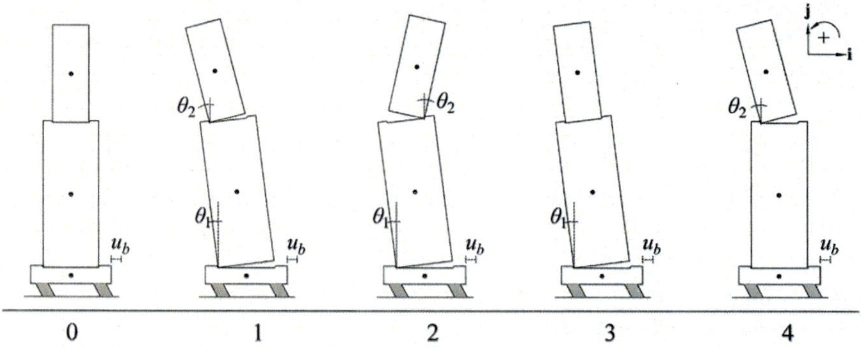

Fig. 1. Possible motion patterns of the two stacked bodies system placed on a base isolation device

The two-block model standing on an isolation system is characterized by five patterns, one more if compared to the case where the system is standing on a rigid foundation [1, 10]. This additional pattern, herein called pattern 0, is characterized by the pure translational motion of the base isolation and by the absence of rocking motion. The other four patterns (Fig. 1) are similar to those found in Spanos et al. [1], with the difference that the rocking motion must take into account the contribution of the isolation device.

The motion of the system can be described using the generalized coordinates. The system has three DOFs, so the motion can be described by the use of three Lagrangian parameters q_i. The first two generalized coordinates are θ_1 and θ_2, the rotations of the lower body and of the upper body respectively. To describe the motion of the base isolation device, the horizontal displacement u_b is used as the third coordinate.

Once the Lagrangian parameters are defined, the equations of motion can be derived using Eq. (1), for each of the coordinates, and repeating the process for every pattern.

$$\frac{d}{dt}\left(\frac{\partial T}{\partial \dot{q}_i}\right) - \frac{\partial T}{\partial q_i} + \frac{\partial D}{\partial \dot{q}_i} + \frac{\partial V}{\partial q_i} = Q_i \tag{1}$$

In order to assemble the equations of motion, the kinetic energy, T, the potential energy, V, the non-conservative forces, Q_i, as well as the dissipation function, D, must be derived as in Spanos et al. [1], but also considering the isolation system contribution. In this case, to consider the energy dissipation due to the isolation device, it is necessary to introduce an appropriate dissipation function, differently to the fixed base system, where the dissipation of energy relies only on impacts.

As already described by Spanos et al. [1] and Prajapati et al. [10], during the motion, the two-block system can change its configuration; this can be due to either impacts or sudden based accelerations. The main difference between the fixed and isolated base systems is in the formulation of the coefficient of restitution due to impacts: in the case of the isolated system it is necessary to consider that when an impact occurs, the isolated base also experiences a change of velocity [4].

3 Sine-Pulse Analysis

The use of sine-pulse analysis, although not exhaustive to understand the complex behavior of a system subjected to an earthquake, can be a powerful tool to study the dynamics of rocking systems. Further, considering the large numbers of parameters describing the system, the sine-pulse analysis can allow to partially nondimensionalize the problem, for example by the use of rocking spectra [11]. Given the many parameters describing the structure, the use of a "reference system" is here assumed. This system has defined geometrical and mechanical characteristics, and during the following analyses some of these parameters will be modified to understand their influence on the dynamics of the entire assembly. The geometrical properties for the reference system are partially based on the values reported in the experimental campaign performed by Fragiadakis et al. [6]. The lower body, here also denoted as the pedestal, is assumed to have a thickness of $2b_1 = 0.4$ m, and a height of $2h_1 = 1.0$ m, while the upper body is assumed to have a thickness of $2b_2 = 0.22$ m, and a height of $2h_1 = 0.75$ m. The seismic isolation device, here assumed linear, has the following mechanical properties: (a) period $T_b = 2.0$ s; (b) ratio between the mass of the base isolation device and the cumulative mass of the two rigid bodies $\chi_b = 0.1$; (c) damping ratio $\xi_b = 0.05$. Furthermore, the amplitude of the sine pulse is named a_p, while the circular frequency and the period of the pulse are ω_p and T_p respectively.

3.1 Collapse Mechanisms

In the case of a single rocking body subjected to a sine-pulse excitation, the overturning can occur in two different ways: (a) without impact (b) after an impact with the foundation [12]. In the case of two stacked rocking bodies, the overturning of the rigid bodies can take place in the mentioned patterns and can involve different types of impacts, for this reason the recurrent collapse mechanisms are investigated (Fig. 2).

Sine-pulse analysis of the reference system revealed four types of principal overturning mechanisms (Fig. 2): (a) without any impact; (b) with an impact between the two blocks, or middle impact; (c) with one or more impacts between the pedestal and the base isolation device; (d) with one impact between the two blocks and one or more impacts between the pedestal and the isolation system.

3.2 Rocking Spectra

In this section, the reference system is used to assess the effectiveness of the base isolation device on the global response. To understand the effect of the isolation system, four rocking spectra are derived for the reference system, each spectrum is characterized by the use of a different period T_b for the base isolation device (Fig. 3). In all the analyses, the response of the upper body alone, placed on a rigid foundation, is used as benchmark to estimate the effect of the different parameters on the global response. Indeed, the upper body can be thought as the statue, or more generally the valuable object to protect using the base isolation technology. For these reasons, the rocking spectra are built normalizing with respect to p_2, the inverted pendulum frequency parameter [2] for the upper body alone, and to $g\tan(\alpha_2)$, where g is the gravity acceleration and $\tan(\alpha_2) = b_2/h_2$. The

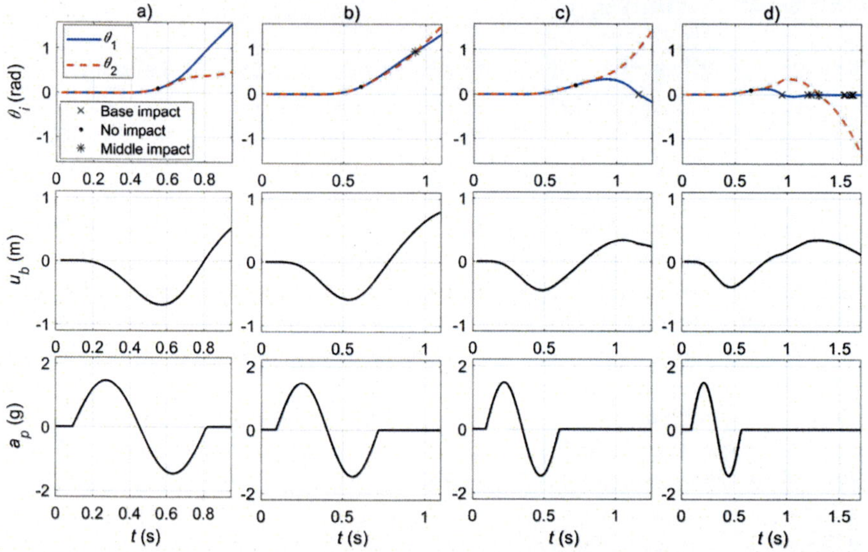

Fig. 2. Sample time histories, in terms of rotation of the bodies, displacement of the isolator and ground excitation, for the recurrent overturning mechanisms: a) no impact b) middle impact c) base impact d) base and middle impacts

response of the upper body alone, is expressed in terms of analytical rocking spectra [12].

The results highlight how the effectiveness of the isolation system strongly depends on the period T_b. As T_b increases the spectra tends to shrink, reducing the overturning area. If compared with the analytical spectrum, obtained for the upper body alone, the results highlight that for short periods T_b the isolation device provides a negative effect (Fig. 3a). As T_b increases, the beneficial effect of the base isolation increases as well (Fig. 3b, c, d), which is evident especially for large values of ω_p or small values of p_2. Further, observing the behavior of the isolated system, regardless of the period T_b, an overturning region below the analytical upper body spectrum is observed. In this area the isolation device decreases the minimum overturning acceleration, if compared to the upper body response, and the safety level of the system decreases. This peculiar response is due to the ground motion amplification caused by the linear isolation system and depends on the ratio (called T_r from now on) between the period of the pulse, T_p, and the period of the linear isolation system, T_b.

To better understand the behavior of the system in this region, four time-history analyses are presented (Fig. 4). The sample analyses are carried out for the same sine-pulse ground motion but for four different isolator periods (point A in Fig. 3a, b, c, d).

The results confirm how increasing the period T_b the safety level of the system is generally enhanced. For the same pulse (Fig. 4c), the system behaves in two different ways (Fig. 4a): (a) overturns for T_b = 1.0 and .02 s; (b) does not rock and the system is safe for T_b =2.5 and 3.0 s. The sample time histories demonstrate how quickly the

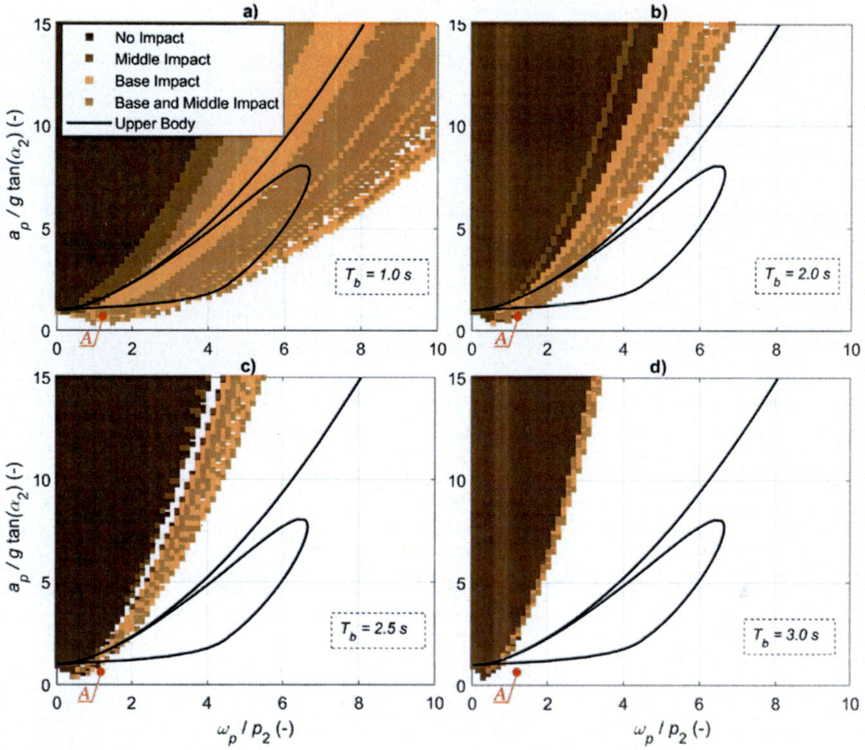

Fig. 3. Rocking spectra of the reference system for different base isolation periods (kept constant during the construction of each spectrum): a) $T_b = 1.0$ s; b) $T_b = 2.0$ s; c) $T_b = 2.5$ s; d) $T_b = 3.0$ s.

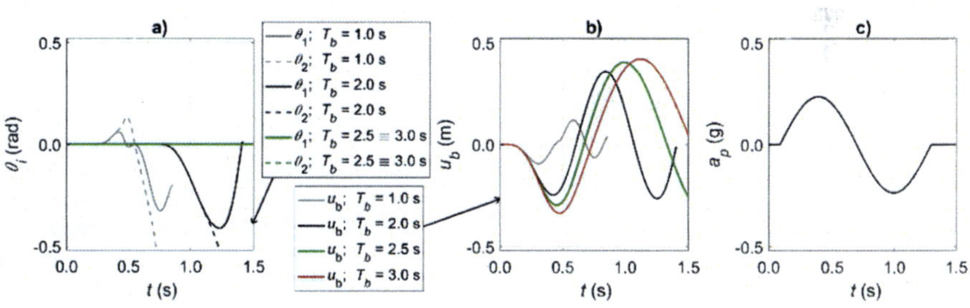

Fig. 4. Sample time histories, for the points A in Fig. 3, in terms of: a) rotation of the bodies; b) displacement of the base isolation; c) ground excitation.

transition from no rocking to overturning condition occurs in this region of the spectrum, emphasizing the importance of a careful design. It is worth mentioning that increasing the period T_b also increases the maximum displacement of the isolation device; the excessive displacement demand could make the base isolation system difficult to design for manufacturing reasons.

4 Conclusions

In this paper, a numerical model able to describe the dynamics of a rocking object on a rocking pedestal placed on a linear base isolation device is presented. The model is used to investigate the highly nonlinear response of the system when subjected to pulse-type ground motion.

By means of a series of rocking spectra, the effectiveness of the base isolation for this kind of systems is investigated, and the results indicate that this protective system can be effective for large-period devices. The study also shows how the response of the isolated system can be amplified for large period pulses or short period devices, when compared to the fixed base system. This effect is due to the amplification introduced by the isolation device in terms of accelerations, for specified ratios between the period of the pulse and that of the isolator.

Acknowledgements. This work was partially carried out under the MONALISA (MONitoraggio Attivo e Isolamento da vibrazioni e Sismi di oggetti d'Arte = Active monitoring and isolation from vibrations and earthquakes of art objects) research project funded by Lazio Regional Government and Italian Ministry of University and Research, project id. 305-2020-35576. The opinions expressed in this publication are those of the authors and are not necessarily endorsed by the funding bodies.

References

1. Spanos, P.D., Roussis, P.C., Politis, N.P.A.: Dynamic analysis of stacked rigid blocks. Soil Dyn. Earthq. Eng. **21**(7), 559–578 (2001)
2. Housner, G.W.: The behavior of inverted pendulum structures during earthquakes. Bull. Seismol. Soc. Am. **53**(2), 403–417 (1963)
3. Caliò, I., Marletta, M.: Passive control of the seismic rocking response of art objects. Eng. Struct. **25**(8), 1009–1018 (2003)
4. Vassiliou, M.F., Makris, N.: Analysis of the rocking response of rigid blocks standing free on a seismically isolated base. Earthq. Eng. Struct. Dynam. **41**(2), 177–196 (2012)
5. Pellecchia, D., Lo Feudo, S., Vaiana, N., Dion, J., Rosati, L.: A procedure to model and design elastomeric-based isolation systems for the seismic protection of rocking art objects. Comput.-Aided Civ. Infrastruct. Eng. 1–18 (2022)
6. Fragiadakis, M., DiSarno, L., Saetta, A., Castellano, M.G., Rocca, I., Diamantopoulos, S., et al.: Experimental seismic assessment and protection of museum artefacts. In: Proceedings of the International Conference on Structural Dynamic, EURODYN 2020, Jan 2020, vol. 2, pp. 3381–3396 (2020)
7. MATLAB 9.7.0.1190202 (R2019b). The MathWorks Inc., Natick, MA (2018)
8. Humar, J.: Dynamics of Structures. Prentice-Hall, Englewood Cliffs, NJ (1990)
9. Meirovitch, L.: Methods of Analytical Dynamics. McGraw-Hill, New York (1970)
10. Prajapati, S., Destro Bisol, G., Alshawa, O., Sorrentino, L.: Non-linear dynamic model of a two-bodies vertical spanning wall elastically restrained at the top. Earthq. Eng. Struct. Dynam. 1–21 (2022)
11. Zhang, J., Makris, N.: Rocking response of free-standing blocks under cycloidal pulses. J. Eng. Mech. **127**(5), 473–483 (2001)
12. Dimitrakopoulos, E.G., DeJong, M.J.: Revisiting the rocking block: closed-form solutions and similarity laws. Proc. Roy. Soc. A: Math. Phys. Eng. Sci. **468**(2144), 2294–2318 (2012)

Seismic Risk Mitigation for Non-Structural Components: From Experimentation to Home Automation

A Shake Table Testing Campaign of Electrical Cabinets

Roberto J. Merino[1], Filippo Dacarro[2], Paolo Dubini[2], Francesco Graziotti[2,3], Luca Grottoli[2], Igor Lanese[2], Daniele Perrone[4](✉), Maria Rota[2], Roberto Nascimbene[1], and Andre Filiatrault[1]

[1] University School for Advanced Studies IUSS Pavia, Pavia, Italy
[2] European Centre for Training and Research in Earthquake Engineering (EUCENTRE), Pavia, Italy
[3] Department of Civil Engineering and Architecture, University of Pavia, Pavia, Italy
[4] University of Salento, Lecce, Italy
daniele.perrone@unisalento.it

Abstract. Non-structural elements such as electrical cabinets play a key role in determining the seismic performance of buildings. Seismic qualification shake table testing of non-structural elements is a relatively simple way to evaluate their expected seismic performance. This paper presents a shake table testing campaign performed on several electrical cabinet specimens using the 6DLAB laboratories at the EUCENTRE Foundation located in Pavia, Italy. These shake table tests were performed as part of the research project Creazione di un Ambiente Domestico Sicuro (CADS). The physical properties of all the tested electrical cabinet archetypes varied considerably including: base-mounted electrical cabinets, wall-mounted electrical cabinets, electrical cabinets with different attachment strengths, and electrical cabinets with different framing structure strengths. The shake table tests were carried out using two seismic testing protocols, one of which is commonly used for the seismic qualification of non-structural elements, whereas the second one was proposed within the scope of the CADS Project. The performance of selected electrical cabinet archetypes during the shake table tests are presented and discussed in this paper.

Keywords: Shake table testing · Non-structural elements · Electrical cabinets · Seismic qualification

1 Introduction

Non-structural elements (NSEs) are not part of a building's load bearing system but are nonetheless subjected to the same dynamic environment as their structural counterparts during an earthquake ground motion. Many recent earthquakes that have occurred in developed countries have demonstrated that preventing buildings from collapsing during an earthquake is a necessary but not a sufficient goal when mitigating overall seismic risk [1–3]. Damage to NSEs can significantly contribute to the seismic risk of a building in three main ways. (1) Damage to NSEs can generate significant economic losses in

G. P. Cimellaro (Ed.): WCSI 2022, LNCE 309, pp. 735–749, 2023.
https://doi.org/10.1007/978-3-031-21187-4_63

buildings. (2) Damage to NSEs can render a building out of function for a significant period of time; this is of special importance in critical facilities such as hospitals and airports [2, 4]. (3) Finally, the collapse of certain NSEs can threaten the lives of building occupants or passer byes and can hinder the exit of the former from the building after the ground motion has subsided. Because of these reasons, consideration of the seismic performance of NSEs has become an integral part of modern performance-based earthquake engineering.

Unfortunately, a lack of research studies and clear and complete design guidelines do not generally allow seismic damage to NSEs to be optimally controlled, and seismic damage to NSEs continues to be a problem for the earthquake engineering community. Even though rational design methodologies for NSEs have been recently proposed in the public scientific literature [5–7], their implementation in practice is still a long way off due to the vast amount of NSE typologies and the accompanying expensive and time-consuming experimental and analytical studies required to bring such methodologies to maturity. One relatively simple and practical way of improving the seismic performance of NSEs in buildings is to perform qualification shake table testing. For example, the Italian Norme Tecniche per le Costruzioni [8] encourages the use of qualification shake table testing for NSEs, although it does not mandate it. In the United States, the current ASCE7-22 standard [9], requires the qualification shake table testing of important NSEs in critical facilities. Additionally, the Department of Health Care Access and Information (HCAI) in California requires, through the California's Hospital Seismic Retrofit Program, that all mechanical and electrical equipment that are installed in hospitals be qualified through shake table testing.

This paper presents a shake table testing campaign of several electrical cabinet specimens that was performed in the 6DLAB in the Centre for Training and Research in Earthquake Engineering (EUCENTRE Foundation) located in Pavia, Italy. The shake table testing campaign was performed as part of the research project: Creazione di un Ambiente Domestico Sicuro (CADS). Electrical cabinets are a typology of NSE that could be subject to the requirements of qualification shake table testing in accordance to NTC [8] or ASCE7-22 [9]. The paper focuses on the main highlights of the testing campaign including a description of the shake table and of the tested specimens, the testing sequences, the instrumentation used to record the response of the specimens, and some preliminary results of the response and damage recorded during the test of three of the electrical cabinet specimens. A discussion comparing the responses of these three specimens is also provided in order to try to identify the typologies of electrical cabinets that are expected to have the best seismic performance.

2 Description of the Shake Table and the Electrical Cabinet Specimens

2.1 Shake Table

The shake table testing campaign of the electrical cabinet specimens was conducted in the 6DLAB shake table in the EUCENTRE Foundation located in Pavia, Italy [10]. The 6DLAB shake table was specifically designed for the qualification shake table testing of

NSEs. The "6D" in the name 6DLAB references the fact that the shake table can move in six degrees of freedom, including the movements in both horizontal and the vertical directions and the rotations along the three principal axes. The 6DLAB shake table is, therefore, ideal to perform multi-axial shake table tests on NSEs.

2.2 Electrical Cabinet Specimens

The shake table testing campaign was conducted using two configurations of electrical cabinets. The first configuration, referred to here as Configuration 1, consisted of four different electrical cabinet specimens installed on the surface of the shake table and tested at the same time. On the other hand, the second configuration, referred to here as Configuration 2, consisted of three additional electrical cabinet specimens. Therefore, a total of seven distinct electrical cabinet specimens were tested as part of the experimental campaign. Table 1 presents a summary of the main properties of the electrical cabinet specimens tested in both Configuration 1 and Configuration 2. The properties presented in Table 1 include the type of fixture (base- or wall-mounted), the length, width, height, elevation of the center of gravity above the base, the seismic weight, and the measured initial fundamental frequencies at both main orthogonal horizontal directions. Figure 1 shows a photograph of the four electrical cabinet specimens tested as part of Configuration 1, while Fig. 2 shows a photograph of the other three electrical cabinet specimens that were tested as part of Configuration 2. The red arrows plotted in Figs. 1 and 2 show the direction of the x and y axes of the shake table in relation to the position of the specimens on the shake table. Note that all the specimens were installed so that their short direction is aligned with the x axis of the shake table. Note also that the electrical cabinet specimens of Configuration 1 are referred to in this study as Specimens 1 to 4, while the electrical cabinet specimens of Configuration 2 are referred to in this study as Specimens 5–7. The seismic weight of most of the specimens includes extra weights that were installed during the test to represent the presence of the electrical equipment to be installed inside the electrical cabinets when they are operational.

Table 1. Main properties of the tested electrical cabinet specimens.

ID	Type of Fixture	Length [mm]	Width [mm]	Height [mm]	Elevation of centre of gravity [mm]	Seismic weight [kN]	Measured Initial fundamental frequency [Hz]	
							x dir	y dir
Configuration 1								
1	Base-mounted	1300	1000	2171	N.A.[a]	6.08	17.0	9.6
2	Base-mounted	1400	420	2143	N.A.[a]	2.70	6.0	17.0
3	Base-mounted	1013	611	2109	1053	9.31	3.3	7.5
4	Base-mounted	1013	811	2109	1076	9.54	3.5	6.7

(continued)

Table 1. (*continued*)

ID	Type of Fixture	Length [mm]	Width [mm]	Height [mm]	Elevation of centre of gravity [mm]	Seismic weight [kN]	Measured Initial fundamental frequency [Hz]	
							x dir	y dir
Configuration 2								
5	Wall-mounted	600	600	800	343	2.45	14.0	8.5
6	Base-mounted	1011	811	2107	1089	9.25	1.4	3.0
7	Base-mounted	819	811	2107	1052	6.11	2.2	4.0

[a] N.A.—Information not available from the manufacturer

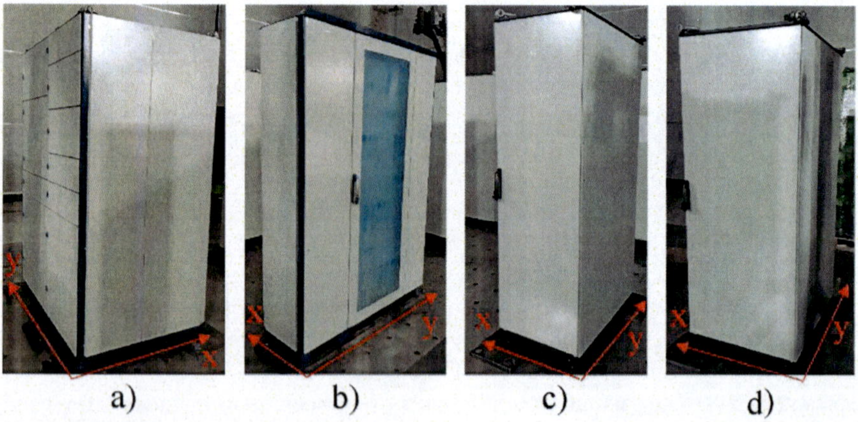

Fig. 1. Electrical cabinet specimens of Configuration 1 (a) Specimen 1; (b) Specimen 2; (c) Specimen 3; (d) Specimen 4.

All the electrical cabinet specimens of Configuration 1 are base-mounted specimens and have a total height of around 2100 mm. All specimens also are connected on top of a base attachment which is in turn anchored to the surface of the shake table. Specimen 1 is the broadest specimen with almost square plan dimensions, while Specimen 2 has the slenderest plan dimensions with a length that is approximately 2.5 times its width. Note that both the elevation of the center of gravity above the base and the internal structure of Specimens 1 and 2 were not revealed by the manufacturer and therefore remain unknown for the purpose of this study. Additionally, the manufacturer did not specify extra weights to be installed in Specimens 1 and 2 to simulate the contained electrical installations. Specimen 1 did have some extra weights and tubes, but Specimen 1 was completely empty for the test. Specimen 3 and 4 have a similar structure but have a different width, as can be observed from Table 1. Both Specimens 3 and 4 have an internal frame structure that supports the weight of the electrical installations. This internal frame structure has pinned connections between the beams and uprights and at the base connection, so it does not have any lateral load resistance. The internal frame is connected by steel channels at mid height and at the top of the specimens to the external structure of the specimens, which includes an external frame and the panel walls; these channels transfer the inertial loads from the electrical installation weights to the lateral force resisting system of the specimens. Both Specimens 3 and 4 have only one door in one of their faces along their shorter direction.

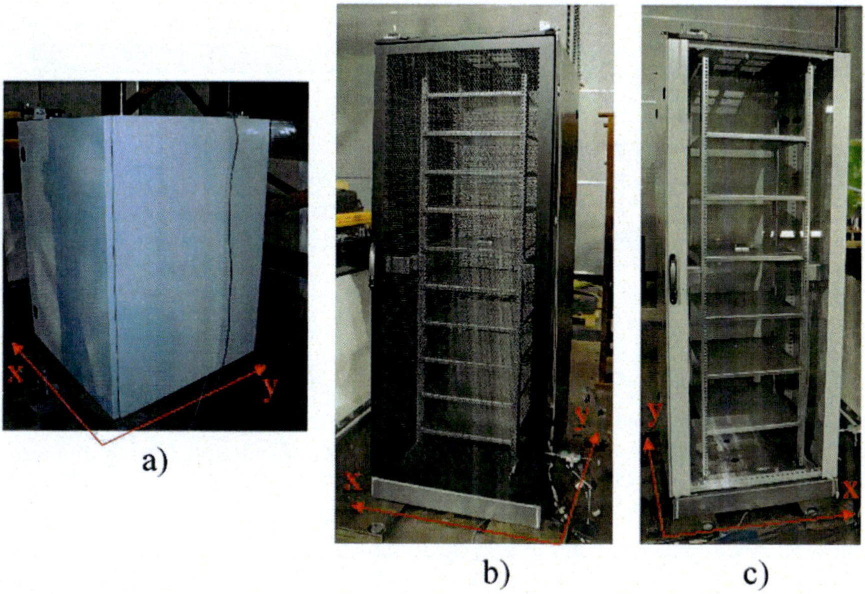

Fig. 2. Electrical cabinet specimens of Configuration 2 (a) Specimen 5; (b) Specimen 6; (c) Specimen 7.

The electrical cabinet specimens of Configuration 2 have the following properties. Specimen 5 is wall-mounted while Specimens 6 and 7 are both base-mounted as the other four specimens of Configuration 1. Given that Specimen 5 is wall-mounted, its length and height are smaller than those of the other base-mounted specimens. As can be observed from Fig. 2 (a), Specimen 5 was installed on a rigid steel frame that was rigidly connected to the base of the shake table. Specimens 6 and 7 have a similar structure as Specimens 3 and 4; they have a laterally weak internal frame structure that is connected to a stronger external frame structure. Both Specimens 6 and 7 have two glass doors, one on each face along their short direction.

Table 1 also presents the initially measured fundamental frequencies of all the specimens along the two principal directions of the shake table. The fundamental frequencies of the specimens were measured using results of dynamic identification white noise tests performed before any of the qualification shake table testing protocols was applied to the specimens. The initial fundamental frequencies of all the specimens vary between 1.4 Hz and 17.0 Hz along the shorter direction (x axis of the shake table) and between 3.0 Hz and 17.0 Hz along the longer direction of the specimens (y axis of the shake table). The stiffest specimens are Specimens 1, 2, and 5, while the most flexible specimens are

Specimens 6 and 7. However, Specimens 3 and 4 are also relatively flexible. One of the main reasons for Specimens 1, 2, and 5 being the stiffest specimens is that their seismic weights are either relatively low. Specimen 1 is only slightly loaded for its size, Specimen 2 has no additional weights beyond its self-weight, and Specimen 5 is relatively small in comparison to all the other specimens, so its seismic weight is small. As can be observed from Table 1, the fundamental frequencies measured along the shorter directions of the specimens tend to be lower than those measured along the longer directions with the exceptions of Specimens 1 and 5. The fundamental frequency of Specimen 1 along its shorter direction could be higher than along its longer direction because the specimen is broad and the flexibility of the elements used to attach the specimen to the base of the shake table could vary in relation to the other specimens. Specimen 5, on the other hand, is a special case since it is a wall-mounted specimen, with the flexibility in the out-of-plane behavior of the specimen being greater than its in-plane flexibility. This is probably due to the flexibility of the attachment structure of the specimen.

3 Instrumentation

This section describes the instrumentation used to measure the dynamic response of the electrical cabinet specimens during the shake table testing campaign. The acceleration response of all the electrical cabinet specimens was measured using an array of accelerometers that were installed as follows. Two accelerometers were installed at the top of each specimen at diagonally opposite corners and one accelerometer was installed as close as possible to the center of gravity. If the center of gravity of the specimen was not known, as is the case with Specimens 1 and 2, then the accelerometer was installed close to the mid-height of the specimen. Additionally to the accelerometers, potentiometers were also installed to measure the displacement response of the electrical cabinet specimens relative to their original position on the shake table. Three horizontal line potentiometers were installed between each of the specimens and rigid braced frames that were built on top of the shake table. For each specimen, two potentiometers were installed along the x axis of the shake table, or along the shorter direction of the specimens, while a third potentiometer was installed diagonally between the specimen and the rigid frame. Figure 3 presents a schematic view of the instrumentation layout for both Configuration 1 and Configuration 2. Two vertical potentiometers were also installed between an elevation of approximately 150 mm on each specimen and the surface of the shake. These potentiometers were used to measure the uplift of the specimens in the case that rocking occurred. Figure 3 presents a schematic view of both testing configurations including the electrical cabinet specimens tested as part of each configuration and the respective instruments used to measure the dynamic responses of the electrical cabinet specimens.

In addition to the accelerometers and potentiometers used to measure the accelerations and displacements of the electrical cabinet specimens, damage to the electrical cabinet specimens was recorded using videos and photographs. In this way the typology of the damage to the specimens can be correctly assessed and the performance of the specimens in terms of, for example, life safety or continuous operation can be determined.

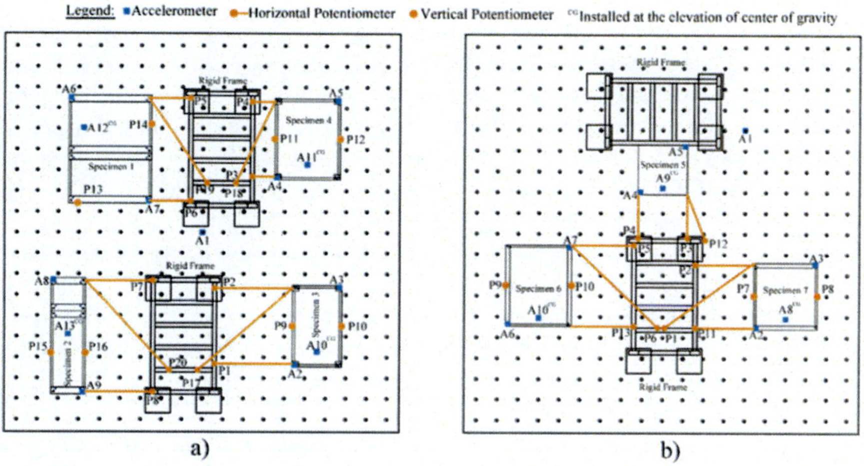

Fig. 3. Layout of electrical cabinet specimens and their instrumentation during the shake table tests: (a) Configuration 1; (b) Configuration 2.

4 Testing Program

This section presents the shake table testing program used for the two configurations of electrical cabinet specimens. The testing program included two different loading protocols as well as dynamic identification white noise tests used to monitor the fundamental frequencies of the electrical cabinet specimens. The loading protocols used to perform the shake table testing were the Telcordia-GR-63-CORE [11] and the protocol from a recently proposed seismic classification procedure for NSEs in Italy developed as part of the CADS project [12].

The Telcordia-GR-63-CORE [11] loading protocol is aimed at electrical NSEs thar are going to be installed in buildings in the United States. The loading protocol consists of a set of predefined Required Response Spectra (RRS) defined for four different zones of the United States. Each electrical NSE should be subjected to a test input motion whose Test Response Spectrum (TRS) is at least equal to the RRS of the zone where the NSE is expected to be installed. The RRS of each zone is defined by various frequency and acceleration parameters. The test input motions are predefined and are given by Telcordia and are different for each one of the four zones. The Telcordia-GR-63-CORE loading protocol should be applied to the three main orthogonal directions of each specimen but not at the same time. In other words, three uniaxial tests, two horizontal and one vertical, should be performed for each tested specimen.

The second loading protocol is part of a seismic classification procedure for NSEs in Italy developed as part of the CADS project [12]. The loading protocol for the seismic classification procedure for NSEs in Italy is based on ISO13033 [13] in combination with the force requirements for NSEs given in Eurocode 8 [14]. The loading protocol developed as part of the CADS project is similar to the loading protocol of Telcordia-GR-63-CORE [11] in that it defines four zones in Italy depending on their expected maximum value of peak floor spectral acceleration at roof level according to the peak ground acceleration seismic hazard map of Italy for a 475-year return period and the equation to estimate the seismic lateral force on NSEs given by Eurocode 8 [14]. Using these parameters, the Italian territory was divided into four zones with a 0.40 g increment in peak floor spectral acceleration with the maximum value being 1.60 g. Each of the four zones is then associated to a horizontal and a vertical RRS whose shape is defined according to ISO13033 (2013) and whose amplitude is given by their corresponding value of peak floor spectral acceleration. Each zone is also associated to class level identified with the nomenclature A, B, C, and D. Therefore, the RRS of Class A has a peak floor spectral acceleration of 1.60 g, the RRS of Class B has one of 1.20 g, the RRS of Class C has one of 0.80 g, and the RRS of Class D has one of 0.40 g. Full triaxial shake table tests are required as part of this loading protocol. The test input motions of the loading protocol are generated according to the procedure specified in ICC-ES [15] and their TRS shall match the RRS using the same criteria.

Table 2 presents a summary of the sequence of the tests performed for both configurations of electrical cabinet specimens. In both cases, the testing sequence began with a white noise test with unloaded electrical cabinet specimens. This was followed by a white noise test with extra weights added to each specimen to simulate the weight of electrical installations. Note that all white noise tests consisted of three separate uniaxial tests performed along the three main axes of the shake table (two horizontal, x and y, and a vertical, z). The testing sequence for Configuration 1 then continued with the implementation of the RRS of Class D for the loading protocol developed for the CADS project. White noise tests were performed after each shake table test for Configuration 1. The testing sequence of Configuration 1 continued with the loading protocol developed for the CADS project up to the RRS of Class B and then proceeded to perform the three uniaxial tests for Zone 1 of Telcordia-GR-63-CORE [11]. After performing the shake table tests for Zone 1 of Telcordia-GR-63-CORE, the sequence continued with the loading protocol developed for the CADS project with the RRS of Class A. After this test, the damage to Specimens 3 and 4 was considerable and they were removed from the shake table. The sequence concluded with the uniaxial tests of Zones 3 and 4 of Telcordia-GR-63-CORE. Note that the vertical uniaxial test (z direction) of the Telcordia-GR-63-CORE Zones 3 and 4 tests could not be carried out due to limitations of the shake table. The sequence of tests of Configuration 2, on the other hand, continued after the initial white noise tests with the full loading sequence of the loading protocol developed for the CADS project, starting with the RRS of Class D and going up to the RRS of Class A. Note that after the test using the RRS of Class C of the loading protocol developed for the CADS project, Specimens 6 and 7 were removed from the shake table due to excessive damage. The testing sequence of Configuration 2 concluded with the performance of Zone 4 of the Telcordia-GR-63-CORE loading protocol. Note also that no white noise tests were performed after every shake table test for Configuration 2 due to the excessive deformations and damage observed in Specimens 6 and 7. It was also deemed unnecessary to perform the Telcordia-GR-63-CORE Zone 3 test since it was already perceived that Specimen 5 was very strong and would suffer no damage even for the higher intensities, as is explained in the next section.

Table 2. Testing sequences of the shake table tests of the two configurations of electrical cabinet specimens.

ID	Test Description	
	Configuration 1	Configuration 2
1	White noise test – unloaded specimens (3 uniaxial, x y z)	White noise test – unloaded specimens (3 uniaxial, x y z)
2	White noise test[a] (3 uniaxial, x y z)	White noise test[a] (3 uniaxial, x y z)
3	CADS Project – Class D (triaxial)	CADS Project – Class D (triaxial)
4	White noise test (3 uniaxial, x y z)	CADS Project – Class C (triaxial)
5	CADS Project – Class C (triaxial)	CADS Project – Class B (triaxial)[b]
6	White noise test (3 uniaxial, x y z)	CADS Project – Class A (triaxial)
7	CADS Project – Class B (triaxial)	Telcordia-GR-63-CORE – Zone 4 (2 uniaxial, x y)[c]
8	White noise test (3 uniaxial, x y z)	
9	Telcordia-GR-63-CORE – Zone 1 (3 uniaxial, x y z)	
10	White noise test (3 uniaxial, x y z)	
11	CADS Project – Class A (triaxial)[3]	
12	White noise test[a] (3 uniaxial, x y z)	
13	Telcordia-GR-63-CORE – Zone 3 (2 uniaxial, x y)[d]	
14	Telcordia-GR-63-CORE – Zone 4 (2 uniaxial, x y)[d]	
15	White noise test (3 uniaxial, x y z)[d]	

[a] From this point on, the electrical cabinet specimens were loaded with extra weights representing electrical equipment
[b] Specimen 6 and 7 were removed from the shake table due to excessive damage.
[c] Specimens 3 and 4 were removed from the shake table due to excessive damage.
[d] Test in vertical direction (z) could not be performed due to limitations of the shake table.

Figure 4 shows an example of the comparison of the TRS with the RRS for both the Telcordia-GR-63-CORE [11] and the loading protocol developed as part of the CADS project [12] for Configuration 1. Figure 4 (a) presents the comparison between the TRS and RRS for Zone 1 of the Telcordia-GR-63-CORE loading protocol for both horizontal directions (x and y axes of the shake table), while Fig. 4 (b) presents the comparison between the TRS and the RRS for the zone of Italy with the highest value of peak floor spectral acceleration (1.60 g) according to the loading protocol developed as part of the CADS project. Note that contrary to the Telcordia-GR-63-CORE (2006) in which the TRS shall be at least equal to the RRS, the requirements of the loading protocol developed as part of the CADS project, which are based on the requirements of ICC-ES (2012), are that the TRS shall be no less than 10% below the RRS and no more than

30% the RRS. Therefore, a zone in which the TRS should fall can be defined and is painted in grey in Fig. 4 (b). Figure 4 (a) shows that the TRS for all three directions (x, y, and z) for the case of the Telcordia-GR-63-CORE loading protocol are above the RRS except for frequencies lower than 2.5 Hz and frequencies between 10 Hz and 15 Hz. The RRS for the loading protocol developed as part of the CADS project, on the other hand, generally falls within the grey zone of Fig. 4 (b) with the exception of the TRS in the y direction which is lower than the limit of 10% below the RRS for a frequency range between 15 Hz and 25 Hz. Note that the horizontal RRS of both loading protocols are relatively similar in amplitude and shape.

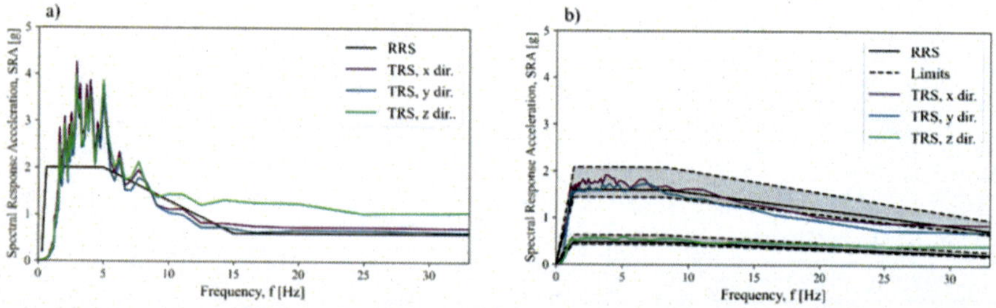

Fig. 4. Examples of comparison of TRS with RRS for the two horizontal directions and the vertical direction for Configuration 1: (a) Telcordia-GR-63-CORE (2006) loading protocol, Zone 1; (b) RRS of Class A for the loading protocol developed as part of the CADS project.

5 Shake Table Testing Results

This section summarizes the overall performance of some representative electrical cabinet specimens during the shake table tests. Figure 5 presents photographs of the main damages to the electrical cabinet specimens that could be observed after the shake table testing of the electrical cabinet specimens of Configuration 1. For example, Specimen 1 suffered no observable damages after the whole sequence of shake table testing. However, it was considered that for the last and strongest test level, which was the Telcordia-GR-63-CORE [11] Zone 4, the specimen may have some probability of overturning, so it was restrained to one of the rigid frames installed on the shake table as can be observed from Fig. 5a. Specimen 4 on the other hand, suffered considerable damage during the sequence of tests. The damage to Specimen 4 started to be noticed for the CADS Project Class B test, for which a slight permanent buckling of one of the lateral panels was observed. The damage to Specimen 4 continued to propagate for the CADS Project Class A test. For this test, the attachment of the specimen failed completely, and significant permanent buckling occurred for the two lateral panels as can be observed for Fig. 5 (b), and (c), respectively. The difference in damage between the two specimens can be explained by the fact that the seismic weight of Specimen 1 was lower than the seismic weight of Specimen 4 even though Specimen 1 is broader than Specimen 4, as can be seen from Table 1. The overall seismic force-resisting system of Specimen 4 consists of an internal frame with a very low lateral strength capacity, since its beam-to-upright connections are very weak, and an external frame that carries most of the lateral forces along with the base attachment. The internal frame is connected to the external frame using strong steel channels that transfer the inertial forces from the additional weights to the external frame and the attachment. Additionally, Specimen 4 was observed to rock slightly during the final test to which it was subjected to (CADS Project – Class A) with a rocking surface that was created between the body of the cabinet specimen and the attachment. No rocking was observed between the attachment and the surface of the shake table. Note that Fig. 4 shows that the amplitudes of the RRS and TRS for the Telcordia-GR-63-CORE, Zone 1 test are higher than that of the RRS and TRS of the CADS Project – Class A test. However, the main damage to Specimen 4 occurred mainly during the CADS Project Class A and B tests and not during the Telcordia-GR-63-CORE, Zone 1 tests. These occurred because the CADS Project loading protocol includes triaxial tests, while the Telcordia-GR-63-CORE [11] includes three separate uniaxial tests.

Clearly, full triaxial shake table tests, which simulate the real conditions during an earthquake ground motion more accurately, are more damaging to NSEs than separate uniaxial tests.

Figure 6 presents photographs of the main damages to the electrical cabinets of Configuration 2 observed during the shake table tests. Specimen 5 did not show any damage throughout the entire testing sequence of Configuration 2, as Fig. 6 (a) shows. This good performance of Specimen 5 was probably due to its small size and weight of the specimen and to the fact that the specimen was wall-mounted rather than base-mounted like all the other specimens. Specimen 6 experienced significant damage after the sequence of shake table tests. The structure of Specimen 6 is very similar to the structure of Specimen 4 with a laterally flexible internal frame that is connected to a stronger external frame. During the CADS Project Class D test, the doors of Specimen 6 flung open and large elastic deformations could be observed in the internal frame. During the CADS Project Class C test, the specimen got severely damaged as the two lateral panels collapsed and large inelastic permanent deformations could be observed in the internal frame after the test, as can be observed from Fig. 6 (b). After that test, a beam-to-upright connection in the external frame failed completely, as can be observed in Fig. 6 (c). The poor performance of Specimen 6 in comparison to Specimen 4 from Configuration 1, can be explained by a weaker and more flexible external frame structure and lateral panel connections in Specimen 6. Note that the lateral panel connections in Specimen 4 were strong enough so that the panel itself buckled inelastically before the connection failed. In the case of Specimen 6, however, this connection was weak and the panel violently disconnected without any permanent damage. The strength of the external frame of Specimen 4 was enough to allow the specimen to rock rigidly above the connection with the attachment producing a failure of the attachment. This did not occur in Specimen 6, for which the external frame yielded and failed instead, producing permanent deformations. Note also that Specimen 4 failed under a much larger seismic intensity than Specimen 6.

6 Conclusions

This paper presented a shake table testing campaign on several typologies of electrical cabinet specimens. The shake table testing campaign was performed using a shake table

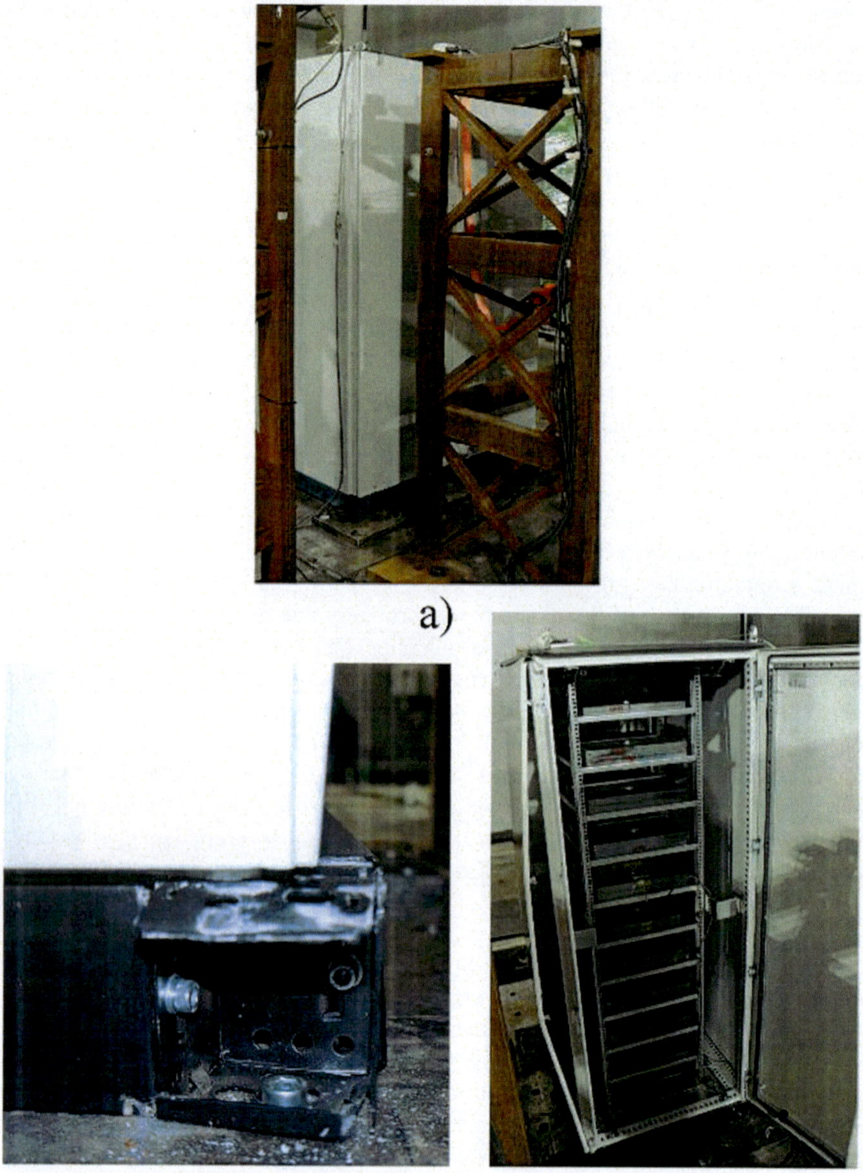

a)

b)

c)

Fig. 5. Damages observed to some of the electrical cabinet specimens of Configuration 1 after shake table testing: (a) Specimen 1 is restrained not to overturn for Telcordia-GR-63-CORE (2006), Zone 4; (b) failure of attachment of Specimen 4 for CADS Project – Class A test; (c) buckling of lateral panels of Specimen 4 for CADS Project – Class A test.

Fig. 6. Damages observed to some of the electrical cabinet specimens of Configuration 2 after shake table testing: (a) Specimen 5 is completely undamaged for Telcordia-GR-63-CORE (2006), Zone 4; (b) excessive permanent lateral deformations and collapse of lateral panels of Specimen 6 for CADS Project – Class C test; (c) failure of beam to upright connection for external frame of Specimen 6 for CADS Project – Class C test.

that allows for movement in six degrees of freedom. Two configurations of electrical cabinet specimens were tested; the first configuration included four electrical cabinet specimens while the second configuration included other three electrical cabinet specimens. Two different loading protocols as well as white noise test for dynamic identification were used to test the electrical cabinet specimens.

Some of the electrical cabinet specimens did not suffer any damage after the whole shake table sequence while others were severely damaged and came close to collapse. The only wall-mounted specimen did not experience any damage throughout the shake table testing campaign. Similarly, a broad base-mounted specimen that was lightly loaded did not experience any damage during the shake table testing campaign neither. Highly loaded base-mounted electrical cabinet specimens that had a structure comprising of a flexible internal frame rigidly connected to an external strong frame were prone to damage during shake table testing. The typology of damage on these specimens depended on the strength of the connections between the external frame and the internal frame and between the external frame and the lateral panels of the specimen. If these connections were strong, the specimen tended to rock and experience inelastic damage in the lateral panels and the attachment while if these connections were weak, the specimens tended to present inelastic damage in the external frame and complete collapse of the lateral panels. The former type of damage is preferable to the latter given that the specimens that experienced this damage were able to sustain higher seismic intensities. Finally, triaxial tests, which simulate better the actual conditions that electrical cabinets experience during the occurrence of earthquake ground motions were shown to produce higher levels of damage to the electrical cabinet specimens than uniaxial tests with higher seismic intensities.

The acceleration and displacement response obtained from the shake table test campaign of electrical cabinets need to be processed to better understand the seismic behavior of this typology of non-structural element. Further, experimental fragility functions can be developed using the damage data presented in this paper lumping together all the different typologies of electrical cabinet specimens tested during this shake table testing campaign.

Acknowledgements. The authors gratefully acknowledge the funding by Italian Ministry of Education, University and Research (PON Grant n.1735/2017, CADS Project). The authors would also like to acknowledged ETA SpA for providing the electrical cabinet specimens used for the qualification shake table testing.

References

1. Dhakal, R.P.: Damage to non-structural components and contents in 2010 Darfield earthquake. Bull. N. Z. Soc. Earthq. Eng. **43**(4), 404–411 (2010)
2. Miranda, E., Mosqueda, G., Retamales, R., Pekcan, G.: Performance of nonstructural components during the 27 February 2010 Chile earthquake. Earthq. Spectra **28**(S1), S453–S471 (2012)
3. Perrone, D., Calvi, P.M., Nascimbene, R., Fischer, E., Magliulo, G.: Seismic performance and damage observation of non-structural elements during the 2016 Central Italy earthquake. Bull. Earthq. Eng. **17**, 5655–5677 (2019)
4. FEMA: Reducing the risks of nonstructural earthquake damage—a practical guide, FEMA E-74. In: Federal Emergency Management Agency and National Earthquake Hazard Reduction Program, USA (2012)
5. Filiatrault, A., Perrone, D., Merino, R.J., Calvi, G.M.: Performance-based seismic design of non-structural building elements. J. Earthq. Eng. **25**(2), 237–269 (2021)
6. Merino, R.J., Perrone, D., Filiatrault, A.: Calibrated equivalent viscous damping for direct displacement-based seismic design of suspended piping trapeze restraint installations. J. Earthq. Eng. (2021). https://doi.org/10.1080/13632469.2021.1987357
7. Perrone, D., Rodriguez, D., Filiatrault, A., Brunesi, E., Beiter, C., Piccinin, R.: A framewaork for the quantification of non-structural seismic performance factors. J. Earthq. Eng. (2022). https://doi.org/10.1080/13632469.2021.1991516

8. NTC: Aggiornamento delle Norme tecniche per le costruzioni. Ministero delle Infrastrutture e dei Trasporti; Roma, Italia (2018)
9. ASCE: ASCE7-22, Minimum design loads for buildings and other structures. American Society of Civil Engineers, Reston, Virginia, USA (2022)
10. EUCENTRE. 2022. 6DLAB Eucentre. https://www.eucentre.it/6dlab-laboratory/?lang=en. Last accessed 25 May 2022
11. Telcordia-GR-63-CORE: NEBSTM Requirements: Physical Protection, Telcordia Technologies Generic Requirements, Piscataway, New Jersey, USA (2006)
12. Merino, R.J., Perrone, D., Filiatrault, A., Nascimbene, R.: A seismic classification procedure for non-structural building elements based on shake-table qualification testing. In: Proceedings of the 3rd European Conference on Earthquake Engineering and Seismology. Bucharest, Romania (2022)
13. ISO13033. 2013. Bases for design of structures—loads, forces and other actions -Seismic actions on nonstructural components for building applications. International Organization for Standardization, Geneva, Switzerland
14. EN 1998-1: Eurocode 8—Design provisions for earthquake resistant structures, EN-1998-1:2011, *Comite Europeen de Normalization*, Brussels, Belgium (2011)
15. ICC-ES: Acceptance criteria for seismic qualification by shake table testing of nonstructural components and systems. *International Code Council Evaluation Service*, International Code Council AC156, Whittier, CA (2012)

Effects of the Vertical and Horizontal Acceleration on the Seismic Response of Piping Networks

Gianni Blasi(✉), Daniele Perrone, and Maria Antonietta Aiello

Department of Engineering for Innovation, University of Salento, Lecce, Italy
Gianni.blasi@unisalento.it

Abstract. Piping networks are complex systems in a building and their seismic response assessment may be challenging. Different types of seismic analyses have been increasingly conducted on such non-structural components, due to their importance from a building serviceability standpoint, as well as the effect of their failure on economic losses. On the other hand, further studies are still required, due to the peculiar geometry of each network. This study aims to analyse the effect of the vertical and horizontal component of the seismic acceleration on the performance of piping networks with irregular geometry. Non-linear dynamic analyses are performed and the response of different types of piping networks is analysed in terms of maximum accelerations and maximum displacements. The dynamic behaviour of the systems is defined accounting for the non-linear response of pipe joints and suspended piping restraint installations. The analysis results are used to provide design optimization and analysis criteria for piping systems.

Keywords: Non-structural components · Piping systems · Irregular networks · Vertical seismic action

1 Introduction

Post-earthquake damage surveys highlighted the vulnerability of non-structural components (NSCs) and their importance at both immediate occupancy and life-safety performance levels [1]. In case of piping systems, hardships in defining their seismic vulnerability are mainly related to their peculiar configurations and high irregularity. Several elements influence the dynamic properties of piping systems, such as type of piping restraints, connections between pipes and presence of multiple pipes connected to a single restraint. Additionally, only gravity restraints (rods or cables) are employed in some cases to connect the pipes to the structures, which may lead to several aspects from a seismic performance view. Firstly, the absence of braced restraints causes high displacements in case of earthquake, leading to possible damage due to impact between pipes. Furthermore, gravity restraints may be simple cables acting as tie elements, leading to high sensitivity to vertical accelerations.

G. P. Cimellaro (Ed.): WCSI 2022, LNCE 309, pp. 750–760, 2023.
https://doi.org/10.1007/978-3-031-21187-4_64

In the latest years, research studies were focused on fire-protection systems in strategic and public buildings, as demonstrated by the increasing number of laboratory tests and numerical analyses conducted to assess their dynamic response. Different piping system configurations were analysed [2, 3] showing failure of pipe joints and piping restraints when subjected to seismic action. Aiming at defining the mechanical response and develop simplified models for global analyses, single components were tested in several experimental campaigns [4, 5]. Different materials, pipe dimensions and pipe connections were analysed to provide a comprehensive characterization of such elements [6, 7].

Modern seismic codes [8, 9], provide formulations to define the seismic demand, depending on the ductility capacity, the dynamic properties and the importance of the considered component. Additionally, several authors proposed simplified procedures accounting for the main parameters influencing the dynamic response of the structure [10, 11]. The "cascade methods" are widely adopted in non-structural analysis and consist in generating the seismic demand on NSCs based on the analysis of the structural system.

Despite the complex configuration of piping layouts may cause interaction between local modes and vertical component of the earthquake, a few studies were performed in the past to assess the effect of the vertical acceleration on the global response of such systems. In this study, the seismic vulnerability of different configurations of fire-fighting piping systems is assessed by mean of non-linear dynamic analysis. The piping systems are modelled using OpenSees [12], simulating the non-linear hysteretic response of seismic and gravity piping restraints and pipe joints to locate damage due to earthquake. According to cascading method, the piping systems were subjected to floor acceleration time-histories generated from the dynamic analysis of a case study infilled RC frame analysed in a previous work [13]. The influence of the vertical acceleration on the seismic fragility of the case study piping systems was assessed. For each piping layout considered, non-linear time-history analyses were conducted either assuming and neglecting the vertical acceleration, comparing the damage to pipe joints and piping restraints.

2 Description of the Case Study Piping Systems

The piping layout configuration is provided in Fig. 1. A main line composed of 2″ steel pipes and two branching lines composed of 1″ pipes were assumed. The total length of the main line and both branching lines was equal to 30.0 m and 33.0 m, respectively. The piping network was restrained to the slab through gravity and seismic piping restraint installations, characterized by simple cables and trapezes with longitudinal and transverse braces, respectively. The spacing between gravity restraints was equal to 3.0 m, while seismic restraints were included every two gravity restraints (i.e. with 6.0 m spacing). The spacing of seismic restraints was defined based on a simplified seismic design according to Italian NTC18 [9], by calculating the equivalent static force, $\boldsymbol{F_{Hr}}$, acting on the single restraint, through to the formulation:

$$F_{Hr} = \frac{W_{ns} S_{ans}}{q_{ns}} \quad (1)$$

In Eq. (1), $\boldsymbol{W_{ns}}$ is the piping weight aliquot assigned to the single seismic restraint, evaluated as the product between the pipeline unit weight and the spacing between restraints. S_{ans} is the spectral acceleration demand of the system and q_{ns} is the behaviour factor of the element considered, assumed equal to 1.0.

In main lines, suspended piping restraints generally host multiple pipes, particularly in strategic buildings (Fig. 2b and c). This feature significantly influences the mass of the system and, consequently, the dynamic behaviour. Therefore, two different layout configurations were assumed in this work, including two and four pipes in the main line, respectively, to evaluate the influence of this aspect on the seismic vulnerability of the system. The diameter of all the pipes in main line was assumed equal to 2″. The mass of the system was computed considering the presence of water in all the pipes in main line and branching lines.

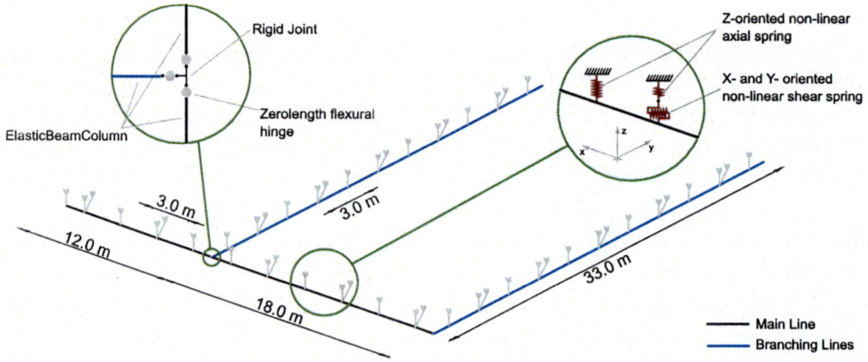

Fig. 1. Numerical model developed in OpenSees for the global analysis of the piping system.

The value of S_{ans} was computed adopting the simplified formulation (2) provided by Italian NTC18 [9]:

$$S_{ans} = PGA \cdot S_s \left\{ \frac{5 \cdot \left(1 + \frac{z}{h}\right)}{\left[1 + 4\left(1 - \frac{T_{ns1}}{0.8 \cdot T_{f1}}\right)^2\right]} \right\} \tag{2}$$

In Eq. (2), PGA and S_s are the peak ground acceleration and the soil coefficient, respectively, z and h are the quote of the piping system and the height of the structure, respectively, T_{ns1} and T_{f1} are the fundamental periods of the piping system and the structure, respectively. S_{ans} was calculated according to a conservative approach, assuming $T_{ns1}/T_{f1} = 1$.

The numerical model was developed by adopting a lumped plasticity approach. Linear elastic beam elements were used to simulate pipes, including zero-length bi-directional flexural springs (Fig. 1) to reproduce the hysteretic moment-rotation behaviour at the joint between multiple pipes. The non-linear response of suspended piping seismic restraints (Fig. 2c) was simulated including non-linear zero-length shear springs. Gravity piping restraints (Fig. 2a and b) were modelled using zero-length axial springs acting in vertical direction.

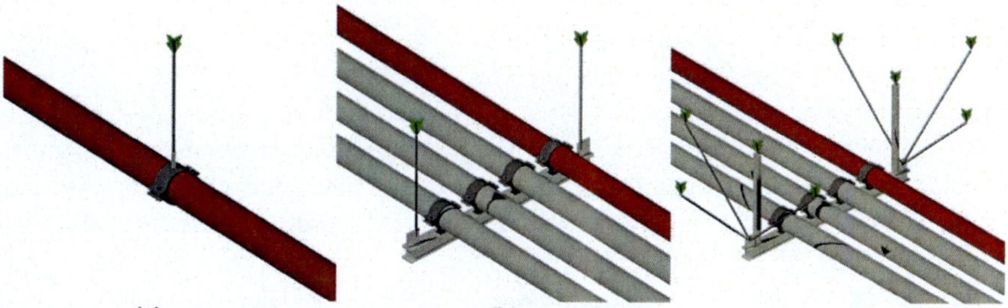

Fig. 2. Illustration of (a) gravity restraints in branching lines (b), gravity restraints in main line and (c) seismic piping restraint in main line.

The mass matrix was defined in the numerical model through a smeared approach, assigning a unit length mass to each linear beam element representing the pipe. Since the seismic demand on a single pipeline is influenced by the mass of all the pipes rigidly connected by the suspended piping restraint, the total mass of the main line was obtained by the sum of the masses of all the pipes (Fig. 2b and c).

The damping of the system was defined adopting the Rayleigh approach [14], by assigning a mass-proportional damping a stiffness-proportional damping to linear elements and lumped hinges, respectively. The damping ratio was assumed equal to 2%, according to the test results obtained by Blasi et al. [7]. A summary of the details of the analysed piping networks is provided in Table 1.

Table 1. Details of the analysed piping networks.

ID	FP_1		FP_2	
	Main line	Branch lines	Main line	Branch lines
Line length [m]	30.0	33.0	30.0	33.0
Pipe diameter [mm]	60.3	33.7	60.3	33.7
Pipe's Young's Modulus [MPa]	194383	194,383	194,383	194,383
Joint type	Threaded	Threaded	Threaded	Threaded
Number of pipes	4	1	8	1

2.1 Description of Nonlinearity Modelling

In the numerical model adopted in OpenSees, the non-linear response of the pipe connections at pipe-joint interface was simulated using the Pinching4 uniaxial material [15]. Two different hysteretic behaviours were defined for 2″ and 1″ pipe joints, respectively (i.e. for pipe connection in main and branching lines, respectively). The parameters defining the moment-rotation (M-θ) backbone curve and the pinching behaviour were calibrated in order to match the curves obtained from laboratory tests on 2″ and 1″ pipe joints [7, 16]. The hysteretic M-θ behaviour defined in OpenSees is provided in Fig. 3a and b for 2″ and 1″ pipe joints, respectively.

Both gravity and seismic restraints were simulated including zero-length non-linear springs, acting in vertical and in the two horizontal directions, respectively, and directly connected to the elements simulating the pipes. The mechanical behaviour of gravity supports was defined assuming a tri-linear elastic-hardening-softening Axial force-deformation (F-d) response in tension only (Fig. 4a). A classical approach was adopted to calibrate the yielding and ultimate point in the F-d bi-linear curve, based on the cross-sectional area of the cables and steel's yielding and ultimate strength, respectively.

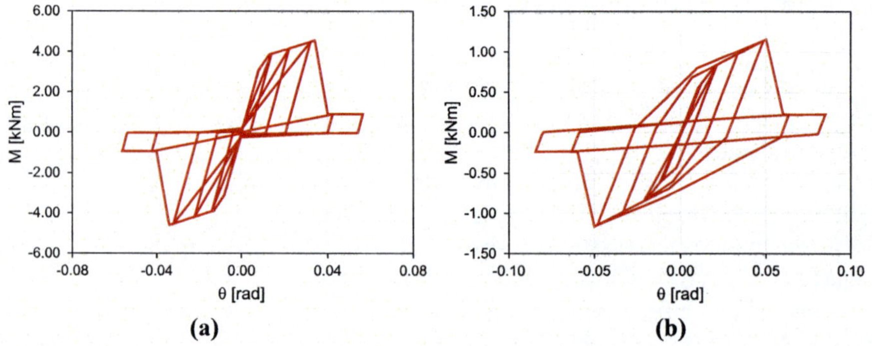

Fig. 3. Hysteretic M-θ behaviour in numerical model for (a) 2″ and (b) 1″ pipe joints.

The softening slope was aimed at simulating tensile failure of the cable, arbitrarily assuming residual strength and deformation values as 10% and 120% of ultimate strength and ultimate deformation, respectively. The F-d behaviour of the lumped springs simulating seismic restraints (Fig. 4b) corresponds to the X- and Y- lateral response of the braced elements. Similarly to pipe joints, the pinching parameters and the backbone curve points in the F-d curve were calibrated based on laboratory tests results [17].

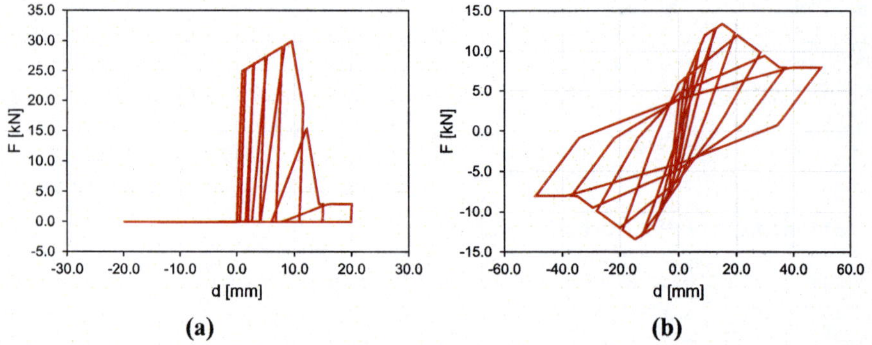

Fig. 4. Hysteretic F-d behaviour in numerical model for (a) gravity restraints and (b) seismic restraints.

2.2 Cascade Analysis

A preliminary non-linear dynamic analysis of an eight-storey RC frame was firstly carried out to obtain the floor acceleration time histories to use for the seismic analysis of the piping networks. The RC moment-resisting frame (Fig. 5) is characterized by nine bays at each floor in both principal directions. The inter-storey height and the bay length are equal to 3.4 m and 4.5 m, respectively. The properties of the RC frame elements were defined according to a simulate seismic design, according Italian NTC 2018, [9]. A high seismic hazard zone (Benevento, Campania), a soil type A and a nominal life of the building equal to 100 years were assumed, as provided in NTC-18 [9] for hospitals. The design peak ground acceleration (PGA) is equal to 0.331 g for a return period equal to 949 years, corresponding to a probability of exceedance equal to 10% in 100 years. More details of the configuration of the building and the modelling approach are available in [13].

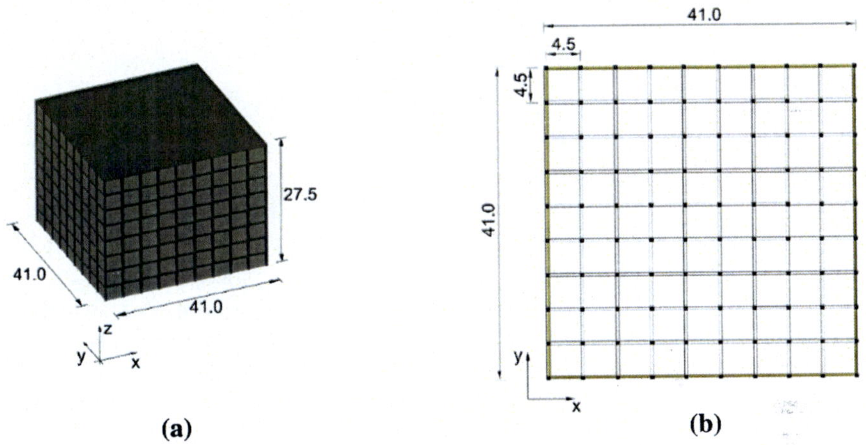

Fig. 5. 3D (a) and plan (b) configuration of the RC building analysed.

The seismic input employed for the analysis of the RC frame is composed of a set of 20 unscaled spectrum-compatible ground motions, selected from the European strong-motion database [18] using REXEL platform [19]. For each ground motion, the X, Y and Z components of the acceleration were applied in the three principal directions of the structure. The selected set was defined assuming 10% upper and lower deviation tolerance of matching between the average and the design spectrum, computed at Life-Safety (LS) performance level. The so-defined criterion meets Eurocode 8 provisions [9, 20]. The design spectrum parameters were set as described in previous section. The 5% damped elastic spectra of the selected ground motions, along with the design spectra and the fundamental period of the frame T_{f1}, are provided in Fig. 6.

Even though the operation of the building is not required at LS performance level, the analysis at this stage might be useful for the damage assessment of the piping system in case of post-elastic response of the structure. Moreover, the stability assessment of NSCs at life-safety performance level is required in modern seismic design codes (e.g. [9]).

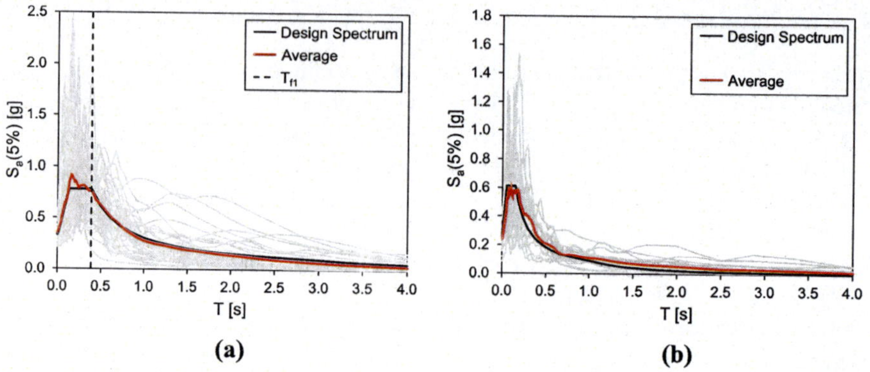

Fig. 6. (a) Horizontal and (b) vertical response spectra of the acceleration time histories considered for cascade analysis.

2.3 Nonlinear Dynamic Analysis Results

The seismic vulnerability of the piping systems analysed was assessed based on the damage of pipe joints and suspended piping seismic restraints, as well as the maximum displacements of the network. The damage rate of pipe joints was expressed as $\theta_{max}/\theta_{lim}$, where θ_{max} is the maximum rotation obtained from the analysis and is the limit rotation θ_{lim}. The value of limit rotation is not necessarily equal to the ultimate rotation capacity of the joint prior failure, because leakage of the joints after the attainment of the yielding rotation, θ_y, was observed by Tian et al. [6]. Since leakage represents loss of operation performance level, the value of θ_{lim} was set equal to θ_y, according to a conservative approach. In case of suspended piping seismic restraints, the damage rate was computed as $\delta_{max}/\delta_{lim}$, being δ_{max} the maximum deformation of the braced restraints obtained from the analysis and δ_{lim} the yielding deformation. Lastly, the maximum displacements (D_{max}) at the nodes connected to suspended piping restraints were monitored.

The analysis results obtained for FP_1 and FP_2, are organized in boxplots showing the median and the variation range among the 20 input motions. It is worth mentioning that the values of $\theta_{max}/\theta_{lim}$, $\delta_{max}/\delta_{lim}$ and D_{max}, discussed in the following refer to the maximum values computed among all the elements in the systems. The boxplots of the $\theta_{max}/\theta_{lim}$ values obtained for the two piping systems analysed in case of 2″ and 1″ pipe joints are provided in Figs. 7 and 8 respectively.

In each chart, the results obtained considering (Z accel ON) and neglecting (Z accel OFF) the vertical component of the acceleration in the analysis are compared. An increase of 2″ pipe joint damage is observed for both FP_1 and FP_2 when considering vertical acceleration (Fig. 7).

In FP_1, the median value of $\theta_{max}/\theta_{lim}$ was equal to 0.16 and 0.09 in case of Z accel ON and OFF, respectively. As expected, significantly higher values were obtained in case of FP_2, due to the higher mass of the main line. A similar difference between median $\theta_{max}/\theta_{lim}$ comparing Z accel OFF and ON is observed comparing FP_2 to FP_1 (− 45% and − 56%, respectively).

It is worth noting that the value of θ_{max} obtained for both FP_1 and FP_2 was lower than the yielding rotation for all the considered floor accelerograms, both in case of

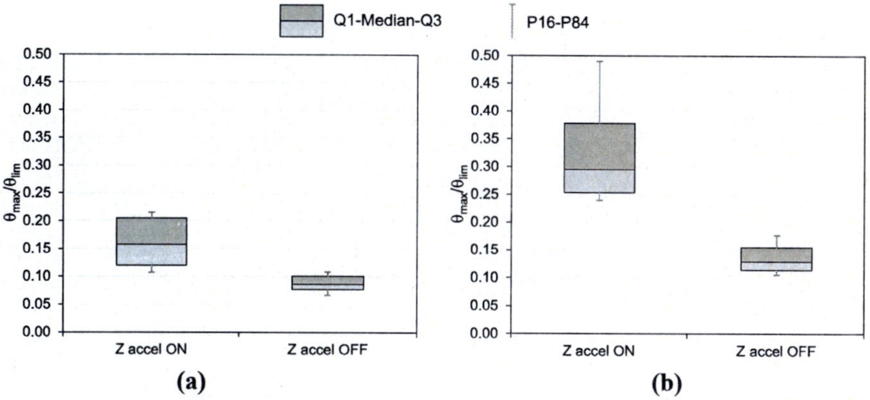

Fig. 7. Boxplots of the $\theta_{max}/\theta_{lim}$ data referred to 2″ pipes for (a) FP_1 and (b) FP_2.

2″ and 1″ pipe joints. Hence, loss of operation of the system was not attained due to absence of yielding. Referring to 1″ joints, a significantly lower influence of the vertical component on the results is observed compared to 2″ joints (Fig. 8). This outcome is mainly caused by the higher mass of the main line with respect to the branching lines, which led to higher seismic demand in the vertical direction on the joints (Fig. 9).

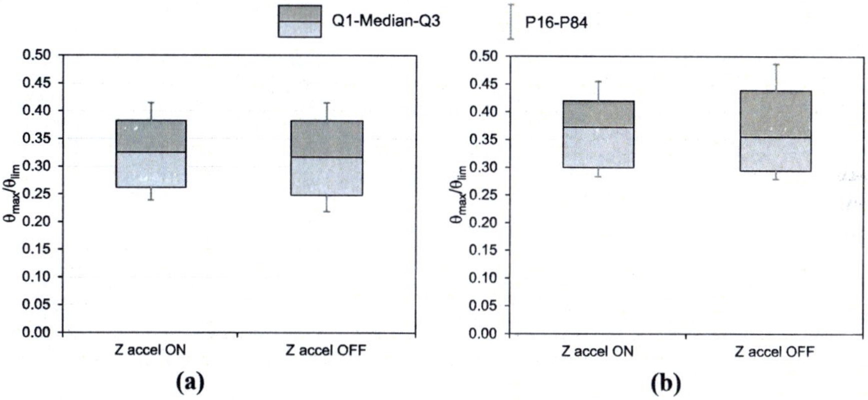

Fig. 8. Boxplots of the $\theta_{max}/\theta_{lim}$ data referred to 1″ pipes for (a) FP_1 and (b) FP_2.

Lastly, the boxplots reporting the maximum displacement D_{max} of the piping systems are provided in Fig. 10. A lower influence of the vertical acceleration on the results is obtained compared to $\theta_{max}/\theta_{lim}$ referred to 2″ pipe joints, both for FP_1 and FP_2. The median value of D_{max} decreased by 13% and 25% neglecting vertical acceleration in case of FP_1 and FP_2, respectively. In this case, the use of cables as vertical supports led to high deformation in the positive verse of vertical direction. This feature had major influence on the vertical displacements of the pipes and led to higher pipe joint damage. The results obtained can suggest a noticeable interaction between vertical motion and

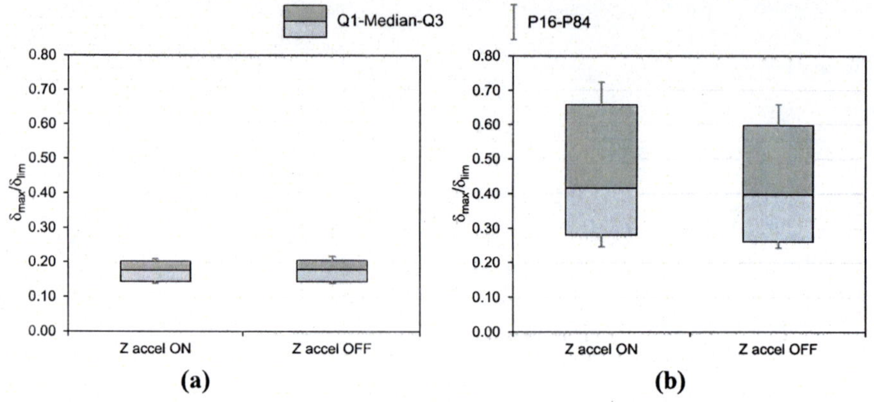

Fig. 9. Boxplots of the $\delta_{max}/\delta_{lim}$ data referred to suspended piping seismic restraints for (a) FP_1 and (b) FP_2.

local modes frequencies, which are highly influenced by the axial stiffness of suspended piping restraints.

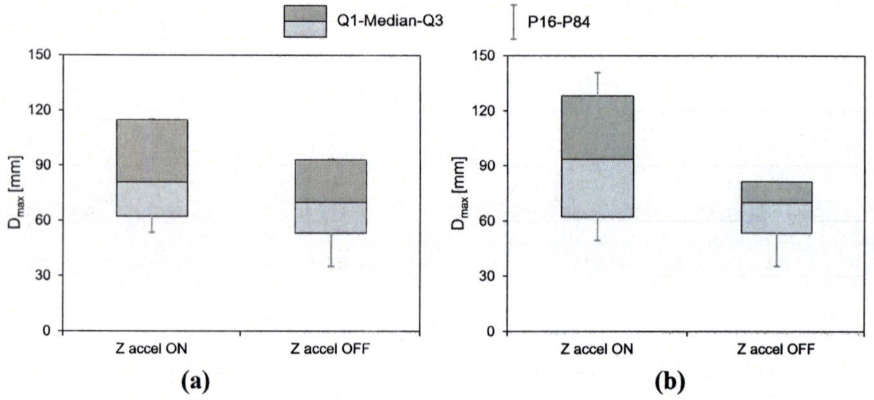

Fig. 10. Boxplots of the D_{max} data for (a) FP_1 and (b) FP_2.

3 Conclusions

The cascade analysis performed in this study was aimed at assessing the influence of vertical seismic acceleration on the vulnerability of two types of piping systems. An accurate modelling approach was adopted to simulate the response of the piping systems, accounting for the non-linear response of suspended piping restraints and pipe connections. The so-defined model allowed to analyse the mechanical damage and the loss of operation of the systems due to earthquake. The two piping systems considered in this study had different number of pipes running along the main line, in order to evaluate the influence of the mass of the network on the seismic vulnerability.

The results of the numerical simulation showed an underestimation of the damage on pipe joints when neglecting the vertical component on the seismic action. Higher joint rotations were obtained, due to the interaction between vertical motion and higher modes. This phenomenon was probably caused by the presence of gravity restraints realised with simple cables, which suffered high displacements. Referring to the suspended piping restraints, no significant influence of the vertical acceleration was detected on the maximum deformation of braced connections. The results obtained in this paper evidence the hardships in analysing the seismic response of piping systems, due to the high variation of the mechanical properties of the elements. An accurate design/assessment procedure should account for vertical acceleration to avoid a significant overestimation of the performance. Some aspects were not considered in this work and need further investigation, such as the presence of different types of gravity restraints rather than simple cables, which may significantly affect maximum displacements due to vertical action.

Acknowledgements. The authors gratefully acknowledge the support of "The Laboratories University Network of Seismic Engineering" (RELUIS) thanks to the research project DPC-RELUIS 2022–2024.

References

1. Perrone, D., Calvi, P.M., Nascimbene, R., Fischer, E.C., Magliulo, G.: Seismic performance of non-structural elements during the 2016 Central Italy earthquake. Bull. Earthq. Eng. **17**(10), 5655–5677 (2018). https://doi.org/10.1007/s10518-018-0361-5
2. Soroushian, S., Zaghi, A.E., Maragakis, E.M., et al.: Seismic fragility study of fire sprinkler piping systems with grooved fit joints. J. Struct. Eng. **141**:04014157 (2015)
3. Tadinada, S.K., Gupta, A.: Structural fragility of T-joint connections in large-scale piping systems using equivalent elastic time-history simulations. Struct. Saf. **65**, 49–59 (2017). https://doi.org/10.1016/j.strusafe.2016.12.003
4. Schneider, S.P.: Flexural capacity of pressurized steel pipies. J. Struct. Eng. **124**, 330–340 (1998)
5. Ju, B.S., Gupta, A.: Seismic fragility of threaded Tee-joint connections in piping systems. Int. J. Press. Vessel. Pip. **132–133**, 106–118 (2015). https://doi.org/10.1016/j.ijpvp.2015.06.001
6. Tian, Y., Filiatrault, A., Mosqueda, G.: Experimental seismic study of pressurized fire sprinkler piping subsystems. Technical Report MCEER-13-0001 (2013)
7. Blasi, G., Aiello, M.A., Maddaloni, G., Pecce, M.R.: Seismic response evaluation of medical gas and fire-protection pipelines' Tee-Joints. Eng. Struct. **173**, 1039–1053 (2018)
8. FEMA 412 (2002) Seismic restraints for mechanical equipment. Federal Emergency Menagement Agency
9. NTC-2018 (2018) Aggiornamento delle «Norme tecniche per le costruzioni». D.M. 17 gennaio 2018, Italy
10. Vukobratović, V., Fajfar, P.: Code-oriented floor acceleration spectra for building structures. Bull. Earthq. Eng. **15**(7), 3013–3026 (2016). https://doi.org/10.1007/s10518-016-0076-4
11. Merino, R.J., Perrone, D., Filiatrault, A.: Consistent floor response spectra for performance-based seismic design of nonstructural elements. Earthq. Eng. Struct. Dyn. **49**, 261–284 (2020). https://doi.org/10.1002/eqe.3236

12. McKenna, F., Fenves, G.L., Scott, M.H., Jeremir, B.: Open System for Earthquake Engineering Simulation. University of Berkeley, OpenSEES (2000)
13. Blasi G, Perrone D, Aiello MA, Pecce MR (2021) Seismic performance assessment of piping systems in bare and infilled RC buildings. Soil. Dyn. Earthq. Eng. **149**.https://doi.org/10.1016/j.soildyn.2021.106897
14. Strutt, J.W.: The Theory of Sound. Macmillan and co., London (UK) (1877)
15. Lowes, L.N., Mitra, N., Altoontash, A.: A Beam-Column Joint Model for Simulating the Earthquake Response of Reinforced Concrete Frames. University of California, Berkeley (2003)
16. Soroushian, S., Zaghi, A.E., Maragakis, E., et al.: Analytical seismic fragility analyses of fire sprinkler piping systems with threaded joints. Earthq. Spectra **31**, 1125–1155 (2015). https://doi.org/10.1193/083112EQS277M
17. Perrone, D., Filiatrault, A., Peloso, S., Brunesi, E., Beiter, C., Piccinin, R.: Experimental seismic response evaluation of suspended piping restraint installations. Bull. Earthq. Eng. **18**(4), 1499–1524 (2019). https://doi.org/10.1007/s10518-019-00755-5
18. Ambraseys, N.N., Smit, P., Douglas, J., et al: Internet site for European strong-motion data. Eur. Comm. Res. Gen. Environ. Clim. Program. (2002). http://www.isesd.hi.is/ESD_Local/frameset.htm
19. Iervolino, I., Galasso, C., Cosenza, E.: REXEL: computer aided record selection for code-based seismic structural analysis. Bull. Earthq. Eng. **8**, 339–362 (2010)
20. EN 1998-1 (2005) Eurocode 8—Design of structures for earthquake resistance—Part 1: General rules, seismic actions and rules for buildings. European Standard

Seismic Vulnerability of Pallet Storage Systems

Giacomo Piredda[1,2], Alberto Zonta[1], Enrico Bernardi[1,3], Marco Donà[1](✉), and Francesca da Porto[1]

[1] Department of Geosciences, University of Padova, via Gradenigo 6, 35131 Padova, Italy
marco.dona.1@unipd.it
[2] Department of Science, Technology and Society, Scuola Universitaria Superiore IUSS Pavia, Piazza Della Vittoria 15, 27100 Pavia, Italy
[3] Guangzhou Higher Education Mega Center, EERTC, Guangzhou University, 230 Wai Huan Xi Road, Guangzhou 510006, China

Abstract. In the last decade, increasing attention has been paid to the seismic vulnerability of steel pallet racks, mainly due to the catastrophic effects recorded in recent earthquakes and the increased use of these systems in the production and e-commerce sectors. For example, in May 2012 the Italian region of Emilia-Romagna was hit by a seismic swarm that affected many industrial buildings, causing significant damage to the structures as well as to the storage systems inside. In this paper, a parametric analysis to assess the vulnerability of these systems is presented. First, a small stock of 6-levels 7-spans pallet racks is generated by varying the upright section's properties and considering the unbraced and braced configurations. Then, the seismic vulnerability of this stock of structure is evaluated through nonlinear time history analyses and a Cloud-like approach. For this purpose, a specific code was developed in the OpenSees framework. The rack modelling includes the main geometric and material non-linearities, both for elements and connections, as well as the pallet-to-beam friction. Fragility curves are derived as a function of the spectral acceleration (S_a) for two engineering demand parameters, significant in representing the damage condition of these structures. Finally, fragility results are presented, discussed, and compared, in order to provide useful information on the vulnerability of these structural systems.

Keywords: Steel racking systems · Braced/unbraced pallet racks · Nonlinear rack modelling · 3D FEM models · Cloud analysis · Seismic vulnerability · Fragility curves

1 Introduction

The increasing demand for the storage of goods, the development of e-commerce and the globalization have increased the production and use of storage systems. The most used structures for this purpose are the adjustable pallet racking, made by assembling cold-formed steel profiles with open thin cross-section. The pallet racks show specific characteristics in the two main directions. In the longitudinal direction (down-aisle), they consist of uprights connected by pallet-racking beams; in the transverse direction

G. P. Cimellaro (Ed.): WCSI 2022, LNCE 309, pp. 761–770, 2023.
https://doi.org/10.1007/978-3-031-21187-4_65

(cross-aisle), they are characterized by a series of trusses, made with horizontal and diagonal elements that connect the uprights to each other.

The slenderness, the use of semi-rigid connections and the high payload, make these structures sensitive to instability problems and therefore very vulnerable to seismic actions. Furthermore, many of the racks currently used have not been designed to withstand horizontal seismic actions in accordance with recent regulations (e.g., [1, 2]).

Their great vulnerability was demonstrated by the earthquakes in Emilia-Romagna in 2012 and in Central Italy in 2016 [3]. These events also highlighted the drastic impact of seismic damage to warehouses and storage systems on production downtime and therefore on the indirect losses of companies, such as to jeopardize their survival [4, 5].

Although there are numerous studies on the seismic behavior of racking systems and their main components (such as [6–8]), specific studies on their seismic fragility are very few. To the authors' knowledge, only Gabbianelli et al. [9] evaluated the fragility for two rack configurations, unbraced and braced, through a multiple-stripe analysis approach [9].

In this paper, the fragility of a small stock of six pallet racks (three unbraced and three braced) is investigated through a Cloud-like approach, based on a 3D rack modeling. For this purpose, a parametric code was developed in the OpenSees framework. Each rack is analyzed in Time-History (TH) by applying 134 bidirectional seismic inputs, representative of the Italian seismicity. The non-linear behaviors of the base and beam-to-column (BTC) connections are modelled, as well as the non-linear flexural and axial responses (including buckling) of the uprights, and the pallet-to-beam friction. The fragility curves are derived conditional on the spectral acceleration associated with the main vibration period of the rack (S_a), and with respect to two representative engineering demand parameters (EDPs), i.e., the maximum rotation of the beam-to-column connections and the maximum drift between load levels.

2 Case Studies and Structural Modelling

2.1 Presentation of the Case Studies

In this study, three pallet racks, designed to support gravitational loads only and which differ in the type of upright, are taken as reference case studies. These racks are analyzed in two configurations, i.e., unbraced (original situation) and with the insertion of a spine or vertical bracing. These structures are double-entry racks, anchored to the industrial floor using base plate connections, which make a semi-rigid joint in the longitudinal direction (down-aisle) and a hinge connection in the transverse direction (cross-aisle).

All the case studies have the same layout in plan, shown in Fig. 1, characterized by spans with a length of 2700 mm and a depth of 1000 mm; the space between the two entries is 300 mm. The racks have a total height of 9000 mm and consist of six load levels with an inter-story height of 1500 mm (see Fig. 2a). As already mentioned, the case studies differ only in the type of upright and in the presence or absence of vertical bracing. The pallet load applied in each structure depends on the type of upright used (as well as the height of the load level and the type of frame).

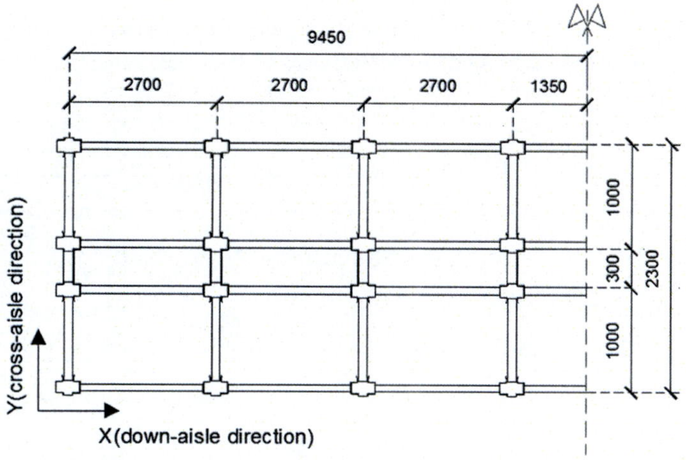

Fig. 1. Plan layout of the case study racks (all measures are in mm).

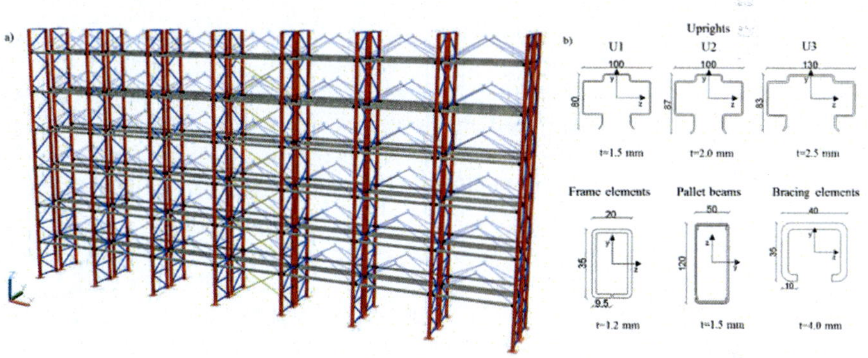

Fig. 2. a) Global view of the generic 7-spans, 6-load-levels case study rack; b) cross sections of uprights, pallet beams, and frame elements.

The uprights, named for simplicity U1, U2, and U3, have different geometry and therefore inertial characteristics (see Fig. 2b). The U1 upright is the one with the smallest section and consequently with the lowest values of resisting moment (in both directions) and maximum axial load. The U2 upright has intermediate inertial characteristics between U1 and U3, while the U3 upright is the one with the greatest inertial and strength properties.

In all the analyzes carried out, the reductions in axial and flexural strengths due to the instability effects are also taken into account, according to [2]. Specifically, for the profiles used, the values of the buckling factor (χ) for the axial stress range from 0.93 to 0.97 depending on the upright, whereas those of the flexural-torsional buckling factor (χ_{LT}) are only slightly lower than 1.

As regards the BTC connections, different values of stiffness ($k_{\varphi,\,Y}$), yield strength ($M_{\varphi y,\,Y}$) and peak strength ($M_{\varphi p,\,Y}$) are assumed according to the uprights, as reported in Table 1. These values derive from experimental evaluations in the availability of the authors. Instead, the stiffness and strength values of the base connections are assumed as in [9].

Table 1. Rotational stiffness of the beam-to-column (BTC) and base connections.

	$k_{\varphi,Y}$ [N·mm/rad]	$M_{\varphi y,Y}$ [10^6 N·mm]	$M_{\varphi p,Y}$ [10^6 N·mm]
BTC joint (U1, U2)	$1.17 \cdot 10^8$	1.13	1.69
BTC joint (U3)	$1.80 \cdot 10^8$	1.34	2.01
Base joint	$4.35 \cdot 10^8$	2.61	5.22

2.2 Presentation of the 3D Finite-Element Models

To effectively analyze the fragility of these structures, a specific procedure is developed in the OpenSees framework [10], in order to automatically generate and analyze, in the time domain, 3D non-linear finite-element (FE) rack models.

The non-linearities of the materials are modeled differently for the various rack elements. In particular, a spread plasticity is assumed for the uprights, for both axial and flexural response, whereas a concentrated plasticity is considered for the base and beam-to-column connections. Pallet beams and spacers, on the other hand, are modelled as linear elastic elements.

The pallet masses are positioned at the top of a pyramidal substructure (dummy substructure), which allows to take into account the actual center of gravity of the pallet, as indicated in [1]. These masses can slide above the pallet beams, in the cross-aisle direction, thanks to the insertion of frictional contact elements.

The uprights are modeled as *Force-Based Beam-Column* elements, with different sectional behaviors for the various degree of freedom. As regards the axial behavior, the *uniaxialMaterial* object *Pinching4* is used (see Fig. 3a), calibrated in such a way as to take into account the buckling in compression, which does not allow the yield strength to be reached in the negative branch. Instead, an elastic-plastic behavior is modeled for the flexural responses in the two main directions. As for the behaviors in the remaining degrees of freedom, i.e., shear and torsion, they are assumed to be linear elastic.

Linear elastic elements with end hinges (*Beam with Hinges* elements) are used for the diagonal and horizontal elements of the cross frame, so as to simulate their truss behavior. Pallet beams, on the other hand, are modeled as simple *Elastic Beam Column* elements, with elastic stiffnesses clearly derived from section properties.

In the case of braced racks, the bracing system is positioned on the central span, and modeled through *Force-Based Beam-Column* elements. Their flexural stiffness is set to zero, making these elements equivalent to trusses, whereas their axial response is simulated using the same shape of hysteretic response as uprights.

The rotational behavior of the base and BTC connections is modeled using *Zero-Length elements* (ZLE). Regarding BTC connections, a hysteretic behavior is implemented in the down-aisle direction through the *uniaxialMaterial* object *Hysteretic*; the shape of this hysteretic law, calibrated with the values set out in Table 1, is provided by [9]. In the other two directions, the behavior of the BTC connections is linear elastic; in particular, the rotational stiffness $k_{\varphi,Y}$ (Table 1) is also assumed for rotations around the Z axis, whereas the rotation around X axis (beam torsion) is set equal to that of the beam (using the *equalDOF* command). Similarly, as regards the base connections, a hysteretic material is implemented in the down-aisle direction, still according to [9]. However, these connections behave like hinges in the cross-aisle direction, and an *equalDOF* is still assumed for torsional rotations.

Lastly, to simulate the sliding of the pallet, a friction model is inserted between the top node of the dummy substructure and the upper node of the pallet mass, as shown in Fig. 3.b, by means of a ZLE.

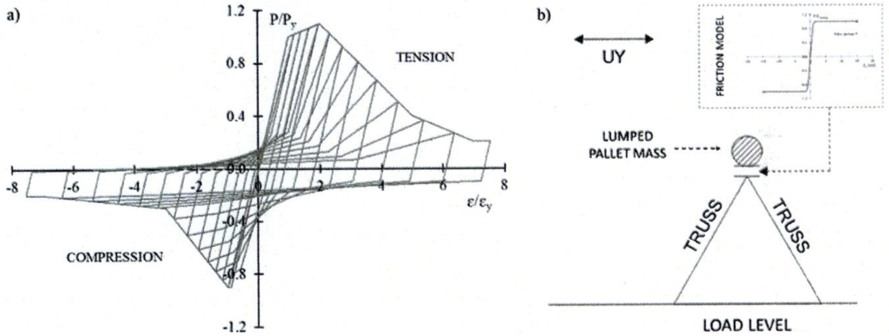

Fig. 3. (a) Non-linear axial behavior of uprights. (b) Dummy substructure and friction model.

3 Methodology

Various methods can be used to derive fragility functions through nonlinear time-history (NLTH) analyses. They are generally based on one of the following analysis procedures: the *Incremental Dynamic Analysis* [11], the *Multiple Stripe Analysis* [12], or the *Cloud Analysis* [13]. Cloud Analysis, unlike the first procedures, makes use of natural records only, and it is selected for this study since a reliable and site-independent database of natural ground motion records was available. Fragility curves are generally expressed as lognormal cumulative density functions (CDFs) and provide the probability of reaching or exceeding a certain damage state (DS_i), conditioned on a certain level of the earthquake intensity measure (IM). Two parameters are needed to define the fragility curve for each DS_i, i.e., the median ($\mu_{DS,i}$) and the logarithmic standard deviation ($\beta_{DS,i}$) of the seismic intensities (IM_j) which result in that DS_i. These parameters are derived by using the moments method [14]. Fragility curves can be computed by means of Eq. (1), where D is the seismic demand and Φ represents the standardized normal CDF.

$$P(D \geq DS_i|IM) = \Phi\left[\frac{1}{\beta_{DS,i}} \cdot ln\left(\frac{IM}{\mu_{DS,i}}\right)\right] \tag{1}$$

The natural records adopted in this study for the NLTH analyses are a subset of a database provided by [15]. In particular, 134 unscaled bidirectional events are selected, which are representative of Italian earthquakes in rigid soils, with return periods from 50 to 5000 years. This dataset was specifically developed within a national project (ReLUIS-DPC 2019–2021, Work Package 4 – MARS) to identify seismic events suitable for obtaining fragility curves with a Cloud-like approach [16].

As already noted by [9], the most suitable IM parameter to evaluate the fragility of such structures is the spectral acceleration (S_a), as it is the actual seismic action on the pallet racking. Indeed, the S_a values could differ significantly from the PGA values, due to the relatively high vibration periods of the racks, which also differ greatly between the two main directions. Hence, the fragility in this study is conditioned on S_a.

Two engineering demands parameters (EDPs) are chosen to describe the damage level of the case study racks, also based on the previous results of [9]. These EDPs are: (*i*) the maximum rotation of the BTC connections in the down-aisle direction, φ_{btc}; (*ii*) the maximum drift between load levels in the down-aisle direction, θ_X. The thresholds for the achievement of three sequential DSs are then set for each EDP, as reported in Table 2 (as a function of the BTC yield rotation, $\varphi_{btc,y}$, and of the yield drift of the uprights, $\theta_{X,\,y}$). The meaning of these three DSs is listed below:

- DS1: elastic range, incipient damage conditions; element integrity and workers' safety are guaranteed
- DS2: inelastic range, post-yield conditions; some elements are damaged, but workers' safety is still guaranteed
- DS3: inelastic range, post-peak conditions; some elements are close to collapse, workers' safety is no longer guaranteed.

Although the achievement of DS3 corresponds to the collapse of a single element, i.e., to a local rather than global collapse, this condition can be conservatively assumed as an incipient global collapse, due to the low level of redundancy and robustness of these structures, which are also sensitive to cascade collapses.

Table 2. Damage state (DS) thresholds, for each engineering demand parameter (EDP).

EDP	Description	DS1	DS2	DS3
φ_{btc}	Maximum rotation of the beam-to-column connections	$\geq 0.8\ \varphi_{btc,y} <$	$\geq 1.0\ \varphi_{btc,y} <$	$\geq \varphi_{btc,p} = 4.0\ \varphi_{btc,y}$
θ_X	Maximum drift between load levels (down-aisle direction)	$\geq 0.8\ \theta_{X,y} <$	$\geq 1.0\ \theta_{X,y} <$	$\geq 1.5\ \theta_{X,y}$

4 Fragility Curves

The fragility curves, obtained from the NLTH analyses with the Cloud-like procedure described above, are shown in Figs. 4 and 5, respectively for the unbraced and braced racking configurations, and for the two EDPs chosen (i.e., φ_{btc} in figures (a), and θ_X in figures (b)). The values of median, $\mu_{DS,i}$, and logarithmic standard deviation, $\beta_{DS,i}$, for the various DSs of each evaluated EDP are reported in Table 3.

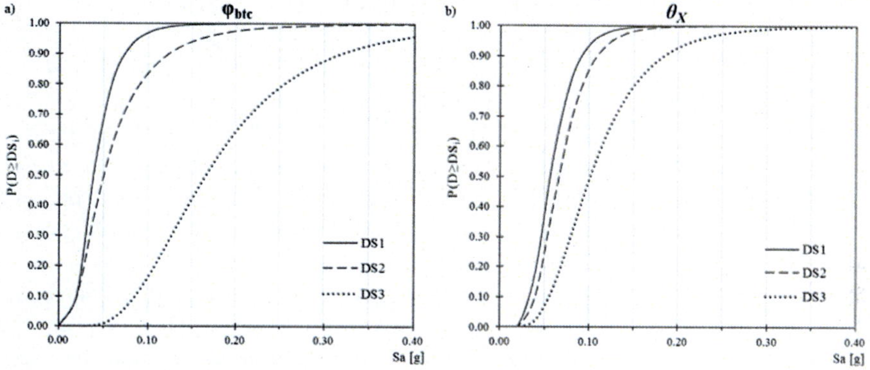

Fig. 4. Fragility curves of unbraced racks: (a) maximum *BTC* joint rotation, φ_{btc}; (b) maximum drift in down-aisle direction, θ_X.

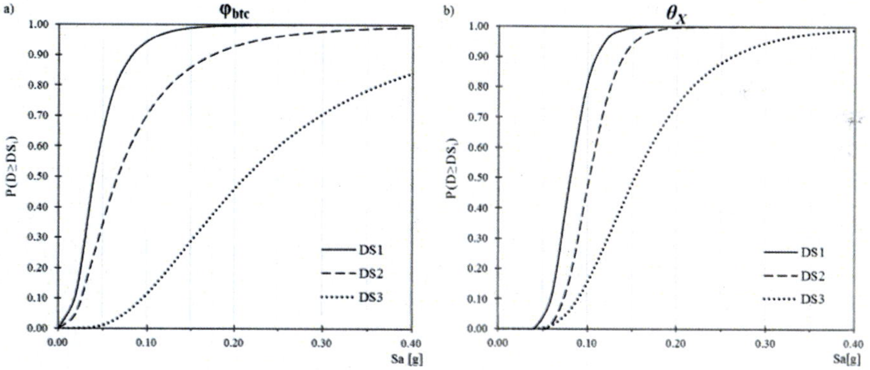

Fig. 5. Fragility curves of braced racks: (a) maximum *BTC* joint rotation, φ_{btc}; (b) maximum drift in down-aisle direction, θ_X.

From Figs. 4 and 5 the following considerations can be drawn.

- In general, for both unbraced and braced rack configurations, the fragility associated with φ_{btc} was slightly higher than that relating to θ_X (i.e., with lower μ_{Sa} values) for the lower DSs (DS1 and DS2), whereas it was lower (i.e., with higher μ_{Sa} values) for the DS3. This is explained by the fact that when some BTC connections exceed

Table 3. Statistical parameters of fragility curves: median, $\mu_{DS,i}$, and logarithmic standard deviation, $\beta_{DS,i}$.

Rack type	EDP	DS1 $\mu_{DS,i}$ ($\beta_{DS,i}$)	DS2 $\mu_{DS,i}$ ($\beta_{DS,i}$)	DS3 $\mu_{DS,i}$ ($\beta_{DS,i}$)
UB	φ_{btc}	0.038 (0.509)	0.050 (0.703)	0.165 (0.511)
		0.056 (0.377)	0.066 (0.392)	0.101 (0.466)
BR	φ_{btc}	0.040 (0.578)	0.067 (0.748)	0.213 (0.630)
		0.080 (0.238)	0.102 (0.247)	0.153 (0.420)

the related DS2, i.e., exceed their yield strain, the inter-story stiffness of the racking in the down-aisle direction is considerably reduced, causing a rapid increase in the values of θ_X. As a consequence, the racks were brought to collapse due to excessive lateral deformation of the uprights rather than the achievement of the peak resistance of the BTC connections. This also demonstrates the correctness of the peak resistance of the BTC connections used.

- Furthermore, considering the EDP θ_X, which governs the rack collapse, it can be seen from both Figs. 4 and 5 that the fragility curves of the various DSs are quite close to each other, i.e., the fragility interval between the conditions of incipient damage (DS1) and incipient collapse (DS3) is quite narrow. In particular, the distance between the curves DS2 and DS3, which defines the response interval of the racking in yield conditions, highlights the low redundancy and ductile capacity of these structural systems, as is well known.
- The outcome that the inter-story drift (θ_X) drives the collapse of these structural systems, and the fact that the distance between fragility curves associated with a post-yield damage state (such as DS2) and a condition of incipient collapse (such as DS3) is relatively small for these systems, also find confirmation in a previous study [9].
- As regards the insertion of the bracing system, the results obtained show that it reduces the seismic fragility of the racks, as expected. In particular, the vulnerability reduction mainly concerns the EDP θ_X, for which the μ_{Sa} values increase for all DSs. As for the EDP φ_{btc}, the influence of the bracing system is observed on the higher DSs. Comparing the curves of Fig. 4 with those of Fig. 5, it is also observed that the presence of the bracing system tends to increase the $\beta_{DS,i}$ values (i.e., the dispersion of the results) for the EDP φ_{btc}, and to reduce them for the EDP θ_X (especially for lower DSs). This would seem to be related to the different vibration modes of the racking in the case of the bracing system, which amplifies the torsional modes.

Lastly, it is worth noting that if the fragility results for racks without bracing system find some confirmations (in the order of magnitude) in previous studies (e.g., [9]), those relating to braced systems must be further investigated, also because they are strongly conditioned by the chosen bracing system (profiles and layout) and by the type of racking in which it is applied (single or double-entry, with various inter-story heights).

5 Conclusions

The seismic fragility of a small stock of steel pallet racks was assessed in this paper. In particular, three unbraced and three braced medium-rise double-entry systems were analyzed, accounting for a variation in the section of the uprights for both types of racking.

The 3-D finite-element models of the racks were created in the OpenSees framework, considering the geometric non-linearities and the main material non-linearities. In particular, base and beam-to-column connections were modelled by using non-linear hinges. Spread non-linearities were assigned to the upright sections, to simulate both axial and flexural non-linear behavior. A friction model was also employed to describe the sliding phenomena of the pallet masses on the pallet beams.

For each of these models, a parametric NLTH analysis was carried out, considering a set of 134 bi-directional natural records, representative of the Italian seismicity. The fragility of these structures was therefore derived through a Cloud-like approach. Two engineering demand parameters (EDPs) were chosen, i.e., the maximum rotation of the BTC connections (φ_{btc}) and the maximum drift between load levels in the down-aisle direction (θ_X), and three damage states (DSs) were defined for each EDP. The fragility curves were derived conditioned to the spectral acceleration (S_a), as it is an appropriate intensity measure for this type of structure.

Overall, the fragility results demonstrated the high vulnerability of these storage systems, as well as the possible reduction of this vulnerability through suitable bracing systems. The EDP that drove the racking to collapse is the inter-story drift (θ_X), and the BTC connections seem to play an important role in determining the damage condition (DS1-DS2), and therefore in delaying/anticipating the achievement of structural collapse. Finally, further investigations are needed in order to broaden the case series and provide results of more general validity.

Acknowledgements. This study was supported by the program "POR FSE VENETO 2014–2020"—project 2105-0031-14632019, title: "*Innovative technologies for the mitigation of the effects of dynamic actions on storage/racks systems*".

References

1. CEN (European Committee for Standardization): EN 15512. Steel Static Storage Systems—Adjustable Pallet Racking Systems—Principles for structural design. Brussels (2020)
2. CEN (European Committee for Standardization): EN 16681. Steel Static Storage Systems—Adjustable Pallet Racking Systems—Principles for seismic design. Brussels (2016)
3. Perrone, D., Calvi, P.M., Nascimbene, R., Fischer, E.C., Magliulo, G.: Seismic performance of non-structural elements during the 2016 Central Italy earthquake. Bull. Earthq. Eng. **17**(10), 5655–5677 (2018). https://doi.org/10.1007/s10518-018-0361-5
4. Donà, M., Bizzaro, L., Carturan, F., da Porto, F.: Effects of business recovery strategies on seismic risk and cost-effectiveness of structural retrofitting for business enterprises. Earthq Spectra **35**(4), 1795–1819 (2019)
5. Brown, C., Stevenson, J., Giovinazzi, S., Seville, E., Vargo, J.: Factors influencing impacts on and recovery trends of organizations: evidence from the 2010/2011 Canterbury earthquakes. Int. J. Disaster Risk Reduction **14**, 56–72 (2015)

6. Baldassino, N., Zandonini, R.: Performance of base-plate connections of steel storage pallet racks. In: Proceedings of Fifth International Conference on Coupled Instabilities in Metal Structures (CIMS2008), pp. 119–130. Gregory J. Hancock Symposium, Sydney, Australia (2008)
7. Padilla-Llano, D.A., Moen, C.D., Eatherton, M.R.: Cyclic axial response and energy dissipation of cold-formed steel framing members. Thin-walled Struct. **78**, 95–107 (2014)
8. Castiglioni, C.A., Drei, A., Kanyilmaz, A., Mouzakis, H.P.: Earthquake-induced pallet sliding in industrial racking systems. J. Build. Eng. **19**, 122–133 (2018)
9. Gabbianelli, G., Francesco, C., Nascimbene, R.: Seismic vulnerability assessment of steel storage pallet racks. Ingegneria Sismica **37**(2) (2020)
10. McKenna, F., Fenves, G.L., Scott, M.H.: Open System for Earthquake Engineering Simulation. University of California, Berkeley, CA (2000)
11. Vamvatsikos, D., Cornell, C.A.: Incremental dynamic analysis. Earthq. Eng. Struct. Dynam. **31**(3), 491–514 (2002)
12. Jalayer, F., Cornell, C.A.: Alternative non-linear demand estimation methods for probability-based seismic assessments. Earthq. Eng. Struct. Dynam. **38**(8), 951–972 (2009)
13. Jalayer, F.: Direct probabilistic seismic analysis: implementing non-linear dynamic assessments. Stanford University (2003)
14. Masi, A., Lagomarsino, S., Dolce, M., Manfredi, V., Ottonelli, D.: Towards the updated Italian seismic risk assessment: exposure and vulnerability modelling. Bull. Earthq. Eng. **19**(8), 3253–3286 (2021). https://doi.org/10.1007/s10518-021-01065-5
15. Paolucci, R., Ozcebe, A. G., Smerzini, C., Masi, A., Manfredi, V.: Selection and spectral matching of recorded ground motions for earthquake engineering analysis. Available at: http://143.225.144.186:5000 (2020)
16. Manfredi, V., Masi, A., Özcebe, A.G., Paolucci, R., Smerzini, C.: Selection and spectral matching of recorded ground motions for seismic fragility analyses. Bull. Earthq. Eng. 1–27 (2022).https://doi.org/10.1007/s10518-022-01393-0

A Simplified Framework to Generate Fragility Functions for In-Plane Behavior of Gypsum Partition Walls

Fabio Longo[1(✉)], Daniele Perrone[1], Emanuele Brunesi[2], Simone Peloso[2], and Maria Antonietta Aiello[2]

[1] Department of Innovation Engineering, University of Salento, via Per Monteroni, 73100 Lecce, Italy
fabio.longo@unisalento.it
[2] Eucentre, via Ferrata, 27100 Pavia, Italy

Abstract. Non-structural components play a significant role in the definition of economic losses produced by seismic events. Gypsum partition walls represent a significant portion of the non-structural components and can manifest considerable damage even at low-intensity seismic events. Nonetheless, the non-structural partition walls are generally not taken into account in the context of traditional analysis and design. This is also a consequence of the lack of knowledge related to a clear definition of performance limit states, owing also to the wide range of partitions' typologies available on the market. To deal with this issue, a database collecting the results of quasi-static experimental tests was developed. The database was used to collect data useful to identify performance parameters and drift levels at which the different limit states are attained. The hysteretic curves collected in the database were used to develop a single-degree-of-freedom model for each partition typology. This paper briefly presents a general framework to calculate fragility functions for internal partitions using the data available in the experimental database. A database of floor displacement time histories evaluated by analyzing a building portfolio composed of reinforced concrete moment resisting frames was used to perform the multi-stripe analyses required to generate the fragility functions. The proposed framework is illustrated by calculating the fragility functions for one of the gypsum partition typologies collected in the experimental database.

Keywords: Non-structural components · Gypsum partitions · Fragility functions · Framework

1 Introduction

The attention to the behavior of non-structural components (NSCs) during earthquake events has steadily grown in the last decades. Despite structures are nowadays designed to be earthquake-resistant, damage can be significant in NSCs, such as partitions, ceiling, floors, furniture, etc. As a consequence, the greatest part of economic losses following seismic events are ascribed to NSCs themselves, [1], and damage to them can potentially interrupt the use of a building, [2–4].

G. P. Cimellaro (Ed.): WCSI 2022, LNCE 309, pp. 771–781, 2023.
https://doi.org/10.1007/978-3-031-21187-4_66

Among NSCs, partition walls are widely adopted worldwide [5]. They can be found nearly in every building and interact with the load-bearing system in case of a seismic event. Beyond the design assumptions, partition walls are subjected to the dynamic response of the building and exposed to drift-imposed damage, [6].

Different types of partition walls can be found among many solutions available in the construction market, implying also that their classification can be based on the construction system and materials involved. This being stated, drywall systems offer the possibility to fasten the installation and reduce the weight of partition elements, which is something designers would favor and opt for. Additionally, different materials can be involved in these systems for both frames (i.e. wood, aluminum, steel) and panels (gypsum, wood, insulating materials, etc.).

Although partition walls are widespread, almost ubiquitous, seismic design codes (e.g. Eurocode 8 and the Italian design code NTC 2018 [7, 8]) do not suggest any specific limit states for their design/assessment. This issue surely relates to the lack of extensive experimental campaigns investigating the seismic performance of the wide range of partition typologies available on the market. In order to define a limit state, in fact, experimental tests are required to simulate the behavior of the partition walls under seismic forces and observe the corresponding damage, correlating it with threshold(s) in terms of one or more engineering demand/capacity parameter. In this context, the development of fragility functions relying upon probabilistic approaches is nowadays of particular relevance in performance-based earthquake engineering [9].

In light of the above, this study deals with the identification of a simplified and general framework to derive fragility curves for partition walls. The fragility curves developed according to the proposed approach can be used in the design and assessment of new and existing buildings equipped with the considered partition wall typology, particularly when the expected annual loss is targeted and taken as the main evaluation metric.

Given that the performance parameters describing the response of partitions were calibrated by using experimental data, a database of experimental tests involving gypsum partition walls was compiled as part of this study, as detailed hereinafter.

2 The Proposed Simplified Framework

In this section, a simplified framework to develop fragility functions for partition walls is described. As already stated, prescriptions are currently lacking in seismic codes, such as [7], when dealing with the definition of limit states for partition walls. The definition of limit states is crucial to calculate fragility functions, regardless of whether numerical and/or experimental data are considered for their derivation. Numerical analyses are generally used to predict the partition displacements induced by drifts of the load-bearing structure. To perform numerical analyses, mechanical component-based finite element models simulating the response of the partition walls could be a viable approach. In this case, however, not only experimental test data of the entire partition would be required, but also those of the components necessary to assemble the partition, in such a way that mechanics-based modelling of the entire partition could be fed with reliable information for all key components. On the other hand, a simplified single-degree-of-freedom (SDOF) modelling approach would only require the hysteretic curves obtained by testing along with the drift levels corresponding to the attainment of the limit states.

The entire proposed process is shown in Fig. 1. Initially, the experimental hysteretic response curves of different specimens are collected and analyzed. Also, a table of drift ratios corresponding to the attainment of the limit states of the partitions is compiled. Subsequently, a simplified numerical model is calibrated. The experimental and numerical hysteretic curves are compared together and the parameters characterizing the numerical model response are modified until a satisfying experimental vs numerical comparison is accomplished. Once the numerical model is calibrated, nonlinear time history analyses are performed. Finally, the outputs of the numerical analyses are postprocessed to compute the parameters required to define the fragility curves. A step-by-step description of the framework procedure is reported in the next paragraphs.

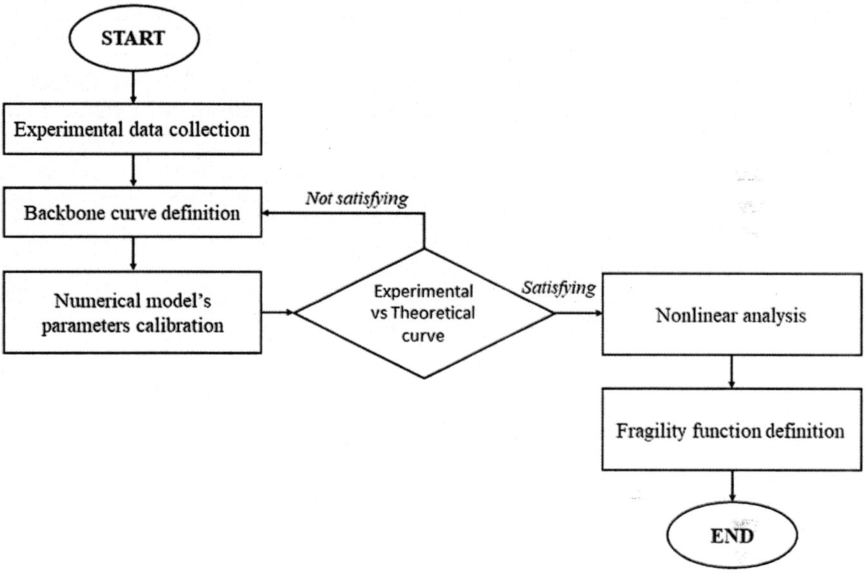

Fig. 1. Flowchart of the proposed framework.

2.1 Experimental Data Collection

Partition walls are in-plane experimentally tested essentially through two different testing procedures. The former consists in quasi-static tests, which involves the use of a hydraulic actuator to impose a cyclic displacement history, simulating an interstory drift applied to the specimen. The latter, instead, consists in shake table dynamic tests, which are generally used for seismic qualification purposes. The database developed in this study collect experimental data from both quasi-static and shaking table tests. Nevertheless, the data analysis was performed only including quasi-static test results. In fact, it is hard to define a hysteretic curve from shake table test results, which for instance do not necessarily provide information to characterize in-cycle degradation.

As already stated, it is essential to collect all the data necessary to calibrate the numerical models and define the drift thresholds at which the different limit states are reached.

The database collected and presented in this work refers to the experimental campaigns carried out in eight different studies, for a total of 19 samples. Kanvinde & Deierlein [10] pointed out that, even when not designed as seismic-resisting elements, gypsum drywall partitions can enhance the strength and stiffness of wood-frame structures. The first experimental quasi-static tests performed on gypsum partition walls are dated back to 2007. Lee et al. [11] examined the effects of cyclic loading on four full-scale drywall partitions with light-gauge steel stud framing. Moreover, the influence of door openings was investigated as well. Different configurations were tested through the years. For example, Restrepo & Bersofsky [12] tested both the door opening's and partial height wallboard's influence through in-plane quasi-static tests.

In-plane quasi-static tests were also presented by Davies et al. [13], who carried out an extensive experimental campaign including 36 different partition wall configurations. The effects of many construction details, such as different materials, connections, and testing protocols, as well as wall dimensions and boundary conditions, on the seismic performance were investigated. Further insights can be found in the work by Petrone et al. [14], wherein plasterboard internal partitions, representing typical partitions in industrial and commercial buildings in the European area, were tested in-plane. Innovative connections aimed to reduce the partition's damage were in-plane quasi-statically tested by Tasligedik et al., Pali et al., Araya-Letelier and Mulligan et al. [15–18].

2.2 Backbone Curve Definition

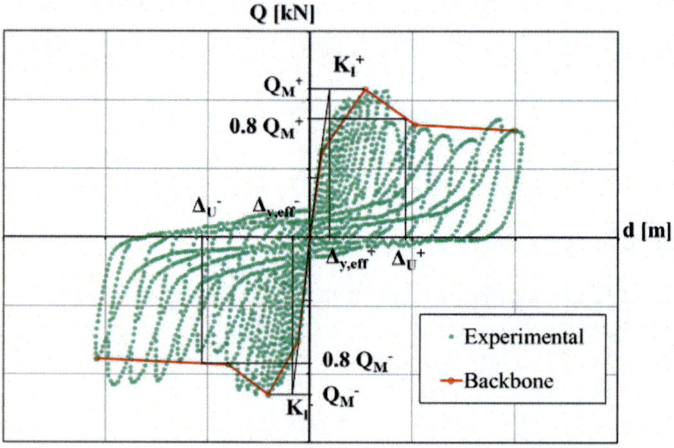

Fig. 2. Example of performance parameters obtained for positive and negative loading directions.

The hysteretic curve data can be treated to define the backbone envelope curve [9], as shown in Fig. 2. The backbone envelope curve represents the nonlinear monotonic response that envelopes the cyclic response and can be described by using some performance parameters. In particular, the main performance parameters evaluated according to FEMA P-795 are: the maximum load, QM, the initial stiffness, KI, the effective yield deformation, Δy,eff, and the ultimate deformation, ΔU. These performance parameters are calculated for both positive and negative load directions (Fig. 2).

The performance parameters are necessary to define the properties of the material used in the numerical model to simulate the behavior of the partition walls, as described in the next paragraph.

2.3 Numerical Model of the Partition Wall

In this study, the finite element software OpenSees [19] was used to implement and calibrate the numerical partition wall model. The hysteretic response of the partition wall is simulated using the Pinching4Material available in OpenSees. The Pinching4 uniaxial material allows reproducing a 'pinched' load-deformation response and exhibits degradation under cyclic loading (see Fig. 3). The parameters governing reloading, unloading, cyclic degradation and maximum energy dissipation are assumed and checked by trial-and-error fitting, as illustrated in Fig. 4. Consequently, the calibration procedure is iterative, as also shown in Fig. 1.

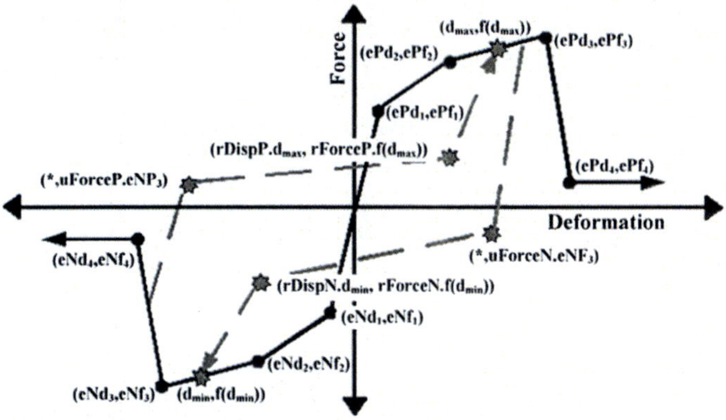

Fig. 3. Parameters included in Pinching4 uniaxial material model definition (source: OpenSees command language manual).

The numerical model involves three zero-length elements in series: one element reproduces the hysteretic behavior of the partition and is placed in the middle of the complete model, while the other two elements are used to account for the stiffness of the partition's connections to the upper and lower slabs. These links are characterized by a linear elastic behavior and are used because different solutions are available in the market to connect the partitions to the supporting structure. It is worth noting that more detailed numerical modelling approaches are available in the literature to simulate the response of the partition walls, e.g. in Rahmanishamsi et al. [20]. However, the level of detail of the numerical model should be defined also based on the scope of the study and the computational efforts underlying it.

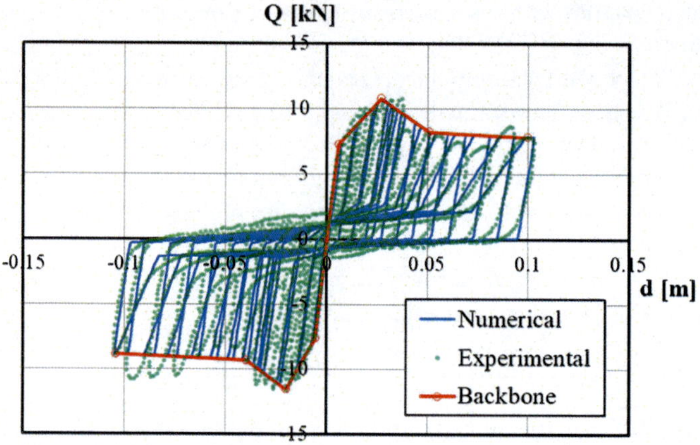

Fig. 4. Example of numerical model calibration.

2.4 Nonlinear Dynamic Analysis

The numerical model is used to determine the maximum displacement to which the partition walls are subjected for increasing levels of seismic intensity. According to the proposed framework, a multi-stripe analysis, involving floor displacement time histories, should be performed. The definition of the seismic demand represents a pivotal step in this framework. The floor displacement time histories should account not only for record-to-record variability but also for the variability related to the dynamic filtering of the supporting structure. To deal with this issue, it should be envisaged to make use of one or more floor interstory displacement time history database. This database should consist of floor displacement time histories recorded during the monitoring of real buildings or analyzing the results of multi-stripe analyses on building portfolios.

2.5 Fragility Function Definition

The drift values computed by nonlinear dynamic analyses of the partitions are compared with drift thresholds corresponding to the attainment of different limit states, thus leading to demand/capacity convolution and derivation of the fragility curve, which provides the probability of reaching or exceeding a limit state as a function of a specific intensity measure (IM). The function can be computed according to the procedure suggested in [21].

3 Illustrative Example

The above-defined framework was applied to a case-study specimen, namely the one labeled "Specimen no. 5" in [14]. This partition is representative of typical partition walls installed in industrial buildings in Europe. Geometrical data (partition dimensions, track and stud spacing, boards thickness, etc.), mass of the sample and mechanical data (frame configuration) were collected, as recommended in Fig. 1. Moreover, the experimental hysteresis curve and the drifts corresponding to limit states were collected. In particular, the first limit state (Damage State 1 – DS1) was achieved at 0.34% drift level due to the superficial damage of the wall (i.e. drop of gypsum dust and/or tape detachment). The second limit state (Damage State 2 – DS2) manifested at 0.87% drift level and was associated with the local damage of gypsum wallboards and/or steel frames (i.e. boards out-of-plane fail and/or frame local plastic deformation). The third limit state (Damage State 3 – DS3) corresponded to the ultimate displacement recorded during the test, equal to 2.78% drift ratio. The DS3 represents severe damage that requires the necessity of replacing the partition wall (i.e. expulsion of partition portion and/or frame buckling).

According to the proposed framework, the experimental hysteretic curve was used to define the backbone curve. After the definition of the backbone curve parameters, the calibration of the numerical hysteresis curve was undertaken, as illustrated in Fig. 5. Once the comparison between numerical and experimental results was deemed satisfying, all the data necessary to run nonlinear analysis were at hand.

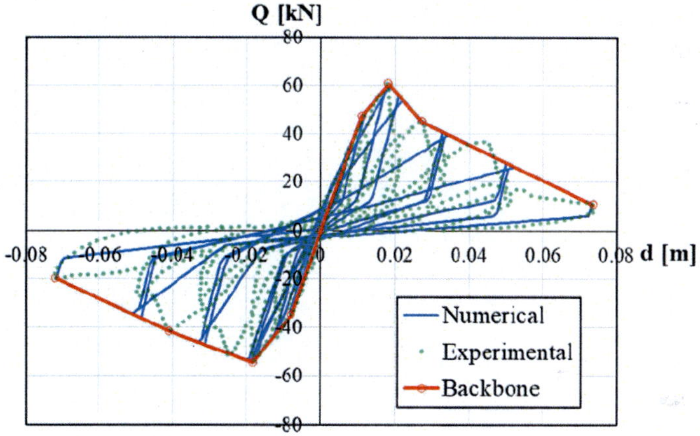

Fig. 5. Numerical hysteresis curve, experimental hysteretic curve and envelope backbone curve of "Specimen no. 1" from [14].

The numerical model previously described has been used in the illustrative example. In this case, the links simulating the stiffness of the partition's connections to the structure were assumed rigid. Despite this assumption is not representative of the wide range of solutions available in the markets, this work is aimed to establish a framework for future analyses and developments. Regarding the seismic input, the database of floor displacement time histories recorded by Perrone et al. [22] has been used. The building portfolio analyzed by Perrone et al. [22] can be considered representative of reinforced concrete moment resisting frames built in Italy and designed according to seismic provisions. The building portfolio, composed by 100 buildings, was analyzed considering ten return periods of the seismic intensity and 20 ground motions for each return period. A total of 32,909 analyses were run to study the performance of the partition wall. The building portfolio was analyzed both considering and neglecting the presence of masonry infills, consequently two sets of floor displacement time histories were recorded.

Finally, the fragility curves were computed as suggested by Baker [23]. Since the fragility curves specify the probability to reach a certain state limit as a function of an IM, a parameter representing the IM itself should be defined. In this case, the interstory drift was assumed as IM.

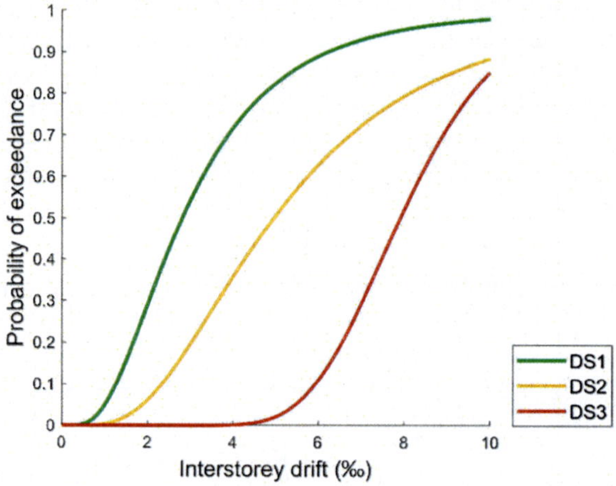

(a)

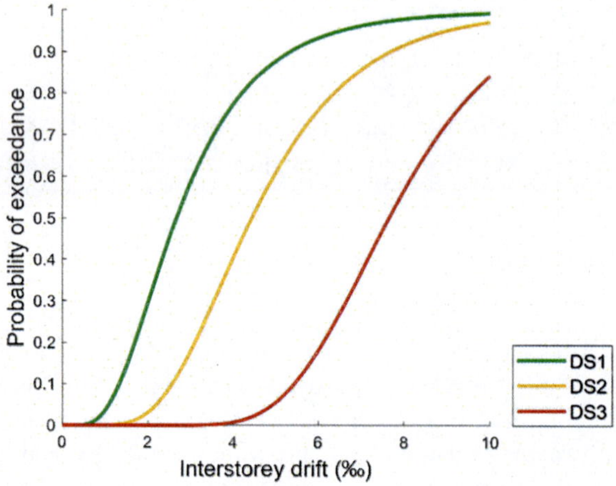

(b)

Fig. 6. Fragility curves of "Specimen no. 5" from Petrone et al. (2015) considering: (a) bare frames and (b) infilled frames.

The fragility curves corresponding to DS1, DS2 and DS3 are plotted in Fig. 6. For each return period used to analyze the response of the building portfolio, the mean interstory drift across all stories and displacement time histories was calculated and considered as IM, then the percentage of analyses exceeding the drifts associated to each limit state was computed to define the fragility curves. The fragility curves were fitted according to Baker [21]. Figure 6a illustrates the results obtained by considering the interstory displacements of bare moment resisting frames, while Fig. 6b illustrates the results pertaining to the infilled frame counterparts. As expected, the probability of reaching any damage state is higher when taking as reference the bare frame building portfolio. In fact, frames with infill walls have higher stiffness than bare frame counterparts and, consequently, result in lower interstory displacements. The analysis' outcomes confirmed the fact that partition walls are prone to exhibit damage even for low-intensity earthquakes. In fact, the DS1 fragility curves exhibited probability values greater than 50% for interstory drifts equal to 3‰, for both bare and infilled frame cases. In contrast, the DS3 curves exhibited almost null probability of occurrence for interstory drifts of up to 4‰.

4 Conclusions

With the advancement of performance-based seismic engineering and seismic risk assessment methodologies, the need for fragility functions for both structural and non-structural elements is becoming more evident. This work proposed a generalized framework to study the seismic behavior/performance of non-structural components through numerical analysis aiming to calculate fragility functions applicable to seismic risk and loss assessment analysis. In particular, the paper has focused on partition walls, which represent a typical non-structural element installed in both residential and industrial buildings. The framework requires the development of database collecting the experimental results available for the analyzed non-structural elements, thereby permitting calibration of simplified – or more complex – numerical models and identification of thresholds corresponding to limit state attainment. Then, nonlinear time-history analyses for increasing levels of seismic shaking should be performed to calculate the fragility functions.

In addition to the description of the proposed framework, an illustrative example was reported to showcase it. A typical partition wall available in the European market was selected and the fragility functions were derived considering two different structural model configurations for the definition of the seismic input acting on the case-study partition wall: bare and infilled reinforced concrete frames. The results of the illustrative example demonstrated the effectiveness of the framework in calculating fragility functions for partition walls.

However, further developments are still required to propose reliable fragility functions for a class of partition walls to be used in loss estimation frameworks. In particular, the proposed framework should be applied to a larger number of samples and different intensity measure should be investigated. Moreover, the effect of the simplified numerical modelling approach should be investigated by adopting more advanced numerical models, in particular when innovative construction details are introduced in the partition walls, or if out-of-plane behavior cannot be neglected.

Acknowledgements. The authors gratefully acknowledge the funding by Italian Ministry of Education, University and Research (PON Grant n. 1735/2017, CADS Project).

References

1. Taghavi, S., Miranda, M.M.: Response assessment of nonstructural building elements. Pacific Earthquake Engineering Research Center (2003)
2. Baird, A., Ferner, H.: Damage to non-structural elements in the 2016 Kaikōura earthquake. Bull. N. Z. Soc. Earthq. Eng. **50**(2), 187–193 (2017)
3. Perrone, D., Calvi, P.M., Nascimbene, R., Fischer, E.C., Magliulo, G.: Seismic performance of non-structural elements during the 2016 Central Italy earthquake. Bull. Earthq. Eng. **17**(10), 5655–5677 (2018). https://doi.org/10.1007/s10518-018-0361-5
4. Sullivan, T.J.: Post-Earthquake reparability of buildings: the role of non-structural elements. Struct. Eng. Int. **30**(2), 217–223 (2020)
5. Petrone, C., Magliulo, G., Manfredi, G.: Mechanical properties of plasterboards: experimental tests and statistical analysis. J. Mater. Civ. Eng. **28**(11) (2016)
6. Sousa, L., Monteiro, R.: Seismic retrofit options for non-structural building partition walls: Impact on loss estimation and cost-benefit analysis. Eng. Struct. **161**, 8–27 (2018)
7. Eurocode 8 (2005). Eurocode 8: design of structures for earthquake resistance-part 1: general rules, seismic actions and rules for buildings. European Committee for Standardization, Brussels
8. D.M. 17/01/2018. Aggiornamento delle "Norme Tecniche per le Costruzioni" (NTC 2018). Supplemento ordinario alla "Gazzetta Ufficiale", n. 42 del 20/02/2018 [in Italian]
9. FEMA. FEMA P-795 Quantification of building system performance and response parameters—component equivalency methodology. Federal Emergency Management Agency, Washington, DC (2011)
10. Kanvinde, A.M., Deierlein, G.G.: Analytical models for the seismic performance of gypsum drywall partitions. Earthq. Spectra **22**(2), 391–411 (2006)
11. Lee, T.H., Kato, M., Matsumiya, T., Suita, K., Nakashima, M.: Seismic performance evaluation of non-structural components: drywall partitions. Earthq. Eng. Struct. Dynam. **36**(3), 367–382 (2007)
12. Restrepo, J.I., Bersofsky, A.M.: Performance characteristics of light gage steel stud partition walls. Thin-Walled Struct. **49**(2), 317–324 (2011)
13. Davies, R., Retamales, R., Mosqueda, G., Filiatrault, A., Allen, D.: Effects of cold-formed steel framed gypsum partition walls on the seismic response of a medical facility (2012)
14. Petrone, C., Magliulo, G., Lopez, P., Manfredi, G.: Seismic fragility of plasterboard partitions via in-plane quasi-static tests. Earthq. Eng. Struct. Dynam. **44**(14), 2589–2606 (2015)
15. Tasligedik, A.S., Pampanin, S., Palermo, A.: Low damage seismic solutions for non-structural drywall partitions. Bull. Earthq. Eng. **13**(4), 1029–1050 (2014). https://doi.org/10.1007/s10518-014-9654-5
16. Pali, T., Bucciero, B., Terracciano, M.T., Macillo, V., Fiorino, L., Landolfo, R.: 11.02: In-plane quasi-static cyclic tests on lightweight steel drywall non-structural partition walls. Ce/papers, **1**(2–3), 2857–2866 (2017)
17. Araya-Letelier, G., Miranda, E., Deierlein, G.: Development and testing of a friction/sliding connection to improve the seismic performance of gypsum partition walls. Earthq. Spectra **35**(2), 653–677 (2019)
18. Mulligan, J., Sullivan, T. J., Dhakal, R.P.: Experimental seismic performance of partly-sliding partition walls. J. Earthq. Eng. 1–26
19. Mazzoni, S., McKenna, F., Scott, M. H., Fenves, G.L.: OpenSees command language manual. Pac. Earthq. Eng. Res. (PEER) Cent. **264**(1), 137–158 (2006)
20. Rahmanishamsi, E., Soroushian, S., Maragakis, E.M.: Analytical model for the in-plane seismic performance of cold-formed steel-framed gypsum partition walls. Earthq. Eng. Struct. Dynam. **45**(4), 619–634 (2016)

21. Baker, J.W.: Efficient analytical fragility function fitting using dynamic structural analysis. Earthq. Spectra **31**(1), 579–599 (2015)
22. Perrone, D., Brunesi, E., Filiatrault, A., Nascimbene, R.: Probabilistic estimation of floor response spectra in masonry infilled reinforced concrete building portfolio. Eng. Struct. **202**, 109842 (2020)
23. Baker, J.W., Cornell, C.A.: A vector-valued ground motion intensity measure consisting of spectral acceleration and epsilon. Earthq. Eng. Struct. Dynam. **34**(10), 1193–1217 (2005)

Dynamic Characterization of Glazed Partition Walls by Operational Modal Analysis Technique

Alessandra De Angelis[1](✉), Giuseppe Maddaloni[1], and Maria Rosaria Pecce[2]

[1] Department of Engineering, University of Sannio, 82100 Benevento, Italy
adeangelis@unisannio.it

[2] Department of Structures for Engineering and Architecture, University of Naples Federico II, 80125 Naples, Italy

Abstract. The vulnerability assessment of glazing systems and partitions under dynamic loads is still an open topic, requiring huge efforts. Especially in terms of dynamic-related effects, the use of Finite Element (FE) software packages can represent an efficient tool for design but special care should be spent for the definition of key input parameters, such as for example the restraints features. Moreover, boundary conditions could also modify, during the lifetime of a given glazed partition, due to external factors or materials degradation, hence resulting in possible unsafe structural performances or premature damage. In this paper, the dynamic characterization of two types of glazed partition is presented based on Operational Modal Analysis (OMA) techniques and Finite Element (FE) numerical simulations. The dynamic in situ tests have been carried out in the framework of the PON research project CADS- "Creating a Safe Home Environment". It is shown, in particular, how the actual restraint condition with respect to ideal boundaries can affect the vibration parameters (frequency, modal shape) of a given glazed partition.

Keywords: Glazed partitions · Dynamic characterization · Boundary conditions · Operational modal analysis (OMA) · Finite element (FE) numerical models

1 Introduction

The seismic vulnerability of non structural elements is a current important topic of the seismic engineering because most losses due to earthquake are related to these components and also the human life safety is threatened by their performance. In this field the assessment of existing non structural components is particularly complex because in situ investigation is necessary to establish the properties of the materials but in many cases the most difficult aspect is to define the boundary conditions, as for infill walls and partitions. However this aspect is a key issue of the component vulnerability that can give its collapse during the seismic event.

An innovative procedure for defining the boundary conditions of infill walls has been applied using the Ambient Vibration Test to calibrate the perimeter restraints trough the

G. P. Cimellaro (Ed.): WCSI 2022, LNCE 309, pp. 782–790, 2023.
https://doi.org/10.1007/978-3-031-21187-4_67

updating of a numerical model considering the experimental frequencies and modes [1]. In this regards, the dynamic in situ tests based on Operational Modal Analysis (OMA) techniques are excellent tools to derive the modal parameters under service conditions [2] and to calibrate reliable numerical models taking into account the real boundary conditions. The results of these works encourage to extend the techniques to glass partitions that are typical non structural elements installed both in homes and offices. One of the greatest advantages of glass partitions is the brightness, i.e. their ability to circulate light from a room to another one in contrast to what a masonry wall would do. However their collapse during an earthquake is much dangerous for human life.

From a technical point of view, glass partitions can be framed or frameless systems even if in both cases a structure, made with aluminium, wooden or other materials, is usually needed to support the glass. Generally, framed glass partitions are systems in which each sheet of glass is restrained along all four sides. Instead, frameless glazed partitions require upper and/or lower rails to support the glass while generally the vertical uprights are only located on the sides of the whole partition. In fact, the vertical joints between two adjacent glass sheets are simply filled with silicone or glue.

The vulnerability and the structural performance assessment of glazing systems and partitions under dynamic loads is still an open topic, requiring huge efforts. Several research studies used the OMA techniques for the dynamic investigation of glazed systems [3–5]. It is worth pointing out that most of the works available in literature are related to the dynamic characterization of structural glass elements, hence suggesting the need for extended investigations on glazed non structural elements such as partition walls.

The study has been carried out in the framework of the PON research project CADS-"Creating a Safe Home Environment" and it aims investigating the effect of the actual restraint conditions on the analysis of the fundamental frequencies and modal shapes. This aspect is particularly significant since boundary conditions could modify, during the lifetime of a given glazed partition, due to external factors or materials degradation, hence resulting in possible unsafe structural performances or premature damage.

In this paper, the first part of the work is reported detailing the outcomes of the output-only vibration experiments carried out on glazed partitions with two different configurations for glazing frame construction type, i.e., framed and frameless. Moreover, the results of the in situ tests are used to validate a numerical model of glazed partitions.

2 Dynamic Identification of Glazed Partitions

Two glazed partitions with a different configurations for the glazing frame construc-tion type were selected as case study. The first one is a typical framed partition in-stalled as main entrance in historical buildings. The selected case ($H = 3.5$ m high $\times$ $L = 6.45$ m wide) consists of a white painted beech wood frame containing the indi-vidual glasses (see Fig. 1a). The second configuration is a frameless partition typically installed in modern offices. The selected case ($H = 4.05$ m high $\times$ $L = 4.8$ m wide) is located in the office building (see Fig. 1b) under construction of the company Edil vetro s.r.l., industrial partner of the CADS project. The glasses are full-height and linked to the structure by upper and lower rails and vertical uprights on the two sides. The joints between the glasses are filled with silicone.

a) b)

Fig. 1. a) The framed and b) frameless glazed partitions selected as case study

2.1 The Preliminary Finite Element Model

In this section, the glazed partitions experimentally tested are modeled numerically with the finite element method using the software SAP2000 [6]. The objective is to find a numerical representation of the partitions able of reproducing their dynamic modal behavior. Thus, the models will be used for the model updating.

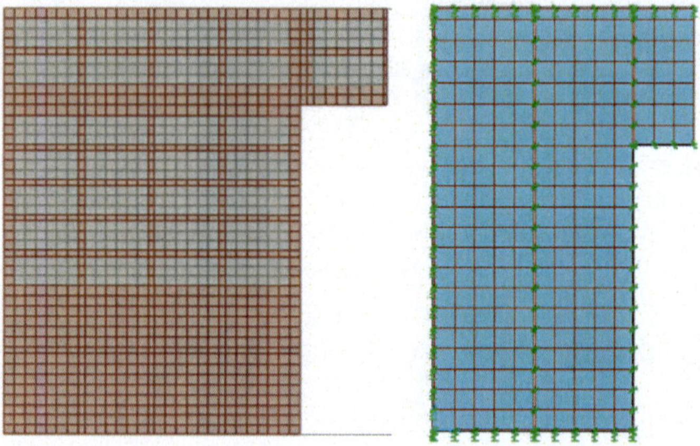

a) b)

Fig. 2. FE models of the framed (a) and frameless (b) glazed partitions.

Due to geometric and restraint symmetry the numerical models of half partition are developed. For the framed glazed partition, both glasses and wood panels are modelled by shell elements (Fig. 2a) with the actual thickness detected on site. At the initial stage, the restraints of the partition are modelled as fully fixed along the four sides. A fairly dense mesh equal to 10 cm × 10 cm is chosen so that the joints are exactly in the same position as the sensors during the experimental test.

Instead for the frameless partition shell elements and frames are adopted for glasses and aluminum profiles, respectively (Fig. 2b). In particular, for the constraint conditions linear and fully fixed link elements are adopted to simulate the contact between the aluminium tracks of the partition and the structural elements of the building, the glass-to-frame attachment and the contact between the glass panels.

The initial value of the material properties are resumed in the fourth column of Tables 3 and 4. In order to obtain the modal parameters and design the layout of the experimental tests, a modal analysis is carried out. Figure 3 shows the modal shapes obtained for the two types of glazed partitions. As can be seen, the greater displacements are concentrated in the central part of the half partition for the framed one and at the side where the door is located for the frameless one.

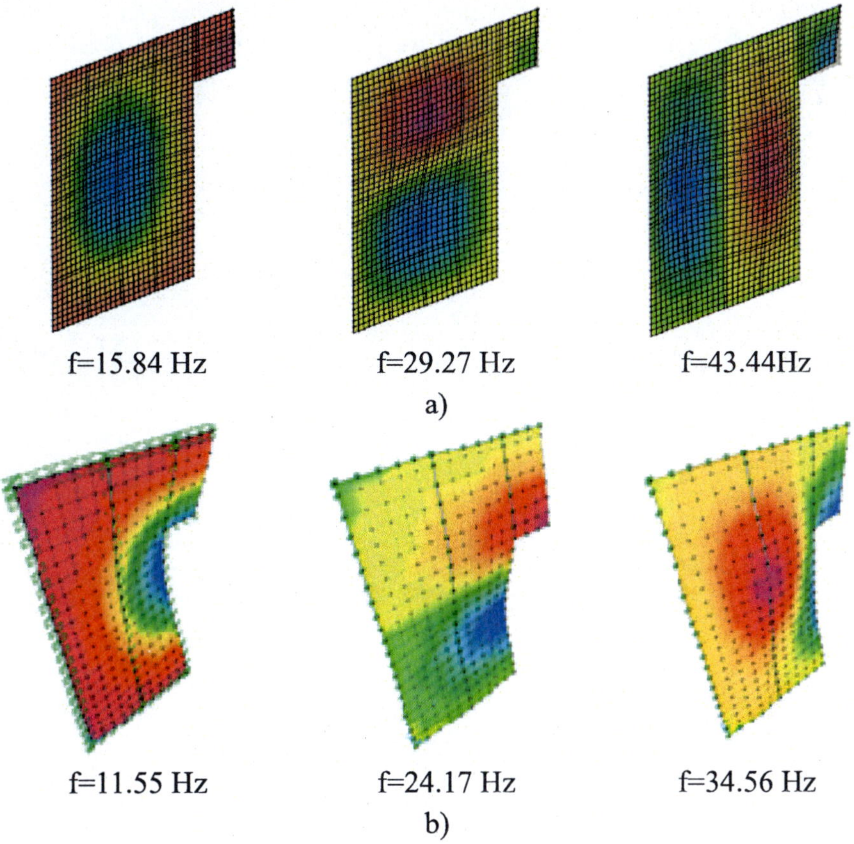

Fig. 3. First three numerical modal shapes for a) the framed and b) frameless partitions

The numerical frequencies obtained are also reported in Fig. 3 for the two cases and they give an idea of the starting parameters to be used for setting the dynamic test on site. It is worth to underline that the first three numerical frequencies of the frameless partition are lower than ones of the framed configuration stressing a greater flexibility of this type of glazed partition.

2.2 The Test Setup and Results

The OMA approach called "output only" is particularly suitable for exploring the dynamic response of systems and components under operating conditions without the knowledge of the imposed excitation. The tests involved the arrangement of a higher number of sensors in order to reliably identify the first three modal shapes. In particular 32 accelerometers were arranged for the framed partition and 33 for the frameless one. However, having only 12 accelerometers available, three setups were defined using two accelerometers as a reference for both tests. The adopted sensors are uniaxial Integrated Circuit Piezoelectric (ICP) accelerometers with the following characteristics: 10 V/g sensitivity, ± 0.5 g measurement range and frequency range from 0.05 to 4000 Hz. The equipment used for the tests also includes signal cables, a 12-channel data acquisition system and a portable computer (Fig. 4).

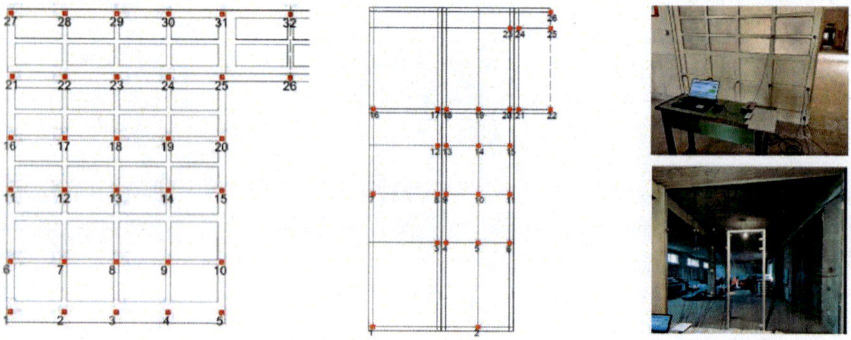

a) b) c)

Fig. 4. Monitored points for the framed (a) and frameless (a) partitions and c) and photo of a single setup

The ambient vibration response of the two glazed partitions was acquired for 600 s at a sampling frequency of 100 Hz. Before the data elaboration, i.e., extracting the modal properties, the time series have been deeply inspected. The OMA technique named Frequency Domain Decomposition (FDD) has been applied [7] using the software ARTeMIS [8]. In this case, three clear modes have been identified for the tested glazed partitions as reported in Fig. 5.

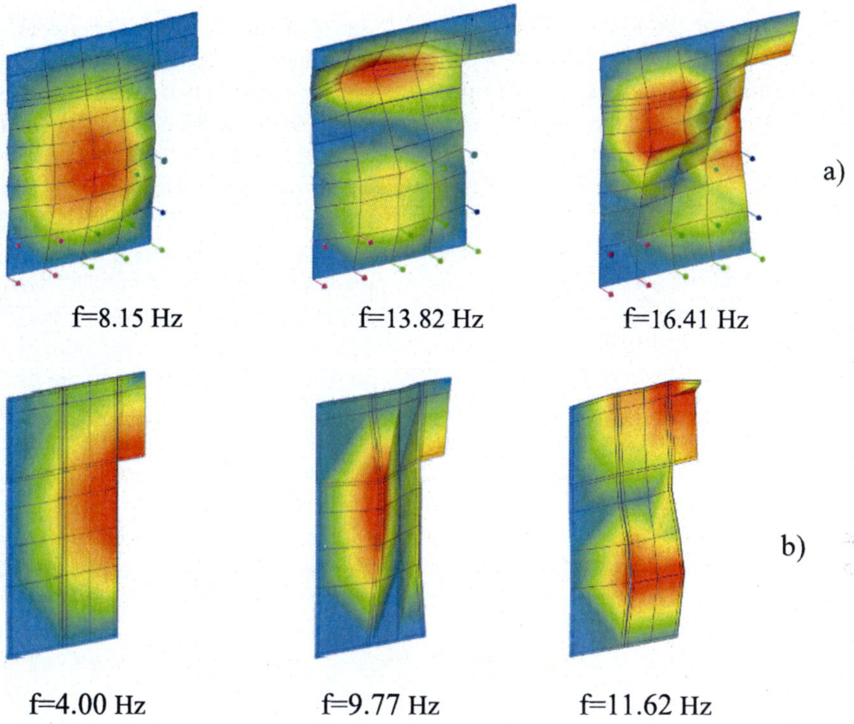

Fig. 5. First three experimental modal shapes for a) the framed and b) frameless partitions

2.3 Model Updating

Dynamic characteristics obtained from field testing generally do not coincide with those of the numerical model, therefore, model updating is required to match the numerical and experimental measurements. As reported in Table 1 the comparison between experimental (f_{exp}) and numerical (f_{num}) frequencies reveals that numerical frequencies are bigger than experimental ones for both cases with a frequency error in percentage (ε) higher than 90%.

Table 1. Comparison before calibration for the framed partition

Framed partition					Frameless partition				
Mode	f_{exp} [Hz]	Mode	f_{num} [Hz]	ε [%]	Mode	f_{exp} [Hz]	Mode	f_{num} [Hz]	ε [%]
1	8.15	1	15.84	94.4	1	4.00	1	11.51	187.8
2	13.82	2	29.27	111.8	2	9.77	3	24.17	–
3	16.41	3	43.44	164.7	3	11.62	2	34.57	–

Moreover, for the frameless partition the order of modes is reversed between the second and the third ones. The above considerations underline that an updating of the numerical model is necessary. The main purpose of the model updating procedure is to minimize the differences between the numerical and experimental dynamic characteristics by changing some uncertain parameters, such as material properties and boundary conditions. In this case, the parameters affected by uncertainties that need to be modified are the elastic modulus and the density of the materials and the boundary condition along the three sides of the half partition. To calibrate the constraint along the sides, the boundary conditions were defined using spring elements. In particular, the out-of-plane translational stiffness is calibrated. Note that the initial values of the spring stiffness (10^5 N/m) correspond to the fully restrained boundary condition while the value 0 corresponds to the free side condition. A total of 7 parameters are set for changing during the manual updating. The starting and updated values of all the parameters together with the range of variation are presented in Tables 2 and 3 for the framed and frameless partitions, respectively. The final values of the stiffness of the springs obtained from the calibration process reveal that for both partitions the degree of constraint along the sides is intermediate between free and fully restrained. Moreover, for the frameless partition it is also found that the degree of constraint of the upper and bottom sides is less effective than vertical ones. The final correlation between the dynamic experimental behaviour and the numerical simulation is developed using the frequency error (ε) and the Modal Assurance Criterion (MAC) [9]. As shown in Table 4, all three relevant natural frequencies are matched within 5% (expect the third mode of the framed partition) and the high MAC values show a good correlation between the experimental and numerical mode shapes.

Table 2. Material properties for the framed partition

Material	Property	Range	Initial value	Optimal value
Glass	Density [kg/m^3]	2200–7200	2400	3000
	Elastic modulus [MPa]	50,000–90,000	70,000	90,000
Wood	Density [kg/m^3]	300–1200	680	800
	Elastic modulus [MPa]	4000–20,000	14,500	5900
Constraint	All sides translational stiffness [kN/m]	0–10^5	10^5	5000

3 Conclusions

In this work, carried out in the framework of the PON research project CADS- "Creating a Safe Home Environment", the modal behavior of glazed partitions with two different configurations for glazing frame construction type, i.e., framed and frameless, is experimentally identified and validated with the finite element method. It is shown that the

Table 3. Material properties for the frameless partition.

Material	Property	Range	Initial value	optimal value
Glass	Density [kg/m^3]	2200–7200	2400	3000
	Elastic modulus [MPa]	50,000–90,000	70,000	70,000
Aluminium	Density [kg/m^3]	2000–3400	2700	2490
	Elastic modulus [MPa]	50,000–80,000	70,000	69,000
Constraint	Upper and lower sides translational stiffness [kN/m]	0–10^5	10^5	100
	Vertical side translational stiffness [kN/m]	0–10^5	10^5	1000

Table 4. Comparison after calibration for the frameless partition

Mode	Framed partition				frameless partition			
	f_{exp} [Hz]	f_{num} [Hz]	ε [%]	MAC [–]	f_{exp} [Hz]	f_{num} [Hz]	ε [%]	MAC [–]
1	8.15	7.86	– 3.6	0.94	4.00	4.13	3.3	0.98
2	13.82	13.83	0.1	0.73	9.77	9.81	0.4	0.78
3	16.41	18.22	11.0	0.88	11.62	11.66	0.3	0.66

dynamic test and the further model updating can provide information about the degree of constraint if the first three mode shapes are considered. Moreover, it is also shown how the vibration parameters of a given type of partition are affected by the actual restraint condition with respect to ideal boundaries. This turns out in possible unsafe structural performances or premature damage, hence suggesting the need for even further dedicated studies.

Acknowledgments. The authors gratefully acknowledge the funding by Italian Ministry of Education, University and Research (PON Grant n.1735/2017, CADS Project).

References

1. De Angelis, A., Pecce, R.: Out-of-plane structural identification of a masonry infill wall inside beam-column RC frames. Eng. Struct. **2018**(173), 546–558 (2018)
2. Rainieri, C., Fabbrocino, G.: Operational modal analysis of civil engineering structures–an introduction and guide for applications. Springer, p. 142 (2014)
3. Lenci, S., Consolini, L., Clementi, F.: On the experimental determination of dynamical properties of laminated glass. Ann. Solid Struct. Mech. 7(1–2), 27–43 (2015). https://doi.org/10.1007/s12356-015-0040-z
4. Pelayo, F., Lopez-Aenlle, M.: Natural frequencies and damping ratios of multi-layered laminated glass beams using a dynamic effective thickness. J. Sandw. Struct. Mater. (2017)

5. Bedon, C., Fasan, M., Amadio, C.: Vibration analysis and dynamic characterization of structural glass elements with different restraints based on operational modal analysis. Buildings **9**(1), 13 (2019)
6. Computers and Structures (2016). SAP2000 version 18, Walnut Creek, CA
7. Brincker, R., Zhang, L., Andersen, P.: Modal identification of output-only systems using frequency domain decomposition. Inst. Phys. Publishing Smart Mater. Struct. **10**, 441–445 (2001)
8. SVS 2010. ARTeMIS extractor 2010 release 5.0. http://www.svibs.com/
9. Allemang, R., Brown, D.: A correlation coefficient for modal vector analysis. In: 1st International Modal Analysis Conference, Orlando, USA (1982)

Dynamic Characterization and Damage Detection of a Fire-Protection Piping System

Alessandra De Angelis[1(✉)], Giuseppe Maddaloni[1], and Maria Rosaria Pecce[2]

[1] Department of Engineering, University of Sannio, 82100 Benevento, Italy
adeangelis@unisannio.it

[2] Department of Structures for Engineering and Architecture, University of Naples Federico II, 80125 Naples, Italy

Abstract. Recent earthquakes worldwide have proven that losses result not only from damage to building structures but also from nonstructural components. Therefore, a correct design of these components can ensure not only the functionality and the life safety but also reduce the economical losses. Among non-structural components fire-protection piping systems play an important role for critical facilities such as hospitals that are required to guarantee the immediate operation after a seismic event. In order to conduct a more accurate seismic analysis or damage assessment of the piping systems, it is necessary to establish reliable numerical models. In this work, the results of an experimental ambient vibration test are reported to determine the dynamic properties of a piping system, such as the natural frequencies and mode shapes, to establish the constraints that are uncertain in case of existing constructions, and upload a numerical model implemented by MIDAS FEA NX software. Most of the numerical natural frequencies achieve a good correlation with the measured frequencies, allowing a calibration of the effective constraints. Finally the preliminary results related to the efficiency of the test layout for damage detection are reported and discussed.

Keywords: Piping system · Ambient vibration test · Dynamic properties · Numerical model · Damage detection

1 Introduction

Non-structural elements, particularly the plant systems, play a fundamental role in ensuring the functionality of a building in post-earthquake. A correct design of these components can ensure not only the functionality and the life safety but also reduce the economical losses. The above aspects are particularly important for fire-protection piping systems in critical facilities such as hospitals that are required to be operative after a seismic event. In general the failure of a fire protection piping system can be linked to one of the following scenarios:

- Failure due to a damage occurred during the seismic event.
- Failure due to a damage occurred before the seismic event caused, for example, by construction imperfections and material shortcomings or deterioration.

G. P. Cimellaro (Ed.): WCSI 2022, LNCE 309, pp. 791–798, 2023.
https://doi.org/10.1007/978-3-031-21187-4_68

In the first case the damage can be avoided by a correct seismic design of piping system itself [1] while in the second case it can be prevented in advance through damage detection. It is well known that damage in a structure produces variations in its geometric and physical properties, which can result in changes in its natural frequencies and mode shapes. In the last years, many damage detection methods based on dynamic parameters have been developed by several researchers [2–4]. However, the most commonly used method of damage detection uses changes of natural frequencies and mode shape directly. In a healthy structure occurrence of damage can locally alter the structure, so that differences appear between the numerical response of FE model and the in situ response of the structure. Therefore an effective manner of structural damage detection can be attained by FE model updating in turn causing changes in structural dynamic characteristics [5]. The above concepts could be applied also to nonstructural components [6–8]. Thus, the updated model of a fire protection piping system based on the results of in situ dynamic tests could be used to the assessment of the numerical model or perform damage detection. It is useful to understand if it is convenient or not to have a monitoring of the piping system as well as to establish the most vulnerable part of the system that should be monitored. Hence, early damage detection to locate incipient damage provides a chance for timely structural maintenance and can guarantee structural reliability and continuing serviceability.

This work, carried out in the framework of the PON research project CADS- "Creating a Safe Home Environment", focuses on the evaluation of the dynamic properties such as natural frequencies and mode shapes of a fire-protection piping system, to assess the numerical model and, as perspective, to design a monitoring system. The results of the ambient vibration test and the subsequent model updating are described in details. Finally the calibrated model is used to reproduce the identification procedure, i.e., the dynamic test is simulated exciting the numerical model, with the aim of evaluating the efficiency of the test layout for damage detection. The first results of these analyses are discussed.

2 Dynamic Identification of a Fire-Protection Piping System

2.1 Description of the Selected Fire-Protection Piping System

In order to analyze the dynamic properties of a typical configuration of fire-protection piping system, the fire surpression piping system with fire extinguishing equipments like Naspo arranged in the laboratories of a University building in Benevento has been selected as case study.

The epoxy powder-coated steel piping system has a main line with a nominal diameter of 3" and a second line with a nominal diameter of 1" ¼.

Great attention has been paid to the points where the piping system connects to the structure and crosses vertical structural and/or nonstructural elements. In particular, it is mainly anchored to the structure with rod hangers (Fig. 1a) to ensure stability for vertical loads. At some points the fire protection system is linked to supports (Fig. 1b), which are also used as supports for other systems, anchored to the ceiling with threaded bars. The selected segment of the piping system crosses both the vertical structural and nonstructural elements (partitions) in four points. In particular, there is a free wall crossing (Fig. 1c) and three constrained one (Fig. 1d).

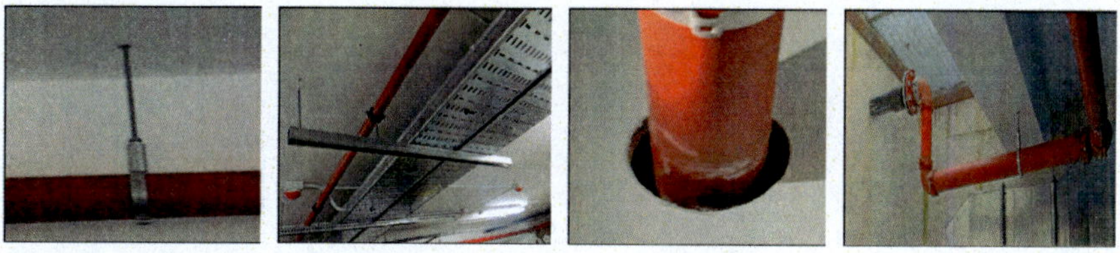

a) b) c) d)

Fig. 1. In-situ constraint conditions: vertical supports (a-b); Free (c) and constrained (d) crossing in walls

2.2 Finite Element Model

A preliminary eigenvalue analysis is carried out to establish the frequency range of interest as well as the associated mode shapes. The 3D finite element model of the selected piping system is developed with the software Midas FEA NX [9] to simulate the elastic response of the piping system. All components of the sample piping system (pipes, elbows, rod hangers) are properly modeled. In particular, the pipes are modelled with frame elements and distributed masses including the weight of the water are assigned to each element. The fittings and or elbows are modeled through rigid link elements. All rod hangers are modeled by means of steel elements with a solid section of diameter equal to 12 mm, modulus of elasticity of 200 GPa and Poisson module of 0.3. For the vertical supports, a pendular connection between the piping and the hanger and a fixed constraint to the ceiling are assumed, respectively. For the boundary conditions, the nodes where the pipes cross structural wall and partitions are fully restrained.

As shown in Fig. 2, the first eight modes involve the two horizontal directions and affect one or more segments of the pipe network. In addition the frequencies range from 2 to 20 Hz.

2.3 Ambient Vibration Test and Model Calibration

The dynamic characteristics of the piping system are assessed through an ambient vibration test (AVT). The accelerations were collected using twelve uniaxial piezoelectric accelerometers with high sensitivity (10 V/g) and low background noise (0.13 $\mu g/\sqrt{Hz}$). According to the outcomes of the modal analysis carried out with the preliminary finite element model, described in the previous section, at least 21 measurements points are necessary. Therefore, in order to study the local dynamic properties of the piping system, five setups were established as shown in Fig. 3, adopting 9 accelerometers as roving sensors and 3 as reference sensors. The latter will play an important role in combining the different setups to obtain the global mode shapes of the piping system.

For each setup, time series of 15 min (900 s) were collected at a sampling frequency of 200 Hz. Then the data were low-pass filtered and downsampled at a rate of 50 Hz. This

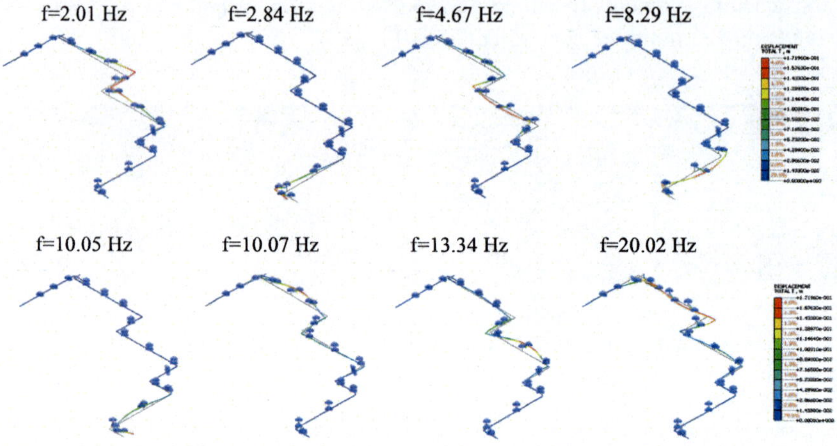

Fig. 2. Numerical frequencies and modal shapes.

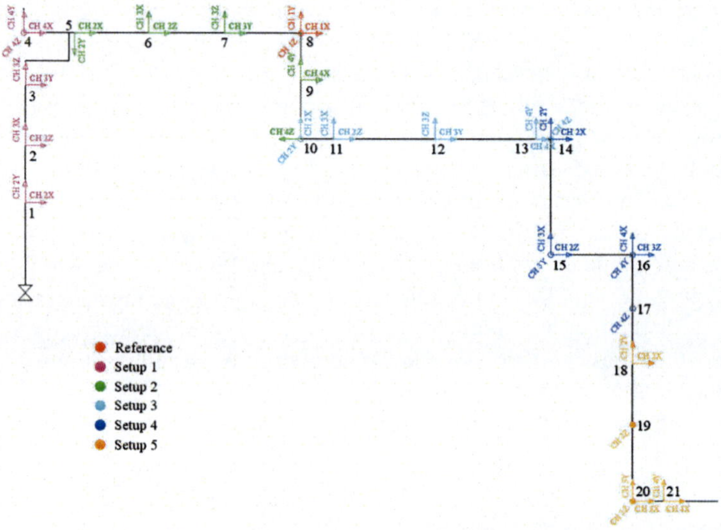

Fig. 3. Measurement points of the ambient vibration tests.

significantly reduces the amount of samples without loss of information in the frequency range of interest. Some pictures of the AVT are reported in Fig. 4. The extraction of the modal parameters (frequencies and modal shapes) is carried out through the software ARTeMIS [10] applying the Frequency Domain Decomposition (FDD) method [11]. Eight clear modes were identified for the tested fire-protection piping system as reported in Fig. 5.

A first comparison with the results of the preliminary numerical model shows that some of the numerical mode shapes correspond with those identified experimentally.

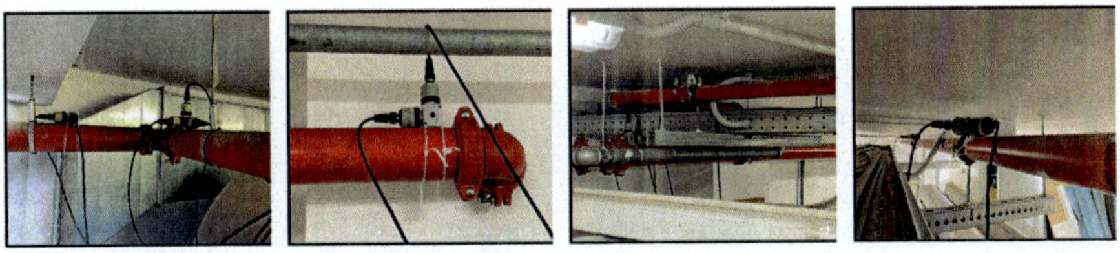

Fig. 4. Pictures of the AVT.

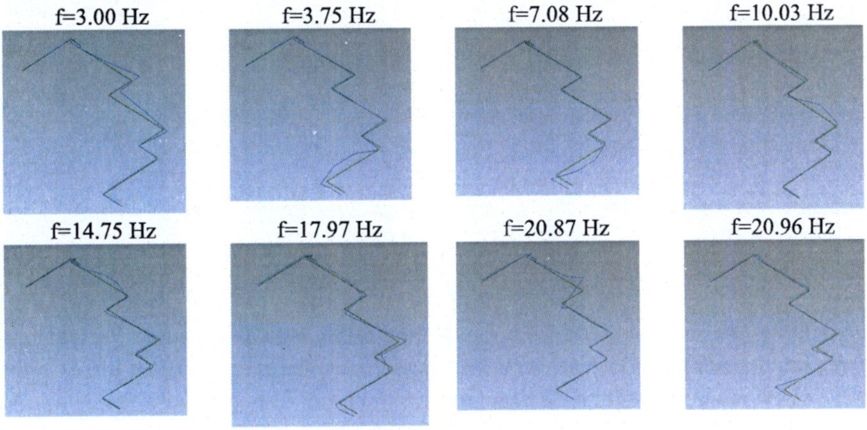

Fig. 5. Experimental frequencies and modal shapes.

However, there is a lack of correspondence in the order of modes and the numerical frequencies are generally lower than the experimental ones.

Therefore, it is necessary an updating of the model modifying in particular the stiffness of the elastic end support between the piping system and the supporting structure with the aim to minimize the differences between the numerical (f_{num}) and experimental (f_{exp}) frequencies. In order to perform the calibration, the fixed restraints arranged in the preliminary model are replaced with springs whose translational and rotational stiffnesses have to be calibrated. The unknown stiffnesses are defined by fine tuning (manual updating) in the simulation model until correlation with experiment data is reached. As reported in Fig. 6, the correlation obtained after calibration is very promising for five experimentally identified modes. In fact it is observed that the calibration of the model leads to have an excellent agreement for the first mode for which a frequency error (Δf) equal to 0,4% is obtained, while for the higher modes it is obtained a good agreement in terms of modal shapes but the frequency error is variable in the range 20–50%.

3 Efficiency of the Test Layout for Damage Detection

In a monitoring system, the accelerometers placed on the piping system will record the dynamic behaviour of the system with a certain time interval, extrapolating from time to time the frequencies of the system that can be compared with the frequencies of

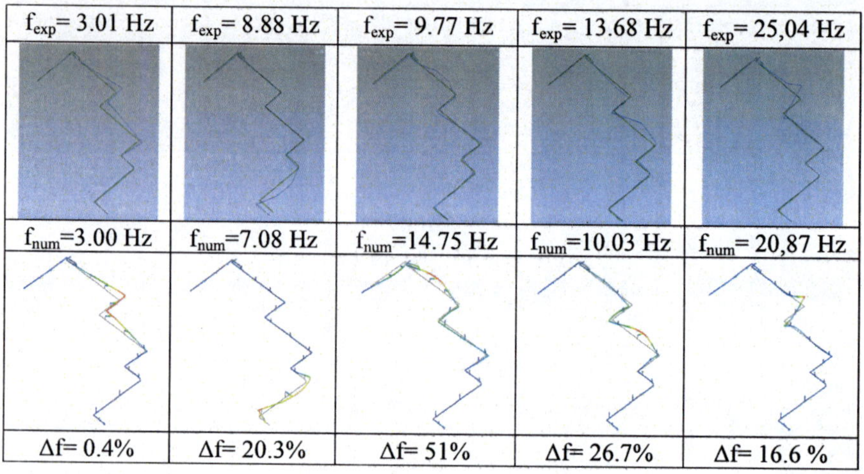

Fig. 6. Experimental and numerical correlation

the health system defined during the installation phase (that is the phase in which the piping system is perfectly assembled and fixed). A damage detection could be performed analyzing the structural response in the time. In fact, if the frequencychanges a warning will be activated allowing to check promptly the network.

In order to use a monitoring system to do damage detection it is necessary to define a series of damage scenarios and see if they would be detected by the system. To assess whether the defined layout of the test is able to detect the presence of damage, in this work two damage scenarios are established and the dynamic test is simulated numerically. Specifically, the calibrated model is excited with white noise simulating the environmental actions and the acceleration responses of the piping system at the points, where the sensors are located, are processed to obtain the modal parameters, in particular the frequency.

The two possible damage scenarios selected are related to the hanger rods supporting the pipes. According to the first mode shape, for which an excellent correlation has been obtained, the points at which there are higher displacements have been identified because these are the points where a failure could cause significant damage. For this reason it is assumed a possible damage in those points, first with the failure of a single hanger (case 1) and then with the failure of two adjacent hangers (case 2). The points involved in the damage scenarios are circled in Fig. 7.

Therefore, the white noise is produced as a sequence of normally distributed random variables using Matlab. The accelerometers arranged on the piping system during the test in situ are numerically simulated as joints in the same locations, referred to as “virtual accelerometers”. Linear time history analyses are performed and at the end of the numerical test the virtual accelerometers histories of acceleration are read and processed using the FDD technique to obtain the natural vibration frequencies and modal shapes. From the results obtained it is observed that in both scenarios there is no variation of the modal shape but the frequencies are highly involved as summarized in Table 1.

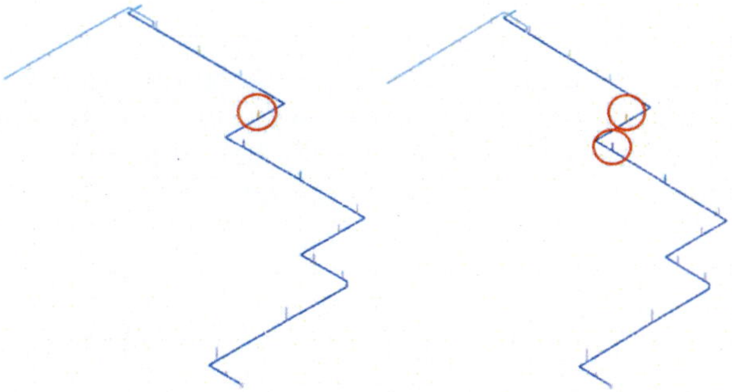

a) b)

Fig. 7. Points involved in the damage scenarios: a) case 1 and b) case 2

Moreover, it is observed that the failure of a single hanger causes a decrease in frequency of 10.96% while the failure of two adjacent hangers leads to double this difference up to 21.93%. Therefore, it can be stated that simulated damage with these two scenarios, i.e., are detectable by a dynamic test with this setup.

Table 1. Results of the damage simulation

Scenario	Before damage	After damage	Comparison
	f_{exp} [Hz]	f_{sim} [Hz]	Δf [%]
Case 1	3.01	2.68	– 10.96
Case 2	3.01	2.35	– 21.93

However, to state that the test setup layout defined for this case study is useful in a monitoring system to perform a damage detection it is necessary to extend these analyses to other damage scenarios and achieve an excellent experimental-numerical correlation for other vibration modes involving other portions of the piping system.

4 Conclusions

The piping systems require an accurate evaluation of their dynamic response in order to establish reliable numerical models. The model of the piping systems are affected by the constraints of the systems used for the connection to the structure, thus the vibration behavior can be useful to explore the elastic response. Therefore, in this paper an experimental modal analysis using an ambient vibration test has been employed to determine the dynamic properties such as natural frequencies and mode shapes of a piping system. The experimental modal parameters have been compared with numerical modal properties simulated by the software Midas FEA NX. Most of the numerical natural frequencies achieve good correlation with the measured frequencies. Moreover, it was found that the accuracy of the vibration behaviour depends on the boundary conditions of the structure, especially vertical supports. Finally the calibrated model is used to reproduce the identification procedure, i.e., the dynamic test is simulated exciting the numerical model, to evaluate the efficiency of the test layout for damage detection. The preliminary results demonstrate that a damage such as the failure of a single or two hangers are detectable by a dynamic test with this setup.

Acknowledgements. The authors gratefully acknowledge the funding by Italian Ministry of Education, University and Research (PON Grant n.1735/2017, CADS Project).

References

1. Blasi, G., Perrone, D., Aiello, M.A., Pecce, M.R.: Seismic performance assessment of piping systems in bare and infilled RC buildings. Soil Dyn. Earthq. Eng. **149** (2021)
2. Fan, W., Qiao, P.Z.: Vibration-based damage identification methods: a review and comparative study. Struct. Health Monit. **10**(1), 83–111 (2011)
3. Messina, A., Williams, E.J., Contursi, T.: Structural damage detection by a sensitivity and statistical-based method. J. Sound Vib. **216**(5), 791–808 (1998)
4. Xu, G.Y., Zhu, W.D., Emory, B.H.: Experimental and numerical investigation of structural damage detection using changes in natural frequencies. J. Vib. Acoust. **129**(6), 686–700 (2007)
5. Wang, D., Tan, Z., Li, Y., Liu, Y.: Review of the application of finite element model updating to civil structures. Key Eng. Mater. **574**, 107–115 (2014)
6. Haapaniemi, H., Luukkanen, P., Nurkkala, P., Rostedt, J.: Correlation analysis of modal analysis results from a pipeline. In: IMAC-XXI: Conference & Exposition on Structural Dynamics (2003)
7. Loh, S.K., Faris, W.F., Hamdi, M., Chin, W.M.: Vibrational characteristics of piping system in air conditioning outdoor unit. Sci. China Technol. Sci. **54**, 1154 (2011)
8. Zhao, L.D., Fouladi, M.H., Haw, Y.K.: Model correlation procedures for air-conditioning copper discharge tube. J. Eng. Sci. Technol. **9**, 40–58 (2014)
9. Midas FEA NX- Midas Engineering Software—Manual
10. SVS 2010. ARTeMIS extractor 2010 release 5.0. http://www.svibs.com/
11. Brincker, R., Zhang, L., Andersen, P.: Modal identification of output-only systems using frequency domain decomposition. Inst. Phys. Publishing Smart Mater. Struct. **10**, 441–445 (2001)

A Rapid Visual Screening Procedure to Evaluate Seismic Risk of Non-structural Elements in Critical Facilities

Alessandra De Angelis[1](✉), Daniele Perrone[2], Giuseppe Maddaloni[1], Maria Rosaria Pecce[3], and Maria Antonietta Aiello[2]

[1] University of Sannio, Benevento, Italy
adeangelis@unisannio.it
[2] University of Salento, Lecce, Italy
daniele.perrone@unisalento.it
[3] University of Naples Federico II, Naples, Italy

Abstract. Non-structural elements (NSEs) represent most of the total construction cost of typical buildings. A significant portion of the total losses in recent earthquakes worldwide has been attributed to damage to NSEs, demonstrating the need for the harmonization of the seismic performance of the NSEs and of the seismic force-resisting systems. Unfortunately, lack of research results on the seismic performance of NSEs available in the public domain has, and continues to hamper, the development of efficient performance-based seismic design methods for buildings. Even though recent research continues to develop new and improved seismic design methodologies for NSEs, significant efforts are still required, particularly for existing building for which seismic mitigation detailing for NSEs are generally not adopted. To deal with this issue, and to reduce non-structural related seismic losses in existing buildings, simplified tools, which allow to quantify the seismic risk at which NSEs are prone, should be developed. This paper presents a rapid visual screening procedure useful to perform a large-scale classification of the seismic vulnerability of NSEs installed in critical facilities. The methodology is based on questionnaire forms that are used to calculate a nonstructural index and to identify the most vulnerable NSE category. The methodology has been applied to some case study buildings focusing the attention on critical facilities.

Keywords: Rapid visual screening · Non-structural elements · Critical facilities · Seismic risk

1 Introduction

The earthquakes that have struck Italy, and more in general densely built regions worldwide, in the twentieth century have shown that losses from damage to non-structural elements (NSEs) far exceed losses from structural damage [1–4]. This is not surprising considering that NSEs account for most of the investment in a typical building [5] and that NSEs typically are installed following prescriptive empirical regulations or expert

G. P. Cimellaro (Ed.): WCSI 2022, LNCE 309, pp. 799–808, 2023.
https://doi.org/10.1007/978-3-031-21187-4_69

judgments rather than according to performance-based seismic design methodologies [6]. The lack of knowledge in this field is mainly related to the fact that, in comparison to structural elements and systems, there is much less experimental data available on the seismic performance of NSEs in order to establish multiple-performance levels for the seismic design of NSEs according to performance-based design/assessment approaches [7]. Based on these considerations engineers are often forced to start almost from scratch: observe what went wrong and try to prevent it from happening again. The earthquake induced losses related to NSEs can be direct and indirect. Direct losses from the earthquake are related to property losses, while indirect losses can be more difficult to quantity. Examples of indirect loss have been widely documented by researchers following past earthquakes. One example is the damage suffered in the Chilean wine industry following the 2010 Maule earthquake [3], as wine production represents one of the major economic activities in Chile. The earthquake caused damage to the fermentation tanks and stacked storage barrels. This damage caused the wine production to shut down for several weeks, with enormous consequences to the Chilean economy. During the same earthquake, the Santiago International Airport was closed for several days due to significant damage suffered by piping and ceiling systems which were related to the improper lateral supports [1]. Four hospitals were closed due to structural and non-structural damage while over 10 lost 75% of their functionality, due to damage to fire sprinklers [1]. Recently, the vulnerability of NSEs was also observed during the 2016 Kaikoura earthquake that struck the north-eastern area of the South Island of New Zealand. The main non-structural damage observed during this earthquake was related to the failure of ceilings, cladding, partitions, building services, and plant equipment [2]. Extensive damage to NSEs has often been observed for less intense earthquakes. For example, Braga et al. [8] reported extensive in-plane and out-of-plane damage to masonry infills in reinforced concrete buildings during the 2009 L'Aquila earthquake. Significant non-structural damage was also observed following the 2012 Emilia earthquake, where storage rack systems in industrial facilities were the most affected components [9] and 2016 Central Italy Earthquake [4].

Based on these considerations, it is worth to be noted not only the importance of improving the seismic design methodologies available in the literature and seismic codes for NSEs installed in new buildings but also the need to develop assessment methodologies which allow to predict the seismic risk at which NSEs installed in existing facilities are prone. This issue is of particular interest also for low-intensity earthquakes, which often do not induce structural damage but could affect the functionality of building due to the damage to NSEs. FEMA E-74 [10] proposed a simplified methodology which allows to catalogue NSEs installed in existing buildings based on the seismic risk at which they are prone and it also provides risk mitigation installation detailing for a wide range of NSE typologies. More recently, De Angelis et al. [11] developed a rapid visual screening methodology for the large-scale assessment of the seismic vulnerability of NSEs in school buildings. This methodology, if used in combination with simplified methodologies for the seismic risk assessment of structures, would allow to better estimate the global performance of structures and would help stakeholders in prioritizing economic investments also focusing on some important aspects that were generally neglected until now.

In this paper, the methodology proposed by De Angelis et al. [11] has been described and modified. The methodology has been applied to some illustrative examples related to police stations built in South of Italy in different seismic regions, in order to verify if different mitigation detailing were adopted to reduce the seismic risk at which NSEs are prone in such critical facilities.

2 The Proposed RVS Procedure for Non-structural Elements

RVS is a qualitative evaluation procedure developed to identify and classify buildings based on their seismic vulnerability. These methodologies are traditionally developed to assess the structural vulnerability and they consist in filling a questionary during a sidewalk survey [12–16]. For this reason, the RVS procedure can be implemented relatively quickly on a large number of buildings. Focusing on structural systems, RVS methodology generally requires the identification of the load-resisting system, the materials of the structure, the respective seismic region, and the inventory of various features that may affect the seismic response of structures. The collected data is then processed to compute a numerical score in order to define a vulnerability trend for all investigated buildings. The results from RVS can be used for a variety of applications, including earthquake disaster risk management programs of a city or a region. The benefit of the application of RVS for preliminary identification of buildings (or building classes) and to assist decision-makers in the implementation of seismic risk reduction strategies is demonstrated by the number of RVS methods developed in the last years and also introduced in national building codes and guidelines [17–19]. Regarding NSEs, the RVS methodologies available in the literature generally include a question about possible hazards related to the vulnerability of NSEs but are not able to provide a specific vulnerability index for those elements in order to estimate their influence on the global seismic performance of the investigated buildings. To deal with this issue, De Angelis et al. [11] proposed a RVS methodology to investigate the seismic vulnerability of NSEs in school buildings.

The methodology proposed in this manuscript slightly modifies the RVS method proposed by De Angelis et al. [11] to extend the building typologies for which it can be applied. The methodology distinguishes between different non-structural categories and comprises 48 questions. The considered categories are: infill wall and partitions, external coating, ceiling and soffits, external appendages, windows/doors/grazing, roofing, electrical/mechanical systems, furniture/contents. The NSEs detailing investigated in the proposed form were established by analyzing the damage observations following post-earthquake inspections, experimental evidence and expert judgment. At the same time, some surveys were carried out in different strategic facilities to identify possible sources of vulnerability. The information required to fill the form account for non-structural typologies, connections to the supporting structure, signs of degradation and possible interaction with other NSEs. For example, Table 1 reports the questions included in the form for ceiling systems. In this case, once the detailing regarding the connections of the ceiling system to the supporting structure is identified, some details about possible deterioration or plaster detachment and possible interaction with piping and lighting systems should be identified.

Table 1. Example of the proposed form for ceiling system.

Question	Vulnerability		
	Low	Medium	High
Are signs of deterioration or plaster detachment present in the plaster ceiling?	0	2	4
Does the suspended ceiling have adequate hangers?	0	8	16
Is the connection system of the ceiling to the walls flexible?	0	4	8
Is lighting restrained independently and not reliant on the ceiling to provide lateral support?	0	4	8
Is the piping located in suspended ceilings restrained independently?	0	4	8

In order to estimate the NSE vulnerability index (I_{NSE}) according to the proposed procedure three components should be considered: the non-structural category, the specific questions and the answer provided to each question. For each question three levels of risk are established: low, medium and high. To each level of risk is associated a numerical value called unit index (*UI*). The *UI* is equal to 0, 2 and 4 for low, medium and high risk, respectively. The level of importance of each question is described by importance factors, p_i, which are equal to 1, 2 and 4 for questions of least, medium and most importance, respectively. Focusing on the questions for ceiling systems reported in Table 1, the question "Does the suspended ceiling have adequate hangers?" assumes the highest importance because it is characterized by a p_i equal to 4, while the question "Are signs of deterioration or plaster detachment present in the plaster ceiling?" is characterized by the lowest importance, consequently the associated p_i is equal to 1. The numerical values reported in Table 1 for each question and level of risk are the combination of p_i with the *UI* associated to each level of risk. For each NSE category, j, a primary vulnerability index I_{PVj} is obtained through the following equation:

$$I_{PVj} = \frac{\sum_{i=1}^{n_j}(UI \cdot p_i)^j}{\sum_{i=1}^{n_j}(MUI \cdot p_i)^j} = \frac{I^j}{MI^j} \tag{1}$$

where *MUI* is the maximum *UI* that can be association to a question.

Before to proceed with the evaluation of the I_{NSE} is it important to define the importance of each non-structural category in the final evaluated. In fact, a different weight, W_j, has been assigned to each non-structural category. The category weight W_j is evaluated by dividing the sum of the highest scores of all the questions in a category by the sum of the highest scores of all categories and multiplying by a modification factor f_j, which normalizes the effect of the different number of questions on the score:

$$W_j = \frac{MI^j}{\sum_{j=1}^{8} MI^j} \cdot \varphi_j \tag{2}$$

where the modification factor f_j of the jth category is obtained by dividing the maximum number of questions of all categories (Mn_j) by its number of questions (n_j):

$$\varphi_j = \frac{Mn_j}{n_j} \tag{3}$$

Table 2 reports the W_j assigned to each non-structural category. De Angelis et al. [11] performed a calibration of the W_j through fragility functions pointing out the effectiveness of the proposed values.

Table 2. Weights of the non-structural categories.

Category	Sum of the highest scores	$\frac{MI^j}{\sum_{j=1}^{8} MI^j}$	n_j	f_j	W_j (%)
Infill walls and partitions	44	13	7	1.43	9
External coatings	20	6	2	5.00	14
Ceiling and soffits	44	13	5	2.00	12
External appendages	88	26	9	1.11	14
Windows, door and glazing	36	10	3	3.33	17
Roofing	32	9	3	3.33	15
Electrical, mechanical and hydraulic components	52	15	10	1.00	7
Furniture and contents	28	8	3	3.33	13
Sum	344	100	42	21	100

Finally, the non-structural index I_{NSE} is obtained by summing the primary vulnerability indices, I_{PVj}, multiplied by the corresponding non-structural category weights, W_j:

$$I_{NSE} = \sum_{j=1}^{8} I_{PVj} \cdot W_j \tag{4}$$

The obtained I_{NSE} allows to classify the NSEs installed in a building in one of the following vulnerability levels:

•Low vulnerability: $I_{NSE} < 0.33$;
•Medium vulnerability: $0.33 \leq I_{NSE} \leq 0.67$.
•High vulnerability: $I_{NSE} > 0.67$.

3 Illustrative Example

The RVS methodology described in Sect. 2 was applied to six reinforced concrete buildings located in the South of Italy. As reported in Fig. 1, the buildings are located in

Puglia, one building for each province of the region was selected. The selected buildings are all used to host Police stations. This choice would allow to investigate if, considering the importance of the buildings, the NSEs are installed according to seismic provisions in order to guaranty the perfect operability of the buildings following seismic events, as also required by the Italian building code [20].

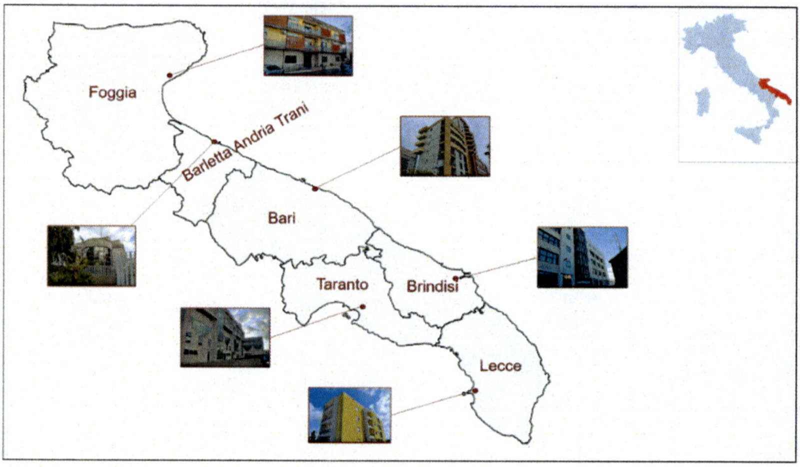

Fig. 1. Case study buildings

The selected buildings were built in different construction periods (varying from 1970 to 2002) and are characterized by a number of floors varying from three to nine. All the considered case study buildings were designed only for gravity loads with a moment resisting frame load-bearing system (Table 3).

Figure 2 reports the results of the RVS methodology applied to all case study buildings. Analyzing the results, it is possible to observe that for the analysed case study building the I_{NSE} assumes values between 0.19 and 0.52; this means that the buildings are characterized by low to medium risk regarding the vulnerability of NSEs. In Fig. 2 the I_{PV} are also reported. The infill walls and partitions are characterized by low to medium risk for all case study buildings because the infill panels are generally built in the supporting frame and the presence of cracking, that can lead to premature detachment of the panels, is not detected. Regarding the external coating, when available, the buildings have a medium risk because the coating are not connected to the structures using seismic detailing; at the same time, the coatings are also present to the upper floors and could hinder the escape routes. Similar considerations can be extended to façades and glazing systems that are generally non installed using supporting systems able to accommodate the interstorey displacements that arise in the building following seismic events.

In all case study buildings, the mechanical, electrical and piping systems, as well as the furniture and contents, are prone to medium risk; this is the consequence of the absence of adequate seismic detailing to prevent excessive lateral displacements and the overturning of rocking NSEs, such as shelves and servers. However, it is worst to be

Table 3. Main features of the selected buildings.

ID	Province	Period of construction	Number of floors	Structural typology
1	Lecce	2002	4	Moment resisting frames
2	Bari	1990	9	Moment resisting frames
3	Barletta-Andria-Trani	1996	4	Moment resisting frames
4	Brindisi	1984	5	Moment resisting frames
5	Foggia	1970	3	Moment resisting frames
6	Taranto	1994	7	Moment resisting frames

noted that in most of the cases the shelves are connected each other and this could help in preventing the overturning, as often observed following previous earthquakes [4]. For what concern the piping systems, it is often observed the absence of lateral restraining and of flexible connections both in the connections with the mechanical equipment and in piping passing through partition and infill walls (Fig. 3). Ceiling systems are generally prone to medium risk due to the inadequate connection to the supporting structure. In fact, the ceiling systems, for all case study buildings, are connected to the slab though vertical wires without the presence of diagonal elements. At the same time, the seismic detailing suggested in seismic risk mitigation guidelines [10] are not followed both regarding the connection of the ceiling system to the perimetral walls and the isolation of lighting and air conditioning systems from the ceiling panels and supporting grids.

4 Conclusions

The damage observed following recent seismic events pointed out the need for seismic risk mitigation measurements dealing with the improvement of both structural and non-structural seismic performance. This is particularly true with the seminal advancement of performance-based seismic design which is leading to a significant step forward in the seismic design of new buildings. Due to the large number of existing buildings requiring preliminary vulnerability evaluations, simplified tools are needed. Rapid Visual Screening procedures have been shown to be an effective and consolidated solution for the classification of structural vulnerability. Based on this consideration, a rapid visual screening procedure for non-structural elements has been recently proposed in the literature. This methodology has been adopted and slightly modified in this study to extend its field of application. The proposed methodology can be applied to define priority lists to be used by local authorities for a better management of available funding

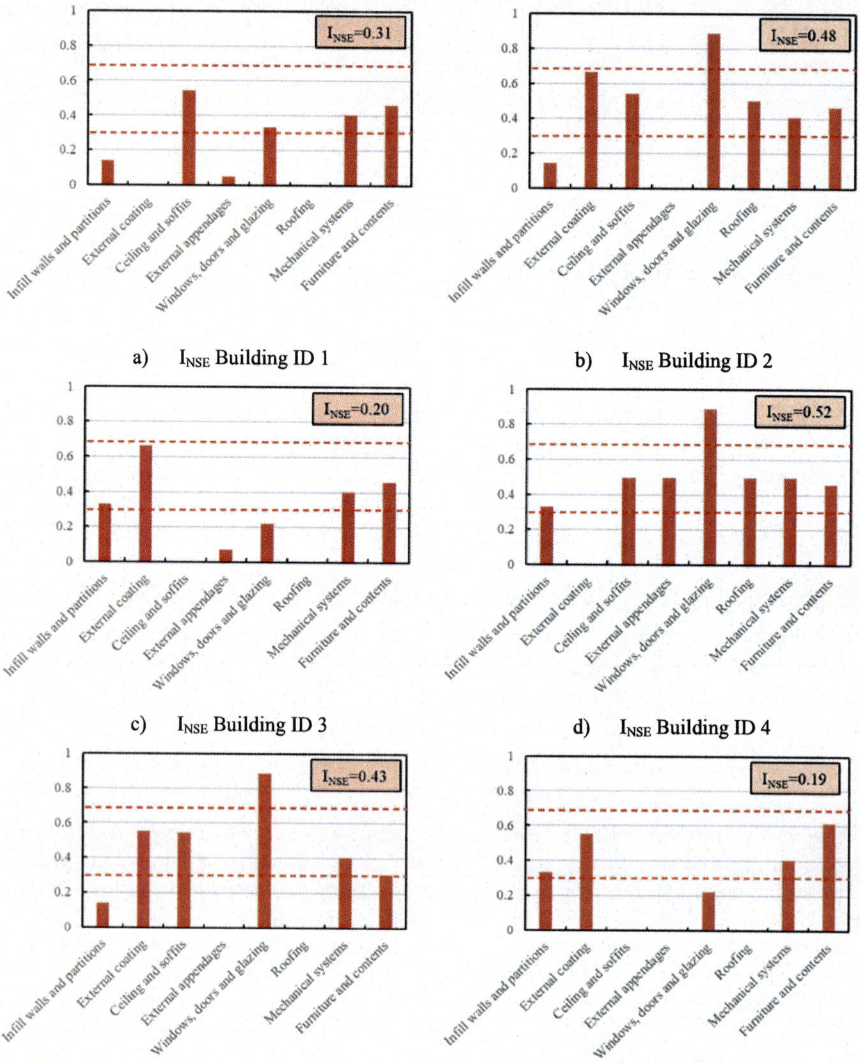

a) I_{NSE} Building ID 1

b) I_{NSE} Building ID 2

c) I_{NSE} Building ID 3

d) I_{NSE} Building ID 4

e) I_{NSE} Building ID 5

f) I_{NSE} Building ID 6

Fig. 2. Results RVS methodology applied to the case study buildings

in the implementation of risk mitigation programs. The proposed methodology has been applied to six police stations located in the South of Italy and allowed to identify the main sources of vulnerability in different non-structural typologies. Although some improvements are still required, this study demonstrated the effectiveness of the methodology and its usefulness to perform a large-scale classification of the seismic vulnerability of non-structural elements installed in critical facilities.

a) Server and air conditioning in Building ID 1

b) Mechanical systems in Building ID 2

c) Ceiling system in Building ID 3

d) Piping system in Building ID 4

e) Partition in Building ID 5

f) Ceiling systems in Building ID 6

Fig. 3. Some details of NSEs in case study buildings

Acknowledgements. The authors gratefully acknowledge the funding by Italian Ministry of Education, University and Research (PON Grant n.1735/2017, CADS Project) and the DPS-DCSTLGP-ReLUIS Project.

References

1. Miranda, E., Mosqueda, G., Retamales, R., Pekcan, G.: Performance of nonstructural components during the 27 February 2010 Chile Earthquake. Earthq. Spectra **28**(S1), S453–S471 (2012)
2. Baird, A., Ferner, H.: Damage to non-structural elements in the 2016 Kaikoura earthquake. Bull. New Zealand Soc. Earthq. Eng. **50**(2), 187–193 (2007)
3. Zareian, F., Sampere, C., Sandoval, V., McCormick, D.L., Moehle, J., Leon, R.: Reconnaissance of the Chilean wine industry affected by the 2010 Chile offshore Maule earthquake. Earthq. Spectra **28**(S1), S503–S512 (2012)

4. Perrone, D., Calvi, P.M., Nascimbene, R., Fischer, E., Magliulo, G.: Seismic performance and damage observation of non-structural elements during the 2016 Central Italy Earthquake. Bull. Earthq. Eng. **17**, 5655–5677 (2018)
5. Miranda, E., Taghavi, S.: Estimation of seismic demands on acceleration-sensitive nonstructural components in critical facilities. In: Proceedings of the seismic design, performance, and retrofit of nonstructural components in critical facilities, pp. 29–2. ATC, Newport Beach, Ca (2003)
6. Filiatrault, A., Sullivan, T.: Performance-based seismic design of nonstructural building components: the next frontier of earthquake engineering. Earthq. Eng. Eng. Vib. **13**(suppl), 17–46 (2014)
7. Filiatrault, A., Perrone, D., Merino, R.J., Calvi, G.M.: Performance-based seismic design of non-structural building elements. J. Earthquake Eng. **25**(2), 237–269 (2021)
8. Braga, F., Manfredi, V., Masi, A., Salvatori, A., Vona, M.: Performance of non-structural elements in RC buildings during the L'Aquila, 2009 Earthquake. Bull. Earthq. Eng. **9**, 307–324 (2011)
9. Ercolino, M., Petrone, C., Coppola, O., Magliulo, G.: Report sui danni registrati a San Felice sul Panaro (Mo) in seguito agli eventi del 20 e 29 maggio 2012–v1.9. http://www.reluis.it (2012)
10. FEMA: Reducing the risks of nonstructural earthquake damage—a practical guide, FEMA E-74. Federal Emergency Management Agency and National Earthquake Hazard Reduction Program, USA (2012)
11. De Angelis, A., Pecce, M.: Seismic nonstructural vulnerability assessment in school buildings. Nat. Hazards **79**(2), 1333–1358 (2015). https://doi.org/10.1007/s11069-015-1907-3
12. Achs, G., Adam, C.: Rapid seismic evaluation of historic brick-masonry buildings in Vienna (Austria) based on visual screening. Bull. Earthq. Eng. **10**, 1833–1856 (2012)
13. Ferreira, T.M., Maio, R., Vicente, R.: Seismic vulnerability assessment of the old city centre of Horta, Azores: calibration and application of a seismic vulnerability index method. Bull. Earthq. Eng. **15**(7), 2879–2899 (2016). https://doi.org/10.1007/s10518-016-0071-9
14. Jain, S.K., Mitra, K., Kumar, M., Shah, M.: A proposed rapid visual screening procedure for seismic evaluation of RC-frame buildings in India. Earthq. Spectra **26**(3), 709–729 (2010)
15. Perrone, D., Aiello, M.A., Pecce, M., Rossi, F.: Rapid visual screening for seismic evaluation of RC hospital buildings. Structures **3**, 57–70 (2015)
16. Yadollai, M., Adnan, A., Zin, R.M.: Seismic vulnerability functional method for rapid visual screening of existing buildings. Arch. Civil Eng. LVIII **3**, 363–377 (2012)
17. JBDPA. Seismic Evaluation and Retrofit, The Japan Building Disaster Prevention Association, Japan (2001)
18. NRCC. Manual for Screening of Buildings for Seismic Investigation, Institute for Research in Construction, National Research Council Canada, Ottawa (1993)
19. NZSEE. The seismic assessment of existing buildings: technical guidelines for engineering assessments., New Zealand society for Earthquake Engineering (2017)
20. NTC. Aggiornamento delle Norme tecniche per le costruzioni. Ministero delle Infrastrutture e dei Trasporti; Roma, Italia (2018)

Comparison of Seismic Losses Associated with Traditional/Innovative Hollow Brick and Plasterboard Internal Partitions

Gennaro Magliulo[1,2], Danilo D'Angela[1(✉)], Pauline Lopez[3], and Gaetano Manfredi[1]

[1] Department of Structures for Engineering and Architecture, University of Naples Federico II, via Claudio 21, 80125 Naples, Italy
danilo.dangela@unina.it

[2] Construction Technologies Institute, National Research Council, 80125 Naples, Italy

[3] Building Physics Innovation and Expertise, Etex Building Performance, Avignon, France

Abstract. Nonstructural elements are typically associated with high seismic risk, regarding functioning interruption, economic losses, and casualties. Architectural elements such as infills and internal partitions are often associated with major construction costs, and, even in case of relatively frequent earthquakes, the post-event repair costs might be critical, due to poor seismic performance. Moreover, damage of architectural elements typically affects functioning and operativity of the facilities. The present study focuses on seismic response of internal partitions, with particular regard to economic losses. Seismic losses associated with both traditional and innovative internal partitions are assessed and compared, considering code-conforming reinforced concrete frame buildings as a case study. Both traditional and innovative partition systems are investigated, considering hollow brick and plasterboard solutions. Seismic demand is assessed through multiple stripe analyses, considering advanced nonlinear modeling and including low-to-high seismicity sites in Italy. Seismic capacity of investigated partition systems is derived from literature experimental data, considering multiple damage states (and performance levels) as a reference. The study quantifies seismic losses associated with the investigated partition systems, shedding lights on their strengths and weaknesses in terms of seismic performance and economic efficiency. Technical insights regarding the efficient use of the investigated partitions are also supplied in the light of the loss comparison analysis.

Keywords: Nonstructural elements · Seismic losses · Internal partitions · Plasterboard · Innovative solutions

1 Introduction

Nonstructural elements are typically associated with an extremely large part of building costs and are particularly critical in terms of seismic losses [1]. Architectural elements include but are not limited to infills, partitions, ceilings, façades, decorative moldings (e.g., plaster), doors/windows, and infills and partitions are among the most critical

G. P. Cimellaro (Ed.): WCSI 2022, LNCE 309, pp. 809–818, 2023.
https://doi.org/10.1007/978-3-031-21187-4_70

regarding seismic losses [2, 3]. Several studies investigated seismic performance of infills and partitions through experimental [3–8] and numerical methods [9–12], and more recently, the research focused on the estimation of seismic losses associated with nonstructural elements, including infills and partitions [2, 13–15]. In many cases, the loss estimations were based on observational data, focused on existing buildings, and did not account for a variety of infill/partition solutions; moreover, past study only considered traditional infill/partition systems.

The present study focuses on partition systems and carries out a comparison of seismic losses considering multiple partition solutions, including traditional and innovative solutions. Low-rise reinforced concrete (RC) frame buildings designed according to Italian building code NTC 2008 [16] considering low-to-high seismicity sites in Italy. Numerical analyses are performed to estimate the seismic demands on the partitions, whereas literature capacity criteria are considered to define capacity thresholds. The seismic losses are estimated considering whole buildings and are correlated to the return period of the reference earthquakes, also with reference to serviceability limit state seismic demands. The present study reports, discusses, and extends the research activities reported in the study by Magliulo et al. [17], which is referred to for technical details.

2 Methodology

2.1 Outline

Seismic losses associated with both traditional and innovative partition systems are investigated in the study. Case study partitions are defined in Sect. 2.2, whereas damage states and considered capacity criteria, estimated according to the literature, are defined in Sect. 2.3. Case study buildings consist in three-story code-conforming RC buildings, located in low-to-high seismicity sites in Italy, i.e., L'Aquila (AQ), Naples (NA), and Milan (MI), illustrated in Sect. 2.4. Seismic demand on partitions was estimated by performing numerical analyses, which are described in Sect. 2.5. The methodology implemented for estimating the global seismic losses is described in Sect. 2.6.

2.2 Case Study Partitions

Three case study partitions were investigated: hollow brick (HB), standard plasterboard (SP), and innovative plasterboard (IP) partitions. HB system consists in traditional hollow bricks (250 × 250 × 80 mm) connected and plastered through mortar [5]. SP partitions consist in traditional plasterboards (double two-layer boards) connected to light steel frames (e.g., [18, 19]), whereas IP system represents an improved plasterboard partition system, developed by Magliulo et al. [18], in order to enhance seismic performance. Figure 1 depicts the investigated case study partitions [17].

2.3 Damage States and Capacity Criteria

Following damage states (DSs) were considered in the study: DS0, DS1, DS2, and DS3, corresponding to null damage, minor damage, moderate damage, major damage. Further

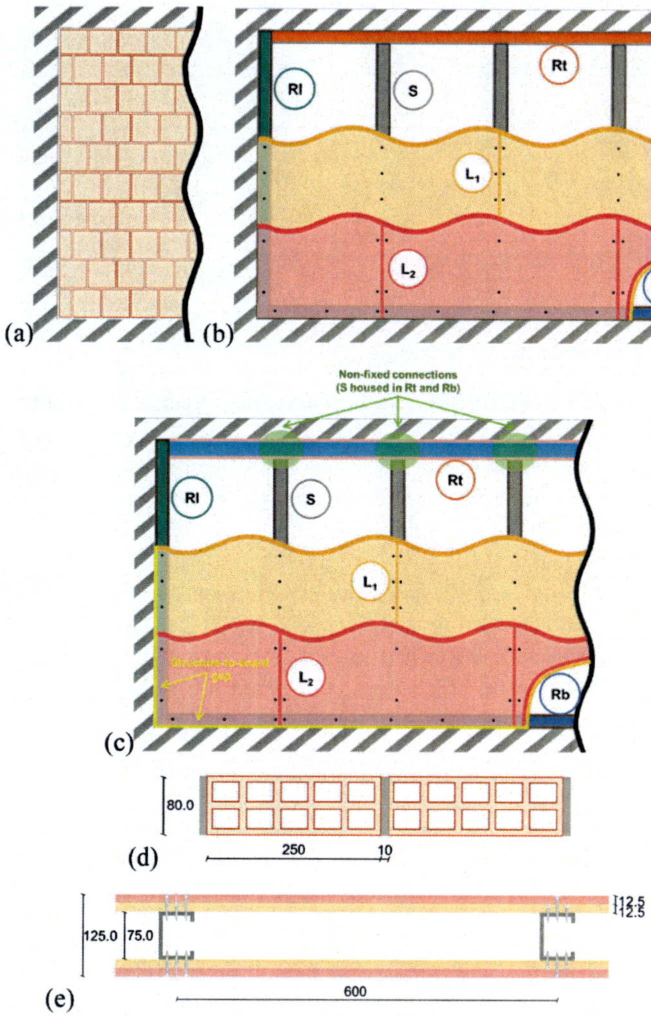

Fig. 1. Geometrical scheme of the investigated partition systems: (a) HB lateral view, (b) SP lateral view, (c) IP lateral view, (d) HB cross-section (mm), and (e) SP/IP cross-section (mm). S, Rt, Rb, Rl, L1, and L2 are representative of vertical stud, top runner, bottom runner, lateral runner, inner board layer, and outer board layer, respectively [17].

details regarding to technical identification of DSs can be found in [18, 20–23]. The damage assessment was implemented only considering in-plane partition damage, i.e., considering interstory drift ratio (IDR) as an engineering demand parameter (EDP). IDR-based partition capacities were estimated according to consolidated literature, favoring the use of robust experimental data [5, 18, 19, 22–24]. Table 1 reports the considered IDR-based capacity thresholds (IDRc) associated with the relevant DSs and the related reference literature.

Table 1. IDR-based capacity thresholds (IDR_C) associated with the relevant DSs.

IDRc [%]		HB	SP	IP
DS	DS1	0.21	0.19	0.58
	DS2	0.34	0.89	0.98
	DS3	0.97	1.49	2.10
References		[5]	[19, 22]	[18, 23, 24]

2.4 Case Study Buildings

Case study buildings consist in residential three-story regular RC frames, designed according to Italian Building code (NTC 2008) [16], which is compatible with Eurocode 8 [25]. Building floor and elevation geometric layout are depicted in Fig. 2a and b, respectively. Three sites with fixed soil conditions (soil type "C") were considered to design the buildings, i.e., L'Aquila (AQ), Naples (NA), and Milan (MI), corresponding to acceleration at bedrock ag equal to 0.104 g (0.261 g), 0.060 g (0.168 g), and 0.024 g (0.049 g) for damage limitation limit state DLS (life safety limit state LSLS), respectively. The buildings were designed considering ductility class "B" ("CDB" in Italian), considering response spectrum analysis for the estimation for the design demands and behavior factor equal to 3.9. Both DLS and LSLS were considered for the design of the buildings and for the structural details.

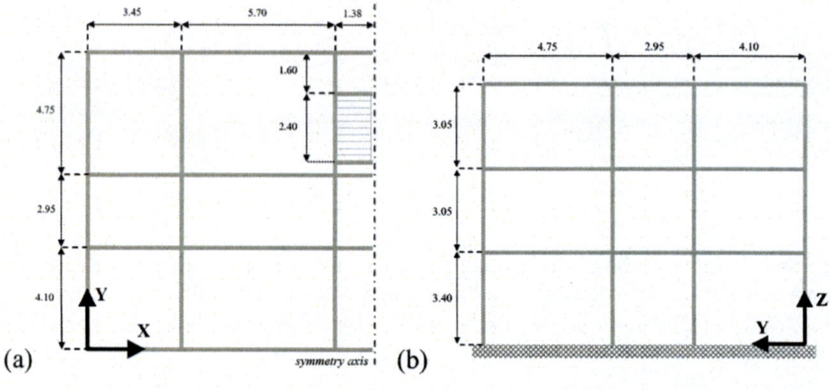

Fig. 2. Building geometrical layout: (a) floor and (b) [17].

2.5 Numerical Analysis and Seismic Demand

Seismic demand was estimated by performing bidirectional multiple stripe analysis (MSA) through the implementation of nonlinear modeling in OpenSees [26]. 3D frame building models were built in OpenSees by considering lumped plasticity approach and assigning well-established Ibarra-Medina-Krawinkler (IMK) response to the inelastic springs [27, 28]. Modeling parameter specification was based on literature studies, and relevant references can be found in [29–33]. Stairwell elements were included within the frame modeling, whereas shear failure and inelastic beam-to-column joint response were not accounted for [29]. P-delta effects (geometric nonlinearities) were implemented in the models. Rayleigh damping was assigned to the models, through a damping ratio equal to 5% [27, 29].

Ten intensity measure (IM) levels were considered to perform MSA, corresponding to return period T_R equal to 10, 50, 100, 250, 500, 1000, 2500, 5000, 10000, and 100000 years. Each analysis stripe (or IM) was associated with 20 pairs of bidirectional records, provided by Iervolino et al. [34] and defined according to conditional spectrum approach [35]. Further details regarding the analysis records can be found in [29, 30].

2.6 Global Loss Analysis

Loss analysis performed in the study is based on the IDR-based comparison between seismic demand on partitions, assessed through MSA, and seismic capacity, derived from the literature, i.e., capacity to demand assessment. Only direct economic costs are accounted for, i.e., initial construction costs and repair costs associated with seismic damage. The DS occurrence of the case study partitions was estimated according to the abovementioned approach, and after the integration of the damage of the whole buildings, the global losses were estimated as a function of the analysis IM levels. Global losses were estimated by applying unitary/nominal costs to the relevant surface areas associated with the damaged partition panels. The unitary costs were estimated in this study, considering the Italian construction price lists (Lazio region) [36] as a reference. Further details regarding the loss assessment assumptions and methodology can be found in [17].

3 Results

3.1 Global Losses

Figure 3 depicts the seismic losses associated expressed in terms of repair costs (Rcs) and Rcs added to initial costs (ICs) (Ics & Rcs), as a function of IM levels, evaluated for (a) AQ, (b) NA, and (c) MI buildings. The most meaningful results are associated with lower intensity IM levels (e.g., IM levels 1–4), since nonstructural elements are designed and assessed considering serviceability limit states. In fact, these elements are generally allowed to be damaged, without potentially causing injuries and casualties, under ultimate limit state earthquakes. However, relatively severe in-plane damage of partition panels (e.g., DS2 or DS3) affects their out-of-plane response, which is typically acceleration-sensitive. Out-of-plane failure might cause severe damage, potentially causing human losses. Therefore, IM levels higher than 3 or 4 can be significant in this context, and they can provide useful information, especially since the study focuses on seismic losses and not on design or limit states assessment.

If only repair costs (Rcs) are taken into account, IP system is the most efficient since it is associated with lower seismic losses, especially over IM levels 1–5 moderate to high seismicity and over all IM levels for low seismicity, whereas HB system is overall the least efficient, even for relatively low IM levels and over all three considered seismicity sites, except for extremely high IM levels (IM levels 8 and 9) and high seismicity. When both initial and repair costs are considered (Ics & Rcs), the results are less straightforward to be evaluated. In particular, for very low to low IM levels, i.e., corresponding to which there is no damage to the partitions, only the initial costs affect global costs; therefore, IP (HB) partition results in least (highest) efficiency, due to higher (lower) initial costs However, when damage to partition become more significant, e.g., from IM level 3 on for medium to high seismicity, both SP and IP result in a more efficient response, compared to HB system. Only for extremely high IM levels and moderate to high seismicity, HB system is found to be more efficient than other systems. As a matter of fact, when significant damage is achieved by higher performing systems, the initial costs again significantly condition the global costs. However, corresponding to these IM levels, severe structural damage is likely to be observed, and the response of nonstructural elements is not meaningful, especially in terms of economic losses.

3.2 Remarks

The estimated economic costs do not account for the potential human losses, and the efficiency of the systems should also account for these latter losses. In other words, economic costs associated with damage of the partition, which might be associated with human losses, especially considering DS2 and DS3, have the same significance of initial costs, which are not associated with potential human losses. Moreover, initial costs are typically less critical than repair costs as economic expenses since the latter are associated with (catastrophic) unforeseen events, whereas the former are typically planned/programmed. Accordingly, is reasonably desirable having higher initial costs rather than higher repair costs, since the latter certainly present additional indirect costs, which might also be associated with human losses. As a final comment, it should be noted that the results cannot be considered to be exhaustive, and further studies should also account for other costs, also including human losses contribution to (un)efficiency.

4 Conclusions

The study investigates the seismic losses associated with alternative partition systems housed in reinforced concrete (RC) code-conforming frame buildings. In particular, traditional hollow brick (HB) and standard plasterboard (SP) partitions are considered, as well as an innovative plasterboard (IP) partition system, developed in the literature. The cost estimation is based on an accurate in-plane demand to capacity assessment, carried out over the whole case study buildings considering interstory drift ratio (IDR) as an engineering demand parameter (EDP). The capacity thresholds are based on robust literature experimental data, whereas the seismic demand is estimated via multiple stripe analysis (MSA), performed considering nonlinear 3D models.

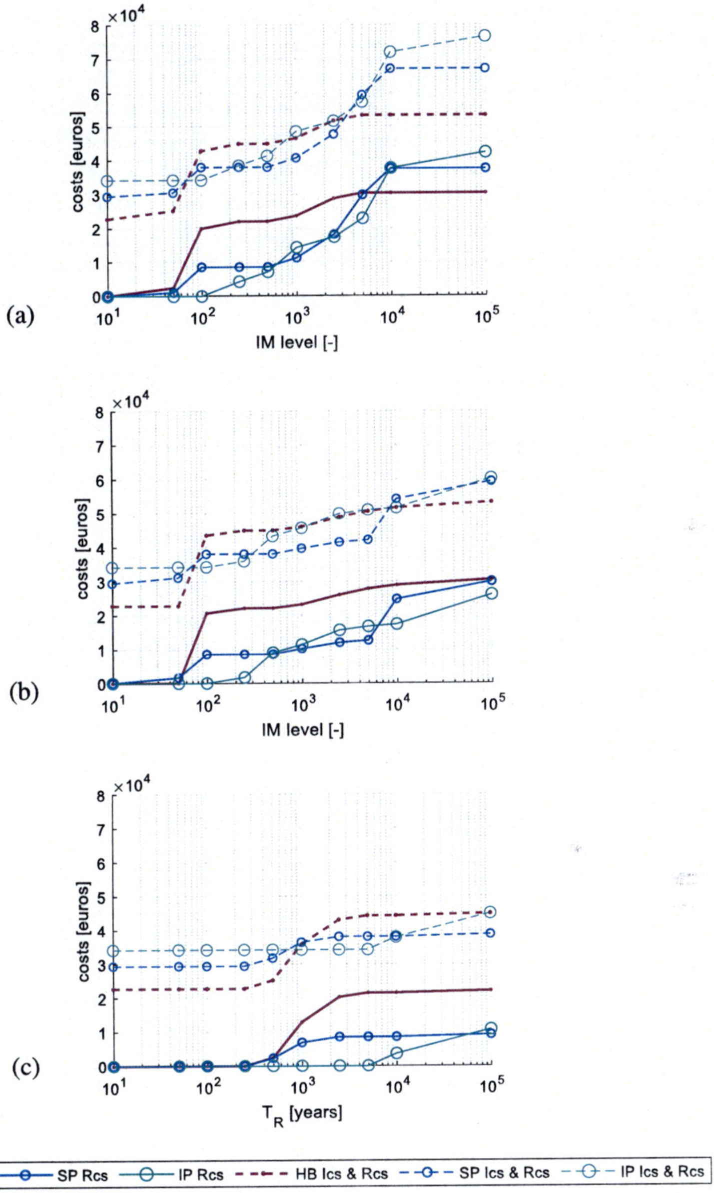

Fig. 3. Seismic losses associated with case study partitions (HB, SP, and IP systems) expressed in terms of repair costs (Rcs) and RCs added to initial costs (ICs) (Ics & Rcs), evaluated for (a) AQ, (b) NA, and (c) MI buildings.

The study sheds lights on efficiency of the investigated partition systems, considering direct economic losses due to extremely low to extremely high severity earthquakes. As a key finding, the study found that high performance plasterboard partition systems might be more efficient that traditional hollow brick system, and in some cases, even than traditional plasterboard system. In fact, when HB results in lower seismic costs, these present higher repair costs, whereas plasterboard partitions present higher initial costs, which are preferred to repair costs (repair costs are potentially associated with human losses). However, the findings cannot be considered to be exhaustive and cannot be generalized or extended to other case studies, without due consideration. Further case studies should be considered, as well as other sources of losses should also be accounted for, to characterize a more comprehensive and robust measure of efficiency.

Acknowledgements. This study was supported by Etex Building Performance International in the framework of the research study agreement ETEX-DIST 2016–2018. The contribution of Eng. Fabiola Napoletano to the numerical analyses is acknowledged.

References

1. Taghavi, S., Miranda, E.: Response assessment of nonstructural building elements. PEER Report 2003/05, Pacific Earthquake Engineering Research Center: University of California at Berkeley, Berkeley, California (2003)
2. Del Vecchio, C., Di Ludovico, M., Pampanin, S., Prota, A.: Repair costs of existing RC buildings damaged by the L'Aquila earthquake and comparison with FEMA P-58 predictions. Earthq. Spectra **34**(1), 237–263 (2018)
3. Restrepo, J.I., Bersofsky, A.M.: Performance characteristics of light gage steel stud partition walls. Thin-Walled Struct. **49**(2), 317–324 (2011)
4. Onat, O., Correia, A.A., Lourenço, P.B., Koçak, A.: Assessment of the combined in-plane and out-of-plane behavior of brick infill walls within reinforced concrete frames under seismic loading. Earthq. Eng. Struct. Dyn. **47**(14), 2821–2839 (2018)
5. Petrone, C., Magliulo, G., Manfredi, G.: Shake table tests for the seismic assessment of hollow brick internal partitions. Eng. Struct. **72**, 203–214 (2014)
6. Tasligedik, A.S., Pampanin, S., Palermo, A.: Low damage seismic solutions for non-structural drywall partitions. Bull. Earthq. Eng. **13**(4), 1029–1050 (2014). https://doi.org/10.1007/s10518-014-9654-5
7. Verderame, G.M., Balsamo, A., Ricci, P., Di Domenico, M., Maddaloni, G.: Experimental assessment of the out-of-plane response of strengthened one-way spanning masonry infill walls. Compos. Struct. **230**, 111503 (2019)
8. Anić, F., Penava, D., Guljaš, I., Sarhosis, V., Abrahamczyk, L.: Out-of-plane cyclic response of masonry infilled RC frames: an experimental study. Eng. Struct. **238**, 112258 (2021)
9. Wang, X., Zhao, W., Kong, J., Zhao, T.: Numerical investigation on the influence of in-plane damage on the out-of-plane behavior of masonry infill walls. Adv. Civil Eng. 6276803 (2020)
10. Petrone, C., Coppola, O., Magliulo, G., Lopez, P., Manfredi, G.: Numerical model for the in-plane seismic capacity evaluation of tall plasterboard internal partitions. Thin-Walled Struct. **122**, 572–584 (2018)
11. Zhai, C.H., Kong, J.C., Wang, X.M., Wang, X.H.: Finite-element analysis of out-of-plane behaviour of masonry infill walls. Proc. Inst. Civil Eng. Struct. Buildings **171**, 203–215 (2018)
12. Preti, M., Bolis, V., Stavridis, A.: Seismic infill–frame interaction of masonry walls partitioned with horizontal sliding joints: analysis and simplified modeling. J. Earthq. Eng. **23**, 1651–1677 (2019)

13. Sousa, L., Monteiro, R.: Seismic retrofit options for non-structural building partition walls: impact on loss estimation and cost-benefit analysis. Eng. Struct. **161**, 8–27 (2018)
14. Del Vecchio, C., Di Ludovico, M., Prota, A.: Repair costs of reinforced concrete building components: from actual data analysis to calibrated consequence functions. Earthq. Spectra **36**, 353–377 (2020)
15. Del Gaudio, C., De Risi, M.T., Ricci, P., Verderame, G.M.: Empirical drift-fragility functions and loss estimation for infills in reinforced concrete frames under seismic loading. Bull. Earthq. Eng. **17**(3), 1285–1330 (2018). https://doi.org/10.1007/s10518-018-0501-y
16. Ministero delle Infrastrutture e dei Trasporti. D.M. del 14/01/2008—Norme tecniche per le Costruzioni 2008. NTC 2008 (in Italian) (2008)
17. Magliulo, G., D'Angela, D., Lopez, P., Manfredi, G.: Nonstructural seismic loss analysis of traditional and innovative partition systems housed in code-conforming RC frame buildings. J. Earthq. Eng. (Ahead-of-print), 1–28 (2021)
18. Magliulo, G., Petrone, C., Capozzi, V., Maddaloni, G., Lopez, P., Manfredi, G.: Seismic performance evaluation of plasterboard partitions via shake table tests. Bull. Earthq. Eng. **12**(4), 1657–1677 (2013). https://doi.org/10.1007/s10518-013-9567-8
19. Landolfo, L., Pali, T., Bucciero, B., Terracciano, M.T., Shakeel, S., Macillo, V., et al.: Seismic response assessment of architectural non-structural LWS drywall components through experimental tests. J. Constr. Steel Res. **162**, 105575 (2019)
20. Fiorino, L., Bucciero, B., Landolfo, R.: Evaluation of seismic dynamic behaviour of drywall partitions, façades and ceilings through shake table testing. Eng. Struct. **180**, 103–123 (2019)
21. Petrone, C., Magliulo, G., Manfredi, G.: Shake table tests on standard and innovative temporary partition walls. Earthq. Eng. Struct. Dyn. **46**, 1599–1624 (2017)
22. Etex Innovation and Technology (EIaT) centre: Test report. In-plane cyclic loading test on partition with double layers of 12.5 mm Standard and Ladura boards with MT75 simple stud @ 600 mm (2018)
23. Petrone, C., Magliulo, G., Lopez, P., Manfredi, G.: Seismic fragility of plasterboard partitions via in-plane quasi-static tests: Seismic fragility of plasterboard partitions. Earthq. Eng. Struct. Dyn. **44**, 2589–2606 (2015)
24. Pali, T., Macillo, V., Terracciano, M.T., Bucciero, B., Fiorino, L., Landolfo, R.: In-plane quasi-static cyclic tests of nonstructural lightweight steel drywall partitions for seismic performance evaluation. Earthq. Eng. Struct. Dyn. **47**, 1566–1588 (2018)
25. CEN. EN 1998–1: Eurocode 8: Design of structures for earthquake resistance—Part 1: general rules, seismic actions and rules for buildings (2004)
26. McKenna, F., Fenves, G.L., Scott, M.H.: OpenSees: Open System for Earthquake Engineering Simulation, Pacific Earthquake Engineering Research Center, University of California, Berkeley, CA. Available at: http://opensees.berkeley.edu (2000)
27. Ibarra, L.F., Krawinkler, H.: Global collapse of frame structures under seismic excitations. In: Stanford CA: The John A. Blume Earthquake Engineering Center, Department of Civil and Environmental Engineering, Stanford University (2005)
28. Ibarra, L.F., Medina, R.A., Krawinkler, H.: Hysteretic models that incorporate strength and stiffness deterioration. Earthq. Eng. Struct. Dyn. **34**, 1489–1511 (2005)
29. Ricci, P., Manfredi, V., Noto, F., Terrenzi, M., Petrone, C., Celano, F., et al.: Modeling and Seismic response analysis of Italian code-conforming reinforced concrete buildings. J. Earthq. Eng. **22**, 105–139 (2018)
30. D'Angela, D., Magliulo, G., Celano, F., Cosenza, E.: Characterization of local and global capacity criteria for collapse assessment of code-conforming RC buildings. Bull. Earthq. Eng. **19**(9), 3701–3743 (2021). https://doi.org/10.1007/s10518-021-01115-y
31. Haselton, C.B., Deierlein, G.G.: Assessing seismic collapse safety of modern reinforced concrete moment frame buildings. In: Stanford CA: The John A. Blume Earthquake Engineering Center, Department of Civil and Environmental Engineering, Stanford University (2007)

32. Haselton, C.B., Liel, A.B., Taylor-Lange, S., Deierlein, G.G.: PEER Report No. 2007/03. Beam-column element model calibrated for predicting flexural response leading to global collapse of RC frame buildings. Berkeley, CA: Pacific Earthquake Engineering Research Center, University of California (2008)
33. Lignos, D.G., Krawinkler, H., Whittaker, A.S.: Prediction and validation of sidesway collapse of two scale models of a 4-story steel moment frame. Earthq. Eng. Struct. Dyn. **40**, 807–825 (2011)
34. Iervolino, I., Spillatura, A., Bazzurro, P.: Seismic reliability of code-conforming Italian buildings. J. Earthq. Eng. **22**, 5–27 (2018)
35. Lin, T., Haselton, C.B., Baker, J.W.: Conditional spectrum-based ground motion selection. Part I: hazard consistency for risk-based assessments: Conditional spectrum-based ground motion selection-I. Earthq. Eng. Struct. Dyn. **42**, 1847–1865 (2013)
36. Ministero delle Infrastrutture e dei Trasporti, Regione Lazio, Dipartimento Sviluppo Infrastrutture e Manutenzione Urbana. Roma Capitale: Bollettino ufficiale della regione Lazio N. 41. Tariffa dei Prezzi 2012. Legge Regionale n.12 del 13 agosto 2011 (2012)

Vibration-Based Test Results for the Investigation of the Infill Masonry Wall Damage

Vanni Nicoletti(✉), Davide Arezzo, Sandro Carbonari, and Fabrizio Gara

Department of Construction, Civil Engineering and Architecture (DICEA), Università Politecnica delle Marche, 60131 Ancona, Italy
v.nicoletti@univpm.it

Abstract. Infills are one of the most common non-structural elements worldwide adopted to build external and internal partition walls in framed structures. Many types of infills can be encountered in practical applications, characterized by different construction typologies and materials, but the most adopted ones are those realized with masonry. It is well known that infill masonry walls suffered lot of damage in the past as a consequence of earthquake excitations and, moreover, they were cause of undesirable effects on the structural seismic performance, such as the activation of soft story mechanisms and shear failure of short columns. For these reasons, the infill contribution on the seismic behaviour of buildings should be investigated both during the design process of new structures and the assessment of the existing ones. The aim of this paper is to investigate the usefulness of vibration data for the damage detection in infilled frame structures starting from the tracking of both the stiffness and modal properties. To this aim, an experimental campaign was performed on a laboratory steel-concrete composite frame with infill masonry walls. The tested mock-up was subjected to stepped-increasing cyclic displacements that produced a progressive in-plane damage of infills, up to their complete damage. In the meantime, vibration-based tests characterized by different level of input excitation provided to the structure, were performed to capture the effects of the infill damage on the mock-up global dynamic response. Results provide useful information for the detection and tracking of damage to non-structural elements during and after low and moderate seismic events, contributing to a better interpretation of data provided by structural health monitoring systems. Finally, a monitoring strategy suitable for new infilled buildings is proposed, with the main aim of providing a contribution in the context of differentiating between structural and non-structural damage, especially in case of blind monitoring.

Keywords: Infill masonry walls · Infill damage detection · Laboratory experimentation · Structural health monitoring

G. P. Cimellaro (Ed.): WCSI 2022, LNCE 309, pp. 819–829, 2023.
https://doi.org/10.1007/978-3-031-21187-4_71

1 Introduction

The assessment of the health status of structures during their life is nowadays a topic of great interest, especially for structures with public functions (e.g., bridges, hospitals, police and fire stations, schools, etc.), for which a Structural Health Monitoring (SHM) is often recommended in case of high hazard exposure or high vulnerability. Indeed, a monitoring system can provide useful information about a building status, in terms of damage on structural and non-structural elements occurred during the building life and after exceptional events, such as earthquakes, hurricanes, blasts, vehicle impacts, etc. Hence, SHM can support and speed up the assessment of the building healthy status, that is commonly performed through on site visual inspections that commonly require long time to be performed. The use of SHM systems on buildings can lead also to gain benefits, such as the reduction of the social hardships and economic losses due to the building inoccupancy, as well as the loss of the structure functionality in case of buildings with strategic functions and whose use is essential for the emergency management.

Among the others, the structural dynamic monitoring is nowadays widely adopted worldwide. The concept beyond this methodology is that the assessment of the healthy status of a structure can be performed by investigating changes in its dynamic properties since a structural or a non-structural damage reflects on changes in mass, stiffness and energy dissipation capabilities of the system, as well as on changes in the boundary conditions. Thus, changes produce modifications of the structure dynamic response that, traced over time, can support the identification of damage activations and/or of damage evolutions. The most adopted dynamic monitoring strategy foresees the adoption of the so called vibration-based SHM (Nicoletti et al. 2022a, Kong et al. 2017), which consists in identifying the structure dynamics based on in-situ vibration measurements (namely, accelerations or velocities).

In the literature, many authors adopted vibration-based tests on real buildings to investigate the damage on structural and non-structural members, adopting a gradual artificially induced damage (Yousefianmoghadam et al. 2020) and using real excitations, such as earthquake strong motions (Uebayashi et al. 2016; Gara et al. 2021). Other authors adopted vibration-based tests to identify the structural and non-structural damage produced by displacements and forces applied to laboratory specimens (Chiellini et al. 2008; Astroza et al. 2016). One of the most investigated non-structural components in framed structures are the infill walls, very often constituted by masonry panels (Gabbianelli et al. 2020; Morandi et al. 2021). In the technical and scientific literature, many works can be found addressing the seismic response of structural and non-structural members (e.g. Scozzese et al. 2017; Scozzese et al. 2018), but only few ones address the use of vibration-based methods for investigating damage on infill walls (Furtado et al. 2019; De Angelis and Pecce 2019). As a consequence, some interesting open issues are still under investigation, including the use of vibration-based SHM for differentiating between damage occurred to structural and non-structural members on buildings, as well as for establishing if a structure can be further used or not after an exceptional event, by eventually repairing the light damage occurred to the non-structural components to prevent risks during aftershocks (Furtado et al. 2018; Mucedero et al. 2020). The former issue becomes even more complicated to solve if the vibration-based tests are performed on existing infilled buildings, where the contribution of the infills to the whole dynamic behaviour is very difficult to understand.

This paper discusses about the usefulness of vibration-based test data for the damage detection in infilled frame structures. The study is based on results of an experimental laboratory campaign performed on an infilled steel-concrete composite mock-up. The experimental campaign includes quasi-static and dynamic tests performed on the laboratory mock-up: quasi-static tests are designed to induce a step-by-step progressive damage to infills, simulating an earthquake damage scenario; meanwhile, different types of dynamic tests producing increasing excitations are performed to evaluate the response of the system at different damage states of infills. The experimental campaign is herein described and the use of dynamic test results is addressed with the main aim of investigating its usefulness in a SHM framework. Moreover, a strategy for the dynamic monitoring of real infilled buildings is proposed in accordance with the main evidences achieved from the laboratory experimentation.

2 Laboratory Experimental Campaign

The laboratory experimental campaign was performed on the steel-concrete composite mock-up shown in Fig. 1a. The mock-up consists in a 3 m high, 2-bay moment resisting frame in the longitudinal direction and 1-bay braced frame in the transverse one, with dimensions of 8.4 m × 2.8 m. The concrete slab is 12 cm thick, casted on collaborating corrugated steel sheets and connected to the beams of the steel frame by means of Nelson's studs. To simulate permanent non-structural and live loads, several concrete blocks are positioned over the floor slab with a symmetric layout. The light infill masonry walls (about 7.5 cm thick, including the gypsum plaster) are built filling one of the two longitudinal spans. More information about the mock-up can be found in Nicoletti et al. (2020).

For the experimental campaign, the mock-up was instrumented with several sensors deployed on the structure, as shown in Fig. 1b: in detail, three linear variable displacement transducers (DT#) were mounted on a fixed external structure measuring the longitudinal displacements produced by the hydraulic actuators used for the quasi-static tests; in addition, six low noise uniaxial piezoelectric accelerometers were positioned over the concrete slab, three measuring in the transverse direction (AC1, AC2, AC3) and three in the longitudinal one (AC4, AC5, AC6). An extensive experimental campaign was undertaken performing quasi-static and dynamic tests (the latter of several types and characterized by different level of excitation provided to the structure) on the bare and on the infilled mock-up, as well as during the progressive damage of the infills. It is worth noting that all the performed tests were designed to produce damage only to the infills, i.e., avoiding the development of the non-linear behaviour of the steel frame structure.

In this paper, only dynamic tests performed on the whole mock-up during the progressive damage of the non-structural components are addressed. In detail, three types of dynamic tests were performed, characterized by different level of excitation (in terms of displacements, velocities and accelerations) provided to the structure: Ambient Vibration Tests (AVTs), Impact Load Tests (ILTs) and Snap-Back Tests (SBTs). These tests were repeated after the quasi-static tests of increasing displacements, aimed to produce damage to infills and to simulate earthquake scenarios. AVTs consisted in measuring the structure accelerations produced by the ambient excitation (mainly induced by microtremors from the surrounding environment for the considered laboratory case study); ILTs foresaw the excitation of the structure through instrumented hammer blows and the recording of both impacts and the produced accelerations. Finally, SBTs were performed by pushing the mock-up in the longitudinal direction up to a predetermined displacement and then by releasing it suddenly. The level of input provided to the mock-up increases with the different dynamic tests, starting from AVTs (where recorded accelerations are of the order of 10–5 g), passing through ILTs (accelerations around 10–2 g) and up to SBTs (accelerations ranging from 0.05 to 0.5 g). The infill progressive damage was produced by longitudinally pushing and pulling the mock-up in a quasi-static manner, starting from a displacement of 1 mm up to 13 mm (with an increase of 0.5 mm at each step). The infill damage was mainly due to the increasing In-Plane (IP) displacements to which the non-structural components were subjected and that produced detaching from the frame and cracking. Correlations between the dynamic parameters of the system, obtained from the three dynamic tests and the damage level experienced by the infills are obtained, discussing about the role of the level of excitation provided to the structure. It is worth noting that ILTs and SBTs are rather difficult to reproduce in real structures; however, the latter are herein considered as different sources of excitation provided to the structure. Such as an example, the system response to SBTs can be compared to that induced by earthquakes of low and moderate intensities, while the system response due to ILTs may be similar to that registered during a seismic swarm or during very strong weather conditions.

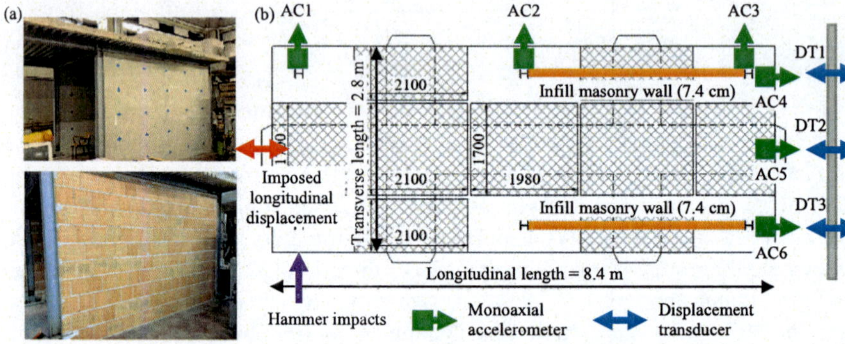

Fig. 1. (a) tested laboratory mock-up; (b) experimental set up.

3 Description of the Infill Damage

Quasi-static cyclic loads are performed by pushing and pulling the mock-up in the longitudinal direction, with the aim of producing an earthquake damage scenario on the non-structural components. Indeed, due to the progressive increase of IP displacements, infills suffered damage that led to the formation and then to the widening of cracks.

The mock-up hysteretic cycles obtained from the plot of the force-displacement graphs during the experimentation are reported in Fig. 2a. Also, the backbone curves are plotted in order to better interpret the results. The latter show sudden drops in correspondence of longitudinal displacements between 6.0 m and 8.0 mm, both in push and pull directions, which represent the activation of an important damage to the infills. To these drops correspond a reduction of both the IP stiffness and strength of the walls. However, the IP strength of infills kept increasing after this drop, while the secant stiffness continued to reduce, tending to the static stiffness of the bare mock-up. These phenomena can be explained by the fact that the resisting mechanism to lateral forces developed by the panels vary as the longitudinal displacements increase. The latter statement is also supported by the crack patterns detected on both walls, which are reported in Fig. 2b. As can be seen, no visible cracks are detected before longitudinal displacements S around 5 mm (drifts around 0.9‰); then, shallow and narrow diagonal cracks occurred at the corners. For displacements S around 6.5 and 7.5 mm (drifts about 1.4–2.0‰) an horizontal crack occurred in both infills following the path of a mortar bed joint. Up to these IP displacements the infill resistant mechanism to lateral forces was governed by the shear response of the wall panel as a whole. Thereafter, strut mechanisms on infills activated, leading to the opening and widening of diagonal stair-stepped cracks. At the end of the cyclic tests, both infills were completely damaged and some brick flakes were also expelled from the walls. Moreover, the secant stiffness calculated for each of the 25 cyclic load steps is reported in Fig. 2c. It is worth highlighting that the term secant stiffness is adopted since the hysteretic cycles of the mock-up (and hence the infills) exhibited a non-linear behaviour from the early steps.

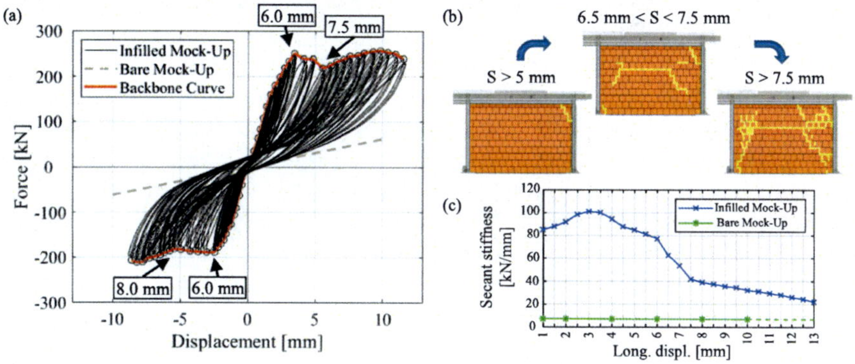

Fig. 2. (a) hysteretic cycles; (b) crack patterns on infills; (c) secant stiffness evolution.

Observing Fig. 2c it is possible to note that, for very low displacements, the secant stiffness increases with the longitudinal displacement; this is due to the fact that a certain level of displacement is necessary to activate the infill resisting mechanisms. After reaching the peak, a decrease is observed, and a sudden drop is well visible for longitudinal displacements between 6.0 and 8.0 mm (drifts from 1.17 and 2.10‰). Finally, an almost constant decrease is noticeably up to the end due to the degradation of the final resisting mechanisms (mainly consisting in friction forces developing between cracks). The final value, although much lower than those relevant to the undamaged infills, is still higher than the static stiffness relevant to the bare mock-up. This proves that, although heavily damaged, the infill walls still contribute to the longitudinal stiffness of the structure.

4 Analysis and Interpretation of Dynamic Test Results

The generalization and the development of a common strategy to be adopted in all buildings for detecting and tracing the damage on infill masonry walls is rather difficult to develop. In fact, the onset and the development of the infill damage strongly depends on both the material and geometrical properties of infills and the surrounding frame, as well as on the construction techniques and the type of stress state mobilized in the panel. However, in this Section, results of performed dynamic tests during the infill progressive damage are analyzed, compared, and discussed trying to correlate their outcomes with the health condition of infills. Only the longitudinal vibration mode is considered since it is the one of greater interest being largely affected by the variation of IP stiffness and strength of infills.

Even if the input level provided to the structure is quite different, AVTs and ILTS can be both classified as low-amplitude tests because, as it will be shown in the sequel, their outcomes (in terms of identified modal parameters) are very similar. This is due to the fact that during both tests the excitation provided to the structure was not enough high to activate dissipative mechanisms. On the contrary, SBTs are classified as high-amplitude tests since during these tests the infilled frame developed dissipative mechanisms, such as frictions at the edges between infills and the frame, and also frictions between cracks.

At first, frequency values identified from the three dynamic test recordings are analyzed, and collected in Fig. 3a as a function of the damage suffered by the infills (expressed in terms of longitudinal displacement of the mock-up). In this figure two different frequency groups can be easily recognized, one on the upper part and one on the bottom part of the graph. The former (at the top) collect frequency data identified from the low-amplitude tests (AVTs and ILTs), whilst the latter refer to data from the high-amplitude tests (SBTs). As can be seen, a frequency decay is always visible and, from the beginning to the end of the experimentation, a frequency reduction of about 30% is detected for both groups. Moreover, the frequency gap between low- and high-amplitude tests proves that only high-amplitude tests are able to activate the non-linear mechanisms, such as frictions between cracks and between the infill and frame, which are responsible for the lower identified frequency values.

However, the frequency decay (i.e. the decay observed by increasing the imposed longitudinal displacement that is responsible of the infill damage) is almost always linear (for both low- and high-amplitude tests). Even if phases in which infills suffered the main damage are difficult to identify, it can be observed that a frequency decay is always visible (for both low- and high-amplitude tests), demonstrating a decrease in the longitudinal stiffness of the mock-up due to the progressive damage of infills. Therefore, it can be asserted that the tracking of frequency values is a useful tool for the monitoring of the health condition of a structure and for supporting the damage detection procedure.

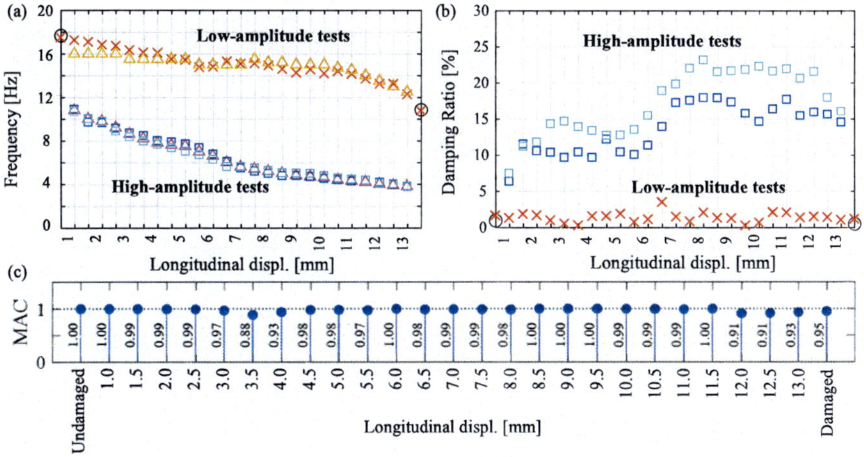

Fig. 3. (a) frequency, (b) damping ratio and (c) MAC evolutions during the progressive damage of infills.

As for the damping ratios, the latter are reported in Fig. 3b as a function of the longitudinal displacement provided to the mock-up (and hence to the damage suffered by infills). As for frequencies, also for damping ratios two different groups are well visible: values on the upper part are identified from high-amplitude tests, while values on the bottom part are obtained from low-amplitude tests. The latter do not show any significant trend and, moreover, damping is always around 1–2%, demonstrating the meaningless of tracing the damping ratios from AVTs and ILTs for the infill damage detection (at least for the case study at hand). Contrarily, the damping identified from high-amplitude tests shows a very interesting trend: initially, damping ratios around 10–15% are identified, then a significant increase is evident (around mock-up longitudinal displacements of about 6 – 8 mm) with values that rapidly increase and then remain almost constant till the end (around 15–25%). This sharp increase in damping occurred in correspondence of the formation of the main damage to infills (development of the horizontal cracks). This demonstrates that the tracing of damping from SBTs (and hence from high-amplitude recordings obtained from SHM systems) is useful not only for identifying the presence of damage in the non-structural elements, but also for capturing steps when the main damage occurs. The almost constant damping ratios after the horizontal and diagonal crack opening (longitudinal mock-up displacements from 8 mm to the end) demonstrates that infills have developed a stable dissipative mechanism that is independent on the amplitude of oscillations. Instead, the initial low values (for longitudinal displacements equal to 1 mm) are consistent with the rather low level of excitation provided to the mock-up, both in terms of displacement and acceleration.

The evolution of the mock-up longitudinal mode shape do not add useful information because the longitudinal mode has almost always the same shape, as demonstrated from high MAC values reported in Fig. 3c, obtained for the different dynamic tests performed. However, it is reasonably to assert that in real infilled buildings also the tracing of the mode shapes can support the task of detecting and tracing the damage to the structural and non-structural elements, especially if the damage is not widespread throughout the building.

5 A Strategy for the Infill Damage Detection in New Infilled Buildings

The outcomes achieved through the laboratory experimental campaign demonstrated that the tracing of modal parameters can be a successful strategy for the damage detection in infilled structures. Anyway, the damage suffered by the mock-up was related only to the infill masonry walls since the experimentation was performed under controlled laboratory protocols that permitted to maintain the steel frame in its elastic range. So, the topic of differentiating between structural and non-structural damage has not been addressed, but this represents a fundamental issue, especially when the dynamic monitoring of buildings is performed without the support of visual inspections, performing the so called data-driven monitoring. To support this task, a better knowledge of the building dynamic behaviour became crucial.

A possible strategy that can support the blind monitoring (data-driven) of an infilled building is proposed in Nicoletti et al. (2022b). In this work it has been suggested to dynamically monitor buildings also during their construction in order to achieve an accurate level of knowledge of the building dynamics and of its evolution over the different construction phases. This strategy is suitable for new buildings, where the construction process can be monitored, or in the restorations of existing ones that involve the complete removal of the non-structural elements. More in detail, authors proposed to perform dynamic tests (as, for instance, AVTs) during the several phases of the bare frame construction, as well as during the several steps in which the infills were built (at each floor of the construction), for identifying the modal parameters that evolves over time. The identified natural frequencies, can be traced over time, and a typical restitution is that of Fig. 4, where the fundamental frequency of a typical low- and mid-rise infilled RC frame building is reported over the evolution of the construction process. As can be seen, from the beginning (point A) to the end of the bare frame construction (point B) the fundamental frequency reduces, while, adding the infills, the fundamental frequency increases up to point C, which is the frequency value at the end of the construction (undamaged new building). So, it is evident that the addition of infills mostly contributes to increase the lateral stiffness of a building.

Differently, from the other points discussed above, point D is numerically determined since it represents the fundamental frequency of the structure when the infills are completely damaged and, hence, when they contribute only in terms of added masses. The latter coincides with the typical assumption that is made when a RC frame building is designed, because, in the practice, infills are commonly modelled only as added masses to the bare frame. The discussed fundamental frequency curve can support the SHM of a building during its life: if the measured fundamental frequency remains almost constant and equal to point C (normalizing environmental effects and operational conditions), then the health status is assessed; if it varies between point C and D, then a possible damage to the non-structural components occurred. The latter consideration

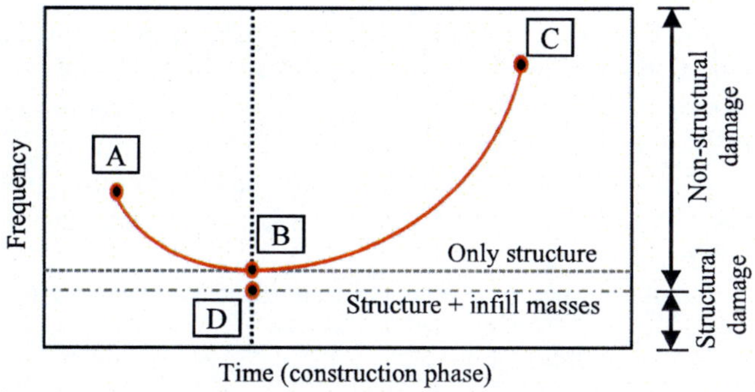

Fig. 4. Typical fundamental frequency evolution during construction of a low- and mid-rise infilled building.

is acceptable for structure built in accordance with modern code prescriptions (including hierarchy principles and seismic detailing), assuring that failure of non-structural components anticipate those of structural ones. Finally, if the fundamental frequency drops below point D, then a possible structural damage (together with the non-structural one) may have occurred. Obviously, for structures not complying with modern codes, a frequency drop from point C may be both due to structural and non-structural damage.

In addition, the discussed monitoring strategy has also many other advantages, supporting in a simple and direct way the assessment of the construction correctness, because anomalous trends in frequency values could be detected during the construction. Thus, the procedure also configures as a kind of proof testing during the building construction. Furthermore, tests on the bare frame can be suitably adopted to validate the design numerical model or to develop a digital-twin model of the building.

6 Conclusions

The presented work aimed at investigating the usefulness of vibration data for the damage detection in infilled frame structures. The study was conducted based on results of an extensive laboratory experimental campaign performed on an infilled steel-concrete composite frame mock-up that was dynamically tested during the progressive damage of infills. In detail, three types of dynamic tests were performed, characterized by different level of excitation (in terms of displacements, velocities and accelerations) provided to the structure: ambient vibration tests, impact load tests and snap-back tests. The infill damage was produced by cyclic quasi-static tests, increasing step-by-step the longitudinal displacements up to the limit relevant to the elastic field of the steel frame. Three different dynamic tests were performed with the main aim of providing correlations between the dynamic parameters of the system and the damage level experienced by the infills, also depending on the level of excitation experienced by the structure.

From the analysis of results achieved from the experimental campaign some considerations can be formulated:

- the tracing of the frequency evolutions during the infill damage from low- and high-amplitude vibration-based tests demonstrated to be an effective tool in the structural health monitoring of the laboratory mock-up, because it allowed the identification of a damage activation through the detection of decay trends;
- the damping ratio evolution achieved from high-amplitude tests (snap-back tests) shown a trend that can be correlated to the beginning and evolution of the progressive damage of infills;
- no useful information are achieved from the analysis of mode shapes; however, it may not be the same in case of real buildings, especially if the damage to the non-structural components is not uniformly distributed within the building.

Results of the laboratory experimental campaign provided useful information that support the growing of knowledge in the topic of damage detection in infilled buildings. A further collection of results, also coming from the technical literature, together with numerical analyses performed on infilled structures, will be able to better address the topic of the assessment of infilled buildings that suffered possible damage after exceptional events.

In the last part of the paper, a monitoring strategy that can support the structural health monitoring of new buildings during their life, has been proposed. Among the others, this procedure aims to address the important topic of differentiating between structural and non-structural damage in cases in which the monitoring is performed without the possibility of performing visual inspections, both for time consuming issues and for the uninhabitability and inaccessibility aspects.

References

Astroza, R., Ebrahimian, H., Conte, J.P., Restrepo, J.I., Hutchinson, T.C.: Influence of the construction process and nonstructural components on the modal properties of a five-story building. Earthq. Eng. Struct. Dyn. **45**(7), 1063–1084 (2016)

Chellini, G., De Roeck, G., Nardini, L., Salvatore, W.: Damage detection of a steel-concrete composite frame by a multilevel approach: Experimental measurements and modal identification. Earthq. Eng. Struct. Dyn. **37**(15), 1763–1783 (2008)

De Angelis, A., Pecce, M.R.: The structural identification of the infill walls contribution in the dynamic response of framed buildings. Struct. Control. Health Monit. **26**(9), e2405 (2019)

Furtado, A., Rodrigues, H., Varum, H., Arede, A.: Mainshock-aftershock damage assessment of infilled RC structures. Eng. Struct. **175**, 645–660 (2018)

Furtado, A., Vila-Pouca, N., Varum, H., Arede, A.: Study of the seismic response on the infill masonry walls of a 15-storey reinforced concrete structure in Nepal. Buildings **9**(2), 39 (2019)

Gabbianelli, G., Perrone, D., Brunesi, E., Monteiro, R.: Seismic acceleration and displacement demand profiles of non-structural elements in hospital buildings. Buildings **10**(12), 243, 1–19 (2020)

Gara, F., Arezzo, D., Nicoletti, V., Carbonari, S.: Monitoring the modal properties of an RC school building during the 2016 Central Italy seismic swarm. J. Struct. Eng. (ASCE) **147**(7), 05021002 (2021)

Kong, X., Cai, C.S., Hu, J.: The state-of-the-art on framework of vibration-based structural damage identification for decision making. Appl. Sci. **7**(5), 497 (2017)

Morandi, P., Hak, S., Milanesi, R.R., Magenes, G.: In-plane/out-of-plane interaction of strong masonry infills: From cyclic tests to out-of-plane verifications. Earthq. Eng. Struct. Dyn (2021)

Mucedero, G., Perrone, D., Brunesi, E., Monteiro, R.: Numerical modelling and validation of the response of masonry infilled RC frames using experimental testing results. Buildings **10**(10), 1–30, 182 (2020)

Nicoletti, V., Arezzo, D., Carbonari, S., Gara, F.: Expeditious methodology for the estimation of infill masonry wall stiffness through in-situ dynamic tests. Constr. Build. Mat. **262**, 120807 (2020)

Nicoletti, V., Arezzo, D., Carbonari, S., Gara, F.: Vibration-based tests and results for the evaluation of infill masonry walls influence on the dynamic behaviour of buildings: a review. Arch. Comput. Methods Eng (2022a)

Nicoletti, V., Arezzo, D., Carbonari, S., Gara, F.: Dynamic monitoring of buildings as a diagnostic tool during construction phases. J. Build. Eng. **46**, 103764 (2022)

Scozzese, F., Terracciano, G., Zona, A., Della Corte, G., Dall'Asta, A.: RINTC Project: nonlinear dynamic analyses of Italian code-conforming steel single-storey buildings for collapse risk assessment. In: Proceedings of the 6th international conference on computational methods in structural dynamics and earthquake engineering (COMPDYN 2017), pp. 15–17, Rhodes Island, Greece (2017)

Scozzese, F., Terracciano, G., Zona, A., Della Corte, G., Dall'Asta, A., Landolfo, R.: Analysis of seismic non-structural damage in single-storey industrial steel buildings. Soil Dyn. Earthq. Eng. **114**, 505–519 (2018)

Uebayashi, H., Nagano, M., Hida, T., Tanuma, T., Yasui, M., Sakai, S.: Evaluation of the structural damage of high-rise reinforced concrete buildings using ambient vibrations recorded before and after damage. Earthq. Eng. Struct. Dyn. **45**, 213–228 (2016)

Yousefianmoghadam, S., et al.: Nonlinear dynamic tests of a reinforced concrete frame building at different damage levels. Earthq. Eng. Struct. Dyn. **49**(9), 924–945 (2020)

Dynamic Properties and Seismic Response of a Museum Display Case with an Art Object

Andrea Prota[1], Martino Zito[1(✉)], Danilo D'Angela[1], Giuseppe Toscano[1], Carla Ceraldi[1], Antimo Fiorillo[1], and Gennaro Magliulo[1,2]

[1] Department of Structures for Engineering and Architecture, University of Naples Federico II, via Claudio 21, 80125 Naples, Italy
martino.zito@unina.it

[2] Construction Technologies Institute, National Research Council, 80125 Naples, Italy

Abstract. The seismic risk associated with the freestanding elements can be high, especially if they are housed within critical buildings (e.g., hospitals and nuclear plants) or have historical/cultural significance (e.g., museum objects or artifacts). In many cases, these systems exhibit a seismic response governed by rigid motion that is typically critical. In this study, an experimental testing campaign is carried out to assess the dynamic properties and seismic performance of museum objects and artifacts. Shake table tests of a typical museum display case containing a representative art object are performed. The specimen was provided by the National Archaeological Museum of Naples (MANN), Italy. Seismic assessment of the test specimen was performed through two types of testing: dynamic identification and seismic performance assessment. The dynamic identification of the display case was carried out, and both natural frequencies and damping ratios of the specimens were evaluated by reliable and consistent methods. The seismic performance tests were conducted by the international seismic certification protocol for nonstructural elements ICC-ES AC156. Both accelerations and displacements of the display case were assessed, as well as the response of the art object was characterized. The study sheds light on the critical behavior of the tested specimen, stressing the need for further studies toward a more comprehensive assessment of freestanding museum objects and artifacts.

Keywords: Nonstructural elements · Art objects · Freestanding · Shake table testing · Rocking

1 Introduction

Museum and art objects are extremely valuable in social, cultural, and economic terms, and they are typically highly vulnerable to dynamic actions and seismic excitations [1]. In many cases, museum and art objects are not mechanically fastened and are not provided with seismic protection devices. In particular, they are often just supported by floors, platforms, or pedestals, and they are likely to behave as freestanding relatively rigid elements under seismic excitation [2, 3]. Despite numerical and theoretical/analytical methods have been widely implemented in the literature to assess the seismic response of

G. P. Cimellaro (Ed.): WCSI 2022, LNCE 309, pp. 830–839, 2023.
https://doi.org/10.1007/978-3-031-21187-4_72

freestanding elements (e.g., [4–6]), experimental assessment through shake table testing represents the most reliable and robust method for assessing freestanding elements, especially for seismic qualification purposes [7].

The literature on experimental assessment of (freestanding) museum and art objects is quite poor if compared with other nonstructural element types. Most of the experimental studies focused on freestanding art objects [8–10] or the display case [11], but very few studies investigated the seismic response of art objects considering their most common configuration, i.e., simply supported on pedestals within freestanding display cases.

The present study aims at assessing the dynamic properties and seismic response of a typical museum arrangement, i.e., the abovementioned configuration. The study was developed in the framework of an agreement between the National Archaeological Museum of Naples (MANN) (Italy) and the Department of Structures for Engineering and Architecture (DIST) (University of Naples Federico II, Italy). The preliminary results of the study were recently published [12]. Figure 1a shows the arrangement of the tested display case (and art object) within the National Archaeological Museum of Naples (MANN), Italy, whereas the specimen mounted on the shake table of the Department of Structures for Engineering and Architecture, University of Naples Federico II, Italy, is depicted in Fig. 1b.

(a) (b)

Fig. 1. Arrangement of the tested display case (a) within the National Archaeological Museum of Naples (MANN), Italy, and (b) on the shake table at the Department of Structures for Engineering and Architecture, University of Naples Federico II, Italy.

2 Methodology

2.1 Specimen

The specimen consists of a museum display case containing a pedestal that supports a ceramic vase. The specimen was provided by MANN and is part of the current exhibition installation of the museum, as depicted in Fig. 1a. Figure 1b shows the arrangement of the display case (and art object) on the shake table at the Department of Structures for Engineering and Architecture, University of Naples Federico II, Italy. Table 1 reports the components of the display case and their main dimensions. The display case was assembled on the shake table by professional workers that typically mount the museum exhibition installations. The case is composed of a base steel frame supported by four corner supports, an internal floor panel, four lateral glass panels (including the opening panel or door) inserted within adjustable slots of the base frame, a top steel frame, supported by the glass panels, and the lighting panel, supported by the top frame (Fig. 1).

Table 1. Main dimensions of display case and components.

Display case	92 × 92 × 244.5 cm (width × length × height) and weight equal to about 1845 N (without art object)
Steel base frame	92 × 92 × 12 cm (width × length × height) and weight of 453 N
Base frame support	1.5 cm (height)
Steel shelter	84 × 84 × 1.8 (width × length × thickness) and weight of 160 N
Glass panel	88.5 × 234 × 1.1 cm (width × height × thickness) and weight of about 540 N
Door runner	92 × 2 cm (length × thickness) and weight of 103 N
Steel top frame	92 × 92 × 12 cm (width × length × height) and weight of 577 N
Pedestal	60 × 60 × 88 cm (width × length × height) and weight of 449 N
Art object	Base diameter and maximum diameter plus handles equal to about 22 and 42 cm, respectively; height equal to about 29 cm; and weight equal to about 69 N

The base steel frame is made of welded box profiles and thin plates (Fig. 2a). The display supports are 1.5 cm height steel threaded rod with a circular base plastic pin, screwed at the four corners of the base frame (Fig. 2b). The floor panel was simply supported by the base frame (no mechanical connection); this is made of a wood plate covered by a cap shelter made of steel. The lateral surfaces of the display case consisted of three fixed and one opening/rotating double-glass panels; each (lateral) panel was inserted within a preexisting slot of the base frame, corresponding to the gap between the steel frame and the think steel runner (Fig. 2c); seals were located at the interface between the panels and the frame/runner. The slot/gap was tightened to fasten the glass panel's bases by friction through fastening screws located at the extremities of the frame/runner. The opening/rotation of the door panel was provided by hinges, connecting base/top frames, and the runners (Figs. 1 and 2d). Both bottom and top runners had locks, that fastened the door panel to the frames. The top frame (Fig. 2e) was made similarly to the bottom one and was connected to the lateral glass panels through the top runners as the bottom frame was. The lighting system was integrated to a think steel sheet, connected superiorly to the top frame (Fig. 2e).

The pedestal was made of welded steel box profiles, welded to steel cover sheets; the artifact consists of a ceramic vase (Fig. 2f). The pedestal was simply supported by the floor panel (freestanding) as well as the vase was simply supported by the pedestal.

2.2 Testing Setup and Instrumentation

The display case was tested as a freestanding element (no connections to the shake table). Triaxial accelerometers were fixed to the most significant points of the specimen, i.e., center and additional points of glass panels (external side), center of lateral surface of base and top frame, at a corner of the floor panel, at a corner of the pedestal's top surface; accelerations of the table were monitored by internal accelerometers. Laser sensors were used to monitor the displacements of the display case in both directions, corresponding to the center of base frame, glass panels, and top frame. In order to prevent the display case from overturning or damage to the instruments, a safety scaffolding structure was built around the display case, with a tolerance distance equal to 25 cm. The tests were recorded by six video cameras. The instrument setup is shown in Fig. 3, where the cardinal sides and global direction axes are also reported. Further details regarding the characteristics and arrangement of the instruments can be found in [12].

2.3 Testing Procedure

The testing procedure consisted of a series of shaking tests, including both dynamic identification tests and incremental seismic performance evaluation tests, according to a consolidated testing and seismic qualification approach [13, 14]. The former tests were aimed at assessing the initial and final dynamic properties of the system and their evolution along the incremental tests, whereas the latter were carried out to assess the seismic capacities of the system as the intensity measure increases. In particular, the dynamic identification tests were performed through monodirectional low-intensity excitation along both horizontal directions, and the seismic performance tests were bidirectional (horizontal). Dynamic identification tests were performed considering initial configuration, between seismic incremental tests, and corresponding to final configuration (after incremental tests).

Dynamic identification inputs corresponded to low-amplitude random vibration acceleration time histories (RAN tests) [15], whereas seismic performance inputs corresponded to scaled AC156 input tests (AC tests) [14]. The testing program consisted of incremental bidirectional AC tests from SDS equal to 0.05 g to 0.45 g, through 0.05 g SDS increments, performing unidirectional RAN tests (along both directions) prior to, between, and after AC tests. Further details regarding the generation of both RAN and AC inputs and main details about the considered signals can be found in [12, 15, 16].

3 Dynamic Identification

The transfer curve method was used to assess the peak frequencies, according to [17–19], whereas damping ratios were estimated by implementing the half-power bandwidth

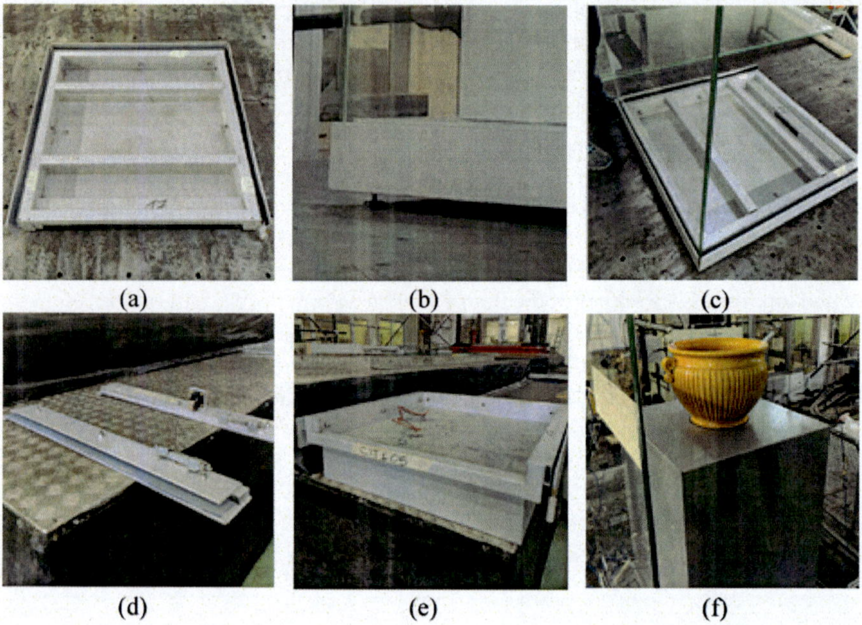

Fig. 2. Details of the tested display case: (a) base frame, (b) detail of the base frame support, (c) top frame, (d) detail of the glass to frame connections, (e) bottom and top door runners, and (f) pedestal and art object.

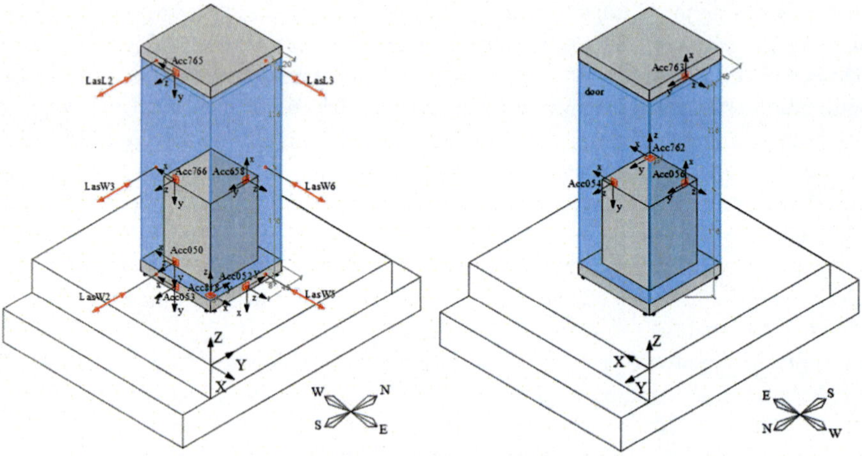

Fig. 3. Schematic views of the specimen and instrumentation.

method [19, 20]. The evolution of the fundamental frequency peaks along the incremental testing procedure is shown in Fig. 4. Along X direction (Fig. 4.a), the peak frequency decreases as the testing intensity increases from an initial frequency just below 8 Hz to about 6 Hz (at the end of the test). An anomaly can be observed corresponding to

RAN1006 test; the most reasonable reason for causing this anomaly could be associated with variation in the arrangement of the specimen assembly due to the previous seismic tests; as a matter of fact, it should be recalled that the specimen consisted in an assembly of several components, which were connected without controlling the connection interfaces, which were significantly conditioned by friction and adjustment phenomena. Similar anomalies were found in the literature, when other freestanding elements, metal cabinets, were tested under a similar arrangement [15, 18].

The peak frequency decrease associated with Y direction (Fig. 4.b) was more regular than along X direction, and the initial (final) peak frequency was just below (above) the peak frequency related to X direction. Therefore, excluding the abovementioned anomaly, the evolution of the peak frequency can be assumed to be quite similar among the horizontal directions. The decrease in frequency along both directions might be associated with the variation of the connection conditions among the assembly components, for example, reduction of gap tolerances or rearrangement of seals.

The evolution of the damping ratios along the incremental testing procedure is depicted in Fig. 5. Along X direction, the damping ratio evolution is more irregular, if compared to the Y direction case, and X direction ratios range within an interval larger and higher (16–25%) than the one associated with Y direction (10–13%). Over the whole testing procedure, damping ratios related to X and Y directions show an increment, which might be potentially associated with higher damping capacities developed in the course of the seismic incremental tests. The estimated ratios are overall reasonably consistent with previous literature, for example, considering experimental tests performed on freestanding hospital metal cabinets [18].

4 Seismic Response

Among the key parameters for the characterization of the seismic response of acceleration-sensitive elements, the component amplification factor (CAF) is reported in the study (Fig. 6). CAF was defined as the ratio between peak component acceleration (PCA) and PFA [15] and was estimated for frame top of the display case, (b) surface top of pedestal, and (c) middle point of glass panels, for both horizontal directions.

Corresponding to lower PFA, CAF along X direction is larger than along Y, and this is more significant considering out of plane response of glasses. As a more general result, it can be observed that the peaks of CAF along PFA correspond to PFA values ranging from 0.10 to 0.20 g, overall closer to 0.10 g.

The seismic motion of the display case was governed by both rigid sliding and rocking, exhibiting a complex response. Figure 7 shows a representative time history response (AC05 test, PFA equal to 0.31 and 0.43 g, along X and Y direction, respectively), i.e., relative displacement, assessed between the top of the display case and the shake table, along both horizontal directions. Two contributions to the global motion can be observed: (a) a relatively reduced oscillation response, probably due to rocking (more) and sliding (less) response, and (b) a clear displacement accumulation trend, approximately linear with time evolution. The residual displacement along Y direction is larger than along X direction, even though the values are comparable.

At the end of the tests, no permanent damage to the display case components and art object was observed. However, both the display case and ceramic vase exhibited

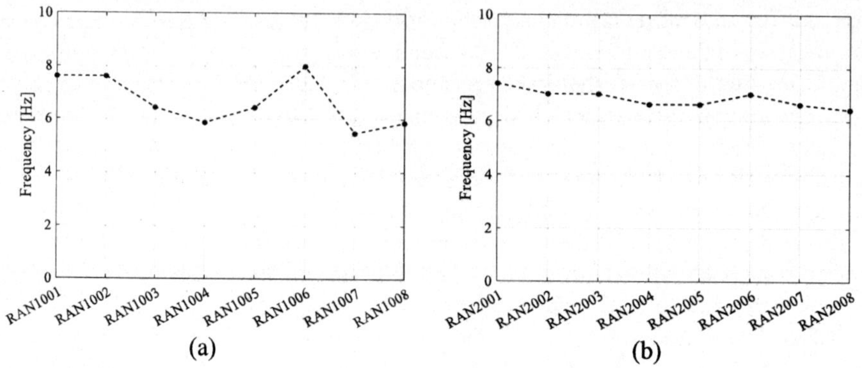

Fig. 4. Evolution of frequency peaks associated with first vibration mode, along (1) X and (2) Y directions.

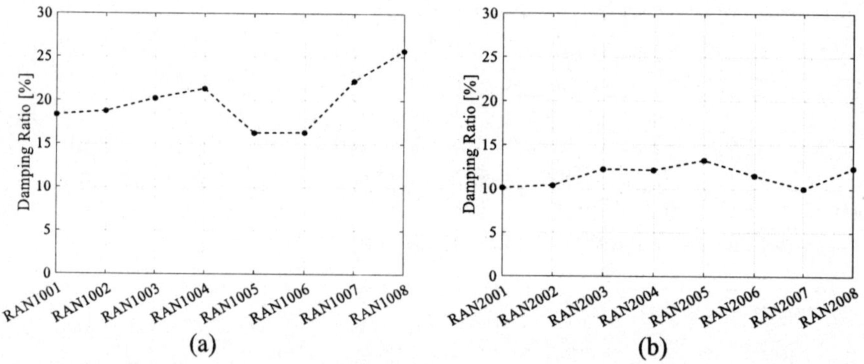

Fig. 5. Evolution of damping associated with first vibration mode, along (1) X and (2) Y directions.

a significant seismic motion. The ceramic vase represented an extremely valuable and vulnerable art object, and its response was considered to characterize the overall achieved damage conditions. In particular, up to test AC04 included, the vase did not exhibit a significant motion/residual displacement, whereas, from test AC05 on, the vase showed a critical sliding response and exhibited a large residual displacement. Accordingly, tentative thresholds of PFA for identifying the potential damage to the ceramic vase can reasonably range from 0.25 to 0.30 g.

5 Conclusions

The study addresses the seismic assessment of museum and art elements by means of shake table testing. A representative display case containing an art object (ceramic vase) was tested under both mono- and bi-directional excitations, according to an incremental procedure. Both dynamic identification tests and seismic performance tests were performed. After estimating the dynamic properties of the specimen, i.e., fundamental frequencies and damping ratio, the seismic response was characterized. In particular, both acceleration and displacement response was assessed, and tentative peak floor acceleration (PFA) thresholds were defined to identify potential damage to the art object.

The study sheds lights on the seismic response of freestanding elements and valuable objects. The estimated dynamic properties can be useful to provide relatively accurate seismic demand estimations, whereas the capacity intensity level ranges identified in the study (i.e., the potential damage of the vase) might be considered to develop more robust capacity criteria and thresholds. The study cannot be considered to be exhaustive and further studies should validate and extend the findings, also considering different case study and performing more advanced results analysis.

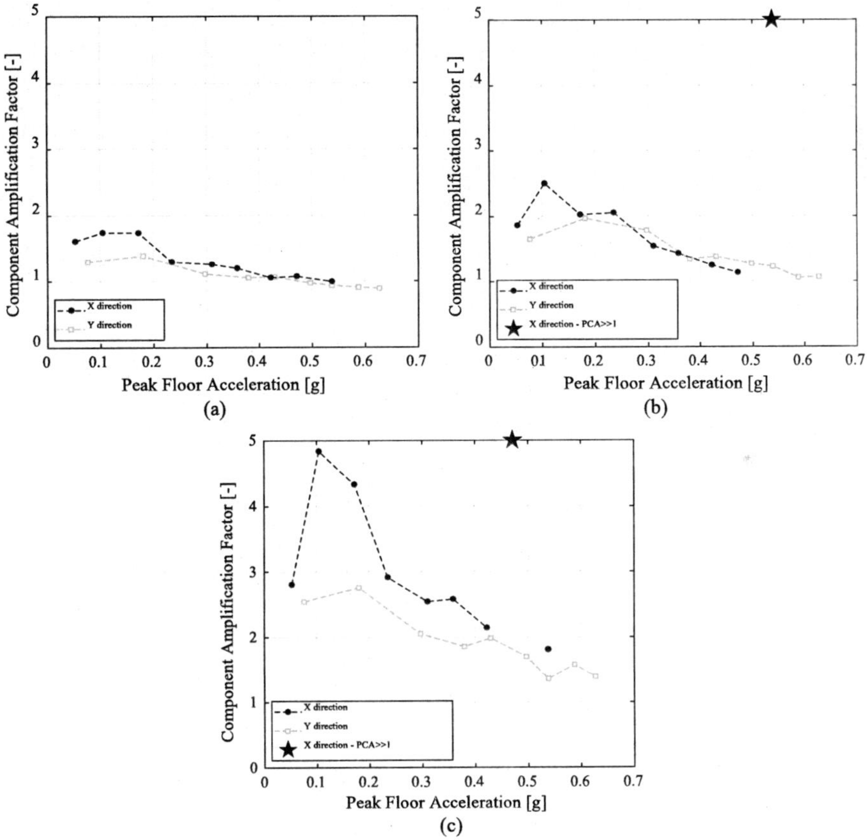

Fig. 6. Component amplification factor (CAF) associated with increasing PFA intensities, estimated considering (a) frame top of display case, (b) surface top of pedestal, and (c) middle point of glass panels (out of plane direction), for both horizontal directions.

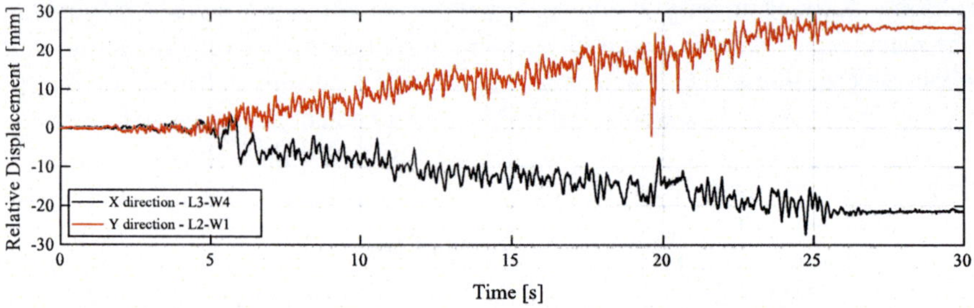

Fig. 7. Relative displacement time history, evaluated between the top of the display case and the shake table, along both horizontal directions.

Acknowledgments. The present study was developed in the framework of the agreement between the National Archaeological Museum of Naples (MANN), Italy, and the Department of Structures for Engineering and Architecture (DIST, University of Naples Federico II), Italy, i.e., *Convenzione MANN-DIST*. The Director of MANN, Dr. Paolo Giulierini, is thanked for his effort regarding the agreement and his support of the related research activities. The technical support of Prof. Giuseppe Maddaloni for performing the tests is fully acknowledged.

References

1. Erdik, M., Durukal, E., Ertürk, N., Sungay, B.: Earthquake risk mitigation in Istanbul museums. Nat. Hazards **53**, 97–108 (2010). https://doi.org/10.1007/s11069-009-9411-2
2. Gesualdo, A., Iannuzzo, A., Minutolo, V., Monaco, M.: Rocking of freestanding objects: theoretical and experimental comparisons. J. Theor. Appl. Mech. **56**, 977–991 (2018). https://doi.org/10.15632/jtam-pl.56.4.977
3. Reinhorn, A.M., Viti, S.: Monumental buildings used as museums: protection or danger for the artifacts? Procedia Struct. Integrity **29**, 40–47 (2020). https://doi.org/10.1016/j.prostr.2020.11.137
4. D'Angela, D., Magliulo, G., Cosenza, E.: Seismic damage assessment of unanchored non-structural components taking into account the building response. Struct. Saf. **93**, 102126 (2021). https://doi.org/10.1016/j.strusafe.2021.102126
5. Dimitrakopoulos, E.G., Paraskeva, T.S.: Dimensionless fragility curves for rocking response to near-fault excitations. Earthq. Eng. Struct. Dyn. **44**, 2015–2033 (2015). https://doi.org/10.1002/eqe.2571
6. Fragiadakis, M., Diamantopoulos, S.: Fragility and risk assessment of freestanding building contents. Earthq. Eng. Struct. Dyn. **49**, 1028–1048 (2020). https://doi.org/10.1002/eqe.3276
7. D'Angela, D., Magliulo, G., Cosenza, E.: Towards a reliable seismic assessment of rocking components. Eng. Struct. **230**, 111673 (2021). https://doi.org/10.1016/j.engstruct.2020.111673
8. Berto, L., Favaretto, T., Saetta, A., Antonelli, F., Lazzarini, L.: Assessment of seismic vulnerability of art objects: the "Galleria dei Prigioni" sculptures at the Accademia Gallery in Florence. J. Cult. Herit. **13**, 7–21 (2012). https://doi.org/10.1016/j.culher.2011.06.005
9. Huang, B., Günay, S., Lu, W.: Seismic assessment of freestanding ceramic vase with shaking table testing and performance-based earthquake engineering. J. Earthq. Eng. 1–23 (2021). https://doi.org/10.1080/13632469.2021.1979132

10. Wittich, C.E., Hutchinson, T.C.: Shake table tests of stiff, unattached, asymmetric structures: shake table tests of stiff, unattached, asymmetric structures. Earthq. Eng. Struct. Dyn. **44**, 2425–2443 (2015). https://doi.org/10.1002/eqe.2589
11. Neurohr, T., McClure, G.: Shake table testing of museum display cases. Can. J. Civ. Eng. **35**, 1353–1364 (2008). https://doi.org/10.1139/L08-084
12. Prota, A., Zito, M., D'Angela, D., Toscano, G., Ceraldi, C., Fiorillo, A., Magliulo, G.: Preliminary results of shake table tests of a typical museum display case containing an art object. Adv. Civ. Eng. 1–18 (2022) https://doi.org/10.1155/2022/3975958
13. Federal Emergency Management Agency (FEMA). Interim protocols for determining seismic performance characteristics of structural and nonstructural components through laboratory testing. Report No. FEMA 461. Washington D.C., USA (2007)
14. International Code Council Evaluation Service (ICC-ES). AC156 acceptance criteria for the seismic qualification of nonstructural components. Brea, California, USA (2012)
15. Di Sarno, L., Magliulo, G., D'Angela, D., Cosenza, E.: Experimental assessment of the seismic performance of hospital cabinets using shake table testing. Earthq. Eng. Struct. Dyn. **48**, 103–123 (2019). https://doi.org/10.1002/eqe.3127
16. Petrone, C., Magliulo, G., Manfredi, G.: Shake table tests on standard and innovative temporary partition walls. Earthq. Eng. Struct. Dyn. **46**, 1599–1624 (2017). https://doi.org/10.1002/eqe.2872
17. Bracci, J.M., Reinhorn, A.M., Mander, J.B.: Seismic resistance of reinforced concrete frame structures designed only for gravity loads: part I—design and properties of a one-third scale model structure. Technical Report NCEER-92-0027. State University of New York at Buffalo, NY, US (1992)
18. Di Sarno, L., Petrone, C., Magliulo, G., Manfredi, G.: Dynamic properties of typical consultation room medical components. Eng. Struct. **100**, 442–454 (2015). https://doi.org/10.1016/j.engstruct.2015.06.036
19. Clough, R.W., Penzien, J.: Dynamics of Structures. McGraw-Hill, New York (1975)
20. Chopra, A.K.: Dynamics of Structures: Theory and Applications to Earthquake Engineering, 3rd edn. Pearson/Prentice Hall, Upper Saddle River, N.J (2007)

Experimental Study on Ceiling Fall Prevention Using New Materials

Takumi Misaki and Osamu Takahashi(✉)

Tokyo University of Science, Katsushika-ku, Tokyo 125-8585, Japan
4121560@ed.tus.ac.jp

Abstract. The Great East Japan Earthquake of 2011 reaffirmed ceilings, which had been regarded as the finishing touch to architectural spaces, as an entity that can cause serious damage to occupants when they fall. The control of damage from falling ceilings was treated as an urgent issue, and in 2013, the "Notice on Specified Ceilings and Specified Ceilings with Structurally Safe Structural Methods (hereinafter referred to as "Ceiling Notice")" was enacted. This defined ceilings that meet certain conditions as "specified ceilings," and required that they be earthquake-resistant and that fall prevention measures be installed in existing buildings as well. Since then, there has been much research on fall prevention measures for ceilings, but there is still an urgent need to develop better materials and construction methods. Against this background, an experimental study was conducted to propose a method of repairing ceiling surfaces that have deteriorated over time and whose safety has not been confirmed. Tensile tests showed that polyurea resin and aramid fiber are effective for repairing suspension bolts inserted in concrete frames. In addition, the ceiling surface (plasterboard) was dropped to confirm its impact resistance against falling impact, and its effectiveness as a ceiling fall prevention measure was confirmed.

Keywords: Ceiling fall prevention · Aramid fiber · Polyurea resin · Safe structural methods

1 Introduction

In the wake of the Great East Japan Earthquake, ceilings, which had been regarded as the finishing touches that set the architectural space, were reaffirmed as entities that could cause serious damage to occupants if they fell. The control of damage from falling ceilings has been treated as an urgent issue, and in 2013, the "Notice on Specified Ceilings and Specified Ceilings with Structurally Safe Structural Methods (hereinafter referred to as the "Ceiling Notice")" was enacted. This defines ceilings that meet certain conditions as "specified ceilings" and requires that they be earthquake resistant and that fall prevention measures be provided in existing buildings. Since then, much research has been conducted on measures to prevent ceilings from falling, but there is still an urgent need to develop better materials and construction methods. It is also important to improve applicability to retrofitting and lower costs. The fail-safe method is important

G. P. Cimellaro (Ed.): WCSI 2022, LNCE 309, pp. 840–852, 2023.
https://doi.org/10.1007/978-3-031-21187-4_73

as it covers the weakness of the ceiling method. It is effective as a lifesaving measure in cases where renewal is difficult due to existing ceilings for which reliable lifesaving protection is not considered, or where only seismic resistance is considered and lifesaving protection is inadequate. Although multiple materials and construction methods can be found as examples, there is a risk of insufficient strength unless design and construction are conducted after a quantitative evaluation of the extent of the fall target. A typical example of a fail-safe is the fall prevention wire shown in Fig. 1. Fall prevention nets are fail-safe as a surface, whereas fall prevention wires are characterized as fail-safe as a point or line. This is important not only for panelized ceiling materials, but also as a fail-safe for equipment, pipes, ducts, etc. Fall prevention wires are easy to use in combination with other measures and can be easily adopted even in crowded ceilings, and can be used for additional retrofitting of existing ceilings. There are also commercially available products that are relatively easy to adopt in terms of cost. It should be noted, however, that the wires are fastened to the ceiling joist and may fall off the ceiling surface under deteriorated conditions of concrete, inserts and suspension bolts, and that the method does not prevent the boards themselves from falling. Polyurea resin and aramid fibers were verified as effective repair methods for these aged and unsafe ceiling surfaces. Figure 2 shows the ceiling surface to be repaired.

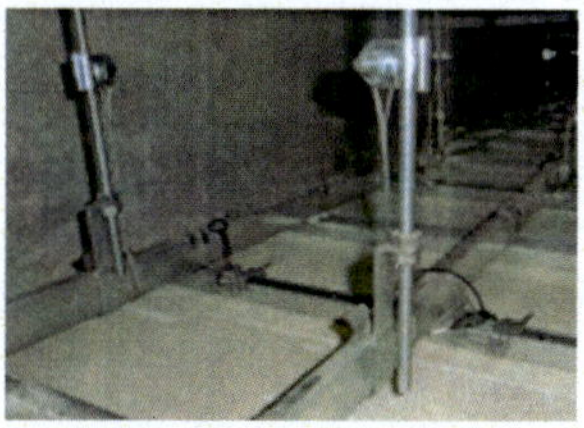

Fig. 1. Ceiling fall prevention wire

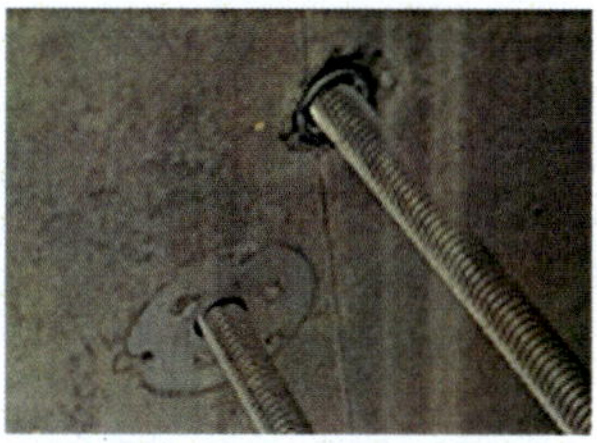

Fig. 2. Surface to be repaired for ceiling

2 Use of New Materials

2.1 Polyurea Resins

Polyurea resins are characterized by excellent tensile and tear strength, elongation, chemical resistance, abrasion resistance, water resistance, explosion resistance, and corrosion resistance. Polyurea resins can be sprayed on with equipment or hand-painted polyurea, which can be used in small quantities by mixing two different materials. In this study, hand-painted polyurea, which can be applied simply, will be used. Polyurea resin coatings can adhere to most materials, and even hard-to-adhere materials can be fixed by applying appropriate surface treatments such as blasting or primer application. Table 1 shows the properties of polyurea resins.

Table 1. Polyurea resin ingredients

Hardness	50 ± 5
Tensile strength (N/mm^2)	19.31–22.06
Tear strength (N/mm^2)	87.56–105.08
Growth rate (%)	400–500
Compressive strength (N/mm^2)	5.52
Full curing time (hour)	24

2.2 Aramid Fiber

Aramid fiber is a high-strength fiber that began industrial production using liquid crystalline polymers, which have extremely strong molecular bonds. This fiber combines low moisture absorption, dimensional stability, and abrasion resistance, and is used as a sling and airship membrane material (Table 2).

Table 2. Aramid fiber composition

Tensile strength (N/mm^2)	3236
Rupture elongation (%)	3.8
Tensile modulus (N/mm^2)	76100
Dry/Wet strength ratio (%)	100

Verify the effectiveness of ceiling repair using polyurea resin and aramid fibers in Figs. 3 and 4.

3 Anchor Tensile Test

3.1 Outline of Experiment

The existing suspension bolts of suspended ceilings inserted into the concrete frame need to be repaired for ceilings where the safety of the suspension bolts has not been confirmed during the on-site survey for seismic retrofit. As a method of repair, the effectiveness of polyurea and Aramid fiber for repair work will be verified by tensile testing.

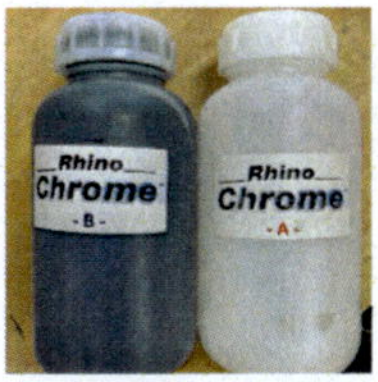

Fig. 3. Hand-painted polyurea

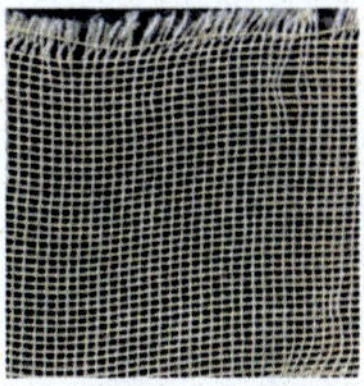

Fig. 4. Aramid fiber

3.2 Outline of Test Specimen

Imaged figures of the test piece are shown in Figs. 5, 6 and 7. A concrete test piece with a round steel of 9Φ diameter, which assumes the target suspension material, is inserted into the end of the test piece, and a M10 chemical anchor is inserted into the other end of the test piece. Hand-painted polyurea is applied to the joint surface of a 9Φ diameter round steel and concrete, and the tensile strength of the polyurea is checked using an Amsler tensile tester. The polyurea depth in the hole was used as a parameter because gaps between the bolts may occur depending on the aged condition of the ceiling concrete and inserts.

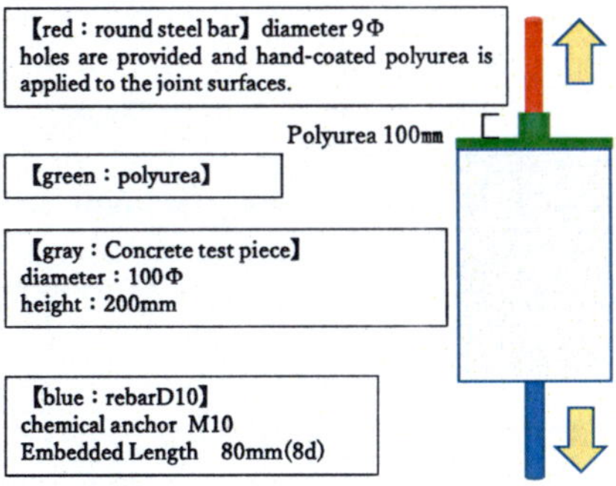

Fig. 5. Imaged figure of the test specimen

3.3 Experimental Results and Discussion

Table 3 shows the list of test specimens and maximum loads. Figures 8 and 9 show the load-displacement relationship with and without aramid fiber as a parameter. It can be

Fig. 6. Without aramid fiber

Fig. 7. Without aramid fiber

confirmed that the maximum bearing capacity varies greatly depending on the presence or absence of aramid fiber. It is believed that the adhesion of aramid fiber to the surface of the specimen and the application of polyurea increased the adhesion to the specimen, resulting in an increase in the bearing capacity. When the specimens were filled with polyurea, a maximum bearing capacity of 4.8 kN with aramid fiber and 1.27 kN without aramid fiber were confirmed. It is thought that the friction between the polyurea and the round steel greatly increased the bearing capacity. Figures 10 and 11 show the load-displacement relationship, which compares the specimens, and Figs. 12 and 13 show the fracture properties of specimens 7 and 8. In specimen 5, the polyurea depth was 20 mm, but the maximum bearing capacity was lower than that of specimens 3 and 4 with a polyurea depth of 10 mm. However, Fig. 13 confirms that test specimen 5 absorbs more energy. In addition, in specimen 7, the round steel and aramid fiber are separated as one piece, whereas in specimen 8, only the round steel is pulled out. Compared to Fig. 14, it can be read that test specimen 7 absorbs more energy.

3.4 Experimental Results and Discussion

The use of polyurea and aramid fiber was shown to provide suspension bolt and insert reinforcement. In addition, depending on the deterioration of the ceiling surface, the gap between the suspension bolts and inserts could be filled with polyurea, which would

Table 3. List of test specimens and maximum load

Specimen	Aramid fiber	Polyurea depth (mm)	Max. capacity (kN)
1	No	0	0.21
2	No	0	0.52
3	No	10	1.3
4	No	10	1.3
5	No	20	1.1
6	Yes	0	0.96
7	Yes	0	0.80
8	Yes	0	0.60
9	Yes	10	3.7
10	Yes	20	4.8

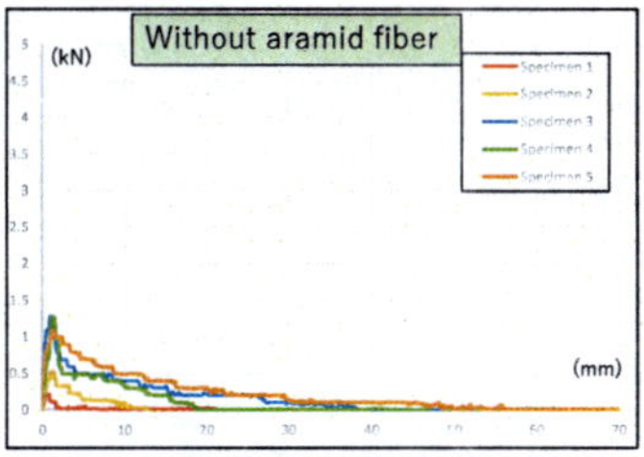

Fig. 8. Load-displacement relationship without aramid fiber

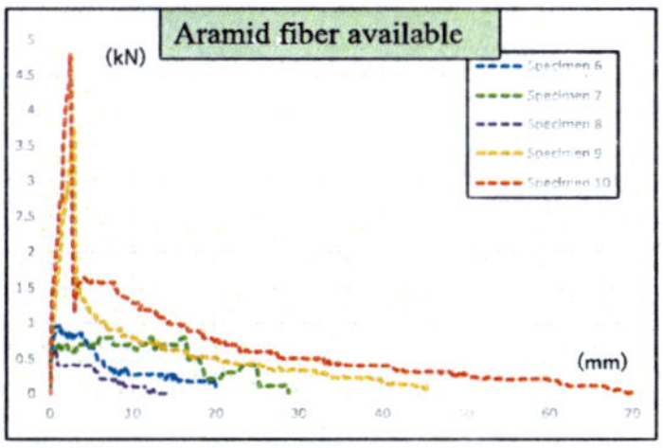

Fig. 9. Load-displacement relationship aramid fiber available

allow for stronger anchorage due to frictional resistance. However, it is not easy to attach aramid fiber to the ceiling surface and apply polyurea, so further study is needed on its workability.

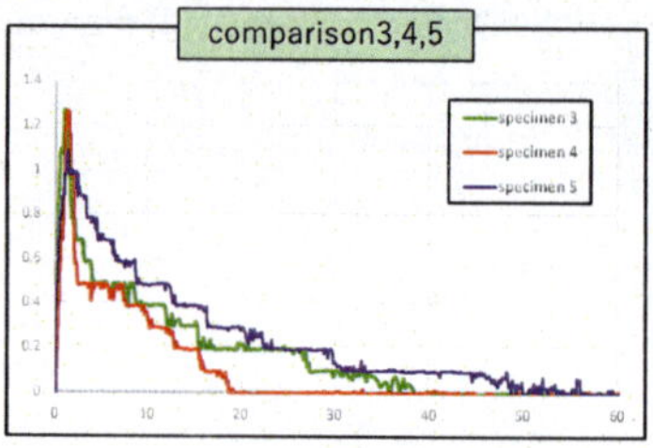

Fig. 10. Load-displacement relationship specimen 3, 4, 5 comparison

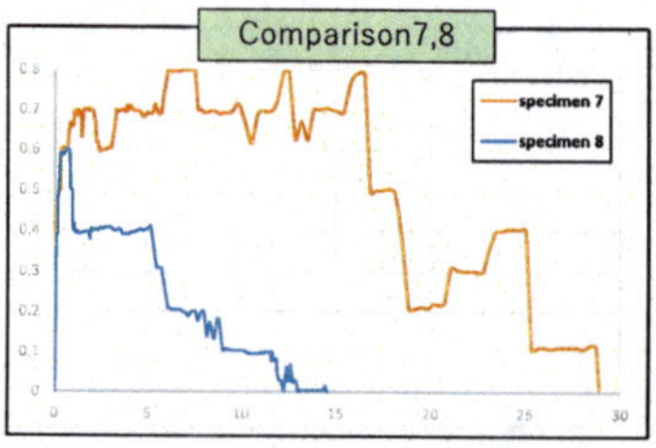

Fig. 11. Load-displacement relationship comparison of test specimens 7 and 8

Fig. 12. Fracture properties specimen 7

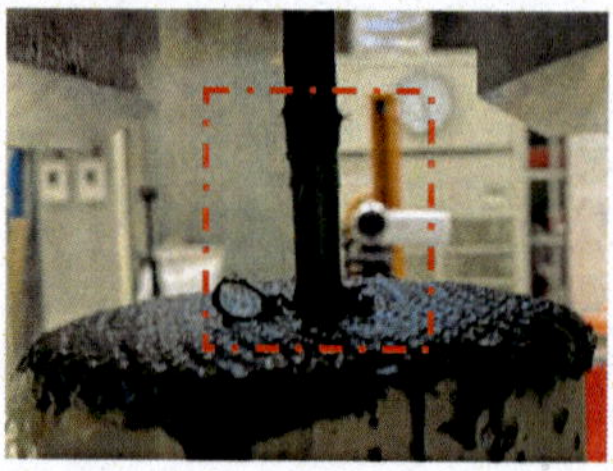

Fig. 13. Fracture properties specimen 8

4 Ceiling Fall Experiment

4.1 Experiment Summary

Prove that fall protection measures are safe on ceiling surfaces using high-strength aramid fibers and hand-painted polyurea. Drop the ceiling surface (plasterboard) to check the impact resistance of the ceiling against falling impact and verify that the ceiling is effective in fall protection measures. The specimen weights were 35, 50, and 70 kg to cover a range from ordinary ceilings to music halls with many finishing materials such as sound-absorbing panels.

4.2 Test Specimen Overview

An image of the specimen is shown in Fig. 14. For the mock-up fabrication of the test specimen, a 910 mm × 1820 mm gypsum board was used. The high-strength aramid fibers are tied in a "moyai knot" (Fig. 15) at the base of the suspension, and the plasterboard and high-strength aramid fibers are connected via a round ring. Figure 16 shows a detailed view of the specimen constituent materials. In addition, a 0.15 m^2 of polyurea is painted on the top to prevent the screws joining ceiling joist and the round ring from pulling out, and is expected to secure ceiling joist and the plasterboard. In the configuration of the specimen, the clip is reinforced and the hanger is unreinforced, as shown in Fig. 17. In recent years, ceilings have been reinforced with clips and other reinforcements. It shall be assumed that the hanger, which is the joint between the hanging bolt and the ceiling joist receiver, or the hanging bolt itself breaks and is damaged, causing the bottom to fall integrally from ceiling joist receiver. The measurement uses a high-speed camera to measure acceleration. Table 4 lists the test specimens. Specimen 1 is a test with a gypsum board weight of 35 kg and polyurea applied to a 0.15 m^2. Specimen 2 is a specimen in which the round rings are 10 cm apart at the center and reinforced with the same material as the ceiling joist receiver (Reinforcement Pattern 1). Specimen 3 is a specimen that is further reinforced at equal intervals on top of the reinforcing pattern 1. Specimen 4 weighs 70 kg and is a specimen with polyurea applied along the ceiling joist (Figs. 18, 19 and 20).

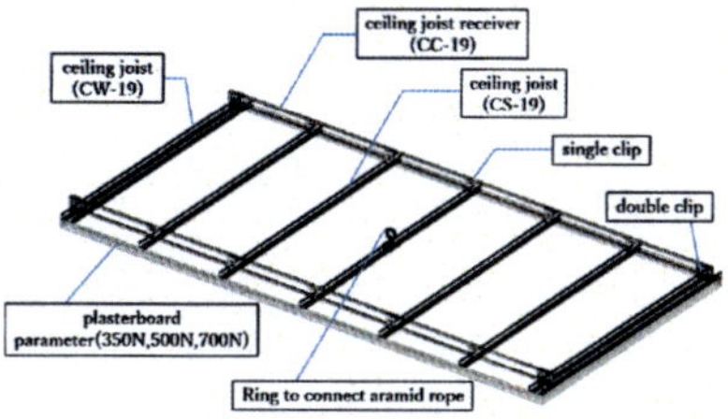

Fig. 14. Imaged figure of the test specimen

4.3 Experimental Results and Discussion

Test specimen 1, weighing 35 kg with polyurea applied to a 0.15 m^2, resulted in bending failure as shown in Fig. 10, but did not fall because the clips on both sides of the ceiling joist were reinforced. No significant damage was observed at a drop height of 0.1 m for specimen 2 with reinforcement pattern 1. At the second drop height of 0.2 m on the same specimen, the polyurea applied to the specimen peeled off the plasterboard, and the

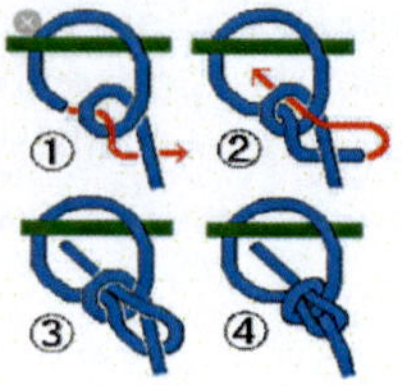

Fig. 15. "moyai knot"

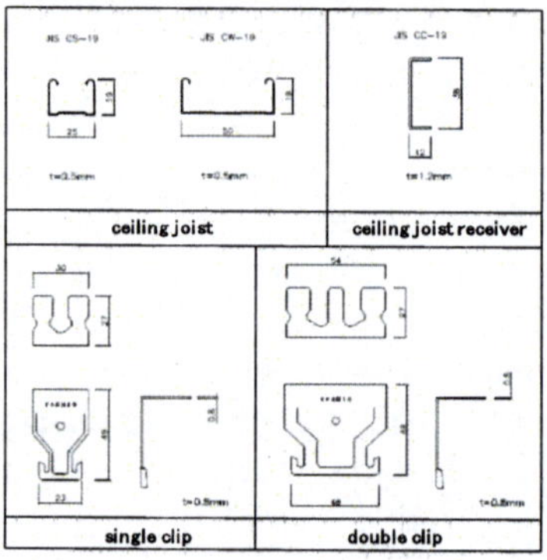

Fig. 16. Specimen constituent materials

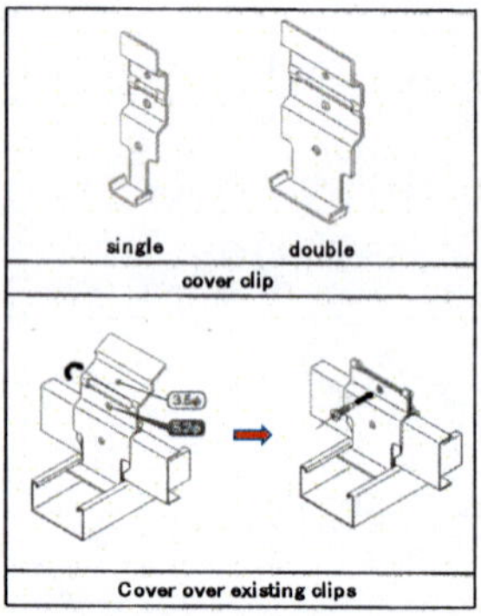

Fig. 17. Cover clip constituent materials

ceiling joist connecting the round ring bent slightly. The third fall height of 0.1 m resulted in bending of the three the ceiling joist receiver by the reinforcement, but none of them fell. Specimen 3 is a specimen with a plasterboard weight of 50 kg and reinforcement pattern 2. At the first drop height of 0.1 m, the applied polyurea peeled off a little, but no other significant damage was observed. The second fall height of 0.2 m resulted in complete separation of the plasterboard and the top from the baseboard, and only the plasterboard fell. It can be inferred that the reinforcement pattern 2 made the ceiling joist more fixed by the two stiffeners, and the impact load of the fall was distributed

Table 4. List of specimens

Specimen number	Weight (kg)	Reinforcement	Fall height (m)	Remarks
1	35	No reinforcement	0.1	
2	35	Reinforcement pattern 1	0.1	The drop was carried out three times in succession
			0.2	
			0.1	
3	50	Reinforcement pattern 2	0.1	
			0.2	
4	70	Reinforcement pattern 2	0.1	Linear application of polyurea
			0.2	

Fig. 18. No reinforcement

Fig. 19. Reinforcement pattern 1

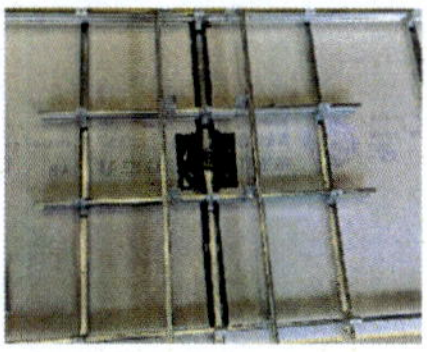

Fig. 20. Reinforcement pattern 2

to the screws connecting ceiling joist and the plasterboard, which may have pulled out the screws. Specimen 4 is a specimen of reinforcement pattern 2 weighing 70 kg, with polyurea applied along the ceiling joist where the round ring is attached. The first drop of 0.1 m did not cause significant deformation as in specimen 4, but the second drop of

0.2 m resulted in the rupture of the ceiling joist at the point where the round ring was connected (Figs. 21, 22, 23 and 24).

Fig. 21. Specimen1 (fall height 0.1 m)

Fig. 22. Specimen2 (fall height 0.2 m). Specimen2 (fall height 0.1 m)

Fig. 23. Specimen3 (fall height 0.2 m)

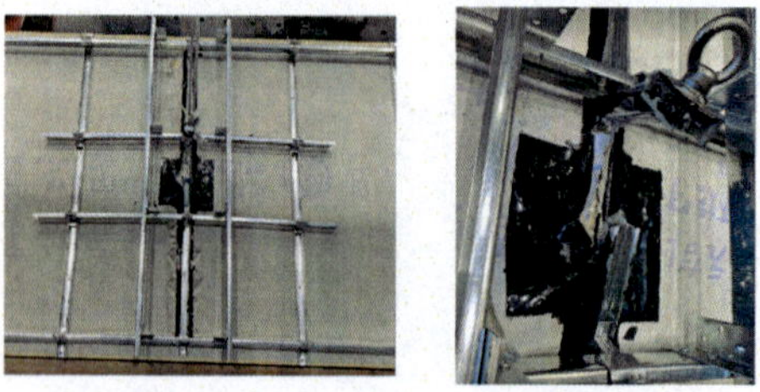

Fig. 24. Specimen4 (fall height 0.2 m)

Next, the results obtained from the high-speed camera are shown in Fig. 25. The maximum instantaneous acceleration was 14.5 (m/sec^2) at a drop height of 0.1 m for specimen 1. The maximum instantaneous accelerations were 45.2 (m/sec^2), 81.5 (m/sec^2), and 41.8 (m/sec^2) for specimens 2, 3, and 4, which were tested at a drop height of 0.2 m, respectively. Specimen 4, which resulted in complete separation of the top from the gypsum board and the ceiling joist, showed nearly twice as much acceleration as the other specimens. This may have been the cause of the separation of the gypsum board from the ceiling joist (Fig. 26).

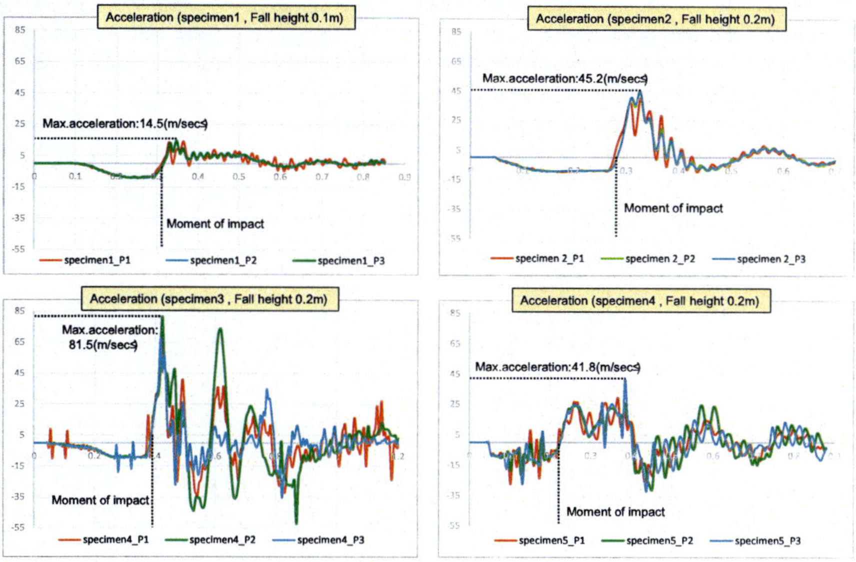

Fig. 25. High-speed camera analysis

specimen number	weight(kg)	reinforcement	Fall Height(m)	Max. acceleration【m/sec^2】	Max. impact load (N)
1	35	No reinforcement	0.1	14.49	507.15
2	35	Reinforcement pattern 1	0.1	34.5	1207.5
			0.2	45.24	1583.4
			0.1	30.83	1079.05
3	50	Reinforcement pattern 2	0.1	32.7	1635
			0.2	81.59	4079.5
4	70	Reinforcement pattern 2	0.1	41.23	2886.1
			0.2	41.81	2926.7

Fig. 26. Acceleration and impact load list

5 Conclusion

The use of polyurea and aramid fibers showed that hanging bolts and insert reinforcement were possible. The ceiling joist was further anchored by the two reinforcements, which showed the danger of falling. In the future, a reinforcement plan should be developed to absorb the energy in a balanced manner.

References

Commentary on technical standards for ceiling fall protection in buildings, 2013.10. National Institute for Land and Infrastructure Management, Ministry of Land, Infrastructure, Transport and Tourism, Building Research Institute, and Association for the Promotion of the New Architectural Professionals System

Guidelines and commentary on safety measures against falls of non-structural members such as ceilings, 2015.01. Architectural Institute of Japan

Yashio Kenzai Kogyo Co. https://yasio.jp/products/ceiling/

Shake-Table Tests on an Industrial Steel Rack Isolated with Innovative Modular Devices

Gabriele Guerrini[1,2(✉)], Francesco Graziotti[1,2], and Andrea Penna[1,2]

[1] Department of Civil Engineering and Architecture, University of Pavia, 27100 Pavia, Italy
gabriele.guerrini@unipv.it
[2] EUCENTRE Foundation, 27100 Pavia, Italy

Abstract. This paper discusses the experimental results from dynamic shake-table tests conducted at the University of Pavia and at the EUCENTRE Foundation (Pavia, Italy) on an industrial steel rack seismically isolated with innovative devices. Despite the undeniable effectiveness of common isolators in reducing seismic demands on superstructures, their application to non-structural systems can be hindered by cost, durability, and mechanical issues. To overcome these drawbacks, an innovative isolator has been patented by Kyneprox S.r.l., based on a multiple articulated quadrilateral mechanism and named "Kinematic Steel Joint (KSJ)". This device can be manufactured from steel sheets, possibly stainless or galvanized to mitigate corrosion issues. The KSJ imposes to the superstructure a pendulum-type motion with self-centering behavior, and its modular nature allows tailoring it to different masses. Friction within pinned joints grants some energy dissipation to the device, and replaceable fuses can be added to act as brakes before reaching the maximum displacement range. KSJ devices were installed at the base of a five-shelf, two-bay industrial steel rack, isolated in the cross-aisle direction and braced in the down-aisle one. Incremental uniaxial shake-table tests were conducted in the isolated direction under three different loading scenarios. The beneficial effects of the isolators on the dynamic response of the rack are demonstrated by the elastic response spectra of the input signal imposed to its base.

Keywords: Industrial steel rack · Kinematic steel joint · Replaceable fuse · Seismic isolation · Self-centering · Shake-table test

1 Introduction

Current seismic design philosophies worldwide focus on life safety, typically through the enforcement of structural ductility: damage is accepted if collapse is prevented. However, protecting buildings from earthquakes requires also to limit the economic and social cost of post-event disruption, repair, and reconstruction. Nonstructural elements, such as architectural, mechanical, or electrical components and contents, account for most of the total investment in a typical building. Furthermore, damage to nonstructural components

G. P. Cimellaro (Ed.): WCSI 2022, LNCE 309, pp. 853–863, 2023.
https://doi.org/10.1007/978-3-031-21187-4_74

occurs at seismic intensities much lower than those required to produce structural damage [1]. Therefore, losses from nonstructural building damage often exceed those from structural damage, and even if the structural performance of a building is good enough to allow immediate occupancy after a seismic event, failure of nonstructural elements can lower the performance level and functionality of the entire building system [2].

After the 2012 Emilia Earthquake sequence, failure of steel racking systems was observed in several industrial buildings, which led to the loss of valuable contents and in many cases to the subsequent collapse of the exterior cladding [3]. Extensive damage to shelves and storage racks was observed also following the 2016 Central Italy earthquake, mainly associated with overturning issues or buckling of the vertical members [4]. The same event proved that bracing and/or proper detailing of heavy storage racks in industrial facilities can limit damage during an earthquake: for instance, no damage was observed where a bracing system was provided along with a special base connection, that allowed relative movement at the floor level.

Seismic isolation acts as a filter for the seismic input transmitted to the superstructure (an entire building, a portion of it, or selected contents) above the isolation layer, reducing acceleration, displacement, and deformation demands, and potential damage to structural and non-structural components. Despite the undeniable effectiveness of common elastomeric and friction-pendulum devices in reducing the seismic accelerations transmitted to isolated structures, they might not be suitable for protecting non-structural systems. In fact, these isolators can become uneconomical because of their material and manufacturing costs. Moreover, they need maintenance or replacement over time, to control elastomer aging, steel corrosion, sliding surface degradation, and other effects that may negatively affect their performance [5, 6].

To address the issues above, Kyneprox S.r.l. has patented a new type of isolator, consisting of a double articulated quadrilateral, named "Kinematic Steel Joint" (KSJ) [7, 8]. This device can be manufactured by simply cutting, folding, and pinning steel sheets, possibly employing stainless or galvanized steel to mitigate corrosion issues. Its modular nature allows tailoring it to different payloads and displacement demands. The device imposes a self-centering pendulum-type motion to the superstructure, as it associates horizontal with upward displacements; as with friction pendulum isolators, the restoring force is proportional to the slope of the trajectory. Friction within the pinned joints grants some energy dissipation to the device, and replaceable fuses can be added to act as brakes before reaching the maximum displacement range.

This paper discusses the main results of a shake-table test conducted at the University of Pavia, Italy, and at the EUCENTRE Foundation laboratories on a five-shelf, two-bay industrial steel rack, isolated with KSJ devices in the cross-aisle direction and braced in the down-aisle one. Incremental uniaxial shake-table tests were conducted in the isolated direction under three different loading scenarios, applying a ground motion recorded in Norcia, Italy, during the 2016 Central Italy earthquake sequence and progressively scaling its acceleration amplitude.

2 Dynamic Shake-Table Test

2.1 Specimen Description and Instrumentation

The five-shelf, two-bay industrial steel rack had the dimensions shown in Fig. 1a.

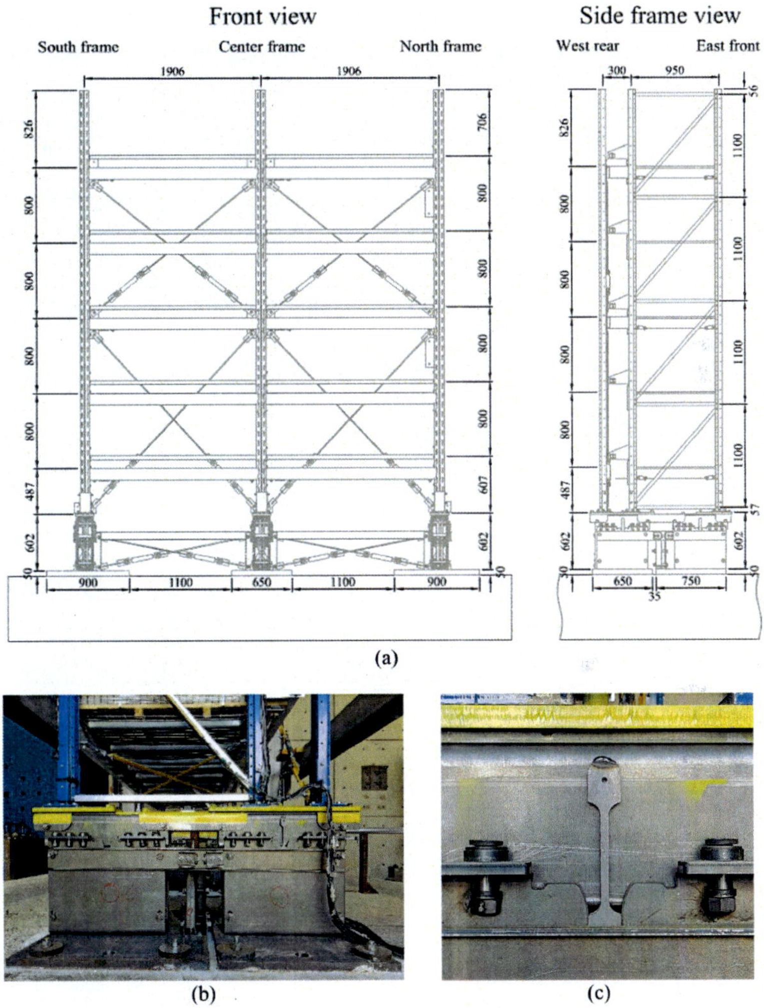

Fig. 1. Test specimen: (a) general dimensions; (b) KSJ isolator pair supporting the north frame; (c) replaceable hysteretic fuse.

Each shelf consisted of two bays, supported by three cross-isle frames. The rack mass was about 575 kg, and any bay could accommodate two 1000 kg pallets. Each frame was supported by a pair of aligned KSJ isolators (Fig. 1a and 1b), with mass of about 400 kg per pair, half of which contributing to the specimen dynamic mass. Each pair of devices was equipped with two or four replaceable hysteretic fuses made of S355 steel (Fig. 1c), with lateral flexural strength of 480 N and stiffness of 150 N/mm. The fuses were engaged at displacements of ± 145 mm to dissipate some kinetic energy by flexural plasticization before reaching the maximum range of ± 180 mm. The KSJ isolators were fastened to 50 mm-thick steel plates, which were in turn connected to the shake-table by four M20 threaded rods.

Three loading configurations were obtained through different pallet distributions on the shelves, to investigate the repeatability and stability of the KSJ isolation response: (i) unloaded configuration, with no pallets, as shown in Fig. 2a; (ii) half-loaded configuration, with two pallets on each bay of shelves 1 and 2, one pallet on each shelf of shelf 3, and no pallets on shelves 4 and 5, as in Fig. 2b; and (iii) fully loaded configuration, with two pallets on both bays of all 5 shelves, as in Fig. 2c. They corresponded to total dynamic masses of 1.175, 11.175, and 21.175 t. Fundamental periods in the cross-isle direction were estimated equal to 0.09 s, 0.38 s, and 0.53 s, respectively.

The dynamic response of rack and isolators was recorded through 27 accelerometers, 13 potentiometers, and 15 wire potentiometers, and 24 strain gages. Figure 3a shows the three potentiometers installed to obtain the trajectory of a KSJ pair, and Fig. 3b the accelerometer to record the acceleration transferred by the isolators to the base of the cross-isle frame. Because the rack was loaded while the data acquisition system was offline, strain-gage recordings did not include gravity load effects and were used only during the test execution to detect possible local buckling of the cold-formed steel columns.

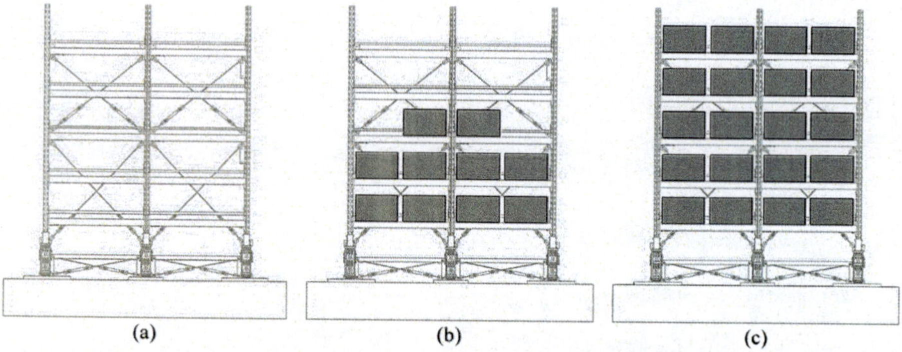

Fig. 2. Loading configurations: (a) unloaded; (b) half-loaded; (c) fully loaded.

2.2 Testing Protocol

The ground motion records were applied through the shake-table only in the cross-isle isolated direction, defined in the following as longitudinal with respect to the imposed table motion. Accordingly, the down-isle direction will be identified as transverse. The input signal was the East-West component of the ground motion recorded during the moment-magnitude (Mw) 6.5 event of October 30th, 2016, at the NRC station in Norcia, Italy, with epicentral distance of 4.6 km and peak ground acceleration (PGA) of 4.76 m/s^2.

(a) (b)

Fig. 3. Sensors for a pair of KSJ isolators: (a) three potentiometers; (b) accelerometer.

A band-stop 4th order Butterworth filter was applied to the original signal over the frequency range 3.2–4.6 Hz. This was intended to avoid resonance issues with the safety steel frame mounted on the shake-table around the specimen, characterized by natural period of about 0.25 s, while minimally affecting the response spectrum for periods longer than 0.4 s, the expected period range for the loaded rack. The target (filtered) acceleration time history, and the corresponding response spectrum for 5% viscous damping ratio, are compared with the original ones in Fig. 4. Filtering the signal at high frequencies resulted in a slight reduction of PGA from 4.76 to 3.57 m/s^2.

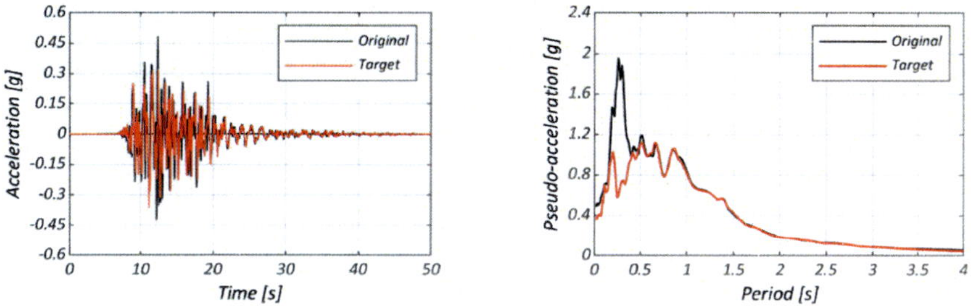

(a) (b)

Fig. 4. Target vs. original input signal: (a) acceleration time history; (b) pseudo-acceleration response spectrum for 5% damping ratio.

The test protocol consisted of applying the target table signal progressively scaling up its acceleration amplitude, with factors 25, 50, 75, 100, and 125%. In all loading configurations the entire test sequence was performed with 2 hysteretic fuses per isolator pair (total 6 fuses); only in the fully loaded configuration the 50, 75, 100, and 125% runs were repeated also with 4 fuses per KSJ pair (total 12 fuses), to investigate the effect of the different number of fuses on the isolator response when approaching the maximum displacement allowance. Low-intensity random noise tests were performed between the seismic runs, for dynamic identification and damage correlation purposes.

3 Experimental Results

3.1 KSJ Isolator Responses

Figure 5 shows the trajectories recorded atop the KSJ isolator pairs during the test runs at 125% of the input acceleration amplitude, and Fig. 6 the corresponding longitudinal force-displacement responses at the isolation level, for the three loading configurations. The variable-curvature trajectories are aligned with the previous analytical and experimental findings [7, 8]. Accordingly, nonlinear restoring force-displacement responses were obtained at the isolation level, with positive/negative stiffness corresponding to positive/negative curvature of the trajectory, zero stiffness at inflection points, and zero force at horizontal-tangent points. Energy dissipation was associated with friction within pinned joints.

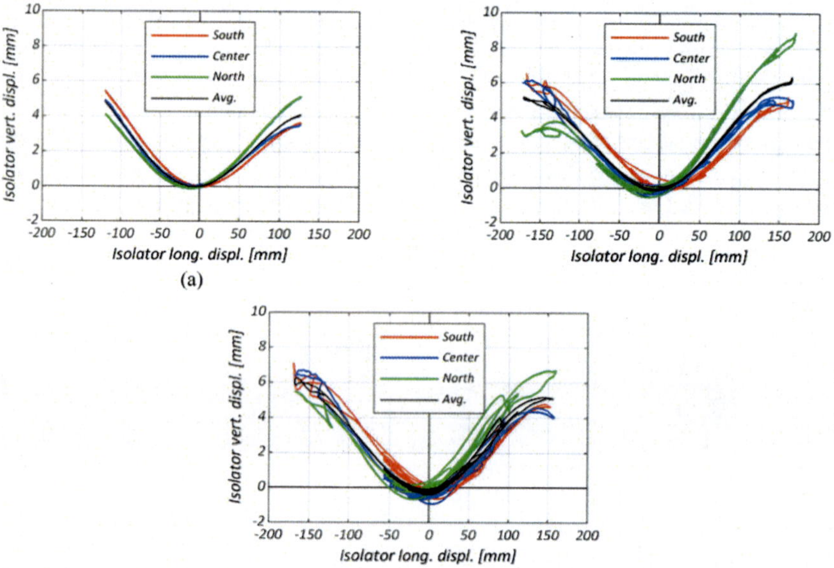

Fig. 5. KSJ isolator pair trajectories from 125%-scaled input: (a) unloaded, (b) half-loaded, and (c) fully loaded configurations.

As expected, the restoring force was proportional to the dynamic mass, reaching a maximum of about 4.5% of the dynamic weight in all loading configurations. Instead, the frictional dissipation included two components: one independent of the supported mass (about 0.3 kN) and one equal to about 3% of the dynamic weight. The overturning moment at the rack base and higher modes of vibration had limited effects on the trajectories and force-displacement curves, more pronounced under higher gravity loads. Minor discrepancies between the trajectories of the three pairs of isolators can also be attributed to some misalignments introduced during installation.

In the unloaded configuration the maximum displacement allowance of the KSJ isolators was never reached. In the half-loaded configuration, it was achieved during

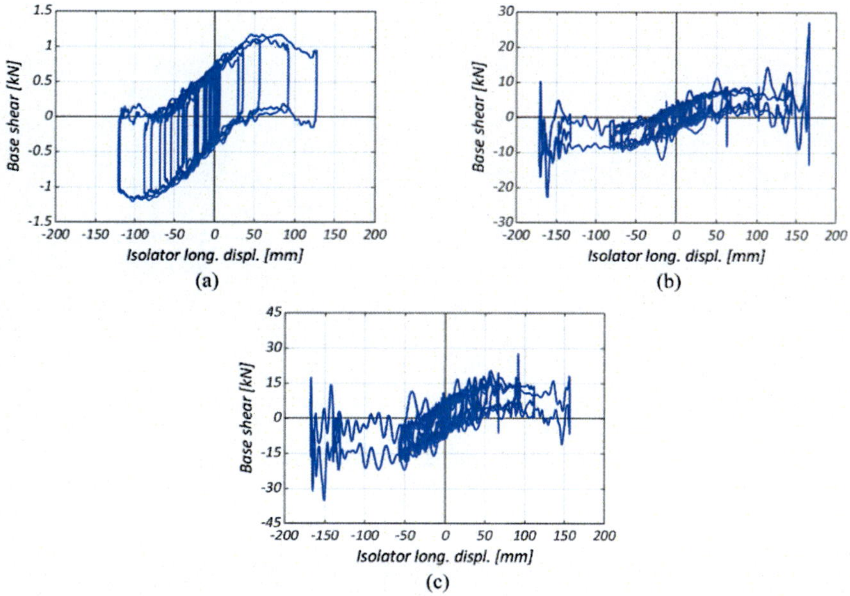

Fig. 6. Longitudinal force-displacement responses from 125%-scaled input: (a) unloaded, (b) half-loaded, and (c) fully loaded configurations.

the test runs scaled to 100% (only positive verse) and 125% (both positive and negative verses). Finally, in the fully loaded case it was reached only under the runs scaled to 125% in both positive and negative verses.

3.2 Isolation Effectiveness

The isolation effectiveness can be appreciated by considering the ratio between the elastic spectral ordinates from the acceleration time series recorded atop the isolators and from the one imposed to the shake-table. In fact, the former acts as input motion for the isolated superstructure. This ratio is plotted in Fig. 7 for the test runs with target signal amplitude scaled to 25%, and in Fig. 8 for those scaled to 125%.

The spectral reduction was more pronounced under higher-intensity input signals. In fact, the superstructure period range over which the ratio was smaller than 1.0 was limited to a period of about 1.25 s in the 25% test runs and extended up to about 2.25 s in the 125% ones. Moreover, in unloaded conditions scaling the input to 25% did not even activate the isolators, with spectral ratios oscillating around 1.0 for any period. These observations can be attributed to two factors: (i) the variable curvature of the KSJ trajectories results in average isolation periods increasing with the displacement demand [7, 8]; and (ii) the KSJ frictional resistance keeps the isolators locked under small lateral accelerations and dynamic masses.

Spikes in the response spectra ratio were recorded at periods of about 0.1 s, corresponding to the actuator oil column period, and around the fixed-base natural period of the rack. The latter was more evident under higher loading, as the interaction between

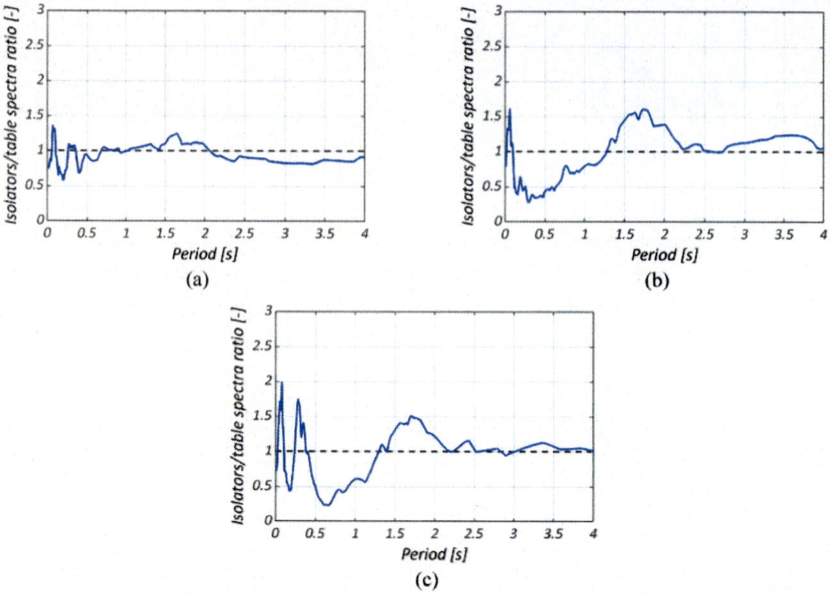

Fig. 7. Response spectra ratio from 25%-scaled input for 5% damping ratio: (a) unloaded, (b) half-loaded, and (c) fully loaded configurations.

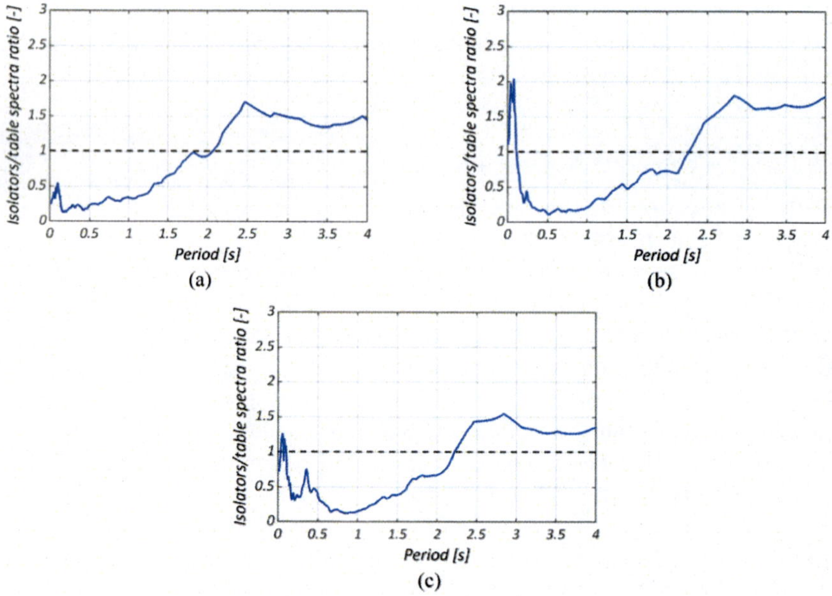

Fig. 8. Response spectra ratio from 125%-scaled input for 5% damping ratio: (a) unloaded, (b) half-loaded, and (c) fully loaded configurations.

superstructure and isolator motion became more pronounced. For superstructure periods approaching or exceeding the average isolation period, undesirable amplification was typically obtained.

3.3 Hysteretic Fuse Number

Repeating the 125%-scaled run in fully loaded configuration with 4 fuses per isolator pair (total 12 fuses) instead of 2 per pair (total 6 fuses) resulted in slightly smaller displacement demand on the isolators and base shear demand upon impact at maximum displacement range (compare Fig. 9a and 9b to Fig. 5c and 6c). The elastic spectra ratio was also slightly affected by this change (Fig. 9c), mainly around the fixed-base natural period of the rack, less excited by the softer impact.

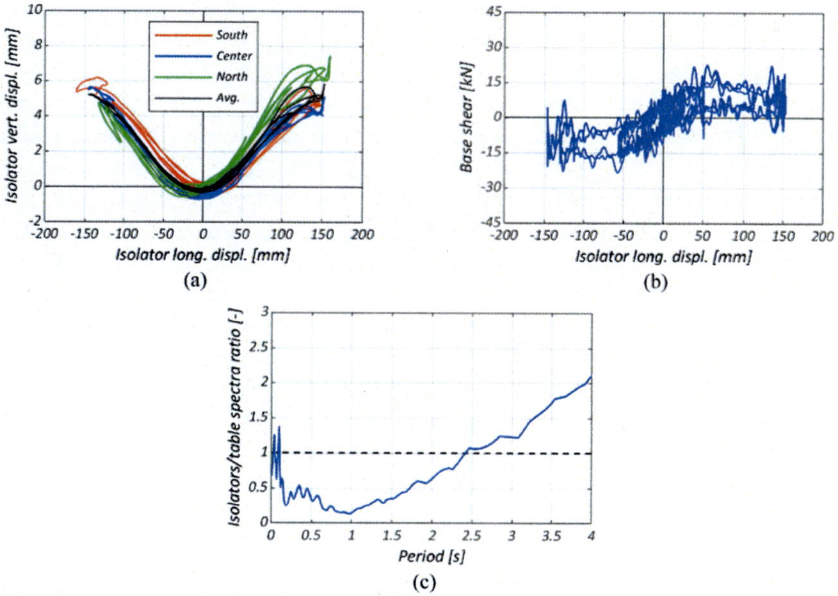

Fig. 9. Results with 12 fuses in fully loaded configuration from 125%-scaled input: (a) KSJ pair trajectories; (b) force-displacement response; (c) response spectra ratio.

4 Conclusions

This paper presented the results of an experimental campaign on an innovative seismic isolation device, applied to an industrial steel rack in the cross-isle direction. The isolator is based on a multiple articulated quadrilateral mechanism and is named "Kinematic Steel Joint" (KSJ). Compared to elastomeric and friction-pendulum isolators currently available on the market, the KSJ offers the advantages of competitive fabrication costs and low-maintenance requirements, if made of galvanized or stainless steel.

Incremental dynamic shake-table tests were performed on the rack in the isolated direction, under three loading scenarios: unloaded, half-loaded, and fully loaded. The outcomes confirmed the behavior of the KSJ devices obtained from previous analytical and experimental studies. The isolators imposed a pendulum-type, self-centering motion to the superstructure base. Restoring forces at the isolation layer were proportional to the slope of the trajectory, and to the mass stored on the rack shelves. KSJ isolators provided some energy dissipation thanks to friction within pinned joints. Replaceable hysteretic fuses were effective at dissipating part of the kinetic energy when approaching the maximum displacement allowance of the isolators, slightly lowering the impact accelerations on the superstructure.

The KSJ solution proved to be effective at reducing the seismic demand on the rack, over a fixed-base period range depending on the displacement demand imposed to the isolators. In fact, the period of the isolation system is related to the curvature of the trajectory, which is not constant with lateral displacement. If the fixed-base period of the superstructure approaches or exceeds the average isolation period calculated at the given displacement demand, undesirable amplification of the effects on the rack are expected as with common isolators. Under low-intensity input motions and small dynamic masses the KSJ devices remained locked, due to frictional resistance within the pinned joints.

The results of this study confirmed the viability of protecting nonstructural components from earthquakes through seismic isolation. They proved the effectiveness of the KSJ technology at isolating industrial racks and encourage further developments and applications of the KSJ isolator. Geometric optimizations will allow to obtain different lateral displacement ranges and trajectory curvatures, compatible with a variety of non-structural component configurations. For instance, ongoing research aims at extending their use in the seismic isolation of museum artworks, electrical cabinets, and server racks.

Acknowledgements. The authors would like to thank Kyneprox S.r.l. for the financial and material support to the experimental campaign. The fundamental contribution of Giovanni Ausenda (Kyneprox S.r.l.) is duly acknowledged, as well as the help provided by the staff of the EUCENTRE Foundation and of the Department of Civil Engineering and Architecture of the University of Pavia.

References

1. Taghavi, S., Miranda, E.: Response assessment of nonstructural building elements. PEER report 2003/05. Pacific Earthquake Engineering Research Center, University of California Berkeley, Berkeley, CA, USA (2003)
2. Filiatrault, A., Sullivan, T.: Performance-based seismic design of nonstructural building components: the next frontier of earthquake engineering. Earthq. Eng. Eng. Vib. **13**(1), 17–46 (2014). https://doi.org/10.1007/s11803-014-0238-9
3. Bournas, D.A., Negro, P., Taucer, F.F.: Performance of industrial buildings during the Emilia earthquakes in Northern Italy and recommendations for their strengthening. Bull. Earthq. Eng. **12**(5), 2383–2404 (2013). https://doi.org/10.1007/s10518-013-9466-z
4. Perrone, D., Calvi, P.M., Nascimbene, R., Fischer, E.C., Magliulo, G.: Seismic performance of non-structural elements during the 2016 Central Italy earthquake. Bull. Earthq. Eng. **17**(10), 5655–5677 (2018). https://doi.org/10.1007/s10518-018-0361-5

5. Clark, P.W., Kelly, J.M., Aiken, I.D.: Aging studies of high-damping rubber and lead-rubber seismic isolators. In: 4th U.S.–Japan Workshop on Earthquake Protective Systems for Bridges; Technical Memorandum no. 3480, pp. 75–89. Public Works Research Institute, Ministry of Construction, Tokyo, Japan (1996)
6. Constantinou, M.C., Whittaker, A.S., Kalpakidis, Y., Fenz, D.M., Warn, G.P.: Performance of seismic isolation hardware under service and seismic loading. Technical report MCEER-07-0012. Multidisciplinary Center for Earthquake Engineering Research, University at Buffalo, The State University of New York, Buffalo, NY, USA (2007)
7. Guerrini, G., Ausenda, G., Graziotti, F., Penna, A.: An innovative seismic isolation device based on multiple articulated quadrilateral mechanisms: analytical study and shake-table test. In: 18th ANIDIS Conference, Ascoli Piceno, Italy (2019)
8. Guerrini, G., Tomassetti, U., Graziotti, F., Rota, M., Penna, A.: Effect of an innovative isolation system on the seismic response of cultural heritage building contents. In: 17th World Conference on Earthquake Engineering, Sendai, Japan (2020)

On the Use of CLT Infills to Improve the Lateral Performance of RC Frames

Li Cavoli Denise[1], Giuseppe D'Arenzo[2], Elisabetta Maria Ruggeri[1], and Marinella Fossetti[1](✉)

[1] Faculty of Engineering and Architecture, Kore University of Enna, Enna, Italy
{denise.licavoli,elisabettamaria.ruggeri}@unikorestudent.it, marinella.fossetti@unikore.it

[2] Timber Structures and Building Rehabilitation, University of Kassel, Kassel, Germany
giuseppe.darenzo@uni-kassel.de

Abstract. Cross-Laminated Timber (CLT) panels allow nowadays to design and realize high-rise timber-based buildings with different construction systems and have shown potential to be used as a sustainable strengthening system for reinforced concrete (RC), steel and masonry structures. Among the various applications of CLT panels in hybrid structures is the very recent interest of the scientific community in using them as infill panels. This paper presents a numerical study aimed at investigating the lateral response of RC frames with traditional masonry and innovative CLT infills. The role of CLT infills and their capability to improve the structural performance of RC structures is investigated. Pushover analyses of RC frames with different number of stories and bays are presented in the study. Results of the analyses show that introducing CLT panel infills in RC frames leads to higher structural performance compared to traditional masonry infills. Given their high in-plane mechanical properties, CLT panel infills allow to reach lower inter-story drifts, resulting in lower structural and non-structural damages and economic losses.

Keywords: CLT infills · Masonry infills · Reinforced concrete structures · Seismic strengthening · Nonlinear analyses

1 Introduction

Although Reinforced Concrete (RC) frame structures infilled with masonry walls are widely spread in many countries, many existing buildings have been designed neglecting the presence of those infills. As many of these countries are seismic prone areas, the seismic interaction between RC frames and masonry infills, which has significant impact on the seismic responses of the structures, has been widely investigated. However, this interaction is not easy to define, since it depends on a plurality of factors, as demonstrated by the vast state of the art on the subject researched by several authors such as Papia et al. [1], Mosalam and Günay [2], Srechai et al. [3].

Many earthquake events have shown that RC structures can suffer unexpected damages or activation of brittle failure mechanisms, such as detachment and collapse of

G. P. Cimellaro (Ed.): WCSI 2022, LNCE 309, pp. 864–874, 2023.
https://doi.org/10.1007/978-3-031-21187-4_75

masonry, shear failure of the columns due to the interaction with the infills, and the activation of dangerous soft-story mechanisms due to the concentration of deformation in a single story, as reported by Reese and Stavridis [4] and Ning et al. [5]. This is especially the case of RC structures in which earthquake resistant design provisions were not applied and which were designed for vertical loads only. The poor performance of buildings exhibiting soft-story mechanisms has led to the development of design procedures that do not allow column side-sway mechanisms that are often associated with a soft-story response. As traditional retrofitting approaches, such as added RC walls or inclined steel braces, pose several obstacles to the architectural functionality of the structures, the increase in strength and stiffness of the columns, while beneficial, would not reduce the expected total damage and losses for the entire structure. Same can be said about increasing the confinement in RC columns, which allows for larger deformation but cannot prevent the collapse caused by the significant P-Delta effects. To this regard relevant are the studies of Beigi et al. [6], who proposed gapped inclined braces as a retrofit system able to take advantage of the positive aspects of the activation of soft story collapse mechanism, while preventing damages caused by P-Delta effects.

Recently, with the aim of reducing the seismic vulnerability of buildings, various authors have developed retrofit interventions based on the use of timber panels. For instance, Stazi et al. [7] analyzed a method in which timber panels are inserted inside the existing RC frames in place of the masonry infills, assuming a perfect bonding at the interfaces between the panel and the frame. Smiroldo et al. [8] instead proposed a system in which the timber panel is inserted into the frame and fixed to a rectangular-based timber subframe through screwed connections with dissipative function. In the method presented by Aloisio et al. [9], Cross Laminated Timber (CLT) panels are added to an existing RC infilled frame via asymmetric friction connections.

The majority of the studies with focus on these hybrid solutions are still at a preliminary stage and further investigation is necessary to grant an adequately solid background for the applicability of timber panels as infills in new buildings and retrofit interventions.

2 Objectives of the Study

The current study aim is to investigate the benefits of CLT panel applications as infill for new and existing RC framed structures, to improve their seismic response. In particular, an infill CLT panel system is proposed, which is conceived to enhance strength and stiffness of the structure and lowering inter-story drifts to prevent structural and non-structural elements damages in occurrence of a seismic events, thus reducing the vulnerability of the structure.

3 Description of the Proposed System

The proposed system is designed to enhance the seismic response of RC frames under horizontal-lateral forces acting in the frame plane by increasing both strength and stiffness of the structure as a whole, while preventing the shear collapse of the columns. This intervention is designed taking into consideration the typical building system for CLT building with typical metal connectors for CLT panels or hybrid connection, and several hybrid building and retrofitting systems such as the ones proposed by Stazi et al. [7], Smiroldo et al. [8] and Aloisio et al. [9].

The intervention presented consist in the replacement of the traditional masonry infill with a CLT panel infill for a RC frame. This panel is inserted with its "strong direction" oriented parallel to the vertical axis, so that it can also contribute to resist vertical loads when the seismic action damages the structural elements. The infill is installed into the frame and connected to the beams above and below the panel. Two different connections have been proposed: the first one consisting of typical angular brackets used at the base of CLT shear walls, see Fig. 1 b, and the second one consisting of a concealed steel plate connected with self-tapping high ductility dowels, see Fig. 1 c. Both these connections are designed to have a dissipative function in the steel-to-timber connection, working mainly under shear loads, and yielding before the collapse of the frame elements. A gap between the CLT infill and the vertical columns is needed and has a double function: it facilitates the installation of the panel into the frame and ensures that, in the first stages of lateral loading, forces are transferred through the metal fasteners, rather than by direct contact between the CLT panel and RC frame.

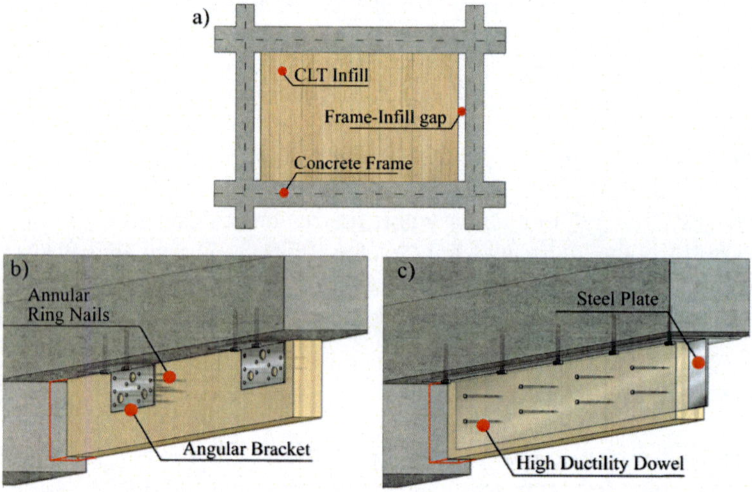

Fig. 1. Proposed system: a) Reinforced concrete frame infilled with CLT panel; b) Installation of the infill with angular brackets; c) Installation of the infill with steel plate and high ductility dowels.

4 Structural Systems and Parametric Study

In this section the structural systems and structural components used to perform a parametric study aimed at evaluating the response of RC frames with different geometries are presented. Three different structural systems were considered: (a) bare RC frame structure, (b) RC frame structure with masonry infills, and (c) RC frame structure with the infill CLT panel system proposed in this study, see Fig. 2.

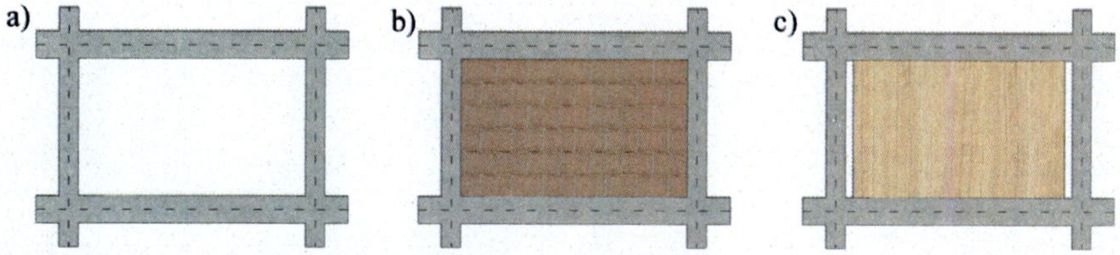

Fig. 2. Structural systems considered: a) bare reinforced concrete frame structure; b) reinforced concrete frame structure with masonry infill; c) reinforced concrete frame structure with CLT panel infill.

In the first step a single-bay single-story RC frame was considered for the analyses. In the second step analyses are extended to multi-story RC frames according to Fig. 3.

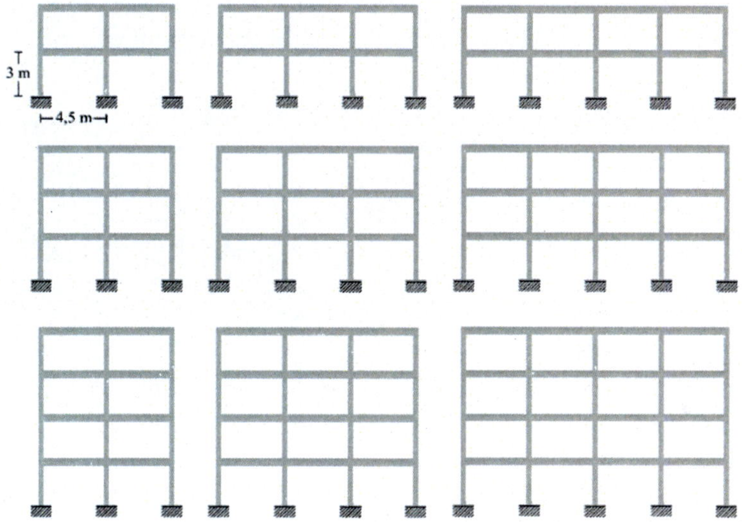

Fig. 3. Multi-bay and multi-story frames configurations investigated.

Each story and each bay retain the geometrical specification used in the single-story single-bay cases, but future steps of the research will include typical problem of infilled structures such as varying width-height ratio of the frames and openings in the infill.

The properties of the bare RC frame systems are reported in Table 1. Tables 2 and 3 provide information of the masonry and CLT infills, respectively. The CLT panels were supposed to be fixed with angular brackets with 11 annular ringed nails according to the experimental tests by Gavric et al. [10], in the first connection solution, and with 7.5 × 95 mm self-drilling high ductility dowels according to the experimental tests by Tuhkanen and Ojamaa [11] in the second connection solution.

Table 1. Geometrical and mechanical characteristics of the bare reinforced concrete frame structure.

Characteristics	Value
Bay width [mm]	4500
Story height [mm]	3000
Beam section [mm]	300 × 500
Beam longitudinal rebars (top side- bottom side)	n2 φ14; n4 φ14
Column section [mm]	300 × 350
Column longitudinal rebars	n3 φ16; n3 φ16
Elastic Modulus of concrete frame [MPa]	31475.8

Table 2. Geometrical and mechanical characteristics of the masonry infill.

Masonry infill thickness [mm]	150
Elastic Modulus of Masonry Infills [MPa]	4600

Table 3. Geometrical and mechanical characteristics of the CLT panel infill and proposed systems.

CLT Panel dimensions [mm]	4050 × 3250 × 85
Lateral gap opening [mm]	50
Elastic Modulus of CLT Panel [MPa] (strong in-plane dir)	7108
Elastic Modulus of CLT Panel [MPa] (weak in-plane dir)	4862
No. of angular brackets used (top side–bottom side)	10–10
No. of high ductility dowels used (top side–bottom side)	16–16

5 Finite Element Analyses

5.1 Modeling of the Structural Systems

The structural systems presented in the previous section were modelled on the commercial software SAP2000, see Fig. 4. The RC frame non-linear behavior has been reproduced through a concentrated-plasticity model, similar to the simplified model proposed by Smiroldo et al. [8], where bending moment deformation-controlled plastic hinges and shear force-controlled plastic hinges were assigned to both extremities of each frame element.

The shear force-controlled plastic hinges were assigned the shear-compression value according to NTC2018 [12]. The behavior of deformation-controlled plastic hinges assigned to beams depends solely on the bending moment, while the plastic hinges

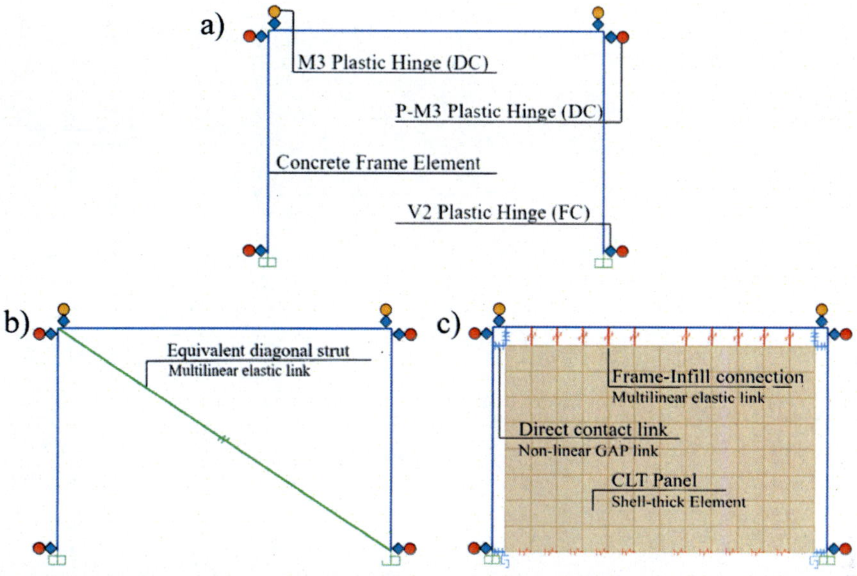

Fig. 4. Mechanical behaviour: a) Bending moment-rotation of plastic hinges in the RC frame b) Force-displacement of equivalent diagonal strut [13]; c) Force-displacement of angular bracket metal connector [10]; d) Force-displacement of dowell connection [11].

assigned to the columns activate based on the interaction between axial load and bending moment. Also in this case, the values used were calculated according to NTC 2018 [12], both for the bending moment-rotation bilinear curve and the axial load-bending moment interaction curve. Figure 5 a depicts the bilinear moment rotation curve used for the bending moment deformation-controlled plastic hinges.

The concrete frame follows these specifications in both the bare frame structure model and the infilled models, regardless of the infill considered.

The masonry infill is modeled by means of equivalent structural macro-model as a multi-linear link element used as diagonal strut, were the force-displacement curve substitute a stress-strain law. The parameters assigned to the link and the modelling approach are derived from the semi-empirical model proposed by Di Trapani et al. [13]. Figure 5 b shows compression only stress-strain curve used for the calculation of the load-displacement curve of the equivalent strut.

The CLT panel is modelled as a shell thin element with an elastic-orthotropic behavior. The in-plane elasticity moduli were calculated as a weighted average between the board layers oriented in the vertical and horizontal direction according to the theory of Blaß and Fellmoser [14]. The equivalent shear modulus was calculated according to Brandner et al. [15]. The interaction between the CLT panel and the RC frame is ensured trough two types of links.

The angular bracket metal connector and the steel plate-dowels connection are modelled as a multi-linear elastic link and have an elasto-plastic behaviour (Fig. 5 c), with values referring to the shear tests of angular brackets investigated by Gavric et al. [10] and the tests carried out by Tuhkanen and Ojamaa [11] for the dowelled connection.

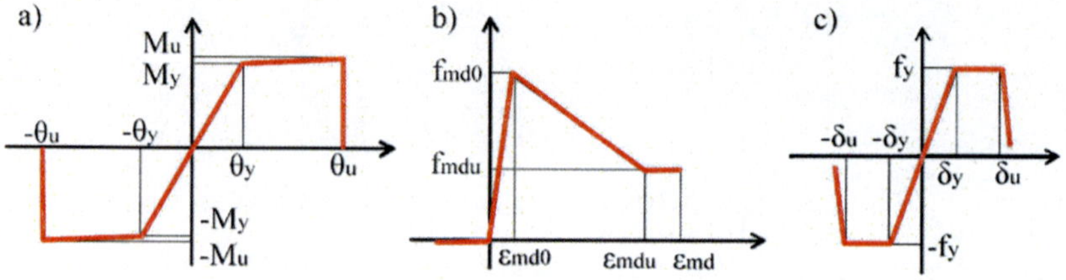

Fig. 5. Finite element models on SAP2000: a) Bare frame; b) Masonry infilled RC frame; c) CLT infilled RC frame.

The contact between the CLT panel and the above and below RC beams was modelled through gap elements.

These elements work only for compressive forces and their stiffness was calculated based on the elastic modulus of the infill and the contact surface between the CLT panel and the RC Frame, according to D'Arenzo et al. [16]. Gap elements from SAP2000 library were also used to model the contact between the CLT panel and the vertical columns, which is triggered when relative displacement between CLT panel and RC frame equals the space-gap between panel and frame.

In all configurations a vertical load of 10 kN/m and a triangular distribution of horizontal loads was considered in the analyses. Nonlinear static analyses were conducted, and pushover curves derived for the different configurations analyzed.

6 Results and Discussion

Figure 6 shows the pushover curves of the single-story single-bay systems considered in the study: bare RC frame, RC frame with masonry infills and RC frame with CLT infills connected with the two connection solutions shown in Fig. 1. It is possible to observe that all systems with infills reach mechanical performance, both at stiffness and load bearing capacity level, higher than those reached by the bare frame system. On the other hand, higher deformation capacities are associated to the bare frame, compared to the solutions with infills.

The masonry-infilled system presented initial stiff response due to the interaction between the diagonal strut and the concrete frame. Through this interaction, stresses are transferred to the column until the maximum capacity is reached, at a displacement of 9 mm. After maximum force is reached a collapse occurs due to exceeding of compressive stresses of the masonry infill.

The CLT-infilled system with angular brackets, is characterized by values of initial elastic stiffness between those of the bare frame model and the masonry-infilled model. This can be explained by the fact that, in the CLT-infilled system, the columns can deform before reaching the contact with the infill, but with an additional resistive contribution provided by the metal connectors.

The backbone curve of the CLT-infilled system presents an increase in load bearing and displacement capacity, with an overall ductile behavior compared to the masonry-infilled model. More specifically, the end of elastic phase, which corresponds to the

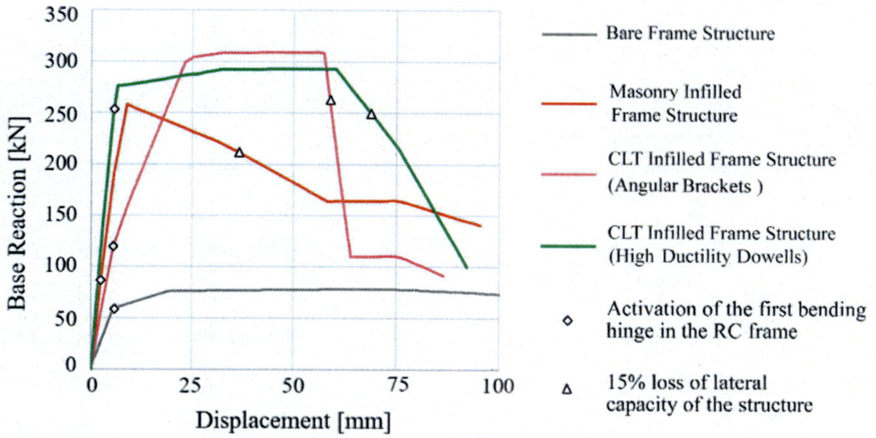

Fig. 6. Single-story single-bay structures models backbone curves: Reinforced concrete bare frame structure; Masonry-infilled reinforced concrete frame structure; CLT-infilled (angular brackets) RC frame structure; CLT-Infilled (high ductility dowels) RC frame structure.

activation of the first bending hinge on the frame, is reached to around 5 mm displacement, and it is followed by the plasticization of the concrete elements. Around 23 mm displacement, the fasteners begin to develop their post-elastic behavior and the curve reach a plastic plateau. The collapse point is reached around 57 mm, with the fasteners failure and the bending moment plastic hinges on the concrete frame fully activated resulting in a kinematic mechanism. The following loss of strength and stiffness indicates a structural collapse.

The CLT-infilled system installed with high ductility dowels presents the higher structural stiffness due to the nature of its concealed connection. Again, the backbone curve presents an increase in load bearing and displacement capacity, with an overall ductile behavior compared to the masonry-infilled model. At 5 mm displacement, the fasteners start to yield and the first bending hinge enters its plasticization phase. The structure engages its post-elastic behavior with a slight increase of its strength. The lateral strength keeps growing until the dowels reach their ultimate displacement, at 60 mm displacement of the system. The dowels failure is followed by the activation of the bending hinges at the base of the columns and the resulting kinematic mechanism.

Figure 7 shows the pushover curves of the multi-story multi-bay systems considered in the study. It is possible to observe that increasing the number of bays enhances the total strength of the structures and increasing the number of stories results in higher displacements. These results also confirm the behavior already analyzed in the single-story single-bay structures, with performance of the CLT infilled structures higher than those of the masonry infilled structures.

In specific terms, masonry infilled frames present lower strength compared to the CLT infilled frames and reach a more brittle failure due to either excessive axial load in the diagonal strut or shear collapse in the column.

On the other hand, CLT infills enhance satisfyingly the structural system stiffness without activating early shear collapse in the column and present the highest results in

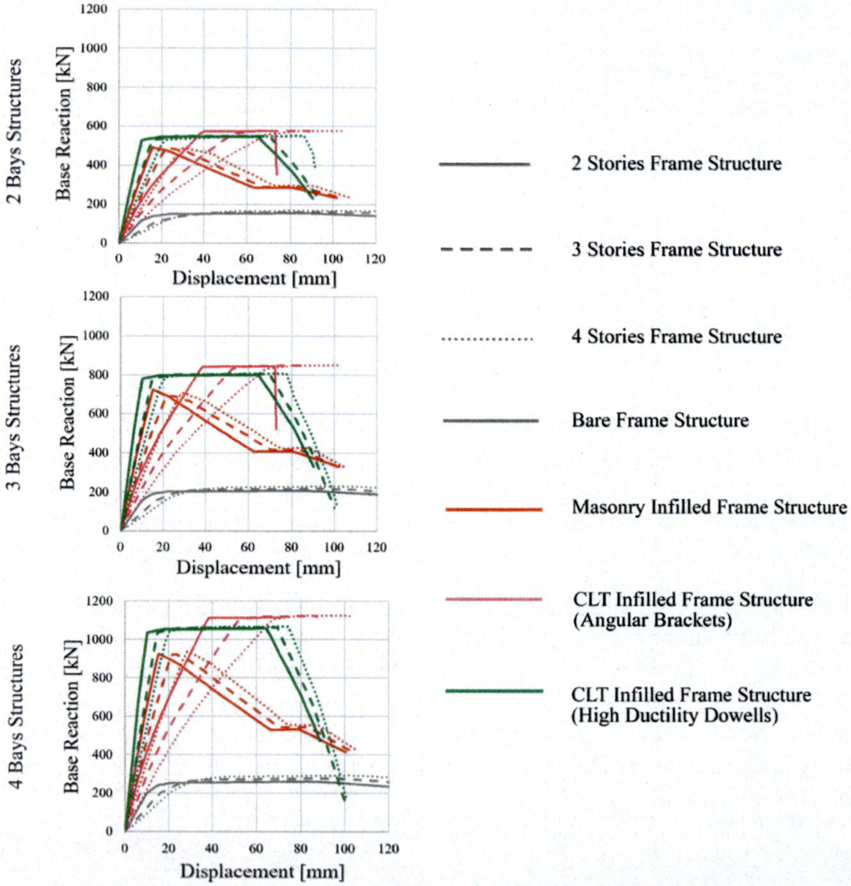

Fig. 7. Multi-story multi-bay structures models backbone curves: reinforced concrete bare frame structure; masonry-infilled reinforced concrete frame structure; CLT-infilled (angular brackets) reinforced concrete frame structure; CLT-infilled (high ductility dowels) reinforced concrete frame structure.

terms of strength. That is due to the fact that the failure of the connections is reached before the contact between CLT infill and the vertical column and to the gap between these two structural components, which prevents increment of shear stress in the columns. In particular, the CLT system with angular brackets have lower stiffness but higher strength while the system with dowelled connection as higher stiffness, lower strength but higher ductility. The dowelled connection, with its higher stiffness and ductility and good strength values better responds to the aim of the present study of reducing inter-story drifts and the seismic response of structures to low and medium intensity earthquakes.

7 Conclusions

Cross-Laminated Timber (CLT) panels allows nowadays to design and realize timber-based buildings with different construction systems and have shown potential to be used as a sustainable strengthening system for reinforced concrete (RC), steel and masonry structures. In this paper a numerical study aimed at investigating the lateral response of RC frames with traditional masonry and innovative CLT infills was presented. Two connection solutions for CLT-infills are proposed, one with traditional angular brackets used to prevent the horizontal sliding of CLT shear walls, and one with concealed steel plate connected with self-tapping high ductility dowels. Structural performances of these systems are compared to those of traditional masonry infills.

The role of CLT infills and their capacity to improve the structural performance of new and existing RC structures was investigated trough pushover analyses of RC frames with different characteristics, in single-bay single-story frames, and in multi-bay multi-story frames. Results of the numerical analyses conducted in this study show that introducing CLT panel infills with the connection solutions proposed in RC frames leads to higher structural performance compared to traditional masonry infills. In the dowelled connection system presented, the high stiffness of the two-shear-planes connection along with the high in-plane mechanical properties of the CLT panel infills allowed to reach lower inter-story drifts, resulting in likely lower structural and non-structural damages and economic losses.

Future aspects to be investigated are the dynamic response of the proposed CLT-infilled systems under earthquake shakings, the width of the gap between CLT infills and the vertical columns to soften the ultimate brittle collapse mechanism, and the effects of the openings in the infill panels.

Acknowledgement. The research is supported from the CADS project (Creazione di un Ambiente Domestico Sicuro), project code: ARS01_00920, funded under the MIUR Announcement "Ricerca Industriale e Sviluppo Sperimentale nelle 12 aree di specializzazione individuate dal PNR 2015–2020" (D.D. 1735/2017), Specialisation Area "Tecnologie per gli Ambienti di Vita".

References

1. Papia, M., Cavaleri, L., Fossetti, M.: Infilled frames: developments in the evaluation of the stiffening effect of infills. Struct. Eng. Mech. **16**(6), 675–693 (2003)
2. Mosalam, K.M., Günay, S.: Progressive collapse analysis of reinforced concrete frames with unreinforced masonry infill walls considering in-plane/out-of-plane interaction. Earthq. Spectra **31**(2), 921–943 (2015)
3. Srechai, J., Leelataviwat, S., Wararuksajja, W., Limkatanyu, S.: Multi-strut and empirical formula-based macro modeling for masonry infilled RC frames. Eng. Struct. **2661**, 114559 (2022)
4. Reese, A., Stavridis, A.: A Simplified method for the estimation of the seismic resistance of RC frames with weak infill panels. In: 10th U.S. National Conference on Earthquake Engineering, Frontiers of Earthquake Engineering. Anchorage, Alaska (2014)
5. Ning, N., Ma, Z., Zhang, P., Yu, D., Wang, J.: Influence of masonry infills on seismic response of RC frames under low frequency cyclic load. Eng. Struct. **183**, 70–82 (2019)

6. Beigi, H.A., Christopoulos, C., Sullivan, T.: Gapped-inclined braces for seismic retrofit of soft-story buildings. J. Struct. Eng. **140**, 1–11 (2014)
7. Stazi, F., Serpili, M., Maracchini, G., Pavone, A.: An experimental and numerical study on CLT panels used as infill shear walls for RC buildings retrofit. Constr. Build. Mater. **211**, 605–616 (2019)
8. Smiroldo, F., Giongo, I., Piazza, M.: Use of timber panels to reduce the seismic vulnerability of concrete frame structures. Eng. Struct. **244**, 112797 (2021)
9. Aloisio, A., Pelliciari, M., Sirotti, S., Boggian, F., Tomasi, R.: Optimization of the structural coupling between RC frames, CLT shear walls and asymmetric friction connections. Bull. Earthq. Eng. **20**(8), 3775–3800 (2021)
10. Gavric, I., Fragiacomo, M., Ceccotti, A.: Cyclic behaviour of typical metal connectors for cross-laminated (CLT) structures. Mater. Struct. **48**(6), 1841–1857 (2014). https://doi.org/10.1617/s11527-014-0278-7
11. Tuhkanen, E., Ojamaa, M.: Early experimental investigations on slotted-in steel plate connections with self-perforating dowels in CLT. Wood Mat. Sci. Eng. **16**(2), 102–109 (2021)
12. Ministero delle Infrastrutture e dei Trasporti: NTC 2018 -Norme Tecniche per le Costruzioni di cui al D.M. 17/01/2018. Gazzetta Ufficiale N.42 del 20/02/2018 (In Italian). Istituto Poligrafico e Zecca dello Stato (2018)
13. Di Trapani, F., Cavaleri, L., Bertagnoli, G., Gino, D.: A semi-empirical stress-strain model for equivalent strut fiber-section modeling of infilled frames. In: 17th ANIDIS Conference, Pistoia, Italy (2017)
14. Blaß, H.J., Fellmoser, P.: Design of solid wood panels with cross layers. In: 8th World Conference Timber Engineering, Lathi, Finland (2004)
15. Brandner, R., Dietsch, P., Dröscher, J., Schulte-Wrede, M., Kreuzinger, H., Sieder, M.: Cross laminated timber (CLT) diaphragms under shear: Test configuration, properties and design. Constr. Build. Mater. **147**, 312–327 (2017)
16. D'Arenzo, G., Casagrande, D., Polastri, A., Fossetti, M., Fragiacomo, M., Seim, W.: CLT shear walls anchored with shear-tension angle brackets: Experimental tests and finite-element modeling. J. Struct. Eng. **147**(7), 1–15 (2021)

Required Response Spectra and Acceleration Loading Histories for Seismic Assessment of Acceleration-Sensitive Nonstructural Elements According to the Italian Building Code

Martino Zito[1], Danilo D'Angela[1], Giuseppe Maddaloni[2], and Gennaro Magliulo[1,3](✉)

[1] Department of Structures for Engineering and Architecture, University of Naples Federico II, Via Claudio 21, 80125 Naples, Italy
{martino.zito,danilo.dangela,gmagliul}@unina.it
[2] Department of Engineering, University of Benevento Sannio, Benevento, Italy
giuseppe.maddaloni@unisannio.it
[3] Construction Technologies Institute, National Research Council, 80125 Naples, Italy

Abstract. The assessment of the seismic performance of nonstructural elements (NEs) is typically required by the current international codes, regulations, and guidelines. Required response spectra (RRS) and acceleration testing inputs are the key parameters for defining shake table testing protocols suitable for both assessment and qualification of acceleration-sensitive NEs. However, international codes and guidelines define RRS and acceleration time histories that are not necessarily consistent with seismic demands associated with Italian and European territories. In this study, novel RRS and acceleration time histories are developed for a consistent shake table assessment of acceleration-sensitive NEs according to the Italian building code. A critical review of existing protocols was carried out to define new approaches for developing code-consistent assessment and qualification protocols. The approach used to develop RRS and testing inputs is widely applicable and easily extendable to different case studies as the process is highly flexible and adaptable. The application of the developed methodology lays the foundations for the development of code-consistent shake table protocols for both assessment and qualification purposes.

Keywords: Nonstructural elements · Testing protocol · Shake table testing · Seismic qualification · Performance-based earthquake engineering

1 Introduction

Nonstructural elements (NEs) are typically extremely critical in terms of seismic response and performance as several post-event surveys found out [1–3]. As a matter of fact, NEs were rarely designed considering seismic actions in the past, and nowadays, even though several regulations prescribe to design/verify NEs considering seismic

G. P. Cimellaro (Ed.): WCSI 2022, LNCE 309, pp. 875–883, 2023.
https://doi.org/10.1007/978-3-031-21187-4_76

actions (such as Italian building code [4]), this is rarely carried out in practice. However, nonstructural elements are typically essential for guaranteeing the functioning of buildings, especially for critical and strategic facilities such as hospitals [5], and they are associated with an extremely large part of building costs [6]. Furthermore, seismic response and damage of NEs can cause injuries and casualties [7, 8], as it can be observed in Fig. 1 that depicts examples of critical seismic damage exhibited by nonstructural elements due to the 2016 Central Italy earthquake.

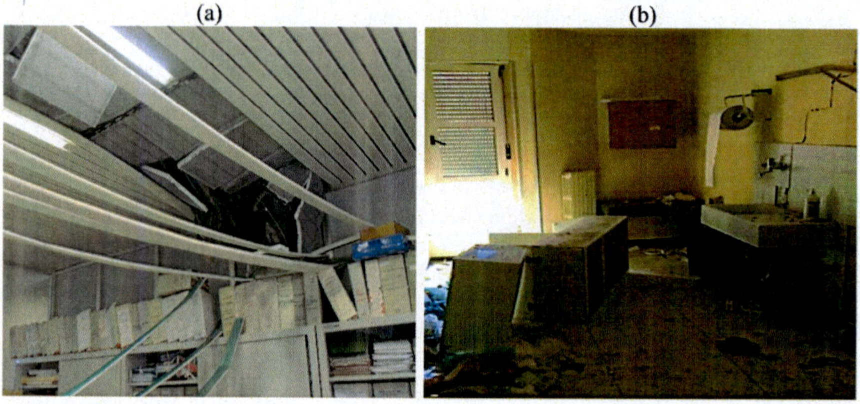

Fig. 1. Critical seismic damage of nonstructural elements due to the 2016 Central Italy earthquake: (a) failure of ceiling system [8] and (b) overturning of hospital contents/furniture (Amandola hospital) [5].

Nonstructural elements are typically categorized in acceleration- and displacement-sensitive elements, even though recent literature pointed out that in several cases they are sensitive to both parameters and that most of elements are (also) sensitive to acceleration. It is quite rare that elements are not sensitive to inertial effects due to acceleration inputs applied at their boundaries, e.g., infill panels, which are significantly sensitive to displacement (drift) within their plane, are sensitive to accelerations out of their plane. Conversely, there are several elements that are sensitive to accelerations but not at displacements/drift, e.g., such as freestanding of base-anchored equipment. Seismic assessment of (acceleration-sensitive) elements can be carried out through different methods, i.e., observational (e.g., post-event), theoretical/analytical, numerical, and experimental. Experimental methods are overall more reliable that other ones, and, differently from civil structures, they can be generally implemented to investigate NEs considering their full-scale dimensions and reproducing seismic loading scenarios in a consistent manner. Shake table testing represents the state-of-art method for assessing seismic response and performance of NEs, especially for acceleration-sensitive elements. Several shake table protocols are defined by regulation, guidelines, and technical documents, allowing the assessment of the capacities and performances of NEs according to standardized and relatively robust procedures. However, recent literature pointed out that existing protocols might not be fully reliable and consistent with actual seismic scenarios associated with NEs (e.g., [2, 9]).

The present study investigates the existing shake table protocols, focusing on the definition of the seismic input and the associated response spectra. In the light of the critical assessment of these protocols, the study proposes a new formulation of response spectra that is compatible with Italian building code [4] and related seismic demand formulation, which was found to be reliable and robust considering accurate estimations of seismic demands on building elements [10, 11]. The formulation is not depending on the building characteristics and is generally applicable given the basic hazard parameters. The provided spectra could be used to define a novel shake table protocol that is consistent with seismic and building scenarios of Italy and Europe, and that is conservative if compared to other protocols.

2 Shake Table Testing and Existing Protocols

Shake table testing represents the most reliable and robust method for assessing seismic response of NEs, especially for acceleration-sensitive elements. As a matter of fact, shake tables are able to reproduce seismic excitations in terms of ground motions and floor records. The selection of the shake table protocol, with particular regard to the seismic inputs to be assigned to the table, is of paramount interest due to the typical record-to-record uncertainty associated with seismic response of NEs [9]. In fact, referring to authoritative shake table protocols allow to arrange, carry out, and elaborate robust and reliable shake table tests, favoring the development of representative and consistent assessment results. Indeed, adherence to reference shake table protocols is mandatory in the case of seismic qualification and certification (e.g., [2, 12, 13]).

Two categories of shake table protocols are typically defined, i.e., (a) protocols for generic elements defined by regulations and guidelines and (b) protocols for specific elements developed by industrial companies and institutions. ICC-ES AC156 [13], FEMA 461 [14], and ISO 13033 [15] protocols belong to the first category; they are defined by international regulations and refer to generic (acceleration-sensitive) elements. In particular, AC156 represents is the only protocol explicitly developed for the purpose of (special) seismic certification [9, 16]. IEEE 344 [17], IEEE 693 [18], IEC 60068-2-57 [19], and GR-63-CORE [20] protocols belong to the second category; they are referred to electric/nuclear/telecommunication equipment and systems and are defined by companies and relevant institutions. Figure 2 shows RRS associated with reference shake table protocols, considering PGA equal to 0.50 g.

The seismic input for shake table testing is an essential feature among the different key aspects and technical characteristics of shake table protocols. In particular, seismic test inputs are typically defined as acceleration time histories, and they are often defined/developed considering artificial procedures/signals and through complicated processing/filtering/matching operations. The consistency the seismic test input and the and compatibility with the reference protocol is often assessed through the analysis of the test response spectra (TRS) and the technical comparison with required response spectra (RRS). Therefore, the severity and the consistency of the test inputs is practically referred to RRS. Accordingly, it can be reasonably be assumed that the robustness and the representativity of shake table protocols is significantly conditioned by the definition of RRS.

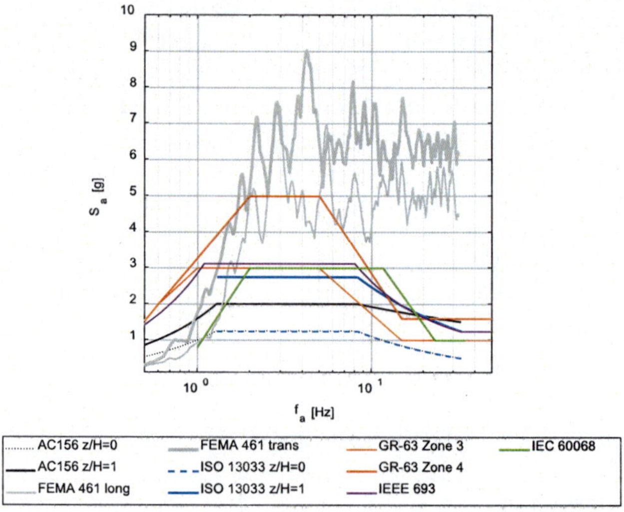

Fig. 2. RRS related to reference protocols (S_a is spectral acceleration and f_a is element frequency), considering PGA equal to 0.50 g.

It should be noted that FEMA 461 protocol does not provide RRS but acceleration signals to be used, which are also depicted in Fig. 1; z/H is the ratio between the installation height of the element and building height. The different protocols are associated with significantly different, and disperse, spectral ordinates, especially over the most amplified frequencies (e.g., within 1–10 Hz). It is worth observing that AC156 protocol is associated with relatively mild RRS, even for the case of z/H equal to one, compared with all other protocols. Conversely, FEMA 461 seismic inputs appear to be extremely severe, especially for medium to high frequencies (e.g., from 2 Hz). The other protocols range between AC156 and FEMA 461 spectral responses and appear to be relatively disperse.

Figure 2 stresses the need for addressing the representativeness of the protocols in terms of RRS, possibly defining new reliable protocols that are aimed to generic elements and that are representative of consistent seismic and seismologic scenarios. As a matter of fact, the most referenced protocol, i.e., AC156, which is the only one aimed at seismic certification, is found to be extremely mild, as other studies also reported [2, 21]. Moreover, FEMA 461 protocol, which does not provide RRS, seems to be extremely severe, possibly determining overconservative and antieconomic results. The other protocols are quite dispersed in terms of RRS. Moreover, they are mostly referred to specific elements, and this arises a question: are they fully applicable for assessing/qualifying different elements.

3 A New Formulation for Required Response Spectra According to Italian Building Code

Several recent studies estimated the seismic demand on nonstructural elements housed in buildings [10, 11, 22–25]. In particular, the simplified formula provided by Italian building code [4, 24, 25], which is applicable for frame buildings, was found to be reliable and relatively conservative if compared with floor responses associated with typical (reinforced concrete) buildings [10, 11]. Equation (1) reports the floor spectra equations provided by Italian building code [4], where α is the design peak ground acceleration at bed rock expressed in g, S is the soil amplification factor, T_a is the fundamental period of the NE, T_1 is the fundamental period of the building, and {a, b, a_p} are factors depending on the building period. Figure 3 depicts the dimensionless floor spectral response associated with Eq. 1, as reported within the building code.

$$\frac{S_a}{\alpha \cdot S} = \begin{cases} \left(1+\frac{z}{H}\right)\left[\frac{a_p}{1+(a_p-1)\left(1-\frac{T_a}{aT_1}\right)^2}\right] \geq 1 \; for\; T_a < aT_1 \\ \left(1+\frac{z}{H}\right)a_p \quad for\; aT_1 \leq T_a < bT_1 \\ \left(1+\frac{z}{H}\right)\left[\frac{a_p}{1+(a_p-1)\left(1-\frac{T_a}{bT_1}\right)^2}\right] \geq 1 \; for\; T_a \geq bT_1 \end{cases} \tag{1}$$

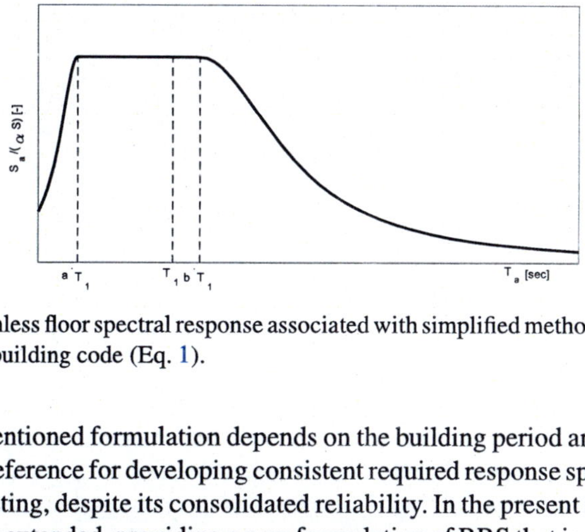

Fig. 3. Dimensionless floor spectral response associated with simplified method (frame buildings) defined in Italian building code (Eq. 1).

The abovementioned formulation depends on the building period and has never been considered as a reference for developing consistent required response spectra for purpose of shake table testing, despite its consolidated reliability. In the present study, Eq. (1) has been analytically extended, providing a new formulation of RRS that is not depending on the building period (step 1). Successively, a set of seismic inputs, i.e., acceleration time histories, has been generated and processed to be compliant with the abovementioned new formulation (step 2).

Step 1 consists in estimating spectral response (Eq. 1) over a reasonable range of building periods, defining an optimal upper bound closed-form equation that is analytically inspired by Eq. (1). Equation (2) reports this upper bound limit, which represents a new formulation of seismic demands on nonstructural elements, that is applicable for a wide and representative range of buildings and facilities (i.e., 0.1–2.0 s building periods). Equation (2) is graphically represented in Fig. 4 (thick line), whereas Eq. (1) curve sets are represented using thin lines.

$$\frac{S_a}{a\ S} = \begin{cases} 4\left(1+\frac{z}{H}\right)+\left[\frac{1+\frac{z}{H}}{f_1-f_0}(f_a-f_0)\right] \geq \alpha S \; for f_a < f_1 \\ 5\left(1+\frac{z}{H}\right) & for f_1 \leq f_a < f_2 \\ \frac{5\left(1+\frac{z}{H}\right)}{1+4\left(1-\frac{f_2}{f_a}\right)} & for f_a \geq f_2 \end{cases} \tag{2}$$

where f_0, f_1, and f_2 are set equal to 1.00, 1.40, and 12.5 Hz, respectively.

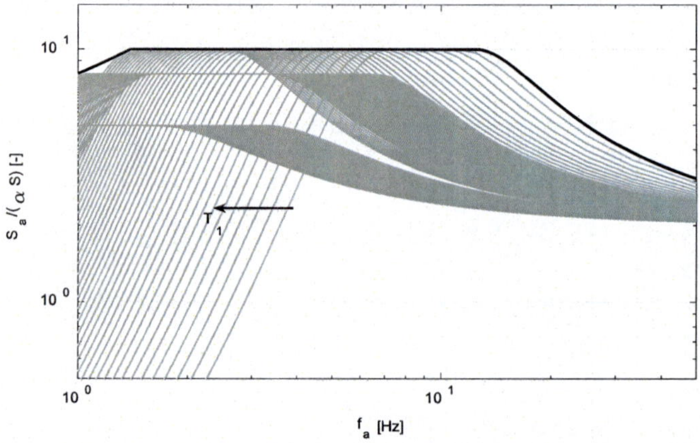

Fig. 4. Dimensionless floor spectral response associated with new formulation (Eq. 2) for required response spectra developed in this study (in thick line) and curve sets related to simplified method defined in Italian building code (Eq. 1).

Step 2 consists in assuming Eq. (2) as the expression of RRS, generating a set of spectrum-compatible seismic inputs by applying the procedure developed in [26]. In particular, in this latter and in other literature studies (e.g., [5, 27], this procedure was used with regard to AC156 RRS formulation, whereas in this study, the procedure was implemented considering Eq. 2 as target RRS, also applying enhancement and optimization procedures. Further details regarding the full procedure can be found in [28]; technical details regarding the input generation are omitted since they were provided in several other studies, such as in [5, 9].

Figure 5 shows (a) two (representative) acceleration loading histories (e.g., longitudinal and transversal signals for shake table testing) and (b) spectral responses of a set of seven loading histories (e.g., a representative set of loading histories to be used to perform numerical analyses) developed according to the abovementioned procedure; in particular RRS (and related inputs) are related to z/h equal to one and α S equal to 0.40 g. The time histories (Fig. 5 a) are macroscopically similar to the seismic signals typically developed according to AC156 protocol, but looking at the spectral responses (Fig. 5 b), it is obvious the significantly more severe spectral response, with particular regard to the amplified frequency region.

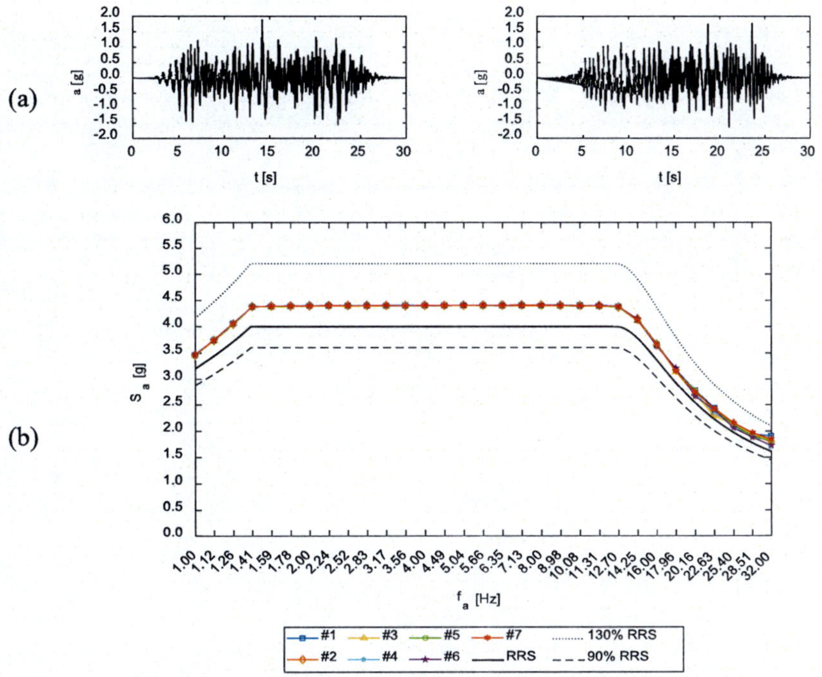

Fig. 5. Seismic signals developed according to the procedure and formulation proposed in this paper: (a) two (representative) acceleration loading histories (e.g., longitudinal and transversal signals for shake table testing) and (b) spectral responses of a set of seven loading histories (e.g., a representative set of loading histories to be used to perform numerical analyses).

4 Concluding Remarks

The study deals with the seismic performance of acceleration-sensitive nonstructural elements, with focus on the definition of seismic demand and the development of shake table testing protocols. In particular, a new formulation of required response spectra (RRS) is developed by extending the seismic demand formulation provided by the Italian building code, applicable for frame buildings. The provided formulation does not depend on the building period and has a general validity, as well as it is likely to be more reliable than exiting RRS formulations, according to latest research studies. Furthermore, representative loading histories are generated and processed according to the developed RRS formulation, providing seismic input signals that are promising for both experimental and numerical applications. Further studies should be carried out to validate the developed procedure and to assess the reliability and representativity of the developed loading histories.

Acknowledgements. The study was supported by the Italian Department of Civil Protection (DCP) in the frame of the national project DPC—ReLUIS 2019–2021 WP 17: “Code contributions for nonstructural elements”.

References

1. Achour, N., Miyajima, M., Kitaura, M., Price, A.: Earthquake-induced structural and non-structural damage in hospitals. Earthq. Spectra **27**, 617–634 (2011). https://doi.org/10.1193/1.3604815
2. Perrone, D, Brunesi, E., Decarro, F., Peloso, S., Filiatrault, A. Seismic assessment and qualification of non-structural elements in Europe: a critical review. In: 4th International Workshop on the Seismic Performance of Non-Structural Elements (SPONSE). (2019) Pavia, Italy. https://doi.org/10.7414/4sponse.ID.10
3. Yön, B., Onat, O., Öncü, M.E.: Earthquake damage to nonstructural elements of reinforced concrete buildings during 2011 van seismic sequence. J. Perform. Constr. Facil. **33**, 04019075 (2019). https://doi.org/10.1061/(ASCE)CF.1943-5509.0001341
4. Consiglio Superiore dei Lavori Pubblici (CS.LL.PP.). Circolare 21 gennaio 2019, n. 7 C.S.LL.PP. Istruzioni per l'applicazione dell'«Aggiornamento delle "Norme tecniche per le costruzioni"» di cui al decreto ministeriale 17 gennaio 2018 (2019)
5. Di Sarno, L., Magliulo, G., D'Angela, D., Cosenza, E.: Experimental assessment of the seismic performance of hospital cabinets using shake table testing. Earthquake Eng. Struct. Dynam. **48**, 103–123 (2019). https://doi.org/10.1002/eqe.3127
6. Taghavi, S., Miranda, E. Response Assessment of Nonstructural Building Elements. PEER Report No. 2003/05. Pacific Earthquake Engineering Research Center: University of California at Berkeley, Berkeley, California (2003)
7. Braga, F., Manfredi, V., Masi, A., Salvatori, A., Vona, M.: Performance of non-structural elements in RC buildings during the L'Aquila, 2009 earthquake. Bull Earthquake Eng **9**, 307–324 (2011). https://doi.org/10.1007/s10518-010-9205-7
8. Perrone, D., Calvi, P.M., Nascimbene, R., Fischer, E.C., Magliulo, G.: Seismic performance of non-structural elements during the 2016 Central Italy earthquake. Bull. Earthq. Eng. **17**(10), 5655–5677 (2018). https://doi.org/10.1007/s10518-018-0361-5
9. D'Angela, D., Magliulo, G., Cosenza, E.: Towards a reliable seismic assessment of rocking components. Eng. Struct. **230**, 111673 (2021). https://doi.org/10.1016/j.engstruct.2020.111673
10. Chichino, B., Peloso, S., Bolognini, D., Moroni, C., Perrone, D., Brunesi, E.: Towards seismic design of nonstructural elements: Italian code-compliant acceleration floor response spectra. Advances in Civil Engineering, 1–18 (2021). https://doi.org/10.1155/2021/4762110
11. Di Domenico, M., Ricci, P., Verderame, G.M.: Floor spectra for bare and infilled reinforced concrete frames designed according to Eurocodes. Earthquake Engng. Struct. Dyn. eqe.3523 (2021). https://doi.org/10.1002/eqe.3523
12. American Society of Civil Engineers. ASCE 7–16. Minimum design loads for buildings and other structures, ASCE standard. (2016) American Society of Civil Engineers: Structural Engineering Institute, Reston, Va
13. International Code Council Evaluation Service (ICC-ES) AC156. Acceptance Criteria for the Seismic Qualification of Nonstructural Components. Brea, California, USA (2012)
14. Federal Emergency Management Agency (FEMA). Interim protocols for determining seismic performance characteristics of structural and nonstructural components through laboratory testing. Report No. FEMA 461. Washington D.C., USA (2007)
15. International Organization for Standardization. ISO 13033:2013. Bases for design of structures—loads, forces and other actions—Seismic actions on nonstructural components for building applications (2013)
16. California Building Standards Commission. California Building Standards Code (CBC) (2016)

17. Institute of Electrical and Electronics Engineers (IEEE). IEEE 344. Seismic Qualification of Equipment for Nuclear Power Generating Stations, New York (2013)
18. Institute of Electrical and Electronics Engineers (IEEE). IEEE 693 Recommended practice for seismic design for substations, New York (2006)
19. International Electrotechnical Commission, International Electrotechnical Commission, Technical Committee 104: Environmental conditions, classification and methods of test. Environmental testing. Part 2–57, Part 2–57 (2013)
20. Network Infrastructure and Operations. NEBS Requirements: Physical Protection. Telcordia Technologies Generic Requirements, GR-63-CORE (No. GR-63-CORE) (2006)
21. D'Angela, D., Magliulo, G., Cosenza, E.: Seismic damage assessment of unanchored non-structural components taking into account the building response. Struct. Saf. **93**, 102126 (2021). https://doi.org/10.1016/j.strusafe.2021.102126
22. Huang, B., Chen, S., Lu, W., Mosalam, K.M.: Seismic demand and experimental evaluation of the nonstructural building curtain wall: a review. Soil Dyn. Earthq. Eng. **100**, 16–33 (2017). https://doi.org/10.1016/j.soildyn.2017.05.025
23. Merino, R.J., Perrone, D., Filiatrault, A.: Consistent floor response spectra for performance-based seismic design of nonstructural elements. Earthquake Engng Struct Dyn **49**, 261–284 (2020). https://doi.org/10.1002/eqe.3236
24. Petrone, C., Magliulo, G., Manfredi, G.: Floor response spectra in RC frame structures designed according to Eurocode 8. Bull. Earthq. Eng. **14**(3), 747–767 (2015). https://doi.org/10.1007/s10518-015-9846-7
25. Petrone, C., Magliulo, G., Manfredi, G.: Seismic demand on light acceleration-sensitive non-structural components in European reinforced concrete buildings. Earthquake Eng. Struct. Dynam. **44**, 1203–1217 (2015). https://doi.org/10.1002/eqe.2508
26. Magliulo, G., Maddaloni, G., Petrone, C.: A procedure to select time-histories for shaking table tests on nonstructural components. In: M. Papadrakakis, V. Papadoupolus, V. Plevris (eds.) Proceedings of the 4th ECCOMAS Thematic Conference on Computational Methods in Structural Dynamics and Earthquake Engineering. Kos Island, Greece (2013). https://doi.org/10.13140/2.1.4693.8881
27. Prota, A., Zito, M., D'Angela, D., Toscano, G., Ceraldi, C., Fiorillo, A., Magliulo, G.: Preliminary results of shake table tests of a typical museum display case containing an art object. Adv. Civil Eng., 1–18 (2022). https://doi.org/10.1155/2022/3975958
28. Zito, M., D'Angela, D., Maddaloni, G., Magliulo, G.: A shake table protocol for seismic assessment and qualification of acceleration-sensitive nonstructural elements. Comput.-Aided Civil Infrastruct. Eng. (Under review). 1–28. https://doi.org/10.1111/mice.12951

Supplemental Energy Dissipation Devices for Passive Structural Control

Multi-EDP Performance Assessment of a Steel BRBF Under Ground Motion Sequences

Fernando Gutiérrez-Urzúa[1](✉), Fabio Freddi[1], and Enrico Tubaldi[2]

[1] University College London, London WC1E 6BT, UK
{f.urzua,f.freddi}@ucl.ac.uk
[2] University of Strathclyde, Glasgow G1 1XJ, UK
enrico.tubaldi@strath.ac.uk

Abstract. The behaviour of Buckling-Restrained Braces (BRBs) is characterised by steady and nearly symmetrical hysteretic loops that provide large energy dissipation capacity. However, their low post-yielding stiffness may result in large residual deformations at the end of the earthquake motion. Moreover, the cumulative ductility demand due to repeated plastic excursions can lead to low-cycle fatigue failure. These two unfavourable conditions could be exacerbated by the occurrence of subsequent earthquakes (e.g., mainshock-aftershock sequences or multiple earthquakes during the design lifetime). Therefore, their assessment requires a framework that considers ground motion (GM) sequences. In this paper, a case study structure subjected to GM sequences is analysed, considering three Engineering Demand Parameters (EDPs) relevant to the performance evaluation of the BRBs and of the frame, namely the maximum ductility demand in the BRBs, the cumulative ductility demand in the BRBs, and the residual inter-storey-drift in the frame. These EDPs are assessed both independently and simultaneously, in order to establish the risk of overpassing any capacity limit that may lead to the collapse or demolition of the structure.

Keywords: Buckling-restrained braces · Ground motion sequences · Cumulative ductility · Residual deformation · Risk assessment

1 Introduction

Buckling-Restrained Braces (BRBs) are dissipative devices that can be used to improve the seismic performance of newly designed and/or existing structures. Unlike other bracing systems, BRBs avoid buckling under compression loads by encasing a steel core in an unbounded restraining sleeve, thus allowing the development of large plastic deformations both in tension and compression with nearly symmetric hysteretic loops [1–4]. However, being characterised by a low post-elastic stiffness, their use may lead to residual deformations, hindering the building's reparability and excessive cumulative ductility demand, possibly compromising the residual capacity of BRBs.

These issues are further emphasised when considering multiple earthquakes. Only a few authors have investigated the performance of BRBs under multiple earthquakes,

G. P. Cimellaro (Ed.): WCSI 2022, LNCE 309, pp. 887–894, 2023.
https://doi.org/10.1007/978-3-031-21187-4_77

particularly by considering mainshock-aftershock sequences [5–7]. However, these studies considered only a single aftershock and no considerations have been made on the effects of multiple earthquakes within the lifespan of the structure nor their effects on the cumulative ductility demand of the BRBs.

The present paper examines the performance of a BRB frame within a probabilistic framework based on hazard-compatible site-specific GM sequences and multiple independent and/or dependent failure modes. The failure modes considered include the overpassing of the maximum and/or the maximum cumulative ductility capacity of the BRB devices and/or the maximum residual storey drifts of the MRF. A three-storey steel Moment Resisting Frame (MRF) retrofitted with BRBs is considered for case study purposes and is numerically analysed in OpenSees [8]. The conclusions are drawn in terms of a risk-based comparison, considering different failure modes, the number of GMs required to reach those failures and the number of GMs expected during the structure's lifetime.

2 Case Study Structure

The case study structure consists of a steel MRF retrofitted by BRBs in an inverted V configuration, as described in Fig. 1. The MRF is based on the North-South external frame of the pre-Northridge Boston 3-storey building designed within the SAC Project [9]. Beams and columns are made of ASTM A572 standard, Grade 50, Group 1 (f_y = 1.1 × 344.74 MPa; E = 199.95 GPa), while the BRB core is made of steel with yielding strength f_y = 322 MPa and Young's modulus E = 199.95 GPa.

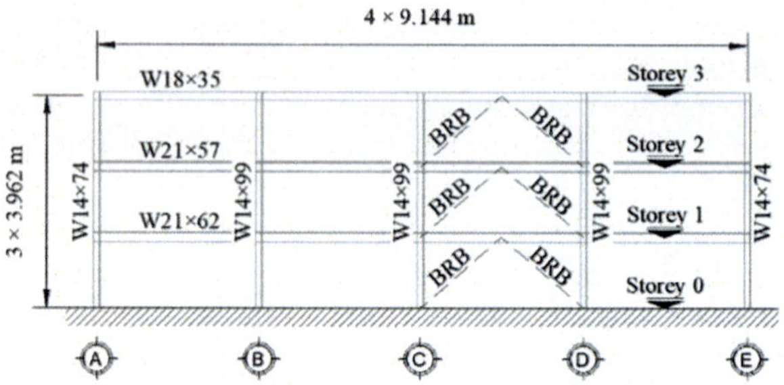

Fig. 1. Elevation view of the case study structure.

The case study structure is modelled in OpenSees [8] by considering a distributed plasticity approach for columns, a concentrated plasticity approach for beams, and a scissors model for panel zones. The BRBs are modelled by considering a series arrangement of the BRB device and of an elastic brace as done in Freddi et al. [10]. Additional details on the case study structure and modelling approach are provided by previous authors [11, 12]. The case study structure's retrofit aims to comply with the demands of a high seismicity site (Los Angeles). The response spectrum for a 5% probability of exceedance in 50 yrs (i.e., T_R = 975 yrs) is considered for the design at the Collapse Prevention (CP) structural performance level, as required by the ASCE 41–17 [13] for buildings within Risk Category II. The retrofit design procedure used [10, 14, 15] aims to maximise the energy dissipation capacity of the BRBs by proportioning their stiffness to match the first mode shape of the bare frame and their strength to induce the simultaneous yielding of the devices in all storeys. The retrofit design is based on a target top storey drift of 0.275 m, which corresponds to a drift at which some of the structural elements in the MRF have overpassed the Immediate Occupancy (IO) [13], but none of the elements has overpassed the Life Safety (LS) structural performance level. The BRB devices are considered to have a maximum ductility capacity of $\mu_{\text{max,cap}} = 20$. The cross-sectional core area of the BRB devices are 3142, 2692 and 1673 mm^2 for each BRB in the 1st, 2nd and 3rd storeys, respectively, while their core length is considered as 2.18, 2.76 and 2.55 m. The cross-sectional area of the elastic braces is 8750, 5046 and 3592 mm^2, respectively.

Figure 2 shows the first-mode proportional pushover curve for the retrofitted frame, including markers at the steps on which the BRBs yield at each storey, as well as the point at which they reach $\mu_{\text{max,cap}}$. As observed, both the yielding of the devices and the overpassing of the $\mu_{\text{max,cap}}$ occur almost simultaneously.

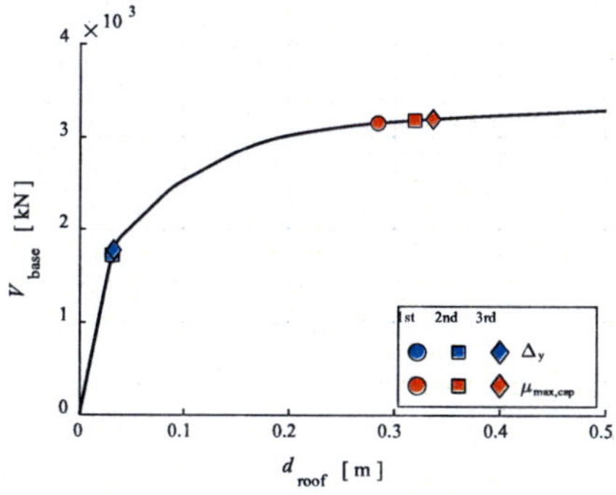

Fig. 2. Pushover curve for the case study structure, including markers at which BRBs yield (Δ_y) and reach their maximum ductility capacity ($\mu_{\text{max,cap}}$) at each floor.

3 Hazard Considerations

The seismic hazard characteristics of the site are obtained through USGS Unified Hazard Tool [16] using the Conterminous US database (v4.2.0). In accordance with the retrofitting design, Los Angeles and soil Class D ($V_{s30} = 259$ m/s) are considered as the site for the definition of the hazard. The spectral acceleration corresponding to the first structural vibration period $S_a(T_1)$ is considered as Intensity Measure (IM). The hazard curve, shown in Fig. 3(a), correlates a given value of IM to its annual rate of exceedance (λ_{im}). Figure 3(b) shows the cumulative distribution function (CDF) for intensities above $im_{\min} = 0.1$g. Assuming earthquakes at this site as independent events and disregarding potential aftershocks, their occurrence follows a Poisson distribution. Therefore, the probability of having n GMs given a lifespan t can be expressed as:

$$P(n|t) = \frac{(\lambda_{\min}t)^n}{n!}e^{-\lambda_{\min}t} \tag{1}$$

where $\lambda_{\min}$ is the annual rate of exceedance for earthquakes with *IM* equal or larger than the minimum *IM* threshold, *i.e.*, $\lambda(IM \geq im_{\min})$. Figure 4 shows the probability mass function (PMF) for the site assuming a lifespan $t = 50$yrs and $im_{\min} = 0.1$g. The maximum probability mass value corresponds to $n = 3$, as $P(n = 3 \mid t, im_{\min}) = 0.22$. The present study considers sequences of 11 GMs, as the probability of having more earthquakes is negligible, *i.e.*, $P(n \leq 11 \mid t, im_{\min}) = 0.9998$.

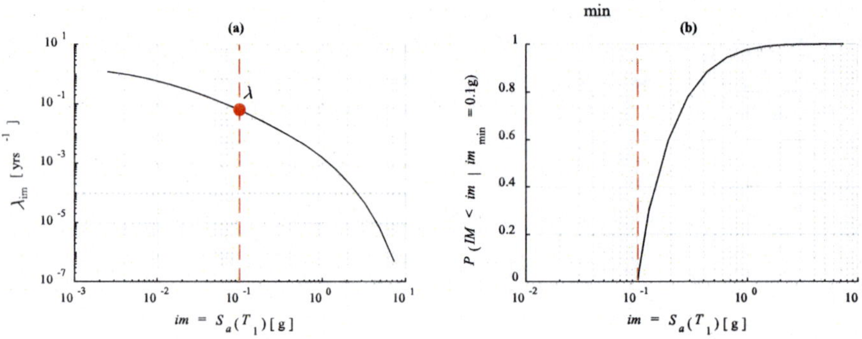

Fig. 3. Hazard characteristics of the site: (a) annual rate of exceedance; (b) intensity distribution above the minimum threshold.

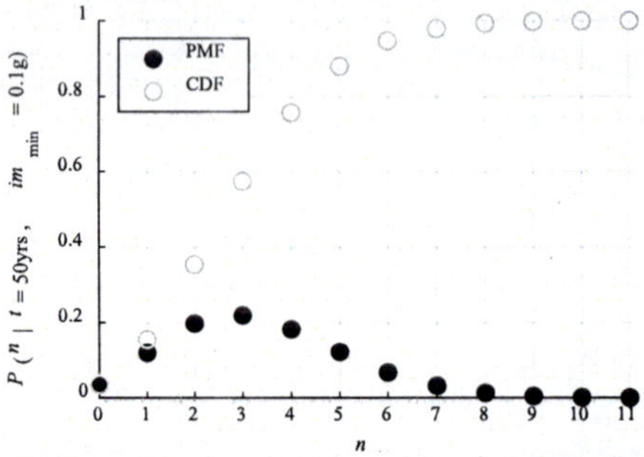

Fig. 4. Probability of n earthquakes occurring in a lifespan t, probability mass function (PMS) and cumulative distribution function (CDF).

A Monte Carlo procedure is used to generate one thousand GM sequences. The *IM* of each GM is selected through random sampling from the CDF in Fig. 3(b). The GM pool used is composed of two pools of GMs of 240 and 150 records [17, 18]. The entire GM pool is classified into bins, delimited by selected *IM* limits. The GM selection at each sequence is carried out through a Latin Hypercube Sampling (LHS) process; thus, each GM record can be selected more than once but not within the same sequence. As the GMs are scaled to match the *IM* obtained through random sampling, the selection of GM records is made by considering only those within the corresponding bin. This procedure allows for avoiding excessive scaling factors.

4 Engineering Demand Parameters and Capacity Limits

Three key EDPs are considered: (1) the maximum ductility demand on the BRBs (i.e., $\mu_{max} = \Delta_{max}/\Delta_y$ where Δ_{max} and Δ_y, are the maximum and yielding deformations), (2) the cumulative ductility demand of the BRBs (i.e., $\mu_c = \Sigma\Delta_{pl}/\Delta_y$. Where $\Sigma\Delta_{pl}$ the sum of plastic deformation in all excursions), and (3) the residual inter-storey drift ratio of the structure (θ_{res}).

These EDPs are monitored throughout the time-history analyses and compared to pre-defined capacity limits. The capacity limit for μ_{max} is considered to be the design value for the devices, $\mu_{max,cap} = 20$. The definition of the capacity limit for μ_c is not a straightforward task, as it is a path-dependent quantity. To overcome this limitation, Andrews et al. [19] developed a model based on extensive experimental results, which considered the peak demand ductility quantities (μ_{max}, μ_c) along with the BRB material and geometric characteristics. For the purpose of this study, the model developed by Andrews et al. [19] is rearranged to derive a deterministic cumulative ductility capacity value for each BRB, as follows:

$$\mu_c(C_{rem} = 0) = \mu_{c,cap} = \frac{C_{tot}\mu_{ult}}{152.9} - \frac{\mu_{max}}{136.5} \tag{2}$$

where $\mu_c(C_{rem} = 0)$ is the cumulative ductility demand at the point on which the remaining ductility capacity (C_{rem}) is zero. C_{tot} represents the total capacity, which given in terms of the geometric and material properties of the BRB device [19], and μ_{ult} is the ultimate ductility capacity of the material, defined as $\varepsilon_u/\varepsilon_y$, where ε_u is assumed to be 0.35. Assuming $\mu_{max} = 20$, the resulting cumulative ductility capacity values, $\mu_{c,cap}$, are 615.16, 582.03 and 473.94 for each BRB in the 1st, 2nd and 3rd storeys, respectively, consistently with values reported by previous authors [20].

Finally, the capacity limit for the residual inter-storey drift ratio ($\theta_{res,cap}$) is assumed as 0.5%, which is associated with repair costs exceeding the costs of demolition/replacement of the structure ([21]), i.e., $\theta_{res,cap} = 0.005$. The re-centring capability of the structure after the BRBs removal is not considered, as the focus is on the failure of the BRB devices.

5 Failure Trends and Risk Assessment

One thousand sequential time-history analyses are performed on the case study structure, each composed of eleven earthquakes. The probabilistic analysis is made first by considering each EDP and corresponding failures independently; therefore, each sequence can register one or more failure modes. If a failure mode is reached at GM *n* of the sequence, the building is considered to have failed for the rest of the sequence, even if subsequent GMs (*e.g.*, $n + 1$, $n + 2$) result in lower deformation demands for the structure.

Figure 5(a) shows the probability of reaching a given failure mode, independently of other failure modes, i.e., failure modes are not mutually exclusive. For all EDPs, the probability of experiencing a failure increases with n. The probability increases for the failure modes related to μ_{max} and θ_{res} follow a nearly linear trend with their intercept close to zero, while the failure mode related to μ_c grows following a cumulative normal distribution. The probability of failure within the lifespan t, is function of the PMF for GMs occurrence (i.e., Eq. (1)) and can be expressed as:

$$P(D \geq C|t) = \sum_{n=1}^{\infty} P(D \geq C|n) \cdot P(n|t) \tag{3}$$

where $P(D \geq C \mid n)$ is the probability of failure after n GMs. The outcomes of Eq. (3) are shown in Fig. 5(b). As a result of the number of earthquakes required to reach the failure mode for μ_c, the probability of this failure mode during the lifetime of the structure t is further reduced.

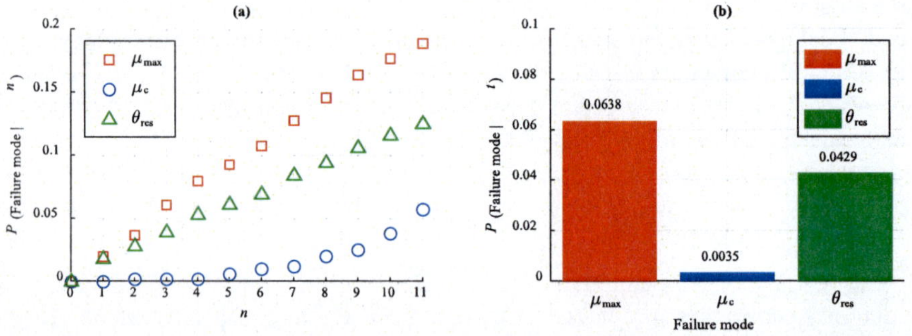

Fig. 5. (a) Number of sequences that exhibit each failure mode after n GMs, independently of other failure modes; (b) Number of sequences that overpass $\mu_{c,cap}$ and reduced values, independently of other failure modes.

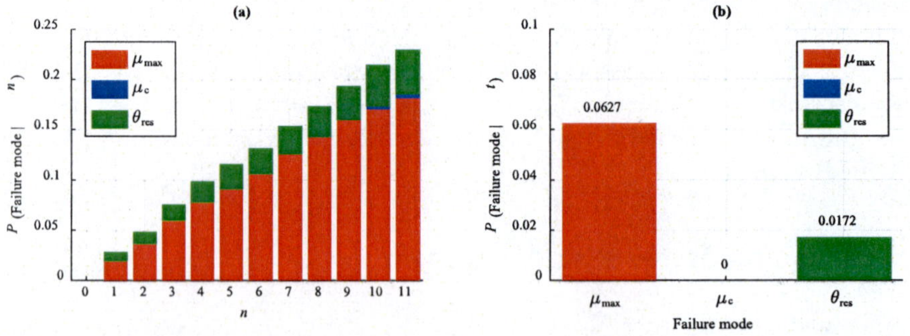

Fig. 6. Probability of each failure type: (a) function of the number of GMs; and (b) considering the probability of having n GMs in lifetime t.

The methodology is successively modified to acknowledge that the occurrence of any failure mode results in a structure that has overpassed its collapse limit or requires demolition, thus, only the first failure mode is registered. Figure 6(a) shows the probability of exhibiting each failure mode, conditional to the number of GMs. As it can be observed, the overall probability of failure increases proportionally with n. The most common failure mode is related to μ_{max}, followed by θ_{res}. On the other hand, the number of cases in which μ_c overpassed its capacity limit is negligible and occurs only after several GMs. Figure 6(b) shows that the probability of failure modes related to μ_c within the lifetime of the structure becomes negligible (i.e., 0.0000069) as this failure mode requires not only several sequential GMs, but also requires that other failure modes are not developed at earlier earthquakes. On the other hand, there is a larger chance of overpassing the capacity limits for μ_{max} and θ_{res} with probability values of 0.0627 and 0.0172, respectively.

6 Conclusions

The seismic performance of a Moment Resisting Frame (MRF) retrofitted with Buckling-Restrained Braces (BRBs) is numerically investigated. The study focuses on the performance under GM sequences and considers multiple Engineering Demand Parameters (EDPs), *i.e.,* maximum (μ_{max}) and cumulative ductility demand (μ_c) of the BRBs, and residual inter-storey drift ratios (θ_{res}) of the MRF. The following conclusions can be drawn for the considered case study structure:

1. The probability of failure related to μ_c is negligible, even when other failure modes are not considered, as it requires several GMs. When other failure typologies were considered, the cumulative ductility failure started to occur only after 10 GMs, which reduces the risk to negligible values.
2. The consideration of the EDPs independently from each other, leads to larger values of risk for each failure typology. Moreover, these values cannot be simply added together to define the collapse or demolition probability, as the occurrence of a failure typology restricts the occurrence of others.
3. The probability of failure related to θ_{res} is non-negligible, as it accounts for 0.0172 probability of failure, on top of the 0.0627 probability of failure due to μ_{max}. At this stage, the re-centring capacity of the MRF after BRB removal is not considered.

It is important to highlight that the considered capacity limits for all the EDPs are established as deterministic values. A more comprehensive stochastic approach requires the consideration of the variability of the capacity.

References

1. Tremblay, R., Poncet, L., Bolduc, P., Neville, R., DeVall, R.: Testing and design of buckling restrained braces for canadian application. In: *13th World Conference on Earthquake Engineering*, Vancouver, Canada (2004)
2. Zona, A., Dall'Asta, A.: Elastoplastic model for steel buckling-restrained braces. J. Construct. Steel Res., **68**(1), 118–125 (2012)

3. Freddi, F., Tubaldi, E., Zona, A., Dall'Asta, A.: Seismic performance of dual systems coupling moment-resisting and buckling-restrained braced frames. Earthquake Eng. Struct. Dyn., **50**(2), 329–353 (2021)
4. Di Sarno, L., Manfredi, G.: Seismic retrofitting with buckling restrained braces: application to an existing non-ductile RC framed building. Soil Dyn. Earthq. Eng. **30**(11), 1279–1297 (2010)
5. Hoveidae, N., Radpour, S.: Performance evaluation of buckling-restrained braced frames under repeated earthquakes. Bull. Earthq. Eng. **19**(1), 241–262 (2020). https://doi.org/10.1007/s10518-020-00983-0
6. Morfuni, F., Freddi, F., Galasso, C.: Seismic performance of dual systems with BRBs under mainshock-aftershock sequences. In: *13th International Conference on Applications of Statistics and Probability in Civil Engineering, ICASP 2019* (2019)
7. Guerrero, H., Ruíz-García, J., Escobar, J.A., Terán-Gilmore, A.: Response to seismic sequences of short-period structures equipped with Buckling-Restrained Braces located on the lakebed zone of Mexico City. J. Constr. Steel Res. **137**(June), 37–51 (2017)
8. McKenna, F., Fenves, G.L., Scott, M.H.: Open system for earthquake engineering simulation (OpenSees) (2000)
9. Gupta, A., Krawinkler, H.: Behavior of ductile SMRFs at various seismic hazard levels. J. Struct. Eng. **126**(1), 98–107 (2000)
10. Freddi, F., Ghosh, J., Kotoky, N., Raghunandan, M.: Device uncertainty propagation in low-ductility RC frames retrofitted with BRBs for seismic risk mitigation. Earthquake Eng. Struct. Dyn., eqe.3456 (2021)
11. Gutiérrez-Urzúa, F., Freddi, F., Di Sarno, L.: Comparative analysis of code-based approaches for seismic assessment of existing steel moment resisting frames. J. Constr. Steel Res. **181**, 106589 (2021)
12. Gutiérrez-Urzúa, F., Freddi, F.: Influence of the design objectives in the seismic performance of steel Moment Resisting Frames retrofitted with Buckling Restrained Braces. Earthquake Eng. Struct. Dyn., (2022)
13. American Society of Civil Engineers. Seismic Evaluation and Retrofit of Existing Buildings. *ASCE/SEI 41–17* (2017)
14. Ragni, L., Zona, A., Dall'Asta, A.: Analytical expressions for preliminary design of dissipative bracing systems in steel frames. J. Construct. Steel Res., **67**(1), 102–113 (2011)
15. Freddi F, Tubaldi E, Ragni L, Dall'Asta A. Probabilistic performance assessment of low-ductility reinforced concrete frames retrofitted with dissipative braces. *Earthquake Engineering and Structural Dynamics* 2013; **42**: 993–1011
16. United States Geological Survey (USGS). Unified Hazard Tool 2021. https://earthquake.usgs.gov/hazards/interactive/.
17. Baker, J.W., Lin, T., Shahu, S.K., Jayaram, N.: *New Ground Motion Selection Procedures and Selected Motions for the PEER Transportation Research Program* (2011)
18. Smerzini, C., Galasso, C., Iervolino, I., Paolucci, R.: Ground motion record selection based on broadband spectral compatibility. Earthq. Spectra **30**(4), 1427–1448 (2014)
19. Andrews, B.M., Fahnestock, L.A., Song, J.: Ductility capacity models for buckling-restrained braces. J. Constr. Steel Res. **65**(8–9), 1712–1720 (2009)
20. Palmer, K.D., Christopulos, A.S., Lehman, D.E., Roeder, C.W.: Experimental evaluation of cyclically loaded, large-scale, planar and 3-d buckling-restrained braced frames. J. Constr. Steel Res. **101**, 415–425 (2014)
21. Iwata, Y., Sugimoto, H., Kuwamura, H.: Reparability limit of steel structural builidings: study on performance based design of steel structural buildings, Part 2. J. Struct. Construct. Eng. **70**(588), 165–172 (2005)

Modern Systems for Wind and Seismic Induced Vibrations

Peter Nawrotzki[1], Daniel Siepe[1](✉), and Fulvio Bottoni[2]

[1] GERB Schwingungsisolierungen GmbH and Co.KG, Berlin, Germany
{peter.nawrotzki,daniel.Siepe}@gerb.de
[2] GERB Italia Srl, Sesto San Giovanni, Italy
fulvio.bottoni@gerb.it

Abstract. This contribution examines the application of modern systems for increasing damping in regard to protect structures against wind and seismic induced vibrations. The Tuned Mass Control System, presented as the first of two systems, consist of an additional mass, elastically connected to the main structure—in general quite similar to the well-known tuned mass systems that are applied at high rise structures against wind-induced effects. Dampers are installed in parallel to the elastic devices to widen the operating frequency band of the system and to reduce the relative displacements between mass and main structure. This passive seismic protection system yields a significant reduction of induced acceleration, displacement levels as well as internal stresses, support reactions and interstory drift. An important advantage is the possibility to use the inside of the building as there is no disturbance by the upgrade activities, if these systems are used as retrofitting systems. Corresponding project examples are presented and results of numerical investigations are described to verify the effectiveness of the mitigation measures. Just as these systems are suitable for new buildings and for existing structures, viscous wall dampers are also appropriate for both cases. They are using the shear resistance force of a highly viscous fluid. The devices consist of an outer steel dashpot, filled with liquid and an internal shearing piston. The extension of a two-story building by seven new floors is presented as corresponding example, where the increase of structural damping was required to protect the existing structure against seismic demands.

Keywords: Tuned mass system · Viscous wall damper · Seismic · Wind

1 Introduction

Seismic mitigation can be obtained through seismic isolation (with or without internal dissipation), energy dissipation using dissipative devices, energy transfer using dynamics absorbers or through the combination of two or more of the previous techniques, as described in [1]. This article focuses on increasing the structural damping. It is a common and well-examined strategy for the protections of structures against vibrations caused by wind and seismic loading. The reduction (e.g. for seismic calculations) of the

G. P. Cimellaro (Ed.): WCSI 2022, LNCE 309, pp. 895–902, 2023.
https://doi.org/10.1007/978-3-031-21187-4_78

induced structural responses by the increase of viscous damping can be taken from different national and international standards. According to these regulations, an increase of structural damping from 5% to 12% provides a correction factor of about 0.75. Thus, this effect causes a further reduction of structural stress, strain, displacement, etc. in a range between 22% and 30%.

The main principles of two eligible systems are described as follows. The mass of the tuned system is usually arranged at the highest level of the structure to yield the optimum effectivity. Typically, reinforced concrete or steel is used as material for the mass block that is connected elastically with the substructure such that the corresponding frequencies are related to specific rules. Viscous dampers are arranged in parallel to reduce the relative displacements between tuned mass and structure and to widen the operating frequency band of the tuned mass system. The development of tuned mass systems for seismic protection (designated as TMCS—Tuned Mass Control Systems) is based on the experience from tuned mass damper systems that work against wind and men-induced vibrations, as presented in [2]. Especially for a seismic event it is important that the protection system works immediately and passively without additional electrical power supply. The TMCS yields a significant reduction of induced acceleration and displacement levels as well as of internal forces and support reactions due to seismic excitation.

A viscous wall damper device is an alternative to other earthquake energy dissipation systems, which can reduce the effect of earthquakes or strong wind excitations on structures and improve the dynamic performance of these structures. These devices are usually installed in buildings and structures connecting different floors. In this position they are reducing interstory drift ratios during wind or seismic excitation. These devices act in one horizontal direction and increase the overall damping of the structure. Reduced interstory drifts result in lower internal forces and moments improving comfort, safety and reliability. The elements consist of a steel housing connected to the lower floor/structure, an inner piston connected to the upper floor, and viscous liquid in-between. During seismic and strong wind excitation, the relative displacement between upper and lower structure causes the piston moving through the viscous liquid. Damping forces are induced and yield kinetic energy dissipation. Several dampers can work in parallel, if required.

This contribution provides practical considerations for the layout and use of systems for increase of structural damping, details of executed projects and results of numerical investigations. It documents the high efficiency of the investigated control systems for the protection of structures against impacts due to wind and seismic excitations.

2 Tower in Romania

The TV transmission tower at Magura Odobesti was designed in 1973 and was executed until the end of 1974. This station is composed of a technical building, a reinforced concrete tower, a steel tower and a pylon. The upper part of the tower structure and one of two tuned mass dampers (TMDs) are shown in Fig. 1. The facility is a high-power transmitter located near the top of Magura Odobesti, at an altitude of approximately 1000 m. The length of the structure on top of the technical building amounts to about

111 m, including the concrete part, steel structure and pylon. More details are discussed in [3].

Fig. 1. TV Transmission Tower (left), layout arrangement and view on top of TMD (right).

This structure had to be protected from the effects of wind in particular. During initial calculations and vibration measurements it was found that the first and second eigenmode have to be damped. Thus, 2 TMDs were designed and installed. The first system, consisting of a mass of about 300 kg is arranged at top of the pylon. The second system, weighting approximately 2850 kg, is installed at a height of about 95 m.

The commissioning of the tuned mass system was finished in 2007, including final vibration measurements. The corresponding results show that the reduction effect reaches about 70–80% for the first eigenmode and up to 90% for the second eigenmode.

3 Hospital in Romania

3.1 General

This section exemplifies the seismic upgrade of an important county hospital in Romania. Slobozia is the capital of Ialomita County and it is located approximately 120 km east of Bucharest. Construction started around 1965 and since the 70s the hospital exhibits more than 700 beds. Its structure consists of reinforced concrete members, like columns, beams and slabs. The total length of the structure amounts to about 110 m with a width of 13 m and a height of 31 m. The total mass of the building could be assessed with approximately 18.000 metric tons. The building consisted of several different parts, separated by seismic gaps. These gaps are very small and not persistent as discussed in [4]. The original design did not consider any load due to seismic excitation. In case of earthquakes hammering effects were expected as well as high interstory drift ratios that exceed the allowable limits of the seismic standard. The corresponding damage would have affected structural members and the required repair work would have been tremendous. Due to the importance of the hospital for the health care in this region of Romania the temporarily closure would not have been acceptable.

Thus, a strategy for the seismic upgrade of the hospital was required. The consolidation steps paid special attention to the strict requirement that important rooms (e.g. emergency rooms) of the hospital must be in operation during the works. Several aspects of the strategy and important design and execution details are presented in the following sections.

3.2 Retrofitting Strategy

The upgrading of the structure consists of conventional strengthening measures and the additional arrangement of a Tuned Mass Control System (TMCS). First, the slabs at the roof are coupled to reduce the relative motion between the separated parts during an earthquake. The expected hammering effects between adjacent parts can be avoided now. During the next step the possibility to arrange new additional shear walls inside the structure to reduce the interstory drift ratio was investigated. The available space for these strengthening measures was quite limited. Thus, it was decided to arrange three new steel frame structures outside the building. These frames are connected rigidly to the existing structure at each floor.

The number and size of the new steel frames are not sufficient to achieve the required improvement of the seismic behavior of the structure. The fulfilment of the requirements of the codes (e.g. limitation of interstory drift ratio) required an additional retrofitting measure. The arrangement of the TMCS at the roof of the building was agreed upon. The new steel frames provided the possibility to implement the TMCS directly inside these structures instead of an extensive and costly direct arrangement/connection to the existing roof. The application of a TMCS for an existing structure without limitations in its use is a very important advantage of this mitigation system as described also in [5].

3.3 Results of Analyses

The analytical investigations consider a seismic input with a peak ground acceleration of 0.28 g. The seismic design spectrum shows that the frequency range with the highest acceleration values starts at 1.0 Hz. A spectral amplification factor of 2.75 must be considered up to 10 Hz. A three-dimensional finite element model of the structure was prepared using the commercial software SAP2000. Three different models (Model A: Existing structure without any mitigation measure/Model B: Structure with new steel frames and coupled slabs at the roof/Model C: Combination of Model B with the arrangement of the TMCS) are prepared. Results of field measurements in terms of eigenfrequencies and corresponding mode shapes are used to adjust the finite element models. After applying the conventional strengthening measures the internal forces and horizontal displacements are already reduced. The main horizontal frequencies are increased slightly due to the stiffening of the structure. As a result, smaller relative displacements occurred. The additional new steel frames bear a part of the seismic forces. The application of the conventional strengthening was not sufficient to keep all structural requirements. Thus, the next consolidation step considered the additional arrangement of the TMCS.

By comparing some important structural results, the optimum parameters of the TMCS are defined. The tuning frequency in y-direction (minor axis) was chosen to about 1.8 Hz with an internal damping of about 26%. For the other horizontal direction (x-direction, major axis) the tuning frequency was defined with 2.5 Hz and a corresponding internal damping of 19%. The TMCS was tuned in combination with such a high damping ratio, because a larger reduction of seismic reactions can be achieved, as discussed in [6], if higher internal damping values of the TMCS itself are applied. It was verified in [5] that a large damping coefficient does not lead necessarily to reduction of the efficiency of the tuned mass system. Due to the subcritical tuning for both directions and the corresponding high viscous damping it is ensured that the operating range of the systems is within approximately 60–98% of the measured main frequencies of the structure. All performed analyses verify that this range covers all dynamic events that realistically could occur. More information about the executed ambient vibration measurements can be taken from [7]. The corresponding efficiency of the proposed measures reaches nearly 50%. As a typical example, the bending moment of an existing corner column can be reduced by more than 20% after considering the conventional strengthening. This value could be further reduced to more than 40% if the combined strategy of strengthening and arranging a TMCS is considered.

It can be concluded that the results of the seismic calculations show that the combination of conventional strengthening and application of a TMCS reduces internal forces, absolute accelerations and displacements. As a consequence, the interstory drift values are reduced also. The seismic retrofit will lead to less damage and a remarkable increase in safety during a seismic event.

3.4 Details of the Tuned Mass Control System and Commissioning

The arrangement of three new steel frames required the division of the total mass of the TMCS into three identical parts, each one weighting approximately 53 metric tons. Reinforced concrete slabs are used as mass blocks, installed at each horizontal bay of the steel frame structures. The mass ratio between these blocks and the total building weight amounts to only 0.9%. The mass blocks are hanging on ropes and are elastically connected in both horizontal directions to the steel structure by elements using helical steel springs. Two viscous dampers are installed between the bottom side of each mass block and the adjacent steel structure. Figure 2 shows a corresponding view of the system.

The commissioning was finalized by adjusting the stiffness of the chosen spring elements according to the defined optimum tuning frequencies, based on ambient vibration measurements, described in [7]. The retrofit of the hospital was successfully finished in 2014 and it was presented to the public by the Prime Minister and District Health Minister of Ialomita County.

Fig. 2. View of tuned mass block within steel frame.

4 Hotel in Bulgaria

4.1 General

This chapter focuses on a project executed in Sofia, Bulgaria. The existing building is a two-story hotel with a reinforced concrete frame structure, designed to be stable only under vertical dead load. It was built in the 1960s and is being upgraded with a seven-story steel structure. As a result of the latest Bulgarian seismic standards, not only the new structure but also the existing structure underneath had to be verified for earthquake loads.

4.2 Retrofitting Strategy and Arrangement of Viscous Wall Dampers

The new steel structure could be easily designed for the seismic forces, but for the existing structure the proof according to the seismic standard was not feasible. Thus, it was required to retrofit the structure to keep the requirements. It was decided to reinforce the structure with vertical steel connections, that are built into the reinforced concrete frames, as presented in [8]. Furthermore, the main columns are reinforced by additional reinforced concrete jackets. Due to the corresponding space requirements for these retrofitting measures it was not feasible to arrange enough connections/jackets to keep the limits.

It was evaluated how to reduce the seismic loads generated by the new structure and as a result it was decided to increase the structural damping. In addition to arranging a tuned mass system, it is also possible to increase the damping by installing elements within the structure. Viscous wall dampers are one typical example for such devices. The devices are suitable for new builds but also applicable for retrofitting measures. Based on seismic analyses, decisions were made to complete the retrofit with the arrangement of 21 viscous wall dampers within the new steel structure.

4.3 Results of Analyses

The dynamic and seismic analyses of the building were performed with the commercial SAP2000 software. A three-dimensional finite element model was prepared, using frame elements (to represent columns and beams), shell elements (for walls and slabs) and so-called link elements (to simulate the viscous wall dampers). The analytical investigations consider a seismic input with a peak ground acceleration of 0.3 g. The seismic design spectrum shows that the frequency range with the highest acceleration values starts at 2.0 Hz.

The initial analyses show that the arrangement of the dampers yields a reduction of seismic shear forces in one existing column (as one typical example) of nearly 50%. Based on these results the detailed design of the structure was performed. Finally, it was decided to slightly increase the numbers of dampers to twenty-one devices. In [8] it was concluded that it was possible to achieve two main objectives—to reduce the seismic relative displacements, and hence avoiding hitting the neighboring buildings and to reduce the effect of seismic impact on existing structural members.

4.4 Construction Works

In addition to the conventional strengthening measures within the existing structure, 21 viscous wall dampers are installed within the new building part. Figure 3 shows how the device is connected to the steel structure.

Fig. 3. View of viscous wall damper during construction works.

The construction works were finished in 2016. Since 2018 the luxurious hotel belongs to the IHG group.

5 Conclusions

The most important conclusion that can be drawn from this paper is that modern systems exist, which could protect structures against wind and seismic loading. Several project examples are presented. The determined results show that the application of tuned mass systems or viscous wall dampers are efficient solutions to reduce occurring wind and seismically induced vibrations. The reduction leads to an increase of comfort and to structural safety. As the topics of retrofitting and sustainability become increasingly important, a particularly noteworthy advantage of the systems presented is that they can be used not only for new structures but also for existing ones.

References

1. AFCEN: PTAN 2018 RCC-CW Seismic Dissipative Devices. 1st edn. afcen, Courbevoie (2018)
2. Meinhardt, C.: Increase of a high rise buildings damping behaviour by applying large scale tuned mass dampers. In: 17th Congress of IABSE, pp. 1–8. IABSE, Zurich. Chicago, USA (2008)
3. Dima, M., Popp, T., Dalmer, F., Nawrotzi, P.: Atenuarea vibratiilor folosind sistemul TMD E-54087—Turnul des transmisie TV Magura Odobesit (in Romanian). AICPS Review **4**, 3–7 (2008)
4. Siepe, D., Nawrotzki, P., Alvarado, M.R.I.: Retrofitting of a hospital using a tuned mass control system. In: 16th World Conference on Earthquake Engineering- WCEE, pp. 1–10. Chilean Association on Seismology and Earthquake Engineering (ACHISINA), Santiago de Chile, Chile (2017)
5. Tributsch, A., Adam, C., Furtmueller, T.: Mitigation of earthquake induced vibrations by Tuned Mass Dampers. In: Structural Dynamics - EURODYN, pp. 1742–1749. K.U. Leuven: Department of Civil Engineering, Leuven, Belgium (2011)
6. Meinhardt, C., Siepe, D., Nawrotzki, P.: Passive damping devices for earthquake protection of bridges and buildings. J. Anti-Seismic Syst. Int. Soc. (ASSISi) **2**(1), 35–55 (2011)
7. Nawrotzki, P., Meinhardt, C., Siepe, D.: Seismic protection of a hospital in Romania with a TMCS. In: 15th D-A-CH-Tagung: Erdbebeningenieurwesen und Baudynamik, pp. 203–212. Bauhaus Universität Weimar, Weimar, Germany (2017)
8. Milev, J., Filatova, D., Milev, N., Toteva, A., Georgiev, G., Dyulgerov, P.: Application of Viscous Wall Dampers for the extension of Radisson Hotel in Sofia (in Bulgarian). In: 1st Scientific Applied Conference with International Participation, Reinforced Concrete and Masonry Structures—Theory and Practice, pp. 1–10. University Sofia, Sofia, Bulgaria (2015)

Enhancing Stiffness and/or Damping in Structural Systems with Cellular Shear Walls

Panagiota Syrimi, Spyridoula-Maria Papathanasiou, and Panos Tsopelas(✉)

Laboratory of Testing and Materials, Department of Mechanics, School of Applied Mathematical and Physical Sciences, National Technical University of Athens, 5, Heroes of Polytechniou Avenue, Theocharis Building, Zografou Campus, 15773 Athens, Greece
tsopelas@central.ntua.gr

Abstract. Throughout the last decades, the idea of substituting conventional steel shear wall structural systems with perforated shear wall panels has been explored. Extending this concept, cellular shear wall systems have been proposed. Such systems demonstrate an enhanced dynamic behavior due to their lightweight properties and their ability to avoid out-of-plane buckling. The existing gaps within the body of the cellular shear wall system, allow for viscoelastic infills to be introduced, providing both stiffness and damping into the structural element. The purpose of this paper is to describe the performance of such cellular shear wall systems with viscoelastic fillers under dynamic loading. Considering the steel matrix of the shear wall system, two types of cell architectures are discussed: (a) 8hexagonal honeycomb cells and (b) chiral cells. A harmonic excitation is imposed on full-scaled shear wall panels of any of the previous cell types, with and without viscoelastic infills. The effectiveness of the proposed shear wall system is verified. Comparative results in terms of stress-strain curves provide better understanding of the proposed concept's behavior.

Keywords: Shear walls · Cellular materials · Viscoelastic fillers · Damping behavior

1 Introduction

Shear walls are vertical structural components designed to resist in-plane lateral forces. They are frequently encountered as parts of structural frames in conventional or high-rise buildings. Due to their load resistance mechanism, shear walls are used in both the construction of new structures and the retrofitting of old ones, for wind and seismic protection purposes. Different types of shear walls include wooden, brick or concrete panels, however, steel plate shear walls are the most popular systems combining lightweight properties and geometrical flexibility along with easy manufacturing and efficiency.

Steel plate shear wall systems (SPSWs) were the subject of many different research endeavors throughout the last few decades. Starting from the simple unstiffened thin steel panels to cold-formed steel shear wall panels, to perforated steel shear wall panels, to steel infill panels with various types of stiffeners, to steel corrugated shear walls (SCSWs)

G. P. Cimellaro (Ed.): WCSI 2022, LNCE 309, pp. 903–910, 2023.
https://doi.org/10.1007/978-3-031-21187-4_79

[1–9]. In view of creating a light-weight shear wall system with adequate stiffness and deformation capacity that, simultaneously, does not exhibit undesirable phenomena, like the out-of-plane buckling or the ineffective usage of material, the concept of cellular shear walls (CeSWs) has been introduced, by Chen [10] and Chen and Tsopelas [11]. They presented a numerical study of the mechanical properties of CeSWs. The efficiency of the proposed system for the mitigation of seismic effects was verified with the analysis of a three-story structural frame. Recently, researchers have published experimental results on scaled shear walls with ring-shaped and auxetic-shaped steel panels [12–14].

The idea of placing infills with damping properties into the voids of cellular architectures to enhance the overall structural dynamic response has been explored by various researchers [15–19]. In most of these cases, viscoelastic materials are selected as fillers. Viscoelastic materials are commonly used in vibration absorption devices for the mitigation of the dynamic response of structures (usually high-rise buildings and bridges) under wind or earthquake loading. However, the previously mentioned idea of combining cellular configurations with viscoelastic fillers has not yet been explored.

2 Cellular-Solid Shear Wall Systems

Cellular-solid shear wall systems are advantageous over classic steel shear wall panels due to: (a) being less vulnerable to out of plane bucking, (b) usually weight less, (c) offering stiffness to the structure attached to, while dissipating energy through plastic straining and (d) offer architectural elegance. Studies of their behavior to certain load cases has revealed that the geometry of the basic cell, the so-called "unit cell", greatly affects their mechanical properties. In the ideal case when it is equiaxed, the cell can be considered isotropic. In case the unit cell is slightly elongated or flattened, the properties of it appear to differ significantly between the two perpendicular directions x and y. In reality the mechanical properties of unit cells are neither isotropic nor homogeneous on their microstructure. Only the macroscopic behavior of cellular solids is considered homogeneous, given that the cell size is relatively small compared to the size of the structural element consisting of the cellular solid. Figure 1a depicts a cellular shear wall with overall dimensions L and H. Figure 1b shows the main geometric variables of honeycomb and chiral cells.

2.1 Modeling of Cellular-Solids

Simple beam elements have been successfully used by many researchers for modeling both the linear and nonlinear mechanical properties of cellular materials [20, 21]. In this work, the ABAQUS [22] software is used for the analyses. The element B21 is used for modeling the walls of the cells, while appropriate meshing was chosen so as both bending and shear effects to be accounted.

The shear loading is applied as lateral deformation (u) of increasing amplitude on the top of the shear wall considered (see Fig. 1a). The ratio of the applied deformation at the top of the wall, u, over its height, H, defines the average shear strain ($\gamma_{aver} =$ u/H). Similarly, the reaction force at the bottom of the wall normalized by the cross-sectional area of the shear wall panel is defined as the average shear stress of the cellular solid ($\tau_{aver} = F/d/L$). This normalization of the lateral displacements and reaction forces makes the results scalable under the assumptions of the analysis and aids the development of "spring" like hysteretic models for full scale cellular shear walls.

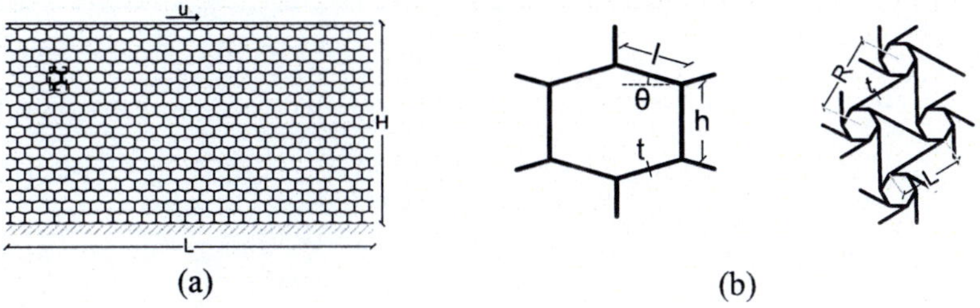

Fig. 1. a) Honeycomb Cellular Shear Wall of dimensions LxH, b) Honeycomb and Chiral unit cells used in this study.

3 Cellular-Solid Shear Wall Systems with Viscoelastic Fillers

3.1 Honeycomb Cellular Geometry

In this section the cellular solid architecture is modified using viscoelastic material to fil the voids, all or portion of them depending on the cell geometries, in order to enhance stiffness and energy dissipation characteristics of a cellular shear wall system. The cellular matrix is made of structural steel, with material properties shown in Table 1. For the viscoelastic material considered, the experimental data for the storage and loss shear moduli as functions of harmonic frequency, are obtained from the work of Shen and Soong [23] and are given in Table 2.

Table 1. Elastic and plastic properties of structural steel

Young's Modulus, E (GPa)	Poisson's Ratio, ν	Yield stress, σ_y (MPa)	Yield strain, ε_y (%)	Ultimate stress, σ_u (MPa)	Ultimate strain, ε_u (%)
200	0.30	275	0.1375	414	25

It should be noted that the analyses of the proposed system in the case of shear walls with cellular matrices of honeycomb architecture has been carried out with the help of representative volume element (RVE) models. Two representative volume elements (RVEs) are considered, each one consisting of seven complete cells. The first RVE model, under the name matrix model, (Fig. 2a) represents the plain steel honeycomb matrix of a cellular wall without any infills. The models are developed in Abaqus [22] environment. The cell walls of the honeycomb are simulated with the B21 beam element, while appropriate meshing was chosen. The second RVE model (Fig. 2b), the filled model, has the same steel honeycomb matrix with each one of the cells filled with the viscoelastic material considered. The viscoelastic filler is simulated through shell elements in full contact with the beam elements of the cell walls. Comparative analysis of these two models shows the viability of the proposed concept and quantifies the contribution of the cell matrix and the viscoelastic filler material.

Table 2. Frequency-dependent properties of viscoelastic material at a temperature of 21 °C [23].

Frequency f (Hz)	Shear storage modulus G' (MPa)	Shear loss modulus G" (MPa)
1.0	1.372	1.786
1.5	1.827	2.248
2.0	2.069	2.724
2.5	2.517	3.192
3.0	2.661	3.358

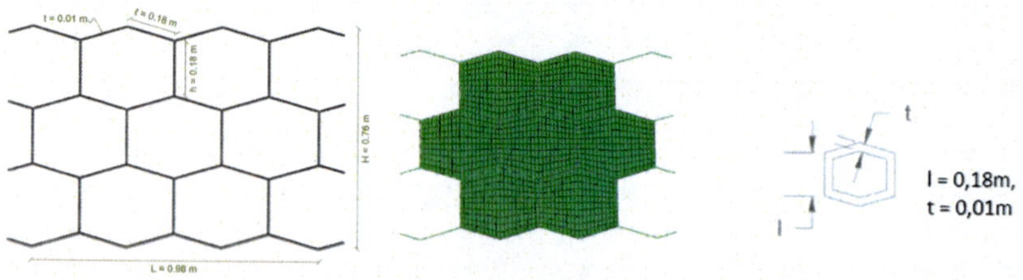

(a) (b) (c)

Fig. 2. Representative volume element (RVE) models: (a) matrix model, (b) filled model. (c) Geometric properties of honeycomb cells

Each one of the aforementioned models is subjected to a harmonic excitation (shear strain amplitude 13%) simulating pure shear cyclic behavior. In Fig. 3a, the dynamic response of the unfilled matrix model under a harmonic excitation of frequency equal to 1 Hz is presented. As expected, due to the bilinear behavior of steel, the results provide a hysteretic stress-strain curve. In Fig. 3b, the dynamic response of the filled model under the same harmonic excitation of frequency equal to 1 Hz is given for three different values of the out-of-plane depth (thickness of the filler) of the viscoelastic infills. It becomes clear that the thicker the viscoelastic infills, the higher the levels of stiffness and damping capacity of the RVE. It is noted that due to the out-of-plane depth of the cellular shear wall the undesirable out-of-plane buckling that steel plate shear wall systems exhibit is avoided.

Subtracting the dynamic response of the matrix model, representing the contribution of the plain steel cellular matrix, from the overall response of the filled model, the behavior of the viscoelastic material and its contribution to the total performance of the system is obtained. The stress-strain curves, describing the dynamic features of the viscoelastic filler, for an out-of-plane depth of 3cm and for three different harmonic excitation frequencies of 1, 2 and 3 Hz, are shown in Fig. 3c.

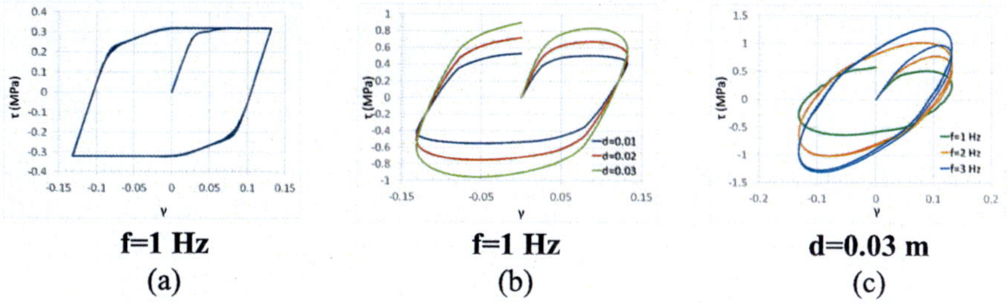

Fig. 3. (a) Stress-strain curve of the unfilled matrix model. (b) Dynamic response of the filled model for three different values of the out-of-plane depth of the viscoelastic infills. All models are subjected to a harmonic excitation of frequency equal to 1Hz. (c) Behavior of the viscoelastic filler under different frequencies.

3.2 Chiral Cellular Geometry

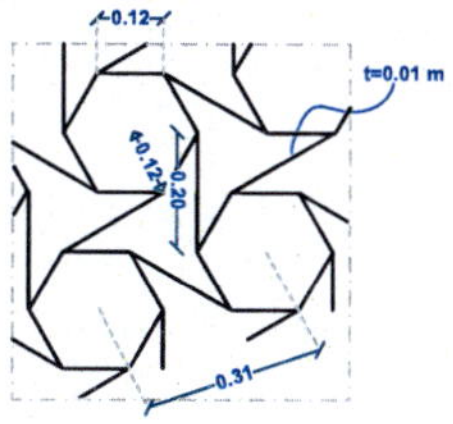

Fig. 4. Size and geometry of the chiral cells.

The contribution of the viscoelastic fillers to the overall performance of the combined cellular-viscoelastic shear wall is also discussed in the case where the cellular matrix is of chiral geometry. The geometry of the chiral cells utilized is presented in Fig. 4. The viscoelastic fillers are introduced only inside the central portion (it has the shape of a honeycomb) of each chiral cell. In order to define the material behavior of the combined steel chiral matrix with viscoelastic fillers, two shear wall models are considered (full shear wall models are considered for the chiral architecture as opposed to the RVE considered in the previous section). The first, matrix, model, (Fig. 5a) represents the plain steel chiral matrix without any infills. The cell walls of the chiral are modeled with beam elements. The second, infilled, model (Fig. 5b) has the same steel chiral matrix with the honey-combs inside the chiral cells (center of chiral) filled with the viscoelastic material. Again, the viscoelastic infill is modeled in Abaqus using shell elements.

Each one of the aforementioned models is subjected to a harmonic excitation simulating pure shear. In Fig. 6, the dynamic response of the unfilled matrix model and of the filled model, under a harmonic excitation of frequency equal to 2 Hz, is presented.

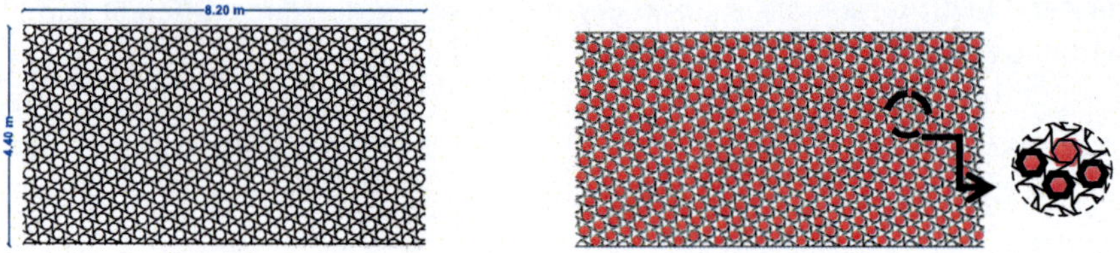

(a) (b)

Fig. 5. Thickness = 0.01 m and Depth = 0.01 m. Shear wall models: (a) matrix model, (c) filled model.

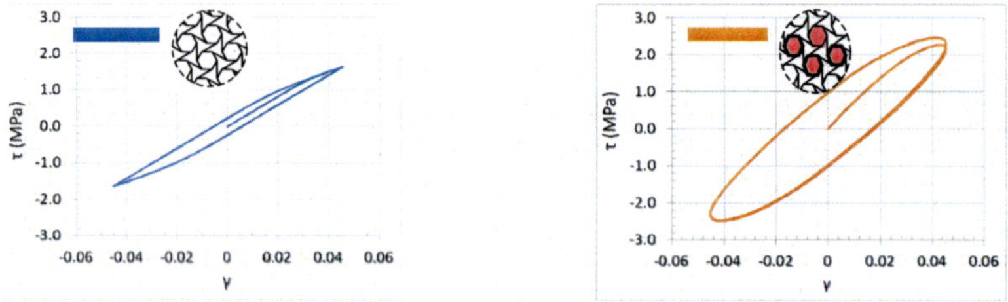

(a) (b)

Fig. 6. (a) Stress-strain curve of the unfilled matrix model. (b) Stress-strain curve of the filled model. All models are subjected to a harmonic excitation of frequency equal to 2Hz.

4 Conclusions

In this paper two types of cellular solid architecture are considered: honeycomb and chiral. After examining the contribution of the viscoelastic inclusions and the overall behavior of the combined cellular steel matrix—viscoelastic infills it is apparent that exhibits an improved dynamic response, in both terms of stiffness and damping. The results obtained from this study lead to the following conclusions: The proposed cellular shear wall with viscoelastic fillers can be a viable "response modification system" for seismic applications providing both stiffness and energy dissipation. Such a system can provide design flexibility, since hysteretic stiffness and damping of the cellular steel matrix are independent from the stiffness and damping of the viscoelastic filler. Thus, the designer can define the proportion between those features based on the requirements of the application, through the proper choice of geometric and material properties. The proposed system is scalable. Both the cell and the overall shear wall dimensions can be scaled to fulfill the needs of each application. Finally, the considered shear wall system prevents out-of-plane buckling under dynamic loading due to the significant out-of-plane depth of the cellular wall with respect to an equivalent steel shear wall panel.

References

1. Thorburn, L.J.: (Lorna J 1958, Montgomery CJ, Kulak GL. Analysis of steel plate shear walls (1983). https://doi.org/10.7939/R3BG2HB64
2. Timler, P.A., Kulak, G.L.: Experimental study of steel plate shear walls (1983)
3. Elgaaly, M.: Thin steel plate shear walls behavior and analysis **32** (1998)
4. Robert Driver, B.G., Kulak, G.L., Laurie Kennedy, D.J., Elwi, A.E.: Cyclic Test of Four-Story Steel Plate Shear Wall
5. Berman, J.W., Bruneau, M.: Experimental Investigation of Light-Gauge Steel Plate Shear Walls for the Seismic Retrofit of Buildings (2003)
6. Berman, J., Bruneau, M.: Experimental investigation of light-gauge steel plate shear walls. J. Struct. Eng. **131**(2), 259–267 (2005). https://doi.org/10.1061/(ASCE)0733-9445(2005)131:2(259)
7. Vian, D., Bruneau, M.: Steel Plate Shear Walls for Seismic Design and Retrofit of Building Structures (2005)
8. Jalilzadeh Afshari, M., Asghari, A., Gholhaki, M.: Shear strength and stiffness enhancement of cross-stiffened steel plate shear walls. Int. J. Adv. Struct. Eng. **11**(2), 179–193 (2019). https://doi.org/10.1007/s40091-019-0224-6
9. Xu, Z., Tong, G., Zhang, L.: Stiffness Demand on Stiffeners of Vertically Stiffened Steel Plate Shear Wall (2020). https://doi.org/10.1061/(ASCE)ST.1943
10. Chen, L.: Seismic Performance of Shear Walls Utilizing Cellular Material. The Catholic University of America, Washington, D.C. (2009)
11. Chen, L., Tsopelas, P.: Cellular-solid shear walls under seismic excitations. In *Proceedings of the 9th U.S. National and 10th Canadian Conference on Earthquake Engineering*, Toronto (2010)
12. Egorova, N., Eatherton, M.R., Maurya, A.: Experimental study of ring-shaped steel plate shear walls. J. Constr. Steel Res. **103**, 179–189 (2014). https://doi.org/10.1016/j.jcsr.2014.09.002
13. Monsef Ahmadi, H., Sheidaii, M.R., Boudaghi, H., de Matteis, G.: Experimental and numerical study on largely perforated steel shear plates with rectangular tube–shaped links. Adv. Struct. Eng. **23**(15), 3307–3322 (2020). https://doi.org/10.1177/1369433220937147
14. Hamed, A.A., Asl, R.B., Rahimzadeh, H.: Experimental and numerical study on the structural performance of auxetic-shaped, ring-shaped and unstiffened steel plate shear walls. J. Build. Eng. **34** (2021). https://doi.org/10.1016/j.jobe.2020.101939
15. Murray, G., Gandhi, F., Hayden, E.: Polymer-filled honeycombs to achieve a structural material with appreciable damping. J. Intell. Mater. Syst. Struct. **23**(6), 703–718 (2012). https://doi.org/10.1177/1045389X12439636
16. Boucher, M.A., Smith, C.W., Scarpa, F., Rajasekaran, R., Evans, K.E.: Effective topologies for vibration damping inserts in honeycomb structures. Compos. Struct. **106**, 1–14 (2013). https://doi.org/10.1016/j.compstruct.2013.05.036
17. Aumjaud, P., Smith, C.W., Evans, K.E.: A novel viscoelastic damping treatment for honeycomb sandwich structures. Compos. Struct. **119**, 322–332 (2015). https://doi.org/10.1016/j.compstruct.2014.09.005
18. Ongaro, F., de Falco, P., Barbieri, E., Pugno, N.M.: Mechanics of filled cellular materials. Mech. Mater. **97**, 26–47 (2016). https://doi.org/10.1016/j.mechmat.2016.01.013
19. Wang, Y.C., Lai, H.W., Ren, X.J.: Enhanced auxetic and viscoelastic properties of filled reentrant honeycomb. Physica Status Solidi (B) Basic Res. **257**(10) (2020). https://doi.org/10.1002/pssb.201900184
20. Papka S, Kyriakides S. In-plane compressive response and crushing of honeycomb. J. Mech. Phys. Solids **42**(10) (1994). https://doi.org/10.1016/0022-5096(94)90085-X

21. Overaker, D., Cuitino, A., Langrana, N.A.: Effects of morphology and orientation on the behavior of two-dimensional hexagonal foams and application in a re-entrant foam anchor model. Mech. Mater. **29**(1) (1998). https://doi.org/10.1016/S0167-6636(98)00004-0
22. ABAQUS/Standard User's Manual, Version 6.14
23. Shen, K., Soong, T.: Modeling of viscoelastic dampers for structural applications. J. Eng. Mech. **121**(6), 694 (1995). https://doi.org/10.1061/(ASCE)0733-9399

Development of New Optimal Passive Non-detuning Mass Dampers

Giampierobruno Grillo[1], Luca Martinelli[1](✉), and Christian Bucher[2]

[1] Department of Civil and Environmental Engineering, Politecnico di Milano, Milan, Italy
luca.martinelli@polimi.it

[2] Department of Civil Engineering, Vienna University of Technology, Vienna, Austria

Abstract. In this paper, derived from a M.Sc. thesis jointly developed at Politecnico di Milano and TU Wien, an innovative design strategy for the well known Tuned Mass Damper (TMD) passive structural control device is presented. Linear TMDs need to stay properly tuned to the principal structure in order to be effective. The newly proposed "NextGenTMD" incarnation aims at overcoming the detuning by implementing a hysteretic version of the TMD, able to remain tuned despite the hysteretic response of the primary structure itself. The usage of Genetic Algorithms introduced in the design process of the NextGenTMD is at the basis of the proposed approach. Within this context, the design phase of the TMD is presented as a multi-objective minimization problem. The proposed design methodology is applied on a reduced numerical model of a four story building, of which full scale seismic pseudo-dynamic test results were available. The calibration of this reduced hysteretic model is carried out as well as a minimization problem exploiting Genetic Algorithms. The capabilities of the NextGenTMD, designed according to the proposed procedure, is assessed with respect to a set of spectrum compatible time histories, according to the Eurocode 8. The very promising results obtained, from numerical simulations, show that the NextGenTMD can be effective in a seismic setting while the simplicity of the installation is promising for its large scale diffusion and, consequently, for a more effective reduction of the seismic vulnerability of the built environment.

Keywords: Seismic design · Passive mass dampers · Nonlinear dynamics · Structural optimization · Genetic algorithms

1 Introduction

In the past two decades, earthquakes were among the deadliest natural disasters, resulting in over 700,000 casualties worldwide and causing several billions of USD damages. Despite much has been done to increase the seismic performance of buildings, many strategies can still be implemented, especially for low rise buildings which are among the most diffused in the European seismic prone areas.

In this paper, which is derived from a M.Sc. thesis work jointly developed at Politecnico di Milano and TU Wien, some new design strategies for a new incarnation of

G. P. Cimellaro (Ed.): WCSI 2022, LNCE 309, pp. 911–920, 2023.
https://doi.org/10.1007/978-3-031-21187-4_80

the well-known Tuned Mass Dampers (TMD) passive structural control devices are presented.

The proposed "NextGen-TMD" is a hysteretic-behavior version of the linear tuned mass damper (L-TMD), a device that, if properly calibrated on the natural frequency of the main building, is able to oscillate with it, adsorbing part of the seismic energy that could otherwise cause structural damage.

As it is well known, L-TMDs need to stay properly tuned to the principal structure in order to be effective. This is problematic for normal buildings, characterized by a nonlinear dynamic response during severe earthquakes.

The newly proposed incarnation aims at overcoming the loss of tuning, and effectiveness by implementing a properly designed hysteretic version of the TMD, broadening the use of TMDs beyond classical applications to high-rise buildings, which behave in a relatively flexible and elastic way under seismic loading.

The use of evolutionary optimization algorithms (genetic algorithms) is at the basis of the proposed design method of the TMD, which, in this paper, is applied to a calibrated reduced model of low-rise representative building. Within this context, also the calibration of the hysteretic reduced building model is presented as a multi-objective minimization problem.

The very promising results, obtained so far from numerical simulations, show that TMDs can be effective in a seismic setting, differently from what has been deemed as possible in some literature sources, and encourage for further extensions of the research efforts towards physical tests. Indeed, if effective, the simplicity of the installation of such an instrument could allow for a large-scale diffusion and a wider reduction of the seismic vulnerability of the built environment.

2 Reduced Model of the Representative Building with the TMD

The "primary" structure, on which the design method of the passive TMD control system presented in this work is focused, is a four-story reinforced concrete building designed according to [1]. The building is a dual structure, with an elevation of 12.5 m, comprising coupled walls and a frame part. The seismic performance of the building was assessed through a pseudo-dynamic test campaign carried out at the ELSA Laboratory of the Joint Research Centre in Ispra (see [2]).

Since the purpose of this work is to introduce a new design possibility in the field of a complex dynamic problem, and to explore its general effectiveness, given the large number of variables involved, some simplifying modelling choices have been made: the combined system composed by the "primary" structure (the one to be controlled) and the damper is modelled as a two degrees of freedom nonlinear system (Fig. 1), in which the mass m_1, representing the primary structure, and the mass m_2, representing the dynamic absorber, are attached to nonlinear springs k_1 and k_2 that follow hysteretic rules.

The nonlinear spring k_1 is calibrated, thanks to a genetic algorithm's optimization procedure, starting from the above mentioned pseudo-dynamic full-scale test of the structure. The nonlinear spring k_2 follows the Bouc-Wen hysteretic law [3].

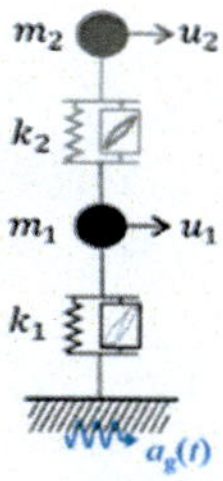

Fig. 1. Reduced model of the building with NextGenTMD

The equations of motion under horizontal seismic input can be written as:

$$\begin{cases} m_1\ddot{u}_1 + c\dot{u}_1 + f_s(u_1) - f_d(u) = -m_s a_g(t) \\ m_2\ddot{u}_2 + f_d(u) = -m_2 a_g(t) \\ u = u_2 - u_1 \end{cases} \quad (1)$$

where u_1 and u_2 are relative displacements to the base of the structure, a_g is the ground acceleration, f_s and f_d are the non-linear restoring forces exerted by the non-linear springs k_1 and k_2, respectively.

The hysteretic model chosen to represent the primary structure behavior is the Takeda model [4], originally developed, basing on experimental observations, to reproduce the nonlinear behavior of Reinforced Concrete (RC) structures under earthquake loading. In this work, the simplest original version with large loops, (see Fig. 2), have been implemented in the Matlab environment. The parameters of the Takeda model (positive and negative yielding strength "$F_y{}^+$" and "$F_y{}^-$", amount of previous yielding that will be maintained upon reloading "β", unloading stiffness parameter "α", post-yielding stiffness parameter "r") used to represent the structure were retrieved via a genetic algorithms-based procedure which relied on the minimization of the error between the measured response of the real structure and the calculated response of the numerical model for the experimental data-set available from [1]. Further details can be found in [6].

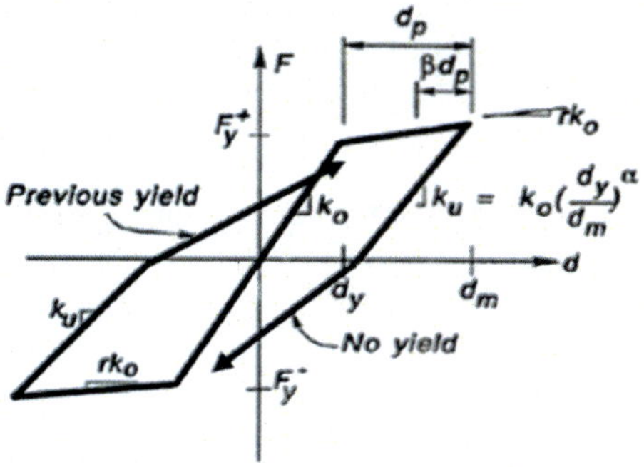

Fig. 2. Takeda hysteretic model (redrawn from [5])

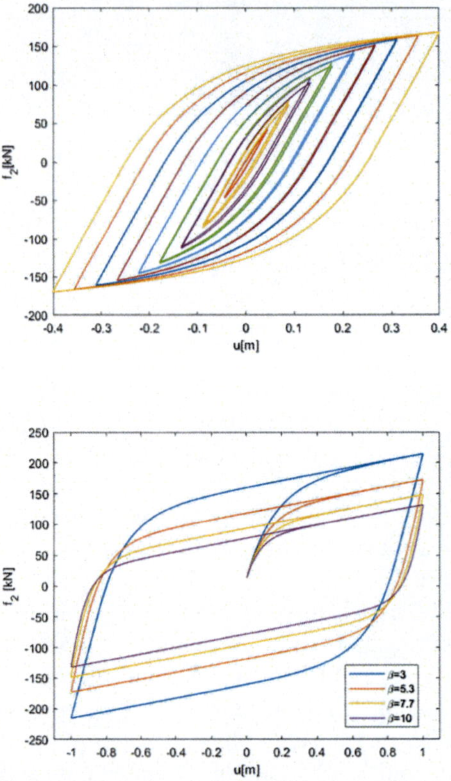

Fig. 3. Features of the Bouc Wen model, hysteresis cycles for n = 1; $\delta_d = 0{:}05$ and: (above) different input peak displacement, (below) variations of $\beta_d = \gamma_d$

Concerning the hysteretic behavior of the mass damper, the modelling assumption of this work is that its non-linear spring will follow the Bouc-Wen hysteresis model. The Bouc-Wen model was selected for its good performance concerning the documented ability to resemble the hysteretic behavior of structural elements (e.g. rubber bearings) [7], and the practical feasibility of a Bouc-Wen hysteretic Mass Damper to control linear viscous-elastic structures as described in [3] and [8].

The Takeda type hysteric cycle is defined in Fig. 2 while the Bouc-Wen model is defined by the set of equations in Eq. (2). A sample of the role of its parameters is shown in Fig. 3, while a full characterization of the model in the context of structural dynamics can be found in [3, 5], and [8].

$$\begin{cases} F_d(u) = k_{e_l}u + z \\ \dot{z} = \left\{k_d - \left[\gamma_d + \beta_d sgn(z\dot{u})\right]|z|^n\right\}\dot{u} \\ k_e l = k_t * \delta_d \\ k_d = k_t - k_{e_l} = k_t(1 - \delta_d) \\ k_d > 0;\ \beta_d > 0;\ \gamma_d > 0; \\ \gamma_d + \beta_d > 0;\ \gamma_d - \beta_d \geq 0 \end{cases} \tag{2}$$

In Eq. (2) u is the displacement across the spring, z is an internal state variable while β_{d}, γ_{d}, δ_d, and k_t, are model parameters.

3 Design Strategy of the TMD

In the design problem here introduced a rather large mass ratio of the TMD in terms of mass of the primary structure is adopted and, differently from most applications, the primary structure is not linear viscous, and the input motion is not as repeatable as customarily assumed. Considering these characteristics, the "non-detuning" damper here presented is also referenced as the NextGen-TMD.

The design strategy presented in this work is based on the research of the set of parameters of the mass damper that will allow to optimize the non-linear earthquake response of the primary structure, possibly also satisfying a limit on the relative displacement $u = u_2 - u_1$ of the TMD mass with respect to the building.

The set of parameters that composes the design space selected for this work consists of the initial elastic tuning frequency, i.e., the ratio between the initial frequency of vibration of the device to that of the primary structure:

$$f = f_{0,d}/f_{0,s} \tag{3}$$

that is a common parameter in the design of elastic mass dampers ([9]), also, the Bouc-Wen model parameter β_d and its hardening coefficient δ_d complete the set of design parameters. If we assumed as the objective function the displacement at the top of the primary structure with respect to the ground, u_1, this turns out to be a highly irregular function in the space of the chosen design parameters (see Fig. 4). Consequently, its differentiability cannot be granted, and a non-gradient based optimization method is to be used.

It will be shown that the TMD design problem can be properly solved by employing soft computing optimization algorithms, such as the Genetic Algorithms (GA), for the minimization of a target response quantity of the structure, generally indicated as "Q".

Whatever the chosen quantity "Q", it is common to define a performance index as:

$$i_Q = 1 - Q_C/Q_{NC} \tag{4}$$

where Q_C and Q_{NC} are the controlled and the non-controlled quantity of interest of the structural response, and the overall index i_Q is an indicator of the effectiveness of the reduction of the structural response. This definition associates to a reduction of the quantity Q a positive value of the performance index. In Fig. 4 a plot of this function, in which the displacement at the top of the structure, u_1, is selected as quantity of interest "Q" is provided. In this case, the performance index is just named as "$i[\%]$".

It will be shown that, despite the number of unknowns and their (difficult to foresee) effect on the nonlinear structural response, genetic algorithms allow to cast the design procedure into a multi objective, constrained, optimization problem, which considers the nonlinear response of structures and the stochastic nature of the input earthquake.

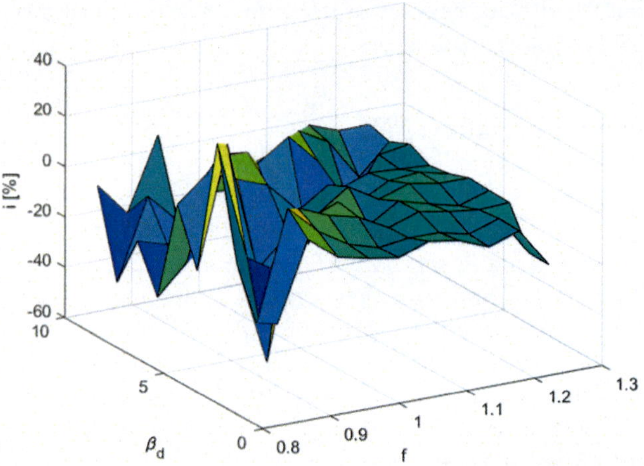

Fig. 4. Effect of the selected design parameters ($\beta_d = \gamma_d, f$) on the displacement performance index "*i*" for a fixed value of $\delta_d = 0.05$, , referring to the Bouc-Wen model defined as in (2) and to the tuning frequency f defined in (3)

4 Genetic Algorithms and Optimization

The Bouc-Wen model is defined by Eq. (2). The number of parameters, and the nonlinearity which characterizes the problem, make it unfeasible to search for reliable solutions in closed form. In addition, it would be a long work to try and optimize such a system with standard gradient-based numerical methods.

Genetic algorithms (GA) ([7, 10, 11]) start from a Random Monte Carlo generation of an initial population of design variables that in the terminology afore mentioned represent the "chromosomes" (see Fig. 4). In our case, the "chromosomes" would either be the parameters of the Takeda model (in the problem of the structural parameters retrieval) or of the tuned mass damper (and of its Bouc-Wen spring) and the task of the minimization is to identify the set of parameters that best minimizes a selected objective function, which, in the case of the experimental data fitting is represented by a suitable error function. For the problem of the design of the mass damper, which is the object of this paper, the behavior of the structure is already considered as known (see [6]) and the objective function is represented by a suitably chosen quantity of interest of the structural response or of a norm of a set of quantities of interest: the flexibility regarding the choice and the regularity of the objective function (see Fig. 4) is, indeed, one of the strong points of the algorithm.

In the design problem, the parameters for which the optimization is performed are those related to the initial tuning of the TMD: the already described frequency ratio between the absorber frequency, which is controlled by the Bouc-Wen model parameters β_d, δ_d and the initial tuning frequency, f.

Starting from such an initial population, a suitable subset is selected as candidate "parents", based on their "fitness", which is related to the "score", i.e., to the value of the target function computed in correspondence of each gene. Penalties on such a score eventually arise if the constraints are violated. Through selection, the genes that best

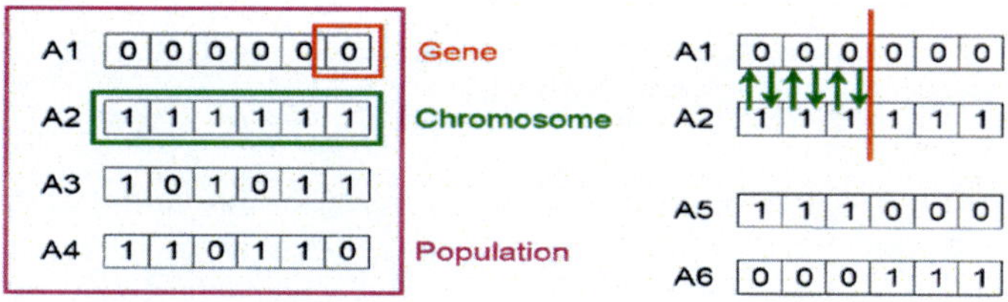

Fig. 5. Genetic algorithms' elements breakdown. (L) population; (R) crossover process and new offspring (from "https://towardsdatascience.com/introduction-to-genetic-algorithms-including-example-code-e396e98d8bf3")

perform on each generation will "survive" from one generation to the next and they are also favored for reproduction ("elitism"). During reproduction genes are recombined through a cross-over operation so to generate offspring (see Figs. 5 and 6). A random "crossover point" is often chosen along the string representing each of the parents and the new generation occurs by exchanging the genes of parents among themselves until the crossover point is reached.

It is also possible to obtain the "child" by using only one starting point through mutation, that allows the genetic algorithms to avoid local minima from being exchanged for global ones.

The so obtained individuals having the best scores are again kept as "parents" for the next reproduction cycle. The population has a fixed size, therefore, when new generations are formed, individuals with least fitness are eliminated, providing space (in the memory) for new offspring.

This sequence of operations is repeated to produce individuals in each new generation which have a better fitness than the previous one, thus allowing the initial population to "evolve" to the result.

In the application here presented, the genetic algorithms main features, as the maximum number of iterations, the generation and mutation criteria and the population size were kept equal to the suggested default optimal ones in Matlab [12].

```
START
Generate the initial population
Compute fitness
REPEAT
    Selection
    Crossover
    Mutation
    Compute fitness
UNTIL population has converged
STOP
```

Fig. 6. Example of a genetic algorithms' "pseudocode"

5 Design Objective Functions Set-Up and Application to the Prototype Building Model

In this work, different design solutions for the problem of the control of the structural response to spectrum compatible input accelerograms, were obtained by setting up the design problem as an optimization one with a single objective function.

Based on the definition of the input time history, three design cases were defined and are here named GA-1 (spectrum-compatible simulated time histories), GA-2 (sine-sweep), GA-3 (white noise).

Indeed, for a hysteretic problem as the one of this work, a single objective function is especially dependent on the input time history used. Consequently, as also prescribed by the EC8, the GA-1 design case was obtained as the set of parameters that optimized, on average, the structural displacement resulting from a set of 8 simulated time histories, generated through the software SIMQKE_GR [13] to be compatible with the design spectrum of the real structure described in [1]. A similar set of simulated design time histories was used for the assessment of the results. The design solutions GA-2 and GA-3 were, instead, obtained by using "analytical" functions that, in analogy with the simulated accelerograms, were calibrated to have a Fourier Transform able to approximate the one of the full scale test design accelerogram (D05) in an effort to avoid the dependence of the method from the specific generated time histories and to reduce the computational time by employing only one input function in the optimization process. Results concerning the main quantities of interest for the structural motion are listed in Table 1.

The best results were obtained with the GA-1 design case that was able to reduce, on average, the structural displacement of $\delta u_1 = 0.039$ m in the positive direction and $\delta u_1 = 0.0355$ m in the negative one, as shown in Table 2. Table 3 reports the response parameters of the damper, while Fig. 7 compares the response of the structure with and without the damper, assessing its effectiveness.

Table 1. TMD parameters as from the different designs

Par	Simulated t.h. (GA-1)	Chirp input (GA-2)	White noise input (GA-3)	Simulated t.h. (GA-4)
δ_d	0.05	0.05	0.10	0.048
f	0.99	0.86	0.86	0.99
β_d	3.51	4.35	4.37	3.54

Table 2. Time histories average results for the controlled structure

Case	$u_{1,max}$	$u_{1,min}$	$u_{1,RMS}$	$A_{1,max}$	$A_{1,min}$	$A_{1,RMS}$
GA-1	0.0751	−0.0834	0.0259	10.5611	−10.7826	2.4892
GA-2	0.1043	−0.1132	0.0344	14.2373	−13.9117	2.6439
GA-3	0.0988	−0.1179	0.0322	13.7877	12.3734	2.4561

Table 3. Average of the peak response of the absorber to the set of "assessment" time-histories

Case	u_{max}	u_{min}	A_{max}	A_{min}
GA-1	0.3524	−0.3651	47.4294	−45.8586
GA-2	0.7293	−0.7798	64.1818	−61.5851
GA-3	0.1637	−0.1719	29.5949	−28.7478

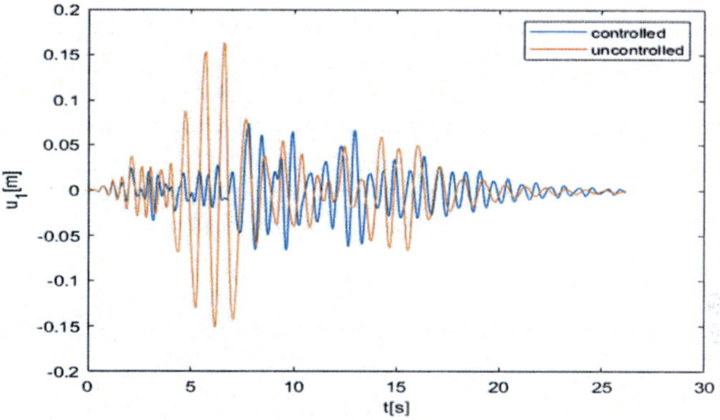

Fig. 7. Displacement comparison between the GA-1 controlled and non-controlled case. The input is the same used in testing the physical structure

6 Conclusions

In this work, which summarizes part of the MSc thesis of the first author conducted under the supervision of the last two authors, the mass dampers class of the passive mass dampers is revisited and cast in a new design strategy that leads to enhanced control performances, to the point that a new name seems justified to identify the final results of the design process: namely the "NextGenTMD" mass damper.

The hysteretic-behaviour on which the NextGentTMD is based on is tailored to the structure on which it will operate on the base of an optimization procedure carried out in the framework of the Genetic Algorithm class as to avoid both the difficulties related to defining gradients when the structural response is non-linear, and to be decoyed by local minima of the parameters' space.

The main structure is in this work represented as a single degree of freedom oscillator on the base of the Takeda model. The Takeda model parameters have been identified as well in the framework of the Genetic Algorithm class by matching the available seismic experimental response of a dual frame-wall multistory reinforced concrete building tested under its design earthquake.

Differently from what has been deemed as possible in some literature sources, excellent results were obtained from the NextGen damper. The simplicity of the installation that such an instrument could allow are promising for its large-scale diffusion and, consequently, for a more effective reduction of the seismic vulnerability of the built environment.

References

1. Panagiotakos, T.B., Fardis, M.N.: A displacement-based seismic design procedure for RC buildings and comparison with EC8. Earthq. Eng. Struct. Des. **30**(10), 1439–1462 (2001)
2. Colombo, A., Fardis, M.N., Molina, J., Negro, P., Panagiotakos, T.B., Tsionis, G.: Pseudo-dynamic response of dual RC structure with different designs. In: 12th European Conference on Earthquake Engineering (2002)
3. Vestroni, F., Casini, P.: Mitigation of structural vibrations by hysteretic oscillators in internal resonance. Nonlinear Dyn. **99**(1), 505–518 (2019). https://doi.org/10.1007/s11071-019-05129-9
4. Takeda, T., Sozen, M.A., Nielsen, N.N.: Reinforced Concrete Response to simulated Earthquakes, Ohbayashi-Gumi Technical Research Report (1971)
5. Carr, A.J.: Ruaumoko 2D, Inelastic Dynamic Analysis. University of Canterbury, Christchurch, Department of Civil Engineering (2004)
6. Grillo, G.: Development and Validation of New Design Strategies for Optimal Passive Non-detuning Mass Dampers, M.Sc. Thesis, Politecnico di Milano (2021)
7. Bucher, C.: Structural Optimization, Lecture Notes, TU-Wien (2020)
8. Morelli, L.: Studio di un Assorbitore Isteretico per la Mitigazione delle Vibrazioni Strutturali, Ph.D. Thesis, Univ. di Roma "La Sapienza" (2019)
9. Den Hartog, J.P.: Mechanical Vibrations, 2nd edn. McGRAW-HILL, Cambridge, MA (1940)
10. https://towardsdatascience.com/introduction-to-genetic-algorithms-including-example-code-e396e98d8bf3, last accessed on 17/11/2021
11. Katoch, S., Chauhan, S.S., Kumar, V.: A review on genetic algorithm: past, present, and future. Multim. Tools Appl. **80**(5), 8091–8126 (2020). https://doi.org/10.1007/s11042-020-10139-6
12. MATLAB: https://www.mathworks.com/help/gads/options-in-genetic-algorithm.html, last accessed on 17/11/2021. The MathWorks, Inc., Natick. Massachusetts, United States
13. Gelfi, P.: https://gelfi.unibs.it/software/simqke/simqke_gr.htm. Last accessed on 17/11/2021

Challenges and Detailing Considerations for the Incorporation of Passive Energy Dissipation Systems in Life Science Occupancies

Sam Richardson(✉), Kurt Lindorfer, and Bryan Lee

PARADIGM Structural Engineers, San Francisco, CA 94111, USA
srichardson@paradigmse.com

Abstract. There is an increased desire for higher levels of building performance during and after seismic events for life science and biotechnology facilities due to the potential for loss of valuable research data, product inventory, or manufacturing capability of critical and high-value products. Methods for increasing building performance include the incorporation of passive energy dissipation systems in the form of viscous damping devices or a seismic base isolation system. Several challenges exist with the structural design of laboratory facilities, including a desire to maintain long spans free of columns for better space planning and tall stories to accommodate demanding ventilation and air filtration requirements associated with clean room spaces. The authors describe these challenges and a set of novel design and detailing solutions implemented to facilitate efficient incorporation of a damped moment frame lateral system for a production laboratory facility in the San Francisco Bay Area in Northern California. The improved structural performance supported the client's campus-wide goals to minimize downtime after a severe seismic event.

Keywords: Passive energy dissipation · Fluid viscous dampers · Life sciences

1 Introduction

Since the discovery of recombinant DNA in 1973 and the founding of Genentech in the late 70s, biotechnology and the life sciences industries have developed into a burgeoning field in the global economy [1]. The field continues to grow with Coldwell Banker Richard Ellis (CBRE) reporting an increase in office stock conversion to laboratories at a rate of 266% from Q1 2021 to Q4 2021 in the San Francisco Bay area [2]. This key fixture of the modern economy has created a significant market for the construction and structural engineering industries. Due to the large capital investments and oftentimes low facility redundancy, there is an increased desire for higher levels of building performance during and after seismic events in life science and biotechnology facilities due to potential for loss of valuable research data, product inventory, or manufacturing capability of critical and high-value products. This paper presents the case study of a life science project in the San Francisco Bay Area that utilizes passive energy dissipation systems to meet these increased structural performance needs.

G. P. Cimellaro (Ed.): WCSI 2022, LNCE 309, pp. 921–928, 2023.
https://doi.org/10.1007/978-3-031-21187-4_81

2 Structural Design Requirements for Lab Occupancy

Life science manufacturing and research processes often have strict requirements on cleanliness and air filtration beyond that of a typical office building. These ventilation and air filtration systems along with other laboratory processes necessitate a high density of overhead distribution systems to be supported by the structure that impact the functionality of the facility. This overhead infrastructure necessitates large floor to floor heights to accommodate all systems as well as provide access for maintenance. This can pose an issue for the seismic design of drift-controlled structural systems such as moment frames which are negatively impacted by large story heights or beam spans. This can also be a challenge for strength-controlled systems due to the need to accommodate large floor to floor differential displacements or drifts, that while still within code limits, can be large in magnitude for many non-structural systems.

Lab space planning is a challenging architectural problem that requires interdisciplinary coordination between conventional design disciplines, lab process engineers, and end-user needs. A not uncommon consequence is development of buildings to accommodate specific tenant needs that can lead to long span, inefficient, or irregular bay layouts. On the subject project this coordination process led to an interior layout that needed to be free of bracing elements and possessed long spans that would pose significant efficiency issues with a moment frame lateral system.

3 Description of Building and Performance Objectives

The subject property is a three-story, 95,000 ft^2 laboratory facility for a private life sciences client in the San Francisco Bay Area. The space program consists of a floor of laboratory occupancy above and below a mechanical support floor. The lateral force resisting system consists of prequalified Post-Northridge WUF-W special moment frames with supplemental damping using non-linear fluid viscous dampers located within select moment frames. Passive energy dissipation was incorporated into the structural design to allow the use of a flexible lateral system that could remain nearly elastic during a design level seismic event to be consistent with the client's campus wide goals of minimal downtime after a seismic event. The dampers were rated for 525 kips (2335 kN) and designed to accommodate ± 8 in. (203 mm) of stroke with a total damper quantity of 60. Damping devices were distributed throughout the height of the building, with 32 dampers on the first floor, 20 dampers on the second floor and 8 dampers on the third floor.

The building was designed to meet ASCE 7-16 [3] Chap. 17 using the non-linear time history analysis (NLTHA) procedure. The NLTHA utilized 11 ground motions scaled using amplitude scaling to a risk targeted maximum rotation spectrum developed in accordance with ASCE 7-16 [3] Ch 21. In addition to the code verification of the building, the structure was analyzed by NLTHA to confirm that the structure met the project performance objectives of immediate occupancy (IO) at the Design Basis Earthquake (DBE) and life safety (LS) at the Maximum Considered (MCEr) event.

The analysis model included concentrated plasticity hinges for moment frame beams and column bases and a non-linear spring panel zone model. Performance acceptance criteria for deformation-controlled components can be seen in Table 1. The acceptance criteria for immediate occupancy for the WUF-W moment frame hinges was taken as the observed deformation before onset of localized buckling from research utilized as part of the connection prequalification [4]. This plastic rotation was found to be more conservative than the allowable rotation listed in ASCE 41-17 [5] for the immediate occupancy limit state. The life safety limit state for the beam hinges was based on ASCE 41-17 [5].

Table 1. Acceptance criteria for different performance metrics.

Performance metric	Acceptance criteria DBE (IO)	Acceptance criteria MCEr (LS)
Drift	1.5% of story height	2.5% of story height
WUF-W hinge plastic rotations	0.015 radians	0.041 radians
Panel zone plastic rotation	ASCE 41-17 IO limit	ASCE 41-17 LS limit
Column hinge plastic rotation	ASCE 41-17 IO limit	ASCE 41-17 LS limit

4 Description of Damped Frame Configuration

As described in previous sections, the interior column bays and story heights were heavily driven by the planned laboratory spaces of the building. This led to long girder and beam spans as well as irregular beam spacing and tall floor heights to allow for the above ceiling distribution. This in turn created challenges for the structural design of a moment frame system by severely impacting the moment frame member sizes. To improve the structural efficiency of the steel frames, the spacing of the perimeter bays were architecturally "decoupled" from the interior bays thereby creating a uniform exterior grid pattern with short bay lengths spaced at approximately 21′ to 6″ (6.55 m) with nearly the entire perimeter of the building following this spacing.

Installing dampers within the interior bays of the structure was not feasible due to the space demands of the laboratory as well as necessity of potential post-earthquake inspection of the dampers given that the lab spaces have validated clean room classifications. This forced the dampers to be located in the same bays as the perimeter moment frame, thus creating challenges due to the tight frame spacing and tall floor to floor heights. The authors considered the implementation of the fluid viscous dampers in four configurations as depicted in Fig. 1 with a comparison of the efficiencies of the four configurations described in Table 2. Configuration Option 1 (Fig. 1a) and Option 2 (Fig. 1b) consist of a chevron brace with horizontal dampers and inclined dampers respectively. Option 3 (Fig. 1c) is a single diagonal brace with an inclined damper. Option 4 (Fig. 1d) is a modification of Option 1 with a lowered work point of the extender to brace intersection. This led to an increase in the efficiency of the supporting structural system as shown in Table 2.

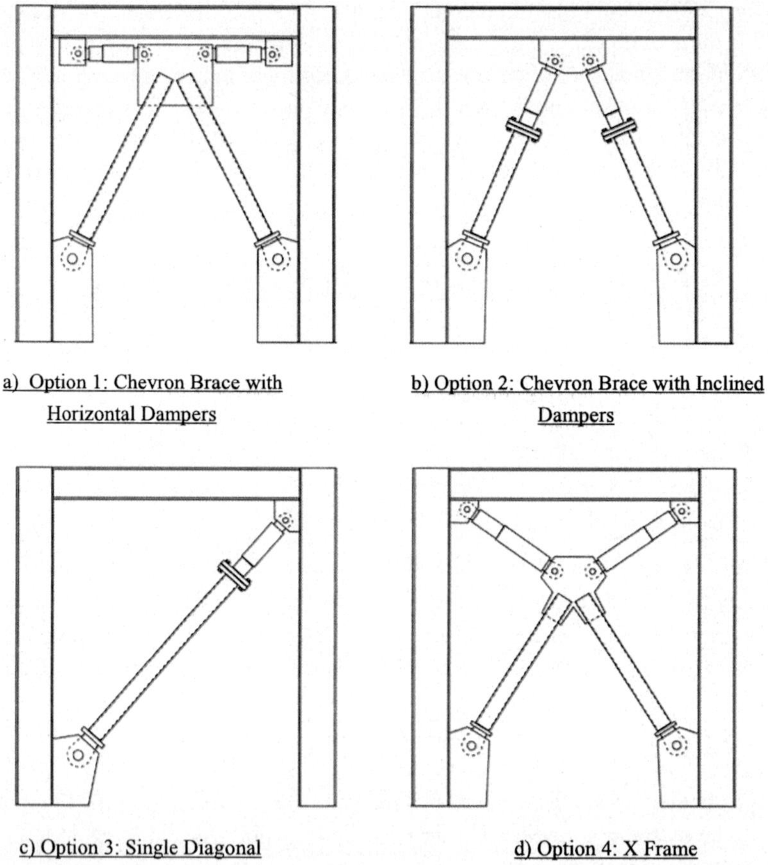

a) Option 1: Chevron Brace with Horizontal Dampers

b) Option 2: Chevron Brace with Inclined Dampers

c) Option 3: Single Diagonal

d) Option 4: X Frame

Fig. 1. Damper frame configurations considered for design implementation.

Table 2. Comparison of damper efficiency and extender brace tonnage for damper frame configurations considered for design implementation.

	Option 1	Option 2	Option 3	Option 4
Damping force efficiency	1.0	0.42	0.67	0.83
Damper stroke reduction	1.0	2.4	1.5	1.2
Extender brace tonnage	1.63 tons (14.5 kN)	1.40 tons (12.5 kN)	1.20 tons (10.7 kN)	1.33 tons (11.8 kN)

Option 1 demonstrates great efficiency in damper forces, and therefore cost; however, the long extender braces became controlled by flexural behavior and required very large members to maintain enough stiffness so as to not impact damping efficiency and therefore negatively impacted cost. The large member sizes would have also created issues with the attachment of exterior cladding. Option 2 has the best stroke efficiency but the lowest force efficiency because of frame geometry and steep damper angle. Due to limitations in the amounts of bays that could receive dampers, this option resulted in considerably higher damper forces. Option 3 was considered for its conventional nature and lowest tonnage per frame but was deemed impractical due to the steep angle and reduced damper efficiency. Also, with the limited number of damper frames and only 1 damper per frame, damper forces are significantly larger resulting in the highest costs per damper of all four options.

The Option 4 "X" frame was ultimately selected due to its balance of force efficiency and steel tonnage reductions compared to the other three configurations. By controlling the work point elevations at each floor, the authors were also able to standardize the damper lengths allowing the manufacturer to fabricate a single structural housing type for the dampers, thus greatly reducing overall costs and improving fabrication speed.

5 Detailing Considerations

A schematic of the "X" frame with a description of members can be seen in Fig. 2. One of the primary concerns of the authors and the project peer reviewers was the stability of the central junction node and how it would behave when subjected to differential movements between the top and bottom of the frame out of plane of the frame. A pair of out of plane braces were provided to kick up to the floor framing above, creating an essentially rigid connection with the floor. Causing the junction node to move with the floor at the top of the damper frames. This does not pose an issue for the damper which is already connected to the same floor level as the junction assembly; however, the connection of the extender braces experiences non-trivial out of plane rotations which could result in hinging of the extenders or more likely, localized damage of the gusseted connections which are attached the moment frame beams. To counteract this effect, the authors worked with the damper manufacturers to utilize a specialized extender brace end connection with a spherical bearing allowing for both in-plane and out of plane rotations thus accommodating the imposed rotations as shown in Fig. 3.

While arguably present in most designs utilizing fluid viscous dampers, the presence of these pin connections highlighted the eccentricities associated with a constantly adjusting brace work point due to out of plane frame movement during a seismic event. These eccentricities result in twisting demands being imposed on the moment frame columns, which were resolved by embedded slab connection plates that allowed development of these forces into the structural diaphragm as seen in Fig. 4.

In addition to the structural detailing considerations associated with the "X" frame, focus was given to the non-structural systems ability to accommodate the differential floor to floor movements so as to enhance the buildings function after a design basis seismic event. Key consideration was given to the performance of the stairs by providing slip joint connections, interfaces between overhead ceilings and distribution systems, as well as incoming above ground utilities.

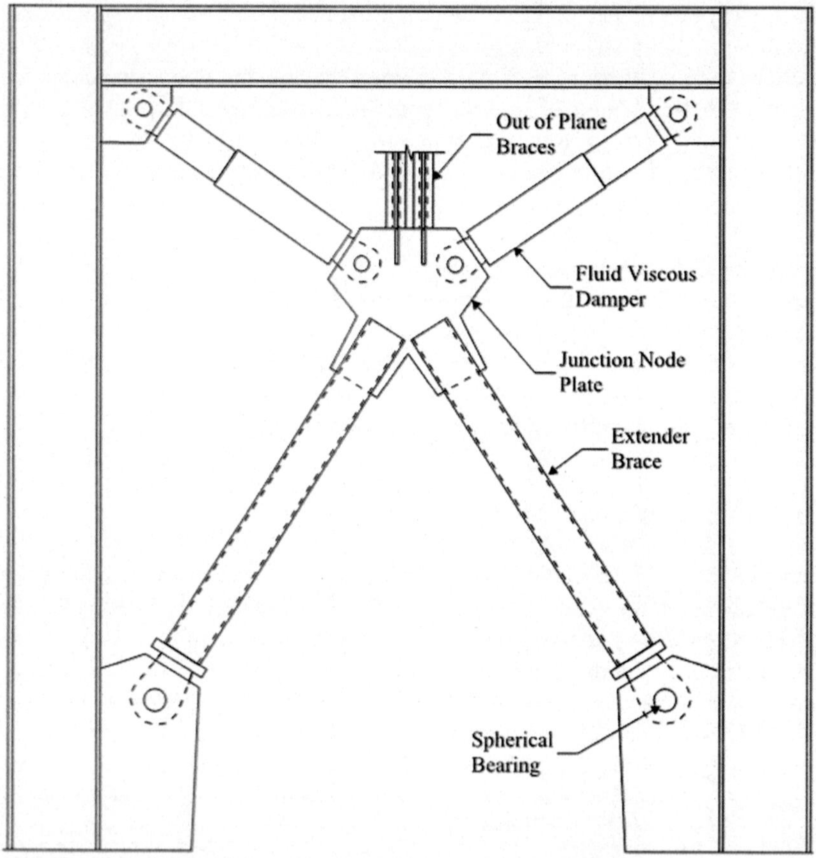

Fig. 2. Option 4 "X" frame schematic elevation.

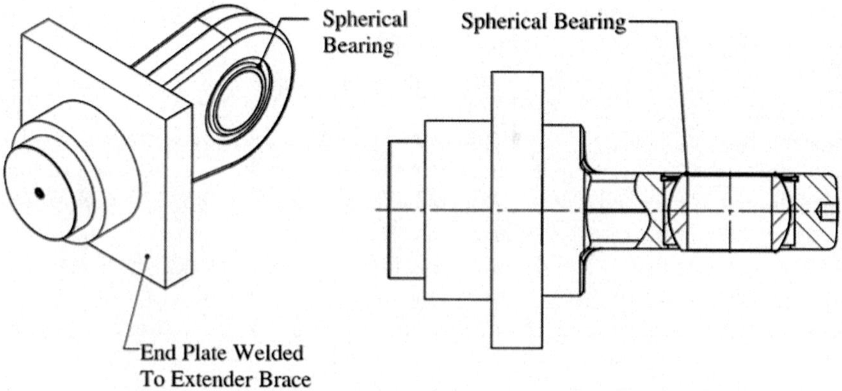

Fig. 3. Isometric and section view of extender brace end connection.

Fig. 4. Embedded slab connection plates (boxed in red) at damper frame column.

6 Conclusions

The rapid growth of the life science and biotechnology industries has increased the desire for higher seismic building performance on laboratory facilities to protect valuable research data and products. This paper presents a case study of a recent life science project located in the San Francisco Bay Area with a special moment frames and supplemental non-linear fluid viscous dampers lateral resisting system. The structure was designed for enhanced performance objectives of immediate occupancy at DBE events and life safety at MCEr events to decrease the post-earthquake recovery time of the structure. A novel "X" frame damper configuration was selected due to its balance between damping force efficiency, damper stroke efficiency, and reductions in supporting steel tonnage. Due to the non-conventional nature of the configuration, extra efforts were put into detailing the elements in the damper frames to ensure deformation compatibility and sufficient robustness could be maintained.

Acknowledgements. The authors would like to acknowledge Taylor Devices Inc. For their partnership and spirit of collaboration throughout the design and construction of the project.

References

1. Bhatia, S.: Introduction to Pharmaceutical Biotechnology, vol. 1. IOP Publishing Ltd. (2018)
2. CBRE Report: San Francisco Bay Area Second Largest Percentage Gain in Life Sciences Conversions (2022). https://www.cbre.com/press-releases/cbre-report-san-francisco-bay-area-second-largest-percentage-gain-in-life-sciences, last accessed 2022/07/13
3. ASCE: Minimum Design Loads and Associated Criteria for Buildings and Other Structures (ASCE7-16). American Society of Civil Engineers, Reston, Virginia (2017a)
4. Ricles, J., Fisher, J., Lu, L., Kaufmann, E.: Development of improved welded moment connections for earthquake-resistant design. J. Constr. Steel Res. **58**, 565–604 (2002)
5. ASCE: Seismic Evaluation and Retrofit of Existing Buildings (ASCE41-17). American Society of Civil Engineers, Reston, Virginia (2017b)

Seismic Isolation in the US Mission Critical Sector

Sam Richardson(✉) and Kurt Lindorfer

PARADIGM Structural Engineers, San Francisco, CA 94111, USA
srichardson@paradigmse.com

Abstract. There is an increased desire for higher levels of building performance during and after seismic events in the mission critical and data center industries due to the costs associated with non-structural damage and uptime losses. Conventional industry-accepted approaches such as voluntary use of a larger importance factor may increase the robustness of the structure itself, but does not result in reduced floor accelerations or reductions of negative impacts on both computer servers and the extensive mechanical and electrical distribution systems associated with data centers. Seismic isolation remains the premier solution for increasing both structural and non-structural performance during a seismic event. The authors describe the project challenges in the design of two base isolated facilities located in California. The first project is located in Northern California and was for a client operating their data center in a "Colocation" format. This project was located on a site with tight space constraints, necessitating the use of fluid viscous dampers as part of the isolation system to limit building displacements to an acceptable level. The second project is located in Southern California and was for a single "Enterprise" client. The larger site for this project allowed the team to design for a larger displacement to maximize the effectiveness of the isolation system and minimize the demands imparted on to the building.

Keywords: Seismic base isolation · Passive energy dissipation · Triple friction pendulum · Fluid viscous dampers · Data centers · Mission critical

1 Introduction

Data centers are physical spaces where information technology infrastructure and computing services are stored. The origins of the modern data center can be traced back to the original dot-com booms of the late 1980s and early 1990s. Recent proliferations of the use of social media, the increased use of streaming media, smart phones and the use of the "cloud" over the past two decades have led to rapid development and large-scale deployment of these facilities. A key component of a successful data center is high speed and reliable access to stored data and computing services. Speed primarily takes the form of geographic proximity to the end user as well as use of new fiber technologies. Reliability takes the form of "uptime" [1] by minimizing interruption of the data center's connection to the outside world through means of redundant power systems.

G. P. Cimellaro (Ed.): WCSI 2022, LNCE 309, pp. 929–936, 2023.
https://doi.org/10.1007/978-3-031-21187-4_82

The key contradiction that puts these goals in competition is that maximization of speed, or reduced latency, necessitates that data centers be located in close proximity to population centers that are often subject to natural hazards that can impact the function of the data center. This creates an increased desire by data center developers, operators, and users for higher levels of building performance during and after seismic hazards to mitigate costs associated with non-structural damage and uptime losses. Conventional industry-accepted approaches such as voluntary use of an importance factor greater than 1.0 may increase the robustness of the structure itself, but it doesn't result in reduced floor accelerations or a reduction in the impacts on both computer servers and the extensive mechanical and electrical distribution systems associated with data centers. Seismic isolation remains the premier solution for increasing both structural and non-structural performance during a seismic event. This paper presents a case study for two seismically isolated data centers located in high seismicity regions of Northern and Southern California, describing observed challenges, lessons learned, and topics for future research.

2 Structural Design Requirements for Data Centers

Data centers represent a challenging structural design problem due to their large concentrations of heavy computer equipment, mechanical cooling equipment, primary and redundant electrical power systems as well as extensive mechanical, electrical, plumbing (MEP) and networking distribution systems. This takes the form of increased weight and seismic mass that data center structures must be designed to resist relative to that of a conventional office building. For the server cabinets, it is common practice to determine an effective live load via weighted average of computer equipment loading and maintenance live loading. Trends toward heavier/denser computer equipment typically results in a minimum effective live load of 250 psf (12 kPa). A portion of this server load should be included in the seismic mass, considering the anticipated in-service cabinet weights. The actual load will be dependent on the user server fill density but can typically be conservatively taken as 75% of the equipment live load absent more detailed information. A second massive contributor to building structural demands is the presence of electrical equipment. The electrical infrastructure of a data center will include primary power systems consisting of transformers and power distribution units as well as heavily redundant backup power systems including generators, batteries, uninterrupted power systems and switchgears. A majority of the electrical architecture is located within the building, with generators and sometimes transformers located outside of the buildings depending on project requirements. It is critical that portions of the structure supporting this equipment are adequately proportioned to resist these heavy pieces of electrical infrastructure. After a major seismic event, the functionality of the data center is highly dependent on the backup power systems which has caused a trend towards specification of seismically certified equipment.

The density of computer equipment generates large heat loads which negatively impact the performance of the data center as a system. As a result, complex cooling schemes are incorporated using air and/or water-cooled equipment to keep computer room temperatures within operating levels. Common cooling strategies involve the use of either rooftop or at-grade air handling units that feed a chilled water loop that can be used to cool the air within the computer room or directly cool servers with a water-cooled system. Impairment of the cooling system after a seismic event would necessitate scaling back the computational and networking abilities of a data center. Key structural considerations for cooling systems include whether the equipment is located at the roof, adding weight and mass to the structure; or at grade, adding exterior foundation elements and additional electrical and chilled water distribution that must enter the building from an exterior service yard.

Networking connections are required for a data center to meet its intended purpose. Exterior telecommunication networks enter the building through an entrance room or meet-me room where connections between the housed computer equipment can be made. These low voltage connections require a massive number of cables which are supported in groups that can lead to significant supported weights. Often times networking distribution is supported by a hanging structural grid that supports layered trapeze elements with trays feeding multiple cables to each server. These networking loads typically necessitate a suspended dead load proportioned based on client needs. Stability of the suspended grid before and after a seismic event is critical to maintaining networking connections between the data center and the outside world.

3 "Importance Factor" as Standard of Practice

As described previously, resiliency to withstand unplanned occurrences is a key component to the goal of maximizing data center uptime and thus reducing business interruption. Both "Colocation" and "Enterprise" clients typically require the design of their facilities for a voluntarily increased risk category or importance factor with the intention of improving the seismic performance of the data center to align with these uptime goals. The practice of this can somewhat be traced to the TIA 942 [1] standard which lays out requirements for different "tiers" of data centers. It is common practice for data center developers to advertise leased space having this increased seismic performance. The authors speculate that the prevalence of voluntarily increasing importance factors in the US data center market is more driven by the perception of increased performance, rather than an actual understanding of how the importance factor increases robustness of the structure. Its also believed by the authors that it is not well understood by developers that the importance factor does not reduce the earthquake demands experienced by the structure and its contents.

As understood by many engineers in the data center industry, the incorporation of an importance factor alone is unable to reduce the large accelerations resulting from ground shaking which potentially damage acceleration sensitive computer equipment, power systems, and overhead distribution. This is somewhat addressed by the tier system laid out in TIA 942 [1] which mentions the use of seismic isolation and passive energy dissipation systems for "Tier 3" and "Tier 4" data centers; however, as noted in Erkus et al. [2] it is somewhat unclear whether this is a requirement and what performance objectives should be targeted when using such systems.

4 Seismic Isolation for Data Centers Key Points

Seismic isolation has gained traction in the engineering community as a premier system for mitigation of the effects of earthquake hazards for building structures. Through the effects of period shift and increased equivalent viscous damping, seismic isolation significantly reduces earthquake induced floor accelerations experienced by a structure. The reduction of acceleration is a key difference between the conventional approaches of the voluntary usage of an increased importance factor. Another notable difference is that the reduction of seismic demands allows for the building structure to be economically designed to remain essentially elastic during major shaking. Comparing back to a fixed based building, this would be akin to designing the structure for an importance factor much greater than the 1.25 or 1.5 typically used for fixed based data center designs. The reduction in floor accelerations and increased performance serves to allow rapid re-occupancy and restoration of function to limit downtime after a seismic event.

While not exclusive to data center occupancies, it is important when utilizing base isolation to develop a clear set of performance objectives. Both structural and non-structural behavior should be considered in the setting of this design basis. An important decision with regards to the non-structural performance is whether to locate backup generators on or off of the isolation plane. While seismically certified options have become more and more available in recent years, it is the authors' opinion that given the likely downtime of local power utilities strong consideration must be given to the added protection of including the generators on the isolation system. When making these decisions, stakeholders should consider the incongruity between the performance of an isolated building, which depending on the design performance objective, may remain functional after a Maximum Considered Earthquake (MCEr) and the performance of a seismically rated generator which is typically rated only to Design Basis Earthquake (DBE) shaking.

It is also important to consider the differential movements that may occur during a seismic event due to the incorporation of seismic isolation. Data centers will tend to have larger quantities of network and power connections than a typical office occupancy. These outside utility lines will require flexible interfaces that accommodate the differential movements at the transition from being "ground supported" to "building supported". In a conventional isolation design, most, if not all, of these transitions will occur in a below building basement level. Careful coordination of the often congested below building MEP distribution needs to be done to ensure all transition elements are capable of accommodating full seismic movement. This coordination should also allow for reasonable access for inspection, maintenance and repair or replacement of the seismic isolation device as well as other below building distribution.

5 Description of a Northern California Data Center

The subject project is a four-story, 160,000 square foot data center located in Northern California for an international client with past experience with seismically isolated data centers. The data center is a "Colocation" facility with the ability to support multiple leased floors of varying critical IT load. The building is constructed with composite floor and roof slabs supported by a steel frame. An above roof platform was used to support mechanical cooling equipment and the backup power system was located on the first floor of the isolation plane. The lateral system consists of buckling restrained braces using the composite slabs as diaphragms for distribution of horizontal loads. The isolation system consisted of Triple Friction Pendulum (TFP) bearings in parallel with Fluid Viscous Dampers (FVD's). The isolators were supported on top of concrete pedestals which were supported by pile caps atop auger cast piles with the pile caps interconnected by grade beams.

Upon careful consideration with the client, the building performance objectives were to meet an immediate occupancy level at a DBE level event as well as damage control level during an MCEr level event. To achieve this would require significant reductions in the seismic demands applicable to the site which requires an isolation system with a very low stiffness. Due to narrow site constraints and the desire to maximize floor area, it was not possible to accommodate the large displacements resulting from the relatively flexible bearings needed to meet the desired performance criteria by using the bearings alone. As a result, (25) FVDs were incorporated into the isolation system, reducing displacements to manageable levels with minimal sacrifice to the reduction in seismic demands experienced by the structure above the isolation plane. In the end, the structure and surrounding "moat" was proportioned to accommodate 32″ (813 mm) of displacement occurring in any direction.

6 Description of a Southern California Data Center

The subject project is a two-story, 60,000 square foot data center located in Southern California for an "Enterprise" client. The building is constructed with composite floor and roof slabs supported by a steel frame. Distributed cooling systems located throughout the building and the backup power system were included on the isolation plane. The lateral system consists of buckling restrained braces using the composite slabs as diaphragms for distribution of horizontal loads. The isolation system consisted of TFP bearings which are supported on top of concrete pedestals and a mat slab.

Client needs dictated that the building performance objectives were to remain operational after a DBE level event and meet immediate occupancy after an MCEr event. The large site suited this requirement by allowing the maximization of seismic movements to greatly minimize floor accelerations. The maximum displacement during an MCEr event is anticipated to be 42″ (1070 mm) with peak data hall accelerations of less than 0.3 g. The large reductions in seismic demands experienced by the building allowed the structure to be designed to remain elastic during an MCEr event.

7 Lessons Learned and Recommended Areas of Study

Accommodation of movement at the interface of a seismically isolated building and the surrounding grade while mitigating foreign matter intrusion into the basement is a critical part of the design process and is often accomplished by extending the basement beyond the building footprint thus creating a "Moat". One approach to mitigating foreign matter intrusion into the basement at the moat is to provide a "moat cover" matching the elevation of surrounding grade that will move with the building during seismic displacements which was done on the Northern California project. A more cost-efficient system that was utilized for the Southern California project involved lowering surrounding grade, allowing the building to "float" over the moat walls. In this scenario walking paths are located further away from the building with site landscaping that would discourage vehicles and pedestrians from being adjacent to the structure where this detail occurs. Since seismic isolation is not common in the United States, let alone the US Data Center sector, architects and clients may not be familiar with the interaction of the building with surroundings and the potential life safety issues that this interaction poses. While these life safety concerns are not typically the purview of the structural engineer of record (SEOR), it is important that we educate our architects and owners about these risks to encourage safe design decisions.

While important for any structure, the functionality of a data center is highly dependent on its utility hookups or interconnection with outside networks, making the design of utility distribution to accommodate the differential movements between the site and the structure a key design consideration for a seismically isolated data center. This presents a challenge for the US data center market where design of MEP distribution supports are often overlooked during the building design phase and not considered until onboarding of trade subcontractors. There may also not be sufficient overlap between companies qualified for data center work and companies with seismic isolation experience. The authors wish to highlight the importance of meeting early and often with the MEP trades at all levels to educate and gain "buy in" with the design approach. For this to work requires cooperation with the trades and general contractor. Experience has shown that one of the significant hurdles is coordination between the trades to eliminate clashes that would restrain the movement of flexible joints or cause impact between a building supported element and a ground supported element. To better plan for these issues, it is encouraged that the SEOR be involved in the design of the flexible transitions and MEP distribution supports. The authors note that the presence of a deeper underfloor moat level on the Southern California project greatly simplified this coordination process due to the extra space and vertical separation this afforded.

With advancements in the structural engineering field promoting broader implementation of performance-based designs targeting reduced downtime, including FEMA P-58 [3] and ATC-138 [4], there should be a focus by researchers and practitioners to investigate the behavior of flexible joints utilized for piping and conduit crossing the "isolation plane". Over the course of the two projects, the authors were unable to find testing data from researchers and manufacturers data was limited. Sizing of these connections appears to be based on analytical procedures or rational engineering judgement; however, the post-earthquake performance is not well understood. The authors managed to perform limited field testing of some concepts with a low degree of technical sophistication, but a thorough investigation of the connection performance is not project feasible from a cost and schedule standpoint. It is encouraged that those involved with code development consider requirements for prototype testing or prequalification standards of these flexible connections. It is also encouraged that manufacturers and researchers work together to develop fragility curves that can be used for FEMA P-58 [3] style loss and business interruption estimation.

The incorporation of FVDs was a pivotal design decision for the Northern California project. In the immediate term, it enabled building design to move forward while maintaining and maximizing building floor area as directed by the owner, without substantial reworking of the original concept and space planning. In the long term, the displacement reductions simplified coordination of "underfloor" distribution and "moat covers". Research on the use of FVD devices as part of the seismic isolation has shown that there can be some performance tradeoffs [5], yet the authors would encourage any designers of a base isolated data center in regions of high seismicity to strongly consider their implementation unless presented with a site possessing more than ample space to accommodate the movements of the building and underbuilding MEP distribution. There are some gaps in ASCE 7 [6] when utilizing supplemental damping devices as part of the seismic isolation system. The language of the code necessitates that most base isolated structures with passive damping systems will require use of the Non-Linear Time History provisions and possess a peer review team. It can be argued that this will effectively resolve gaps that may exist in the code; however, it leaves critical aspects of the design such as testing and minimum design forces up to the interpretation of the design professionals and the design review team. The authors encourage the code committee responsible for Chaps. 17 and 18 to develop language that more smoothly integrates the use of velocity dependent damping devices as well as a practical set of minimum testing requirements.

A principal driving force to the selection of seismic isolation for a data center is the ability of this passive system to limit risk exposure from a seismic event caused by damage to building contents and business interruption. There is a void of information geared towards the data center market to help quantify these benefits. The authors feel that those involved in the data center industry should explore building off of the "Functional Recovery" framework models posed by ATC-138 [4] and FEMA P-58 [3], to develop a functional recovery framework for mission critical facilities. The goal of this work would be to arm stakeholders with information about potential business downtime to better inform decision making about incorporating seismic isolation, or other high performing structural systems, into new data centers designs. A secondary component of this risk analysis is better understanding the performance of server cabinets during seismic movements which is an under-researched topic.

8 Conclusions

Seismic isolation is a premier structural design solution for increasing the post-earthquake performance of data centers in regions of high seismicity. The authors' experience on two data centers located in California note some important lessons learned and recommendations on future developments to increase the ease of utilizing this innovative technology in the US data center market.

References

1. TIA: Telecommunications Infrastructure Standard for Data Centers (ANSI/TIA-942-A-2012). Telecommunications Industry Association: Standards and Technology Department, Arlington, Virginia (2012)

2. Erkus, B., Polat, S., Darama, H.: Seismic design of data centers for Tier III and Tier IV resilience: basis of design. In: 11th U.S. National Conference on Earthquake Engineering, Los Angeles, CA (2018)
3. ATC: Seismic Performance Assessment of Buildings (FEMA P-58). Applied Technology Council, Redwood City, CA (2018)
4. ATC: Seismic Performance Assessment of Buildings, vol. 8 (ATC 138-3). Applied Technology Council, Redwood City, California (2021)
5. Wolff, E., Ipek, C., Constantinou, M.: Effect of viscous damping devices on the response of seismically isolated structures. Earthquake Eng. Struct. Dynam. **44**, 185–198 (2014)
6. ASCE: Minimum Design Loads and Associated Criteria for Buildings and Other Structures (ASCE7-16). American Society of Civil Engineers, Reston, Virginia (2017)

The Brittle Failure of Fluid Viscous Dampers and the Related Consequences on the Reliability of a Medium-Rise Steel Building

Laura Gioiella[1](✉), Fabrizio Scozzese[1], Enrico Tubaldi[2], Laura Ragni[3], and Andrea Dall'Asta[1]

[1] School of Architecture and Design, University of Camerino, Camerino, Italy
laura.gioiella@unicam.it

[2] Department of Civil and Environmental Engineering, University of Strathclyde, Glasgow, UK

[3] Department of Civil and Building Engineering and Architecture, Università Politecnica Delle Marche, Ancona, Italy

Abstract. Damper failures are generally brittle mechanisms, which can compromise the capacity of the structure to withstand the seismic action, leading to a lack of robustness of the overall system. The brittle failure is due to the attainment of the maximum force capacity, because of end-stroke impacts, which causes an arising of the forces, or to an excessive velocity of the piston (over-velocity). Moreover, anti-seismic devices need to be designed with proper safety margins against their failure in order to reach a target safety level. Seismic standards generally prescribe safety factors (reliability factors), that in the case of Fluid Viscous Dampers (FVDs) are applied to stroke and velocity, with values that are not homogenous among seismic codes. The effect of damper failure and different reliability factors on both the fragility and the seismic risk of the structural system is investigated by performing multiple-stripe analysis and monitoring different global and local demand parameters of a medium-rise steel moment-resisting frame building, widely studied in literature, thanks to an advanced model implemented in OpenSees, which considers the brittle mechanism. Moreover, the problem of damper failure is also analysed in terms of fragility functions, providing information about the dependency of the probability of failure with the seismic intensity.

Keywords: Fluid viscous dampers · Brittle failure · Reliability factors · Seismic risk assessment · Damper fragility functions

1 Introduction

Fluid Viscous Dampers (FVDs) are devices widely used for seismic passive protection of both new and existing structures, with the aim of reducing both displacements and drift demands. Such devices are usually installed within the frame of the building to protect [1–4], while recently their use is becoming more and more frequent in external

G. P. Cimellaro (Ed.): WCSI 2022, LNCE 309, pp. 937–947, 2023.
https://doi.org/10.1007/978-3-031-21187-4_83

configurations [5, 6]. These latter may involve localized external structures or exoskeletons that realize with an appropriate and integrated design, not only the upgrading of the image of the existing building, but especially provide an enhancement of the structural and energy-saving capacities, as also testified by some Horizon 2020 research projects (e.g. [7, 8]).

Several approaches are to date available for designing both size and location of viscous dampers in many different configurations, based on direct procedures [9, 10] or optimization methods [11, 12]. These design approaches generally control the seismic performance of buildings under the design seismic intensity level, also weighting the variability of the mechanical properties of the FVDs [13], as prescribed by the European Code on anti-seismic devices (EN 15129). However, the reliability under extreme earthquake events may be characterized by low robustness and inadequate safety levels due to dampers that usually exhibit a brittle collapse behaviour, which may lead to the collapse of the whole structural system [14, 15]. Consequently, the choice of adequate safety factors for the design of dampers is very important for gaining a satisfactory performance under strong actions and for controlling the probability of failure [16].

A model able to describe the brittle failure of FVDs is proposed and used in the present paper to evaluate the probability of collapse of structures equipped with such devices. More in detail, the proposed model assumes that the collapse of the damper is due to the attainment of its force capacity, caused by the over-velocity experienced by the damper or by the achievement of the end-stroke. An insight on the ability of the model in describing the consequences on both the devices and the frame to protect is provided first with reference to a 3-storey steel building subjected to an incremental sinusoidal input. In detail, results in terms of global and local parameters are presented. The formers are, respectively floor displacements (u) and floor absolute accelerations (A), while the latter are the forces (F_{di}) and the strokes (Δ_{di}) of the FVDs. Successively, the seismic risk of a medium-rise benchmark steel building, widely studied in literature [17], consisting in a 9-storey moment-resisting frame equipped with FVDs, is evaluated by means of a probabilistic approach. The probabilistic response of the building is assessed by performing Multi Stripe Analysis (MSA) [18] at different intensity measure (IM) levels, while results are given in terms of demand hazard curves of the maximum interstorey drift (IDR) and focusing on the effect of the amplification factors on the sequence of dampers failure. Both linear and nonlinear devices have been considered and their capacity (stroke and strength) has been designed according to a seismic action having a Mean Annual Frequency (MAF) of exceeding equal to 2×10^{-3} and by assuming five combinations of amplification factors (also called γ-factors) relevant to damper stroke (γ_Δ) and velocity (γ_v) that is strength, in accordance with the main international codes (EN 15129 [19], Eurocode 8 [20], ASCE/SEI 7-10 [21]). The case without dampers and the one in which the damper failure is disregarded are also considered for comparison purposes. Finally, fragility functions are also presented aiming at providing information about the dependency of the probability of failure with the seismic intensity of linear and nonlinear dampers placed at different floors, for the different γ-factors analysed.

2 Fluid Viscous Dampers Modelling

The relationship that describes the constitutive law of FVD device, Constantinou and Symans [1], is described by the following expression

$$F_d(v) = c|v|^{\alpha} sgn(v) \tag{1}$$

where c and α are the constitutive parameters of the device representing, respectively, an amplification factor and the extent of non-linearity in the behaviour of the device; v is the relative velocity between the ends of the device; $|v|$ is the absolute value of v; sgn is the sign operator and F_d is the damper resisting force. Regarding the behaviour of the device, it is worth to remember that FVDs can be produced with values of the constitutive parameter α from 0.1 up to 2. Among them those having $\alpha = 1$ behaves linearly, while the range $0.3 \leq \alpha \leq 1$ is the most widespread, due to their ability in dissipating seismic input energy.

As previously introduced, a dissipative device may undergo brittle failure due to the attainment of its strength capacity, $F_{d,max}$, which in this study is assumed to be the same both in compression (maximum shortening of the device) and in traction (maximum elongation of the device). Two mechanisms may lead to the capacity attainment: (*i*) over-velocity; (*ii*) end-stroke. The first one is related to an excessive velocity of the piston, with respect to the design one that leads to a leak of fluid or to the damage of mechanical components of the device. The latter, instead, is related to forces that arise when the device exploits all its available stroke, $\pm\ \Delta_{d,max}$, resulting in an impact. The attainment of the impact, however, does not strictly imply the damper failure because the impact force may be lower than the device strength. It is noteworthy that once the maximum capacity is attained, the resulting failure mechanisms is brittle, thus making the device ineffective, with no residual ability to sustain loads or dissipate energy.

Figure 1 depicts the model proposed and used in this paper for accounting for the brittle failure of the device. It is composed of three elements: a dashpot, describing the dissipative behaviour; a hook and gap element, set in parallel to the dissipative device, which simulate the impact due to either excessive shortening ($-\Delta_{d,max}$) or elongation ($+\Delta_{d,max}$); and a third element, set in series with the others, simulating the failure due to the attainment of the force capacity. The failure occurs when the modulus of damper force attains the limit value $F_{d,max}$. The model is implemented in OpenSees [22] using two-node link elements simulating each of the three components and various material properties to describe the different behaviours; further details can be found in Scozzese et al. [16].

2.1 The Efficiency of the Model in Reproducing the Brittle Failure Mechanism

This section provides a further insight on the model ability in simulating both the end-stroke mechanism and the over-velocity one, highlighting their consequences in terms of both local and global EDPs. More in detail the force-displacement loops of two FVDs behaving linearly and installed, respectively, at the first and last elevation of a 3-storey moment-resisting steel frame building subjected to an incremental sinusoidal input are reported in Fig. 2. While Fig. 3 shows the time-histories of floor displacements and floor

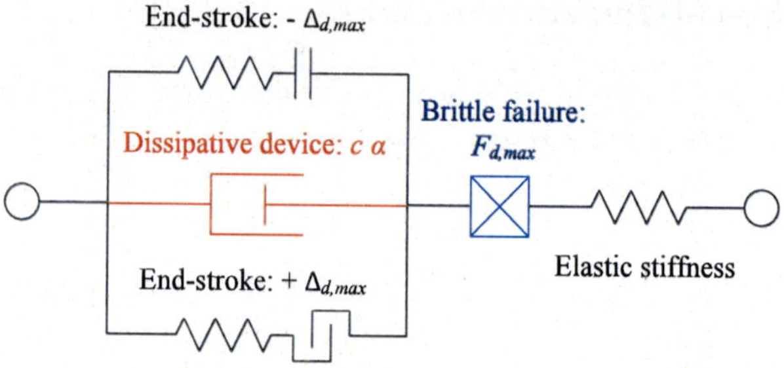

Fig. 1. FVD model encompassing the brittle failure mechanism.

absolute acceleration from which it is possible to better understand the consequences of the dampers' failure on the frame structure to protect.

By observing the stroke-force relationships of Fig. 2a, b it is evident that at the beginning of the third increment of the input motion, both the monitored dampers experience the end-stroke attainment without failure, with impacts that are more evident for the top-storey device. After these impacts, occurred without consequences, the FVDs restore their behaviour as pure dissipative devices. Few instants later, suddenly, the damper located at the ground floor fails due to over-velocity, triggering the damper failure at the upper elevation. The sequence of failure is highlighted by a series of ripples in the stroke-force relationship of the top-storey device. The ripples begin when the first device fails and last until the upper device fails too for over-velocity.

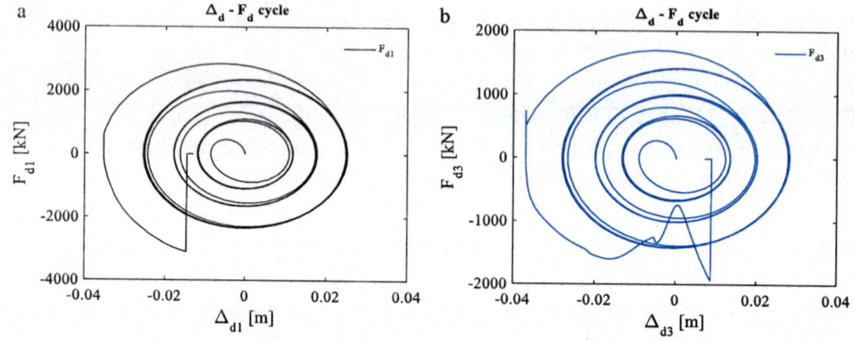

Fig. 2. Force-displacement cycles of linear FVDs subjected to sinusoidal incremental input: (a) first elevation; (b) third elevation.

Figure 3a, b show the time-histories of the parameters strictly related to the frame response, highlighting the consequences of the damper failures on the frame itself. It is useful to say that the incremental input lasts nearly 23 s and few more seconds are shown to understand how the frame restore its initial rest condition. Generally, the responses in terms of displacements (u) and absolute accelerations (A) are significantly amplified by the impacts occurring in the dampers and by their failure. The absolute accelerations are more affected than the displacements. It is worth to note that the peaks in terms of absolute accelerations, recorded between 14 and 15 s, that is the instants where the dampers' failure occurs, are mainly related to the impacts experimented by the devices before their failure.

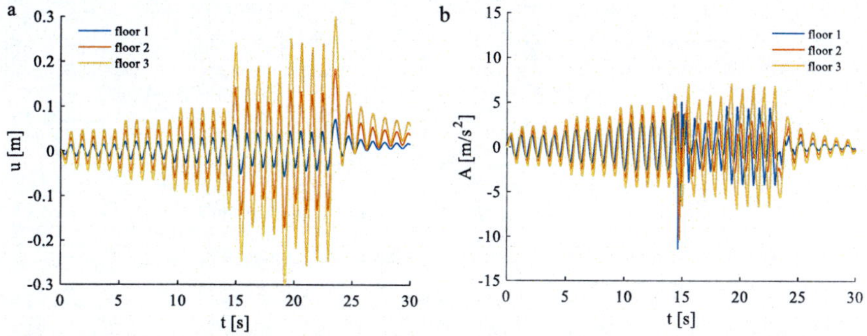

Fig. 3. Frame response in terms of: (a) floor displacements; (b) floor absolute accelerations.

3 Case Study and Method

In this section, the case study analysed is introduced together with the considered seismic scenario and the design properties of both linear and non-linear FVDs ($\alpha = 1.0$ and $\alpha = 0.3$). Moreover, a brief introduction to the probabilistic framework applied together with the amplification factors used for the design of the FVDs are presented.

3.1 Benchmark Structure and Hazard Scenario

The case study considered is a 9-storey benchmark steel building, consisting of perimeter moment-resisting frames and internal gravity frames with shear connections (Fig. 4a). The period of the first vibration mode of the building is $T_1 = 2.225$ s, the frame has its own damping coefficient $\xi = 2\%$ and the seismic hazard of the site in terms of spectral pseudo-acceleration $S_a(T_1)$ is reported in Fig. 4b. The IM levels at which MSA are performed are 20 and highlighted by circles in Fig. 4b, the red ones refer to the main limit states suggested by codes.

3.2 Damping System Design Properties

The design of the FVDs is carried out to enhance the building performance under a seismic scenario with a 10% probability of exceedance in 50 years (ULS scenario according to Eurocode 8). To this aim, a target value $\xi_{add} = 30\%$ has been chosen for supplemental damping. Linear dampers have been designed so that the viscous constants are distributed proportionally to the storey shear force of the first mode of the bare frame, while the coefficients of the non-linear ones are derived based on the equivalence of the energies dissipated by the linear and nonlinear FVDs, for a given value of α. Further

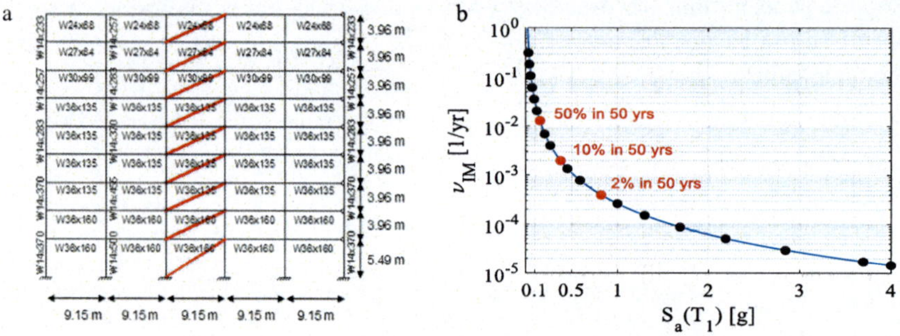

Fig. 4. (a) The 9-storey building; (b) IM hazard curve.

details regarding the design process can be found in Scozzese et al. [16]. Table 1 reports the FVDs constitutive parameters, while Table 2 shows the values of mean displacement $\Delta_{d,j}$, force $F_{d,j}$ and velocity v_j demand of the dampers, evaluated at the design condition, with unitary amplification factors (i.e. $\gamma_v = \gamma_\Delta = 1$).

Table 1. Damping properties for different levels of dampers nonlinearity.

α	Floor 1	Floor 2	Floor 3	Floor 4	Floor 5	Floor 6	Floor 7	Floor 8	Floor 9
	c_1	c_2	c_3	c_4	c_5	c_6	c_7	c_8	c_9
	[kNs$^\alpha$/m$^\alpha$]								
1	48,103	46,834	44,578	41,282	36,918	31,534	25,199	17,903	9675
0.3	8133	7899	7518	6962	6226	5318	4250	3019	1632

Table 2. Damper design parameters at the design condition.

		Floor 1	Floor 2	Floor 3	Floor 4	Floor 5	Floor 6	Floor 7	Floor 8	Floor 9
$\alpha = 1$	$\Delta_{d,j}$ [mm]	38.4	32.7	31.9	31.9	29.5	27.8	28.7	29.9	24.3
	$F_{d,j}$ [kN]	7781	6060	5416	5075	4379	3930	3559	3011	1614
	v_j [m/s]	0.16	0.13	0.12	0.12	0.12	0.12	0.14	0.17	0.17
$\alpha = 0.3$	$\Delta_{d,j}$ [mm]	38.7	32.4	31.3	30.8	28.7	26.1	29.1	31.0	25.8
	$F_{d,j}$ [kN]	5244	4688	4345	4014	3546	2952	2507	1920	1062
	v_j [m/s]	0.23	0.18	0.16	0.16	0.15	0.14	0.17	0.22	0.24

3.3 Amplification Factors

A probabilistic framework based on a conditional probabilistic approach is used to build the demand hazard curves, expressing the MAF (ν_{EDP}) of exceedance of the monitored EDPs. The probabilistic procedure, described in detail in Scozzese et al. [16], has been applied by considering different amplification factors relevant to velocity (γ_v) and stroke (γ_Δ). Such values are partially derived from the main international seismic Standards, both European and American [20–22], moreover further values are proposed in this study to further investigate their effectiveness in ensuring the reliability of both the FVDs and the overall structural system and the influence of this design choice. Consequently, values ranging from 1 to 3 have been considered. Moreover, two more limit cases are evaluated for comparison purposes: "No Failure" that is the case where no dampers' failures are permitted (i.e. $\gamma_v = \gamma_\Delta = \infty$), and "Bare Model", which represents the frame without FVDs.

4 Results

In this section, few results in terms of demand hazard curves related to IDR and on the effect of the amplification factors on the sequence of dampers failure among different storeys are presented first. Successively, the problem of damper failure is analysed also in terms of fragility functions.

4.1 Demand Hazard Curves and Failure Probabilities

Figure 5 shows the demand hazard curves of IDR, with respect to the mean annual rate of exceedance ν_D, for each damper typology ($\alpha = 1.0$, $\alpha = 0.3$). In the figure the coloured solid lines refer to the different γ-factors, while the two horizontal dotted lines identify, respectively, the design hazard level 0.0021 yr^{-1} (grey one) corresponding to 10th IM level and the target risk level desired for the structural systems, 2×10^{-4} yr^{-1}, (red dotted line). It can be observed that the case without amplification factors (blue curve, $\gamma_v = \gamma_\Delta = \gamma = 1$) deviates from the "no failure" case (red curve) at $\nu = 10^{-2}$ yr^{-1} because the damper design is carried out based on the first mode response approximation, which is less accurate for the medium and high-rise buildings, whose response is significantly influenced by higher-order modes. The values of $\gamma = 3$ lead to the achievement of the desired target risk only in the case of linear FVDs. Moreover, the efficiency of the added dampers reduces for decreasing MAF of exceedances, for both the cases of linear and nonlinear dampers. The nonlinear behaviour of the frame (and consequent period-elongation), indeed, has a significant contribution and affects the dampers performance and their efficiency. It is also observed that the beneficial effect in terms of IDR reduction reduces for increasing levels of nonlinearity of the dampers (i.e. lower alpha values).

Regarding the effect of the amplification factors on the sequence of dampers failure among different storeys, Fig. 6 shows the average trends of the $F_d/F_{d,max}$ ratios (together with the 16th and 84th response percentiles by dotted lines), for different intensity levels and for different building storeys (i.e., floors n. 1, 3, 6, 9). It is worth to remember that the FVDs design has been conducted at the tenth IM level. With reference to Fig. 6,

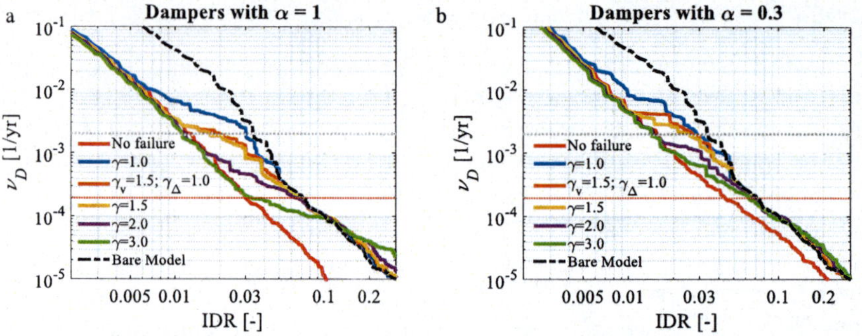

Fig. 5. Demand hazard curves of the IDR parameter for different damper's amplification factors. Case of building with (a) linear dampers $\alpha = 1$; and (b) $\alpha = 0.3$ nonlinear dampers.

it is observed that when one device fails, all the other devices fail too, even though at different times. When no amplification factors are used ($\gamma_v = \gamma_\Delta = 1$), the curves are almost perfectly overlapped (Fig. 6a), thus there are not cases in which some devices remain active while others fail. The response changes if higher γ-factors are used, as shown in Fig. 6b ($\gamma_v = \gamma_\Delta = 3$). Indeed, beside the curve shifting towards higher IMs, curves of dampers belonging to the upper floors slightly deviate at the highest seismic intensities, by testifying a lower average rate of failures, and thus a concentration of collapse cases in the dampers at the lower floors. It is noteworthy that the damage concentration on the structural elements at the storeys with failed (inactive) dampers may result in a building performance worse than the one of the bare frame case, this result can be also related to the design method used for the FVDs, which disregards higher order modes.

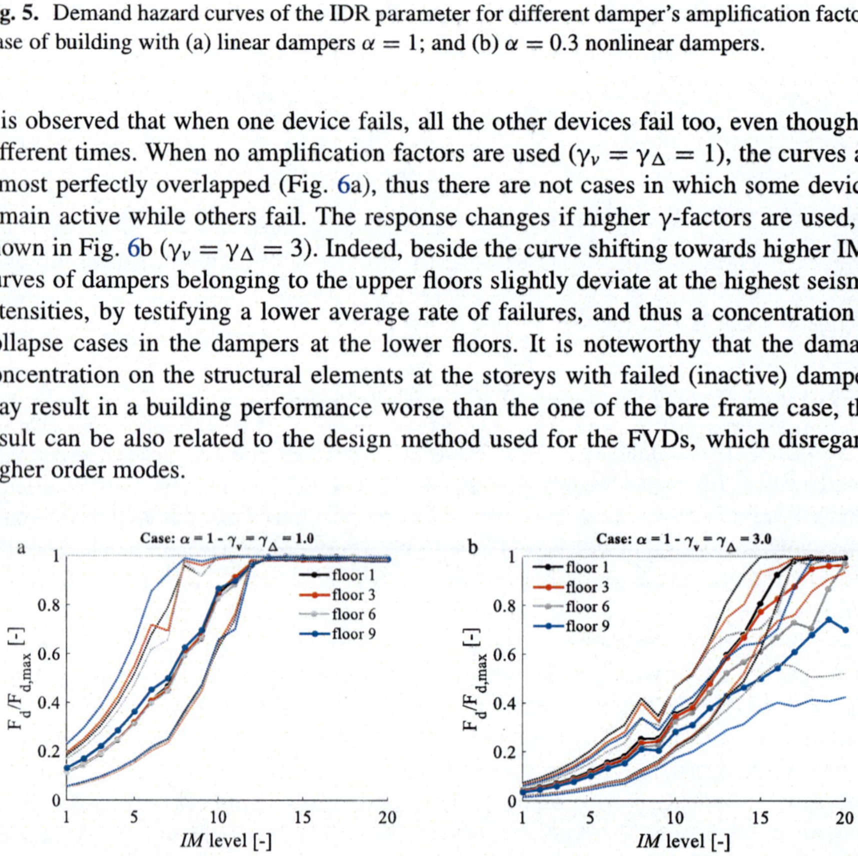

Fig. 6. Average and 16th and 84th response percentiles of the $F_d/F_{d,max}$ – IM trends at different floors for γ factors (a) $\gamma_v = \gamma_\Delta = 1$; (b) $\gamma_v = \gamma_\Delta = 3$

4.2 Fragility Functions

The damper reliability is finally analysed by showing in Fig. 7 the fragility functions for, respectively, the linear and nonlinear dampers placed at the first and last floors, for the different γ-factors analysed. It can be observed that the absence of amplification factors leads to high damper failure probabilities (>50%) at seismic intensities lower than the design level (i.e., IM = 0.38g), and from IM = 1.0g a 100% probability of damper failure is obtained. The beneficial effect of γ-factors larger than 1 is testified by the shifting of the fragility curves towards higher seismic intensities. Moreover, failure probabilities also reduce by moving from floor 1 to floor 9, especially for linear devices, although no differences are observed among the floors for the case without amplification ($\gamma_v = \gamma_\Delta = 1$).

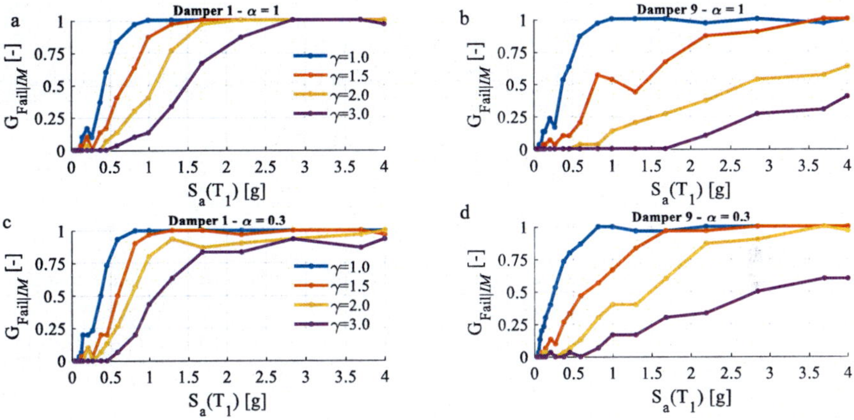

Fig. 7. Damper collapse fragility at: (a) 1st floor with linear FVDs; (b) 9th floor with linear FVDs; (c) 1st floor with non-linear FVDs; (d) 9th floor with non-linear FVDs.

5 Conclusions

In the present paper a parametric investigation is carried out to explore the influence of the safety factors (γ-factors) on the seismic risk of a benchmark structure equipped with linear and nonlinear FVDs. The brittle failure of dampers, occurring when their internal force attains the device strength, has been considered in the probabilistic analyses. Based on the outcomes of the present study, the following conclusions can be drawn:

- The likelihood of the damper failure as well as the "rapidity" of the response transition from damped to bare (or partially damped) structural system are governed by the magnitude of the two amplification factors (γ_Δ, γ_v) adopted for damper stroke and velocity.
- If no amplification is provided ($\gamma_v = \gamma_\Delta = 1.0$), the dampers probability of failure is higher than the design hazard level (assumed equal to 0.0021 yr^{-1} in this work), thus, dampers experience failure at intensity levels lower than the design one, while the use of amplification factors higher than 1.0 allows attaining lower failure probabilities, and this beneficial effect is more significant for larger γ-factors.

- Nonlinear dampers exhibit higher failure probabilities than the linear ones, thus the safety coefficient should depend on the degree of nonlinearity of the adopted dampers.

References

1. Constantinou, M.C., Symans, M.D.: Experimental and Analytical Investigation of Seismic Response of Structures with Supplemental Fluid Viscous Dampers. National Center for earthquake engineering research, Buffalo NY (1992)
2. Pavlou, E., Constantinou, M.C.: Response of nonstructural components in structures with damping systems. J. Struct. Eng. **132**(7), 1108–1117 (2006)
3. Lavan, O., Dargush, G.F.: Multi-objective evolutionary seismic design with passive energy dissipation systems. J. Earthquake Eng. **13**(6), 758–790 (2009)
4. Pavia, A., Scozzese, F., Petrucci, E., Zona, A.: Seismic upgrading of a historical Masonry Bell Tower through an internal dissipative steel structure. Buildings **11**(1), 24 (2021)
5. Gioiella, L., Tubaldi, E., Gara, F., Dezi, L., Dall'Asta, A.: Modal properties and seismic behaviour of buildings equipped with external dissipative pinned rocking braced frames. Eng. Struct. **172** (2018a)
6. Gioiella, L., Tubaldi, E., Gara, F., Dezi, L., Dall'Asta, A.: Stochastic seismic analysis and comparison of alternative external dissipative systems. Shock Vib. **47** (2018b)
7. H2020-LC-SC3-EE-2019: e-SAFE, Energy and Seismic AFfordable rEnovation solutions (2020–2024)
8. H2020-EE-2016-PPP: Pro-GET-OnE, proactive synergy of inteGrated efficient technologies on buildings. Envelopes (2017–2021)
9. Hwang, J.S., Lin, W.C., Wu, N.J.: Comparison of distribution methods for viscous damping coefficients to building. Struct. Infrastruct. Eng.: Maintenance, Manag., Life-Cycle Des. Perform. **9**(1), 28–41 (2010)
10. Tubaldi, E., Gioiella, L., Scozzese, F., Ragni, L., Dall'Asta, A.: A design method for viscous dampers connecting adjacent structures. Front. Built Environ. **6**, 25 (2020)
11. Altieri, D., Tubaldi, E., De Angelis, M., Patelli, E., Dall'Asta, A.: Reliability-based optimal design of nonlinear viscous dampers for the seismic protection of structural systems. Bull. Earthq. Eng. **16**(2), 963–982 (2018)
12. Pollini, N., Lavan, O., Amir, O.: Optimization-based minimum-cost seismic retrofitting of hysteretic frames with nonlinear fluid viscous dampers. Earthquake Eng. Struct. Dynam. **47**(15), 2985–3005 (2018)
13. Scozzese, F., Dall'Asta, A., Tubaldi, E.: Seismic risk sensitivity of structures equipped with anti-seismic devices with uncertain properties. Struct. Saf. **77**, 30–47 (2019)
14. Miyamoto, H.K., Gilani, A.S., Wada, A., Ariyaratana, C.: Identifying the collapse hazard of steel special moment-frame buildings with viscous dampers using the FEMA P695 methodology. Earthq. Spectra **27**(4), 1147–1168 (2011)
15. Caprili, S., Panzera, I., Salvatore, W.: Resilience-based methodologies for design of steel structures equipped with dissipative devices. Eng. Struct. **228** (2021)
16. Scozzese, F., Gioiella, L., Dall'Asta, A., Ragni, L., Tubaldi, E.: Influence of viscous dampers ultimate capacity on the seismic reliability of building structures. Struct. Saf. (2021)
17. Barroso, L.R., Winterstein, S.: Probabilistic seismic demand analysis of controlled steel moment-resisting frame structures. Earthquake Eng. Struct. Dynam. **31**(12), 2049–2066 (2002)
18. Scozzese, F., Tubaldi, E., Dall'Asta, A.: Assessment of the effectiveness of multiple-stripe analysis by using a stochastic earthquake input model. Bull. Earthq. Eng. 1–37 (2020)

19. European Committee for Standardization: EN 15129:2010—Antiseismic Devices. Brussels, Belgium (2010)
20. European Committee for Standardization: Eurocode 8-Design of Structures for Earthquake Resistance. Part 1: General Rules, Seismic Actions and Rules for Buildings, Brussels, Belgium (2004)
21. ASCE (American Society of Civil Engineers): Minimum Design Loads for Buildings and Other Structures. Standard ASCE/SEI 7-10 (2010)
22. McKenna, F.: OpenSees: a framework for earthquake engineering simulation. Comput. Sci. Eng. **13**(4), 58–66 (2011)

Preliminary Numerical Analysis of the Response of Base-Isolated SDOF Systems Constrained by Two Deformable Devices Under Seismic Excitations

Giuseppe Perna(✉), Maurizio De Angelis, and Ugo Andreaus

Department of Structural and Geotechnical Engineering, Sapienza University of Rome, Via Eudossiana 18, 00148 Rome, Italy
{giuseppe.perna,maurizio.deangelis,ugo.andreaus}@uniroma1.it

Abstract. Seismic isolation is a passive control methodology widely used for the mitigation of the dynamic response of structures. However, significant seismic actions can induce considerable displacements in these structures, which in turn can cause damage to the isolation system or impacts against adjacent structures in the event of insufficient seismic gap. A technique to prevent the occurrence of these negative phenomena can be obtained by interposing dissipative and deformable devices, called bumpers between the structure and the adjacent ones. Therefore, in this work the response, from a numerical point of view, of single-degree-of-freedom, SDOF, system isolated at the base subjected to known seismic actions and whose displacements are limited by bumpers has been studied. The system was designed according to an optimality relationship that relates the mechanical parameters of the isolator with those of the bumper, and is independent of the gap.

The SDOF system investigated is representative of a typical base-isolated structure, therefore, the system exhibits a damping ratio $\xi = 10\%$ and a system natural period $T_n = 3.00$ s. The dynamic response of the system obtained by impact with an appropriately designed bumper is compared with the response of the same system obtained without the impact and with an impact whit rigid bumper. Finally, the influence of the stiffness of the bumper, designed with the optimality relationship, and the gap is evaluated. The results show a considerable reduction in displacements compared to a modest increase in accelerations with respect to the no-impact condition, evidence that highlights the effectiveness of the optimality relationship used.

Keywords: Vibro-impact systems · Dynamic response control · Gap · Bumper · Numerical modelling · Seismic excitations

1 Introduction

Seismic isolation is a widely used strategy for controlling the dynamic response of bridges, works of art, equipment and nonstructural components as well as both new and existing buildings. The concept on which the base-isolation system is based is to

G. P. Cimellaro (Ed.): WCSI 2022, LNCE 309, pp. 948–965, 2023.
https://doi.org/10.1007/978-3-031-21187-4_84

reduce the natural frequency of the structure to values far from the frequencies of seismic excitation that have strong energy contributions.

However, reducing the natural frequency of the system, results in greater flexibility of the structure which will experience greater displacements of the isolated floor. In the presence of pulse-like earthquakes (near-fault), these displacements may be excessively large [1, 2] exceeding the design seismic displacement or gap. This phenomenon can result in damage to the isolation devices on the one hand, as they exceed the limit deformation, and on the other hand, pounding of the structure with surrounding structures. The pounding has been considerably studied in the literature [3–5]; it is associated with considerable increases in dynamic response as increases in the maximum value of acceleration and inter-story displacement. The occurrence of increases in accelerations, caused by the impact, can cause not only structural damage but also damage to nonstructural elements such as equipment and facilities.

One strategy for mitigating the adverse effects due to this event is the interposition of deformable and dissipative devices, called bumpers. This strategy has been extensively covered by authors who have carried out many studies of both numerical and experimental nature [6–14] on single-degree-of-freedom, SDOF, systems subjected to a stepwise sine sweep base excitation, highlighting a remarkable potential of these systems as devices for control of dynamic response. The design of these control systems, derived from these studies, makes it possible to reduce the parameters governing the impact from three (gap, stiffness and damping ratio of the bumper) to two, because the formulation of an optimality relationship links stiffness to the damping ratio of the bumper [14].

This work aims to extend the studies carried out by the authors on SDOF systems subjected to sine base excitation, to the case of seismic excitations. Therefore, different scenarios obtained from numerical analyses on SDOF systems subjected to seismic actions of two earthquakes known in the literature (Irpinia-Sturno, Cape Mendocino-Petrolia) are compared here, varying the parameters governing the impact with the bumpers (such as namely bumper stiffness and initial gap).

The work is developed as follows: Sect. 2 reports the parameters governing the model and the equations governing motion; Sect. 3 introduces the optimality relationship obtained for sine base excitation by the authors [14]; Sect. 4 compares the scenarios obtained with three different values of bumper stiffness and four values of the initial gap with the scenarios that characterize the two limit cases, free flight (no-impact) and so-called rigid impact; finally, Sect. 5 reports the conclusions and possible developments of future work.

2 Model and Equations of Motion

The numerical analyses conducted refer to the system in the Fig. 1. This is single-degree-of-freedom, SDOF, system consisting of a mass M and a damper D that is constrained by two deformable and dissipative bumpers, arranged symmetrically on both sides of the mass with an initial gap G_0. The devices are referred to as B_j, where $j = R$ refers to the right bumper, and $j = L$ refers to the left bumper.

The damper is modeled as a linear elastic spring, with stiffness K, in parallel with a dashpot of damping coefficient C. In analogy, the bumpers are modeled as a linear elastic element, of stiffness K_j, and with a linear viscosity element, of damping coefficient C_j ($j = R, L$).

The system is excited by an accelerogram $A_t(t)$, which in a first analysis is the accelerogram of the Irpinia earthquake (1980, Sturno), and in a second analysis is the accelerogram of the Cape Mendocino earthquake (1992, Petrolia).

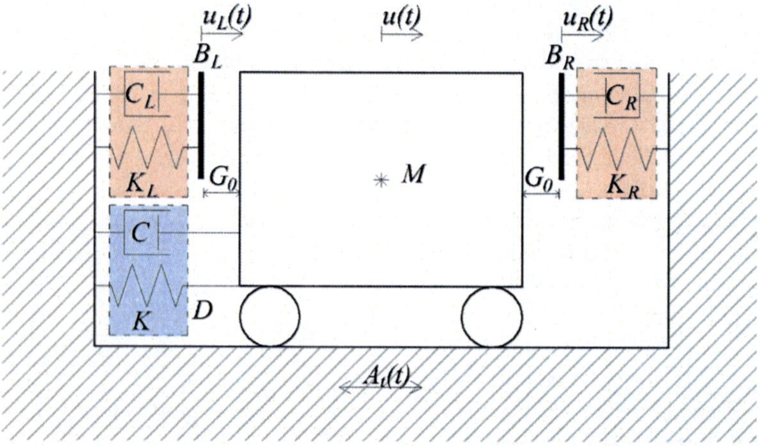

Fig. 1. Model of the system: the cyan box contains the mechanical elements that characterize the damper, and the red boxes contain the mechanical elements that characterize the bumpers.

During the motion, the mass may or may not be in contact with one of the two bumpers. Introducing gap functions $G_j(t)$, which represent the size of the right ($j = R$) and the left ($j = L$) gap in time, and the Heaviside functions ψ_k, which govern the contact ($k = 1$) and separation ($k = 2$), yields the following definitions:

$$G_R(t) = G_0 + [u_R(t) - u(t)];\ G_L(t) = G_0 + [-u_L(t) + u(t)] \tag{1}$$

$$Contact\quad \psi_1\left[G_j(t)\right] = \begin{cases} 0,\ G_j(t) > 0 \\ 1,\ G_j(t) = 0 \end{cases};\ (j = R, L) \tag{2}$$

$$Separation\quad \psi_2\left[F_j(t)\right] = \begin{cases} 0,\ F_R(t) \le 0 or F_L(t) \ge 0 \\ 1,\ F_R(t) > 0 or F_L(t) < 0 \end{cases} \tag{3}$$

where $F_j(t)$ is the contact force occurring during the contact time with the jth bumper and is equal to $F_j(t) = C_j\dot{u}_j(t) + K_ju_j(t)$ ($j = R, L$); the equations of motion can be written in the following compact form:

$$M\ddot{u}(t) + C\dot{u}(t) + Ku(t) + \psi_1\left[G_j(t)\right]\cdot\psi_2\left[F_j(t)\right]\cdot F_j(t) = -MA(t) \tag{4}$$

$$F_i(t) = 0;\ (if\ j = R,\ then\ i = L;\quad if\ j = L,\ then\ i = R) \tag{5}$$

where $u(t)$ and $u_j(t)$ ($j = R, L$) are the relative displacement (with respect to the ground) of the mass and of the bumpers respectively; the single dot ($\dot{}$) and double dot ($\ddot{}$) represent the first and second derivative respectively with respect to time t.

3 Optimal Design of Bumpers Parameters

The optimality relationship obtained in [14] is recalled here. In that work, the effect of the presence of deformable and dissipative bumpers on the dynamic response of a SDOF system subjected to sine excitation was studied. It was observed that the presence of bumpers is not always unfavorable compared to no-impact (free flight conditions, FF) if bumper parameters are selected appropriately.

Two dimensionless parameters were introduced as follows:

$$\gamma_j = \frac{C_j}{C} \tag{6}$$

$$\lambda_j = \frac{K_j}{K} \tag{7}$$

where γ_j is the ratio between the viscous damping coefficient C_j of the jth bumper and the viscous damping coefficient C of the damper; e λ_j is the ratio between the stiffness K_j of the jth bumper and the stiffness K of the damper.

Through parametric numerical analysis, for each investigated gap value G_0, and for fixed values of the system damping ratio $\xi = C/2\sqrt{MK}$ and γ_j, a relationship was obtained that defines the optimal value of the stiffness ratio λ_j that minimizes the peaks of the absolute acceleration of the mass in the primary resonance. This λ_j value coincides with the full utilization of the bumpers because they have time to completely recover their deformation, totally dissipating the deformation energy accumulated until then, and they do not remain inactive because the next impact occurs practically immediately after complete recovery. The optimality relationship obtained was as follows:

$$\frac{\gamma_j}{\lambda_j} = \frac{1}{2\xi} \tag{8}$$

This relationship turns out to be independent of the gap and makes it possible to reduce the number of parameters characterizing the impact, from three $(G_0, \lambda_j, \gamma_j)$ to two (G_0, λ_j), since it links λ_j and γ_j. Other considerations have to be made with reference to the value of λ_j to be considered in the analyses: based on research [14, 15], it is considered preferable to adopt values of bumper stiffness K_j comparable to the damper stiffness K, that is $\lambda_j \simeq 1$.

In this work the values $\lambda_R = \lambda_L = \lambda$, $\gamma_R = \gamma_L = \gamma$, as well as an equal initial gap value G_0 for both sides have been assumed; moreover, the bumper design refers to the optimality relationship (8) to a value of the stiffness ratio $\lambda = 1$ to define the impact case with a bumper considered preferable (PB).

4 Numerical Analysis of Response Scenarios

Based on the model given in Sect. 2 and depicted in Fig. 1, the effect of introducing deformable and dissipative bumpers on the dynamic response of an SDOF system is studied through numerical analysis.

The analyses are conducted on a system characterized by damping ratio $\xi = 0.1$, natural period $T_n = 3.00$ s, and with mass $M = 550$ Kg. The examined responses are:

- Absolute o total acceleration of mass $a(t) = \ddot{u}(t) + A(t)$;
- relative displacement of mass $u(t)$;
- force acting on the mass $F(t) = C\dot{u}(t) + Ku(t) + \psi_1[G_j(t)] \cdot \psi_2[F_j(t)] \cdot F_j(t), (j = R, L)$;
- bumper contact force $F_j(t) = C_j\dot{u}_j(t) + K_j u_j(t), (j = R, L)$;
- bumper deformation $u_j(t)(j = R, L)$;
- bumper displacement, that is the bumper deformations increased by the initial gap, $v_R(t) = u_R(t) + G_0, v_L(t) = u_L(t) - G_0$.

In a first set of analyses, whose results are given in Sect. 4.1, three situations are compared. The first case presents no-impact and is defined as Free Flight (FF). In the other two cases, instead, two symmetrical obstacles placed at an equal initial distance $G_0 = 0.20$ m are introduced [16, 17]: the rigid bumper (RB), is characterized by $\lambda = 100$ and $\gamma = 3.5$, and the preferable bumper (PB), whose stiffness is comparable to that of the damper, characterized by $\lambda = 1$ and $\gamma = 5$ which satisfies the optimality relationship (8).

In Sect. 4.2, the effect of varying the stiffness of the bumper, designed by adopting γ that satisfies the optimality relationship (8), on the dynamic response is analyzed. The three cases examined are: (1) bumper less stiffs than damper $\lambda = 0.25, \gamma = 1.25$; (2) bumper stiffer than damper $\lambda = 2, \gamma = 10$; (3) bumper stiffness equal to damper stiffness (PB) $\lambda = 1, \gamma = 5$.

Section 4.3 explores the influence of varying the distance between mass and obstacle, gap $G_{0i}(i = 1, 2, 3, 4)$. The adopted bumper is the preferred bumper (PB), while the investigated gaps are $G_{01} = 0.20$ m, $G_{02} = 0.10$ m, $G_{03} = 0.05$ m, $G_{04} = 0.00$ m.

While in the first three Sects. 4.1, 4.2, and 4.3, the SDOF system is subjected to the accelerogram of the Irpinia earthquake, in Sect. 4.4, scenarios related to the different gaps G_{01}, G_{02}, G_{03} and G_{04} (which were previously defined) under the excitation of the accelerogram of the Cape Mendocino earthquake are compared.

4.1 Comparison of Free Flight (FF), Rigid Bumper Impact (RB) and Preferable Bumper Impact (PB)

The effect of interposing a particular so-called "preferable" bumper, designed according to the optimality relationship (8) and having stiffness equal to that of the damper ($\lambda = 1, \gamma = 5$), on the dynamic response is studied. The scenario obtained in the PB case is compared with the response of both the system without impact (FF) and with impact with rigid bumper (RB), defined by $\lambda = 100$ e $\gamma = 3.5$.

Figure 2 shows the response in terms of total mass acceleration $a(t)$ as a function of time t. Free flight reports lower values of peak acceleration than the other two scenarios. The RB scenario shows the highest peak accelerations, with an absolute maximum that is 507% of FF; the scenario defined by the PB, records the maximum value of +45% with respect to free flight, while, when compared to the RB, it experiences significantly lower accelerations, reduced by 71%.

Figure 3 shows the response in terms of relative mass displacement with respect to the ground $u(t)$ as a function of time t. The scenario obtained in the RB case is characterized by the lowest values of peak displacements, with a maximum of 0.22 m.

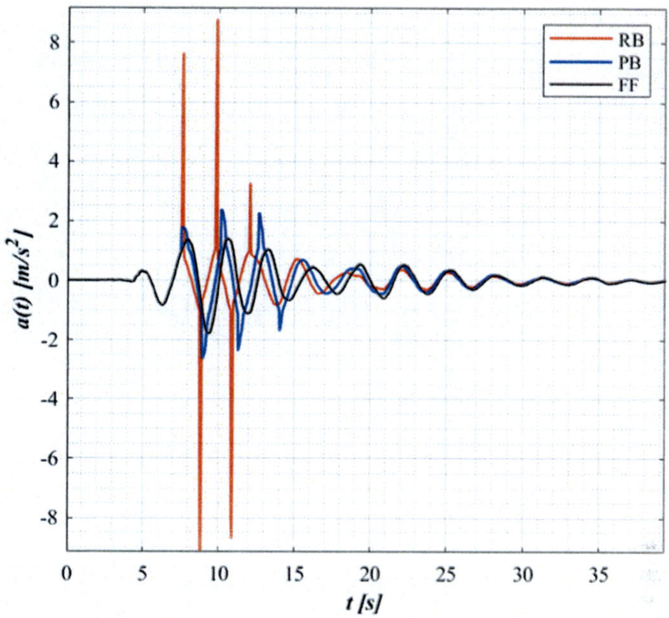

Fig. 2. Time history of the absolute acceleration of the mass *a(t)*, due to Irpinia earthquake and for the three cases analyzed: in red, the response obtained by impact with rigid bumper (RB, $\lambda = 100$ and $\gamma = 3.5$); in black, the response obtained without impact (FF); and in blue, the response obtained by impact with preferable bumper (PB, $\lambda = 1$ and $\gamma = 5$).

Free flight, on the contrary, reports larger displacements than the other two conditions, with a maximum peak of 0.40 m. Finally, the scenario obtained in the PB case falls between the two extreme scenarios and shows, compared to free flight, a 23% reduction in maximum displacement, which is worth 0.31 m, and a 41% increase, with respect to the RB case.

Figure 4 represents the force-displacement cycles of the system, a graph that summarizes the information provided in the previous graphs, i.e., Figs. 2 and 3. Free flight also highlights the linear behavior of the system, as well as reporting greater displacements and lower forces than the other scenarios, which instead show nonlinearity due to the presence of the impact. The impact with rigid bumper indeed reports an evident change in the slope of the curve thus presenting high reductions in displacement (−45% compared to FF) but in turn important increases in force (+200% compared to free flight). The scenario obtained with the preferable bumper impact, on the contrary, shows a more dissipative behavior and therefore a hysteresis cycle characterized by a larger area. In this scenario, the force acting on the system undergoes a maximum increase, already found in the acceleration response study, of +45% against a displacement reduction of −23% with respect to the free flight response.

After studying the mass response, Figs. 5 and 6 show the bumper responses under the two impact conditions considered. The trend of contact force $F_j(t)$ can be analyzed from the graph of Fig. 5, which displays significantly smaller peaks of force for preferable bumpers than for rigid bumpers by reporting a 79% reduction.

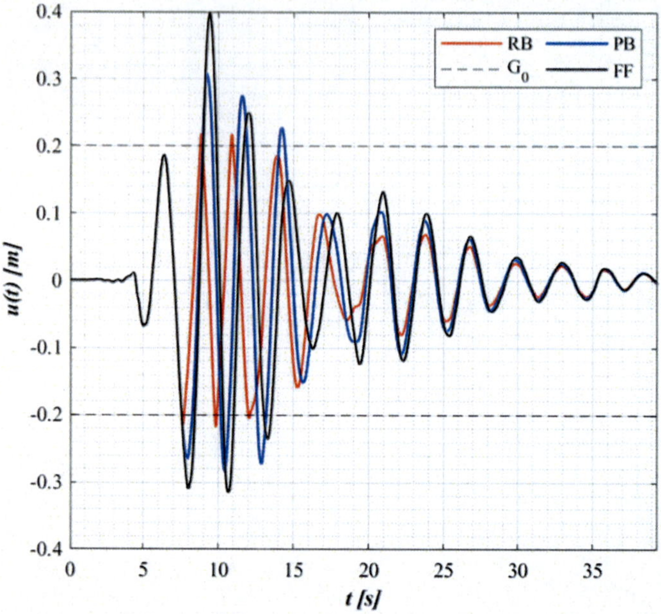

Fig. 3. Time history of the relative displacement of the mass *u(t)*, due to Irpinia earthquake and for the three different cases analyzed: in red the response obtained by impact with rigid bumper (RB, $\lambda = 100$ and $\gamma = 3.5$); in black the response obtained without impact (FF); in blue the response obtained by impact with preferable bumper (PB, $\lambda = 1$ and $\gamma = 5$). The initial condition of gap $G_0 = 0.20$ m is also shown with black horizontal bashed line.

Regarding the bumper deformation increased by the initial gap $v_j(t)$, shown in Fig. 6, it can be seen that the trend of the curves representing the two scenarios is reversed: the preferable bumper reports higher v_j deformation peaks than the rigid bumper by 41% (it turns out to be 475% when considering only the u_j deformation, net of the initial gap G_0). Impact stiffness in the RB case is evidenced by the presence of low-intensity peaks that increase the initial G_0 by a maximum of 9.5%, in contrast to the PB scenario where the increase turns out to be 55%. In addition, the preferred bumpers show complete recovery of deformation before being deformed again by the next impact. This behavior is due to the design of the preferable bumper made according to the optimality relationship (8). From the graphs in Figs. 5 and 6, we can also count the number of impacts of the mass occurring which turns out to be greater in the case with preferable bumper than in the case with rigid bumper.

The response of the bumpers, in the two different cases examined (RB and PB), is summarized by the graph in Fig. 7 representing the bumpers force-displacement cycles. The cycles of preferable bumpers are larger than those of rigid bumpers, which dissipate less energy than the former ones. When the mass shows a displacement $u(t)$ between $\pm G_0$ the mass is in free-flight conditions, so the contact force in the bumpers F_j is zero.

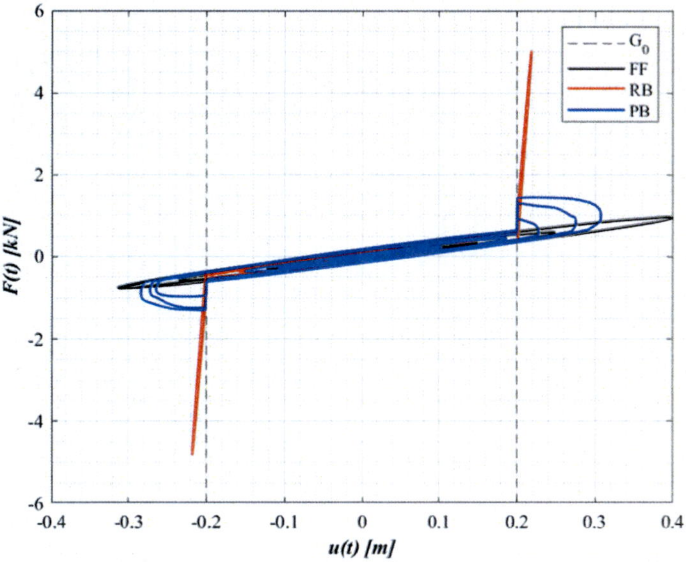

Fig. 4. Force-displacement cycles of the system *F(t)−u(t)*, due to Irpinia earthquake and for the three different cases analyzed: in red the response obtained by impact with rigid bumper (RB, $\lambda = 100$ and $\gamma = 3.5$); in black the response obtained without impact (FF); in blue the response obtained by impact with preferable bumper (PB, $\lambda = 1$ and $\gamma = 5$). The initial condition of gap $G_0 = 0.20$ m is also shown with black vertical bashed line.

4.2 Comparison of Bumper with Different Values $\lambda - \gamma$

As mentioned above, this paragraph studies the effects on the system response of varying the stiffness of bumpers designed with the optimality relationship (8), keeping fixed the gap $G_{01} = 0.20$ m. Three different stiffnesses are considered: the least rigid bumper is characterized by $K_j = 0.25K$; the preferred bumper, of intermediate stiffness, by $K_j = K$; and finally, the most rigid bumper by $K_j = 2K$. These scenarios are also compared with those for the limit cases of rigid bumper impact and free flight.

The response of the system can be interpreted and commented on through the graphs in Fig. 8 showing the force-displacement cycles of the system for the respective cases investigated. All three bumper impact curves designed with the optimality relationship (8) exhibit large force-displacement hysteresis cycles commensurate with energy dissipation. The most rigid optimal bumper reports a 32% reduction in displacement with respect to free flight (FF) and a 55% reduction in acceleration with respect to rigid bumper impact (RB); the least rigid optimal bumper reports a −9.5% reduction in displacement with respect to the FF scenario and a −80% reduction in acceleration with respect to the RB scenario. In contrast, the bumper with stiffness equal to the damper stiffness reports a 23% reduction in displacement and a 71% reduction in acceleration with respect to FF and RB, respectively. Evaluating the increases in acceleration relative to free flight and displacement relative to rigid bumper impact, we see that the stiffer optimal bumper shows increases of 128% and 23%, respectively; the less rigid optimal bumper, on the other hand, shows +8% in acceleration and +68% in displacement; finally, the optimal

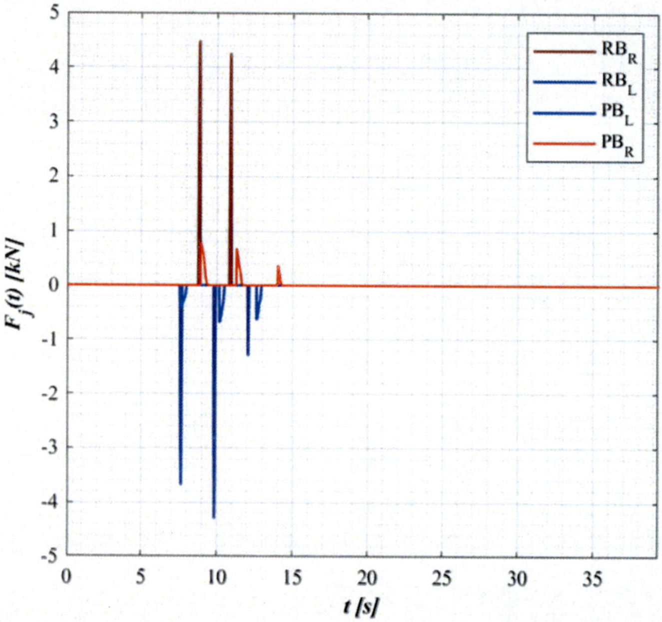

Fig. 5. Time history of the contact force in the bumpers $F_j(t)$ ($j = R,L$), due to Irpinia earthquake and for the two cases with impact: in dark red (right bumper, R) and dark blue (left bumper, L) the response obtained for impact with rigid bumper (RB, $\lambda = 100$ and $\gamma = 3.5$); in red (right bumper, R) and blue (left bumper, L) the response obtained for impact with preferable bumper (PB, $\lambda = 1$ and $\gamma = 5$).

bumper with $\lambda = 1$ reports an increase in acceleration relative to the FF case of 55% and in displacement relative to the RB case of 41%.

Analysis of Fig. 9 which represents the force-displacement cycles of the bumpers, shows a reduction in the amplitude of hysteresis cycles as the bumper stiffness decreases. Moreover, displacement increases as stiffness decreases, reporting, compared with the RB case, an increase for $\lambda = 0.25$ of 55%, for $\lambda = 1$ of 41% and for $\lambda = 2$ of 27%. The contact force, on the opposite, decreases as the stiffness decreases. Thus, compared to the rigid bumper impact, for smaller λ there is a reduction of 95%, for intermediate λ a reduction of 82% and finally, for larger λ a reduction of 62%.

4.3 Comparison of Different Gaps

The effects obtained by interposing bumpers (designed with the optimality relationship and with stiffness equal to that of the damper) between mass and obstacle for different initial gaps G_0, are evaluated in this subsection. Four different cases defined by four distinct G_0 are examined: a relatively large seismic gap (0.20 m), two small gaps (0.10 m, 0.05 m) and null gap, i.e. bumper adjacent to the mass (0.00 m). These four cases are always compared with respect to free flight and rigid bumper impact for gap of 0.20 m.

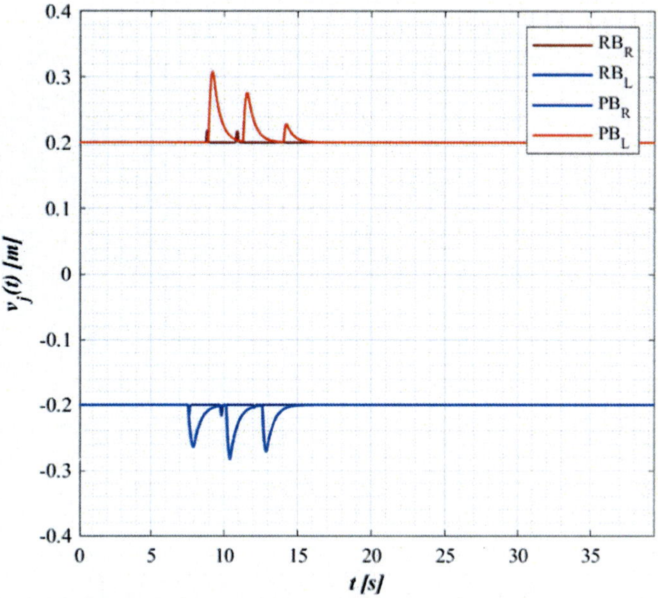

Fig. 6. Time history of the bumper displacement $v_j(t)$ ($j = R,L$), due to Irpinia earthquake and for the two cases with impact: in dark red (right bumper, R) and dark blue (left bumper, L) the response obtained for impact with rigid bumper (RB, $\lambda = 100$ and $\gamma = 3.5$); in red (right bumper, R) and in blue (left bumper, L) the response obtained for impact with preferable bumper (PB, $\lambda = 1$ and $\gamma = 5$).

Observing of Fig. 10 shows a decrease in both maximum displacement and total force acting on the system in going from $G_{01} = 0.20$ m to $G_{04} = 0.00$ m. Comparing these results with free flight shows a decrease in displacement that is for G_{01} of 23%, for G_{02} of 45%, for G_{03} of 55% and for G_{04} of 67%. Comparison in terms of forces, with respect to free flight, shows for G_{01} an increase of 45%, for G_{02} of 30%, for G_{03} of 7% while for G_{04} there is a reduction of 16%.

The null gap scenario obtained with the preferred bumpers, designed with the optimality relationship (8) and with $\lambda = 1$, is therefore also found to be favorable in terms of acceleration compared to free flight, as well as in terms of displacement, where it exhibits the greatest reduction compared to the other gaps.

Comparison of the same scenarios with rigid impact shows that only in scenario G_{01} there is an increase in displacement (+41%), while for G_{02} we obtain an equal mass displacement and for scenarios G_{03} and G_{04} a reduction in displacement of 18% and 41% respectively is found. Comparing the forces acting on the system, on the other hand, there is an increasing decrease ranging from −71% for G_{01} to −83% in the G_{04} case.

Figure 11 shows a reduction in bumper deformation, u_j, as the gap decreases. This deformation, with respect to the RB scenario, reports increases of 475% for G_{01}, of 600% for G_{02}, of 650% for G_{03} and of 660% for G_{04}. Referring to the bumper deformation increased by the initial gap, we find a reduction in the maximum value of v_j as the gap decreases, which, compared with the RB case, reports an increase of 41% in the largest

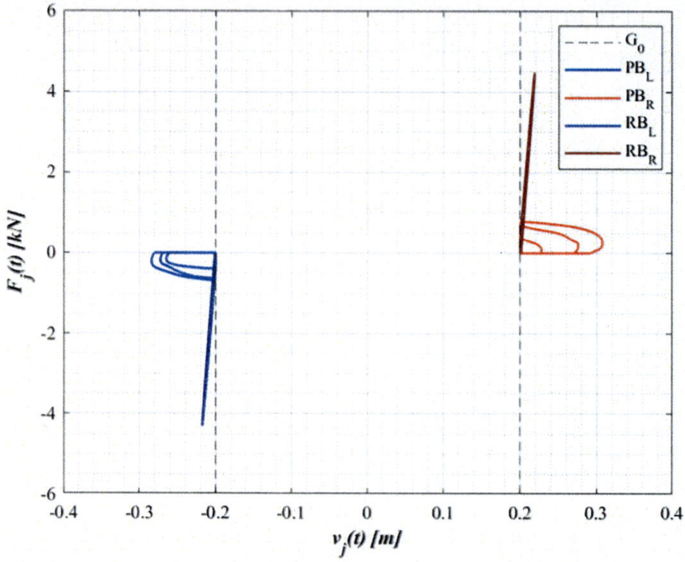

Fig. 7. Force-displacement cycles of the bumpers $F_j(t)$-$v_j(t)$ ($j = R,L$), due to Irpinia earthquake and for the two cases with impact: in dark red (right bumper, R) and dark blue (left bumper, L) the response obtained for impact with rigid bumper (RB, $\lambda = 100$ and $\gamma = 3.5$); in red (right bumper, R) and in blue (left bumper, L) the response obtained for impact with preferable bumper (PB, $\lambda = 1$ and $\gamma = 5$). The initial gap condition, $G_0 = 0.20$m, is shown with black vertical bashed line.

gap case, to a maximum value equal to that of rigid bumper impact in the 0.10 m gap case, until a reduction of 18 and 41% is obtained for the 0.05 m gap and the zero gap, respectively.

The contact force decreases as the gap decreases reporting decreases of 82% in the case of larger gap, 83% for gap equal to 0.10 m, 85% for gap 0.05 m and 89% for null gap.

Hysteresis cycles in addition, as the gap decreases, show greater energy dissipation commensurate with more area enclosed by the cycle itself.

4.4 Response Scenarios Under the Excitation Due to the Cape Mendocino Earthquake

In this Paragraph, the different G_{0i} gap scenarios ($i = 1, 2, 3, 4$) obtained under the Irpinia earthquake are compared with those obtained under the Cape Mendocino earthquake action.

Figure 12 shows the force-displacement cycles of the system, while Fig. 13 shows the same cycles relative to the bumpers. As the gap increases, the maximum force acting on the system does not show a monotonic trend as could be recognized for the Irpinia earthquake (Fig. 10). The maximum force turns out to be always greater than the maximum force in free flight: for G_{01}, the maximum force is increased of 131%; for G_{02} of 118%; for G_{03}, instead of showing a reduction in the percentage increase compared

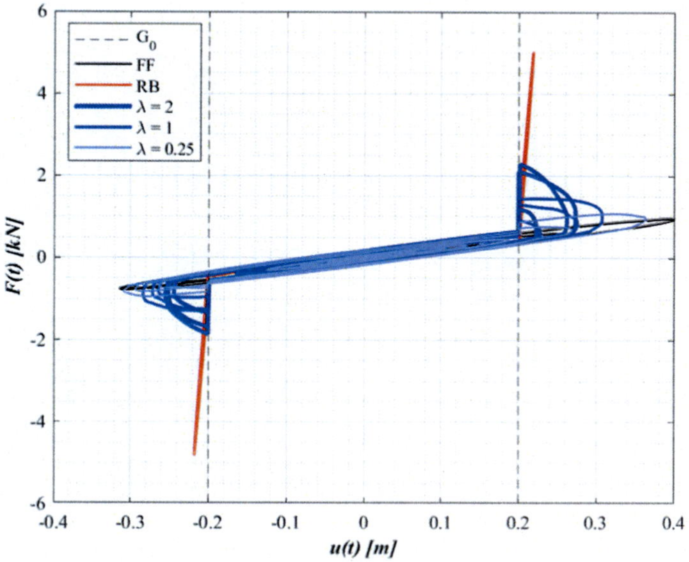

Fig. 8. Force-displacement cycles of the system $F(t)-u(t)$, due to Irpinia earthquake and for three different bumper stiffnesses analyzed, comparing them with FF and RB. The red color represents the response obtained by impact with rigid bumper (RB, $\lambda = 100$ and $\gamma = 3.5$), the black represents the response obtained without impact (FF), while the blue color, with different thicknesses and shades, represents the responses obtained by impact with bumpers designed according to the optimality relationship. Less thickness and lighter shades refer to less rigid bumper ($\lambda = 0.25$ and $\gamma = 1.25$); intermediate thickness and shades refer to bumper with stiffness equal to damper stiffness ($\lambda = 1$ and $\gamma = 5$); greater thickness and darker shades refer to bumper with greater stiffness ($\lambda = 2$ and $\gamma = 10$). The initial gap condition, $G_0 = 0.20$m, is shown with black vertical bashed line.

to the FF case, the maximum force shows an increase of 128%; finally, for the G_{04} case, an increase of 100% is observed. This analysis shows that the null gap condition results in a lower total force acting on the system than the scenarios offered by the other G_{0i}, nevertheless the maximum force turns out to be higher than in free flight unlike in the Irpinia earthquake where it was reduced. Compared to the RB case, the force undergoes a decrease in all scenarios that is comprised between 82 and 84%.

Regarding displacement, a decrease in the peak is noted as the gap decreases. In contrast to the Irpinia earthquake, which shows the maximum displacements all in the same direction (toward the side defined as right), in the Cape Mendocino case the impact obtained for the various G_{0i} changes this trend: below G_{01} the largest displacements of the mass are found at the opposite side with respect to free flight. The reduction in the maximum displacement, compared to the FF case, turns out to be as follows for the different G_{0i} scenarios: G_{01} results in a reduction of 18.5%; G_{02} of 23%; G_{03} of 34%; finally, the null gap a decrease of 42%. Making a comparison with respect to the RB case, the only scenario that reports a smaller peak displacement is the scenario defined by the mass adjacent to the bumper, which experiences a decrease of 8%. The other

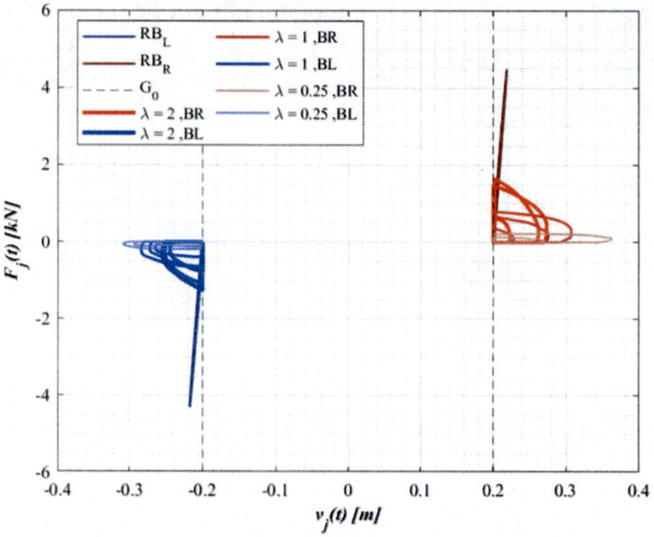

Fig. 9. Force-displacement cycles of the bumpers $F_j(t)$-$v_j(t)$ ($j = R,L$), due to Irpinia earthquake and for three different bumper stiffnesses analyzed, comparing them with FF and RB. The red color represents the response obtained by impact with rigid bumper (RB, $\lambda = 100$ and $\gamma = 3.5$), the black represents the response obtained without impact (FF), while the blue color, with different thicknesses and shades, represents the responses obtained by impact with bumpers designed according to the optimality relationship. Less thickness and lighter shades refer to less rigid bumper ($\lambda = 0.25$ and $\gamma = 1.25$); intermediate thickness and shades refer to bumper with stiffness equal to damper stiffness ($\lambda = 1$ and $\gamma = 5$); greater thickness and darker shades refer to bumper with greater stiffness ($\lambda = 2$ and $\gamma = 10$). The initial gap condition, $G_0 = 0.20$ m, is shown with black vertical bashed line.

scenarios report percentage increases that as the gap increases turn out to be 4% (G_{03}), 20% (G_{02}), finally, for a gap equal to $G_{01} = 0.20$ m, 29%, respectively.

From the study of the force-displacement cycles of the bumpers in Fig. 13, it is observed that maximum v_j decreases with G_0, while the force remains for all scenarios at values that are in the range of 88%, not showing sharp reductions as the gap decreases, as - on the contrary - was observed in Fig. 11 in the case of Irpinia's earthquake.

5 Conclusions and Future Developments

This paper presents a preliminary numerical analysis of the dynamic effects due to interposition of deformable and dissipative bumpers on an SDOF system with typical dynamic parameters of a base-isolated structure and subjected to two different seismic excitations known in the literature. The case studies analyzed involve the design of the bumpers according to the optimality relationship (8) obtained in [14] through the parametric study of the dynamic response of a base-isolated SDOF system and subjected to a stepwise sine sweep base excitation. This formulation makes it possible to minimize the absolute acceleration peaks of the mass in the primary resonance by designing the

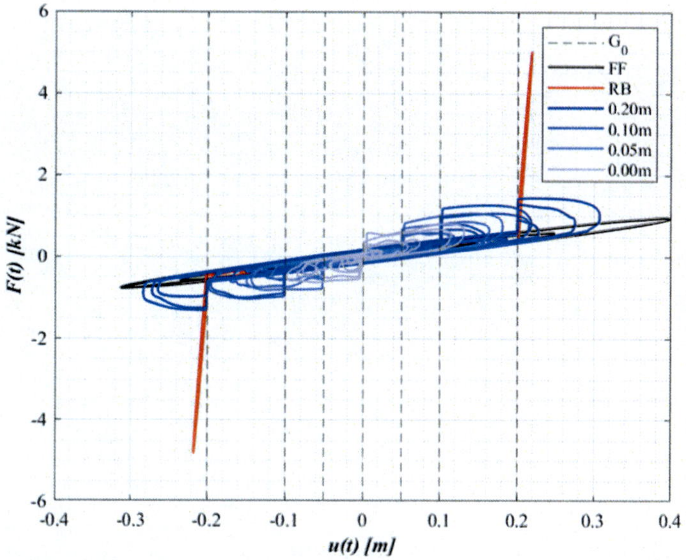

Fig. 10. Force-displacement cycles of the system *F(t)-u(t)*, due to Irpinia earthquake and for four different gap G_0 analyzed comparing them with FF and RB. The red color represents the response obtained by impact with rigid bumper (RB, $\lambda = 100$ and $\gamma = 3.5$), the black color represents the response obtained without impact (FF), while the blue color, with different shades, represents the responses obtained with different initial gaps G_0 for impact with preferable bumper ($\lambda = 1$ e $\gamma = 5$): the reduction in the intensity of blue color results in a halving of the gap G_0, which decreases from 0.20m, to 0.10m, to 0.05m until it take the value 0.00m. The lines representing the gaps G_0 are shown with black vertical bashed line.

bumper in such a way that it is fully exploited: it totally recovers the deformation before the next impact, which occurs immediately after the recovery preventing the bumper from remaining inactive (thus completely dissipating the accumulated deformation energy). The reported studies are characterized by three different values of the stiffness ratio λ and four different values of the gap G_0 that are compared with two limit scenarios: system response without impact (FF) and system response with impact to rigid bumper (RB). The response quantities analyzed are absolute mass acceleration, relative mass displacement, force acting on the mass and the bumper's contact force, deformation and displacement. The main results of this study are the following:

- Using deformable and dissipative bumpers, designed through the optimality relationship (8) and assuming a value of the stiffness ratio $\lambda = 1$, PB, allows a 71% reduction in acceleration, against a 41% increase in displacement, with respect to a defined rigid impact, in the case studied in which the expected earthquake imparts to the system a displacement at the base greater than a possible seismic gap of 0.20 m. In reference to the free-flight (FF) scenario, on the other hand, it allows a maximum displacement reduction of 23% against an acceleration increase of 44%, bearing in mind, however,

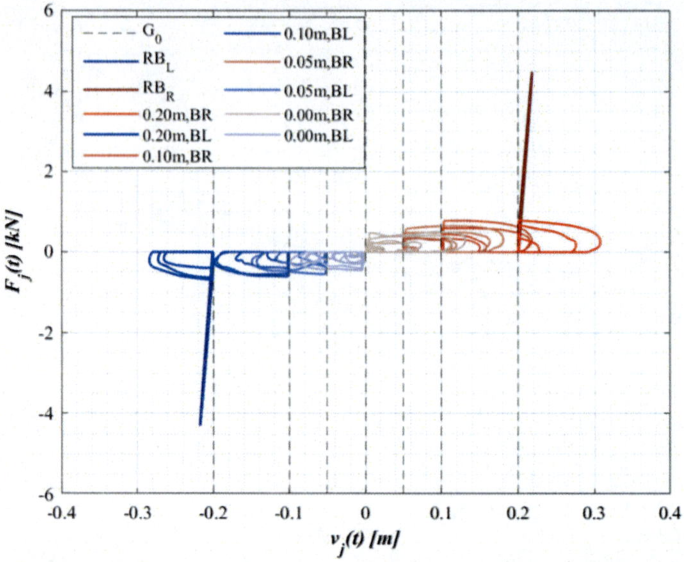

Fig. 11. Force-displacement cycles of the bumpers $F_j(t)$-$v_j(t)$ ($j = R,L$), due to Irpinia earthquake and for four different gap G_0 analyzed comparing them with FF and RB. The red color represents the response obtained by impact with rigid bumper (RB, $\lambda = 100$ and $\gamma = 3.5$), the black color represents the response obtained without impact (FF), while the blue color, with different shades, represents the responses obtained with different initial gaps G_0 for impact with preferable bumper ($\lambda = 1$ e $\gamma = 5$): the reduction in the intensity of blue color results in a halving of the gap G_0, which decreases from 0.20m, to 0.10m, to 0.05m until it take the value 0.00m. The lines representing the gaps G_0 are shown with black vertical bashed line.

that for tipical periods of isolation, the system is subjected to low intensity accelerations and therefore high percentage increases do not necessarily correspond to high absolute acceleration values.

- Comparison with optimal bumpers designed with different values of λ and with a gap $G_0 = 0.20$ m shows that too low stiffness values do not allow large reductions in displacement (−9.5%), while encountering small increases in acceleration (+8.0%), compared to free flight; while high values of λ show greater reductions in displacement (−32%), but high increases in acceleration (+128%), always related to free flight. Comparison with the rigid bumper (RB) case shows, as λ increases, decreasing reductions in acceleration, from least rigid to most rigid by 80%, 71% and 55%, respectively, and decreasing increases in displacement: 68% for $\lambda = 0.25$, 41% for $\lambda = 1$, and 23% for $\lambda = 2$.
- The analysis of the response obtained from different initial gaps, demonstrated that gap is an important design parameter to get the maximum performance on controlling the dynamic response in terms of force and displacement of the system. From the study of the scenarios, we find the possibility that, under a certain threshold of the gap value, we have reductions also in the system force and thus in the total mass acceleration in addition to the reduction of the mass displacement compared to free-flight conditions, as obtained for the Irpinia earthquake under conditions of mass adjacent to the bumper

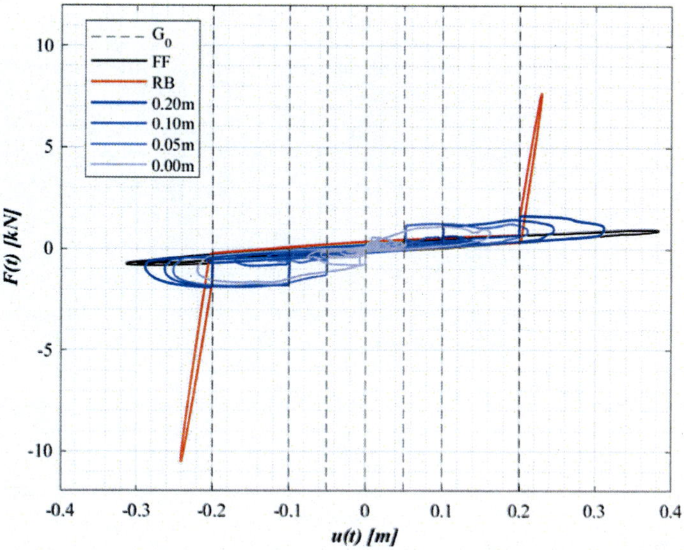

Fig. 12. Force-displacement cycles of the system *F(t)-u(t)*, due to Cape Mendocino earthquake and for four different gap G_0 analyzed comparing them with FF and RB. The red color represents the response obtained by impact with rigid bumper (RB, $\lambda = 100$ and $\gamma = 3.5$), the black color represents the response obtained without impact (FF), while the blue color, with different shades, represents the responses obtained with different initial gaps G_0 for impact with preferable bumper ($\lambda = 1$ e $\gamma = 5$): the reduction in the intensity of blue color results in a halving of the gap G_0, which decreases from 0.20 m, to 0.10 m, to 0.05 m until it take the value 0.00 m. The lines representing the gaps G_0 are shown with black vertical bashed line.

and for a system natural period of $T_n = 3.00$ s. This result contrasts with the general trend that shows increases in acceleration against decreases in displacement for low system frequencies (far from the primary resonance, in the range of isolation periods), as can be observed in the case of the Cape Mendocino earthquake by comparing the scenarios obtained with the various gaps compared to the free flight case.

- Looking at the bumper response, for larger gains on the control of dynamic mass response (in terms of force and displacement reduction), there are larger increases in deformation and contact force, and therefore greater wear of the bumpers.
- The preliminary study conducted herein led to conclude that appropriate deformable (λ) and dissipative (γ) bumpers, properly placed (G_0) are an effective strategy for seismic response mitigation.

The results obtained also point to the need for a parametric investigation to be carried out on a larger population of earthquakes, groupable into specific classes, to study and evaluate the effects of these devices on the dynamic response of the system.

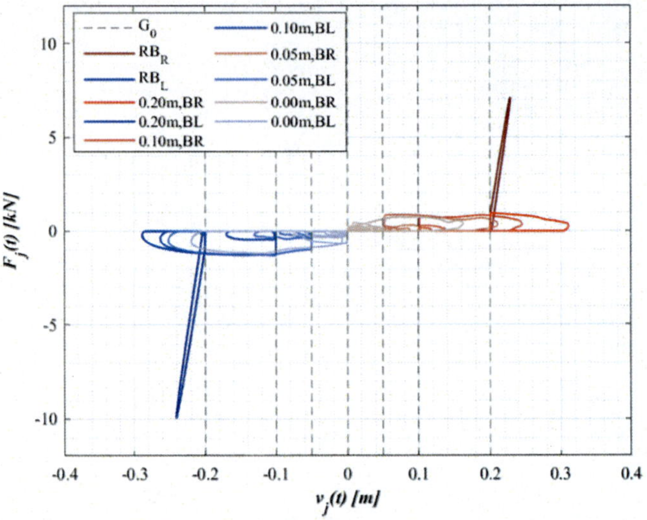

Fig. 13. Force-displacement cycles of the bumpers $F_j(t)$-$v_j(t)$ ($j = R,L$), due to Cape Mendocino earthquake and for four different gap G_0 analyzed comparing them with FF and RB. The red color represents the response obtained by impact with rigid bumper (RB, $\lambda = 100$ and $\gamma = 3.5$), the black represents the response obtained without impact (FF), while the blue color, with different shades, represents the responses obtained with different initial gaps G_0 for impact with preferable bumper ($\lambda = 1$ e $\gamma = 5$): the reduction in the intensity of blue results in a halving of the gap G_0, which decreases from 0.20 m, to 0.10 m, to 0.05 m until it take the value 0.00 m. The lines representing the gaps G_0 are shown with black vertical bashed line.

References

1. Ismail, M., Rodellar, J., Pozo, F.: Passive and hybrid mitigation of potential near-fault inner pounding of a self-braking seismic isolator. Soil Dyn. Earthq. Eng. **69**, 233–250 (2015). https://doi.org/10.1016/j.soildyn.2014.10.019
2. Jangid, R.S., Kelly, J.M.: Base isolation for near-fault motions. Earthq. Eng. Struct. Dyn. **30**(5), 691–707 (2001). https://doi.org/10.1002/eqe.31
3. Polycarpou, P.C., Komodromos, P.: On poundings of a seismically isolated building with adjacent structures during strong earthquakes. Earthq Eng. Struct Dyn **39**(8), 933–940 (2010)
4. Mavronicola, E.A., Polycarpou, P.C., Komodromos, P.: Effect of planar impact modeling on the pounding response of base-isolated buildings. Front. Built Environ. **2**(11), 1–16 (2016)
5. Komodromos, P., Polycarpou, P.C., Papaloizou, L., Phocas, M.C.: Response of seismically isolated buildings considering poundings. Earthq. Eng. Struct. Dyn. **36**(12), 1605–1622 (2007)
6. Andreaus, U., De Angelis, M.: Influence of the characteristics of isolation and mitigation devices on the response of single-degree-of-freedom vibro-impact systems with two-sided bumpers and gaps via shaking table tests. Struct. Control Heal. Monit. **27**(5), 1–21 (2020). https://doi.org/10.1002/stc.2517
7. Andreaus, U., De Angelis, M.: Nonlinear dynamic response of a base-excited SDOF oscillator with double-side unilateral constraints. Nonlinear Dyn. **84**(3), 1447–1467 (2016). https://doi.org/10.1007/s11071-015-2581-4

8. Andreaus, U., Baragatti, P., De Angelis, M., Perno, S.: A preliminary experimental study about two-sided impacting SDOF oscillator under harmonic excitation. J. Comput. Nonlinear Dyn. **12**(6), 061010 (2017). https://doi.org/10.1115/1.4036816
9. Andreaus, U., Baragatti, P., De Angelis, M., Perno, S.: Shaking table tests and numerical investigation of two-sided damping constraint for end-stop impact protection. Nonlinear Dyn. **90**(4), 2387–2421 (2017). https://doi.org/10.1007/s11071-017-3810-9
10. Andreaus, U., De Angelis, M.: Experimental and numerical dynamic response of a SDOF vibro-impact system with double gaps and bumpers under harmonic excitation. Int. J. Dyn. Control **7**(4), 1278–1292 (2019). https://doi.org/10.1007/s40435-019-00532-x
11. Stefani, G., De Angelis, M., Andreaus, U.: Scenarios in the experimental response of a vibro-impact single-degree-of-freedom system and numerical simulations. Nonlinear Dyn. **103**(4), 3465–3488 (2020). https://doi.org/10.1007/s11071-020-05791-4
12. Stefani, G., De Angelis, M., Andreaus, U.: Numerical study on the response scenarios in a vibro-impact single-degree-of-freedom oscillator with two unilateral dissipative and deformable constraints. Commun. Nonlinear Sci. Numer. Simul. **99**, 105818 (2021). https://doi.org/10.1016/j.cnsns.2021.105818
13. Stefani, G., De Angelis, M., Andreaus, U.: Influence of the gap size on the response of a single-degree-of-freedom vibro-impact system with two-sided constraints: experimental tests and numerical modeling. Int. J. Mech. Sci. **206**, 106617 (2021). https://doi.org/10.1016/j.ijmecsci.2021.106617
14. Stefani, G., De Angelis, M., Andreaus, U.: The effect of the presence of obstacles on the dynamic response of single-degree-of-freedom systems: Study of the scenarios aimed at vibration control. J. Sound Vib. **531**(116949), 1–21 (2022). https://doi.org/10.1016/j.jsv.2022.116949
15. Du, H., Wang, Y., Han, M., Ibarra, L.F.: Experimental seismic performance of a base-isolated building with displacement limiters. Eng. Struct. **2441**, 112811 (2021). https://doi.org/10.1016/j.engstruct.2021.112811
16. Folhento, P., Barros, R., Braz-César, M.: Mitigation of earthquake-induced structural pounding between adjoining buildings—state-of-the-art. In: Gonçalves, J.A., Braz-César, M., Coelho, J.P. (eds) CONTROLO 2020, Lecture Notes in Electrical Engineering, vol. 695, pp. 752–761. Springer, Cham. (2020). https://doi.org/10.1007/978-3-030-58653-9_72
17. Brown, T., Elshaer, A.: Pounding of structures at proximity: a state-of-the-art review. J Building Eng. **48**, 103991 (2022). https://doi.org/10.1016/j.jobe.2022.103991

Seismic Vulnerability Assessment and Mitigation of Existing Buildings

Seismic Retrofit of r.c. Buildings with Base Isolation

Giacomo Buffarini[1], Paolo Clemente[1], Andrea De Flaviis[2], Chiara Ormando[3](✉), and Antonello Salvatori[2]

[1] ENEA, Casaccia Research Centre, 00123 Rome, Italy
{giacomo.buffarini,paolo.clemente}@enea.it

[2] Department of Civil, Construction-Architectural and Environmental Engineering, University of L'Aquila, 67100 L'Aquila, Italy
antonello.salvatori@univaq.it

[3] Department of Civil and Computer Science Engineering, University of Tor Vergata, 00133 Rome, Italy
chiara.ormando@uniroma2.it

Abstract. Seismic isolation is today largely used as a passive mitigation system for seismic actions. While this technique can be easily implemented in new structures, more difficult, whenever possible, is the implementation in existing buildings, which can become a real challenge especially in countries where maintenance of old structures, sometimes of historical importance, is an important issue, like Italy. The use of seismic isolation in existing buildings presents interesting features. Using a base isolation system, the period of vibration can be chosen in order to allow input of low spectral amplitude, so that the superstructure can substantially remain in the elastic range. This possibility allows for limiting or even avoiding the conventional retrofit intervention in the superstructure. In fact, the intervention can be concentrated at the lowest level, while no intervention is necessary in the superstructure, so it is not necessary to demolish non-structural elements. As a result, an "operational seismic retrofit" would be possible, that is the superstructure could remain operative during the retrofit works. In order to do that, at least a safety degree not lower than the existing one should be guaranteed. This can be obtained by using temporary fixing systems for the isolation devices or temporary structures for the columns. After a brief overview of the available research in this field, the paper proposes and discusses different solutions for "operational seismic retrofit", pointing out also the convenience in terms of costs. Furthermore, a case study is presented, in which some solutions are implemented.

Keywords: Seismic isolation · Existing buildings · Seismic retrofit · Rubber bearings · Retrofit interventions

1 Introduction

It is well known that base isolation consists in the insertion of particular devices between the structure in elevation and its foundations, with the aim of increasing the oscillation

G. P. Cimellaro (Ed.): WCSI 2022, LNCE 309, pp. 969–977, 2023.
https://doi.org/10.1007/978-3-031-21187-4_85

period and thus obtaining the "decoupling" of the motion of the building with respect to that of the ground. The seismic actions on the superstructure can be reduced by 70–80% compared to those of a fixed-base building and it is possible to design a structure in the elastic range, i.e. so that it does not suffer significant damage even under the design earthquake [1, 2]. The good performance of seismically isolated buildings under strong earthquakes was demonstrated in several cases [3]. The observed behaviour under low intensity events pointed out the necessity of an accurate design also for these events for seismic isolation systems composed of HDRBs [4–9] and SDs [10].

Seismic isolation is almost always applicable in new buildings (it may not be in the cases of low frequency amplifying subsoils or very deformable buildings), for which it has also been shown that it does not involve increases in the construction cost, at least for areas with high or medium seismicity [11, 12]. Indeed, if the system is well designed [13–15], significant savings can be achieved. For existing buildings, however, the adoption of seismic isolation is not yet widespread despite the numerous applications following the seismic events that have occurred since the L'Aquila earthquake in 2009 [16–18] as well as interesting proposals [19–21]. To the advantage of the adoption of isolation in existing buildings two aspects must be remembered:

- as for new buildings, it guarantees a level of safety that cannot be pursued with traditional techniques and should always be used for a seismic retrofit, making the building new from the point of view of structural safety; acting on the reduction of the action rather than on the strengthening of the structure, it also allows to limit or avoid reinforcement interventions on the superstructure;
- in many cases it does not require the emptying of the building and, therefore, the abandonment by the inhabitants; this feature makes it particularly interesting in prevention works for which tax incentives are provided.

The first applications of seismic isolation in the retrofit of existing buildings date back to the 80s of the twentieth century. There were many applications in Japan and California, where following the strong 1989 Loma Prieta earthquake, many public buildings, often of historical interest, were adapted by seismic isolation. Among the main applications, we recall the City Hall of San Francisco, where a cut was made at the level of the extrados of the foundations, temporarily entrusting the task of supporting the weight of the superstructure to hydraulic jacks. Under each column, one or more seismic isolators were placed. Above the insulation system a rigid slab was created in its floor.

In Italy, the first applications concerned two reinforced concrete structures, the building in via Latini in Fabriano, where the devices were inserted between the existing foundation and a new foundation below the previous one, and the Multifunctional Center in Rione Traiano in Naples, in which the isolators were inserted along the columns of the first level after cutting them. Several interesting applications have been made in the reconstruction following the 2009 L'Aquila earthquake on reinforced concrete buildings.

With reference to the aforementioned second solution, in this paper, some techniques useful to keep the building operative during the retrofit works are studied. Keeping the building operative during retrofit works leads to economical savings and to a higher level of comfort for those who live in the building. Obviously, it is necessary to guarantee an adequate safety degree through the use of some temporary devices.

2 Retrofit of Reinforced Concrete Buildings with Seismic Isolation

The most used technique for the seismic isolation of existing reinforced concrete buildings consists in cutting and eliminating a portion of the columns (and the walls, if any), and successively inserting the isolators. As in the case of new buildings, the best solution, when possible, is to insert the devices at the top of the columns of the lowest floor (the underground floor, if any). Furthermore, it is advisable to place the isolation devices at the top of the columns or walls of the chosen floor, so as not to interfere with the functionality of the floor itself. In this way, the floor above the isolators guarantees the horizontal stiffness required, and the portions of the columns under the isolation devices can be enlarged to obtain the stiffness needed or just to support the isolators [22].

Usually, the installation of the isolators requires the following steps for each column (Fig. 1):

- enlargement of the column at the basement, if necessary,
- installation of metal brackets by means of threaded bars,
- placement of hydraulic jacks to unload the part of the column to be removed,
- diamond wire cutting and removal of the segment of the column,
- insertion of the isolator and subsequent grouting of the anchors and the collar with anti-shrinkage cement mortar,
- removal of jacks and loading of the isolator.

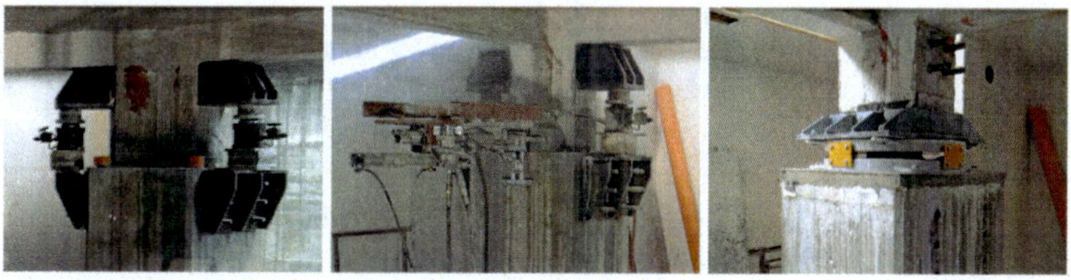

Fig. 1. Installation of an isolation device in a column (courtesy of FIP Industriale).

One of the issues that characterises the intervention on existing buildings is the transfer of the load from the existing situation to the final one. This must take place through the use of temporary elements, to which the load can be transferred temporarily.

The load is temporarily transferred from the structural element immediately above to the one immediately below the portion to be cut, through a suitable number of hydraulic jacks, put into loading in order to discharge totally the structural element to be interrupted or only its part to be cut.

In the usual case of cutting the columns on the first level above the foundations, the jacks can be placed between the beams of the first deck and the foundations (Fig. 2).

The loading of the isolation devices is usually done by means of flat jacks inserted underneath the isolator and injected with resin until the device is loaded.

Fig. 2. Insertion of an isolation device in a column (courtesy of G&P Intech).

3 Operational Intervention Techniques and Safety During the Works

In order to keep the building functional during the retrofit works, it should be guaranteed an adequate safety degree. This should be between the minimum value, which characterizes the structure in its condition before the intervention, and the maximum one, relative to the final condition after the intervention. Usually, the minimum value can be accepted, taking into account that this choice guarantees to keep the same safety degree as the previous situation for a limited time.

The first step is to carry out a vulnerability analysis of the building in order to find out the seismic event that brings it to the ultimate limit state.

The present safety can be guaranteed by ensuring the present strength for each column, with reference to all the internal forces, or by ensuring that each column will be able to support the actual internal forces during the quake. Usually, the first condition turns out to be too expensive or even impossible to realize and the second one is accepted. In any case, it should be clear that, to guarantee an adequate level of safety during the works, it is necessary to keep the structure locked during the works. This can be done by acting on the devices or on the columns.

One example is a technique typically used in Japan, in which external r.c. slabs are built around the first floor just above the isolation interface. These slabs are connected to r.c. walls, with the aim of increasing the restraint degree and absorbing most of the seismic action. The slabs are cut at the end of the intervention.

Another example is a technique used by Salvatori and Luciano in the retrofit of a building damaged by the 2009 L'Aquila earthquake. In this case, the rubber bearings were locked using some steel profiles, vertically bolted between the upper and the lower steel plates. This building was not operational during the works.

A third example is a technique used to keep the building fixed at the base, locking the columns, without touching the isolators [23]. In this case, the building was operational during the works.

The two last techniques will be discussed in the following.

3.1 Locking of the Isolation Devices

An isolator can be locked by means of steel profiles between the upper and the lower plate of the device. A seismic action in x direction causes shear and bending moment. These generate shear tension in the profiles parallel to the seismic direction and traction and compression tensions in the profiles orthogonal to the seismic direction. Analogously for a seismic action in y direction. Obviously, for a generic seismic action, the two cases are combined and both shear and normal tensions will be in all the profiles.

The steel profiles can be connected to the isolator plates by means of vertical (Fig. 3) or horizontal bolts (Fig. 4). In the first case, the horizontal size of the plates could be increased, while in the second case the thickness of the plates should be increased. It should be noted that the length of the profiles, the number of bolts and the class of steel should be carefully verified. They should be compatible with the device sizes and allow a simple and effective placement.

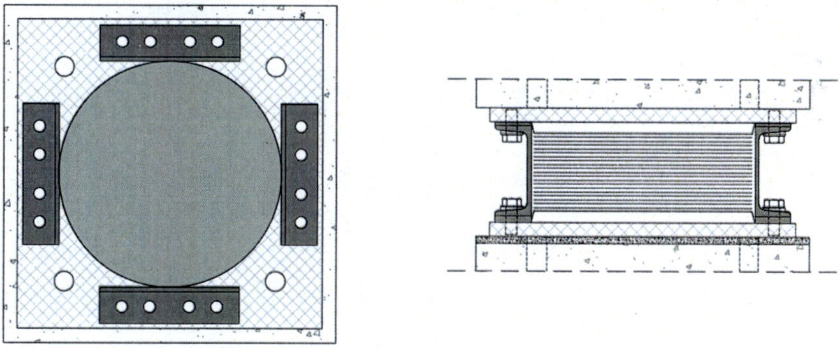

Fig. 3. Locking devices with vertical bolts.

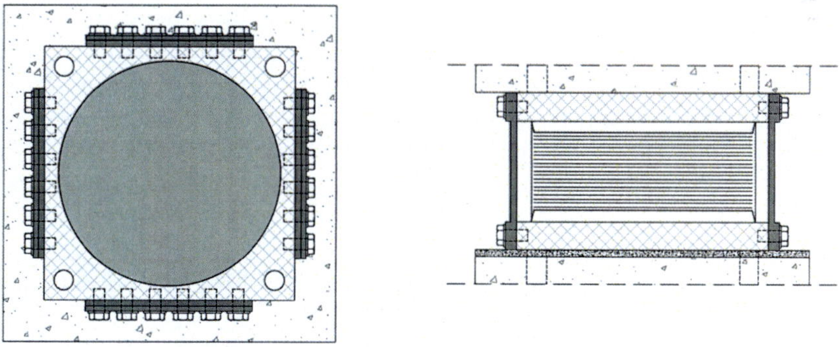

Fig. 4. Locking devices with horizontal bolts.

Furthermore, if the goal is to ensure the present strength of the column, the isolation device could not be able to transmit the bending moments from the upper to the lower part of the column, due to the limited resistance to the extraction of the steel clamps.

3.2 Locking of Columns

In this case (Fig. 5), the profiles entitled to prevent the movement of the isolation device are bolted to other profiles, connected to the column through threaded bars and that could also be used to support the hydraulic jacks to discharge the column before its cut.

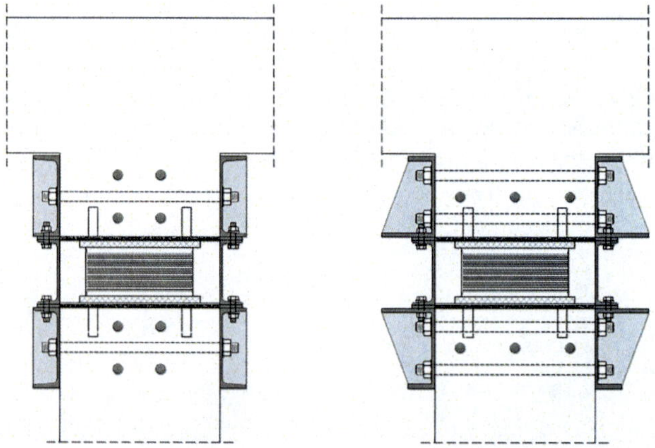

Fig. 5. Column section with elastomeric isolators and locking profiles.

In this case, also the bending moments can be transmitted from the upper to the lower part of the column and the isolation devices are subject only to vertical loads during the works.

4 Costs and Timetable

The techniques described in the previous paragraph have been applied to a typical building, designed with reference to lower seismic action and to be retrofitted.

The building used as case study has a regular shape. It has one underground floor and four floors above the ground. The dimensions in plan are 24 × 12 m, and the total height is about 14 m. The structure is a r.c. frame, with 24 columns (30 × 60 cm) and beams (30 × 50 cm or 100 × 20 cm). It is worth pointing out that this building, having several columns with a low distance from each other, is not very appropriate for the insertion of isolation devices in retrofit.

Costs and times of intervention are evaluated and normalized with reference to the demolition and reconstruction of the same building.

In terms of time, both the techniques can be completed in about 180 days (taking into account also Saturdays and Sundays), considering all the phases (intervention on the superstructure, preliminary phases, intervention on foundation and columns of the substructure, isolation, and final phases), that means 7.5 days per column. In particular, the isolation phase takes 2 days per column. The increase of time compared to the traditional isolation technique is negligible; the intervention lasts about 33% of the total time of an intervention of demolition and reconstruction.

In terms of cost, for the locking of isolators, the extra cost, compared to classic isolation, is about 4%, due to the increase in the size of the steel plates and to the presence of the steel profiles used to lock the isolators. For the locking of columns, the extra cost due to the steel profiles used to lock them is about 2%, so the increase in cost is very low.

The total cost of the two techniques compared to demolition and reconstruction is about 21% for the former and about 23% for the latter.

Even if a comparison between the cost of a traditional retrofit and the pro-posed interventions was not realized, it is worth noting that in the second case people must not leave their homes during the works, if doesn't for a very limited time necessary for some minor interventions in the superstructure, saving money and reducing the discomfort for the residents.

5 Conclusions

The seismic retrofit of existing buildings could be very hard to achieve with traditional technologies. Actually, in several cases, the structure has been designed to support lower or nil seismic actions and therefore almost all the structural elements should be reinforced. In these cases, the demolition and reconstruction of the building often turn out to be convenient, both from an economic and a technical point of view. In any case, the building must be evacuated during the works.

On the other hand, seismic isolation, which is based on the increase of the natural period of vibration that leads to a reduction of the seismic effects in the structure, is certainly a suitable technique for the complete retrofit of a structure.

In this paper, two main techniques are described in order to keep the building functional during the retrofit works. In fact, by locking the isolators or the columns, it is possible to guarantee at least the same previous seismic safety and so to operate on the substructure without the necessity for people to leave the building during the intervention. The proposed interventions entail only a negligible increase in cost with respect to a traditional seismic isolation implementation.

Also, the suitability of seismic isolation has been demonstrated from an economic point of view with respect to demolition and reconstruction of buildings. In fact, the two proposed interventions were designed for the retrofit of an actual building. It was estimated that the retrofit requires only a few months and that the cost was under the 25% of the cost required for demolition and reconstruction.

References

1. Clemente, P., Martelli, A.: Seismically isolated buildings in Italy: state-of-the-art review and applications. Soil Dyn. Earthq. Eng. **119**, 471–487 (2019). https://doi.org/10.1016/j.soildyn.2017.12.029
2. Clemente, P.: Seismic isolation: past, present and the importance of SHM for the future. J. Civ. Struct. Heal. Monit. **7**(2), 217–231 (2017). https://doi.org/10.1007/s13349-017-0219-6

3. Clemente, P., Martelli, A.: Recent applications of seismic isolation in Italy. In: Proc. of the 16th World Conference on Seismic Isolation, Energy Dissipation and Active Vibration Control of Structures (16WCSI, St. Petersburg, July 1–6), Keynote lecture, pp. 21–36 (2017). https://doi.org/10.37153/2686-7974-2019-16-21-36
4. Clemente, P., Bongiovanni, G., Buffarini, G., Saitta, F., Castellano, M.G., Scafati, F.: Effectiveness of HDRB isolation systems under low energy earthquakes. Soil Dyn. Earthq. Eng. **118**, 207–220 (2019). https://doi.org/10.1016/j.soildyn.2018.12.018
5. Clemente, P., Bongiovanni, G., Buffarini, G., Saitta, F., Scafati, F.: Monitored seismic behavior of base isolated buildings in Italy. In: Limongelli, M., Çelebi, M. (eds), Seismic Structural Health Monitoring, Springer Tracts in Civil Engineering, pp. 115–137 (2019) https://doi.org/10.1007/978-3-030-13976-6_5
6. Salvatori, A., Di Cicco, A., Clemente, P.: Monitoring of base isolated building subjected to far fault earthquake. In: Rainieri, C., Fabbrocino, G., Caterino, N., Ceroni, F., Notarangelo, M.A. (eds.) CSHM 2021. LNCE, vol. 156, pp. 311–338. Springer, Cham (2021). https://doi.org/10.1007/978-3-030-74258-4_21
7. Salvatori, A., Di Cicco, A., Clemente, P.: Seismic monitoring of buildings with base isolation. In: Papadrakakis, M., Fragiadakis, M. (eds) Computational Methods in Structural Dynamics and Earthquake Engineering (7th ECCOMAS Thematic Conf., COMPDYN 2019, Crete Island, Greece, 24–26 June 2019), ID 19221, Institute for Structural Analysis and Antiseismic Research, National Technical University of Athens (2019). https://2019.compdyn.org/proceedings/
8. Clemente, P., Di Cicco, A., Saitta, F., Salvatori, A.: Seismic behaviour of base isolated civil protection operative centre in Foligno, Italy. J. Perform. Constructed Facil. **35**(4), 04021027 (2021). https://doi.org/10.1061/(ASCE)CF.1943-5509.0001589
9. Salvatori, A., Bongiovanni, G., Clemente, P., Ormando, C., Saitta, F., Scafati, F.: Observed seismic behaviour of a HDRB and SD isolation system under far fault earthquakes. Infrastructures **7**(2), 13 (2022). https://doi.org/10.3390/infrastructures7020013
10. Saitta, F., Clemente, P., Bongiovanni, G., Buffarini, G., Salvatori, A., Grossi, C.: Base isolation of buildings with curved surface sliders: basic design criteria and critical issues. Adv. Civ. Eng. **2018**(1569683) 14, Hindawi (2018). https://doi.org/10.1155/2018/1569683
11. Clemente, P., Buffarini, G.: Base isolation: design and optimization criteria. Seismic Isolation Prot. Syst. **1**(1), 17–40 (2010). https://doi.org/10.2140/siaps.2010.1.17
12. Clemente, P., Bontempi, F., Boccamazzo, A.: Seismic Isolation in masonry buildings: technological and economic issues. In: Modena, C., da Porto, F., Valluzzi, M.R. (eds) Brick and Block Masonry: Trends, Innovation and Challenges (Proc. 6th International Conference IB2MAC, 26–30 Jun 2016, Padua), pp. 2207–2215, Taylor & Francis Group, London, UK (2016). https://doi.org/10.1201/b21889-273
13. Tripepi, C., Clemente, P.: Graphic procedure for the optimum design of elastomeric isolators. Pract. Periodical Struct. Des. Constr. **26**(1), 04020058-1_9 (2021). https://doi.org/10.1061/(ASCE)SC.1943-5576.0000547
14. Ormando, C., Clemente, P., Ianniruberto, U., Scafati, F.: Onset of motion of curved surface sliders used in seismic isolation systems. Pract. Period. Struct. Des. Constr. **26**(3), 04021015 (2021). https://doi.org/10.1061/(ASCE)SC.1943-5576.0000580
15. Scafati, F., Ormando, C., Clemente, P., Bongiovanni, G.: Observed behavior of buildings seismically isolated with CSSs under a low energy earthquake. J. Civ. Struct. Heal. Monit. 12(2), 225–243 (2022). https://doi.org/10.1007/s13349-021-00539-z
16. Bongiovanni, G., Buffarini, G., Clemente, P., Saitta, F., Tripepi, C.: Retrofit of existing buildings with seismic isolation: design issues and applications. In: Proc. of the 16th World Conference on Seismic Isolation, Energy Dissipation and Active Vibration Control of Structures (16WCSI, St. Petersburg, July 1–6), pp. 111–122, (2019). https://doi.org/10.37153/2686-7974-2019-16-111-122

17. Buffarini, G., Clemente, P., Mancinelli, G.: Experimental analysis and seismic rehabilitation of an earthquake damaged building. In: Proc. of the First Middle East Conference on Smart Monitoring, Assessment and Rehabilitation of Civil Structures (Smar2011, Dubai, 8–10 Feb 2011), Paper 171 (2011)
18. Saitta, F., Clemente, P., Buffarini, G., Bongiovanni, G.: Vulnerability analysis and seismic retrofit of a strategic building. J. of Perform. Constr. Facil. **31**(2), 04016085 (2017). https://doi.org/10.1061/(ASCE)CF.1943-5509.0000948
19. Clemente, P.; De Stefano, A.: Application of seismic isolation in the retrofit of historical buildings. In: Brebbia, C.A., Maugeri M. (eds) 8th World Conference on Earthquake Resistant Engineering Structures, ERES 2011 (Sept. 7–9, Chianciano, Italy), pp. 41–52, WIT Press, Southampton, UK, 120, 1746–4498 (2011). https://doi.org/10.2495/ERES110041
20. De Stefano, A., Clemente, P., Invernizzi, S., Matta, E., Quattrone, A.: Innovative technique for the base isolation of existing buildings. In: Papadrakakis, M., Papadopoulos, V., Plevris, V. (eds) Computtional Methods in Structural Dynamics and Earthquake Engineering (5th International Conference COMPDYN 2015, Crete Island, Greece, 25–27 May 2015) 1, pp. 377–387. NTUA, Athens, Greece (2015). ISBN: 978-960-99994-7-2. https://doi.org/10.7712/120115.3403.1825
21. Mezzi, M., Vetturini, R., Cecchini, W.: Actual applications of base isolation to heritage buildings after the 2009 L'Aquila earthquake. In: Proc. of the 14th World Conference on Seismic Isolation. Energy Dissipation and Active Vibration Control of Structures (14WCSI, San Diego, 9–11 Sept) (2015)
22. Castellano, M.G.: Italian experience in seismic retrofit of buildings through seismic isolation or energy dissipation. Proceedings of the Third Conference on Smart Monitoring, Assessment and Rehabilitation of Civil Structures (SMAR 2015, Turkey, Antalya) (2015)
23. Cardone, D., Gesualdi, G.: Seismic rehabilitation of existing reinforced concrete buildings with seismic isolation: a case study. ACADEMIA, Earthq. Spectra **30**(4), 1–24 (2014)

Advanced Constitutive Laws for Nonlinear Static Analyses of Masonry Structures

Adalgisa Zirpoli and Stefano Farina(✉)

Harpaceas s.r.l., Milano, Italy
{zirpoli,farina}@harpaceas.it

Abstract. In recent years, the use of static non-linear analysis, the so-called *pushover* analysis, is increasingly common when dealing with seismic vulnerability assessment and mitigation of existing buildings.

The *pushover* analysis has the final goal to determine the relation between the base shear and the displacement of a control node to determine the capacity curve of the structure, together with the evaluation of the maximum displacement reached after a seismic event. Finally, by post-processing such information, it is possible to perform comparisons in terms of displacements or acceleration to evaluate vulnerability index of the structure.

Many Finite Element software include this type of analysis among their various features. Therefore, computational algorithms are becoming useful tools to deal with seismic vulnerability issues, thanks to their robustness.

Non-linear static analyses are usually performed on masonry and reinforced concrete structures, by assigning plastic material properties to the main elements of the structure.

In this paper, an advanced constitutive law is presented, the Concrete Damage Plasticity model, highlighting the main features and the parameters involved in its implementation.

Finally, the application of this constitutive model is presented on a real existing concrete-masonry structure, by means of the software Midas Gen.

Keywords: Masonry structures · Non-Linear static analysis · Midas gen · Concrete damage plasticity · Pushover analysis · Risk assessment · Existing bridges

1 Suitable Non-linear Constitutive Laws for Masonry Material

In structural modeling, there are many possible sources of non-linearity that can be considered, for example tension-only braces in steel structures, compression-only springs used for representing soil-structure interaction and elastoplastic constitutive laws assigned to the materials.

Non-linear constitutive laws are used to represent in a more accurate way the structural response of the material. In general, the definition of additional parameters is required to accurately represent the stress-strain relationship.

G. P. Cimellaro (Ed.): WCSI 2022, LNCE 309, pp. 978–997, 2023.
https://doi.org/10.1007/978-3-031-21187-4_86

Table 1. Non-linear constitutive laws available in midas gen.

Constitutive law	Application	Input	Failure criterion	Hardening	Softening
Strumas	Irregular weaving of masonry (Existing structures)	ft, $E_{El\text{-}Pl}$	σ_{max}	Yes	No
Mohr-coulomb	Regular weaving of masonry (New Structures)	c, φ	τ (confinement)	Yes	No
Drucker-prager	Regular weaving of masonry (New structures)	c, φ	τ (confinement)	Yes	No
Concrete damage plasticity	Irregular weaving of masonry (existing structures)	f_t, f_c, $\varepsilon_{in,t}$, $\varepsilon_{in,c}$	$\sigma_{max}/\sigma_{min}$	Yes	Yes

In Midas Gen it is possible to choose between four different constitutive laws when dealing with masonry structures (Table 1):

The correct choice of the model to be used is based first on the typology of masonry we are dealing with.

In fact, shear failure can occur with different cracking mechanisms, depending on the realization process of the masonry itself. Two principal failure modes can be distinguished:

- Diagonal Cracking
- Shear-Sliding

Both failure modes affect the new structures, but, due to the regular weaving of the masonry, diagonal cracking will occur in mortar joints between bricks.

On the other hand, in existing structures only diagonal cracking can develop. In fact, the weaving of masonry for existing buildings is usually not regular and therefore the diagonal cracking involves both mortar joints and bricks. Moreover, there is not a preferential plane in which the cracks can develop.

Therefore, when dealing with existing masonry building, Strumas or Concrete Damage Plasticity should be set for plastic material properties.

The following hypothesis is made: failure occur when principal macroscopic tensile stress exceed a fixed limit, assumed equal to the tensile resistance of masonry.

1.1 Strumas Constitutive Law

In the Strumas constitutive law just few parameters have to be set for the material:

- f_t: maximum admissible tensile strength
- ν: Poisson's ratio
- E: Young's modulus
- Stiffness reduction factor: parameter related to the possible presence of hardening.

These parameters can be set for all the three components of the masonry material: bricks, head and bed mortar joints (Fig. 1).

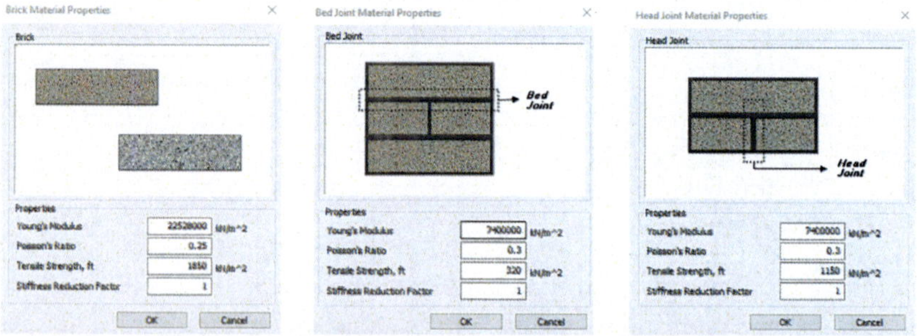

Fig. 1. Example of the parameters to be inserted for the constituents of masonry in Strumas model.

Then, the software performs a homogenization of masonry's elements (first horizontal homogenization, then vertical), and converts them into an equivalent material. After the analysis, the results are traced back to the original components.

As a result, it is possible to visualize, step by step, the Gauss points where the tensile principal stress exceeds the limit of the material (Fig. 2).

Fig. 2. Example of yield point diagonal distribution using Strumas constitutive law.

The principal limitation of Strumas constitutive model is related to the compressive behaviour of materials. In fact, the resistance is limited just in tension, but not in compression, for which the material will remain indefinitely elastic.

2 Concrete Damage Plasticity

Concrete Damage Plasticity is a constitutive law able to represent the asymmetrical behaviour of concrete or masonry, by separating its behaviour in tension and compression. Moreover, it allows softening development in post-peak behaviour.

In addition, with Concrete Damage Plasticity it is also possible to consider the progressive damaging of material under cyclic loading.

It is suitable for quasi-brittle material such as concrete, masonry and some kind of soils in very specific conditions.

2.1 Possible Applications

Concrete Damage Plasticity can be used both for non-linear static and dynamic analyses. In fact, it may be used to study the damaging of the material when cyclic loads are considered. By setting a relationship between plastic strains and damage (expressed through a specific function), it is possible to perform non-linear dynamic analysis. In this case, time forcing functions are considered by means of Time History Analysis.

2.2 Parameters Involved in Concrete Damage Plasticity

Dilation Angle The first parameter to be set is the dilation angle. This parameter has a clear physical meaning, since it controls the amount of plastic volumetric strain that develops during the plastic shearing. In fact, compacted dense granular materials tend to expand volumetrically when subjected to shear loading. Due to their compact state, the grains are interlocking and don't have the possibility to move each other relatively.

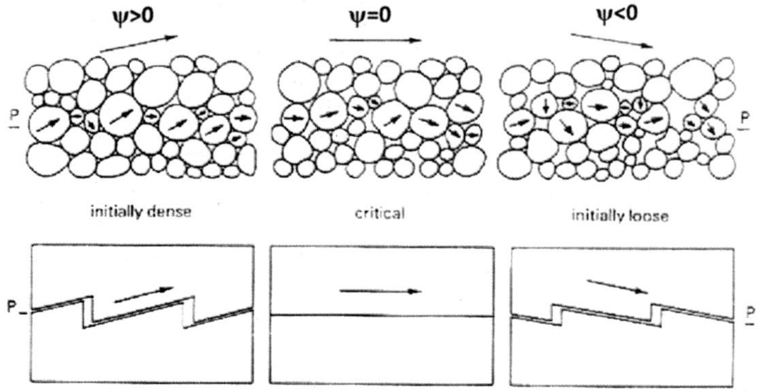

Fig. 3. Graphical representation of dilation angle ψ.

The dilation angle can vary between zero and the friction angle of the material. A dilation angle equal to zero ($\psi = 0$) means that the material preserves its volume when shear deformation arises. On the contrary, if $\psi = \emptyset$ the material swells the same quantity as it slides, increasing non realistically its volume (Fig. 3).

Eccentricity This parameter defines the rate at which the flow potential function approaches the asymptote (see Figs. 5 and 6). In fact, the flow potential G (Fig. 4) tends to a straight line as the eccentricity tends to zero. The flow potential ensures that the flow direction is always uniquely defined.

Default flow potential value is 0.1 and implies that the material has almost the same dilation angle over a wide range of confinement pressure stress values. Increasing the value of the eccentricity provides more curvature to the flow potential, implying also that the dilation angle increases more rapidly as the confining pressure decreases. Values of eccentricity that are significantly less than the default value may leads to convergence problem if the material is subjected to low confining pressure [1].

$$G = \sqrt{(\epsilon \sigma_{t0} \tan \psi)^2 + \bar{q}^2} - \bar{p} \tan \psi,$$

Fig. 4. Flow potential function.

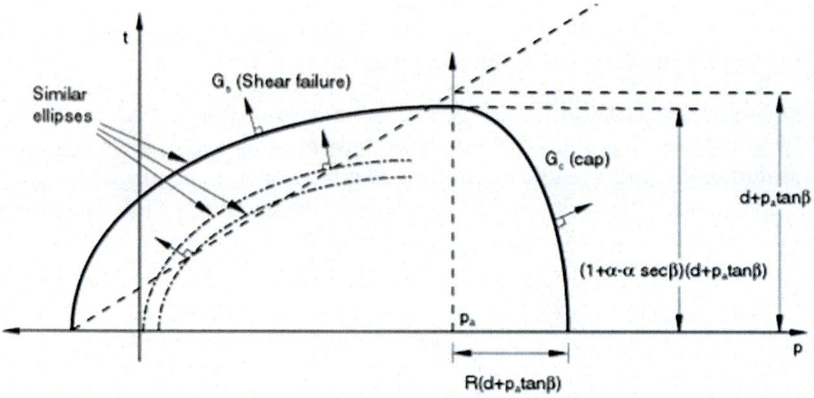

Fig. 5. Graphical representation of the flow potential function.

Ratio f_{b0}/f_{c0} The third parameter to be set is the ratio between two values f_{b0} and f_{c0} for the material. They are defined as follows:

- f_{b0}: elastic limit in a compressive biaxial test
- f_{c0}: elastic limit in a compressive uniaxial test

Typical values of this ratio for concrete are in the range 1.10–1.16 [1].

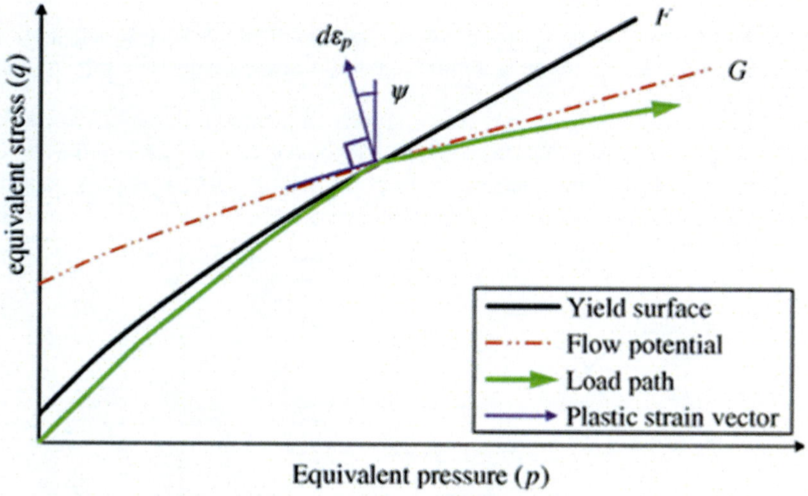

Fig. 6. Influence of dilation angle on the evolution of the Yield Surface.

K This parameter is the ratio of the second stress invariant on the tensile meridian to that on compressive meridian at initial yield for any given value of the pressure invariant, such that the maximum principal stress is negative.

It must satisfy the condition $0.5 < K < 1.0$ and the default value is 0.666. Therefore, K affects the shape of the yield surface in the deviatoric plane (Fig. 7).

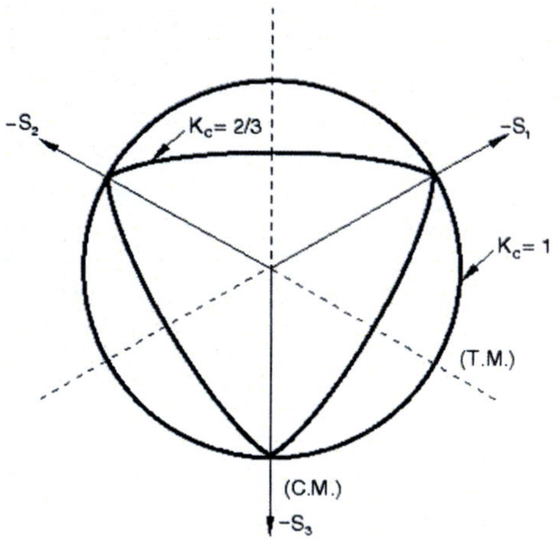

Fig. 7. Yield surface in the deviatoric plane, corresponding to different values of K_C.

Viscosity Parameter Dealing with materials that exhibit softening behaviour and stiffness degradation often lead to convergence problems. A common technique to overcome these problems is using a regularization of the constitutive equations, which causes the consistent tangent stiffness of the softening material to become positive for sufficiently small-time increments.

We are basically admitting that points, which represent the stress state of the material, can go beyond the elastic limit, outside of the plasticization surface (Fig. 8).

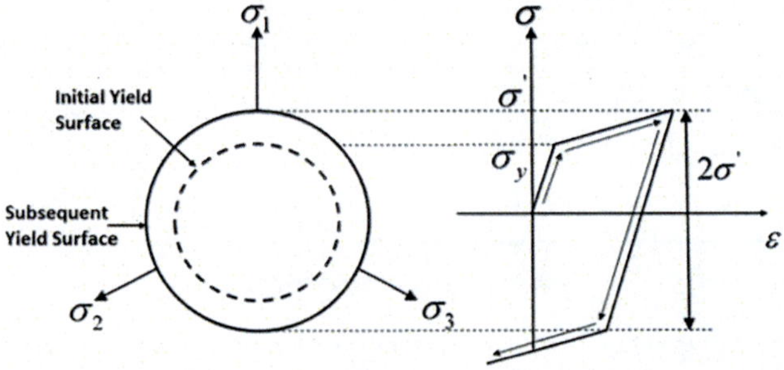

Fig. 8. Graphical representation of the yield surface evolution because of viscosity parameter.

This parameter improves the rate of convergence of the model in the softening regime, without compromising results.

Inelastic Strain—Yield Stress The last parameters to be set regard the stress-strain relations, in tension and in compression, to be assigned to the material (Fig. 9).

In particular, the constitutive law must be input in terms of inelastic strain and yield stress. Therefore, the elastic branch of the constitutive law is not present, the plastic strains start from zero, but the yield stress will start from a value different from zero (Fig. 10).

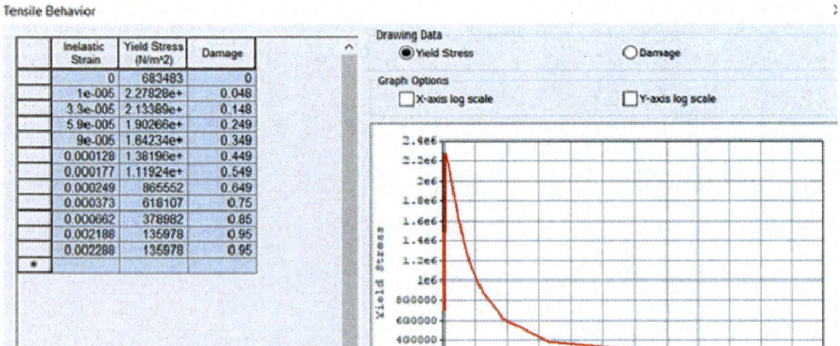

Fig. 9. Example of a constitutive law in tension.

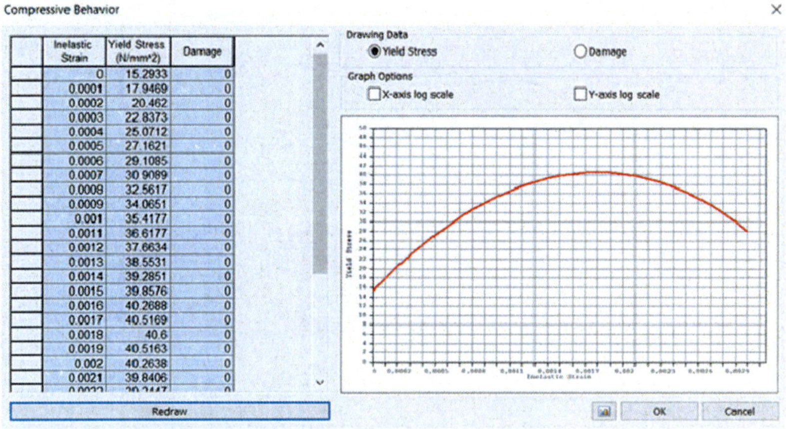

Fig. 10. Example of a constitutive law in compression.

In this way, it is possible to assign different constitutive laws in tension or in compression, setting an unsymmetrical plastic behaviour (Fig. 11).

Moreover, it can be noted that these constitutive laws exhibit a softening branch that would also lead to possible softening structural behaviour.

Damage It is possible to define the damage function for the material to study the behaviour of structures under cyclic loadings.

The damage is assumed to be isotropic and therefore it can be described by means of just one scalar parameter. Two damage variables are defined for the material, one in tension and one in compression. These variables can vary from zero to one (See Figs. 12 and 13).

It is recommended to avoid using values of the damage variables above 0.99, because it corresponds to a 99% reduction of the stiffness (Fig. 11).

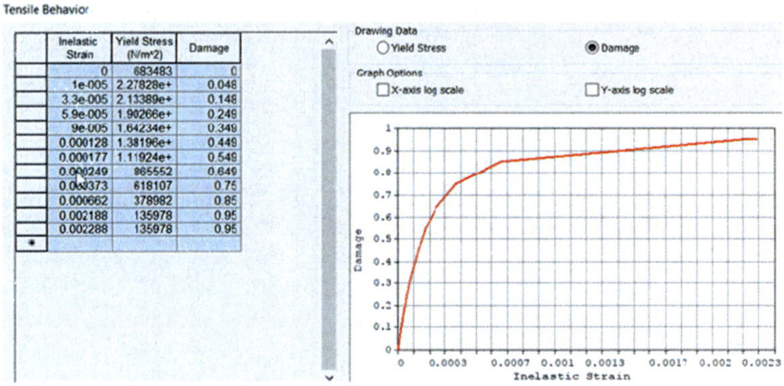

Fig. 11. Example of damage function in tension.

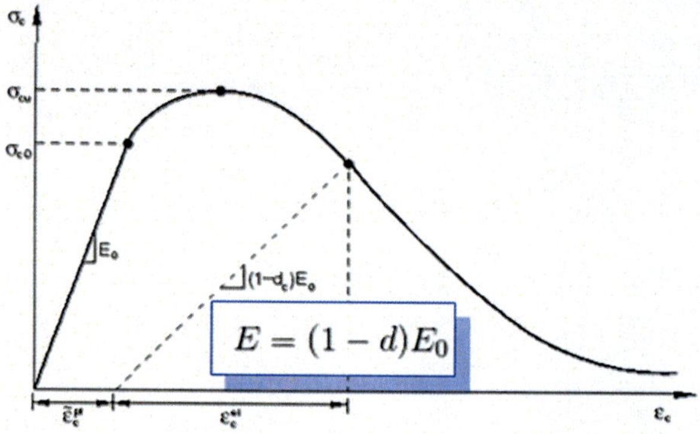

Fig. 12. Graphical representation of the damage in terms of constitutive law.

$$\bar{\sigma}_t = \frac{\sigma_t}{(1-d_t)} = E_0(\varepsilon_t - \tilde{\varepsilon}_t^{pl})$$

$$\bar{\sigma}_c = \frac{\sigma_c}{(1-d_c)} = E_0(\varepsilon_c - \tilde{\varepsilon}_c^{pl})$$

Fig. 13. Definition of effective tensile and compressive cohesion stresses.

3 Case Study: Masonry Rail Bridge

In this section the description of the analysis of an existing rail bridge is reported. The work has been taken by ITEC Engineering s.r.l. and has been analysed in Midas Gen.

3.1 Description of the Bridge

The structural work is composed of a concrete arch bridge positioned in the railway section between Florence and Bologna, in Cantagallo.

The bridge is 105 m long and consists of three oblique segmental concrete arches spans, together with masonry abutments and piers.

This structural work allows the two-way railway line Florence-Bologna to cross Bisenzio River (Fig. 14)

The structure has been analysed to evaluate the seismic vulnerability in relation to the improvement project that has been requested, due to the high importance of the structure itself.

3.2 Description of the Numerical Model

The numerical model of the bridge is composed of thirty thousand solid tridimensional elements created by extrusion of the section of the bridge. As it can be seen in Fig. 15 each colour corresponds to a different material.

Plastic properties have been assigned to just some materials:

Fig. 14. Picture of the existing bridge.

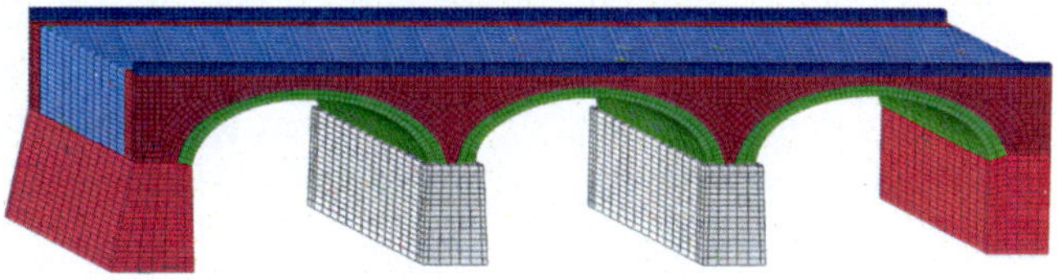

Fig. 15. Numerical model.

- Arches
- Abutments
- Piers

The other materials remained elastic during the analysis:

- Lateral sides
- Upper side
- Reinforced concrete coat of the piers
- Soil filling

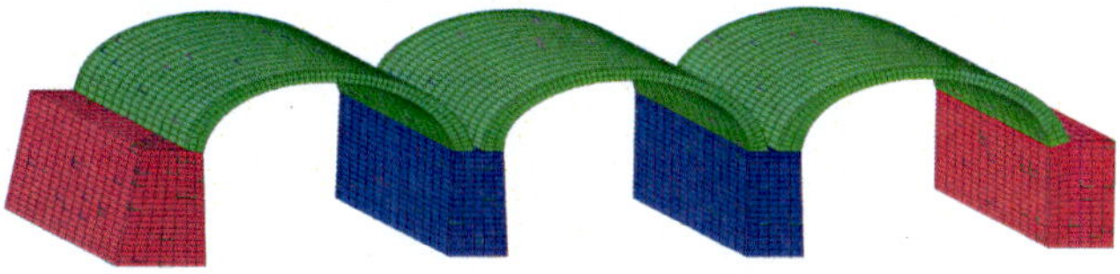

Fig. 16. Plastic materials.

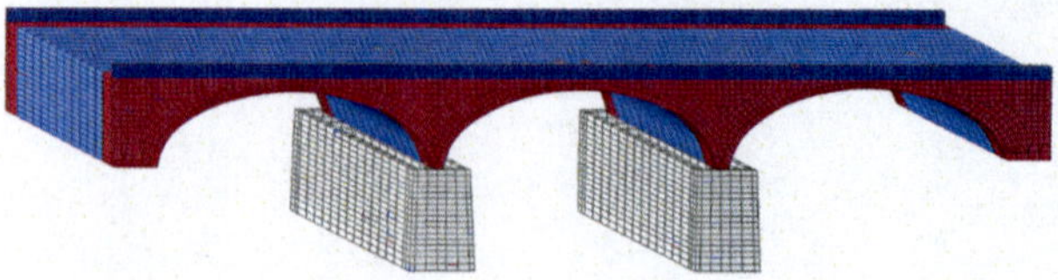

Fig. 17. Elastic materials.

3.3 Eigenvalue Analysis

The first analysis that has been carried out is eigenvalue analysis. In fact, eigenvalue analysis is essential to perform diagnostics of the model. The results of this type of analysis are useful to verify the possible presence of errors and to check the correctness of elastic moduli, boundary conditions and so forth (Figs. 16 and 17).

The principal mode shapes in longitudinal and transversal direction were inspected. The vibration frequencies, modal participation masses and mode representation are reported in the following (Table 2).

Table 2. Results of eigenvalue analysis.

Direction	Frequency [Hz]	Modal participation mass [%]
X	4.21	33.22
Y	2.03	53.21

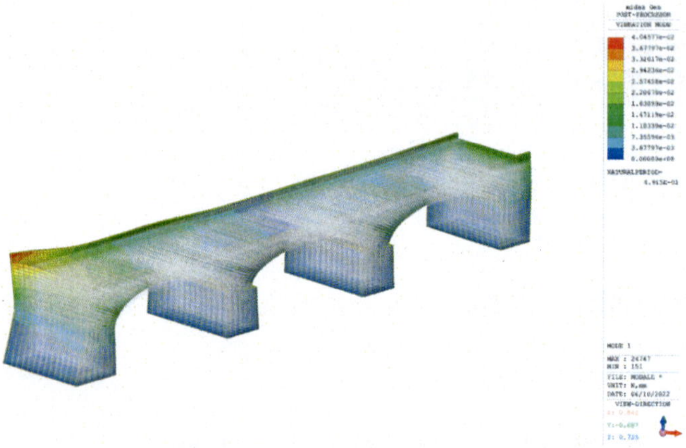

Fig. 18. Principal mode shape in X direction.

3.4 Concrete Damage Plasticity setting

After the preliminary analysis, plastic material's parameters to be assigned to each material have been set (Figs. 18 and 19).

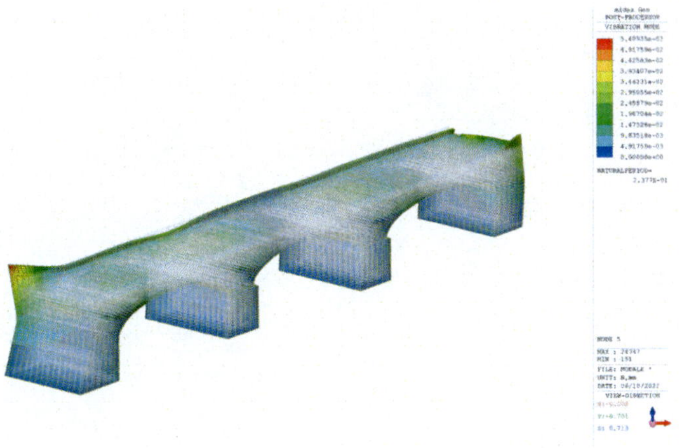

Fig. 19. Principal mode shape in Y direction.

Complete constitutive laws for the material were initially obtained. Then, the elastic branch of the relation was identified and the deformation corresponding to the elastic limit observed. Finally, the complete constitutive law was translated towards the origin of the axes so to represent just the stress-plastic strain relationship (Figs. 20, 21 and 22).

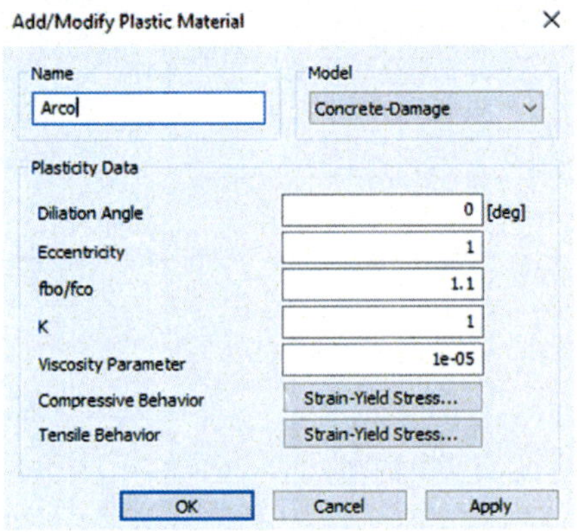

Fig. 20. Plastic material parameters for the arches.

In the final part of the stress-plastic strain relationship, a residual stress has been assigned for convergence reasons.

Due to the presence of the post-peak softening, the Arc Length algorithm [2] has been used, in order to get the convergence of the solver. Figure 23 shows the settings of the method used for these analyses.

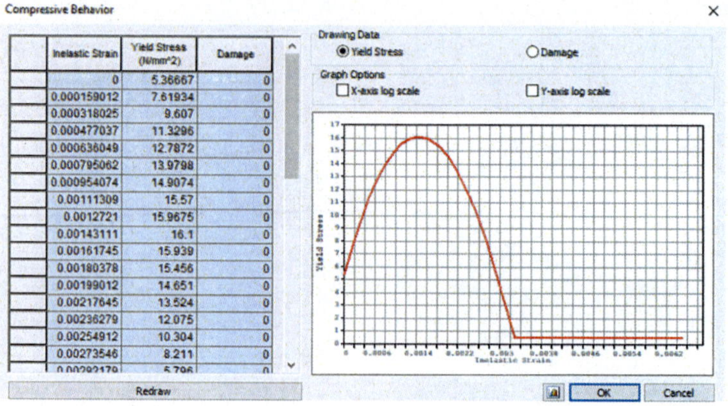

	Inelastic Strain	Yield Stress (N/mm^2)	Damage
	0	5.36667	0
	0.000159012	7.61934	0
	0.000318025	9.607	0
	0.000477037	11.3296	0
	0.000636049	12.7872	0
	0.000795062	13.9798	0
	0.000954074	14.9074	0
	0.00111309	15.57	0
	0.0012721	15.9675	0
	0.00143111	16.1	0
	0.00161745	15.939	0
	0.00180378	15.456	0
	0.00199012	14.651	0
	0.00217645	13.524	0
	0.00236279	12.075	0
	0.00254912	10.304	0
	0.00273546	8.211	0
	0.00292179	5.796	0

Fig. 21. Compressive plastic constitutive law for the arches.

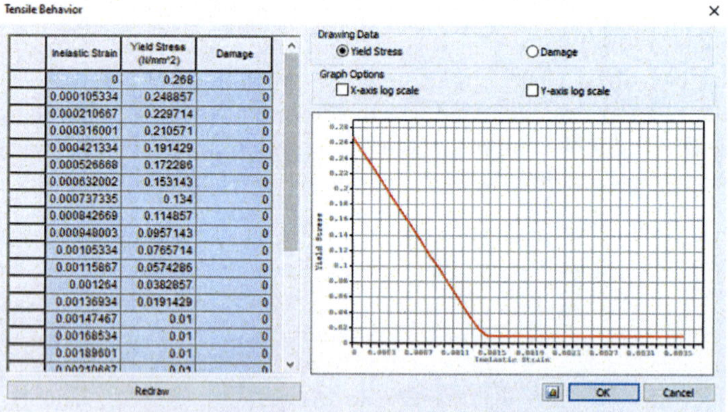

	Inelastic Strain	Yield Stress (N/mm^2)	Damage
	0	0.268	0
	0.000105334	0.248857	0
	0.000210667	0.229714	0
	0.000316001	0.210571	0
	0.000421334	0.191429	0
	0.000526668	0.172286	0
	0.000632002	0.153143	0
	0.000737335	0.134	0
	0.000842669	0.114857	0
	0.000948003	0.0957143	0
	0.00105334	0.0765714	0
	0.00115867	0.0574286	0
	0.001264	0.0382857	0
	0.00136934	0.0191429	0
	0.00147467	0.01	0
	0.00168534	0.01	0
	0.00189601	0.01	0
	0.00210667	0.01	0

Fig. 22. Tensile constitutive law of the arches.

3.5 ULS Loading

After the setting of plastic materials, vertical loading analysis has been performed.

An Ultimate Limit State combination has been used to study the behaviour of the structure under vertical loading. The following load cases were considered:

- Self-Weight
- Ballast
- Static earth pressure
- Symmetrical train loading

The combination coefficients were 1.35 for permanent loads (self-weight, static earth pressure and ballast), 1.45 for variable traffic loads (Fig. 24).

Results In the following, principal tensile stress at 10% (Fig. 25) and 25% (Fig. 26) of the total application of loading are reported. It can be noticed that the highest tensile

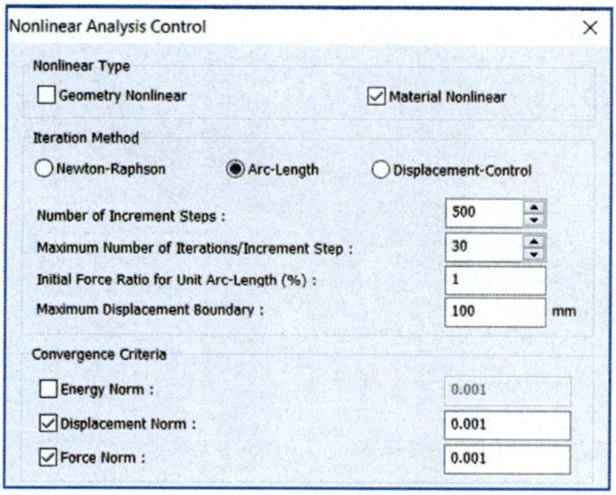

Fig. 23. Arc length settings.

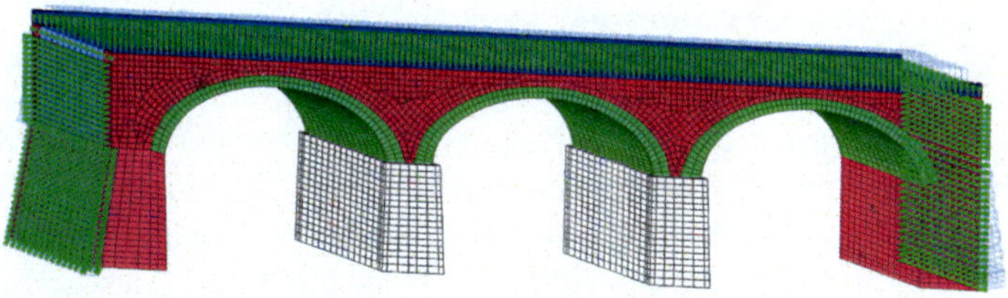

Fig. 24. Representation of the Ultimate Limit State combination.

stresses develop in the central part of the arches. These central parts of the arches are responsible for the analysis stop before reaching the complete loading combination.

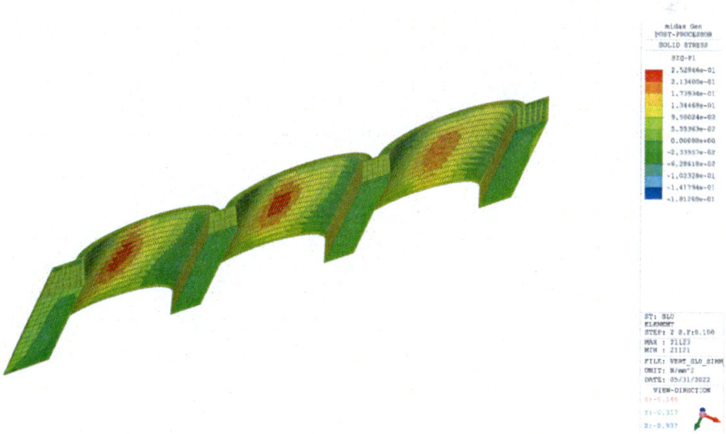

Fig. 25. Principal tensile stress at 10% of loading.

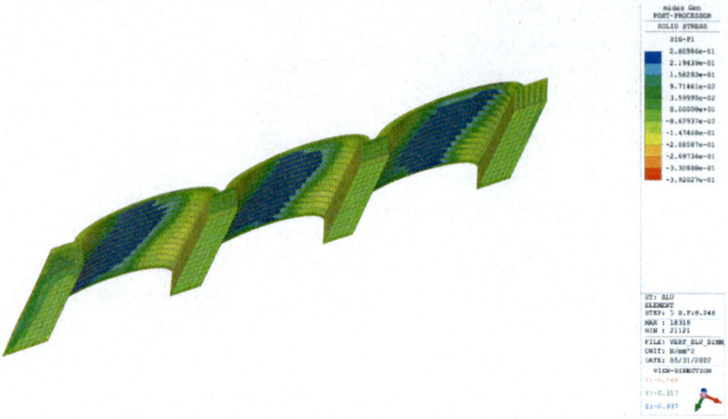

Fig. 26. Principal tensile stress at 25% of loading.

3.6 Pushover Load Case Proportional to Mode Shape

This is the first pushover load case that has been performed. It is proportional to mode shapes analyzed with the eigenvalue analysis, as specified in the Italian Building Code. This loading distribution can be derived from an eigenvalue analysis with response spectrum function (Fig. 27). All the modes, such that the total modal participation mass reach at least the 85% of the total, were considered and combined. The combination has been performed with the Complete Quadratic Combination.

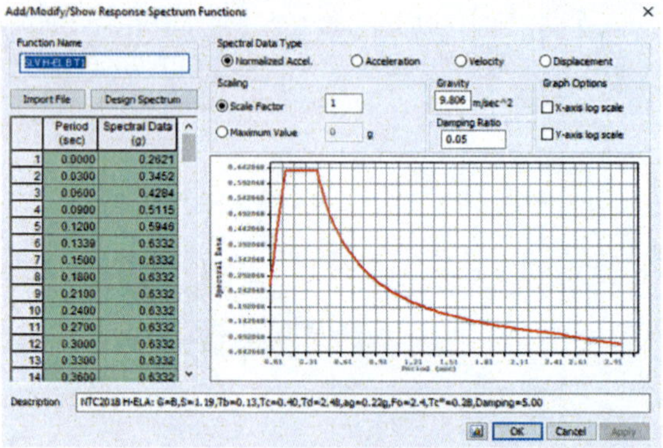

Fig. 27. Response spectrum function.

Finally, thanks to the useful command "Nodal Results of RS", it has been possible to convert inertia forces resulting from a linear dynamic analysis into external nodal forces applied to the structure. Therefore, in this way, the output has become an input and the original dynamic analysis was turned into an equivalent static analysis.

Of course, in addition to this distribution, also the vertical quasi-permanent load combination was applied.

Results In the following the results for the pushover load case proportional to mode shapes in transverse direction are presented.

First, displacement contour results near collapse are reported (Fig. 28).

Fig. 28. Displacement contour at collapse for the pushover loading proportional to mode shapes in transverse direction (top view).

Then, also principal tensile stresses near collapse (Fig. 29) are reported. The lower bound in the legend has been set equal to zero, so to visualize just the element in which tensile stresses are present.

Fig. 29. Principal tensile stress contour at collapse for the pushover loading proportional to mode shapes in transverse direction (bottom view).

Pushover curve After the analysis, the capacity curve in transverse direction of the structure has been derived. First, the control node was selected, according to the first mode shape (Fig. 30).

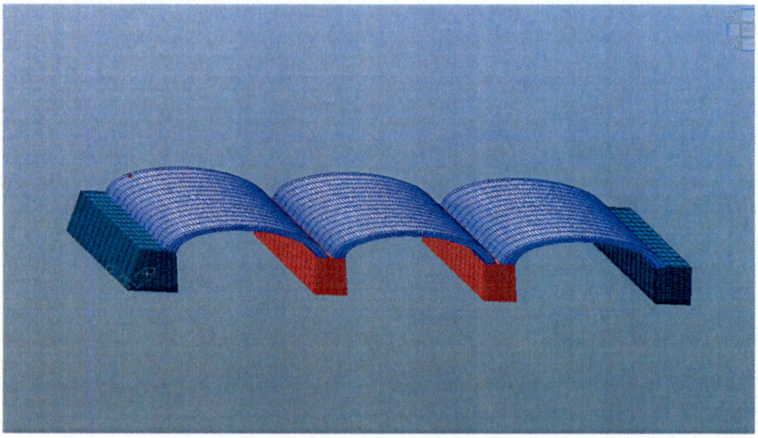

Fig. 30. Control node chosen for the bridge.

Then its displacement was monitored during all the steps of the analysis. In Fig. 30 the control node displacement is plotted together with the load factor. The load factor represents the percentage of the total loading applied.

For this pushover loading it is possible to notice some local snap-through phenomena. Thanks to the possibility to represent softening branch with Concrete Damage Plasticity it was possible to capture this phenomenon, even if globally there was no softening behavior (Fig. 31).

3.7 Pushover Load Case Proportional to Masses

The second pushover load case presented is proportional to masses, according to Italian Building Code.

In this case, its implementation in the software was easier. In fact, the used software offers a specific command called "Nodal Body Force", useful to apply loads according to the mass distribution in each node.

In this case, the quasi-permanent load combination was also applied.

Results In the following the results for the pushover load case proportional to masses in transverse direction is presented.

Like in the previous section, displacement contour results near collapse are first reported (Fig. 32).

Once again, also principal tensile stresses near collapse (Fig. 33) are reported. The lower bound in the legend has been set again equal to zero.

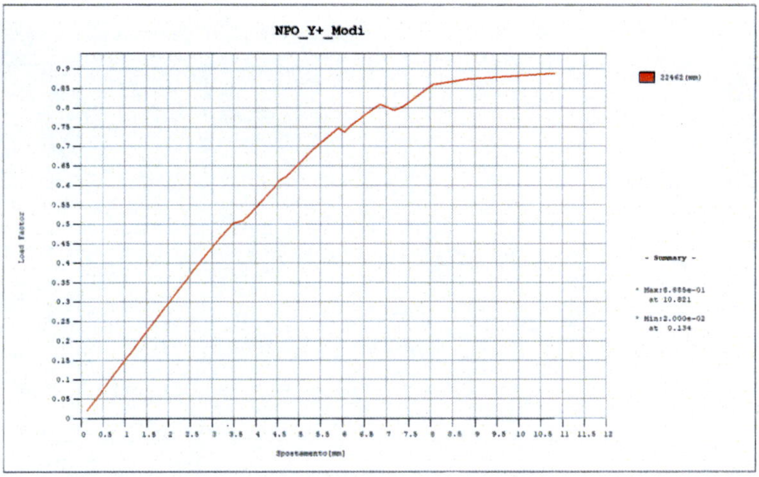

Fig. 31. Capacity curve for loading proportional to mode shapes in transverse direction.

Fig. 32. Displacement contour at collapse for the pushover loading proportional to masses in transverse direction (top view).

Pushover curve After the analysis, the capacity curve in transverse direction of the structure has been derived. The same control node of the previous load case has been inspected (Fig. 30). Its displacement was monitored during all the steps of the analysis. In Fig. 34 displacement of control node is plotted together with the load factor. For this pushover loading no softening behavior has been captured.

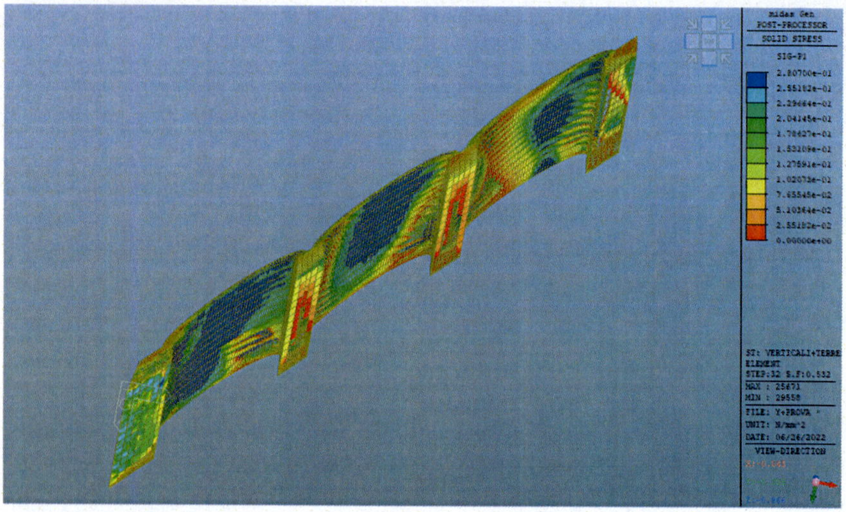

Fig. 33. Principal tensile stress contour at collapse for the pushover loading proportional to masses in transverse direction (top view).

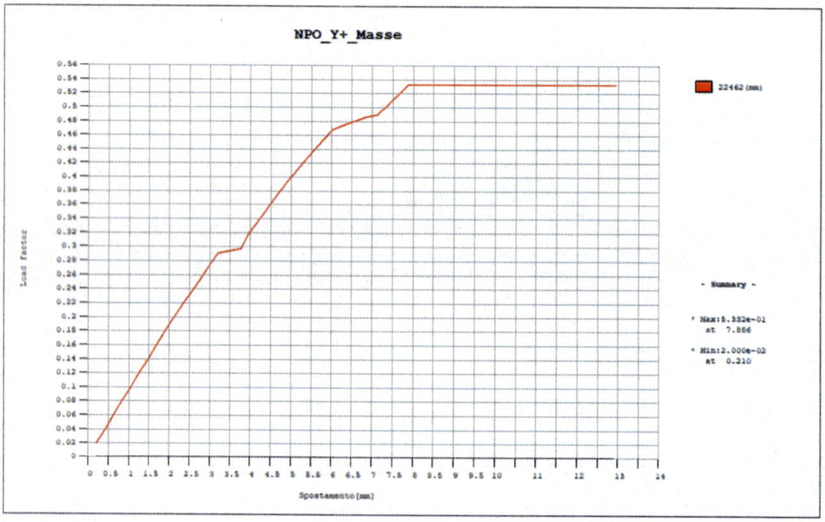

Fig. 34. Capacity curve for loading proportional to masses in transverse direction.

4 Conclusions

In the first part of this paper the constitutive law Concrete Damage Plasticity, as implemented in Midas Gen, has been presented. Despite of its nomenclature and the materials it was developed for, the Authors have proved that it can be usefully adopted for masonry structures as well.

A case history, concerning an historical bridge in the north of Italy, has been presented. Some results, part of a wider analysis campaign, aimed to assess the seismic risk index, have been shown. Specifically, cracking state and capacity curves have been proposed.

Acknowledgments. We would like to thank ITEC engineering S.r.l. for kindly providing us the material necessary to perform the analyses presented in this paper.

We want to thank Ing. Vallarino, C.E.O. of the company, and Ing. Antonelli, Ing. Benedetti, Ing. Puca and Ing. Sguazzino.

References

1. Lubliner, J., Oliver, J.: A plastic-damage model for concrete. Int. J. Solids Struct 299–326 (1989)
2. Zirpoli, A.: Il metodo dell'Arc Length per analisi FEM in presenza di non linearità di materiale. Ingenio

Seismic Response Spectra of the 24th August 2016 Amatrice Earthquake

Melissa De Iuliis[1], Francesco Potenza[2](✉), and Vincenzo Gattulli[1]

[1] Sapienza University of Rome, Via Eudossiana 18, 00184 Rome, Italy
{melissa.deiuliis,vincenzo.gattulli}@uniroma1.it
[2] University G. d'Annunzio of Chieti-Pescara, Viale Pindaro 42, 65127 Pescara, Italy
francesco.potenza@unich.it

Abstract. The severe seismic events affecting urban areas may have disastrous consequences if buildings are designed without appropriate seismic standards. On 24 August 2016, an earthquake of a magnitude Mw 6.0 caused severe damages, casualties, and collapses in four regions of Central Italy. Due to the high level of observed ground motion and not regular damage distribution, in-depth studies have been conducted on the Amatrice earthquake. In this context, the paper investigates the seismic demand affecting the buildings of the Amatrice during the earthquake focusing on evaluating if the registered seismic actions exceed design spectra. Processing both accelerations acquired by the National Accelerometric Network (RAN) stations and *Osservatorio Sismico delle Strutture* (OSS) operated by the Italian *Dipartimento della Protezione Civile* (DPC) and considering available results on site amplification, conclusions on the return period of the seismic event are derived. Moreover, an interpolation method has been introduced to define an Amplification function (*Af*) to use the information contained in microzoning maps evidencing the enormous amplified elastic demand induced by the peculiar site of Amatrice. Finally, the computed elastic response spectra were compared with the elastic demand spectra provided by the Italian Building Standard (DMIT 2018) in three different manners: (i) taking the response spectra of the registration components directly at AMT and AMTS, (ii) evaluating through realistic de-amplification effects these components at the bedrock, and (iii) considering a realistic average of the two registration AMT and AMTS corrected by possible underestimation errors occurred at AMTS.

Keywords: Ground motion · Amatrice earthquake · Seismic response · Amplification factors · Microzoning maps

1 Introduction

During the last decade, Italy has experienced severe seismic sequences, i.e., L'Aquila (2009), Emilia (2012), and Central Italy (2016–2017), whose largest-magnitude earthquakes were Mw 6.3, Mw 6.1, and Mw 6.5, respectively [1]. These earthquakes have caused considerable damage to constructions due to the cumulative effect of being subjected to repeated strong-motion shocks and the shaking features close to the seismic

G. P. Cimellaro (Ed.): WCSI 2022, LNCE 309, pp. 998–1009, 2023.
https://doi.org/10.1007/978-3-031-21187-4_87

rupture [2]. The 24 August 2016 Amatrice (Central Italy) Mw 6.0 earthquake occurred in an area with the highest seismic hazard in Italy, characterized by prevalent normal faulting focal mechanisms [3]. The epicenter was located below the village of Accumuli, in the Tronto River valley (Lazio Region) (Latitude 42.70°, Longitude 13.23°) with a depth of 8.1 km. Therefore, the seismic event can be classified as a swallow earthquake and near-fault. Near-fault seismic ground motions are frequently characterized by extreme velocity and displacement pulses of a relatively long period that distinguish them from typical far-fault ground motions. The Italian seismic code uses a probabilistic approach to evaluate the seismic spectra and return period based on the expected earthquake magnitude [4]. Recent near-fault earthquakes, e.g., 2009 L'Aquila and 2012 Emilia, have highlighted that the magnitude is correlated neither to the damage level nor the exceptional nature of the event [5, 6].

The seismic event has devastated mainly the localities located south of the epicenter, namely the small town of Amatrice, Arquata Del Tronto, and Accumoli, with about 300 fatalities and several hundreds of people injured [7, 8]. During the main shock, the accelerometric Amatrice station (code: AMT) of the Italian Accelerometric Network (RAN) [9], at 10 km from the epicenter on soil type B (Shear-wave velocity at the surface ($V_{S,30}$) between 360 and 800 m/s) according to [10], recorded the highest values of the ground motion (Peak Ground Acceleration of the east component reached 0.87 g) and vastly exceeded the Italian code spectrum [11–13].

Due to the high levels of observed ground motion and not regular damage distribution, several authors have investigated the seismic demand of the Amatrice earthquake, focusing on the leading causes of the registered seismic actions exceeding design spectra. A possible underestimation of the seismic input analyzed via the Probabilistic Seismic Hazard Analysis (PSHA) was accounted for by Panza et al. [14]. They discussed the limits in the formulation and use of PSHA to obtain the reference hazard maps for the Italian seismic code. In their work, they claimed that within PSHA, the assessment of seismic hazards and associated risks are based on subjective and unrealistic assumptions about seismic recurrence. Meletti et al. [3] highlighted a possible underestimation of the Italian seismic hazard by the Italian Seismic Hazard Map 2004 (MPS04), to which seismic actions are usually referred. Later, the Italian reference seismic hazard model was under re-evaluation, considering state-of-the-art methodological approaches, input data, and computational codes. A first presentation of the results of the review activity promised in 2016 can be found in [15], where the main results of the new MPS19 developed by the INGV Seismic Hazard Centre (CPS) are described. MPS19 includes many new data and methods that have appeared since the latest version of MPS and has been developed according to the most advanced international seismic risk standards [16].

On the other hand, near-source effects have been considered relevant for the resulting ground motion [17–19]. It should be noted that from a geological point of view, Amatrice is built on an alluvial terrace extended in the NW-SE direction with a length of about 2000 m and a maximum width of about 600 m. Soon after the first main event, seismic microzonation (SM) studies were conducted to identify stable zones prone to local amplification of seismic movement and zones prone to instability [20, 21]. It has been found that the amplification of ground motion reaches its maximum values in the historic center with a resonant frequency of about 2.0–2.5 Hz, in agreement with the distribution of earthquake damages [22, 23].

The present paper aims to investigate further the seismic structural response through accelerations recorded at different stations in Amatrice, accounting for site amplification phenomena described by seismic microzoning maps. An interpolation method has been introduced to define an Amplification function (*Af*). The M_W earthquake of 24 August has been investigated through accelerations acquired by the National Accelerometric Network (RAN, AMT station) available in the ITACA database (http://itaca.mi.ingv.it/ItacaNet/) and "Osservatorio Sismico delle Strutture" managed by the Italian Department of Civil Protection (OSS, AMTS station) [24, 25]. Then, the computed elastic response spectra were compared with the elastic demand spectra provided by the Italian seismic codes. Finally, different Intensity Measures (IMs) have also been estimated and analyzed.

2 Ground Motion Data Set

The first mainshock of the central Italy seismic sequence was recorded on 24 August 2016 by 350 seismic stations belonging to the Italian Strong-Motion Network managed by the Department of Civil Protection (RAN), to the Italian Seismic Network (Rete Sismica Nazionale, RSN) managed by the Istituto Nazionale di Geofisica e Vulcanologia (INGV), Osservatorio Sismico delle Strutture, managed by the Italian Department of Civil Protection (OSS), and to other local stations (see Fig. 1). The recorded data were downloaded from the ITACA database (Italian Accelerometric Archive, https://itaca.mi.ingv.it/ItacaNet_30/#/home).

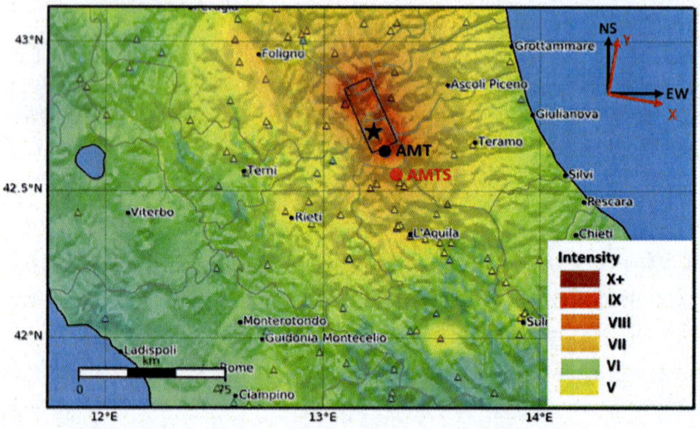

Fig. 1. Intensity map of the study area retrieved from data available on the website: htpps://shakemap.rm.ingv.it/shake/index.htms. White circles represent the AMT and AMTS station

Intensity measures (Ims) of the horizontal components of the triaxial signals recorded at the AMT and AMTS stations are listed in Table 1. Note that the E-W direction of AMT is close to x direction of AMTS, and the N-S direction of AMT is close to the y direction of AMTS station (see Fig. 1). The rotation angle between the directions of AMT and AMTS stations is estimated at 5 degrees.

In addition to PGA, PGV, and PGD, other parameters have been computed to interpret the damage potential of ground motion:

- Housner Intensity (IH), defined as the area below the elastic pseudo-velocity spectrum between the period range of 0.1–2.5 s: $I_H = \int_{0.1}^{2.5} S_{pv}(T, \xi = 0.05)\mathrm{d}T = \frac{1}{2\pi}\int_{0.1}^{2.5} S_{pa}(T, \xi = 0.05)T\mathrm{d}T$; where S_{pv} is the pseudo-velocity at the undamped natural period T and damping ratio $\xi = 0.05$, and S_{pa} is the pseudo-acceleration at the undamped natural period T and damping ratio $\xi = 0.05$ [26].
- Arias Intensity (IA), which represents the sum of the total energies, per unit mass, stored at the end of the earthquake ground motion, in a population of undamped linear oscillators and dimensionally is a velocity [27]: $I_A = \frac{\pi}{2g}\int_0^{t_0} a_g^2(t)dt$.

Table 1. PGA, PGV, PGD, Housner Intensity (I_H), and Arias Intensity (I_A) parameter values for signals recorded at AMT and AMTS stations

Station	Component	PGA (g)	PGV (cm/s)	PGD (cm)	I_H (cm)	I_A (cm/s)
AMT	NS	0.38	40.5	7.2	113.4	73.2
	EW	0.89	43.3	3.0	74.8	190.2
	UP	0.40	26.2	6.6	86.7	57.6
AMTS	NS	0.66	70.6	9.3	236.1	331.9
	EW	0.46	20.2	4.7	279.0	125.9
	UP	0.51	36.3	9.5	123.2	128.3

The highest PGA values for the event were recorded at the AMT station and range between 0.40 g and 0.88 g. Compared to the PGA, the velocity-based parameters provide a better approach to the definition of the destructive potential of earthquakes [28]. The highest PGV values were measured in the Amatrice range between 26.2 and 45.0 cm/s. By looking at the I_H parameter, some interesting points arise. For the August 24th event, it is possible to observe the highest values recorded at the AMTS station, i.e., I_H = 236.1 cm for the NS component and I_H = 279.0 for the EW component. High values of I_H indicate that the acceleration spectrum is shifted towards low frequencies, as will be highlighted by acceleration and energy spectra. It also indicates the presence of long-duration pulses due to the directivity or soil amplification effects. The Arias intensity is accounted for as it is needed to evaluate the significant duration of the earthquake. The largest Arias Intensity I_A, equal to 331.9 cm/s, was achieved for the NS component registered at the AMTS station.

The 5% damped elastic pseudo-acceleration spectra for the horizontal components of AMT and AMTS are shown in Fig. 2.

Peak values of the August 24th event occur in the period 0.1–0.5 s, with a fast decrease for higher periods. Maximum spectral acceleration values are close to or even

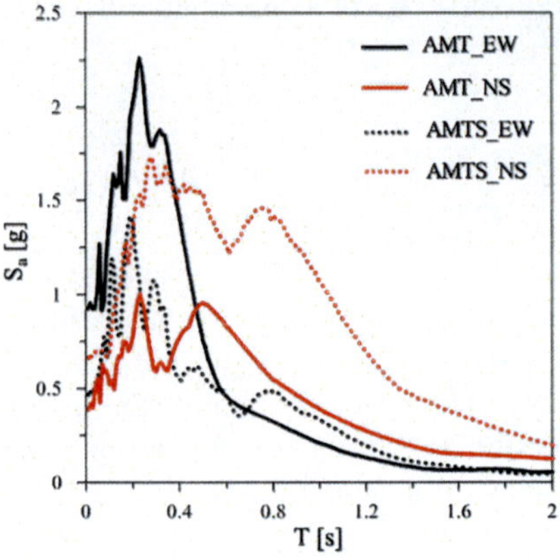

Fig. 2. Comparison of the pseudo-acceleration spectra at 5% damping ratio for two horizontal components for the main event of the sequence registered at the stations AMT and AMTS

greater than 2.0 g for AMT station. This can be justified as a small bilateral source rupture towards to south.

3 Microzoning Maps and Amplification Factors

After the hazardous event, the Italian Government funded structural and non-structural actions to reconstruct the area damaged by the seismic sequence [29]. Seismic Microzonation (from now on, SM) plays a crucial role in reconstruction, earthquake risk reduction, and management as it identifies the subsoil local response in each area in terms of selected ground shaking intensity parameters [30–32]. The Italian regulations establish an SM study consisting of three grades of approach to zonation, from the preliminary (SM1) to the advanced (SM3) level:

– Level 1 (SM1) relies on collecting existing data processed to classify the analyzed area into zones.
– Level 2 (SM2) integrates existing data and introduces the quantitative assessment of local seismic hazards through simplified methods. Results are expressed in terms of expected amplification factors (*Afs*).
– Level 3 (SM3) produces maps from the results of numerical analyses. This is done for areas characterized by high seismic hazards and subsoil complexity.

The *Afs* are computed as the ratio between the value of a given motion parameter at the study site (e.g., a ground surface) and the value of the corresponding parameter at the reference site (e.g., outcropping bedrock). *Afs* are defined for three different range of period: $T_1 = 0.1$–0.5 s; $T_2 = 0.4$–0.8 s, and $T_3 = 0.7$–1.1 s selected by considering the typical fundamental periods of buildings in the studied municipality [33]. The seismic microzonation map of Amatrice municipality related to the spectral interval of 0.1–0.5 s, which commonly includes the fundamental periods of different building typologies, is plotted in Fig. 3. Beyond AMT and AMTS stations, two different places where a high level of damage occurred have been investigated, i.e., A and B. In Fig. 3, the selected places are indicated by red circles. Moreover, it is worth highlighting that although the AMT station is just out the microzonation map edge, it has been scientifically reasonable to assume the closer *Af* (see Table 2).

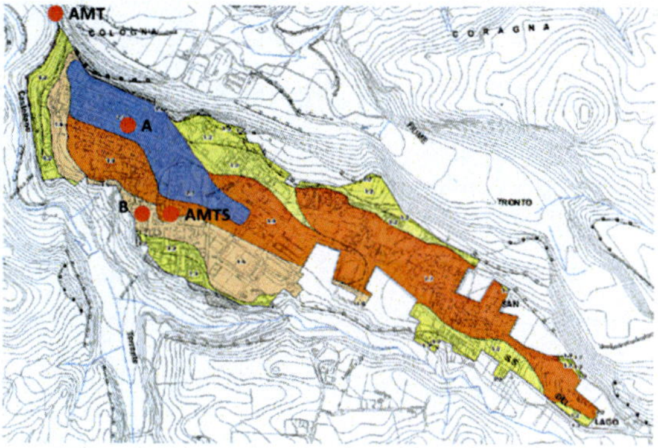

Fig. 3. Seismic Microzonation map of Amatrice municipality for period range of 0.1–0.5 s from [33]

Afs resulted from [33] for each period range are summarized in Table 2.

Table 2. *Afs* for the analyzed places in Amatrice municipality

Station/Place	Period range [s]		
	0.1–0.5	0.4–0.8	0.7–1.1
AMT	1.2	1	1
AMTS	1.8	1.1	1.5
A	2.3	1.1	1.7
B	1.6	1	1.1

Since the ranges of the period used to define the *Afs* in SM3 overlap (i.e., 0.4–0.5 s and 0.7–0.8 s), a linear interpolation method has been implemented to compute the *Af* in the ranges 0.1–0.4 s, 0.5–0.7 s, and 0.8–1.0 s, thus avoiding the overlapping. To simulate the seismic response in the points A and B, the measurements recorded at AMT and AMTS stations are multiplied by the normalized *NAf*. As an example, the *Naf* of point A from the AMT station in the period range of 0.1–0.4 s is computed as:

$$NAf_{x,y} = \frac{Af_x}{Af_y} \tag{1}$$

where x is the analyzed point (e.g., A) and y is the source point (e.g., AMT).

The resulted *Naf* is illustrated in Fig. 4 and evaluated in given period in Table 3.

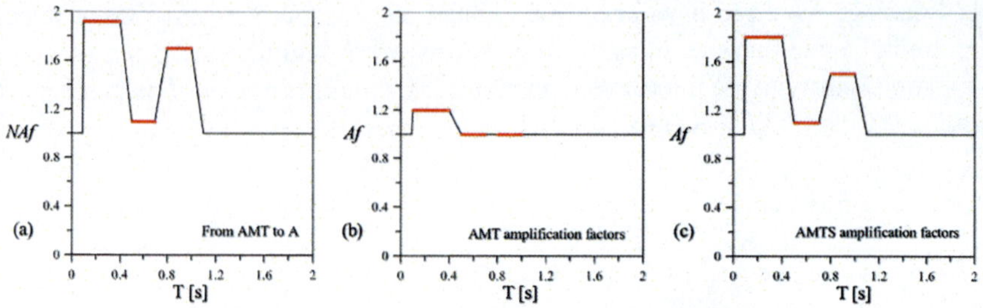

Fig. 4. *Naf* and *Af* graphical representation of: (a) point A from point AMT, (b) bedrock from point AMT, and (c) bedrock from point AMTS. The period ranges 0.1–0.4 s, 0.5–0.7 s, and 0.8–1.0 s have been highlighted in bold red lines.

Table 3. Amplification factors *Nafs* of points A and B in not overlapped period ranges

Station/Place	Period range [s]		
	0.1–0.4	0.5–0.7	0.8–1.0
A from AMT	1.92	1.10	1.70
A from AMTS	1.28	1.00	1.13
B from AMT	1.33	1.00	1.10
B from AMTS	0.89	0.91	0.74

The *Naf* are used to simulate the pseudo-acceleration response spectra for two horizontal components for the main event of the sequence registered at the AMT station. Figure 5 shows the response spectra simulated in points A and B (see Fig. 3) and amplified with the microzonation map from the sequence registered at the AMT station.

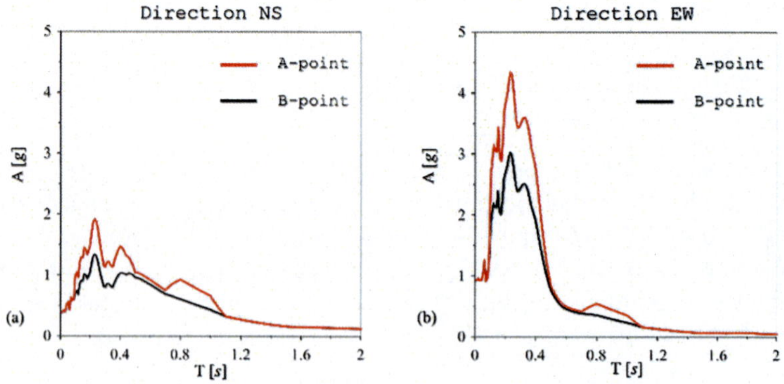

Fig. 5. Pseudo-acceleration response spectra of (a) NS and (b) EW horizontal ground motions simulated in the points A and B through the signal recorded at the AMT station

4 Observations on AMTS Source

The registration at AMTS station shows shaking data underestimated compared to those experienced in the registration site. First, substantial evidence is given by the fact that the installed accelerometer had a full scale of 0.5 g and was, therefore, able to measure with precision accelerations up to this value with the possibility to grasp acceleration values up to 30% higher than the full scale, i.e., up to 0.65 g. Todrani and Cultrera [13] confirmed that the seismic measure provided by the AMTS station underestimates the real action (in terms of PGA) by a factor of about 1.45. Therefore, the elastic response spectra developed from AMTS must be correct, i.e., amplified by a corrective factor of 1.45. According to the third-level seismic microzonation study (MS3), the elastic response spectra for AMTS obtained from AMT must be amplified by an *NAf* equal to 1.5. Figure 6 depicts the geometric direction averages of the elastic spectra for AMTS developed from (i) amplified accelerations recorded at the AMT station and (ii) corrected accelerations recorded at the AMTS station. The geometric directional average is obtained as follows:

$$S_{a_{average}} = \sqrt{S_{a_{NS}}^2 + S_{a_{EW}}^2} \quad (2)$$

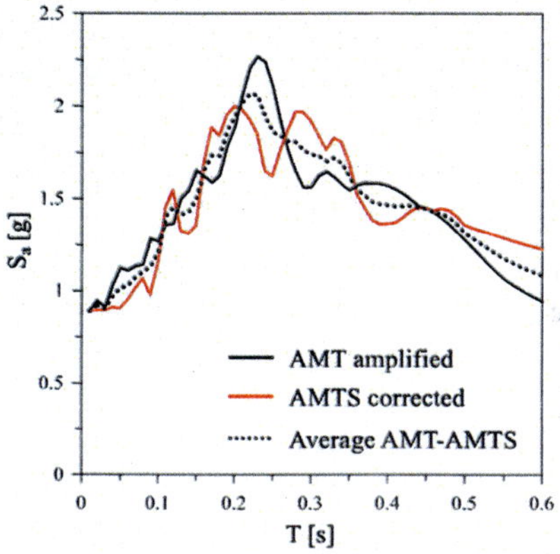

Fig. 6. Geometric directional averages of elastic spectra in AMTS station obtained from amplified AMT (black line) and corrected AMTS (red line). In dash line is reported the average between the two geometric directional averages.

Figure 5 shows the similarity between the geometric directional averages of the horizontal components of AMT and AMTS. Such similarity minimizes the importance of the orientation of the single component. The red curve (obtained from corrected AMTS) and the black curve (obtained from amplified AMT) overlap substantially. Another parameter used to check the similarity between the geometric directional averages of the horizontal components of AMT and AMTS is the difference between the area computed in the period range of 0–0.6s illustrated in Fig. 6. The percentage difference is equal to 0.36%.

4.1 Comparison with Italian Seismic Codes

This section investigates the ground motion intensities recorded during the sequence at AMT and AMTS stations and the values used for design based on the Italian seismic code (NTC18). It is worth mentioning that since 2008, the expected maximum acceleration at the site is no longer computed by dividing the territory into seismic zones but as a function of geographic coordinates of the site, return period, and different percentiles of the spectral ordinates' confidence level (i.e., 16th, 50th, and 84th). Furthermore, it can be observed that Amatrice municipality, affected by the seismic sequence, was classified before the 1980s; thus, a lower seismic hazard was assigned compared to the current one [8]. The observed pseudo acceleration, at 5% of critical damping, is compared with the elastic design spectra provided by the DMIT 2018 for three different return periods (TR), i.e., 475, 975, and 2475 years, corresponding to probabilities of exceedance in 50 years equal to 10%, 5%, and 2%, respectively. Comparisons are depicted in Fig. 7, where, for each station, the spectrum of the horizontal components is reported. The DMIT 2018 spectra are computed for the EC8 soil categories, i.e., soil class B for AMT and AMTS (non-corrected) and topographic category equal to T1. Such a comparison can discuss some valuable results. First, as illustrated in Fig. 7, in the 0–0.5 s period of interest for buildings, records registered at AMTS_NS (non-corrected) exceed the design spectra even for TR = 2475. Moreover, the AMT_EW spectra vastly exceed, in the plateau range of periods, the 2475-year return period spectrum. It is also interesting to focus on the average of the geometric directional average between AMT amplified and AMTS corrected (Fig. 7, blue line). Indeed, the average overcame the TR = 2475 design spectra in the full considered period range (0–2 s).

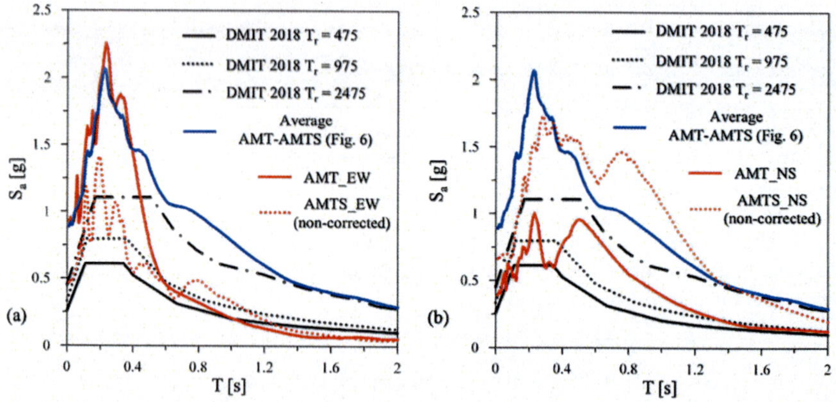

Fig. 7. Pseudo-acceleration response spectra of (a) EW and (b) NS horizontal ground motions recording during the August 24th event (AMT and AMTS stations) compared to the elastic spectra provided by the current Italian Building Standard (DMIT 2018).

Figure 8 compares the average AMT-AMTS reduced by the *Af* provided in Fig. 4c and the elastic spectra computed at bedrock following the current Italian Building Standard (DMIT 2018). The design of the elastic spectra for three different TR has been carried out by setting the soil type to A (bedrock). The comparison aims at identifying the elastic spectra that best fit the reduced AMTS. Indeed, from Fig. 8, it is possible to conclude that the elastic spectrum computed for TR = 1424 is the spectrum that, on average (in the range between 0 and 0.5 s), better estimates the average AMT-AMTS de-amplified.

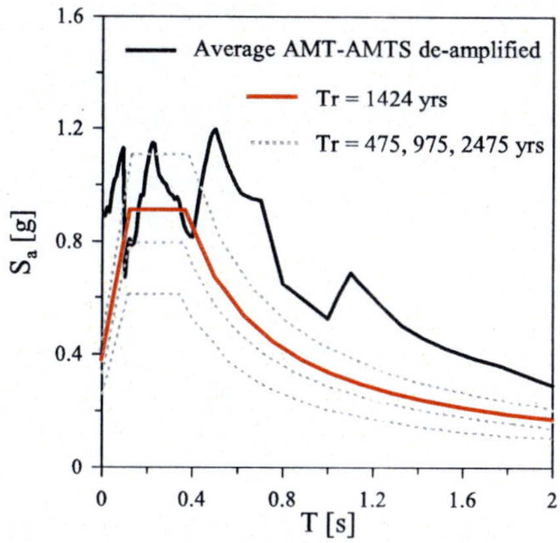

Fig. 8. Comparison between average AMT-AMTS and the elastic spectra computed at bedrock through the current Italian Building Standard (DMIT 2018).

5 Conclusions

A study of the seismic demand at Amatrice municipality during the Central Italy seismic sequence was carried out through experimental and computational approaches. The accelerogram damage potential registered during the sequence's strongest events was evaluated. Different intensity measures related to the ground motion records have been computed, e.g., elastic spectral demands in terms of pseudo-accelerations, energies, etc. A linear interpolation method has been proposed to account for the microzonation effects through the introduced function *NAf* used to amplify the response spectra. The spectral demands are compared with those of Italian seismic codes. The main conclusions of the study are summarized as follows:

- Pseudo-acceleration spectra showed high demands in the low periods (up to 0.6 s). High accelerations recorded at AMT could be due to pulse-like motions in the near field. A similar effect was also observed for the L'Aquila 2009 earthquake.
- The calculation of intensity measures, such as the Housner Intensity, shows that the highest values were attained at AMTS station during the main event. Such values indicate that the acceleration spectrum is shifted towards low frequencies and the presence of long-duration pulses due to soil amplification effects or directivity. The correlation between damage and Housner Intensity may explain the extensive damage distribution in Amatrice.

- The comparison between the acceleration spectra and the design code spectra highlighted that in Amatrice during the August 24th event, the records' spectra could be approximate with a 1424 years return period design spectrum in the plateau range of periods (0.1–0.4 s).
- Based on the results obtained in this study, the procedure proposed in the Italian building code (DMIT 2018) seems to be conservative at medium-to-long periods, while it can lead to significative underestimation of amplification factors (and therefore of spectral accelerations) in lower period range (i.e., 0.1–0.5 s).

References

1. Iervolino, I., Giorgio, M.: È possibile evitare il superamento delle azioni di progetto nell'area epicentrale di terremoti forti? Progettazione Sismica **3** (2017)
2. Iervolino, I., Baltzopoulos, G., Chioccarelli, E., Suzuki, A.: Seismic actions on structures in the near-source region of the 2016 central Italy sequence. Bull. Earthq. Eng. **17**(10), 5429–5447 (2017). https://doi.org/10.1007/s10518-017-0295-3
3. Meletti, C. et al.: Seismic hazard in central Italy and the 2016 Amatrice earthquake. Ann. Geophys. (2016)
4. Albarello, D. et al.: Pericolosità sismica del territorio nazionale: carte di consenso GNDT e SSN. Le Ricerche de GNDT nel Campo della Pericolosità Sismica (1996–1999), pp. 265–273 (2000)
5. Falsone, G., Alderucci, T., Spinella, N.: I recenti terremoti poco profondi in Italia: caratteristiche peculiari trascurate dalle norme (2018)
6. Ceci, A.M., Gattulli, V., Potenza, F.: Serviceability and damage scenario in irregular RC structures: post-earthquake observations and modeling predictions. J. Perform. Constr. Facil. **27**(1), 98–115 (2013)
7. Arcoraci, L. et al.: QUEST—Rilievo macrosismico per i terremoti nell'Italia central. Aggiornamento dopo le scosse del 26 e 30 ottobre 2016. INGV Technical Report (2016)
8. Mollaioli, F., AlShawa, O., Liberatore, L., Liberatore, D., Sorrentino, L.: Seismic demand of the 2016–2017 Central Italy earthquakes. Bull. Earthq. Eng. **17**(10), 5399–5427 (2018). https://doi.org/10.1007/s10518-018-0449-y
9. Albarello, D., et al.: Seismic characterization and reconstruction of reference ground motion at accelerometric sites of the Italian national accelerometric network (RAN). Nat. Hazards **86**(2), 401–416 (2016). https://doi.org/10.1007/s11069-016-2310-4
10. Ministero Infrastrutture e Trasporti, D.M., NTC (2008) Norme tecniche per le costruzioni. 2008, Roma
11. Luzi, L., et al.: The central Italy seismic sequence between August and December 2016: Analysis of strong-motion observations. Seismol. Res. Lett. **88**(5), 1219–1231 (2017)
12. D'Amico, M. et al. Italian Accelerometric Archive v 3.1. Istituto Nazionale di Geofisica e Vulcanologia, Dipartimento della Protezione Civile Nazionale: Rome, Italy (2020)
13. Todrani, A., Cultrera, G.: Near-source simulation of strong ground motion in Amatrice downtown including site effects. Geosciences **11**(5), 186 (2021)
14. Panza, G., et al.: Why are the standard probabilistic methods of estimating seismic hazard and risks too often wrong. In: Earthquake hazard, risk and disasters, pp. 309–357. Elsevier (2014)
15. Meletti, C. et al.: The new Italian seismic hazard model (MPS19). Ann. Geophys. **64**(1) (2021)
16. Gerstenberger, M.C. et al.: Probabilistic seismic hazard analysis at regional and national scales: State of the art and future challenges. Rev. Geophys. **58**(2), e2019RG000653 (2020)

17. Gaudiosi, I. et al.: Evaluation of site effects at Amatrice (central Italy) after the August 24th, 2016, Mw 6.0 earthquake. Soil Dyn. Earthq. Eng. **144**, 106699 (2021)
18. Panzera, F., et al.: Correlation between earthquake damage and seismic site effects: the study case of Lentini and Carlentini Italy. Eng. Geol. **240**, 149–162 (2018)
19. Sextos, A., et al.: Local site effects and incremental damage of buildings during the 2016 Central Italy earthquake sequence. Earthq. Spectra **34**(4), 1639–1669 (2018)
20. SM Working Group. Guidelines for seismic microzonation (2015)
21. Moretti, M. et al.: SISMIKO: Emergency network deployment and data sharing for the 2016 central Italy seismic sequence. Ann. Geophys. (2016)
22. Milana, G., et al.: Local site effects estimation at Amatrice (Central Italy) through seismological methods. Bull. Earthq. Eng. **18**(12), 5713–5739 (2019). https://doi.org/10.1007/s10518-019-00587-3
23. Cara, F., et al.: Temporary dense seismic network during the 2016 Central Italy seismic emergency for microzonation studies. Sci. Data **6**(1), 1–15 (2019)
24. Dolce, M., Nicoletti, M., De Sortis, A., Marchesini, S., Spina, D., Talanas, F.: Osservatorio sismico delle strutture: the Italian structural seismic monitoring network. Bull. Earthq. Eng. **15**(2), 621–641 (2015). https://doi.org/10.1007/s10518-015-9738-x
25. De Sortis, A., et al.: Analysis of the recorded response of a school building heavily damaged by the 2016 central italy earthquake. Buildings **12**(7), 907 (2022)
26. Housner, G.W.: Spectrum intensities of strong-motion earthquakes (1952)
27. Arias, A.: Measure of Earthquake Intensity. Massachusetts Inst. of Tech., Cambridge. Univ. of Chile, Santiago de Chile (1970)
28. Decanini, L., Liberatore, L., Mollaioli, F.: Damage potential of the 2009 L'Aquila, Italy, earthquake. J. Earthq. Tsunami **6**(03), 1250032 (2012)
29. Dolce, M. Di Bucci, D. The 2016–2017 central apennines seismic sequence: analogies and differences with recent Italian earthquakes. In: European Conference on Earthquake Engineering Thessaloniki. Springer, Greece (2018)
30. Pergalani, F., et al.: Seismic microzoning map: approaches, results and applications after the 2016–2017 Central Italy seismic sequence. Bull. Earthq. Eng. **18**(12), 5595–5629 (2019). https://doi.org/10.1007/s10518-019-00640-1
31. Celikbilek, A., Sapmaz, G.: Risk management and microzonation in urban planning: an analysis for Istanbul. Disaster Sci. Eng. **2**(2), 59–66 (2016)
32. Aversa, S., Crespellani, T.: Seismic microzonation: an essential tool for urban planning in seismic areas. UPLanD-J. Urban Plan. Landscape Environ. Des. **1**(1), 121 (2016)
33. Microzonazione sismica, Relazione Illustrativa (2017) [cited 2022; Available from: https://www.amicidisaletta.it/wpcontent/uploads/2018/04 Amatrice_MS3_Relazione-Illustrativa.pdf

Assessment of the Acceleration Floor Spectra Through Dynamic Identification: The Museum of Bargello in Florence

Riccardo M. Azzara[1], Daniele Pellegrini[2], Vieri Cardinali[3](✉), Stefania Viti[3], and Marco Tanganelli[3]

[1] Istituto Nazionale di Geofisica e Vulcanologia, Arezzo, Osservatorio Sismologico, Italy
riccardo.azzara@ingv.it

[2] Istituto di Scienze e Tecnologie dell'Informazione "A. Faedo", CNR-ISTI, Pisa, Italy
daniele.pellegrini@isti.cnr.it

[3] Department of Architecture (DIDA), University of Florence, Florence, Italy
vieri.cardinali@unifi.it

Abstract. Artworks represent a priceless asset to the economic and cultural features of communities. Most art collections are hosted in Museums, which can be new buildings, appositely made for expositive purposes, or monumental buildings, whose high artistic and historical value enhances the exposed art pieces. In this latter case, however, the Museums can disregard the seismic safety requirements provided for new constructions, becoming the main source of hazard for the precious contents they should preserve.

In this paper, the dynamic behavior of the National Museum of Bargello in Florence is studied by means of a dynamic identification, focusing the attention on the "*Sala Donatello*". An experimental campaign was performed by simultaneously installing two sets of three seismometric stations in the mentioned room and in the "*Sala Michelangelo*", placed at the lower level. Analysis of the recorded data via Operational Modal Analysis techniques has furnished the structure's natural frequencies, damping ratio and mode shapes allowing the assessment of the amplification of the seismic acceleration experienced by the art works exposed in "*Sala Donatello*". The effect of the seismic acceleration on the artifacts has been checked on a case-study, i.e. the masterpiece "*Marzocco*". It is the statue of the lion considered the symbol of Florence, realized by Donatello in 1420, placed on a marble pedestal made by *Benedetto da Maiano* in 1480, which is a work of art as well. The assessment has been made by performing a simplified rigid-block analysis. The geometrical data of Marzocco has been stated based on a detailed photogrammetric survey, which provided a reliable representation of the mass distribution.

Keywords: Seismic vulnerability · Dynamic identification · Works of Art · Sculptures · Bargello Museum · Lion of Marzocco

G. P. Cimellaro (Ed.): WCSI 2022, LNCE 309, pp. 1010–1021, 2023.
https://doi.org/10.1007/978-3-031-21187-4_88

1 Introduction

The problems related to the conservation of cultural assets and their protection are collecting an increasing interest by researchers and academics [1–6]. In Italy, as in the other countries affected by seismic risk, many losses of immovable and movable cultural goods occurred as a consequence of earthquakes [7, 8]. Recent seismic events such as Assisi (1997), L'Aquila (2009), Emilia (2012) and the last one in Central Italy (2016), confirm the need to strengthen prevention activities. The definition of the vulnerability or seismic security of movable property contained within museums or present in archaeological areas is of vital importance. The safety of the works of art exhibited within the museums requires to check the dynamic response both of the "container" (building-museum) and of the single manufacts [9].

In this frame, the first step should be oriented at modeling the seismic response of the buildings where the artifacts are contained. Dealing, generally, with buildings of cultural or historical interest, the experimental practice should be absolutely not invasive and as possible do not interfere with the use of the structure. These are the principal operative characteristics of the seismometric monitoring of built structures, a practice that is increasingly developing in recent times, in particular when cultural heritage is involved [10–17].

This work has been developed within the research framework "Resimus" [18], which is aimed at assessing the seismic safety of artworks through a comprehensive approach. Such procedure, briefly presented in the Sect. 2, joints a careful assessment of the seismic acceleration occurring at the artwork (taking into account the role of the exposition building in the accelerations transmission) and its seismic response. This study is focused on the seismic safety of art collections exhibited in the Museum of Bargello, in Florence, and – more precisely – of the famous marble sculpture "Lion of Marzocco", placed in the *Sala Donatello.*

A special attention has been paid to the assessment of the seismic acceleration faced by the artwork at the occurring of a ground motion compatible with the seismic hazard of the area. Indeed, the seismic intensity occurring at the *Sala Donatello* after the assumed ground motion has been found by performing a dynamic monitoring of the building, i.e. by placing proper seismic stations. The waveforms recorded during the test have been analyzed in the frequency and time domains. The dynamic behavior of the building has been compared to the one of the *Volognana* tower, which belongs to the same building and is adjacent to *Sala Donatello.* The collected information has been used to perform a simplified analysis of the seismic performance of the Lion of Marzocco. The obtained safety level of the statue has been compared to the one found according to the conventional approach provided by the current Italian Technical Code. The comparison has shown that the Code procedure induces unconservative assessment of the seismic performance of artworks.

2 The Performed Analysis: The International Museum of Bargello, the Donatello Room and the Lion of Marzocco

The current work is focused on the seismic performance of the famous statue "Lions of Marzocco", the famous symbol of Florence [19–21] shown in Fig. 1. It was commissioned to Donatello by the Republic of Florence around 1420, in order to decorate the papal apartments of Santa Maria Novella. Because of its state of preservation, it was later moved to the Uffizi, and finally to the Museum of Bargello, first at ground floor, in the stone sculpture room, then in the *Sala Donatello* at the first floor (Fig. 2a,b). It is located above a marble base, made around 1480 and recently attributed to *Benedetto da Maiano.* It is a parallelepiped element consisting of a molded foot, a central body completed by four banisters and an entablature composed of architrave, frieze and frame, all enriched with heraldic symbols. It was moved to the Bargello Museum only in 1999, thus rediscovering its legitimate function [22].

Quantity	Unit	Single component		
Material	-	stone	marble	marble
Mass	kg	561	269	2264
Height	cm	133.0	15.5	130.0
X-coordinate of MC	cm	32.3	33.2	32.8
X-coordinate of MC	cm	86.2	93.0	93.8
X-coordinate of MC	cm	200	137.6	69.8

Fig. 1. View and main information (material, mass, height and Mass Center) on the Lion of Marzocco.

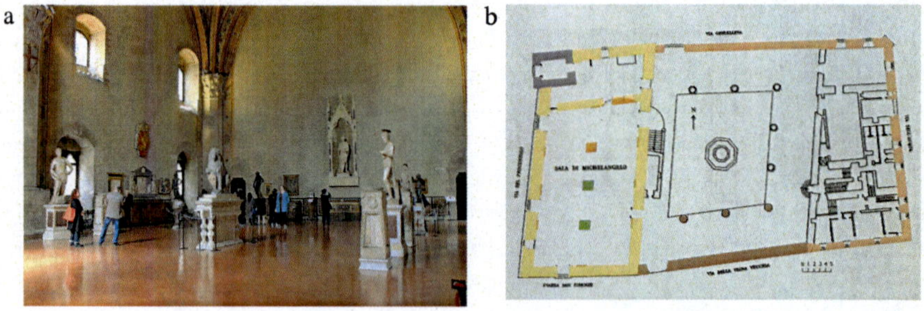

Fig. 2. a. View of the "*Sala di Donatello*"; b. Plan of "*Sala Donatello*" (from [23]).

The presented analysis has been made within the methodological approach developed by the research project "Resimus". It provides a comprehensive frame of operational steps, briefly represented in the flowchart shown in Fig. 3. The seismic performance of an artwork is assessed by comparing its vulnerability to the seismic hazard suffered as a consequence of the occurring earthquake. Both vulnerability and hazard, in turn, are assessed after proper analyses, which can be analytical and/or empirical.

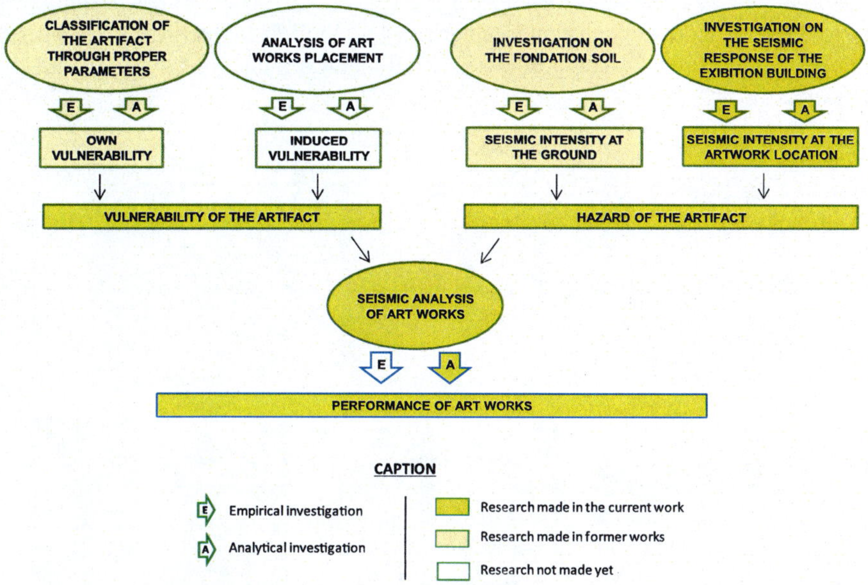

Fig. 3. Flowchart of the research activity on the seismic safety of artifacts

The vulnerability of an artwork can be distinguished as "own", i.e. related to its own features, of "induced", i.e. related to its position within the exhibition (and to the possible collapse of the exhibition building itself or other artworks). In both cases, the vulnerability sources can be found through proper classification procedures [18]. In Fig. 3, the research steps presented in this work have been evidenced with a yellow label, to distinguish them from the ones developed in former steps of the research program. These latter results can be found in other papers made by the authors [24–26].

In this work only the "own vulnerability", neglecting any other induced ones, has been investigated. The collapse mechanism has been assumed related to the overturning, based on former studies focused on the vulnerability classification of artworks. The seismic hazard assumed for the assessment has been quantified after a careful analysis, aimed at investigating the seismic propagation of the base acceleration form the foundation to the exhibition room. The seismic intensity at the soil foundation has been defined according to the seismic seismicity of the area and the soil class, which has been object of former investigations [27]. A special attention has been paid to quantify the effect of the dynamic response of the building on the amount of seismic acceleration arising at the statue base at the occurring of the assumed ground motion. The dynamic response of the Museum of Bargello has been found by performing a dynamic identification of the building, and by validating the obtained results through a numerical modal analysis.

3 The Dynamic Response of the Building

The experimental measurements were performed in 2018, October 29th. Six seismometric stations were installed for about 3 h inside the Museum. Three stations were deployed on each floor aligned along a diagonal direction (Fig. 4), in order to reconstruct the response along a horizontal profile for each floor and along the vertical alignment at each point.

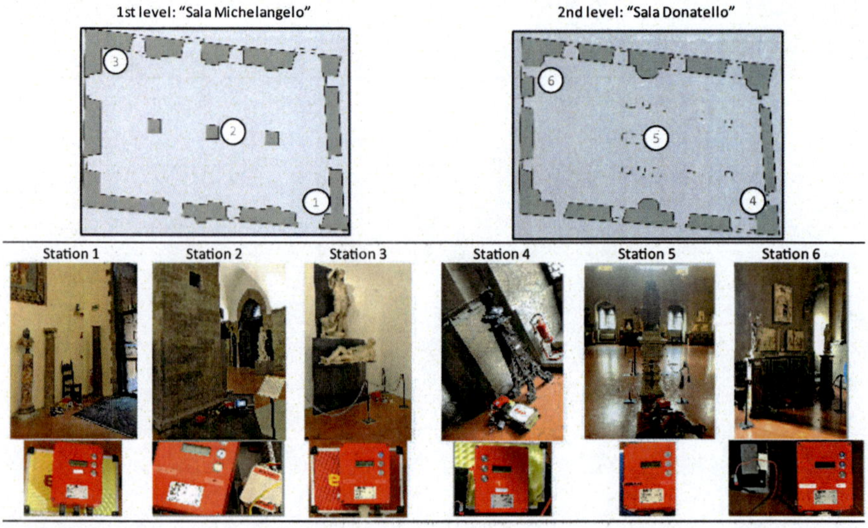

Fig. 4. Distribution of the seismic stations inside the Bargello Museum, on the left, ground floor, *Sala Michelangelo*, on the right, first floor, *Sala Donatello* where the *Marzocco* monument is exposed.

Each seismic station was equipped with an SS20 three axial seismometers (velocity transducer, eigenfrequency 2 Hz, useful band 0.1–40 Hz) coupled to a Digital Acquisition System (DAS) SL06 (A/D converter 24 bit) produced by SARA electronic instruments. The six seismic stations have jointly recorded for about 3 h at a sampling rate of 100 Hz. Unfortunately, three of them (#2 and #5, located at the center of the halls and # 3 at the corner of Michelangelo Hall), were too far from a window or other opening, so that it was not possible to connect the GPS antenna, therefore the recorded data were manually synchronized identifying common signals occurred on the floors and on the vertical. The remaining stations were synchronized through an external GPS. The sensors were oriented along longitudinal (Y) and transverse (X) directions. Before proceeding with the analysis, the velocity data were corrected for the instrumental response of the seismometers and integrated, in order to obtain displacement. Table 1 shows the values of the spectral peaks detected from the analysis that are common to all the stations, while Fig. 5 shows the displacement average spectra computed for the data recorded inside the *Sala Donatello* and *Michelangelo*, respectively at the top and at the bottom. In each panel, the spectra for X (black), Y (blue) and Z (red) directions and the code of the station are shown.

Table 1. Experimental frequencies of the normal modes of oscillation.

Station number	1	2	3	4	5	6
Frequency [Hz]	1.00	1.52	2.66	2.99	3.64	3.91

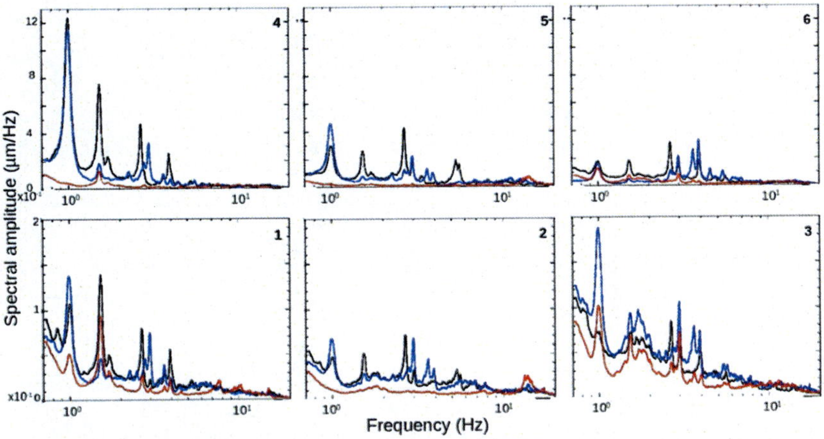

Fig. 5. Displacement average spectra provided by the seismic stations along the directions X (black), Y (blue) and Z (red).

The obtained results have been compared to the ones obtained from experiments previously performed inside the *Volognana* Tower, that is part of the complex of the Bargello Museum and adjacent to the monitored section. Indeed, the Tower is firmly connected to the "*Palazzo del Bargello*" up to the height of 28 m. The Tower, 45 m high, was instrumented through five seismic stations installed at different heights.

Figure 6 shows the comparison between the spectrum found for the *Sala Donatello* (station #4) and the ones referred to the *Volognana* Tower. Namely, the comparison shown in the left panel refers to the spectrum provided by the seismic section placed at the top of the Tower (blue curve), whilst the right panel refers to the one placed at 9 m of height (red curve). In order to better compare the trends, the vertical scale has been normalized to the maximum of each curve. The first two spectral peaks, at 1.00 Hz and 1.52 Hz, represent the frequencies of the first two normal modes respectively for the longitudinal and the transversal component. The comparison shows that the first two frequencies correspond to a cross-effect produced by the free oscillation of the tower. As a result, it is possible conclude that the first two frequencies listed in Table 1 are due to the free oscillation of the *Volognana* Tower and that the subsequent two values are surely ascribable to the first mode of oscillation of the monitored section: 2.66 Hz for the X direction, 3.0 Hz for the Y one.

For each of these frequencies the time trend of the maximum amplitude has been evaluated. After filtering the signal in a narrow band centered at the identified frequency, the time pattern for the average maximum amplitude of the displacement has been evaluated for all the measurement points. The distribution of the maximum amplitudes has been computed, evaluating the maximum displacements on subsequent intervals of 10 s. Evaluating the ratio between the average maxima of vertical corresponding stations, it is possible to estimate an amplification factor for each measuring point at the first floor relative to each mode. At the end, the damping factors have been computed for each measurement points in the same frequency band use for evaluating the maxima distribution.

Table 2 lists the values of amplification and damping estimated for each measurement point. Thanks to the dynamic identification it was possible to determine the fundamental modes of the structural unit. This information is crucial for several reasons: *(i)* calibrate suitable structural models in order to perform seismic analysis; *(ii)* determine the acceleration response at the *Sala Donatello* level, which the statues are subjected to.

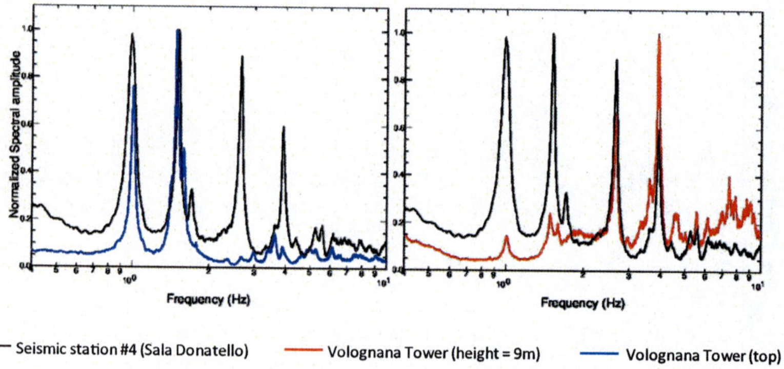

Fig. 6. Comparison between the frequency histories at *Volognana* Tower and *Sala Donatello.*

Table. 2 Estimated values of amplification and damping factor at the first floor (*Sala Donatello*)

Stations	2.6 Hz		3.0 Hz	
	Amplification	Damping [%]	Amplification	Damping [%]
4	5.40	0.56	4.44	0.74
5	5.56	0.57	3.39	0.73
6	3.20	0.57	2.09	0.74

The dynamic identification has been adopted to calibrate a FE model of the *Sala Donatello.* The model consists of a simplified 2-d structure made by 32592 plates and 26970 nodes realized in Strand7. The model represents a first modelling step, as it does not involve the openings and it only considers the massive bearing walls and the vaulted structures. To simulate the boundary conditions of the unit, i.e. the connections to the other aggregations of the Bargello Museum (*Volognana* Tower, courtyard etc.), as a first approximation they have been accounted by modelling vertical carriages orthogonal to the direction of the structural elements. This simplified model allowed obtaining the first modes of the unit by defining an Elastic Young's modulus of the masonry equal to 2150 MPa, with a density of 20 kN/m^2. In particular, the model led to a first mode equal to 2.66 Hz along the X direction, while frequencies equal to 2.96 and 4.08 Hz were found in the Y direction. The latter will be adopted in further phases of the research to execute seismic analysis on the museum content.

4 Simplified Analysis of the Lion of Marzocco

The obtained value of the first vibrational mode has been used to calibrate the simplified analysis performed to check the seismic performance of the Lion of Marzocco; such analysis has been carried out according to NTC2018 [28], through the floor spectra, defined on the basis on the fundamental Period, T_0, of the structure. In this paper, the floor spectrum found by assuming the proper value of T_0 has been compared to the one corresponding to the simplified approach provided by NTC 2008 ($T_0 = 0.05 * H^{3/4}$, where H is the building's height) [29]. The analysis has been performed by assuming two limit states (i.e. the *damage* and the *life safety* limit states), with a return period equal to 712 years (nominal life of the building equal to 50 years and important class III), and a class soil B, which is the most probable for the area. Figure 7 shows the floor spectra found according to these assumptions, together with the soil spectrum.

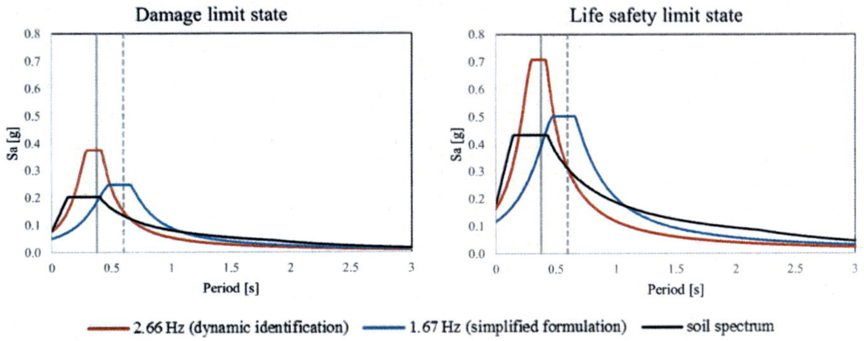

Fig. 7. Comparison between the floor spectra for simplified and experimental evaluation of the fundamental period.

In this preliminary investigation, the seismic analyses have been performed considering three distinct overturning mechanisms, assumed alternatively based on the epistemic uncertainties that concern the statue. As the relationships within statue, basement and support are unknown, three different overturning systems have been accounted: *(i)* statue only; *(ii)* statue + support; *(iii)* statue + support + pedestal. The out-of-plane mechanisms have been considered rather than other possible behavior (e.g. sliding) because of the slender dimension of the composition. Other studies conducted on different statuesque configurations, together with researches on the friction properties of statues over different supports and analytical FE investigations prove the reliability of this assumption [30, 31]. The overturning analysis was made according to the principle of virtual work, by assuming a linear behavior of the system. The computation of the multiplier activating each mechanism has been later converted into PGA values according to MIT2019; in addition, a confidence factor equal to 1.35 was considered. The capacity of the structure has been then compared with the seismic demand presented in Fig. 6 for both damage and life safety limit state. For the latter, a simplified *q-factor* equal to 2.0 has been accounted according to MIT2019. As linear kinematic analyses have been considered, the demand was considered with respect of $T = 0\ s$. Within the three considered mechanisms, the one involving the statue, the support and the pedestal resulted to be the one with a lower multiplier, hence, given this level of knowledge, it represents the most likely phenomenon. In Fig. 8, for the mentioned out-of-plane mechanisms a comparison in terms of safety indexes (accounted as the ratio within seismic capacity and seismic demand) is shown.

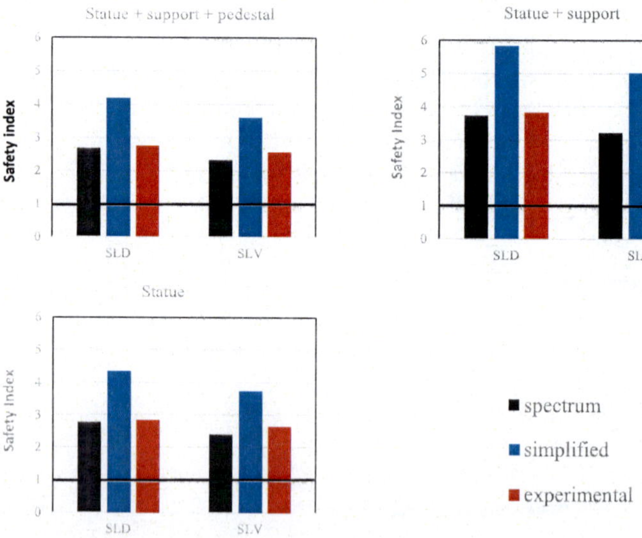

Fig. 8. Comparison between the different safety indexes.

The research shows the variability of the safety indexes of the statue based on its museal location and the accounted considerations. Based on the seismic floor spectra developed according to NTC2008, the *z/h* ratio of the *Sala Donatello* with respect of the total height of the museum leads to decreasing seismic values. In this sense, the localization of the statue in the Donatello room, rather than in the Michelangelo's one, leads to a decrease of the demand with a consequent increase of the safety indexes. For the life safety limit state, the safety index increases of the 10% when the floor spectrum is plotted accounting on the dynamic identification results, instead of increasing of 55% as suggested by the simplified formulation.

5 Conclusions

The present work illustrates the preliminary results of a research concerning the seismic safety of the artworks exhibited in museums. Namely, a simplified rigid-block analysis of the marble statue "Lion of Marzocco", currently located in the International Museum of Bargello in Florence, has been made, based on the floor seismic spectra provided by the National Technical Code (NTC 2018). Such spectra have been found as a function of the Fundamental Period (T_0) of the building, found through an experimental investigation based on the lecture of six seismic stations, which provided the amplification factor of the base-acceleration at the various floors of the Museum.

The results of the experimental test made on the building have been compared to the ones provided by a solid FE model, which confirmed their reliability. The T_0 value provided by the dynamic investigation has been used to set the seismic floor spectra of the building; they have been compared to the ones found through the T_0 value as provided by the simplified approach of the Code, i.e., as a function of the building height. The analysis showed that the conventional procedure suggested by the Code provides very unconservative values of the seismic vulnerability of the statue; indeed, the safety index found after the Code procedure is about 40% higher than the value found by using the effective T_0, provided by the dynamic identification.

The presented results refer to the first analyses performed on the *Lions of Marzocco* within the research project; the investigation has to be extended, in order to include more detailed and comprehensive Finite Element models to represent the seismic response of the building, and to check the dynamic response of the case-study to the assumed seismic acceleration.

References

1. Agbabian, M.S., Ginell, W.S., Masri, S.F., Nigbor, R.L.: Evaluation of earthquake damage mitigation methods for museum objects. Stud. Conserv. **36**(2), 111–120 (1991)
2. Agbadian, M.S., Masri, S.F., Nigbor, R.L., Ginell, W.S.: Seismic damage mitigation concepts for art objects in museums. In: Proc. IX World Conference on Earthquake Engineering, Tokyo-Kyoto, Japan (1988)
3. Lowry, M.K., Farrar, B.J., Armendariz, D., Podany, J.: Protecting Collections in the J. Paul Getty Museum from Earthquake Damage. WAAC Newsletter **29** (2007)
4. Podany, J.: Advances in the reassembly of large stone sculpture at the J. Paul Getty Museum. In: Recent Advances in the Conservation and Analysis of Artifacts, pp. 375–381. Institute of Archaeology, London (1987)
5. Spyrakos, C.C., Maniatakis, C.A., Taflampas, I.M.: Assessment of seismic risk for museum artifacts. The 14th World Conference on Earthquake Engineering. October 12–17, 2008, Beijing, China (2008)
6. Spyrakos, C.C., Maniatakis, C.A., Taflampas, I.M.: Application of predictive models to assess failure of museum artifacts under seismic loads. JCH 23, 11–21.Author, F., Author, S.: Title of a proceedings paper. In: Editor, F., Editor, S. (eds) CONFERENCE 2016, LNCS, vol. 9999, pp. 1–13. Springer, Heidelberg (2016)
7. Parisi, F., Augenti, N.: Earthquake damages to cultural heritage constructions and simplified assessment of artworks. Eng. Fail. Anal. **34**, 735–760 (2013)
8. Liberatore, D., Spera, G., Claps, A., Larotonda, A.: Tipologie dei Beni archeologici. In: "Vulnerabilità dei beni archeologici e degli oggetti esibiti nei musei" (A cura di Liberatore D.), CNR – Gruppo Nazionale per la difesa dai terremoti – Roma (2000). ISBN 88-900449-4-2
9. Viti, S., Reinhorn, A.M.: Monumental buildings used as museums: protection or danger for the artifacts?, Procedia Struct. Integrity **29**: 40–47 (2020). ISSN: 2452–3216. https://doi.org/10.1016/j.prostr.2020.11.137

10. Azzara, R., De Falco, A., Girardi, M., Pellegrini, D.: Ambient vibration recording on the Maddalena Bridge in Borgo a Mozzano (Italy): data analysis. Ann. Geophys., **60**(4), S0441 (2017). https://doi.org/10.4401/ag-7159
11. Azzara, R.M., De Roeck, G., Girardi, M., Padovani, C., Pellegrini, D., Reynders, E.: The influence of environmental parameters on the dynamic behaviour of the San Frediano bell tower in Lucca. Eng. Struct. **156**, 175–187 (2018)
12. Azzara, R.M., De Roeck, G., Girardi, M., Padovani, C., Pellegrini, D., Reynders, E.: Assessment of the dynamic behaviour of an ancient masonry tower in Lucca via ambient vibrations. In: Van Balen, K., Verstrynge, E. (eds) Structural Analysis of Historical Constructions. Anamnesis, diagnosis, therapy, controls. Taylor & Francis Group, 2016, Print ISBN: 978-1-138-02951-4 (2016). ISBN: 978-1-317-20662-0
13. Azzara, R.M., Girardi, M., Padovani, C., Pellegrini, D.: Experimental and numerical investigations on the seismic behaviour of the San Frediano bell tower in Lucca. Ann. Geophys. **61**(2018) (2019). https://doi.org/10.4401/ag-8025
14. Azzara, R.M., Zaccarelli, L., Morelli, A., Trombetti, T., Dallavalle, G., Cavaliere, A., Danesi, S.: Seismic monitoring of the Asinelli and Garisenda medioeval towers in Bologna (Italy), an instrumental contribution to the engineering modeling direct to their protection. In: Proc. 2nd International Conference on Protection of Historical Constructions (PROHITECH 2014), Istanbul, May 7–9 (2014)
15. Baraccani, S., Palermo, M., Azzara, R.M., Gasparini, G., Silvestri, S., Trombetti, T.: Structural interpretation of data from static and dynamic structural health monitoring of monumental buildings. Key Eng. Mater. **747**, 431–439 (2017)
16. Ubertini, F., Cavalagli, N., Kita, A., Comanducci, G.: Assessment of a monumental masonry bell-tower after2016 Central Italy seismic sequence by long-term SHM. Bull. Earthq. Eng. **16**(2), 775–801 (2018)
17. Gentile, C., Saisi, A.: Ambient vibration testing of historic masonry towers for structural identification and damage assessment. Constr. Build. Mater. **21**(6), 1311–1321 (2007)
18. Viti, S., Tanganelli, M.: Resimus: a research project on the seismic vulnerability of museums' collections. In: 7th International Conference on Computational Methods in Structural Dynamics and Earthquake Engineering, COMPDYN 2019, vol. 2, pp. 2819–2829. National Technical University of Athens (2019)
19. Gaeta Bertelà, G.: Museo Nazionale del Bargello. La guida ufficiale. Firenze, Giunti Gruppo Editoriale, 2000, ISBN 9788809217430 (in Italian) (2000)
20. Paolozzi Strozzi, B.: Museo Nazionale del Bargello, Guida ufficiale., Firenze, Giunti Editore S.p.a., 2014 (5), ISBN 978–8809759626 (in Italian) (2014)
21. Lumachi, F.: Firenze, nuova guida illustrata, storica, artistica, aneddotica della città e dintorni, Firenze, Società Editrice Fiorentina 1929 (in Italian) (1929)
22. Manescalchi, R.: Il Marzocco/The lion of Florence. In collaborazione con Maria Carchio, Alessandro del Meglio, ISBN 978-88-95450-23-0
23. Giorgi, L., Matracchi, P.: Il Bargello a Firenze. Da Palazzo del Podestà a Museo Nazionale. In: "Santa Maria del Fiore. Teorie e storie dell'archeologia e del restauro nella città delle fabbriche arnolfiane", pp. 125–174, Alinea, Firenze (in Italian) (2006)
24. Tanganelli, M., Coli, M., Cimellaro, G.P., Marasco, S., Cardoni, A., Noori, A.Z., Viti, S.: Dynamic analysis of artifacts: experimental tests for the validation of numerical models. In: Proceedings of the 7th International Conference on Computational Methods on Structural Dynamics and Earthquake Engineering. vol. 2, p. 2865–2877, M. Papadrakakis, M. Fragiadakis, ISBN: 978-618-82844-7-0, Crete, Greece, 24–26/6/2019 (2019)
25. Tanganelli, M., Galassi, S., Viti, S.: Simplified analyses for the model setting of sculptures: the "Oceano" by Giambologna. Proc. 8 th ECCOMAS Thematic Conference on Computational Methods in Structural Dynamics and Earthquake Engineering. pp. 1–10, COMPDYN 2021, ATENE, 27–30 June (2021)

26. Tanganelli, M., Viti, S., Cimellaro, G.P., Domaneschi, M.: Developing a laboratory facility to assess friction coefficients of standing samples. PROCEDIA STRUCTURAL INTEGRITY, vol. 29, pp. 142–148 (2020). ISSN: 2452-3216. https://doi.org/10.1016/j.prostr.2020.11.150
27. Sapia, V., Materni, V., Giannattasio, F., Marchetti, M.: Esplorazione geofisica del sottosuolo: primi risultati nel centro storico di Firenze. In: "RESIMUS Un progetto rivolto alla vulnerabilità sismica delle opere museali", a cura di Stefania Viti, DIDAPRESS, Firenze (2019)
28. Norme Tecniche per le Costruzioni 2018, Decreto del Ministro delle Infrastrutture del 17 gennaio 2018, Nuove norme tecniche per le costruzioni, (in Italian) (2018)
29. Norme Tecniche per le Costruzioni 2008. Norme tecniche per le costruzioni. D.M. Ministero Infrastrutture e Trasporti 14 gennaio 2008, G.U.R.I. 4 Febbraio 2008, Roma (in Italian)
30. Cerri, G., Pirazzoli, G., Verdiani, G., Rotunno, T., Pintucchi, B., Viti, S., Tanganelli, M.: Seismic Assessment of Artefacts: The Case of Juno's Fountain of The National Museum of Bargello. In: Florence Heri-Tech 2018 - The Future of Heritage Science and Technologies, Florence (Italy), 16–18 Maggio 2018, vol. 36. IOP publishing (2018)
31. Viti, S., Pintucchi, B., Rotunno, T., Tanganelli, M.: The seismic analysis of Cerere at the Museum of Bargello. Bull. Earthq. Eng. **18**(6), 2635–2656 (2020). https://doi.org/10.1007/s10518-020-00802-6

Experimental Evaluation of the Cyclic Behaviour of Different Smooth Rebar Anchoring Layouts

Simone Pelucco(✉), Anthony Paderno, and Marco Preti

University of Brescia, D.I.C.A.T.A.M., Via Branze 43, 25123 Brescia , BS, Italy
{s.pelucco,a.paderno007,marco.preti}@unibs.it

Abstract. The existing reinforced concrete frame structures built before the 1970s were designed only for gravity loads, and they were mainly built using smooth steel rebars. The behaviour of these kinds of structures is different with respect to that exhibited by structures reinforced with deformed rebars. Low bond capacity and the associated slip of the anchorages characterise a more flexible structural response and a lower energy dissipation of the elements. The contribution of rebar slip to structural deformability is often not considered in traditional lumped plasticity models used for the assessment of existing structures. An innovative modelling approach has been recently proposed to directly account for smooth rebar slip, so the characterization of the cyclic behaviour of different types of anchorages is of primary importance. A first experimental campaign was conducted on straight, hooked-end, and bent anchorages to evaluate their hysteretic behaviour and their strength capacity. Some aspects highlighted by the first campaign were investigated in the second one, the main results are here presented. The influence of the bond conditions and of the embedded length on the anchorage performance of plain rebars was investigated, and the mean bond strength was thus evaluated. Finally, the cyclic behaviour of bent anchorages for different bond conditions is analysed.

Keywords: Existing RC buildings · Plain reinforcement · Bar slip · Experimental test · Cyclic behaviour · Rebar anchorage

1 Introduction

A large amount of existing Reinforced Concrete (RC) frame structures was designed only for gravity loads; seismic considerations and details were not considered. In the period between World War II and the 1970's, smooth rebars were used as longitudinal and transversal reinforcement. Large longitudinal rebar slips due to the low bond capacity characterise a different flexural behaviour and a different crack pattern of the members with respect to that exhibited by structures realised using deformed rebars. Experimental results on clomuns [Di Ludovico et al. 2014; Melo et al. 2015; Verderame et al. 2008a; b] and beam-column joints [Beschi et al. 2015; Braga et al. 2009] show that the deformation is mainly concentrated at the member ends where few cracks open. An added contribution

G. P. Cimellaro (Ed.): WCSI 2022, LNCE 309, pp. 1022–1031, 2023.
https://doi.org/10.1007/978-3-031-21187-4_89

to the deformation is given by the rotation at the member ends due to the rebar slip. Consequently, a more flexible response and lower energy dissipation are associated. A new modelling approach has been proposed to directly account for the rebar slip [Paderno, Pelucco, Metelli, et al. 2021a], so the mechanical characterization of the cyclic stress-slip behaviour of the different types of anchorages is essential.

The behaviour of smooth rebars has been studied for short straight anchor lengths and hooked rebars [Fabbrocino et al. 2005; Verderame, De Carlo, et al. 2009a; Verderame, Ricci, et al. 2009b]. A previous experimental campaign was conducted to investigate the cyclic axial stress-slip behaviour of long straight anchors, hooked-end anchors and bent anchors [Paderno, Pelucco, and Preti 2021b]. Here are presented the main results of a second experimental campaign conducted to investigate some aspects highlighted by the first one.

The influence of the bond conditions on the anchor strength performance was investigated by means of two specimens. Straight smooth rebars were placed at different heights and the variability of the mean bond strength was evaluated. The two specimens differ in length, so the influence of the embedded length is also considered.

The cyclic behaviour of bent anchors was analysed through four specimens. The bond conditions vary along the length of this type of anchor, so three specimens were casted with the loading side located at the top of the block and a specimen with the loading side at the bottom of the block, so the influence of the bond conditions is also investigated.

2 Description of the Specimens

The experimental campaign consists of six concrete blocks. Specifically, two blocks (named SM1 and SM2) are intended to investigate the influence of the bond conditions and of the anchoring length on the performance of straight anchorages in terms of strength; the other four blocks (BS1, BS2, BS3 and BS4) are used to characterise the behaviour of bent anchorages and to study the influence of the bond conditions.

2.1 Geometry of the Specimens SM1 and SM2

Each block contains six straight 12 mm diameter smooth rebars. They are located at different heights to study the influence of the bond conditions on the anchorage strength. Each bar protrudes from the block side by 75 mm to allow the clamping of the bar head by the loading machine. To allow a simpler application of the load to each single bar, the rebar heads protrude alternately from the block on two opposite sides, and the rebar tail end simply emerges from the block side to allow the monitoring of the rebar end slip. A concrete cover of 100 mm with respect to the external faces of the block and a distance of 125 mm between each rebar are guaranteed for all the rebars.

The two blocks SM1 and SM2 differ in their length, which is equal to 1000 mm and 760 mm, respectively. So, the influence of the anchoring length for the same bond condition is investigated. These specific embedded lengths are chosen for direct comparison of the results with the previous recalled experimental campaign [Paderno, Pelucco, and Preti 2021b].

The geometry of the specimens is represented in Fig. 1.

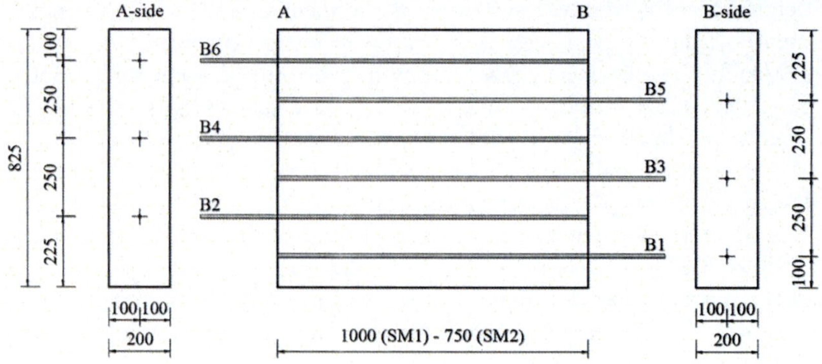

Fig. 1. Geometry of the specimens SM1 and SM2

2.2 Geometry of the Specimens BS1 to BS4

As for the bent anchor specimens, a single anchor bar is located in a concrete block. The 12 mm diameter smooth rebars are characterised by two 45° bends with a straight inclined length in between. The rebar ends with two horizontal straight lengths, with the head protruding from the block to allow clamping for the application of the load. Unlike the specimens of the previous experimental campaign, the block is as long as the anchor, so the tail end of the rebar emerges from the block to allow monitoring of its displacement.

The specimens BS1, BS2 and BS3 were casted with the loading side at the bottom of the concrete block, in good bond conditions. Specimen BS4 was casted in the opposite direction, with the bar head protruding from the top part of the block side, so in bad bond conditions. Note that for the bent rebar anchors, the bond conditions progressively vary along the inclined length. The specimens BS3 and BS4 were instrumented with a double strain gauge located at the mid-length of the inclined length to evaluate the different strength contibution offered by the two bends to the anchor and to monitor the propagation of the slip along the rebar.

In Fig. 2, it is possible to see the geometry of the anchorages BS1 to BS3; the specimen BS4 is similar.

2.3 Materials

The smooth rebars were realised with steel of grade S275JR [EN 10025 2019]. The mechanical characterization of the tensile behaviour was made according to [UNI EN 10080 2005]. Two different steel batches were used, respectively, for the straight specimens and the bent ones. The mean values of yielding and ultimate strength are reported in Table 1 for each group of specimens.

Each specimen was realised with a different cast of the same mix design. The mean cylindrical strength and the mean elastic modulus are reported in Table 1, as well. The mechanical properties of concrete are designed to be consistent with those of existing RC construction which adopts smooth rebars.

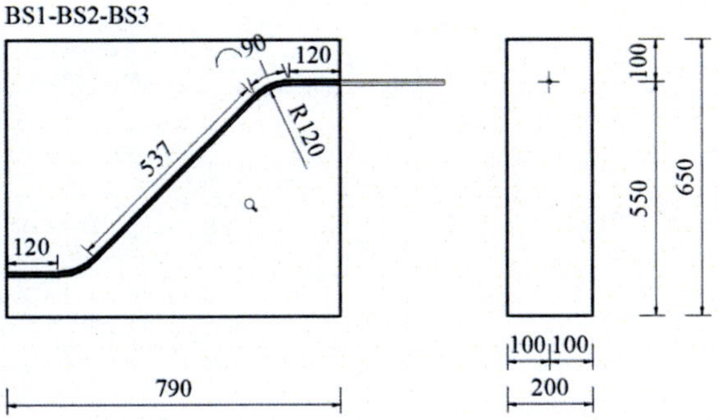

Fig. 2. Geometry of the specimens BS1, BS2 and BS3

Table 1. Mechanical properties of the material used.

Specimen	Steel		Concrete	
	f_y [Mpa]	f_u [Mpa]	f_c [Mpa]	E_c [Mpa]
SM1	461.24	549.63	20.59	26485
SM2			19.61	25942
BS1	361.17	518.07	22.37	25885
BS2			18.72	23933
BS3			18.46	22796
BS4			18.94	23930

3 Test Set-Up

3.1 Loading Frame and Instrumentation

The steel frame used to test the anchorages and the instrumentation system are those described in [Paderno, Pelucco, and Preti 2021b]. It is a self-balanced steel system (Fig. 3) which allows to impose a cyclic pull and push force at the head of the rebar anchor. The clamping of the rebar is obtained by means of a head coupler.

Tensile and compressive loads are measured using a load cell, while head and tail slip are measured using three LVDTs located on the loading frame and one LVDT located on the backside of the specimen.

3.2 Loading Protocol

The anchor specimens were tested under cyclic loading. The loading protocols are graphically reported in Fig. 4.

For straight rebar anchors, the tensile load was gradually increased during the first cycle until the peak strength was reached. Then, the slip is increased up to twice the value

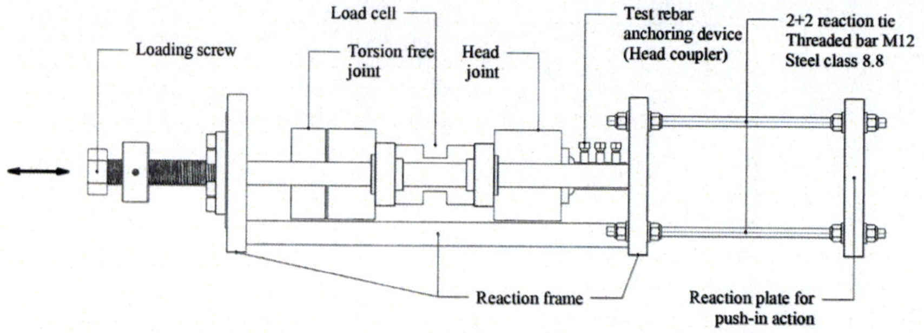

Fig. 3. Representation of the loading frame, [Paderno, Pelucco, and Preti 2021b].

reached at the peak strength before unloading and reversing the load in compression till the full recovery of the cumulated slip. In the following cycles, the magnitude of the slip is progressively increased.

For bent rebars, the load was progressively increased up to yielding, then cycles followed in slip control, similar to what was explained for straight anchorages. The unloading and reverse loading were pushed to the full recovery of the slip, unless buckling occurred in compression.

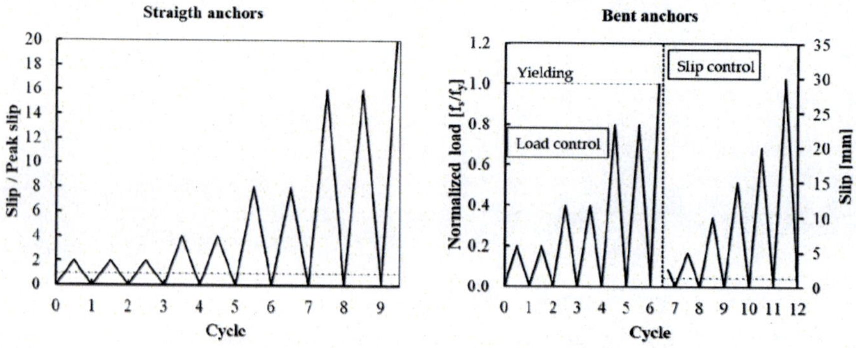

Fig. 4. Loading protocol for the two types of anchors.

4 Experimental Results

4.1 Straight Anchors

In Fig. 5, the results of the tests of the SM1 specimen are reported. From the bottom to the top rebar, the rebars are numbered in ascending order. The overall behaviour is the same as highlighted in the previous experimental campaign. After the reaching of the yielding or of the peak strength, there is a degradation in strength. A partial recovery of the slip occurs at the release of the load. After that, the full recovery of the slip is possible by the application of a compression unless rebar buckling occurs. At the reloading in tension, a lower strength plateau is reached for slip amplitude larger than already explored in the previous cycle. A further strength reduction occurs for repeated cycles in the previously explored amplitude range, until a slip of larger amplitude is reached. Only the bottom anchor was able to reach the rebar yielding; the peak strength, the slip corresponding to the peak strength, and the residual strength progressively reduce from the bottom rebar to the top one. The SM2 specimen shows similar results, which are here not reported for conciseness. In this case, the reduced embedded length of the anchor rebar did not allow any anchor specimen to reach the yielding strength of the rebar.

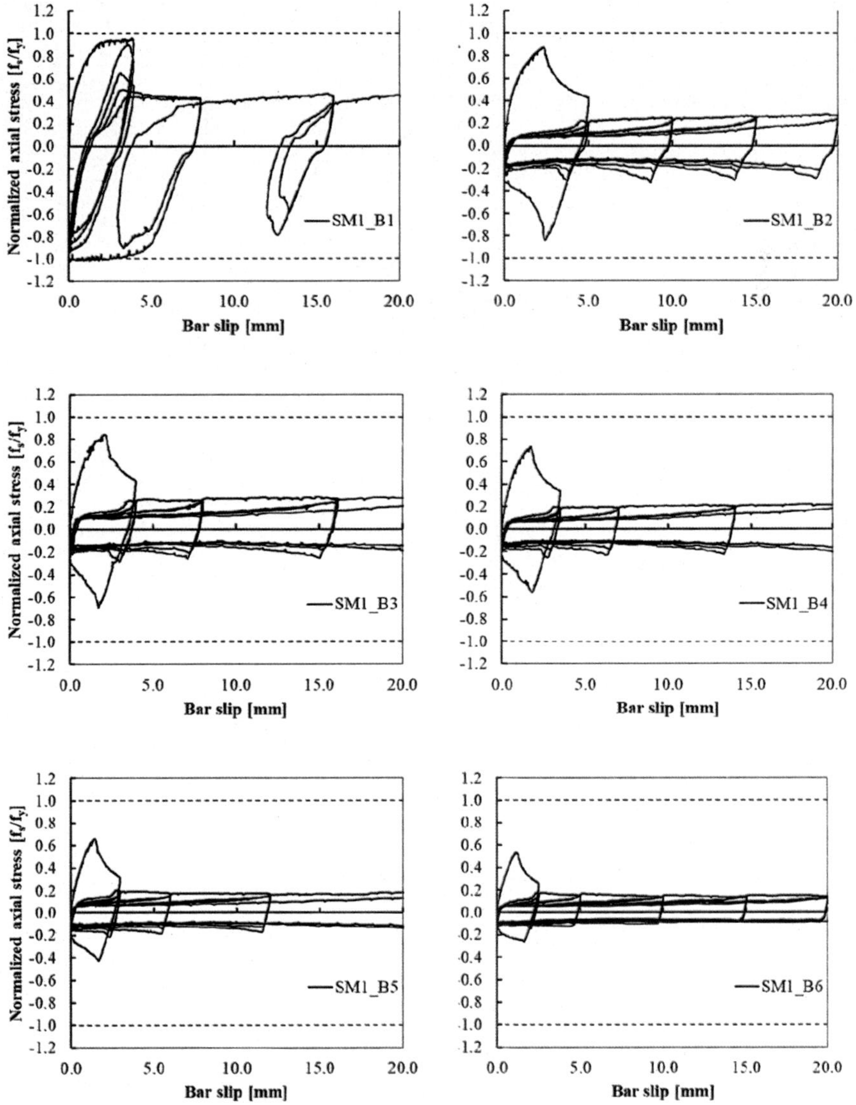

Fig. 5. Results of the SM1 specimen.

In Fig. 6, for each test, is represented a focus on the positive branch of the first cycle. The comparison shows that both the strength and stiffness decrease from the bottom rebar to the top one. The tests on the SM2 (Fig. 7) specimen show a little variation to this trend for the rebars located at mid-height.

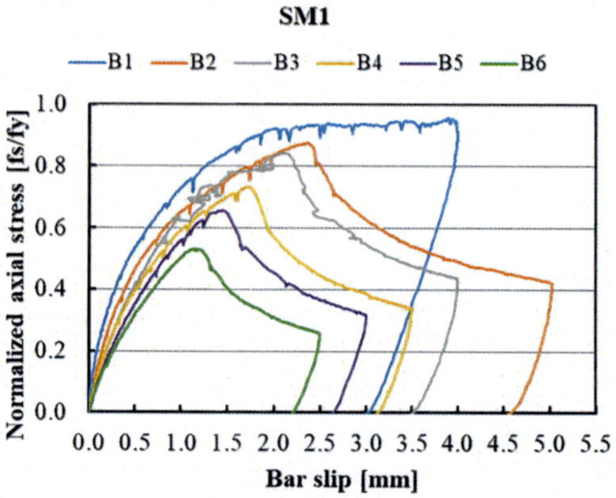

Fig. 6. Comparison of the first positive branch for each test of the SM1 specimen.

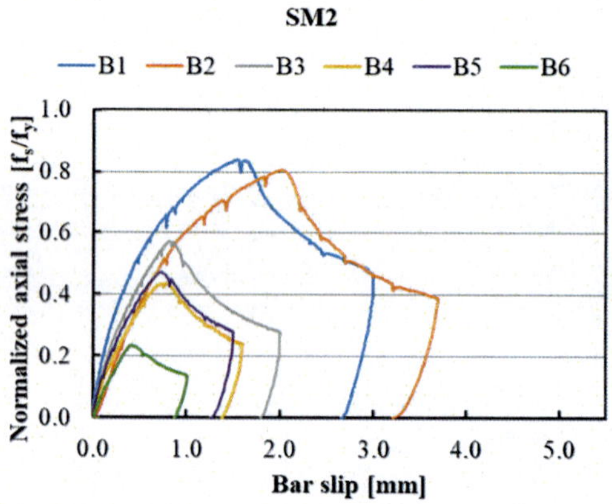

Fig. 7. Comparison of the first positive branch for each test of the SM2 specimen.

4.2 Bent Anchors

The bent anchor is able to equilibrate the yielding and the hardening of the rebar both with the loading side in good and bad condition. At the reloading in tension, the behaviour is secant to the maximum slip explored in the previous cycle. In compression, the yielding is reached unless the buckling of the rebar occurs. The specimen with the loading side in good bond condition (BS4) shows a stiffer initial branch. In Fig. 8, the results of a specimen in bad bond conditions (BS1) and of the specimen in good bond conditions (BS4) are reported.

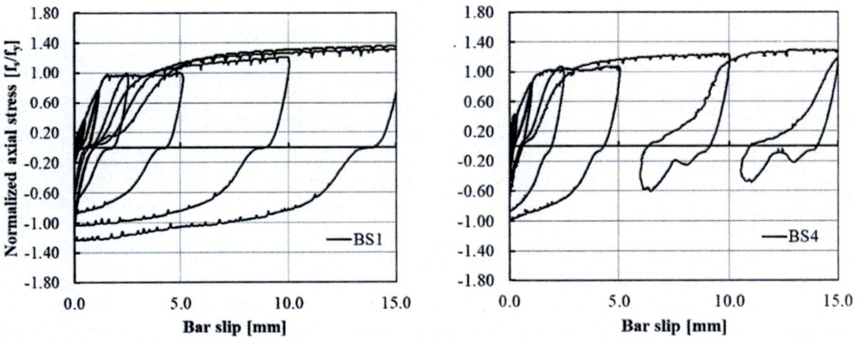

Fig. 8. Results of a specimen in bad bond conditions (BS1) and of the specimen in good bond conditions (BS4).

5 Discussion of the Results of the Traight Anchors

5.1 Influence of the Bond Conditions

Assuming a constant distribution of bond stresses along the anchor length, a mean bond-stress (τ_b) magnitude is evaluated by dividing the axial load (N) by the rebar lateral surface. At the peak load and at the cyclic branch, all the length of the rebar (l_b) is subjected to slip. These two conditions are taken as a reference.

$$\tau_b = \frac{N}{\pi \phi l_b} \tag{1}$$

For comparison with previous works and indications [CEB-FIP Model Code 2010; Paderno, Pelucco, and Preti 2021b; Verderame, De Carlo, et al. 2009a; Verderame, Ricci, et al. 2009b], the ratio C between the bond stress and the square root of the concrete compressive stress is considered.

$$\tau_b = C\sqrt{f_c} \tag{2}$$

For the sake of simplicity, the four lower rebars are considered in good bond conditions, while the other two are considered in bad bond conditions. The mean value of the bond coefficient C for the two specimens and for the two bond conditions is reported in Table 2; they agree with the indication of the references. It is worth noting that the lower rebar of the SM1 specimen achieves the yielding without the activation of the rebar tail, so the bond strength is not completely developed, and the bond coefficient is underestimated. The results show that the length of the anchorage is not so relevant to the bond strength value. The strength of the anchors in bad bond conditions is a little higher than that proposed by [CEB-FIP Model Code 2010].

Table 2. Mean bond coefficient C evaluated by the experimental tests and comparison with other works.

Specimen	Bond coefficient C	
	Peak	Cyclic
SM1 Good	0.26	0.04
SM2 Good	0.27	0.05
SM1 Bad	0.18	0.03
SM2 Bad	0.15	0.02
[CEB-FIP Model Code 2010], Good	0.30	NA
[CEB-FIP Model Code 2010], Bad	0.15	NA
[Paderno, Pelucco, and Preti 2021b], Good	0.25	0.06
[Verderame, De Carlo, et al. 2009a], Good	0.31	0.05

6 Conclusions

The results of an experimental campaign on the cyclic behaviour of smooth rebar anchors are here reported.

Two specimens of straight anchors characterised by two different lengths are considered; each of them contains six smooth rebars located at different heights. The influence of the bond conditions and of the embedded length on the anchor performance is investigated. The mean bond coefficient is evaluated at the peak load and at the cyclic branch for each rebar; both the specimens show a reduction of both the bond strengths along the member height. Instead, the embedded length is not so relevant to the coefficient.

The cyclic behaviour of bent anchors is investigated by means of four specimens; three of them were casted with the loading side placed at the bottom of the block, in good bond conditions; the fourth was casted with the loading side at the top. The anchors equilibrate the yielding and the hardening of the rebars. An initial stiffer behaviour is shown by the specimen with the loading side in good bond condition.

Acknowledgements. The experimental campaign presented in this work has been supported within the framework of DPC/ReLUIS project 2019–2021, WP5.

References

Beschi, C., Riva, P., Metelli, G., Meda, A.: HPFRC jacketing of non seismically detailed RC corner joints. J. Earthquake Eng. **19**(1), 25–47 (2015)

Braga, F., Gigliotti, R., Laterza, M.: R/C existing structures with smooth reinforcing bars: experimental behaviour of beam-column joints subject to cyclic lateral loads. Open Constr. Build. Technol. J. **3**(1), 52–67 (2009)

CEB-FIP Model Code. "fib Model Code for Concrete Structures 2010," p. 436 (2010)

Di Ludovico, M., Verderame, G.M., Prota, A., Manfredi, G., Cosenza, E.: Cyclic behavior of nonconforming full-scale RC columns. J. Struct. Eng. **140**(5), 04013107 (2014)

EN 10025. EN 10025 Hot Rolled products of structural steels (2019)
Fabbrocino, G., Verderame, G.M., Manfredi, G.: Experimental behaviour of anchored smooth rebars in old type reinforced concrete buildings. Eng. Struct. **27**(10), 1575–1585 (2005)
Melo, J., Varum, H., Rossetto, T.: Experimental cyclic behaviour of RC columns with plain bars and proposal for Eurocode 8 formula improvement. Eng. Struct. **88**, 22–36 (2015)
Paderno, A., Pelucco, S., Metelli, G., Preti, M.: Analytical and numerical modelling of existing RC frames with smooth rebars. Eng. Struct. **249**, 113160 (2021)
Paderno, A., Pelucco, S., Preti, M.: Experimental investigation on anchorage performance of embedded smooth rebars subjected to cycling loading, pp. 1201–1211. Greece, Athens (2021)
UNI EN 10080. UNI EN 10080 Steel for the Reinforcement of Concrete—Weldable Reinforcing Steel (2005)
Verderame, G.M., De Carlo, G., Ricci, P., Fabbrocino, G.: Cyclic bond behaviour of plain bars. Part II: analytical investigation. Constr. Build. Mater. **23**(12), 3512–3522 (2009)
Verderame, G.M., Fabbrocino, G., Manfredi, G.: Seismic response of r.c. columns with smooth reinforcement. Part I: Monotonic tests. Eng. Struct. **30**(9), 2277–2288 (2008)
Verderame, G.M., Fabbrocino, G., Manfredi, G.: Seismic response of r.c. columns with smooth reinforcement. Part II: Cyclic tests. Eng. Struct. **30**(9), 2289–2300 (2008)
Verderame, G.M., Ricci, P., Carlo, G.D., Manfredi, G.: Cyclic bond behaviour of plain bars. Part I: experimental investigation. Constr. Build. Mater. **23**(12), 3499–3511 (2009)

Dynamic Characterisation and Seismic Vulnerability Assessment of Existing Masonry Port Structures

Raffaele Tarantini[1], Alessandro Cardoni[1], Sebastiano Marasco[1], Jacopo Merlin[1], Enrico Pribaz[2], Gianluca Rupolo[2], Marco Domaneschi[1(✉)], and Gian Paolo Cimellaro[1]

[1] Department of Structural, Geotechnical and Building Engineering, Politecnico di Torino, Corso Duca degli Abruzzi 24, 10129 Turin, TO, Italy
{alessandro.cardoni,sebastiano.marasco,marco.domaneschi, gianpaolo.cimellaro}@polito.it, jacopo.merlin@studenti.polito.it

[2] Autorità di Sistema Portuale del Mar Tirreno Settentrionale, Calata Carrara 1, 57123 Livorno , LI, Italy

Abstract. This paper focuses on the dynamic identification and the seismic vulnerability assessment of an existing masonry port structure located in Livorno (Italy), which is a former railway station bombed and reconstructed after WWII. To this end, a historical-critical analysis, and an investigation and testing campaign has been carried out, along with a number of Ambient Vibration Tests (AVT). According to the Operational Modal Analysis (OMA), a Frequency Domain Decomposition (FDD) algorithm has been leveraged for dynamic identification. The severe noise around the harbour area and the poor soil quality make it difficult to identify the modal parameters of the structure, even if specific signal processing techniques were applied to remove the interferences. Although a ground sensor is not needed for output-only identification, it has been useful to identify a local phenomenon that it would have been arduous to figure out without its support. The investigations and the test outputs have been the baseline to create a finite element model, although not reproducing the real structural behaviour of some simplifying assumptions. However, the dynamic Structural Health Monitoring (SHM) outcomes have not been leveraged to calibrate it. A Modal Response Spectrum Analysis (MRSA) and a non-linear static (Pushover) analysis (POA) have been carried out to determine the structure capacity, and thus the Seismic Vulnerability Index (SVI) of the simplified model.

Keywords: Existing structures · Masonry buildings · Structural health monitoring · Ambient vibration test · Operational modal analysis · Dynamic identification of structures · Seismic vulnerability assessment

1 Introduction

In recent years, the masonry construction sector has shown a growing vitality, both from a cultural point of view and from that of the building activity. This recovery has

G. P. Cimellaro (Ed.): WCSI 2022, LNCE 309, pp. 1032–1042, 2023.
https://doi.org/10.1007/978-3-031-21187-4_90

certainly contributed to the pressing interest on the part of the scientific community in the rehabilitation, restoration and seismic retrofitting of the enormous heritage built in Italy.

It was therefore taken note of a significant problem that afflicts our country, the preservation and reuse of existing buildings. In any intervention concerning the building heritage it is necessary, in addition to the knowledge of the state of conservation of the masonry, to investigate the design criteria, the construction techniques and the rule of art adopted during the construction of the artifact. It is also of fundamental importance to define the mechanical behaviour of both recently built and historical masonries.

The modelling and seismic analysis of masonry buildings topic is therefore absolutely topical. In fact, the assessment of the seismic vulnerability of existing masonry structures represents one of the most critical aspects that the civil engineer and, specifically, the structural engineer must face. A valid support on which the professional can rely is represented by Structural Health Monitoring (SHM), a tool of fundamental importance that allows to observe and estimate the state of health of civil works for the purpose of assessing their safety.

One of the key steps of the seismic vulnerability assessment lies in the design of a structural model on which to carry out static and dynamic analyses. Often, due to the lack of information on the structure, the model created on the basis of historical-critical analysis and classical investigation techniques does not constitute a perfect digital twin of the real structure and, consequently, a sufficiently faithful representation of the real dynamic behaviour of the latter. In order to match the behaviour of the model with that of the real structure, the methodology of dynamic identification can be used. The purpose of the latter is to estimate the modal parameters of the building, on the basis of which it is possible to calibrate the finite element model.

In the next chapters the seismic vulnerability assessment of the former railway station "Livorno Porto Vecchio" will be analysed, starting from the historical-critical analysis, ending up to the numerical modelling, as well as the Structural Health Monitoring (SHM) and the dynamic identification. We believe that the present work might be a useful contribution for the enhancement of the scientific background regarding the Mediterranean and European built heritage.

2 General Overview

Located in Piazza del Portuale n°4 (Calata Carrara), 57123, Livorno, the building under study is part of the building complex constituting the former Livorno Porto Vecchio railway station, until 1939 called Stazione Marittima, today a branch of the Port System Authority of the Northern Tyrrhenian Sea.

The building, almost symmetrical with respect to one of the main directions, can be virtually divided into 5 blocks (Fig. 1). Blocks 1,2,4 and 5 are divided into 2 levels, while block 3 is divided into a single level. The first floor slabs of blocks 1,2,4,5 are located at the same height (3.50 m), while the roof slab of block 3 is located at a higher altitude (5.20 m). Finally, the roofing floors of blocks 1,2,4,5 stand at a height of 7.50 m. Between the buildings 1,2 and 4,5 there is a structural joint that will be hypothesized to be effective also from the seismic point of view (Fig. 1).

All blocks are regular in plan, while only blocks 1 and 5 are regular even in elevation. This will affect the calculation of the behaviour factor q (NTC2018).

The load-bearing structure consists of masonry walls of various thicknesses; the perimeter ones have a regular distribution of the openings in height. The horizons are one-way concrete slabs.

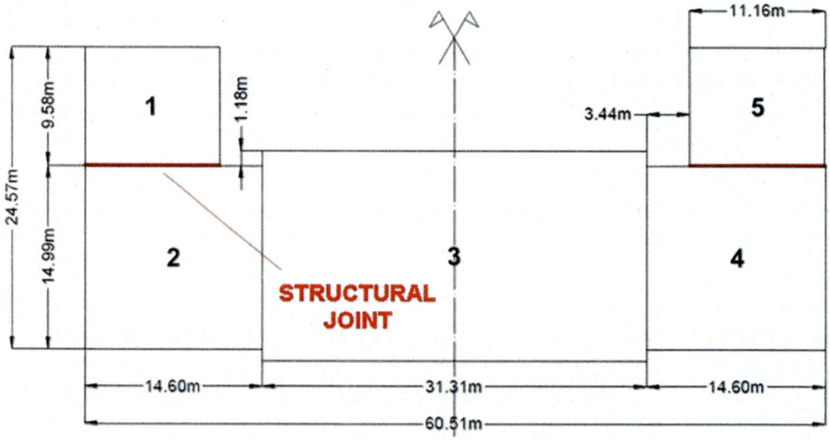

Fig. 1. Structure block schematisation.

3 Historical-Critical Analysis

The building in question is a former goods yard located in the port area of Livorno, now a branch of the Port System Authority of the Northern Tyrrhenian Sea, which allowed to connect the Port of Livorno to the railway. An important work built on land taken from the sea.

The works of the railway yard, intended exclusively for freight service, began in 1856 and were completed in 1858. A vast dock was created, flanked by railway tracks and sheds used for the storage of merchandise. In addition, a building for office use and a small building were erected for the Guardia di Finanza, the latter in the middle of the water.

The area suffered extensive damage during World War II (1939–1945). The main building of the station was destroyed by bombing and promptly rebuilt in the post-war period, although in different forms (Fig. 2).

Over the years, the traffic centre of gravity was moved to the docks of the New Port of Livorno. The station gradually lost importance and the new building and sheds were used for other uses; moreover, the dock facing the station was filled.

In the course of time, the structure has undergone changes in the distribution of the interiors, which have led to the demolition and reconstruction of some partitions.

The historical-critical analysis has also allowed us to detect the legislation in force at the time and the available existing documentation.

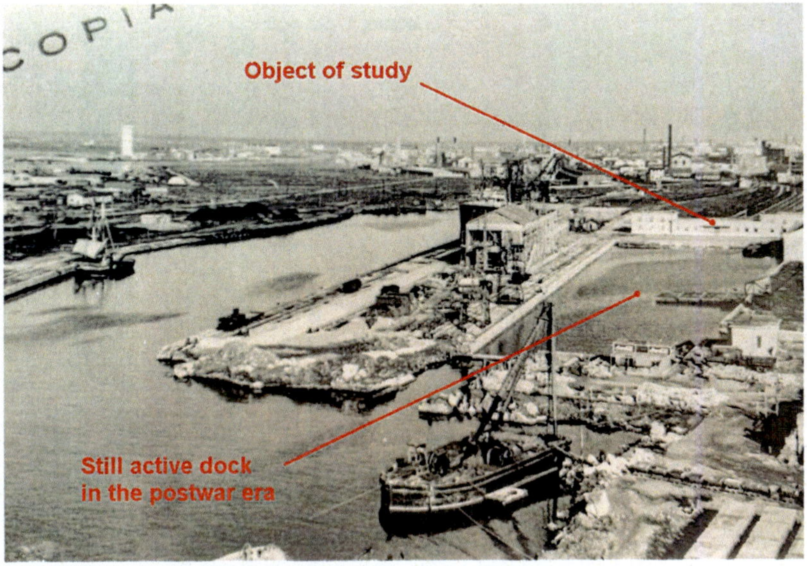

Fig. 2. Rebuilding after WWII (www.gflivorno.it)

The following summarizes the main results of the historical-critical analysis and highlights the first critical issues and the main observations:

- Soil: There is no geological report referring to the area on which the building insists, but it was possible to find one relating to an area located about 300 m away. According to what is reported inside, taking into account the geophysical surveys in a down hole type hole carried out, a soil category C was assigned (NTC2018);
- Geometry: from the existing documentation it was possible to obtain the thicknesses of the structural elements and the interstory heights;
- Slabs and loads: from the certificate of static suitability found it was possible to obtain the stratigraphy of the horizons and the related permanent loads.

No information was found regarding the foundation apparatus, the degree of interlocking of the walls, the presence of curbs and lintels, the warping of the floors, and so on.

In order to verify the correctness of the information obtained from the historical-critical analysis and obtain the missing information, a campaign of investigations and tests has been designed, as reported in the following paragraphs.

4 On-Site Testing and Investigation Campaign

In the preliminary phase, all the available documentation was analysed. This analysis highlighted the lack of useful information for accurate structural modelling. Therefore, it was necessary to provide for an extensive plan of inspections and tests.

In the first place, visual investigations were carried out that made it possible to examine the wall texture and its degree of interlocking, to analyse the cracking pattern and to evaluate the state of general damage to the structure. Based on these investigations, several tests, including non-destructive (thermography, sclerometer, cover meter) and minor-destructive ones (single and double flat jacks, shove test, penetrometer), were planned in order to assess the mechanical properties of the constituent materials, as shown in Fig. 3 [1–3].

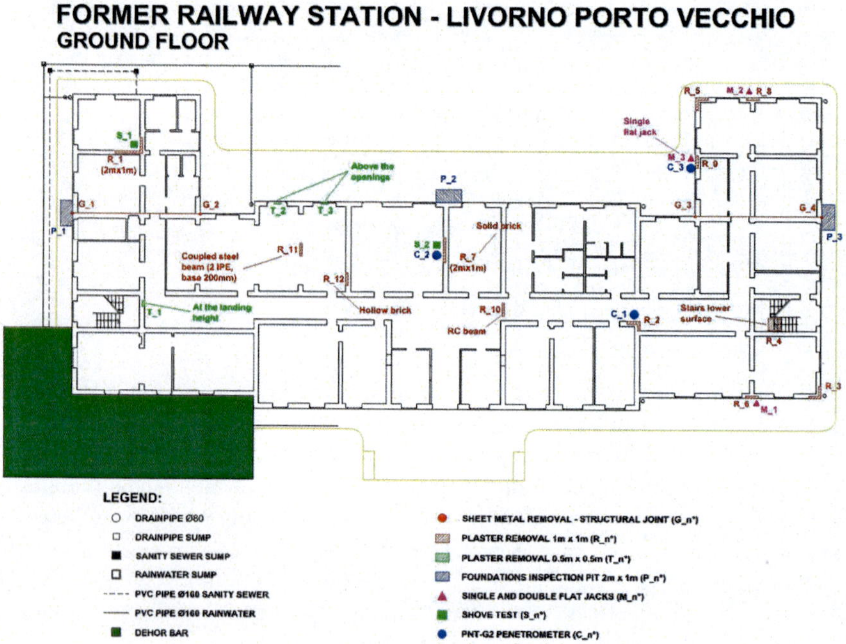

Fig. 3. Testing and investigation plan.

5 Dynamic Identification

5.1 SHM Acquisition System

The LUNITEK data acquisition and processing system model TRITON was used. The system is equipped with an integrated Force Balance triaxial accelerometric sensor. The 24-bit resolution A/D conversion card was programmed with a sampling rate of 250 samples/second per channel with synchronous sampling. The system is equipped with a microSD memory that allows you to manage a ring-buffer for long continuous recordings and a GPS receiver for synchronization in absolute time. The connection to the instrument can be ensured via local network connection (cable or WiFi) or via UMTS / HSPA modem that allows you to control its operation remotely. The system is powered by the external network and integrates an internal battery (LiPo) that guarantees continuous operation for over 30 h. The system was used for multi-channel monitoring by installing multiple units [1–3].

5.2 SEnsor's Setup

All accelerometers in the various configurations have been oriented with respect to a common global reference system (N,E,Z). Figure 4. Setup example.

Shows a configuration example adopted during the dynamic characterization tests of the building.

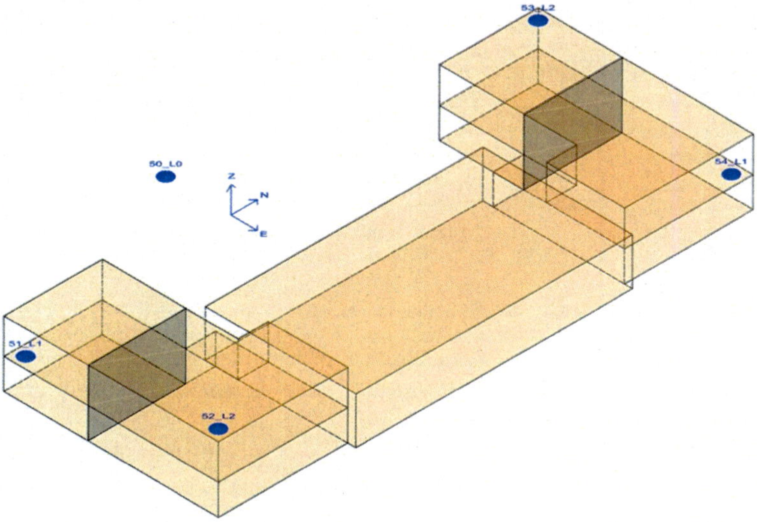

Fig. 4. Setup example.

5.3 Ambient Vibration Test (AVT)

According to operational modal analysis, the building was subjected to Ambient Vibration Tests (AVT), in order to determine its modal characteristics, such as natural frequencies and modal shapes [1–4].

The data acquired by the sensors are acceleration time histories, the actual duration of which was approximately 15min per single setup. Within this recording window, a "clean" sub-window was then identified, that is, free of anomalous peaks due for example to bumps, and therefore identifiable as white noise Fig. 5.

5.4 Filtering

By means of a MATLAB script, appropriate signal theory techniques were applied. In particular, we applied a low-pass filter, with a cut-off frequency equal to 30 Hz, and a band-stop filter to reduce frequency peaks attributable to external noise sources (Fig. 6).

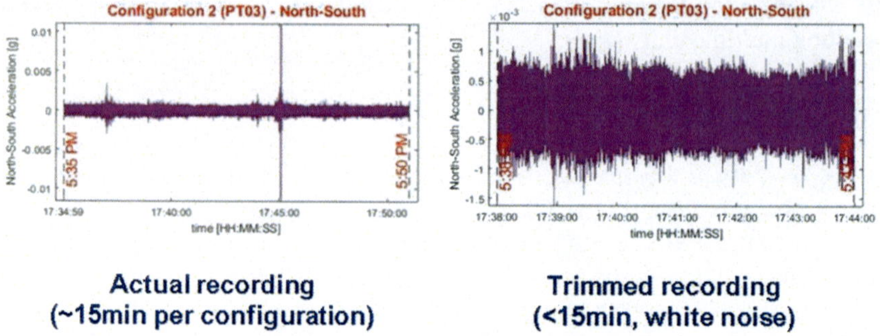

Fig. 5. Example of acceleration time history.

Fig. 6. Filtering.

5.5 Frequency Domain Decomposition (FDD)

Subsequently, we applied the dynamic identification technique called Frequency Domain Decomposition (FDD), based on the singular value decomposition algorithm of the Power Spectral Density (PSD) matrix.

The singular values are returned in output, from whose analysis it is possible to identify the spectral density eigenfunctions corresponding to each mode of the system and, therefore, to estimate the modal shape at each peak [1–4].

5.6 Results

In most of the results it is possible to observe a frequency response peak located around 3.5 Hz, which could be attributed to the structure. However, on closer inspection, a serious mistake would be made.

Although not necessary in the case of dynamic output-only identification, a ground sensor was also installed during the tests. By analysing the PSDs of the signals from individual accelerometers, an unexpected discovery was made. The 3.5 Hz tone is simultaneously present both for the accelerometers positioned on the structure and on the ground (Fig. 7), and in all 3 instrumental directions. This reasonably means that it is not referred to the structure but depends on the frequency component of an external input source. In light of what was discovered, it was not possible to identify the modal parameters of the structure and therefore to calibrate the subsequent structural model.

6 Numerical Modelling

At the present stage of analysis, the following simplifications in the preparation of the finite element model of the structure have been assumed.

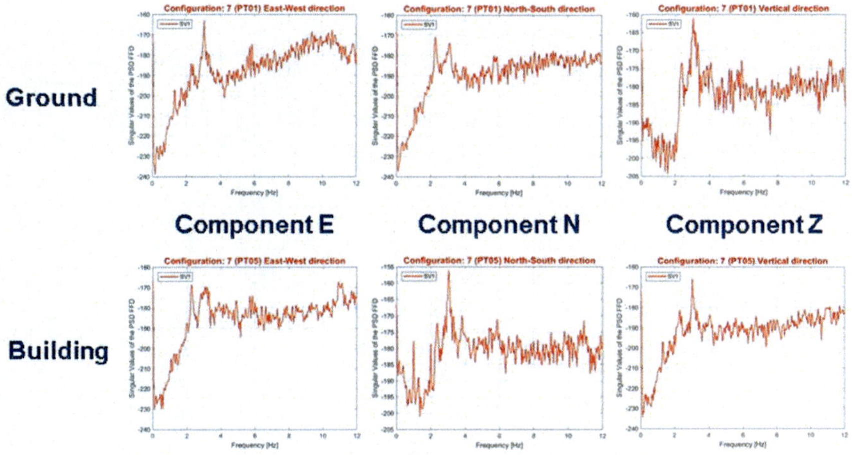

Fig. 7. Ground versus building accelerometers.

- The actual settlement (rigid rotation) of blocks 1 and 5 was not considered;
- The degree of interlocking between orthogonal walls was neglected;
- The structural seismic joints between blocks were considered as perfectly effective. Therefore, the possible interaction (hammering) between adjacent blocks during the seismic event has been neglected and the blocks have been modelled separately;
- Soil-structure interaction has not been considered.

The basic hypotheses formulated above have made it possible to model a structure that differs from the real one, but have not at all compromised the methodology for assessing seismic vulnerability, which remains unchanged.

In the beginning a 3D frame in AutoCAD environment has been created, representing the fixed wires of the structural elements. This was then imported into the SAP2000 FE code in which each line represents a frame type element. Then, the fixed constraints to the base, the rigid floor diaphragms, defined the materials, the sections and assigned the loads have been assigned [5].

The building has been modelled following the "equivalent frame" approach (Fig. 8). Therefore, by means of the "rigid end offset" command and on the basis of Dolce's theory, the zones with rigid behaviour in the masonry walls and in the floor strips have been defined.

SAP2000 does not provide for the automatic creation of plastic hinges for masonry elements. Therefore, to take into account the non-linear response of the structure, ad hoc plastic hinges have been designed, using an elasto-plastic shear-displacement capacity model, in which the ultimate shear is the minimum among those resulting from the 3 typical collapse mechanisms of wall panels: rocking and toe crushing, diagonal shear, and sliding shear failure (Fig. 9).

After model preparation, Modal analysis, Linear dynamic analysis ad Pushover analysis can be considered. Specifically, in addition to the vertical loads, two distributions of inertia forces have been assigned: those called Group 1, proportional to the shear forces at each storey, and those called Group 2, proportional to the floor mass. The two

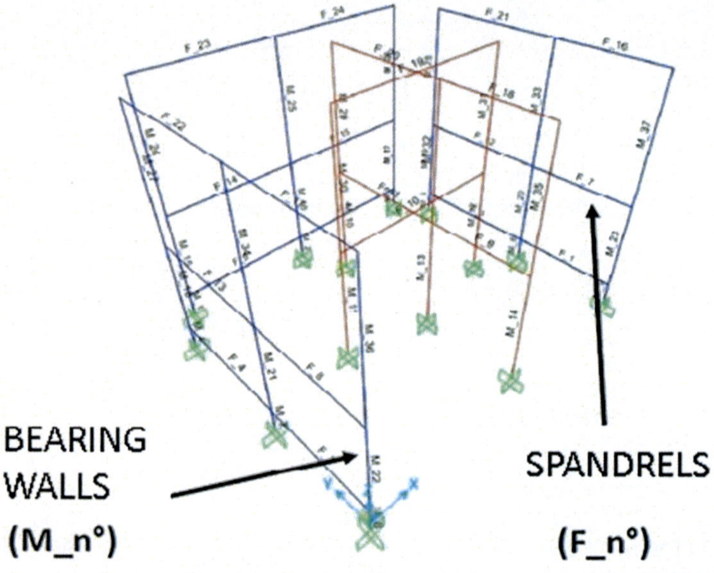

Fig. 8. Equivalent frame model.

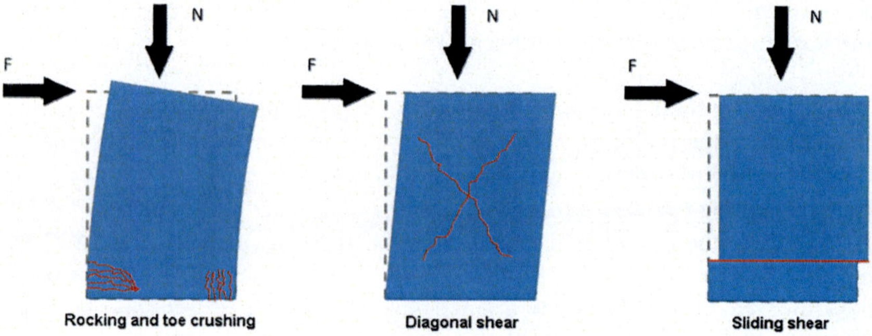

Fig. 9. Failure mechanisms

distributions were applied in the 2 main directions, both with positive and negative signs, for a total of 8 pushover analyses for each block.

In Fig. 10, an example of capacity curve of the MDOF system is depicted, expressed as a function of the shear at the base and the displacement of one or more control points, with a representation of the activated plastic hinges.

After that, the capacity curves provided by SAP2000 are reduced to equivalent SDOF systems, scaling them by the modal participation factor Γ referred to the specific load profile. Finally, the equivalent bilinear curves have been evaluated using the equal energy rule. Obviously, among all the curves, the one that represents the capacity of the structural system is that one with the lowest maximum shear.

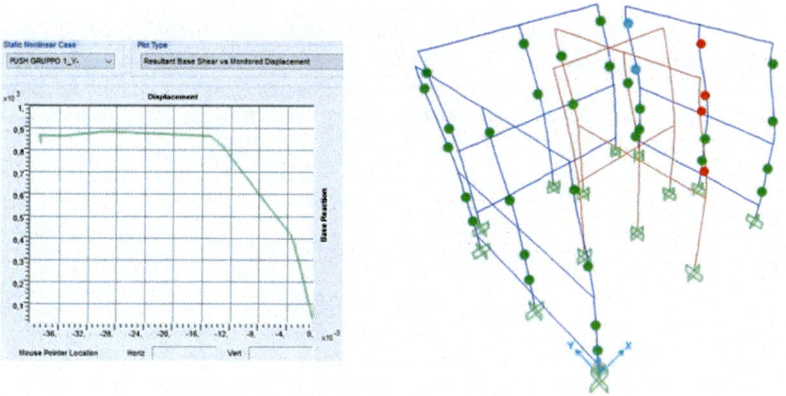

Fig. 10. MDOF capacity curve and plastic hinges evolution.

7 Seismic Vulnerability Assessment

The ζ_E seismic vulnerability index, in agreement with the Italian code, can be expressed as the ratio between capacity and demand in terms of Peak Ground Acceleration [2, 3].

The compute indices of all the blocks result higher than unit: i.e., 1.06 for Blocks 1 and 5, and 1.20 for Blocks 2,3,4; with a Demand PGA of 0.1814 g. Thus, the structural model complies with the safety requirements set by the NTC2018 with regard to seismic actions.

8 Conclusions

The Structural Health Monitoring performed on the structure allowed to understand the importance of the soil-structure interaction on buildings with high stiffness and low ductility.

Recording the accelerations at the ground and comparing them with the ones recorded in the building it was possible to understand that the structure behaves as a rigid body.

This work lends itself to being a starting point for further studies, which involve the development of more refined structural models that include the soil-structure interaction (e.g., Winkler soil model).

References

1. Cimellaro, G.P., Piantà, S., De Stefano, A.: Output-only modal identification of ancient L'Aquila city hall and civic tower. J. Struct. Eng. **138**(4) (2012)
2. Marasco, S., Zamani Noori, A., Domaneschi, M., Cimellaro, G.P.: Seismic vulnerability assessment indices for buildings: proposals, comparisons and methodologies at collapse limit states. Int. J. Disaster Risk Reduction **63**, 102466 (2021)
3. Domaneschi, M., Zamani Noori, A., Pietropinto, M.M.V., Cimellaro, G.P.: Seismic vulnerability assessment of existing school buildings. Comput. Struct. **248**, 106522 (2021)

4. Brincker, R., Zhang, L., Andersen, P.: Modal identification of output only systems using frequency domain decomposition. Smart Mater. Struct. **10**(3) (2001)
5. Pasticier, L., Amadio, C., Fragiacomo, M.: Non-linear seismic analysis and vulnerability evaluation of a masonry building by means of the SAP2000 V 10 code. Earthquake Eng. Struct. Dynam. **37**(3), 467–485 (2007)

Horizontal and Vertical BIM Interoperability Aimed at Seismic Vulnerability Assessment

Francesca Maria Ugliotti(✉), Marco Domaneschi, Anna Osello, Salvatore Tuccitto, and Salvatore Monforte

Politecnico di Torino, Torino, Italia
{francesca.ugliotti,marco.domaneschi,anna.osello}@polito.it, s263247@studenti.polito.it, salvo.tuccitto@hotmail.it

Abstract. A vital task of this millennium is to protect the existing heritage, also through the adoption of resilient management systems. In this framework, the organization of knowledge remains one of the critical points. For this reason, new methodologies and cross-disciplinary technologies are increasingly being chosen to optimize resources toward more sustainable interventions. Therefore, the ability to model the building geometry and behavior must be maximized through interoperable processes between Building Information Modeling and Finite Element Modeling methods aimed at the seismic vulnerability assessment. Setting up an integrated digitalization process is undoubtedly challenging initially but returns more significant benefits during the infrastructure life cycle. The interoperability tests' bi-directionality is essential for constantly evaluating activities to update data following facilities' modifications. The Modal Assurance Criterion indicator is used to assess the coherence of the models after possible simplifications introduced for non-linear state analyses.

Keywords: Building information modeling · Finite element modeling · Interoperability · Seismic vulnerability · Level of knowledge

1 Introduction

Nowadays, built environment management is increasingly evolving towards systematizing knowledge through integrated systems based on three-dimensional parametric representation. The Building Information Modeling (BIM) methodology aims to create a service model where processes, tools and people can interact in the best way. This approach is increasingly deepened and declined in the Italian and European contexts as the most significant benefits are primarily observed in the operation phase. The reference existing heritage requires considerable care from the point of view of maintenance, monitoring and preservation, even concerning catastrophic natural events such as landslides or earthquakes. Moreover, buildings with historical and artistic value often host strategic functions fundamental to the community, such as public offices, hospitals, schools, and power systems. The interruption or limitation of these buildings' usability can strongly impact the social activities, logistics, and economy. For these reasons, the adoption of

G. P. Cimellaro (Ed.): WCSI 2022, LNCE 309, pp. 1043–1062, 2023.
https://doi.org/10.1007/978-3-031-21187-4_91

streamlined processes for the organization and retrieval of available documentation plays a key role whenever it is necessary to carry out actions on these infrastructures. The Italian NTC 2018 [1] standard and its Circular [2], in line with the Eurocodes, outlined the procedure for assessing structure knowledge. BIM tools are evaluated to standardize and speed up that procedure so that more resources can be invested in evaluating design interventions than returning models. Professionals can set the seismic vulnerability assessment from a digital model conceived as a collector of information updated over time, valuable to return a coherent representation of the state of the art and to map the level of knowledge of the asset in an appropriate and shared way. The testing of integrated BIM-FEM (Finite Element Modeling) processes for modeling, management, and structural analysis establishes a multidisciplinary research field that enriches traditional investigation practices. Currently, lots of research focuses on specific aspects that integrate a particular process moment. A central element is using a physical model in the context of calculation software [3–5], also starting from laser scanner surveys in the case of interest in accurately rendering construction irregularities [6]. Conversely, the export of analytical models in the BIM environment is evaluated [7]. Within this framework, this study primary aims to check the horizontal and vertical interoperability between Model Authoring and Structural Calculation software to optimize the workflow and guarantee a correct information transfer.

2 Methodology

The study's objective is to define an optimized workflow to manage the knowledge processes and seismic vulnerability assessment of a point-type infrastructure employing a three-dimensional model associated with an information database. This approach effectively addresses the steps defined in the NTC 2018 [1], including historical-critical analysis, geometric-structural survey, mechanical characterization of materials, knowledge levels and confidence factors, actions and structural analysis. Figure 1 summarizes the procedural workflow.

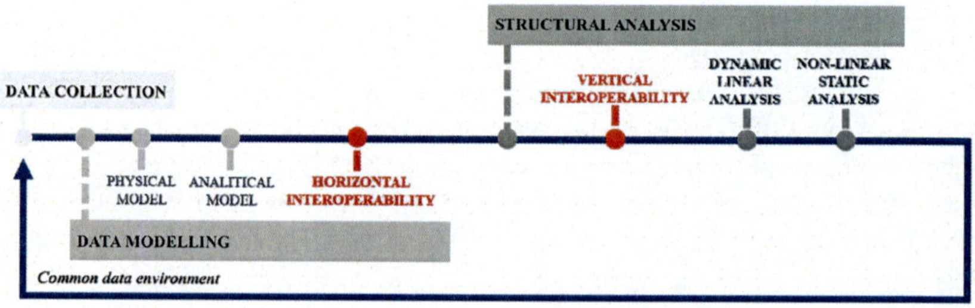

Fig. 1. Seismic vulnerability assessment methodological workflow

Within this framework, the concept of interoperability represents one of the critical issues. For this reason, the transfer of information between Model Authoring and calculation software is tested to identify the most efficient path. The FEM model resulting from this activity is made available for specialized simulations.

The approach is validated by applying it to a real case study [8, 9]: a hydroelectric power plant of the early 1900s in the center of Italy. It is a multi-storey reinforced concrete structure spread over around 2000 $m^2$of three adjacent parts built at different times. The maximum height is 40 m.

2.1 Data Collection

A rigorous data collection process is essential to provide a coherent summary of the information contained in the digital models. An expeditious field survey must follow the drawings drawn up during construction time to verify the information. The geometric check of the visible elements constitutes the first step from which to recreate the model according to the actual state. The restitution of the structural component requires a thermographic analysis to identify the elements and diagnostic tests to assess the type of materials used and their mechanical characteristics. In the case under study, the following tests were carried out: pacometric analysis, scarification and removal of reinforcement, analysis and hardness of the reinforcement, sclerometric analysis, concrete sampling and carbonation depth analysis. BIM environment can be exploited both for (i) set up an optimal survey plan depending on the components that can be investigated on-site, (ii) map destructive tests, (iii) gather the data collected during the survey [10].

2.2 Data Modeling

The BIM model must be designed in the most practical and versatile way to be helpful in the building life cycle. For this reason, it is essential to carefully project the model's architecture at an early stage to handle its overall coordination and facilitate the work of individual professionals. It is therefore considered appropriate to set up a federated model (Fig. 2) that combines the various disciplines' contributions with the most appropriate tools for geometric restitution and simulation activities. This integration makes it possible to detect possible inconsistencies and optimize the assessment phase (Fig. 3).

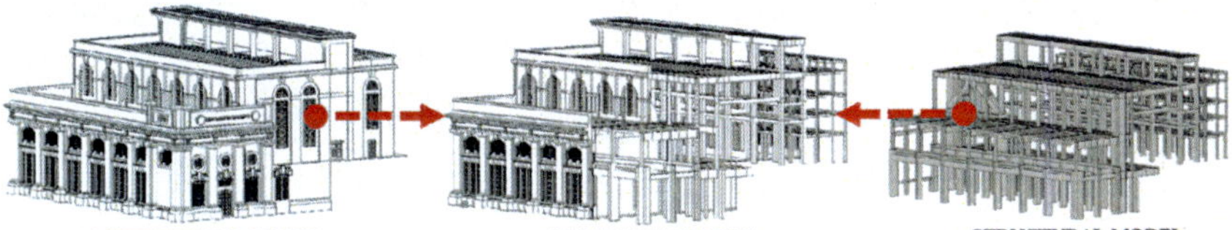

Fig. 2. BIM federated model

Concerning the objectives of the study, Autodesk Revit and Tekla Structures were selected as Model Authoring software, while Midas Structure Midas Gen for Finite Element analysis tool. Autodesk Revit is one of the world's most widely used parametric programs. It is suitable for modeling in different disciplines and integrated with the visual programming application Dynamo, which extends its ability to manipulate data and generate complex geometries. Tekla Structures adopt the truly constructible BIM process making it possible to model any structure with the highest Level of Detail

(LOD) regardless of its size or material. Midas Structure Midas Gen has the highest market share in the seismic region. It offers advanced seismic analyses such as pushover analysis, nonlinear time history analysis, nonlinear boundary analysis and fiber analysis, and conventional seismic analysis such as static seismic and response spectrum analysis. The object modeling method and the most appropriate environment to do this are derived from the interoperability texts discussed below.

Through the use of a Model Authoring software, the physical and analytical modeling of the components is accompanied by the automatic setting up of a database of information which, if carefully entered, can accompany the assessments of the building throughout its life cycle. At this stage, it is beneficial to set up schedules and thematic views so that the information in the digital model can be easily accessed, even by those who have not done the modeling. Geometrical information on structural components can therefore be accompanied by information on materials, reinforcement, loads, as well as images or links to external documents or platforms. In general, a LOD 350 [11] has been evaluated for the seismic vulnerability assessment of an existing structure. The introduction of additional parameters associated with objects makes it possible to extend the range of data that can be managed and to provide information on the use of the model. For example, defining the level of knowledge of the asset under study (Fig. 4) is crucial to identifying the most appropriate structural analysis for assessing seismic vulnerability as required by Italian regulations.

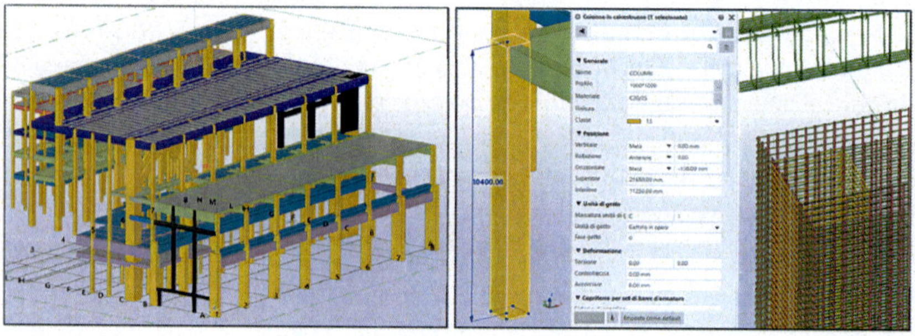

Fig. 3. Tekla structures BIM model

2.3 Interoperability Tests

Interoperability is the principle around which the BIM methodology is hinged, i.e. the possibility of exchanging data without loss of information throughout the building life-cycle. Horizontal interoperability makes it possible to bring together representations obtained through software programs that perform the same functions from different vendors. An example is the integration of different disciplines - architectural, structural and systems - whose three-dimensional modeling can be managed with different software. On the other hand, vertical interoperability concerns complementary tools such as simulation or calculation. Data exchange can occur via open or proprietary formats in both cases. Interoperability tests were carried out to optimize the workflow and to

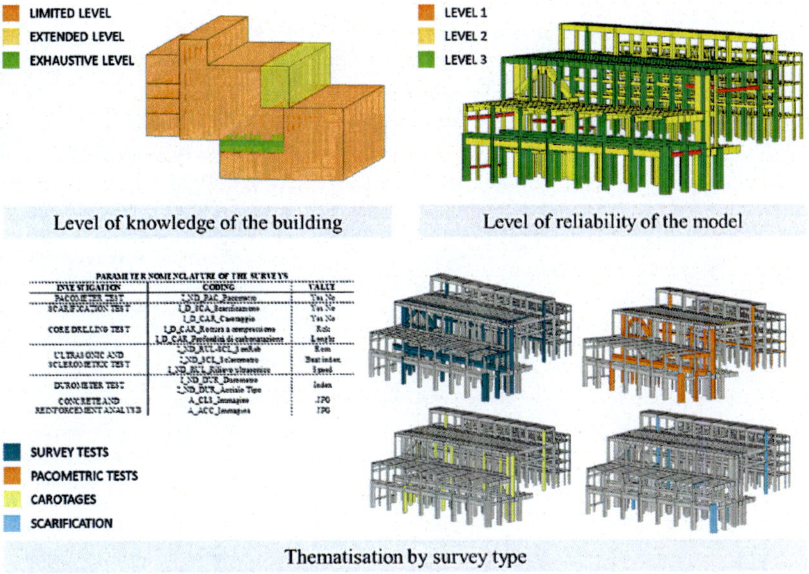

Fig. 4. Accessibility of the information in the Autodesk Revit BIM model

limit as far as possible the loss of data or the transfer of incorrect information, which are difficult to detect when the model is used by different users than those who generated it. The consequences are errors, even significant ones, in the subsequent analysis and evaluation stages. Therefore, this phase aims to identify the most efficient path using different software and exchange formats within an integrated process between disciplines and professionals. As shown in Fig. 5, interoperability is assessed from one Model Authoring software to another (Tekla Structure and Autodesk Revit) and from the Model Authoring software (Tekla Structure and Autodesk Revit) to the Finite Element software (Midas Gen) in a cross- and bi-directional manner (Tables 1 and 2).

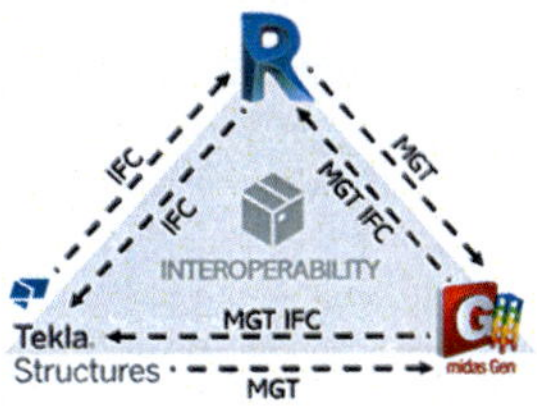

Fig. 5. Interoperability tests carried out.

Table 1. Horizontal interoperability overview tests.

Input model	Output model	Exchange format	Issue
Tekla Structures 2020	Autodesk Revit 2020	IFC2x3 Coordination view 2.0	*Issue* Adequate mapping of IFC classes is required to ensure proper association of objects to Revit categories
			Outcome All elements were imported in the proper position and with the correct attributes assigned. The only exception is a personalized section
Autodesk Revit 2020	Tekla Structures 2020	IFC2x3 Coordination view 2.0	*Issue* Adequate mapping of IFC classes is required to ensure proper association of Revit categories to objects. The import of reinforcement is allowed by changing the IFC class from *IFCReinforcingBar* to *IfcBuildingElementProxy*. This makes it possible to recognize the presence of reinforcement within a structural object and its quantity, even if this involves associating the Revit category with a generic IFC object. The imported model must undergo a conversion to transform the objects into native elements of the output software. Verification of material and section recognition settings is required
			Outcome All elements were exported correctly, except for some critical issues such as: the lack of some customized sections, the rotation of some elements according to their barycentric axis, and some elements incorrectly imported as extrusion solids. The reinforcement is exported but is recognized as *Proxy construction elements* and not as *Reinforcement*. Longitudinal bars are detected by the software as *Beam* elements, with the material set during the mapping process. In contrast, cross bars are correctly assigned to the reinforcement bar templates but lose their material information

Table 2. Vertical interoperability overview tests.

Input	Output	Exchange	Issue and Outcome
Tekla Structures 2020	Midas GEN 2021	Bidirectional Plug-in (MGT)	*Issue* The plug-in converts all structural model objects into analytical elements in the MGT format. A text file is generated to inform about problems that occurred during the export. Subsequently, the model must be cleaned, i.e. all information not useful for structural analysis must be removed. *Outcome* Section geometries were exported correctly, except for customized sections that retained their mechanical properties. Reinforcement was not exported.
Midas GEN 2021	Tekla Structures 2020	Bidirectional Plug-in (MGT)	*Issue* The plug-in updates the Tekla model in terms of: material, sections, position. Once the file has been imported, Tekla starts converting all analytical elements into native elements. *Outcome* All elements were imported in the proper position and with the correct section geometries and material. Importing via MGT files is not suitable when custom sections are to be exported, as the software cannot find a match with those in its own catalogue. Reinforcement was not exported.
		IFC	*Issue* Imported elements are initially recognized as IFC elements, after a conversion to native Tekla elements is possible. Material and sections can be mapped, so that this conversion can take place correctly. *Outcome* Many elements have not been transferred, although the sections are correctly located, including the customized ones. Overall, the positioning of the elements is correct, with the exception of one section that is rotated with respect to its center of gravity. Reinforcement was not imported.

(*continued*)

Table 2. (*continued*)

Input	Output	Exchange	Issue and Outcome
Autodesk Revit 2020	Midas GEN 2021	Plug-in (MGT)	*Issue* The plug-in generates an MGT format file containing all valuable information for creating the analytical model, identifying materials, sections, the orientation of elements, and coordinates of start and end nodes. A dedicated interface makes it possible to map the elements by controlling the symbologies managed by both software. Once the parameter association has been defined for a section, the plug-in can recognize all similar sections and apply the parameter association. For Midas Gen, the presence of an offset corresponds to the insertion of a rigid link between an element's axis passing through the section's geometric center of gravity and the calculation axis. Whereas Midas Gen offsets are considered element section properties, Revit defines them through constraints or geometric position. Offsets in pillars can only be defined as constraints, considering a change in the start and end node height leaving the analytical axis passing through the center of gravity of the section unchanged. On Midas Gen, this will correspond to a Z-axis translation of the node. As far as beams are concerned, *offsets* can be defined as a constraint or a geometric position. In the first case, constraints are vertical and define how far the element is from the level to which it belongs and the position of the analytical axis to the level. It will result in Midas Gen translating the analytical axis to a different height from the real one, with the consequent division of the columns at the beginning and end of the beam. In the second case, Revit sets by default the analytical axis in the extrados of the beam. This aspect will be recognized as a property of the section. *Outcome* Objects are only transferred if the elements match the predefined Midas Gen sections. Custom sections are therefore not exported. In the case of rectangular sections, the plugin fails to associate the corresponding parameters for the use of capital letters in Midas Gen. All elements are assigned to the correct type, with the exception of *Plate* elements for which a mesh is recreated. It is better to define offsets in the geometric position in Revit to have a clearer view of the beams' positioning and avoid errors. Reinforcement was not exported.

(*continued*)

Table 2. (*continued*)

Input	Output	Exchange	Issue and Outcome
Midas GEN 2021	Autodesk Revit 2020	Plug-in (MGT)	*Issue* It is not possible to export non-standard geometric shapes. Unlike Midas Gen, the Revit software platform is not based on standards for defining the material of elements. Because of this, the material is not assigned correctly. Supporting schedules can be used for the reassignment task. *Outcome* Some elements, such as shear walls and everything connected to them, are not imported. The presence of custom sections creates problems and it is necessary to remove them at the export stage. Furthermore, a visual analysis of the analytical model shows that elements are assigned to parametric families with an inversion of the base and height parameters. The unit of measurement is incorrect.
		IFC	*Issue* Midas Gen does not provide the possibility of selecting and managing the elements to be exported. Therefore, it is not possible to assign IFC classes to analytical objects (as is the case when exporting a Tekla Structures model in IFC format). The IFC map of elements is done directly from Revit. *Outcome* All elements, including custom sections, are transferred and assigned to the correct Revit categories. However, the imported components do not result as parametric families. In addition, there is no center of gravity axis and the associated elevations. All objects are connected at the same level: their position is only defined via the "base offset" parameter. No structural material is maintained. For these reasons, the resulting model is not suitable for the structural discipline, although it can be a reference for the architectural component. For example, if you want to use the analytical model, you have to modify each beam manually and sometimes reshape them. Columns must be replaced with parametric families, such as the default Revit ones or those created by the user, so the exact position of the center of gravity can be defined. Missing information requires a not inconsiderable amount of time.

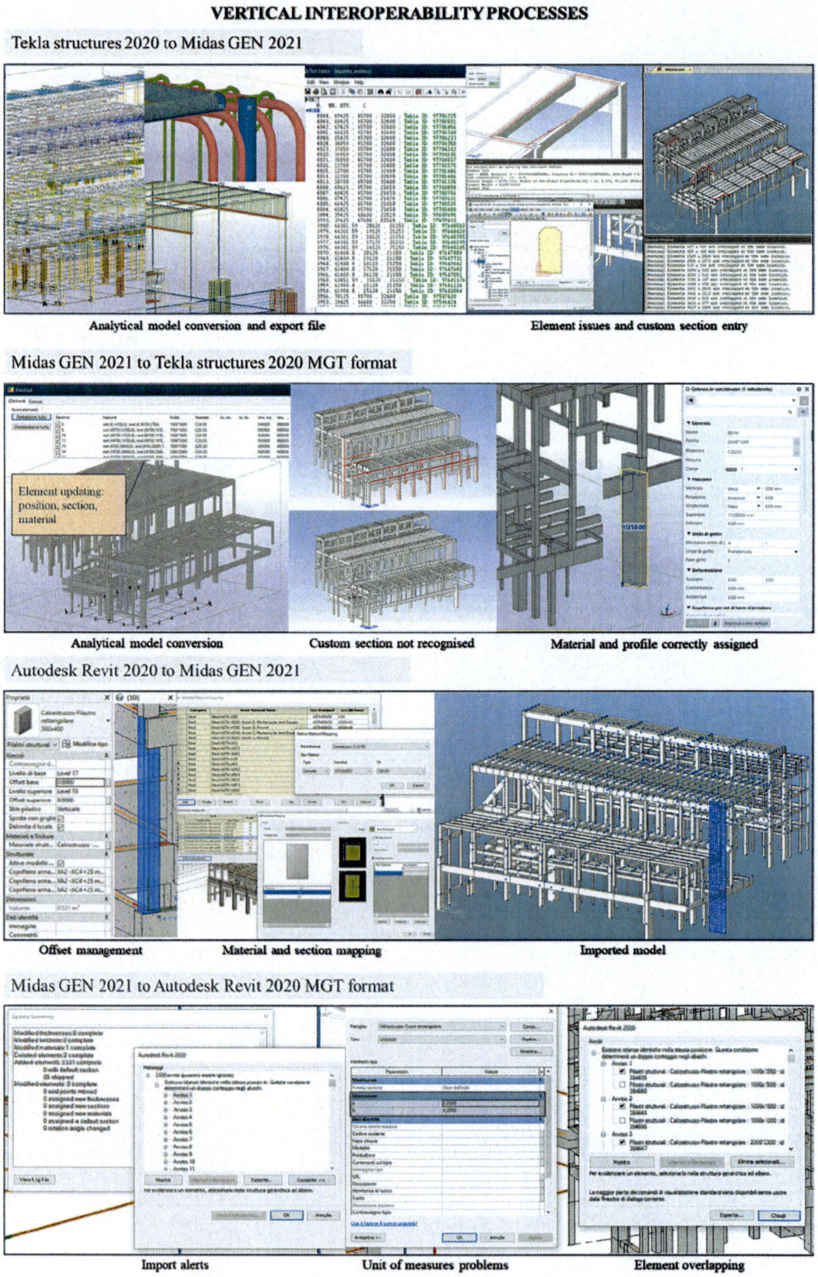

Fig. 6. Vertical interoperability procedures.

The interoperability tests have been evaluated through the formulation of comparison parameters. A score from 0 to 3 is assigned based on the goodness and correctness of the information transferred. Table 3 shows the categories of information considered and the weights assigned according to the effort required for their correction or re-input. Table 4 shows the results obtained for the interoperability processes considered. In general, it can be noted that the horizontal interoperability flow via open IFC format scored best. However, the same consideration does not apply in the case of vertical interoperability, i.e. in the case of the connection between BIM-FEM models. Although a substantial research line is oriented towards the correct mapping of fields according to the open Industry Foundation Classes (IFC) data model, the exchange process between different applications is not yet regulated by a perfectly codified workflow. Dedicated plug-ins made on the basis of proprietary formats, e.g. MGT, ensure a sufficient level of interoperability. A very critical aspect lies in the management of reinforcement. These, which are essential for structural analyses, are not maintained in the exchange processes between the different software. Only in the case of horizontal interoperability between Tekla Structure and Autodesk Revit have desirable results been achieved. This undoubtedly entails a considerable amount of data re-entry work, even if supported by the models produced (Fig. 6).

Table 3. Comparison parameters for interoperability score evaluation

Comparison parameters	Weight [%]
Material	10
Elements	20
Reinforcement	15
Generic section	12
Customized section	20
Position	15
Unit of measures	8

2.4 Structural Analysis

FEM Modeling The structural analysis is performed by preparing a suitable structural model from the results of BIM-FEM interoperability. Based on the performed interoperability tests, the Midas Gen - Tekla structures software association has been chosen. The starting analytical model from Tekla Structures 2020 was then imported into the Midas GEN 21 structural analysis software (Fig. 7).

The pushover analysis conducted on the refined model is definitely complex due to the amount of details that affect the analysis result and significantly increase the computational burden. Therefore, it has been decided to reasonably simplify and replace the custom sections with equivalent elements (Fig. 8). Then, by merging the three separated buildings into a single structural body by removing the rigid links that had been previously inserted (Fig. 8). To further reduce the computational burden, additional information inherent to secondary elements that were not useful for structural analysis has

Table 4. Interoperability tests score.

Comparison parameters	IFC				MGT			
	Tekla Structures → R	Tekla Structures ← R	Tekla Structures ← Gen	R ← Gen	Tekla Structures → Gen	R → Gen	Tekla Structures ← Gen	R ← Gen
Material	3	3	3	0	3	2.5	3	1
Elements	3	2.5	2	3	2.5	3	2.5	2.5
Reinforcement	3	1	0	0	0	0	0	0
Generic section	2.5	3	3	2	3	2.5	3	2
Customized section	2.5	2.5	3	3	1.5	0	0	0
Position	3	3	2.5	3	2	3	3	3
Unit of measures	3	3	3	3	3	3	3	0
Score	**2.8**	**2.5**	**2.3**	**2.1**	**2.0**	**1.8**	**1.9**	**1.3**

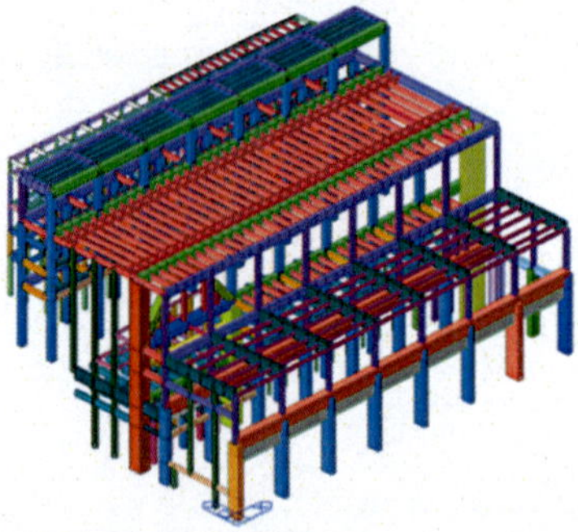

Fig. 7. Imported Tekla structures model in Midas GEN for structural analysis

been removed (Fig. 8). These simplifications have been introduced one by one, verifying the equivalence of the results in terms of static and dynamic response between the model before and after the specific simplification adopted.

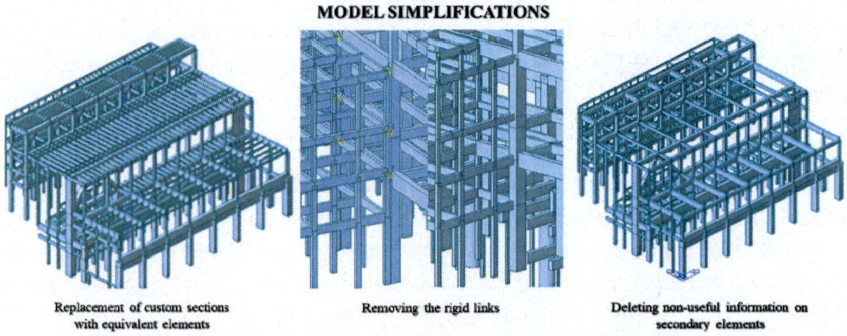

Fig. 8. Model simplifications for structural analysis

A linear dynamic analysis has been performed by evaluating the consistency in modal shapes and eigenfrequencies, employing the Modal Assurance Criterion (MAC) [12] to compare the dynamic range. Modal analysis associated with the design spectrum (linear dynamic analysis) has also been carried out, considering a sufficient number of vibration modes for 85% of the participating mass to be reached in each direction. The natural modes obtained for the analytical model have been compared in Table 5. A reasonable coherency of the mode shapes between the two models is highlighted. Then the MAC was used for a quantitative assessment of the equivalence.

Table 5. Comparison between refined model and simplified model

N° mode	Frequencies [rad/sec]		Period [s]	
	MR	MS	MR	MS
1	6.185	6.421	1.02	0.98
2	6.858	7.078	0.92	0.89
3	8.358	8.726	0.75	0.72
4	15.157	14.837	0.41	0.42
5	18.552	19.397	0.34	0.32
6	19.846	20.815	0.32	0.30
7	21.150	21.216	0.30	0.30
8	21.764	21.723	0.29	0.29
9	22.047	22.331	0.29	0.28
10	22.736	23.041	0.28	0.27

The MAC allows to evaluate the correlation between two vectors of real or complex elements. For the present study, the MAC is applied to two eigenvectors, $\{\varphi_r\}$ and $\{\varphi_s\}$, representing the mode shapes associated with two different FEM models with different levels of refinement. In particular, the eigenvectors associated with the MR and MS models have been compared in their respective vibration modes and directions. MAC is evaluated as:

$$MAC(r, s) = \frac{|\{\varphi_r\}^T \cdot \{\varphi_s\}|^2}{\left(\{\varphi_r\}^T \cdot \{\varphi_r\}\right)\left(\{\varphi_s\}^T \cdot \{\varphi_s\}\right)}$$

where $\{\varphi r\}$ is the eigenvector representing the mode shape of the considered vibration mode, for the refined model MR; $\{\varphi s\}$ is the associated eigenvector for the simplified model MS. The computational process has been implemented in Matlab.

Since the MAC values returned for the pairs of eigenvectors of interest are always higher than 0.8 and higher than 0.9 for the first modes, the two models can be considered reasonable equivalent, which implies that the results performed on the simplified model do not differ essentially from the detailed one. Thus, the simplified FEM model is adopted for seismic capacity verifications.

Seismic vulnerability assessment The Pushover analysis is used for the vulnerability assessment of the building to compute the performance point. It is done by exploiting the capacity curve (Fig. 9). It is bilinearized according to the procedures defined in the NTC-2018 as method A - also known as method N2, based on the identification of the inelastic demand through the equal displacements or equal energy rule, and method B - also known as method CSM, based on the construction of a capacity spectrum.

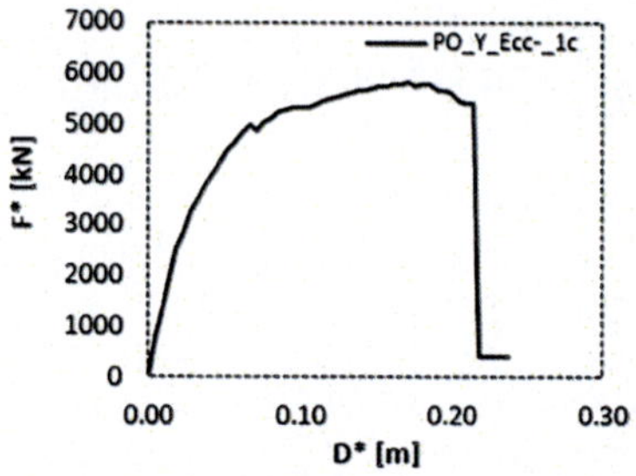

Fig. 9. Capacity curve

Bilinearization results in transformation of the capacity curve into an equivalent elasto-plastic system, where the performance point of the structure is determined. The Italian regulations do not give guidance on what values to use for evaluating the seismic vulnerability index (ζ_E). For this reason, it has been evaluated through three formulations in terms of: (i) base shear; (ii) displacements; (iii) PGA.

(i) *Base shear*. The first formulation is based on the N2 and CSM methods provided by the NTCs, bilinearizing the capacity curve and identifying performance point to evaluate the demand, i.e., the maximum base shear. While the capacity is the maximum force, i.e., the collapse shear force (Fig. 11). The results vulnerability indices for the N2 and CSM methods are $\zeta_E = 1,07$ and $\zeta_E = 1,11$, respectively.

$$\zeta_E = \frac{F^*_{COLL}}{F^*_{MAX}} \tag{1}$$

(ii) *Displacements*. The second formulation is based on N2 and CSM methods again but considers the maximum displacement demand, while the capacity is the displacement at the collapse state (Fig. 11). The computed results are the following ones: $\zeta_E = 1,78$ and $\zeta_E = 1,54$, respectively.

$$\zeta_E = \frac{S^*_{COLL}}{S^*_{MAX}} \tag{2}$$

(iii) *PGA*. In this case the demand is evaluated as the PGA at the reference site, while the capacity is that one obtained by a scaling procedure so as the elastic displacement at the reference period of the structure coincides with the ultimate displacement evaluated on the bilinearized curve by the N2 method (Fig. 10). The results show $\zeta_E = 1,78$.

$$\zeta_E = \frac{PGA_{CAP}}{PGA_{SITE}} \tag{3}$$

An alternative to the (3) formulation considers the vulnerability index as the ratio of the maximum force of the bilinearized capacity curve, divided by the mass of the equivalent single degree of freedom system, and the design PGA at the site, i.e., the spectrum value at the null period.

Results analysis The resulting capacity curves allow to investigate the structural behavior through different aspects, evaluating demand and capacity. First and foremost is the ratio of the collapse base shear to the maximum shear demand, which indicates the structure's strength. Another aspect is related to the results in displacements, which indicates the structure's capacity to absorb the seismic demand in terms of ductility. Indeed, very high values of the displacement ratio indicate a high dissipative capacity.

These NTC-2018 seismic vulnerability indices are based on finding a performance point on the capacity curve of the structure to be compared with the capacity of the system represented. In the case study, by the collapse base shear, as well as the highest value from the capacity curve. However, such approaches (i and ii) do not allow obtaining values of ζ_E less than 1. The only exception is the limit case, and the dangerous one, because not verified, with very high seismic demand and thus the absence of the performance point. The described approaches have been compared with an incremental dynamic analysis [13] performed on an example FEM model (Fig. 11) to give validity to the last consideration.

The different scenarios in terms of ζ_E as a function of the seismic hazard at three different sites have been considered (CO, ME, SR). Accelerograms compatible with the spectra at the sites have been selected (Fig. 12).

According to the literature [14], the maximum inter-story drift has been chosen to evaluate the system capacity and vulnerability index evaluated by exploiting the formulation of method one from [15].

$$\zeta_E = \frac{PGA_{CAP}}{PGA_{SITE}}$$

where the PGA_{SITE} is the seismic demand at the reference site, while PGA_{CAP} is the capacity PGA, that one that scales the compatible spectrum accelerograms until the predetermined inter-story displacement is reached (Fig. 13).

Tables 6 and 7 show the comparison of vulnerability indices between the incremental dynamic analysis (IDA, nonlinear analysis, the most accurate) and those computed through the Pushover analysis in terms of forces and displacements, respectively. It is worth noticing that when the vulnerability index computed through IDA is less than the unity, the performance point in the Pushover analysis can not be identified (*) with both N2 and CSM. This is because the performance point would be beyond the collapse where there is no longer a capacity curve. It highlights how the simplified and straightforward

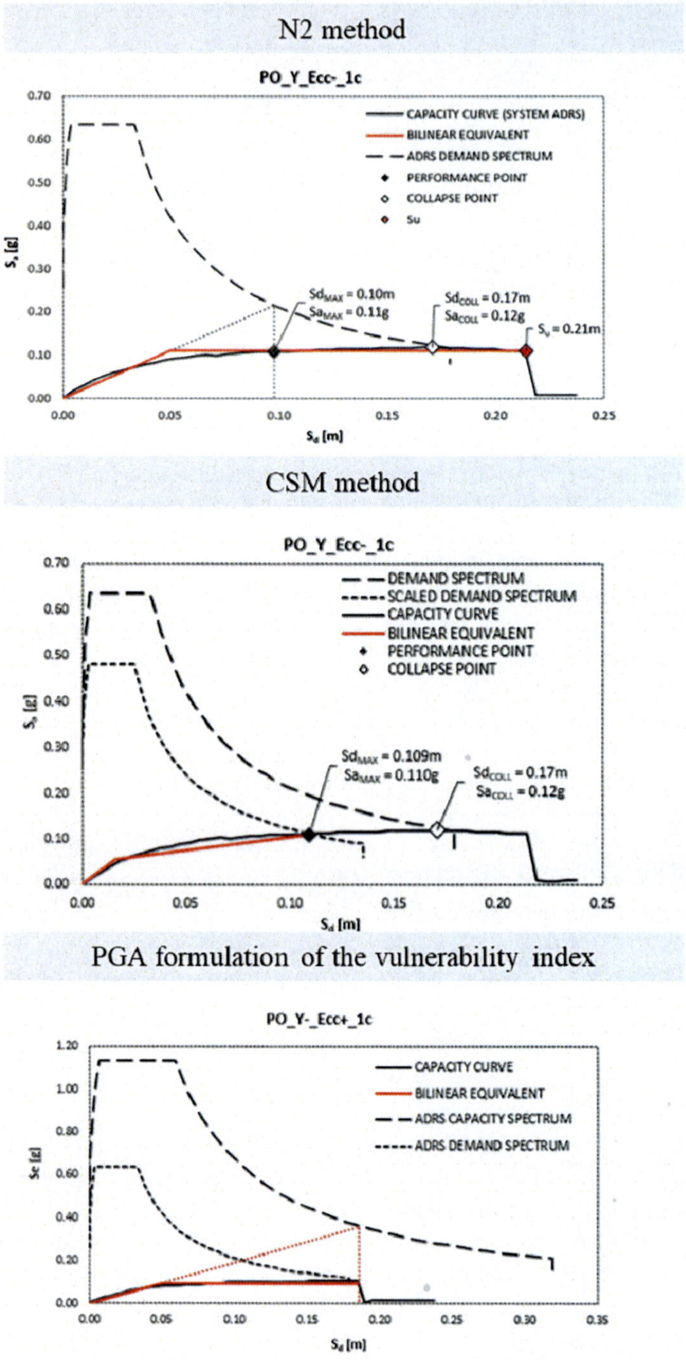

Fig. 10. Evaluation of the seismic vulnerability index methods

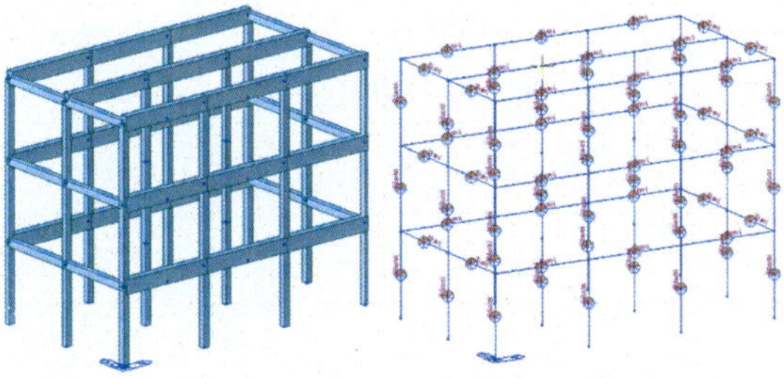

Fig. 11. Example model

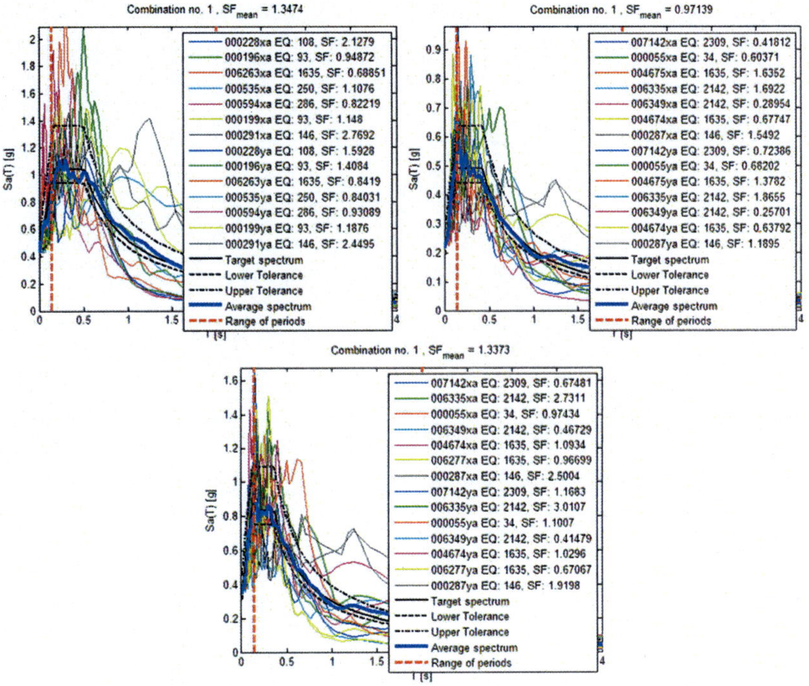

Fig. 12. Spectrum compatible accelerograms (CO, ME, SR)

approaches based on the Pushover analysis are not able to provide sub-unit values (contemplated by NTC-2018). Therefore, the nonlinear dynamic analysis would have to be used to accurately assess ζ_E.

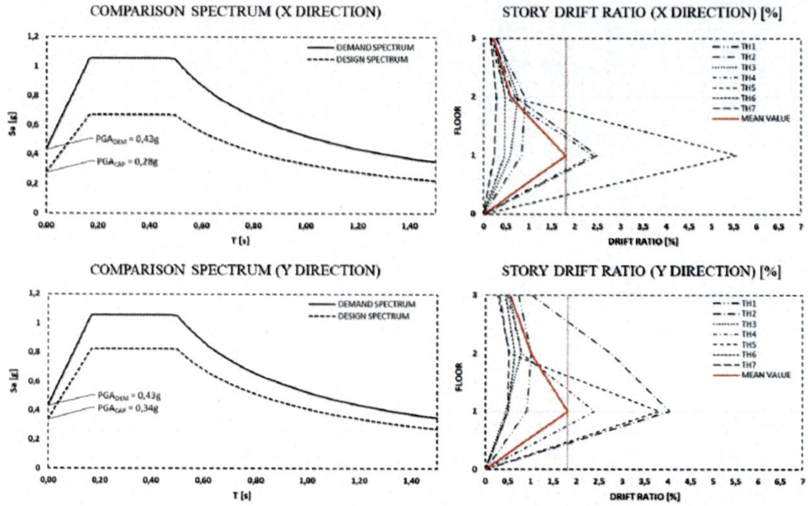

Fig. 13. Demand spectrum scaling until the target inter-floor shift is reached

Table 6. Comparison of results: Pushover indices in terms of forces

Collapse Analysis		Case $\zeta_E < 1$ (CO)	Case $\zeta_E \simeq 1$ (ME)	Case $\zeta_E > 1$ (SR)
		ζ_E	ζ_E	ζ_E
Incremental Dynamic Analysis		0.60	0.99	1.98
Pushover	N2 (force)	(*)	1.02	1.02
	CSM (force)	(*)	1.00	1.00
	PGA	0.96	1.09	1.69

Table 7. Comparison of results: Pushover indices in terms of displacements

Collapse Analysis		Case $\zeta_E < 1$ (CO)	Case $\zeta_E \simeq 1$ (ME)	Case $\zeta_E > 1$ (SR)
		ζ_E	ζ_E	ζ_E
Incremental Dynamic Analysis		0.60	0.99	1.98
Pushover	N2 (spost.)	(*)	1.08	1.69
	CSM (spost.)	(*)	1.13	2.30
	PGA	0.96	1.09	1.69

3 Conclusion

For existing and historical heritage, retrieving critical information to carry out seismic safety assessments is usually very time-consuming. The study proposes the setting up of a BIM digitalization process to organize a dynamic archive of information that can be updated over time. Therefore, the digital model's realization must be optimized given not only a specific intervention but also considering the possible survey, simulation, management and maintenance activities that may occur during the building's life cycle. In order to ensure an optimal workflow, much attention must be paid to interoperable processes between professionals, disciplines and software. In this study, interoperability tests were carried out to create a structural database and perform structural analyses. It is essential to set up a working environment, realized in this framework through a federated model, in which individual professionals can access the knowledge in the most appropriate format for specific purposes.

A case study has been analyzed to test the proposed approach in terms of horizontal modeling (using different software but consistent in the type of application) and vertical modeling (using different software by switching from BIM modeling to structural FEM modeling).

The process of analyzing complex structures also requires a high computational burden. In this regard, simplifications to be introduced into the analytical model were evaluated by verifying congruence with a reference model through an accurate sensitivity analysis, comparing the modal forms obtained from a linear dynamic analysis for the reference model with those obtained for the simplified model through the Modal Assurance Criterion. The linear dynamic analysis made it possible to understand whether modeling errors were present, compare the capacity and demand for each structural element of the building following current regulations, and identify the force distribution required for the subsequent non-linear static analysis. Based on the work rates obtained, it was possible to design local interventions to solve the seismic problem. The seismic vulnerability coefficient of the case study was calculated using a simplified method (non-linear static analysis, Pushover). This computational methodology has finally been evaluated against a more accurate methodology type, "incremental dynamic analysis", on a simple model adopted as an example and placed in different sites on the Italian territory, characterized by different seismic hazard levels. It can be observed that the adopted methods based on Pushover analysis can only give a preliminary assessment of the seismic vulnerability index. It is left to the non-linear dynamic analysis to more accurately assess the seismic vulnerability index through the relationship between the demand and capacity PGAs. It should be pointed out that this is a preliminary study to be repeated on structures characterized by different structural schemes.

The study, therefore, aims to evaluate a complete process from the data collection phase on a structure to seismic vulnerability assessment. In this framework, the horizontal and vertical interoperability analysis is intended as an iterative process where modeling must be optimized to maximize the information transfer. Furthermore, any inadequacies observed in seismic vulnerability would be resolved by repeating the process and introducing the necessary interventions and modifications to the existing structure by taking advantage of the support of this procedure.

Author Contribution. Abstract: all authors; Introduction: A.O; Methodology: F.M.U, M.D.; Data Collection: F.M.U, S.T; Data Modeling: F.M.U, S.M; Interoperability Tests: F.M.U, S.T, S.M., Structural Analysis: M.D, S.T, S.M., Conclusion: F.M.U, M.D.

References

1. Ministero delle Infrastrutture e dei Trasporti: Aggiornamento delle "Norme Tecniche per le Costruzioni). Roma (2018)
2. Ministero delle Infrastrutture e dei Trasporti: Circolare 21 gennaio 2019. Istruzioni per l'applicazione dell'«Aggiornamento delle "Norme tecniche per le costruzioni"» di cui al decreto ministeriale 17 gennaio 2018. Roma (2019)
3. Fernández-Mora, V., Navarro, I.J., Yepes, V.: Integration of the structural project into the BIM paradigm: aA literature review. In: J. Building Eng. **53** (2022)
4. Beirnaert, F., Lippens, A.: Analysis of the interoperability from BIM to FEM. Master Thesis in Construction Engineering. Tampere University of Applied Sciences (2018)
5. Crespi P., Franchi, A., Ronca, P., Giordano, N., Scamardo, M., Gusmeroli, G., Schiantarelli, G.: From BIM to FEM: the analysis of an historical masonry building. WIT Trans. Built Environ. **149**, 581–592 (2015)
6. Barazzetti, L., Banfi, F., Brumana, R., Gusmeroli, G., Previtali, M., Schiantarelli, G.: Cloud-to-BIM-to-FEM: Structural simulation with accurate historic BIM from laser scans. Simul. Pract. Theory **57**, 71–87 (2015)
7. Toska, K., Zanchetta, C.: Computational design per il BIM strutturale. In: Ingenio. (2018)
8. Monforte, S.: Modellazione e Interoperabilità BIM-FEM di un edificio esistente. Analisi statica non lineare e approcci per la valutazione della vulnerabilità sismica. Master Thesis in Civil Engineering. Politecnico di Torino (2021)
9. Tuccitto, S.: Modellazione e Interoperabilità BIM-FEM di un edificio esistente. Analisi dinamica lineare e confronto forme modali tramite Modal Assurance Criterion. Master Thesis in Civil Engineering. Politecnico di Torino (2021)
10. Ugliotti, F.M., Osello, A., Rizzo, C., Muratore, L.: BIM-based structural survey design. In: Struct. Integrity Procedia **18**(2019), 809–815. Elsevier, Amsterdam (2019)
11. BIM Forum: 2021 Level of Development (LOD) Specification. For Building Information Models. Part I, Guide, & Commentary. https://bimforum.org/wp-content/uploads/2022/02/LOD-Spec-2021-Part-I-FINAL-2021-12-28.pdf. Accessed 2022/06/17
12. Domaneschi, M., Martinelli, L.: Optimal passive and semi-active control of a wind excited suspension bridge. Struct. Infrastruct. Eng. **9**(3), 242–259. Taylor & Francis (2013)
13. El Khoudri, M., Ben Allal, L., Himi, M., Adak, D.: Seismic vulnerability assessment of reinforced concrete buildings using Incremental Dynamic Analysis IDA. J. Mater Environ. Sci. **7**, 481–487 (2016)
14. Ghobarah, A.: On drift limits associated with different damage levels. In: Paper Read at International Workshop on Performance-Based Seismic Design (2004)
15. Marasco, S., Noori, A.Z., Domaneschi, M., Cimellaro, G.P.: Seismic vulnerability assessment indices for buildings: proposals, comparisons and methodologies at collapse limit states. Int. J. Disaster Risk Reduction **63**(2021), 102466. Elsevier, Amsterdam (2021)

Automating the Frequency Domain Decomposition Technique Using the Modal Assurance Criterion

Amir Reza Elahi, Alessandro Cardoni(✉), Marco Domaneschi, and Gian Paolo Cimellaro

Department of Structural, Building and Geotechnical Engineering, Politecnico Di Torino, 10129 Torino, Italy
s274041@studenti.polito.it, {alessandro.cardoni, marco.domaneschi,gianpaolo.cimellaro}@polito.it

Abstract. Infrastructures are degrading rapidly, which necessitates the application of automated dynamic characterization in damage identification schemes. The Operational Modal Analysis (OMA) is an effective tool to process monitoring data of structural assets. In this study, the Frequency Domain Decomposition (FDD) technique is automated based on two different criteria, the Discrimination Factor (DF) and Modal Assurance Criterion (MAC). The recorded ambient vibration response of a cable-stayed bridge was used as input to the proposed method to extract its modal properties automatically. The method was also tested on a calibrated finite element model of the cable-stayed bridge for validation purposes. The optimal automation performance is achieved by establishing the best range of each criterion.

Keywords: Operational modal analysis · Frequency domain decomposition · Cable-stayed bridges · Modal assurance criterion · Discrimination factor

1 Introduction

As Structural Health Monitoring (SHM) broadens by introducing numerous disciplines, OMA approaches are incredibly recognized to fulfill the monitoring requisites without interrupting the service of the civil assets. In OMA approaches, two main categories of time and frequency domain exist. Despite the advent of various time-domain techniques such as Stochastic Subspace Identification (SSI) [1] and Random Decrement Technique (RDT) [2], the frequency domain approaches are still sought due to their interaction with users and ease of employment. However, the drawback of using such techniques is the risk of not exciting all modes by ambient vibration. Also, these approaches are prone to mistakes made by the user in the mode selection.

In the frequency domain, the FDD technique was first suggested by Brincker et al. to estimate the modal properties of a system without the knowledge of input vibration [3]. The method relied on constructing the Power Spectral Density (PSD) matrix from the ambient vibration response. Consequently, the peaks of the first Singular Values

G. P. Cimellaro (Ed.): WCSI 2022, LNCE 309, pp. 1063–1073, 2023.
https://doi.org/10.1007/978-3-031-21187-4_92

(SV) corresponding to the decomposed PSD matrix were considered as the modal frequencies. In addition, the related singular vectors represented the system's mode shapes. The approach accurately estimated the modal properties of a two-story building [3, 4]. Moreover, it was further developed to estimate damping ratios even though the results were problematic in many cases [5, 6].

The introduction of new identification methods based on ambient vibration is still an ongoing field of study due to their essential role in damage identification schemes. In a recent study, Di Matteo et al. [7] proposed a novel identification method by decomposing the outputs using a proper Butterworth filter to obtain mono-component signals corresponding to an SDOF system. Then, the Analytical Signal (AS) and Hilbert Transform were applied to the decomposed response to extract modal parameters. The method resulted in the successful modal identification of Chiaramonte Palace in Palermo.

As stated, access to automatic modal identification approaches is essential to incorporate into damage detection algorithms. In this regard, the first effort to introduce an automated version was based on the least square complex frequency. In this method, a fuzzy clustering approach was applied based on different criteria to select the physical poles from a high-order model. The drawback of the method was the high computational effort and complexity in the application [8].

Since the manual peak selection in the FDD is a clumsy task with the risk of error, an automatic version was later proposed to enhance its performance. In order to increase the accuracy of automatic peak picking, they established the modal coherence criterion to identify the modal domain around each peak of the SV plot. The limits associated with the modal coherence were obtained based on the standard deviation of correlation between mode shapes and the number of measurement channels [9].

In another attempt to apply the automated OMA approaches, Cheynet et al. [10] used the algorithms pertaining to automatic covariance-driven SSI and AFDD to determine the modal properties from both numerical and experimental data of a suspension bridge in Norway. In automating the FDD technique, the peak picking algorithm proposed by Liutkus [11] was employed to identify the peaks in the SV spectrum. The study recommends the use of time-domain approaches rather than the frequency ones for suspension bridges to obtain more accurate damping ratios.

To the author's knowledge, the existing algorithms for the AFDD are insufficient in extracting the modal properties of signals with high noise. In detail, it was observed that the noisy SV spectrum resulting from amplified ambient vibration gives rise to an inaccurate selection of peaks without any physical interpretation. Therefore, to deal with this challenge, the MAC and DF are used as distinguishing factors between the sinusoidal mode shapes while improving the accuracy of peak identifications. In this paper, the next section is dedicated to a brief description of the FDD methodology and the proposed method. Then, it is applied to the ambient vibration response of the cable-stayed bridge. Finally, the results are compared with a calibrated finite element (FE) model and manual FDD outputs to evaluate their accuracy.

2 Methodology

Based on the theories related to stochastic processes, the relation between two realizations can be defined by a Correlation Function (CF) matrix comprised of diagonal elements as the autocorrelations and non-diagonals as cross-correlations. For the Gaussian signals or close ones, the correlation function contains significant properties if and only if they are stationary, and ergodic, with a mean value equal to zero. In the vibrational analysis, the components of the CF matrix for two random signals of $x(t)$ and $y(t)$ are defined as:

$$\begin{aligned} R_x(\tau) &= E[x(t)x(t+\tau)] \\ R_y(\tau) &= E[y(t)y(t+\tau)] \\ R_{xy}(\tau) &= E[x(t)y(t+\tau)] \end{aligned} \tag{1}$$

where E represents the expectation and using the time averaging technique, it has the integral form of:

$$\begin{aligned} R_x(\tau) &= \frac{1}{T}\int_0^T x(t)x(t+\tau)dt \\ R_y(\tau) &= \frac{1}{T}\int_0^T y(t)y(t+\tau)dt \\ R_{xy}(\tau) &= \frac{1}{T}\int_0^T x(t)y(t+\tau)dt \end{aligned} \tag{2}$$

In the above relations, the parameter T is the time interval. Since it is assumed that the signal is stationary, the shift in the integral's time bounds is allowed. This leads to the derivation of the convolution integral presented below:

$$\begin{aligned} R_x(\tau) &= \frac{1}{T}\int_{-T-\tau}^{-\tau} x(-\theta-\tau)x(-\theta)d\theta = \frac{1}{T}\int_0^T x(-\theta)x(\tau-\theta)d\theta = x(-t)*x(t) \\ R_y(\tau) &= \frac{1}{T}\int_{-T-\tau}^{-\tau} y(-\theta-\tau)y(-\theta)d\theta = \frac{1}{T}\int_0^T y(-\theta)y(\tau-\theta)d\theta = y(-t)*y(t) \\ R_{xy}(\tau) &= \frac{1}{T}\int_{-T-\tau}^{-\tau} x(-\theta-\tau)y(-\theta)d\theta = \frac{1}{T}\int_0^T x(-\theta)y(\tau-\theta)d\theta = x(-t)*y(t) \end{aligned} \tag{3}$$

In the derived relations, the integral variable is changed. At this point, It is proven that the correlation function is the convolution of the two signals. The autocorrelation represents the free decay of motion for an SDOF system as well.

To investigate the power distribution in the frequency domain, the Fourier transform of the correlation function is proposed. As a result, the PSD matrix will consist of diagonal and non-diagonal elements which are calculated by the Fourier transform of auto and cross-correlations respectively. Thus, the PSD matrix elements have the form of:

$$\begin{aligned} G_{xx}(\omega) &= X(-\omega)X(\omega) = X^*(\omega)X(\omega) \\ G_{yy}(\omega) &= Y(-\omega)Y(\omega) = Y^*(\omega)Y(\omega) \\ G_{yx}(\omega) &= Y(-\omega)X(\omega) = Y^*(\omega)X(\omega) \end{aligned} \tag{4}$$

It is worthwhile to state that in the derivation process, the reversal property of Fourier transform is used in which the $X(-\omega)$ becomes $X^*(-\omega)$ (complex conjugate). Referring to the classical structural dynamics, the Fourier transform of the response is equal to the Fourier transform of input acceleration multiplied by the Frequency Response Function (FRF). This is indicated in Eq. (5) where H represents FRF. Thus, by replacing the Fourier transform of the response in Eq. (4), the PSD of output would be:

$$Y(\omega) = X(\omega)H(i\omega)$$

$$G_{yy}(\omega) = Y^*(\omega)Y(\omega) = X^*(\omega)X(\omega)H^*(i\omega)H(i\omega) = G_{xx}(\omega)H^*(i\omega)H(i\omega) = G_{xx}(\omega)|H(i\omega)|^2 \tag{5}$$

In addition, by recalling the integral form of the convolutions, the same relation can be established for the PSD of the output. The above relation is the fundamental theorem for stationary signals where the PSD of output is a function of the input counterpart.

The problem formulation is approximated based on the following assumptions: the input of the ambient vibration is accepted as white noise; the modal coordinates are uncorrelated. In theory, the FRF can be written as the summation of:

$$H(i\omega) = \sum_{k=1}^{n}\left(\frac{R_k}{(i\omega - \lambda_k)} + \frac{R_k^*}{(i\omega - \lambda_k^*)}\right) \tag{6}$$

In which n, λ_k, R_k and R_k^* are the numbers of modes, pole, residue, and its complex conjugate respectively. As stated, the simplification is made by assuming the input as a white noise signal. This assumption results in considering the correlation of the input signal (R_{xx}) as Dirac's delta while its Fourier transform (G_{xx}) would be a scalar value (C). Therefore, Eq. (5) becomes:

$$G_{yy}(i\omega) = \sum_{k=1}^{n}\sum_{s=1}^{n}\left[\frac{R_k}{(i\omega - \lambda_k)} + \frac{R_k^*}{(i\omega - \lambda_k^*)}\right]C\left[\frac{R_s}{(i\omega - \lambda_s)} + \frac{R_s^*}{(i\omega - \lambda_s^*)}\right] \tag{7}$$

Conducting a detailed mathematical manipulation as reported in [12], while assuming that the *kth* residue of the output PSD is proportional to the mode shape vector in the lightly damped systems, the PSD of the output will be written as:

$$G_{yy}(i\omega) = \sum_{k \in sub(\omega)}\left(\frac{d_k\phi_k\phi_k^T}{i\omega - \lambda_k} + \frac{d_k^*\phi_k^*\phi_k^H}{i\omega - \lambda_k^*}\right) \tag{8}$$

where d is the scalar constant, ϕ is the mode shape vector, and ϕ_k^H is the complex conjugate of its transpose. The above relation is the approximate solution for the PSD of the output. In the power spectral density of the outputs, each peak represents the dominance of a mode or its combination with close ones. If only one mode dominates, the summation indicated in Eq. (8) has only one term [5, 12]. To obtain the modal properties from that term near the peak, it is suggested to use the Singular Value Decomposition (SVD) technique to decompose the PSD. SVD can be considered a powerful technique in comparison with Eigen Value Decomposition (EVD). For instance, to obtain orthogonal mode shapes of matrix A using EVD, several mathematical properties such as being symmetrical and rectangular must be satisfied which is not necessary for the SVD. To elaborate the method, the decomposition involves finding the best fitting subspace for a set of data scattered in the k dimension. In other words, assuming a set of data in 2 dimensions, the best fitting line with the least vertical distance from the data can be obtained by maximizing the projection length or minimizing the square distance. Then, the algorithm of this best-fitting line with the least square distance is further developed for the k lines to build the subspace fitting the data in the k dimension. Assuming $G_{yy}(iw_i)$ as the power spectral density at discrete frequencies, its SVD decomposition has the form of:

$$G_{yy}(i\omega_i) = U_i S_i V_i^H \tag{9}$$

where S is the ordered first singular values and U, and V are the corresponding first singular vectors.

The above methodology is the starting point of the proposed AFDD that is presented in Fig. 1.

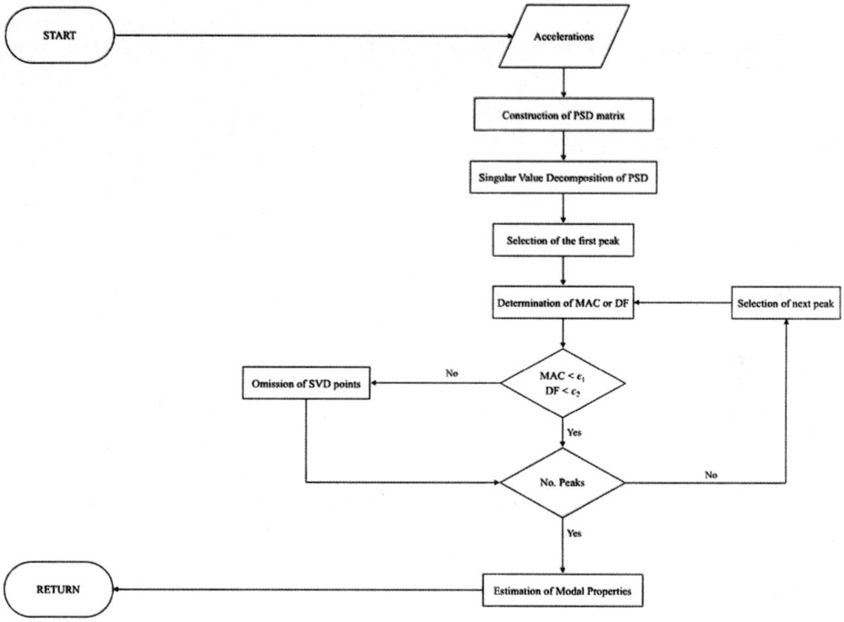

Fig. 1. The flowchart of the AFDD algorithm.

In the proposed methodology, the MAC and DF between two vectors can be determined by the following relations:

$$\mathrm{MAC(a, b)} = \frac{|\mathrm{a^H b}|^2}{(\mathrm{a^H a})(\mathrm{b^H b})} \tag{10}$$

$$\mathrm{DF(a, b)} = \mathrm{a^H b} \tag{11}$$

3 Case Study: Yonghe Cable Stayed Bridge

The Yonghe cable-stayed bridge, located in the port of Tianjin, Hebei province, China, is a benchmark dedicated to the SHM monitoring of cable-stayed bridges. The structure is 510 m long and 11 m wide. It is comprised of one 210 m main and two 125 m secondary spans casting adjacent to it (**Error! Reference source not found.**.a). The bridge consists of prestressed main girders connected with the cast-in-place joints. In addition, 88 pairs of 5 mm cables are supporting the deck while being attached to the towers. The towers are formed by two transverse beams and are 60.5 m high, using sliding formwork technology for their construction. Being utilized in December 1987, the bridge was retrofitted in 2005–2007. During the reconstruction stage, a comprehensive monitoring system designed at Harbin University of Science and Technology was instrumented for further monitoring and assessments [13].

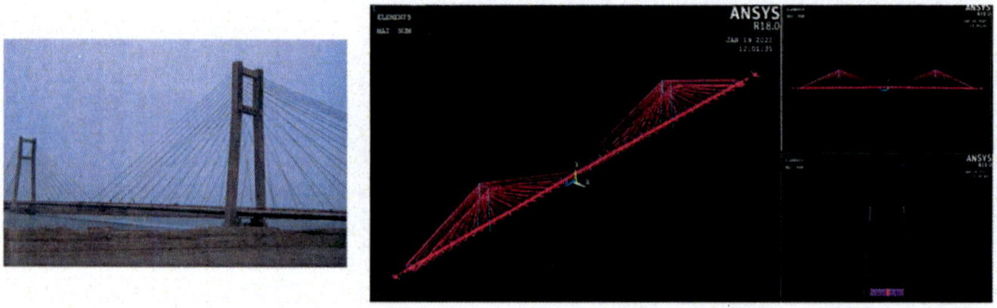

(a) (b)

Fig. 2. The Yonghe cable-stayed bridge: (a) picture of the actual bridge and (b) FE model.

As indicated in Fig. 2b, a 3D FE model is prepared to perform modal analysis and derive the reference frequencies and mode shapes. The FE model was developed based on the documented drawings in the ANSYS Mechanical APDL [14]. The model includes defining the main girders and towers using the three-dimensional beam element (BEAM 44). Moreover, the transverse beams were assigned mass elements with a spacing of 2.9 m. The cables were defined as linear elastic links (LINK 10). All of the boundary conditions, such as floating the main girder on the fixed towers and modeling the rubber supports were defined properly. The model was further updated by changing the mechanical properties based on comparing the initial and extracted modal properties using the FDD method [13].

The SHM system includes more than 150 sensors from which 14 uniaxial accelerometers were instrumented on the deck. Additionally, an anemoscope and temperature sensors were mounted to the southern tower to measure wind velocity and temperature change, and Fiber Bragg Grating (FBG) sensors were embedded in the girders to document strains.

This study intends to identify the modal properties of the cable-stayed bridge using the acceleration records from the 14 uniaxial accelerometers (Fig. 3). To this aim, the records related to hour 4, January 17, are selected. A Butterworth bandpass filter with the low and high frequency cut of 0.1 and 30 Hz is applied to the raw data. Then, they are inserted as the input of the AFDD method. Following the procedure presented in the methodology section, a script in MATLAB [15] was prepared and used for conducting the analysis. In the analysis, the frequency range is fixed within 0–1.3 Hz, and the number of required peaks is to equal 6. A sensitivity analysis is performed for two criteria where the MAC variation is between 0 and 1 with the step of 0.01. The same range is proposed to the DF with the variation step of 0.1. The optimum range of MAC is defined within 0.18–0.21, while the best performance of DF is 0.8.

The selected peaks are presented in Fig. 4 when the distinguishing factor is MAC.

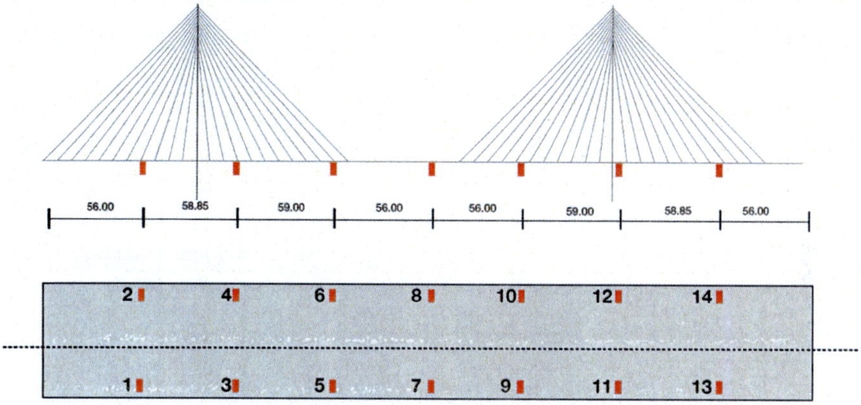

Fig. 3. The layout of the instrumented SHM system.

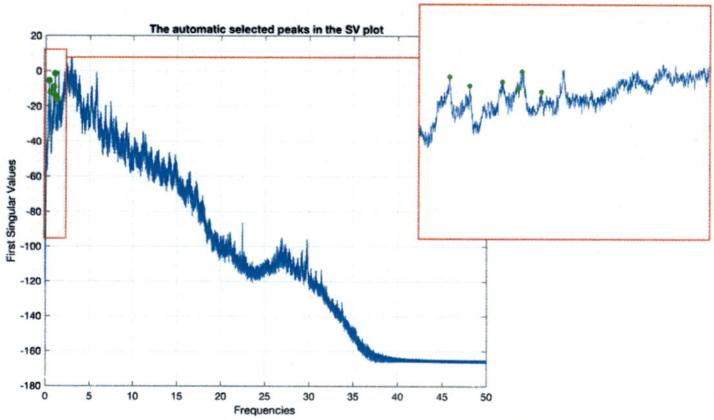

Fig. 4. The plot of first singular values and selected peaks.

In Fig. 5 the distribution of MAC and DF during the selection of the first peak in the SVD plot is illustrated.

To have a better comparison, the outputs from FDD, modal analysis of the FE model, and the AFDD of 7 h acceleration records are obtained. The frequencies are compared in Table 1.

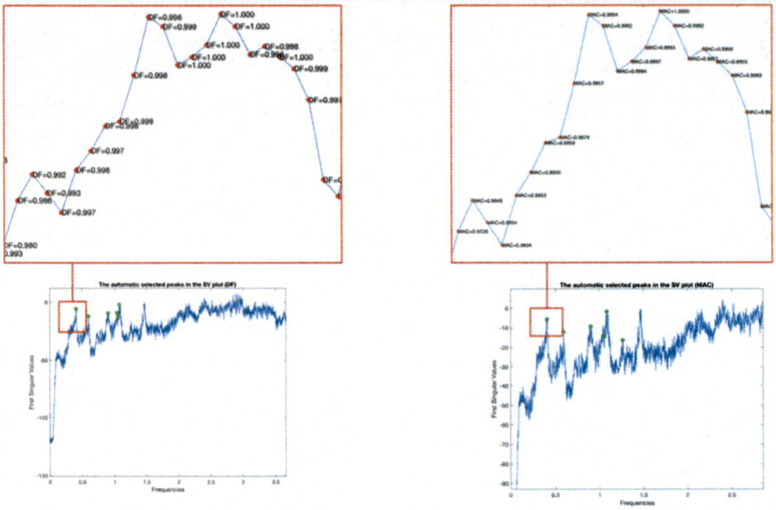

Fig. 5. The distribution of MAC and DF during the selection of the first peak.

Table 1: The comparison between the derived frequencies.

Method	FE model	FDD	AFDD with MAC (1 h)	AFDD with MAC (7 h)	AFDD with DF (1 h)
Identified frequencies (Hz)	0.4175	0.4171	0.4059	0.4040	0.4059
	0.5937	0.5873	0.5928	0.5929	0.5928
	0.8774	0.8913	0.9010	0.8973	0.9010
	1.0442	1.0377	1.04141	1.0412	1.0384
	1.0892	1.1014	1.0803	1.0957	1.0414
	1.2145	1.2708	1.2611	1.2609	1.083

It was observed that the AFDD with MAC results in a more accurate estimation of frequencies compared with the DF counterpart. In fact, the AFDD with DF is not capable of distinguishing two closely spaced peaks, and the fourth peak is selected incorrectly.

To further investigate the method's validity, the mode shapes derived from AFDD with MAC for one hour of acceleration records are plotted using MATLAB. In deformed configurations, due to the low number of sensors, the mode shapes are smoothed using spline interpolation. They are scaled and normalized as well. The deformed states of the bridge connected with the derived mode shapes are presented in Fig. 6.

To plot the 2D mode shapes and comparing with the FE counterpart, the normalized mode shapes from two rows of accelerometers are considered. The rows are separated as the even and odd sensors based on the numbering indicated in Fig. 3. The consistency of acquired mode shapes is illustrated in the following figure:

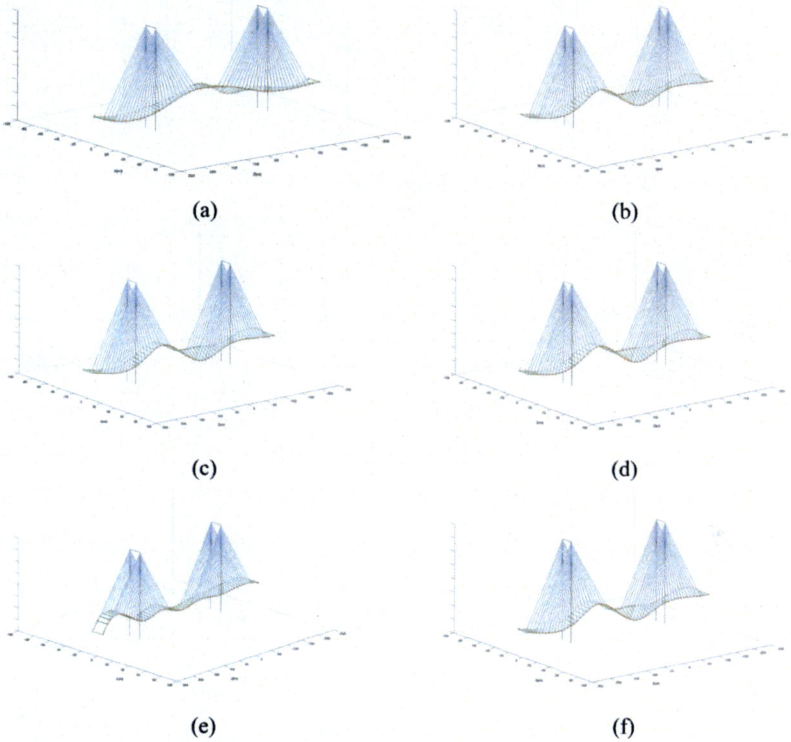

Fig. 6. The mode shapes obtained from AFDD approach: 1st mode shape (f = 0.4059), 2nd mode shape (f = 0.5928), 3rd mode shape (f = 0.901), 4th mode shape (f = 1.04141), 5th mode shape (f = 1.0803), 6th mode shape (f = 1.2611).

Despite good coherence between most derived mode shapes, there is an inaccuracy in some plots since the data related to sensor No.10 at $z = -56$ m were erroneous. It was deduced that the sensor is defective, which aligns with previous findings.

4 Conclusion

In this study, the AFDD methods based on the MAC and DF are proposed to derive the modal properties. To evaluate their accuracy, records from the ambient vibration response of a cable-stayed bridge were considered as the input. Subsequently, the modal properties were extracted and compared with the data related to the existing modal identification techniques. The proposed methodology with MAC can capture the properties more accurately in comparison with the case where DF is a distinguishing factor. Furthermore, it is more efficient than the manual FDD technique since there is no user intervention. The best range of MAC was 0.18–0.2. However, it is necessary to further investigate the effect of factors such as noise, frequency of sampling, duration of records, sensor numbers and layout, the number of required peaks, and the type of structure on the optimum range of MAC. This will be covered in future studies on different infrastructures.

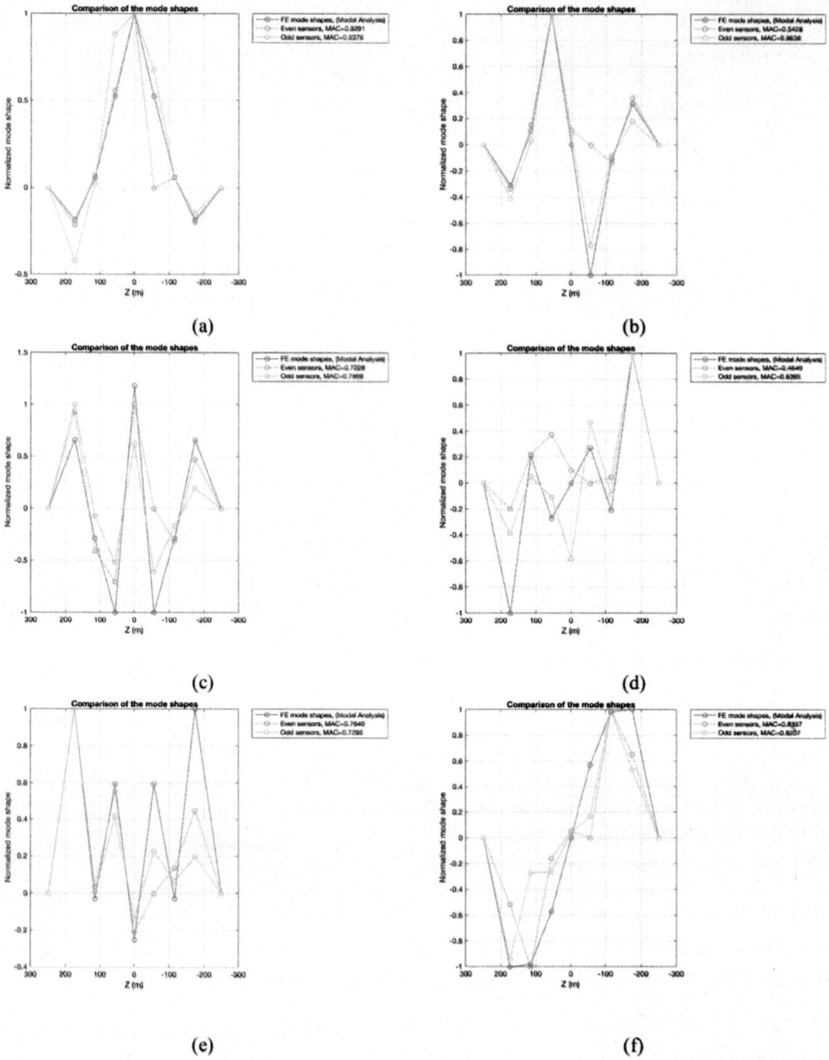

Fig. 7. Comparison between the mode shapes obtained from AFDD and FE model: (a) 1st mode shape, (b) 2nd mode shape, (c) 3rd mode shape, (d) 4th mode shape, (e) 5th mode shape, (f) 6th mode shape.

References

1. Overschee, P.: Subspace identification for linear systems. Kluwer Academic Publishers
2. Asmussen, J.C., Ibrahim, R., Brincker, R.: Random decrement: identification of structures subjected to Ambient Excitation. In: Proceedings of the 16th International Modal Analysis Conference, vol. 2, pp. 914–921. Society for Experimental Mechanics, Santa Barbara, California, USA, Feb. 1998 (1998)

3. Brincker, R., Zhang, L., Andersen, P.: Modal identification from ambient responses using frequency domain decomposition. In: Proceedings of the International Modal Analysis Conference (IMAC), pp. 625–630, San Antonio, Texas, USA 10 (2000)
4. Brincker, R., Zhang, L., Andersen, P.: Modal identification of output-only systems using frequency domain decomposition. Smart Mater. Struct. **10**(3), 441–445 (2001). https://doi.org/10.1088/0964-1726/10/3/303
5. Bajric, A., Brincker, R., Thöns, S: Evaluation of damping estimates in the presence of closely spaced modes using operational modal analysis techniques. In: Proceedings of the 6th International Operational Modal Analysis Conference (2015)
6. Magalhães, F., Cunha, Á., Caetano, E., Brincker, R.: Damping estimation using free decays and ambient vibration tests. Mech. Syst. Signal Process. **24**(5), 1274–1290 (2010). https://doi.org/10.1016/j.ymssp.2009.02.011
7. Di. Matteo, A., Masnata, C., Russotto, S., Bilello, C., Pirrotta, A.: A novel identification procedure from ambient vibration data. Meccanica **56**(4), 797–812 (2020). https://doi.org/10.1007/s11012-020-01273-4
8. Vanlanduit, S., Verboven, P., Guillaume, P., Schoukens, J.: An automatic frequency domain modal parameter estimation algorithm. J. Sound Vib. **265**(3), 647–661 (2003). https://doi.org/10.1016/S0022-460X(02)01461-X
9. Brincker, R., Andersen, P., Jacobsen, N-J.: Automated frequency domain decomposition for operational modal analysis. In: Conference Proceedings: IMAC-XXIV: A Conference & Exposition on Structural Dynamics Society for Experimental Mechanics (2007)
10. Cheynet, E., Jakobsen, J.B., Snæbjörnsson, J.: Damping estimation of large wind-sensitive structures. Procedia Eng. **199**, 2047–2053 (2017). https://doi.org/10.1016/j.proeng.2017.09.471
11. Liutkus, A.: Scale-space peak picking. Research Rep. Inria Nancy-Grand Est. Villers-les-Nancy, France: Inria Nancy—Grand Est (2015)
12. Brincker, R., Ventura, C.: Introduction to operational modal analysis, Wiley (2015)
13. Li, S., Li, H., Liu, Y., Lan, C., Zhou, W., Ou, J.: SMC structural health monitoring benchmark problem using monitored data from an actual cable-stayed bridge: SMC structural health monitoring benchmark problem. Struct. Control Health Monit. **21**(2), 156–172 (2014). https://doi.org/10.1002/stc.1559
14. ANSYS, Academic Research Mechanical, Release 18.1
15. MATLAB R2021, last accessed 2016/11/21